MULTIVARIATE ANALYSIS
WITH APPLICATIONS IN EDUCATION AND PSYCHOLOGY

BROOKS/COLE SERIES IN STATISTICS
Roger E. Kirk, Editor

METHODS IN THE STUDY OF HUMAN BEHAVIOR

Vernon Ellingstad and Norman W. Heimstra, The University of South Dakota

EXPERIMENTAL DESIGN: PROCEDURES FOR THE BEHAVIORAL SCIENCES

Roger E. Kirk, Baylor University

STATISTICAL ISSUES: A READER FOR THE BEHAVIORAL SCIENCES

Roger E. Kirk, Baylor University

THE PRACTICAL STATISTICIAN: SIMPLIFIED HANDBOOK FOR STATISTICS

Marigold Linton, The University of Utah, and
Philip S. Gallo, Jr., San Diego State University

MULTIVARIATE ANALYSIS WITH APPLICATIONS IN EDUCATION AND PSYCHOLOGY

Neil H. Timm, University of Pittsburgh

MULTIVARIATE ANALYSIS
WITH APPLICATIONS IN EDUCATION AND PSYCHOLOGY

NEIL H. TIMM
University of Pittsburgh

BROOKS/COLE PUBLISHING COMPANY
Monterey, California
A Division of Wadsworth Publishing Company, Inc.

ISBN: 0-8185-0096-4
L.C. Catalog Card No.: 74-83250
Printed in the United States of America
10 9 8 7 6 5 4 3 2 1

PREFACE

Multivariate analysis is a branch of statistics dealing with procedures for summarizing, representing, and analyzing multiple quantitative measurements obtained on a number of individuals or objects. *Multivariate Analysis* is designed to be used both as a text for students of multivariate analysis and as a complete reference for persons engaged in research requiring multivariate methods. Though concerned primarily with applications, with a focus in education and psychology, the book contains the necessary theory for the multivariate procedures developed. The material presented is intended for individuals with facility in introductory material in univariate analysis of variance and in regression analysis. However, because many persons who have a need for multivariate techniques do not have sufficient training in univariate analysis, each multivariate procedure is preceded, when possible, by the corresponding univariate method. As a text, the organization and coverage of the topics allow the instructor to treat the material at either an applied level or at an intermediate to advanced mathematical level, depending on the ability of the students.

The development of many of the statistical procedures in this text depend on the linear model; unlike other books using the linear model, this text integrates the four standard approaches to the analysis of experimental designs—side conditions, reparameterization, generalized inverses, and geometry. This text contains important material usually omitted from other texts, such as growth-curve analysis, the extended linear model, the analysis of heterogeneous data, partial and bi-partial canonical correlation analysis, the Behrens-Fisher problem, the analysis of nonorthogonal designs, and more. Extensive consideration is given to multivariate analysis of variance and covariance analysis and simultaneous inference using several multivariate criteria.

Because this text does not depend on a particular statistical program, the applications used to illustrate the multivariate procedures are closely associated with statistical theory in such a manner that students and researchers should be able to interpret and understand the origin of summary statistics generated by any standard computer program designed to do multivariate data analysis. When feasible, computational formulas for use on a desk calculator have been included and, because complete data sets for each example are contained in the text for all multivariate procedures illustrated, computations are easily verified. This approach should allow the reader to take advantage of any computer program as well as the desk calculator.

The book is divided into six chapters, an answer section, and an appendix containing statistical tables. Chapter 1 presents an introduction to the elements of matrix algebra, vector spaces, and derivatives. Readers familiar with these topics may want to skim this chapter. Those not familiar with matrix algebra may have to consult additional sources.

Chapter 2 is concerned with distribution theory and quadratic forms. This chapter is at a somewhat higher level than the remaining chapters in the book. Those readers concerned mainly with application and not with theory may wish to skip this chapter.

In Chapters 3, 4, and 5, the linear model approach to experimentation for one- and two-sample problems, regression and correlation analysis, and analysis-of-variance designs are discussed and illustrated. These chapters are concerned mainly with application; however, theory is also included for a deeper understanding of the multivariate techniques developed.

Chapter 6 is primarily concerned with principal-component analysis. Although this book does not emphasize discriminant analysis or factor analysis, sufficient information on these topics has been included for completeness.

The univariate review preceding the development of each multivariate technique, the inclusion of a chapter on matrix algebra, and the complete set of statistical tables for multivariate methods, make this book self-contained both as a reference for univariate and multivariate methods and as an introductory text in applied multivariate analysis.

The text provides enough material in Chapters 1 and 2, Chapters 3 and 4, and Chapters 5 and 6 for a three-semester course in multivariate data analysis. For students with some mathematical training, the material can be covered in two-semesters if Chapters 1 and 6 are covered only briefly. For students with some mathematical statistics training, Chapters 2 and 3 and parts of 4, 5, and 6 provide enough material for a very complete one-semester course. With Chapter 2 deleted, brief coverage of the material in Chapters 1 and 6 and detailed coverage of Chapters 3, 4, and 5 are sufficient for a two-semester course primarily concerned with applications; suggested reading for a one-semester course would include portions of Chapters 1, 3, 4, and 5.

The preparation of this text evolved from lectures given at training sessions on multivariate statistical analysis in 1970 and 1971, under the sponsorship of the American Educational Research Association and from class notes in the teaching of a three-semester course on multivariate methods at the Department of Educational Research, University of Pittsburgh. I am indebted to the more than 300 students and training session participants who provided valuable suggestions and criticisms of the material presented. Thanks must also be extended to the other instructors of the training sessions, Professors M. I. Charles E. Woodson, Darrell R. Bock, Jerry D. Finn, Joel R. Levin, and Robert M. Pruzek, for their encouragement during the initial development stages of this manuscript. My deepest appreciation and thanks are given to my colleague James E. Carlson at the University of Pittsburgh. He provided many thoughtful readings and suggestions during the development of the manuscript. I also wish to thank the reviewers of the manuscript: Bert Green, Jr., Johns Hopkins University; David T. Hughs, University of Florida; and Roger E. Kirk, Baylor University.

Grateful acknowledgement and appreciation are extended to the following professors at the University of California, Berkeley, for stimulating my interest in this field and guiding my academic program while I was a graduate student: David R. Brillinger, Gus W. Haggstrom, and Erich L. Lehmann of the Department of Statistics; Henry F. Kaiser and Leonard A. Marascuilo of the School of Education. Dr. Marascuilo was also helpful in critiquing the final version of the manuscript.

The material selected for discussion is dependent on the original work of C. R. Rao, T. W. Anderson, S. N. Roy, Henry Scheffé, and Karl Jöreskog, among others. I wish to acknowledge a deep debt of appreciation to these statisticians; without their original contributions and books, many of the theoretical foundations of statistics would not be available to applied researchers, and this book might never have been written.

It is not possible to mention all persons who have made this book a reality; but thanks must be given to Ellen BeGole, Richard Fellers, John Murphy, Dale Perinetti and Wee Kiat Tan for proofreading and checking the examples in the text. The manuscript was typed and retyped several times. The preparation of the whole manuscript and every revision was performed with great care and patience by Sherrie Harbert, to whom I am most grateful.

Neil H. Timm

CONTENTS

To Margaret, Bill, and Stephen

CHAPTER 1

FUNDAMENTALS OF VECTOR AND MATRIX ALGEBRA

1.1 INTRODUCTION

Fundamental to the study of multivariate data analysis is a collection of p selected variables:

$$[y_1, y_2, \ldots, y_p]$$

Such a set (or p-tuple) of scores is termed a *p-variate response* or *multivariate vector-valued observation.* This vector of variables represents a collection of measurements on an individual subject. Typical vectors might be a person's scores on p specified ability tests or the total number of error responses to a stimulus given at each of p trials. In the first case, the variables are usually different, whereas in the second situation they are very similar in nature.

The study of multivariate observations is seldom limited to one subject. In general, an $N \times p$ array is required to represent the p scores on N subjects; that is,

$$\underset{(N \times p)}{\mathbf{Y}} = \begin{bmatrix} y_{11} & y_{12} & \cdots & y_{1p} \\ y_{21} & y_{22} & \cdots & y_{2p} \\ \vdots & \vdots & \vdots\vdots\vdots & \vdots \\ y_{N1} & y_{N2} & \cdots & y_{Np} \end{bmatrix}$$

where the first subscript indicates the subject and the second subscript denotes the variable. Such an array of scores is called a *data matrix.*

To facilitate the manipulation of vector-valued observations and data matrices, familiarity with the rules of vector and matrix algebra is essential. The following sections of this chapter serve as a review of the theory of vectors and matrices. For further details, you may find it necessary to refer to the references.

1.2 VECTORS AND VECTOR SPACES

Vectors. An *n-vector* (written $\mathbf{y}$) is an ordered *n*-tuple of real* numbers arranged in a column;

(1.2.1)
$$\mathbf{y} = \begin{bmatrix} y_1 \\ y_2 \\ \vdots \\ y_n \end{bmatrix}$$

From a geometric viewpoint, the elements y_i $(i = 1, 2, \ldots, n)$ of a multivariate observation vector $\mathbf{y}$ may be regarded as coordinates of a point in an *n*-dimensional Euclidean space where $\mathbf{y}$ represents the line segment joining the origin with the point $[y_1, \ldots, y_n]$. Such a directed line segment is called a *position vector.* In a three-dimensional Euclidean space, $\mathbf{y}$ denotes a vector from the origin to the point $[y_1, y_2, y_3]$ illustrated in Figure 1.2.1.

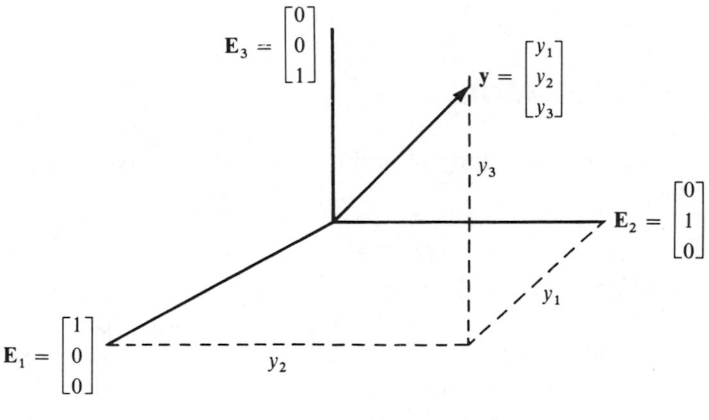

FIGURE 1.2.1. Vector represented in Euclidean three-dimensional space.

Addition and Scalar Multiplication of Vectors. Two *n*-vectors $\mathbf{x}$ and $\mathbf{y}$ are *equal* if and only if x_i equals y_i for every element in the two vectors:

(1.2.2)
$$\mathbf{x} = \mathbf{y} \quad \text{if and only if} \quad x_i = y_i \qquad (i = 1, \ldots, n)$$

* All vectors and matrices in this text are assumed to be real valued unless stated otherwise.

Vectors, like real numbers, obey a set of rules or axioms (see Halmos, 1958, p. 2). Fundamental to these rules are the operations of vector addition and multiplication of a vector by a *real number*, or *scalar*.

The *sum* of two vectors $\mathbf{x}$ and $\mathbf{y}$ with the same number of elements is obtained by adding their corresponding components; thus

(1.2.3) $$\mathbf{z} = \mathbf{x} + \mathbf{y} \quad \text{if and only if} \quad z_i = x_i + y_i \qquad (i = 1, \ldots, n)$$

To *multiply* a vector by a scalar s, multiply each element of the vector by the scalar; thus

(1.2.4) $$\mathbf{x} = s\mathbf{y} \quad \text{if and only if} \quad x_i = sy_i \qquad (i = 1, \ldots, n)$$

Geometrically, in two dimensions, we observe that vector addition obeys the parallelogram law (see Figure 1.2.2), so that the vector sum is the diagonal of a parallelogram. Scalar multiplication may change the length as well as the direction of a vector, as shown in Figure 1.2.3; the same is true in general for n dimensions.

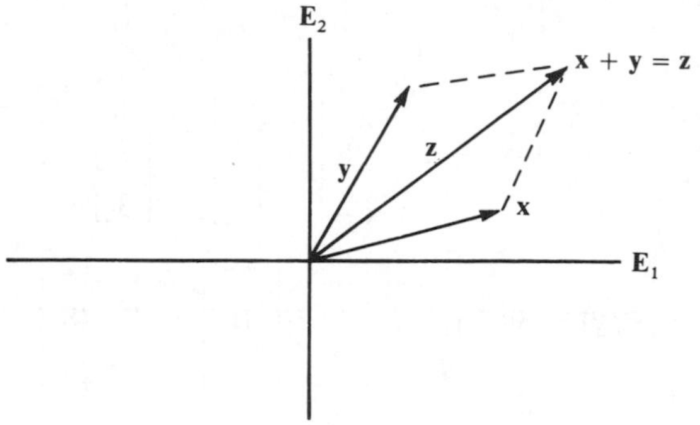

FIGURE 1.2.2. Parallelogram law for two dimensions.

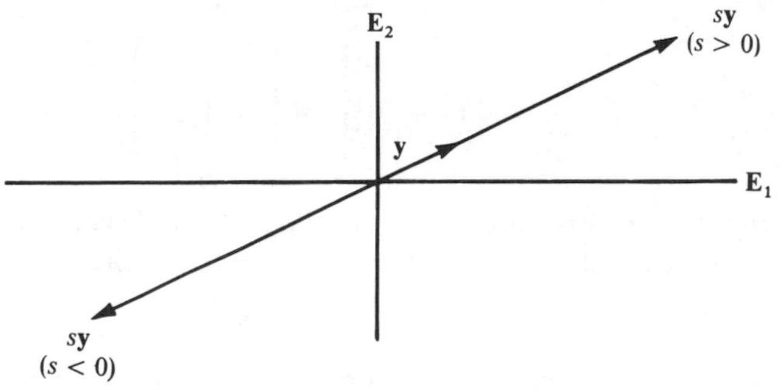

FIGURE 1.2.3. Multiplication of a vector by a scalar.

From equations (1.2.3) and (1.2.4) and the properties of real numbers, or from the geometry of vectors, it follows that

(1.2.5) $$\mathbf{x} + \mathbf{y} = \mathbf{y} + \mathbf{x} \qquad \text{(commutative law)}$$

(1.2.6) $$(\mathbf{x} + \mathbf{y}) + \mathbf{z} = \mathbf{x} + (\mathbf{y} + \mathbf{z}) \qquad \text{(associative law)}$$

(1.2.7) $s(c\mathbf{y}) = (sc)\mathbf{y} = (cs)\mathbf{y} = c(s\mathbf{y})$ (associative law)

(1.2.8) $s(\mathbf{x} + \mathbf{y}) = s\mathbf{x} + s\mathbf{y}$ (distributive law for vectors)

(1.2.9) $(s + c)\mathbf{y} = s\mathbf{y} + c\mathbf{y}$ (distributive law for scalars)

for vectors **x**, **y**, and **z** and real numbers s and c.

EXAMPLE 1.2.1. Let $s = 3$, $c = 4$,

$$\mathbf{x} = \begin{bmatrix} 1 \\ 2 \\ 0 \end{bmatrix}, \quad \mathbf{y} = \begin{bmatrix} -1 \\ 4 \\ 2 \end{bmatrix}, \quad \text{and} \quad \mathbf{z} = \begin{bmatrix} 1 \\ 1 \\ 1 \end{bmatrix}$$

Then

$$\mathbf{x} + \mathbf{y} = \begin{bmatrix} 1 \\ 2 \\ 0 \end{bmatrix} + \begin{bmatrix} -1 \\ 4 \\ 2 \end{bmatrix} = \begin{bmatrix} 1 + (-1) \\ 2 + 4 \\ 0 + 2 \end{bmatrix} = \begin{bmatrix} 0 \\ 6 \\ 2 \end{bmatrix}$$

$$(\mathbf{x} + \mathbf{y}) + \mathbf{z} = \begin{bmatrix} 0 \\ 6 \\ 2 \end{bmatrix} + \begin{bmatrix} 1 \\ 1 \\ 1 \end{bmatrix} = \begin{bmatrix} 1 \\ 7 \\ 3 \end{bmatrix}$$

$$s(c\mathbf{y}) = 3\left\{4\begin{bmatrix} -1 \\ 4 \\ 2 \end{bmatrix}\right\} = 3\begin{bmatrix} -4 \\ 16 \\ 8 \end{bmatrix} = \begin{bmatrix} -12 \\ 48 \\ 24 \end{bmatrix}$$

$$s(\mathbf{x} + \mathbf{y}) = 3\begin{bmatrix} 0 \\ 6 \\ 2 \end{bmatrix} = \begin{bmatrix} 0 \\ 18 \\ 6 \end{bmatrix}$$

$$(s + c)\mathbf{y} = (4 + 3)\begin{bmatrix} -1 \\ 4 \\ 2 \end{bmatrix} = \begin{bmatrix} -7 \\ 28 \\ 14 \end{bmatrix}$$

By extending the operations of vector addition and scalar multiplication, we define a linear combination of a set of vectors. A vector of the form

(1.2.10) $$\mathbf{y} = s_1\mathbf{y}_1 + \cdots + s_k\mathbf{y}_k$$

is called a *linear combination* of the vectors $\mathbf{y}_1, \mathbf{y}_2, \ldots, \mathbf{y}_k$, where $s_1, s_2, \ldots, s_k$ are arbitrary real numbers.

Vector Spaces. In general let $\mathbf{y}_1, \mathbf{y}_2, \ldots, \mathbf{y}_k$ be a finite, nonempty set of real n-vectors and $s_1, s_2, \ldots, s_k$ be an arbitrary set of scalars. Then the set V of all vectors of the form

$$V = \{\mathbf{v} \mid \mathbf{v} = s_1\mathbf{y}_1 + \cdots + s_k\mathbf{y}_k\}$$

is called a *real vector space*. The vectors $\mathbf{y}_1, \mathbf{y}_2, \ldots, \mathbf{y}_k$ are said to *span* (or generate) V.

EXAMPLE 1.2.2. Let

$$\mathbf{y}_1 = \begin{bmatrix} 1 \\ 0 \\ 0 \end{bmatrix} \quad \text{and} \quad \mathbf{y}_2 = \begin{bmatrix} 0 \\ 1 \\ 0 \end{bmatrix}$$

The set of all vectors V of the form $\mathbf{y} = s_1\mathbf{y}_1 + s_2\mathbf{y}_2$ span a plane in three-dimensional space. Any vector $\mathbf{y}$ in the two-dimensional subspace can be represented as a linear combination of the vectors $\mathbf{y}_1$ and $\mathbf{y}_2$.

The vectors $\mathbf{y}_1, \mathbf{y}_2, \ldots, \mathbf{y}_m$ of a vector space are *linearly dependent* if there exist real numbers $s_1, s_2, \ldots, s_m$ not all zero such that

(1.2.11) $$s_1\mathbf{y}_1 + \cdots + s_m\mathbf{y}_m = \mathbf{0}$$

Otherwise the set of vectors are *linearly independent.*

EXAMPLE 1.2.3. Let

$$\mathbf{y}_1 = \begin{bmatrix} 1 \\ 1 \\ 1 \end{bmatrix}, \quad \mathbf{y}_2 = \begin{bmatrix} 0 \\ 1 \\ -1 \end{bmatrix}, \quad \text{and} \quad \mathbf{y}_3 = \begin{bmatrix} 1 \\ 4 \\ -2 \end{bmatrix}$$

To determine whether the vectors $\mathbf{y}_1, \mathbf{y}_2$, and $\mathbf{y}_3$ are linearly dependent or independent, the preceding definition is employed. Consider the linear combination

$$s_1\mathbf{y}_1 + s_2\mathbf{y}_2 + s_3\mathbf{y}_3 = \mathbf{0}$$

Apply equations (1.2.3) and (1.2.4) for $\mathbf{y}_1$, $\mathbf{y}_2$, and $\mathbf{y}_3$ defined above, so that

$$s_1 \begin{bmatrix} 1 \\ 1 \\ 1 \end{bmatrix} + s_2 \begin{bmatrix} 0 \\ 1 \\ -1 \end{bmatrix} + s_3 \begin{bmatrix} 1 \\ 4 \\ -2 \end{bmatrix} = \begin{bmatrix} 0 \\ 0 \\ 0 \end{bmatrix}$$

$$\begin{bmatrix} s_1 \\ s_1 \\ s_1 \end{bmatrix} + \begin{bmatrix} 0 \\ s_2 \\ -s_2 \end{bmatrix} + \begin{bmatrix} s_3 \\ 4s_3 \\ -2s_3 \end{bmatrix} = \begin{bmatrix} 0 \\ 0 \\ 0 \end{bmatrix}$$

We thus obtain a system of three equations and three unknowns:

(1) $s_1 + \quad\quad s_3 = 0$
(2) $s_1 + s_2 + 4s_3 = 0$
(3) $s_1 - s_2 - 2s_3 = 0$

From equation (1), $s_1 = -s_3$. By substituting for s_1 in equation (2), $s_2 = -3s_3$. If s_1 and s_2 are defined in terms of s_3, equation (3) is satisfied. If $s_3 \neq 0$, there exist real numbers s_1, s_2, and s_3 not all equal to zero such that

$$s_1\mathbf{y}_1 + s_2\mathbf{y}_2 + s_3\mathbf{y}_3 = \mathbf{0}$$

Thus $\mathbf{y}_1, \mathbf{y}_2$, and $\mathbf{y}_3$ are seen to be linearly dependent. For example, $\mathbf{y}_1 + 3\mathbf{y}_2 - \mathbf{y}_3 = \mathbf{0}$.

EXAMPLE 1.2.4. As an example of a set of linearly independent vectors, let

$$\mathbf{y}_1 = \begin{bmatrix} 0 \\ 1 \\ 1 \end{bmatrix}, \quad \mathbf{y}_2 = \begin{bmatrix} 1 \\ 1 \\ -2 \end{bmatrix}, \quad \text{and} \quad \mathbf{y}_3 = \begin{bmatrix} 3 \\ 4 \\ 1 \end{bmatrix}$$

Using the definition

$$s_1\mathbf{y}_1 + s_2\mathbf{y}_2 + s_3\mathbf{y}_3 = \mathbf{0}$$

$$s_1 \begin{bmatrix} 0 \\ 1 \\ 1 \end{bmatrix} + s_2 \begin{bmatrix} 1 \\ 1 \\ -2 \end{bmatrix} + s_3 \begin{bmatrix} 3 \\ 4 \\ 1 \end{bmatrix} = \begin{bmatrix} 0 \\ 0 \\ 0 \end{bmatrix}$$

a system of simultaneous equations is obtained:

$$\begin{array}{ll} (1) & s_2 + 3s_3 = 0 \\ (2) & s_1 + s_2 + 4s_3 = 0 \\ (3) & s_1 - 2s_2 + s_3 = 0 \end{array}$$

From equation (1), $s_2 = -3s_3$. By substituting $-3s_3$ for s_2 in equation (2), $s_1 = -s_3$; by substituting for s_1 and s_2 in equation (3), $s_3 = 0$. Thus the only solution to the system is $s_1 = s_2 = s_3 = 0$, or $\mathbf{y}_1, \mathbf{y}_2$, and $\mathbf{y}_3$ are found to be linearly independent.

Linearly independent and linearly dependent vectors are fundamental to vector spaces and the study of multivariate analysis. As an example, let us suppose that a test is administered to N students where scores for k subtests are recorded. If the set of vectors $\mathbf{y}_1, \mathbf{y}_2, \ldots, \mathbf{y}_k$ is linearly independent and a vector

$$\mathbf{y}_{k+1} = \sum_{i=1}^{k} \mathbf{y}_i$$

(which represents a vector of total scores for the N subjects) is included with the set of independent vectors, the set of vectors now becomes dependent.

A set of linearly independent vectors that spans a vector space V is said to form a *basis* for the vector space V. The number of nonzero vectors in any basis is called the *dimension* or *rank* of the vector space. The space V of vectors consisting of only the $\mathbf{0}$ vector is by definition zero dimensional. The vectors

$$\mathbf{y}_1 = \begin{bmatrix} 1 \\ 0 \\ 0 \end{bmatrix}, \quad \mathbf{y}_2 = \begin{bmatrix} 0 \\ 1 \\ 0 \end{bmatrix}, \quad \text{and} \quad \mathbf{y}_3 = \begin{bmatrix} 0 \\ 0 \\ 1 \end{bmatrix}$$

form a basis for Euclidean three-dimensional space, and every vector in the space can be represented as a linear combination of the basis vectors. To indicate that the vectors $\mathbf{y}_1, \mathbf{y}_2$, and $\mathbf{y}_3$ span a three-dimensional Euclidean space, a subscript on V is usually employed to denote the dimension of the vector space. That is, $\mathbf{y}_1, \mathbf{y}_2$, and $\mathbf{y}_3$ span V_3. A subscript on V will also be used in this text to indicate that a vector $\mathbf{y}$ is an element of a vector space. In Example 1.2.2, the vectors $\mathbf{y}_1$ and $\mathbf{y}_2$ span V_2, a plane that is a two-dimensional vector space, but $\mathbf{y}_1$ and $\mathbf{y}_2$ are vectors in V_3. Since V_2 is contained in V_3, we write this as $V_2 \subset V_3$.

The following results (Halmos, 1958, pp. 11 and 13), establish the existence and uniqueness of a basic for a vector space.

(1.2.12) Every vector space has a basis.

(1.2.13) Every vector in a vector space has a unique representation as a linear combination of a given basis.

(1.2.14) Any two bases for a vector space have the same number of vectors.

A nonempty subset of vectors, V_r, of a real vector space, V_n, is a *subspace* or *linear manifold* of V_n if, for every pair of vectors $\mathbf{x}$ and $\mathbf{y}$ in V_r and scalars s_1 and s_2, $s_1\mathbf{x} + s_2\mathbf{y}$ is also in V_r. Thus every subspace of V_n is itself a vector space, and hence all results applied to V_n hold for subspaces. Any subspace V_r, where $0 < r < n$, is called a *proper* subspace. The subset of vectors containing only the zero vector and the subset containing the whole space are extreme examples of subspaces and are termed *improper* subspaces. V_2 is a proper subspace of V_3; that is, $V_2 \subset V_3$.

EXERCISES 1.2

1. Given the vectors

$$\mathbf{y}_1 = \begin{bmatrix} 1 \\ 1 \\ 1 \end{bmatrix} \quad \text{and} \quad \mathbf{y}_2 = \begin{bmatrix} 2 \\ 0 \\ -1 \end{bmatrix}$$

find the vectors

a. $2\mathbf{y}_1 + 3\mathbf{y}_2$

b. $s_1\mathbf{y}_1 + s_2\mathbf{y}_2$

c. $\mathbf{y}_3$ such that $3\mathbf{y}_1 - 2\mathbf{y}_2 + 4\mathbf{y}_3 = 0$

2. For the vectors and scalars defined in Example 1.2.1, verify equations (1.2.5) through (1.2.9).

3. Prove results (1.2.12) through (1.2.14).

4. Prove that

a. Any set of vectors containing $\mathbf{0}$ is linearly dependent.

b. Any subset of a linearly independent set is also linearly independent.

c. In a linearly dependent set of vectors, at least one of the vectors is a linear combination of the remaining vectors.

d. Any set of $n + 1$ vectors in a n-dimensional vector space is linearly dependent.

5. Show that the four vectors given below are linearly dependent.

$$\mathbf{y}_1 = \begin{bmatrix} 1 \\ 0 \\ 0 \end{bmatrix}, \quad \mathbf{y}_2 = \begin{bmatrix} 2 \\ 3 \\ 5 \end{bmatrix}, \quad \mathbf{y}_3 = \begin{bmatrix} 1 \\ 0 \\ 1 \end{bmatrix}, \quad \text{and} \quad \mathbf{y}_4 = \begin{bmatrix} 0 \\ 4 \\ 6 \end{bmatrix}$$

6. Are the following vectors linearly dependent or linearly independent?

$$\mathbf{y}_1 = \begin{bmatrix} 1 \\ 1 \\ 1 \end{bmatrix}, \quad \mathbf{y}_2 = \begin{bmatrix} 1 \\ 2 \\ 3 \end{bmatrix}, \quad \text{and} \quad \mathbf{y}_3 = \begin{bmatrix} 2 \\ 2 \\ 3 \end{bmatrix}$$

7. Do the vectors

$$\mathbf{y}_1 = \begin{bmatrix} 2 \\ 4 \\ 2 \end{bmatrix}, \quad \mathbf{y}_2 = \begin{bmatrix} 1 \\ 2 \\ 3 \end{bmatrix}, \quad \text{and} \quad \mathbf{y}_3 = \begin{bmatrix} 6 \\ 12 \\ 10 \end{bmatrix}$$

span the same space as the vectors

$$\mathbf{x}_1 = \begin{bmatrix} 0 \\ 0 \\ 2 \end{bmatrix} \quad \text{and} \quad \mathbf{x}_2 = \begin{bmatrix} 2 \\ 4 \\ 10 \end{bmatrix}$$

1.3 ORTHOGONALIZATION OF VECTOR SPACES

Inner Product. For Euclidean n-dimensional space, there is yet another fundamental operation for vectors called the inner (or scalar or dot) product of two vectors. The *inner product* of two vectors is the scalar quantity

(1.3.1) $$x_1y_1 + x_2y_2 + \cdots + x_ny_n = \sum_{i=1}^{n} x_iy_i = (\mathbf{x}, \mathbf{y})$$

The scalar product so defined is seen to satisfy the following equations for vectors $\mathbf{x}$, $\mathbf{y}$, and $\mathbf{z}$ and scalars s and c.

(1.3.2) $(\mathbf{x}, \mathbf{y}) = (\mathbf{y}, \mathbf{x})$ (commutative)
(1.3.3) $(\mathbf{x}, \mathbf{x}) \geq 0$ (positiveness)
(1.3.4) $(s\mathbf{x}, c\mathbf{y}) = sc(\mathbf{x}, \mathbf{y})$ (associative)
(1.3.5) $(\mathbf{x} + \mathbf{y}, \mathbf{z}) = (\mathbf{x}, \mathbf{z}) + (\mathbf{y}, \mathbf{z})$ (distributive)
(1.3.6) $(\mathbf{x} + \mathbf{y}, \mathbf{w} + \mathbf{z}) = (\mathbf{x}, \mathbf{w} + \mathbf{z}) + (\mathbf{y}, \mathbf{w} + \mathbf{z})$ (linear)

EXAMPLE 1.3.1. Let

$$\mathbf{x} = \begin{bmatrix} -1 \\ 3 \\ 2 \end{bmatrix}, \quad \mathbf{y} = \begin{bmatrix} 1 \\ 2 \\ 0 \end{bmatrix}, \quad \text{and} \quad \mathbf{z} = \begin{bmatrix} 1 \\ 1 \\ -2 \end{bmatrix}$$

and let $s = 2$ and $c = 3$. Then $(\mathbf{x}, \mathbf{y}) = (-1)(1) + (3)(2) + (2)(0) = 5$. We can also demonstrate the results of equations (1.3.2) through (1.3.6) in a similar manner.

Vector spaces with an inner product are termed *inner-product spaces.* Associated with the definition of an inner product is the discussion of lengths, distances, and angles in Euclidean n-dimensional space.

Lengths, Distances, and Angles. The Pythagorean theorem may be generalized to n dimensions. Figure 1.3.1 illustrates $d^2 = y_1^2 + y_2^2 = (\mathbf{y}, \mathbf{y})$ for two-dimensional space. In three-dimensional space, $d^2 = y_1^2 + y_2^2 + y_3^2 = (\mathbf{y}, \mathbf{y})$; see Figure 1.3.2.

In general it is evident that the square of the length of any vector $\mathbf{y}$ is the inner product of $\mathbf{y}$ with itself so that the *length* of $\mathbf{y}$, denoted by $\|\mathbf{y}\|$, is the positive square

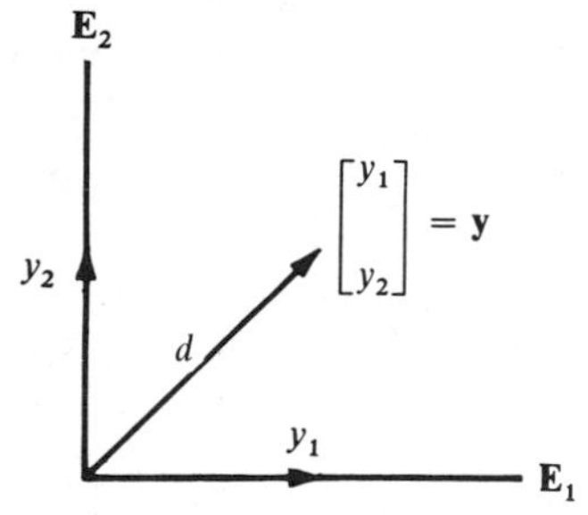

FIGURE 1.3.1. Pythagorean theorem in two-dimensional space.

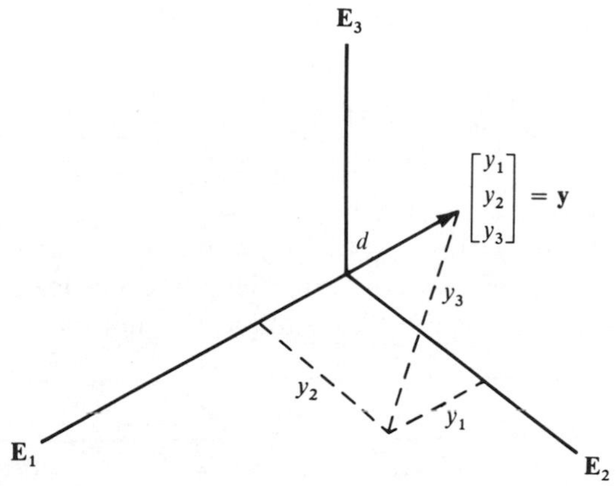

FIGURE 1.3.2. Pythagorean theorem in three-dimensional space.

root of its own inner product:

(1.3.7) $$\|\mathbf{y}\| = \sqrt{(\mathbf{y}, \mathbf{y})}$$

The length, or *distance*, between two vectors is given by

(1.3.8) $$\|\mathbf{x} - \mathbf{y}\| = \sqrt{(\mathbf{x} - \mathbf{y}, \mathbf{x} - \mathbf{y})}$$

Diagrammatically, equation (1.3.8) is represented in Figure 1.3.3.

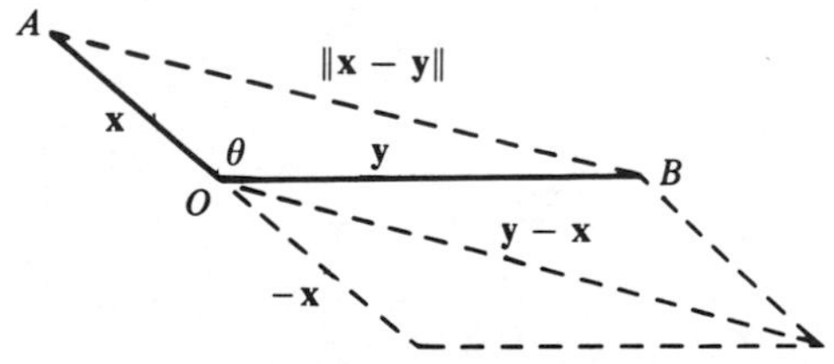

FIGURE 1.3.3. Distance between two vectors.

For determining the squared distance between **y** and **x** in terms of the squared lengths of **y** and **x**, the law of cosines is applied to the triangle AOB in Figure 1.3.3;

$$\|\mathbf{y} - \mathbf{x}\|^2 = \|\mathbf{x}\|^2 + \|\mathbf{y}\|^2 - 2\|\mathbf{x}\|\,\|\mathbf{y}\| \cos \theta$$

Equating this with the definition of distance in (1.3.8), the expression for the *cosine of the angle* between the two nonzero vectors $\mathbf{x}$ and $\mathbf{y}$ is obtained:

$$\text{(1.3.9)} \qquad \cos\theta = \frac{(\mathbf{x}, \mathbf{y})}{\|\mathbf{x}\| \, \|\mathbf{y}\|} \qquad 0° \le \theta \le 180°$$

EXAMPLE 1.3.2. Let

$$\mathbf{x} = \begin{bmatrix} -1 \\ 1 \\ 2 \end{bmatrix} \quad \text{and} \quad \mathbf{y} = \begin{bmatrix} 1 \\ 0 \\ -1 \end{bmatrix}$$

The distance between $\mathbf{x}$ and $\mathbf{y}$ is then $\|\mathbf{x} - \mathbf{y}\| = \sqrt{(\mathbf{x} - \mathbf{y}, \mathbf{x} - \mathbf{y})} = \sqrt{14}$ and the cosine of the angle between $\mathbf{x}$ and $\mathbf{y}$ is

$$\cos\theta = \frac{(\mathbf{x}, \mathbf{y})}{\|\mathbf{x}\| \, \|\mathbf{y}\|} = \frac{-3}{\sqrt{6}\sqrt{2}} = \frac{-\sqrt{3}}{2}$$

so that the angle between $\mathbf{x}$ and $\mathbf{y}$ is $\theta = \cos^{-1}(-\sqrt{3}/2) = 150°$. If the vectors $\mathbf{x}$ and $\mathbf{y}$ in (1.3.9) have *unit length*, so that $\|\mathbf{x}\| = \|\mathbf{y}\| = 1$, $\cos\theta$ is just the inner product of the vectors $\mathbf{x}$ and $\mathbf{y}$. Any n-vector $\mathbf{y}$ not of unit length can be made so if each element of $\mathbf{y}$ is divided by the length of $\mathbf{y}$. For this example, the unit vectors in the direction of $\mathbf{x}$ and $\mathbf{y}$ are

$$\mathbf{x}^* = \frac{\mathbf{x}}{\|\mathbf{x}\|} = \begin{bmatrix} -1/\sqrt{6} \\ 1/\sqrt{6} \\ 2/\sqrt{6} \end{bmatrix} \quad \text{and} \quad \mathbf{y}^* = \frac{\mathbf{y}}{\|\mathbf{y}\|} = \begin{bmatrix} 1/\sqrt{2} \\ 0/\sqrt{2} \\ -1/\sqrt{2} \end{bmatrix}$$

since $(\mathbf{x}^*, \mathbf{x}^*) = (\mathbf{y}^*, \mathbf{y}^*) = 1$ and $\cos\theta = (\mathbf{x}^*, \mathbf{y}^*) = -\sqrt{3/2}$. Alternatively, we may express $\cos\theta$ in terms of direction cosines. If $\mathbf{y}$ is a vector in an n-dimensional space, the *direction cosines* of the vector $\mathbf{y}$ are the cosines of the angles made between $\mathbf{y}$ and each of its reference axes. If the vector $\mathbf{y}$ is of unit length and the reference axes are orthogonal, the direction cosines are the coordinates of the vector $\mathbf{y}$. In Figure 1.3.4, the angles associated with the direction cosines are $\cos\alpha_1$, $\cos\alpha_2$, and $\cos\alpha_3$; they are seen to satisfy the relationship $\cos^2\alpha_1 + \cos^2\alpha_2 + \cos^2\alpha_3 = 1$. Using (1.3.9) and the definition of direction cosines, the direction cosines of $\mathbf{y}$ in Figure 1.3.4 are

$$\cos\alpha_1 = \frac{y_1}{\|\mathbf{y}\|}$$

$$\cos\alpha_2 = \frac{y_2}{\|\mathbf{y}\|}$$

$$\cos\alpha_3 = \frac{y_3}{\|\mathbf{y}\|}$$

Similarly, for two vectors,

$$\cos\theta = \frac{(\mathbf{y}, \mathbf{x})}{\|\mathbf{x}\| \, \|\mathbf{y}\|} = \frac{y_1 x_1}{\|\mathbf{x}\| \, \|\mathbf{y}\|} + \frac{y_2 x_2}{\|\mathbf{x}\| \, \|\mathbf{y}\|} + \frac{y_3 x_3}{\|\mathbf{x}\| \, \|\mathbf{y}\|}$$

$$= \cos\alpha_1 \cos\beta_1 + \cos\alpha_2 \cos\beta_2 + \cos\alpha_3 \cos\beta_3$$

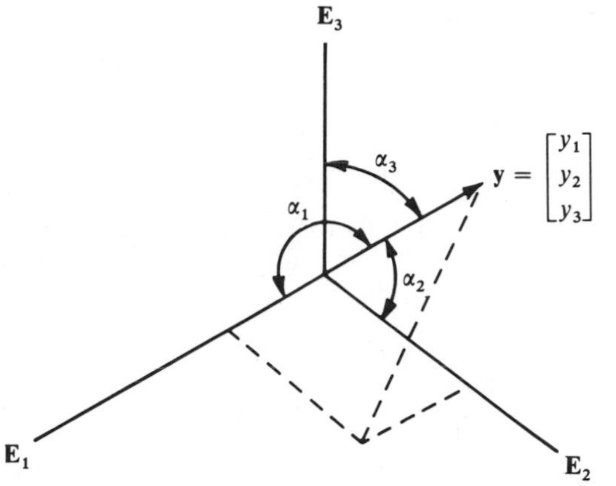

FIGURE 1.3.4. Direction cosines in three-dimensional space.

as indicated in Figure 1.3.5. In n dimensions, the formula for $\cos\theta$, in terms of direction cosines, now becomes immediate:

$$\cos\theta = \sum_{i=1}^{n} \cos\alpha_i \cos\beta_i \tag{1.3.10}$$

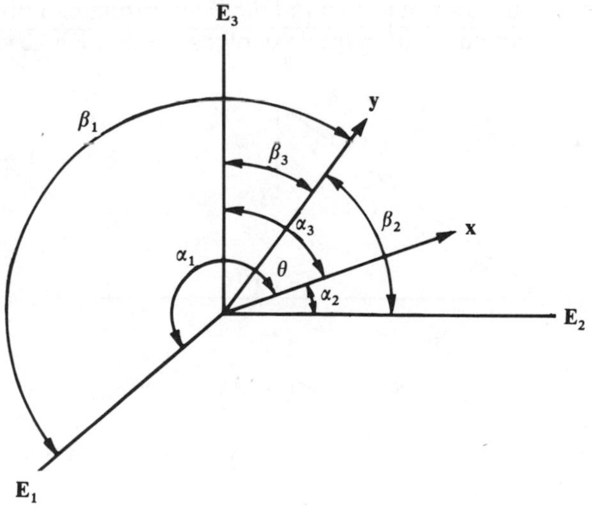

FIGURE 1.3.5. Direction cosines for two vectors.

Orthogonal Vectors. Two nonzero vectors **x** and **y** are *orthogonal*, written as ($\mathbf{x} \perp \mathbf{y}$), if and only if their inner product is zero; that is,

$$(\mathbf{x} \perp \mathbf{y}) \quad \text{if and only if} \quad (\mathbf{x}, \mathbf{y}) = 0 \tag{1.3.11}$$

If each vector has unit length ($\|\mathbf{x}\| = \|\mathbf{y}\| = 1$), the vectors are said to be *orthonormal.*

EXAMPLE 1.3.3. Let

$$\mathbf{x} = \begin{bmatrix} -1 \\ 2 \\ -4 \end{bmatrix} \quad \text{and} \quad \mathbf{y} = \begin{bmatrix} -4 \\ 0 \\ 1 \end{bmatrix}$$

Then $(\mathbf{x}, \mathbf{y}) = (-1)(-4) + (2)(0) + (-4)(1) = 0$; however, the vectors are not of unit length. That is, $\|\mathbf{x}\| = [(-1)^2 + (2)^2 + (-4)^2]^{1/2} = \sqrt{21}$ and $\|\mathbf{y}\| = \sqrt{17}$. The vectors

$$\mathbf{x}^* = \frac{1}{\sqrt{21}} \begin{bmatrix} -1 \\ 2 \\ -4 \end{bmatrix} \quad \text{and} \quad \mathbf{y}^* = \frac{1}{\sqrt{17}} \begin{bmatrix} -4 \\ 0 \\ 1 \end{bmatrix}$$

are of unit length. The vectors $\mathbf{x}$ and $\mathbf{y}$ are often called *normalized* vectors when they are represented by $\mathbf{x}^*$ and $\mathbf{y}^*$.

A basis for a vector space V is called an *orthogonal basis* when every pair of vectors in the set is found to be pairwise orthogonal; the basis is considered an *orthonormal basis* if each vector also has unit length.

The Gram-Schmidt Process. Fundamental to the concept of orthogonal vectors is the *orthogonal projection* of a vector. In two-dimensional space, let us consider the vectors $\mathbf{y}$ and $\mathbf{x}$ given in Figure 1.3.6. The orthogonal projection of $\mathbf{y}$ on $\mathbf{x}$, denoted by $P_{\mathbf{x}}\mathbf{y}$, is some constant multiple $s\mathbf{x}$ of $\mathbf{x}$ such that $P_{\mathbf{x}}\mathbf{y} \perp (\mathbf{y} - P_{\mathbf{x}}\mathbf{y})$. Thus,

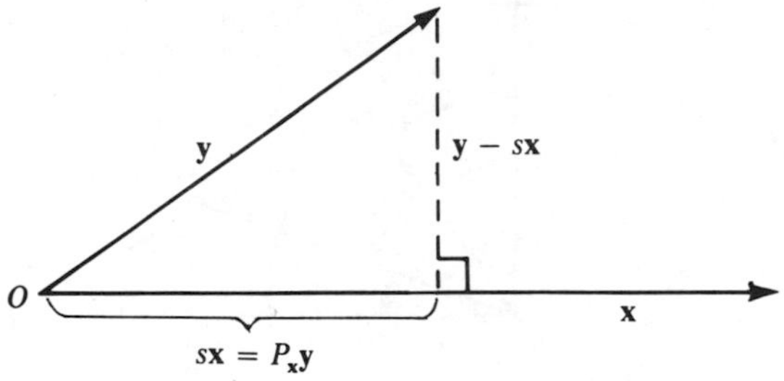

FIGURE 1.3.6. Orthogonal projection of $\mathbf{y}$ on $\mathbf{x}$, where $P_{\mathbf{x}}\mathbf{y} = s\mathbf{x}$.

if $(P_{\mathbf{x}}\mathbf{y}, \mathbf{y} - P_{\mathbf{x}}\mathbf{y}) = 0$ or $(s\mathbf{x}, \mathbf{y} - s\mathbf{x}) = 0$, $s = (\mathbf{y}, \mathbf{x})/(\mathbf{x}, \mathbf{x}) = (\mathbf{y}, \mathbf{x})/\|\mathbf{x}\|^2$; so that the projection of $\mathbf{y}$ on $\mathbf{x}$ is

$$P_{\mathbf{x}}\mathbf{y} = s\mathbf{x} = \frac{(\mathbf{y}, \mathbf{x})\mathbf{x}}{\|\mathbf{x}\|^2} \tag{1.3.12}$$

EXAMPLE 1.3.4. Let

$$\mathbf{x} = \begin{bmatrix} 1 \\ 1 \\ 1 \end{bmatrix} \quad \text{and} \quad \mathbf{y} = \begin{bmatrix} 1 \\ 4 \\ 2 \end{bmatrix}$$

Then

$$P_{\mathbf{x}}\mathbf{y} = \frac{(\mathbf{y}, \mathbf{x})\mathbf{x}}{\|\mathbf{x}\|^2} = \frac{7}{3}\begin{bmatrix} 1 \\ 1 \\ 1 \end{bmatrix}$$

With this example, it is important to note that the coefficient $(\mathbf{y}, \mathbf{x})/\|\mathbf{x}\|^2$ is just the average of the elements of $\mathbf{y}$. This is always the case when projecting onto a vector of all 1s—that is, onto an *equiangular vector*.

To obtain an orthogonal basis for any vector space V, spanned by any set of vectors $\mathbf{x}_1, \mathbf{x}_2, \ldots, \mathbf{x}_k$, the preceding projection process is employed. It is now outlined sequentially. Set $\mathbf{y}_1 = \mathbf{x}_1$, then

$$\mathbf{y}_2 = \mathbf{x}_2 - P_{\mathbf{y}_1}\mathbf{x}_2 = \mathbf{x}_2 - \frac{(\mathbf{x}_2, \mathbf{y}_1)\mathbf{y}_1}{\|\mathbf{y}_1\|^2} \qquad \mathbf{y}_2 \perp \mathbf{y}_1$$

$$\mathbf{y}_3 = \mathbf{x}_3 - P_{\mathbf{y}_1}\mathbf{x}_3 - P_{\mathbf{y}_2}\mathbf{x}_3$$

$$= \mathbf{x}_3 - \frac{(\mathbf{x}_3, \mathbf{y}_1)\mathbf{y}_1}{\|\mathbf{y}_1\|^2} - \frac{(\mathbf{x}_3, \mathbf{y}_2)\mathbf{y}_2}{\|\mathbf{y}_2\|^2} \qquad \mathbf{y}_3 \perp \mathbf{y}_2 \perp \mathbf{y}_1$$

or, generally,

$$\mathbf{y}_i = \mathbf{x}_i - \sum_{j=1}^{i-1} c_{ij}\mathbf{y}_j \qquad \text{where } c_{ij} = \frac{(\mathbf{x}_i, \mathbf{y}_j)}{\|\mathbf{y}_j\|^2}$$

deleting those $\mathbf{y}_i$ for which $\mathbf{y}_i = \mathbf{0}$.

This process of finding an orthogonal basis $\mathbf{y}_1, \mathbf{y}_2, \ldots, \mathbf{y}_n$ for any given set of vectors $\mathbf{x}_1, \mathbf{x}_2, \ldots, \mathbf{x}_k$ is called the *Gram-Schmidt orthogonalization* process. To find an orthonormal basis, the orthogonal basis must be normalized.

EXAMPLE 1.3.5. Let V be spanned by

$$\mathbf{x}_1 = \begin{bmatrix} 1 \\ -1 \\ 1 \\ 0 \\ 1 \end{bmatrix}, \quad \mathbf{x}_2 = \begin{bmatrix} 2 \\ 0 \\ 4 \\ 1 \\ 2 \end{bmatrix}, \quad \mathbf{x}_3 = \begin{bmatrix} 1 \\ 1 \\ 3 \\ 1 \\ 1 \end{bmatrix}, \quad \text{and} \quad \mathbf{x}_4 = \begin{bmatrix} 6 \\ 2 \\ 3 \\ -1 \\ 1 \end{bmatrix}$$

To find an orthonormal basis, the Gram-Schmidt process is used as follows:

$$\mathbf{y}_1 = \mathbf{x}_1 = \begin{bmatrix} 1 \\ -1 \\ 1 \\ 0 \\ 1 \end{bmatrix} \quad \text{and} \quad \mathbf{y}_2 = \mathbf{x}_2 - \frac{(\mathbf{x}_2, \mathbf{y}_1)\mathbf{y}_1}{\|\mathbf{y}_1\|^2}$$

$$\mathbf{y}_2 = \begin{bmatrix} 2 \\ 0 \\ 4 \\ 1 \\ 2 \end{bmatrix} - \frac{\left\{ \begin{bmatrix} 2 \\ 0 \\ 4 \\ 1 \\ 2 \end{bmatrix}, \begin{bmatrix} 1 \\ -1 \\ 1 \\ 0 \\ 1 \end{bmatrix} \right\}}{\|[1, -1, 1, 0, 1]\|^2} \begin{bmatrix} 1 \\ -1 \\ 1 \\ 0 \\ 1 \end{bmatrix}$$

$$= \begin{bmatrix} 2 \\ 0 \\ 4 \\ 1 \\ 2 \end{bmatrix} - \frac{8}{4} \begin{bmatrix} 1 \\ -1 \\ 1 \\ 0 \\ 1 \end{bmatrix}$$

$$= \begin{bmatrix} 0 \\ 2 \\ 2 \\ 1 \\ 0 \end{bmatrix}$$

$$\mathbf{y}_3 = \mathbf{x}_3 - \frac{(\mathbf{x}_3, \mathbf{y}_1)\mathbf{y}_1}{\|\mathbf{y}_1\|^2} - \frac{(\mathbf{x}_3, \mathbf{y}_2)\mathbf{y}_2}{\|\mathbf{y}_2\|^2} = \begin{bmatrix} 0 \\ 0 \\ 0 \\ 0 \\ 0 \end{bmatrix}$$

so delete $\mathbf{y}_3$:

$$\mathbf{y}_4 = \begin{bmatrix} 6 \\ 2 \\ 3 \\ -1 \\ 1 \end{bmatrix} - \frac{(\mathbf{x}_4, \mathbf{y}_1)\mathbf{y}_1}{\|\mathbf{y}_1\|^2} - \frac{(\mathbf{x}_4, \mathbf{y}_2)\mathbf{y}_2}{\|\mathbf{y}_2\|^2}$$

$$= \begin{bmatrix} 6 \\ 2 \\ 3 \\ -1 \\ 1 \end{bmatrix} - \frac{8}{4} \begin{bmatrix} 1 \\ -1 \\ 1 \\ 0 \\ 1 \end{bmatrix} - \frac{9}{9} \begin{bmatrix} 0 \\ 2 \\ 2 \\ 1 \\ 0 \end{bmatrix} = \begin{bmatrix} 4 \\ 2 \\ -1 \\ -2 \\ -1 \end{bmatrix}$$

An orthogonal basis for V is $\{\mathbf{y}_1, \mathbf{y}_2, \mathbf{y}_4\}$ and an orthonormal basis is $\mathbf{y}_1^* = \mathbf{y}_1/\sqrt{4}$, $\mathbf{y}_2^* = \mathbf{y}_2/3$, $\mathbf{y}_3^* = \mathbf{y}_3/\sqrt{26}$. From the example, we can conclude that the dimension of V is three.

By employing the Gram-Schmidt process, we can extend a basis for any subspace $V_r \subset V_n$ to an orthonormal basis for V_n.

Orthogonal Spaces. Having defined orthogonality of vectors, we move into a discussion of vectors orthogonal to vector spaces and of the orthogonality of vector spaces. Let us suppose that V is a subspace in V_n. A vector $\mathbf{y} \in V_n$ is *orthogonal to the subspace* V, written as $(\mathbf{y} \perp V)$, if and only if $\mathbf{y}$ is orthogonal to every vector spanning V. Thus, if $\mathbf{x}_1, \mathbf{x}_2, \ldots, \mathbf{x}_k$ span V,

(1.3.13) $$(\mathbf{y} \perp V) \quad \text{if and only if} \quad \mathbf{y} \perp \mathbf{x}_i \qquad (i = 1, \ldots, k)$$

By extending orthogonality to spaces, the set of all vectors orthogonal to a given subspace $V \subset V_n$ is the subspace $V^\perp$. $V^\perp$ is called the *orthocomplement* of V in the space of V_n. Hence

(1.3.14) $$V^\perp = \{\mathbf{x}^* \in V_n \mid (\mathbf{x}^*, \mathbf{x}) = 0 \quad \text{for all } \mathbf{x} \in V\}$$

The geometry of the orthocomplement is seen in Figure 1.3.7. Every vector $\mathbf{x}^* \in V^\perp$ is orthogonal to every vector $\mathbf{x} \in V$. $V^\perp$ is clearly seen to be a subspace of V_n since, for any pair of vectors $\mathbf{x}_1^*$ and $\mathbf{x}_2^*$ in $V^\perp$, $(s_1\mathbf{x}_1^* + s_2\mathbf{x}_2^*, \mathbf{x}) = 0$ for any $\mathbf{x} \in V$, indicating that $s_1^*\mathbf{x}_1^* + s_2\mathbf{x}_2^* \in V^\perp$, or $V^\perp$ is a subspace.

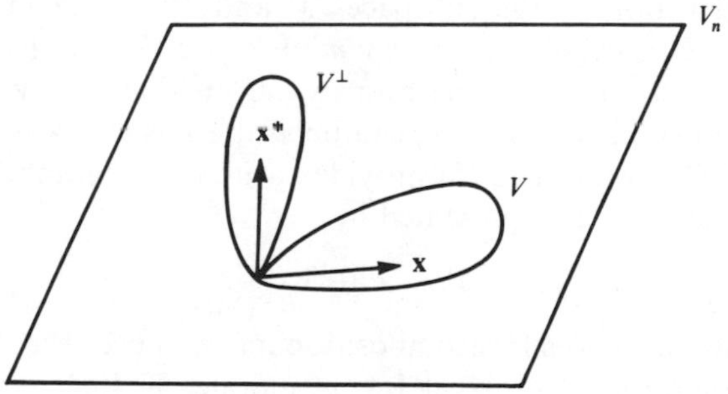

FIGURE 1.3.7. The orthocomplement of V relative to V_n, the subspace $V^\perp$.

The subsequent discussion allows us to illustrate how the decomposition of a vector into unique orthogonal components is performed.

(1.3.15) For any vector $\mathbf{y} \in V_n$, there exist unique vectors $\mathbf{x} \in V$ and $\mathbf{x}^* \in V^\perp$ such that $\mathbf{y} = \mathbf{x} + \mathbf{x}^*$. Furthermore, the vector $\mathbf{x}$, called the *orthogonal projection of* $\mathbf{y}$ *onto* V and denoted by $P_V\mathbf{y}$, minimizes $\|\mathbf{y} - \mathbf{z}\|$ among all vectors $\mathbf{z} \in V$.

The proof of this theorem follows that of Scheffé (1959, p. 383). Let $\mathbf{x}_1, \mathbf{x}_2, \ldots, \mathbf{x}_r$ be an orthonormal basis for V. Let

$$\mathbf{x} = \sum_{i=1}^{r} (\mathbf{y}, \mathbf{x}_i)\mathbf{x}_i$$

denote the projection of $\mathbf{y}$ onto V and let $\mathbf{x}^* = \mathbf{y} - \mathbf{x}$. Then $\mathbf{x} \in V$, since it is a linear combination of the basis vectors of V, and $\mathbf{x}^* \in V^\perp$. To show that $\mathbf{x}^* \in V^\perp$, where $V^\perp = \{\mathbf{v} \in V_n \mid \mathbf{v} \perp V\}$ and $V^\perp = \{\mathbf{v} \in V_n \mid \mathbf{v} \perp \mathbf{x}_i\}$ $(i = 1, \ldots, r)$, consider any $\mathbf{x}_j \in V$,

$$(\mathbf{x}_j, \mathbf{x}^*) = (\mathbf{x}_j, \mathbf{y}) - \sum_{i=1}^{r} (\mathbf{y}, \mathbf{x}_i)\mathbf{x}_i = (\mathbf{x}_j, \mathbf{y}) - (\mathbf{x}_j, \mathbf{y}) = 0$$

since, for $i \neq j$, $(\mathbf{x}_j, \mathbf{x}_i) = 0$. Hence $\mathbf{x}^* \perp V$ or $\mathbf{x}^* \in V^\perp$ since $\mathbf{x}^*$ is orthogonal to all basis vectors $\mathbf{x}_j$. This completes the proof of existence.

To illustrate uniqueness, suppose $\mathbf{y} = \mathbf{x} + \mathbf{x}^* = \mathbf{w} + \mathbf{w}^*$, where $\mathbf{x}$ and $\mathbf{w}$ are in V and $\mathbf{x}^*$ and $\mathbf{w}^*$ are in $V^\perp$. Then $\mathbf{x} - \mathbf{w} = \mathbf{w}^* - \mathbf{x}^*$. Since V and $V^\perp$ are subspaces, $(\mathbf{x} - \mathbf{w}) \in V$ and $(\mathbf{w}^* - \mathbf{x}^*) \in V^\perp$. The only self-orthogonal vector is the zero vector, therefore $(\mathbf{x} - \mathbf{w}) = \mathbf{0}$ and $(\mathbf{w}^* - \mathbf{x}^*) = \mathbf{0}$, or $\mathbf{x} = \mathbf{w}$ and $\mathbf{w}^* = \mathbf{x}^*$, thus completing the proof of uniqueness.

To indicate minimization, we must illustrate that the obtained $\mathbf{x}$ is closest to $\mathbf{y}$. Consider $\|\mathbf{y} - \mathbf{z}\|^2$ for any $\mathbf{z} \in V$, the $\|\mathbf{y} - \mathbf{z}\|^2 = \|(\mathbf{y} - \mathbf{x}) + (\mathbf{x} - \mathbf{z})\|^2 = \|\mathbf{y} - \mathbf{x}\|^2 + \|\mathbf{x} - \mathbf{z}\|^2 + 2[(\mathbf{x} - \mathbf{z}), (\mathbf{y} - \mathbf{x})]$. However, since $(\mathbf{x} - \mathbf{z}) \in V$ and $(\mathbf{y} - \mathbf{x}) \in V^\perp$, to minimize $\|\mathbf{y} - \mathbf{z}\|^2$, $\mathbf{z} = \mathbf{x}$ must be chosen. This completes the proof.

In summary, for any $\mathbf{y} \in V_n$, there exist unique vectors $\mathbf{x} \in V$ and $\mathbf{x}^* \in V^\perp$ where $V \subset V_n$ such that $\mathbf{y} = \mathbf{x} + \mathbf{x}^*$ and the $P_V\mathbf{y} = \mathbf{x}$ is such that the $\|\mathbf{y} - \mathbf{x}\|$ is minimal. For such a decomposition, $\mathbf{y} = \mathbf{x} + \mathbf{x}^*$ for unique vectors $\mathbf{x} \in V$ and $\mathbf{x}^* \in V^\perp$. More generally, if V and W are any two subspaces of V_n, the subspace of vectors

$$U = \{\mathbf{y} \mid \mathbf{y} = \mathbf{v} + \mathbf{w}; \mathbf{v} \in V \text{ and } \mathbf{w} \in W\}$$

is called the sum of the vector spaces V and W and is written $U = V + W$. If, in addition, the intersection of the subspaces V and W is the null space, so that $W \cap V = \{\mathbf{0}\}$, then U is called the *direct sum* of V and W, denoted by $U = V \oplus W$. Although $V \cap W = \{\mathbf{0}\}$ does not necessarily imply that $V \perp W$, if $V \perp W$, then $V \cap W = \{\mathbf{0}\}$. From (1.3.15), any $\mathbf{y} \in V_n$ can be written as $\mathbf{y} = \mathbf{x} + \mathbf{x}^*$, for $\mathbf{x} \in V$ and $\mathbf{x}^* \in V^\perp$; but $V \perp V^\perp$, so that (1.3.15) provides a unique decomposition of V_n for a given vector subspace $V \subset V_n$ represented by

(1.3.16) $$V_n = V \oplus V^\perp$$

which is called an orthogonal decomposition of V_n since $V \perp V^\perp$. Furthermore, given V, the decomposition of V_n and $\mathbf{y}$ is unique. If $V_n = \{\mathbf{v}_1, \mathbf{v}_2, \ldots, \mathbf{v}_n\}$ and $V_r = \{\mathbf{v}_1, \mathbf{v}_2, \ldots, \mathbf{v}_r\}$ are orthonormal bases for V_n and V_r, where $V_r \subset V_n$,

$$\mathbf{y} = P_V\mathbf{y} + P_{V^\perp}\mathbf{y} = \sum_{i=1}^{r} (\mathbf{y}, \mathbf{v}_i)\mathbf{v}_i + \sum_{i=r+1}^{n} (\mathbf{y}, \mathbf{v}_i)\mathbf{v}_i$$

where the dimension of $V^\perp$ is $n - r$ and the dimension of V is r. By using the projection operation, a vector $\mathbf{y}$ may be constructed from a component in an r-dimensional subspace "close" to $\mathbf{y}$ as measured by the residual component $P_{V^\perp}\mathbf{y} = \mathbf{x}^*$ (see Figure 1.3.8).

EXAMPLE 1.3.6. Let

$$V = \begin{bmatrix} -1 \\ 1 \\ 2 \end{bmatrix} \quad \text{and} \quad \mathbf{y} = \begin{bmatrix} 1 \\ 0 \\ -1 \end{bmatrix}$$

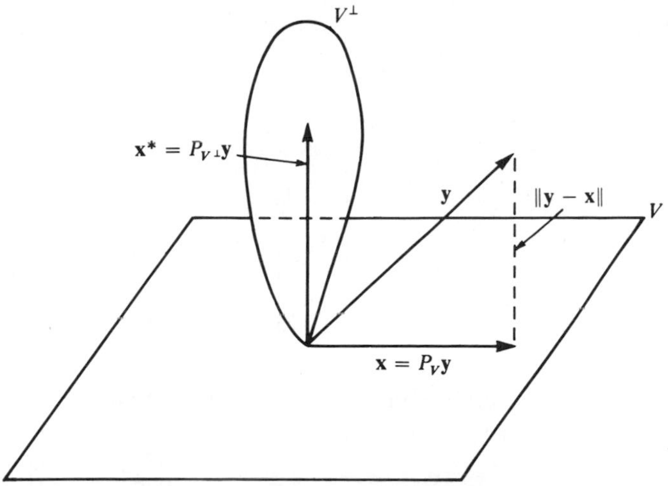

FIGURE 1.3.8. The unique decomposition of **y**.

To express $\mathbf{y}$ as $\mathbf{y} = \mathbf{x} + \mathbf{x}^*$, where $\mathbf{x} \in V$ and $\mathbf{x}^* \in V^\perp$, (1.3.15) is employed.

$$\mathbf{x}^* = \mathbf{y} - P_V\mathbf{y} = \begin{bmatrix} 1 \\ 0 \\ -1 \end{bmatrix} - \begin{bmatrix} 1/2 \\ -1/2 \\ -1 \end{bmatrix} = \begin{bmatrix} 1/2 \\ 1/2 \\ 0 \end{bmatrix}$$

so

$$\mathbf{y} = \mathbf{x} + \mathbf{x}^* = \begin{bmatrix} 1/2 \\ -1/2 \\ -1 \end{bmatrix} + \begin{bmatrix} 1/2 \\ 1/2 \\ 0 \end{bmatrix} = \begin{bmatrix} 1 \\ 0 \\ -1 \end{bmatrix}$$

and $\mathbf{x} \in V$ and $\mathbf{x}^* \in V^\perp$. Furthermore, the vector $\mathbf{y} \in V_3$, so the dimension of $V^\perp$ is two.

To find an orthogonal basis for $V^\perp$ let

$$V^\perp = \left\{ \begin{bmatrix} 1/2 \\ 1/2 \\ 0 \end{bmatrix} \quad \begin{bmatrix} v_1 \\ v_2 \\ v_3 \end{bmatrix} \right\}$$

The vector $\mathbf{v} \in V^\perp$ must be orthogonal to the vectors $\mathbf{x}^* \in V^\perp$ and $\mathbf{x} \in V$. Thus $(1/2)v_1 - (1/2)v_2 - v_3 = 0$ and $(1/2)v_1 + (1/2)v_2 = 0$, or $v_2 = -v_1$ and $v_3 = v_1$. By letting $v_1 = 1$, $v_2 = -1$, and $v_3 = 1$, a basis for $V^\perp$ is

$$V^\perp = \left\{ \begin{bmatrix} 1/2 \\ 1/2 \\ 0 \end{bmatrix} \quad \begin{bmatrix} 1 \\ -1 \\ 1 \end{bmatrix} \right\}$$

EXAMPLE 1.3.7. Let

$$\mathbf{y} \in V_3 \quad \text{and} \quad V = \left\{ \begin{bmatrix} 1 \\ 0 \\ -1 \end{bmatrix} \quad \begin{bmatrix} 0 \\ 1 \\ -1 \end{bmatrix} \right\} = \{\mathbf{x}_1, \mathbf{x}_2\}$$

To find $V^\perp$ where

$$\begin{aligned} V^\perp &= \{\mathbf{v} \in V_3 \mid (\mathbf{v}, \mathbf{x}) = 0 \quad \text{for all } \mathbf{x} \in V_2\} \\ &= \{\mathbf{v} \in V_3 \mid \mathbf{v} \perp V\} \\ &= \{\mathbf{v} \in V_3 \mid \mathbf{v} \perp \mathbf{x}_i\} \qquad (i = 1, 2) \end{aligned}$$

a vector

$$\mathbf{v} = \begin{bmatrix} v_1 \\ v_2 \\ v_3 \end{bmatrix}$$

such that $(\mathbf{v}, \mathbf{x}_1) = 0$ and $(\mathbf{v}, \mathbf{x}_2) = 0$ must be found. This implies that $v_1 - v_3 = 0$, or $v_1 = v_3$ and $v_2 - v_3 = 0$, or $v_2 = v_3$. Thus

$$V^\perp = \begin{bmatrix} 1 \\ 1 \\ 1 \end{bmatrix} = \{\mathbf{1}\} \qquad \text{where } V_3 = V^\perp \oplus V$$

Now the

$$P_{\mathbf{1}}\mathbf{y} = \begin{bmatrix} \bar{y} \\ \bar{y} \\ \bar{y} \end{bmatrix} \quad \text{and} \quad P_V\mathbf{y} = \mathbf{y} - P_{\mathbf{1}}\mathbf{y} = \begin{bmatrix} y_1 - \bar{y} \\ y_2 - \bar{y} \\ y_3 - \bar{y} \end{bmatrix} = \frac{1}{3}\begin{bmatrix} 2y_1 - y_2 - y_3 \\ 2y_2 - y_1 - y_3 \\ 2y_3 - y_1 - y_2 \end{bmatrix}$$

Alternatively, an orthogonal basis for V is

$$\left\{\begin{bmatrix} 1 \\ 0 \\ -1 \end{bmatrix} \begin{bmatrix} -1/2 \\ 1 \\ -1/2 \end{bmatrix}\right\} = \{\mathbf{v}_1^*, \mathbf{v}_2^*\}$$

and

$$\begin{aligned} P_V\mathbf{y} = P_{\mathbf{v}_1^*}\mathbf{y} + P_{\mathbf{v}_2^*}\mathbf{y} &= \frac{y_1 - y_3}{2}\begin{bmatrix} 1 \\ 0 \\ -1 \end{bmatrix} + \frac{2y_2 - y_1 - y_3}{3}\begin{bmatrix} -1/2 \\ 1 \\ -1/2 \end{bmatrix} \\ &= \frac{1}{3}\begin{bmatrix} 2y_1 - y_2 - y_3 \\ 2y_2 - y_1 - y_3 \\ 2y_3 - y_1 - y_2 \end{bmatrix} = \begin{bmatrix} y_1 - \bar{y} \\ y_2 - \bar{y} \\ y_3 - \bar{y} \end{bmatrix} \end{aligned}$$

Hedce $\mathbf{y} = P_{\mathbf{1}}\mathbf{y} + P_V\mathbf{y}$ as stated in the theorem. For this example, represent $\mathbf{y}$, $V^\perp$, V, the $P_V\mathbf{y}$, and the $P_{V^\perp}\mathbf{y}$ geometrically.

Given a subspace $V \subset V_n$, Examples 1.3.7 and 1.3.8 indicate how $V^\perp$, the orthocomplement of V relative to the whole space V_n, is constructed and used to determine a unique orthogonal decomposition of a vector $\mathbf{y}$. Another important vector-space concept is the orthocomplement of a subspace relative to another subspace rather than to the whole space. That is, suppose that $W \subset V \subset V_n$, then all vectors in V orthogonal to W (denoted by V/W) constitute the orthocomplement (subspace) of W relative to the subspace V, so that $V = (V/W) \oplus W$. The orthocomplement of V relative to the whole space is $V^{\perp}$, and $V^{\perp} \oplus (V/W) \oplus W = V_n$. If the dimension of V is r and the dimension of W is k, the dimension of $V^{\perp}$ is $n - r$ and the dimension of V/W is $r - k$, so that $(n - r) + (r - k) + k = n$. In Figure 1.3.9, $W \oplus (V/W) = V$ and $V \oplus V^{\perp} = V_n$, or $W \oplus (V/W) \oplus V^{\perp} = V_n$.

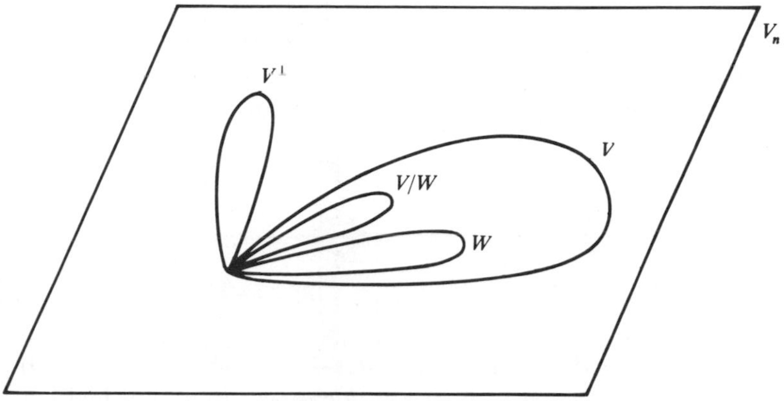

FIGURE 1.3.9. The orthocomplement of W relative to V, V/W.

The preceding discussion is important in analysis of variance where an observation vector $\mathbf{y}$ is projected onto orthogonal subspaces. By using the Pythagorean theorem, the sums of squares for hypotheses are the squared lengths of the projections of the observation vector onto the subspaces and the degrees of freedom are the dimensions of the associated subspaces.

EXAMPLE 1.3.8. As an example of the concept of projecting onto orthogonal subspaces, let the vector space V be spanned by

$$V = \left\{ \begin{bmatrix} 1 \\ 1 \\ 1 \\ 1 \end{bmatrix} \begin{bmatrix} 1 \\ 1 \\ 0 \\ 0 \end{bmatrix} \begin{bmatrix} 0 \\ 0 \\ 1 \\ 1 \end{bmatrix} \right\} = \{\mathbf{1}, \mathbf{a}_1, \mathbf{a}_2\}$$

where $\mathbf{1}$, $\mathbf{a}_1$, and $\mathbf{a}_2$ are elements of V_4. The vectors in V are linearly dependent. Let A denote a basis for V where

$$A = \left\{ \begin{bmatrix} 1 \\ 1 \\ 0 \\ 0 \end{bmatrix} \begin{bmatrix} 0 \\ 0 \\ 1 \\ 1 \end{bmatrix} \right\} = \{\mathbf{a}_1, \mathbf{a}_2\}$$

The vector $\mathbf{1} \subset A$. Thus the orthocomplement of the subspace $\{\mathbf{1}\} \equiv \mathbf{1}$ relative to A, denoted by $A/\mathbf{1}$. is given by

$$A/\mathbf{1} = \{\mathbf{a}_1 - P_{\mathbf{1}}\mathbf{a}_1, \mathbf{a}_2 - P_{\mathbf{1}}\mathbf{a}_2\}$$

$$= \left\{ \begin{bmatrix} 1/2 \\ 1/2 \\ -1/2 \\ -1/2 \end{bmatrix} \begin{bmatrix} -1/2 \\ -1/2 \\ 1/2 \\ 1/2 \end{bmatrix} \right\}$$

The vectors in $A/\mathbf{1}$ span the space; however, a basis for $A/\mathbf{1}$ is given by

$$A/\mathbf{1} = \begin{bmatrix} 1 \\ 1 \\ -1 \\ -1 \end{bmatrix}$$

where $A/\mathbf{1} \oplus \mathbf{1} = A$ and $A \subset V_4$. Thus $A/\mathbf{1} \oplus \mathbf{1} \oplus A^{\perp} = V_4$. Geometrically, as shown in Figure 1.3.10, the vector space $V \equiv A$ has been partitioned into two orthogonal subspaces such that $A = (A/\mathbf{1}) \oplus \mathbf{1}$, where $A/\mathbf{1}$ is the orthocomplement of $\mathbf{1}$

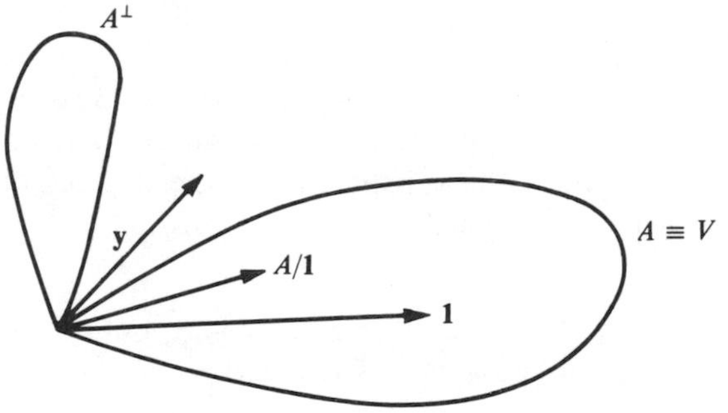

FIGURE 1.3.10. The orthogonal decomposition of V_N for a one-way layout.

relative to A, and $A \oplus A^{\perp} = V_4$. Thus an observation vector $\mathbf{y} \in V_4$ may be thought of as a vector with components in various orthogonal subspaces. The projection of the vector

$$\mathbf{y} = \begin{bmatrix} y_{11} \\ y_{12} \\ y_{21} \\ y_{22} \end{bmatrix}$$

onto the orthogonal subspaces of A, $P_A\mathbf{y} = P_{\mathbf{1}}\mathbf{y} + P_{A/\mathbf{1}}\mathbf{y}$, is now obtained.

$$P_{\mathbf{1}}\mathbf{y} = \bar{y}\begin{bmatrix}1\\1\\1\\1\end{bmatrix} = \hat{\mu}\begin{bmatrix}1\\1\\1\\1\end{bmatrix}$$

$$P_{A/\mathbf{1}}\mathbf{y} = \frac{y_{11} + y_{12} - y_{21} - y_{22}}{4}\begin{bmatrix}1\\1\\-1\\-1\end{bmatrix}$$

$$= \left(\frac{y_{11} + y_{12}}{2} - \frac{y_{11} + y_{12} + y_{21} + y_{22}}{4}\right)\begin{bmatrix}1\\1\\-1\\-1\end{bmatrix}$$

$$= (\bar{y}_1 - \bar{y})\begin{bmatrix}1\\1\\-1\\-1\end{bmatrix} = \hat{\alpha}_1\begin{bmatrix}1\\1\\-1\\-1\end{bmatrix}$$

where $\bar{y} = \hat{\mu}$ and $\hat{\alpha}_1 = \bar{y}_1 - \bar{y}$. Alternatively, since $A/\mathbf{1} \perp \mathbf{1}$, $P_{A/\mathbf{1}}\mathbf{y}$ may be written as

$$\begin{aligned} P_{A/\mathbf{1}}\mathbf{y} &= P_A\mathbf{y} - P_{\mathbf{1}}\mathbf{y} \\ &= \frac{(\mathbf{y}, \mathbf{a}_1)\mathbf{a}_1}{\|\mathbf{a}_1\|^2} + \frac{(\mathbf{y}, \mathbf{a}_2)\mathbf{a}_2}{\|\mathbf{a}_2\|^2} - \frac{(\mathbf{y}, \mathbf{1})\mathbf{1}}{\|\mathbf{1}\|^2} \\ &= \sum_{i=1}^{2}\left[\frac{(\mathbf{y}, \mathbf{a}_i)}{\|\mathbf{a}_i\|^2} - \frac{(\mathbf{y}, \mathbf{1})}{\|\mathbf{1}\|^2}\right]\mathbf{a}_i \\ &= \sum_{i=1}^{2}(\bar{y}_i - \bar{y})\mathbf{a}_i \\ &= \sum_{i=1}^{2}\hat{\alpha}_i\mathbf{a}_i \end{aligned}$$

As an exercise, find the projection of $\mathbf{y}$ onto $A^{\perp}$ and the $\|P_{A/\mathbf{1}}\mathbf{y}\|^2$.

From the analysis of variance, the coefficients of the basis vectors for $\mathbf{1}$ and $A/\mathbf{1}$ yield the estimators for the overall effect μ and all treatment effects α_i for a one-way layout, with $I = 2$ treatment conditions employing side conditions. Furthermore, the total sum of squares, $\|\mathbf{y}\|^2$, is the sum of squared lengths of the projections of $\mathbf{y}$ onto each subspace, $\|\mathbf{y}\|^2 = \|P_{\mathbf{1}}\mathbf{y}\|^2 + \|P_{A/\mathbf{1}}\mathbf{y}\|^2 + \|P_{A^{\perp}}\mathbf{y}\|^2$. The dimensions of the subspaces corresponding to the decomposition of $\mathbf{y}$ that satisfy the relationship $N = 1 + (I - 1) + (N - I)$, or the degrees of freedom of the usual analysis-of-variance sum of squares, are the dimensions of the orthogonal subspaces that can be found by using the vector-space concept of orthocomplement.

EXERCISES 1.3

1. For the vectors and scalars defined in Example 1.3.1, verify equations (1.3.2) through (1.3.6).

2. Derive equation (1.3.9) by employing (1.3.8) and the law of cosines.

3. Given the vectors

$$\mathbf{y}_1 = \begin{bmatrix} 2 \\ -2 \\ 1 \end{bmatrix} \quad \text{and} \quad \mathbf{y}_2 = \begin{bmatrix} 3 \\ 0 \\ -1 \end{bmatrix}$$

   a. Find their lengths, the distance between them, and the angle between the vectors.

   b. Find a vector of length 3 with direction cosines $[1/\sqrt{2}, -1/\sqrt{2}]$.

4. Find the projection of the vector $\mathbf{y}' = [1, 9, -7]$ onto the vector space spanned by

$$\mathbf{v}_1 = \begin{bmatrix} 2 \\ 3 \\ -1 \end{bmatrix} \quad \text{and} \quad \mathbf{v}_2 = \begin{bmatrix} 5 \\ 0 \\ 4 \end{bmatrix}$$

   a. Interpret the result.

   b. In general, if $\mathbf{y} \perp V$, what can be said about the $P_V\mathbf{y}$?

5. Use the Gram-Schmidt orthogonalization process to find an orthogonal basis for the set of vectors in Exercise 6 (Section 1.2).

6. a. Fill in the steps in the solution for s leading to equation (1.3.12).

   b. In Example 1.3.8, find the $P_{V_\perp}\mathbf{y}$.

   c. Show that the projection of a vector $\mathbf{y}$ onto a vector $\mathbf{x}$ having all components equal to 1 is a vector in which each component is the mean of the components of $\mathbf{y}$.

7. The vectors

$$\mathbf{v}_1 = \begin{bmatrix} 1 \\ 2 \\ -1 \end{bmatrix} \quad \text{and} \quad \mathbf{v}_2 = \begin{bmatrix} 2 \\ 3 \\ 0 \end{bmatrix}$$

   span a plane in Euclidean three-dimensional space.

   a. Find an orthogonal basis for the plane.

   b. Find the orthocomplement of the plane.

   c. From (a) and (b), obtain an orthonormal basis for three-dimensional space.

8. Find an orthonormal basis for Euclidean three-dimensional space that includes the vector $\mathbf{y}' = [-1/\sqrt{3}, 1/\sqrt{3}, -1/\sqrt{3}]$.

9. a. Find the orthocomplement of the space spanned by $\mathbf{v}' = [4, 2, 1]$ relative to Euclidean three-dimensional space.

   b. Find the orthocomplement of the space spanned by $\mathbf{v}' = [4, 2, 1]$ relative to the space spanned by $\mathbf{v}_1' = [1, 1, 1]$ and $\mathbf{v}_2' = [2, 0, -1]$.

   c. Find the orthocomplement of the space spanned by $\mathbf{v}_1' = [1, 1, 1]$ and $\mathbf{v}_2' = [2, 0, -1]$ relative to Euclidean three-dimensional space.

   d. Write Euclidean three-dimensional space as a direct sum of the relevant spaces in (a), (b), and (c) in all possible ways.

10. Let V be the space spanned by the orthonormal basis

$$\mathbf{v}_1 = \begin{bmatrix} 1/\sqrt{2} \\ 0 \\ 1/\sqrt{2} \\ 0 \end{bmatrix} \quad \text{and} \quad \mathbf{v}_2 = \begin{bmatrix} 0 \\ -1/\sqrt{2} \\ 0 \\ -1/\sqrt{2} \end{bmatrix}$$

a. Express $\mathbf{x}' = [0, 1, 1, 1]$ as $\mathbf{x} = \mathbf{x}_1 + \mathbf{x}_2$, where $\mathbf{x}_1 \in V$ and $\mathbf{x}_2 \in V^{\perp}$.

b. Verify that $\|P_V\mathbf{x}\|^2 = \|P_{\mathbf{y}_1}\mathbf{v}\|^2 + \|P_{\mathbf{y}_2}\mathbf{v}\|^2$.

c. Which vector $\mathbf{y} \in V$ is "closest to" $\mathbf{x}$? Calculate this minimum distance.

11. Find the dimension of the vector space generated by

$$\begin{matrix} \mathbf{v}_1 & \mathbf{v}_2 & \mathbf{v}_3 & \mathbf{v}_4 & \mathbf{v}_5 \\ \begin{bmatrix}1\\1\\1\\1\end{bmatrix} & \begin{bmatrix}1\\1\\0\\0\end{bmatrix} & \begin{bmatrix}0\\0\\1\\1\end{bmatrix} & \begin{bmatrix}1\\0\\1\\0\end{bmatrix} & \begin{bmatrix}0\\1\\0\\1\end{bmatrix} \end{matrix}$$

12. Prove that $(\mathbf{x}, \mathbf{y})^2 \le (\mathbf{x}, \mathbf{x})(\mathbf{y}, \mathbf{y})$. (This is the Cauchy-Schwartz inequality.)

13. Let the n-vector $\mathbf{y}$ denote an observation vector in V_n, and let V be the vector space spanned by the vector $\mathbf{1}$, a vector of n 1s.

a. Find the projection of $\mathbf{y}$ onto $V^{\perp}$, the orthocomplement of V relative to V_n.

b. Represent $\mathbf{y}$ as $\mathbf{y} = \mathbf{x} + \mathbf{x}^*$, where $\mathbf{x} \in V$ and $\mathbf{x}^* \in V^{\perp}$. What are the dimensions of V and $V^{\perp}$?

c. Since $\|\mathbf{y}\|^2 = \|\mathbf{x}\|^2 + \|\mathbf{x}^*\|^2 = \|P_V\mathbf{y}\|^2 + \|P_{V^{\perp}}\mathbf{y}\|^2$, determine a general expression for the $\|\mathbf{y}\|^2$ in terms of its components. Divide $\|P_{V^{\perp}}\mathbf{y}\|^2$ by the dimension of $V^{\perp}$. What can you conclude in general about the ratio $\|P_{V^{\perp}}\mathbf{y}\|^2/\dim V^{\perp}$?

14. Let the n-vector $\mathbf{y}$ denote an observation vector, and let $\{\mathbf{1}, \mathbf{x}\}$ be a basis for V, where $\mathbf{x}$ is an n-vector of real numbers and $\mathbf{1}$ is a vector of n 1s.

a. Find the orthocomplement of $\mathbf{1}$ relative to the space spanned by the V (that is, $V/\mathbf{1}$) so that $\mathbf{1} \oplus (V/\mathbf{1}) = V$. What is the dimension of $V/\mathbf{1}$?

b. Find the projection of the vector $\mathbf{y}$ onto $V/\mathbf{1}$ and the projection of $\mathbf{y}$ onto the vector $\mathbf{1}$. Employing the simple linear regression model $y = \alpha + \beta(x - \bar{x})$, what can you conclude about the coefficient s in (1.3.12) of the projections for this exercise?

c. Find $\mathbf{y} - P_V\mathbf{y}$ and $\|\mathbf{y} - P_V\mathbf{y}\|^2$. How are these quantities related to the simple linear regression model.

d. In Example 1.3.8, evaluate the squared lengths $\|\mathbf{y} - P_{\mathbf{1}}\mathbf{y}\|^2$, $\|P_{A/\mathbf{1}}\mathbf{y}\|^2$, and $\|P_{A^{\perp}}\mathbf{y}\|^2$. Use Figure 1.3.10 to represent these quantities geometrically and associate with each a label commonly used in one-way, analysis-of-variance designs.

15. Let the vector space V be spanned by

$$V = \left\{ \begin{matrix} \mathbf{v}_1 & & \mathbf{v}_2 & \mathbf{v}_3 & & \mathbf{v}_4 & \mathbf{v}_5 & & \mathbf{v}_6 & \mathbf{v}_7 & \mathbf{v}_8 & \mathbf{v}_9 \\ \begin{bmatrix}1\\1\\1\\1\\1\\1\\1\\1\end{bmatrix} & , & \begin{bmatrix}1\\1\\1\\1\\0\\0\\0\\0\end{bmatrix} & \begin{bmatrix}0\\0\\0\\0\\1\\1\\1\\1\end{bmatrix} & , & \begin{bmatrix}1\\1\\0\\0\\1\\1\\0\\0\end{bmatrix} & \begin{bmatrix}0\\0\\1\\1\\0\\0\\1\\1\end{bmatrix} & , & \begin{bmatrix}1\\1\\0\\0\\0\\0\\0\\0\end{bmatrix} & \begin{bmatrix}0\\0\\1\\1\\0\\0\\0\\0\end{bmatrix} & \begin{bmatrix}0\\0\\0\\0\\1\\1\\0\\0\end{bmatrix} & \begin{bmatrix}0\\0\\0\\0\\0\\0\\1\\1\end{bmatrix} \end{matrix} \right\}$$

$$= \{\ \mathbf{1},\quad A,\quad B,\quad AB\ \}$$

a. Find the space $A + B = \mathbf{1} \oplus (A/\mathbf{1}) \oplus (B/\mathbf{1})$ and the space $AB/(A + B)$ so that $V = \mathbf{1} \oplus (A/1) \oplus (B/\mathbf{1}) \oplus [AB/(A + B)]$. What are the dimensions of each of these spaces?

b. Find the projection of the observation vector $\mathbf{y} = [y_{111}, y_{112}, y_{211}, y_{212}, y_{311}, y_{312}, y_{411}, y_{412}]$ in V_8 onto each space in the direct sum decomposition of the vector space V and the squared lengths of the projections. Represent these quantities geometrically and compare the squared lengths with a 2×2 factorial design.

c. Summarize your findings.

1.4 FUNDAMENTALS OF MATRIX ALGEBRA

An $n \times m$ *matrix* is a rectangular or square array of real numbers arranged in n rows and m columns:

$$\underset{(n \times m)}{\mathbf{Y}} = \begin{bmatrix} y_{11} & y_{12} & \cdots & y_{1m} \\ y_{21} & y_{22} & \cdots & y_{2m} \\ \vdots & \vdots & \vdots & \vdots \\ y_{n1} & y_{n2} & \cdots & y_{nm} \end{bmatrix} \tag{1.4.1}$$

The matrix is of *order* $n \times m$ with elements y_{ij} $(i = 1, \ldots, n; j = 1, \ldots, m)$. A convenient shorthand for the matrix $\mathbf{Y}$ is

$$\mathbf{Y} = [y_{ij}] \tag{1.4.2}$$

Alternatively, a matrix can be represented in terms of its columns or rows:

$$\underset{(n \times m)}{\mathbf{Y}} = [\mathbf{u}_1 \quad \mathbf{u}_2 \quad \cdots \quad \mathbf{u}_m] \quad \text{and} \quad \mathbf{u}_i \in V_n \tag{1.4.3}$$

or

$$\underset{(n \times m)}{\mathbf{Y}} = \begin{bmatrix} \mathbf{y}_1' \\ \mathbf{y}_2' \\ \vdots \\ \mathbf{y}_n' \end{bmatrix} \quad \text{and} \quad \mathbf{y}_i' \in V_m \tag{1.4.4}$$

The prime notation on vectors, used in (1.4.4) will be made clear shortly. If $\mathbf{Y}$ is an individual-by-variable matrix, then a column vector of $\mathbf{Y}$ is an element of the individual space and a row vector of $\mathbf{Y}$ is contained in the variable space. Methods of multivariate analysis have simple descriptions in terms of either or both spaces.

Equality, Addition, and Multiplication of Matrices. Two matrices of the same order are equal if and only if there is element-by-element equality:

$$\mathbf{A} = \mathbf{B} \quad \text{if and only if} \quad \mathbf{a}_{ij} = \mathbf{b}_{ij} \qquad (i = 1, \ldots, n; j = 1, \ldots, m) \tag{1.4.5}$$

Given equality, we will look at some fundamental operations with matrices. To *add* two matrices of the same order, their corresponding elements are added:

$$\mathbf{A} + \mathbf{B} = \mathbf{C} \quad \text{if and only if} \quad \mathbf{c}_{ij} = a_{ij} + b_{ij} \qquad (i = 1, \ldots, n; j = 1, \ldots, m) \tag{1.4.6}$$

To multiply a matrix by a scalar s, every element in the matrix is multiplied by the scalar; thus

$$s\mathbf{A} = [sa_{ij}] \tag{1.4.7}$$

From these definitions and the properties of real numbers, we can see that matrices satisfy the following laws. For matrices **A**, **B**, and **C** and scalars s and c,

(1.4.8) $\mathbf{A} + \mathbf{B} = \mathbf{B} + \mathbf{A}$ (commutative law)
(1.4.9) $(\mathbf{A} + \mathbf{B}) + \mathbf{C} = \mathbf{A} + (\mathbf{B} + \mathbf{C})$ (associative law)
(1.4.10) $s(c\mathbf{A}) = (sc)\mathbf{A} = (cs)\mathbf{A} = c(s\mathbf{A})$ (associative law)
(1.4.11) $s(\mathbf{A} + \mathbf{B}) = s\mathbf{A} + s\mathbf{B}$ (distributive law for matrices)
(1.4.12) $(s + c)\mathbf{A} = s\mathbf{A} + c\mathbf{A}$ (distributive law for scalars)

EXAMPLE 1.4.1. Let

$$\mathbf{A} = \begin{bmatrix} 1 & 2 \\ 3 & 7 \\ -4 & 8 \end{bmatrix} \quad \text{and} \quad \mathbf{B} = \begin{bmatrix} 2 & 2 \\ 7 & 5 \\ 3 & 1 \end{bmatrix}$$

Then

$$\mathbf{A} + \mathbf{B} = \begin{bmatrix} 1 & 2 \\ 3 & 7 \\ -4 & 8 \end{bmatrix} + \begin{bmatrix} 2 & 2 \\ 7 & 5 \\ 3 & 1 \end{bmatrix} = \begin{bmatrix} 1+2 & 2+2 \\ 3+7 & 7+5 \\ -4+3 & 8+1 \end{bmatrix} = \begin{bmatrix} 3 & 4 \\ 10 & 12 \\ -1 & 9 \end{bmatrix}$$

$$5(\mathbf{A} + \mathbf{B}) = \begin{bmatrix} 15 & 20 \\ 50 & 60 \\ -5 & 45 \end{bmatrix}$$

The next fundamental matrix operation is the *multiplication* of matrices. The product of an $n \times m$ matrix **A** by an $m \times p$ matrix **B** is an $n \times p$ matrix **C** whose elements are the sum of products of the ith row of **A** with the elements in the jth column of **B**; thus

(1.4.13) $$\underset{(n\times m)}{\mathbf{A}}\ \underset{(m\times n)}{\mathbf{B}} = \underset{(n\times p)}{\mathbf{C}} \quad \text{if and only if all} \quad \mathbf{c}_{ij} = \sum_{k=1}^{m} a_{ik}b_{kj}$$

For matrix multiplication as specified in (1.4.13) to be defined, the number of columns in **A** must be the same as the number of rows in **B**. The matrices **A** and **B** are then said to be *conformable* for matrix multiplication.

EXAMPLE 1.4.2.

$$\mathbf{A} = \begin{bmatrix} -1 & 2 & 3 \\ 5 & 1 & 0 \end{bmatrix} \quad \text{and} \quad \mathbf{B} = \begin{bmatrix} 1 & 2 & 1 \\ 1 & 2 & 0 \\ 1 & 2 & -1 \end{bmatrix}$$

Then

$$\mathbf{AB} = \begin{bmatrix} (-1)(1) + 2(1) + 3(1) & -1(2) + 2(2) + 3(2) & -1(1) + 2(0) + 3(-1) \\ 5(1) + 1(1) + 0(1) & 5(2) + 1(2) + 0(2) & 5(1) + 1(0) + 0(-1) \end{bmatrix}$$

$$= \begin{bmatrix} 4 & 8 & -4 \\ 6 & 12 & 5 \end{bmatrix}$$

From (1.4.13) and the properties of real numbers, matrices satisfy the following rules:

(1.4.14) $(\mathbf{AB})\mathbf{C} = \mathbf{A}(\mathbf{BC})$ (associative law)

(1.4.15) $(\mathbf{A} + \mathbf{B})(\mathbf{C} + \mathbf{D}) = \mathbf{A}(\mathbf{C} + \mathbf{D}) + \mathbf{B}(\mathbf{C} + \mathbf{D}) = \mathbf{AC} + \mathbf{AD} + \mathbf{BC} + \mathbf{BD}$ (distributive law)

In general, matrix multiplication is not commutative; that is, $\mathbf{AB} \neq \mathbf{BA}$. In fact, if **AB** is defined, **BA** may not be.

Special Matrices. The *transpose* of a matrix, indicated by a prime, is a matrix obtained by interchanging the rows and columns of the original matrix; thus, given any $n \times m$ matrix **A**, the $m \times n$ matrix **A**′ will be

(1.4.16)
$$\mathbf{A}' = \begin{bmatrix} a_{11} & a_{21} & \cdots & a_{n1} \\ a_{12} & a_{22} & \cdots & a_{n2} \\ \vdots & \vdots & \vdots & \vdots \\ a_{1m} & a_{2m} & \cdots & a_{nm} \end{bmatrix}$$

which is the transpose of **A**. That is,

$$\mathbf{A}' = [a_{ji}] \quad \text{when} \quad \mathbf{A} = [a_{ij}] \qquad (i = 1, \ldots, n; j = 1, \ldots, m)$$

Some properties of matrix transposition follow:

(1.4.17) $$(\mathbf{AB})' = \mathbf{B}'\mathbf{A}'$$

(1.4.18) $$(\mathbf{A} + \mathbf{B})' = \mathbf{A}' + \mathbf{B}'$$

(1.4.19) $$(\mathbf{A}')' = \mathbf{A}$$

(1.4.20) $$(\mathbf{ABC})' = \mathbf{C}'\mathbf{B}'\mathbf{A}'$$

EXAMPLE 1.4.3. Let

$$\mathbf{A} = \begin{bmatrix} 1 & 3 \\ -1 & 4 \end{bmatrix} \quad \text{and} \quad \mathbf{B} = \begin{bmatrix} 2 & 1 \\ 1 & 1 \end{bmatrix}$$

Then

$$\mathbf{A}' = \begin{bmatrix} 1 & -1 \\ 3 & 4 \end{bmatrix} \quad \text{and} \quad \mathbf{B}' = \begin{bmatrix} 2 & 1 \\ 1 & 1 \end{bmatrix}$$

$$\mathbf{AB} = \begin{bmatrix} 5 & 4 \\ 2 & 3 \end{bmatrix} \quad \text{and} \quad (\mathbf{AB})' = \begin{bmatrix} 5 & 2 \\ 4 & 3 \end{bmatrix} = \mathbf{B}'\mathbf{A}'$$

$$(\mathbf{A} + \mathbf{B})' = \begin{bmatrix} 3 & 0 \\ 4 & 5 \end{bmatrix} = \mathbf{A}' + \mathbf{B}'$$

A matrix **A** is said to be *symmetric* if and only if $\mathbf{A} = \mathbf{A}'$; that is, if $[a_{ij}] = [a_{ji}]$, for all i and j. In Example 1.4.3, the matrix **B** is symmetric. Transposes are used in making symmetric matrices. Given any matrix **A**, **AA**′ is symmetric, as is **A**′**A**. However, $\mathbf{AA}' \neq \mathbf{A}'\mathbf{A}$. The transpose of a column vector of a matrix is a row vector.

By convention, vectors without primes are taken to be column vectors. Having introduced the transpose of a matrix, we will now look at an alternative definition of matrix multiplication. First, write the $n \times m$ matrix $\mathbf{A}$ in column form and the $m \times p$ matrix $\mathbf{B}$ in row form, then the product $\mathbf{AB}$ is the sum of the *outer products* of the columns of $\mathbf{A}$ with the rows of $\mathbf{B}$. Thus, if

$$\mathbf{A} = [\mathbf{a}_1 \quad \mathbf{a}_2 \quad \cdots \quad \mathbf{a}_m] \quad \text{and} \quad \mathbf{B} = \begin{bmatrix} \mathbf{b}'_1 \\ \mathbf{b}'_2 \\ \vdots \\ \mathbf{b}'_m \end{bmatrix}$$

the product $\mathbf{AB}$ is defined as

$$\mathbf{AB} = \sum_{i=1}^{m} \mathbf{a}_i \mathbf{b}'_i \tag{1.4.21}$$

The inner vector product of two vectors $(\mathbf{a}_i, \mathbf{b}_i) = \mathbf{a}'_i \mathbf{b}_i$ is a scalar; however, the outer product is a matrix.

EXAMPLE 1.4.4. Let

$$\mathbf{A} = \begin{bmatrix} -1 & 2 & 3 \\ 5 & 1 & 0 \end{bmatrix} \quad \text{and} \quad \mathbf{B} = \begin{bmatrix} 1 & 2 & 1 \\ 1 & 2 & 0 \\ 1 & 2 & -1 \end{bmatrix}$$

Then

$$\mathbf{a}_1 = \begin{bmatrix} -1 \\ 5 \end{bmatrix}, \quad \mathbf{a}_2 = \begin{bmatrix} 2 \\ 1 \end{bmatrix}, \quad \mathbf{a}_3 = \begin{bmatrix} 3 \\ 0 \end{bmatrix}$$

and

$$\mathbf{b}'_1 = [1, 2, 1], \quad \mathbf{b}'_2 = [1, 2, 0], \quad \mathbf{b}'_3 = [1, 2, -1]$$

Thus

$$\sum_{i=1}^{3} \mathbf{a}_i \mathbf{b}'_i = \begin{bmatrix} -1 & -2 & -1 \\ 5 & 10 & 5 \end{bmatrix} + \begin{bmatrix} 2 & 4 & 0 \\ 1 & 2 & 0 \end{bmatrix} + \begin{bmatrix} 3 & 6 & -3 \\ 0 & 0 & 0 \end{bmatrix}$$

$$= \begin{bmatrix} 4 & 8 & -4 \\ 6 & 12 & 5 \end{bmatrix} = \mathbf{AB}$$

the same result as illustrated in Example 1.4.2.

A *square* matrix (a matrix of order $n \times n$) whose diagonal elements are all 1s and off-diagonal elements are all 0s, is called the *identity matrix* and is written as

$$\mathbf{I} = \begin{bmatrix} 1 & 0 & \cdots & 0 \\ 0 & 1 & \cdots & 0 \\ \vdots & \vdots & \vdots & \vdots \\ 0 & 0 & \cdots & 1 \end{bmatrix} \tag{1.4.22}$$

The order of an identity matrix is often written as $\mathbf{I}_n$. If all the elements in a matrix are 0s, the matrix is termed a *null matrix* and denoted by **0**. When the products and sums are defined,

$$\mathbf{IA} = \mathbf{A} = \mathbf{AI} \tag{1.4.23}$$

$$\mathbf{A} + \mathbf{0} = \mathbf{A} \tag{1.4.24}$$

An important operation for square matrices is the *trace* operator. The trace of a square matrix is the sum of the diagonal elements of the matrix; thus, for the $n \times n$ matrix $\mathbf{A}$,

$$\text{Tr}(\mathbf{A}) = a_{11} + a_{22} + \cdots + a_{nn} = \sum_{i=1}^{n} a_{ii} \tag{1.4.25}$$

It follows directly for square matrices that

$$\text{Tr}(\mathbf{A} + \mathbf{B}) = \text{Tr}(\mathbf{A}) + \text{Tr}(\mathbf{B}) \tag{1.4.26}$$

$$\text{Tr}(\mathbf{AB}) = \text{Tr}(\mathbf{BA}) \tag{1.4.27}$$

where matrix multiplication and addition are defined.

Any square matrix whose off-diagonal elements are 0s is called a *diagonal matrix*. Diagonal matrices are written as

$$\mathbf{D} = \begin{bmatrix} d_{11} & 0 & \cdots & 0 \\ 0 & d_{22} & \cdots & 0 \\ \vdots & \vdots & \vdots & \vdots \\ 0 & 0 & \cdots & d_{nn} \end{bmatrix} \tag{1.4.28}$$

If $\mathbf{A}$ is any square matrix, we denote the diagonal of $\mathbf{A}$ by Dia $\mathbf{A}$ or by Dia $[a_{ii}]$.

Pre- and postmultiplication of any rectangular matrix by a diagonal matrix have important consequences. Premultiplication of $\mathbf{A}$ by a diagonal matrix $\mathbf{DA}$ multiplies each element in the ith row of $\mathbf{A}$ by d_{ii}; postmultiplication by a diagonal matrix $\mathbf{AD}$ multiplies each element in the jth column of $\mathbf{A}$ by d_{jj}. For example,

$$\underset{(n \times n)}{\mathbf{D}} \; \underset{(n \times m)}{\mathbf{A}} = \begin{bmatrix} d_{11}a_{11} & \cdots & d_{11}a_{1m} \\ d_{22}a_{21} & \cdots & d_{22}a_{2m} \\ \vdots & \vdots & \vdots \\ d_{nn}a_{n1} & \cdots & d_{nn}a_{nm} \end{bmatrix}$$

and

$$\underset{(n \times m)}{\mathbf{A}} \; \underset{(m \times m)}{\mathbf{D}} = \begin{bmatrix} d_{11}a_{11} & \cdots & d_{mm}a_{1m} \\ d_{11}a_{21} & \cdots & d_{mm}a_{2m} \\ \vdots & \vdots & \vdots \\ d_{11}a_{n1} & \cdots & d_{mm}a_{nm} \end{bmatrix}$$

A square matrix whose elements above (or below) the principal or main diagonal are 0s is called a *lower* (*or upper*) *triangular matrix*. If the elements on the diagonal are 1s, the matrix is called a *unit lower* (*or unit upper*) *triangular matrix*.

EXAMPLE 1.4.5.

$$\text{lower } \mathbf{L} = \begin{bmatrix} 2 & 0 & 0 \\ 6 & 5 & 0 \\ 3 & 1 & 7 \end{bmatrix} \qquad \text{unit lower} = \begin{bmatrix} 1 & 0 & 0 \\ 6 & 1 & 0 \\ 3 & 5 & 1 \end{bmatrix}$$

$$\text{upper } \mathbf{U} = \begin{bmatrix} 2 & 6 & 8 \\ 0 & 1 & 3 \\ 0 & 0 & 6 \end{bmatrix} \qquad \text{unit upper} = \begin{bmatrix} 1 & 6 & 8 \\ 0 & 1 & 3 \\ 0 & 0 & 1 \end{bmatrix}$$

Partitioned Matrices. It is convenient for us to partition a matrix into *submatrices*. For illustration, an $n \times m$ matrix $\mathbf{A}$ may be partitioned as

$$\mathbf{A} = [\mathbf{A}_1 \quad \mathbf{A}_2] \tag{1.4.29}$$

where $\mathbf{A}_1$ is of dimensionality $n \times m_1$, $\mathbf{A}_2$ is $n \times m_2$, and $m_1 + m_2 = m$. More generally, $\mathbf{A}$ may be partitioned into

$$\mathbf{A} = \begin{bmatrix} \mathbf{A}_{11} & \mathbf{A}_{12} & \cdots & \mathbf{A}_{1m} \\ \mathbf{A}_{21} & \mathbf{A}_{22} & \cdots & \mathbf{A}_{2m} \\ \vdots & \vdots & \vdots & \vdots \\ \mathbf{A}_{n1} & \mathbf{A}_{n2} & \cdots & \mathbf{A}_{nm} \end{bmatrix}. \tag{1.4.30}$$

where $\mathbf{A}_{ij}$ has n_i rows and m_j columns and where $\Sigma_j m_j = m$ and $\Sigma_i n_i = n$. Thus the elements of the partitioned matrix are submatrices. Transposition, addition, and multiplication of partitioned matrices adhere to the definitions outlined for matrices whose elements are scalars. The transpose of the partitioned matrix $\mathbf{A}$

$$\mathbf{A} = \begin{bmatrix} \mathbf{A}_{11} & \mathbf{A}_{12} \\ \mathbf{A}_{21} & \mathbf{A}_{22} \end{bmatrix}$$

is

$$\mathbf{A}' = \begin{bmatrix} \mathbf{A}'_{11} & \mathbf{A}'_{21} \\ \mathbf{A}'_{12} & \mathbf{A}'_{22} \end{bmatrix} \tag{1.4.31}$$

The sum of the partitioned matrix $\mathbf{B}$

$$\mathbf{B} = \begin{bmatrix} \mathbf{B}_{11} & \mathbf{B}_{12} \\ \mathbf{B}_{21} & \mathbf{B}_{22} \end{bmatrix}$$

and the partitioned matrix $\mathbf{A}$ is

$$\mathbf{A} + \mathbf{B} = \begin{bmatrix} \mathbf{A}_{11} + \mathbf{B}_{11} & \mathbf{A}_{12} + \mathbf{B}_{12} \\ \mathbf{A}_{21} + \mathbf{B}_{21} & \mathbf{A}_{22} + \mathbf{B}_{22} \end{bmatrix} \tag{1.4.32}$$

only if the submatrices have conformable dimensions.

Finally, the product of two partitioned matrices **AB** is

$$\mathbf{AB} = \begin{bmatrix} \mathbf{A}_{11} & \mathbf{A}_{12} \\ \mathbf{A}_{21} & \mathbf{A}_{22} \end{bmatrix} \begin{bmatrix} \mathbf{B}_{11} & \mathbf{B}_{12} \\ \mathbf{B}_{21} & \mathbf{B}_{22} \end{bmatrix}$$

$$(1.4.33) \qquad = \begin{bmatrix} \mathbf{A}_{11}\mathbf{B}_{11} + \mathbf{A}_{12}\mathbf{B}_{21} & \mathbf{A}_{11}\mathbf{B}_{12} + \mathbf{A}_{12}\mathbf{B}_{22} \\ \mathbf{A}_{21}\mathbf{B}_{11} + \mathbf{A}_{22}\mathbf{B}_{21} & \mathbf{A}_{21}\mathbf{B}_{12} + \mathbf{A}_{22}\mathbf{B}_{22} \end{bmatrix}$$

only if the submatrices have conformable dimensions.

EXAMPLE 1.4.6. Let

$$\mathbf{A} = \left[\begin{array}{cc:cc} 1 & 2 & 0 & 1 \\ \hdashline 1 & -1 & 3 & 1 \\ 2 & 3 & 2 & -1 \end{array}\right] = \begin{bmatrix} \mathbf{A}_{11} & \mathbf{A}_{12} \\ \mathbf{A}_{21} & \mathbf{A}_{22} \end{bmatrix}$$

$$\mathbf{B} = \left[\begin{array}{c:c} 1 & 1 \\ 1 & -1 \\ \hdashline 2 & 0 \\ 0 & 5 \end{array}\right] = \begin{bmatrix} \mathbf{B}_{11} & \mathbf{B}_{12} \\ \mathbf{B}_{21} & \mathbf{B}_{22} \end{bmatrix}$$

Then

$$\mathbf{A}_{11}\mathbf{B}_{11} + \mathbf{A}_{12}\mathbf{B}_{21} = [1, 2]\begin{bmatrix} 1 \\ 1 \end{bmatrix} + [0, 1]\begin{bmatrix} 2 \\ 0 \end{bmatrix} = 3$$

$$\mathbf{A}_{11}\mathbf{B}_{12} + \mathbf{A}_{12}\mathbf{B}_{22} = [1, 2]\begin{bmatrix} 1 \\ -1 \end{bmatrix} + [0, 1]\begin{bmatrix} 0 \\ 5 \end{bmatrix} = 4$$

$$\mathbf{A}_{21}\mathbf{B}_{11} + \mathbf{A}_{22}\mathbf{B}_{21} = \begin{bmatrix} 1 & -1 \\ 2 & 3 \end{bmatrix}\begin{bmatrix} 1 \\ 1 \end{bmatrix} + \begin{bmatrix} 3 & 1 \\ 2 & -1 \end{bmatrix}\begin{bmatrix} 2 \\ 0 \end{bmatrix} = \begin{bmatrix} 6 \\ 9 \end{bmatrix}$$

$$\mathbf{A}_{21}\mathbf{B}_{12} + \mathbf{A}_{22}\mathbf{B}_{22} = \begin{bmatrix} 1 & -1 \\ 2 & 3 \end{bmatrix}\begin{bmatrix} 1 \\ -1 \end{bmatrix} + \begin{bmatrix} 3 & 1 \\ 2 & -1 \end{bmatrix}\begin{bmatrix} 0 \\ 5 \end{bmatrix} = \begin{bmatrix} 7 \\ -6 \end{bmatrix}$$

so that

$$\mathbf{AB} = \left[\begin{array}{c:c} 3 & 4 \\ \hdashline 6 & 7 \\ 9 & -6 \end{array}\right]$$

Direct Matrix Products. We conclude this section by defining the *direct* (*or Kronecker*) *product* of two matrices. Let **A** be an $n \times m$ matrix and **B** be a $p \times q$ matrix. Then the direct (or Kronecker) product of **A** and **B** (written $\mathbf{A} \otimes \mathbf{B}$) is defined as the $np \times mq$ matrix:

$$(1.4.34) \qquad \mathbf{A} \otimes \mathbf{B} = \begin{bmatrix} a_{11}\mathbf{B} & \cdots & a_{1m}\mathbf{B} \\ \vdots & \vdots\vdots\vdots & \vdots \\ a_{n1}\mathbf{B} & \cdots & a_{nm}\mathbf{B} \end{bmatrix}$$

As an example, if

$$\mathbf{A} = \begin{bmatrix} 1 & 2 \\ 3 & 4 \end{bmatrix} \quad \text{and} \quad \mathbf{B} = [1, 2]$$

then

$$\mathbf{A} \otimes \mathbf{B} = \begin{bmatrix} 1\mathbf{B} & 2\mathbf{B} \\ 3\mathbf{B} & 4\mathbf{B} \end{bmatrix} = \begin{bmatrix} 1 & 2 & 2 & 4 \\ 3 & 6 & 4 & 8 \end{bmatrix}$$

The following properties of Kronecker products are a direct derivation of the definition (for example, see Bellman, 1960).

(1.4.35) For vectors $\mathbf{x}$ and $\mathbf{y}$,

$$\mathbf{x}' \otimes \mathbf{y} = \mathbf{y}\mathbf{x}' = \mathbf{y} \otimes \mathbf{x}'$$

(1.4.36) For a scalar s,

$$s\mathbf{A} \otimes \mathbf{B} = \mathbf{A} \otimes s\mathbf{B} = s(\mathbf{A} \otimes \mathbf{B})$$

(1.4.37)
$$(\mathbf{A} \otimes \mathbf{B}) \otimes \mathbf{C} = \mathbf{A} \otimes (\mathbf{B} \otimes \mathbf{C})$$

(1.4.38) Provided matrix sums are defined,

$$(\mathbf{A} + \mathbf{B}) \otimes \mathbf{C} = (\mathbf{A} \otimes \mathbf{C}) + (\mathbf{B} \otimes \mathbf{C})$$

(1.4.39) Provided ordinary matrix multiplication is defined,

$$(\mathbf{A} \otimes \mathbf{B})(\mathbf{C} \otimes \mathbf{D}) = (\mathbf{AC} \otimes \mathbf{BD})$$

(1.4.40) For square matrices $\mathbf{A}$ and $\mathbf{B}$,

$$\text{Tr}(\mathbf{A} \otimes \mathbf{B}) = \text{Tr}(\mathbf{A})\,\text{Tr}(\mathbf{B})$$

(1.4.41)
$$(\mathbf{A} \otimes \mathbf{B})' = \mathbf{A}' \otimes \mathbf{B}'$$

(1.4.42) In general,

$$\mathbf{A} \otimes \mathbf{B} \neq \mathbf{B} \otimes \mathbf{A}$$

(1.4.43)
$$\mathbf{I} \otimes \mathbf{A} = \begin{bmatrix} \mathbf{A} & \mathbf{0} & \cdots & \mathbf{0} \\ \mathbf{0} & \mathbf{A} & \cdots & \mathbf{0} \\ \vdots & \vdots & & \vdots \\ \mathbf{0} & \mathbf{0} & \cdots & \mathbf{A} \end{bmatrix}$$

EXERCISES 1.4

1. Given

$$\mathbf{A} = \begin{bmatrix} 1 & 2 \\ 0 & -1 \\ 4 & 5 \end{bmatrix}, \quad \mathbf{B} = \begin{bmatrix} 3 & 0 \\ -1 & 1 \\ 2 & 7 \end{bmatrix}, \quad \mathbf{C} = \begin{bmatrix} 1 & 2 \\ -3 & 5 \\ 0 & -1 \end{bmatrix}, \quad \mathbf{D} = \begin{bmatrix} 1 & 1 \\ -1 & 2 \\ 6 & 0 \end{bmatrix}$$

$s = 2$, and $c = 3$, verify equations (1.4.8) through (1.4.12), (1.4.14), (1.4.15), and (1.4.17) through (1.4.20).

2. For

$$\mathbf{A} = \begin{bmatrix} 1 & -2 & 3 \\ 0 & 4 & 2 \\ 1 & 2 & 1 \end{bmatrix} \quad \text{and} \quad \mathbf{B} = \begin{bmatrix} 1 & 1 & 2 \\ 0 & 0 & 4 \\ 2 & -1 & 3 \end{bmatrix}$$

show that

a. $\mathbf{AB} \neq \mathbf{BA}$.

b. $\mathbf{AA}'$ and $\mathbf{A}'\mathbf{A}$ are both symmetric but are not equal.

3. a. Prove that the definition of outer product, equation (1.4.21), is equivalent to the definition (1.4.13).

b. If $\mathbf{X} = [\mathbf{x}_1, \mathbf{x}_2, \ldots, \mathbf{x}_p]$ and $\mathbf{x}$ and $\boldsymbol{\varepsilon}$ are $n \times 1$ vectors while $\boldsymbol{\beta}$ is a $p \times 1$ vector, show that $\mathbf{y} = \mathbf{X}\boldsymbol{\beta} + \boldsymbol{\varepsilon}$ may be written as $\mathbf{y} = \beta_1\mathbf{x}_1 + \beta_2\mathbf{x}_2 + \cdots + \beta_p\mathbf{x}_p + \boldsymbol{\varepsilon}$.

4. Prove equations (1.4.26) and (1.4.27).

5. Given

$$\mathbf{A} = \begin{bmatrix} \sigma_1^2 & \sigma_{12} \\ \sigma_{21} & \sigma_2^2 \end{bmatrix} \quad \text{and} \quad \mathbf{B} = \begin{bmatrix} 1/\sigma_1 & 0 \\ 0 & 1/\sigma_2 \end{bmatrix}$$

form **BAB**. Interpret this statistically.

6. a. In Example 1.4.5, show that the product of the two lower triangular matrices is lower triangular.

b. By considering transposes, obtain the corresponding result for upper triangular matrices.

c. Show that

$$\begin{bmatrix} 4 & 12 & 16 \\ 12 & 41 & 63 \\ 6 & 19 & 69 \end{bmatrix} = \begin{bmatrix} 2 & 0 & 0 \\ 6 & 5 & 0 \\ 3 & 1 & 7 \end{bmatrix} \begin{bmatrix} 2 & 6 & 8 \\ 0 & 1 & 3 \\ 0 & 0 & 6 \end{bmatrix}$$

Infer a factorization theorem from this example. Is the factorization unique?

7. a. In Example 1.4.6, recompute **AB**, where **A** is partitioned between its second and third rows rather than between its first and second rows.

b. Verify equations (1.4.34) through (1.4.43).

1.5 RANK, INVERSE, AND DETERMINANT OF A MATRIX

The Rank of a Matrix. An $n \times m$ matrix **A** may be represented in row or column form. The m column n-vectors of **A** span the column space of **A**, and the n row m-vectors of **A** generate the row space of **A**. The *column* (*or row*) *rank* of the matrix **A** is the number of linearly independent vectors or the dimensionality of the column (or row) space of **A**. It will be shown in the subsection, finding the rank of a matrix, that these two ranks are always equal. This unique number r is called the *rank* of the matrix **A**. If $\mathbf{A} = \mathbf{0}$, the rank of **A**, denoted by $R(\mathbf{A})$, is equal to zero; otherwise, $0 < R(\mathbf{A}) \leq \min(n, m)$. If $m \leq n$, $R(\mathbf{A})$ cannot exceed m, and if $R(\mathbf{A}) = r = m$, **A** is of *full rank*. If **A** is not of full rank, then there are $m - r$ dependent column vectors in **A**. Conversely, if $n \leq m$, there are $n - r$ dependent row vectors in **A**.

Some listed properties of rank follow:

(1.5.1) $$R(\mathbf{A}) = R(\mathbf{A}')$$

(1.5.2) $$R(\mathbf{AB}) \leq \min\,[R(\mathbf{A}), R(\mathbf{B})]$$

(1.5.3) $$R(\mathbf{AA}') = R(\mathbf{A}) = R(\mathbf{A}'\mathbf{A})$$

The first statement is the result of the equivalence of row and column ranks. To prove (1.5.2), note that the columns of $\mathbf{C} = \mathbf{AB}$ are linear combinations of the columns of **A** since the jth column of **C** is of the form

$$\mathbf{c}_j = [\mathbf{a}_1 \quad \mathbf{a}_2 \quad \cdots \quad \mathbf{a}_m] \begin{bmatrix} b_{1j} \\ b_{2j} \\ \vdots \\ b_{mj} \end{bmatrix}$$

Similarly, the ith row of **C** is a linear combination of the rows of **B**. Thus the rows of **C** are linear combinations of the *rows* of **B**, and the number of linearly independent rows of **C** is less than or equal to the number in **B**. Hence $R(\mathbf{AB}) \leq R(\mathbf{B})$. Similarly, employing columns, $R(\mathbf{AB}) \leq R(\mathbf{A})$. Result (1.5.2) is thus obtained. To prove (1.5.3), let V_A denote the vector space spanned by the columns of **A**. Suppose that there are vectors **v** such that $\mathbf{v} \perp \mathbf{A}$ or $\mathbf{v}'\mathbf{A} = \mathbf{0}'$. Then $\mathbf{v}'(\mathbf{AA}') = \mathbf{0}'$. Conversely, if $\mathbf{v}'(\mathbf{AA}') = \mathbf{0}'$, then $\mathbf{v}'\mathbf{A}(\mathbf{v}'\mathbf{A})' = \mathbf{0}$ or $\|\mathbf{v}'\mathbf{A}\|^2 = 0$, and thus $\mathbf{v}'\mathbf{A} = \mathbf{0}$. Every vector orthogonal to **A** is orthogonal to $\mathbf{AA}'$, and conversely. Therefore the dimension of V_A is the same as the dimension of $V_{AA'}$, or $R(\mathbf{AA}') = R(\mathbf{A})$. Replacing **A** by $\mathbf{A}'$ everywhere and observing (1.5.1), $R(\mathbf{A}) = R(\mathbf{A}'\mathbf{A})$.

The Inverse of a Matrix. The *inverse* of a square matrix of order n is an $n \times n$ matrix **B** such that pre- or postmultiplication of **A** by **B** yields the identity matrix. The inverse of the square matrix **A** is written as $\mathbf{A}^{-1}$; thus

(1.5.4) $$\mathbf{B} = \mathbf{A}^{-1} \quad \text{if and only if} \quad \mathbf{BA} = \mathbf{I} = \mathbf{AB}$$

A square matrix **A** is *nonsingular* if an inverse exists for **A**; otherwise the matrix **A** is *singular*. The inverse of **A** is clearly unique. To prove uniqueness, suppose $\mathbf{A}_1^{-1}$ is the inverse of **A**, and $\mathbf{A}_2^{-1}$ is another inverse of **A**. Then $\mathbf{AA}_1^{-1} = \mathbf{A}_1^{-1}\mathbf{A} = \mathbf{I}$ and $\mathbf{AA}_2^{-1} = \mathbf{A}_2^{-1}\mathbf{A} = \mathbf{I}$, so that $\mathbf{A}_2^{-1}\mathbf{AA}_1^{-1} = \mathbf{A}_1^{-1}$ or $\mathbf{A}_2^{-1} = \mathbf{A}_1^{-1}$.

Some useful properties of inverses follow:

(1.5.5) $$(\mathbf{AB})^{-1} = \mathbf{B}^{-1}\mathbf{A}^{-1}$$

(1.5.6) $$(\mathbf{A}')^{-1} = (\mathbf{A}^{-1})'$$

(1.5.7) The inverse of a symmetric matrix is symmetric.

(1.5.8) $$(\mathbf{A}^{-1})^{-1} = \mathbf{A}$$

Statement (1.5.5) follows from the definition since $\mathbf{B}^{-1}\mathbf{A}^{-1}\mathbf{AB} = \mathbf{I}$ and $\mathbf{ABB}^{-1}\mathbf{A}^{-1} = \mathbf{I}$, $\mathbf{B}^{-1}\mathbf{A}^{-1} = (\mathbf{AB})^{-1}$. Similarly, to prove (1.5.6), $(\mathbf{AA}^{-1})' = (\mathbf{A}^{-1})'\mathbf{A}' = \mathbf{I}$ and $(\mathbf{A}^{-1}\mathbf{A})' = \mathbf{A}'(\mathbf{A}^{-1})' = \mathbf{I}$, or $(\mathbf{A}')^{-1} = (\mathbf{A}^{-1})'$. If **A** is symmetric, $\mathbf{A} = \mathbf{A}'$, so

$\mathbf{A}^{-1} = (\mathbf{A}')^{-1}$, and by (1.5.7) it equals $(\mathbf{A}^{-1})'$, or $\mathbf{A}^{-1} = (\mathbf{A}^{-1})'$, so that $\mathbf{A}^{-1}$ is also symmetric given that $\mathbf{A}$ is symmetric. Premultiplication of $\mathbf{A}^{-1}(\mathbf{A}^{-1})^{-1} = \mathbf{I}$ by $\mathbf{A}$ establishes (1.5.8).

The Determinant of a Matrix. Associated with any $n \times n$ matrix $\mathbf{A}$ is a unique scalar function of the elements of $\mathbf{A}$ termed the *determinant* of $\mathbf{A}$ and written $|\mathbf{A}|$. Formally, the determinant of a square matrix $\mathbf{A}$ is the real-valued function defined by

(1.5.9) $$|\mathbf{A}| = \sum^{n} (-1)^k a_{1i_1} a_{2i_2} \cdots a_{ni_n}$$

where the summation is taken over all $n!$ permutations of the elements of $\mathbf{A}$ such that each product contains only one element from each row and each column of $\mathbf{A}$. The first subscript is always in its natural order and the second subscripts are equal to $1, 2, \ldots, n$ taken in some order. The exponent k represents the necessary number of interchanges of two successive elements in the sequence so that the second subscripts can be placed in their natural order, $1, 2, \ldots, n$. For example, if

$$\mathbf{A} = \begin{bmatrix} a_{11} & a_{12} \\ a_{21} & a_{22} \end{bmatrix}$$

$$|\mathbf{A}| = (-1)^k a_{11}a_{22} + (-1)^k a_{12}a_{21}$$

Now the value of k for the first term is 0 and for the second 1, so that

$$|\mathbf{A}| = a_{11}a_{22} - a_{12}a_{21}$$

If

$$\mathbf{A} = \begin{bmatrix} a_{11} & a_{12} & a_{13} \\ a_{21} & a_{22} & a_{23} \\ a_{31} & a_{32} & a_{33} \end{bmatrix}$$

$$\begin{aligned} |\mathbf{A}| &= (-1)^k a_{11}a_{22}a_{23} + (-1)^k a_{12}a_{23}a_{31} + (-1)^k a_{13}a_{21}a_{32} + (-1)^k a_{11}a_{23}a_{32} \\ &\quad + (-1)^k a_{12}a_{21}a_{33} + (-1)^k a_{13}a_{22}a_{31} \\ &= a_{11}a_{22}a_{33} + a_{12}a_{23}a_{31} + a_{13}a_{21}a_{32} - a_{11}a_{23}a_{32} - a_{12}a_{21}a_{33} - a_{13}a_{22}a_{31} \end{aligned}$$

The above definition of a determinant is called the *row expansion* of $\mathbf{A}$ since the first subscripts remained in their natural order; alternatively, the *column expansion* of $\mathbf{A}$ is

(1.5.10) $$|\mathbf{A}| = \sum^{n!} (-1)^k a_{i_1 1} a_{i_2 2} \cdots a_{i_n n}$$

The row expansion of $\mathbf{A}$ is the same as the column expansion of $\mathbf{A}$, so that

(1.5.11) $$|\mathbf{A}| = |\mathbf{A}'|$$

Finding the Determinant of a Matrix. The definitions for rank, inverse, and determinant of a matrix do not enable us to determine these quantities effectively for a given matrix $\mathbf{A}$. Evaluating determinants by employing the formal definition would be quite burdensome if n were large. We therefore need to examine more systematic procedures.

The determinant of a matrix **A** obtained by deleting the ith row and jth column of **A** is called the *minor* of a_{ij} and is denoted by $|\mathbf{M}_{ij}|$. The *cofactor* of a_{ij} (denoted by A_{ij}) is $-1^{i+j}|\mathbf{M}_{ij}|$, so that a cofactor is a *signed minor.* A determinant of **A** of order n may be conveniently expressed in terms of determinants of order $n - 1$ by either of the following formulas:

(1.5.12a) $$|\mathbf{A}| = \sum_{j=1}^{n} a_{ij}A_{ij} \qquad \text{for any } i$$

(1.5.12b) $$|\mathbf{A}| = \sum_{i=1}^{n} a_{ij}A_{ij} \qquad \text{for any } j$$

This method for determining the $|\mathbf{A}|$ is called finding the $|\mathbf{A}|$ by the *expansion of cofactors.*

EXAMPLE 1.5.1. Let

$$\mathbf{A} = \begin{bmatrix} a_{11} & a_{12} \\ a_{21} & a_{22} \end{bmatrix}$$

Expanding by the first row of **A**, the determinant of **A** is

$$|\mathbf{A}| = a_{11}(-1)^2|a_{22}| + a_{12}(-1)^3|a_{21}| = a_{11}a_{22} - a_{12} - a_{12}a_{21}$$

or, if

$$\mathbf{A} = \begin{bmatrix} a_{11} & a_{12} & a_{13} \\ a_{21} & a_{22} & a_{23} \\ a_{31} & a_{32} & a_{33} \end{bmatrix}$$

$$|\mathbf{A}| = a_{11}(-1)^2\begin{vmatrix} a_{22} & a_{23} \\ a_{32} & a_{33} \end{vmatrix} + a_{12}(-1)^3\begin{vmatrix} a_{21} & a_{23} \\ a_{31} & a_{33} \end{vmatrix} + a_{13}(-1)^4\begin{vmatrix} a_{21} & a_{22} \\ a_{31} & a_{32} \end{vmatrix}$$
$$= a_{11}a_{22}a_{33} - a_{11}a_{32}a_{23} - a_{12}a_{21}a_{33} + a_{12}a_{31}a_{23} + a_{13}a_{21}a_{32} - a_{13}a_{31}a_{22}$$

Observation of these expansions clarifies the equivalence of the formal definition with the cofactor technique. The cofactor procedure, although systematic, involves numerous calculations to determine the value of the $|\mathbf{A}|$. The following properties of determinants are useful in finding the $|\mathbf{A}|$; they are obtainable from the formal definition of the $|\mathbf{A}|$ (see Aitken, 1956).

(1.5.13) The value of a determinant remains unaltered if corresponding rows and columns of **A** are interchanged: $|\mathbf{A}| = |\mathbf{A}'|$.

(1.5.14) If any two rows (or columns) of **A** are interchanged, the determinant of **A** changes sign: $|\mathbf{A}| = -|\mathbf{A}|$, for an interchange.

(1.5.15) If any two rows (or columns) of **A** are equal, $|\mathbf{A}| = -|\mathbf{A}|$, so that $|\mathbf{A}| = 0$.

(1.5.16) If any row (column) of **A** is all 0s, the $|\mathbf{A}| = 0$.

(1.5.17) If every element of a row (or column) is multiplied by a scalar s, the value of the $|\mathbf{A}|$ is multiplied by s.

(1.5.18) If every element in **A** is multiplied by a scalar s, the determinant of **A** is multiplied by s^n.

(1.5.19) The value of the $|\mathbf{A}|$ remains unchanged if every element of a row (or column) is increased by a scalar multiple of the corresponding elements of another row (or column).

(1.5.20) If $\mathbf{A}$ is a triangular matrix, the $|\mathbf{A}|$ is equal to the product of its diagonal elements.

Employing these properties in a systematic manner facilitates the evaluation of the determinant of a matrix.

EXAMPLE 1.5.2. Let

$$\mathbf{A} = \begin{bmatrix} 6 & 1 & 0 \\ 3 & -1 & 2 \\ 4 & 0 & -1 \end{bmatrix}$$

Then, by property (1.5.17), factor 6 from row one:

$$|\mathbf{A}| = 6 \begin{vmatrix} 1 & 1/6 & 0 \\ 3 & -1 & 2 \\ 4 & 0 & -1 \end{vmatrix}$$

By property (1.5.19), multiply row one by -3 and add to row two. Multiply row one by -4 and add to row three:

$$|\mathbf{A}| = 6 \begin{vmatrix} 1 & 1/6 & 0 \\ 0 & -3/2 & 2 \\ 0 & -2/3 & -1 \end{vmatrix}$$

By property (1.5.17), factor $-3/2$ from row two:

$$|\mathbf{A}| = 6\left(\frac{-3}{2}\right) \begin{vmatrix} 1 & 1/6 & 0 \\ 0 & 1 & -4/3 \\ 0 & -2/3 & -1 \end{vmatrix}$$

By property (1.5.19), multiply row two by 2/3 and add to row three:

$$|\mathbf{A}| = 6\left(\frac{-3}{2}\right) \begin{vmatrix} 1 & 1/6 & 0 \\ 0 & 1 & -4/3 \\ 0 & 0 & -17/9 \end{vmatrix}$$

By property (1.5.17), factor $-17/9$ from row three:

$$|\mathbf{A}| = 6\left(\frac{-3}{2}\right)\left(\frac{-17}{9}\right) \begin{vmatrix} 1 & 1/6 & 0 \\ 0 & 1 & -4/3 \\ 0 & 0 & 1 \end{vmatrix}$$

By property (1.5.20),

$$|\mathbf{A}| = 6\left(\frac{-3}{2}\right)\left(\frac{-17}{9}\right) = 17$$

The process of performing these operations on rows (or columns) is known as the *sweep-out process* by triangularization.

A significant result for determinants is the property that establishes the determinant of a product of two square matrices equal to the product of the determinants of each matrix:

(1.5.21) $$|\mathbf{AB}| = |\mathbf{A}|\,|\mathbf{B}|$$

In evaluating the determinant of a partitioned matrix, if the matrix $\mathbf{A}_{11}$ is nonsingular and $\mathbf{A}_{22}$ is square, then

(1.5.22) $$|\mathbf{A}| = \begin{vmatrix} \mathbf{A}_{11} & \mathbf{A}_{12} \\ \mathbf{A}_{21} & \mathbf{A}_{22} \end{vmatrix} = |\mathbf{A}_{11}|\,|\mathbf{A}_{22} - \mathbf{A}_{21}\mathbf{A}_{11}^{-1}\mathbf{A}_{12}|$$

Finally, the determinant of the Kronecker product of two square matrices of order n and m, respectively, is the product of their individual determinants raised to the power of their orders:

(1.5.23) $$|\mathbf{A} \otimes \mathbf{B}| = |\mathbf{A}|^n\,|\mathbf{B}|^m$$

The Geometry of a Determinant. The $|\mathbf{A}|$ has a convenient geometric interpretation if we consider the triangle B_1OB_2 in Figure 1.5.1, where $\mathbf{a}_1$ and $\mathbf{a}_2$ are columns of a 2×2 matrix $\mathbf{A}$.

$$\begin{aligned}
\text{area of } B_1OB_2 &= (\text{base})(\text{height}) \\
&= \frac{1}{2}\|\mathbf{a}_2\|\,\|\mathbf{a}_1 - P_{\mathbf{a}_2}\mathbf{a}_1\| \\
&= \frac{1}{2}\sqrt{a_{12}^2 + a_{22}^2}\left\|\begin{matrix} \dfrac{a_{11}a_{22}^2 - a_{21}a_{22}a_{12}}{a_{12}^2 + a_{22}^2} \\[2ex] \dfrac{a_{21}a_{12}^2 - a_{11}a_{12}a_{22}}{a_{12}^2 + a_{22}^2} \end{matrix}\right\| \\
&= \frac{1}{2}\sqrt{\frac{(a_{12}^2 + a_{12}^2)(a_{11}a_{22} - a_{21}a_{12})^2}{a_{12}^2 + a_{22}^2}} \\
&= \frac{1}{2}(a_{11}a_{22} - a_{21}a_{12}) = \frac{1}{2}|\mathbf{A}|
\end{aligned}$$

The area of the parallelogram B_1OB_2O' is $|\mathbf{A}|$. By extending this result to higher dimensions, the determinant of $\mathbf{A}$ is equal to the volume of a parallelotope. This result is important in multivariate analysis in its relation to the *generalized variance* of a set of multivariate observations.

Finding the Inverse of a Matrix. Associated with a square matrix $\mathbf{A}$ is an *adjoint* (*or adjugate*) *matrix*. If A_{ij} denotes the cofactor of an element a_{ij} in the matrix

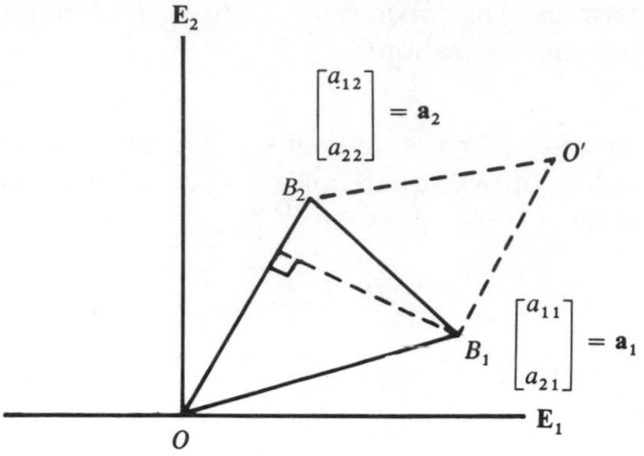

FIGURE 1.5.1. Triangle in two-dimensional space.

A, the adjoint of **A** is the matrix

(1.5.24) $$\mathbf{C} = [A_{ij}]' = \begin{bmatrix} A_{11} & A_{21} & \cdots & A_{n1} \\ A_{12} & A_{22} & \cdots & A_{n2} \\ \vdots & \vdots & \vdots & \vdots \\ A_{1n} & A_{2n} & \cdots & A_{nn} \end{bmatrix}$$

For example, if

$$\mathbf{A} = \begin{bmatrix} 1 & 2 & 3 \\ 4 & 5 & 6 \\ 7 & 8 & 10 \end{bmatrix}$$

the cofactors

$$A_{11} = (-1)^{1+1}\begin{vmatrix} 5 & 6 \\ 8 & 10 \end{vmatrix} = 2$$

$$A_{12} = (-1)^{1+2}\begin{vmatrix} 4 & 6 \\ 7 & 10 \end{vmatrix} = 2$$

$$A_{13} = (-1)^{1+3}\begin{vmatrix} 4 & 5 \\ 7 & 8 \end{vmatrix} = -3$$

Continuing,

$$\mathbf{C} = \begin{bmatrix} 2 & 4 & -3 \\ 2 & -11 & 6 \\ -3 & 6 & -3 \end{bmatrix} = [\mathbf{c}_1 \quad \mathbf{c}_2 \quad \mathbf{c}_3]$$

Representing **A** as

$$\mathbf{A} = \begin{bmatrix} \mathbf{a}_1' \\ \mathbf{a}_2' \\ \vdots \\ \mathbf{a}_n' \end{bmatrix}$$

an important property of the columns of $\mathbf{C}$ is that the inner products $(\mathbf{a}_i, \mathbf{c}_j) = 0$, for $i \neq j$, and $(\mathbf{a}_i, \mathbf{c}_j) = |\mathbf{A}|$, for $i = j$. Thus

$$\mathbf{AC} = \begin{bmatrix} |\mathbf{A}| & \cdots & \mathbf{0} \\ \vdots & \vdots & \vdots \\ \mathbf{0} & \cdots & |\mathbf{A}| \end{bmatrix} = \mathbf{CA}$$

or

$$\frac{\mathbf{AC}}{|\mathbf{A}|} = \mathbf{I} = \frac{\mathbf{C}}{|\mathbf{A}|}\mathbf{A}$$

so that $\mathbf{A}^{-1} = \mathbf{C}/|\mathbf{A}|$ if and only if $|\mathbf{A}| \neq 0$. If the determinant of $\mathbf{A}$ is 0, no inverse exists for $\mathbf{A}$, and $\mathbf{A}$ is singular. One computational scheme for finding the inverse of a matrix $\mathbf{A}$ is given by

(1.5.25) $$\mathbf{A}^{-1} = \frac{\mathbf{C}}{|\mathbf{A}|} \qquad \text{where } \mathbf{C} \text{ is the adjoint of } \mathbf{A}$$

EXAMPLE 1.5.3.

$$\mathbf{A} = \begin{bmatrix} 1 & 2 & 3 \\ 4 & 5 & 6 \\ 7 & 8 & 10 \end{bmatrix}$$

The inverse of $\mathbf{A}$, using (1.5.25), is

$$\mathbf{A}^{-1} = \frac{-1}{3} \begin{bmatrix} 2 & 4 & -3 \\ 2 & -11 & 6 \\ -3 & 6 & -3 \end{bmatrix}$$

Some further useful properties of inverses are summarized.

(1.5.26) The matrix $\mathbf{A}^{-1}$ is nonsingular: $|\mathbf{A}^{-1}| = 1/|\mathbf{A}|$.

(1.5.27) The inverse of a diagonal matrix $\mathbf{D}_n$ is

$$\mathbf{D}_n^{-1} = \begin{bmatrix} 1/d_{11} & 0 & \cdots & 0 \\ 0 & 1/d_{22} & \cdots & 0 \\ \vdots & \vdots & \vdots & \vdots \\ 0 & 0 & \cdots & 1/d_{nn} \end{bmatrix}$$

(1.5.28) The inverse of a partitioned matrix (see Searle, 1966, p. 210),

$$\mathbf{A} = \begin{bmatrix} \mathbf{A}_{11} & \mathbf{A}_{12} \\ \mathbf{A}_{21} & \mathbf{A}_{22} \end{bmatrix} \qquad \text{if } |\mathbf{A}_{11}| \neq 0, \quad |\mathbf{A}_{22}| \neq 0, \quad \text{and} \quad |\mathbf{A}| \neq 0$$

is

$$\begin{bmatrix} \mathbf{B}^{-1} & -\mathbf{B}^{-1}\mathbf{A}_{12}\mathbf{A}_{22}^{-1} \\ -\mathbf{A}_{22}^{-1}\mathbf{A}_{21}\mathbf{B}^{-1} & \mathbf{A}_{22}^{-1} + \mathbf{A}_{22}^{-1}\mathbf{A}_{21}\mathbf{B}^{-1}\mathbf{A}_{12}\mathbf{A}_{22}^{-1} \end{bmatrix}$$

or

$$\begin{bmatrix} \mathbf{A}_{11}^{-1} + \mathbf{A}_{11}\mathbf{A}_{12}\mathbf{C}^{-1}\mathbf{A}_{21}\mathbf{A}_{11}^{-1} & -\mathbf{A}_{11}^{-1}\mathbf{A}_{12}\mathbf{C}^{-1} \\ -\mathbf{C}^{-1}\mathbf{A}_{21}\mathbf{A}_{11}^{-1} & \mathbf{C}^{-1} \end{bmatrix}$$

where

$$\mathbf{B} = \mathbf{A}_{11} - \mathbf{A}_{12}\mathbf{A}_{22}^{-1}\mathbf{A}_{21} \quad \text{and} \quad \mathbf{C} = \mathbf{A}_{22} - \mathbf{A}_{21}\mathbf{A}_{11}^{-1}\mathbf{A}_{12}$$

Furthermore, from (1.5.22),

$$|\mathbf{A}| = |\mathbf{B}|\,|\mathbf{A}_{22}| = |\mathbf{C}|\,|\mathbf{A}_{11}|$$

To find the inverse of a Kronecker product of two matrices, the following result is employed

(1.5.29) $$(\mathbf{A} \otimes \mathbf{B})^{-1} = \mathbf{A}^{-1} \otimes \mathbf{B}^{-1}$$

Finding the Rank of a Matrix. To determine the rank of a matrix, it is necessary to define some elementary matrix operations. There are three basic types:

(1.5.30)
(a) In type $\mathbf{E}_1$, any two rows (or columns) are interchanged.
(b) In type $\mathbf{E}_2$, any row (or column) is multiplied by some scalar.
(c) In type $\mathbf{E}_3$, a row (or column) is replaced by adding to the replaced row (or column) a scalar multiple of another row (or column).

For example, $\mathbf{E}_1\mathbf{A}$, $\mathbf{E}_2\mathbf{A}$, and $\mathbf{E}_3\mathbf{A}$ where

$$\mathbf{E}_1 = \begin{bmatrix} 0 & 1 & 0 & \cdots & 0 \\ 1 & 0 & 0 & \cdots & 0 \\ 0 & 0 & 1 & \cdots & 0 \\ \vdots & \vdots & \vdots & \vdots & \vdots \\ 0 & 0 & 0 & \cdots & 1 \end{bmatrix}$$

$\mathbf{E}_1$ interchanges row one with row two in $\mathbf{A}$.

$$\mathbf{E}_2 = \begin{bmatrix} 1 & 0 & 0 & \cdots & 0 \\ 0 & s & 0 & \cdots & 0 \\ 0 & 0 & 1 & \cdots & 0 \\ \vdots & \vdots & \vdots & \vdots & \vdots \\ 0 & 0 & 0 & \cdots & 1 \end{bmatrix}$$

Row two of $\mathbf{A}$ is multiplied by s.

$$\mathbf{E}_3 = \begin{bmatrix} 1 & 0 & 0 & \cdots & 0 \\ s & 1 & 0 & \cdots & 0 \\ 0 & 0 & 1 & \cdots & 0 \\ \vdots & \vdots & \vdots & \vdots & \vdots \\ 0 & 0 & 0 & \cdots & 1 \end{bmatrix}$$

Row two of **A** is replaced by s times row one of **A** added to the elements of row two of **A**. Postmultiplication by the transpose of these matrices would operate on the columns of **A**.

Each elementary matrix is nonsingular. With this and the fact that pre- or postmultiplication of any matrix by a square, nonsingular matrix preserves the rank of the original matrix,

(1.5.31) $$R(\mathbf{AQ}) = R(\mathbf{A}) = R(\mathbf{PA}) \text{ if } \mathbf{P} \text{ and } \mathbf{Q} \text{ are square and nonsingular}$$

We may determine the rank of any $n \times m$ matrix **A**. To prove (1.5.31), observe property (1.5.2) of the rank of a matrix product, $R(\mathbf{AQ}) \leq \min [R(\mathbf{A}), R(\mathbf{Q})]$, and that $\mathbf{A} = \mathbf{AQQ}^{-1}$. Thus $R(\mathbf{A}) = R(\mathbf{AQQ}^{-1}) \leq R(\mathbf{AQ}) \leq R(\mathbf{A})$ or $R(\mathbf{AQ}) = R(\mathbf{A})$. Similarly, $R(\mathbf{A}) = R(\mathbf{PA})$ because of (1.5.2) and the fact that $\mathbf{A} = \mathbf{P}^{-1}\mathbf{PA}$.

By successively premultiplying a matrix **A** by elementary matrices, the matrix **A** is reduced to *row echelon form.* The row echelon form of **A** is an $n \times m$ matrix **C** with an $r \times r$ upper unit triangular submatrix followed by $n - r$ rows of 0 and $m - r$ columns of additional elements, thus

(1.5.32) $$\mathbf{E}_s \cdots \mathbf{E}_2\mathbf{E}_1\mathbf{A} = \mathbf{PA} = \begin{bmatrix} \underbrace{\mathbf{U}_r}_{r} & \underbrace{\mathbf{C}_1}_{m-r} \\ \mathbf{0} & \mathbf{0} \end{bmatrix} \begin{matrix} \}r \\ \}n-r \end{matrix} = \underset{(n \times m)}{\mathbf{C}}$$

The rank of **A** is r since the r rows of **C** constitute a linearly independent set of row vectors. Operating next on the columns of **C**, **C** may be reduced to a matrix of the following form

(1.5.33) $$\mathbf{E}_s \cdots \mathbf{E}_2\mathbf{E}_1\mathbf{AE}_{s+1} \cdots \mathbf{E}_t = \mathbf{PAQ} = \begin{bmatrix} \mathbf{I}_r & \mathbf{0} \\ \mathbf{0} & \mathbf{0} \end{bmatrix}$$

This procedure of reducing **A** to the matrix

$$\begin{bmatrix} \mathbf{I}_r & \mathbf{0} \\ \mathbf{0} & \mathbf{0} \end{bmatrix}$$

is called reducing **A** to *canonical form.* Alternatively, **A** may be reduced to *diagonal form,* where the diagonal form of **A** is represented by

(1.5.34) $$\mathbf{PAQ} = \begin{bmatrix} \mathbf{D}_r & \mathbf{0} \\ \mathbf{0} & \mathbf{0} \end{bmatrix} = \mathbf{\Lambda}$$

for some sequence of elementary operations. By using (1.5.33) or (1.5.34) and observing that the $R(\mathbf{PAQ}) = R(\mathbf{A})$ for any nonsingular matrices **P** and **Q**, we see that the row rank and the column rank of **A** are equal to r, which is the unique rank of the matrix **A**.

EXAMPLE 1.5.4. Let

$$\mathbf{A} = \begin{bmatrix} 1 & 2 \\ 3 & 9 \\ 5 & 6 \end{bmatrix}, \quad \mathbf{E}_1 = \begin{bmatrix} 1 & 0 & 0 \\ -3 & 1 & 0 \\ 0 & 0 & 1 \end{bmatrix}, \quad \mathbf{E}_2 = \begin{bmatrix} 1 & 0 & 0 \\ 0 & 1 & 0 \\ -5 & 0 & 1 \end{bmatrix}$$

$$\mathbf{E}_3 = \begin{bmatrix} 1 & 0 & 0 \\ 0 & 1 & 0 \\ 0 & 4/3 & 1 \end{bmatrix}, \quad \mathbf{E}_4 = \begin{bmatrix} 1 & 0 & 0 \\ 0 & 1/3 & 0 \\ 0 & 0 & 1 \end{bmatrix}, \quad \mathbf{E}_5 = \begin{bmatrix} 1 & -2 \\ 0 & 1 \end{bmatrix}$$

Then

$$\mathbf{E}_4\mathbf{E}_3\mathbf{E}_2\mathbf{E}_1\mathbf{A}\mathbf{E}_5 = \begin{bmatrix} 1 & 0 \\ 0 & 1 \\ 0 & 0 \end{bmatrix} \quad \text{or} \quad \begin{bmatrix} 1 & 0 & 0 \\ -1 & 1/3 & 0 \\ -1 & 4/3 & 1 \end{bmatrix} \mathbf{A} \begin{bmatrix} 1 & -2 \\ 0 & 1 \end{bmatrix} = \begin{bmatrix} 1 & 0 \\ 0 & 1 \\ 0 & 0 \end{bmatrix}$$

so that the rank of $\mathbf{A}$ is 2. Alternatively, using the reduction to diagonal form,

$$\mathbf{E}_3\mathbf{E}_2\mathbf{E}_1\mathbf{A}\mathbf{E}_5 = \begin{bmatrix} 1 & 0 \\ 0 & 3 \\ 0 & 0 \end{bmatrix} \quad \text{or} \quad \begin{bmatrix} 1 & 0 & 0 \\ -3 & 1 & 0 \\ -9 & 4/3 & 1 \end{bmatrix} \mathbf{A} \begin{bmatrix} 1 & -2 \\ 0 & 1 \end{bmatrix} = \begin{bmatrix} 1 & 0 \\ 0 & 3 \\ 0 & 0 \end{bmatrix}$$

which also illustrates that the rank of $\mathbf{A}$ is 2.

The reduction of $\mathbf{A}$ to canonical form leads to the following factorization theorem:

(1.5.35) If $\mathbf{A}$ is any $n \times m$ matrix of rank r, there exist nonsingular square matrices $\mathbf{P}$ and $\mathbf{Q}$ of order n and m, respectively, such that

$$\mathbf{PAQ} = \begin{bmatrix} \mathbf{I}_r & \mathbf{0} \\ \mathbf{0} & \mathbf{0} \end{bmatrix}$$

Also, $\mathbf{A}$ may be factored into

$$\mathbf{A} = \mathbf{P}^{-1} \begin{bmatrix} \mathbf{I}_r & \mathbf{0} \\ \mathbf{0} & \mathbf{0} \end{bmatrix} \mathbf{Q}^{-1} = \mathbf{P}_1\mathbf{Q}_1$$

where $\mathbf{P}_1$ and $\mathbf{Q}_1$ are $n \times r$ and $r \times m$ matrices of rank r.

When $\mathbf{A}$ is a square symmetric matrix, the following theorem results:

(1.5.36) If $\mathbf{A}$ is a square symmetric matrix of order n and rank r, there exists a nonsingular matrix $\mathbf{P}$ of order n such that

$$\mathbf{PAP}' = \begin{bmatrix} \mathbf{D}_r & \mathbf{0} \\ \mathbf{0} & \mathbf{0} \end{bmatrix} = \mathbf{\Lambda}$$

and $\mathbf{A}$ may be represented as

$$\mathbf{A} = \mathbf{P}^{-1}\mathbf{\Lambda}(\mathbf{P}')^{-1}$$

If the rank $\mathbf{A}$ is of full rank, then $\mathbf{PAP}' = \mathbf{D}_n$ or $\mathbf{A} = \mathbf{P}^{-1}\mathbf{D}_n(\mathbf{P}')^{-1}$.

Given any square $n \times n$ matrix $\mathbf{A}$ of full rank, elementary row operations are used to determine $\mathbf{A}^{-1}$. For clarity, observe that $\mathbf{PA} = \mathbf{U}_n$ and that only $n(n-1)/2$ row operations are needed to reduce $\mathbf{U}_n$ to $\mathbf{I}_n$. Thus, $\mathbf{P}^*\mathbf{PA} = \mathbf{I}_n$. Further, $\mathbf{A} = \mathbf{P}^{-1}(\mathbf{P}^*)^{-1}\mathbf{I}_n$ or $\mathbf{A}^{-1} = \mathbf{P}^*\mathbf{PI}_n$. Hence, if one uses the same elementary row operations on $\mathbf{I}_n$ and

A until **A** is reduced to $\mathbf{I}_n$, then $\mathbf{I}_n$ becomes $\mathbf{A}^{-1}$. This procedure of inversion is known as *Gauss' matrix inversion technique.*

EXAMPLE 1.5.5. Let

$$\mathbf{A} = \begin{bmatrix} 2 & 3 & 1 \\ 1 & 2 & 3 \\ 3 & 1 & 2 \end{bmatrix}$$

To find $\mathbf{A}^{-1}$, write

$$(\mathbf{A} \mid \mathbf{I} \parallel \text{row totals}) = \left[\begin{array}{ccc|ccc||c} 2 & 3 & 1 & 1 & 0 & 0 & 7 \\ 1 & 2 & 3 & 0 & 1 & 0 & 7 \\ 3 & 1 & 2 & 0 & 0 & 1 & 7 \end{array}\right]$$

Multiply row one by 1/2, and subtract row two from row one. Multiply row three by 1/3, and subtract row three from row one.

$$\left[\begin{array}{ccc|ccc||c} 1 & 3/2 & 1/2 & 1/2 & 0 & 0 & 7/2 \\ 0 & 1/2 & 5/2 & -1/2 & 1 & 0 & 7/2 \\ 0 & -7/6 & 1/6 & -1/2 & 0 & 1/3 & -7/6 \end{array}\right]$$

Multiply row two by 2 and row three by $-6/7$. Then subtract row three from row two. Multiply row three by $-7/36$.

$$\left[\begin{array}{ccc|ccc||c} 1 & 3/2 & 1/2 & 1/2 & 0 & 0 & 7/2 \\ 0 & 1 & 5 & -1 & 2 & 0 & 7 \\ 0 & 0 & 1 & -5/18 & 7/18 & 1/18 & 7/6 \end{array}\right]$$

Multiply row three by -5, and add to row two. Then multiply row three by $-1/2$, and add to row one.

$$\left[\begin{array}{ccc|ccc||c} 1 & 3/2 & 0 & 23/26 & -7/36 & -1/36 & 105/36 \\ 0 & 1 & 0 & 7/18 & 1/18 & -5/18 & 7/6 \\ 0 & 0 & 1 & -5/18 & 7/18 & 1/18 & 7/6 \end{array}\right]$$

Multiply row two by $-3/2$, and add to row one.

$$\left[\begin{array}{ccc|ccc||c} 1 & 0 & 0 & 1/18 & -5/18 & 7/18 & 7/6 \\ 0 & 1 & 0 & 7/18 & 1/18 & -5/18 & 7/6 \\ 0 & 0 & 1 & -5/18 & 7/18 & 1/18 & 7/6 \end{array}\right] = (\mathbf{I} \mid \mathbf{A}^{-1} \parallel \text{row totals})$$

so that

$$\mathbf{A}^{-1} = \frac{1}{18}\begin{bmatrix} 1 & -5 & 7 \\ 7 & 1 & -5 \\ -5 & 7 & 1 \end{bmatrix}$$

The operations applied to the rows are systematically applied to the totals in order that calculations may be checked at each stage. The sum of the elements in each row of the first two partitions must equal the total. Gauss' method is well suited for a desk calculator and is computationally more efficient than (1.5.25). However, matrix inversion on an electronic computer usually makes use of the Gauss-Jordan method, a compact version of Gauss' procedure (see Householder, 1964, or Faddeeva, 1959).

EXERCISES 1.5

1. Find two matrices **A** and **B** such that $R(\mathbf{AB})$ is strictly less than $\min[R(\mathbf{A}), R(\mathbf{B})]$.

2. For

$$\mathbf{A} = \begin{bmatrix} 2 & 1 & -1 \\ 0 & 2 & 3 \\ 1 & 1 & 1 \end{bmatrix}$$

find the $|\mathbf{A}|$

a. by the expansion of cofactors.

b. by the sweep-out process.

3. Verify properties (1.5.13) through (1.5.20) by constructing appropriate matrices.

4. a. Verify (1.5.22) for

$$\mathbf{A} = \left[\begin{array}{cccc} 1 & 2 & 1 & 5 \\ -3 & 4 & 2 & 0 \\ \hdashline 0 & 3 & 1 & 2 \\ 2 & 0 & 2 & 4 \end{array}\right]$$

b. Verify (1.5.23) for

$$\mathbf{A} = \begin{bmatrix} 10 \\ 25 \end{bmatrix} \quad \text{and} \quad \mathbf{B} = \begin{bmatrix} 1 & 0 & 1 \\ 0 & 2 & 0 \\ 3 & 0 & 2 \end{bmatrix}$$

5. Find the area of the parallelogram whose vertices are the points [0, 0], [2, 3], [1, 4], and [3, 7].

6. Find the inverse of

$$\mathbf{A} = \begin{bmatrix} 1 & 2 & 3 \\ 1 & 0 & 0 \\ 2 & 1 & -1 \end{bmatrix}$$

a. by using the method of (1.5.25).

b. by using Gauss' matrix inversion technique.

7. If

$$\mathbf{A} = \begin{bmatrix} 1 & 0 & 0 & 1 \\ 3 & 5 & 1 & 2 \\ 1 & -1 & 2 & -1 \\ 0 & 3 & 4 & 1 \end{bmatrix}$$

find $\mathbf{A}^{-1}$ by appropriately partitioning **A** and applying (1.5.28), first verifying that $|\mathbf{A}| \neq 0$ by (1.5.22) and also by showing that $R(\mathbf{A}) = 4$ by reducing **A** to canonical form.

8. Given

$$\mathbf{A} = \begin{bmatrix} 1 & 0 & 2 \\ 3 & 1 & 5 \\ 5 & 2 & 8 \\ 0 & 0 & 1 \end{bmatrix}$$

of rank r, employ (1.3.35) to obtain the factorization

$$\underset{(4 \times 3)}{\mathbf{A}} = \underset{(4 \times r)}{\mathbf{P}_1} \underset{(r \times 3)}{\mathbf{Q}_1}$$

where $R(\mathbf{P}_1) = R(\mathbf{Q}_1) = r$.

b. If

$$\mathbf{A} = \begin{bmatrix} 2 & 1 & 2 \\ 1 & 0 & 4 \\ 2 & 4 & -16 \end{bmatrix}$$

with $R(\mathbf{A}) = r$, employ (1.5.36) to obtain the factorization

$$\mathbf{A} = \underset{(3 \times r)}{\mathbf{P}_1} \underset{(r \times 3)}{\mathbf{P}_1'}$$

where $R(\mathbf{P}_1) = R(\mathbf{P}_1') = r$.

1.6 SIMULTANEOUS LINEAR EQUATIONS AND LINEAR TRANSFORMATIONS

The inverse of a matrix is most frequently used in the solution of a system of n simultaneous equations in m unknowns:

(1.6.1)
$$\begin{aligned} a_{11}x_1 + a_{12}x_2 + \cdots + a_{1m}x_m &= y_1 \\ a_{21}x_1 + a_{22}x_2 + \cdots + a_{2m}x_m &= y_2 \\ \vdots \qquad \vdots \qquad \vdots \qquad \vdots \quad &\quad \vdots \\ a_{n1}x_1 + a_{n2}x_2 + \cdots + a_{nm}x_m &= y_n \end{aligned}$$

Or, employing matrix notation,

(1.6.2)
$$\underset{(n \times m)}{\mathbf{A}} \underset{(m \times 1)}{\mathbf{x}} = \underset{(n \times 1)}{\mathbf{y}}$$

where the vector $\mathbf{x}$ is a vector of unknowns, the vector $\mathbf{y}$ is a known nonnull vector of scalars, and the $R(\mathbf{A}) = r$.

Consistent Systems. Before considering the solution of the general system, some special cases of (1.6.2) need to be considered. When the number of equations is the same as the number of unknowns ($n = m$) and the matrix $\mathbf{A}$ is of full rank, $R(\mathbf{A}) = m$, a unique solution to (1.6.2) exists in terms of $\mathbf{A}^{-1}$; that is, $\mathbf{x} = \mathbf{A}^{-1}\mathbf{y}$. In many statistical applications, $R(\mathbf{A}) = m$, but $m < n$. This is still the case of full rank; however, $\mathbf{x} = (\mathbf{A}'\mathbf{A})^{-1}\mathbf{A}'\mathbf{y}$ is the unique solution. In general, whenever the matrix $\mathbf{A}$ is of full rank, the solution to (1.6.2) is unique. If (1.6.2) admits a solution, the system is said to be *consistent*; otherwise, (1.6.2) is *inconsistent*. A unique solution for (1.6.2) does not have to exist to allow a consistent system. There may be an infinite number of solutions to a consistent system.

To determine whether the system $\mathbf{Ax} = \mathbf{y}$ is consistent or not, the following theorem is employed:

(1.6.3) The system $\mathbf{Ax} = \mathbf{y}$ is consistent if and only if the rank of the matrix $[\mathbf{A} \quad \mathbf{y}]$ is equal to the $R(\mathbf{A}) = r$ (that is, if $\mathbf{y}$ is in the column space of $\mathbf{A}$).

If the $R(\mathbf{A} \quad \mathbf{y}) > R(\mathbf{A})$, the system is inconsistent; if the $R(\mathbf{A} \quad \mathbf{y}) = R(\mathbf{A}) = m$, the system has a unique solution. To prove (1.6.3), let $\mathbf{x}$ be a solution to (1.6.2). Then, since the $R(\mathbf{A}) = r$, (1.6.2) may be written as

$$\begin{matrix} r\{ \\ n-r\{ \end{matrix}\begin{bmatrix} \mathbf{A}_1 \\ \mathbf{CA}_1 \end{bmatrix}\mathbf{x} = \begin{bmatrix} \mathbf{y}_1 \\ \mathbf{y}_2 \end{bmatrix} \quad \text{or} \quad \begin{matrix} \mathbf{A}_1\mathbf{x} = \mathbf{y}_1 \\ \mathbf{CA}_1\mathbf{x} = \mathbf{y}_2 \end{matrix} \tag{1.6.4}$$

By (1.6.4), $\mathbf{CA}_1\mathbf{x} = \mathbf{Cy}_1$, or $\mathbf{Cy}_1 = \mathbf{y}_2$, hence $R(\mathbf{A} \quad \mathbf{y}) = R(\mathbf{A}) = r$. Conversely, if $R(\mathbf{A} \quad \mathbf{y}) = R(\mathbf{A}) = r$, $\mathbf{y}$ must be a linear combination of r columns of $\mathbf{A}$. Thus, for some scalars $x_1, x_2, \ldots, x_r$,

$$\mathbf{y} = \sum_{i=1}^{r} \mathbf{a}_i x_i$$

where $\mathbf{a}_1, \mathbf{a}_2, \ldots, \mathbf{a}_r$ is a column basis for $\mathbf{A}$. Letting $x_i = 0$ (for $i = r + 1, \ldots, m$), $\mathbf{y} = \mathbf{Ax}$, so that a solution exists. Establishing a solution for (1.6.2), given (1.6.3), observe that if $\mathbf{x}$ solves (1.6.2), it need only satisfy $\mathbf{A}_1\mathbf{x} = \mathbf{y}_1$ in (1.6.4). To solve (1.6.4), partition $\mathbf{A}_1 = [\mathbf{A}_{11} \quad \mathbf{A}_{12}]$ such that the $R(\mathbf{A}_{11}) = r$, where $\mathbf{A}_{11}$ is $r \times r$. Then

$$[\mathbf{A}_{11} \quad \mathbf{A}_{12}]\begin{bmatrix} \mathbf{x}_1 \\ \mathbf{x}_2 \end{bmatrix} = \mathbf{y}_1 \quad \text{or} \quad \mathbf{A}_{11}\mathbf{x}_1 + \mathbf{A}_{12}\mathbf{x}_2 = \mathbf{y}_1 \tag{1.6.5}$$

and

$$\mathbf{x}_1 = \mathbf{A}_{11}^{-1}\mathbf{y}_1 - \mathbf{A}_{11}^{-1}\mathbf{A}_{12}\mathbf{x}_2$$

A solution for (1.6.2) is then

$$\mathbf{x} = \begin{bmatrix} \mathbf{x}_1 \\ \mathbf{x}_2 \end{bmatrix} = \begin{bmatrix} \mathbf{A}_{11}^{-1}\mathbf{y}_1 - \mathbf{A}_{11}^{-1}\mathbf{A}_{12}\mathbf{x}_2 \\ \mathbf{x}_2 \end{bmatrix} \tag{1.6.6}$$

The solution vector admits an infinity of solutions to (1.6.2) because $\mathbf{x}_2$ is arbitrary. Therefore, given (1.6.3), we may find a solution for (1.6.2) by selecting any r linearly independent equations and solving for r of the unknowns in terms of elements of $\mathbf{y}_1$ and $m - r$ of the remaining $\mathbf{x}_i$'s.

The solution to (1.6.2) would be tedious if equation (1.6.6) were employed. *Gauss' elimination procedure* is more systematic and efficient. The method is similar to the method employed in finding $\mathbf{A}^{-1}$.

EXAMPLE 1.6.1. Consider the system

$$\begin{aligned} x_1 + 3x_2 - x_3 &= 4 \\ 2x_1 + x_2 + x_3 &= 7 \\ x_1 - 2x_2 + 2x_3 &= 3 \\ 3x_1 + 4x_2 \phantom{{}+2x_3} &= 11 \end{aligned}$$

where the $R(\mathbf{A}\quad\mathbf{y}) = R(\mathbf{A}) = 2$. Form the matrix

$$\begin{array}{c} \text{Totals} \\ \left[\begin{array}{ccc|c||c} 1 & 3 & -1 & 4 & 7 \\ 2 & 1 & 1 & 7 & 11 \\ 1 & -2 & 2 & 3 & 4 \\ 3 & 4 & 0 & 11 & 18 \end{array}\right] \end{array}$$

Multiply row one by -2, -1, and -3; add to rows two, three, and four, respectively.

$$\left[\begin{array}{ccc|c||c} 1 & 3 & -1 & 4 & 7 \\ 0 & -5 & 3 & -1 & -3 \\ 0 & -5 & 3 & -1 & -3 \\ 0 & -5 & 3 & -1 & -3 \end{array}\right]$$

Multiply row two by $-1/5$; then multiply row two by 5, and add to row three. Multiply row two by 5, and add to row four.

$$\left[\begin{array}{ccc|c||c} 1 & 3 & -1 & 4 & 7 \\ 0 & 1 & -3/5 & 1/5 & 3/5 \\ 0 & 0 & 0 & 0 & 0 \\ 0 & 0 & 0 & 0 & 0 \end{array}\right]$$

Thus

$$x_1 - 3x_2 - x_3 = 4 \quad \text{and} \quad x_2 - \frac{3x_3}{5} = \frac{1}{5}$$

or

$$x_2 = \frac{3x_3}{5} + \frac{1}{5} \quad \text{and} \quad x_1 = \frac{17}{5} - \frac{4x_3}{5}$$

where x_3 is arbitrary. Other procedures commonly used on computers are outlined in Fadeeva (1959). The Gauss-Doolittle method is often employed.

Theorem (1.6.3) applies to the *nonhomogeneous system* (1.6.2). The *homogeneous* system of equations

$$\underset{(n \times m)}{\mathbf{A}} \; \underset{(m \times 1)}{\mathbf{x}} = \underset{(n \times 1)}{\mathbf{0}} \tag{1.6.7}$$

is always consistent since the null vector $\mathbf{x}' = [0, \ldots, 0]$ satisfies the system. Moreover, if the $R(\mathbf{A}) = m$, the only solution to (1.6.7) is the trivial solution $\mathbf{x} = \mathbf{0}$. Hence, for (1.6.7) to have a nontrivial solution, the $R(\mathbf{A}) = r$ must be less than m. For a square matrix $\mathbf{A}$, this implies that the $|\mathbf{A}| = 0$.

Three procedures often used in statistical work to solve systems such as (1.6.2) and (1.6.7) are: (1) solution by *restricting the number of unknowns*, (2) solution by *reparameterization*, and (3) solution by use of a *generalized inverse*. Detailed coverage of these methods is given in Scheffé (1959, p. 17), Graybill (1961, p. 235) and Rao (1965a, p. 26 and Chap. 4).

Restricting the Number of Unknowns. This method adds a sufficient number of independent equations—side conditions—to the original consistent system to remove its rank deficiency. The new system is then of full rank and a unique solution is obtained with the imposed side conditions. If

$$\underset{(n\times m)}{\mathbf{A}}\ \underset{(m\times 1)}{\mathbf{x}} = \underset{(n\times 1)}{\mathbf{y}} \qquad \text{where } n > m \text{ and } R(\mathbf{A}) = r < m$$

represents the original consistent system, then side conditions of the form

$$\underset{(m-r)\times m}{\mathbf{R}}\ \underset{(m\times 1)}{\mathbf{x}} = \underset{(m-r)\times 1}{\boldsymbol{\theta}} \tag{1.6.8}$$

are added to $\mathbf{Ax} = \mathbf{y}$ so that the $R(\mathbf{R}) = m - r$ and the $R(\mathbf{A}'\ \ \mathbf{R}') = r + (m - r) = m$, where the rows in $\mathbf{R}$ are not linearly dependent on the rows in $\mathbf{A}$. The new consistent system

$$\begin{bmatrix} \mathbf{A} \\ \mathbf{R} \end{bmatrix} \mathbf{x} = \begin{bmatrix} \mathbf{y} \\ \boldsymbol{\theta} \end{bmatrix} \tag{1.6.9}$$

now has a unique solution:

$$\mathbf{x} = (\mathbf{A}'\mathbf{A} + \mathbf{R}'\mathbf{R})^{-1}(\mathbf{A}'\mathbf{y} + \mathbf{R}'\boldsymbol{\theta}) \tag{1.6.10}$$

EXAMPLE 1.6.2. Consider the system

$$\begin{bmatrix} 1 & 1 & 0 \\ 1 & 1 & 0 \\ 1 & 0 & 1 \\ 1 & 0 & 1 \end{bmatrix} \begin{bmatrix} \mu \\ \alpha_1 \\ \alpha_2 \end{bmatrix} = \begin{bmatrix} y_{11} \\ y_{12} \\ y_{21} \\ y_{22} \end{bmatrix}$$

where the $R(\mathbf{A}\ \ \mathbf{y}) = R(\mathbf{A}) = 2$. One restriction must be imposed to bring $\mathbf{A}$ up to full rank. Choose $\alpha_1 + \alpha_2 = 0$. Then $\mathbf{R} = [0, 1, 1]$ and $\boldsymbol{\theta} = 0$. The new system with side conditions becomes

$$\begin{bmatrix} 1 & 1 & 0 \\ 1 & 1 & 0 \\ 1 & 0 & 1 \\ 1 & 0 & 1 \\ \hline 0 & 1 & 1 \end{bmatrix} \begin{bmatrix} \mu \\ \alpha_1 \\ \alpha_2 \end{bmatrix} = \begin{bmatrix} y_{11} \\ y_{12} \\ y_{21} \\ y_{22} \\ \hline 0 \end{bmatrix}$$

or, using (1.6.10),

$$\begin{bmatrix} \mu \\ \alpha_1 \\ \alpha_2 \end{bmatrix} = \begin{bmatrix} 4 & 2 & 2 \\ 2 & 3 & 1 \\ 2 & 1 & 3 \end{bmatrix}^{-1} \begin{bmatrix} 4y_{..} \\ 2y_{1.} \\ 2y_{2.} \end{bmatrix}$$

where

$$y_{..} = \frac{y_{11} + y_{12} + y_{21} + y_{22}}{4}$$

$$y_{1.} = \frac{y_{11} + y_{12}}{2}$$

$$y_{2.} = \frac{y_{21} + y_{22}}{2}$$

Thus

$$\begin{bmatrix} \mu \\ \alpha_1 \\ \alpha_2 \end{bmatrix} = \frac{1}{16} \begin{bmatrix} 8 & -4 & -4 \\ -4 & 8 & 0 \\ -4 & 0 & 8 \end{bmatrix} \begin{bmatrix} 4y_{..} \\ 2y_{1.} \\ 2y_{2.} \end{bmatrix}$$

$$= \begin{bmatrix} 1/2 & -1/4 & -1/4 \\ -1/4 & 1/2 & 0 \\ -1/4 & 0 & 1/2 \end{bmatrix} \begin{bmatrix} 4y_{..} \\ 2y_{1.} \\ 2y_{2.} \end{bmatrix}$$

$$= \begin{bmatrix} 2y_{..} - (y_{1.}/2) - (y_{2.}/2) \\ y_{1.} - y_{..} \\ y_{2.} - y_{..} \end{bmatrix}$$

or

$$\mu = 2y_{..} - \frac{y_{1.}}{2} - \frac{y_{2.}}{2} = y_{..}$$

$$\alpha_1 = y_{1.} - y_{..}$$

$$\alpha_2 = y_{2.} - y_{..}$$

Alternatively, Gauss' elimination procedure may be employed to solve the system:

$$\begin{bmatrix} 4 & 2 & 2 \\ 2 & 3 & 1 \\ 2 & 1 & 3 \end{bmatrix} \begin{bmatrix} \mu \\ \alpha_1 \\ \alpha_2 \end{bmatrix} = \begin{bmatrix} 4y_{..} \\ 2y_{1.} \\ 2y_{2.} \end{bmatrix}$$

or

$$\begin{bmatrix} 1 & 1/2 & 1/2 \\ 1 & 3/2 & 1/2 \\ 1 & 1/2 & 3/2 \end{bmatrix} \begin{bmatrix} \mu \\ \alpha_1 \\ \alpha_2 \end{bmatrix} = \begin{bmatrix} y_{..} \\ y_{1.} \\ y_{2.} \end{bmatrix}$$

or

$$\begin{bmatrix} 1 & 1/2 & 1/2 \\ 0 & 1 & 0 \\ 0 & 0 & 1 \end{bmatrix} \begin{bmatrix} \mu \\ \alpha_1 \\ \alpha_2 \end{bmatrix} = \begin{bmatrix} y_{..} \\ y_{1.} - y_{..} \\ y_{2.} - y_{..} \end{bmatrix}$$

so

$$\mu = y_{..}$$
$$\alpha_1 = y_{1.} - y_{..}$$
$$\alpha_2 = y_{2.} - y_{..}$$

Reparameterization. In the second procedure, we again consider (1.6.2), where the $R(\mathbf{A} \quad \mathbf{y}) = R(\mathbf{A}) = r \leq m$. In using the method of reparameterization, the system is reduced to full rank r by solving for only r linear functions of the unknowns. The approach is to introduce a new set of unknowns $\mathbf{x}^* = \mathbf{Cx}$, where $\mathbf{x}^*$ is an $r \times 1$ vector, $\mathbf{C}$ is an $r \times m$ matrix of rank r, and $\mathbf{x}$ is the $m \times 1$ original vector of unknowns. From (1.5.35) there exist matrices $\mathbf{B}$ and $\mathbf{C}$ both of rank r such that

$$\underset{(n \times m)}{\mathbf{A}} = \underset{(n \times r)}{\mathbf{B}} \; \underset{(r \times m)}{\mathbf{C}}$$

Substitution of this into (1.6.2) provides a unique solution:

$$\begin{aligned} \mathbf{Ax} &= \mathbf{y} \\ \mathbf{BCx} &= \mathbf{y} \\ (\mathbf{B'B})\mathbf{Cx} &= \mathbf{B'y} \\ \mathbf{Cx} &= (\mathbf{B'B})^{-1}\mathbf{B'y} \\ \mathbf{x}^* &= (\mathbf{B'B})^{-1}\mathbf{B'y} \end{aligned} \tag{1.6.11}$$

The quantity $\mathbf{x}^*$ is the reparameterized vector of unknowns. To find the matrix $\mathbf{B}$ in (1.6.11), $\mathbf{C}$ must be selected so that the rows of $\mathbf{C}$ are in the row space of $\mathbf{A}$, since $\mathbf{A} = \mathbf{BC}$, or so that the

$$R\begin{bmatrix} \mathbf{A} \\ \mathbf{C} \end{bmatrix} = R(\mathbf{A}) = R(\mathbf{C}) = r \tag{1.6.12}$$

Given $\mathbf{C}$, $\mathbf{B}$ is determinable in terms of the matrices $\mathbf{A}$ and $\mathbf{C}$:

$$\mathbf{B} = \mathbf{B}(\mathbf{CC'})(\mathbf{CC'})^{-1} = \mathbf{AC'}(\mathbf{CC'})^{-1} \tag{1.6.13}$$

Substitution of (1.6.13) into (1.6.11) provides a unique solution $\mathbf{x}^*$ for the reparameterized system.

In statistical work, the rows of $\mathbf{C}$ are usually chosen in such a manner that the vector of new variables $\mathbf{x}^*$ has a useful interpretation.

EXAMPLE 1.6.3. Let

$$\begin{bmatrix} 1 & 1 & 0 \\ 1 & 1 & 0 \\ 1 & 0 & 1 \\ 1 & 0 & 1 \end{bmatrix} \begin{bmatrix} \mu \\ \alpha_1 \\ \alpha_2 \end{bmatrix} = \begin{bmatrix} y_{11} \\ y_{12} \\ y_{21} \\ y_{22} \end{bmatrix}$$

where the $R(\mathbf{A}) = R(\mathbf{A} \quad \mathbf{y}) = 2$. $\mathbf{C}$ must be chosen such that $R(\mathbf{C}) = R(\mathbf{A}' \quad \mathbf{C}') = 2$. Thus let

$$\mathbf{C} = \begin{bmatrix} 1 & 1/2 & 1/2 \\ 0 & 1 & -1 \end{bmatrix}$$

then

$$(\mathbf{CC}') = \begin{bmatrix} 3/2 & 0 \\ 0 & 2 \end{bmatrix}, \qquad (\mathbf{CC}')^{-1} = \begin{bmatrix} 2/3 & 0 \\ 0 & 1/2 \end{bmatrix}$$

$$\mathbf{B} = \mathbf{AC}'(\mathbf{CC}')^{-1} = \begin{bmatrix} 1 & 1/2 \\ 1 & 1/2 \\ 1 & -1/2 \\ 1 & -1/2 \end{bmatrix}$$

so

$$\begin{aligned} \mathbf{x}^* &= (\mathbf{BB}')^{-1}\mathbf{B}'\mathbf{y} \\ &= \begin{bmatrix} 4 & 0 \\ 0 & 1 \end{bmatrix}^{-1} \begin{bmatrix} 4y_{..} \\ y_{1.} - y_{2.} \end{bmatrix} \\ &= \begin{bmatrix} y_{..} \\ y_{1.} - y_{2.} \end{bmatrix} \end{aligned}$$

or

$$\mathbf{x}^* = \mathbf{Cx} = \begin{bmatrix} 1 & 1/2 & 1/2 \\ 0 & 1 & -1 \end{bmatrix} \begin{bmatrix} \mu \\ \alpha_1 \\ \alpha_2 \end{bmatrix}$$

and

$$\mu + \frac{\alpha_1 + \alpha_2}{2} = y_{..} \qquad \alpha_1 - \alpha_2 = y_{1.} - y_{2.}$$

This is consistent with the previous results since, if $\alpha_1 = y_{1.} - y_{..}$ and $\alpha_2 = y_{2.} - y_{..}$, then $\alpha_1 - \alpha_2 = y_{1.} - y_{2.}$. However, $\mu = y_{..}$ only if $\alpha_1 + \alpha_2 = 0$. As a special case, $\mu = y_{..}$ when $\alpha_1 = \alpha_2 = 0$.

Generalized Inverse. The last procedure (the use of a generalized inverse) is a method designed to determine a solution to a consistent system of equations. It is a general technique allowing us to solve (1.6.2) when a solution exists, without specifying explicitly the "appropriate" side conditions required in the first method or choosing a matrix $\mathbf{C}$ such that the rows of $\mathbf{C}$ are in the row space of $\mathbf{A}$, a requirement of the reparameterization technique.

If the matrix $\mathbf{A}$ in (1.6.2) is square of order m and full rank, the solution is $\mathbf{x} = \mathbf{A}^{-1}\mathbf{y}$. For a rectangular matrix $\mathbf{A}$ of full column rank m, the solution is $\mathbf{x} = (\mathbf{A}'\mathbf{A})^{-1}\mathbf{A}'\mathbf{y}$. The matrix $(\mathbf{A}'\mathbf{A})^{-1}\mathbf{A}'$ is a type of "inverse" matrix in the sense that $[(\mathbf{A}'\mathbf{A})^{-1}\mathbf{A}']\mathbf{A} = \mathbf{I}_m$; however, $\mathbf{A}$ in this case is not square and $\mathbf{A}[(\mathbf{A}'\mathbf{A})^{-1}\mathbf{A}'] \neq \mathbf{I}_n$. In general, it would be desirable to have a matrix $\mathbf{A}^+$ that behaves in some way as the usual inverse of **A**, so that a solution to (1.6.2) would have representation similar to the full-rank case $\mathbf{x} = \mathbf{A}^+\mathbf{y}$, when the $R(\mathbf{A}) = r \le m \le n$. Such a matrix may be termed a *generalized inverse* for a matrix **A** of order $n \times m$.

Penrose (1955) was the first to define and use the generalized inverse in the solution of (1.6.2). He defined the generalized inverse of a matrix **A** as the unique matrix $\mathbf{A}^+$ satisfying the following four conditions:

$$\mathbf{A}\mathbf{A}^+\mathbf{A} = \mathbf{A} \qquad (\mathbf{A}\mathbf{A}^+)' = \mathbf{A}\mathbf{A}^+$$

$$\mathbf{A}^+\mathbf{A}\mathbf{A}^+ = \mathbf{A}^+ \qquad (\mathbf{A}^+\mathbf{A})' = \mathbf{A}^+\mathbf{A}$$

With $\mathbf{A}^+$ defined by $\mathbf{A}^+ = (\mathbf{A}'\mathbf{A})^{-1}\mathbf{A}'$ for the full-rank case, $\mathbf{A}^+$ is seen to satisfy the four conditions defined above. However, Rao (1962, 1965a, 1966b) found that all of Penrose's conditions were not necessary to obtain a solution for a consistent system of equations $\mathbf{A}\mathbf{x} = \mathbf{y}$. For Rao and this text, a *generalized inverse* (*or g-inverse*) of an $n \times m$ matrix **A** is any matrix satisfying only the first of Penrose's conditions. That is, using Rao's notation, a generalized inverse is any matrix $\mathbf{A}^-$ satisfying the condition

$$\mathbf{A}\mathbf{A}^-\mathbf{A} = \mathbf{A} \tag{1.6.14}$$

The symbol $\mathbf{A}^-$ is used for a *g*-inverse to distinguish it from Penrose's *g*-inverse.

In the literature, the term *g*-inverse is not used exclusively for Rao's definition. Graybill (1969, p. 119) calls it a *conditional inverse* and even Rao (1955a) has called it a *pseudo-inverse*. Many authors have also used the term pseudo-inverse for Penrose's generalized inverse. Several other variants of Penrose's inverse have been suggested and are referred to as some type of generalized inverse. For an extensive review of generalized inverses, see Boullion and Odell (1971) and Rao and Mitra (1971).

Rao (1962, 1965a, 1966b) presents several procedures for determining $\mathbf{A}^-$ for any $n \times m$ matrix **A**. We now consider two. From (1.5.36) there exist nonsingular matrices **P** and **Q** of order n and m, respectively, such that $\mathbf{PAQ} = \mathbf{\Lambda}$ where $\mathbf{\Lambda}$ is of the form

$$\mathbf{\Lambda} = \begin{bmatrix} \mathbf{D}_r & \mathbf{0} \\ \mathbf{0} & \mathbf{0} \end{bmatrix}$$

for any $n \times m$ matrix **A** of rank r, so that $\mathbf{A} = \mathbf{P}^{-1}\mathbf{\Lambda}\mathbf{Q}^{-1}$. Let $\mathbf{\Lambda}^-$ denote the transpose of $\mathbf{\Lambda}$ with its nonzero elements replaced by their reciprocals. Since $\mathbf{\Lambda}\mathbf{\Lambda}^-\mathbf{\Lambda} = \mathbf{\Lambda}$, $\mathbf{\Lambda}^-$ is a *g*-inverse of $\mathbf{\Lambda}$. By setting $\mathbf{A}^- = \mathbf{Q}\mathbf{\Lambda}^-\mathbf{P}$, $\mathbf{A}\mathbf{A}^-\mathbf{A} = \mathbf{P}^{-1}\mathbf{\Lambda}\mathbf{Q}^{-1}\mathbf{Q}\mathbf{\Lambda}^-\mathbf{P}\mathbf{P}^{-1}\mathbf{\Lambda}\mathbf{Q}^{-1} = \mathbf{P}^{-1}\mathbf{\Lambda}\mathbf{\Lambda}^-\mathbf{\Lambda}\mathbf{Q}^{-1} = \mathbf{P}^{-1}\mathbf{\Lambda}\mathbf{Q}^{-1} = \mathbf{A}$. Thus one procedure for determining $\mathbf{A}^-$ is to form the product

$$\mathbf{A}^- = \mathbf{Q}\mathbf{\Lambda}^-\mathbf{P} \tag{1.6.15}$$

As **P** and **Q** are not unique, $\mathbf{A}^-$ is not necessarily unique. However, if **A** is of full rank and square, so that $\mathbf{A}^{-1}$ exists for **A**, as defined in (1.5.4), $\mathbf{A}\mathbf{A}^{-1}\mathbf{A} = \mathbf{A}$, or $\mathbf{A}^- = \mathbf{A}^{-1}$, and $\mathbf{A}^-$ is unique. $\mathbf{A}^-$ is also unique if it satisfies all of Penrose's conditions.

EXAMPLE 1.6.4. Let

$$\mathbf{A} = \begin{bmatrix} 2 & 4 \\ 2 & 2 \\ -2 & 0 \end{bmatrix}$$

Then, with

$$\mathbf{P} = \begin{bmatrix} 1 & 0 & 0 \\ -1 & 1 & 0 \\ -1 & 2 & 1 \end{bmatrix} \quad \text{and} \quad \mathbf{Q} = \begin{bmatrix} 1 & -2 \\ 0 & 1 \end{bmatrix}$$

$$\mathbf{PAQ} = \begin{bmatrix} 2 & 0 \\ 0 & -2 \\ 0 & 0 \end{bmatrix}$$

Thus

$$\mathbf{\Lambda}^- = \begin{bmatrix} 1/2 & 0 & 0 \\ 0 & -1/2 & 0 \end{bmatrix} \quad \text{and} \quad \mathbf{A}^- = \mathbf{Q\Lambda}^-\mathbf{P} = \begin{bmatrix} -1/2 & 0 & 0 \\ 1/2 & -1/2 & 0 \end{bmatrix}$$

EXAMPLE 1.6.5. Let

$$\mathbf{A} = \begin{bmatrix} 4 & 2 & 2 \\ 2 & 2 & 0 \\ 2 & 0 & 2 \end{bmatrix}$$

Choose

$$\mathbf{P} = \begin{bmatrix} 1/4 & 0 & 0 \\ -1/2 & 1 & 0 \\ -1 & 1 & 1 \end{bmatrix}, \qquad \mathbf{Q} = \begin{bmatrix} 1/4 & -1/2 & -1 \\ 0 & 1 & 1 \\ 0 & 0 & 1 \end{bmatrix}$$

and

$$\mathbf{\Lambda}^- = \begin{bmatrix} 4 & 0 & 0 \\ 0 & 1 & 0 \\ 0 & 0 & 0 \end{bmatrix}$$

Then

$$\mathbf{A}^- = \begin{bmatrix} 1/2 & -1/2 & 0 \\ -1/2 & 1 & 0 \\ 0 & 0 & 0 \end{bmatrix}$$

From Examples 1.6.4 and 1.6.5, another method for computing $\mathbf{A}^-$ becomes evident if the rank of $\mathbf{A}$ is known. If the $R(\mathbf{A}) = r$ and $\mathbf{A}$ is partitioned so that

$$\underset{(n \times m)}{\mathbf{A}} = \begin{bmatrix} \mathbf{A}_{11} & \mathbf{A}_{12} \\ \mathbf{A}_{21} & \mathbf{A}_{22} \end{bmatrix}$$

where the $r \times r$ matrix $\mathbf{A}_{11}$ is of rank r, then a generalized inverse of $\mathbf{A}$ is

$$\underset{(m \times n)}{\mathbf{A}^-} = \begin{bmatrix} \mathbf{A}_{11}^{-1} & \mathbf{0} \\ \mathbf{0} & \mathbf{0} \end{bmatrix} \tag{1.6.16}$$

To prove that $\mathbf{A}^-$ as defined in (1.6.16) is a g-inverse, we see that

$$\mathbf{AA}^-\mathbf{A} = \begin{bmatrix} \mathbf{A}_{11} & \mathbf{A}_{11} \\ \mathbf{A}_{21} & \mathbf{A}_{21}\mathbf{A}_{11}^{-1}\mathbf{A}_{12} \end{bmatrix} = \begin{bmatrix} \mathbf{A}_{11} & \mathbf{A}_{12} \\ \mathbf{A}_{21} & \mathbf{A}_{22} \end{bmatrix} = \mathbf{A}$$

since $\mathbf{A}_{21}\mathbf{A}_{11}^{-1}[\mathbf{A}_{11} \quad \mathbf{A}_{12}] = [\mathbf{A}_{21} \quad \mathbf{A}_{22}]$ by the way $\mathbf{A}$ was partitioned.

It is not necessary that $\mathbf{A}_{11}$ be in the first r rows and r columns of $\mathbf{A}$. Suppose it is not; then let $\mathbf{R}$ and $\mathbf{C}$ denote the product of elementary matrices, which interchange rows and columns. This allows $\mathbf{A}$ to be partitioned as

$$\mathbf{RAC} = \hat{\mathbf{A}} = \begin{bmatrix} \hat{\mathbf{A}}_{11} & \hat{\mathbf{A}}_{12} \\ \hat{\mathbf{A}}_{21} & \hat{\mathbf{A}}_{22} \end{bmatrix}$$

where the rank and order of $\hat{\mathbf{A}}_{11}$ is r. Then

(1.6.17) $$\mathbf{A}^- = \mathbf{C}\begin{bmatrix} \mathbf{A}_{11}^{-1} & \mathbf{0} \\ \mathbf{0} & \mathbf{0} \end{bmatrix}\mathbf{R}$$

EXAMPLE 1.6.5 (continued). Let $\mathbf{A}$ be defined as in Example 1.6.5 where the $R(\mathbf{A}) = 2$. Choose

$$\mathbf{C} = \begin{bmatrix} 0 & 0 & 1 \\ 0 & 1 & 0 \\ 1 & 0 & 0 \end{bmatrix} \quad \text{and} \quad \mathbf{R} = \begin{bmatrix} 0 & 0 & 1 \\ 0 & 1 & 0 \\ 1 & 0 & 0 \end{bmatrix}$$

so that

$$\mathbf{RAC} = \begin{bmatrix} 2 & 0 & 2 \\ 0 & 2 & 2 \\ 2 & 2 & 4 \end{bmatrix}$$

Thus

$$\begin{bmatrix} \hat{\mathbf{A}}_{11}^{-1} & \mathbf{0} \\ \mathbf{0} & \mathbf{0} \end{bmatrix} = \begin{bmatrix} 1/2 & 0 & 0 \\ 0 & 1/2 & 0 \\ 0 & 0 & 0 \end{bmatrix}$$

and

$$\mathbf{A}^- = \mathbf{C}\begin{bmatrix} 1/2 & 0 & 0 \\ 0 & 1/2 & 0 \\ 0 & 0 & 0 \end{bmatrix}\mathbf{R} = \begin{bmatrix} 0 & 0 & 0 \\ 0 & 1/2 & 0 \\ 0 & 0 & 1/2 \end{bmatrix}$$

Idempotent Matrices and Properties of a g-inverse. An application of a generalized inverse is found in the solution of a system of simultaneous equations. Before illustrating its use in the solution of (1.6.2), a definition of an idempotent matrix is required as well as a discussion of some important properties of $\mathbf{A}^-$.

(1.6.18) A square matrix $\mathbf{A}$ is *idempotent* if $\mathbf{A}^2 = \mathbf{A}$.

Graybill (1969, p. 339) discusses the properties of idempotent matrices. A select few are now summarized.

(1.6.19) For any square matrix of rank r such that $\mathbf{A}^2 = \mathbf{A}$, the $\text{Tr}(\mathbf{A}) = R(\mathbf{A}) = r$.
(1.6.20) If $\mathbf{A}$ is idempotent, $\mathbf{I} - \mathbf{A}$ is also idempotent.
(1.6.21) For any idempotent matrix $\mathbf{A}$ of order n and rank r, $R(\mathbf{A}) + R(\mathbf{I} - \mathbf{A}) = n$.

The importance of idempotency will become more obvious in Chapter 2, where idempotent matrices are used in the distribution theory of quadratic forms (see also Section 1.7). The proof of properties (1.6.19) through (1.6.21) follows immediately. From (1.5.35) there exist matrices $\mathbf{P}_1$ and $\mathbf{Q}_1$, each of rank r, such that $\mathbf{A} = \mathbf{P}_1\mathbf{Q}_1$. Since $\mathbf{A}^2 = \mathbf{A}$, $(\mathbf{P}_1\mathbf{Q}_1)(\mathbf{P}_1\mathbf{Q}_1) = \mathbf{P}_1\mathbf{Q}_1$, or $\mathbf{Q}_1\mathbf{P}_1 = \mathbf{I}_r$; hence $\text{Tr}(\mathbf{A}) = \text{Tr}(\mathbf{P}_1\mathbf{Q}_1) = \text{Tr}(\mathbf{Q}_1\mathbf{P}_1) = \text{Tr}(\mathbf{I}_r) = r$, proving (1.6.19). If $\mathbf{A}$ is idempotent, $(\mathbf{I} - \mathbf{A})^2 = \mathbf{I} - 2\mathbf{A} + \mathbf{A}^2 = \mathbf{I} - \mathbf{A}$, establishing (1.6.20). Finally, (1.6.19) and (1.6.20) are easily combined to prove (1.6.21).

It is now possible to summarize some general properties of any g-inverse and show how it is used to find a solution to a consistent system of equations—for example, $\mathbf{Ax} = \mathbf{y}$ (Rao, 1965a, 1966b). Let $\mathbf{A}^-$ be any g-inverse of $\mathbf{A}$, $\mathbf{AA}^-\mathbf{A} = \mathbf{A}$, and let $\mathbf{H} = \mathbf{A}^-\mathbf{A}$. Then

(1.6.22) $$\mathbf{H}^2 = \mathbf{H}$$

(1.6.23) $$R(\mathbf{A}) = R(\mathbf{H}) = \text{Tr}(\mathbf{H}) \le R(\mathbf{A}^-)$$

(1.6.24) A necessary and sufficient condition for the system (1.6.2) to be consistent (have a solution) is that $\mathbf{AA}^-\mathbf{y} = \mathbf{y}$.
(1.6.25) A general solution to a consistent system (1.6.2) is given by $\mathbf{A}^-\mathbf{y} + (\mathbf{I} - \mathbf{H})\mathbf{z}$ for arbitrary vectors $\mathbf{z}$ and every solution has this form.
(1.6.26) A general solution to (1.6.7) is given by $(\mathbf{I} - \mathbf{H})\mathbf{z}$ for arbitrary vectors $\mathbf{z}$ and every solution has this form.
(1.6.27) If $(\mathbf{A}'\mathbf{A})^-$ is any g-inverse of $\mathbf{A}'\mathbf{A}$, then $(\mathbf{A}'\mathbf{A})^-\mathbf{A}'$ is a g-inverse of $\mathbf{A}$ and the matrix $\mathbf{A}(\mathbf{A}'\mathbf{A})^-\mathbf{A}'$ is unique and symmetric.

To prove (1.6.22) notice that $\mathbf{H}^2 = (\mathbf{A}^-\mathbf{A})(\mathbf{A}^-\mathbf{A}) = \mathbf{A}^-(\mathbf{AA}^-\mathbf{A}) = \mathbf{A}^-\mathbf{A} = \mathbf{H}$. Property (1.6.23) is immediately proved from (1.6.22) with the following observations. The $R(\mathbf{A}) = R(\mathbf{AA}^-\mathbf{A}) \le R(\mathbf{A}^-\mathbf{A}) = R(\mathbf{H}) = \text{Tr}(\mathbf{H}) \le R(\mathbf{A})$; hence $R(\mathbf{A}) = R(\mathbf{H}) = \text{Tr}(\mathbf{H})$. Furthermore, $R(\mathbf{A}) \le R(\mathbf{AA}^-) \le R(\mathbf{A}^-)$, proving (1.6.23). To prove (1.6.24), suppose that $\hat{\mathbf{x}}$ is a solution to (1.6.2), so that $\mathbf{A}\hat{\mathbf{x}} = \mathbf{y}$. Multiply both sides of the equality by $\mathbf{AA}^-$, $\mathbf{AA}^-\mathbf{A}\hat{\mathbf{x}} = \mathbf{AA}^-\mathbf{y}$ so that $\mathbf{A}\hat{\mathbf{x}} = \mathbf{AA}^-\mathbf{y}$, which implies that $\mathbf{AA}^-\mathbf{y} = \mathbf{y}$. Conversely, if $\mathbf{AA}^-\mathbf{y} = \mathbf{y}$, $\mathbf{y} = \mathbf{Az}$, for a vector $\mathbf{z}$, and the system is consistent since $\mathbf{y}$ is in the column space of $\mathbf{A}$. Furthermore, $\hat{\mathbf{x}} = \mathbf{A}^-\mathbf{y}$ is a solution. Thus, for a consistent system, $\mathbf{A}^-\mathbf{y}$ is a solution for any g-inverse $\mathbf{A}$. However, as (1.6.25) shows, $\mathbf{A}^-\mathbf{y}$ is not the only solution. To prove (1.6.25), let $\hat{\mathbf{x}} = \mathbf{A}^-\mathbf{y} + (\mathbf{I} - \mathbf{H})\mathbf{z}$. Then $\mathbf{A}\hat{\mathbf{x}} = \mathbf{AA}^-\mathbf{y} + \mathbf{A}(\mathbf{I} - \mathbf{H})\mathbf{z} = \mathbf{y} + (\mathbf{A} - \mathbf{AA}^-\mathbf{A})\mathbf{z} = \mathbf{y}$, so $\hat{\mathbf{x}}$ satisfies $\mathbf{Ax} = \mathbf{y}$. For any other solution $\mathbf{x}^*$ of (1.6.2), let $\mathbf{z} = (\mathbf{I} - \mathbf{H})\mathbf{x}^*$. Then $\hat{\mathbf{x}} = \mathbf{A}^-\mathbf{y} + (\mathbf{I} - \mathbf{H})(\mathbf{I} - \mathbf{H})\mathbf{x}^* = \mathbf{A}^-\mathbf{y} + (\mathbf{I} - \mathbf{H})\mathbf{x}^* = \mathbf{A}^-\mathbf{y} + \mathbf{x}^* - \mathbf{A}^-\mathbf{A}\mathbf{x}^* = \mathbf{x}^*$. Hence every solution to (1.6.2) can be written in the form given by (1.6.25) for the proper choice of $\mathbf{z}$. From (1.6.25), (1.6.26) is immediate. To prove (1.6.27), first note that, for matrices $\mathbf{A}$ and $\mathbf{B}$, if $\mathbf{AB} = \mathbf{0}$, it does not necessarily imply that $\mathbf{A} = \mathbf{0}$ or $\mathbf{B} = \mathbf{0}$. Find an example for which $\mathbf{AB} = \mathbf{0}$ and $\mathbf{A} \ne \mathbf{0}$ and $\mathbf{B} \ne \mathbf{0}$. For a symmetric matrix $\mathbf{A}'\mathbf{A}$, if $\mathbf{A}'\mathbf{A} = \mathbf{0}$, then $\mathbf{A} = \mathbf{0}$. Letting $\mathbf{W} = \mathbf{A}[\mathbf{I} - (\mathbf{A}'\mathbf{A})^-(\mathbf{A}'\mathbf{A})]$, $\mathbf{W}'\mathbf{W} = \mathbf{0}$, so that $\mathbf{A} = \mathbf{A}(\mathbf{A}'\mathbf{A})^-\mathbf{A}'\mathbf{A} = \mathbf{A}[(\mathbf{A}'\mathbf{A})^-\mathbf{A}']\mathbf{A}$, or

$(\mathbf{A}'\mathbf{A})^-\mathbf{A}'$ is a g-inverse of $\mathbf{A}$ given $(\mathbf{A}'\mathbf{A})^-$ is a g-inverse of $\mathbf{A}'\mathbf{A}$. To prove the uniqueness of $\mathbf{A}(\mathbf{A}'\mathbf{A})^-\mathbf{A}'$, let $(\mathbf{A}'\mathbf{A})^*$ denote another g-inverse of $\mathbf{A}'\mathbf{A}$. Verify that $[\mathbf{A}(\mathbf{A}'\mathbf{A})^-\mathbf{A}' - \mathbf{A}(\mathbf{A}'\mathbf{A})^*\mathbf{A}'][\mathbf{A}(\mathbf{A}'\mathbf{A})^-\mathbf{A}' - \mathbf{A}(\mathbf{A}'\mathbf{A})^*\mathbf{A}']' = \mathbf{0}$ by using the fact that $\mathbf{A} = \mathbf{A}(\mathbf{A}'\mathbf{A})^-\mathbf{A}'\mathbf{A} = \mathbf{A}(\mathbf{A}'\mathbf{A})^*\mathbf{A}'\mathbf{A}$; that is, $\mathbf{A}(\mathbf{A}'\mathbf{A})^-\mathbf{A}' = \mathbf{A}(\mathbf{A}'\mathbf{A})^*\mathbf{A}'$. That $\mathbf{A}(\mathbf{A}'\mathbf{A})^-\mathbf{A}'$ is symmetric, whether $(\mathbf{A}'\mathbf{A})^-$ is or not, follows from the uniqueness property. Given a symmetric g-inverse of $\mathbf{A}'\mathbf{A}$, denoted by $(\mathbf{A}'\mathbf{A})^*$, $\mathbf{A}(\mathbf{A}'\mathbf{A})^*\mathbf{A}$ is symmetric. However, $\mathbf{A}(\mathbf{A}'\mathbf{A})^*\mathbf{A}' = \mathbf{A}(\mathbf{A}'\mathbf{A})^-\mathbf{A}'$ for a nonsymmetric g-inverse $(\mathbf{A}'\mathbf{A})^-$, so $\mathbf{A}(\mathbf{A}'\mathbf{A})^-\mathbf{A}'$ is symmetric. This completes the proof of (1.6.27).

EXAMPLE 1.6.6. Consider the system

$$\begin{bmatrix} 1 & 1 & 0 \\ 1 & 1 & 0 \\ 1 & 0 & 1 \\ 1 & 0 & 1 \end{bmatrix} \begin{bmatrix} \mu \\ \alpha_1 \\ \alpha_2 \end{bmatrix} = \begin{bmatrix} y_{11} \\ y_{12} \\ y_{21} \\ y_{22} \end{bmatrix}$$

Using (1.6.27), $(\mathbf{A}'\mathbf{A})^-\mathbf{A}'$ is a g-inverse of $\mathbf{A}$ if $(\mathbf{A}'\mathbf{A})^-$ is a g-inverse of $\mathbf{A}'\mathbf{A}$. From Example 1.6.5,

$$\mathbf{A}'\mathbf{A} = \begin{bmatrix} 4 & 2 & 2 \\ 2 & 2 & 0 \\ 2 & 0 & 2 \end{bmatrix} \quad \text{and} \quad (\mathbf{A}'\mathbf{A})^- = \begin{bmatrix} 1/2 & -1/2 & 0 \\ -1/2 & 1 & 0 \\ 0 & 0 & 0 \end{bmatrix}$$

so

$$\mathbf{A}^-\mathbf{y} = (\mathbf{A}'\mathbf{A})^-\mathbf{A}'\mathbf{y} = \begin{bmatrix} y_{2.} \\ y_{1.} - y_{2.} \\ 0 \end{bmatrix}$$

Since

$$\mathbf{H} = \mathbf{A}^-\mathbf{A} = (\mathbf{A}'\mathbf{A})^-\mathbf{A}'\mathbf{A} = \begin{bmatrix} 1 & 0 & 1 \\ 0 & 1 & -1 \\ 0 & 0 & 0 \end{bmatrix}$$

$$\mathbf{I} - \mathbf{H} = \begin{bmatrix} 0 & 0 & -1 \\ 0 & 0 & 1 \\ 0 & 0 & 1 \end{bmatrix}$$

and a general solution to the system is

$$\begin{bmatrix} \mu \\ \alpha_1 \\ \alpha_2 \end{bmatrix} = \begin{bmatrix} y_{2.} \\ y_{1.} - y_{2.} \\ 0 \end{bmatrix} + (\mathbf{I} - \mathbf{H})\mathbf{z}$$

$$= \begin{bmatrix} y_{2.} \\ y_{1.} - y_{2.} \\ 0 \end{bmatrix} + \begin{bmatrix} -z_3 \\ z_3 \\ z_3 \end{bmatrix}$$

Choosing $z_3 = y_{2.} - y_{..}$, the solution

$$\begin{bmatrix} \mu \\ \alpha_1 \\ \alpha_2 \end{bmatrix} = \begin{bmatrix} y_{..} \\ y_{1.} - y_{..} \\ y_{2.} - y_{..} \end{bmatrix}$$

is consistent with the procedure of restricting the number of unknowns (see Example 1.6.2).

EXAMPLE 1.6.7. To demonstrate that the construction of a g-inverse of $\mathbf{A}'\mathbf{A}$ does not change the set of solution vectors, let

$$(\mathbf{A}'\mathbf{A})^- = \begin{bmatrix} 0 & 0 & 0 \\ 0 & 1/2 & 0 \\ 0 & 0 & 1/2 \end{bmatrix}$$

be a g-inverse of $\mathbf{A}'\mathbf{A}$ (as in Example 1.6.5). Then

$$\mathbf{H} = (\mathbf{A}'\mathbf{A})^-\mathbf{A}'\mathbf{A} = \begin{bmatrix} 0 & 0 & 0 \\ 1 & 1 & 0 \\ 1 & 0 & 1 \end{bmatrix}$$

$$\mathbf{I} - \mathbf{H} = \begin{bmatrix} 1 & 0 & 0 \\ -1 & 0 & 0 \\ -1 & 0 & 0 \end{bmatrix}$$

and

$$\begin{bmatrix} \mu \\ \alpha_1 \\ \alpha_2 \end{bmatrix} = (\mathbf{A}'\mathbf{A})^-\mathbf{A}'\mathbf{y} + (\mathbf{I} - \mathbf{H})\mathbf{z}$$

$$= \begin{bmatrix} 0 \\ y_{1.} \\ y_{2.} \end{bmatrix} + \begin{bmatrix} z_1 \\ -z_1 \\ -z_1 \end{bmatrix}$$

Selecting $z_1 = y_{..}$,

$$\begin{bmatrix} \mu \\ \alpha_1 \\ \alpha_2 \end{bmatrix} = \begin{bmatrix} y_{..} \\ y_{1.} - y_{..} \\ y_{2.} - y_{..} \end{bmatrix}$$

which is consistent with Example 1.6.6. Furthermore, for $z_1 = y_{2.} - z_3$, the general solution in this example can be reduced to the general solution given in Example 1.6.6.

Property (1.6.25) only indicates the general form of solutions to (1.6.2) for a consistent system. Rao (1962; 1965a, p. 182) has proved that certain linear combinations of the unknowns in a consistent system $\mathbf{Ax} = \mathbf{y}$ are unique no matter what solution to (1.6.2) is employed.

(1.6.28) With $\mathbf{AA^-A} = \mathbf{A}$ and $\mathbf{H} = \mathbf{A^-A}$, linear combinations $\mathbf{p'x}$ of the unknowns of the consistent system $\mathbf{Ax} = \mathbf{y}$ have a unique solution in terms of a solution $\hat{\mathbf{x}}$ of (1.6.2) if and only if $\mathbf{p'H} = \mathbf{p'}$.

To prove (1.6.28), let $\hat{\mathbf{x}} = \mathbf{A^-y} + (\mathbf{I} - \mathbf{H})\mathbf{z}$. Then $\mathbf{p'}\hat{\mathbf{x}} = \mathbf{p'A^-y}$ is independent of $\mathbf{z}$ if and only if $\mathbf{p'H} = \mathbf{p'}$. If $\mathbf{x}^*$ is any other solution to (1.6.2), let $\mathbf{z} = (\mathbf{I} - \mathbf{H})\mathbf{x}^*$ so that $\mathbf{x} = \mathbf{x}^*$ by the proof of (1.6.25) and $\mathbf{p'}\hat{\mathbf{x}} = \mathbf{p'x}^*$, if $\mathbf{p'H} = \mathbf{p'}$.

By the idempotent property of $\mathbf{H}$, if $\mathbf{p'} = \mathbf{t'H}$, then $\mathbf{p'H} = \mathbf{p'}$ for arbitrary vectors $\mathbf{t}$. Given that the $R(\mathbf{A}) = r$, there are only r linearly independent vectors $\mathbf{p'} = \mathbf{t'H}$ for which $\mathbf{p'H} = \mathbf{p'}$ since $R(\mathbf{H}) = R(\mathbf{A}) = r$. Restating (1.6.28), the condition for uniqueness becomes as follows.

(1.6.29) The linear combinations of unknowns $\mathbf{p'x}$ of a consistent system $\mathbf{Ax} = \mathbf{y}$ have a unique solution if and only if $\mathbf{p'H} = \mathbf{p'}$. The solutions are given by $\mathbf{p'}\hat{\mathbf{x}} = \mathbf{t'HA^-y}$ for r linearly independent vectors $\mathbf{p'} = \mathbf{t'H}$ and arbitrary vectors $\mathbf{t'}$.

If, for $\mathbf{A}^-$ in (1.6.29), the g-inverse of $\mathbf{A}$ is given by $\mathbf{A}^- = (\mathbf{A'A})^-\mathbf{A'}$, where $(\mathbf{A'A})^-$ is a g-inverse of $\mathbf{A'A}$, $\mathbf{p'}\hat{\mathbf{x}}$ in (1.6.29) becomes $\mathbf{p'}\hat{\mathbf{x}} = \mathbf{t'}[(\mathbf{A'A})^-\mathbf{A'A}](\mathbf{A'A})^-\mathbf{A'y}$.

EXAMPLE 1.6.8. Apply (1.6.29) to the consistent system

$$\begin{bmatrix} 1 & 1 & 0 \\ 1 & 1 & 0 \\ 1 & 0 & 1 \\ 1 & 0 & 1 \end{bmatrix} \begin{bmatrix} \mu \\ \alpha_1 \\ \alpha_2 \end{bmatrix} = \begin{bmatrix} y_{11} \\ y_{12} \\ y_{21} \\ y_{22} \end{bmatrix}$$

to determine if unique solutions for the linear combinations of the unknowns $\alpha_1 - \alpha_2$ and $\mu + (\alpha_1 + \alpha_2)/2$ can be found. To check that unique solutions exist, verify the condition that $\mathbf{p'H} = \mathbf{p'}$. For $\alpha_1 - \alpha_2$, $\mathbf{p'} = [0, 1, -1]$ and

$$\mathbf{p'H} = [0, 1, -1] \begin{bmatrix} 1 & 0 & 1 \\ 0 & 1 & -1 \\ 0 & 0 & 0 \end{bmatrix} = [0, 1, -1] = \mathbf{p'}$$

For $\mu + (\alpha_1 + \alpha_2)/2$, $\mathbf{p'} = [1, 1/2, 1/2]$ and

$$\mathbf{p'H} = [1, 1/2, 1/2] \begin{bmatrix} 1 & 0 & 1 \\ 0 & 1 & -1 \\ 0 & 0 & 0 \end{bmatrix} = [1, 1/2, 1/2] = \mathbf{p'}$$

Thus solutions exist and are unique. To find solutions in general, observe that

$$\begin{aligned} \mathbf{p'x} = \mathbf{t'Hx} &= [t_0, t_1, t_2] \begin{bmatrix} 1 & 0 & 0 \\ 0 & 1 & -1 \\ 0 & 0 & 0 \end{bmatrix} \begin{bmatrix} \mu \\ \alpha_1 \\ \alpha_2 \end{bmatrix} \\ &= t_0(\mu + \alpha_2) + t_1(\alpha_1 - \alpha_2) \end{aligned}$$

is the general form for linear combinations of the unknowns in terms of an arbitrary vector $\mathbf{t}$; the unique solution $\mathbf{p}'\hat{\mathbf{x}}$ is, from Example 1.6.6,

$$\mathbf{p}'\hat{\mathbf{x}} = \mathbf{t}'\mathbf{H}\mathbf{A}^{-}\mathbf{y} = \mathbf{t}'[(\mathbf{A}'\mathbf{A})^{-}\mathbf{A}'\mathbf{A}](\mathbf{A}'\mathbf{A})^{-}\mathbf{A}'\mathbf{y}$$

$$= \mathbf{t}'\begin{bmatrix} 1 & 0 & 1 \\ 0 & 1 & -1 \\ 0 & 0 & 0 \end{bmatrix}\begin{bmatrix} y_{2.} \\ y_{1.} - y_{2.} \\ 0 \end{bmatrix}$$

$$= t_0 y_{2.} + t_1(y_{1.} - y_{2.})$$

Setting $t_0 = 0$ and $t_1 = 1$, $\mathbf{p}'\mathbf{x} = \alpha_1 - \alpha_2$ and $\mathbf{p}'\hat{\mathbf{x}} = y_{1.} - y_{2.}$. Thus $\alpha_1 - \alpha_2 = y_{1.} - y_{2.}$. For $t_0 = 1$ and $t_1 = 1/2$,

$$\mu + \frac{1}{2}(\alpha_1 + \alpha_2) = \frac{y_{1.} + y_{2.}}{2} = y_{..}$$

These results agree with the solution obtained by using the reparameterization technique. Here, however, we are not explicitly checking that the matrix $\mathbf{C}$ is in the row space of $\mathbf{A}$, but only that $\mathbf{p}'\mathbf{H} = \mathbf{p}'$. Furthermore, by using the side condition $\alpha_1 + \alpha_2 = 0$, the solution found by restricting the unknowns is demonstrated.

EXAMPLE 1.6.9. Expanding on the concept that unique solutions for certain linear combinations of the unknowns do not depend on the construction of a g-inverse, let

$$(\mathbf{A}'\mathbf{A})^{-} = \begin{bmatrix} 0 & 0 & 0 \\ 0 & 1/2 & 0 \\ 0 & 0 & 1/2 \end{bmatrix}$$

as in Example 1.6.7. Then

$$\mathbf{p}'\mathbf{x} = \mathbf{t}'\mathbf{H}\mathbf{x} = [t_0, t_1, t_2]\begin{bmatrix} 0 & 0 & 0 \\ 1 & 1 & 0 \\ 1 & 0 & 1 \end{bmatrix}\begin{bmatrix} \mu \\ \alpha_1 \\ \alpha_2 \end{bmatrix}$$

$$= (t_1 + t_2)\mu + t_1\alpha_1 + t_2\alpha_2$$

Solutions to $\mathbf{p}'\mathbf{x}$ are presented by

$$\mathbf{p}'\hat{\mathbf{x}} = \mathbf{t}'\mathbf{H}\mathbf{A}^{-}\mathbf{y} = \mathbf{t}'\begin{bmatrix} 0 & 0 & 0 \\ 1 & 1 & 0 \\ 1 & 0 & 1 \end{bmatrix}\begin{bmatrix} 0 \\ y_{1.} \\ y_{2.} \end{bmatrix}$$

$$= t_1 y_{1.} + t_2 y_{2.}$$

Choosing $\mathbf{t}' = [0, 1, -1]$ and $\mathbf{t}' = [0, 1/2, 1/2]$, unique solutions become $\alpha_1 - \alpha_2 = y_{1.} - y_{2.}$ and

$$\mu + \frac{1}{2}(\alpha_1 + \alpha_2) = \frac{y_{1.} + y_{2.}}{2} = y_{..}$$

which are consistent with previous findings.

Theorem (1.6.29), concerning linear combinations of unknowns, and the concept of a generalized inverse are important to the estimation theory of linear statistical models (to be discussed in Chapter 3).

Linear Transformations. The system of equations $\mathbf{Ax} = \mathbf{y}$ is typically conceived of as a *linear transformation.* The $m \times 1$ vector $\mathbf{x}$ is operated on by the $n \times m$ matrix $\mathbf{A}$, yielding the $n \times 1$ image vector $\mathbf{y}$. A transformation is linear if, in carrying $\mathbf{x}_1$ into $\mathbf{y}_1$ and $\mathbf{x}_2$ into $\mathbf{y}_2$, it transforms the vector $s_1\mathbf{x}_1 + s_2\mathbf{x}_2$ into $s_1\mathbf{y}_1 + s_2\mathbf{y}_2$, for every pair of scalars s_1 and s_2. That is, the m-vector $\mathbf{x}$ is mapped into the n-vector $\mathbf{y} \in V_r$ if the rank of $\mathbf{A}$ is r.

Suppose that $\mathbf{Ax}_1 = \mathbf{y}_1$ and $\mathbf{By}_1 = \mathbf{z}$, then these two transformations may be replaced by a single transformation $\mathbf{BAx}_1 = \mathbf{z}$ involving a new matrix, which is the product of the two individual matrices.

Of particular interest in statistical applications are linear transformations that map vectors of a space onto vectors of the same space. The matrix $\mathbf{A}$ associated with this transformation is now of order n.

(1.6.30)
$$\underset{(n \times n)}{\mathbf{A}} \ \underset{(n \times 1)}{\mathbf{x}} = \underset{(n \times 1)}{\mathbf{y}}$$

The linear transformation is nonsingular if and only if $|\mathbf{A}| \neq 0$. In such a case, the transformation is one to one and there exists a unique transformation $\mathbf{A}^{-1}$ that yields the original vector

(1.6.31)
$$\mathbf{x} = \mathbf{A}^{-1}\mathbf{y}$$

If $\mathbf{A}$ is less than full rank, the transformation is singular and many to one. These transformations map vectors into subspaces.

EXAMPLE 1.6.10. As a simple example of a nonsingular linear transformation in Euclidean two-dimensional space, consider the square formed by the vectors $\mathbf{x}_1' = [1, 0]$, $\mathbf{x}_2' = [0, 1]$, and $\mathbf{x}_1 + \mathbf{x}_2 = [1, 1]$ under the transformation

$$\mathbf{A} = \begin{bmatrix} 1 & 4 \\ 0 & 1 \end{bmatrix}$$

Then

$$\mathbf{Ax}_1 = \mathbf{y}_1 = \begin{bmatrix} 1 \\ 0 \end{bmatrix}$$

$$\mathbf{Ax}_2 = \mathbf{y}_2 = \begin{bmatrix} 4 \\ 1 \end{bmatrix}$$

$$\mathbf{A}(\mathbf{x}_1 + \mathbf{x}_2) = \mathbf{y}_1 + \mathbf{y}_2 = \begin{bmatrix} 5 \\ 1 \end{bmatrix}$$

Geometrically, observe that the parallel segments $\{[0, 1], [1, 1]\}$ and $\{[0, 0], [1, 0]\}$ are transformed into the parallel segments $\{[4, 1], [5, 1]\}$ and $\{[0, 0], [1, 0]\}$ as are the other sides. However, some lengths, angles, and hence distances of the original figure changed under the transformation.

A linear nonsingular transformation that preserves these quantities is called an *orthogonal transformation.*

(1.6.32) A linear transformation $\mathbf{Tx} = \mathbf{y}$ is said to be orthogonal if $\mathbf{TT}' = \mathbf{I}$.

The condition $\mathbf{TT}' = \mathbf{I}$ implies that the row vectors of $\mathbf{T}$ are of unit length and pairwise orthogonal so that they form an orthonormal basis for V_n. The square matrix $\mathbf{T}$ associated with the transformation is called an *orthogonal matrix.*

(1.6.33) An orthogonal matrix is a square matrix of order n such that

$$\mathbf{TT}' = \mathbf{I} = \mathbf{T}'\mathbf{T}$$

This definition implies that $\mathbf{T}' = \mathbf{T}^{-1}$, and, since $\mathbf{T}^{-1}\mathbf{T} = \mathbf{I}$, $\mathbf{T}'\mathbf{T} = \mathbf{I}$. Thus the columns of an orthogonal matrix form an orthonormal basis. Furthermore, since the $|\mathbf{TT}'| = |\mathbf{T}||\mathbf{T}'| = |\mathbf{I}| = 1$ and the $|\mathbf{T}| = |\mathbf{T}'|$, the $|\mathbf{T}| = \pm 1$. That is, any orthogonal matrix has the property that the determinant is 1 or -1.

To verify that lengths, distances, angles, and areas are invariant under orthogonal transformations, the aforementioned properties are employed. For the transformation $\mathbf{T}$, let $\mathbf{x} \to \mathbf{y}$. Then the length of $\mathbf{y}$ squared is

$$\|\mathbf{y}\|^2 = \mathbf{y}'\mathbf{y} = (\mathbf{x}'\mathbf{T}')(\mathbf{Tx}) = \mathbf{x}'\mathbf{x} = \|\mathbf{x}\|^2$$

For distance, let $\mathbf{x}_1 \to \mathbf{y}_1$ and $\mathbf{x}_2 \to \mathbf{y}_2$. Then the distance squared between $\mathbf{y}_1$ and $\mathbf{y}_2$ is

$$\begin{aligned}\|(\mathbf{y}_1 - \mathbf{y}_2)\|^2 &= (\mathbf{y}_1 - \mathbf{y}_2)'(\mathbf{y}_1 - \mathbf{y}_2)\\ &= (\mathbf{Tx}_1 - \mathbf{Tx}_2)'(\mathbf{Tx}_1 - \mathbf{Tx}_2)\\ &= (\mathbf{x}_1 - \mathbf{x}_2)'(\mathbf{T}'\mathbf{T})(\mathbf{x}_1 - \mathbf{x}_2)\\ &= \|\mathbf{x}_1 - \mathbf{x}_2\|^2\end{aligned}$$

For angles, let $\mathbf{x}_1 \to \mathbf{y}_1$ and $\mathbf{x}_2 \to \mathbf{y}_2$. Then the cosine of the angle between $\mathbf{y}_1$ and $\mathbf{y}_2$ is

$$\begin{aligned}\cos\theta &= \frac{\mathbf{y}_1'\mathbf{y}_2}{\sqrt{(\mathbf{y}_1'\mathbf{y}_1)(\mathbf{y}_2'\mathbf{y}_2)}} = \frac{\mathbf{x}_1'(\mathbf{T}'\mathbf{T})\mathbf{x}_1}{\sqrt{(\mathbf{x}_1'\mathbf{T}'\mathbf{T}\mathbf{x}_1)(\mathbf{x}_2'\mathbf{T}'\mathbf{T}\mathbf{x}_2)}}\\ &= \frac{\mathbf{x}_1'\mathbf{x}_2}{\sqrt{(\mathbf{x}_1'\mathbf{x}_1)(\mathbf{x}_2'\mathbf{x}_2)}}\end{aligned}$$

For area, let $\mathbf{x}_1 \to \mathbf{y}_1$ and $\mathbf{x}_2 \to \mathbf{y}_2$. Then the area of the parallelogram determined by $\mathbf{y}_1$ and $\mathbf{y}_2$ is the

$$\begin{vmatrix} y_{11} & y_{12} \\ y_{21} & y_{22} \end{vmatrix} = |\mathbf{T}| \begin{vmatrix} x_{11} & x_{12} \\ x_{21} & x_{22} \end{vmatrix} = \begin{vmatrix} x_{11} & x_{12} \\ x_{21} & x_{22} \end{vmatrix}$$

so that all these quantities are preserved under orthogonal transformations.

Some fundamental properties of orthogonal matrices are worth remembering.

(1.6.34) An orthogonal transformation is nonsingular.
(1.6.35) If $\mathbf{T}$ is orthogonal, $|\mathbf{T}| = \pm 1$.
(1.6.36) If $\mathbf{A}$ is any square nonsingular matrix, $|\mathbf{T}'\mathbf{AT}| = |\mathbf{A}|$.
(1.6.37) The product of a finite number of orthogonal matrices is itself orthogonal.

Transformations in the above context operate on vectors within some vector space relative to one given coordinate system. It is sometimes more convenient to discuss transformations with respect to different coordinate systems where vectors remain fixed and reference axes are transformed. For example, if $\mathbf{x} \to \mathbf{y}$ under $\mathbf{A}$ with respect to a specific system, Figure 1.6.1 results.

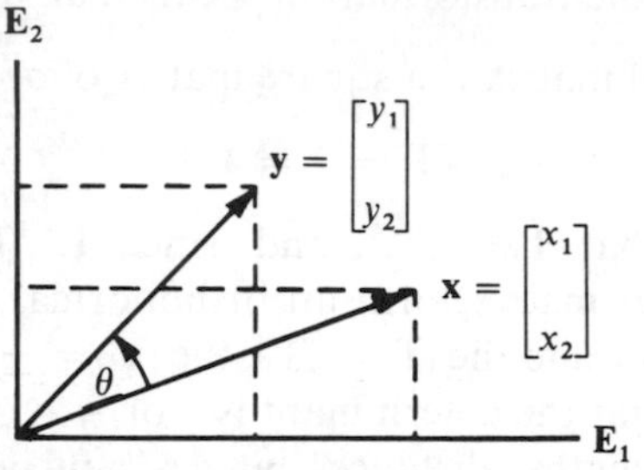

FIGURE 1.6.1. Fixed-system transformation.

Alternatively, selection may be made to fix a vector $\mathbf{v}$ and rotate the system under $\mathbf{A}$, so that Figure 1.6.2 results. In this case, the coordinates of $\mathbf{v}$ are discussed with respect to the old and new reference systems. To determine the relationship between

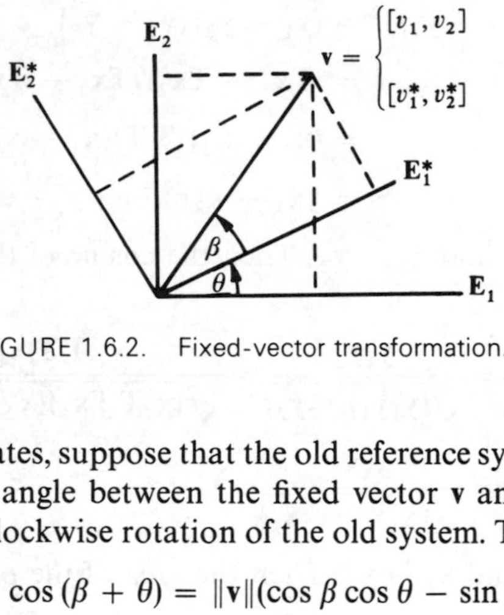

FIGURE 1.6.2. Fixed-vector transformation.

the old and new coordinates, suppose that the old reference system is rotated through an angle θ and β is the angle between the fixed vector $\mathbf{v}$ and the new system that resulted from a counterclockwise rotation of the old system. Then, from Figure 1.6.2,

$$v_1 = \|\mathbf{v}\| \cos(\beta + \theta) = \|\mathbf{v}\|(\cos\beta\cos\theta - \sin\beta\sin\theta)$$

$$v_2 = \|\mathbf{v}\| \sin(\beta + \theta) = \|\mathbf{v}\|(\sin\beta\cos\theta + \cos\beta\sin\theta)$$

and

$$v_1^* = \|\mathbf{v}\| \cos\beta$$

$$v_2^* = \|\mathbf{v}\| \sin\beta$$

so that

$$v_1 = v_1^* \cos\theta - v_2^* \sin\theta$$

$$v_2 = v_1^* \sin\theta + v_2^* \cos\theta$$

or

$$\mathbf{v} = \begin{bmatrix} \cos\theta & -\sin\theta \\ \sin\theta & \cos\theta \end{bmatrix} \mathbf{v}^* \tag{1.6.38}$$

and

$$\mathbf{v}^* = \begin{bmatrix} \cos\theta & \sin\theta \\ -\sin\theta & \cos\theta \end{bmatrix} \mathbf{v} \tag{1.6.39}$$

where $\mathbf{v}$ represents the old coordinates and $\mathbf{v}^*$ represents the new coordinates.

The rotation of the old axes E_1 and E_2 to the new axes E_1^* and E_2^* in Figure 1.6.2 represents a rigid rotation of axes, since the angle between each corresponding axis is fixed and equal to θ. The orthogonal transformation matrix that accomplished this was shown to be

$$\mathbf{T}_1' = \begin{bmatrix} \cos\theta & \sin\theta \\ -\sin\theta & \cos\theta \end{bmatrix}$$

Instead of rotating both axes as in Figure 1.6.2, suppose that the axis E_2^* is reflected following a rigid rotation, as shown in Figure 1.6.3. Following the derivation of (1.6.38) and (1.6.39), the relationship between $\mathbf{v}$ and $\mathbf{v}^*$ is seen to be that

$$\mathbf{v}^* = \begin{bmatrix} \cos\theta & \sin\theta \\ \sin\theta & -\cos\theta \end{bmatrix} \mathbf{v}$$

In this case the orthogonal transformation matrix is

$$\mathbf{T}_2' = \begin{bmatrix} \cos\theta & \sin\theta \\ \sin\theta & -\cos\theta \end{bmatrix}$$

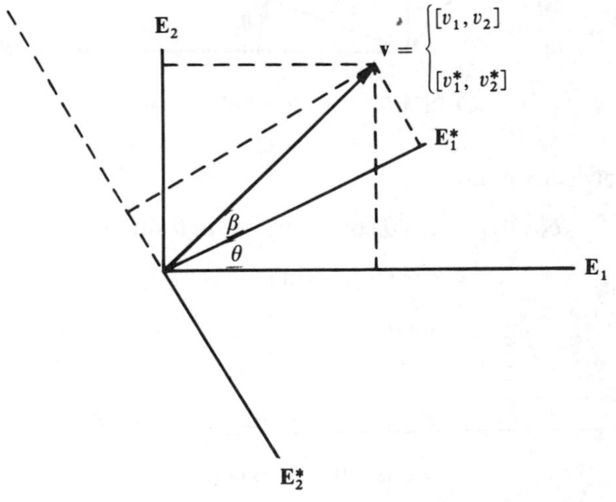

Figure 1.6.3. Fixed-vector rotation and reflection.

Both matrices $\mathbf{T}'_1$ and $\mathbf{T}'_2$ are orthogonal; however, the $|\mathbf{T}'_1| = 1$ and the $|\mathbf{T}'_2| = -1$. That is, an orthogonal matrix transformation can be interpreted as a rigid rotation or as a rigid rotation followed by a reflection of coordinate axes, depending on the value of its determinant. Any transformation involving an orthogonal matrix will be simply referred to as an *orthogonal* transformation in this text since the new reference axes are orthogonal. A transformation altering an orthogonal system to a nonorthogonal system is called an *oblique* transformation. Oblique transformations rotate axes, changing both angles and lengths.

The transformation of coordinates in relation to an oblique transformation is best described in terms of direction cosines, $\cos \theta_{ij}$, where θ_{ij} is taken as the angle between the ith old axes and the jth new axis. The transformation (1.6.39) becomes

$$\mathbf{v}^* = \begin{bmatrix} \cos \theta_{11} & \cos \theta_{21} \\ \cos \theta_{12} & \cos \theta_{22} \end{bmatrix} \mathbf{v} \tag{1.6.40}$$

where the elements of $\mathbf{v}$ and $\mathbf{v}^*$ contain the coordinates of a fixed vector with respect to the old and new system, respectively.

The relationship between the angles specified in (1.6.40) and θ in (1.6.39) for a rigid rotation is best seen in Figure 1.6.4. The correspondence between Figures 1.6.4 and 1.6.2 is summarized as follows:

$$\theta_{11} = \theta \qquad \theta_{21} = \theta - 90^\circ$$

$$\theta_{12} = \theta + 90^\circ \qquad \theta_{22} = \theta$$

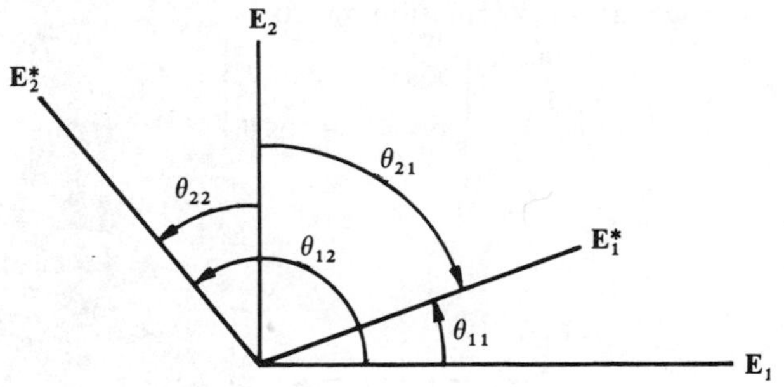

FIGURE 1.6.4. Rigid rotation of axes.

Using trigonometric formula,

$$\begin{aligned} \cos \theta_{21} &= \cos \theta \cos -90^\circ - \sin \theta \sin -90^\circ \\ &= \cos \theta(0) - \sin \theta(-1) \\ &= \sin \theta \end{aligned}$$

and

$$\begin{aligned} \cos \theta_{12} &= \cos \theta \cos 90^\circ - \sin \theta \sin 90^\circ \\ &= \cos \theta(0) - \sin \theta(1) \\ &= -\sin \theta \end{aligned}$$

By substituting these into (1.6.40), the general result is seen to include (1.6.39) as a special case. In addition, the extension of (1.6.40) to n dimensions becomes obvious. For three dimensions, the transformation is

$$\begin{bmatrix} v_1^* \\ v_2^* \\ v_3^* \end{bmatrix} = \begin{bmatrix} \cos\theta_{11} & \cos\theta_{21} & \cos\theta_{31} \\ \cos\theta_{12} & \cos\theta_{22} & \cos\theta_{32} \\ \cos\theta_{13} & \cos\theta_{23} & \cos\theta_{33} \end{bmatrix} \begin{bmatrix} v_1 \\ v_2 \\ v_3 \end{bmatrix}$$

where $\cos\theta_{ij}$ is the direction cosine of the ith old axis with the jth new axis, using the new coordinate axes as the reference system.

Projection Transformations. A linear transformation that maps vectors of a given space onto a subspace is called a *projection.* This transformation is of particular interest in the class of all singular transformations useful in statistical analysis. As seen in Section 1.3, the projection operation allows decomposing a vector $\mathbf{y} \in V_n$ into orthogonal components

(1.6.41) $$\mathbf{y} = P_{V_r}\mathbf{y} + P_{V_{n-r}^{\perp}}\mathbf{y}$$

so that one component of $\mathbf{y}$ is in r-dimensional space and the other is in $(n - r)$-dimensional space. The projection process is illustrated by projection matrices, making the geometry of the projection process algebraic.

Let $\mathbf{P}_V$ denote a projection matrix that projects a vector $\mathbf{y}$ onto a subspace V.

(1.6.42) The matrix $\mathbf{P}_V$ is an orthogonal projection if and only if $\mathbf{P}_V$ is symmetric, $\mathbf{P}_V = \mathbf{P}'_V$, and idempotent, $\mathbf{P}_V^2 = \mathbf{P}_V$.

To prove this, let $\mathbf{P}_V$ denote such a projection matrix $\mathbf{P}_V\mathbf{y} = \mathbf{x}$, where $\mathbf{y} \in V_n$ and $\mathbf{x} \in V_r$ and $V_r \subset V_n$. Clearly, $\mathbf{P}_V\mathbf{x} = \mathbf{x}$, so that $\mathbf{P}_V\mathbf{y} = \mathbf{x} = \mathbf{P}_V\mathbf{x} = \mathbf{P}_V\mathbf{P}_V\mathbf{y}$. Thus $(\mathbf{P}_V - \mathbf{P}_V^2)\mathbf{y} = \mathbf{0}$ for any $\mathbf{y}$, or $\mathbf{P}_V^2 = \mathbf{P}_V$, so that $\mathbf{P}_V$ is idempotent. To show symmetry, let $\mathbf{y}$ and $\mathbf{z}$ be elements of V_n. Then $\mathbf{y}'\mathbf{P}_V\mathbf{z} = (\mathbf{P}_V\mathbf{y} + \mathbf{P}_{V^\perp}\mathbf{y})'\mathbf{P}_V\mathbf{z} = (\mathbf{P}_V\mathbf{y})'\mathbf{P}_V\mathbf{z} + (\mathbf{P}_{V^\perp}\mathbf{y})'\mathbf{P}_V\mathbf{z} = (\mathbf{P}_V\mathbf{y})'\mathbf{P}_V\mathbf{z}$. Similarly, $(\mathbf{P}_V\mathbf{y})'\mathbf{z} = (\mathbf{P}_V\mathbf{y})'\mathbf{P}_V\mathbf{z}$. Thus $\mathbf{y}'(\mathbf{P}_V\mathbf{z}) = (\mathbf{P}_V\mathbf{y})'\mathbf{z} = \mathbf{y}'\mathbf{P}'_V\mathbf{z}$, so that $\mathbf{P}_V = \mathbf{P}'_V$, or $\mathbf{P}_V$ is symmetrical. Conversely, suppose $\mathbf{P}_V$ is symmetrical and idempotent, where $\mathbf{P}_V\mathbf{y} = \mathbf{x}$, $\mathbf{x} \in V_n$, and $\mathbf{y} \in V_n$. Let $\mathbf{z} \in V$. Then there exists a vector $\mathbf{y}^*$ such that $\mathbf{P}_V\mathbf{y}^* = \mathbf{z}$. Thus it must be shown that $(\mathbf{y} - \mathbf{P}_V\mathbf{y}) \perp \mathbf{z}$. Consider $(\mathbf{y} - \mathbf{P}_V\mathbf{y})'\mathbf{z} = (\mathbf{y} - \mathbf{P}_V\mathbf{y})'\mathbf{P}_V\mathbf{y}^* = \mathbf{y}'\mathbf{P}_V\mathbf{y}^* - \mathbf{y}'\mathbf{P}'_V\mathbf{P}_V\mathbf{y}^*$. But $\mathbf{P}'_V = \mathbf{P}_V$ and $\mathbf{P}'_V\mathbf{P}_V = \mathbf{P}_V^2 = \mathbf{P}_V$, so that $(\mathbf{y} - \mathbf{P}_V\mathbf{y})'\mathbf{z} = 0$, for any $\mathbf{z} \in V_r$. Thus $\mathbf{P}_V$ is an orthogonal projection matrix.

(1.6.43) If $\mathbf{P}_V$ is an orthogonal projection matrix, so is $\mathbf{I} - \mathbf{P}_V$.

Since $\mathbf{P}_V$ is idempotent and symmetrical, so is the matrix $\mathbf{I} - \mathbf{P}_V$. By (1.6.42), $\mathbf{I} - \mathbf{P}_V$ is a projection transformation. $\mathbf{I} - \mathbf{P}_V$ projects $\mathbf{y}$ onto $V^\perp$, the orthocomplement of V relative to V_n. Since $\mathbf{I} - \mathbf{P}_V$ projects vectors onto the space $V^\perp$, and $V_n = V_r \oplus V_{n-r}$, where $V_{n-r} = V_r^\perp$, it can be inferred that the rank of a projection matrix is just the dimension of the space projected onto:

(1.6.44) $$R(\mathbf{P}_V) = \dim V = r$$

(1.6.45) $$R(\mathbf{I} - \mathbf{P}_V) = \dim V^\perp = n - r$$

Multiplying the matrix $\mathbf{I} - \mathbf{P}_V$ on the right by $\mathbf{P}_V$, $(\mathbf{I} - \mathbf{P}_V)\mathbf{P}_V = \mathbf{P}_V - \mathbf{P}_V = 0$. Similarly, $\mathbf{P}_V(\mathbf{I} - \mathbf{P}_V) = 0$. This implies that the matrices $\mathbf{I} - \mathbf{P}_V$ and $\mathbf{P}_V$ are *disjoint.*

Given a vector $\mathbf{y} \in V_n$, $\mathbf{x} = \mathbf{P}_V\mathbf{y} \in V$. Operating on $\mathbf{x}$ with $\mathbf{I} - \mathbf{P}_V$, $(\mathbf{I} - \mathbf{P}_V)\mathbf{x} = (\mathbf{I} - \mathbf{P}_V)\mathbf{P}_V\mathbf{y} = \mathbf{0}$. Similarly, given a vector $\boldsymbol{\varepsilon} = (\mathbf{I} - \mathbf{P}_V)\mathbf{y}$, and operating on this vector with $\mathbf{P}_V$, $\mathbf{P}_V\boldsymbol{\varepsilon} = \mathbf{P}_V(\mathbf{I} - \mathbf{P}_V)\mathbf{y} = \mathbf{0}$, so that $\boldsymbol{\varepsilon}$ and $\mathbf{x}$ are orthogonal by the disjoint property of $(\mathbf{I} - \mathbf{P}_V)$ and $\mathbf{P}_V$. The unique decomposition of $\mathbf{y}$ given in 1.3.15 in terms of projection matrices is now evident.

$$\mathbf{y} = \mathbf{P}_V\mathbf{y} + (\mathbf{I} - \mathbf{P}_V)\mathbf{y} = \mathbf{x} + \boldsymbol{\varepsilon} \tag{1.6.46}$$

Using (1.6.46), the length squared of the observation vector $\mathbf{y}$ is written as

$$\|\mathbf{y}\|^2 = \|\mathbf{P}_V\mathbf{y}\|^2 + \|(\mathbf{I} - \mathbf{P}_V)\mathbf{y}\|^2 \tag{1.6.47}$$

The subscript V on $\mathbf{P}$ was a reminder that the matrix $\mathbf{P}_V$ projected vectors of V_n onto the subspace V. This subscript will now be removed from $\mathbf{P}_V$ for simplicity. To illustrate the use of a projection matrix, let $\mathbf{A}$ be an $n \times r$ matrix of rank r such that the columns span an r-dimensional subspace of V_n. Consider the matrix

$$\mathbf{A}(\mathbf{A}'\mathbf{A})^{-1}\mathbf{A}' = \mathbf{P} \tag{1.6.48}$$

This matrix is a projection matrix since it is idempotent, $[\mathbf{A}(\mathbf{A}'\mathbf{A})^{-1}\mathbf{A}'][\mathbf{A}(\mathbf{A}'\mathbf{A})^{-1}\mathbf{A}'] = \mathbf{A}(\mathbf{A}'\mathbf{A})^{-1}\mathbf{A}'$, symmetrical, and of rank r. Using the projection matrix defined in (1.6.48), any vector $\mathbf{y} \in V_n$ may be constructed from two orthogonal components such that one component is an r-dimensional subspace V_r and the other is an $(n - r)$-dimensional subspace $V^{\perp}$. This is accomplished by using (1.6.46), which avoids having to find an orthogonal basis for the space generated by the columns of $\mathbf{A}$. That is,

$$\begin{aligned}\mathbf{y} &= P_V\mathbf{y} + P_{V^{\perp}}\mathbf{y} \\ &= \mathbf{P}\mathbf{y} + (\mathbf{I} - \mathbf{P})\mathbf{y} \\ &= \mathbf{A}(\mathbf{A}'\mathbf{A})^{-1}\mathbf{A}'\mathbf{y} + [\mathbf{I} - \mathbf{A}(\mathbf{A}'\mathbf{A})^{-1}\mathbf{A}']\mathbf{y}\end{aligned}$$

Furthermore, the length squared of $\mathbf{y}$ is easily partitioned into orthogonal sums of squares:

$$\begin{aligned}\|\mathbf{y}\|^2 &= \|\mathbf{P}\mathbf{y}\|^2 + \|(\mathbf{I} - \mathbf{P})\mathbf{y}\|^2 \\ \mathbf{y}'\mathbf{y} &= \mathbf{y}'\mathbf{P}\mathbf{y} + \mathbf{y}'(\mathbf{I} - \mathbf{P})\mathbf{y} \\ &= \mathbf{y}'\mathbf{A}(\mathbf{A}'\mathbf{A})^{-1}\mathbf{A}'\mathbf{y} + \mathbf{y}'[\mathbf{I} - \mathbf{A}(\mathbf{A}'\mathbf{A})^{-1}\mathbf{A}']\mathbf{y}\end{aligned}$$

Suppose $\mathbf{A}$ is an $n \times m$ matrix and not of full rank, where the $R(\mathbf{A}) = r < m \leq n$. In this case, $\mathbf{P}$ becomes

$$\mathbf{P} = \mathbf{A}(\mathbf{A}'\mathbf{A})^{-}\mathbf{A}' \tag{1.6.49}$$

By applying the findings that $(\mathbf{A}'\mathbf{A})^{-}\mathbf{A}'$ is a g-inverse of $\mathbf{A}$ if $(\mathbf{A}'\mathbf{A})^{-}$ is a g-inverse of $\mathbf{A}'\mathbf{A}$, $\mathbf{A} = \mathbf{A}[(\mathbf{A}'\mathbf{A})^{-}\mathbf{A}']\mathbf{A}$, and it is seen that $\mathbf{P}$ is idempotent, as is $(\mathbf{I} - \mathbf{P})$. Furthermore, $\mathbf{P}$ is symmetrical and unique by (1.6.27). To prove that the $R(\mathbf{P}) = r$, observe that the $R(\mathbf{P}) = R[\mathbf{A}(\mathbf{A}'\mathbf{A})^{-}\mathbf{A}'] \leq R(\mathbf{A}) = R[\mathbf{A}(\mathbf{A}'\mathbf{A})^{-}\mathbf{A}'\mathbf{A}] \leq R[\mathbf{A}(\mathbf{A}'\mathbf{A})^{-}\mathbf{A}'] = R(\mathbf{P})$; hence the $R(\mathbf{P}) = R(\mathbf{A}) = r$. General expressions for the decomposition of $\mathbf{y}$ and the length squared of $\mathbf{y}$ are now evident:

$$\begin{aligned}\mathbf{y} &= \mathbf{A}(\mathbf{A}'\mathbf{A})^{-}\mathbf{A}'\mathbf{y} + [\mathbf{I} - \mathbf{A}(\mathbf{A}'\mathbf{A})^{-}\mathbf{A}']\mathbf{y} \\ \|\mathbf{y}\|^2 &= \mathbf{y}'\mathbf{A}(\mathbf{A}'\mathbf{A})^{-}\mathbf{A}'\mathbf{y} + \mathbf{y}'[\mathbf{I} - \mathbf{A}(\mathbf{A}'\mathbf{A})^{-}\mathbf{A}']\mathbf{y}\end{aligned}$$

To illustrate the use of the projection matrix, Example 1.3.8 is again considered.

EXAMPLE 1.6.11. Let

$$\mathbf{V} = \begin{bmatrix} 1 & 1 & 0 \\ 1 & 1 & 0 \\ 1 & 0 & 1 \\ 1 & 0 & 1 \end{bmatrix} \quad \text{and} \quad \mathbf{y} = \begin{bmatrix} y_{11} \\ y_{12} \\ y_{21} \\ y_{22} \end{bmatrix}$$

where

$$(\mathbf{V}'\mathbf{V})^- = \begin{bmatrix} 0 & 0 & 0 \\ 0 & 1/2 & 0 \\ 0 & 0 & 1/2 \end{bmatrix}$$

Determine the projection of $\mathbf{y}$ onto the space V (the column space of $\mathbf{V}$), using the methods of Section 1.3, by projecting $\mathbf{y}$ onto a set of orthogonal basis vectors that generate the space V. Form the projection matrix

$$\mathbf{P} = \mathbf{V}(\mathbf{V}'\mathbf{V})^-\mathbf{V}' = \begin{bmatrix} 1/2 & 1/2 & 0 & 0 \\ 1/2 & 1/2 & 0 & 0 \\ 0 & 0 & 1/2 & 1/2 \\ 0 & 0 & 1/2 & 1/2 \end{bmatrix}$$

then the projection of $\mathbf{y}$ onto V is

$$\mathbf{x} = \mathbf{V}(\mathbf{V}'\mathbf{V})^-\mathbf{V}'\mathbf{y} = \begin{bmatrix} y_{1.} \\ y_{1.} \\ y_{2.} \\ y_{2.} \end{bmatrix}$$

By identifying the space A in Example 1.3.8 with V as used here, it is seen that both procedures lead to the same result:

$$\begin{aligned} \mathbf{x} &= \mathbf{V}(\mathbf{V}'\mathbf{V})^-\mathbf{V}'\mathbf{y} = P_V\mathbf{y} \\ &= P_1\mathbf{y} + P_{A/1}\mathbf{y} \\ &= \bar{y}\begin{bmatrix} 1 \\ 1 \\ 1 \\ 1 \end{bmatrix} + (y_{1.} - \bar{y})\begin{bmatrix} 1 \\ 1 \\ -1 \\ -1 \end{bmatrix} \\ &= \begin{bmatrix} y_{1.} \\ y_{1.} \\ y_{2.} \\ y_{2.} \end{bmatrix} \end{aligned}$$

By obtaining the projection of $\mathbf{y}$ onto $V^{\perp}$, the orthocomplement of V relative to V_n, the matrix $\mathbf{I} - \mathbf{P}$ is constructed. For this example,

$$\mathbf{I} - \mathbf{P} = \mathbf{I} - \mathbf{V}(\mathbf{V}'\mathbf{V})^{-}\mathbf{V}' = \begin{bmatrix} 1/2 & -1/2 & 0 & 0 \\ -1/2 & 1/2 & 0 & 0 \\ 0 & 0 & 1/2 & -1/2 \\ 0 & 0 & -1/2 & 1/2 \end{bmatrix}$$

so that

$$\boldsymbol{\varepsilon} = [\mathbf{I} - \mathbf{V}(\mathbf{V}'\mathbf{V})^{-}\mathbf{V}']\mathbf{y} = \begin{bmatrix} (y_{11} - y_{12})/2 \\ (y_{12} - y_{11})/2 \\ (y_{21} - y_{22})/2 \\ (y_{22} - y_{21})/2 \end{bmatrix} = \begin{bmatrix} y_{11} - y_{1.} \\ y_{21} - y_{1.} \\ y_{21} - y_{2.} \\ y_{22} - y_{2.} \end{bmatrix}$$

is the projection of $\mathbf{y}$ onto $V^{\perp}$. Alternatively, we could obtain the vector $\boldsymbol{\varepsilon} \in V^{\perp}$ by using (1.6.46), $\mathbf{y} - \mathbf{x} = \boldsymbol{\varepsilon}$.

In Example 1.3.8 the vector spaces $\mathbf{1}$ and $A/\mathbf{1}$ were determined such that $V \equiv A = A/\mathbf{1} \oplus \mathbf{1}$. The vector space $A/\mathbf{1}$ is the orthocomplement of $\mathbf{1}$ relative to A, and the vector space $\mathbf{1}$ is generated by a vector of unities. Having determined these subspaces, the vector $\mathbf{y}$ was projected onto each subspace. For larger matrices this could be a tedious process. However, by using the projection matrix approach, the amount of work is reduced. To establish projection matrices for these projection operations, let the matrix $\mathbf{V}$ in this example be represented by $\mathbf{V} = \{\mathbf{1} \quad \mathbf{A}\}$ where

$$\mathbf{1} = \begin{bmatrix} 1 \\ 1 \\ 1 \\ 1 \end{bmatrix} \quad \text{and} \quad \mathbf{A} = \begin{bmatrix} 1 & 0 \\ 1 & 0 \\ 0 & 1 \\ 0 & 1 \end{bmatrix}$$

Next, define the matrices $\mathbf{P}_1$, $\mathbf{P}_2$, and $\mathbf{P}_3$ as follows:

$$\mathbf{P}_1 = \mathbf{1}(\mathbf{1}'\mathbf{1})^{-}\mathbf{1}'$$
$$\mathbf{P}_2 = \mathbf{V}(\mathbf{V}'\mathbf{V})^{-}\mathbf{V}' - \mathbf{1}(\mathbf{1}'\mathbf{1})^{-}\mathbf{1}'$$
$$\mathbf{P}_3 = \mathbf{I} - \mathbf{V}(\mathbf{V}'\mathbf{V})^{-}\mathbf{V}'$$

so that

$$\mathbf{I} = \mathbf{P}_1 + \mathbf{P}_2 + \mathbf{P}_3$$

The matrices $\mathbf{P}_1$, $\mathbf{P}_2$, and $\mathbf{P}_3$ are projection matrices since each is symmetric and idempotent. We have just seen that the matrix $\mathbf{V}$ is used to project $\mathbf{y}$ onto V so that $\mathbf{x} = \mathbf{V}(\mathbf{V}'\mathbf{V})^{-}\mathbf{V}'\mathbf{y} \in V$. The matrix $\mathbf{P}_1$, defined by

$$\mathbf{P}_1 = \frac{1}{4}\begin{bmatrix} 1 & 1 & 1 & 1 \\ 1 & 1 & 1 & 1 \\ 1 & 1 & 1 & 1 \\ 1 & 1 & 1 & 1 \end{bmatrix}$$

projects $\mathbf{y}$ onto the column space of $\mathbf{1}$. Since $\mathbf{1} \subseteq V$, the vector

$$\mathbf{m} = \mathbf{P}_1\mathbf{y} = \bar{y}\begin{bmatrix} 1 \\ 1 \\ 1 \\ 1 \end{bmatrix}$$

is an element of V and is seen to be the $P_1\mathbf{y}$. The vector $\mathbf{z} = \mathbf{P}_2\mathbf{y} = \mathbf{P}\mathbf{y} - \mathbf{P}_1\mathbf{y} = \mathbf{x} - \mathbf{m}$ is an element of V. The matrix $\mathbf{P}_2$ is given by

$$\mathbf{P}_2 = \mathbf{P} - \mathbf{P}_1$$

$$= \begin{bmatrix} 1/4 & 1/4 & -1/4 & -1/4 \\ 1/4 & 1/4 & -1/4 & -1/4 \\ -1/4 & -1/4 & 1/4 & 1/4 \\ -1/4 & -1/4 & 1/4 & 1/4 \end{bmatrix}$$

so that

$$\mathbf{z} = \mathbf{P}_2\mathbf{y} = \begin{bmatrix} (y_{11} + y_{12} - y_{21} - y_{22})/4 \\ (y_{11} + y_{12} - y_{21} - y_{22})/4 \\ (y_{21} + y_{22} - y_{11} - y_{12})/4 \\ (y_{21} + y_{22} - y_{11} - y_{12})/4 \end{bmatrix}$$

$$= \begin{bmatrix} y_{1.} - \bar{y} \\ y_{1.} - \bar{y} \\ y_{2.} - \bar{y} \\ y_{2.} - \bar{y} \end{bmatrix} = (y_{1.} - \bar{y}) \begin{bmatrix} 1 \\ 1 \\ -1 \\ -1 \end{bmatrix}$$

That is, $\mathbf{P}_2\mathbf{y} = \mathbf{z}$ is the projection of $\mathbf{y}$ onto $V/\mathbf{1}$ or $A/\mathbf{1}$ using the notation in Example 1.3.8. This follows from the fact that $\mathbf{P}_2\mathbf{P}_1 = \mathbf{P}_1\mathbf{P}_2 = \mathbf{0}$, which implies that $\mathbf{m}$ and $\mathbf{z}$ are orthogonal since $\mathbf{P}_2\mathbf{m} = \mathbf{P}_2\mathbf{P}_1\mathbf{y} = \mathbf{0}$ and $\mathbf{P}_1\mathbf{z} = \mathbf{P}_1\mathbf{P}_2\mathbf{y} = \mathbf{0}$. The vector $\mathbf{z}$ was obtained by operating on $\mathbf{y}$ with $\mathbf{P}_2$. Since $\mathbf{z} = \mathbf{P}_2\mathbf{y} = (\mathbf{P} - \mathbf{P}_1)\mathbf{y} = \mathbf{x} - \mathbf{m}$, it is seen that $\mathbf{P}_1\mathbf{x} = \mathbf{P}_1\mathbf{m} = \mathbf{P}_1(\mathbf{P}_1\mathbf{y}) = \mathbf{P}_1\mathbf{y} = \mathbf{m}$ because $\mathbf{P}_1\mathbf{z} = \mathbf{0}$. Hence $\mathbf{m}$ is acquired by using either $\mathbf{P}_1$ directly on $\mathbf{y}$ or by first projecting $\mathbf{y}$ onto V to obtain $\mathbf{x}$ and then operating on $\mathbf{x}$ with $\mathbf{P}_1$ to determine $\mathbf{m}$. One could, of course, locate $\mathbf{z}$ in a similar manner by using $\mathbf{P}_2$ in place of $\mathbf{P}_1$. These results are easily seen in Figure 1.6.5. Compare Figures 1.6.5 and 1.3.10.

By employing various projection matrices in Example 1.6.11, we have indicated how we may easily project a vector $\mathbf{y}$ onto orthogonal subspaces. These results have important implications in statistical theory when partitioning a sum of squares. They are discussed further in Section 2.6.

EXERCISES 1.6

1. Write each of the following systems of simultaneous linear equations in the matrix form (1.6.2), determine whether the system is consistent by using (1.6.3), determine whether the consistent systems have unique or infinite solutions sets, and, finally, find the general solution

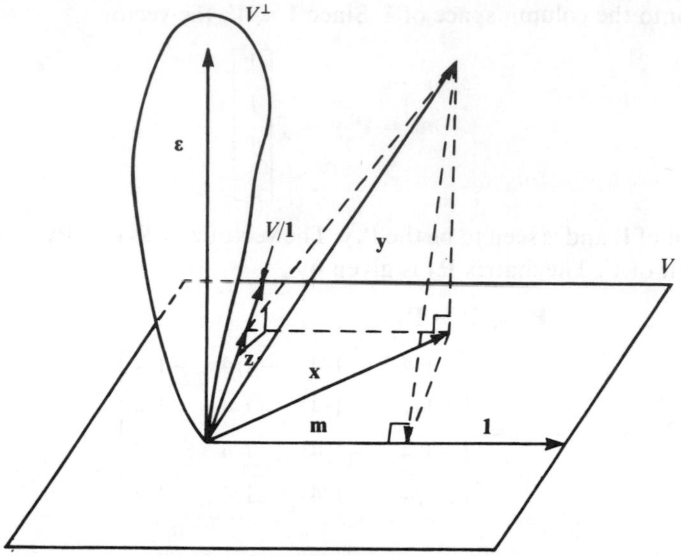

FIGURE 1.6.5. The orthogonal decomposition of V_n.

to each consistent system by Gauss' elimination procedure.

a. $\begin{aligned} 2x - 3y + z &= 5 \\ 6x - 9y + 3z &= 10 \end{aligned}$

b. $\begin{aligned} x + y &= 6 \\ x - z &= -2 \\ y + z &= 8 \\ x + z &= 0 \end{aligned}$

c. $\begin{aligned} x + y &= 0 \\ 2x - 3y &= 0 \end{aligned}$

d. $\begin{aligned} x + y - z &= 0 \\ 2x - y + z &= 0 \\ x + 4y - 4z &= 0 \end{aligned}$

e. $\begin{aligned} x + y + z &= 2 \\ x - y + z &= 6 \\ x + 2y + z &= 0 \\ 3x - y + 3z &= 14 \end{aligned}$

2. Rework Example 1.6.2 with the side conditions

   a. $\alpha_2 = 0$ b. $\alpha_1 - \alpha_2 = 0$

3. Solve the system of equations in Exercise 1e using the side-condition approach with the restrictions

   a. $x - z = 0$ b. $x + z = 0$ c. $y = 0$ d. $x = 0$

4. Rework Example 1.6.3 by the reparameterization method for the set of new variables

   a. $\mu + \alpha_1$ and $\mu + \alpha_2$ b. μ and $\alpha_1 + \alpha_2$

   c. $\mu + \alpha_1$ and $\alpha_1 - \alpha_2$ d. $\mu + \alpha_1$ and α_2

5. Solve the system of equations in Exercise 1e by reparameterizing to the new variables

   a. y and $x + z$ b. $x + y$ and z

6. Find the g-inverses for

   a. $\begin{bmatrix} 1 \\ 2 \\ 0 \end{bmatrix}$ b. $[0, 1, 2]$ c. $\begin{bmatrix} 1 & 2 \\ 0 & -1 \\ 1 & 0 \end{bmatrix}$

d. $\begin{bmatrix} 8 & 4 & 4 \\ 4 & 4 & 0 \\ 4 & 0 & 4 \end{bmatrix}$ e. $\begin{bmatrix} 0 & 0 & 0 & 2 \\ 0 & 1 & 2 & 3 \\ 0 & 4 & 5 & 6 \\ 0 & 7 & 8 & 9 \end{bmatrix}$

7. Prove (1.6.21).

8. Solve the system in Exercise 1e by the method of generalized inverses, and show that the solution agrees with those obtained in Exercises 3 and 4.

9. Solve the following system of equations, using the g-inverse approach:

$$\begin{bmatrix} 1 & 1 & 0 & 1 & 0 \\ 1 & 1 & 0 & 1 & 0 \\ 1 & 1 & 0 & 0 & 1 \\ 1 & 1 & 0 & 0 & 1 \\ 1 & 0 & 1 & 1 & 0 \\ 1 & 0 & 1 & 1 & 0 \\ 1 & 0 & 1 & 0 & 1 \\ 1 & 0 & 1 & 0 & 1 \end{bmatrix} \begin{bmatrix} \mu \\ \alpha_1 \\ \alpha_2 \\ \beta_1 \\ \beta_2 \end{bmatrix} = \begin{bmatrix} y_{111} \\ y_{112} \\ y_{121} \\ y_{122} \\ y_{211} \\ y_{212} \\ y_{221} \\ y_{222} \end{bmatrix}$$

For what linear combinations of $\mu, \alpha_1, \alpha_2, \beta_1$, and β_2 do unique solutions exist? What is the form of the unique solutions?

10. In Example 1.6.8 determine whether unique solutions exist for the following linear combinations of the unknowns and, if they do, find them.

a. $\mu + \alpha_1$ b. $\mu + \alpha_1 + \alpha_2$ c. $\alpha_1 - (\alpha_2/2)$ d. α_1

11. Prove (1.6.34) through (1.6.37).

12. a. Work through the analysis of Example 1.6.11 for the system given in Exercise 9.

b. Letting

$$\mathbf{1} = \begin{Bmatrix} 1 \\ 1 \\ 1 \\ 1 \\ 1 \\ 1 \\ 1 \\ 1 \end{Bmatrix}, \quad A = \begin{Bmatrix} 1 & 0 \\ 1 & 0 \\ 1 & 0 \\ 1 & 0 \\ 0 & 1 \\ 0 & 1 \\ 0 & 1 \\ 0 & 1 \end{Bmatrix}, \quad B = \begin{Bmatrix} 1 & 0 \\ 1 & 0 \\ 0 & 1 \\ 0 & 1 \\ 1 & 0 \\ 1 & 0 \\ 0 & 1 \\ 0 & 1 \end{Bmatrix}, \quad X = \begin{Bmatrix} 1 & 1 & 0 & 1 & 0 \\ 1 & 1 & 0 & 1 & 0 \\ 1 & 1 & 0 & 0 & 1 \\ 1 & 1 & 0 & 0 & 1 \\ 1 & 0 & 1 & 1 & 0 \\ 1 & 0 & 1 & 1 & 0 \\ 1 & 0 & 1 & 0 & 1 \\ 1 & 0 & 1 & 0 & 1 \end{Bmatrix}$$

where $A/\mathbf{1}$ and $B/\mathbf{1}$ denote vector spaces, determine the projection of y onto $\mathbf{1}$, $A/\mathbf{1}$, and $B/\mathbf{1}$, using projection matrices.

c. What are your conclusions from your analysis of parts a and b?

d. For part b, determine the length squared of projections. How might you relate these to your results obtained in part a?

1.7 QUADRATIC FORMS AND CHARACTERISTIC EQUATIONS

In Section 1.6 it was observed that the length of **y** squared could be decomposed into the sum of two products, each of the form $\mathbf{y}'\mathbf{A}\mathbf{y} = Q$. Q is called a *quadratic form.* A quadratic form is a quadratic function of the form

$$Q = \mathbf{y}'\mathbf{A}\mathbf{y} = \sum_{i=1}^{n} \sum_{j=1}^{n} a_{ij}y_iy_j = a_{11}y_1^2 + (a_{12} + a_{21})y_1y_2 + a_{22}y_2^2 + \cdots \tag{1.7.1}$$

where **y** is an $n \times 1$ vector and **A** is a symmetrical matrix of order n. The matrix **A** is called the *matrix of the quadratic form.*

Following Rao (1965a, p. 31), a (real) quadratic form is said to be *positive definite* (p.d.) if $\mathbf{y}'\mathbf{A}\mathbf{y} > 0$ for all nonzero vectors **y** and $\mathbf{y}'\mathbf{A}\mathbf{y} = 0$ only for $\mathbf{y} = \mathbf{0}$ and is said to be *positive semidefinite* (p.s.d.) if $\mathbf{y}'\mathbf{A}\mathbf{y} \geq 0$ for all $\mathbf{y} \neq \mathbf{0}$, with $\mathbf{y}'\mathbf{A}\mathbf{y} = 0$ for at least one $\mathbf{y} \neq \mathbf{0}$. Similar definitions apply for negative definite and negative semidefinite quadratic forms. The symmetric matrix **A** of the quadratic form is given the same definiteness as the form; thus, if $\mathbf{y}'\mathbf{A}\mathbf{y}$ is p.d., **A** is also said to be p.d. In statistical applications, p.d. and p.s.d. matrices arise frequently and are usually called *Gramian* or *nonnegative definite* (n.n.d.) matrices.

To simplify the form of (1.7.1), every quadratic form may be reduced to a sum of squares. From (1.5.36) there exists a nonsingular matrix **P** such that $\mathbf{P}'\mathbf{A}\mathbf{P} = \mathbf{\Lambda}$, where $\mathbf{\Lambda}$ is a diagonal matrix of order n. Letting $\mathbf{y} = \mathbf{P}\mathbf{x}$, (1.7.1) becomes

$$\mathbf{y}'\mathbf{A}\mathbf{y} = \mathbf{x}'\mathbf{P}'\mathbf{A}\mathbf{P}\mathbf{x} = \mathbf{x}'\mathbf{\Lambda}\mathbf{x} = \sum_{i=1}^{n} \lambda_i x_i^2 .$$

(1.7.2) Every quadratic form may be simplified to

$$\mathbf{y}'\mathbf{A}\mathbf{y} = \sum_{i=1}^{n} \lambda_i x_i^2$$

by a nonsingular linear transformation.

That is, every quadratic form has a diagonal representation. From *Sylvester's law of inertia*, the number of positive, negative, and zero coefficients λ_i is uniquely determined by the matrix **A** of the quadratic form (Hohn, 1964, p. 330). The number of x_i^2 terms with nonzero λ_i is called the *rank* of the form, and the difference between the number of positive λ_i and negative λ_i is called the *signature* of the quadratic form.

TABLE 1.7.1. Summary of Quadratic Forms.

λ_i Positive	Zero	Negative	Type of Form	Matrix of Form
√			positive definite	nonsingular
√	√		positive semidefinite	singular
√		√	indefinite	nonsingular
√	√	√	indefinite	singular
		√	negative definite	nonsingular
	√	√	negative semidefinite	singular
	√		null	singular

Relationships among positive, negative, and zero λ_i in (1.7.2), types of quadratic forms, and singularity are summarized in Table 1.7.1.

In (1.7.2), if each $\lambda_i > 0$, $\mathbf{A}$ is p.d. and there exists a further simplification by letting $y_i = z_i/\sqrt{\lambda_i}$:

$$\text{(1.7.3)} \qquad \mathbf{y}'\mathbf{A}\mathbf{y} = \mathbf{z}'\mathbf{z} = \sum_{i=1}^{n} z_i^2$$

To show this, let $\mathbf{y} = \mathbf{P}\mathbf{\Lambda}^{-1/2}\mathbf{z}$ where

$$\mathbf{\Lambda}^{-1/2} = \begin{bmatrix} 1/\sqrt{\lambda_1} & 0 & \cdots & 0 \\ 0 & 1/\sqrt{\lambda_2} & \cdots & 0 \\ \vdots & \vdots & \vdots & \vdots \\ 0 & 0 & \cdots & 1/\sqrt{\lambda_n} \end{bmatrix}$$

so that $\mathbf{y}'\mathbf{A}\mathbf{y} = \mathbf{z}'\mathbf{\Lambda}^{-1/2}\mathbf{P}'\mathbf{A}\mathbf{P}\mathbf{\Lambda}^{-1/2}\mathbf{z} = \mathbf{z}'\mathbf{z}$. $\mathbf{A}$ may also be written as $\mathbf{A} = \mathbf{F}\mathbf{F}'$ by letting $\mathbf{F}' = \mathbf{\Lambda}^{1/2}\mathbf{P}^{-1}$. In a similar fashion, if $\mathbf{A}$ is p.s.d. and the $R(\mathbf{A}) = r < n$, a nonsingular transformation exists such that

$$\text{(1.7.4)} \qquad \mathbf{y}'\mathbf{A}\mathbf{y} = \sum_{i=1}^{r} \lambda_i x_i^2 = \sum_{i=1}^{r} z_i^2$$

where $\lambda_{r+1} = \cdots = \lambda_n = 0$.

Results (1.7.3) and (1.7.4) motivate a theorem that allows us to determine whether a symmetric matrix $\mathbf{A}$ is Gramian (p.d. or p.s.d.).

(1.7.5) A (real) symmetric matrix $\mathbf{A}$ is p.d. if and only if there exists a nonsingular matrix $\mathbf{F}$ such that $\mathbf{A} = \mathbf{F}\mathbf{F}'$. A (real) symmetric matrix $\mathbf{A}$ of rank r is p.s.d. if and only if there exists a matrix $\mathbf{F}$ of rank r such that $\mathbf{A} = \mathbf{F}\mathbf{F}'$.

Given that $\mathbf{A}$ is Gramian, (1.5.36) shows that $\mathbf{A}$ is factorable and that the factor matrices are of proper order and rank. Furthermore, if $\mathbf{A} = \mathbf{F}\mathbf{F}'$, where the rank of $\mathbf{F}$ is n, $\mathbf{y}'\mathbf{F}\mathbf{F}'\mathbf{y} = \mathbf{z}'\mathbf{z} > 0$, for all $\mathbf{z} \neq \mathbf{0}$ or $\mathbf{y} \neq \mathbf{0}$, and is equal to 0 only if $\mathbf{y} = \mathbf{0}$, so that $\mathbf{A}$ is p.d. If the $R(\mathbf{F}) = r < n$ and $\mathbf{A} = \mathbf{F}\mathbf{F}'$, $\mathbf{F}'\mathbf{y} = \mathbf{0}$, for some $\mathbf{y} \neq \mathbf{0}$, so $\mathbf{A}$ is p.s.d. If $|\mathbf{F}\mathbf{F}'| = 0$, $\mathbf{A}$ is p.s.d.; if $\mathbf{A}$ is p.d., $\mathbf{F}$ is $n \times n$. So $\mathbf{A}^{-1} = (\mathbf{F}^{-1})'\mathbf{F}^{-1}$, and $\mathbf{A}^{-1}$ is p.d.

Cholesky Factoring. Theorem (1.7.5) indicates the existence of a matrix $\mathbf{F}$ such that $\mathbf{A} = \mathbf{F}\mathbf{F}'$. To determine a matrix $\mathbf{F}$, use one of many factorization schemes available in Forsythe (1951). One very fundamental procedure will be illustrated for a symmetric p.d. matrix $\mathbf{A}$. A matrix product may be defined, from (1.4.21), as

$$\text{(1.7.6)} \qquad \underset{(n \times n)}{\mathbf{A}} = \underset{(n \times n)}{\mathbf{T}}\ \underset{(n \times n)}{\mathbf{T}'} = \sum_{i=1}^{n} \mathbf{t}_i\mathbf{t}_i'$$

so that

$$\mathbf{A} - \sum_{i=1}^{n} \mathbf{t}_i\mathbf{t}_i' = \mathbf{0}$$

The *Cholesky* or *square root factorization* procedure constructs vectors $\mathbf{t}_i'$ in a systematic manner until (1.7.6) is obtained.

EXAMPLE 1.7.1. Let

$$\mathbf{A} = \begin{bmatrix} 2 & 0 & -2 \\ 0 & 4 & 2 \\ -2 & 2 & 4 \end{bmatrix} = \begin{bmatrix} a_{11} & a_{12} & a_{13} \\ a_{21} & a_{22} & a_{23} \\ a_{31} & a_{32} & a_{33} \end{bmatrix} = \begin{bmatrix} \mathbf{a}_1' \\ \mathbf{a}_2' \\ \mathbf{a}_3' \end{bmatrix}$$

Set $\mathbf{t}_1' = [\sqrt{2}, 0, -\sqrt{2}] = \mathbf{a}_1'/\sqrt{a_{11}}$ and form the matrix

$$\mathbf{B} = \mathbf{A} - \mathbf{t}_1\mathbf{t}_1' = \begin{bmatrix} 2 & 0 & -2 \\ 0 & 4 & 2 \\ -2 & 2 & 4 \end{bmatrix} - \begin{bmatrix} 2 & 0 & -2 \\ 0 & 0 & 0 \\ -2 & 0 & 2 \end{bmatrix}$$

$$= \begin{bmatrix} 0 & 0 & 0 \\ 0 & 4 & 2 \\ 0 & 2 & 2 \end{bmatrix} = \begin{bmatrix} \mathbf{b}_1' \\ \mathbf{b}_2' \\ \mathbf{b}_3' \end{bmatrix}$$

Then set $\mathbf{t}_2' = [0, 2, 1] = \mathbf{b}_2'/\sqrt{b_{22}}$ and form

$$\mathbf{t}_2\mathbf{t}_2' = \begin{bmatrix} 0 & 0 & 0 \\ 0 & 4 & 2 \\ 0 & 2 & 1 \end{bmatrix}$$

so that

$$\mathbf{C} = \mathbf{A} - \mathbf{t}_1\mathbf{t}_1' - \mathbf{t}_2\mathbf{t}_2' = \begin{bmatrix} 0 & 0 & 0 \\ 0 & 0 & 0 \\ 0 & 0 & 0 \end{bmatrix} = \begin{bmatrix} \mathbf{c}_1' \\ \mathbf{c}_2' \\ \mathbf{c}_3' \end{bmatrix}$$

Setting $\mathbf{t}_3' = [0, 0, 1] = \mathbf{c}_3'/\sqrt{c_{33}}$

$$\mathbf{A} - \sum_{i=1}^{3} \mathbf{t}_i\mathbf{t}_i' = 0$$

Thus

$$\mathbf{T}' = \begin{bmatrix} \mathbf{t}_1' \\ \mathbf{t}_2' \\ \mathbf{t}_3' \end{bmatrix} = \begin{bmatrix} \sqrt{2} & 0 & -\sqrt{2} \\ 0 & 2 & 1 \\ 0 & 0 & 1 \end{bmatrix} \quad \text{and} \quad \mathbf{A} = \mathbf{TT}' = ◺\,◹$$

is a square root factorization of the symmetric p.d. matrix $\mathbf{A}$, where $\mathbf{T}'$ is an upper triangular matrix.

Following Example 1.7.1, an algorithm for factoring any p.d. symmetric matrix $\mathbf{A}$ is evident. Let

$$\underset{(n \times n)}{\mathbf{T}'} = [t_{ij}] = \begin{bmatrix} t_{11} & t_{12} & t_{13} & \cdots & t_{1n} \\ 0 & t_{22} & t_{23} & \cdots & t_{2n} \\ 0 & 0 & t_{33} & \cdots & t_{3n} \\ 0 & 0 & 0 & \cdots & t_{nn} \end{bmatrix} \quad \text{and} \quad \underset{(n \times n)}{\mathbf{A}} = [a_{ij}]$$

then

$$t_{11} = a_{11}^{1/2} \qquad t_{1j} = a_{1j}t_{11}^{-1} \qquad (1 \le j \le n)$$

$$t_{ii} = \left(a_{ii} - \sum_{k=1}^{i=1} t_{ki}^2\right)^{1/2} \qquad (1 < i \le n)$$

$$t_{ij} = \left(a_{ij} - \sum_{k=1}^{i-1} t_{ki}t_{kj}\right)t_{ii}^{-1} \qquad (1 < i < j \le n)$$

$$t_{ij} = 0 \qquad (1 < j < i \le n)$$

so that $\mathbf{A} = \mathbf{TT}'$. By partitioning $\mathbf{T}$ so that $\mathbf{T} = \mathbf{L\Lambda}^{1/2}$, where $\mathbf{L}$ is a unit lower triangular matrix, $\mathbf{A}$ has the form $\mathbf{A} = \mathbf{L\Lambda}^{1/2}\mathbf{\Lambda}^{1/2}\mathbf{L}' = \mathbf{L\Lambda U}$, where $\mathbf{U}$ is a unit upper triangular matrix.

(1.7.7) Every (real) symmetric p.d. matrix $\mathbf{A}$ may be expressed as a unique product $\mathbf{L\Lambda L}' = \mathbf{TT}'$, where $\mathbf{T}$ is a lower triangular matrix.

This is a special case of the **LDU** *decomposition theorem* (see Forsythe, 1951, and Marcus, 1960).

The Cholesky factorization of a symmetric p.d. matrix $\mathbf{A}$ also provides a very efficient method for finding the inverse of any symmetric p.d. matrix:

$$\mathbf{A}^{-1} = (\mathbf{T}^{-1})'\mathbf{T}^{-1} = \triangledown\ \triangle$$

Given $\mathbf{T}$, a lower triangular matrix defined by

$$\mathbf{T} = \begin{bmatrix} t_{11} & 0 & 0 & \cdots & 0 \\ t_{12} & t_{22} & 0 & \cdots & 0 \\ t_{13} & t_{23} & t_{33} & \cdots & 0 \\ \vdots & \vdots & \vdots & \vdots & \vdots \\ t_{1n} & t_{2n} & t_{3n} & \cdots & t_{nn} \end{bmatrix} = [t_{ij}]$$

let

$$\mathbf{T}^{-1} = \begin{bmatrix} t^{11} & 0 & 0 & \cdots & 0 \\ t^{21} & t^{22} & 0 & \cdots & 0 \\ t^{31} & t^{32} & t^{33} & \cdots & 0 \\ \vdots & \vdots & \vdots & \vdots & \vdots \\ t^{n1} & t^{n2} & t^{n3} & \cdots & t^{nn} \end{bmatrix} = [t^{ij}]$$

Then an algorithm for finding $\mathbf{T}^{-1}$ from $\mathbf{T}$ is shown to be

$$t^{ii} = t_{ii}^{-1} \qquad (i = j)$$

$$t^{ij} = -t_{ii}^{-1}\left(\sum_{k=j}^{i-1} t_{ki}t^{kj}\right) \qquad (1 < j < i \le n)$$

$$t^{ij} = 0 \qquad (1 < i < j \le n)$$

EXAMPLE 1.7.2. Let

$$\mathbf{A} = \begin{bmatrix} 2 & 0 & -2 \\ 0 & 4 & 2 \\ -2 & 2 & 4 \end{bmatrix}$$

then

$$\mathbf{A} = \begin{bmatrix} \sqrt{2} & 0 & 0 \\ 0 & 2 & 0 \\ -\sqrt{2} & 1 & 1 \end{bmatrix} \begin{bmatrix} \sqrt{2} & 0 & -\sqrt{2} \\ 0 & 2 & 1 \\ 0 & 0 & 1 \end{bmatrix} = \mathbf{TT}'$$

From Example 1.7.1,

$$\mathbf{T} = \begin{bmatrix} t_{11} & 0 & 0 \\ t_{12} & t_{22} & 0 \\ t_{13} & t_{23} & t_{33} \end{bmatrix} = \begin{bmatrix} \sqrt{2} & 0 & 0 \\ 0 & 2 & 0 \\ -\sqrt{2} & 1 & 1 \end{bmatrix}$$

Letting

$$\mathbf{T}^{-1} = \begin{bmatrix} t^{11} & 0 & 0 \\ t^{21} & t^{22} & 0 \\ t^{31} & t^{32} & t^{33} \end{bmatrix}$$

where

$$t^{11} = t_{11}^{-1} = \frac{1}{\sqrt{2}} \qquad t^{22} = t_{22}^{-1} = \frac{1}{2} \qquad t^{33} = t_{33}^{-1} = 1$$

$$t^{21} = -t_{22}^{-1}\left(\sum_{k=1}^{1} t_{k2}t^{k1}\right) = -\frac{1}{2}\left[(0)\left(\frac{1}{2}\right)\right] = 0$$

$$\begin{aligned} t^{31} &= -t_{33}^{-1}\left(\sum_{k=1}^{2} t_{k3}t^{k1}\right) = -t_{33}^{-1}(t_{13}t^{11} + t_{23}t^{21}) \\ &= -1\left[(-\sqrt{2})\left(\frac{1}{\sqrt{2}}\right) + (1)(0)\right] = 1 \\ t^{32} &= -t_{33}^{-1}\left(\sum_{k=2}^{2} t_{k3}t^{k2}\right) = -t_{33}^{-1}(t_{23}t^{22}) = -1\left[(1)\left(\frac{1}{2}\right)\right] = -\frac{1}{2} \end{aligned}$$

and thus

$$\mathbf{T}^{-1} = \begin{bmatrix} 1/\sqrt{2} & 0 & 0 \\ 0 & 1/2 & 0 \\ 1 & -1/2 & 1 \end{bmatrix}$$

Forming the product $\mathbf{A}^{-1} = (\mathbf{T}^{-1})'\mathbf{T}^{-1}$,

$$\mathbf{A}^{-1} = \begin{bmatrix} 3/2 & -1/2 & 1 \\ -1/2 & 1/2 & -1/2 \\ 1 & -1/2 & 1 \end{bmatrix}$$

If $\mathbf{A}$ is not symmetric, the algorithm for the inverse may still be used since $\mathbf{A}^{-1} = (\mathbf{A}'\mathbf{A})^{-1}\mathbf{A}'$ when $\mathbf{A}$ is square, nonsingular, and not symmetric, provided $\mathbf{A}'\mathbf{A}$ is p.d. However,

(1.7.8) $\quad\mathbf{A}'\mathbf{A}$ is p.d. if $\mathbf{A}$ has full column rank.

To prove (1.7.8), observe that $\mathbf{x}'\mathbf{A}'\mathbf{A}\mathbf{x} = (\mathbf{A}\mathbf{x})'\mathbf{A}\mathbf{x} > 0$, for all $\mathbf{x} \neq \mathbf{0}$, and is equal to 0 only if $\mathbf{x} = \mathbf{0}$ since $\mathbf{A}$ has full column rank. Thus $\mathbf{x}'\mathbf{A}'\mathbf{A}\mathbf{x}$ is p.d., or $\mathbf{A}'\mathbf{A}$ is p.d. If $\mathbf{A}$ is less than full column rank, then $\mathbf{A}\mathbf{x} = \mathbf{0}$, for some $\mathbf{x} \neq \mathbf{0}$, or $\mathbf{A}'\mathbf{A}$ is p.s.d. As a corollary to (1.7.8) or from (1.7.5), we immediately see that $\mathbf{A}\mathbf{A}'$ is p.d. if $\mathbf{A}$ has full row rank; otherwise $\mathbf{A}\mathbf{A}'$ is p.s.d.

If $\mathbf{A}$ is p.s.d., the Cholesky factoring procedure is still applicable; however, when $t_{ii} = 0$, the entire row is replaced by 0s. More important, this procedure is conveniently used to find $\mathbf{A}^-$ when $\mathbf{A}$ is p.s.d. and symmetric. If the $R(\mathbf{A}) = r$, then $n - r$ rows of the matrix $\mathbf{T}'$ in a Cholesky factoring procedure will be 0. Given $\mathbf{T}'$, (1.6.17) is employed to find a g-inverse of $\mathbf{T}'$. That is, delete the rows of $\mathbf{T}'$ that contain all 0s and the corresponding columns, and find the inverse of the reduced matrix $\mathbf{T}_{11}$. Next, restore the inverse obtained to its original order with 0s in the rows and columns originally deleted. Denote this matrix by $(\mathbf{T}')^-$. Then $\mathbf{A}^- = (\mathbf{T}')^-(\mathbf{T}^-)$.

EXAMPLE 1.7.3. Let

$$\mathbf{A} = \begin{bmatrix} 4 & 4 & -2 \\ 4 & 4 & -2 \\ -2 & -2 & 10 \end{bmatrix}$$

Using the Cholesky procedure,

$$\mathbf{T}' = \begin{bmatrix} 2 & 2 & -1 \\ 0 & 0 & 0 \\ 0 & 0 & 3 \end{bmatrix}$$

Delete row two and column two to form the matrix $\mathbf{T}_{11}$,

$$\mathbf{T}_{11} = \begin{bmatrix} 2 & -1 \\ 0 & 3 \end{bmatrix} \quad \text{and} \quad \mathbf{T}_{11}^{-1} = \frac{1}{6}\begin{bmatrix} 3 & 1 \\ 0 & 2 \end{bmatrix}$$

so

$$(\mathbf{T}')^- = \frac{1}{6}\begin{bmatrix} 3 & 0 & 1 \\ 0 & 0 & 0 \\ 0 & 0 & 2 \end{bmatrix}$$

and

$$\mathbf{A}^- = (\mathbf{T}')^-(\mathbf{T}^-) = \left\{\frac{1}{6}\begin{bmatrix} 3 & 0 & 1 \\ 0 & 0 & 0 \\ 0 & 0 & 2 \end{bmatrix}\right\}\left\{\frac{1}{6}\begin{bmatrix} 3 & 0 & 0 \\ 0 & 0 & 0 \\ 1 & 0 & 2 \end{bmatrix}\right\}$$

$$= \frac{1}{36}\begin{bmatrix} 10 & 0 & 2 \\ 0 & 0 & 0 \\ 2 & 0 & 4 \end{bmatrix}$$

The Geometry of a Quadratic Form. Geometrically, the equation $\mathbf{y}'\mathbf{A}\mathbf{y} = Q$ specifies an ellipsoid in an n-dimensional space with its center at the origin and $Q > 0$. The matrix $\mathbf{A}$ of order n determines the form of the ellipsoid. For $\mathbf{A} = \mathbf{I}$, the ellipsoid becomes spherical. As an example, in two dimensions, let

$$\mathbf{A} = \begin{bmatrix} 1 & 1/2 \\ 1/2 & 1 \end{bmatrix}$$

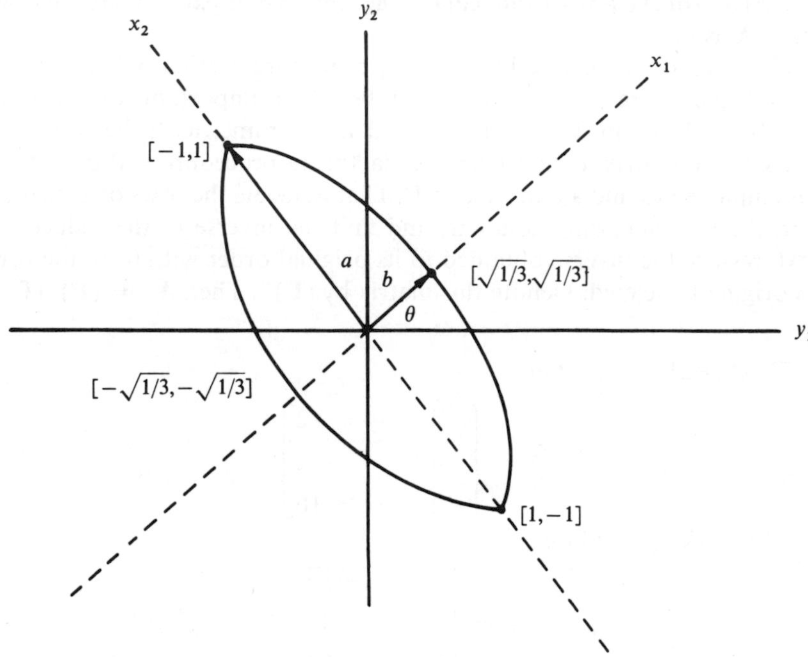

FIGURE 1.7.1. Geometric representation of $\mathbf{y}'\mathbf{A}\mathbf{y} = y_1^2 + y_1y_2 + y_2^2 = 1$.

then $Q = \mathbf{y}'\mathbf{A}\mathbf{y} = y_1^2 + y_1^2 + y_1y_2 + y_2^2$ is the equation of an ellipse. In Figure 1.7.1, the form of the ellipse is represented geometrically in the plane. Rotating $\mathbf{y}_1 \to \mathbf{x}_1$ and $\mathbf{y}_2 \to \mathbf{x}_2$, where the angle $\theta = 45°$, the transformation matrix from (1.6.40) is

$$\mathbf{P}' = \begin{bmatrix} \cos 45° & \sin 45° \\ -\sin 45° & \cos 45° \end{bmatrix} = \begin{bmatrix} 1/\sqrt{2} & 1/\sqrt{2} \\ -1/\sqrt{2} & 1/\sqrt{2} \end{bmatrix}$$

which is a matrix of direction cosines of the ith old axis with the jth new axis. Since $\mathbf{PP}' = \mathbf{P}'\mathbf{P} = \mathbf{I}$, $\mathbf{P}'$ is an orthogonal matrix, and the equation for the ellipse becomes, for $\mathbf{x} = \mathbf{P}'\mathbf{y}$,

$$y_1^2 + y_1y_2 + y_2^2 = \mathbf{y}'\mathbf{A}\mathbf{y} = \frac{3}{2}x_1^2 + \frac{1}{2}x_2^2 = \mathbf{x}'\mathbf{\Lambda}\mathbf{x}$$

$$= \lambda_1 x_1^2 + \lambda_2 x_2^2 \tag{1.7.9}$$

Through a rotation transformation, the equation of the ellipse in the new system appears in a simpler form than the one given in the old system. The cross-product term $y_1 y_2$ has been eliminated. The original equation has been reduced to diagonal form by the orthogonal matrix $\mathbf{P}'$.

The diagonal elements of Λ, $\lambda_1 = 3/2$ and $\lambda_2 = 1/2$, are related to the lengths of the *semimajor* and *semiminor axes* of the ellipse. From analytic geometry, the equation of an ellipse, for $Q = 1$, is given by

$$\left(\frac{1}{b}\right)^2 x_1^2 + \left(\frac{1}{a}\right)^2 x_2^2 = 1 \tag{1.7.10}$$

where a and b represent the lengths of the semimajor and semiminor axes, respectively.

When comparing expressions (1.7.9) and (1.7.10), note that $a^2 = 1/\lambda_2 = 2$ and $b^2 = 1/\lambda_1 = 2/3$; thus the diagonal elements of Λ, λ_1 and λ_2, are proportional to the inverses of the squared lengths of the axes of the ellipse. As Q varies, concentric ellipsoids are generated so that $a^2 = \lambda_1^* Q$ and $b^2 = \lambda_2^* Q$. From (1.7.19), λ_1^* and λ_2^* are the roots of $\mathbf{A}^{-1}$.

Characteristic Equations. From the above example, note that the column vectors $\mathbf{p}_i$ of $\mathbf{P}$ are orthogonal, $\mathbf{PP}' = \mathbf{P}'\mathbf{P} = \mathbf{I}$, and that $\mathbf{P}'$ is employed to reduce the matrix to a diagonal matrix $\mathbf{\Lambda}$. In general, the following theorem, the *spectral decomposition theorem* (see Rao, 1965a, p. 36, or Nobel, 1969, p. 321), may be used to reduce a symmetric matrix $\mathbf{A}$ of order n to diagonal form.

(1.7.11) If $\mathbf{A}$ is a (real) symmetric matrix of order n, there exists an orthogonal matrix $\mathbf{P}$ of order n with column vectors $\mathbf{p}_i$ such that

$$\mathbf{P}'\mathbf{AP} = \mathbf{\Lambda}, \qquad \mathbf{AP} = \mathbf{P\Lambda},$$

$$\mathbf{PP}' = \mathbf{I} = \sum_i \mathbf{p}_i\mathbf{p}_i', \quad \text{and} \quad \mathbf{A} = \mathbf{P\Lambda P}' = \sum_i \lambda_i \mathbf{p}_i\mathbf{p}_i'$$

where $\mathbf{\Lambda}$ is a diagonal matrix with diagonal elements

$$\lambda_1 \geq \lambda_2 \geq \cdots \geq \lambda_n$$

If the $R(\mathbf{A}) = r \leq n$, then there are r nonzero elements on the diagonal of $\mathbf{\Lambda}$. This is concluded from the fact that $\mathbf{P}$ is nonsingular and $R(\mathbf{A}) = R(\mathbf{\Lambda}) = r$. Further, if $\mathbf{A}$ is Gramian, r of the $\lambda_i > 0$, and $n - r$ of the $\lambda_i = 0$.

Also, from the example, the column vectors $\mathbf{p}_i$ of $\mathbf{P}$ satisfy the homogeneous system of equations

$$(\mathbf{A} - \lambda_i \mathbf{I})\mathbf{p}_i = \mathbf{0} \qquad \text{for } i = 1, 2, \ldots, n \tag{1.7.12}$$

where $\mathbf{A}$ is a square symmetric matrix of order n associated with the quadratic form. As seen in (1.6.7), (1.7.12) has a nontrivial solution if and only if the $R(\mathbf{A} - \lambda\mathbf{I}) = r < n$ or the

$$|\mathbf{A} - \lambda\mathbf{I}| = 0 \tag{1.7.13}$$

Equation (1.7.13) is called the *characteristic* or *eigenequation* of the square matrix $\mathbf{A}$ and is an n-degree polynomial in λ with *characteristic roots* $\lambda_1, \lambda_2, \ldots, \lambda_n$. The vectors $\mathbf{p}_i$ satisfying (1.7.12) are called *characteristic vectors* of $\mathbf{A}$. The λ_i's and $\mathbf{p}_i$'s are also known in the literature as *eigenvalues* and *eigenvectors*, or *latent roots* and *latent vectors*. In this text, discussion of (1.7.12) and (1.7.13) will be restricted to (real) symmetric matrices since this is the usual form of **A** in multivariate analysis. For a general elementary treatment of these equations, see Searle (1966, Chap. 7).

EXAMPLE 1.7.4. As an example of (1.7.13), let

$$\mathbf{A} = \begin{bmatrix} 1 & 1/2 \\ 1/2 & 1 \end{bmatrix}$$

then

$$|\mathbf{A} - \lambda\mathbf{I}| = \left|\begin{bmatrix} 1-\lambda & 1/2 \\ 1/2 & 1-\lambda \end{bmatrix}\right| = 0$$

$$= (1-\lambda)^2 - \frac{1}{4} = 0$$

$$= \lambda^2 - 2\lambda + \frac{3}{4} = 0$$

$$= \left(\lambda - \frac{3}{2}\right)\left(\lambda - \frac{1}{2}\right) = 0$$

Or, $\lambda_1 = 3/2$ and $\lambda_2 = 1/2$.

Some useful properties of the eigenvalues of (1.7.13) are used in statistical applications.

(1.7.14) $$\operatorname{Tr}(\mathbf{A}) = \sum_i \lambda_i$$

(1.7.15) $$|\mathbf{A}| = \prod_i \lambda_i$$

(1.7.16) The eigenvalues of **A** remain unchanged under orthogonal transformations.

(1.7.17) $R(\mathbf{A})$ equals the number of nonzero eigenvalues of **A**.

(1.7.18) The eigenvalues of a Gramian matrix are positive or zero.

(1.7.19) If λ is an eigenvalue of the nonsingular matrix $\mathbf{\Lambda}$, then $1/\lambda$ is an eigenvalue of $\mathbf{A}^{-1}$.

To prove these properties, first show that the eigenvalues of **A** in (1.7.14) are the elements of the matrix $\mathbf{\Lambda}$ given in (1.7.11). By (1.7.11), $\mathbf{P}'\mathbf{A}\mathbf{P} = \mathbf{\Lambda}$ or $\mathbf{P}'(\mathbf{A} - \lambda\mathbf{I})\mathbf{P} = \mathbf{\Lambda} - \lambda\mathbf{I}$, so that

$$|\mathbf{P}'(\mathbf{A} - \lambda\mathbf{I})\mathbf{P}| = |\mathbf{A} - \lambda\mathbf{I}| = |\mathbf{\Lambda} - \lambda\mathbf{I}| = (-1)^n \prod_{i=1}^{n} (\lambda - \lambda_i)$$

since $|\mathbf{P}| = |\mathbf{P}'| = \pm 1$. Thus the eigenvalues of $|\mathbf{A} - \lambda\mathbf{I}| = 0$ are the same as the elements of $\mathbf{\Lambda}$ in (1.7.11) denoted by $\lambda_i, i = 1, \ldots, n$. Using this relationship, properties (1.7.14) through (1.7.18) follow immediately. To prove (1.7.19), observe from (1.7.12) that $\mathbf{A}^k\mathbf{p}_i = \lambda^k\mathbf{p}_i$, for any positive integer k, if **A** is singular, and equals any nonzero integer otherwise. Hence, by setting $k = -1$, (1.7.19) follows.

The solution to (1.7.12), for an eigenvector corresponding to the eigenvalue λ_i, is found by utilizing property (1.6.26).

EXAMPLE 1.7.5. Let

$$\mathbf{A} = \begin{bmatrix} 1 & 1/2 \\ 1/2 & 1 \end{bmatrix}$$

with $\lambda_1 = 3/2$. Thus

$$\begin{aligned} \hat{\mathbf{x}} &= (\mathbf{I} - \mathbf{H})\mathbf{z} \\ &= [\mathbf{I} - (\mathbf{A} - \lambda_1\mathbf{I})^-(\mathbf{A} - \lambda_1\mathbf{I})]\mathbf{z} \\ &= \left\{\begin{bmatrix} 1 & 0 \\ 0 & 1 \end{bmatrix} - \begin{bmatrix} 0 & 0 \\ 0 & -2 \end{bmatrix}\begin{bmatrix} -1/2 & 1/2 \\ 1/2 & -1/2 \end{bmatrix}\right\}\mathbf{z} \\ &= \begin{bmatrix} 1 & 0 \\ 1 & 0 \end{bmatrix}\mathbf{z} \\ &= \begin{bmatrix} z_1 \\ z_1 \end{bmatrix} \end{aligned}$$

so by letting $z_1 = 1$, $\hat{\mathbf{x}}'_1 = [1, 1]$. In a similar manner, with $\lambda_2 = 1/2$, $\hat{\mathbf{x}}'_2 = [z_1, -z_1]$. By setting $z_1 = 1$, $\hat{\mathbf{x}}'_2 = [1, -1]$. Now form the matrix

$$\mathbf{P}_0 = [\hat{\mathbf{x}}_1 \quad \hat{\mathbf{x}}_2] = \begin{bmatrix} 1 & 1 \\ 1 & -1 \end{bmatrix}$$

The matrix $\mathbf{P}_0$ is not orthogonal. By normalizing each column vector in $\mathbf{P}_0$, the orthogonal matrix becomes

$$\mathbf{P}_1 = \begin{bmatrix} 1/\sqrt{2} & 1/\sqrt{2} \\ 1/\sqrt{2} & -1/\sqrt{2} \end{bmatrix}$$

However, the determinant of this matrix is not 1 but -1. For a "pure" rotation, the $|\mathbf{P}_1| = 1$. By changing the signs of the second column vector in $\mathbf{P}_1$, the desired transformation matrix is

$$\mathbf{P} = \begin{bmatrix} 1/\sqrt{2} & -1/\sqrt{2} \\ 1/\sqrt{2} & 1/\sqrt{2} \end{bmatrix} \quad \text{with} \quad \mathbf{P}' = \begin{bmatrix} 1/\sqrt{2} & 1/\sqrt{2} \\ -1/\sqrt{2} & 1/\sqrt{2} \end{bmatrix}$$

and the $|\mathbf{P}| = 1$. Alternatively, we could change the signs of $\mathbf{P}_1$ as follows:

$$\mathbf{P}_1 = \begin{bmatrix} -1/\sqrt{2} & 1/\sqrt{2} \\ -1/\sqrt{2} & -1/\sqrt{2} \end{bmatrix}$$

The $|\mathbf{P}_1| = 1$ and the transformation matrix is then

$$\mathbf{P}'_1 = \begin{bmatrix} -1/\sqrt{2} & -1/\sqrt{2} \\ 1/\sqrt{2} & -1/\sqrt{2} \end{bmatrix}$$

This transformation matrix changes the direction of the axes in Figure 1.7.1. The angle of rotation is $\theta' = \theta + 180° = 225°$. This illustrates that the vectors of $\mathbf{P}$ are uniquely defined only up to the multiplication of a scalar factor.

Transformation matrices constructed from equation (1.7.13) are especially important in giving a geometric interpretation to principal component analysis, where axes are rotated by employing a maximum variance criterion. This method will be discussed in Chapter 6.

Because the determination of roots and vectors of the characteristic equation given by (1.7.13) for real symmetric matrices is instrumental to multivariate statistical analysis, it is necessary that an efficient method be employed. For n as large as 4, the procedure used in the solution of roots and vectors of (1.7.13), as performed in our simple example, becomes unrealistic and laborious. Iterated methods suited to computers have been proposed by Householder (1964) and Wilkinson (1965). Extremely relevant references are Givins (1957), Goldstine et al. (1959), Ortega (1960), and Wilkinson (1962). In addition, the work of Barth et al. (1967), Rutishauser (1966), and Welch (1967) should be reviewed. Among the methods proposed, Jacobi's procedure is probably the most widely used since it is the easiest to program. However, tridiagonalization methods such as the QR procedure appear to be superior, in general, and faster for finding all roots and vectors.

When interest in statistical problems is for a few distinct roots and vectors, a common procedure, first used in statistics by Hotelling (1936b), known as the *power method* may be employed. By iteration, roots and vectors are obtained one at a time beginning with the largest.

To initiate the process, selection is made of a vector $\mathbf{x}_0$ consisting of ± 1's corresponding to the signs of the elements in the first column of the matrix $\mathbf{A}$. The vector $\mathbf{x}_0$ is then powered for k iterations by forming the Krylov sequence $\mathbf{Ax}_i = \mathbf{x}_{i+1}$, $i = 0, 1, 2, \ldots, k-1$:

$$\begin{aligned} \mathbf{Ax}_0 &= \mathbf{x}_1 \\ \mathbf{Ax}_1 &= \mathbf{x}_2 \\ &\vdots \\ \mathbf{Ax}_{k-1} &= \mathbf{x}_k \end{aligned} \qquad \text{or} \qquad \begin{aligned} \mathbf{Ax}_0 &= \mathbf{x}_1 \\ \mathbf{A}^2\mathbf{x}_0 &= \mathbf{x}_2 \\ &\vdots \\ \mathbf{A}^k\mathbf{x}_0 &= \mathbf{x}_k \end{aligned}$$

At each step in the sequence, a constant of proportionality is removed so that the modified iteration scheme becomes $\mathbf{Ax}_i = s_{i+1}\mathbf{x}_{i+1}$, for $i = 0, 1, \ldots, k-1$. As k increases, the sequence converges to the first characteristic vector $\mathbf{p}_1$ upon normalization.

To prove convergence, let $\mathbf{x}_0 = s_1\mathbf{p}_1 + s_2\mathbf{p}_2 + \cdots + s_n\mathbf{p}_n$, where $\mathbf{p}_i$ represents the ith characteristic vector of $\mathbf{A}$. Then

$$\begin{aligned} \mathbf{Ax}_0 &= s_1\mathbf{Ap}_1 + s_2\mathbf{Ap}_2 + \cdots + s_n\mathbf{Ap}_n \\ \mathbf{x}_1 &= s_1\lambda_1\mathbf{p}_1 + s_2\lambda_2\mathbf{p}_2 + \cdots + s_n\lambda_n\mathbf{p}_n \end{aligned}$$

After the kth iteration,

$$\mathbf{x}_k = s_1\lambda_1^k\mathbf{p}_1 + s_2\lambda_2^k\mathbf{p}_2 + \cdots + s_n\lambda_n^k\mathbf{p}_n$$

or

$$\frac{\mathbf{x}_k}{\lambda_1} = s_1\mathbf{p}_1 + s_2\left(\frac{\lambda_2}{\lambda_1}\right)^k\mathbf{p}_2 + \cdots + s_n\left(\frac{\lambda_n}{\lambda_1}\right)^k\mathbf{p}_n$$

As $k \to \infty$, provided $\lambda_1 > \lambda_2 > \cdots > \lambda_n$, the terms involving $(\lambda_i/\lambda_1)^k \to 0$, and $\mathbf{x}_k \to s^*\mathbf{p}_1$. By normalizing the limiting vector, $\mathbf{x}_k$ is seen to converge to $\mathbf{p}_1$.

Having obtained $\mathbf{p}_1$, the corresponding root is determined from the fact that $\mathbf{p}'_1\mathbf{A}\mathbf{p}_1 \to \lambda_1$. This follows from the fact that $\mathbf{p}_1$ satisfies the equation $(\mathbf{A} - \lambda_1\mathbf{I})\mathbf{p}_1 = \mathbf{0}$. To find the next root and vector, the residual matrix $\mathbf{A}_1 = \mathbf{A} - \lambda_1\mathbf{p}_1\mathbf{p}'_1$ is utilized and the process is repeated. This method should be used only when the roots are very distinct; otherwise, convergence is slow.

EXAMPLE 1.7.6. As an example of the process, let

$$\mathbf{A} = \begin{bmatrix} 2 & -1 & 0 \\ -1 & 2 & -1 \\ 0 & -1 & 2 \end{bmatrix}$$

By observation of the first column of **A**, set

$$\mathbf{x}_0 = \begin{bmatrix} 1 \\ -1 \\ 1 \end{bmatrix}, \quad \mathbf{A}\mathbf{x}_0 = \begin{bmatrix} 3 \\ -4 \\ 3 \end{bmatrix} = \mathbf{x}_1, \quad \mathbf{A}\mathbf{x}_1 = \begin{bmatrix} 10 \\ -14 \\ 10 \end{bmatrix} = \mathbf{x}_2$$

$$\mathbf{x}_3 = \begin{bmatrix} 34 \\ -48 \\ 34 \end{bmatrix}, \quad \mathbf{x}_4 = \begin{bmatrix} 116 \\ -164 \\ 116 \end{bmatrix}, \quad \mathbf{x}_5 = \begin{bmatrix} 396 \\ -560 \\ 396 \end{bmatrix}$$

$$\mathbf{x}_6 = \begin{bmatrix} 1352 \\ -1912 \\ 1352 \end{bmatrix}, \quad \mathbf{x}_7 = \begin{bmatrix} 4616 \\ -6528 \\ 4616 \end{bmatrix}, \quad \mathbf{x}_8 = \begin{bmatrix} 15{,}760 \\ -22{,}288 \\ 15{,}760 \end{bmatrix}$$

Using as the proportionality constant the largest element in each vector,

$$\begin{aligned}
\mathbf{x}'_0 &= [1.0, -1.0, 1.0] \\
\mathbf{x}'_1 &= [.75, -1.0, .75] \\
\mathbf{x}'_2 &= [.7142857, -1.0, .7142857] \\
\mathbf{x}'_3 &= [.7083333, -1.0, .7083333] \\
\mathbf{x}'_4 &= [.7073171, -1.0, .7073171] \\
\mathbf{x}'_5 &= [.7071429, -1.0, .7071429] \\
\mathbf{x}'_6 &= [.7071129, -1.0, .7071129] \\
\mathbf{x}'_7 &= [.7071078, -1.0, .7071078] \\
\mathbf{x}'_8 &= [.7071069, -1.0, .7071069]
\end{aligned}$$

and

$$\mathbf{x}_k \to \begin{bmatrix} \sqrt{2}/2 \\ -1.0 \\ \sqrt{2}/2 \end{bmatrix}$$

Normalizing $\mathbf{x}_8$,

$$\mathbf{p}_1 = \begin{bmatrix} .5000000 \\ -.7071069 \\ .5000000 \end{bmatrix} \to \begin{bmatrix} 1/2 \\ -\sqrt{2}/2 \\ 1/2 \end{bmatrix}$$

so that $\mathbf{p}_1'\mathbf{A}\mathbf{p}_1 = \lambda_1 = 3.4142132 \to (2 + \sqrt{2})$. By locating the second root and vector, the matrix $\mathbf{A}_1 = \mathbf{A} - \lambda_1\mathbf{p}_1\mathbf{p}_1'$ is formed and the process is repeated using $\mathbf{A}_1$. Then verify that

$$\mathbf{p}_2 \to \begin{bmatrix} 1/\sqrt{2} \\ 0 \\ -1/\sqrt{2} \end{bmatrix} \quad \text{and} \quad \lambda_2 \to 2$$

To expedite convergence of the power method, Aitken's (1937) δ^2 procedure is often employed. For this technique, three consecutive scalar vectors $\mathbf{x}_{i-1}$, $\mathbf{x}_i$, and $\mathbf{x}_{i+1}$ are used and new modified vectors are formed. If $\mathbf{y}^{(1)}, \mathbf{y}^{(2)}, \ldots, \mathbf{y}^{(k)}$ denotes Aitken's modified vectors, then the jth element of the vector is obtained by employing the formula

$$y_j^{(i-1)} = \frac{x_{i-1}^{(j)}x_{i+1}^{(j)} - (x_i^{(j)})^2}{x_{i+1}^{(j)} - 2x_i^{(j)} + x_{i-1}^{(j)}} \qquad \text{for } j = 1, \ldots, n \text{ and } i \geq 2$$

$$y_j^{(i-1)} = \pm 1 \qquad \text{if } x_{i-1} = x_{i+1} = x_i = \pm 1$$

where $x_i^{(j)}$ denotes the jth element in the vector $\mathbf{x}_i$ and $y_j^{(i-1)}$ denotes the jth element in the vector $\mathbf{y}^{(i-1)}$ in Aitken's sequence. As $k \to \infty$, $\mathbf{y}^{(k)} \to s_1^*\mathbf{p}_1$, and normalizing $\mathbf{y}_k \to \mathbf{p}_1$. For a further discussion of this method, read Faddeeva (1959).

In theory,

$$\mathbf{A} = \sum_i \lambda_i \mathbf{p}_i \mathbf{p}_i'$$

However, since decimals are being used in the calculations of λ_i and $\mathbf{p}_i$, there may be rounding error. To investigate the "closeness" of the decomposition of $\mathbf{A}$ by iteration procedures, the *matrix norm* may be employed. If $\mathbf{A}$ is an $n \times m$ matrix, the norm squared of the matrix $\mathbf{A}$ is defined by

$$\|\mathbf{A}\|^2 = \mathrm{Tr}(\mathbf{A}\mathbf{A}') = \mathrm{Tr}(\mathbf{A}'\mathbf{A}) = \sum_{i,j} a_{ij}^2 \tag{1.7.20}$$

For two matrices, the norm squared is

$$\|\mathbf{A} - \mathbf{B}\|^2 = \mathrm{Tr}[(\mathbf{A} - \mathbf{B})(\mathbf{A} - \mathbf{B})'] = \sum_{i,j} (a_{ij} - b_{ij})^2 \tag{1.7.21}$$

This norm is an immediate generalization of the norm squared of a vector.

In a similar manner, the inner product of a vector $\mathbf{y}$ may be generalized to a space where the axes of the space are nonorthogonal. Such a space is termed an *affine space* and usually arises in statistics as a result of oblique transformations. Recall that the inner product of an $n \times 1$ vector $\mathbf{y}$ is $\mathbf{y}'\mathbf{y} = \mathbf{y}'\mathbf{I}\mathbf{y}$ for an n-dimensional Euclidean space. The matrix $\mathbf{I}$ inserted into the inner product is there to remind us that the coordinate axes for a Euclidean space are orthogonal and form an orthonormal basis for the space. Instead of using an orthonormal basis for the space, suppose an arbitrary basis is employed. That is, suppose we operate on $\mathbf{y}$ with the $n \times n$ nonsingular matrix $\mathbf{T}$ such that $\mathbf{Ty} = \mathbf{x}$, so that $\mathbf{y} = \mathbf{T}^{-1}\mathbf{x}$. Then $\mathbf{y}'\mathbf{y} = \mathbf{x}'(\mathbf{T}')^{-1}\mathbf{T}^{-1}\mathbf{x} = \mathbf{x}'(\mathbf{TT}')^{-1}\mathbf{x} = \mathbf{x}'\mathbf{Ax}$. The quantity $\mathbf{x}'\mathbf{Ax}$ is a generalization of the definition of an inner product for an affine space and is called the *inner product in the metric* of $\mathbf{A}$. The length of $\mathbf{x}'\mathbf{Ax}$ is $\|\mathbf{x}'\mathbf{Ax}\| = \sqrt{\mathbf{x}'\mathbf{Ax}}$. The symmetric matrix $\mathbf{A}$ in the quadratic form $\mathbf{x}'\mathbf{Ax}$ is only a metric matrix if it preserves positiveness of length, $\mathbf{x}'\mathbf{Ax} > 0$ for all nonzero vectors $\mathbf{x}$, and the triangular inequality of distance, $\|\mathbf{x} - \mathbf{z}\| \leq \|\mathbf{x} - \mathbf{y}\| + \|\mathbf{y} - \mathbf{z}\|$. Thus a matrix $\mathbf{A}$ forms a metric if and only if $\mathbf{A}$ is p.d.

It is clear from the definitions of inner product and length that the cosine between two vectors in a general metric is defined by $\cos\theta = \mathbf{x}'\mathbf{Ay}/\|\mathbf{x}'\mathbf{Ax}\|\,\|\mathbf{y}'\mathbf{Ay}\|$. To obtain generalizations in an affine space, the metric $\mathbf{I}$ of the Euclidean space need only be replaced with a metric matrix. For a more complete discussion of affine spaces and affine geometry, Dempster (1969) is suggested reading. The concepts summarized above are sufficient for this text.

It is also possible to generalize the characteristic equation (1.7.13) in an arbitrary metric $\mathbf{B}$ where $\mathbf{B}$ is a (real) $n \times n$ symmetric p.d. matrix and $\mathbf{A}$ is a symmetric matrix of order n:

(1.7.22) $$|\mathbf{A} - \lambda\mathbf{B}| = 0$$

The homogeneous system of equations

(1.7.23) $$(\mathbf{A} - \lambda_i\mathbf{B})\mathbf{q}_i = 0 \qquad (i = 1, \ldots, n)$$

has a nontrivial solution if and only if (1.7.22) is satisfied. The quantities λ_i and $\mathbf{q}_i$ are called the eigenvalues and eigenvectors of $\mathbf{A}$ in the metric of $\mathbf{B}$. As in an orthogonal space, a decomposition theorem corresponding to (1.7.11) exists for affine spaces (see Rao, 1965a, p. 37).

(1.7.24) If $\mathbf{A}$ and $\mathbf{B}$ are (real) symmetric matrices of order n and $\mathbf{B}$ is p.d., then there exists a nonsingular matrix $\mathbf{Q}$ of order n with column vectors $\mathbf{q}_i$ such that

$$\mathbf{Q}'\mathbf{AQ} = \mathbf{\Lambda} \quad \text{and} \quad \mathbf{Q}'\mathbf{BQ} = \mathbf{I}$$

$$\mathbf{A} = (\mathbf{Q}')^{-1}\mathbf{\Lambda}\mathbf{Q}^{-1} \quad \text{and} \quad (\mathbf{Q}')^{-1}\mathbf{Q}^{-1} = \mathbf{B}$$

$$\mathbf{A} = \sum_{i=1}^{n} \lambda_i\mathbf{x}_i\mathbf{x}_i' \quad \text{and} \quad \mathbf{B} = \sum_{i=1}^{n} \mathbf{x}_i\mathbf{x}_i'$$

where $\mathbf{x}_i$ is the ith column vector of $(\mathbf{Q}')^{-1}$ and $\mathbf{\Lambda}$ is a diagonal matrix with eigenvalues λ_i from the equation $|\mathbf{A} - \lambda\mathbf{B}| = 0$, for $i = 1, 2, \ldots, n$.

Geometrically, (1.7.24) is employed to reduce two ellipsoids in an affine space to a sphere and an ellipsoid in a Euclidean space. Thus, through the transformation $\mathbf{x} = \mathbf{Q}^{-1}\mathbf{y}$,

$$\mathbf{y}'\mathbf{Ay} \rightarrow \sum_{i=1}^{n} \lambda_i x_i^2 \quad \text{and} \quad \mathbf{y}'\mathbf{By} \rightarrow \sum_{i=1}^{n} x_i^2$$

Solution to the two matrix characteristic equations is achieved by reducing them to a single characteristic equation. Given the system

(1.7.25) $$(\mathbf{A} - \lambda_i\mathbf{B})\mathbf{q}_i = 0 \qquad (i = 1, \ldots, n)$$

with

$$|\mathbf{A} - \lambda\mathbf{B}| = 0$$

recall that the matrix **B** may be factored by the square root method into $\mathbf{B} = \mathbf{TT}'$, where $\mathbf{T}^{-1}\mathbf{B}(\mathbf{T}')^{-1} = \mathbf{I}$. Using this and the transformation $\mathbf{q}_i = (\mathbf{T}')^{-1}\mathbf{x}_i$, (1.7.25) reduces to

(1.7.26) $$[\mathbf{T}^{-1}\mathbf{A}(\mathbf{T}')^{-1} - \lambda_i\mathbf{I}]\mathbf{x}_i = \mathbf{0} \qquad (i = 1, \ldots, n)$$

so that the $|\mathbf{T}^{-1}\mathbf{A}(\mathbf{T}')^{-1} - \lambda\mathbf{I}| = 0$, where $\mathbf{T}^{-1}\mathbf{A}(\mathbf{T}')^{-1}$ is symmetrical. Thus the roots of (1.7.26) are equal to the roots of (1.7.25) since nonsingular transformations preserve the values of the roots. The vectors of (1.7.25) are obtainable through the transformation $\mathbf{q}_i = (\mathbf{T}')^{-1}\mathbf{x}_i$.

Alternatively, the transformation $\mathbf{q}_i = \mathbf{B}^{-1}\mathbf{x}_i$ could have been employed. In this case (1.7.25) reduces to

(1.7.27) $$(\mathbf{AB}^{-1} - \lambda_i\mathbf{I})\mathbf{x}_i = 0$$

and $|\mathbf{AB}^{-1} - \lambda\mathbf{I}| = 0$. However, the matrix $\mathbf{AB}^{-1}$ is not necessarily symmetrical, and special iterative procedures must be used to find its roots and vectors (see Householder, 1964).

The eigenvalues $\lambda_1, \lambda_2, \ldots, \lambda_n$ of the characteristic equation $|\mathbf{A} - \lambda\mathbf{B}| = 0$ are essential to the study of multivariate analysis. Therefore it is important to relate the roots of various characteristic equations:

(1.7.28) The roots of v_i of the characteristic equation $|\mathbf{B} - v(\mathbf{B} + \mathbf{A})| = 0$ are related to the roots of $|\mathbf{A} - \lambda\mathbf{B}| = 0$ by the formula

$$\lambda_i = \frac{1 - v_i}{v_i} \quad \text{or} \quad v_i = \frac{1}{1 + \lambda_i}$$

(1.7.29) The roots θ_i of the characteristic equation $|\mathbf{A} - \theta(\mathbf{A} + \mathbf{B})| = 0$ are related to the roots of $|\mathbf{A} - \lambda\mathbf{B}| = 0$ by the formula

$$\lambda_i = \frac{\theta_i}{1 - \theta_i} \quad \text{or} \quad \theta_i = \frac{\lambda_i}{1 + \lambda_i}$$

and to the roots of $|\mathbf{B} - v(\mathbf{B} + \mathbf{A})| = 0$ by the relationship $v_i = (1 - \theta_i)$.

To prove (1.7.28), let $[\mathbf{B} - v(\mathbf{B} + \mathbf{A})]\mathbf{x} = \mathbf{0}$. Then $\mathbf{Bx} - v\mathbf{Bx} - v\mathbf{Ax} = \mathbf{0}$, or $(1 - v)\mathbf{Bx} - v\mathbf{Ax} = \mathbf{0}$, so that $v\mathbf{Ax} - (1 - v)\mathbf{Bx} = \mathbf{0}$. Hence

$$\left[\mathbf{A} - \left(\frac{1 - v}{v}\right)\mathbf{B}\right]\mathbf{x} = \mathbf{0} \quad \text{or} \quad \left|\mathbf{A} - \left(\frac{1 - v}{v}\right)\mathbf{B}\right| = 0$$

and (1.7.28) is proven. The relationship given in (1.7.29) follows similarly.

EXERCISES 1.7

1. Given the quadratic forms

$$3y_1^2 + y_2^2 + 2y_3^2 + 2y_1y_3 \quad \text{and} \quad y_1^2 + 5y_2^2 + y_3^2 + 2y_1y_2 - 4y_2y_3$$

a. Find the matrices associated with each form.

b. Transform both to the form $\Sigma\lambda_i x_i^2$.

c. Determine whether the forms are positive definite, positive semidefinite, or neither, and find their ranks.

d. Factor both matrices in part a as $\mathbf{FF}'$.

e. If the form is positive definite, write it as Σz_i^2; find its Cholesky factorization and hence its inverse.

2. Use the procedure of Example 1.7.3 to find a g-inverse of

$$\mathbf{A} = \begin{bmatrix} 1 & 1 & 0 \\ 1 & 5 & -2 \\ 0 & -2 & 1 \end{bmatrix}$$

3. Given the ellipse defined by

$$2y_1^2 + y_1 y_2 + 2y_2^2 = 3$$

a. Write the equation in terms of new variables x_1 and x_2, so that no cross-product term appears.

b. Find the lengths of the major and minor axes.

c. What are the eigenvalues of the associated matrices $\mathbf{A}$ and $\mathbf{A}^{-1}$?

4. For each of the $n \times n$ matrices

$$\mathbf{A} = \begin{bmatrix} 2 & 1 \\ 1 & 2 \end{bmatrix} \quad \text{and} \quad \mathbf{B} = \begin{bmatrix} 1 & 1 & 0 \\ 1 & 5 & -2 \\ 0 & -2 & 1 \end{bmatrix}$$

a. Find eigenvalues.

b. Find n mutually orthogonal eigenvectors and write the matrix as $\mathbf{P\Lambda P}'$ using (1.7.11).

5. Prove (1.7.14) through (1.7.18) and (1.7.30).

6. Employ the "power method" of Example 1.7.6 to find the largest eigenvalue of the matrix $\mathbf{B}$ of Exercise 4.

7. Determine the eigenvalues and eigenvectors for the $n \times n$ matrix $\mathbf{R} = [r_{ij}]$, where $r_{ij} = 1$, for $i \neq j$, and $r_{ij} = r \neq 0$, for $i \neq j$.

8. For the symmetric matrices $\mathbf{A}$ and $\mathbf{B}$ defined by

$$\mathbf{A} = \begin{bmatrix} 498.807 & \\ 426.757 & 374.657 \end{bmatrix} \quad \text{and} \quad \mathbf{B} = \begin{bmatrix} 1830.25 & \\ -334.750 & 12{,}555.25 \end{bmatrix}$$

solve (1.7.25) for λ_i and $\mathbf{q}_i$ by using $|\mathbf{A} - \lambda\mathbf{B}| = 0$.

1.8 DERIVATIVES, MAXIMA, AND MINIMA

Some fundamentals of differential calculus are helpful in understanding statistical least-squares estimation theory. To make this text complete at an elementary level, a brief review of derivatives of algebraic functions, vectors, and matrices is included.

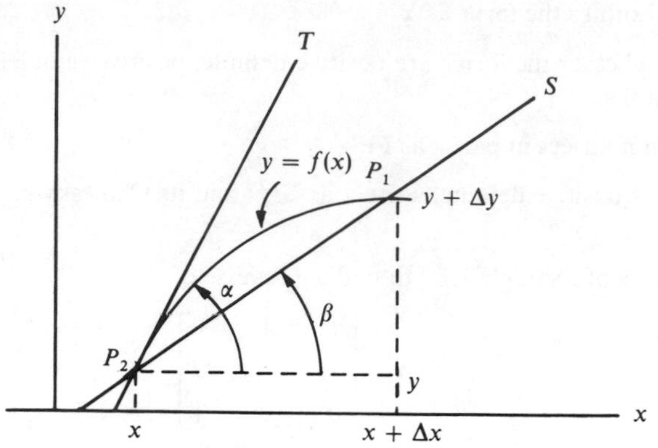

FIGURE 1.8.1. Geometric representation of the derivative.

Consider a function of x represented by the curve $y = f(x)$ in Figure 1.8.1. The derivation of y with respect to x is defined by the relationship

$$f'(x) = \lim_{\Delta x \to 0} \frac{f(x + \Delta x) - f(x)}{\Delta x}$$

(1.8.1)

$$\frac{dy}{dx} = \lim_{\Delta x \to 0} \frac{\Delta y}{\Delta x}$$

The function $f(x)$ (more briefly, f) is said to be differentiable at x if (1.8.1) exists. Geometrically, the point $P_1 \to P_2$ as $\Delta x \to 0$ and the secant line S approaches the tangent line T. Or the $\tan \beta = \Delta y/\Delta x \to \tan \alpha = dy/dx$ as $\Delta x \to 0$, so that the derivative represents the slope of the tangent line P_2T.

Employing definition (1.8.1), let $y = f(x) = cx^n$, where $n > 0$ and c is a constant; then

$$\begin{aligned} f(x + \Delta x) &= c(x + \Delta x)^n \\ &= c\left[x^n + nx^{n-1}(\Delta x) + \frac{n(n-1)}{2!}x^{n-2}(\Delta x)^2 + \cdots + (\Delta x)^n\right] \end{aligned}$$

and

$$\Delta y = f(x + \Delta x) - f(x) = c[nx^{n-1}(\Delta x) + \cdots + (\Delta x)^n]$$

so that

$$\frac{\Delta y}{\Delta x} = c\left[nx^{n-1} + \frac{n(n-1)}{2!}x^{n-2}(\Delta x) + \cdots + (\Delta x)^{n-1}\right]$$

or the derivative of y with respect to x is

$$\frac{dy}{dx} = \lim_{\Delta x \to 0} \frac{\Delta y}{\Delta x} = cnx^{n-1}$$

Hence, if $y = cx^n$, the

(1.8.2)

$$\frac{dy}{dx} = cnx^{n-1}$$

if $y = c$, where c is a constant, $dy/dx = 0$. Some other rules and examples are summarized below. For proofs, see Thomas (1960, Chap. 2).

If $u = f(x)$, $y = g(x)$, and $y = u + v$, the

(1.8.3) $$\frac{dy}{dx} = \frac{du}{dx} + \frac{dv}{dx}$$

EXAMPLE 1.8.1. Let $y = x^2 + 2x + 4$. Then

$$\frac{dy}{dx} = 2x + 2$$

If $u = f(x)$, $y = g(x)$, and $y = uv$, the

(1.8.4) $$\frac{dy}{dx} = u\frac{dv}{dx} + \frac{du}{dx}v$$

If $u = f(x)$, $y = g(x)$, and $y = u/v$, the

(1.8.5) $$\frac{dy}{dx} = \frac{v(du/dx) - u(dv/dx)}{v^2}$$

If $y = f(u)$ and $u = g(x)$, the

(1.8.6) $$\frac{dy}{dx} = \frac{dy}{du}\frac{du}{dx}$$

Rule (1.8.6) is commonly referred to as the *chain rule*; letting $y = u^n$, the

$$\frac{dy}{dx} = nu^{n-1}\frac{du}{dx}$$

EXAMPLE 1.8.2. Let $y = x^{-2} = 1/x^2$ by (1.8.5); then

$$\frac{dy}{dx} = \frac{x^2(d/dx)(1) - 1(d/dx)(x^2)}{x^4}$$

$$= \frac{-2x}{x^4} = \frac{-2}{x^3} = -2x^{-3}$$

so that (1.8.2) applies for negative integers. Letting $y = (x^2 + 1)^3x^2$ by (1.8.4),

$$\frac{dy}{dx} = (x^2 + 1)^3\frac{d}{dx}(x^2) + x^2\frac{d}{dx}(x^2 + 1)^3$$

$$= (x^2 + 1)^3 2x + x^2\frac{d}{dx}(x^2 + 1)^3$$

by (1.8.6),

$$\frac{dy}{dx} = 2x(x^2 + 1)^3 + x^2 3(x^2 + 1)^2\frac{d}{dx}(x^2 + 1)$$

$$= 2x(x^2 + 1)^3 + 3x^2(x^2 + 1)^2 2x$$

$$= 2x(x^2 + 1)^3 + 6x^3(x^2 + 1)^2$$

Rules for differentiating natural logarithmic, exponential, sine, and cosine functions follow:

If $y = \log u$, $u = f(x)$,

$$\frac{dy}{dx} = \frac{1}{u}\frac{du}{dx} \tag{1.8.7}$$

If $y = e^u$, $u = f(x)$,

$$\frac{dy}{dx} = e^u\frac{du}{dx} \tag{1.8.8}$$

If $y = \sin u$, $u = f(x)$,

$$\frac{dy}{dx} = \cos u\frac{du}{dx} \tag{1.8.9}$$

If $y = \cos u$, $u = f(x)$,

$$\frac{dy}{dx} = -\sin u\frac{du}{dx} \tag{1.8.10}$$

Maxima and Minima. If the first derivative of y, with respect to x evaluated at x_0, is equal to zero, $f'(x_0) = 0$, x_0 is called a *critical point* of the function $f(x)$. The value of y at x_0, $y = f(x_0)$, is called a *critical value of the function.*

EXAMPLE 1.8.3. Let $y = f(x) = x(4 - x) = 4x - x^2$. The first derivation of y with respect to x is $dy/dx = f'(x) = 4 - 2x$. By equating this to zero, $f'(x) = 4 - 2x = 0$; by solving for x, we find that $x_0 = 2$ is critical point of $f(x)$. The critical value of the function at $x_0 = 2$ is $y = f(2) = 4$. The slope of the tangent line at $x_0 = 2$ is zero. At a point $x < x_0$, say $x = 1$, the slope of the tangent line is positive since $dy/dx = 2$ at $x = 1$. At a point $x > x_0$, say $x = 3$, the slope of the tangent line is negative, and $dy/dx = -2$ at $x = 3$. This indicates that $y = f(x_0) = 4$ is a maximum of the function $y = 4x - x^2$ since the slope changed signs as x passed through x_0. Now graph $f(x) = x(4 - x)$.

In general, the critical value of a function at the critical point x_0 is a *maximum* at x_0 if

$$f'(x) \text{ is } \begin{cases} > 0 & \text{for } x < x_0 \\ 0 & \text{for } x = x_0 \\ < 0 & \text{for } x > x_0 \end{cases} \tag{1.8.11}$$

and a *minimum* at x_0 if

$$f'(x) \text{ is } \begin{cases} < 0 & \text{for } x < x_0 \\ 0 & \text{for } x = x_0 \\ > 0 & \text{for } x > x_0 \end{cases} \tag{1.8.12}$$

If neither (1.8.11) nor (1.8.12) holds, but $f'(x_0) = 0$, then the critical point x_0 yields neither a maximum nor a minimum value for y. Such a point is called a *point of inflection.* The slope of the tangent line T at the point x_0 is zero at a maximum or minimum. The use of (1.8.11) and (1.8.12) to find maximum or minimum points of a function is termed the *first-derivative test.* Alternatively, the *second-derivative test* may be employed to find a maximum or a minimum.

Since the derivative of y with respect to x is a function of x, its derivative may also be determined. The *second derivative* of $y = f(x)$ with respect to x is defined by a number of equivalent expressions.

(1.8.13)
$$\frac{d}{dx}\left(\frac{dy}{dx}\right) = \frac{d}{dx}(f'(x)) = \frac{d^2y}{dx^2} = \frac{d^2f}{dx^2} = f''(x)$$

The second-derivative test for finding extreme values proceeds as follows:

(1.8.14) Given that $f'(x_0) = 0$, the second derivative is tested for positiveness, negativeness, and nullness. If $f''(x_0) < 0$, then $f(x_0)$ is a maximum; if $f''(x_0) > 0$, then $f(x_0)$ is a minimum. If $f''(x_0) = 0$, the test fails and $f(x_0)$ may or may not be an extreme value.

EXAMPLE 1.8.4. Using Example 1.8.3, let $y = 4x - x^2$. Then $f''(x_0) = -2$ and $f''(x_0) < 0$, so that $f(x_0) = 4$ is a maximum by the second-derivative test. Points where the first derivative changes its sign are called *points of inflection.* They may be found among the roots of $f''(x) = 0$. However, not every root will be a point of inflection. For a detailed introduction to maxima and minima problems, see Apostol (1961, Chap. 8).

The preceding discussion treated the one-variable problem. We now consider z as a function of two independent variables x and y, so that $z = f(x, y)$. Two variables x and y are said to be (mathematically) *independent* if they are not functionally dependent, so that $y \neq f(x)$. By fixing one independent variable y, the *partial derivative* of z with respect to x is defined as

(1.8.15)
$$\frac{\partial z}{\partial x} = \frac{\partial f}{\partial x} = \lim_{\Delta x \to 0} \frac{f(x + \Delta x, y) - f(x, y)}{\Delta x}$$

For fixed x, the partial derivative of z with respect to y is

(1.8.16)
$$\frac{\partial z}{\partial y} = \frac{\partial f}{\partial y} = \lim_{\Delta y \to 0} \frac{f(x, y + \Delta y) - f(x, y)}{\Delta x}$$

The *curl operator,* ∂, is used to distinguish this differentiation from ordinary single-variable differentiation; however, the rules for single-variable differentiation remain applicable since the other variable is always held constant.

Geometrically, the function $f(x, y)$ may be thought of as some surface in three dimensions. The partial derivative of $z = f(x, y)$, with respect to x, represents the slope of the tangent line T_x of the curve cut from the surface by a plane for a constant value of y. Similarly, the $\partial z/\partial y$ is the slope of the tangent line T_y of the curve cut from the surface by a plane for a constant value of x (see Figure 1.8.2).

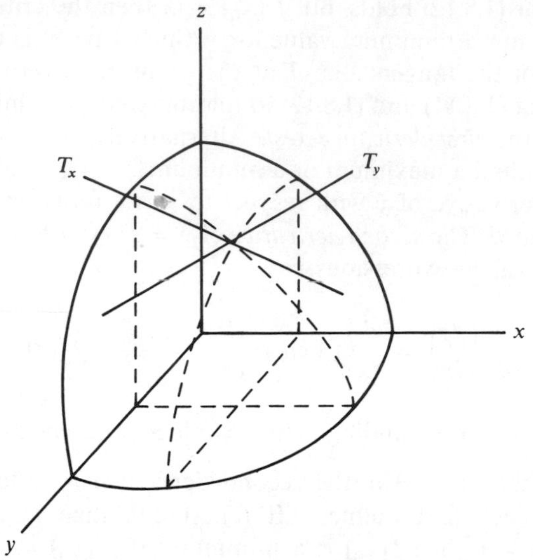

FIGURE 1.8.2. Geometric representation of the partial derivative.

The *second partial derivatives* of $z = f(x, y)$ are defined by the relations

(1.8.17)
$$\frac{\partial^2 z}{\partial x^2} = \frac{\partial^2 f}{\partial x^2} = \frac{\partial}{\partial x}\left(\frac{\partial f}{\partial x}\right)$$

(1.8.18)
$$\frac{\partial^2 z}{\partial y^2} = \frac{\partial^2 f}{\partial y^2} = \frac{\partial}{\partial y}\left(\frac{\partial f}{\partial y}\right)$$

(1.8.19)
$$\frac{\partial^2 z}{\partial x\,\partial y} = \frac{\partial^2 f}{\partial x\,\partial y} = \frac{\partial}{\partial x}\left(\frac{\partial f}{\partial y}\right) = \frac{\partial}{\partial y}\left(\frac{\partial f}{\partial x}\right)$$

To find maxima and minima for functions of two variables, critical points (x_0, y_0) are found by solving the homogeneous system of *normal equations*

$$\frac{\partial f}{\partial x} = 0 \quad \text{and} \quad \frac{\partial f}{\partial y} = 0$$

Having obtained a solution to the system (x_0, y_0), the expression

$$\delta = \left(\frac{\partial^2 f}{\partial x^2}\right)\left(\frac{\partial^2 f}{\partial y^2}\right) - \left(\frac{\partial^2 f}{\partial x\,\partial y}\right)^2$$

is evaluated at (x_0, y_0). The value of $z = f(x_0, y_0)$ is a maximum, minimum, or neither according to the following conditions (see Apostol, 1962, p. 205):

(1.8.20)
$$\text{Maximum} \quad \text{if } \delta > 0 \quad \frac{\partial^2 f}{\partial x^2} < 0$$

$$\text{Minimum} \quad \text{if } \delta > 0 \quad \frac{\partial^2 f}{\partial x^2} > 0$$

$$\text{Neither} \quad \text{if } \delta < 0 \qquad \text{Test fails} \quad \text{if } \delta = 0$$

Geometrically, location of a plane tangent to the surface at $z = f(x_0, y_0)$ is sought.

By representing δ in matrix form, and evaluating the matrix Δ at (x_0, y_0) where

(1.8.21) $$\Delta = \begin{bmatrix} \partial^2 f/\partial x^2 & \partial^2 f/\partial x\, \partial y \\ \partial^2 f/\partial y\, \partial x & \partial^2 f/\partial y^2 \end{bmatrix}_{(x_0, y_0)}$$

Bellman (1960, p. 3) shows that $z = f(x_0, y_0)$ is a minimum if Δ is positive definite or a maximum if $-\Delta$ is positive definite.

As an example of (1.8.21), use the method of least squares to fit a straight line to a set of data points. Let $[x_i, y_i]$, $i = 1, \ldots, n$, indicate pairs of observed data points plotted in the (x, y) plane (see Figure 1.8.3), and let $y = \alpha + \beta x$ represent the theoretical line that best fits the data in the sense that the sum of the squares of deviations of the observed y_i's, at each of the points x_i, from the line are minimal.

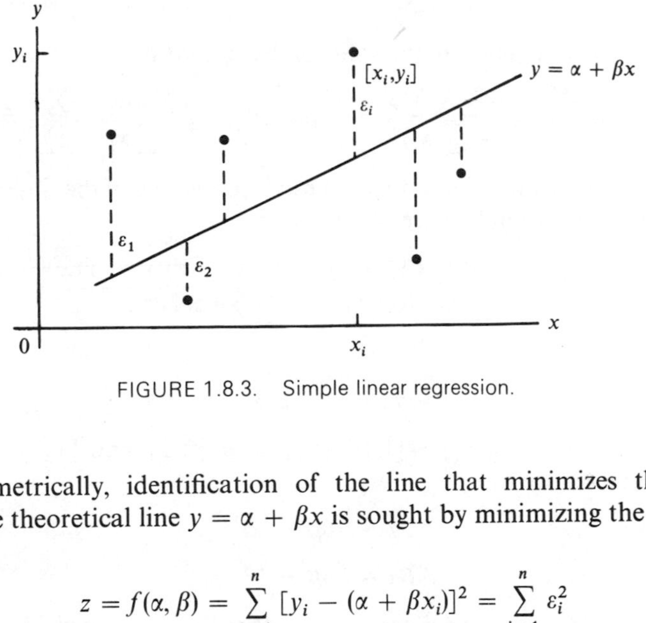

FIGURE 1.8.3. Simple linear regression.

Geometrically, identification of the line that minimizes the deviations $\Sigma_i \varepsilon_i^2$ from the theoretical line $y = \alpha + \beta x$ is sought by minimizing the function

$$z = f(\alpha, \beta) = \sum_{i=1}^{n} [y_i - (\alpha + \beta x_i)]^2 = \sum_{i=1}^{n} \varepsilon_i^2$$

In three-dimensional space, the point $f(\alpha_0, \beta_0)$ is determined such that the function $z = f(\alpha, \beta)$ is a minimum (see Figure 1.8.4). To accomplish this, the *normal equations*

$$\frac{\partial f}{\partial \alpha} = 0 \quad \text{and} \quad \frac{\partial f}{\partial \beta} = 0$$

must be solved for α_0 and β_0. These become, for this problem,

$$\frac{\partial f}{\partial \alpha} = 2 \sum_{i=1}^{n} (y_i - \alpha - \beta x_i)(-1) = 0 \quad \text{and} \quad \frac{\partial f}{\partial \beta} = 2 \sum_{i=1}^{n} (y_i - \alpha - \beta x_i)(-x_i) = 0$$

or

$$n\alpha + (\sum x_i)\beta = \sum y_i \quad \text{and} \quad \sum x_i + (\sum x_i^2)\beta = \sum x_i y_i$$

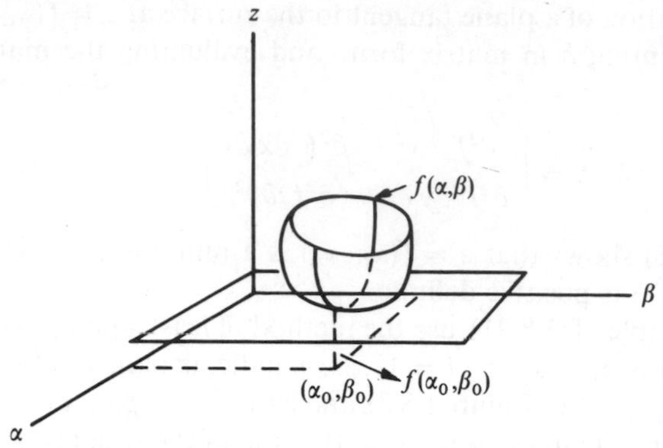

FIGURE 1.8.4. Function minimized under linear regression.

Solving for α and β and denoting the solution by α_0 and β_0

$$\alpha_0 = \frac{\sum y_i \sum x_i^2 - \sum x_i \sum x_i y_i}{n\sum x_i^2 - (\sum x_i)^2} \quad \text{and} \quad \beta_0 = \frac{n\sum x_i y_i - \sum x_i \sum y_i}{n\sum x_i^2 - (\sum x_i)^2}$$

Simplifying the expressions for α_0 and β_0, the familiar values of the slope and intercept for the estimated regression line are obtained

$$\beta_0 = \frac{(1/n)\sum x_i y_i - \overline{xy}}{\sum (x_i - \bar{x})^2/n} = \frac{\sum (x_i - \bar{x})(y_i - \bar{y})}{\sum (x_i - \bar{x})^2}$$

(1.8.22)

$$\alpha_0 = \bar{y} - \beta_0 \bar{x}$$

For the observed pairs $[x_1, y_1] = [1, 0]$, $[x_2, y_2] = [2, 2]$, and $[x_3, y_3] = [3, 3]$, the normal equations reduce to

$$3\alpha + 6\beta = 5$$

$$6\alpha + 14\beta = 13$$

so that $\alpha_0 = -4/3$, $\beta_0 = 3/2$. To determine whether these values yield a minimum for the function $f(\alpha, \beta)$, the matrix

$$\mathbf{\Delta} = \begin{bmatrix} \dfrac{\partial^2 f}{\partial \alpha^2} & \dfrac{\partial^2 f}{\partial \alpha\, \partial \beta} \\ \dfrac{\partial^2 f}{\partial \beta\, \partial \alpha} & \dfrac{\partial^2 f}{\partial \beta^2} \end{bmatrix}_{(\alpha_0, \beta_0)}$$

is investigated. The values of the first-order partials are

$$\frac{\partial f}{\partial \alpha} = -\ \ 5 + 3\alpha + \ 6\beta$$

$$\frac{\partial f}{\partial \beta} = -13 + 6\alpha + 14\beta$$

so that

$$\frac{\partial^2 f}{\partial \alpha^2} = 3, \quad \frac{\partial^2 f}{\partial \beta^2} = 14, \quad \frac{\partial^2 f}{\partial \alpha \, \partial \beta} = 6$$

and the matrix

$$\mathbf{\Delta} = \begin{bmatrix} 3 & 6 \\ 6 & 14 \end{bmatrix}$$

is p.d. since $\lambda_1 > \lambda_2 > 0$. Thus $z = f(\alpha_0, \beta_0) = 1/6$ is the minimal value of the function $f(\alpha, \beta)$. The line represented by $\hat{y} = \alpha_0 + \beta_0 x$ is the line of best fit to the data in a least-squares sense—that is, the best approximation to the theoretical line $y = \alpha + \beta x$.

When the variables x and y of $z = f(x, y)$ are not independent, (1.8.22) may no longer be employed to find extreme values. For example, we might be interested in minimizing $z = f(x, y)$ subject to the constraint that $x^2 + y^2 = 1$. To find an extreme value under a constraint, we may solve for one variable in terms of the other and substitute the result into z so that the two-variable problem is reduced to a one-variable extremal problem. However, with a complicated function, this becomes burdensome. An alternative method is to introduce a *Lagrange multiplier* (see Apostol, 1957, p. 152). This technique is discussed later in this section.

The case of n independent variables is an immediate extension of the two-variable problem. Suppose y is a function of n independent variables so that $y = f(x_1, x_2, \ldots, x_n)$. Fixing any one variable, the partial derivative of y with respect to x_i is defined as

$$\text{(1.8.23)} \quad \frac{\partial y}{\partial x_i} = \frac{\partial f}{\partial x_i} = \lim_{\Delta x \to 0} \frac{f(x_1, x_2, \ldots, x_i + \Delta x_i, \ldots, x_n) - f(x_1, \ldots, x_n)}{\Delta x_i}$$

By establishing extreme values for functions in n variables, the homogeneous system of normal equations is solved for critical points $(x_{01}, x_{02}, \ldots, x_{0n})$:

$$\frac{\partial f}{\partial x_1} = 0$$
$$\frac{\partial f}{\partial x_2} = 0$$
$$\vdots$$
$$\frac{\partial f}{\partial x_n} = 0$$

Obtaining a solution, $(x_{01}, \ldots, x_{0n})$, an extended version of the second-derivative test in the two-variable situation is applied. By letting

$$\mathbf{\Delta} = \begin{bmatrix} \partial^2 f/\partial x_1^2 & \partial^2 f/\partial x_1 \, \partial x_2 & \cdots & \partial^2 f/\partial x_1 \, \partial x_n \\ \partial^2 f/\partial x_2 \, \partial x_1 & \partial^2 f/\partial x_2^2 & \cdots & \partial^2 f/\partial x_2 \, \partial x_n \\ \vdots & \vdots & \ddots & \vdots \\ \partial^2 f/\partial x_n \, \partial x_1 & \partial^2 f/\partial x_n \, \partial x_2 & \cdots & \partial^2 f/\partial x_n^2 \end{bmatrix}$$

and evaluating it at $(x_{01}, \ldots, x_{0n})$, the value of the function $y = f(x_{01}, \ldots, x_{0n})$ is a

$$\text{(1.8.24)} \qquad \begin{array}{ll} \text{Minimum} & \text{if} \quad \boldsymbol{\Delta} \text{ is p.d.} \\ \text{Maximum} & \text{if} \; -\boldsymbol{\Delta} \text{ is p.d.} \end{array}$$

Vector and Matrix Derivatives. Throughout this text, situations will occur when it is necessary to determine the derivatives of functions involving vectors and matrices. Although it is always possible to use the familiar rules developed in the preceding sections element by element, vector and matrix differentiation rules will facilitate the ability to work with vector and matrix expressions directly. Let $\mathbf{x}$ be an $n \times 1$ vector with elements x_i and $y = f(\mathbf{x})$, then the *partial derivative of a function f with respect to the vector* $\mathbf{x}$, which is a function of n independent variables $x_1, \ldots, x_n$, is defined by

$$\text{(1.8.25)} \qquad \frac{\partial y}{\partial \mathbf{x}} = \frac{\partial f}{\partial \mathbf{x}} = \begin{bmatrix} \partial f/\partial x_1 \\ \partial f/\partial x_2 \\ \vdots \\ \partial f/\partial x_n \end{bmatrix}$$

EXAMPLE 1.8.5. Let $y = f(\mathbf{x}) = 2x_1 + 3x_2^2 + 2x_3^5$. Then

$$\frac{\partial y}{\partial \mathbf{x}} = \frac{\partial f}{\partial \mathbf{x}} = \begin{bmatrix} \partial y/\partial x_1 \\ \partial y/\partial x_2 \\ \partial y/\partial x_3 \end{bmatrix} = \begin{bmatrix} 2 \\ 6x_2 \\ 10x_3^4 \end{bmatrix}$$

If $y = f(\mathbf{x}) = a$, where a is a constant, the

$$\text{(1.8.26)} \qquad \frac{\partial f}{\partial \mathbf{x}} = \mathbf{0}$$

If $y = f(\mathbf{x}) = \mathbf{a}'\mathbf{x}$, or $y = f(\mathbf{x}) = \mathbf{x}'\mathbf{a}$, where $\mathbf{a}$ is a constant vector, the

$$\text{(1.8.27)} \qquad \frac{\partial f}{\partial \mathbf{x}} = \mathbf{a}$$

Extending (1.8.25) a row at a time, the derivative of a vector function $\mathbf{y} = f(\mathbf{x})$ with respect to the vector $\mathbf{x}$ is defined. Let $\mathbf{y}$ be an $m \times 1$ vector and $\mathbf{x}$ be an $n \times 1$ vector such that $y_i = f(\mathbf{x})$, then the *partial derivative of a vector function f with respect to the vector* $\mathbf{x}$, which is a function of n independent variables $x_1, \ldots, x_n$, is given by the $n \times m$ matrix of partial derivatives defined as

$$\text{(1.8.28)} \qquad \frac{\partial \mathbf{y}}{\partial \mathbf{x}} = \frac{\partial f}{\partial \mathbf{x}} = \begin{bmatrix} \partial y_1/\partial x_1 & \partial y_2/\partial x_1 & \cdots & \partial y_m/\partial x_1 \\ \partial y_1/\partial x_2 & \partial y_2/\partial x_2 & \cdots & \partial y_m/\partial x_2 \\ \vdots & \vdots & \vdots & \vdots \\ \partial y_1/\partial x_n & \partial y_2/\partial x_n & \cdots & \partial y_m/\partial x_n \end{bmatrix}$$

Suppose that

$$\mathbf{y} = \mathbf{A}\mathbf{x} = \begin{bmatrix} 2 & 3 & 4 \\ 1 & 2 & 6 \\ 5 & 7 & 2 \end{bmatrix} \begin{bmatrix} x_1 \\ x_2 \\ x_3 \end{bmatrix} = \begin{bmatrix} 2x_1 + 3x_2 + 4x_3 \\ x_1 + 2x_2 + 6x_3 \\ 5x_1 + 7x_2 + 2x_3 \end{bmatrix} = \begin{bmatrix} y_1 \\ y_2 \\ y_3 \end{bmatrix}$$

Then, by applying definition (1.8.28) a row at a time, the

$$\frac{\partial \mathbf{y}}{\partial \mathbf{x}} = \frac{\partial(\mathbf{A}\mathbf{x})}{\partial \mathbf{x}} = \begin{bmatrix} \partial y_1/\partial x_1 & \partial y_2/\partial x_1 & \partial y_3/\partial x_1 \\ \partial y_1/\partial x_2 & \partial y_2/\partial x_2 & \partial y_3/\partial x_2 \\ \partial y_1/\partial x_3 & \partial y_2/\partial x_3 & \partial y_3/\partial x_3 \end{bmatrix} = \mathbf{A}'$$

If $\mathbf{y} = f(\mathbf{x}) = \mathbf{A}\mathbf{x}$, where $\mathbf{A}$ is a matrix of constants, the

(1.8.29) $$\frac{\partial f}{\partial \mathbf{x}} = \mathbf{A}'$$

If $\mathbf{y} = f(\mathbf{x}) = \mathbf{A}\mathbf{x}$, where $\mathbf{A}$ is a matrix of constants, the

(1.8.30) $$\frac{\partial f}{\partial \mathbf{x}'} = \mathbf{A}$$

If $y = f(\mathbf{x}) = \mathbf{x}'\mathbf{A}\mathbf{x}$, where $\mathbf{A}$ is a matrix of constants, the

(1.8.31) $$\frac{\partial f}{\partial \mathbf{x}} = (\mathbf{A} + \mathbf{A}')\mathbf{x} \quad \text{and} \quad \frac{\partial f}{\partial \mathbf{x}'} = \mathbf{x}'(\mathbf{A} + \mathbf{A}')$$

When $\mathbf{A}$ is symmetric, $\mathbf{A} = \mathbf{A}'$, the

$$\frac{\partial f}{\partial \mathbf{x}} = 2\mathbf{A}\mathbf{x}$$

As an example of these formulas, the method of least squares for the model $y = \alpha + \beta x$ will be illustrated by using vector and matrix notation to find α_0 and β_0. Each observation y_i has the form $y_i = \alpha + \beta x_i + \varepsilon_i$, where $\alpha + \beta x_i$ is on the theoretical line and ε_i is the deviation. Using matrix notation, the n observations are represented as

$$\begin{bmatrix} y_1 \\ y_2 \\ \vdots \\ y_n \end{bmatrix} = \begin{bmatrix} 1 & x_1 \\ 1 & x_2 \\ \vdots & \vdots \\ 1 & x_n \end{bmatrix} \begin{bmatrix} \alpha \\ \beta \end{bmatrix} + \begin{bmatrix} \varepsilon_1 \\ \varepsilon_2 \\ \vdots \\ \varepsilon_n \end{bmatrix}$$

$$\underset{(n \times 1)}{\mathbf{y}} = \underset{(n \times 2)}{\mathbf{X}} \ \underset{(2 \times 1)}{\boldsymbol{\eta}} + \underset{(n \times 1)}{\boldsymbol{\varepsilon}}$$

The line of best fit in the least-squares sense minimizes the quantity $\Sigma_{i=1}^{n} \varepsilon_i^2$ obtained by minimizing the function

$$f(\alpha, \beta) = f(\boldsymbol{\eta}) = \sum_{i=1}^{n} \varepsilon_i^2 = \boldsymbol{\varepsilon}'\boldsymbol{\varepsilon} \quad \text{or} \quad f(\boldsymbol{\eta}) = (\mathbf{y} - \mathbf{X}\boldsymbol{\eta})'(\mathbf{y} - \mathbf{X}\boldsymbol{\eta})$$

Taking the partial derivative of $f(\boldsymbol{\eta})$ with respect to $\boldsymbol{\eta}$, using rules (1.8.30), (1.8.29), and (1.8.31), the

$$\begin{aligned}\frac{\partial f}{\partial \boldsymbol{\eta}} &= \frac{\partial}{\partial \boldsymbol{\eta}}(\mathbf{y}'\mathbf{y} - \boldsymbol{\eta}'\mathbf{X}'\mathbf{y} - \mathbf{y}'\mathbf{X}\boldsymbol{\eta} + \boldsymbol{\eta}'\mathbf{X}'\mathbf{X}\boldsymbol{\eta}) \\ &= -\mathbf{X}'\mathbf{y} - (\mathbf{y}'\mathbf{X})' + 2(\mathbf{X}'\mathbf{X})\boldsymbol{\eta} \\ &= -2\mathbf{X}'\mathbf{y} + 2(\mathbf{X}'\mathbf{X})\boldsymbol{\eta}\end{aligned}$$

Equating the partial derivative of the function $f(\boldsymbol{\eta})$ to zero yields the normal equations $(\mathbf{X}'\mathbf{X})\boldsymbol{\eta} = \mathbf{X}'\mathbf{y}$. If $\mathbf{X}'\mathbf{X}$ is not of full rank, a solution to the normal equations is $\boldsymbol{\eta}_0 = (\mathbf{X}'\mathbf{X})^-\mathbf{X}'\mathbf{y} + (\mathbf{I} - \mathbf{H})\mathbf{z}$, where $\mathbf{H} = (\mathbf{X}'\mathbf{X})^-\mathbf{X}'\mathbf{X}$ and $\mathbf{z}$ is an arbitrary vector. If the $R(\mathbf{X}) = R(\mathbf{X}'\mathbf{X}) = 2$, so that $\mathbf{X}'\mathbf{X}$ is of full rank, a unique solution exists for the normal equations since $(\mathbf{X}'\mathbf{X})^- = (\mathbf{X}'\mathbf{X})^{-1}$. Then, solving for the critical value $\boldsymbol{\eta}_0$ for the full-rank case,

$$\mathbf{X}'\mathbf{X} = \begin{bmatrix} n & \sum x_i \\ \sum x_i & \sum x_i^2 \end{bmatrix}$$

$$(\mathbf{X}'\mathbf{X})^{-1} = \frac{\text{adjoint}\,(\mathbf{X}'\mathbf{X})}{|\mathbf{X}'\mathbf{X}|} = \frac{1}{n\sum x_i^2 - (\sum x_i)^2}\begin{bmatrix} \sum x_i^2 & -\sum x_i \\ -\sum x_i & n \end{bmatrix}$$

$$\mathbf{X}'\mathbf{y} = \begin{bmatrix} \sum y_i \\ \sum x_i y_i \end{bmatrix}$$

and

$$\boldsymbol{\eta}_0 = \begin{bmatrix} \dfrac{\sum y_i \sum x_i^2 - \sum x_i \sum x_i y_i}{n\sum x_i^2 - (\sum x_i)^2} \\[2ex] \dfrac{n\sum x_i y_i - \sum x_i \sum y_i}{n\sum x_i^2 - (\sum x_i)^2} \end{bmatrix} = \begin{bmatrix} \alpha_0 \\ \beta_0 \end{bmatrix}$$

which is in agreement with (1.8.22). To verify that $\boldsymbol{\eta}_0$ yields a minimum, the second derivative of the function $f(\boldsymbol{\eta})$ with respect to $\boldsymbol{\eta}$ is obtained. The

$$\frac{\partial^2 f}{\partial \boldsymbol{\eta}^2} = \mathbf{X}'\mathbf{X}$$

Associating $\boldsymbol{\Delta}$ with $\mathbf{X}'\mathbf{X}$ in (1.8.21), $\boldsymbol{\eta}_0$ yields a minimum if $\mathbf{X}'\mathbf{X}$ is p.d. In this case, the $|\mathbf{X}'\mathbf{X}| \neq 0$.

The advantage of the vector differentiation approach over the scalar element-by-element method is that it is easily generalized for more than one independent variable. For

$$\underset{(n\times q)}{\mathbf{X}} = \begin{bmatrix} 1 & x_{11} & \cdots & x_{1k} \\ 1 & x_{21} & \cdots & x_{2k} \\ \vdots & \vdots & \vdots & \vdots \\ 1 & x_{n1} & \cdots & x_{nk} \end{bmatrix} \quad \text{and} \quad \underset{(q\times 1)}{\boldsymbol{\eta}} = \begin{bmatrix} \alpha \\ \beta_1 \\ \vdots \\ \beta_k \end{bmatrix}$$

the above results are immediately applicable. Instead of minimizing $f(\boldsymbol{\eta}) = \boldsymbol{\varepsilon}'\boldsymbol{\varepsilon}$, the $\text{Tr}(\boldsymbol{\varepsilon}\boldsymbol{\varepsilon}')$ could be equivalently minimized since the

$$\text{Tr}(\boldsymbol{\varepsilon}\boldsymbol{\varepsilon}') = \sum_{i=1}^{n} \varepsilon_i^2$$

If f is a function of nm independent variables $x_{11}, x_{12}, \ldots, x_{nm}$, represented by the matrix

$$\underset{(n \times m)}{\mathbf{X}} = \begin{bmatrix} x_{11} & x_{12} & \cdots & x_{1m} \\ x_{21} & x_{22} & \cdots & x_{2m} \\ \vdots & \vdots & \vdots\vdots\vdots & \vdots \\ x_{n1} & x_{n2} & \cdots & x_{nm} \end{bmatrix}$$

so that $y = f(\mathbf{X})$, then the *partial derivative of a function f with respect to the matrix* $\mathbf{X}$ is defined by the $n \times m$ matrix

$$\frac{\partial f}{\partial \mathbf{X}} = \left[\frac{\partial f}{\partial \mathbf{x}_{ij}}\right] \tag{1.8.32}$$

which is a matrix of partial derivatives.

EXAMPLE 1.8.6. Let

$$y = x_{11} + x_{21} + x_{22}^2$$

Then

$$\frac{\partial f}{\partial \mathbf{X}} = \begin{bmatrix} \partial f/\partial x_{11} & \partial f/\partial x_{12} \\ \partial f/\partial x_{21} & \partial f/\partial x_{22} \end{bmatrix} = \begin{bmatrix} 1 & 0 \\ 1 & 2x_{22} \end{bmatrix}$$

EXAMPLE 1.8.7. Let

$$y = [a_1, a_2]\begin{bmatrix} x_{11} & x_{12} \\ x_{21} & x_{22} \end{bmatrix}\begin{bmatrix} a_1 \\ a_2 \end{bmatrix}$$

$$= a_1^2 x_{11} + a_1 a_2 x_{21} + a_1 a_2 x_{12} + a_2^2 x_{22}$$

Then

$$\frac{\partial y}{\partial \mathbf{X}} = \begin{bmatrix} a_1^2 & a_1 a_2 \\ a_1 a_2 & a_2^2 \end{bmatrix} = \mathbf{a}\mathbf{a}'$$

If $\mathbf{X}$ is symmetric,

$$\frac{\partial y}{\partial \mathbf{X}} = \begin{bmatrix} a_1^2 & a_1 a_2 + a_1 a_2 \\ a_1 a_2 + a_1 a_2 & a_2^2 \end{bmatrix}$$

$$= 2\mathbf{a}\mathbf{a}' - \text{Dia}\, \mathbf{a}\mathbf{a}'$$

Definition (1.8.32) can be applied as follows. If $\mathbf{X}$ is an $n \times n$ matrix and $\mathbf{A}$ is a conformable matrix of constants, then

$$\frac{\partial}{\partial \mathbf{X}}[\text{Tr}(\mathbf{X})] = \frac{\partial}{\partial \mathbf{X}'}[\text{Tr}(\mathbf{X}')] = \mathbf{I}_n \tag{1.8.33a}$$

$$\text{(1.8.33b)} \quad \frac{\partial}{\partial \mathbf{X}}[\mathrm{Tr}(\mathbf{AX})] = \frac{\partial}{\partial \mathbf{X}'}[\mathrm{Tr}(\mathbf{XA})] = \mathbf{A}'$$

$$\text{(1.8.33c)} \quad \frac{\partial}{\partial \mathbf{X}'}[\mathrm{Tr}(\mathbf{AX})] = \frac{\partial}{\partial \mathbf{X}'}[\mathrm{Tr}(\mathbf{XA})] = \mathbf{A}$$

$$\text{(1.8.33d)} \quad \frac{\partial}{\partial \mathbf{X}}[\mathrm{Tr}(\mathbf{X}'\mathbf{AX})] = (\mathbf{A} + \mathbf{A}')\mathbf{X} \quad \text{and} \quad \frac{\partial}{\partial \mathbf{X}'}[\mathrm{Tr}(\mathbf{X}'\mathbf{AX})] = \mathbf{X}'(\mathbf{A}' + \mathbf{A})$$

If $\mathbf{A}$ is symmetric, $\mathbf{A} = \mathbf{A}'$ and

$$\text{(1.8.33e)} \quad \frac{\partial}{\partial \mathbf{X}}[\mathrm{Tr}(\mathbf{X}'\mathbf{AX})] = 2\mathbf{AX}$$

The proof follows from the definition of the $\mathrm{Tr}(\mathbf{X}) = \Sigma_{i=1}^{n} x_{ii}$. For example, to prove (1.8.33a),

$$\frac{\partial\left(\sum_i x_{ii}\right)}{\partial x_{ij}} = 0 \qquad (i \neq j)$$

$$\frac{\partial\left(\sum_j x_{ii}\right)}{\partial x_{ij}} = 1 \qquad (i = 1, \ldots, n)$$

As an example of these operations, consider minimizing the function

$$f(\mathbf{B}) = \mathrm{Tr}[(\mathbf{Y} - \mathbf{XB})'(\mathbf{Y} - \mathbf{XB})]$$

where $\mathbf{Y}$ is an $n \times p$ constant matrix, $\mathbf{X}$ is an $n \times q$ constant matrix, and $\mathbf{B}$ is a $q \times p$ matrix of independent variables. Then the

$$\begin{aligned}\frac{\partial f}{\partial \mathbf{B}} &= \frac{\partial}{\partial \mathbf{B}}[\mathrm{Tr}(\mathbf{Y}'\mathbf{Y} - \mathbf{Y}'\mathbf{XB} - \mathbf{B}'\mathbf{X}'\mathbf{Y} + \mathbf{B}'\mathbf{X}'\mathbf{XB})] \\ &= \frac{\partial}{\partial \mathbf{B}}[\mathrm{Tr}(\mathbf{Y}'\mathbf{Y}) - \mathrm{Tr}(\mathbf{Y}'\mathbf{XB}) - \mathrm{Tr}(\mathbf{B}'\mathbf{X}'\mathbf{Y}) + \mathrm{Tr}(\mathbf{B}'\mathbf{X}'\mathbf{XB})] \\ &= -(\mathbf{Y}'\mathbf{X})' - (\mathbf{X}'\mathbf{Y}) + 2\mathbf{X}'\mathbf{XB}\end{aligned}$$

Equating the $\partial f/\partial \mathbf{B}$ to zero yields the normal equations

$$\text{(1.8.34)} \qquad (\mathbf{X}'\mathbf{X})\mathbf{B} = \mathbf{X}'\mathbf{Y}$$

A general solution to this system is given by $\mathbf{B}_0 = (\mathbf{X}'\mathbf{X})^{-}\mathbf{X}'\mathbf{Y} + (\mathbf{I} - \mathbf{H})\mathbf{Z}$, where $\mathbf{H} = (\mathbf{X}'\mathbf{X})^{-}\mathbf{X}'\mathbf{X}$ and $\mathbf{Z}$ is an arbitrary matrix. This result follows from (1.6.25).

Differentiation of a function with respect to a vector (1.8.25), of a vector function with respect to a vector (1.8.28), and of a function with respect to a matrix (1.8.32) have been defined. Consideration of the differentiation of a matrix function with respect to a scalar is now performed. Let $\mathbf{Y}$ be a $k \times p$ matrix such that each element of $\mathbf{Y}$ is a function of an $n \times m$ matrix $\mathbf{X}$ of independent variables x_{ij}. The *partial derivative of the matrix* **Y** *with respect to the variables* x_{ij} is defined by the matrix

$$\frac{\partial \mathbf{Y}}{\partial x_{ij}} = \begin{bmatrix} \partial y_{11}/x_{ij} & \partial y_{12}/\partial x_{ij} & \cdots & \partial y_{1p}/\partial x_{ij} \\ \partial y_{21}/x_{ij} & \partial y_{22}/\partial x_{ij} & \cdots & \partial y_{2p}/\partial x_{ij} \\ \vdots & \vdots & \vdots\vdots\vdots & \vdots \\ \partial y_{k1}/\partial x_{ij} & \partial y_{k2}/\partial x_{ij} & \cdots & \partial y_{kp}/\partial x_{ij} \end{bmatrix} \tag{1.8.35}$$

Some important rules for the differentiation of matrices with respect to a scalar are

$$\begin{aligned} \frac{\partial(\mathbf{YZ})}{x_{ij}} &= \mathbf{Y}\left(\frac{\partial \mathbf{Z}}{\partial x_{ij}}\right) + \left(\frac{\partial \mathbf{Y}}{\partial x_{ij}}\right)\mathbf{Z} \\ \frac{\partial \mathbf{X}^{-1}}{\partial x_{ij}} &= -\mathbf{X}^{-1}\left(\frac{\partial \mathbf{X}}{\partial x_{ij}}\right)\mathbf{X}^{-1} \qquad \text{if } |\mathbf{X}| \neq 0 \end{aligned} \tag{1.8.36}$$

where **Y** and **Z** are functions of x_{ij}. The first expression in (1.8.36) is immediate from scalar differentiation. To prove the second, let $\mathbf{I} = \mathbf{XX}^{-1}$. By applying the first result, the

$$\frac{\partial \mathbf{I}}{\partial x_{ij}} = \mathbf{X}\frac{\partial \mathbf{X}^{-1}}{\partial x_{ij}} + \left(\frac{\partial \mathbf{X}}{\partial x_{ij}}\right)\mathbf{X}^{-1} = \mathbf{0}$$

and the second result follows.

We now examine taking the partial derivative of a determinant with respect to its elements. Let **X** be an $n \times n$ matrix of independent variables x_{ij}. The determinant of **X**, using the ith row for expansion, is $|\mathbf{X}| = x_{i1}X_{i1} + x_{i2}X_{i2} + \cdots + x_{in}X_{in}$, where X_{ij} is the cofactor of x_{ij}. Since no X_{ij} involves x_{ij}, the

$$\frac{\partial |\mathbf{X}|}{\partial x_{ij}} = X_{ij}$$

Hence the partial derivative of the determinant of **X** with respect to its elements is

$$\frac{\partial |\mathbf{X}|}{\partial \mathbf{X}} = [X_{ij}] = \text{adjoint } \mathbf{X}' \qquad \text{for } |\mathbf{X}| \neq 0 \tag{1.8.37}$$

When **X** is symmetric, expression (1.8.37) may not be used since every cofactor X_{ij} contains the element x_{ij} in the form x_{ji}. The form of the partial derivative is now more complicated. Before considering the general case, a 3×3 example is examined. Let

$$\mathbf{X} = \begin{bmatrix} x_{11} & x_{12} & x_{13} \\ x_{12} & x_{22} & x_{23} \\ x_{13} & x_{32} & x_{33} \end{bmatrix}$$

The determinant of **X** is

$$\begin{aligned} |\mathbf{X}| &= x_{11}\begin{vmatrix} x_{22} & x_{23} \\ x_{23} & x_{33} \end{vmatrix} - x_{12}\begin{vmatrix} x_{12} & x_{23} \\ x_{13} & x_{33} \end{vmatrix} + x_{13}\begin{vmatrix} x_{12} & x_{22} \\ x_{13} & x_{23} \end{vmatrix} \\ &= x_{11}X_{11} + x_{12}X_{12} + x_{13}X_{13} \end{aligned}$$

Taking the partial derivative of the X with respect to x_{13}, the

$$\frac{\partial|\mathbf{X}|}{\partial x_{13}} = (-x_{12})\frac{\partial}{\partial x_{13}}(x_{12}x_{33} - x_{13}x_{23}) + \frac{\partial}{\partial x_{13}}[x_{13}(x_{12}x_{23} - x_{13}x_{22})]$$

$$= 2(x_{12}x_{23} - x_{22}x_{13})$$

$$= 2\begin{vmatrix} x_{12} & x_{22} \\ x_{13} & x_{23} \end{vmatrix}$$

$$= 2X_{13}$$

For the element x_{11}, the partial derivative is

$$\frac{\partial|\mathbf{X}|}{\partial x_{11}} = X_{11}$$

More generally, let

$$\underset{(n \times n)}{\mathbf{Y}} = \begin{bmatrix} y_{11} & y_{12} & \cdots & y_{1n} \\ y_{21} & y_{22} & \cdots & y_{2n} \\ \vdots & \vdots & \vdots & \vdots \\ y_{n1} & y_{n2} & \cdots & y_{nn} \end{bmatrix}$$

where $y_{pq} = f_{pq}(x_{11}, x_{12}, \ldots, x_{nn})$. Following Graybill (1969, p. 266), the chain rule (1.8.6) is employed. If $u = f(y)$ and $y = g(\mathbf{X})$, then

$$\frac{\partial f}{\partial \mathbf{X}} = \frac{\partial u}{\partial y}\frac{\partial y}{\partial \mathbf{X}}$$

Thus

(1.8.38)
$$\frac{\partial|\mathbf{Y}|}{\partial x_{ij}} = \sum_{q=1}^{n}\sum_{p=1}^{n}\frac{\partial|\mathbf{Y}|}{\partial y_{pq}}\left(\frac{\partial y_{pq}}{\partial x_{ij}}\right) = \sum_{p=1}^{n}\sum_{q=1}^{n} Y_{pq}\left(\frac{\partial y_{pq}}{\partial x_{ij}}\right)$$

where Y_{pq} is the cofactor of y_{pq} in $\mathbf{Y}$. Substituting the matrix $\mathbf{X}$ for the matrix $\mathbf{Y}$ in (1.8.38), the

$$\frac{\partial|\mathbf{X}|}{\partial x_{ij}} = X_{ij} + X_{ji}$$

For a symmetric matrix $\mathbf{X}$, the

$$\frac{\partial|\mathbf{X}|}{\partial x_{ij}} = 2X_{ij} \qquad \text{for } i \neq j$$

$$\frac{\partial|\mathbf{X}|}{\partial x_{ij}} = X_{ii} \qquad \text{for } i = j$$

Thus, if $\mathbf{X}$ is an $n \times n$ symmetric matrix of independent variables, the

(1.8.39)
$$\frac{\partial|\mathbf{X}|}{\partial \mathbf{X}} = 2[X_{ij}] - \text{Dia}\,[X_{ij}]$$

Suppose $y = \log|\mathbf{X}|$. Then the

(1.8.40) $$\frac{\partial}{\partial \mathbf{X}}(\log|\mathbf{X}|) = \frac{1}{|\mathbf{X}|}\frac{\partial}{\partial \mathbf{X}}|\mathbf{X}| = (\mathbf{X}^{-1})'$$

by the chain rule. For a symmetric matrix $\mathbf{X}$, (1.8.40) becomes

(1.8.41) $$\frac{\partial(\log|\mathbf{X}|)}{\partial \mathbf{X}} = 2\mathbf{X}^{-1} - \text{Dia}\,\mathbf{X}^{-1}$$

Other formulas for matrix differentiation are summarized by Dwyer (1967). More recently, McDonald and Swaminathan (1971) have developed a simplified matrix calculus. Their procedures are particularly useful in factor analysis and covariance structure analysis. However, we shall use the more standard procedures as outlined above.

Lagrange Multipliers. It often occurs in multivariate analysis that functions must be minimized or maximized subject to a set of constraints, so that the variables are no longer independent. To investigate such problems, *Lagrange multipliers* are employed.

In the two-variable case, in order to determine extreme values of a function $f(x, y)$ by the method of Lagrange multipliers, where the variables x and y are subjected to a constraint $g(x, y) = 0$, the procedure includes the following steps.

(1.8.42) Form the function $F = f(x, y) - \lambda g(x, y)$, where λ is a Lagrange multiplier. Then determine the values of λ, x, and y that solve the homogeneous system of equations

$$\frac{\partial F}{\partial x} = 0, \quad \frac{\partial F}{\partial y} = 0, \quad \text{and} \quad \frac{\partial F}{\partial \lambda} = 0$$

for its critical points.

Provided that the partial derivatives given in (1.8.42) exist and that the partial derivatives $\partial g/\partial x$ and $\partial g/\partial y$ do not vanish at the critical points of the function $f(x, y)$, extreme values may be obtained. No simple tests exist to help determine whether a maximum or minimum has been achieved; thus reliance is usually on the geometry of the problem.

As an example of this procedure, suppose the maximum and minimum distances from the origin to the ellipse in (1.8.9), $x^2 + xy + y^2 = 1$, are desired. The problem is to establish the extreme values of $f(x, y) = x^2 + y^2$, subject to the constraint $g(x, y) = x^2 + xy + y^2 - 1 = 0$. Geometrically, the constraint and circles take the form given in Figure 1.8.5. The equations given in (1.8.42) for $F = x^2 + y^2 - (x^2 + xy + y^2 - 1)$ become

(1.8.43) $$\begin{aligned} \frac{\partial F}{\partial x} &= 2x - \lambda(2x + y) = 0 \\ \frac{\partial F}{\partial y} &= 2y - \lambda(x + 2y) = 0 \\ \frac{\partial F}{\partial \lambda} &= x^2 + xy + y^2 - 1 = 0 \end{aligned}$$

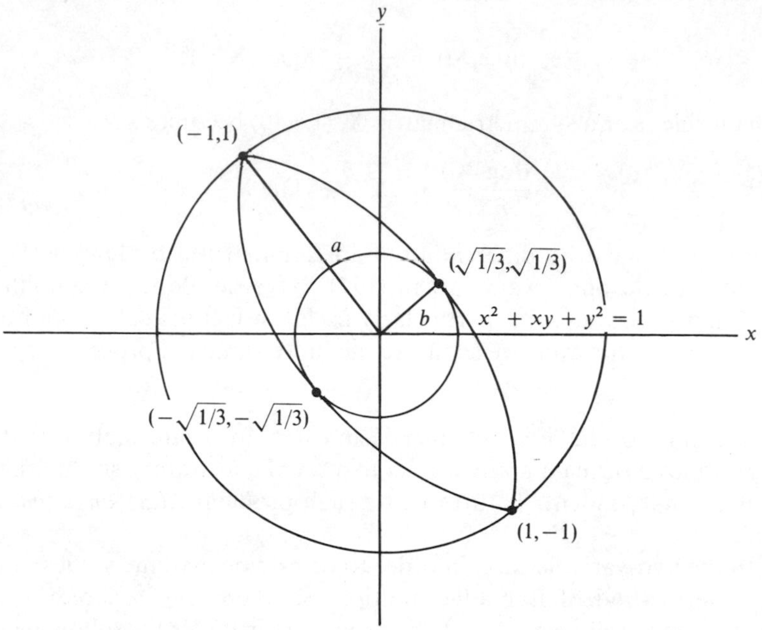

FIGURE 1.8.5. Maximize $f(x, y) = x^2 + y^2$, where $x^2 + xy + y^2 = 1$.

Multiplying the first equation by y, the second by x, and subtracting yield $\lambda(y^2 - x^2) = 0$, so that $y^2 = x^2$ or $y = \pm x$. Substitution of y into the constraining equations determines x, $3x^2 = 1$ and $x^2 = 1$. Solutions to the system are, for the orientation given in Figure 1.8.5, $[x_0, y_0] = [\sqrt{1/3}, \sqrt{1/3}]$, $[x_0, y_0] = [-\sqrt{1/3}, -\sqrt{1/3}]$, $\lambda = 2/3$, and $[x_0, y_0] = [-1, 1]$, $[x_0, y_0] = [1, -1]$, and $\lambda = 2$. For $x^2 = 1/3$ and $y^2 = 1/3$, $f(\sqrt{1/3}, \sqrt{1/3}) = f(-\sqrt{1/3}, -\sqrt{1/3}) = 2/3 = \lambda_2$; for $x^2 + y^2 = 1$, $f(-1, 1) = f(1, -1) = 2 = \lambda_1$. The maximum is obviously 2 and the minimum is 2/3. Hence, in Figure 1.8.5, $a^2 = 2$ and $b^2 = 2/3$.

Equations (1.8.43) for the $F/\partial x$ and the $\partial F/\partial y$ can be written in terms of λ:

$$\left(1 - \frac{1}{\lambda}\right)x + \frac{y}{2} = 0 \quad \text{and} \quad \frac{x}{2} + \left(1 - \frac{1}{\lambda}\right)y = 0$$

These equations have a nontrivial solution only if the determinant

$$\begin{vmatrix} 1 - \dfrac{1}{\lambda} & 1/2 \\ 1/2 & 1 - \dfrac{1}{\lambda} \end{vmatrix} = 0 \tag{1.8.44}$$

The roots of (1.8.44) are $\lambda_1 = 2$ and $\lambda_2 = 2/3$, which we found to be the maximum and minimum values of the function $f(x, y) = x^2 + y^2$, with the restriction that $x^2 + xy + y^2 = 1$.

Alternatively, suppose extreme values of the function $f(x, y) = x^2 + xy + y^2$ are sought subject to the constraint $g(x, y) = x^2 + y^2 - 1 = 0$. Thus instead of allowing a view of all values of the function $f(x, y) = x^2 + xy + y^2$, attention is restricted to unit vectors since $x^2 + y^2 = 1$. Employing (1.8.41), $F = x^2 + xy + y^2 - \lambda(x^2 + y^2 - 1)$ and

$$\begin{aligned} \frac{\partial F}{\partial x} &= 2x + y - 2\lambda x = 0 \\ \frac{\partial F}{\partial y} &= 2y + x - 2\lambda y = 0 \\ \frac{\partial F}{\partial \lambda} &= x^2 + y^2 - 1 = 0 \end{aligned} \tag{1.8.45}$$

or, upon simplification, these become

$$\begin{aligned} (1 - \lambda)x + \frac{y}{2} &= 0 \\ \frac{x}{2} + (1 - \lambda)y &= 0 \\ x^2 + y^2 &= 1 \end{aligned}$$

A solution to the first two equations exists only if the determinant

$$\begin{vmatrix} 1 - \lambda & 1/2 \\ 1/2 & 1 - \lambda \end{vmatrix} = 0 \tag{1.8.46}$$

The roots of this characteristic equation are $\lambda_1^* = 3/2$ and $\lambda_2^* = 1/2$. Using (1.8.45), x and y are related by the expressions $x - y = 0$ and $x + y = 0$, so that $y = \pm x$. Employing the constraint with $x = y$, $2x^2 = 1$, and with $y = -x$, $2x^2 = 1$. Thus critical points are $[x_0, y_0] = [1/\sqrt{2}, 1/\sqrt{2}]$ and $[x_0, y_0] = [-1/\sqrt{2}, 1/\sqrt{2}]$, which in matrix form denote the columns of the orthogonal matrix $\mathbf{P}$ obtained in (1.6.9). By substituting these points into $f(x, y) = x^2 + xy + y^2$, the maximum and minimum are $\lambda_1^* = 3/2$ and $\lambda_2^* = 1/2$, respectively, which were the largest and smallest roots of the characteristic equation (1.8.46) and are the reciprocals of the roots of (1.8.44). Relating a^2 and b^2 to the roots λ_1^* and λ_2^*, $a^2 = 1/\lambda_2^*$ and $b^2 = 1/\lambda_1^*$, which was designated in (1.7.9).

The two problems are "inverse" of one another due to the relation of the roots. Locating the maximum in one problem immediately yields the minimum in the other. The reversal of this concept is also true. If establishing extreme values is of interest, points need not be determined as the result of the relationship of the roots to the maxima and minima.

Extending the technique to n variables,

(1.8.47) if extreme values of the function $y = f(x_1, x_2, \ldots, x_n)$ are to be formed subject to m constraints $g_i(x_1, x_2, \ldots, x_n)$, for $i = 1, \ldots, m$, where $m < n$, the function

$$F = f(x_1, x_2, \ldots, x_n) - \sum_{i=1}^{m} \lambda_i g_i(x_1, x_2, \ldots, x_n)$$

is formed, where λ_i are m Lagrange multipliers. This function then solves the system of equations

$$\frac{\partial F}{\partial x_i} = 0 \qquad \text{for all } i$$

$$\frac{\partial F}{\partial \lambda_i} = 0 \qquad \text{for all } i$$

for critical points.

Again, partial derivatives given in (1.8.47) must exist, and not all partial derivatives $\partial F/\partial \lambda_i$ vanish at extreme values.

As an example of this process, suppose the quadratic form $\mathbf{y} = \mathbf{x}'\mathbf{A}\mathbf{x}$ is to be maximized subject to the constraint that $\mathbf{x}'\mathbf{x} = 1$, where $\mathbf{A}$ is a constant matrix of order n and $\mathbf{x}$ is an $n \times 1$ vector. Employing matrix notation, $F = \mathbf{x}'\mathbf{A}\mathbf{x} - \lambda(\mathbf{x}'\mathbf{x} - 1)$. The partial derivative of F with respect to the vector $\mathbf{x}$ and the scalar λ is

$$\frac{\partial F}{\partial \mathbf{x}} = 2\mathbf{A}\mathbf{x} - 2\lambda\mathbf{x}$$

$$\frac{\partial F}{\partial \lambda} = \mathbf{x}'\mathbf{x} - 1$$

Equating the partial derivatives to zero, $(\mathbf{A} - \lambda\mathbf{I})\mathbf{x} = 0; \mathbf{x}'\mathbf{x} = 1$. To obtain a solution, λ must satisfy the characteristic equation $|\mathbf{A} - \lambda\mathbf{I}| = 0$. By multiplying $(\mathbf{A} - \lambda\mathbf{I})\mathbf{x} = \mathbf{0}$ on the left by $\mathbf{x}'$ and using the constraint, $\mathbf{x}'\mathbf{A}\mathbf{x} = \lambda\mathbf{x}'\mathbf{x} = \lambda$, the maximum of $\mathbf{x}'\mathbf{A}\mathbf{x}$ subject to constraint $\mathbf{x}'\mathbf{x} = 1$ is λ_1, the largest root of $|\mathbf{A} - \lambda\mathbf{I}|$ is 0, and the smallest root, λ_n, is the minimum. These values are acquired by selecting the eigenvectors associated with the roots.

(1.8.48) $$\max_{\mathbf{x}} \frac{\mathbf{x}'\mathbf{A}\mathbf{x}}{\mathbf{x}'\mathbf{x}} = \lambda_1 \quad \text{and} \quad \min_{\mathbf{x}} \frac{\mathbf{x}'\mathbf{A}\mathbf{x}}{\mathbf{x}'\mathbf{x}} = \lambda_n$$

Suppose the extremal values of the quadratic form $\mathbf{x}'\mathbf{A}\mathbf{x}$, subject to the constraint $\mathbf{x}'\mathbf{B}\mathbf{x} = 1$, where $\mathbf{A}$ is a symmetric matrix and $\mathbf{B}$ is a p.d. matrix. Letting $F = \mathbf{x}'\mathbf{A}\mathbf{x} - \lambda(\mathbf{x}'\mathbf{B}\mathbf{x} - 1)$, and taking partial derivatives and equating them to zero, the

$$\frac{\partial F}{\partial \mathbf{x}} = 2\mathbf{A}\mathbf{x} - 2\lambda\mathbf{B}\mathbf{x} = 0$$

$$\frac{\partial F}{\partial \lambda} = \mathbf{x}'\mathbf{B}\mathbf{x} - 1 = 0$$

Proceeding as before, the roots and vectors of $(\mathbf{A} - \lambda\mathbf{B})\mathbf{x} = \mathbf{0}$ are found. Again the eigenvalues must satisfy the characteristic equation $|\mathbf{A} - \lambda\mathbf{B}| = 0$, with roots $\lambda_1, \lambda_2, \ldots, \lambda_n$.

(1.8.49)
$$\max_{\mathbf{x}} \frac{\mathbf{x}'\mathbf{A}\mathbf{x}}{\mathbf{x}'\mathbf{B}\mathbf{x}} = \lambda_1 \quad \text{and} \quad \min_{\mathbf{x}} \frac{\mathbf{x}'\mathbf{A}\mathbf{x}}{\mathbf{x}'\mathbf{B}\mathbf{x}} = \lambda_n$$

The procedures summarized in this section are fundamental to the development of principal component and canonical correlation analysis as well as to estimation and hypothesis test theory. The specific implications will be brought out in the following chapters.

EXERCISES 1.8

1. Find the dy/dx for the following:

   a. $y = 3x^5 - \cos x$ b. $y = e^x \sin x$ c. $y = \dfrac{3x + 1}{x - 1}$

   d. $y = \sin e^{x^2}$ e. $y = \log(\cos x)$

   f. Evaluate dy/dx in parts a and b at $x = 0$, and interpret the results geometrically.

2. a. Find all zeros, maxima, minima, and points of inflection of the function $y = x^3 + x^2$.

   b. Using the information in part a, carefully sketch the graph of the function.

3. a. Show that $\beta_0 = \hat{\rho}_{xy}(\hat{\sigma}_y/\hat{\sigma}_x)$ in (1.8.22).

   b. Prove (1.8.31), (1.8.33a to e), and (1.8.36).

   c. Derive (1.8.41) from (1.8.40).

4. a. If

$$\mathbf{x} = \begin{bmatrix} x_1 \\ x_2 \\ x_3 \end{bmatrix} \quad \text{and} \quad \mathbf{y} = \begin{bmatrix} x_1 + x_3 \\ x_1 x_2 \end{bmatrix}$$

   find the $\partial\mathbf{y}/\partial\mathbf{x}$.

   b. The quadratic form $f(\mathbf{x}) = x_1^2 + 4x_1x_2 + x_2^2$ may be written

$$f(\mathbf{x}) = \mathbf{x}'\mathbf{A}\mathbf{x} = \mathbf{x}'\begin{bmatrix} 1 & 2 \\ 2 & 1 \end{bmatrix}\mathbf{x}$$

   as well as

$$f(\mathbf{x}) = \mathbf{x}'\mathbf{B}\mathbf{x} = \mathbf{x}'\begin{bmatrix} 1 & 3 \\ 1 & 1 \end{bmatrix}\mathbf{x}$$

   Compute the $\partial f/\partial\mathbf{x}$ in two different ways by using (1.8.31), and show that they agree. Compute $\partial f/\partial\mathbf{A}$ and $\partial f/\partial\mathbf{B}$ (see Example 1.8.7).

   c. If $\mathbf{A}$ is not symmetric, show that the $\partial/\partial\mathbf{A}|\mathbf{A}|^{-1/2} = (-1/2)|\mathbf{A}|^{-1/2}(\mathbf{A}^{-1})'$. What is the formula if $\mathbf{A}$ is symmetric?

5. a. Find the extreme values of $f(x_1, x_2) = x_1^2 x_2^3$, subject to the constraint $x_1 + x_2 = 1$.

   b. Find the extreme values of $f(x_1, x_2) = x_1 + x_2$, subject to the constraint $x_1^2 x_2^3 = 1$.

6. Determine the vector **x** such that the quadratic form **x′Ax** for an $n \times n$ matrix **A** is minimal, subject to the condition that the $\Sigma_{i=1}^{n} x_i = 1$.

CHAPTER 2

MULTIVARIATE SAMPLING THEORY

2.1 INTRODUCTION

Fundamental to the study of multivariate analysis is the multivariate normal distribution, the estimation of its parameters, and the algebra of expectations for vector- and matrix-valued random variables. These topics, as well as other aspects of distribution theory essential to multivariate data analysis, are discussed in this chapter.

2.2 RANDOM MATRICES AND VECTORS

As indicated in Section 1.1, multivariate analysis is concerned with the study of random $N \times p$ data matrices of the form

$$\underset{(N \times p)}{Y} = \begin{bmatrix} Y_{11} & Y_{12} & \cdots & Y_{1p} \\ Y_{21} & Y_{22} & \cdots & Y_{2p} \\ \vdots & \vdots & \vdots\vdots\vdots & \vdots \\ Y_{N1} & Y_{N2} & \cdots & Y_{Np} \end{bmatrix}$$

where $Y_{11}, Y_{12}, \ldots, Y_{Np}$ are random variables. The expected value of the random $N \times p$ matrix $\mathbf{Y}$ is the matrix of expectations of its elements,

$$E(\mathbf{Y}) = [E(Y_{ij})] \tag{2.2.1}$$

Similarly, the expectation of a random vector is the vector of the expected values of its elements. From the operations of univariate random variables, the following operations occur for random matrices $\mathbf{X}$ and $\mathbf{Y}$ and constant matrices $\mathbf{A}$, $\mathbf{B}$, and $\mathbf{C}$:

$$E(\mathbf{X} + \mathbf{Y}) = E(\mathbf{X}) + E(\mathbf{Y}) \tag{2.2.2}$$

$$E(\mathbf{AX}) = \mathbf{A}E(\mathbf{X}) \tag{2.2.3}$$

$$E(\mathbf{XB}) = E(\mathbf{X})\mathbf{B} \tag{2.2.4}$$

$$E(\mathbf{C}) = \mathbf{C} \tag{2.2.5}$$

Mean, Covariance, and Correlation Matrices. The *mean vector* of a random vector $\mathbf{Y}' = [Y_1, Y_2, \ldots, Y_p]$ is denoted by

$$\boldsymbol{\mu} = E(\mathbf{Y}) = \begin{bmatrix} E(Y_1) \\ E(Y_2) \\ \vdots \\ E(Y_p) \end{bmatrix} = \begin{bmatrix} \mu_1 \\ \mu_2 \\ \vdots \\ \mu_p \end{bmatrix}$$

where the mean of each univariate random variable is μ_i. To extend the notion of a variance to vectors, let σ_{ij} denote the covariance between the elements Y_i and Y_j of the random p-vector $\mathbf{Y}$,

$$\sigma_{ij} = \operatorname{cov}(Y_i, Y_j) = E\{[Y_i - E(Y_i)][Y_i - E(Y_j)]\}.$$

Then the variance of the random p-vector $\mathbf{Y}$ is the *variance covariance* or *dispersion matrix* defined by

$$\begin{aligned} V(\mathbf{Y}) &= E\{[\mathbf{Y} - E(\mathbf{Y})][\mathbf{Y} - E(\mathbf{Y})]'\} \\ &= \begin{bmatrix} \sigma_{11} & \sigma_{12} & \cdots & \sigma_{1p} \\ \sigma_{21} & \sigma_{22} & \cdots & \sigma_{2p} \\ \vdots & \vdots & \vdots & \vdots \\ \sigma_{p1} & \sigma_{p2} & \cdots & \sigma_{pp} \end{bmatrix} \\ &= \boldsymbol{\Sigma} \end{aligned} \tag{2.2.6}$$

where $\sigma_{ii} = \sigma_i^2 = V(Y_i)$.

Two properties of the variance operator for vectors follow from (2.2.6). For a random p-vector $\mathbf{Y}$ and scalar vector $\mathbf{a}$, the

$$V(\mathbf{Y} + \mathbf{a}) = V(\mathbf{Y}) \tag{2.2.7}$$

and for a scalar matrix **A**, the

$$V(\mathbf{AY}) = \mathbf{A\Sigma A}' \tag{2.2.8}$$

To prove (2.2.8). observe that

$$\begin{aligned} V(\mathbf{AY}) &= E\{[\mathbf{AY} - E(\mathbf{AY})][\mathbf{AY} - E(\mathbf{AY})]'\} \\ &= E\{\mathbf{A}[\mathbf{Y} - E(\mathbf{Y})][\mathbf{Y} - E(\mathbf{Y})]'\mathbf{A}'\} \\ &= \mathbf{A\Sigma A}' \end{aligned}$$

As a special case of (2.2.8), the variance of a linear combination of random variables $\mathbf{a}'\mathbf{Y} = a_1 Y_1 + \cdots + a_p Y_p$, where **a** is a constant vector, is $\mathbf{a}'\mathbf{\Sigma a}$; that is,

$$V(\mathbf{a}'\mathbf{Y}) = \mathbf{a}'\mathbf{\Sigma a} = \sum_{i=1}^{p} \sum_{j=1}^{p} a_i a_j \sigma_{ij} \tag{2.2.9}$$

By extending (2.2.6), the dispersion matrix of two random vectors **X** and **Y** is defined by

$$\operatorname{cov}(\mathbf{X}, \mathbf{Y}) = E\{[\mathbf{X} - E(\mathbf{X})][\mathbf{Y} - E(\mathbf{Y})]'\} = \mathbf{\Sigma}_{\mathbf{xy}} \tag{2.2.10}$$

From (2.2.9),

$$\operatorname{cov}(\mathbf{a}'\mathbf{X}, \mathbf{b}'\mathbf{Y}) = \mathbf{a}'\mathbf{\Sigma}_{\mathbf{xy}}\mathbf{b} \tag{2.2.11}$$

For random vectors **X** and **Y**, scalar matrices **A** and **B**, and scalar vector **a**, the following rules for the covariance operator are obtained:

$$\operatorname{cov}(\mathbf{Y}, \mathbf{Y}) = V(\mathbf{Y}) \tag{2.2.12}$$

$$\operatorname{cov}(\mathbf{X}, \mathbf{Y}) = \operatorname{cov}(\mathbf{Y}, \mathbf{X}) \tag{2.2.13}$$

$$\operatorname{cov}(\mathbf{a} + \mathbf{AX}, \mathbf{Y}) = \mathbf{A} \operatorname{cov}(\mathbf{X}, \mathbf{Y}) \tag{2.2.14}$$

$$\operatorname{cov}(\mathbf{X}, \mathbf{a} + \mathbf{BY}) = \operatorname{cov}(\mathbf{X}, \mathbf{Y})\mathbf{B}' \tag{2.2.15}$$

The correlation between the random variables Y_i and Y_j is given by

$$\rho_{ij} = \frac{\sigma_{ij}}{\sigma_i \sigma_j} = \frac{\operatorname{cov}(Y_i, Y_j)}{\sqrt{V(Y_i)V(Y_j)}} \qquad \text{where } -1 \le \rho_{ij} \le 1$$

The *correlation matrix* for the p-vector-valued random variable **Y** is

$$\mathbf{P} = \begin{bmatrix} 1 & \rho_{12} & \cdots & \rho_{1p} \\ \rho_{21} & 1 & \cdots & \rho_{2p} \\ \vdots & \vdots & \vdots & \vdots \\ \rho_{p1} & \rho_{p2} & \cdots & 1 \end{bmatrix} \tag{2.2.16}$$

Letting Dia $[\sigma_i]$ denote the diagonal matrix with diagonal elements equal to the square root of the diagonal elements of $\mathbf{\Sigma}$, we are able to determine the relationship between $\mathbf{P}$ and $\boldsymbol{\Sigma}$ by using matrix notation:

$$\mathbf{P} = \text{Dia}\,[\sigma_i^{-1}]\boldsymbol{\Sigma}\,\text{Dia}\,[\sigma_i^{-1}]$$

$$\boldsymbol{\Sigma} = \text{Dia}\,[\sigma_i]\mathbf{P}\,\text{Dia}\,[\sigma_i] \tag{2.2.17}$$

The matrices $\boldsymbol{\Sigma}$ and $\mathbf{P}$ occur frequently in the study of multivariate data. To invoke the theory developed in Chapter 1, observe that $\boldsymbol{\Sigma}$ and $\mathbf{P}$ are symmetric and Gramian:

(2.2.18) Matrices $\boldsymbol{\Sigma}$ and $\mathbf{P}$ are symmetric and Gramian.

Symmetry is obvious; the Gramian property follows from (2.2.9). If the $|\boldsymbol{\Sigma}| = 0$, there exists a linear combination of the random variables that is constant, with probability 1.

By partitioning the p-vector $\mathbf{Y}$ into distinct subsets

$$\mathbf{Y} = \begin{bmatrix} \mathbf{Y}_1 \\ \mathbf{Y}_2 \end{bmatrix}$$

the

$$V(\mathbf{Y}) = \begin{bmatrix} \boldsymbol{\Sigma}_{11} & \boldsymbol{\Sigma}_{12} \\ \boldsymbol{\Sigma}_{21} & \boldsymbol{\Sigma}_{22} \end{bmatrix} = \begin{bmatrix} V(\mathbf{Y}_1) & \text{cov}(\mathbf{Y}_1, \mathbf{Y}_2) \\ \text{cov}(\mathbf{Y}_2, \mathbf{Y}_1) & V(\mathbf{Y}_2) \end{bmatrix}$$

where the $\boldsymbol{\Sigma}_{ij}$ denote dispersion matrices. Using this notation,

(2.2.19) $\mathbf{Y}_1$ and $\mathbf{Y}_2$ are uncorrelated if and only if $\boldsymbol{\Sigma}_{12} = 0$ and the individual components of $\mathbf{Y}_i$ are uncorrelated if and only if $\boldsymbol{\Sigma}_{ii}$ is a diagonal matrix.

EXERCISES 2.2

1. a. Employ (2.2.1) to prove (2.2.2) through (2.2.5).
   b. Use (2.2.6) and (2.2.8) to prove (2.2.12) through (2.2.15).
   c. Prove (2.2.18).
2. For $E(\mathbf{X}) = \boldsymbol{\mu}$, the $V(\mathbf{X}) = \boldsymbol{\Sigma}$, and a constant matrix $\mathbf{A}$, for a random vector $\mathbf{X}$, prove that
   a. $E(\mathbf{X}'\mathbf{A}\mathbf{X}) = \text{Tr}(\mathbf{A}\boldsymbol{\Sigma}) + \boldsymbol{\mu}'\mathbf{A}\boldsymbol{\mu}$
   b. $\text{cov}(\mathbf{X}', \mathbf{X}'\mathbf{A}\mathbf{X}) = 2\boldsymbol{\Sigma}\mathbf{A}\boldsymbol{\mu}$
   c. $V(\mathbf{X}'\mathbf{A}\mathbf{X}) = 2\,\text{Tr}(\mathbf{A}\boldsymbol{\Sigma})^2 + 4\boldsymbol{\mu}'\mathbf{A}\boldsymbol{\Sigma}\mathbf{A}\boldsymbol{\mu}$

   [*Hint*: $E(\mathbf{X}\mathbf{X}') = \boldsymbol{\Sigma} + \boldsymbol{\mu}\boldsymbol{\mu}'$, and the $\text{Tr}(\mathbf{A}\mathbf{X}\mathbf{X}') = \mathbf{X}'\mathbf{A}\mathbf{X}$.]
3. Show that the $E[(\mathbf{Y} - \mathbf{c})(\mathbf{Y} - \mathbf{c})']$, for a constant vector $\mathbf{c}$, is minimal when $\mathbf{c} = \boldsymbol{\mu} = E(\mathbf{Y})$.

2.3 THE UNIVARIATE NORMAL DISTRIBUTION

The *density function* for a normal variable Y is defined by

$$f(y) = \frac{1}{\sigma\sqrt{2\pi}} e^{(-1/2)(y-\mu)^2/\sigma^2} \quad -\infty < y < \infty \tag{2.3.1}$$

with mean $\mu = E(Y)$ and variance $\sigma^2 = V(Y)$. To indicate that Y is distributed normally with mean μ and variance σ^2, use the notation $Y \sim N(\mu, \sigma^2)$.

To estimate the parameters μ and σ^2 in (2.3.1), suppose a random sample of observations $Y_1, Y_2, \ldots, Y_N$ is obtained from a normal population. Denoting the observed values of $Y_1, \ldots, Y_N$ by $y_1, \ldots, y_N$, the *maximum-likelihood estimators* of μ and σ^2 are

$$\begin{aligned} \hat{\mu} &= \frac{1}{N}\sum_{i=1}^{N} y_i = \bar{y} \\ \hat{\sigma}^2 &= \frac{1}{N}\sum_{i=1}^{N} (y_i - \bar{y})^2 \end{aligned} \tag{2.3.2}$$

We will now review the method of obtaining maximum-likelihood estimators. The likelihood function for a random sample of N observations is given by

$$L(y_1, y_2, \ldots, y_N \mid \boldsymbol{\theta}) = \sum_{i=1}^{N} f(y_i \mid \boldsymbol{\theta}) \tag{2.3.3}$$

where $\boldsymbol{\theta}$ represents a set of unknown parameters and $f(y_i \mid \boldsymbol{\theta})$ is the density function of the ith observation (see Mood and Graybill, 1963). The principle of maximum likelihood allows, for a given sample, the location of values of the parameters most likely to yield sample values near the observed sample point. Thus the maximum-likelihood estimator of $\boldsymbol{\theta}$ is the statistic $\hat{\boldsymbol{\theta}}$ that maximizes (2.3.3) or the equivalently monotonic functions of (2.3.3). Since many functions in statistics contain exponential terms, it is more convenient to work with the natural logarithm of the likelihood rather than with the likelihood directly. Applying this procedure in the normal case,

$$\begin{aligned} L = L(y_1, y_2, \ldots, y_N \mid \mu, \sigma^2) &= \prod_{i=1}^{N} \frac{1}{\sigma\sqrt{2\pi}} \exp - \frac{(y_i - \mu)^2}{2\sigma^2} \\ &= \left(\frac{1}{2\pi\sigma^2}\right)^{N/2} \exp - \frac{\sum_{i=1}^{N} (y_i - \mu)^2}{2\sigma^2} \end{aligned}$$

By the techniques of Section 1.8, the $\log L$ is maximized. Taking logarithms,

$$L^* = \log L = -\frac{N}{2}\log 2\pi - \frac{N}{2}\log \sigma^2 - \frac{1}{2\sigma^2}\sum_i (y_i - \mu)^2$$

and differentiating, the normal equations become

$$\begin{aligned} \frac{\partial L^*}{\partial \mu} &= \frac{1}{\sigma^2}\sum_i (y_i - \mu) \\ \frac{\partial L^*}{\partial \sigma^2} &= \frac{-N}{2\sigma^2} + \frac{1}{2\sigma^4}\sum_i (y_i - \mu)^2 \end{aligned}$$

By equating these to zero and solving for μ and σ^2, (2.3.2) follows. Using the procedures in Section 1.8, L^* is maximal and therefore so is L.

To determine whether the maximum-likelihood estimators are unbiased for the population parameters μ and σ^2, we must show that $E(\hat{\theta}) = \theta$ for the parameter θ. Clearly,

$$E(\hat{\mu}) = \mu$$

(2.3.4)
$$E(\hat{\sigma}^2) = \left(\frac{N-1}{N}\right)\sigma^2$$

so that $\hat{\mu} = \bar{y}$ is unbiased for μ, but $\hat{\sigma}^2$ is biased for σ^2. Acquiring an unbiased estimator, we set

(2.3.5)
$$s^2 = \left(\frac{N}{N-1}\right)\hat{\sigma}^2 = \frac{\sum_{i=1}^{N}(y_i - \bar{y})^2}{N-1}$$

Then $E(s^2) = \sigma^2$ (see Hays, 1963, p. 203).

An important property of independent normal random variables is that, if $Y_1, Y_2, \ldots, Y_N$ are normally distributed with mean μ_i and variance σ_i^2 for $i = 1, \ldots, N$ represented by $Y_i \sim IN(\mu_i, \sigma_i^2)$, then

(2.3.6)
$$X = \sum_{i=1}^{N} a_i Y_i \sim N\left(\sum_i a_i\mu_i, \sum_i a_i^2\sigma_i^2\right)$$

That is, linear combinations of normally distributed random variables are also normally distributed. As a special case of (2.3.6), let $Y_i \sim IN(\mu, \sigma^2)$; then

(2.3.7)
$$\bar{Y} = \frac{\sum_i Y_i}{N} \sim N\left(\mu, \frac{\sigma^2}{N}\right)$$

EXERCISES 2.3

1. a. Suppose that $Y \sim N(\mu, \sigma^2)$. Find parameters μ and σ^2 so that the $P(Y \leq 2) = 1/2$.

   b. Suppose that the random variable Y is normally distributed with mean 4 and that the $E(Y^2) = 36$. Find the $P(|Y - 3| > 1)$.

2. If $X \sim N(1, 4)$, $Y \sim N(2, 8)$, and X and Y are independent, find the distribution of

   a. $X + Y$ b. $X - Y$ c. $3X - 2Y$

3. Show that maximizing the log L^* of the likelihood function is equivalent to maximizing L.

2.4 THE MULTIVARIATE NORMAL DISTRIBUTION

Knowledge of integral calculus, transformation of variables, and Jacobians is intrinsic to the derivation of the multivariate normal distribution. These procedures are beyond the scope of this text. For a development of the multivariate normal distribution using these tools, see Anderson (1958) or Rao (1965a). A development by analogy, beginning with a standard zero-one random variable and proceeding to the general case, is adequate for the purposes of this text. A random variable Z is called a *standard zero-one* random variable if it has a normal density function with mean 0 and variance 1.

Suppose $Z_i \sim IN(0, 1)$ so that $E(Z_i) = 0$ and the $V(Z_i) = 1$. The density function of the standard zero-one random variable Z_i is

$$f(z_i) = \frac{1}{\sqrt{2\pi}} e^{-z_i^2/2} \quad -\infty < z_i < \infty \tag{2.4.1}$$

By letting $\mathbf{Z}' = [Z_1, Z_2, \ldots, Z_p]$, a vector of independent random variables, the joint density function of $\mathbf{Z}$ is the product of the individual density functions

$$f(z_1, z_2, \ldots, z_p) = f(\mathbf{z}) = \prod_{i=1}^{p} f(z_i) \tag{2.4.2}$$

For the standard normal case, (2.4.2) becomes

$$f(\mathbf{z}) = \prod_{i=1}^{p} f(z_i) = (2\pi)^{-p/2} \exp - \frac{\sum_{i=1}^{p} z_i^2}{2}$$

$$= (2\pi)^{-p/2} e^{-\mathbf{z}'\mathbf{z}/2} \quad -\infty < z_i < \infty \tag{2.4.3}$$

Given (2.4.3), the random p-vector $\mathbf{Z}$ is said to have a *standard multivariate normal distribution* with mean $\boldsymbol{\mu} = \mathbf{0}$ and variance-covariance matrix $\boldsymbol{\Sigma} = \mathbf{I}$, denoted by $\mathbf{Z} \sim N_p(\mathbf{0}, \mathbf{I})$.

Suppose that random variables $Y_i \sim IN(\mu_i, \sigma^2)$, so that $E(Y_i) = \mu_i$ and $V(Y_i) = \sigma^2$, for $i = 1, 2, \ldots, N$. By (2.4.2), the density function of the random vector $\mathbf{Y}' = [Y_1, Y_2, \ldots, Y_p]$ is

$$\begin{aligned} f(\mathbf{y}) &= \prod_{i=1}^{p} f(y_i) \\ &= \prod_{i=1}^{p} \frac{1}{\sigma\sqrt{2\pi}} \exp - \frac{(y_i - \mu_i)^2}{2\sigma^2} \\ &= (2\pi)^{-p/2} \left(\frac{1}{\sigma^p}\right) \exp - \frac{\sum_{i=1}^{n} (y_i - \mu_i)^2}{2\sigma^2} \\ &= (2\pi)^{-p/2} |\sigma^2 I|^{-1/2} \exp - \frac{(\mathbf{y} - \boldsymbol{\mu})'(\sigma^2\mathbf{I})^{-1}(\mathbf{y} - \boldsymbol{\mu})}{2} \end{aligned} \tag{2.4.4}$$

The vector $\mathbf{Y}$ has an *independent multivariate normal distribution*, written $\mathbf{Y} \sim N_p(\boldsymbol{\mu}, \sigma^2\mathbf{I})$, with mean vector and variance-covariance matrix

$$E(\mathbf{Y}) = \boldsymbol{\mu} = \begin{bmatrix} \mu_1 \\ \mu_2 \\ \vdots \\ \mu_p \end{bmatrix} \quad \text{and} \quad V(\mathbf{Y}) = \begin{bmatrix} \sigma^2 & 0 & \cdots & 0 \\ 0 & \sigma^2 & \cdots & 0 \\ \vdots & \vdots & \ddots & \vdots \\ 0 & 0 & \cdots & \sigma^2 \end{bmatrix} = \sigma^2\mathbf{I}_p$$

respectively.

More generally, the multivariate normal distribution for the random vector **Y** with mean and variance-covariance matrix

$$E(\mathbf{Y}) = \boldsymbol{\mu} = \begin{bmatrix} \mu_1 \\ \mu_2 \\ \vdots \\ \mu_p \end{bmatrix} \quad \text{and} \quad V(\mathbf{Y}) = \begin{bmatrix} \sigma_{11} & \sigma_{12} & \cdots & \sigma_{1p} \\ \sigma_{21} & \sigma_{22} & \cdots & \sigma_{2p} \\ \vdots & \vdots & \vdots & \vdots \\ \sigma_{p1} & \sigma_{p2} & \cdots & \sigma_{pp} \end{bmatrix} = \boldsymbol{\Sigma}$$

respectively, is given by

$$(2.4.5) \qquad f(\mathbf{y}) = (2\pi)^{-p/2}|\boldsymbol{\Sigma}|^{-1/2} \exp - \frac{(\mathbf{y}-\boldsymbol{\mu})'\boldsymbol{\Sigma}^{-1}(\mathbf{y}-\boldsymbol{\mu})}{2} \qquad -\infty < y_i < \infty$$

written $\mathbf{Y} \sim N_p(\boldsymbol{\mu}, \boldsymbol{\Sigma})$. Notice that $\mathbf{x}'\mathbf{A}\mathbf{x} = \mathrm{Tr}(\mathbf{A}\mathbf{x}\mathbf{x}')$, so that (2.4.5) may be represented as

$$(2.4.6) \qquad f(\mathbf{y}) = (2\pi)^{-p/2}|\boldsymbol{\Sigma}|^{-1/2} \exp - \frac{\mathrm{Tr}[\boldsymbol{\Sigma}^{-1}(\mathbf{y}-\boldsymbol{\mu})(\mathbf{y}-\boldsymbol{\mu})']}{2}$$

In the bivariate case,

$$\mathbf{y} = \begin{bmatrix} y_1 \\ y_2 \end{bmatrix}, \quad \boldsymbol{\mu} = \begin{bmatrix} \mu_1 \\ \mu_2 \end{bmatrix}, \quad \boldsymbol{\Sigma} = \begin{bmatrix} \sigma_{11} & \sigma_{12} \\ \sigma_{21} & \sigma_{22} \end{bmatrix} = \begin{bmatrix} \sigma_1^2 & \rho\sigma_1\sigma_2 \\ \rho\sigma_1\sigma_2 & \sigma_2^2 \end{bmatrix}$$

and (2.4.5) reduces to

$$f(\mathbf{y}) = \frac{\exp\left\{\dfrac{-1}{2(1-\rho^2)}\left[\left(\dfrac{y_1-\mu_1}{\sigma_1}\right)^2 - 2\rho\left(\dfrac{y_1-\mu_1}{\sigma_1}\right)\left(\dfrac{y_2-\mu_2}{\sigma_2}\right) + \left(\dfrac{y_2-\mu_2}{\sigma_2}\right)^2\right]\right\}}{2\pi\sigma_1\sigma_2\sqrt{1-\rho^2}}$$

(2.4.7)

where ρ is the correlation between Y_1 and Y_2. The density function in (2.4.7) is called the *bivariate normal density function.*

From analytic geometry, the quadratic form of the exponent in (2.4.7),

$$\frac{-1}{2(1-\rho^2)}\left[\frac{(y_1-\mu_1)^2}{\sigma_1^2} - 2\rho\left(\frac{y_1-\mu_1}{\sigma_1}\right)\left(\frac{y_2-\mu_2}{\sigma_2}\right) + \frac{(y_2-\mu_2)^2}{\sigma_2^2}\right] = Q$$

represents an ellipse with center at the point $[\mu_1, \mu_2]$, called the *centroid of the distribution.* The shape of the ellipse in the y_1y_2 plane depends on the value of ρ and the ratio of σ_1 to σ_2. For $\sigma_1 = \sigma_2$ and $\rho = 0$, the ellipse becomes a circle. For $\sigma_1 = \sigma_2$ and $\rho > 0$, the orientation of the ellipse is as in Figure 2.4.1 for various values of Q.

By intersecting the three-dimensional, bell-shaped surface in Figure 2.4.2 with planes parallel to the y_1y_2 plane and projecting these intersections onto the y_1y_2 plane generates the set of concentric ellipses given in Figure 2.4.1. The surface in Figure 2.4.2 depicts a bivariate normal density function.

Letting

$$z_1 = \frac{y_1-\mu_1}{\sigma_1} \quad \text{and} \quad z_2 = \frac{y_2-\mu_2}{\sigma_2}$$

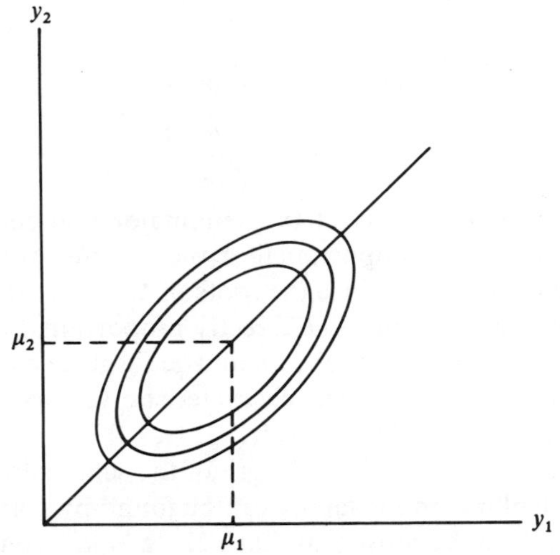

FIGURE 2.4.1. Constant-probability density ellipses.

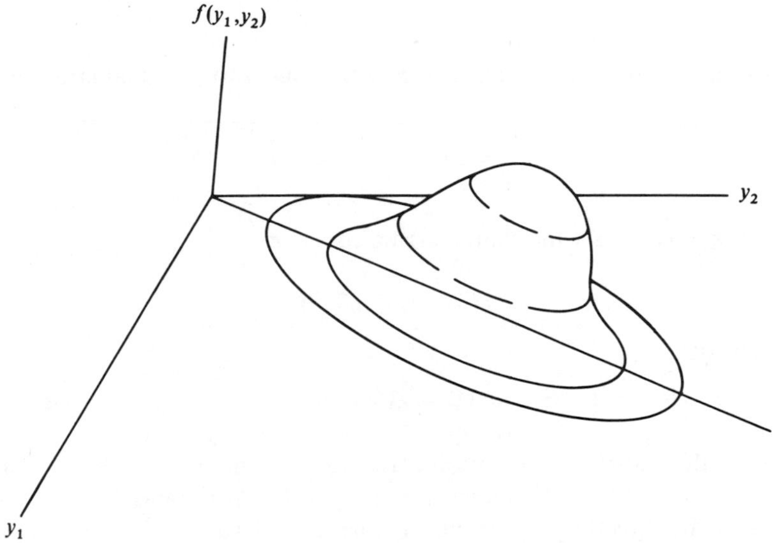

FIGURE 2.4.2. A bivariate normal density function.

the density function for the bivariate normal becomes

$$f(\mathbf{z}) = \frac{\exp\left[\dfrac{-1}{2(1-\rho^2)}(z_1^2 - 2\rho z_1 z_2 + z_2^2)\right]}{2\pi\sqrt{1-\rho^2}}$$

The ellipses associated with the exponent of $f(\mathbf{z})$ are

$$\frac{z_1^2 - 2\rho z_1 z_2 + z_2^2}{1-\rho^2} = Q = [z_1, z_2]\mathbf{\Sigma}^{-1}\begin{bmatrix} z_1 \\ z_2 \end{bmatrix}$$

where

$$\boldsymbol{\Sigma}^{-1} = \begin{bmatrix} 1/(1-\rho^2) & -\rho/(1-\rho^2) \\ -\rho/(1-\rho^2) & 1/(1-\rho^2) \end{bmatrix} \quad \text{and} \quad \boldsymbol{\Sigma} = \begin{bmatrix} 1 & \rho \\ \rho & 1 \end{bmatrix}$$

From (1.7.10), the squared lengths of the semimajor and semiminor axes of the ellipses associated with Q are proportional to the inverses of the roots of $\boldsymbol{\Sigma}^{-1}$, the matrix of the quadratic form. Or, since thc roots of $\boldsymbol{\Sigma}^{-1}$ are the inverses of the roots of $\boldsymbol{\Sigma}$, the squared lengths are seen to be directly proportional to the roots of $\boldsymbol{\Sigma}$, the inverse of the matrix associated with the quadratic form. The characteristic roots of $\boldsymbol{\Sigma}$ are $\lambda_1 = 1 + \rho$ and $\lambda_2 = 1 - \rho$. The characteristic vectors associated with these roots are $\mathbf{p}_1' = [1/\sqrt{2}, 1/\sqrt{2}]$ and $\mathbf{p}_2' = [-1/\sqrt{2}, 1/\sqrt{2}]$.

Alternatively, from (1.8.43), the length of the semimajor axis, called the *first principal axis*, of each ellipse in the family can be found by maximizing the function $f(\mathbf{z}) = z_1^2 + z_2^2$, subject to the constraint that $\mathbf{z}'\boldsymbol{\Sigma}^{-1}\mathbf{z} = k$, a constant. That is, values of the function

$$F = \mathbf{z}'\mathbf{z} - \lambda(\mathbf{z}'\boldsymbol{\Sigma}^{-1}\mathbf{z} - k)$$

are sought such that F is maximal.

Taking the partial derivative of F with respect to $\mathbf{z}$ and equating it to zero,

$$\frac{\partial F}{\partial \mathbf{z}} = 2\mathbf{z} - 2\lambda\boldsymbol{\Sigma}^{-1}\mathbf{z} = \mathbf{0}$$

The value of $\mathbf{z}$ must satisfy the characteristic equation

$$(\mathbf{I} - \lambda\boldsymbol{\Sigma}^{-1})\mathbf{z} = \mathbf{0}$$

or, equivalently,

$$(\boldsymbol{\Sigma} - \lambda\mathbf{I})\mathbf{z} = \mathbf{0}$$

The vectors $\mathbf{z}_i$ that satisfy this equation correspond to the roots of the determinantal equation $|\boldsymbol{\Sigma} - \lambda\mathbf{I}| = 0$. Since the largest root is $\lambda_1 = 1 + \rho$, the length of the semimajor axis is seen to be directly proportional to the largest root of $\boldsymbol{\Sigma}$. Furthermore, the coordinates of the axis are proportional to the elements of the first eigenvector of $\boldsymbol{\Sigma}$.

Estimation of Parameters. Maximum-likelihood estimators are obtained in order to estimate the parameters $\boldsymbol{\mu}$ and $\boldsymbol{\Sigma}$ in (2.4.5). Suppose that a random sample of N p-vectors $\mathbf{Y}_1, \mathbf{Y}_2, \ldots, \mathbf{Y}_N$, represented by rows in the matrix

$$\underset{(N \times p)}{\mathbf{Y}} = \begin{bmatrix} \mathbf{Y}_1' \\ \mathbf{Y}_2' \\ \vdots \\ \mathbf{Y}_N' \end{bmatrix} = \begin{bmatrix} Y_{11} & Y_{12} & \cdots & Y_{1p} \\ Y_{21} & Y_{22} & \cdots & Y_{2p} \\ \vdots & \vdots & \vdots & \vdots \\ Y_{N1} & Y_{N2} & \cdots & Y_{Np} \end{bmatrix} \tag{2.4.8}$$

are obtained from a multivariate normal population with parameters $\boldsymbol{\mu}$ and $\boldsymbol{\Sigma}$, where $N > p$. The maximum-likelihood estimators of $\boldsymbol{\mu}$ and $\boldsymbol{\Sigma}$ are

$$\hat{\boldsymbol{\mu}} = \frac{1}{N}\sum_{i=1}^{N} y_i = \bar{\mathbf{y}}$$

(2.4.9)

$$\hat{\boldsymbol{\Sigma}} = \frac{1}{N}\sum_{i=1}^{N} (\mathbf{y}_i - \bar{\mathbf{y}})(\mathbf{y}_i - \bar{\mathbf{y}})'$$

For proof, the likelihood function

$$L = L(\mathbf{y} \mid \boldsymbol{\mu}, \boldsymbol{\Sigma}) = \prod_{i=1}^{N} f(\mathbf{y}_i) = (2\pi)^{-Np/2}|\boldsymbol{\Sigma}|^{-N/2} \exp -\frac{1}{2}\sum_{i=1}^{N} (\mathbf{y}_i - \boldsymbol{\mu})'\boldsymbol{\Sigma}^{-1}(\mathbf{y}_i - \boldsymbol{\mu})$$

is utilized. Taking logarithms, with $\mathbf{V} = \boldsymbol{\Sigma}^{-1}$,

$$L^* = \log L = -\frac{1}{2}Np \log 2\pi + \frac{1}{2}N \log|\mathbf{V}| - \frac{1}{2}\sum_{i=1}^{N} (\mathbf{y}_i - \boldsymbol{\mu})'\mathbf{V}(\mathbf{y}_i - \boldsymbol{\mu})$$

To maximize L^*, the partial derivatives, with respect to $\boldsymbol{\mu}$ and $\boldsymbol{\Sigma}$, are set equal to zero. By (1.8.31), the partial derivative of L^* with respect to $\boldsymbol{\mu}$ becomes

$$\begin{aligned}\frac{\partial L^*}{\partial \boldsymbol{\mu}} &= -\frac{1}{2}\sum_{i=1}^{N} \frac{\partial}{\partial \boldsymbol{\mu}}[(\mathbf{y}_i - \boldsymbol{\mu})'\mathbf{V}(\mathbf{y}_i - \boldsymbol{\mu})] \\ &= -\frac{1}{2}\sum_{i=1}^{N} [-2\mathbf{V}(\mathbf{y}_i - \boldsymbol{\mu})] \\ &= \mathbf{V}\sum_{i=1}^{N} (\mathbf{y}_i - \boldsymbol{\mu})\end{aligned}$$

Equating this to zero,

$$\mathbf{V}\sum_{i=1}^{N} (\mathbf{y}_i - \boldsymbol{\mu}) = \mathbf{0}$$

Since $\mathbf{V}$ is nonsingular,

$$\hat{\boldsymbol{\mu}} = \frac{1}{N}\sum_{i=1}^{N} \mathbf{y}_i = \bar{\mathbf{y}}$$

is the maximum-likelihood estimator for $\boldsymbol{\mu}$.

To utilize the partial derivative of L^* with respect to the symmetric matrix $\boldsymbol{\Sigma}$, (1.8.39), (1.8.41), and Example 1.8.7 are used. The partial derivative of $(N/2) \log|\mathbf{V}|$, with respect to $\mathbf{V}$, is

$$\frac{\partial}{\partial \mathbf{V}}\left(\frac{N}{2}\log|\mathbf{V}|\right) = \frac{N}{2}[2\mathbf{V}^{-1} - \text{Dia } \mathbf{V}^{-1}]$$

The partial derivative of

$$F = -\frac{1}{2}\sum_{i=1}^{N} (\mathbf{y}_i - \boldsymbol{\mu})'\mathbf{V}(\mathbf{y}_i - \boldsymbol{\mu})$$

with respect to $\mathbf{V}$ is

$$\frac{\partial F}{\partial \mathbf{V}} = -\frac{1}{2}\sum_{i=1}^{N} [2(\mathbf{y}_i - \boldsymbol{\mu})(\mathbf{y}_i - \boldsymbol{\mu})' - \text{Dia}(\mathbf{y}_i - \boldsymbol{\mu})(\mathbf{y}_i - \boldsymbol{\mu})']$$

By substituting $\bar{\mathbf{y}}$ for $\boldsymbol{\mu}$ and equating the sum of the partial derivatives to the null matrix, the partial derivative of L^* with respect to $\mathbf{V}$ becomes

$$\frac{\partial L^*}{\partial \mathbf{V}} = \frac{N}{2}(2\mathbf{V}^{-1} - \text{Dia } \mathbf{V}^{-1}) - \frac{N}{2}(2\hat{\boldsymbol{\Sigma}} - \text{Dia } \hat{\boldsymbol{\Sigma}}) = \mathbf{0}$$

Multiplying through by $2/N$, observe that the elements of $2\mathbf{V}^{-1} - \text{Dia } \mathbf{V}^{-1}$ and $2\hat{\boldsymbol{\Sigma}} - \text{Dia } \hat{\boldsymbol{\Sigma}}$ are identical. That is, with v^{ij} and $\hat{\sigma}_{ij}$ denoting the elements of $\mathbf{V}^{-1}$ and $\hat{\boldsymbol{\Sigma}}$, respectively, $v^{ij} = \hat{\sigma}_{ij}$, for $i = j$, and $2v^{ij} = 2\hat{\sigma}_{ij}$, for $i \neq j$. Since $\mathbf{V}^{-1} = (\boldsymbol{\Sigma}^{-1})^{-1} = \boldsymbol{\Sigma}$, the maximum-likelihood estimator of $\boldsymbol{\Sigma}$ is $\hat{\boldsymbol{\Sigma}}$ as required (see Anderson, 1958, pp. 45–49).

As in the univariate case, $\hat{\boldsymbol{\mu}}$ is an unbiased estimator for $\boldsymbol{\mu}$, and $\hat{\boldsymbol{\Sigma}}$ is a biased estimator of $\boldsymbol{\Sigma}$:

$$E(\hat{\boldsymbol{\mu}}) = E(\bar{Y}) = \frac{1}{N}E\left(\sum_i \mathbf{Y}_i\right) = \frac{1}{N}N\boldsymbol{\mu} = \boldsymbol{\mu}$$

$$E(\hat{\boldsymbol{\Sigma}}) = E\left[\frac{\sum_{i=1}^{N} (\mathbf{Y}_i - \bar{\mathbf{Y}})(\mathbf{Y}_i - \bar{\mathbf{Y}})'}{N}\right]$$

$$= E\left[\frac{\sum_{i=1}^{N} (\mathbf{Y}_i - \boldsymbol{\mu})(\mathbf{Y}_i - \boldsymbol{\mu})'}{N} - (\bar{\mathbf{Y}} - \boldsymbol{\mu})(\bar{\mathbf{Y}} - \boldsymbol{\mu})'\right]$$

but

$$E[(\bar{\mathbf{Y}} - \boldsymbol{\mu})(\bar{\mathbf{Y}} - \boldsymbol{\mu})'] = E\left\{\left[\frac{1}{N}\sum_{i=1}^{N} (\mathbf{Y}_i - \boldsymbol{\mu})\right]\left[\frac{1}{N}\sum_{i=1}^{N} (\mathbf{Y}_i - \boldsymbol{\mu})\right]'\right\}$$

$$= \frac{\sum_{i=1}^{N} E[(\mathbf{Y}_i - \boldsymbol{\mu})(\mathbf{Y}_i - \boldsymbol{\mu})']}{N^2}$$

$$= \frac{N\boldsymbol{\Sigma}}{N^2}$$

$$= \frac{\boldsymbol{\Sigma}}{N}$$

since

$$E\left[\sum_{i=1}^{N} (\mathbf{Y}_i - \boldsymbol{\mu})(\mathbf{Y}_i - \boldsymbol{\mu})'\right] = N\boldsymbol{\Sigma}$$

Thus

$$E(\hat{\mathbf{\Sigma}}) = \frac{N\mathbf{\Sigma} - \mathbf{\Sigma}}{N} = \left(\frac{N-1}{N}\right)\mathbf{\Sigma}$$

By setting

$$\mathbf{S} = \frac{N\hat{\mathbf{\Sigma}}}{N-1} = \frac{\sum_{i=1}^{N} (\mathbf{Y}_i - \overline{\mathbf{Y}})(\mathbf{Y}_i - \overline{\mathbf{Y}})'}{N-1} \tag{2.4.10}$$

an unbiased estimator for $\mathbf{\Sigma}$ is obtained. S is then computed from the $N \times p$ data matrix $\mathbf{Y}$. For illustration,

$$\mathbf{S} = \frac{\sum_{i=1}^{N} (\mathbf{y}_i - \overline{\mathbf{y}})(\mathbf{y}_i - \overline{\mathbf{y}})'}{N-1}$$

$$\mathbf{S} = \frac{\sum_{i=1}^{N} \mathbf{y}_i\mathbf{y}_i' - N\overline{\mathbf{y}\mathbf{y}'}}{N-1}$$

$$\mathbf{S} = \frac{\mathbf{Y}'\mathbf{Y} - N\overline{\mathbf{y}\mathbf{y}'}}{N-1} \tag{2.4.11}$$

$$\mathbf{S} = \frac{\mathbf{Y}'\mathbf{Y} - \mathbf{Y}'\mathbf{1}(\mathbf{1}'\mathbf{1})^{-1}\mathbf{1}'\mathbf{Y}}{N-1}$$

$$\mathbf{S} = \frac{\mathbf{Y}'[\mathbf{I} - \mathbf{1}(\mathbf{1}'\mathbf{1})^{-1}\mathbf{1}']\mathbf{Y}}{N-1}$$

where **1** denotes a vector of N 1s.

As in univariate analysis, linear combinations of multivariate random variables are again normally distributed. In particular:

If $\mathbf{Y}_i \sim IN_p(\boldsymbol{\mu}, \mathbf{\Sigma})$, for $i = 1, \ldots, N$, then

$$\mathbf{Y} = \sum_i a_i\mathbf{Y}_i \sim N_p\left(\sum_i a_i\boldsymbol{\mu}, \sum_i a_i^2\mathbf{\Sigma}\right) \tag{2.4.12}$$

for fixed constants a_i.

If $\mathbf{Y} = \mathbf{A}\mathbf{X} + \mathbf{b}$, where $\mathbf{X} \sim N_p(\boldsymbol{\mu}, \mathbf{\Sigma})$, $\mathbf{A}$ is a $q \times p$ constant matrix $(q \leq p)$ of rank q, and $\mathbf{b}$ is a constant q-vector, then

$$\mathbf{Y} \sim N_q(\mathbf{A}\boldsymbol{\mu} + \mathbf{b}, \mathbf{A}\mathbf{\Sigma}\mathbf{A}') \tag{2.4.13}$$

As an application of (2.4.12), let $a_i = 1/N$; then

$$\overline{\mathbf{Y}} = \frac{\sum_i \mathbf{Y}_i}{N} \sim N_p\left(\boldsymbol{\mu}, \frac{\mathbf{\Sigma}}{N}\right) \tag{2.4.14}$$

Illustrating (2.4.13), suppose $\mathbf{Y} \sim N_p(\boldsymbol{\mu}, \boldsymbol{\Sigma})$, and Dia $[\sigma_i]$ is a diagonal matrix whose diagonal elements are the square roots of the diagonal elements of $\boldsymbol{\Sigma}$. By setting

$$\mathbf{Z} = \text{Dia}\,[\sigma_i^{-1}](\mathbf{Y} - \boldsymbol{\mu}) \tag{2.4.15}$$

a vector of standardized variables is obtained. The mean and variance-covariance matrices of $\mathbf{Z}$ are

$$E(\mathbf{Z}) = \mathbf{0}$$
$$V(\mathbf{Z}) = \text{Dia}\,[\sigma_i^{-1}]V(\mathbf{Y})\,\text{Dia}\,[\sigma_i^{-1}] = \mathbf{P}$$

By (2.4.13), the random vector of standardized variables $\mathbf{Z}$ has a multivariate normal distribution with mean vector $\mathbf{0}$ and variance-covariance matrix equal to a correlation matrix $\mathbf{P}$. That is,

$$f(\mathbf{z}) = (2\pi)^{-p/2}|\mathbf{P}|^{-1/2}e^{-\mathbf{z}'\mathbf{P}^{-1}\mathbf{z}/2} \qquad -\infty < z_i < \infty \tag{2.4.16}$$

Employing the techniques used to find the maximum-likelihood estimator of $\boldsymbol{\Sigma}$, the maximum-likelihood estimator of $\mathbf{P}$ is

$$\begin{aligned}\hat{\mathbf{P}} &= \frac{1}{N}\sum_{i=1}^{N} \mathbf{z}_i\mathbf{z}_i' \\ &= \frac{1}{N}\sum_{i=1}^{N}\left(\frac{y_{ij} - \mu_j}{\sigma_j}\right)\left(\frac{y_{ik} - \mu_k}{\sigma_k}\right) \\ &= [\hat{\rho}_{ij}]\end{aligned} \tag{2.4.17}$$

such that

$$\frac{1}{N}\sum_{i=1}^{N}\left(\frac{y_{ij} - \mu_j}{\sigma_j}\right)^2 = 1 \qquad \text{for } j = 1, \ldots, p$$

The maximum-likelihood estimator $\hat{\rho}_{ij}$ of ρ_{ij} is not the same as the *sample correlation coefficient*

$$\hat{\rho}_{ij} = r_{ij} = \frac{\hat{\sigma}_{ij}}{\sqrt{\hat{\sigma}_{ii}\hat{\sigma}_{jj}}}$$

which is the maximum-likelihood estimator of ρ_{ij}, assuming all elements of $\boldsymbol{\mu}$ and the diagonal elements of $\boldsymbol{\Sigma}$ are unknown (see Kendall and Stuart, 1961, p. 294). A convenient method for obtaining $\hat{\mathbf{P}} = \mathbf{R}$, when $\boldsymbol{\mu}$ and $\boldsymbol{\Sigma}$ are unknown, is with the formula

$$\begin{aligned}\mathbf{R} &= \text{Dia}\,[s_i^{-1}]\mathbf{S}\,\text{Dia}\,[s_i^{-1}] = [r_{ij}] \\ &= (\text{Dia}\,\mathbf{S})^{-1/2}\,\mathbf{S}\,(\text{Dia}\,\mathbf{S})^{-1/2}\end{aligned} \tag{2.4.18}$$

where $\mathbf{S}$ is defined as in (2.4.10).

EXERCISES 2.4

1. Suppose $\mathbf{Y} \sim N_4(\boldsymbol{\mu}, \boldsymbol{\Sigma})$, where

$$\boldsymbol{\mu} = \begin{bmatrix} 1 \\ 2 \\ 3 \\ 4 \end{bmatrix} \quad \text{and} \quad \boldsymbol{\Sigma} = \begin{bmatrix} 3 & 1 & 0 & 0 \\ 1 & 4 & 0 & 0 \\ 0 & 0 & 1 & 4 \\ 0 & 0 & 4 & 20 \end{bmatrix}$$

   a. Find the joint distribution of Y_1 and Y_2 and of Y_2 and Y_4.

   b. Determine ρ_{12} and ρ_{24}.

   c. Find the length of the semimajor axis of the ellipse associated with the multivariate normal random variable $\mathbf{Y}$ for a constant $k = 100$.

2. Determine the normal density function associated with the quadratic form

$$Q = 2y_1^2 + y_2^2 + 3y_3^2 + 2y_1y_2 + 2y_1y_3$$

3. Rework Exercise 2 for X and Y dependent and $\rho = .25$.

4. For the bivariate normal density function (2.4.7), graph the ellipse of the exponent for $\mu_1 = \mu_2 = 0$, $\sigma_1^2 = \sigma_2^2 = 1$, $Q = 2$, and $\rho = 0$, .5, and .9.

5. a. Give an example of two univariate normal random variables that have a nonnormal bivariate distribution.

   b. Give an example of two uncorrelated random variables that are dependent.

2.5 CONDITIONAL DISTRIBUTIONS AND CORRELATION

In this section we will review some results of conditional distributions as they relate to the theory of correlation and regression. The implications of the theory presented here and in the other sections of this chapter are relevant to the applications discussed in the remaining chapters of the text. However, if you are mainly interested in applications, you may skip this material without loss of continuity and return to it as needed throughout the text.

The Bivariate Normal. In the bivariate case,

$$\mathbf{U} = \begin{bmatrix} Y \\ X \end{bmatrix} \sim N_2\left\{\begin{bmatrix} \mu_y \\ \mu_x \end{bmatrix}, \begin{bmatrix} \sigma_y^2 & \rho\sigma_y\sigma_x \\ \rho\sigma_y\sigma_x & \sigma_x^2 \end{bmatrix}\right\} \tag{2.5.1}$$

so that

$$f(y, x) = \frac{\exp\left\{\dfrac{-1}{2(1-\rho^2)}\left[\left(\dfrac{y-\mu_y}{\sigma_y}\right)^2 - 2\rho\left(\dfrac{y-\mu_y}{\sigma_y}\right)\left(\dfrac{x-\mu_x}{\sigma_x}\right) + \left(\dfrac{x-\mu_x}{\sigma_x}\right)^2\right]\right\}}{2\pi\sigma_y\sigma_x\sqrt{1-\rho^2}}$$

The *conditional* density function of Y, for fixed values of $X = x$, is defined by $f(y|x) = f(x, y)/f(x)$. However, by using (2.4.13), you can see that the marginal distributions of a bivariate normal distribution are each normal, so that

$$f(x) = \frac{1}{\sigma_x\sqrt{2\pi}} e^{-(x-\mu_x)^2/2\sigma^2} \tag{2.5.2}$$

Letting $z_x = (x - \mu_x)/\sigma_x$ and $z_y = (y - \mu_y)/\sigma_y$,

$$f(y|x) = \frac{\dfrac{\exp - \dfrac{1}{2(1-\rho)^2}(z_y^2 - 2\rho z_y z_x + z_x^2)}{2\pi\sigma_y\sigma_x\sqrt{1-\rho^2}}}{\dfrac{\exp - z_x^2/2}{\sigma_x\sqrt{2\pi}}}$$

$$= \frac{\exp -(1/2)[(z_y - \rho z_x)/\sqrt{1-\rho^2}]^2}{\sigma_y\sqrt{2\pi}\sqrt{1-\rho^2}}$$

$$= \frac{\exp -(1/2)\{[y - \mu_y - \rho(\sigma_y/\sigma_x)(x - \mu_x)]/\sigma_y\sqrt{(1-\rho^2)}\}^2}{\sigma_y\sqrt{2\pi}\sqrt{1-\rho^2}}$$

From (2.5.1) and (2.5.2), the conditional distribution of Y, given $X = x$, is normal with mean and variance, respectively,

$$\begin{aligned} \mu_{y\cdot x} &= \mu_y + \rho\left(\frac{\sigma_y}{\sigma_x}\right)(x - \mu_x) \\ \sigma_{y\cdot x}^2 &= \sigma_y^2(1 - \rho^2) \end{aligned} \tag{2.5.3}$$

The mean of the conditional normal distribution, $\mu_{y\cdot x}$, is seen to be linear in x, and is called the *regression function* of y on x. The variance of the conditional normal distribution, $\sigma_{y\cdot x}^2$, is independent of x, and is called the *partial* (*residual*) *variance* of y, given x. By the symmetry of the problem, a similar expression for the regression function and partial variance hold for y and x interchanged. Assuming bivariate normality, the roles of Y and X can be interchanged, and two linear regression functions result.

Using (2.5.3) and the fact that Y and X have a bivariate normal density function, the value of Y, for fixed $X = x$, can be predicted by using $\mu_{y\cdot x}$ if the parameters associated with the regression function are replaced by their maximum-likelihood estimates. That is,

$$\begin{aligned} \hat{\mu}_{y\cdot x} &= \hat{\mu}_y + \hat{\rho}\left(\frac{\hat{\sigma}_y}{\hat{\sigma}_x}\right)(x - \hat{\mu}_x) \\ &= \left[\hat{\mu}_y - \hat{\rho}\left(\frac{\hat{\sigma}_y}{\hat{\sigma}_x}\right)\hat{\mu}_x\right] + \hat{\rho}\left(\frac{\hat{\sigma}_y}{\hat{\sigma}_x}\right)x \\ &= \hat{\alpha} + \hat{\beta}x \end{aligned} \tag{2.5.4}$$

which is an estimate of the mean of the conditional distribution of Y for fixed X.

In practice it is unusual to assume that Y and X have a bivariate normal distribution for regression studies, so that (2.5.4) is not applicable. The values of

X are usually selected before an experiment is conducted. Let $x_1, x_2, \ldots, x_N$ denote the predetermined values of X. Corresponding to the values of X, let $Y_1, Y_2, \ldots, Y_N$ be conditional independent normally distributed random variables such that the regression function is linear in x with mean $E(Y_i) = \alpha + \beta x_i$ $(i = 1, \ldots, N)$ and common variance σ^2. Then the conditional density function of Y_i, given x_i, is

$$f(y_i|x_i) = \frac{1}{\sigma\sqrt{2\pi}} e^{-(y_i - \alpha - \beta x_i)^2/2\sigma^2}$$

By letting $\mu_{y \cdot x} = \alpha + \beta x$, the problem of predicting y from x is reduced to estimating the parameters α and β. The maximum-likelihood estimates for α and β agree with the estimates given in (2.5.4). This correspondence is seen to depend on linearity and normality. Another method for securing estimates of α and β is the method of least squares as illustrated in (1.8.22). However, since this procedure does not assume anything about the joint distribution of X and Y for a general distribution, no relationship need exist between (2.5.4) and the estimates obtained for α and β by employing the least-squares theory.

The Multivariate Normal. Extending these results to the multivariate case becomes more involved. Employing the notation used in Section 2.2, let

$$\mathbf{U} = \begin{bmatrix} \mathbf{Y} \\ \mathbf{X} \end{bmatrix} \sim N_{p+q}\left\{\begin{bmatrix} \boldsymbol{\mu}_1 \\ \boldsymbol{\mu}_2 \end{bmatrix}, \quad \boldsymbol{\Sigma} = \begin{bmatrix} \boldsymbol{\Sigma}_{11} & \boldsymbol{\Sigma}_{12} \\ \boldsymbol{\Sigma}_{21} & \boldsymbol{\Sigma}_{22} \end{bmatrix}\right\}$$

where $\mathbf{Y}$ is a random p-vector, $\mathbf{X}$ is a random q-vector and $|\boldsymbol{\Sigma}_{22}| \neq 0$. Then the conditional distribution of $\mathbf{Y}$, given $\mathbf{X} = \mathbf{x}$, is multivariate normal with mean and variance-covariance matrix

$$\begin{aligned} \boldsymbol{\mu}_{1\cdot 2} &= \boldsymbol{\mu}_1 + \boldsymbol{\Sigma}_{12}\boldsymbol{\Sigma}_{22}^{-1}(\mathbf{x} - \boldsymbol{\mu}_2) \\ \boldsymbol{\Sigma}_{1\cdot 2} &= \boldsymbol{\Sigma}_{11} - \boldsymbol{\Sigma}_{12}\boldsymbol{\Sigma}_{22}^{-1}\boldsymbol{\Sigma}_{21} \end{aligned} \tag{2.5.5}$$

respectively.

To prove (2.5.5), two results of partitioned matrices are invoked—expressions (1.5.22) and (1.5.28). Following the univariate proof exactly with

$$\begin{aligned} f(\mathbf{y}, \mathbf{x}) &= (2\pi)^{-(p+q)/2}\begin{vmatrix} \boldsymbol{\Sigma}_{11} & \boldsymbol{\Sigma}_{12} \\ \boldsymbol{\Sigma}_{21} & \boldsymbol{\Sigma}_{22} \end{vmatrix}^{-1/2} \exp -\frac{1}{2}\begin{bmatrix} \mathbf{y} - \boldsymbol{\mu}_1 \\ \mathbf{x} - \boldsymbol{\mu}_2 \end{bmatrix}'\begin{bmatrix} \boldsymbol{\Sigma}_{11} & \boldsymbol{\Sigma}_{12} \\ \boldsymbol{\Sigma}_{21} & \boldsymbol{\Sigma}_{22} \end{bmatrix}^{-1}\begin{bmatrix} \mathbf{y} - \boldsymbol{\mu}_1 \\ \mathbf{x} - \boldsymbol{\mu}_2 \end{bmatrix} \\ &= (2\pi)^{-(p+q)/2}|\boldsymbol{\Sigma}_{22}|^{-1/2}|\mathbf{A}|^{-1/2} \exp -\frac{1}{2}\begin{bmatrix} \mathbf{y} - \boldsymbol{\mu}_1 \\ \mathbf{x} - \boldsymbol{\mu}_2 \end{bmatrix}'\boldsymbol{\Sigma}^{-1}\begin{bmatrix} \mathbf{y} - \boldsymbol{\mu}_1 \\ \mathbf{x} - \boldsymbol{\mu}_2 \end{bmatrix} \end{aligned}$$

with

$$\mathbf{A} = \boldsymbol{\Sigma}_{11} - \boldsymbol{\Sigma}_{12}\boldsymbol{\Sigma}_{22}^{-1}\boldsymbol{\Sigma}_{21}$$

$$\boldsymbol{\Sigma}^{-1} = \begin{bmatrix} \mathbf{A}^{-1} & -\mathbf{A}^{-1}\boldsymbol{\Sigma}_{12}\boldsymbol{\Sigma}_{22}^{-1} \\ -\boldsymbol{\Sigma}_{22}^{-1}\boldsymbol{\Sigma}_{21}\mathbf{A}^{-1} & \boldsymbol{\Sigma}_{22}^{-1} + \boldsymbol{\Sigma}_{22}^{-1}\boldsymbol{\Sigma}_{21}\mathbf{A}^{-1}\boldsymbol{\Sigma}_{12}\boldsymbol{\Sigma}_{22}^{-1} \end{bmatrix}$$

and

$$f(\mathbf{x}) = (2\pi)^{-q/2}|\boldsymbol{\Sigma}_{22}| \exp -[(\mathbf{x} - \boldsymbol{\mu}_2)'\boldsymbol{\Sigma}_{22}^{-1}(\mathbf{x} - \boldsymbol{\mu}_2)]/2$$

the conditional distribution of **Y**, given **X** = **x**, becomes

$$f(\mathbf{y}|\mathbf{x}) = (2\pi)^{-p/2}|\mathbf{A}|^{-1/2}\, e^{-[(\mathbf{y}-\mathbf{a})'\mathbf{A}^{-1}(\mathbf{y}-\mathbf{a})]/2}$$

where

$$\mathbf{a} = \mu_1 + \boldsymbol{\Sigma}_{12}\boldsymbol{\Sigma}_{22}^{-1}(\mathbf{x} - \boldsymbol{\mu}_2)$$

proving (2.5.5) if $\mathbf{a}$ is set equal to $\boldsymbol{\mu}_{1\cdot 2}$ and $\mathbf{A} = \boldsymbol{\Sigma}_{1\cdot 2}$.

The derivation of (2.5.5) did not depend on selecting a particular set of q variables for the random vector **X**. Because of the joint multivariate normality assumption, result (2.5.5) holds for any subset of p variables in **Y** and the remaining q variables in **X**. In the simple bivariate case, this amounts to being able to regress **Y** on **X** and **X** on **Y**.

The $p \times q$ matrix $\mathbf{B}' = \boldsymbol{\Sigma}_{12}\boldsymbol{\Sigma}_{22}^{-1}$ is called the *matrix of regression coefficients* of **Y** on **X** and reduces to $\beta = \rho(\sigma_1/\sigma_2)$ in the bivariate case. If **Y** contains a single element,

$$\boldsymbol{\Sigma} = \begin{bmatrix} \boldsymbol{\sigma}_{11} & \boldsymbol{\sigma}'_{12} \\ \boldsymbol{\sigma}_{21} & \boldsymbol{\Sigma}_{22} \end{bmatrix}$$

and the matrix **B** reduces to a vector $\boldsymbol{\beta} = \boldsymbol{\Sigma}_{22}^{-1}\boldsymbol{\sigma}_{21}$, called the vector of *multiple regression coefficients*. The quantity $\boldsymbol{\mu}_1 - \boldsymbol{\Sigma}_{12}\boldsymbol{\Sigma}_{22}^{-1}(\mathbf{x} - \boldsymbol{\mu}_2)$ is known as the general *multivariate regression function* and has the form $\alpha + \beta_1 x_1 + \cdots + \beta_k x_k$ in the multiple regression case. Again, the linearity of regression in this case depends on the joint multinormal assumption, which may be unrealistic in many applied problems (see Chapter 4).

Partial Correlation. The matrix $\boldsymbol{\Sigma}_{1\cdot 2} = \boldsymbol{\Sigma}_{11} - \boldsymbol{\Sigma}_{12}\boldsymbol{\Sigma}_{22}^{-1}\boldsymbol{\Sigma}_{21}$ in (2.5.5) is called the *matrix of partial variances and covariances*. Letting $\sigma_{ij\cdot p+1,\ldots,p+q}$ denote the ijth element in $\boldsymbol{\Sigma}_{1\cdot 2}$, the matrix defined by the elements

$$\rho_{ij\cdot p+1,\ldots,p+q} = \frac{\sigma_{ij\cdot p+1,\ldots,p+q}}{\sqrt{\sigma_{ii\cdot p+1,\ldots,p+q}}\;\sqrt{\sigma_{jj\cdot p+1,\ldots,p+q}}} \qquad (2.5.6)$$

is the *matrix of partial correlations*, where $\rho_{ij\cdot p+1,\ldots,p+q}$ is the partial correlation between Y_i and Y_j holding the variables in the **X** set fixed, held constant. The partial-correlation coefficient allows us to measure the linear dependence of any two variables in the first set by removing the linear association of the variables in the second set with the variables in the first set.

EXAMPLE 2.5.1. Suppose

$$\mathbf{Y} = \begin{bmatrix} y_1 \\ y_2 \end{bmatrix} \qquad \mathbf{X} = x_3$$

Then

$$\rho_{12\cdot 3} = \frac{\rho_{12} - \rho_{13}\rho_{23}}{\sqrt{1 - \rho_{13}^2}\sqrt{1 - \rho_{23}^2}}$$

In this case,

$$\boldsymbol{\Sigma} = \begin{bmatrix} \boldsymbol{\Sigma}_{11} & \boldsymbol{\Sigma}_{12} \\ \boldsymbol{\Sigma}_{21} & \boldsymbol{\Sigma}_{22} \end{bmatrix} = \left[\begin{array}{cc|c} \sigma_{11} & \sigma_{12} & \sigma_{13} \\ \sigma_{21} & \sigma_{22} & \sigma_{23} \\ \hline \sigma_{31} & \sigma_{32} & \sigma_{33} \end{array}\right]$$

and

$$\boldsymbol{\Sigma}_{11} - \boldsymbol{\Sigma}_{12}\boldsymbol{\Sigma}_{22}^{-1}\boldsymbol{\Sigma}_{21} = \begin{bmatrix} \sigma_{11} - \sigma_{13}\sigma_{33}^{-1}\sigma_{31} & \sigma_{12} - \sigma_{13}\sigma_{33}^{-1}\sigma_{32} \\ \sigma_{21} - \sigma_{23}\sigma_{33}^{-1}\sigma_{31} & \sigma_{22} - \sigma_{23}\sigma_{33}^{-1}\sigma_{32} \end{bmatrix}$$

so

$$\begin{aligned} \rho_{12\cdot 3} &= \frac{\sigma_{12\cdot 3}}{\sqrt{\sigma_{11\cdot 3}}\sqrt{\sigma_{22\cdot 3}}} = \frac{\sigma_{12} - \sigma_{13}\sigma_{33}^{-1}\sigma_{32}}{\sqrt{\sigma_{11} - \sigma_{13}\sigma_{33}^{-1}\sigma_{31}}\sqrt{\sigma_{22} - \sigma_{23}\sigma_{33}^{-1}\sigma_{32}}} \\ &= \frac{\sigma_{12} - \sigma_{13}\sigma_{33}^{-1}\sigma_{32}}{\sqrt{\sigma_{11}\sigma_{22}}\sqrt{1 - \dfrac{\sigma_{13}\sigma_{33}^{-1}\sigma_{31}}{\sigma_{11}}}\sqrt{1 - \dfrac{\sigma_{23}\sigma_{33}^{-1}\sigma_{32}}{\sigma_{22}}}} \\ &= \frac{\dfrac{\sigma_{12}}{\sqrt{\sigma_{11}\sigma_{22}}} - \dfrac{\sigma_{13}\sigma_{33}^{-1}\sigma_{31}}{\sqrt{\sigma_{11}\sigma_{22}}}}{\sqrt{1 - \rho_{13}^2}\sqrt{1 - \rho_{23}^2}} = \frac{\rho_{12} - \rho_{13}\rho_{23}}{\sqrt{1 - \rho_{13}^2}\sqrt{1 - \rho_{23}^2}} \end{aligned}$$

In general, Anderson (1958, p. 34) shows that the

$$\text{(2.5.7)} \quad \rho_{ij\cdot p+1,\ldots,p+q} = \frac{\rho_{ij\cdot p+1,\cdots,p+q-1} - \rho_{i(p+q)\cdot p+1,\cdots,p+q-1}\rho_{j(p+q)\cdot p+1,\cdots,p+q-1}}{\sqrt{1 - \rho_{i(p+q).p+1,\ldots,p+q-1}^2}\sqrt{1 - \rho_{j(p+q)\cdot p+1,\ldots,p+q-1}^2}}$$

For example, let $\mathbf{Y}' = [y_1, y_2]$ and $\mathbf{X}' = [x_3, x_4, x_5]$. Then the

$$\rho_{12\cdot 345} = \frac{\rho_{12\cdot 34} - \rho_{15\cdot 34}\rho_{25\cdot 34}}{\sqrt{1 - \rho_{15\cdot 34}^2}\sqrt{1 - \rho_{25\cdot 34}^2}}$$

$$\rho_{12\cdot 34} = \frac{\rho_{12\cdot 3} - \rho_{14\cdot 3}\rho_{24\cdot 3}}{\sqrt{1 - \rho_{14\cdot 3}^2}\sqrt{1 - \rho_{24\cdot 3}^2}}$$

$$\rho_{14\cdot 3} = \frac{\rho_{14} - \rho_{13}\rho_{43}}{\sqrt{1 - \rho_{13}^2}\sqrt{1 - \rho_{43}^2}}$$

so that, working in a recursive manner, any desired partial correlation can be obtained.

When an experimenter can assume joint multinormality, the mean of the conditional distribution may be used for prediction. The difference between the elements of $\mathbf{Y}$ and the values predicted by using the conditional mean vector $\mathbf{W}$, where

$$\text{(2.5.8)} \qquad \mathbf{W} = \mathbf{Y} - \boldsymbol{\mu}_1 - \boldsymbol{\Sigma}_{12}\boldsymbol{\Sigma}_{22}^{-1}(\mathbf{X} - \boldsymbol{\mu}_2) = \mathbf{Y} - \boldsymbol{\mu}_{1\cdot 2}$$

is called the *residual vector*. It represents the errors made in using $\boldsymbol{\mu}_{1\cdot 2}$ to predict $\mathbf{Y}$. In the simple bivariate case, $W = Y - \mu_1 - \rho(\sigma_1/\sigma_2)(X - \mu_2)$. The covariance between Y and W is

$$\begin{aligned}\operatorname{cov}(Y, W) &= E\{[Y - E(Y)][W - E(W)]\} \\ &= E[(Y - \mu_1)W] \\ &= E(YW) \\ &= \sigma_1^2 - \rho\left(\frac{\sigma_1}{\sigma_2}\right)\sigma_{12} \\ &= \sigma_1^2(1 - \rho^2) \\ &= \sigma_{1\cdot 2}^2\end{aligned}$$

Thus the covariance between Y and the residual W is just the variance of the conditional distribution. As $\sigma_{1\cdot 2}^2$ goes to zero ($\sigma_{1\cdot 2}^2 \to 0$), the error in linear prediction is less, since the centroids of the bivariate distribution are long and narrow—cigar shaped. Alternatively, $\sigma_{1\cdot 2}^2 = \sigma_1^2(1 - \rho^2)$, so that, as the measure of error $\sigma_{1\cdot 2}^2 \to 0$, $\rho^2 = (1 - \sigma_{1\cdot 2}^2)/\sigma_1^2 \to 1$. The square of the correlation coefficient, ρ^2, provides a measure of linear association between the random variables Y and X. The covariance between W and X is zero. That is,

$$\begin{aligned}\operatorname{cov}(X, W) &= E\{[X - E(X)][W - E(W)]\} \\ &= E(XW) - E(X)E(W) \\ &= E(XW) \\ &= [E(XY) - E(X)E(Y)] - \rho\left(\frac{\sigma_1}{\sigma_2}\right)\{E(X^2) - [E(X)]^2\} \\ &= \sigma_{12} - \rho\left(\frac{\sigma_1}{\sigma_2}\right)\sigma_2^2 \\ &= 0\end{aligned}$$

This indicates that the residual W and the variate X are uncorrelated or independently distributed because of the normality assumption.

For the multivariate case,

$$\begin{aligned}\operatorname{cov}(\mathbf{Y}, \mathbf{W}) &= \boldsymbol{\Sigma}_{11} - \boldsymbol{\Sigma}_{12}\boldsymbol{\Sigma}_{22}^{-1}\boldsymbol{\Sigma}_{21} = \boldsymbol{\Sigma}_{1\cdot 2} \\ \operatorname{cov}(\mathbf{X}, \mathbf{W}) &= \mathbf{0}\end{aligned} \tag{2.5.9}$$

(The proof of (2.5.9), employing the operators defined in Section 2.2, is left to the reader.) When the random vector $\mathbf{Y}$ contains only one variate, while $\mathbf{X}$ is a random q-vector, $\boldsymbol{\Sigma}_{1\cdot 2}$ reduces to

$$\sigma_{1\cdot 2}^2 = \sigma_1^2 - \boldsymbol{\sigma}_{12}'\boldsymbol{\Sigma}_{22}^{-1}\boldsymbol{\sigma}_{21} = \frac{|\boldsymbol{\Sigma}|}{|\boldsymbol{\Sigma}_{22}|}$$

As above,

$$\begin{aligned}\sigma_{1\cdot 2}^2 &= \sigma_1^2\frac{1 - \boldsymbol{\sigma}_{12}'\boldsymbol{\Sigma}_{22}^{-1}\boldsymbol{\sigma}_{21}}{\sigma_1^2} \\ &= \sigma_1^2(1 - \rho_{0(1,\ldots,q)}^2)\end{aligned}$$

so that $\rho^2_{0(1,\ldots,q)} \to 1$ as $\sigma^2_{1\cdot 2} \to 0$. The quantity $\rho^2_{0(1,\ldots,q)}$ is termed the *square of the multiple correlation* between Y and $\mathbf{X}$. It is the square of the maximum correlation between Y and the linear combination $\boldsymbol{\beta}'\mathbf{X}$. Thus

$$\rho^2_{0(1,\ldots,q)} = \max_{\boldsymbol{\beta}} \operatorname{cor}(Y, \boldsymbol{\beta}'\mathbf{X})^2 \tag{2.5.10}$$

or

$$\rho_{0(1,\ldots,q)} = \max_{\boldsymbol{\beta}} \frac{\boldsymbol{\beta}'\boldsymbol{\sigma}_{21}}{\sigma_1\sqrt{\boldsymbol{\beta}'\boldsymbol{\Sigma}_{22}\boldsymbol{\beta}}}$$

where $\rho_{0(1,\ldots,q)}$ is the *multiple correlation coefficient.*

Multiple Correlation Coefficient. Since the correlation between Y and $\boldsymbol{\beta}'\mathbf{X}$ is the same as that between Y and vectors with unit variance, instead of maximizing over all $\boldsymbol{\beta}$, maximization is restricted to normalized vectors $\boldsymbol{\beta}$ satisfying the constraint that $\boldsymbol{\beta}'\boldsymbol{\Sigma}_{22}\boldsymbol{\beta} = 1$. Hence the following expression may equivalently be maximized:

$$\max_{\boldsymbol{\beta}} \frac{\boldsymbol{\beta}'\sigma_{21}}{\sigma_1} \qquad \text{with } \boldsymbol{\beta}'\boldsymbol{\Sigma}_{22}\boldsymbol{\beta} = 1$$

Employing the procedures of Section 1.8, let

$$F(\boldsymbol{\beta}) = \frac{\boldsymbol{\beta}'\boldsymbol{\sigma}_{21}}{\sigma_1} - \lambda(\boldsymbol{\beta}'\boldsymbol{\Sigma}_{22}\boldsymbol{\beta} - 1)$$

then

$$\frac{\partial F}{\partial \boldsymbol{\beta}} = \frac{\boldsymbol{\sigma}_{21}}{\sigma_1} - 2\lambda\boldsymbol{\Sigma}_{22}\boldsymbol{\beta} \quad \text{and} \quad \frac{\partial F}{\partial \lambda} = \boldsymbol{\beta}'\boldsymbol{\Sigma}_{22}\boldsymbol{\beta} - 1$$

Equating each of the partial derivatives to zero,

$$\boldsymbol{\beta} = \frac{1}{2\lambda\sigma_1}\boldsymbol{\Sigma}_{22}^{-1}\boldsymbol{\sigma}_{21} \quad \text{and} \quad \boldsymbol{\beta}'\boldsymbol{\Sigma}_{22}\boldsymbol{\beta} = 1$$

Substituting $\boldsymbol{\beta} = (1/2\lambda\sigma_1)\boldsymbol{\Sigma}_{22}^{-1}\boldsymbol{\sigma}_{21}$ into the side condition $\boldsymbol{\beta}'\boldsymbol{\Sigma}_{22}\boldsymbol{\beta} = 1$,

$$\frac{\boldsymbol{\sigma}'_{21}\boldsymbol{\Sigma}_{22}^{-1}}{2\lambda\sigma_1}\boldsymbol{\Sigma}_{22}\frac{\boldsymbol{\Sigma}_{22}^{-1}\boldsymbol{\sigma}_{21}}{2\lambda\sigma_1} = 1$$

so that

$$\lambda = \frac{1}{2\sigma_1}\sqrt{\boldsymbol{\sigma}'_{21}\boldsymbol{\Sigma}_{22}^{-1}\boldsymbol{\sigma}_{21}} \quad \text{and} \quad \boldsymbol{\beta} = \frac{\boldsymbol{\Sigma}_{22}^{-1}\boldsymbol{\sigma}_{21}}{\sqrt{\boldsymbol{\sigma}'_{21}\boldsymbol{\Sigma}_{22}^{-1}\boldsymbol{\sigma}_{21}}}$$

such that $\boldsymbol{\beta}'\boldsymbol{\Sigma}_{22}\boldsymbol{\beta} = 1$. Thus

$$\rho_{0(1,\ldots,q)} = \max_{\boldsymbol{\beta}} \frac{\boldsymbol{\beta}'\boldsymbol{\sigma}_{21}}{\sigma_1} = \frac{\boldsymbol{\sigma}'_{21}\boldsymbol{\Sigma}_{22}^{-1}\boldsymbol{\sigma}_{21}}{\sigma_1\sqrt{\boldsymbol{\sigma}'_{21}\boldsymbol{\Sigma}_{22}^{-1}\boldsymbol{\sigma}_{21}}} = \frac{\sqrt{\boldsymbol{\sigma}_{21}\boldsymbol{\Sigma}_{22}^{-1}\boldsymbol{\sigma}_{21}}}{\sigma_1}$$

or

$$\rho^2_{0(1,\ldots,q)} = \frac{\boldsymbol{\sigma}'_{12}\boldsymbol{\Sigma}_{22}^{-1}\boldsymbol{\sigma}_{21}}{\sigma_1^2}$$

as claimed. The condition $\boldsymbol{\beta}'\boldsymbol{\Sigma}_{22}\boldsymbol{\beta} = 1$ was employed to maximize (2.5.10). To verify that no loss of generality was imposed by the constraint, remove the normalization factor $\sqrt{\boldsymbol{\sigma}_{21}'\boldsymbol{\Sigma}_{22}^{-1}\boldsymbol{\sigma}_{21}}$ from $\boldsymbol{\beta}$, then $\boldsymbol{\beta} = \boldsymbol{\Sigma}_{22}^{-1}\boldsymbol{\sigma}_{21}$ and

$$\rho_{0(1,\ldots,q)} \frac{\boldsymbol{\beta}'\boldsymbol{\sigma}_{21}}{\sigma_1\sqrt{\boldsymbol{\beta}'\boldsymbol{\Sigma}_{22}\boldsymbol{\beta}}} = \frac{\boldsymbol{\sigma}_{21}'\boldsymbol{\Sigma}_{22}^{-1}\boldsymbol{\sigma}_{21}}{\sigma_1\sqrt{\boldsymbol{\sigma}_{21}'\boldsymbol{\Sigma}_{22}^{-1}\boldsymbol{\Sigma}_{22}\boldsymbol{\Sigma}_{22}^{-1}\boldsymbol{\sigma}_{21}}} = \frac{\sqrt{\boldsymbol{\sigma}_{21}'\boldsymbol{\Sigma}_{22}^{-1}\boldsymbol{\sigma}_{21}}}{\sigma_1}$$

so that the value of $\rho_{0(1,\ldots,q)}^2$ remains unchanged. By substituting the set of regression coefficients $\boldsymbol{\beta}$, defined by the regression coefficient matrix in the expression $\boldsymbol{\beta}'\mathbf{x}$, the multiple correlation between $\mathbf{Y}$ and $\boldsymbol{\beta}'\mathbf{X}$ is maximized.

Generalizations of the multivariate case are discussed in Chapter 4.

EXERCISES 2.5

1. Suppose that the random vector $\mathbf{Y} \sim N(\mathbf{0}, \boldsymbol{\Sigma})$, where

$$\boldsymbol{\Sigma} = \begin{bmatrix} 2 & 1 & 1 \\ 1 & 1 & 0 \\ 1 & 0 & 3 \end{bmatrix}$$

Find all first-order partials.

2. For the random vector $\mathbf{Y}$ in Exercise 1, Section 2.4, find the following:
   a. The conditional mean and variance of Y_1, given y_2, y_3, and y_4.
   b. The partial-correlation coefficient $\rho_{13.2}$.
   c. The multiple correlation between Y_1 and (Y_2, Y_3, and Y_4).
3. Let

$$\mathbf{U} = \begin{bmatrix} \mathbf{Y} \\ \mathbf{X} \\ \mathbf{Z} \end{bmatrix} \sim N_{p+q+r}\left\{\begin{bmatrix} \boldsymbol{\mu}_1 \\ \boldsymbol{\mu}_2 \\ \boldsymbol{\mu}_3 \end{bmatrix}, \; \boldsymbol{\Sigma} = \begin{bmatrix} \boldsymbol{\Sigma}_{11} & \boldsymbol{\Sigma}_{12} & \boldsymbol{\Sigma}_{13} \\ \boldsymbol{\Sigma}_{21} & \boldsymbol{\Sigma}_{22} & \boldsymbol{\Sigma}_{23} \\ \boldsymbol{\Sigma}_{31} & \boldsymbol{\Sigma}_{32} & \boldsymbol{\Sigma}_{33} \end{bmatrix}\right\}$$

where $\mathbf{Y}$, $\mathbf{X}$, and $\mathbf{Z}$ are $p \times 1$, $q \times 1$, and $r \times 1$ random vectors, respectively. Show that the conditional mean and variance-covariance matrix of $\mathbf{Y}$ and $\mathbf{X}$, given $\mathbf{z}$, are

$$\boldsymbol{\mu}_{.3} = \begin{bmatrix} \boldsymbol{\mu}_1 \\ \boldsymbol{\mu}_2 \end{bmatrix} + \begin{bmatrix} \boldsymbol{\Sigma}_{13}\boldsymbol{\Sigma}_{33}^{-1} \\ \boldsymbol{\Sigma}_{23}\boldsymbol{\Sigma}_{33}^{-1} \end{bmatrix}(\mathbf{z} - \boldsymbol{\mu}_3)$$

$$\boldsymbol{\Sigma}_{.3} = \begin{bmatrix} \boldsymbol{\Sigma}_{11} - \boldsymbol{\Sigma}_{13}\boldsymbol{\Sigma}_{33}^{-1}\boldsymbol{\Sigma}_{31} & \boldsymbol{\Sigma}_{12} - \boldsymbol{\Sigma}_{13}\boldsymbol{\Sigma}_{33}^{-1}\boldsymbol{\Sigma}_{31} \\ \boldsymbol{\Sigma}_{21} - \boldsymbol{\Sigma}_{23}\boldsymbol{\Sigma}_{33}^{-1}\boldsymbol{\Sigma}_{32} & \boldsymbol{\Sigma}_{22} - \boldsymbol{\Sigma}_{23}\boldsymbol{\Sigma}_{33}^{-1}\boldsymbol{\Sigma}_{32} \end{bmatrix}$$

2.6 THE CHI-SQUARE AND WISHART DISTRIBUTIONS

The *chi-square distribution* is obtained from a sum of squares of independent normal zero-one random variables and is fundamental to the study of analysis of variance. In this section we will review the chi-square distribution and generalize the results, in an intuitive manner, to its multivariate analogue.

The Central Chi-Square Distribution. If $Y_1, Y_2, \ldots, Y_p$ are independent normal random variables with mean 0 and variance 1, $Y_i \sim N(0, 1)$, or, employing vector notation, $\mathbf{Y} \sim N_p(\mathbf{0}, \mathbf{I})$, then

$$W = \mathbf{Y}'\mathbf{Y} = \sum_{i=1}^{p} Y_i^2 \sim \chi^2(p) \qquad 0 < W < \infty \tag{2.6.1}$$

That is, $\mathbf{Y}'\mathbf{Y}$ has a central χ^2 distribution with p degrees of freedom.

The Noncentral Chi-Square Distribution. If the random p-vector $\mathbf{Y} \sim N_p(\boldsymbol{\mu}, \sigma^2\mathbf{I})[Y_i \sim IN(\mu_i, \sigma^2)$ for $i = 1, \ldots, p]$, then

$$\frac{\mathbf{Y}'\mathbf{Y}}{\sigma^2} = \frac{\sum_{i=1}^{p} Y_i^2}{\sigma^2} \sim \chi^2(p, \delta) \tag{2.6.2}$$

That is, $\mathbf{Y}'\mathbf{Y}/\sigma^2$ has a noncentral χ^2 distribution with p degrees of freedom and noncentrality parameter $\delta = \boldsymbol{\mu}'\boldsymbol{\mu}/\sigma^2$ that reduces to a central χ^2 distribution, if $\boldsymbol{\mu} = \mathbf{0}$. To obtain the noncentrality parameter from $\mathbf{Y}'\mathbf{Y}/\sigma^2$, substitute for $\mathbf{Y}$ its expected value, $\delta = E(\mathbf{Y}')E(\mathbf{Y})/\sigma^2 = \boldsymbol{\mu}'\boldsymbol{\mu}/\sigma^2$. Special cases of (2.6.2) follow:

(2.6.3) If $\mathbf{Y} \sim N_p(\mathbf{0}, \mathbf{I})$, then $\mathbf{Y}'\mathbf{Y} \sim \chi^2(p)$.
(2.6.4) If $\mathbf{Y} \sim N_p(\boldsymbol{\mu}, \mathbf{I})$, then $\mathbf{Y}'\mathbf{Y} \sim \chi^2(p, \delta)$, with $\delta = \boldsymbol{\mu}'\boldsymbol{\mu}$.

Distribution of Quadratic Forms. The inner product $\mathbf{Y}'\mathbf{Y}$ in (2.6.3) and (2.6.4) is a special case of a quadratic form $\mathbf{Y}'\mathbf{A}\mathbf{Y}$, with $\mathbf{A} = \mathbf{I}$. In general, there is the following important result for the distribution of quadratic forms:

(2.6.5) Let $\mathbf{Y} \sim N_p(\boldsymbol{\mu}, \sigma^2\mathbf{I})$ and $\mathbf{A}$ be a symmetric matrix of rank r, then $\mathbf{Y}'\mathbf{A}\mathbf{Y}/\sigma^2 \sim \chi^2(r, \delta)$, where $\delta = \boldsymbol{\mu}'\mathbf{A}\boldsymbol{\mu}/\sigma^2$, if and only if $\mathbf{A} = \mathbf{A}^2$.

To prove sufficiency, if $\mathbf{A}^2 = \mathbf{A}$, there exists an orthogonal matrix $\mathbf{P}$ from (1.7.11) such that $(\mathbf{P}'\mathbf{A}\mathbf{P})(\mathbf{P}'\mathbf{A}\mathbf{P}) = \mathbf{P}'\mathbf{A}\mathbf{P}$, so that the values λ_i in the decomposition are either 0 or 1. Thus

$$\mathbf{P}'\mathbf{A}\mathbf{P} = \begin{bmatrix} \mathbf{I}_r & \mathbf{0} \\ \mathbf{0} & \mathbf{0} \end{bmatrix} \quad \text{and} \quad \mathbf{A} = \mathbf{P}_1\mathbf{P}_1'$$

where $\mathbf{P}_1$ is a $p \times r$ matrix of rank r. Setting $\mathbf{X} = \mathbf{P}'\mathbf{Y}$,

$$\mathbf{Y}'\mathbf{A}\mathbf{Y} = \mathbf{X}'\mathbf{P}'\mathbf{A}\mathbf{P}\mathbf{X} = \sum_{i=1}^{r} X_i^2$$

By (2.4.13), $\mathbf{X} \sim N_p(\mathbf{P}'\boldsymbol{\mu}, \sigma^2\mathbf{I})$, and by (2.6.2),

$$\frac{\sum_{i=1}^{r} X_i^2}{\sigma^2} \sim \chi^2(r, \delta)$$

where $\delta = \boldsymbol{\mu}'\mathbf{P}_1\mathbf{P}_1'\boldsymbol{\mu}/\sigma^2 = \boldsymbol{\mu}'\mathbf{A}\boldsymbol{\mu}/\sigma^2$. The proof of necessity is given by Rao (1965a, p. 150) and involves moment generating or characteristic functions, which are beyond the scope of this text.

EXAMPLE 2.6.1. As an example of Theorem (2.6.5), let $Y_i \sim IN(\mu, \sigma^2)$, so that $\mathbf{Y} \sim N_p(\boldsymbol{\mu}, \sigma^2\mathbf{I})$, where $\boldsymbol{\mu}' = [\mu, \mu, \ldots, \mu]$ is a p-vector of μ's. Then

$$\frac{(p-1)s^2}{\sigma^2} = \frac{\sum_{i=1}^{p} (Y_i - \bar{Y})^2}{\sigma^2} = \frac{\mathbf{Y}'[\mathbf{I} - \mathbf{1}(\mathbf{1}'\mathbf{1})^{-1}\mathbf{1}']\mathbf{Y}}{\sigma^2}$$

where $\mathbf{1}$ denotes a p-vector of 1s. By letting $\mathbf{C} = \mathbf{1}(\mathbf{1}'\mathbf{1})^{-1}\mathbf{1}' = [1/p]$, the matrix $\mathbf{C}$ is an averaging projection matrix.

$$\mathbf{CC} = [\mathbf{1}(\mathbf{1}'\mathbf{1})^{-1}\mathbf{1}'][\mathbf{1}(\mathbf{1}'\mathbf{1})^{-1}\mathbf{1}'] = \mathbf{1}(\mathbf{1}'\mathbf{1})^{-1}\mathbf{1}' = \mathbf{C}$$

and $\mathbf{C}' = \mathbf{C}$. For an idempotent matrix, the $R(\mathbf{C}) = \mathrm{Tr}(\mathbf{C})$. For $\mathbf{C}$ as defined above, the $\mathrm{Tr}(\mathbf{C}) = 1$, or the $R(\mathbf{C}) = 1$. With $\mathbf{A} = \mathbf{I}_p - \mathbf{C}$, $\mathbf{A}^2 = \mathbf{A}$, and the $R(\mathbf{A}) = \mathrm{Tr}(\mathbf{A}) = p - 1$. By (2.6.5),

$$\frac{(p-1)s^2}{\sigma^2} \sim \chi^2(p-1, \delta)$$

The noncentrality parameter is obtained by evaluating

$$\begin{aligned}
\delta &= \frac{E(\mathbf{Y}')(\mathbf{I} - \mathbf{C})E(\mathbf{Y})}{\sigma^2} \\
&= \frac{\boldsymbol{\mu}'\boldsymbol{\mu} - \left(\sum_i \mu\right)p^{-1}\left(\sum_i \mu\right)}{\sigma^2} \\
&= \frac{p\mu - (p\mu)p^{-1}(p\mu)}{\sigma^2} \\
&= \frac{p\mu^2 - p\mu^2}{\sigma^2} \\
&= 0
\end{aligned}$$

or $(p-1)s^2/\sigma^2$ has a central χ^2 distribution with $p - 1$ degrees of freedom:

$$\frac{(p-1)s^2}{\sigma^2} \sim \chi^2(p-1, \delta = 0)$$

The generalization of (2.6.5) to random vectors with dependent elements extends in a natural manner as follows:

(2.6.6) If $\mathbf{Y} \sim N_p(\boldsymbol{\mu}, \boldsymbol{\Sigma})$, then the quadratic form $\mathbf{Y}'\mathbf{AY} \sim \chi^2(r, \delta)$, where $\delta = \boldsymbol{\mu}'\mathbf{A}\boldsymbol{\mu}$ and the $R(\mathbf{A}) = r$, if and only if $\mathbf{A\Sigma A} = \mathbf{A}$ or if $\mathbf{A\Sigma}$ is idempotent.

To prove this, let $\mathbf{X} = \mathbf{F}^{-1}\mathbf{Y}$ and $\mathbf{FF}' = \boldsymbol{\Sigma}$. Then

$$\mathbf{X} \sim N_p[\mathbf{F}^{-1}\boldsymbol{\mu}, \mathbf{F}^{-1}\boldsymbol{\Sigma}(\mathbf{F}')^{-1} = \mathbf{I}] \quad \text{and} \quad \mathbf{Y}'\mathbf{AY} = \mathbf{X}'\mathbf{F}'\mathbf{AFX} \sim \chi^2(r, \delta)$$

by (2.6.5) if and only if $(\mathbf{F}'\mathbf{AF})(\mathbf{F}'\mathbf{AF}) = \mathbf{F}'\mathbf{AF}$ or $\mathbf{A\Sigma A} = \mathbf{A}$.

EXAMPLE 2.6.2 One important illustration of (2.6.6) is the determination of the distribution of $X^2 = N(\bar{\mathbf{Y}} - \boldsymbol{\mu})'\boldsymbol{\Sigma}^{-1}(\bar{\mathbf{Y}} - \boldsymbol{\mu})$. Suppose $\mathbf{Y} \sim N_p(\boldsymbol{\mu}, \boldsymbol{\Sigma})$, then $\bar{\mathbf{Y}} \sim N_p(\boldsymbol{\mu}, \boldsymbol{\Sigma}/N)$ and $\sqrt{N}(\bar{\mathbf{Y}} - \boldsymbol{\mu}) \sim N_p(\mathbf{0}, \boldsymbol{\Sigma})$. Hence, by (2.6.6),

$$X^2 = N(\bar{\mathbf{Y}} - \boldsymbol{\mu})'\boldsymbol{\Sigma}^{-1}(\bar{\mathbf{Y}} - \boldsymbol{\mu}) \sim \chi^2(p, \delta = 0)$$

since $\boldsymbol{\Sigma}^{-1}\boldsymbol{\Sigma}\boldsymbol{\Sigma}^{-1} = \boldsymbol{\Sigma}$. The statistic X^2 can be modified to test the hypothesis that $\boldsymbol{\mu} = \boldsymbol{\mu}_0$, when $\boldsymbol{\Sigma}$ is known and $\boldsymbol{\mu}_0$ is a specified vector. This follows from the fact that

$$X_0^2 = N(\bar{\mathbf{Y}} - \boldsymbol{\mu}_0)'\boldsymbol{\Sigma}^{-1}(\bar{\mathbf{Y}} - \boldsymbol{\mu}_0) \sim \chi^2(p, \delta)$$

with noncentrality parameter $\delta = N(\boldsymbol{\mu} - \boldsymbol{\mu}_0)'\boldsymbol{\Sigma}^{-1}(\boldsymbol{\mu} - \boldsymbol{\mu}_0)$.

Some special cases of (2.6.5) and (2.6.6) are summarized:

(2.6.7a) If $\mathbf{Y} \sim N_p(\mathbf{0}, \mathbf{I})$, then $\mathbf{Y}'\mathbf{A}\mathbf{Y} \sim \chi^2(r, \delta = 0)$ if and only if $\mathbf{A}^2 = \mathbf{A}$ and the $R(\mathbf{A}) = r$.

(2.6.7b) If $\mathbf{Y} \sim N_p(\mathbf{0}, \boldsymbol{\Sigma})$, then $\mathbf{Y}'\mathbf{A}\mathbf{Y} \sim \chi^2(r, \delta = 0)$ if and only if $\mathbf{A}\boldsymbol{\Sigma}$ is idempotent and the $R(\mathbf{A}) = r$.

(2.6.7c) If $\mathbf{Y} \sim N_p(\boldsymbol{\mu}, \mathbf{I})$, then $\mathbf{Y}'\mathbf{A}\mathbf{Y} \sim \chi^2(r, \boldsymbol{\mu}'\mathbf{A}\boldsymbol{\mu})$ if and only if $\mathbf{A}^2 = \mathbf{A}$ and the $R(\mathbf{A}) = r$.

Before discussing the independence of quadratic forms, we will look at a general theorem to determine the expectation of a quadratic form:

(2.6.8) If $\mathbf{Y} \sim N_p(\boldsymbol{\mu}, \boldsymbol{\Sigma})$, then $E(\mathbf{Y}'\mathbf{A}\mathbf{Y}) = \mathrm{Tr}(\mathbf{A}\boldsymbol{\Sigma}) + \boldsymbol{\mu}'\mathbf{A}\boldsymbol{\mu}$.

To prove (2.6.8), notice that the

$$\begin{aligned} E(\mathbf{Y}'\mathbf{A}\mathbf{Y}) &= E[\mathrm{Tr}(\mathbf{A}\mathbf{Y}\mathbf{Y}')] \\ &= \mathrm{Tr}[\mathbf{A}E(\mathbf{Y}\mathbf{Y}')] \\ &= \mathrm{Tr}[\mathbf{A}(\boldsymbol{\Sigma} + \boldsymbol{\mu}\boldsymbol{\mu}')] \\ &= \mathrm{Tr}(\mathbf{A}\boldsymbol{\Sigma}) + \mathrm{Tr}(\mathbf{A}\boldsymbol{\mu}\boldsymbol{\mu}') \\ &= \mathrm{Tr}(\mathbf{A}\boldsymbol{\Sigma}) + \boldsymbol{\mu}'\mathbf{A}\boldsymbol{\mu} \end{aligned}$$

From the proof of (2.6.8), we see that $\mathbf{Y}$ need not be normally distributed. Furthermore, by setting $\mathbf{X} = \mathbf{Y} - \boldsymbol{\mu}$, $\mathbf{X} \sim N_p(\mathbf{0}, \boldsymbol{\Sigma})$ and $E(\mathbf{X}'\mathbf{A}\mathbf{X}) = \mathrm{Tr}(\mathbf{A}\boldsymbol{\Sigma})$, $V(\mathbf{X}'\mathbf{A}\mathbf{X}) = 2\,\mathrm{Tr}[(\mathbf{A}\boldsymbol{\Sigma})^2]$.

Independence of Quadratic Forms. The preceding theorems are applied to a single quadratic form; however, in analysis of variance, one quadratic form is usually expressed as a sum of independent quadratic forms. Fundamental theorems for the sum of independent quadratic forms are therefore summarized. These theorems allow for a geometric interpretation of the analysis of variance. The first is Cochran's theorem (Cochran, 1934), which was generalized by Madow (1940).

(2.6.9) If $\mathbf{Y} \sim N_p(\boldsymbol{\mu}, \mathbf{I})$ and

$$\mathbf{Y}'\mathbf{Y} = \sum_{i=1}^{k} \mathbf{Y}'\mathbf{A}_i\mathbf{Y} \qquad \text{where } R(\mathbf{A}_i) = r_i \quad \text{and} \quad \sum_{i=1}^{k} \mathbf{A}_i = \mathbf{I}_p$$

then the quadratic forms $\mathbf{Y}'\mathbf{A}_i\mathbf{Y} \sim \chi^2(r_i, \delta_i)$, where $\delta_i = \boldsymbol{\mu}'\mathbf{A}_i\boldsymbol{\mu}$ are statistically independent for all i if and only if

$$\sum_{i=1}^{k} r_i = p \quad \text{or} \quad \sum_i R(\mathbf{A}_i) = R\left(\sum_i \mathbf{A}_i\right)$$

Graybill and Marsaglia (1957), Hogg and Craig (1958), Banarjee (1964), and Loynes (1966) extend Cochran's theorem to the following:

(2.6.10) If $\mathbf{Y} \sim N_p(\boldsymbol{\mu}, \boldsymbol{\Sigma})$,

$$\mathbf{Y}'\mathbf{A}\mathbf{Y} = \sum_{i=1}^{k} \mathbf{Y}'\mathbf{A}_i\mathbf{Y}$$

such that

$$\mathbf{A} = \sum_{i=1}^{k} \mathbf{A}_i$$

the $R(\mathbf{A}) = r$, and the $R(\mathbf{A}_i) = r_i$, then

(1) $\mathbf{Y}'\mathbf{A}_i\mathbf{Y} \sim \chi^2(r_i, \boldsymbol{\mu}'\mathbf{A}_i\boldsymbol{\mu})$,
(2) $\mathbf{Y}'\mathbf{A}_i\mathbf{Y}$ and $\mathbf{Y}'\mathbf{A}_j\mathbf{Y}$ are statistically independent if $i \neq j$, and
(3) $\mathbf{Y}'\mathbf{A}\mathbf{Y} \sim \chi^2(r, \boldsymbol{\mu}'\mathbf{A}\boldsymbol{\mu})$ if and only if any two of the following three conditions hold
  (a) each $\mathbf{A}_i\boldsymbol{\Sigma}$ is idempotent or $\mathbf{A}_i\boldsymbol{\Sigma}\mathbf{A}_i = \mathbf{A}_i$,
  (b) $\mathbf{A}_i\boldsymbol{\Sigma}\mathbf{A}_j = \mathbf{0}$ for $i \neq j$, and
  (c) $\mathbf{A}\boldsymbol{\Sigma}$ is idempotent or $\mathbf{A}\boldsymbol{\Sigma}\mathbf{A} = \mathbf{A}$, *or* $\mathbf{Y}'\mathbf{A}\mathbf{Y} \sim \chi^2(r, \boldsymbol{\mu}'\mathbf{A}\boldsymbol{\mu})$ if and only if
  (d) $r = \sum_{i=1}^{k} r_i$ and $\mathbf{A}\boldsymbol{\Sigma}$ is idempotent or $\mathbf{A}\boldsymbol{\Sigma}\mathbf{A} = \mathbf{A}$.

The proof of (2.6.10) is involved. The interested reader is referred to Loynes (1966). For further detail on the distribution of quadratic forms, read Graybill (1961, Chap. 4).

EXAMPLE 2.6.3. As an example of (2.6.10), we can easily show that the sample mean and variance of a random sample from a normal population are independent. Let $\mathbf{Y} \sim N_p(\mu\mathbf{1}, \sigma^2\mathbf{I})$, and let $\mathbf{C} = [1/p]$ denote the averaging projection matrix. Consider

$$\frac{\mathbf{Y}'\mathbf{I}\mathbf{Y}}{\sigma^2} = \frac{\mathbf{Y}'\mathbf{C}\mathbf{Y}}{\sigma^2} + \frac{\mathbf{Y}'(\mathbf{I}-\mathbf{C})\mathbf{Y}}{\sigma^2} \quad \text{or} \quad \frac{\sum_{i=1}^{p} Y_i^2}{\sigma^2} = \frac{p\bar{Y}^2}{\sigma^2} + \frac{(p-1)s^2}{\sigma^2}$$

$\bar{Y}$ and s^2 are statistically independent since $R(\mathbf{I}) = p = R(\mathbf{C}) + R(\mathbf{I}-\mathbf{C}) = 1 + (p-1)$ and $\mathbf{I} = \mathbf{C} + (\mathbf{I} - \mathbf{C})$. Furthermore, the distribution of the sum of squares is evident; that is,

$$\chi^2\left(p, \frac{p\mu^2}{\sigma^2}\right) = \chi^2\left(1, \frac{p\mu^2}{\sigma^2}\right) + \chi^2(p-1, 0)$$

EXAMPLE 2.6.4. Let $\mathbf{y} \sim N_4(\boldsymbol{\mu}, \sigma^2\mathbf{I})$,

$$\mathbf{A} = \begin{bmatrix} 1 & 1 & 0 \\ 1 & 1 & 0 \\ 1 & 0 & 1 \\ 1 & 0 & 1 \end{bmatrix} = [\mathbf{A}_1 \quad \mathbf{A}_2], \quad \text{and} \quad \mathbf{y} = \begin{bmatrix} y_{11} \\ y_{12} \\ y_{21} \\ y_{22} \end{bmatrix}$$

where

$$\mathbf{A}_1 = \begin{bmatrix} 1 \\ 1 \\ 1 \\ 1 \end{bmatrix} \quad \text{and} \quad \mathbf{A}_2 = \begin{bmatrix} 1 & 0 \\ 1 & 0 \\ 0 & 1 \\ 0 & 1 \end{bmatrix}$$

In Example 1.6.11, projection matrices of the form

$$\mathbf{P}_1 = \mathbf{A}_1(\mathbf{A}_1'\mathbf{A}_1)^-\mathbf{A}_1$$
$$\mathbf{P}_2 = \mathbf{A}(\mathbf{A}'\mathbf{A})^-\mathbf{A}' - \mathbf{A}_1(\mathbf{A}_1'\mathbf{A}_1)^-\mathbf{A}_1'$$
$$\mathbf{P}_3 = \mathbf{I} - \mathbf{A}(\mathbf{A}'\mathbf{A})^-\mathbf{A}'$$

were employed to project the observation vector $\mathbf{y}$ onto orthogonal subspaces (see Example 1.3.8). In Section 1.5 it was indicated that the length squared of an observation vector could be partitioned into orthogonal sums of squares by using projection matrices where each length squared was a quadratic form. To determine the distribution of the squared lengths, (2.6.10) is used.

The projection matrices $\mathbf{P}_1$, $\mathbf{P}_2$, and $\mathbf{P}_3$ are idempotent and $\mathbf{I} = \mathbf{P}_1 + \mathbf{P}_2 + \mathbf{P}_3$. Furthermore, the $R(\mathbf{P}_1) = 1$, the $R(\mathbf{P}_2) = 1$, and the $R(\mathbf{P}_3) = 2$, so that $R(\mathbf{I}) = R(\mathbf{P}_1) + R(\mathbf{P}_2) + R(\mathbf{P}_3)$. Forming the equation for the quadratic forms,

$$\mathbf{y}'\mathbf{y} = \mathbf{y}'\mathbf{P}_1\mathbf{y} + \mathbf{y}'\mathbf{P}_2\mathbf{y} + \mathbf{y}'\mathbf{P}_3\mathbf{y}$$

or

$$\|\mathbf{y}\|^2 = \|\mathbf{P}_1\mathbf{y}\|^2 + \|\mathbf{P}_2\mathbf{y}\|^2 + \|\mathbf{P}_3\mathbf{y}\|^2$$

since $\mathbf{P}_i\mathbf{P}_j = \mathbf{0}$, for $i \neq j$. The quadratic forms that are constructed are familiar. From Example 1.6.11,

$$\mathbf{A}(\mathbf{A}'\mathbf{A})^-\mathbf{A}' = \begin{bmatrix} 1/2 & 1/2 & 0 & 0 \\ 1/2 & 1/2 & 0 & 0 \\ 0 & 0 & 1/2 & 1/2 \\ 0 & 0 & 1/2 & 1/2 \end{bmatrix}$$

and

$$\mathbf{A}(\mathbf{A}'\mathbf{A})^-\mathbf{A}'\mathbf{y} = \begin{bmatrix} y_{1.} \\ y_{1.} \\ y_{2.} \\ y_{2.} \end{bmatrix}$$

so that

$$\begin{aligned}
\mathbf{y}'\mathbf{A}(\mathbf{A}'\mathbf{A})^{-}\mathbf{A}'\mathbf{y} &= y_{11}y_{1.} + y_{12}y_{1.} + y_{21}y_{2.} + y_{22}y_{2.} \\
&= y_{1.}(y_{11} + y_{12}) + y_{2.}(y_{21} + y_{22}) \\
&= 2y_{1.}^2 + 2y_{2.}^2 \\
&= \sum_i 2y_{i.}^2
\end{aligned}$$

Since $\mathbf{y}'\mathbf{I}\mathbf{y} = \Sigma_i \Sigma_j y_{ij}^2$, the

$$\|\mathbf{P}_3\mathbf{y}\|^2 = \mathbf{y}'\mathbf{P}_3\mathbf{y} = \sum_i \sum_j (y_{ij} - y_{i.})^2$$

From Example 1.6.11,

$$\mathbf{A}_1(\mathbf{A}_1'\mathbf{A}_1)^{-}\mathbf{A}_1' = \frac{1}{4}\begin{bmatrix} 1 & 1 & 1 & 1 \\ 1 & 1 & 1 & 1 \\ 1 & 1 & 1 & 1 \\ 1 & 1 & 1 & 1 \end{bmatrix}$$

so that

$$\|\mathbf{P}_1\mathbf{y}\|^2 = \mathbf{y}'\mathbf{P}_1\mathbf{y} = 4y_{..}^2$$

Furthermore, the

$$\|\mathbf{P}_2\mathbf{y}\|^2 = \mathbf{y}'\mathbf{P}_2\mathbf{y} = \sum_i 2y_{i.}^2 - 4y_{..}^2 = \sum_i 2(y_{i.} - y_{..})^2$$

Thus

$$\mathbf{y}'\mathbf{y} = \mathbf{y}'\mathbf{P}_1\mathbf{y} + \mathbf{y}'\mathbf{P}_2\mathbf{y} + \mathbf{y}'\mathbf{P}_3\mathbf{y}$$

reduces to

$$\sum_i \sum_j y_{ij}^2 = 4y_{..}^2 + \sum_i 2(y_{i.} - y_{..})^2 + \sum_i \sum_j (y_{ij} - y_{i.})^2$$

To compare this result with the one-way, analysis-of-variance design with two groups and two observations per cell, note that the total sum of squares has been partitioned into a sum of squares due to the overall mean plus a sum of squares for between groups plus a sum of squares for within groups, by using orthogonal projection matrices $\mathbf{P}_1$, $\mathbf{P}_2$, and $\mathbf{P}_3$. By dividing each sum of squares by σ^2 and applying (2.6.10), the total sum of squares is partitioned in a sum of independent chi-squares where the degrees of freedom are just the rank of the projection matrices.

In the usual analysis-of-variance table, the quantity $\mathbf{y}'(\mathbf{I} - \mathbf{P}_1)\mathbf{y}$ is termed the *total sum of squares about the mean.* If $\mathbf{y}$ has N observations, the $R(\mathbf{I} - \mathbf{P}_1) = N - 1$. For I groups, the rank of $\mathbf{A}(\mathbf{A}'\mathbf{A})^{-}\mathbf{A}'$ is I, so that the $R(\mathbf{P}_2) = I - 1$ and the rank of $\mathbf{P}_3 = N - I$.

Cochran's theorem and its extension allow us to partition a sum of squares into orthogonal components with projection matrices that add geometrical meaning to the procedure. In Chapter 3 we will learn to form ratios of appropriate sums of squares to test hypotheses.

The Wishart Distribution. The multivariate generalization of the χ^2 distribution is the *Wishart distribution.* The mathematical derivation of the Wishart density function is given by Anderson (1958, Chap. 7) and Wishart (1928). Rao (1965a, Chap. 10) discusses in detail many of its properties. Following Rao (1965a), let

$$\underset{(N \times p)}{\mathbf{Y}} = \begin{bmatrix} \mathbf{Y}'_1 \\ \mathbf{Y}'_2 \\ \vdots \\ \mathbf{Y}'_N \end{bmatrix} = \begin{bmatrix} Y_{11} & Y_{12} & \cdots & Y_{1p} \\ Y_{21} & Y_{22} & \cdots & Y_{2p} \\ \vdots & \vdots & \vdots & \vdots \\ Y_{N1} & Y_{N2} & \cdots & Y_{Np} \end{bmatrix} = [\mathbf{U}_1 \quad \mathbf{U}_2 \quad \cdots \quad \mathbf{U}_p] \tag{2.6.11}$$

where

$$\mathbf{W} = \mathbf{Y}'\mathbf{Y} = \sum_{i=1}^{N} \mathbf{Y}_i\mathbf{Y}'_i = \begin{bmatrix} \mathbf{U}'_1\mathbf{U}_1 & \mathbf{U}'_1\mathbf{U}_2 & \cdots & \mathbf{U}'_1\mathbf{U}_p \\ \mathbf{U}'_2\mathbf{U}_1 & \mathbf{U}'_2\mathbf{U}_2 & \cdots & \mathbf{U}'_2\mathbf{U}_p \\ \vdots & \vdots & \vdots & \vdots \\ \mathbf{U}'_p\mathbf{U}_1 & \mathbf{U}'_p\mathbf{U}_2 & \cdots & \mathbf{U}'_p\mathbf{U}_p \end{bmatrix} \tag{2.6.12}$$

The matrix $\mathbf{W}$ is the generalization of W given in (2.6.1). Similarly, a "quadratic form" $\mathbf{Y}'\mathbf{A}\mathbf{Y}$ for a symmetric matrix $\mathbf{A}$ in the multivariate case becomes

$$\mathbf{Y}'\mathbf{A}\mathbf{Y} = \begin{bmatrix} \mathbf{U}'_1\mathbf{A}\mathbf{U}_1 & \mathbf{U}'_1\mathbf{A}\mathbf{U}_2 & \cdots & \mathbf{U}'_1\mathbf{A}\mathbf{U}_p \\ \mathbf{U}'_2\mathbf{A}\mathbf{U}_1 & \mathbf{U}'_2\mathbf{A}\mathbf{U}_2 & \cdots & \mathbf{U}'_2\mathbf{A}\mathbf{U}_p \\ \vdots & \vdots & \vdots & \vdots \\ \mathbf{U}'_p\mathbf{A}\mathbf{U}_1 & \mathbf{U}'_p\mathbf{A}\mathbf{U}_2 & \cdots & \mathbf{U}'_p\mathbf{A}\mathbf{U}_p \end{bmatrix} \tag{2.6.13}$$

The diagonal elements of $\mathbf{Y}'\mathbf{A}\mathbf{Y}$ are familiar quadratic forms; however, the off-diagonal elements $\mathbf{U}'_i\mathbf{A}\mathbf{U}_j$ are bilinear forms.

Following (2.6.1) in the univariate case, the *central Wishart distribution* is defined by using the notation in (2.6.12):

(2.6.14) If $\mathbf{Y}_i \sim IN_p(\mathbf{0}, \boldsymbol{\Sigma})$, for $i = 1, 2, \ldots, N$, then

$$\mathbf{W} = \mathbf{Y}'\mathbf{Y} = \sum_{i=1}^{N} \mathbf{Y}_i\mathbf{Y}'_i \sim W_p(N, \boldsymbol{\Sigma})$$

$\mathbf{W} = \mathbf{Y}'\mathbf{Y}$ is said to have a central Wishart distribution with N degrees of freedom. In general, if $\mathbf{Y}_i \sim IN_p(\boldsymbol{\mu}_i, \boldsymbol{\Sigma})$, $i = 1, \ldots, N$, then

$$\mathbf{W} = \mathbf{Y}'\mathbf{Y} = \sum_{i=1}^{N} \mathbf{Y}_i\mathbf{Y}'_i \sim W_p(N, \boldsymbol{\Sigma}, \boldsymbol{\Gamma}) \tag{2.6.15}$$

has a *noncentral Wishart distribution* with N degrees of freedom and noncentrality parameter $\boldsymbol{\Gamma} = E(\mathbf{Y}')E(\mathbf{Y})$ (see James, 1964). If $N < p$, the density function of the Wishart distribution does not exist. When $p = 1$, $\mathbf{W}$ reduces to

$$W = \sum_{i=1}^{N} Y_i^2 \sim \sigma^2\chi^2(N, \delta)$$

where $\delta = \boldsymbol{\mu}'\boldsymbol{\mu}/\sigma^2$ as expected.

Wishart-distributed matrices have many of the properties of χ^2 variables discussed in univariate analysis. The multivariate analogue of (2.6.5) is as follows:

(2.6.16) Let $\mathbf{Y}_i \sim IN_p(\boldsymbol{\mu}_i, \boldsymbol{\Sigma})$, for $i = 1, \ldots, N$, and let $\mathbf{A}$ be a symmetric matrix of rank r. Then $\mathbf{Y}'\mathbf{A}\mathbf{Y} \sim W_p(r, \boldsymbol{\Sigma}, \Gamma)$, where $\Gamma = E(\mathbf{Y}')\mathbf{A}E(\mathbf{Y})$ if and only if $\mathbf{A} = \mathbf{A}^2$.

EXAMPLE 2.6.5. As an example of (2.6.16), the distribution of the sample variance-covariance matrix $\mathbf{S}$ will be determined. Let $\mathbf{Y}_i \sim IN_p(\boldsymbol{\mu}, \boldsymbol{\Sigma})$, for $i = 1, \ldots, N$ and $N \geq p$, then

$$\begin{aligned}(N-1)\mathbf{S} &= \sum_{i=1}^{N} (\mathbf{Y}_i - \overline{\mathbf{Y}})(\mathbf{Y}_i - \overline{\mathbf{Y}})' \\ &= \mathbf{Y}'[\mathbf{I} - \mathbf{1}(\mathbf{1}'\mathbf{1})^{-1}\mathbf{1}']\mathbf{Y}\end{aligned}$$

where $\mathbf{Y}$ is defined by (2.6.11) and $\mathbf{1}$ is a vector of N 1s, has a central Wishart distribution $(N-1)\mathbf{S} \sim W_p(N-1, \boldsymbol{\Sigma})$ or $\mathbf{S} \sim W_p[N-1, \boldsymbol{\Sigma}/(N-1)]$.

In multivariate analysis, the sum of independent Wishart distributions, although following the same rules as in the univariate case, is not used as often as ratios of independently distributed Wishart matrices or, more specifically, their determinants. Extending independence to its multivariate analogue, a theorem for Wishart distributions is obtained:

(2.6.17) If $\mathbf{Y}_i \sim IN_p(\boldsymbol{\mu}_i, \boldsymbol{\Sigma})$, for $i = 1, \ldots, N$, then $\mathbf{Y}'\mathbf{A}_1\mathbf{Y} \sim W_p(r_1, \boldsymbol{\Sigma}, \boldsymbol{\Gamma}_1)$ and $\mathbf{Y}'\mathbf{A}_2\mathbf{Y} \sim W_p(r_2, \boldsymbol{\Sigma}, \boldsymbol{\Gamma}_2)$, where the matrices $\mathbf{A}_i$ are symmetric and of rank r_i, are statistically independent if and only if $\mathbf{A}_1\mathbf{A}_2 = \mathbf{0}$.

Cochran's theorem for Wishart matrices is now stated for the central case:

(2.6.18) If $\mathbf{Y}_i \sim IN_p(\mathbf{0}, \boldsymbol{\Sigma})$ and

$$\mathbf{Y}'\mathbf{Y} = \sum_{i=1}^{k} \mathbf{Y}'\mathbf{A}_i\mathbf{Y} \qquad \text{where } R(\mathbf{A}_i) = r_i \quad \text{and} \quad \sum_{i=1}^{k} \mathbf{A}_i = \mathbf{I}_N$$

the forms $\mathbf{Y}'\mathbf{A}_i\mathbf{Y} \sim W_p(r_i, \boldsymbol{\Sigma})$ are statistically independent for all i if and only if

$$\sum_{i=1}^{k} r_i = N$$

If $r_i < p$, the Wishart density function does not exist.

EXAMPLE 2.6.6. Suppose $\mathbf{Y}_i \sim IN_p(\boldsymbol{\mu}, \boldsymbol{\Sigma})$, $i = 1, \ldots, N$. Then $\mathbf{Y}'[\mathbf{I} - \mathbf{1}(\mathbf{1}'\mathbf{1})^{-1}\mathbf{1}']\mathbf{Y} \sim W_p(N-1, \boldsymbol{\Sigma})$ and $\mathbf{Y}'[\mathbf{1}(\mathbf{1}'\mathbf{1})^{-1}\mathbf{1}']\mathbf{Y} \sim W_p(1, \boldsymbol{\Sigma}, \Gamma_2)$ are independent. Furthermore,

$$\begin{aligned}\mathbf{Y}'\mathbf{Y} &= \mathbf{Y}'[\mathbf{I} - \mathbf{1}(\mathbf{1}'\mathbf{1})^{-1}\mathbf{1}']\mathbf{Y} + \mathbf{Y}'[\mathbf{1}(\mathbf{1}'\mathbf{1})^{-1}\mathbf{1}']\mathbf{Y} \\ \mathbf{Y}'\mathbf{I}\mathbf{Y} &= \mathbf{Y}'\mathbf{A}_1\mathbf{Y} + \mathbf{Y}'\mathbf{A}_2\mathbf{Y} \\ \mathbf{Y}'\mathbf{Y} &= (N-1)\mathbf{S} + N\overline{\mathbf{Y}}\overline{\mathbf{Y}}'\end{aligned}$$

or

$$W_p(N, \mathbf{\Sigma}, \mathbf{\Gamma}) = W_p(N - 1, \mathbf{\Sigma}, \mathbf{\Gamma}_1 = \mathbf{0}) + W_p(1, \mathbf{\Sigma}, \mathbf{\Gamma}_2)$$

where $\mathbf{\Gamma} = \mathbf{\Gamma}_1 + \mathbf{\Gamma}_2$, so that $\mathbf{S}$ and $\bar{\mathbf{Y}}$ are independently distributed.

The multivariate analogue of the variance σ^2 is the variance-covariance matrix $\mathbf{\Sigma}$. In Chapter 1, the determinant of a matrix was shown to be related to the concept of volume. Wilks (1932) calls the $|\mathbf{\Sigma}|$ the *generalized variance* of a multivariate normal distribution. In a sample, the generalized variance becomes the $|\mathbf{S}|$, where $\mathbf{S}$ is the sample variance-covariance matrix and is used as a measure of the variability of the set of vectors $\mathbf{Y}_i \sim IN_p(\boldsymbol{\mu}, \mathbf{\Sigma})$. Since the generalized sample variance often appears in multivariate hypothesis testing, the distribution of the $|\mathbf{S}|$ is obtained.

Fundamental to finding the distribution of the $|\mathbf{S}|$ is *Bartlett's decomposition* of $\mathbf{S}$:

(2.6.19) If $\mathbf{W} \sim W_p(N, \mathbf{I})$, then there exists a $p \times p$ lower triangular matrix $\mathbf{T}$ such that $\mathbf{TT}' = \mathbf{W}$, where $T_{ii} \sim \chi_{N-i-1}$; that is, where T_{ii} has a chi-square distribution, and $T_{ij} \sim N(0, 1)$, for $i > j$, where all entries are statistically independent.

The proof of (2.6.19) follows Kshirsagar (1959). Let

$$\underset{(N \times p)}{\mathbf{Y}} = \begin{bmatrix} \mathbf{Y}_1' \\ \mathbf{Y}_2' \\ \vdots \\ \mathbf{Y}_N' \end{bmatrix} = \begin{bmatrix} Y_{11} & Y_{12} & \cdots & Y_{1p} \\ Y_{21} & Y_{22} & \cdots & Y_{2p} \\ \vdots & \vdots & \vdots & \vdots \\ Y_{N1} & Y_{N2} & \cdots & Y_{Np} \end{bmatrix} = [\mathbf{U}_1 \quad \mathbf{U}_2 \quad \cdots \quad \mathbf{U}_p]$$

then

$$\mathbf{W} = \mathbf{Y}'\mathbf{Y} = \sum_{i=1}^{N} \mathbf{Y}_i\mathbf{Y}_i' = \begin{bmatrix} \mathbf{U}_1'\mathbf{U}_1 & \mathbf{U}_1'\mathbf{U}_2 & \cdots & \mathbf{U}_1'\mathbf{U}_p \\ \mathbf{U}_2'\mathbf{U}_1 & \mathbf{U}_2'\mathbf{U}_2 & \cdots & \mathbf{U}_2'\mathbf{U}_p \\ \vdots & \vdots & \vdots & \vdots \\ \mathbf{U}_p'\mathbf{U}_1 & \mathbf{U}_p'\mathbf{U}_2 & \cdots & \mathbf{U}_p'\mathbf{U}_p \end{bmatrix}$$

where $\mathbf{Y}_i \sim IN_p(\mathbf{0}, \mathbf{I})$, so that $\mathbf{W} \sim W_p(\mathbf{N}, \mathbf{I})$. Orthogonalizing the columns of $\mathbf{Y}$ by the Gram-Schmidt orthogonalization process,

$$\mathbf{Z}_i = \mathbf{U}_i - \sum_{j=1}^{i-1} c_{ij}\mathbf{Z}_j \qquad \text{where } c_{ij} = \frac{\mathbf{U}_i'\mathbf{Z}_j}{\|\mathbf{Z}_j\|^2}$$

Letting

$$b_{pj} = \frac{\mathbf{U}_p'\mathbf{Z}_j}{\|\mathbf{Z}_j\|}$$

$$\begin{aligned} \mathbf{U}_p'\mathbf{U}_p &= \mathbf{Z}_p'\mathbf{Z}_p + c_{p1}^2\mathbf{Z}_1'\mathbf{Z}_1 + \cdots + c_{p,p-1}^2\mathbf{Z}_{p-1}'\mathbf{Z}_{p-1} \\ &= \mathbf{Z}_p'\mathbf{Z}_p + \sum_{k=1}^{p-1} b_{pk}^2 \end{aligned}$$

or, in general, if $b_{ij} = \mathbf{U}_i'\mathbf{Z}_j/\|\mathbf{Z}_j\|$, since $\mathbf{Z}_i'\mathbf{Z}_j = 0$, for $i \neq j$, then

$$\mathbf{U}_i'\mathbf{U}_i = \mathbf{Z}_i'\mathbf{Z}_i + \sum_{k=1}^{i-1} b_{ik}^2$$

Also notice that, for $j > i$,

$$\mathbf{U}_j = \mathbf{Z}_j + c_{j1}\mathbf{Z}_1 + \cdots + c_{j,j-1}\mathbf{Z}_{j-1} \quad \text{and} \quad \mathbf{U}_i = \mathbf{Z}_i + \cdots + c_{i,i-1}\mathbf{Z}_{i-1}$$

and that, for $i \neq j$,

$$\begin{aligned}\mathbf{U}_i'\mathbf{U}_j &= c_{ji}\mathbf{Z}_i'\mathbf{Z}_i + c_{i1}c_{j1}\mathbf{Z}_1'\mathbf{Z}_1 + \cdots + c_{i,i-1}c_{j,j-1}\mathbf{Z}_{i-1}'\mathbf{Z}_{i-1} \\ &= \sum_{k=1}^{i-1} b_{ik}b_{jk} + b_{ji}\|\mathbf{Z}_i\|\end{aligned}$$

hence $\mathbf{Y}'\mathbf{Y} = \mathbf{T}\mathbf{T}' = \mathbf{W}$ if

$$\mathbf{T} = \begin{bmatrix} (\mathbf{Z}_1'\mathbf{Z}_1)^{1/2} & & & & \\ b_{21} & (\mathbf{Z}_2'\mathbf{Z}_2)^{1/2} & & & \\ b_{31} & b_{32} & (\mathbf{Z}_3'\mathbf{Z}_3)^{1/2} & & \\ \vdots & \vdots & \vdots & & \\ b_{p1} & b_{p2} & b_{p3} & \cdots & (\mathbf{Z}_p'\mathbf{Z}_p)^{1/2} \end{bmatrix}$$

yields the appropriate representation. $\mathbf{Y}_i \sim IN_p(\mathbf{0}, \mathbf{I})$, so that $\mathbf{U}_i \sim IN(\mathbf{0}, \mathbf{I})$. Since the b_{ij}'s ($i = 1, \ldots, j-1$) are orthogonal (linear in the entries of $\mathbf{U}_i$ with mean 0 and variance 1), the b_{ij}'s are independent, $N(0, 1)$. Also, the conditional distribution of $\mathbf{U}_j$, given $\mathbf{U}_1, \ldots, \mathbf{U}_{j-1}$, does not depend on $\mathbf{U}_1, \ldots, \mathbf{U}_{j-1}$ since $\mathbf{U}_i \sim IN(\mathbf{0}, \mathbf{I})$; since the b_{ij}'s are linear in the entries of $\mathbf{U}_i$, the b_{ij}'s are independent of all that has preceded, which implies independence of $\mathbf{Z}_i$. Furthermore,

$$\mathbf{Z}_i'\mathbf{Z}_i = \sum_{i=1}^{N} Z_i^2 = \sum_{i=1}^{N} (U_i - b_{i1}Z_1 - \cdots - b_{i,i-1}Z_{i-1})^2$$

is the residual sum of squares after regression, so that $\mathbf{Z}_i'\mathbf{Z}_i \sim \chi^2[N-(i-1)]$. Hence

$$|\mathbf{Y}'\mathbf{Y}| = |\mathbf{T}||\mathbf{T}'| = \prod_{i=1}^{p} T_{ii}^2 \sim \sum_{i=1}^{p} \chi^2(N-i+1) \tag{2.6.20}$$

In order to determine the distribution of the $|\mathbf{S}|$, use Example 2.6.5. That is, since $(N-1)\mathbf{S} \sim W_p(N-1, \mathbf{\Sigma})$, there exist vectors $\mathbf{V}_i \sim IN_p(\mathbf{0}, \mathbf{\Sigma})$ such that

$$(N-1)\mathbf{S} = \sum_{i=1}^{N-1} \mathbf{V}_i\mathbf{V}_i' = \mathbf{V}'\mathbf{V}$$

Let $\mathbf{W}_i = \mathbf{\Sigma}^{-1/2}\mathbf{V}_i$, where $\mathbf{V}_i \sim IN_p(\mathbf{0}, \mathbf{\Sigma})$. Then $\mathbf{W}_i \sim IN_p(\mathbf{0}, \mathbf{I})$. Replacing each row of $\mathbf{Y}$ in (2.6.20) by $\mathbf{W}_i$, for $i = 1, \ldots, N-1$,

$$|\mathbf{W}'\mathbf{W}| \sim \prod_{i=1}^{p} \chi^2(N-i)$$

so that

$$|\mathbf{W}'\mathbf{W}| = |\mathbf{V}'\mathbf{\Sigma}^{-1/2}\mathbf{\Sigma}^{-1/2}\mathbf{V}|$$
$$= \frac{(N-1)^p|\mathbf{S}|}{|\mathbf{\Sigma}|} \sim \prod_{i=1}^{p} \chi^2(N-i)$$

Hence, if $\mathbf{Y}_i \sim IN_p(\boldsymbol{\mu}, \mathbf{\Sigma})$, for $i = 1, \ldots, N$, then the $|\mathbf{S}|$ are distributed as the $|\mathbf{\Sigma}|/(N-1)^p$ times a product of independent χ^2 variates:

(2.6.21) $$|\mathbf{S}| \sim \frac{|\mathbf{\Sigma}|}{(N-1)^p} \prod_{i=1}^{p} \chi^2(N-i)$$

For an alternative proof of (2.6.21), see Anderson (1958, p. 171). Distributions of the ratio of determinants for independent random Wishart matrices are discussed in the next section.

EXERCISES 2.6

1. If $\mathbf{Y} \sim N_p(\boldsymbol{\mu}, \mathbf{\Sigma})$, prove that $(\mathbf{Y} - \boldsymbol{\mu})'\mathbf{\Sigma}^{-1}(\mathbf{Y} - \boldsymbol{\mu}) \sim \chi^2(p)$.
2. If $\mathbf{Y} \sim N_p(\mathbf{0}, \mathbf{\Sigma})$, show that

$$\mathbf{Y}'\mathbf{A}\mathbf{Y} = \sum_{j=1}^{p} \lambda_j z_j^2$$

where the λ_j are the roots of the determinantal equation $|\mathbf{\Sigma}^{1/2}\mathbf{A}\mathbf{\Sigma}^{1/2} - \mathbf{I}| = 0$, where $\mathbf{A}$ is a symmetric matrix and $Z_i \sim IN(0, 1)$.
3. If $\mathbf{Y} \sim N_p(\mathbf{0}, \mathbf{P})$, where $\mathbf{P}$ is a projection matrix, show that the $\|\mathbf{Y}\|^2 \sim \chi^2(p)$.
4. Use (2.6.10) to write the distribution of $\mathbf{y}'\mathbf{P}_i\mathbf{y}/\sigma^2$, and evaluate the noncentrality parameters for Example 2.6.4.
5. If

$$\begin{bmatrix} \mathbf{Y}'\mathbf{Y} & \mathbf{Y}'\mathbf{X} \\ \mathbf{X}'\mathbf{Y} & \mathbf{X}'\mathbf{X} \end{bmatrix} \sim W_{p+q}(N, \mathbf{\Sigma})$$

where

$$\mathbf{\Sigma} = \begin{bmatrix} \mathbf{\Sigma}_{11} & \mathbf{\Sigma}_{12} \\ \mathbf{\Sigma}_{21} & \mathbf{\Sigma}_{22} \end{bmatrix}$$

and $\mathbf{Y}'\mathbf{Y}$, $\mathbf{Y}'\mathbf{X}$, and $\mathbf{X}'\mathbf{X}$ are $p \times p$, $p \times q$, and $q \times q$ matrices, respectively, show that $\mathbf{X}'\mathbf{X} - \mathbf{X}'\mathbf{Y}(\mathbf{Y}'\mathbf{Y})^{-1}\mathbf{Y}'\mathbf{X} \sim W_q(N - p, \mathbf{\Sigma}_{22} - \mathbf{\Sigma}_{21}\mathbf{\Sigma}_{11}^{-1}\mathbf{\Sigma}_{12})$.

2.7 THE t, F, BETA, T^2, AND U DISTRIBUTIONS

For testing hypotheses, two distributions frequently employed in univariate analysis are the t and F distributions.

The t Distribution. The definition of *Student's t distribution* is as follows:

(2.7.1) If X and Y are independent random variables such that $X \sim N(\mu, \sigma^2)$ and $Y \sim \chi^2(v, \delta)$, then $t = X/\sqrt{Y/v} \sim t(v, \delta)$, $-\infty < t < \infty$.

That is, $X/\sqrt{Y/v}$ has a noncentral t distribution with v degrees of freedom and noncentrality parameter $\delta = \mu/\sigma$. If $\mu = 0$, the noncentral t distribution reduces to the familiar central t distribution known as Student's t distribution.

EXAMPLE 2.7.1. Suppose that a random sample of N observations with observed values $y_1, y_2, \ldots, y_N$ is selected from a normal population mean μ and unknown variance σ^2, where $Y_i \sim IN(\mu, \sigma^2)$, and suppose that the null hypothesis $H_0: \mu = \mu_0$ is to be tested against the alternative hypothesis $H_1: \mu \neq \mu_0$. The sample mean and variance of the observations are given by

$$\bar{y} = \frac{\sum_{i=1}^{N} y_i}{N} \quad \text{and} \quad s^2 = \frac{\sum_{i=1}^{N} (y_i - \bar{y})^2}{N - 1}$$

Since $\bar{y}$ and s^2 are statistically independent, $(N - 1)s^2/\sigma^2 \sim \chi^2(N - 1)$; under the null hypothesis, $N(\bar{y} - \mu_0)/\sigma \sim N(0, 1)$, the ratio

$$\frac{\sqrt{N}(\bar{y} - \mu_0)/\sigma}{\sqrt{(N - 1)s^2/(N - 1)\sigma^2}} = \frac{\sqrt{N}(\bar{y} - \mu_0)}{s} \sim t(N - 1)$$

In general,

$$\frac{N(\bar{y} - \mu_0)}{s} \sim t\left(N - 1, \delta = \frac{|\mu - \mu_0|}{\sigma/\sqrt{N}}\right)$$

The F Distribution. A special case of the noncentral F distribution, fundamental to power calculations in univariate analysis and to some multivariate procedures, is *Snedecor's central F distribution* (named in honor of R. A. Fisher):

(2.7.2) If X and Y are independent random variables such that $X \sim \chi^2(v_h, \delta)$ and $Y \sim \chi^2(v_e, 0)$, then

$$F = \frac{X/v_h}{Y/v_e} \sim F(v_h, v_e, \delta) \qquad 0 \leq F < \infty$$

The notation $F(v_h, v_e, \delta)$ denotes a *noncentral* F distribution with degrees of freedom v_h and v_e, and noncentrality parameter δ, where δ is obtained from the noncentral χ^2 distribution. The noncentrality parameter is determined by replacing the sample observations with their expected values. In this text, the symbols v_h and v_e are used to denote the numerator and denominator degrees of freedom of the χ^2 random variables in forming the F ratio, instead of the more familiar symbols v_1 and v_2. The symbols v_h and v_e stress the fact that the numerator of the F ratio is obtained from a hypothesis sum of squares while the denominator is obtained from an error sum of squares in regression and analysis-of-variance applications.

The Beta Distribution. A distribution closely related to the F distribution is the *beta distribution*:

(2.7.3) If X and Y are independent random variables such that $X \sim \chi^2(\nu_h, \delta)$ and $Y \sim \chi^2(\nu_e, 0)$, then

$$B = \frac{X}{X + Y} \sim \beta(\nu_h, \nu_e, \delta) \qquad 0 \leq B \leq 1$$

That is, $X/(X + Y)$ has a noncentral beta distribution with parameters ν_h and ν_e and noncentrality parameter δ.

The relationship between B and F is that $B = \nu_h F/(\nu_e + \nu_h F)$. To illustrate,

$$F = \frac{X/\nu_h}{Y/\nu_e}$$

$$\frac{\nu_h F}{\nu_e} = \frac{X}{Y}$$

since

$$B = \frac{X}{X + Y} = \frac{X/Y}{(X/Y) + 1}$$

$$= \frac{\nu_h F/\nu_e}{(\nu_h F/\nu_e) + 1}$$

$$= \frac{\nu_h F}{\nu_e + \nu_h F}$$

The relationship between B and F was used by Tang (1938) in the construction of the Tang tables for power calculations in analysis-of-variance applications. B is a monotonic increasing function of F. Alternatively, the beta distribution can also be related to a monotonic decreasing function of F as tabled by K. Pearson (1934) and reproduced in E. S. Pearson and H. O. Hartley (1966, pp. 150–167). Both relations are summarized as follows:

(2.7.4) If $F \sim F(\nu_h, \nu_e)$, then the random variables

$$B = \frac{\nu_h F}{\nu_e + \nu_h F} = \frac{X}{X + Y} \sim B(\nu_h, \nu_e)$$

and

$$B' = \frac{1}{1 + \nu_h F/\nu_e} = \frac{Y}{X + Y} \sim B'(\nu_e, \nu_h)$$

where F represents a ratio of independent chi-square random variables and the notation $B'(\nu_e, \nu_h)$ indicates a beta distribution with the degrees of freedom interchanged.

To obtain a percentage point in B' from one in B, we merely form $1 - B(\nu_h, \nu_e) = B'(\nu_e, \nu_h)$. From (2.7.4), the relation between the beta and t distributions is summarized as follows:

(2.7.5) If t is a random variable such that $t \sim t(v_e)$, then

$$B' = \frac{1}{1 + (t^2/v_e)} \sim B'(v_e, 1)$$

EXAMPLE 2.7.2 (Central Case). Suppose $Y_i \sim IN(\mu, \sigma^2)$, for $i = 1, \ldots, N$, and the null hypothesis $H_0 : \mu = 0$ is to be tested against the alternative hypothesis $H_1 : \mu \neq 0$. The sample mean and variance of the observations are $\bar{y}$ and s^2, respectively, and as defined in Example 2.7.1. Since $(N - 1)s^2/\sigma^2 \sim \chi^2(N - 1)$, $N\bar{y}^2/\sigma^2 \sim \chi^2(1)$ under the null hypothesis, and since $\bar{y}$ and s^2 are independent,

$$F = \frac{N\bar{y}^2/\sigma^2}{(N - 1)s^2/(N - 1)\sigma^2} = \frac{N\bar{y}^2}{s^2} \sim F(1, N - 1)$$

has a central F distribution under the null hypothesis with degrees of freedom $v_h = 1$ and $v_e = N - 1$.

Selecting the type I error $\alpha = .05$, with a sample size of $N = 9$, the hypothesis would be rejected if the observed F value is larger than the upper .05 percentage point of the central F distribution obtained from the F table (Table IV in the Appendix): $F(v_h, v_e) = F^{.05}_{(1,8)} = 5.32$. The type I error α is the probability of rejecting H_0, given that H_0 is true, and is called the *size* of the test.

EXAMPLE 2.7.3 (Central Case, *continued*). Employing the central beta distribution using relationship (2.7.4),

$$\begin{aligned} B &= \frac{N\bar{y}^2/\sigma^2}{(N\bar{y}^2/\sigma^2) + [(N - 1)s^2/\sigma^2]} \\ &= \frac{N\bar{y}^2/(N - 1)s^2}{[N\bar{y}^2/(N - 1)s^2] + 1} \\ &= \frac{N\bar{y}^2/s^2}{(N\bar{y}^2/s^2) + (N - 1)} \\ &= \frac{F}{(N - 1) + F} \sim B(1, N - 1) \end{aligned}$$

Since B is a monotonic increasing function of F, Tang's table could be employed to test the null hypothesis. For $\alpha = .05$, H_0 would be rejected if

$$B = \frac{F}{(N - 1) + F} > B^{.05}_{(1,8)} = .399$$

By employing the Pearson and Hartley tables (1966, p. 156) and (2.7.4), H_0 is rejected at the $\alpha = .05$ level if

$$B' = \frac{v_e}{v_e + v_h F} = \frac{N - 1}{(N - 1) + F} < B'^{.05}_{(8,1)} = .60071$$

since B' is a monotonic decreasing function of F. This value is obtained from Table IX in the Appendix, with $p = 1$, $\nu_h = q = 1$, and $n = \nu_e = 8$, since $B'(\nu_e, \nu_h) = U(1, \nu_h, \nu_e)$. The value given is $U^{.05}_{(1,1,8)} = .600708$, which demonstrates the relationship noted earlier that $1 - B(\nu_h, \nu_e) = B'(\nu_e, \nu_h)$. We shall examine the relation between B' and U within this chapter.

Finally, Student's t distribution could be employed to test the null hypothesis. In this case,

$$t = \frac{\sqrt{N}\,\bar{y}}{s} \sim t(N - 1)$$

For $N = 9$ and $\alpha = .05$, the hypothesis is rejected if

$$|t| > t^{\alpha/2}_{(8)} = t^{.025}_{(8)} = 2.306$$

which is obtained from the t table (Table III in the Appendix) and is equal to the square root of the upper α percentage point of the central F distribution.

The techniques illustrated in the preceding example review the one-sample t test and the relationship among the central t, F, and beta distributions, which are often excluded from elementary statistical texts. Further details of the one-sample t test are given in Chapter 3, which reviews fundamentals included in texts such as Kirk (1968) and Winer (1971).

The *power* of a size α test is equal to $1 - \beta$, where β denotes the probability of a type II error—the probability of accepting a false hypothesis. Power $= 1 - \beta$ is the probability of rejecting a null hypothesis when the parameters under study have specified values indicated by an alternative hypothesis. To calculate the power of the test identified in Example 2.7.2 with $\alpha = .05$, the noncentral F distribution is employed. When H_0 is not true,

$$\frac{N\bar{y}^2}{s^2} \sim F(1, N - 1, \delta)$$

has a noncentral F distribution with noncentrality parameter $\delta = N\mu^2/\sigma^2$. Since δ involves both μ and σ^2, values of these parameters are needed to determine the power of the test from the noncentral F distribution or from a noncentral beta distribution. Tang (1938) and Titku (1967) compiled tables, whereas Pearson and Hartley (1951) constructed charts. (The charts are included in the Appendix.) To use either the charts or the tables, ϕ and not δ is required, as well as $\nu_1 = \nu_h$ and $\nu_2 = \nu_e$ (the degrees of freedom associated with the numerator and the denominator, respectively, of the F ratio). The quantity ϕ is defined by

$$\phi = \left(\frac{\delta}{\nu_h + 1}\right)^{1/2} \tag{2.7.6}$$

EXAMPLE 2.7.4. With $\alpha = .05$ and $N = 9$, suppose that $\sigma^2 = 2$ and that an alternative value of μ, $\mu = 2$, is specified. By substituting the values of σ^2 and μ

in the formula for ϕ,

$$\begin{aligned}\phi &= \left(\frac{\delta}{v_h + 1}\right)^{1/2}\\ &= \left(\frac{N\mu^2/\sigma^2}{1 + 1}\right)^{1/2}\\ &= \mu\frac{\sqrt{N}/\sigma}{\sqrt{2}}\\ &= \frac{2\sqrt{9}}{\sqrt{2}\sqrt{2}}\\ &= 3\end{aligned}$$

Using the chart labeled $v_1 = 1(v_h = v_1)$ in Table V in the Appendix, the value $\phi = 3$ for $\alpha = .05$ is located. Since $v_2 = 8(v_2 = v_e)$, we proceed down the power curve labeled 8 until it intersects the vertical line indicated by $\phi = 3$. Continuing horizontally, the value for the power of the test is read from the left margin as power $= 1 - \beta = .96$.

The previous distributions and techniques should be familiar to researchers using univariate statistical techniques. Further detail is postponed until Chapter 3. Generalization of these distributions to their multivariate analogues is now considered.

The T^2 Distribution. The multivariate extension of Student's central t ratio is *Hotelling's T^2 statistic* (see Hotelling, 1931). Bose and Roy (1938) and Hsu (1938) obtained the noncentral distribution of T^2. The generalization of t follows:

(2.7.7) Let $\mathbf{Y}$ and $\mathbf{W}$ be independent so that $\mathbf{Y} \sim N_p(\boldsymbol{\mu}, \boldsymbol{\Sigma})$ and $n\mathbf{W} \sim W_p(n, \boldsymbol{\Sigma})$; then

$$T^2 = \mathbf{Y}'\mathbf{W}^{-1}\mathbf{Y} \sim T^2(p, n, \delta) \qquad 0 \leq T^2 < \infty$$

That is, T^2 has a noncentral T^2 distribution with noncentrality parameter $\delta = \boldsymbol{\mu}'\boldsymbol{\Sigma}^{-1}\boldsymbol{\mu}$ and degrees of freedom p and n.

Wijsman (1957) and Bowker (1960) related T^2 to the ratio of independent χ^2 random variables. Their results are summarized in the following theorem:

(2.7.8) Let $T^2 = \mathbf{Y}'\mathbf{W}^{-1}\mathbf{Y}$, where $\mathbf{Y} \sim N_p(\boldsymbol{\mu}, \boldsymbol{\Sigma})$ and $n\mathbf{W} \sim W_p(n, \boldsymbol{\Sigma})$, and let $\mathbf{Y}$ and $\mathbf{W}$ be independent; then

$$\frac{n - p + 1}{P}\frac{T^2}{n} \sim F(p, n - p + 1, \delta)$$

has a noncentral F distribution with degrees of freedom p and $n - p + 1$ and noncentrality parameter $\delta = \boldsymbol{\mu}'\boldsymbol{\Sigma}^{-1}\boldsymbol{\mu}$.

To use the statistic T^2 in multivariate applications, either the F or T^2 percentage points may be employed. Tables prepared by Jensen and Howe (1968) give the upper percentage points of the T^2 distribution and are in the T^2 Table VI included in the Appendix. Application of the tables is postponed until Chapter 3. Observe, however, that both $T^2(1, n)$ and $t^2(n)$ are equal when the number of variables $p = 1$. This follows from (2.7.8).

EXAMPLE 2.7.5. One important form of Theorem (2.7.7) is now illustrated. Suppose that $\mathbf{Y}_1, \mathbf{Y}_2, \ldots, \mathbf{Y}_N$ is a random sample from a multivariate normal population, $\mathbf{Y}_i \sim IN_p(\boldsymbol{\mu}, \boldsymbol{\Sigma})$. Then $\overline{\mathbf{Y}} \sim N_p(\boldsymbol{\mu}, \boldsymbol{\Sigma}/N)$, $(N-1)\mathbf{S} \sim W_p(N-1, \boldsymbol{\Sigma})$, and $\overline{\mathbf{Y}}$ and $\mathbf{S}$ are independent. Hence $T^2 = N\overline{\mathbf{Y}}'\mathbf{S}^{-1}\overline{\mathbf{Y}} \sim T^2(p, N-1, \delta)$, where $\delta = N\boldsymbol{\mu}'\boldsymbol{\Sigma}^{-1}\boldsymbol{\mu}$. Let $\mathbf{d} = \overline{\mathbf{Y}} - \boldsymbol{\mu}$; then $\sqrt{N}\mathbf{d} \sim N_p(\mathbf{0}, \boldsymbol{\Sigma})$ and

$$T^2 = N(\overline{\mathbf{Y}} - \boldsymbol{\mu}_0)'\mathbf{S}^{-1}(\overline{\mathbf{Y}} - \boldsymbol{\mu}_0) \sim T^2(p, N-1, \delta),$$

where $\delta = N(\boldsymbol{\mu} - \boldsymbol{\mu}_0)'\boldsymbol{\Sigma}^{-1}(\boldsymbol{\mu} - \boldsymbol{\mu}_0)$. If $\mu = \mu_0$, then the distribution of T^2 is central, or, by (2.7.8),

$$\frac{N-p}{p}\,\frac{T^2}{N-1} \sim F(p, N-p)$$

The U Distribution. The extension of the beta distribution to its multivariate analogue is represented by the following result (see Wilks, 1932):

(2.7.9) If $\mathbf{W}_1 \sim W_p(v_e, \boldsymbol{\Sigma})$ and $\mathbf{W}_2 \sim W_p(v_h, \boldsymbol{\Sigma})$ are independent matrices with degrees of freedom v_e and v_h, respectively, and if $v_e \geq p$, the distribution of

$$\Lambda = \frac{|\mathbf{W}_1|}{|\mathbf{W}_1 + \mathbf{W}_2|} \sim U(p, v_h, v_e) \qquad 0 \leq \Lambda \leq 1$$

Λ is distributed as a product of independent beta variables with parameters $v_e - i + 1$ and v_h, for $i = 1, \ldots, p$. Anderson (1958, Chap. 8) discusses in detail the distribution of the Λ statistic. The statistic Λ is known as the *Wilks' Λ criterion.* Illustration of its use is included in Chapter 3.

In the special case that $p = 1$, (2.7.9) is seen to reduce to $B'(v_e, v_h)$, so that

$$U(1, v_h, v_e) = B'(v_e, v_h) = \frac{1}{1 + (v_h F/v_e)}$$

as indicated earlier. The percentage points of the U distribution are included in Table IX of the Appendix (see also, Wall, 1968). Before the construction of Wall's table, considerable work was conducted to obtain approximations to the distribution of Λ and then to relate Λ to familiar univariate tabulated distributions. Rao (1952, p. 261) relates Λ to the F distribution and Bartlett (1947) relates it to the χ^2 distribution. For Bartlett's approximation,

$$X_B^2 = -\left[v_e - \frac{1}{2}(p - v_h + 1)\right] \log \Lambda \mathrel{\dot{\sim}} \chi^2(pv_h) \tag{2.7.10}$$

That is, X_B^2 has approximately a χ^2 distribution with degrees of freedom pv_h (see Anderson, 1958, p. 208), when N is large. Rao (1965a, p. 471) summarizes several other approximation procedures with references and exact relationships when they exist, between the null distribution of Λ and F.

Other Distributions. As an alternative development of (2.7.9), Λ is related to the roots of three frequently used characteristic equations in multivariate analysis. The results are based on the "invariance principle," which simply means that changes in the origin and units of the original observations do not alter the value of Λ. The only invariants of $\mathbf{W}_1$ and $\mathbf{W}_2$, where $\mathbf{W}_1 \sim W_p(v_e, \boldsymbol{\Sigma})$ and $\mathbf{W}_2 \sim W_p(v_h, \boldsymbol{\Sigma})$ are independently distributed, are the eigenvalues or roots of the characteristic equation

(2.7.11) $$|\mathbf{W}_2 - \lambda \mathbf{W}_1| = 0$$

(See Anderson, 1958, p. 222.) By employing result (1.7.29), the roots of three characteristic equations are related to Λ:

(2.7.12) Let λ_i, v_i, and θ_i denote the roots of

(a) $|\mathbf{W}_2 - \lambda \mathbf{W}_1| = 0$
(b) $|\mathbf{W}_1 - v(\mathbf{W}_1 + \mathbf{W}_2)| = 0$
(c) $|\mathbf{W}_2 - \theta(\mathbf{W}_2 + \mathbf{W}_1)| = 0$

respectively, where $\mathbf{W}_1 \sim W_p(v_e, \boldsymbol{\Sigma})$ independently of $\mathbf{W}_2 \sim W_p(v_h, \boldsymbol{\Sigma})$ with $v_e \geq p$. Then

(d) $$\Lambda = |\mathbf{W}_1(\mathbf{W}_1 + \mathbf{W}_2)^{-1}| = \prod_i v_i = \prod_i (1 + \lambda_i)^{-1} = \prod_i (1 - \theta_i)$$

Also, the

(e) $$|\mathbf{W}_2 \mathbf{W}_1^{-1}| = \prod_i \lambda_i = \prod_i \left(\frac{1 - v_i}{v_i}\right) = \prod_i \left(\frac{\theta_i}{1 - \theta_i}\right)$$

(f) $$|\mathbf{W}_2(\mathbf{W}_2 + \mathbf{W}_1)^{-1}| = \prod_i \theta_i = \prod_i \left(\frac{\lambda_i}{1 + \lambda_i}\right) = \prod_i (1 - v_i)$$

for the nonzero ordered roots from largest to smallest, and $i = 1, 2, \ldots, s = \min(v_h, p)$.

Thus Λ is equal to a product of the roots of (2.7.11).

The central distribution of the nonzero roots of (2.7.12) was determined independently by several statisticians: Fisher (1939), Girshick (1939a), Hsu (1939), and Roy (1939).

Numerous test statistics proposed in the literature, with important applications in multivariate analysis, have been tabulated and are functions of the roots of (2.7.12). The distributions are complex mathematical expressions, and their forms are not directly relevant to warrant a review of all original derivations in a text primarily concerned with applications. Instead, available sources of tables, notation, and use in the literature is of concern. Bibliographies in the cited references mention journals that contain the forms of the density functions.

Roy (1957) suggested the largest root λ_1 of (2.7.12a), *Roy's criterion*, as a multivariate test statistic and derived the null distribution of

(2.7.13) $$\theta_s = \frac{\lambda_1}{1 + \lambda_1}$$

which is the largest root of (2.7.12c). Pillai (1960, 1965, 1967) prepared tables for θ_s, and Heck (1960) prepared charts. A portion of Pillai's tables and Heck's charts are reproduced in Table VII of the Appendix.

Lawley (1938) and Hotelling (1951) proposed the statistic

$$T_0^2 = \nu_e \operatorname{Tr}(\mathbf{W}_2\mathbf{W}_1^{-1}) = \nu_e \sum_{i=1}^{s} \lambda_i \tag{2.7.14}$$

as a test statistic, known as the *Lawley-Hotelling trace criterion* or *Hotelling's generalized T_0^2 statistic*. Pillai (1960) tabled percentage points of the distribution of the sum of the roots of (2.7.12a):

$$U^{(s)} = \sum_{i=1}^{s} \lambda_i \tag{2.7.15}$$

Davis (1970) also tabulated percentage points for T_0^2/ν_h. For Davis, $T_0^2/\nu_h = (\nu_e/\nu_h)U^{(s)}$. In the literature, (2.7.15) is also referred to as the *trace criterion* or *Hotelling's trace criterion*. Table XI in the Appendix contains percentage points for $U^{(s)}$.

Pillai (1960) suggested a trace criterion known as *Pillai's trace criterion*, which is the sum of the roots of (2.7.12c). A table of percentage points for his test statistic

$$V^{(s)} = \operatorname{Tr}[\mathbf{W}_2(\mathbf{W}_2 + \mathbf{W}_1)^{-1}] = \sum_{i=1}^{s} \theta_i \tag{2.7.16}$$

is given in Table XI of the Appendix. We shall consider table usage later in the text.

Except for Hotelling's T^2 statistic, only central or null percentage points for the preceding statistics are given. James (1964) derived the noncentral density of the roots of (2.7.12a); however, there is a dearth of information on the noncentral distributions of the proposed test statistics, except in special cases such as Roy (1966), Ito (1962, 1969), and Pillai (1970). As a result, material addressed to the comparative power of the various criteria is unavailable. Gabriel (1968, 1969), Schatzoff (1966), Posten and Bargmann (1964), and others have begun to investigate this specific area. See, especially, Olson (1973). A discussion on power and robustness of the criteria will be provided when we apply test statistics to specific problems.

In addition to test statistics dependent on the distribution of the roots of (2.7.12), Krishnaiah (1969) reviewed proposed test statistics that depend on the multivariate χ^2, t, and F distributions. Roy (1958), Krishnaiah (1965, 1969), and Das Gupta (1970) also proposed step-down procedures applicable to multivariate data analysis. Gabriel (1969) reviewed multivariate test procedures and multivariate distributions. For a complete bibliography on multivariate analysis, see Anderson, Das Gupta, and Styan (1972).

EXERCISES 2.7

1. If $\mathbf{W}_1 \sim W_p(\nu_e, \mathbf{\Sigma})$ and $\mathbf{W}_2 \sim W_p(\nu_h, \mathbf{\Sigma})$, and if $\mathbf{W}_1$ and $\mathbf{W}_2$ are statistically independent, show that $|\mathbf{W}_1|/|\mathbf{W}_1 + \mathbf{W}_2|$ is distributed as a product of independent beta variables for $\nu_e \geq p$. *Hint:* see Rao (1965, p. 457).
2. For $\mathbf{Y}_1, \ldots, \mathbf{Y}_N \sim IN_p(\boldsymbol{\mu}, \mathbf{\Sigma})$ and $T^2 = N(\bar{\mathbf{Y}} - \boldsymbol{\mu})'\mathbf{S}^{-1}(\bar{\mathbf{Y}} - \boldsymbol{\mu})$, show that
$$T^2 = N \max_{\mathbf{a}} \frac{[\mathbf{a}'(\bar{\mathbf{Y}} - \boldsymbol{\mu})]^2}{\mathbf{a}'\mathbf{S}\mathbf{a}}$$
for arbitrary vectors a and that
$$\Lambda = \frac{1}{1 + T^2/(N-1)}.$$

CHAPTER 3

ONE- AND TWO-SAMPLE TESTS ON MEANS AND THE GENERAL LINEAR MODEL

3.1 INTRODUCTION

Traditionally, a discussion of univariate statistical inference is begun with the study of statistical tests of hypotheses regarding means by assuming that population variances are known (see Hays, 1963). However, known population variances are seldom available in the behavioral sciences. This chapter is limited to the generalization of familiar univariate tests on means in one- and two-normal-sample cases to their multivariate analogues, when variances are unknown. Multivariate extensions of univariate tests with known variances are presented by Anderson (1958).

Tests on the means of one or two univariate populations are easily obtained from the general linear univariate model or the Gauss-Markoff setup. Integrating this approach with more traditional approaches is considered in this chapter. With a firm understanding of the univariate procedures, an extension to the multivariate situation will be made.

Reliance on univariate procedures for extension by analogy will provide the justification for some of the multivariate methods developed, since many mathematical proofs of general statistical theories are not fully reproduced in this text.

3.2 ONE-SAMPLE PROBLEM—UNIVARIATE

To integrate the theory of the general linear model with estimation and hypothesis test theory, familiar to users of univariate methods, we review univariate analysis through the one-sample problem.

Estimation Theory. To provide inferences about a random variable Y, a random sample of size N of N random variables $Y_1, Y_2, \ldots, Y_N$, which reflect the characteristics of the distribution of Y, is drawn from a distribution $f(y|\theta_1, \theta_2, \ldots, \theta_m)$ dependent on some unknown parameters $\{\theta_i\}$. The *sample data* or *observed sample values* of the random variables $Y_1, Y_2, \ldots, Y_N$, denoted by $y_1, y_2, \ldots, y_N$, are used to find estimates for the population parameters and to test hypotheses about them. In this text, capital letters will generally be used to denote random variables, and lowercase letters will denote the value of the random variable.

The family of functions for a random variable Y depends on a set of parameters $\{\theta_i\}$ restricted to the *parameter space* Ω, which defines all admissible values of θ_i. Let $\hat{\theta}_i$ denote the estimate of the population parameter θ_i. Strictly speaking, a distinction should be made between an estimate and an estimator. An *estimator* is a random variable that is a function of $Y_1, Y_2, \ldots, Y_N$. An *estimate* is the value of the estimator and is a function of $y_1, y_2, \ldots, y_N$. In this text, the same symbol will at times denote both, since an estimator is obtained with the replacement of sample-data values by random variables.

(3.2.1) It is desirable that estimators have the properties of being (1) unbiased, (2) minimum-variance unbiased, (3) consistent, and (4) best asymptotically normal (BAN).

Consistent and BAN estimators are properties that depend on large samples. The other characteristics are satisfied by estimators obtained for any size sample.

(3.2.2)

(a) An estimator $\hat{\theta}$ is called an *unbiased* estimator for θ if $E(\hat{\theta}) = \theta$.

(b) If $\hat{\theta}$ is unbiased for θ and has the smallest variance among all such estimators, $\hat{\theta}$ is said to be a *minimum-variance unbiased estimator.*

(c) If $\hat{\theta}_1, \hat{\theta}_2, \ldots, \hat{\theta}_N$ are estimators for θ and depend on $Y_1, Y_2, \ldots, Y_N$ (random variables) so that $\hat{\theta}_N$ converges in probability to θ as N increases

$$\lim_{N\to\infty} P(|\hat{\theta}_N - \theta| \geq \delta) = 0 \qquad \text{for every } \delta > 0$$

then $\hat{\theta}_N$ is *consistent* for θ.

(d) A consistent estimator $\hat{\theta}_N$ is BAN if the asymptotic distribution of $\sqrt{N}\,(\hat{\theta}_N - \theta)$ is normal with mean 0 and the variance $\sigma^2(\theta)$ has the least possible value.

Numerous procedures have been proposed for obtaining "best" estimators having some of the properties specified in (3.2.1). However, the two most frequently used methods in statistical analysis are those of maximum likelihood and least squares, discussed in Sections 2.3 and 1.8.

To investigate both procedures for the one-sample problem, suppose that a random sample of N observations with values $y_1, y_2, \ldots, y_N$ is obtained from a univariate normal population with mean μ and unknown variance σ^2 so that

$Y_i \sim IN(\mu, \sigma^2)$. An alternative way to state these assumptions is as follows:

$$Y_i = \mu + \varepsilon_i \qquad i = 1, \ldots, N \tag{3.2.3}$$

$$\varepsilon_i \sim IN(0, \sigma^2) \tag{3.2.4}$$

The Y_i's and ε_i's have the same variance because the two differ by a constant and, since $E(Y_i) = \mu$, $E(\varepsilon_i) = 0$. The random variable ε_i denotes the random error component of Y_i; by (3.2.4), the errors are uncorrelated with common variance σ^2.

Employing matrix notation and the multivariate normal density function, let $\mathbf{Y}$ denote the $N \times 1$ random vector, $\boldsymbol{\mu}$ the $N \times 1$ vector of means with elements μ, and $\sigma^2 \mathbf{I}_N$ the variance-covariance matrix of the vector $\mathbf{Y}$. Then $Y_i \sim IN(\mu, \sigma^2)$, for $i = 1, \ldots, N$, is represented by $\mathbf{Y} \sim N_N(\mu, \sigma^2 I_N)$. Using this notation, (3.2.3) and (3.2.4) become

$$\underset{(N \times 1)}{\mathbf{Y}} = \underset{(N \times 1)}{\mathbf{X}} \; \underset{(1 \times 1)}{\boldsymbol{\beta}} + \underset{(N \times 1)}{\boldsymbol{\varepsilon}}$$

$$\begin{bmatrix} Y_1 \\ Y_2 \\ \vdots \\ Y_N \end{bmatrix} = \begin{bmatrix} 1 \\ 1 \\ \vdots \\ 1 \end{bmatrix} \mu + \begin{bmatrix} \varepsilon_1 \\ \varepsilon_2 \\ \vdots \\ \varepsilon_N \end{bmatrix} \tag{3.2.5}$$

$$\boldsymbol{\varepsilon} \sim N_N(\mathbf{0}, \sigma^2 \mathbf{I}_N) \tag{3.2.6}$$

Representation (3.2.5), with the condition that

$$\begin{aligned} E(\mathbf{Y}) &= \mathbf{X}\boldsymbol{\beta} = \mu \quad \text{or} \quad E(\boldsymbol{\varepsilon}) = \mathbf{0} \\ V(\mathbf{Y}) &= \sigma^2 \mathbf{I}_N \quad \text{or} \quad V(\boldsymbol{\varepsilon}) = \sigma^2 \mathbf{I}_N \end{aligned} \tag{3.2.7}$$

is a special case of the general linear univariate model or the univariate Gauss-Markoff setup. The expected value of Y_i is linear in the elements of β and the random errors ε_i are uncorrelated with common variance σ^2. Condition (3.2.6) is usually not considered part of the Gauss-Markoff setup; however, it is needed to test hypotheses about μ as well as to obtain maximum-likelihood estimators for μ and σ^2.

Maximum-likelihood estimators for μ and σ^2 can only be obtained if the density function for the random variables Y_i is known. Given that $Y_i \sim IN(\mu, \sigma^2)$, we demonstrated in Section 2.3 that

$$\hat{\mu} = \bar{y} \quad \text{and} \quad \hat{\sigma}^2 = \frac{\sum_{i=1}^{N} (y_i - \bar{y})^2}{N}$$

were the maximum-likelihood estimators for the parameters μ and σ^2. In general, maximum-likelihood estimators are not unbiased but they are consistent and are usually BAN estimators.

To obtain the least-squares estimator for μ, the error or residual sum of squares

$$Q = \sum_{i=1}^{N} \varepsilon_i^2 = \sum_{i=1}^{N} (y_i - \mu)^2$$

is minimized. Taking the partial derivative of Q with respect to μ and equating it to 0, we find that the least-squares estimator for μ is the same as the maximum-likelihood estimator; that is,

$$\frac{\partial Q}{\partial \mu} = 2 \sum_{i=1}^{N} (y_i - \mu)(-1) = 0$$

$$N\mu - \sum_{i=1}^{N} y_i = 0$$

$$\hat{\mu} = \bar{y}$$

The least-squares procedure does not yield an estimator for σ^2 directly; however, in terms of the estimator for μ, an estimator obtained from Q, which is unbiased for σ^2, is shown to be

$$s^2 = \frac{\sum_{i=1}^{N} (y_i - \hat{\mu})^2}{N - 1}$$

Under the normality assumption, the least-squares estimator for μ and the unbiased estimator for σ^2, expressed as a function of $\hat{\mu}$, are consistent and minimum-variance unbiased estimators.

Estimating μ and more generally $\boldsymbol{\beta}$, when the density function of $\mathbf{Y}$ is unknown, the univariate Gauss-Markoff theorem is employed and will be discussed in Section 3.5. For this case, we need to define a new class of restricted minimum-variance estimators.

Hypothesis Testing. Statistical hypothesis testing involves assertions about the parameters of a density function $f(y|\theta_1, \theta_2, \ldots, \theta_m)$ of a random variable Y. Partitioning the parameter space Ω into disjoint regions ω and $\Omega - \omega$ such that $\Omega = \omega \cup \Omega - \omega$ and $\omega \cap \Omega - \omega = \phi$, a statistical hypothesis called the *null hypothesis*, denoted by H_0, states that the parameters of a density function are elements of ω. The statement that the parameters are elements of the space $\Omega - \omega$ is termed the *alternative hypothesis* and is written as H_1. If a hypothesis specifies the values of all the parameters of a density function, it is called a *simple hypothesis*; otherwise, it is called a *composite hypothesis.*

Testing the null hypothesis H_0 against the alternative hypothesis H_1, a random sample of N observations is obtained from the distribution of the random variable Y. The N-tuple of observed values of the random variable Y is

$$\mathbf{y}' = [y_1, y_2, \ldots, y_N]$$

Let S denote the sample space of all possible observed N-tuples $\mathbf{y}$. The problem in test theory is to partition the sample space S into two regions S_ω and $S_{\Omega-\omega}$, where $S_\omega \cup S_{\Omega-\omega} = S$ and $S_\omega \cap S_{\Omega-\omega} = \phi$. Thus, if a sample point $\mathbf{y} \in S_\omega$, the null hypothesis is favored; whereas, if $\mathbf{y} \in S_{\Omega-\omega}$, the null hypothesis is not favored and the alternative hypothesis is tenable. The region $S_{\Omega-\omega}$ of the sample space is termed the *critical region* for the test.

In testing any hypothesis, there are two types of errors an experimenter may make. Given that H_0 represents the true state of nature, he may reject H_0 in favor of H_1. This is called a *type I error*. The probability associated with this error is denoted by α and is called the *significance level* or *size* of the test:

$$\alpha = P(\mathbf{y} \in S_{\Omega-\omega}|\theta \in \omega) = P\,(\text{rejecting } H_0|H_0 \text{ is true})$$

Alternatively, if H_1 specifies the true state of nature, he may reject H_1 in favor of H_0. That is, he may accept H_0 when H_1 is really true. This is termed a *type II error*. The probability of this happening is denoted by β:

$$\beta = P(\mathbf{y} \in S_\omega | \theta \in \Omega - \omega) = P\,(\text{accepting } H_0 | H_1 \text{ is true})$$

The other alternatives an experimenter may make are *correct decisions.* Table 3.2.1 is constructed to represent the aforementioned decisions and their corresponding probabilities. The complement $1 - \beta$ of a type II error probability is the *power* of the test and the complement $1 - \alpha$ of a type I error probability is the *confidence* of the test.

TABLE 3.2.1

Decision	H_0 is true	H_1 is true
reject H_0	type I error (α)	no error ($1 - \beta$)
accept H_0	no error ($1 - \alpha$)	type II error (β)

In principle, an experimenter would like to have both errors as small as possible. However, for a fixed sample size, the size of the type II error will usually increase as the size of the type I error decreases. In the spirit of classical test theory, for a fixed sample size, a value for the type I error that one can tolerate is chosen; then, among all tests with a fixed level of significance, a rejection region is selected that minimizes β. This fundamental idea underlies test theory and was suggested by Neyman and Pearson (1933). It is known as the *fundamental lemma of Neyman and Pearson* and was originally proposed to test a simple null hypothesis against a simple alternative hypothesis.

In review of the Neyman-Pearson lemma, suppose that a simple hypothesis is tested against a simple alternative:

$$\begin{aligned} H_0 &: \theta = \theta_0 \\ H_1 &: \theta = \theta_1 \end{aligned} \qquad (\theta_1 > \theta_0)$$

for a random variable Y with a probability density function $f(y|\theta)$ that depends, for convenience, on only one parameter. Let $\mathbf{y} \in S$; then the value of the likelihood function is a measure of the "likelihood" of observing $\mathbf{y}$ when θ is the "true" parameter. Let

$$\lambda = \frac{\prod_i f(y_i|\theta_0)}{\prod_i f(y_i|\theta_1)} = \frac{L(\mathbf{y}|\theta_0)}{L(\mathbf{y}|\theta_1)}$$

denote the ratio of the value of the likelihood function under H_0, divided by the value of the likelihood function under H_1. If $L(y|\theta_0) < L(\mathbf{y}|\theta_1)$, the tendency would be to reject H_0 in favor of H_1 for small values of λ such that

$$P(\lambda < \lambda_0) = \alpha$$

which specifies the best critical region for the test. This procedure is shown to yield a most powerful test that minimizes β or maximizes power. For a proof, see Hoel (1971) or Mood and Graybill (1963). Treatment of the topic at a more advanced level may be found in Lehmann's (1959) text.

The Neyman-Pearson lemma was developed for simple hypotheses. For composite hypotheses, there are situations in which the procedure yields a best test; however, the nature of the composite hypotheses and of the density function greatly restricts their use in practice without placing further restrictions on the class of tests. An alternative procedure for constructing tests involving composite hypotheses employs the *generalized-likelihood-ratio criterion*, which does not usually yield best tests in terms of maximizing power. Such a test is called a *likelihood-ratio test* (LRT).

The construction of an LRT is similar to the procedure outlined by the Neyman-Pearson lemma except that the alternative hypothesis is not specifically considered. Let $Y \sim f(y|\theta_1, \ldots, \theta_m)$ and $\mathbf{y} \in S$. To test the hypothesis that the parameters are in ω against the alternative hypothesis that they are in $\Omega - \omega$, two likelihood functions are again considered. Let

$$L = L(\mathbf{y} \mid \theta_1, \ldots, \theta_m) = \prod_{i=1}^{N} f(y_i \mid \theta_1, \ldots, \theta_m)$$

denote the likelihood function, $L(\hat{\omega})$ the value of the likelihood function maximized under the null hypothesis H_0, and $L(\hat{\Omega})$ the value of the likelihood function maximized over the whole parameter space Ω. Defining λ as the ratio of the likelihoods, the likelihood ratio (LR)

$$\lambda = \frac{L(\hat{\omega})}{L(\hat{\Omega})} \tag{3.2.8}$$

is constructed. Since $\omega \subset \Omega$, $L(\hat{\omega}) \leq L(\hat{\Omega})$, so that $0 \leq \lambda \leq 1$. $L(\hat{\omega})$ is a measure of the "likelihood" that the hypothesis is "true"; it would not be increased significantly by allowing the parameters to assume values in Ω if λ is near 1. Thus, when λ is near 1, the null hypothesis is favored. If λ is near 0, the values of the parameters, other than those restricted by H_0, would more likely be implying that the null hypothesis is not tenable or that H_0 would be rejected for small values of λ. A critical region for an LRT is constructed by using λ so that the null hypothesis is rejected if $\lambda < \lambda_0$, where the $P(\lambda < \lambda_0) = \alpha$. For large sample sizes and under very general conditions, Wald (1943) showed that $-2 \log \lambda \mathrel{\dot\sim} \chi^2(v)$ when H_0 is true as $N \to \infty$, where the degrees of freedom v are equal to the number of independent parameters estimated in Ω minus the number of independent parameters estimated under ω.

To illustrate the LRT procedure, the familiar one-sample case is utilized. Suppose a random sample of N observations with observed values $y_1, \ldots, y_N$ is obtained from a normal distribution with mean μ and unknown variance σ^2, where $Y_i \sim IN(\mu, \sigma^2)$. To test the hypothesis

$$\begin{aligned} H_0&: \mu = \mu_0 \qquad \text{(specified)} \\ H_1&: \mu \neq \mu_0 \end{aligned} \tag{3.2.9}$$

Student's t distribution is employed (Section 2.7), where t is defined by

$$t = \frac{\sqrt{N}(\bar{y} - \mu_0)}{s} \sim t(N-1) \tag{3.2.10}$$

when the null hypothesis is true.

Alternatively, let $d = (\bar{y} - \mu_0)$ represent the distance between the sample mean and the hypothesized population mean. Then (3.2.10) becomes

$$t = \frac{d\sqrt{N}}{s} = D\sqrt{N} \tag{3.2.11}$$

where D denotes the studentized distance. The relationship among t, D, and F is related by the formula $t^2 = ND^2 = F$.

To derive the test criterion in (3.2.10) to test (3.2.9), the LR in (3.2.8) is used. Here $\Omega = \{\mu, \sigma^2\}$ and $\omega = \{\mu = \mu_0, \sigma^2\}$. Since the value of σ^2 is not specified by the null hypothesis, the null hypothesis is composite. Furthermore, $Y \sim N(\mu, \sigma^2)$ so that the likelihood function is

$$L = L(\mathbf{y} \mid \mu, \sigma^2) = (2\pi\sigma^2)^{-N/2} \exp - \frac{\sum_{i=1}^{N} (y_i - \mu)^2}{2\sigma^2}$$

The estimators that maximize L in Ω are

$$\hat{\mu}_\Omega = \bar{y} \quad \text{and} \quad \hat{\sigma}^2_\Omega = \frac{\sum_{i=1}^{N} (y_i - \bar{y})^2}{N}$$

Substituting these values into L,

$$\begin{aligned} L(\hat{\Omega}) &= \left(\frac{1}{2\pi\sigma^2_\Omega}\right)^{N/2} \exp - \frac{\sum_{i=1}^{N} (y_i - \bar{y})^2}{2\hat{\sigma}^2_\Omega} \\ &= \left(\frac{1}{2\pi\hat{\sigma}^2_\Omega}\right)^{N/2} e^{-N/2} \end{aligned}$$

Maximizing L under ω, $\mu = \mu_0$ and the estimator for σ^2 is shown to be

$$\hat{\sigma}^2_\omega = \frac{\sum_{i=1}^{N} (y_i - \mu_0)^2}{N}$$

so that

$$\begin{aligned} L(\hat{\omega}) &= \left(\frac{1}{2\pi\hat{\sigma}^2_\omega}\right)^{N/2} \exp - \frac{\sum_{i=1}^{N} (y_i - \mu_0)^2}{2\hat{\sigma}^2_\omega} \\ &= \left(\frac{1}{2\pi\hat{\sigma}^2_\omega}\right)^{N/2} e^{-N/2} \end{aligned}$$

Thus

$$\lambda = \frac{L(\hat{\omega})}{L(\hat{\Omega})} = \left(\frac{\hat{\sigma}_\Omega^2}{\hat{\sigma}_\omega^2}\right)^{N/2}$$

$$= \left[\frac{\sum_{i=1}^{N} (y_i - \bar{y})^2}{\sum_{i=1}^{N} (y_i - \mu_0)^2}\right]^{N/2}$$

But

$$\sum_{i=1}^{N} (y_i - \mu_0)^2 = \sum_{i=1}^{N} (y_i - \bar{y} + \bar{y} - \mu_0)^2 = \sum_{i=1}^{N} [(y_i - \bar{y}) + (\bar{y} - \mu_0)]^2$$

$$= \sum_{i=1}^{N} (y_i - \bar{y})^2 + N(\bar{y} - \mu_0)^2$$

so that

$$\lambda = \left[\frac{1}{1 + N(\bar{y} - \mu_0)^2 \Big/ \sum_{i=1}^{N} (y_i - \bar{y})^2}\right]^{N/2} = \left[\frac{1}{1 + t^2/(N-1)}\right]^{N/2}$$

It is not necessary to determine the distribution of λ in this case since if $t^2 = 0, \lambda = 1$, and as $t^2 \to \infty, \lambda \to 0$. Thus rejecting small values of λ is equivalent to rejecting large values of t^2. Testing (3.2.9) at the significance level α, H_0 is rejected if

$$|t| > t^{\alpha/2}(N - 1) \tag{3.2.12}$$

where $t^{\alpha/2}(N - 1)$ denotes the upper $\alpha/2$ percentage point of the Student's t distribution with $N - 1$ degrees of freedom.

Confidence Intervals. A *confidence interval* (set) for the parameter μ is obtained by using a general method (see Neyman, 1937). Let $\mathbf{y} \in S$ and $\theta \in \Omega$, where $Y \sim f(y \mid \theta)$. Suppose a mapping is defined that assigns to each $\mathbf{y} \in S$ a set $t(\mathbf{y})$ in Ω. This mapping defines level $1 - \alpha$ confidence intervals (sets) for θ if the $P[\theta \in t(\mathbf{y}) \mid \theta] = 1 - \alpha$ for all $\theta \in \Omega$. With $S_{\Omega-\omega}$, the critical region of size α for testing that θ is the true parameter, the $P(\mathbf{y} \in S_{\Omega-\omega} \mid \theta) = \alpha$, for all $\theta \in \omega$. Define

$$t(\mathbf{y}) = \{\theta \mid \mathbf{y} \in S_\omega\} \tag{3.2.13}$$

then the $P[\theta \in t(\mathbf{y}) \mid \theta] = P(\mathbf{y} \in S_\omega \mid \theta) = 1 - \alpha$, since $P(\mathbf{y} \notin S_\omega \mid \theta) = \alpha$. The set $t(\mathbf{y})$ is a $100(1 - \alpha)\%$ *confidence-interval (set) estimator* of θ for a given $\mathbf{y}$. That is, the probability that the interval (set) includes the true value of θ is $1 - \alpha$.

Applying these principles to the one-sample problem,

$$S_{\Omega-\omega} = \{\mathbf{y} \mid |t| > t^{\alpha/2}(N - 1)\}$$

$$= \left\{\mathbf{y} \,\middle|\, \left|\frac{\sqrt{N}(\bar{y} - \mu_0)}{s}\right| > t^{\alpha/2}(N - 1)\right\}$$

so

$$t(\mathbf{y}) = \{\mu \mid \mathbf{y} \in S_\omega\}$$
$$= \left\{\mu_0 - \frac{t^{\alpha/2}(N-1)s}{\sqrt{N}} \leq \bar{y} \leq \mu_0 + \frac{t^{\alpha/2}(N-1)s}{\sqrt{N}}\right\}$$
$$= \left\{\mu \mid \bar{y} - \frac{st^{\alpha/2}(N-1)}{\sqrt{N}} \leq \mu \leq \bar{y} + \frac{t^{\alpha/2}(N-1)s}{\sqrt{N}}\right\}$$

is the confidence interval (set) for μ defined by the points

$$\left(\bar{y} - \frac{st^{\alpha/2}(N-1)}{\sqrt{N}}, \bar{y} + \frac{st^{\alpha/2}(N-1)}{\sqrt{N}}\right)$$

A $100(1-\alpha)\%$ confidence interval for the parameter μ is determined by evaluating the expression

$$\bar{y} - \frac{s}{\sqrt{N}} t^{\alpha/2}(N-1) \leq \mu \leq \bar{y} + \frac{s}{\sqrt{N}} t^{\alpha/2}(N-1) \tag{3.2.14}$$

derived by using the test statistic and the complement of the rejection region when testing the null hypothesis.

EXERCISES 3.2

1. Verify the expression for $\hat{\sigma}^2_\omega$ (the estimate of σ^2 obtained under the hypothesis) for the one-sample t test.
2. Suppose that a random sample of size N is obtained from a normal distribution with mean μ and known variance σ^2, then obtain the LRT of size α for the hypothesis $H_0 : \mu = \mu_0$ against the alternative hypothesis $H_1 : \mu \neq \mu_0$. Also, determine a $100(1-\alpha)\%$ confidence interval for the parameter μ.
3. If $Y_i \sim IN(\mu, \sigma^2)$, for $i = 1, \ldots, N$, obtain the LRT of size α for the hypothesis $H_0 : \sigma^2 = \sigma_0^2$ against the alternative hypothesis $H_1 : \sigma^2 \neq \sigma_0^2$; also, obtain a $100(1-\alpha)\%$ confidence interval for σ^2.

3.3 ONE-SAMPLE PROBLEM—MULTIVARIATE

Rather than observing one random variable, studies in the behavioral sciences are primarily concerned with random samples of p-vector-valued random response variables. Suppose we obtain a random sample of N p-vector random variables Y_i:

$$\begin{aligned} \mathbf{Y}_1' &= [Y_{11}, Y_{12}, \cdots, Y_{1p}] \\ \mathbf{Y}_2' &= [Y_{21}, Y_{22}, \ldots, Y_{2p}] \\ &\vdots \qquad \vdots \\ \mathbf{Y}_N' &= [Y_{N1}, Y_{N2}, \ldots, Y_{Np}] \end{aligned} \tag{3.3.1}$$

or, as a matrix,

(3.3.2)
$$\underset{(N\times p)}{\mathbf{Y}} = \begin{bmatrix} Y_{11} & Y_{12} & \cdots & Y_{1p} \\ Y_{21} & Y_{22} & \cdots & Y_{2p} \\ \vdots & \vdots & \vdots\vdots\vdots & \vdots \\ Y_{N1} & Y_{N2} & \cdots & Y_{Np} \end{bmatrix}$$

Further, suppose that each row of $\mathbf{Y}$ is derived from a multivariate normal distribution with mean vector $\boldsymbol{\mu}$ and variance-covariance matrix $\boldsymbol{\Sigma}$ so that $\mathbf{Y}_i \sim IN_p(\boldsymbol{\mu}, \boldsymbol{\Sigma})$.

By employing the general linear multivariate model, the observations are represented in the following linear model form:

$$\underset{(N\times p)}{\mathbf{Y}} = \underset{(N\times 1)}{\mathbf{X}} \quad \underset{(1\times p)}{\mathbf{B}} + \underset{(N\times p)}{\mathbf{E}_0}$$

$$\begin{bmatrix} Y_{11} & Y_{12} & \cdots & Y_{1p} \\ Y_{21} & Y_{22} & \cdots & Y_{2p} \\ \vdots & \vdots & \vdots\vdots\vdots & \vdots \\ Y_{N1} & Y_{N2} & \cdots & Y_{Np} \end{bmatrix} = \begin{bmatrix} 1 \\ 1 \\ \vdots \\ 1 \end{bmatrix} [\mu_1, \mu_2, \ldots, \mu_p] + \begin{bmatrix} \varepsilon_{11} & \varepsilon_{12} & \cdots & \varepsilon_{1p} \\ \varepsilon_{21} & \varepsilon_{22} & \cdots & \varepsilon_{2p} \\ \vdots & \vdots & \vdots\vdots\vdots & \vdots \\ \varepsilon_{N1} & \varepsilon_{N2} & \cdots & \varepsilon_{Np} \end{bmatrix}$$

(3.3.3)
$$E(\mathbf{Y}) = \mathbf{XB} = \begin{bmatrix} \mu_1 & \mu_2 & \cdots & \mu_p \\ \mu_1 & \mu_2 & \cdots & \mu_p \\ \vdots & \vdots & \vdots\vdots\vdots & \vdots \\ \mu_1 & \mu_2 & \cdots & \mu_p \end{bmatrix}$$

$$V(\mathbf{Y}) = \mathbf{I}_N \otimes \boldsymbol{\Sigma} = \begin{bmatrix} \boldsymbol{\Sigma} & \cdot & \cdots & \mathbf{0} \\ \mathbf{0} & \boldsymbol{\Sigma} & \cdots & \mathbf{0} \\ \vdots & \vdots & \vdots\vdots\vdots & \vdots \\ \mathbf{0} & \mathbf{0} & \cdots & \boldsymbol{\Sigma} \end{bmatrix}$$

where $\boldsymbol{\Sigma}$ is the common variance-covariance matrix of any row of $\mathbf{Y}$. Observation of the variance-covariance matrix of the data matrix $\mathbf{Y}$ indicates that, for $i \neq i'$, each row of $\mathbf{Y}$ is uncorrelated. Representation (3.3.3) is a special case of the multivariate Gauss-Markoff setup.

Each column of $\mathbf{Y}$ follows a general univariate model. Let $\boldsymbol{\Sigma} = [\sigma_{ij}]$, $\mathbf{Y} = [\mathbf{U}_1 \quad \mathbf{U}_2 \quad \cdots \quad \mathbf{U}_p]$, and $\boldsymbol{\mu}' = [\mu_1, \mu_2, \ldots, \mu_p]$, where $\mathbf{U}_j$ and μ_j are, respectively, the jth columns of $\mathbf{Y}$ and $\boldsymbol{\mu}$; then the above representation becomes

(3.3.4)
$$E(\mathbf{U}_j) = \begin{bmatrix} 1 \\ 1 \\ \vdots \\ 1 \end{bmatrix} \mu_j$$

$$\text{cov}(\mathbf{U}_j, \mathbf{U}_{j'}) = \sigma_{jj'}\mathbf{I}_N$$

In testing the hypothesis that the means for all variables are specified

$$H_0: \begin{bmatrix} \mu_1 \\ \mu_2 \\ \vdots \\ \mu_p \end{bmatrix} = \begin{bmatrix} \mu_{01} \\ \mu_{02} \\ \vdots \\ \mu_{0p} \end{bmatrix} \quad \text{(specified)}$$

$$H_1: \begin{bmatrix} \mu_1 \\ \mu_2 \\ \vdots \\ \mu_p \end{bmatrix} \neq \begin{bmatrix} \mu_{01} \\ \mu_{02} \\ \vdots \\ \mu_{0p} \end{bmatrix} \quad \text{(say)}$$

or

$$\begin{aligned} &H_0: \boldsymbol{\mu} = \boldsymbol{\mu}_0 \qquad \text{(specified)} \\ &H_1: \boldsymbol{\mu} \neq \boldsymbol{\mu}_0 \end{aligned} \tag{3.3.5}$$

we again consider the t statistic. Representing t^2 as

$$\begin{aligned} t^2 &= \frac{N(\bar{y} - \mu_0)^2}{s^2} \\ &= N(\bar{y} - \mu_0)(s^2)^{-1}(\bar{y} - \mu_0) \end{aligned} \tag{3.3.6}$$

where $(s^2)^{-1}$ is the inverse of the estimate of the common variance σ^2, and replacing $\bar{y}$ and s^2 by their multivariate analogues,

$$\bar{\mathbf{y}} = \frac{1}{N} \sum_{i=1}^{N} y_i$$

and

$$\mathbf{S} = \frac{1}{N-1} \sum_{i=1}^{N} (\mathbf{y}_i - \bar{\mathbf{y}})(\mathbf{y}_i - \bar{\mathbf{y}})'$$

where $\mathbf{y}_i$ denotes the ith observed row of $\mathbf{Y}$, t^2 becomes

$$T^2 = N(\bar{\mathbf{y}} - \boldsymbol{\mu}_0)'\mathbf{S}^{-1}(\bar{\mathbf{y}} - \boldsymbol{\mu}_0) \tag{3.3.7}$$

This quadratic form is the single-sample Hotelling's T^2 statistic used to test (3.3.5).

As in the univariate case,

$$T^2 = ND^2 \qquad \text{where } D^2 = (\bar{\mathbf{y}} - \boldsymbol{\mu}_0)'\mathbf{S}^{-1}(\bar{\mathbf{y}} - \boldsymbol{\mu}_0) \tag{3.3.8}$$

The quantity D^2 is termed *Mahalanobis' D^2 statistic* (see Mahalanobis, 1936). For T^2 to have an F distribution, (2.7.8) is used so that

$$\frac{(N-p)T^2}{(N-1)p} = \frac{(N-p)ND^2}{(N-1)p} \sim F(p, N-p) \tag{3.3.9}$$

when H_0 is true.

The test of H_0 is rejected at the significance level α if

$$\frac{(N-p)T^2}{p(N-1)} > F^\alpha(p, N-p) \tag{3.3.10}$$

where $F^\alpha(p, N-p)$ is the upper α percentage point of the F distribution with degrees of freedom p and $N-p$ found in Table IV of the Appendix.

From Table VI in the Appendix, H_0 is rejected at the significance level α if

$$T^2 > T^\alpha(p, N-1) \tag{3.3.11}$$

where $T^\alpha(p, N-1)$ denotes the upper α percentage point of Hotelling's T^2 distribution.

The critical region given in (3.3.11) to test (3.3.5) was established by an extension, in an intuitive manner, of the result obtained for the univariate case. To prove (3.3.11), the LRT for testing (3.3.5) is derived. The likelihood function is

$$L = L(\mathbf{y} \mid \boldsymbol{\mu}, \boldsymbol{\Sigma}) = (2\pi)^{-Np/2}|\boldsymbol{\Sigma}|^{-N/2} \exp -\frac{1}{2}\sum_{i=1}^{N} (\mathbf{y}_i - \boldsymbol{\mu})'\boldsymbol{\Sigma}^{-1}(\mathbf{y}_i - \boldsymbol{\mu}) \tag{3.3.12}$$

where $\Omega = \{\boldsymbol{\mu}, \boldsymbol{\Sigma}\}$ and $\omega = \{\boldsymbol{\mu} = \boldsymbol{\mu}_0, \boldsymbol{\Sigma}\}$. Maximizing L in Ω, the estimators for $\boldsymbol{\mu}$ and $\boldsymbol{\Sigma}$ are the maximum-likelihood estimators

$$\hat{\boldsymbol{\mu}}_\Omega = \bar{\mathbf{y}} \quad \text{and} \quad \hat{\boldsymbol{\Sigma}}_\Omega = \frac{\sum_{i=1}^{N} (\mathbf{y}_i - \bar{\mathbf{y}})(\mathbf{y}_i - \bar{\mathbf{y}})'}{N}$$

Under the hypothesis,

$$\hat{\boldsymbol{\mu}}_\omega = \boldsymbol{\mu}_0 \quad \text{and} \quad \hat{\boldsymbol{\Sigma}}_\omega = \frac{\sum_{i=1}^{N} (\mathbf{y}_i - \boldsymbol{\mu}_0)(\mathbf{y}_i - \boldsymbol{\mu}_0)'}{N}$$

the likelihood-ratio criterion is

$$\lambda = \frac{L(\hat{\omega})}{L(\hat{\Omega})} = \frac{(2\pi)^{-Np/2}|\hat{\boldsymbol{\Sigma}}_\omega|^{-N/2}\, e^{-Np/2}}{(2\pi)^{-Np/2}|\hat{\boldsymbol{\Sigma}}_\Omega|^{-N/2}\, e^{-Np/2}}$$

$$= \frac{|\hat{\boldsymbol{\Sigma}}_\Omega|^{N/2}}{|\hat{\boldsymbol{\Sigma}}_\omega|^{N/2}} = \left[\frac{\left|\sum_{i=1}^{N} (\mathbf{y}_i - \bar{\mathbf{y}})(\mathbf{y}_i - \bar{\mathbf{y}})'\right|}{\left|\sum_{i=1}^{N} (\mathbf{y}_i - \boldsymbol{\mu}_0)(\mathbf{y}_i - \boldsymbol{\mu}_0)'\right|}\right]^{N/2}$$

Letting

$$\mathbf{Q}_e = \sum_{i=1}^{N} (\mathbf{y}_i - \bar{\mathbf{y}})(\mathbf{y}_i - \bar{\mathbf{y}})' = \mathbf{Y}'[\mathbf{I}_N - \mathbf{X}(\mathbf{X}'\mathbf{X})^{-1}\mathbf{X}']\mathbf{Y} \tag{3.3.13}$$

and noting that

$$\sum_{i=1}^{N} (\mathbf{y}_i - \boldsymbol{\mu}_0)(\mathbf{y}_i - \boldsymbol{\mu}_0)' = \mathbf{Q}_e + N(\bar{\mathbf{y}} - \boldsymbol{\mu}_0)(\bar{\mathbf{y}} - \boldsymbol{\mu}_0)'$$

we have

$$\Lambda = \lambda^{2/N} = \frac{|\mathbf{Q}_e|}{|\mathbf{Q}_e + N(\bar{\mathbf{y}} - \boldsymbol{\mu}_0)(\bar{\mathbf{y}} - \boldsymbol{\mu}_0)'|}$$

However, by applying (1.5.28) twice,

$$\begin{aligned}\Lambda &= \frac{|\mathbf{Q}_e|}{\begin{vmatrix} 1 & \sqrt{N}(\bar{\mathbf{y}} - \boldsymbol{\mu}_0)' \\ -\sqrt{N}(\bar{\mathbf{y}} - \boldsymbol{\mu}_0) & \mathbf{Q}_e \end{vmatrix}} \\ &= \frac{|\mathbf{Q}_e|}{|\mathbf{Q}_e|\,|1 + N(\bar{\mathbf{y}} - \boldsymbol{\mu}_0)'\mathbf{Q}_e^{-1}(\bar{\mathbf{y}} - \boldsymbol{\mu}_0)|} \\ &= \frac{1}{1 + [T^2/(N-1)]}\end{aligned}$$

where $\mathbf{S} = \mathbf{Q}_e/(N - 1)$ and

$$\begin{aligned}T^2 &= N(\bar{\mathbf{y}} - \boldsymbol{\mu}_0)'\mathbf{S}^{-1}(\bar{\mathbf{y}} - \boldsymbol{\mu}_0) \\ &= (N-1)\left[\frac{|\mathbf{Q}_e + N(\bar{\mathbf{y}} - \boldsymbol{\mu}_0)(\bar{\mathbf{y}} - \boldsymbol{\mu}_0)'|}{|\mathbf{Q}_e|} - 1\right]\end{aligned}$$

As in the univariate case, it is not necessary to determine the distribution of Λ. As $T^2 \to \infty$, $\Lambda \to 0$ and (3.3.11) follows.

Relating Λ to the F distribution,

$$\Lambda = \frac{|\mathbf{Q}_e|}{|\mathbf{Q}_e + N(\bar{\mathbf{y}} - \boldsymbol{\mu}_0)(\bar{\mathbf{y}} - \boldsymbol{\mu}_0)'|} = \frac{1}{1 + [T^2/(N-1)]} = \frac{1}{1 + [p/(N-1)]F} \tag{3.3.14}$$

From this relationship, the following statement may be made for the one-sample problem:

$$\frac{1 - \Lambda}{\Lambda}\,\frac{N - p}{p} \sim F(p, N - p) \tag{3.3.15}$$

The statistic Λ is Wilks' criterion and is related to the general U distribution. By employing Λ, H_0 is rejected at the significance level α if

$$\frac{1 - \Lambda}{\Lambda}\,\frac{N - p}{p} > F^{\alpha}(p, N - p) \quad \text{or} \quad \Lambda < U^{\alpha}(p, 1, N - 1) \tag{3.3.16}$$

where $U^{\alpha}(p, 1, N - 1)$ is used to denote the lower α percentage point of the U distribution. Tabled values of U are included in Table IX in the Appendix.

Returning for a moment to the more familiar univariate situation, the statistic $N(\bar{\mathbf{y}} - \boldsymbol{\mu}_0)(\bar{\mathbf{y}} - \boldsymbol{\mu}_0)'$ corresponds to $N(\bar{y} - \mu_0)^2$ and if the hypothesis $\mu = 0$ is being tested, $N(\bar{y} - \mu_0)^2 = N\bar{y}^2$ when H_0 is true. The quantity $N\bar{y}^2$ is just the sum of squares for the overall mean in a univariate ANOVA table or the hypothesis sum of squares for testing $\mu = 0$ with one degree of freedom. With this as motivation, let

$$\Lambda = \frac{|\mathbf{Q}_e|}{|\mathbf{Q}_e + \mathbf{Q}_h|} \tag{3.3.17}$$

where

$$\mathbf{Q}_h = N(\bar{\mathbf{y}} - \boldsymbol{\mu}_0)(\bar{\mathbf{y}} - \boldsymbol{\mu}_0)' \tag{3.3.18}$$
$$\mathbf{Q}_e = \mathbf{Y}'[\mathbf{I} - \mathbf{X}(\mathbf{X}'\mathbf{X})^{-1}\mathbf{X}']\mathbf{Y}$$

Later in this chapter, we shall show that $\mathbf{Q}_h$ and $\mathbf{Q}_e$ are distributed as independent Wishart matrices; thus, by (2.7.12),

$$\Lambda = \prod_{i=1}^{s} v_i = \prod_{i=1}^{s} (1 + \lambda_i)^{-1} \tag{3.3.19}$$

where the v_i are the roots of $|\mathbf{Q}_e - v(\mathbf{Q}_e + \mathbf{Q}_h)| = 0$ and the λ_i are the roots of $|\mathbf{Q}_h - \lambda\mathbf{Q}_e| = 0$. For testing (3.3.5), s is just 1 so that $\Lambda = 1/(1 + \lambda_1)$, where λ_1 is the largest root of $|\mathbf{Q}_h - \lambda\mathbf{Q}_e| = 0$.

Roy (1953, 1957) used only the largest root λ_1, from among the s roots of $|\mathbf{Q}_h - \lambda\mathbf{Q}_e| = 0$, as a test statistic to investigate more general hypotheses about means. Tables are not provided for Roy's test statistic; however, Heck (1960) tabulated the distribution of the largest root of $|\mathbf{Q}_h - \theta(\mathbf{Q}_h + \mathbf{Q}_e)| = 0$ (see Tables VII and VIII in the Appendix). This may be used to test (3.3.5) since, by (2.7.12), $\theta_1 = \lambda_1/(1 + \lambda_1)$.

Employing Roy's criterion, the one-sample hypothesis given by (3.3.5) is rejected at the significance level α if

$$\theta_s = \frac{\lambda_1}{1 + \lambda_1} > \theta^{\alpha}(s, m, n) \tag{3.3.20}$$

where for this hypothesis $s = 1$, $m = (|p - 1| - 1)/2$, and $n = (N - p - 2)/2$. General expressions for s, m, and n will be given later.

Lawley (1938) and Hotelling (1951) generalized the concept of the T^2 statistic for testing more general hypotheses about means. They suggested that the sum of the roots of $|\mathbf{Q}_h - \lambda\mathbf{Q}_e| = 0$,

$$T_0^2 = \nu_e \operatorname{Tr}(\mathbf{Q}_h\mathbf{Q}_e^{-1}) = \nu_e \sum_{i=1}^{s} \lambda_i \tag{3.3.21}$$

be used as a test statistic called the *Lawley-Hotelling trace criterion.* Pillai (1960) tabled percentage points of the sum of the roots of $|\mathbf{Q}_h - \lambda\mathbf{Q}_e| = 0$ so that

$$U^{(s)} = \sum_{i=1}^{s} \lambda_i = T_0^2/\nu_e \tag{3.3.22}$$

where ν_e is the degree of freedom associated with $\mathbf{Q}_e$.

For the one-sample hypothesis being considered, $U^{(s)}$ is equal to λ_1, and, since $\nu_e = N - 1$, $T_0^2 = (N - 1)\lambda_1$. However, for the Λ criterion, $(1 - \Lambda)/\Lambda = T^2/(N - 1)$. Since $\Lambda = (1 + \lambda_1)^{-1}$, $T^2 = (N - 1)\lambda_1$, so that $T^2 = T_0^2$ in this case.

Employing the Lawley-Hotelling trace criterion and Pillai's tables, the null hypothesis $H_0: \boldsymbol{\mu} = \boldsymbol{\mu}_0$ would be rejected at the significance level α if

$$U^{(s)} = \frac{T_0^2}{(N - 1)} > U_0^{\alpha}(s, m, n) \tag{3.3.23}$$

where s, m, and n maintain their previous definitions and $U_0^{\alpha}(s, m, n)$ represents the upper α percentage point of Pillai's table (see Table X in the Appendix).

Pillai's trace criterion $V^{(s)}$ could also be used to test (3.3.5), where

$$V^{(s)} = \sum_{i=1}^{s} \theta_i \tag{3.3.24}$$

and the θ_i are the roots of $|\mathbf{Q}_h - \theta(\mathbf{Q}_h + \mathbf{Q}_e)| = 0$. However, in this case, Pillai's criterion and Roy's criterion are identical.

The preceding discussion has established relationships existing among frequently employed test statistics used to test more general hypotheses about means and on which future multivariate techniques will depend. This development has also shown that each test statistic for the one-sample problem is related to the familiar F distribution when the degrees of freedom for the hypothesis are $v_h = 1$ and the degrees of freedom for error are $v_e = N - 1$. In summary, the hypothesis

$$H_0: \boldsymbol{\mu} = \boldsymbol{\mu}_0 \quad \text{(specified)}$$

$$H_1: \boldsymbol{\mu} \neq \boldsymbol{\mu}_0$$

is rejected at the significance level α if

(3.3.25) (a) Hotelling:

$$T^2 > T^\alpha(p, N-1) \quad \text{or} \quad \frac{(N-p)T^2}{p(N-1)} > F^\alpha(p, N-p)$$

(b) Wilks:

$$\Lambda < U^\alpha(p, 1, N-1) \quad \text{or} \quad \frac{N-p}{p}\frac{1-\Lambda}{\Lambda} > F^\alpha(p, N-p)$$

(c) Roy:

$$\theta_s = \frac{\lambda_1}{1+\lambda_1} > \theta^\alpha(s, m, n) \quad \text{or} \quad \frac{N-p}{p}\frac{\theta_s}{1-\theta_s} > F^\alpha(p, N-p)$$

(d) Lawley-Hotelling:

$$U^{(s)} = \frac{T_0^2}{N-1} = \lambda_1 > U_0^\alpha(s, m, n) \quad \text{or} \quad \frac{N-p}{p}\frac{T_0^2}{N-1} > F^\alpha(p, N-p)$$

Employing any of the above test statistics, the same conclusion would be reached concerning the acceptance of H_0 since the rank of the noncentrality matrix is 1. This is not always the case when testing more general hypotheses about means (see Chapter 5).

As in the univariate case, T^2 may be converted to give a confidence set for the population mean vector $\boldsymbol{\mu}$. In univariate analysis, the complement of the rejection region for testing H_0 is used to construct a $100(1-\alpha)\%$ simultaneous confidence interval (set). By analogy, a $100(1-\alpha)\%$ confidence ellipsoid for $\boldsymbol{\mu}$ consists of all vectors that satisfy the inequality

$$N(\bar{\mathbf{y}} - \boldsymbol{\mu})'\mathbf{S}^{-1}(\bar{\mathbf{y}} - \boldsymbol{\mu}) \leq T^\alpha(p, N-1) \tag{3.3.26}$$

The boundary of this region is an ellipsoid with its center at the point $\mathbf{y}' = [\bar{y}_1, \bar{y}_2, \ldots \bar{y}_p]$. On rejection of H_0, the ellipsoid yields those tuples of the parameter vector $\boldsymbol{\mu}$ outside the ellipsoid that are unlikely to occur; however, the practical usefulness of this region is of limited value when p is large.

Continuing,

$$T^2 = \max_{\mathbf{a}} \frac{N[\mathbf{a}'(\bar{\mathbf{y}} - \boldsymbol{\mu}_0)]^2}{\mathbf{a}'\mathbf{S}\mathbf{a}} \tag{3.3.27}$$

for arbitrary vectors **a**. To prove (3.3.27), simply observe that

$$\mathbf{a}'\mathbf{Q}_h\mathbf{a} = N\mathbf{a}'(\bar{\mathbf{y}} - \boldsymbol{\mu}_0)(\bar{\mathbf{y}} - \boldsymbol{\mu}_0)'\mathbf{a} \quad \text{and} \quad \mathbf{a}'\mathbf{S}\mathbf{a} = \frac{\mathbf{a}'\mathbf{Q}_e\mathbf{a}}{N-1}$$

Then the

$$\max_{\mathbf{a}} \frac{N[\mathbf{a}'(\bar{\mathbf{y}} - \boldsymbol{\mu}_0)]^2}{\mathbf{a}'\mathbf{S}\mathbf{a}} = \max_{\mathbf{a}} \frac{\mathbf{a}'\mathbf{Q}_h\mathbf{a}}{\mathbf{a}'\mathbf{Q}_e\mathbf{a}/(N-1)} = \tau$$

where τ is the largest root of $|(\mathbf{Q}_h - \tau\mathbf{Q}_e)/(N-1)| = 0$. But since $\lambda_1(N-1) = \tau = T^2$, where λ_1 is the largest root of $|\mathbf{Q}_h - \lambda\mathbf{Q}_e| = 0$, (3.3.27) is proven. Hence

$$\frac{N[\mathbf{a}'(\bar{\mathbf{y}} - \boldsymbol{\mu})]^2}{\mathbf{a}'\mathbf{S}\mathbf{a}} \leq N(\bar{\mathbf{y}} - \boldsymbol{\mu})'\mathbf{S}^{-1}(\bar{\mathbf{y}} - \boldsymbol{\mu}) \tag{3.3.28}$$

From the distribution of T^2, the

$$P\{N[\mathbf{a}'(\bar{\mathbf{y}} - \boldsymbol{\mu})]^2 \leq \mathbf{a}'\mathbf{S}\mathbf{a}T^\alpha(p, N-1) \quad \text{for all } \mathbf{a}\} = 1 - \alpha \tag{3.3.29}$$

For a given arbitrary vector **a**, the

$$P\{N[\mathbf{a}'(\bar{\mathbf{y}} - \boldsymbol{\mu})]^2 \leq \mathbf{a}'\mathbf{S}\mathbf{a}T^\alpha(p, N-1) \quad \text{for any } \mathbf{a}\} \geq 1 - \alpha \tag{3.3.30}$$

By expanding (3.3.29), $100(1 - \alpha)\%$ simultaneous confidence intervals for an arbitrary vector **a** are

$$\mathbf{a}'\bar{\mathbf{y}} - c_0\sqrt{\frac{\mathbf{a}'\mathbf{S}\mathbf{a}}{N}} \leq \mathbf{a}'\boldsymbol{\mu} \leq \mathbf{a}'\bar{\mathbf{y}} + c_0\sqrt{\frac{\mathbf{a}'\mathbf{S}\mathbf{a}}{N}} \tag{3.3.31}$$

where $c_0^2 = T^\alpha(p, N-1)$.

The inequality given in (3.3.31) yields the set of $100(1 - \alpha)\%$ simultaneous confidence intervals for linear combinations of the elements of $\boldsymbol{\mu}$ as determined by various choices of **a**. Rejecting the multivariate hypothesis implies that there exists at least one significant linear combination of the means involving $\boldsymbol{\mu}$ that differs from $\boldsymbol{\mu}_0$. In constructing the intervals, restriction to intervals involving means of the individual variables is not necessarily indicated when the overall multivariate hypothesis is rejected. Investigation of meaningful linear combinations of means across variables may have to be considered to find significance.

The simultaneous intervals obtained in (3.3.31) correspond to Scheffé's (1953) S method used in univariate analysis for data snooping or a posteriori comparisons. Scheffé's procedure is a special case of a general simultaneous inference technique proposed by Roy and Bose (1953) that uses the heuristic union-intersection principle for test construction (see also Roy, 1953).

For a few comparisons, the confidence coefficient may be considerably greater than $1 - \alpha$. When a researcher is interested in only making comparisons involving the p individual means, an alternative procedure using planned Bonferroni intervals will usually be shorter (see Miller, 1966). With probability at least $(1 - \alpha)$, the p statements

$$\bar{y}_i - \frac{c_0 s_i}{\sqrt{N}} \leq \mu_i \leq \bar{y}_i + \frac{c_0 s_i}{\sqrt{N}} \qquad i = 1, \ldots, p \tag{3.3.32}$$

hold simultaneously. The constant $c_0 = t^{\alpha/2p}(N - 1)$ in (3.3.32) is the upper $\alpha/2p$ percentage point of the central t distribution, and s_i is the square root of the ith diagonal element of the variance-covariance matrix **S**. Tables of the t distribution for the critical points $t^{\alpha/2k}(v_e)$, for k comparisons, have been prepared by Dayton and Schafer (1973) and are reproduced in Table XII of the Appendix. This procedure does not require that the overall multivariate test be performed. The multivariate hypothesis is rejected if at least one of the a priori or planned comparisons is significant. The overall level of significance is less than or equal to the predetermined α level. This technique also allows the researcher to distribute his α level unequally among the p comparisons, assigning probability α_i to each, so that

$$\sum_{i=1}^{p} \alpha_i = \alpha$$

Cramer and Bock (1966) suggest another approach, which is an extension of Fisher's (1949) least significant difference (LSD) test and is not a simultaneous test procedure. This technique combines the overall multivariate test and univariate inference procedures. Following a significant overall multivariate test of size α, it is recommended that univariate F tests be calculated, a variable at a time and at the same α level, to locate significant dependent variables. Hummel and Sligo (1971) show that this method is less conservative, with respect to the probability of errors, than the previously mentioned procedures with an experimental error rate near the nominal chosen α level. Miller (1966, p. 93), however, criticizes the LSD test, for which the procedure proposed by Cramer and Bock is a generalization: "The preliminary F test guards against falsely rejecting the null hypothesis when the null hypothesis is true. However, when, in fact, the null hypothesis is false and likely to be rejected, the second stage of the LSD gives no increased protection to that part (if any) of the null hypothesis which still remains true."

Furthermore, rejection of the multivariate test does not guarantee that there exists at least one significant univariate F ratio. For a given set of data, the significant comparison may involve some linear combination of the elements of the vector $\boldsymbol{\mu}$. One advantage of this procedure over the Bonferroni method is that it leads to a smaller probability of accepting a false null hypothesis and hence to a smaller type II error. In this sense, the method is similar to doing multiple F tests; however, multiple F tests do not have the protection of the overall test. Multiple F tests are extensions of the familiar multiple t tests in univariate analysis that do not require an overall test. Furthermore, the larger the number of comparisons the greater the probability of a type I error.

Combining the advantages of Fisher's procedure with the Bonferroni method, one might follow the multivariate test by using Bonferroni t's in a stepwise manner. That is, perform each univariate test at the significance level α_i such that the sum of the α_i's add to α. This is similar to conducting pseudo step-down F tests (see Roy, 1958).

EXERCISES 3.3

1. Using equation (3.3.7) for Hotelling's T^2 statistic, show that T^2 is invariant under all affine transformations $\mathbf{x} = \mathbf{A}\mathbf{y} + \mathbf{c}$ of the observations, where $\mathbf{A}$ is a nonsingular matrix and $\mathbf{c}$ is a vector of constants.

2. a. For a p-vector $\mathbf{a}$, containing a 1 in the ith position and 0 elsewhere, evaluate (3.3.31).

   b. Setting $p = 2, 5, 10$, and 20, compare the result in part a with (3.3.32) if $N = 41$ and $\alpha = .05$.

3.4 EXAMPLE—ONE-SAMPLE MULTIVARIATE CASE

A random number of N students enroll in an algebra II course annually. It has been the instructor's practice to administer an exam at the beginning of the course to test the students' basic mathematics preparation and skill in dealing with word problems. The instructor hopes to find out from the examination whether his students' preparation differs from the school district's standard, which shows scores in these areas to be $\mu_{01} = 80$ and $\mu_{02} = 50$. For a new term, the instructor has a class enrollment of $N = 28$ students and obtains the scores summarized in Table 3.4.1 for each

TABLE 3.4.1. Sample Data: Hotelling's One-Sample T^2.

BM, WP	BM, WP	BM, WP	BM, WP
72, 66	42, 43	91, 79	32, 30
60, 53	37, 40	56, 68	60, 50
56, 57	33, 29	79, 65	35, 37
41, 29	32, 30	81, 80	39, 36
32, 32	63, 45	78, 55	50, 34
30, 35	54, 46	46, 38	43, 37
39, 39	47, 51	39, 35	48, 54

student. The first score measures basic mathematics (BM) preparation and the second indicates the students' ability to solve word problems (WP). The question to be answered is "Does the students' preparation in basic mathematics and word problems differ from the district's standard?" If so, On which variables?

$$\text{Test } H_0: \begin{bmatrix} \mu_1 \\ \mu_2 \end{bmatrix} = \begin{bmatrix} 80 \\ 50 \end{bmatrix}$$

$$H_1: \begin{bmatrix} \mu_1 \\ \mu_2 \end{bmatrix} \neq \begin{bmatrix} 50 \\ 80 \end{bmatrix}$$

with a significance level $\alpha = .05$.

Before solving this problem, we review the possible test statistics that may be employed to test H_0:

(1) Hotelling: From Table VI, reject H_0 if

$$T^2 = N(\bar{\mathbf{y}} - \boldsymbol{\mu}_0)'\mathbf{S}^{-1}(\bar{\mathbf{y}} - \boldsymbol{\mu}_0) > T^{\alpha}(p, N-1) = T^{.05}(2, 27) = 6.997$$

(2) Wilks: From Table IX, reject H_0 if

$$\Lambda = \frac{|\mathbf{Q}_e|}{|\mathbf{Q}_e + \mathbf{Q}_h|} < U^{\alpha}(p, 1, N-1) = U^{.05}(2, 1, 27) = .794192$$

where

$$\mathbf{Q}_e = \sum_{i=1}^{N} (\mathbf{y}_i - \bar{\mathbf{y}})(\mathbf{y}_i - \bar{\mathbf{y}})' \quad \text{and} \quad \mathbf{Q}_h = N(\bar{\mathbf{y}} - \boldsymbol{\mu}_0)(\bar{\mathbf{y}} - \boldsymbol{\mu}_0)'$$

(3) Roy: Heck's charts do not exist for $s = 1$; instead, the tables for the U distribution may be employed. H_0 is rejected if

$$\theta_s = \frac{\lambda_1}{1 + \lambda_1} > \theta^{\alpha}(s, m, n) = \theta^{.05}(1, 0, 12) = 1 - U^{.05}(2, 1, 27)$$
$$= .205808$$

(4) Lawley-Hotelling: Pillai's (1960) tables do not exist for $s = 1$; however, since $T^{\alpha}(p, N-1) = (N-1)U_0^{\alpha}(s, m, n)$ in this case, H_0 is rejected if

$$U^{(s)} = \frac{T_0^2}{N-1} = \lambda_1 > U_0^{.05}(s, m, n) = U_0^{.05}(1, 0, 12) = .259148$$

Using the above critical values, the following relationships may be verified;

$$\frac{p(N-1)}{N-p} F^{\alpha}(p, N-p) = T^{\alpha}(p, N-1)$$

(3.4.1)
$$U^{\alpha}(p, 1, N-1) = \frac{1}{1 + [T^{\alpha}(p, N-1)/(N-1)]} = \frac{1}{1 + [p/(N-p)]F^{\alpha}(p, N-p)}$$
$$\theta^{\alpha}(s, m, n) = \frac{pF^{\alpha}(p, N-p)}{(N-p) + pF^{\alpha}(p, N-p)} = 1 - U^{\alpha}(p, 1, N-1)$$
$$U_0^{\alpha}(s, m, n) = \frac{T^{\alpha}(p, N-1)}{N-1} = \frac{pF^{\alpha}(p, N-p)}{N-p}$$

For the one-sample hypothesis, any one of the procedures may be used to test H_0. The value of F to be substituted into the above expressions is $F^{.05}_{(2,26)} = 3.369$.

Calculations will be carried out by using Hotelling's criterion where

$$T^2 = N(\bar{\mathbf{y}} - \boldsymbol{\mu}_0)'\mathbf{S}^{-1}(\bar{\mathbf{y}} - \boldsymbol{\mu}_0)$$

For this problem, the data matrix $\mathbf{Y}$ is of the form

$$\mathbf{Y} = \begin{bmatrix} y_{11} & y_{12} \\ y_{21} & y_{22} \\ \vdots & \vdots \\ y_{28,1} & y_{28,2} \end{bmatrix}$$

so that

$$\bar{\mathbf{y}} = \begin{bmatrix} \sum_{j=1}^{28} y_{j1}/28 \\ \sum_{j=1}^{28} y_{j2}/28 \end{bmatrix} = \begin{bmatrix} 50.536 \\ 46.179 \end{bmatrix}$$

Now calculate the error sum of squares and cross-products matrix $\mathbf{Q}_e$,

$$\begin{aligned} \mathbf{Q}_e &= \mathbf{Y}'[\mathbf{I} - \mathbf{X}(\mathbf{X}'\mathbf{X})^{-1}\mathbf{X}']\mathbf{Y} \\ &= \mathbf{Y}'[\mathbf{I} - \mathbf{1}(\mathbf{1}'\mathbf{1})^{-1}\mathbf{1}']\mathbf{Y} \\ &= \mathbf{Y}'\mathbf{Y} - \mathbf{Y}'\mathbf{1}(\mathbf{1}'\mathbf{1})^{-1}\mathbf{1}'\mathbf{Y} \\ &= \begin{bmatrix} 79{,}349.0 & \\ 71{,}384.0 & 65{,}647.0 \end{bmatrix} - \begin{bmatrix} 1415.0 \\ 1293.0 \end{bmatrix} \left(\frac{1}{28}\right) [1415.0 \quad 1293.0] \end{aligned}$$

$$= \begin{bmatrix} 7840.96 & \\ 6041.32 & 5938.11 \end{bmatrix} \tag{3.4.2}$$

From $\mathbf{Q}_e$, obtain the unbiased estimate $\mathbf{S}$ of $\mathbf{\Sigma}$,

$$\mathbf{S} = \frac{\mathbf{Q}_e}{N-1} = \begin{bmatrix} 290.406 & \\ 223.753 & 219.930 \end{bmatrix} \tag{3.4.3}$$

so that

$$s_1^2 = 290.406 \qquad s_1 = 17.04$$
$$s_2^2 = 219.930 \qquad s_2 = 14.83$$

Computing $\mathbf{S}^{-1}$,

$$\mathbf{S}^{-1} = \begin{bmatrix} .015932 & \\ -.016209 & .021038 \end{bmatrix} \tag{3.4.4}$$

and $T^2 = 293.69$. Since this is greater than the critical value of $T^{.05}_{(2,27)} = 6.997$, the hypothesis is rejected. Employing any of the other criteria, the same result would have been obtained since

(1) Wilks: $\Lambda = .0842 < .794192$;
(2) Roy: $\theta_s = .9158 > .205808$; and
(3) Lawley-Hotelling: $U^{(s)} = 10.88 > .259148$.

The instructor may thus conclude that review is necessary for the class. However, clarification does not exist as to which area needs to be reviewed. To determine whether one or both variables contributed to the rejection of H_0, the following expression is evaluated:

$$\mathbf{a}'\bar{\mathbf{y}} - c_0\sqrt{\frac{\mathbf{a}'\mathbf{S}\mathbf{a}}{N}} \le \mathbf{a}'\boldsymbol{\mu} \le \mathbf{a}'\bar{\mathbf{y}} + c_0\sqrt{\frac{\mathbf{a}'\mathbf{S}\mathbf{a}}{N}} \tag{3.4.5}$$

$$\text{for } c_0 = \sqrt{T^{.05}_{(2,27)}} = \sqrt{6.997} = 2.645$$

Investigating the first variable, let $\mathbf{a}' = [1, 0]$, then (3.4.5) becomes

$$50.536 - 2.645\sqrt{\frac{290.406}{28}} \leq \mu_1 \leq 50.536 + 2.645\sqrt{\frac{290.406}{28}}$$

$$42.02 \leq \mu_1 \leq 59.05$$

Since this interval does not include $\mu_{01} = 80$, we may conclude that variable 1, basic mathematic skills, contributed to the rejection of H_0. For $\mathbf{a}' = [0, 1]$, the confidence interval for μ_2 is

$$46.179 - 2.645\sqrt{\frac{219.930}{29}} \leq \mu_2 \leq 46.179 + 2.645\sqrt{\frac{219.930}{28}}$$

$$38.77 \leq \mu_2 \leq 53.59$$

Since this interval includes the parameter $\mu_{02} = 50$, it may be stated that variable 2, word problem skills, did not significantly contribute to the rejection of H_0.

Alternatively, by performing the experiment on an a priori basis where only intervals involving single means are of interest, the Bonferroni procedure (3.3.32) yields significance with shorter intervals. Using this method, the number of comparisons $k = 2$, $c_0 = t_{(27)}^{.05/2k} = 2.375$, and the intervals become

$$50.536 - 2.375\sqrt{\frac{290.406}{29}} \leq \mu_1 \leq 50.536 + 2.375\sqrt{\frac{290.406}{28}}$$

$$42.88 \leq \mu_1 \leq 58.18$$

and

$$46.179 - 2.375\sqrt{\frac{219.930}{28}} \leq \mu_2 \leq 46.179 + 2.375\sqrt{\frac{219.930}{28}}$$

$$39.52 \leq \mu_2 \leq 52.84$$

where again only μ_1 is significant.

Since the overall multivariate test is significant, following Fisher, simple t intervals might be constructed for each element of $\boldsymbol{\mu}$. For this method, the constant c_0 in (3.4.5) is $c_0 = t^{\alpha/2}(v_e) = t_{(27)}^{.025} = 2.052$. Intervals for each element in $\boldsymbol{\mu}$ become

$$43.93 \leq \mu_1 \leq 57.14$$

$$40.43 \leq \mu_2 \leq 51.93$$

However, the probability of a type I error may be larger than the nominal level α. The Bonferroni procedure and the simultaneous intervals constructed by using (3.4.5) guarantee that the probability of a type I error will be less than or equal to α.

EXERCISES 3.4

1. In a pilot study designed to investigate a new training program in grammar usage (G), reading skills (R), and spelling (S), the mean performance of 11 randomly selected students in the three areas was hypothesized to be $\mu_G = 80$, $\mu_R = 75$, and $\mu_S = 70$ at the end of the first week of instruction.

Subject	G	R	S
1	31	12	24
2	52	64	32
3	57	42	21
4	63	19	54
5	42	12	41
6	71	79	64
7	65	38	52
8	60	14	57
9	54	75	58
10	67	22	69
11	70	34	24

a. Using $\alpha = .05$, test the overall hypothesis $H: \boldsymbol{\mu} = \boldsymbol{\mu}_0$ with $\boldsymbol{\mu}_0' = [80, 75, 70]$ by using the criteria discussed for the one-sample multivariate problem.

b. Use the Bonferroni procedure to test the hypothesis $H: \boldsymbol{\mu} = \boldsymbol{\mu}_0$ a variable at a time. Do your conclusions for this method of analysis differ from those obtained from part a?

c. Summarize your findings.

3.5 THE GENERAL LINEAR MODEL—UNIVARIATE

It was indicated previously that tests on means could be obtained from the Gauss-Markoff setup. This is accomplished by applying some fundamental results of least-squares theory. We will indicate the manner in which this might be accomplished and integrate more traditional approaches with the general method by first reviewing the univariate case (see also Rao, 1965a, Chap. 4, and Graybill, 1961, Chaps. 5 and 6).

Consider a set of observable variables $[y_i, x_{i1}, x_{i2}, \ldots, x_{ik}]$, for $i = 1, \ldots, N$, such that y_i is related to the x_{ij}'s by

$$y_i = \beta_0 + x_{i1}\beta_1 + \cdots + x_{ik}\beta_k + \varepsilon_i \tag{3.5.1}$$

where the ε_i's are random errors, the β_j's are unknown parameters, and the observations $x_{ij} = x_j$, for $i = 1, \ldots, N$, denote N observations on variables $x_{j=1,\ldots,k}$. Using matrix notation, let

$$\underset{(N \times 1)}{\mathbf{y}} = \begin{bmatrix} y_1 \\ y_2 \\ \vdots \\ y_N \end{bmatrix}, \qquad \underset{(q \times 1)}{\boldsymbol{\beta}} = \begin{bmatrix} \beta_0 \\ \beta_1 \\ \vdots \\ \beta_k \end{bmatrix}, \qquad \underset{(N \times 1)}{\boldsymbol{\varepsilon}} = \begin{bmatrix} \varepsilon_1 \\ \varepsilon_2 \\ \vdots \\ \varepsilon_N \end{bmatrix}$$

and the matrix of x_{ij}'s

$$\underset{(N \times q)}{\mathbf{X}} = \begin{array}{c} \begin{matrix} & x_1 & x_2 & \cdots & x_k \end{matrix} \\ \begin{bmatrix} 1 & x_{11} & x_{12} & \cdots & x_{1k} \\ 1 & x_{21} & x_{22} & \cdots & x_{2k} \\ \vdots & \vdots & \vdots & \vdots & \vdots \\ 1 & x_{N1} & x_{N2} & \cdots & x_{Nk} \end{bmatrix} \end{array}$$

Then, in matrix form, (3.5.1) is written as

$$\underset{(N\times 1)}{\mathbf{y}} = \underset{(N\times q)}{\mathbf{X}}\ \underset{(q\times 1)}{\boldsymbol{\beta}} + \underset{(N\times 1)}{\boldsymbol{\varepsilon}} \tag{3.5.2}$$

This equation represents the fundamental linear model used in numerous univariate applications depending on the relationships among $\mathbf{y}$, $\mathbf{X}$, $\boldsymbol{\beta}$, and $\boldsymbol{\varepsilon}$. The model is called a *linear model* because it is linear in the parameters, as seen in (3.5.1).

For *functionally related models*, the vector of observations $\mathbf{y}$ contains N independent observations on an observable random variable Y assumed to be observed without measurement error. The matrix $\mathbf{X}$ is a matrix of known constants, not random variables but predetermined observed mathematical variables. The random error vector $\boldsymbol{\varepsilon}$ has N unobservable elements containing measurement errors and other unexplained variables not included in $\mathbf{X}$. The parameter vector $\boldsymbol{\beta}$ is assumed to be nonrandom. An important special case of functionally related models is the *experimental design model* where the matrix $\mathbf{X}$ contains only 0s and 1s.

Another important class of models that can be examined by using (3.5.2) are *conditional regression models*. For these models, Y and $X_1, X_2, \ldots, X_k$ are random variables. However, the data in the observation vector $\mathbf{y}$ is observed by holding $X_1 = x_1, X_2 = x_2, \ldots, X_k = x_k$ fixed. Y is observed conditioned on $X_1, X_2, \ldots, X_k$ to obtain the N observed values of $\mathbf{y}$. In this case, (3.5.2) is sometimes written as $\mathbf{y} \mid \mathbf{X} = \mathbf{X}\boldsymbol{\beta} + \boldsymbol{\varepsilon}$. However, the assumptions made for $\boldsymbol{\varepsilon}$ and $\boldsymbol{\beta}$ are the same as for functionally related models. A mixture of experimental design models and functionally related models or conditional regression models is known as the *covariance model*.

For functionally related models and conditional regression models, the vector $\boldsymbol{\beta}$ of parameters is nonrandom and the matrix $\mathbf{X}$ contains known, nonrandom elements. When the elements of $\mathbf{X}$ are preselected variables, $\mathbf{X}$ is called a *design matrix*. If $\boldsymbol{\beta}$ is random and independent of $\boldsymbol{\varepsilon}$, we have what are commonly called *components-of-variance models* or, in analysis-of-variance terminology, *random-effects models*. A mixture of components-of-variance models and experimental design models is called a *mixed model*.

The above classification of models indicates the general importance of (3.5.2) in univariate analysis. In all these models, the $E(\boldsymbol{\varepsilon})$ is assumed to be $\mathbf{0}$. For statistical inference, it is usually assumed that $\boldsymbol{\varepsilon}$ follows a multivariate normal distribution where the $V(\boldsymbol{\varepsilon}) = \boldsymbol{\Sigma}$ are unknown. In many models it is further assumed that $\boldsymbol{\Sigma} = \sigma^2\mathbf{I}$, where σ^2 is an unknown variance. This implies that the errors are uncorrelated and have equal variance. The assumption of equal variance is termed *homoscedasticity* of errors. The linear model in (3.5.2) with

$$\begin{aligned} E(\boldsymbol{\varepsilon}) &= \mathbf{0} \\ V(\boldsymbol{\varepsilon}) &= \sigma^2\mathbf{I} \end{aligned} \quad \text{or} \quad \begin{aligned} E(\mathbf{Y}) &= \mathbf{X}\boldsymbol{\beta} \\ V(\mathbf{Y}) &= \sigma^2\mathbf{I} \end{aligned} \tag{3.5.3}$$

is known as the *univariate Gauss-Markoff setup*, where

$\mathbf{y}$ = observation vector for random N-vector $\mathbf{Y}$,
$\mathbf{X}$ = known design matrix of constants given by the design whose rank $r \le q \le N$,
$\boldsymbol{\beta}$ = column vector of q unknown, nonrandom parameters, and
$\boldsymbol{\varepsilon}$ = column vector of random errors.

All models summarized above satisfy (3.5.2); however, only functional regression models and conditional regression models with a variance-covariance matrix $\boldsymbol{\Sigma}$ of errors of the form $\boldsymbol{\Sigma} = \sigma^2\mathbf{I}$ satisfy (3.5.3). For components of variance models and mixed models, the vector $\boldsymbol{\beta}$ contains random elements. To fit these models into (3.5.3), $\boldsymbol{\beta}$ would have to be considered as nonrandom.

There are other models that occur in the analysis of univariate data and may have the form of (3.5.2) but may not be considered under (3.5.3). This would occur if $\mathbf{X}$ were a random matrix or if $\mathbf{y}$ and $\mathbf{X}$ were unobservable. These are called *random regression models* and *error-in-variable models.* Models that do not satisfy (3.5.2) are called *nonlinear models.*

In this text we will only be concerned with linear models that can be easily modified to satisfy the Gauss-Markoff setup.

As observed in Section 3.2, if $\mathbf{X} = \mathbf{1}$ (a vector of N 1s) and $\boldsymbol{\beta} = \mu$ (a parameter "vector"), then the one-sample problem is just a very special case of the more general univariate Gauss-Markoff setup given in (3.5.3).

Estimation Theory. To estimate the elements of $\boldsymbol{\beta}$ in (3.5.2) under (3.5.3), the method of least squares is employed so that the error sum of squares

(3.5.4)
$$\sum_{i=1}^{N} \varepsilon_i^2 = \text{Tr}[(\mathbf{y} - \mathbf{X}\boldsymbol{\beta})'(\mathbf{y} - \mathbf{X}\boldsymbol{\beta})]$$

is minimized. As found in Section 1.8, minimization yields the *normal equations*

(3.5.5)
$$(\mathbf{X}'\mathbf{X})\boldsymbol{\beta} = \mathbf{X}'\mathbf{y}$$

From (1.6.25), the least-squares solution to (3.5.5) is

(3.5.6)
$$\hat{\boldsymbol{\beta}} = (\mathbf{X}'\mathbf{X})^{-}\mathbf{X}'\mathbf{y} + (\mathbf{I} - \mathbf{H})\mathbf{z}$$

where $\mathbf{H} = (\mathbf{X}'\mathbf{X})^{-}\mathbf{X}'\mathbf{X}$ and $\mathbf{z}$ is an arbitrary vector. The vector $\hat{\boldsymbol{\beta}}$ always exists but need not be unique. A unique solution exists if the rank of $\mathbf{X}$ $(r = q)$ is such that $(\mathbf{X}'\mathbf{X})^{-} = (\mathbf{X}'\mathbf{X})^{-1}$.

Although $\hat{\boldsymbol{\beta}}$, as given by (3.5.6), is not generally a unique solution to (3.5.5), certain linear combinations of the elements of $\boldsymbol{\beta}$, often called *parametric functions*, are unique and independent of the solution $\hat{\boldsymbol{\beta}}$.

(3.5.7) A parametric function $\mathbf{c}'\boldsymbol{\beta}$ is said to be *estimable* if there exists a vector $\mathbf{a}$ such that $E(\mathbf{a}'\mathbf{Y}) = \mathbf{c}'\boldsymbol{\beta}$

That is, $\mathbf{c}'\boldsymbol{\beta}$ has a linear unbiased estimate given by $\mathbf{a}'\mathbf{Y}$. Hence $\mathbf{c}'\boldsymbol{\beta}$ is an estimable function if and only if $\mathbf{c}' = \mathbf{a}'\mathbf{X}$—that is, if $\mathbf{c}$ is a linear combination of the rows of $\mathbf{X}$ or if $\mathbf{c}$ is a vector in the space spanned by the columns of $\mathbf{X}'$, written as $\mathbf{c} \in V(\mathbf{X}')$ or, equivalently, $\mathbf{c} \in V(\mathbf{X}'\mathbf{X})$. If the $R(\mathbf{X}) = r$, there are only r linearly independent estimable functions. However, for a known vector $\mathbf{c}$, obtaining a vector $\mathbf{a}$ such that $\mathbf{c}' = \mathbf{a}'\mathbf{X}$ is not necessarily easy when the $R(\mathbf{X})$ is large.

Alternatively, if $\mathbf{c}' = \mathbf{a}'\mathbf{X}$, then $\mathbf{c}'\mathbf{H} = \mathbf{a}'\mathbf{X}\mathbf{H} = \mathbf{a}'\mathbf{X}(\mathbf{X}'\mathbf{X})^{-}\mathbf{X}'\mathbf{X} = \mathbf{a}'\mathbf{X} = \mathbf{c}'$; if $\mathbf{c}'\mathbf{H} = \mathbf{c}'$, then $\mathbf{c}' = \mathbf{c}'(\mathbf{X}'\mathbf{X})^{-}\mathbf{X}'\mathbf{X} = \mathbf{a}'\mathbf{X}$ by setting $\mathbf{a}' = \mathbf{c}'(\mathbf{X}'\mathbf{X})^{-}\mathbf{X}'$. A necessary and sufficient condition that $\mathbf{c}'\boldsymbol{\beta}$ be uniquely estimable is that $\mathbf{c}' = \mathbf{c}'\mathbf{H}$ since, for any $\hat{\boldsymbol{\beta}}$, $\mathbf{c}'\hat{\boldsymbol{\beta}} = \mathbf{c}'\mathbf{H}\hat{\boldsymbol{\beta}} = \mathbf{c}'(\mathbf{X}'\mathbf{X})^{-}\mathbf{X}'\mathbf{y} = \mathbf{a}'\mathbf{X}(\mathbf{X}'\mathbf{X})^{-}\mathbf{X}'\mathbf{y} = \mathbf{a}^{*\prime}\mathbf{y}$ is unique when $\mathbf{c}'\boldsymbol{\beta}$ is estimable.

Furthermore, $\mathbf{c}'\hat{\boldsymbol{\beta}}$ is unbiased for $\mathbf{c}'\boldsymbol{\beta}$ since

$$\begin{aligned} E(\mathbf{c}'\hat{\boldsymbol{\beta}}) &= E(\mathbf{c}'(\mathbf{X}'\mathbf{X})^{-}\mathbf{X}'\mathbf{Y}) + \mathbf{c}'(\mathbf{I} - \mathbf{H})\mathbf{z} \\ &= \mathbf{c}'(\mathbf{X}'\mathbf{X})^{-}\mathbf{X}'E(\mathbf{Y}) \\ &= \mathbf{c}'(\mathbf{X}'\mathbf{X})^{-}\mathbf{X}'\mathbf{X}\boldsymbol{\beta} \\ &= \mathbf{c}'\mathbf{H}\boldsymbol{\beta} \\ &= \mathbf{c}'\boldsymbol{\beta} \end{aligned}$$

if and only if $\mathbf{c}' = \mathbf{c}'\mathbf{H}$. Since $\mathbf{c}' = \mathbf{c}'\mathbf{H}$ if and only if $\mathbf{c}' = \mathbf{t}'\mathbf{H}$ for arbitrary vectors $\mathbf{t}$ by the idempotency property of $\mathbf{H}$, the following theorem, which is a restatement of (1.6.29), results (see Searle, 1965):

(3.5.8) The parametric function $\mathbf{c}'\boldsymbol{\beta}$, where $\boldsymbol{\beta}$ is such that $(\mathbf{X}'\mathbf{X})\boldsymbol{\beta} = \mathbf{X}'\mathbf{y}$, has a unique linear unbiased estimator given by $\mathbf{c}'\hat{\boldsymbol{\beta}} = \mathbf{c}'(\mathbf{X}'\mathbf{X})^{-}\mathbf{X}'\mathbf{y}$ if and only if $\mathbf{c}'\mathbf{H} = \mathbf{c}'$, where in general, for arbitrary vectors $\mathbf{t}$ such that $\mathbf{c}' = \mathbf{t}'\mathbf{H}$,

$$\mathbf{c}'\hat{\boldsymbol{\beta}} = \mathbf{t}'(\mathbf{X}'\mathbf{X})^{-}\mathbf{X}'\mathbf{X}(\mathbf{X}'\mathbf{X})^{-}\mathbf{X}'\mathbf{y} = \mathbf{t}'\mathbf{H}\hat{\boldsymbol{\beta}}$$

If $(\mathbf{X}'\mathbf{X})^{-}$ is constructed such that $(\mathbf{X}'\mathbf{X})^{-}\mathbf{X}'\mathbf{X}(\mathbf{X}'\mathbf{X})^{-} = (\mathbf{X}'\mathbf{X})^{-}$, then $\mathbf{c}'\hat{\boldsymbol{\beta}} = \mathbf{t}'(\mathbf{X}'\mathbf{X})^{-}\mathbf{X}'\mathbf{y}$.

The variance of the estimator $\mathbf{c}'\hat{\boldsymbol{\beta}}$ of $\mathbf{c}'\boldsymbol{\beta}$ is

$$V(\mathbf{c}'\hat{\boldsymbol{\beta}}) = \sigma^2\mathbf{c}'(\mathbf{X}'\mathbf{X})^{-}\mathbf{c} \tag{3.5.9}$$

To prove (3.5.9), we recall from (1.6.27) that $\mathbf{X} = \mathbf{X}(\mathbf{X}'\mathbf{X})^{-}\mathbf{X}'\mathbf{X}$ and, if $\mathbf{c}'\boldsymbol{\beta}$ is estimable, $\mathbf{c} = \mathbf{X}'\mathbf{a}$. Thus

$$\begin{aligned} V(\mathbf{c}'\hat{\boldsymbol{\beta}}) &= \mathbf{c}'(\mathbf{X}'\mathbf{X})^{-}\mathbf{X}'(\sigma^2\mathbf{I})\mathbf{X}(\mathbf{X}'\mathbf{X})^{-}\mathbf{c} \\ &= \sigma^2\mathbf{c}'(\mathbf{X}'\mathbf{X})^{-}\mathbf{X}'\mathbf{X}(\mathbf{X}'\mathbf{X})^{-}\mathbf{c} \\ &= \sigma^2\mathbf{a}'\mathbf{X}(\mathbf{X}'\mathbf{X})^{-}\mathbf{X}'\mathbf{X}(\mathbf{X}'\mathbf{X})^{-}\mathbf{X}'\mathbf{a} \\ &= \sigma^2\mathbf{a}'\mathbf{X}(\mathbf{X}'\mathbf{X})^{-}\mathbf{X}'\mathbf{a} \\ &= \sigma^2\mathbf{c}'(\mathbf{X}'\mathbf{X})^{-}\mathbf{c} \end{aligned}$$

To establish an estimator for the common error variance σ^2 in (3.5.9), the error sum of squares is minimized by replacing $\boldsymbol{\beta}$ with $\hat{\boldsymbol{\beta}}$:

$$\sum_{i=1}^{N} e_i^2 = \mathbf{e}'\mathbf{e} = (\mathbf{y} - \mathbf{X}\hat{\boldsymbol{\beta}})'(\mathbf{y} - \mathbf{X}\hat{\boldsymbol{\beta}}) = Q_e \tag{3.5.10}$$

The term $\mathbf{X}\hat{\boldsymbol{\beta}} = \hat{\mathbf{y}}$ is the best estimate of $\mathbf{y}$ obtained by minimizing (3.5.4) and is called the *best estimate* of $\mathbf{y}$ or the *fitted value*. The difference $\mathbf{y} - \mathbf{X}\hat{\boldsymbol{\beta}} = \mathbf{e}$ is termed the *residual vector*, the error made in estimating $\mathbf{y}$ by $\hat{\mathbf{y}}$. To estimate σ^2, (3.5.10) is employed. Recalling (1.6.27), $\mathbf{X} = \mathbf{X}(\mathbf{X}'\mathbf{X})^{-}\mathbf{X}'\mathbf{X}$; thus

$$\begin{aligned} Q_e &= \mathbf{y}'\mathbf{y} - \mathbf{y}'\mathbf{X}\hat{\boldsymbol{\beta}} - \hat{\boldsymbol{\beta}}'\mathbf{X}'\mathbf{y} + \hat{\boldsymbol{\beta}}'\mathbf{X}'\mathbf{X}\hat{\boldsymbol{\beta}} \\ &= \mathbf{y}'\mathbf{y} - \mathbf{y}'\mathbf{X}\hat{\boldsymbol{\beta}} \\ &= \mathbf{y}'\mathbf{y} - \hat{\boldsymbol{\beta}}'\mathbf{X}'\mathbf{X}\hat{\boldsymbol{\beta}} \\ &= \mathbf{y}'\mathbf{y} - \mathbf{y}'\mathbf{X}(\mathbf{X}'\mathbf{X})^{-}\mathbf{X}'\mathbf{y} \\ &= \mathbf{y}'[\mathbf{I} - \mathbf{X}(\mathbf{X}'\mathbf{X})^{-}\mathbf{X}']\mathbf{y} \qquad (3.5.11) \end{aligned}$$

By applying (2.6.8), $E(\mathbf{Y}'\mathbf{A}\mathbf{Y}) = \text{Tr}(\mathbf{A}\boldsymbol{\Sigma}) + \boldsymbol{\mu}'\mathbf{A}\boldsymbol{\mu}$, and the expected value of Q_e is determined. Since $\mathbf{A} = \mathbf{I} - \mathbf{X}(\mathbf{X}'\mathbf{X})^-\mathbf{X}'$, $\boldsymbol{\Sigma} = \sigma^2\mathbf{I}$, and $\boldsymbol{\mu} = \mathbf{X}\boldsymbol{\beta}$, the

$$\begin{aligned}
E(Q_e) &= \text{Tr}\{\sigma^2[\mathbf{I} - \mathbf{X}(\mathbf{X}'\mathbf{X})^-\mathbf{X}']\} + \boldsymbol{\beta}'\mathbf{X}'[\mathbf{I} - \mathbf{X}(\mathbf{X}'\mathbf{X})^-\mathbf{X}']\mathbf{X}\boldsymbol{\beta} \\
&= \text{Tr}\{\sigma^2[\mathbf{I} - \mathbf{X}(\mathbf{X}'\mathbf{X})^-\mathbf{X}']\} + \boldsymbol{\beta}'\mathbf{X}'\mathbf{X}\boldsymbol{\beta} - \boldsymbol{\beta}'\mathbf{X}'\mathbf{X}(\mathbf{X}'\mathbf{X})^-\mathbf{X}'\mathbf{X}\boldsymbol{\beta} \\
&= \sigma^2\text{Tr}[\mathbf{I} - \mathbf{X}(\mathbf{X}'\mathbf{X})^-\mathbf{X}'] \\
&= (N - r)\sigma^2 \qquad (3.5.12)
\end{aligned}$$

Because $\mathbf{I} - \mathbf{X}(\mathbf{X}'\mathbf{X})^-\mathbf{X}'$ is idempotent, the $\text{Tr}[\mathbf{I} - \mathbf{X}(\mathbf{X}'\mathbf{X})^-\mathbf{X}']$ equals the rank of $\mathbf{I} - \mathbf{X}(\mathbf{X}'\mathbf{X})^-\mathbf{X}$ and equals $N - r$. Hence an unbiased estimate of σ^2 is given by

$$s^2 = \frac{\mathbf{y}'[\mathbf{I} - \mathbf{X}(\mathbf{X}'\mathbf{X})^-\mathbf{X}']\mathbf{y}}{N - r} \qquad (3.5.13)$$

where r is the $R(\mathbf{X})$.

By applying results (3.5.8) and (3.5.13) to the one-sample univariate problem, estimates $\hat{\mu}$ and s^2 for μ and σ^2 are obtained. Since $\mathbf{X} = \mathbf{1}$ is of full rank, $r = 1$,

$$\begin{aligned}
\hat{\mu} &= (\mathbf{X}'\mathbf{X})^{-1}\mathbf{X}'\mathbf{y} \\
&= (\mathbf{1}'\mathbf{1})^{-1}\mathbf{1}'\mathbf{y} \\
&= \frac{1}{N}\sum_{i=1}^{N} y_i \\
&= \bar{y}
\end{aligned}$$

and

$$\begin{aligned}
(N - 1)s^2 &= (\mathbf{y} - \mathbf{1}\hat{\boldsymbol{\beta}})'(\mathbf{y} - \mathbf{1}\hat{\boldsymbol{\beta}}) \\
&= \mathbf{y}'[(\mathbf{I} - \mathbf{1}(\mathbf{1}'\mathbf{1})^{-1}\mathbf{1}']\mathbf{y} \\
&= \mathbf{y}'\mathbf{y} - N\bar{y}^2 \\
&= \sum_{i=1}^{N} y_i^2 - N\bar{y}^2
\end{aligned}$$

or

$$s^2 = \sum_{i=1}^{N} \frac{(y_i - \bar{y})^2}{N - 1}$$

which are the "traditional" unbiased estimators of μ and σ^2 obtained from normal theory. However, in using least-squares theory, we did not require any distribution theory to find desirable estimators.

One more important result must be proved before stating the univariate Gauss-Markoff theorem. Among all linear unbiased estimators of $\mathbf{c}'\boldsymbol{\beta}$ when $\mathbf{c}'\boldsymbol{\beta}$ is estimable, $\mathbf{c}'\hat{\boldsymbol{\beta}}$ has minimum variance. Following the proof of Rao (1965a, p. 182), let $\mathbf{a}'\mathbf{Y}$ be any unbiased estimator of $\mathbf{c}'\boldsymbol{\beta}$ so that $\mathbf{a}'\mathbf{X} = \mathbf{c}'$. Then the $V(\mathbf{c}'\hat{\boldsymbol{\beta}} - \mathbf{a}'\mathbf{Y}) = V(\mathbf{c}'\hat{\boldsymbol{\beta}}) + V(\mathbf{a}'\mathbf{Y}) - 2\,\text{cov}(\mathbf{c}'\hat{\boldsymbol{\beta}}, \mathbf{a}'\mathbf{Y})$. But $\text{cov}(\mathbf{c}'\hat{\boldsymbol{\beta}}, \mathbf{a}'\mathbf{Y}) = \text{cov}[\mathbf{c}'(\mathbf{X}'\mathbf{X})^-\mathbf{X}'\mathbf{Y}, \mathbf{a}'\mathbf{Y}] = \mathbf{c}'(\mathbf{X}'\mathbf{X})^-\mathbf{X}'(\sigma^2\mathbf{I})\mathbf{a} = \sigma^2\mathbf{c}'(\mathbf{X}'\mathbf{X})^-\mathbf{X}'\mathbf{a} = \sigma^2\mathbf{c}'(\mathbf{X}'\mathbf{X})^-\mathbf{c} = V(\mathbf{c}'\hat{\boldsymbol{\beta}})$. Thus $V(\mathbf{c}'\hat{\boldsymbol{\beta}} - \mathbf{a}'\mathbf{Y}) = V(\mathbf{a}'\mathbf{Y}) - V(\mathbf{c}'\hat{\boldsymbol{\beta}}) \geq 0$ and $V(\mathbf{a}'\mathbf{Y}) \geq V(\mathbf{c}'\hat{\boldsymbol{\beta}})$, proving the result.

The univariate *Gauss-Markoff theorem*, on which much of univariate estimation theory is based, is as follows:

(3.5.14) Let $\mathbf{y} = \mathbf{X\beta} + \boldsymbol{\varepsilon}$ such that the $E(\mathbf{Y}) = \mathbf{X\beta}$ and the $V(\mathbf{Y}) = \sigma^2\mathbf{I}$. If $\psi = \mathbf{c}'\boldsymbol{\beta}$ is estimable and $\hat{\boldsymbol{\beta}} = (\mathbf{X}'\mathbf{X})^-\mathbf{X}'\mathbf{y}$ is a least-squares estimator of $\boldsymbol{\beta}$, then
(a) $\hat{\psi} = \mathbf{c}'\hat{\boldsymbol{\beta}}$ is the unique linear unbiased estimate of ψ that has minimum variance among all unbiased linear estimates (that is, $\hat{\psi}$ is a best linear unbiased estimate),
(b) $V(\hat{\psi}) = \sigma^2\mathbf{c}'(\mathbf{X}'\mathbf{X})^-\mathbf{c}$, and
(c) an unbiased estimate of σ^2 is given by

$$s^2 = \frac{(\mathbf{y} - \mathbf{X}\hat{\boldsymbol{\beta}})'(\mathbf{y} - \mathbf{X}\hat{\boldsymbol{\beta}})}{N - r}$$

$$= \frac{\mathbf{y}'[\mathbf{I} - \mathbf{X}(\mathbf{X}'\mathbf{X})^-\mathbf{X}']\mathbf{y}}{N - r}$$

The estimate $\hat{\psi}$ is often termed the *Gauss-Markoff estimator*.

For a geometric interpretation of the theory of least squares, let the observation vector $\mathbf{y}$ be an element of V_N and let V_r denote the space spanned by the columns of $\mathbf{X}$. Then $\mathbf{X}\hat{\boldsymbol{\beta}} = \hat{\mathbf{y}}$ may be interpreted as the vector that comes closest to $\mathbf{y}$ as possible; that is, $\mathbf{X}\hat{\boldsymbol{\beta}}$ is the projection of $\mathbf{y}$ on V_r, so that $P_{V_r}\mathbf{y} = \mathbf{X}\hat{\boldsymbol{\beta}}$. From (1.6.51),

(3.5.15) $$P_{V_r}\mathbf{y} = \mathbf{X}(\mathbf{X}'\mathbf{X})^-\mathbf{X}'\mathbf{y} = \mathbf{X}\hat{\boldsymbol{\beta}} \quad \text{and} \quad P_{V_r^\perp}\mathbf{y} = [\mathbf{I} - \mathbf{X}(\mathbf{X}'\mathbf{X})^-\mathbf{X}']\mathbf{y}$$

Furthermore, $(N - r)s^2 = \mathbf{y}'[\mathbf{I} - \mathbf{X}(\mathbf{X}'\mathbf{X})^-\mathbf{X}']\mathbf{y} = \|P_{V_r^\perp}\mathbf{y}\|^2$, where $N - r$ denotes the dimension of $V_r^\perp$. This is illustrated in Figure 3.5.1. From the orthogonality of the vectors,

(3.5.16) $$\|\mathbf{y}\|^2 = \|\mathbf{X}\hat{\boldsymbol{\beta}}\|^2 + \|\mathbf{y} - \mathbf{X}\hat{\boldsymbol{\beta}}\|^2$$

The vector $\mathbf{y} - \mathbf{X}\hat{\boldsymbol{\beta}}$ measures the error made in estimating $\mathbf{y}$ by $\mathbf{X}\hat{\boldsymbol{\beta}}$.

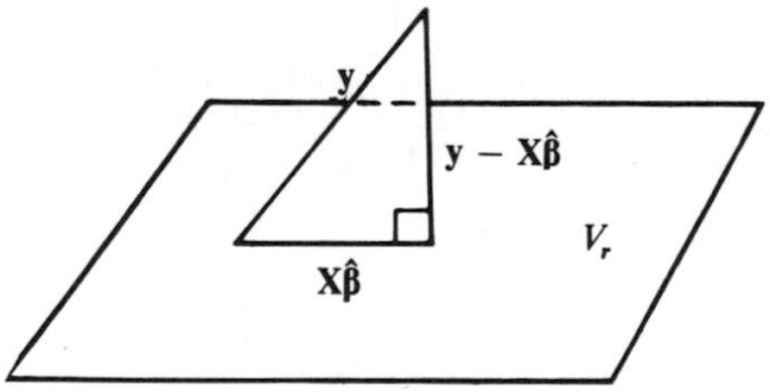

FIGURE 3.5.1. Geometric interpretation of least-squares estimation.

Expression (3.5.14) may also be thought of geometrically (see Scheffé, 1959, p. 14). Suppose an unbiased estimate for $\psi = \mathbf{c}'\boldsymbol{\beta}$ is located, so that $E(\mathbf{a}'\mathbf{Y}) = \mathbf{c}'\boldsymbol{\beta}$. To find the "best" one, observe that, by substituting $\mathbf{a}^*$ for $\mathbf{a}$, where $\mathbf{a}^* = P_V\mathbf{a} = \mathbf{X}(\mathbf{X}'\mathbf{X})^-\mathbf{X}'\mathbf{a}$

(3.5.17) $$\mathbf{a}^{*\prime}\mathbf{y} = \mathbf{a}'\mathbf{X}(\mathbf{X}'\mathbf{X})^-\mathbf{X}'\mathbf{y} = \mathbf{c}'\hat{\boldsymbol{\beta}}$$

Thus, given that $\mathbf{c} \in V(\mathbf{X}')$ among the unbiased estimators $\mathbf{a}'\mathbf{y}$ of ψ, $\hat{\psi}$ is the only one such that the vector $\mathbf{a} \in V(\mathbf{X})$. Hence $\hat{\psi}$ may be guessed at and, if $\mathbf{a} \in V(\mathbf{X})$, the Gauss-Markoff estimate, $\hat{\psi}$, is obtained. For example, in the one-sample problem, let $c = 1$

so that $c \in V(\mathbf{X}')$ and μ is estimable. Try y_1, then $E(Y_1) = \mu$, where here $\mathbf{a}' = [1, 0, \ldots, 0]$; however, $\mathbf{a} \notin V(\mathbf{X})$. To obtain the unique Gauss-Markoff estimate, set

$$\mathbf{a}^* = P_V\mathbf{a} = \mathbf{X}(\mathbf{X}'\mathbf{X})^{-1}\mathbf{X}'\mathbf{a} = \begin{bmatrix} 1/N \\ 1/N \\ \vdots \\ 1/N \end{bmatrix}$$

an element of $V(\mathbf{X})$, so that $\hat{\mu} = \mathbf{c}'\hat{\boldsymbol{\beta}} = \mathbf{a}^*\mathbf{y} = \bar{y}$ is the Gauss-Markoff estimator.

The other procedures—reparameterization and side conditions—discussed in Section 1.6 may also be employed to find "best" estimates of $\mathbf{c}'\boldsymbol{\beta}$. These will be compared to the present two methods using the g-inverse and geometry when the two-sample case is considered.

Hypothesis Testing. Under the *Gauss-Markoff* setup, $\mathbf{y} = \mathbf{X}\boldsymbol{\beta} + \boldsymbol{\varepsilon}$, where $\boldsymbol{\theta} = \mathbf{X}\boldsymbol{\beta} \in \Omega$ and Ω is an r-dimensional subspace $V_r \subset V_N$. Consider the problem of testing the hypothesis $\boldsymbol{\theta} \in \omega$, where ω is an $(r - g)$-dimensional subspace of V_r. For implementation, it is assumed that the random vector $\mathbf{Y}$ has a multivariate normal distribution with mean $\boldsymbol{\eta} = \mathbf{X}\boldsymbol{\beta}$ and variance $\sigma^2\mathbf{I}$, so that $\mathbf{Y} \sim N_N(\boldsymbol{\eta}, \sigma^2\mathbf{I})$. The *fundamental univariate least-squares theorem* is now applied to test hypotheses:

(3.5.18) Under the Gauss-Markoff setup, assume that $\mathbf{Y} \sim N_N(\mathbf{X}\boldsymbol{\beta}, \sigma^2\mathbf{I})$. Let $\mathbf{C}$ be a matrix of order $g \times q$ of rank $g \leq r$, where r is the rank of $\mathbf{X}$ such that the g linearly independent parametric functions $\mathbf{C}\boldsymbol{\beta}$ are individually estimable and $Q_t = \min (\mathbf{y} - \mathbf{X}\boldsymbol{\beta})'(\mathbf{y} - \mathbf{X}\boldsymbol{\beta})$, subject to the condition that $\mathbf{C}\boldsymbol{\beta} = \boldsymbol{\xi}$ (specified). Then

(a) $Q_e = \min\limits_{\boldsymbol{\beta}} (\mathbf{y} - \mathbf{X}\boldsymbol{\beta})'(\mathbf{y} - \mathbf{X}\boldsymbol{\beta}) = \mathbf{y}'[\mathbf{I} - \mathbf{X}(\mathbf{X}'\mathbf{X})^{-}\mathbf{X}']\mathbf{y}$ and $Q_t - Q_e = (\mathbf{C}\hat{\boldsymbol{\beta}} - \boldsymbol{\xi})'[\mathbf{C}(\mathbf{X}'\mathbf{X})^{-}\mathbf{C}']^{-1}(\mathbf{C}\hat{\boldsymbol{\beta}} - \boldsymbol{\xi})$ are independently distributed;

(b) $Q_e \sim \sigma^2\chi^2(N - r)$ and $Q_t - Q_e \sim \sigma^2\chi^2(g, \delta)$; and,

(c) if the hypothesis $H_0\colon \mathbf{C}\boldsymbol{\beta} = \boldsymbol{\xi}$ is true, then $\dfrac{(Q_t - Q_e)/v_h}{Q_e/v_e} \sim F(v_h, v_e)$

where $g = v_h$ and $N - r = v_e$, or $\dfrac{Q_e}{Q_e + (Q_t - Q_e)} = \dfrac{Q_e}{Q_t} = \dfrac{1}{1 + v_h F/v_e}$ $\sim B'(v_e, v_h)$.

Rao (1965a, p. 155) gives a detailed proof of (3.5.18); however, he does not determine the form of $Q_t - Q_e$. We shall now determine the form of $Q_t - Q_e$ and prove the theorem (see also, Searle, 1965).

To minimize Q_t, where

$$Q_t = \min_{\mathbf{C}\boldsymbol{\beta} = \boldsymbol{\xi}} (\mathbf{y} - \mathbf{X}\boldsymbol{\beta})'(\mathbf{y} - \mathbf{X}\boldsymbol{\beta})$$

a vector of Lagrange multipliers is introduced so that

(3.5.19) $$F = (\mathbf{y} - \mathbf{X}\boldsymbol{\beta})'(\mathbf{y} - \mathbf{X}\boldsymbol{\beta}) + 2\boldsymbol{\lambda}'(\mathbf{C}\boldsymbol{\beta} - \boldsymbol{\xi})$$

must be minimized. Taking partial derivatives with respect to $\boldsymbol{\beta}$ and $\boldsymbol{\lambda}$ and equating them to zero, the resulting equations are

$$(3.5.20) \qquad \begin{aligned} (\mathbf{X}'\mathbf{X})\boldsymbol{\beta}_\omega + \mathbf{C}'\boldsymbol{\lambda} &= \mathbf{X}'\mathbf{y} \\ \mathbf{C}\boldsymbol{\beta}_\omega &= \boldsymbol{\xi} \end{aligned}$$

Using the first equation, $\boldsymbol{\beta}_\omega = \hat{\boldsymbol{\beta}} - (\mathbf{X}'\mathbf{X})^-\mathbf{C}'\boldsymbol{\lambda}$. Multiplying by $\mathbf{C}$,

$$\boldsymbol{\lambda} = [\mathbf{C}(\mathbf{X}'\mathbf{X})^-\mathbf{C}']^{-1}(\mathbf{C}\hat{\boldsymbol{\beta}} - \boldsymbol{\xi})$$

so that the best estimate of $\boldsymbol{\beta}$ under the hypothesis is

$$(3.5.21) \qquad \hat{\boldsymbol{\beta}}_\omega = \hat{\boldsymbol{\beta}} - (\mathbf{X}'\mathbf{X})^-\mathbf{C}'[\mathbf{C}(\mathbf{X}'\mathbf{X})^-\mathbf{C}']^{-1}(\mathbf{C}\hat{\boldsymbol{\beta}} - \boldsymbol{\xi})$$

and

$$(3.5.22) \qquad Q_t = (\mathbf{y} - \mathbf{X}\hat{\boldsymbol{\beta}}_\omega)'(\mathbf{y} - \mathbf{X}\hat{\boldsymbol{\beta}}_\omega) = Q_e + \boldsymbol{\lambda}'[\mathbf{C}(\mathbf{X}'\mathbf{X})^-\mathbf{C}']\boldsymbol{\lambda}$$

or

$$(3.5.23) \qquad Q_h = Q_t - Q_e = (\mathbf{C}\hat{\boldsymbol{\beta}} - \boldsymbol{\xi})'[\mathbf{C}(\mathbf{X}'\mathbf{X})^-\mathbf{C}']^{-1}(\mathbf{C}\boldsymbol{\beta} - \boldsymbol{\xi})$$

That $[\mathbf{C}(\mathbf{X}'\mathbf{X})^-\mathbf{C}']^{-1}$ exists follows from the fact that it is a $g \times g$ matrix of rank g. This establishes the form of $Q_t - Q_e$ in (3.5.18a).

To prove (3.5.18b), use the fact that $\mathbf{Y} \sim N_N(\mathbf{X}\boldsymbol{\beta}, \sigma^2\mathbf{I})$. Since $\hat{\boldsymbol{\beta}} = (\mathbf{X}'\mathbf{X})^-\mathbf{X}'\mathbf{y}$,

$$\mathbf{C}\hat{\boldsymbol{\beta}} - \boldsymbol{\xi} \sim N_g[\mathbf{C}\boldsymbol{\beta} - \boldsymbol{\xi}, \mathbf{C}(\mathbf{X}'\mathbf{X})^-\mathbf{C}'\sigma^2]$$

Applying (2.6.6),

$$\frac{(\mathbf{C}\hat{\boldsymbol{\beta}} - \boldsymbol{\xi})'[\mathbf{C}(\mathbf{X}'\mathbf{X})^-\mathbf{C}']^{-1}(\mathbf{C}\hat{\boldsymbol{\beta}} - \boldsymbol{\xi})}{\sigma^2} \sim \chi^2(g, \delta)$$

and

$$\delta = \frac{(\mathbf{C}\boldsymbol{\beta} - \boldsymbol{\xi})'[\mathbf{C}(\mathbf{X}'\mathbf{X})^-\mathbf{C}']^{-1}(\mathbf{C}\boldsymbol{\beta} - \boldsymbol{\xi})}{\sigma^2}$$

Furthermore, by (2.6.8), with $\mathbf{A} = [\mathbf{C}(\mathbf{X}'\mathbf{X})^-\mathbf{C}']^{-1}$ and $\boldsymbol{\Sigma} = \mathbf{C}(\mathbf{X}'\mathbf{X})^-\mathbf{C}'\sigma^2$, the expected value of the sum of squares Q_h due to the hypothesis is

$$\begin{aligned} E(Q_h) &= \mathrm{Tr}\{[\mathbf{C}(\mathbf{X}'\mathbf{X})^-\mathbf{C}']^{-1}[\mathbf{C}(\mathbf{X}'\mathbf{X})^-\mathbf{C}']\sigma^2\} + \delta\sigma^2 \\ &= \mathrm{Tr}(\sigma^2\mathbf{I}_g) + \delta\sigma^2 \\ &= g\sigma^2 + \delta\sigma^2 \end{aligned}$$

Defining the mean square (MS) of a sum of squares (SS) as the SS divided by its degrees of freedom, the expected value of a MS associated with any hypothesis under the Gauss-Markoff setup is obtained by using the formula

$$\begin{aligned} E\left(\frac{Q_h}{g}\right) &= E(\mathrm{MS}_h) = \sigma^2 + \frac{\delta\sigma^2}{g} \\ (3.5.24) \qquad &= \sigma^2 + \frac{(\mathbf{C}\boldsymbol{\beta} - \boldsymbol{\xi})'[\mathbf{C}(\mathbf{X}'\mathbf{X})^-\mathbf{C}']^{-1}(\mathbf{C}\boldsymbol{\beta} - \boldsymbol{\xi})}{g} \end{aligned}$$

Following Example 2.6.1,

$$\frac{\mathbf{y}'[\mathbf{I} - \mathbf{X}(\mathbf{X}'\mathbf{X})^{-}\mathbf{X}']\mathbf{y}}{\sigma^2} \sim \chi^2(N - r, \delta = 0)$$

Proving independence is more involved since the quadratic forms are not written in terms of the same normally distributed random variables. In order to establish independence, observe that

$$Q_e = [\mathbf{y} - \mathbf{X}\mathbf{C}'(\mathbf{C}\mathbf{C}')^{-1}\boldsymbol{\xi}]'[\mathbf{I} - \mathbf{X}(\mathbf{X}'\mathbf{X})^{-}\mathbf{X}'][\mathbf{y} - \mathbf{X}\mathbf{C}'(\mathbf{C}\mathbf{C}')^{-1}\boldsymbol{\xi}]$$

$$= \mathbf{y}'[\mathbf{I} - \mathbf{X}(\mathbf{X}'\mathbf{X})^{-}\mathbf{X}']\mathbf{y} \tag{3.5.25}$$

since $\mathbf{X} = \mathbf{X}(\mathbf{X}'\mathbf{X})^{-}\mathbf{X}'\mathbf{X}$ and $\mathbf{X}' = \mathbf{X}'\mathbf{X}(\mathbf{X}'\mathbf{X})^{-}\mathbf{X}'$. Furthermore,

$$Q_t - Q_e = (\mathbf{C}\hat{\boldsymbol{\beta}} - \boldsymbol{\xi})'[\mathbf{C}(\mathbf{X}'\mathbf{X})^{-}\mathbf{C}']^{-1}(\mathbf{C}\hat{\boldsymbol{\beta}} - \boldsymbol{\xi})$$

$$= [\mathbf{C}(\mathbf{X}'\mathbf{X})^{-}\mathbf{X}'\mathbf{y} - \boldsymbol{\xi}]'[\mathbf{C}(\mathbf{X}'\mathbf{X})^{-}\mathbf{C}']^{-1}[\mathbf{C}(\mathbf{X}'\mathbf{X})^{-}\mathbf{X}'\mathbf{y} - \boldsymbol{\xi}]$$

$$= \{\mathbf{y}'\mathbf{X}[(\mathbf{X}'\mathbf{X})^{-}]'\mathbf{C}' - \boldsymbol{\xi}'\}[\mathbf{C}(\mathbf{X}'\mathbf{X})^{-}\mathbf{C}']^{-1}[\mathbf{C}(\mathbf{X}'\mathbf{X})^{-}\mathbf{X}'\mathbf{y} - \boldsymbol{\xi}]$$

$$= [\mathbf{y} - \mathbf{X}\mathbf{C}'(\mathbf{C}\mathbf{C}')^{-1}\boldsymbol{\xi}]'\mathbf{X}[(\mathbf{X}'\mathbf{X})^{-}]'\mathbf{C}'[\mathbf{C}(\mathbf{X}'\mathbf{X})^{-}\mathbf{C}']^{-1}\mathbf{C}(\mathbf{X}'\mathbf{X})^{-}\mathbf{X}' \times [\mathbf{y} - \mathbf{X}\mathbf{C}'(\mathbf{C}\mathbf{C}')^{-1}\boldsymbol{\xi}] \tag{3.5.26}$$

since $\mathbf{C} = \mathbf{A}\mathbf{X}$, for some matrix $\mathbf{A}$, and

$$[\mathbf{C}(\mathbf{X}'\mathbf{X})^{-}\mathbf{X}'][\mathbf{X}\mathbf{C}'(\mathbf{C}\mathbf{C}')^{-1}] = \mathbf{A}\mathbf{X}\mathbf{C}'(\mathbf{C}\mathbf{C}')^{-1} = \mathbf{I}_g$$

The quadratic forms Q_e and $Q_t - Q_e$ in (3.5.25) and (3.5.26) now have the same form. But, since $\mathbf{X} = \mathbf{X}(\mathbf{X}'\mathbf{X})^{-}\mathbf{X}'\mathbf{X}$,

$$[\mathbf{I} - \mathbf{X}(\mathbf{X}'\mathbf{X})^{-}\mathbf{X}']\mathbf{X}[(\mathbf{X}'\mathbf{X})^{-}]'\mathbf{C}'[\mathbf{C}(\mathbf{X}'\mathbf{X})^{-}\mathbf{C}']^{-1}\mathbf{C}(\mathbf{X}'\mathbf{X})^{-}\mathbf{X}' = \mathbf{0}$$

and, by (2.6.10), (3.5.25) and (3.5.26) are statistically independent. Both have chi-square distributions since the matrices in each quadratic form are idempotent. Thus $Q_e \sim \sigma^2\chi^2(N - r)$ and $Q_t - Q_e \sim \sigma^2\chi^2(g, \delta)$. The degrees of freedom are just the ranks of the matrices in the quadratic forms given in (3.5.25) and (3.5.26). The ratio of two statistically independent χ^2 random variables is by definition an F ratio, proving the fundamental univariate least-squares theorem.

EXAMPLE 3.5.1. Applying (3.5.18) to test the single-sample hypothesis $H_0: \mu = \mu_0$,

$$\begin{aligned} Q_t &= \min_{\mathbf{C}\boldsymbol{\beta}=\boldsymbol{\xi}} (\mathbf{y} - \mathbf{X}\boldsymbol{\beta})'(\mathbf{y} - \mathbf{X}\boldsymbol{\beta}) \\ &= (\mathbf{y} - \mathbf{1}\mu_0)'(\mathbf{y} - \mathbf{1}\mu_0) \\ &= \sum_{i=1}^{N} y_i^2 - 2\mu_0 \sum_{i=1}^{N} y_i + N\mu_0^2 \\ &= \sum_{i=1}^{N} (y_i - \mu_0)^2 \end{aligned}$$

and

$$Q_e = \sum_{i=1}^{N} (y_i - \bar{y})^2$$

so that

$$Q_t - Q_e = \sum_{i=1}^{N} (\bar{y} - \mu_0)^2$$

Alternatively, we could have used (3.5.23) directly to determine $Q_t - Q_e$. Further, $Q_t = (Q_t - Q_e) + Q_e = Q_h + Q_e$, which could also be obtained by partitioning a sum of squares in the "traditional sense" by writing $y_i - \mu_0 = (\bar{y} - \mu_0) + (y_i - \bar{y})$, squaring, and summing:

$$\sum_{i=1}^{N} (y_i - \mu_0)^2 = \sum_{i=1}^{N} (\bar{y} - \mu_0)^2 + \sum_{i=1}^{N} (y_i - \bar{y})^2$$

so that the total sum of squares equals the hypothesis sum of squares plus the within sum of squares as expected. By (2.6.6), the

$$\sum_{i=1}^{N} (y_i - \bar{y})^2 = (N - 1)s^2 \sim \sigma^2\chi^2(N - 1, \delta = 0)$$

and

$$\sum_{i=1}^{N} \frac{(y_i - \mu_0)^2}{\sigma^2} = \sum_{i=1}^{N} z_i^2 \sim \chi^2(N, \delta)$$

since $z_i \sim IN(0, 1)$, or

$$\sum_{i=1}^{N} (\bar{y} - \mu_0)^2 \sim \sigma^2\chi^2(N, \delta)$$

Also, by Cochran's theorem,

$$\sum_{i=1}^{N} (\bar{y} - \mu_0)^2 \sim \sigma^2\chi^2(N, \delta)$$

where

$$\delta = \frac{\sum_{i=1}^{N} (\mu - \mu_0)^2}{\sigma^2}$$

Thus, using (3.5.18) or Cochran's theorem,

$$\frac{(Q_t - Q_e)/1}{Q_e/(N - 1)} = \frac{\sum_{i=1}^{N} (\bar{y} - \mu_0)^2}{s^2} = \frac{N(\bar{y} - \mu_0)^2}{s^2} \sim F(1, N - 1)$$

when H_0 is true, or

$$t = \frac{N(\bar{y} - \mu_0)}{s} \sim t(N - 1)$$

may be used to test H_0.

For a geometric interpretation of (3.5.18), let V_r denote the space spanned by the columns of $\mathbf{X}$, V_{r-g} the hypothesis subspace of V_r, and V_g the orthocomplement of V_{r-g} relative to V_r. In minimizing Q_e, the estimate of $\mathbf{X\beta}$ is $\mathbf{X\hat{\beta}} = \hat{\boldsymbol{\theta}}_\Omega$, obtained under the general model Ω. In minimizing Q_t, the estimate of $\mathbf{X\beta}$, $\mathbf{X}\hat{\boldsymbol{\beta}}_\omega$, is acquired under the constraint imposed by the hypothesis, so let $\hat{\boldsymbol{\theta}}_\omega = \mathbf{X}\hat{\boldsymbol{\beta}}_\omega$. By using the geometry of least-squares estimation, $\hat{\boldsymbol{\theta}}_\Omega$ is the projection of $\mathbf{y}$ on V_r and $\hat{\boldsymbol{\theta}}_\omega$ is the projection of $\mathbf{y}$ on V_{r-g}, so that

$$Q_t = \|\mathbf{y} - \hat{\boldsymbol{\theta}}_\omega\|^2 \quad \text{and} \quad Q_e = \|\mathbf{y} - \hat{\boldsymbol{\theta}}_\Omega\|^2$$

and

$$Q_h = Q_t - Q_e = \|\hat{\boldsymbol{\theta}}_\Omega - \hat{\boldsymbol{\theta}}_\omega\|^2$$

by the orthogonality of the projections. Pictorially, the relations between Q_t and Q_e are seen in Figure 3.5.2, where we see that $Q_t = (Q_t - Q_e) + Q_e$. Thus $Q_e \in V_r^\perp$ of

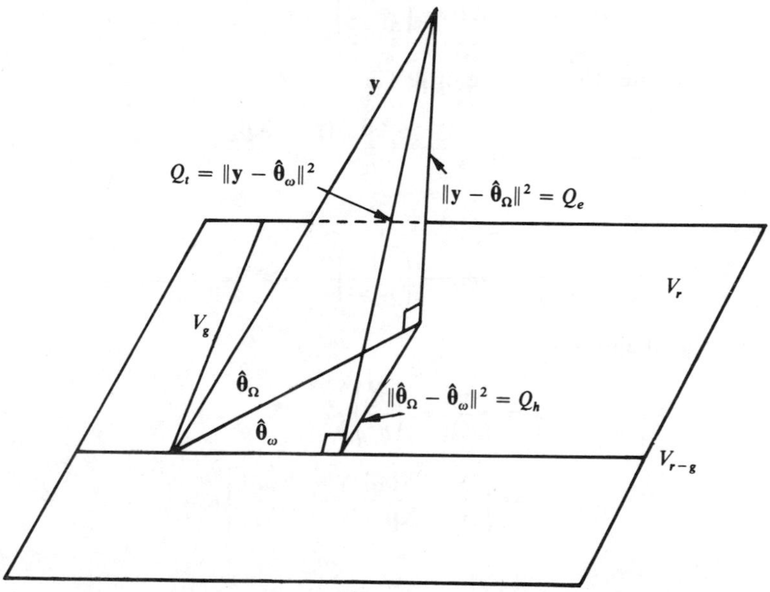

FIGURE 3.5.2. Geometric interpretation of the *F* statistic.

dimension $N - r$, $Q_t \in V_{N-r+g}$ of dimension $N - r + g$, and $Q_t - Q_e \in V_g$ of dimension g. The degrees of freedom in the F statistic correspond to the dimensions of the subspaces. From (3.5.18), $Q_t - Q_e$ and Q_e are independent. Independence also follows from the geometric interpretation of $Q_t - Q_e$ and Q_e using projection operators (see Seber, 1966, p. 25).

The geometry of the fundamental theorem gives a visual interpretation of the F statistic. As $\hat{\boldsymbol{\theta}}_\Omega$ and $\hat{\boldsymbol{\theta}}_\omega$ differ, the best estimates of $\boldsymbol{\theta}$ under Ω and ω (the hypothesis), $Q_t - Q_e$, increase so that the value of the F statistic becomes large, yielding rejection of the null hypothesis.

As Example 3.5.1 and the geometry of the fundamental least-squares theorem indicate, the test of the hypothesis H_0: $\mathbf{C\beta} = \boldsymbol{\xi}$ is a likelihood-ratio test. With

$\mathbf{Y} \sim N(\mathbf{X\beta}, \sigma^2\mathbf{I})$, the likelihood function is

$$L = L(\mathbf{y} \mid \boldsymbol{\beta}, \sigma^2) = (2\pi)^{-N/2}|\sigma^2\mathbf{I}|^{-1/2} \exp - \frac{(\mathbf{y} - \mathbf{X\beta})'(\sigma^2\mathbf{I})^{-1}(\mathbf{y} - \mathbf{X\beta})}{2}$$

$$= (2\pi\sigma^2)^{-N/2} \exp - \frac{(\mathbf{y} - \mathbf{X\beta})'(\mathbf{y} - \mathbf{X\beta})}{2\sigma^2}$$

Maximizing L with respect to $\boldsymbol{\beta}$ and σ^2 is equivalent to maximizing $L^* = \log L$. Taking partial derivatives and equating them to 0 under Ω,

$$\hat{\boldsymbol{\beta}}_\Omega = (\mathbf{X}'\mathbf{X})^-\mathbf{X}'\mathbf{y} \quad \text{and} \quad \hat{\sigma}^2_\Omega = \frac{(\mathbf{y} - \mathbf{X}\hat{\boldsymbol{\beta}}_\Omega)'(\mathbf{y} - \mathbf{X}\hat{\boldsymbol{\beta}}_\Omega)}{N}$$

so that

$$L(\hat{\Omega}) = \left(\frac{1}{2\pi\hat{\sigma}^2_\Omega}\right)^{N/2} e^{-N/2}$$

Under ω, $\hat{\boldsymbol{\beta}}_\omega$ denotes the estimate of $\boldsymbol{\beta}$. Then

$$\hat{\sigma}^2_\omega = \frac{(\mathbf{y} - \mathbf{X}\hat{\boldsymbol{\beta}}_\omega)'(\mathbf{y} - \mathbf{X}\hat{\boldsymbol{\beta}}_\omega)}{N}$$

and

$$L(\hat{\omega}) = \left(\frac{1}{2\pi\hat{\sigma}^2}\right)^{N/2} e^{-N/2}$$

By forming the likelihood ratio

$$\lambda = \frac{L(\hat{\omega})}{L(\hat{\Omega})} = \left(\frac{\hat{\sigma}^2_\Omega}{\hat{\sigma}^2_\omega}\right)^{N/2}$$

$$= \left[\frac{(\mathbf{y} - \mathbf{X}\hat{\boldsymbol{\beta}}_\Omega)'(\mathbf{y} - \mathbf{X}\hat{\boldsymbol{\beta}}_\Omega)}{(\mathbf{y} - \mathbf{X}\hat{\boldsymbol{\beta}}_\omega)'(\mathbf{y} - \mathbf{X}\hat{\boldsymbol{\beta}}_\omega)}\right]^{N/2}$$

or

$$\Lambda = \lambda^{2/N} = \frac{Q_e}{Q_t} = \frac{Q_e}{Q_e + (Q_t - Q_e)} = \frac{1}{1 + (v_h F/v_e)}$$

the result follows. Reject the null hypothesis at the significance level α if

$$F = \frac{(Q_t - Q_e)/v_h}{Q_e/v_e} > F^\alpha(v_h, v_e)$$

Provided the g linearly independent functions $\mathbf{C\beta}$ are individually estimable, a hypothesis of the form $H_0: \mathbf{C\beta} = \boldsymbol{\xi}$ can be tested, using the central F distribution as derived above (that is, using the likelihood-ratio criterion). In forming a matrix $\mathbf{C}$ to test a specific hypothesis represented in terms of the model parameters, it is important to realize that the matrix $\mathbf{C}$ is not unique. The matrix $\mathbf{C}$ is only determinable up to a nonsingular transformation. However, since the hypothesis sum of squares $Q_h = Q_t - Q_e$ in the F ratio is invariant under nonsingular transformations $\mathbf{T}$, testing $H_0: \mathbf{C\beta} = \boldsymbol{\xi}$ and $H'_0: \mathbf{TC\beta} = \mathbf{T}\boldsymbol{\xi}$ shows that they are equivalent. Observe that the hypothesis sum of squares for testing H_0' is identical to that used to test H_0:

$$\begin{aligned} Q_h &= (\mathbf{TC}\hat{\boldsymbol{\beta}} - \mathbf{T}\boldsymbol{\xi})'[\mathbf{TC}(\mathbf{X}'\mathbf{X})^{-}\mathbf{C}'\mathbf{T}']^{-1}(\mathbf{TC}\hat{\boldsymbol{\beta}} - \mathbf{T}\boldsymbol{\xi}) \\ &= (\mathbf{C}\hat{\boldsymbol{\beta}} - \boldsymbol{\xi})'\mathbf{T}'(\mathbf{T}')^{-1}[\mathbf{C}(\mathbf{X}'\mathbf{X})^{-}\mathbf{C}']^{-1}\mathbf{T}^{-1}\mathbf{T}(\mathbf{C}\hat{\boldsymbol{\beta}} - \boldsymbol{\xi}) \\ &= (\mathbf{C}\hat{\boldsymbol{\beta}} - \boldsymbol{\xi})'[\mathbf{C}(\mathbf{X}'\mathbf{X})^{-}\mathbf{C}']^{-1}(\mathbf{C}\hat{\boldsymbol{\beta}} - \boldsymbol{\xi}) \end{aligned}$$

General Gauss-Markoff Setup. A generalization of the Gauss-Markoff setup defined in (3.5.3) is to allow *heteroscedasticity of errors*; that is,

$$\boldsymbol{\Sigma} = \sigma^2\mathbf{V} = \begin{bmatrix} \sigma_1^2 & 0 & \cdots & 0 \\ 0 & \sigma_2^2 & \cdots & 0 \\ \vdots & \vdots & \vdots\vdots\vdots & \vdots \\ 0 & \cdot & \cdots & \sigma_n^2 \end{bmatrix}$$

where $\mathbf{V}$ is a known diagonal matrix. More generally, $\mathbf{V}$ may be any known p.d. matrix so that the errors may be correlated, heteroscedastic, or both. The setup adopted in this case is that

(3.5.27)
$$\begin{aligned} E(\boldsymbol{\varepsilon}) &= \mathbf{0} \qquad & E(\mathbf{Y}) &= \mathbf{X}\boldsymbol{\beta} \\ & \qquad \text{or} & & \\ V(\boldsymbol{\varepsilon}) &= \sigma^2\mathbf{V} \qquad & V(\mathbf{Y}) &= \sigma^2\mathbf{V} \end{aligned}$$

The linear model is

(3.5.28)
$$\mathbf{y} = \mathbf{X}\boldsymbol{\beta} + \boldsymbol{\varepsilon}$$

where $E(\boldsymbol{\varepsilon}) = \mathbf{0}$ and $V(\boldsymbol{\varepsilon}) = \sigma^2\mathbf{V}$.

Although we have a linear model, representation (3.5.27) does not satisfy the assumption of the Gauss-Markoff theorem. However, by use of the transformation $\mathbf{y}_0 = \mathbf{V}^{-1/2}\mathbf{y}$, (3.5.27) reduces to (3.5.3) since

(3.5.29)
$$\begin{aligned} E(\mathbf{Y}_0) &= \mathbf{V}^{-1/2}\mathbf{X}\boldsymbol{\beta} = \mathbf{X}_0\boldsymbol{\beta} \\ V(\mathbf{Y}_0) &= \mathbf{V}^{-1/2}\sigma^2\mathbf{V}\mathbf{V}^{-1/2} = \sigma^2\mathbf{I} \end{aligned}$$

The linear model in (3.5.28) becomes

(3.5.30)
$$\mathbf{y}_0 = \mathbf{X}_0\boldsymbol{\beta} + \boldsymbol{\varepsilon}_0$$

by setting $\mathbf{X}_0 = \mathbf{V}^{-1/2}\mathbf{X}, \mathbf{y}_0 = \mathbf{V}^{-1/2}\mathbf{y}$, and $\boldsymbol{\varepsilon}_0 = \mathbf{V}^{-1/2}\boldsymbol{\varepsilon}$, so that $E(\boldsymbol{\varepsilon}_0) = \mathbf{0}$ and $V(\boldsymbol{\varepsilon}_0) = \sigma^2\mathbf{I}$; hence Gauss-Markoff theory is now applicable using (3.5.30). An estimator for $\boldsymbol{\beta}$ using (3.5.30) is

(3.5.31)
$$\begin{aligned} \hat{\boldsymbol{\beta}} &= (\mathbf{X}_0'\mathbf{X}_0)^{-}\mathbf{X}_0'\mathbf{y}_0 \\ &= (\mathbf{X}'\mathbf{V}^{-1}\mathbf{X})^{-}\mathbf{X}'\mathbf{V}^{-1}\mathbf{y} \end{aligned}$$

which is obtained by minimizing

$$\begin{aligned} \boldsymbol{\varepsilon}_0'\boldsymbol{\varepsilon}_0 &= (\mathbf{y}_0 - \mathbf{X}_0\boldsymbol{\beta})'(\mathbf{y}_0 - \mathbf{X}_0\boldsymbol{\beta}) \\ &= (\mathbf{y} - \mathbf{X}\boldsymbol{\beta})'\mathbf{V}^{-1}(\mathbf{y} - \mathbf{X}\boldsymbol{\beta}) \end{aligned}$$

This procedure is called the method of *generalized least squares* (see Aitken, 1935). When $\mathbf{V}$ is diagonal, we have what is often called the *weighted regression model.* The best linear unbiased estimate of $\psi = \mathbf{c}'\boldsymbol{\beta}$ is

$$\hat{\psi} = \mathbf{c}'(\mathbf{X}'\mathbf{V}^{-1}\mathbf{X})^{-}\mathbf{X}'\mathbf{V}^{-1}\mathbf{y} \tag{3.5.32}$$

with variance given by

$$V(\hat{\psi}) = \sigma^2\mathbf{c}'(\mathbf{X}'\mathbf{V}^{-1}\mathbf{X})^{-}\mathbf{c}' \tag{3.5.33}$$

The model classifications discussed in the univariate case each have multivariate analogues. We will be primarily concerned with multivariate functionally related models and multivariate conditional regression models.

EXERCISES 3.5

1. Let $E(Y_i) = \beta_0 + \beta_1 x_i + \beta_2 x_i^2 + \beta_3 x_i^3$, for $x_1 = -1, x_2 = 0, x_3 = 1$, and $E(Y_i) = \sigma^2$.
   a. Solve the normal equations.
   b. Show that β_0 and β_2 are estimable but that β_1 and β_3 are not estimable.
   c. Is $\beta_0 + \beta_2$ or $\beta_1 + \beta_3$ estimable? Why?
   d. Determine the least-squares estimate for the estimable functions in parts b and c.
   e. Find the estimate for the common variance σ^2.
2. Consider the model $E(Y_i) = \mu + \beta x$, for $i = 1, \ldots, N$, where all x's are equal and the $V(Y_i) = \sigma^2$.
   a. Solve the normal equations.
   b. Does a unique solution for μ or β exist?
   c. For what functions of the parameters can one always find a unique solution?
   d. Find the variance of the estimate $\hat{\psi}$ for $\psi = \mu + \beta x$.
3. For the model $E(Y_{ij}) = \mu + (i - 1)\beta$ and the $V(Y_{ij}) = \sigma^2$, for $i = 0, 1, 2$ and $j = 1, \ldots, J$, consider the hypothesis $H_0: \beta = 0$ for testing.
   a. Show that the hypothesis H_0 is a testable hypothesis.
   b. Construct an ANOVA table for the analysis of the hypothesis and include Q_h, Q_e, degrees of freedom, mean squares, and expected mean squares.
   c. Find the formula for the noncentrality parameter δ.
   d. Evaluate each of the expressions obtained in part a by using the following data:

	j = 1	2	3	4
$y_{0,j}$	3	4	6	3
$y_{1,j}$	3	4	5	5
$y_{2,j}$	1	5	3	6

   e. Using $\alpha = .05$, test the null hypothesis $H_0: \beta = 0$ and determine a $100(1 - \alpha)\%$ confidence interval for β.
4. Prove the results summarized in (3.5.32) and (3.5.33) by following the proof of (3.5.14).
5. Assume that $Y_i = \beta_0 + \beta_1 x_i + \varepsilon_i$, where $E(\varepsilon_i) = 0$, and $\boldsymbol{\Sigma} = \sigma^2\mathbf{V}$, where $\mathbf{V}$ is a diagonal matrix with elements $1/w_i$.

a. Determine the "weighted" least-square estimates of β_0 and β_1.

b. Show that

$$Q_e = \sum_i w_i y_i^2 - \hat{\beta}_0 \sum_i w_i y_i - \hat{\beta}_1 \sum_i w_i x_i y_i$$

c. Derive a 95% confidence interval for β_1.

3.6 THE GENERAL LINEAR MODEL—MULTIVARIATE

To extend the results of Section 3.5 to the multivariate situation, consider a random sample of N p-variate vectors $\mathbf{Y}_i$ such that

$$\underset{(N \times p)}{\mathbf{Y}} = \underset{(N \times q)}{\mathbf{X}} \; \underset{(q \times p)}{\mathbf{B}} + \underset{(N \times p)}{\mathbf{E}_0} \tag{3.6.1}$$

where

$\mathbf{y}_i' =$ ith row of $\mathbf{Y}$, the observed data matrix;
$\mathbf{X} =$ known design matrix as in the univariate case of rank $r \leq q \leq N$;
$\mathbf{B} =$ matrix of unknown nonrandom parameters;
$\mathbf{E}_0 =$ matrix of random errors such that $E(\mathbf{E}_0) = \mathbf{0}$; and the
$\mathbf{V}(\mathbf{E}_0) = \mathbf{I}_N \otimes \mathbf{\Sigma}$, where $p \leq (N - r)$, which insures that $\mathbf{\Sigma}$ is p.d.

Thus

$$\begin{aligned} E(\mathbf{Y}) &= \mathbf{XB} \\ V(\mathbf{Y}) &= \mathbf{I} \otimes \mathbf{\Sigma} \end{aligned} \tag{3.6.2}$$

where (3.6.1) represents the *multivariate linear model* and (3.6.2) denotes the *multivariate Gauss-Markoff setup.*

Estimation Theory. As in the univariate case, the estimation of the elements of $\mathbf{B}$ by least squares is desired. To accomplish this, the

$$\text{Tr}[(\mathbf{Y} - \mathbf{XB})'(\mathbf{Y} - \mathbf{XB})] \tag{3.6.3}$$

is minimized. From Section 1.8, the *normal equations* for this case are

$$(\mathbf{X}'\mathbf{X})\mathbf{B} = \mathbf{X}'\mathbf{Y} \tag{3.6.4}$$

and the least-squares solution to (3.6.4) is

$$\hat{\mathbf{B}} = (\mathbf{X}'\mathbf{X})^{-}\mathbf{X}'\mathbf{Y} + (\mathbf{I} - \mathbf{H})\mathbf{Z} \tag{3.6.5}$$

where $\mathbf{H} = (\mathbf{X}'\mathbf{X})^{-}\mathbf{X}'\mathbf{X}$ and $\mathbf{Z}$ is an arbitrary matrix. A unique solution to (3.6.5) exists if $(\mathbf{X}'\mathbf{X})^{-} = (\mathbf{X}'\mathbf{X})^{-1}$, and, in this case,

$$\hat{\mathbf{B}} = (\mathbf{X}'\mathbf{X})^{-1}\mathbf{X}'\mathbf{Y} \quad \text{or} \quad \hat{\mathbf{B}}' = \mathbf{Y}'\mathbf{X}(\mathbf{X}'\mathbf{X})^{-1} \tag{3.6.6}$$

Representing $\hat{\mathbf{B}}$ as

$$\hat{\mathbf{B}} = (\mathbf{X}'\mathbf{X})^{-}\mathbf{X}'[\mathbf{U}_1 \quad \mathbf{U}_2 \quad \cdots \quad \mathbf{U}_p] \tag{3.6.7}$$

where $\mathbf{U}_j$ denotes the jth column of the data matrix $\mathbf{Y}$,

$$\hat{\mathbf{B}} = [\hat{\boldsymbol{\beta}}_1 \quad \hat{\boldsymbol{\beta}}_2 \quad \cdots \quad \hat{\boldsymbol{\beta}}_p] \tag{3.6.8}$$

where $\hat{\boldsymbol{\beta}}_j = (\mathbf{X}'\mathbf{X})^{-}\mathbf{X}'\mathbf{U}_j$ is the usual univariate least-squares estimate if we consider each column of $\mathbf{Y}$ a variable at a time.

Geometrically then, each column of $\mathbf{Y}$ is being projected onto the space spanned by the columns of $\mathbf{X}$. Pictorially, $\mathbf{Y}$ denotes a subspace and $\mathbf{X}\hat{\boldsymbol{\beta}}$ is a subspace of best fit to the vectors $\mathbf{U}_j$ in $\mathbf{Y}$ (see Figure 3.6.1).

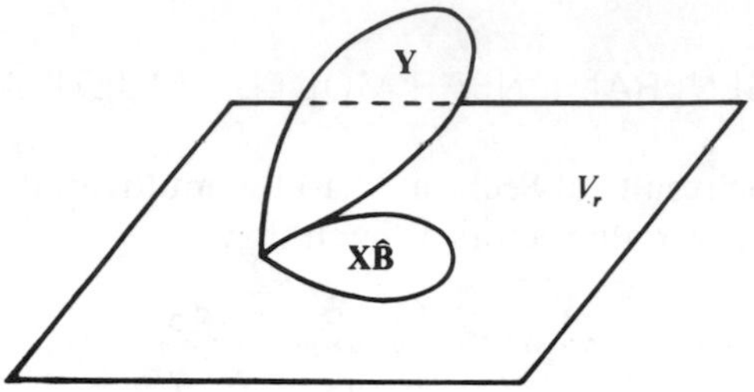

FIGURE 3.6.1. Geometric interpretation of multivariate least-squares estimator.

Again, $\mathbf{Y} - \mathbf{X}\hat{\boldsymbol{\beta}}$ is the error made in determining $\mathbf{Y}$ by $\mathbf{X}\hat{\boldsymbol{\beta}} = \hat{\mathbf{Y}}$, and $\mathbf{Y} - \mathbf{X}\hat{\boldsymbol{\beta}}$ is the error matrix. An estimate of this error matrix is given by the

$$\mathbf{Q}_e = (\mathbf{Y} - \mathbf{X}\hat{\boldsymbol{\beta}})'(\mathbf{Y} - \mathbf{X}\hat{\boldsymbol{\beta}}) \tag{3.6.9}$$

from which an unbiased estimate of $\boldsymbol{\Sigma}$ is obtained:

$$\mathbf{S} = \frac{(\mathbf{Y} - \mathbf{X}\hat{\mathbf{B}})'(\mathbf{Y} - \mathbf{X}\hat{\mathbf{B}})}{N - r}$$

$$= \frac{\mathbf{Y}'[\mathbf{I}_N - \mathbf{X}(\mathbf{X}'\mathbf{X})^{-}\mathbf{X}']\mathbf{Y}}{N - r} \tag{3.6.10}$$

To prove (3.6.10), observe that $\mathbf{S}$ may be written as

$$\mathbf{S} = \frac{1}{N - r}[(\mathbf{U}_j - \mathbf{X}\hat{\boldsymbol{\beta}}_j)'(\mathbf{U}_{j'} - \mathbf{X}\hat{\boldsymbol{\beta}}_{j'})]$$

where $j = j'$ on the diagonal and $j = 1, \ldots, p$. Furthermore, from (3.3.4), the $\text{cov}(\mathbf{U}_j\mathbf{U}_{j'}) = \sigma_{jj'}^2 I$, and the result follows.

Applying these results to the one-sample problem where $\mathbf{X} = \mathbf{1}$ is of full rank, the unbiased least-squares estimate of $\boldsymbol{\mu}$ and an unbiased estimate for $\boldsymbol{\Sigma}$ are obtained:

$$\hat{\boldsymbol{\mu}}' = (\mathbf{1}'\mathbf{1})^{-1}\mathbf{1}'\mathbf{Y} = \bar{\mathbf{y}}'$$

and

$$\mathbf{S} = \frac{\mathbf{Y}'[\mathbf{I}_N - \mathbf{1}(\mathbf{1}'\mathbf{1})^{-1}\mathbf{1}']\mathbf{Y}}{N - 1}$$

$$= \frac{\mathbf{Y}'\mathbf{Y} - N\overline{\mathbf{Y}}\overline{\mathbf{Y}}'}{N - 1}$$

$$= \frac{\sum_{i=1}^{N} (\mathbf{y}_i - \bar{\mathbf{y}})(\mathbf{y}_i - \bar{\mathbf{y}})'}{N - 1}$$

Before stating the multivariate Gauss-Markoff theorem, it is convenient to introduce the following notation: Let

$$\begin{aligned} \mathbf{Y} &= [\mathbf{U}_1 \quad \mathbf{U}_2 \quad \cdots \quad \mathbf{U}_p] \\ \mathbf{B} &= [\boldsymbol{\beta}_1 \quad \boldsymbol{\beta}_2 \quad \cdots \quad \boldsymbol{\beta}_p] \\ \mathbf{E}_0 &= [\boldsymbol{\varepsilon}_1 \quad \boldsymbol{\varepsilon}_2 \quad \cdots \quad \boldsymbol{\varepsilon}_p] \end{aligned} \tag{3.6.11}$$

where each $\mathbf{U}_j$, $\boldsymbol{\beta}_j$, and $\boldsymbol{\varepsilon}_j$ is a column of $\mathbf{Y}$, $\mathbf{B}$, and $\mathbf{E}_0$, respectively. Rolling out each of these matrices,

$$\underset{(Np \times 1)}{\mathbf{y}^*} = \begin{bmatrix} \mathbf{U}_1 \\ \mathbf{U}_2 \\ \vdots \\ \mathbf{U}_p \end{bmatrix}, \quad \underset{(qp \times 1)}{\boldsymbol{\beta}^*} = \begin{bmatrix} \boldsymbol{\beta}_1 \\ \boldsymbol{\beta}_2 \\ \vdots \\ \boldsymbol{\beta}_p \end{bmatrix}, \quad \text{and} \quad \underset{(Np \times 1)}{\boldsymbol{\varepsilon}^*} = \begin{bmatrix} \boldsymbol{\varepsilon}_1 \\ \boldsymbol{\varepsilon}_2 \\ \vdots \\ \boldsymbol{\varepsilon}_p \end{bmatrix} \tag{3.6.12}$$

Employing this notation, (3.6.1) becomes

$$\underset{(Np \times 1)}{\mathbf{y}^*} = \underset{(Np \times qp)}{(\mathbf{I}_p \otimes \mathbf{X})} \underset{(qp \times 1)}{\boldsymbol{\beta}^*} + \underset{(Np \times 1)}{\boldsymbol{\varepsilon}^*} \tag{3.6.13}$$

and (3.6.2) becomes

$$\begin{aligned} E(\mathbf{Y}^*) &= (\mathbf{I}_p \otimes \mathbf{X})\boldsymbol{\beta}^* \\ V(\mathbf{Y}^*) &= \boldsymbol{\Sigma} \otimes \mathbf{I}_N \end{aligned} \tag{3.6.14}$$

Finding unique linear unbiased estimates in the multivariate situation is only notationally involved. The univariate theory may be applied almost directly to justify the multivariate results.

In multivariate analysis, parametric functions of the elements of $\mathbf{B}$ are represented as

$$\psi = \mathbf{c}'\mathbf{B}\mathbf{a} \tag{3.6.15}$$

where the $(1 \times q)$ vector $\mathbf{c}'$ is operating within variables and the $(p \times 1)$ vector $\mathbf{a}$ is used "between" variables. Alternatively, by using (3.6.12),

$$\psi = \mathbf{c}'\mathbf{B}\mathbf{a} = \sum_{k=1}^{p} a_k(\mathbf{c}'\boldsymbol{\beta}_k) = (\mathbf{a} \otimes \mathbf{c})'\boldsymbol{\beta}^* = \mathbf{r}^{*\prime}\boldsymbol{\beta}^* \tag{3.6.16}$$

The *multivariate Gauss-Markoff theorem* is now stated as follows:

(3.6.17) Let $\mathbf{Y} = \mathbf{XB} + \mathbf{E}_0$ such that $E(\mathbf{Y}) = \mathbf{XB}$ and the $V(\mathbf{Y}) = \mathbf{I} \otimes \boldsymbol{\Sigma}$. If $\psi = \mathbf{c}'\hat{\mathbf{B}}\mathbf{a}$ is estimable and $\hat{\mathbf{B}} = (\mathbf{X}'\mathbf{X})^{-}\mathbf{X}'\mathbf{Y}$ is a least-squares estimator of $\mathbf{B}$. Then

(a) $\hat{\psi} = \mathbf{c}'\hat{\mathbf{B}}\mathbf{a}$ is the unique linear unbiased estimate of ψ and has minimum variance among all linear unbiased estimates (that is, ψ is a best linear unbiased estimate);

(b) $V(\hat{\psi}) = \mathbf{a}'\boldsymbol{\Sigma}\mathbf{a}[\mathbf{c}'(\mathbf{X}'\mathbf{X})^{-}\mathbf{c}]$; and

(c) an unbiased estimate of $\boldsymbol{\Sigma}$ is given by

$$\begin{aligned} \mathbf{S} &= \frac{(\mathbf{Y} - \mathbf{X}\hat{\mathbf{B}})'(\mathbf{Y} - \mathbf{X}\hat{\mathbf{B}})}{N - r} \\ &= \frac{\mathbf{Y}'[\mathbf{I} - \mathbf{X}(\mathbf{X}'\mathbf{X})^{-}\mathbf{X}']\mathbf{Y}}{N - r} \end{aligned}$$

The proof of (3.6.17) follows Neudecker (1968). Using (3.5.32), with $\sigma^2\mathbf{V} = \mathbf{\Sigma} \otimes \mathbf{I}$, the best linear unbiased estimate of $\psi = \mathbf{r}^{*\prime}\boldsymbol{\beta}^*$ is

$$\begin{aligned}
\hat{\psi} &= \mathbf{r}^{*\prime}\{[(\mathbf{I} \otimes \mathbf{X}')(\mathbf{\Sigma} \otimes \mathbf{I})^{-1}(\mathbf{I} \otimes \mathbf{X})]^{-}(\mathbf{I} \otimes \mathbf{X})'(\mathbf{\Sigma} \otimes \mathbf{I})^{-1}\}\mathbf{y}^* \\
&= \mathbf{r}^{*\prime}\{[(\mathbf{I} \otimes \mathbf{X}')(\mathbf{\Sigma}^{-1} \otimes \mathbf{I})(\mathbf{I} \otimes \mathbf{X})]^{-}(\mathbf{I} \otimes \mathbf{X}')(\mathbf{\Sigma}^{-1} \otimes \mathbf{I})\}\mathbf{y}^* \\
&= \mathbf{r}^{*\prime}[\mathbf{\Sigma}^{-1} \otimes \mathbf{X}'\mathbf{X}]^{-}(\mathbf{\Sigma}^{-1} \otimes \mathbf{X}')\mathbf{y}^* \\
&= \mathbf{r}^{*\prime}[\mathbf{\Sigma} \otimes (\mathbf{X}'\mathbf{X})^{-}](\mathbf{\Sigma}^{-1} \otimes \mathbf{X}')\mathbf{y}^* \\
&= \mathbf{r}^{*\prime}[\mathbf{I} \otimes (\mathbf{X}'\mathbf{X})^{-}\mathbf{X}']\mathbf{y}^* \\
&= \mathbf{r}^{*\prime}[(\mathbf{X}'\mathbf{X})^{-}\mathbf{X}'\mathbf{Y}]^* \\
&= \mathbf{r}^{*\prime}\hat{\boldsymbol{\beta}}^* \\
&= (\mathbf{a} \otimes \mathbf{c})'\hat{\boldsymbol{\beta}}^* \\
&= \sum_{k=1}^{p} a_k(\mathbf{c}'\hat{\boldsymbol{\beta}}_k) \\
&= \mathbf{c}'\hat{\mathbf{B}}\mathbf{a}
\end{aligned}$$

which is unbiased for ψ. For the parametric function $\psi = \mathbf{c}'\mathbf{B}\mathbf{a}$ to be uniquely estimable, the condition that $\mathbf{c}'\mathbf{H} = \mathbf{c}'$ where $\mathbf{H} = (\mathbf{X}'\mathbf{X})^{-}(\mathbf{X}'\mathbf{X})$ is necessary and sufficient. This follows from the fact that linear combinations of estimable functions are also estimable.

To find the variance of $\hat{\psi} = \mathbf{r}^{*\prime}(\hat{\boldsymbol{\beta}}^*)$,

$$\begin{aligned}
V(\hat{\psi}) &= \mathbf{r}^{*\prime}V(\hat{\boldsymbol{\beta}}^*)\mathbf{r}^* \\
&= \mathbf{r}^{*\prime}V\{[\mathbf{I} \otimes (\mathbf{X}'\mathbf{X})^{-}\mathbf{X}']\mathbf{y}^*\}\mathbf{r}^* \\
&= \mathbf{r}^{*\prime}[\mathbf{I} \otimes (\mathbf{X}'\mathbf{X})^{-}\mathbf{X}'](\mathbf{\Sigma} \otimes \mathbf{I})[\mathbf{I} \otimes (\mathbf{X}'\mathbf{X})^{-}\mathbf{X}']'\mathbf{r}^* \\
&= \mathbf{a}'\mathbf{\Sigma}\mathbf{a} \otimes \mathbf{c}'(\mathbf{X}'\mathbf{X})^{-}(\mathbf{X}'\mathbf{X})(\mathbf{X}'\mathbf{X})^{-}\mathbf{c} \\
&= \mathbf{a}'\mathbf{\Sigma}\mathbf{a} \otimes \mathbf{a}_0'\mathbf{X}(\mathbf{X}'\mathbf{X})^{-}(\mathbf{X}'\mathbf{X})(\mathbf{X}'\mathbf{X})^{-}\mathbf{X}'\mathbf{a}_0 \\
&= \mathbf{a}'\mathbf{\Sigma}\mathbf{a} \otimes \mathbf{c}'(\mathbf{X}'\mathbf{X})^{-}\mathbf{c} \\
&= \mathbf{a}'\mathbf{\Sigma}\mathbf{a}[\mathbf{c}'(\mathbf{X}'\mathbf{X})^{-}\mathbf{c}]
\end{aligned}$$

since, if ψ is estimable, $\mathbf{c} \in V(\mathbf{X}')$.

The estimates $\hat{\psi}$ obtained in the multivariate situation are just arbitrary linear combinations of univariate estimable functions. To clarify, observe that $\mathbf{c}'\hat{\mathbf{B}} = \mathbf{c}'[\hat{\boldsymbol{\beta}}_1 \quad \hat{\boldsymbol{\beta}}_2 \quad \cdots \quad \hat{\boldsymbol{\beta}}_p]$, where each column is of the form $\mathbf{c}'\hat{\boldsymbol{\beta}}_i$, for $i = 1, \ldots, p$. The vector $\mathbf{a}$ is combining estimable functions across variables. Thus, for each variable, the estimators of the columns of $\mathbf{B}$ are what would have been obtained in univariate analysis individually.

The difference between univariate analysis and multivariate analysis is not in estimation theory, but in hypothesis testing.

Hypothesis Testing. Assume that each $\mathbf{Y}_i$ possesses a multivariate normal distribution with the appropriate mean, given implicitly in the Gauss-Markoff setup, and variance-covariance matrix $\mathbf{\Sigma}$. The *fundamental multivariate least-squares theorem* is now stated to test multivariate hypotheses (see also, Wilks, 1932).

(3.6.18) Under the Gauss-Markoff setup, assume that each row $\mathbf{Y}_i'$ of $\mathbf{Y}$ has a multivariate normal distribution such that the $E(\mathbf{Y}) = \mathbf{XB}$ and the $V(\mathbf{Y}) = \mathbf{I}_N \otimes \mathbf{\Sigma}$. Let $\mathbf{C}$ be a matrix of order $g \times q$ of rank $g \leq r$ such that each element of $\mathbf{CB}$ is individually estimable and

$$Q = \min \operatorname{Tr}[(\mathbf{Y} - \mathbf{XB})'(\mathbf{Y} - \mathbf{XB})]$$

subject to the condition that $\mathbf{CB} = \mathbf{\Gamma}$ (specified) where $\mathbf{Q}_t = (\mathbf{Y} - \mathbf{X}\hat{\mathbf{B}}_\omega)'(\mathbf{Y} - \mathbf{X}\hat{\mathbf{B}}_\omega)$ under the hypothesis $\mathbf{CB} = \mathbf{\Gamma}$. Then

(a) $\mathbf{Q}_e = \mathbf{Y}'[\mathbf{I} - \mathbf{X}(\mathbf{X}'\mathbf{X})^-\mathbf{X}']\mathbf{Y}$ and

$$\mathbf{Q}_t - \mathbf{Q}_e = (\mathbf{C}\hat{\mathbf{B}} - \mathbf{\Gamma})'[\mathbf{C}(\mathbf{X}'\mathbf{X})^-\mathbf{C}']^{-1}(\mathbf{C}\hat{\mathbf{B}} - \mathbf{\Gamma})$$

are independently distributed;

(b) $\mathbf{Q}_e \sim W_p(N - r, \mathbf{\Sigma})$ and $\mathbf{Q}_t - \mathbf{Q}_e \sim W_p(g, \mathbf{\Sigma}, \mathbf{\Gamma}_h)$; and

(c) if the hypothesis $H_0: \mathbf{CB} = \mathbf{\Gamma}$ is true, then

$$\Lambda = \frac{|\mathbf{Q}_e|}{|\mathbf{Q}_e + \mathbf{Q}_t - \mathbf{Q}_e|} \sim U(p, v_h, v_e)$$

where $g = v_h$ and $v_e = N - r \geq p$, the number of variables.

To prove (3.6.18), it is first shown that $\mathbf{Q}_h = \mathbf{Q}_t - \mathbf{Q}_e$ has the forms specified in (3.6.18a). The function Q is minimized by introducing a matrix $\mathbf{\Theta}$ of Lagrange multipliers so that $F = \operatorname{Tr}[(\mathbf{Y} - \mathbf{XB})'(\mathbf{Y} - \mathbf{XB})] + 2\operatorname{Tr}[\mathbf{\Theta}'(\mathbf{CB} - \mathbf{\Gamma})]$ must be minimized. Using result (1.8.34) and formula (1.8.33b), the partial derivative of F with respect to $\mathbf{B}$ and $\mathbf{\Theta}$ in similar to the univariate case (3.5.20), when equated to zero. That is, the equations become

$$(\mathbf{X}'\mathbf{X})\mathbf{B}_\omega + \mathbf{C}'\mathbf{\Theta} = \mathbf{X}'\mathbf{Y}$$
$$\mathbf{CB}_\omega = \mathbf{\Gamma}$$

Furthermore,

$$\hat{\mathbf{B}}_\omega = \hat{\mathbf{B}} - (\mathbf{X}'\mathbf{X})^-\mathbf{C}'[\mathbf{C}(\mathbf{X}'\mathbf{X})^-\mathbf{C}']^{-1}(\mathbf{C}\hat{\mathbf{B}} - \mathbf{\Gamma})$$
$$\mathbf{Q}_t = (\mathbf{Y} - \mathbf{X}\hat{\mathbf{B}}_\omega)'(\mathbf{Y} - \mathbf{X}\hat{\mathbf{B}}_\omega) = \mathbf{Q}_e + \mathbf{\Theta}'[\mathbf{C}(\mathbf{X}'\mathbf{X})^-\mathbf{C}']\mathbf{\Theta}$$
$$\mathbf{Q}_h = \mathbf{Q}_t - \mathbf{Q}_e = (\mathbf{C}\hat{\mathbf{B}} - \mathbf{\Gamma})'[\mathbf{C}(\mathbf{X}'\mathbf{X})^-\mathbf{C}']^{-1}(\mathbf{C}\hat{\mathbf{B}} - \mathbf{\Gamma})$$

as claimed. To show that $\mathbf{Q}_h$ has a Wishart distribution, $\mathbf{Q}_h$ is written as

$$\mathbf{Q}_h = [\mathbf{Y} - \mathbf{XC}'(\mathbf{CC}')^{-1}\mathbf{\Gamma}]'\mathbf{X}[(\mathbf{X}'\mathbf{X})^-]'\mathbf{C}'[\mathbf{C}(\mathbf{X}'\mathbf{X})^-\mathbf{C}']^{-1}\mathbf{C}(\mathbf{X}'\mathbf{X})^-\mathbf{X}'[\mathbf{Y} - \mathbf{XC}'(\mathbf{CC}')^{-1}\mathbf{\Gamma}]$$

since for some matrix $\mathbf{A}$, $\mathbf{C} = \mathbf{AX}$ and $\mathbf{X} = \mathbf{X}(\mathbf{X}'\mathbf{X})^-(\mathbf{X}'\mathbf{X})$. Since the matrix $\mathbf{X}[(\mathbf{X}'\mathbf{X})^-\mathbf{X}]'\mathbf{X}'\mathbf{A}[\mathbf{AX}(\mathbf{X}'\mathbf{X})^-\mathbf{X}'\mathbf{A}]^{-1}\mathbf{AX}(\mathbf{X}'\mathbf{X})^-\mathbf{X}'$ is idempotent and of rank g, (2.16.16) is invoked to establish that $\mathbf{Q}_h \sim W_p(g, \mathbf{\Sigma}, \mathbf{\Gamma}_h)$. Furthermore, by Example 2.6.6, $\mathbf{Q}_e \sim W_p(N - r, \mathbf{\Sigma})$. Following the univariate proof and employing (2.6.7), $\mathbf{Q}_e$ and $\mathbf{Q}_h$ are statistically independent. Thus, by (2.7.9), (3.6.18c) follows and (3.6.18) is proven.

Deriving the expected value of $\mathbf{Q}_h$ in the multivariate case, the jj' term of $\mathbf{Q}_h$, $\mathbf{Q}_h^{jj'}$, is investigated by using (3.3.4). That is,

$$E(\mathbf{Q}_h^{jj'}) = \text{Tr}\{[\mathbf{C}(\mathbf{X}'\mathbf{X})^-\mathbf{C}']^{-1}[\mathbf{C}(\mathbf{X}'\mathbf{X})^-\mathbf{C}']\sigma_{jj'}\} + \delta_{jj'}\sigma_{jj'}$$
$$= g\sigma_{jj'} + \delta_{jj'}\sigma_{jj'}$$

where

$$\delta_{jj'} = (\mathbf{C}\boldsymbol{\beta}_j - \boldsymbol{\xi}_j)'[\mathbf{C}(\mathbf{X}'\mathbf{X})^-\mathbf{C}']^{-1}(\mathbf{C}\boldsymbol{\beta}_{j'} - \boldsymbol{\xi}_{j'})$$

and $\boldsymbol{\beta}_j$ and $\boldsymbol{\xi}_j$ are column vectors of $\mathbf{B}$ and $\boldsymbol{\Gamma}$. Transforming to matrix notation and defining the mean square in the multivariate situation as $\text{MS}_h = \mathbf{Q}_h/g$,

$$E\left(\frac{\mathbf{Q}_h}{g}\right) = \boldsymbol{\Sigma} + \frac{(\mathbf{CB} - \boldsymbol{\Gamma})'[\mathbf{C}(\mathbf{X}'\mathbf{X})^-\mathbf{C}']^{-1}(\mathbf{CB} - \boldsymbol{\Gamma})}{g}$$

as expected from the univariate result.

To test the null hypothesis H_0: $\mathbf{CB} = \boldsymbol{\Gamma}$, the likelihood-ratio criterion is employed. Following the univariate proof,

$$\lambda = \left(\frac{|\hat{\boldsymbol{\Sigma}}_\Omega|}{|\hat{\boldsymbol{\Sigma}}_\omega|}\right)^{N/2} = \left[\frac{|(\mathbf{Y} - \mathbf{X}\hat{\mathbf{B}}_\Omega)'(\mathbf{Y} - \mathbf{X}\hat{\mathbf{B}}_\Omega)|}{|(\mathbf{Y} - \mathbf{X}\hat{\mathbf{B}}_\omega)'(\mathbf{Y} - \mathbf{X}\mathbf{B}_\omega)|}\right]^{N/2}$$

or

$$\Lambda = \lambda^{2/N} = \frac{|\mathbf{Q}_e|}{|\mathbf{Q}_t|} = \prod_i v_i$$

where the v_i are the roots of the determinantal equation $|\mathbf{Q}_e - v\mathbf{Q}_t| = 0$. The null hypothesis as stated in (3.6.18c) is therefore rejected at the significance level α if

$$\Lambda = \prod_i^s v_i = \prod_i^s (1 + \lambda_i)^{-1} < U^\alpha(p, v_h, v_e)$$

where the roots λ_i are the eigenvalues of the characteristic equation $|\mathbf{Q}_h - \lambda\mathbf{Q}_e| = 0$ and

$$\Lambda = \frac{|\mathbf{Q}_e|}{|\mathbf{Q}_e + \mathbf{Q}_h|} = \prod_i^s (1 + \lambda_i)^{-1}$$

where $s = \min(p, v_h)$.

Employing this general theorem to test $H: \boldsymbol{\mu} = \boldsymbol{\mu}_0$ in the one-sample case,

$$\mathbf{Q}_t - \mathbf{Q}_e = \mathbf{Q}_h = (\mathbf{C}\hat{\mathbf{B}} - \boldsymbol{\Gamma})'[\mathbf{C}(\mathbf{X}'\mathbf{X})^-\mathbf{C}']^{-1}(\mathbf{C}\hat{\mathbf{B}} - \boldsymbol{\Gamma})$$
$$= (\bar{\mathbf{y}}' - \boldsymbol{\mu}_0')'N(\bar{\mathbf{y}} - \boldsymbol{\mu}_0')$$
$$= N(\bar{\mathbf{y}} - \boldsymbol{\mu}_0)(\bar{\mathbf{y}} - \boldsymbol{\mu}_0)'$$

or

$$\Lambda = \frac{|\mathbf{Q}_e|}{|\mathbf{Q}_e + \mathbf{Q}_h|} = \frac{1}{1 + [T^2/(N-1)]} \sim U(p, 1, N-1)$$

so that

$$T^2 = (N-1)\frac{1-\Lambda}{\Lambda} = N(\bar{\mathbf{y}} - \boldsymbol{\mu}_0)'\mathbf{S}^{-1}(\bar{\mathbf{y}} - \boldsymbol{\mu}_0)$$

as previously obtained.

The hypothesis test matrix **C** refers only to hypotheses on the elements within given columns of the parameter matrix **B**. This does not allow the researcher to generate hypotheses among the different response parameters. To accomplish this, the *general fundamental least-squares theorem* is stated (see also Roy, 1957):

(3.6.19) Under the multivariate Gauss-Markoff setup, $E(\mathbf{Y}) = \mathbf{XB}$ and $V(\mathbf{Y}) = \mathbf{I}_N \otimes \boldsymbol{\Sigma}$, consider the $g \times q$ matrix **C** of rank $g \leq r$ such that each element of **CB** is individually estimable. Let **A** be any $p \times u$ matrix of rank $u \leq p \leq (N-r)$ and $Q = \min \operatorname{Tr}[(\mathbf{YA} - \mathbf{XBA})'(\mathbf{YA} - \mathbf{XBA})]$, subject to the condition $\mathbf{CBA} = \boldsymbol{\Gamma}$ (specified). Then

(a) $\mathbf{Q}_e = \mathbf{A}'\mathbf{Y}'[\mathbf{I} - \mathbf{X}(\mathbf{X}'\mathbf{X})^{-}\mathbf{X}']\mathbf{YA}$ and

$$\mathbf{Q}_h = (\mathbf{C}\hat{\mathbf{B}}\mathbf{A} - \boldsymbol{\Gamma})'[\mathbf{C}(\mathbf{X}'\mathbf{X})^{-}\mathbf{C}']^{-1}(\mathbf{C}\hat{\mathbf{B}}\mathbf{A} - \boldsymbol{\Gamma})];$$

(b) $\mathbf{Q}_e \sim W_u(N - r, \mathbf{A\Sigma A}')$ and $\mathbf{Q}_h \sim W_u(g, \mathbf{A\Sigma A}', \boldsymbol{\Gamma}_h)$; and

(c) if the hypothesis $H_0: \mathbf{CBA} = \boldsymbol{\Gamma}$ is true, then

$$\Lambda = \frac{|\mathbf{Q}_e|}{|\mathbf{Q}_e + \mathbf{Q}_h|} \sim U(u, v_h, v_e)$$

where $v_h = g$, $v_e = N - r$, and $u = R(\mathbf{A})$.

This general theorem reduces to (3.6.18) if $\mathbf{A} = \mathbf{I}_p$.

Tests of (3.6.18c) and (3.6.19c) are derived by using the likelihood-ratio-test criterion. To sketch Roy's procedure for (3.6.18c) with $\boldsymbol{\Gamma} = \mathbf{0}$, his union-intersection principle is employed. For Roy, the null multivariate hypothesis H_0 is expressed as an intersection of composite univariate hypotheses, while the alternative hypothesis is a union of corresponding alternative hypotheses. The multivariate null and alternative hypotheses

$$H_0: \mathbf{CB} = \mathbf{0}$$

$$H_1: \mathbf{CB} \neq \mathbf{0}$$

become, employing the union-intersection principle,

$$H_0: \bigcap_{\mathbf{a}} (\mathbf{CBa} = \mathbf{0})$$

$$H_1: \bigcup_{\mathbf{a}} (\mathbf{CBa} = \mathbf{0})$$

for nonnull arbitrary vectors **a**. The multivariate hypothesis is true if and only if, for all **a**, every univariate hypothesis is true and is rejected if at least one univariate hypothesis is rejected. The region of acceptance for H_0 is the intersection of the acceptance regions of all univariate regions. This region is

$$\bigcap_{\mathbf{a}} \frac{\mathbf{a}'\mathbf{Q}_h\mathbf{a}}{\mathbf{a}'\mathbf{Q}_e\mathbf{a}} \leq \frac{g}{N-r} F^{\alpha^*}(g, N-r)$$

Since the

$$\max_{\mathbf{a}} \frac{\mathbf{a}'\mathbf{Q}_h\mathbf{a}}{\mathbf{a}'\mathbf{Q}_e\mathbf{a}} = \lambda_1$$

where λ_1 is the largest root of $|\mathbf{Q}_h - \lambda\mathbf{Q}_e| = 0$, for arbitrary vectors **a**, the test is referred to as the largest-root criterion. Letting $\theta_s = \lambda_1/(1 + \lambda_1)$, the hypothesis is rejected if the $P\{\theta_s > \theta^\alpha(s, m, n)|H_0\} = \alpha$, for which Heck's (1960) charts are used. If H_0 is rejected at the level α, then for at least one **a**, a univariate hypothesis is rejected at the level $\alpha^* < \alpha$. It should also be noticed that even if H_0 is accepted using Roy's criterion, there may still be several univariate hypotheses rejected at the significance level α.

Under (3.6.19c), with $\mathbf{\Gamma} = \mathbf{0}$, the null and alternative hypotheses become

$$H_0\colon \bigcap_{\mathbf{a}} (\mathbf{CBAa} = \mathbf{0})$$

$$H_1\colon \bigcup_{\mathbf{a}} (\mathbf{CBAa} = \mathbf{0})$$

where

$$\max_{\mathbf{a}} \frac{\mathbf{a}'\mathbf{Q}_h\mathbf{a}}{\mathbf{a}'\mathbf{Q}_e\mathbf{a}} = \lambda_1$$

is used with $\mathbf{Q}_h$ and $\mathbf{Q}_e$ as defined in (3.6.19). Defining the alternative hypothesis not as a union but as an intersection, the smallest root of $|\mathbf{Q}_h - \lambda\mathbf{Q}_e| = 0$ would be used.

Multivariate techniques are dependent, in general, on some multipurpose statistical routine. Although there exist, at almost every institution, special-purpose multivariate routines, there are a wide variety of general routines available to researchers. Some of the more widely used general-purpose routines developed for multivariate data analysis are due to the work of Bargmann (1967), Clyde (1969), and Finn (1972). Special-purpose routines have been developed by Cooley and Lohnes (1971), Dixon (1969), and Fowlkes and Lee (1971).

Rather than having a text dependent on one routine, which is subject to change, this text is organized to be independent of routines used to do multivariate data analysis. It is hoped that the theory and examples discussed will help the researcher to better interpret and comprehend the summary statistics produced by any general-purpose routine.

EXERCISES 3.6

1. a. In the proof of (3.6.18), derive the expressions given there for $\hat{\mathbf{B}}_\omega$, $\mathbf{Q}_t$, and $\mathbf{Q}_h$.

   b. Following the univariate proof and using (2.6.7), show that $\mathbf{Q}_e$ and $\mathbf{Q}_h$ are statistically independent. What can you conclude about $\mathbf{Q}_e$ and $\mathbf{Q}_e + \mathbf{Q}_h$; about $\mathbf{X}\hat{\mathbf{B}}_\omega$ and $\mathbf{Q}_t$?

2. Let the data matrix

$$\mathbf{Y} = \begin{bmatrix} y_{11} & y_{12} \\ y_{21} & y_{22} \\ y_{31} & y_{32} \\ y_{41} & y_{42} \\ y_{51} & y_{52} \\ y_{61} & y_{62} \end{bmatrix}$$

where $E(\mathbf{Y}) = \mathbf{XB}$, the $V(\mathbf{Y}) = \mathbf{I} \otimes \boldsymbol{\Sigma}$, and each row of $\mathbf{Y}$ follows a bivariate normal distribution. Suppose the design matrix $\mathbf{X}$ and parameter matrix $\mathbf{B}$ are as follows:

$$\mathbf{X} = \begin{bmatrix} 1 & 1 & 0 & 0 & 1 & 0 \\ 1 & 1 & 0 & 0 & 0 & 1 \\ 1 & 0 & 1 & 0 & 1 & 0 \\ 1 & 0 & 1 & 0 & 0 & 1 \\ 1 & 0 & 0 & 1 & 1 & 0 \\ 1 & 0 & 0 & 1 & 0 & 1 \end{bmatrix} \qquad \mathbf{B} = \begin{bmatrix} \beta_{11} & \beta_{12} \\ \beta_{21} & \beta_{22} \\ \beta_{31} & \beta_{32} \\ \beta_{41} & \beta_{42} \\ \beta_{51} & \beta_{52} \\ \beta_{61} & \beta_{62} \end{bmatrix}.$$

a. Solve the normal equations.

b. Are $\psi = \beta_{21} - \beta_{31}$, $\psi = \beta_{22}$, $\psi = \beta_{11} + \beta_{51}$, and $\psi = \beta_{52} - \beta_{62} + \beta_{32} - \beta_{22}$ estimable?

c. Determine the least-square estimates of the estimable function(s) in part b and the estimated variance(s) of the estimable function(s).

d. For each estimable function in part b, test the hypothesis that $\psi = 0$ by employing the Roy criterion and the likelihood-ratio criterion using $\alpha = .05$ and the data matrix

$$\mathbf{Y} = \begin{bmatrix} 9 & 7 \\ 6 & 14 \\ 16 & 21 \\ 23 & 28 \\ 30 & 34 \\ 41 & 47 \end{bmatrix}$$

e. For part d, determine $100(1 - \alpha)\%$ confidence intervals for ψ using both criteria.

3.7 TWO-SAMPLE PROBLEM—UNIVARIATE

Two random samples of sizes N_1 and N_2 have been independently drawn from normal populations with different means μ_1 and μ_2 common variance σ^2, so that $Y_{11}, \ldots, Y_{1N_1} \sim IN(\mu_1, \sigma^2)$ and $Y_{21}, \ldots, Y_{2N_2} \sim IN(\mu_2, \sigma^2)$. Given two independent random samples from two normal populations with common variance, the next hypothesis encountered in univariate analysis is $H_0\colon \mu_1 = \mu_2$.

Employing the general linear model, the observations are naturally represented as

$$\underset{(N\times 1)}{\mathbf{y}} = \underset{(N\times 2)}{\mathbf{X}}\ \underset{(2\times 1)}{\boldsymbol{\beta}} + \underset{(N\times 1)}{\boldsymbol{\varepsilon}}$$

$$\begin{bmatrix} y_{11} \\ y_{12} \\ \vdots \\ y_{1N_1} \\ y_{21} \\ y_{22} \\ \vdots \\ y_{2N_2} \end{bmatrix} = \begin{bmatrix} 1 & 0 \\ 1 & 0 \\ \vdots & \vdots \\ 1 & 0 \\ 0 & 1 \\ 0 & 1 \\ \vdots & \vdots \\ 0 & 1 \end{bmatrix} \begin{bmatrix} \mu_1 \\ \mu_2 \end{bmatrix} + \begin{bmatrix} \varepsilon_{11} \\ \varepsilon_{12} \\ \vdots \\ \varepsilon_{1N_1} \\ \varepsilon_{21} \\ \varepsilon_{22} \\ \vdots \\ \varepsilon_{2N_2} \end{bmatrix} \tag{3.7.1}$$

where $N = N_1 + N_2$,

$$E(\mathbf{Y}) = \mathbf{X}\boldsymbol{\beta} = \begin{bmatrix} \mu_1 \\ \vdots \\ \mu_1 \\ \mu_2 \\ \vdots \\ \mu_2 \end{bmatrix}, \quad \text{and} \quad V(\mathbf{Y}) = \sigma^2\mathbf{I} \tag{3.7.2}$$

To test the hypothesis $H_0\colon \mu_1 = \mu_2$ against the alternative hypothesis $H_1\colon \mu_1 \neq \mu_2$ (say), the Student t statistic

$$t = \frac{y_{1.} - y_{2.}}{s_p\sqrt{(N_1 + N_2)/N_1N_2}} \sim t(N_1 + N_2 - 2) \tag{3.7.3}$$

when H_0 is true is employed where

$$s_p^2 = \frac{(N_1 - 1)s_1^2 + (N_2 - 1)s_2^2}{N_1 + N_2 - 2} = \frac{\sum_{i=1}^{N_1} (y_{1i} - y_{1.})^2 + \sum_{i=1}^{N_2} (y_{2i} - y_{2.})^2}{N_1 + N_2 - 2}$$

and

$$y_{1.} = \frac{\sum_{i=1}^{N_1} y_{1i.}}{N_1}$$

is the sample mean for the first sample,

$$y_{2.} = \frac{\sum_{i=1}^{N_2} y_{2i}}{N_2}$$

is the sample mean for the second sample,

$$s_i^2 = \frac{\sum_{j=1}^{N_i} (y_{ij} - y_{i.})^2}{N_i - 1}$$

is the sample variance of the ith sample, and s_p^2 is the pooled-within, unbiased estimate of the common variance σ^2. Now by applying the least-squares theory,

$$\hat{\boldsymbol{\beta}} = \begin{bmatrix} \hat{\mu}_1 \\ \hat{\mu}_2 \end{bmatrix} = (\mathbf{X}'\mathbf{X})^{-1}\mathbf{X}'\mathbf{y}$$

$$= \begin{bmatrix} 1/N_1 & 0 \\ 0 & 1/N_2 \end{bmatrix} \begin{bmatrix} \sum_i y_{1i} \\ \sum_i y_{2i} \end{bmatrix}$$

$$= \begin{bmatrix} y_{1.} \\ y_{2.} \end{bmatrix} \tag{3.7.4}$$

$$s^2 = \frac{(\mathbf{y} - \mathbf{X}\hat{\boldsymbol{\beta}})'(\mathbf{y} - \mathbf{X}\hat{\boldsymbol{\beta}})}{N_1 + N_2 - 2}$$

$$= \frac{\sum_i (y_{1i} - y_{1.})^2 + \sum_i (y_{2i} - y_{2.})^2}{N_1 + N_2 - 2}$$

$$= s_p^2 \tag{3.7.5}$$

where s_p^2 is an unbiased estimate of σ^2. Using the distance concept, define $d = y_{1.} - y_{2.}$. Then

$$t = \frac{d\sqrt{N_1 N_2/(N_1 + N_2)}}{s_p} = D\sqrt{\frac{N_1 N_2}{N_1 + N_2}} \tag{3.7.6}$$

where D is the studentized distance. It will be remembered that $t^2(N_1 + N_2 - 2) = F(1, N_1 + N_2 - 2)$; so

$$t^2 = \left(\frac{N_1 N_2}{N_1 + N_2}\right) D^2 = F \tag{3.7.7}$$

If the LRT of H_0 is performed at the significance level α, H_0 is rejected if

$$|t| > t^{\alpha/2}(N_1 + N_2 - 2) \tag{3.7.8}$$

where $t^{\alpha/2}(N_1 + N_2 - 2)$ is the upper $\alpha/2$ percentage point of the t distribution, with degrees of freedom $N_1 + N_2 - 2$. A $100(1 - \alpha)\%$ confidence interval for the difference $\mu_1 - \mu_2$ is obtained by evaluating the expression

$$y_{1.} - y_{2.} - c_0 s_p\sqrt{\frac{N_1 + N_2}{N_1 N_2}} \leq \mu_1 - \mu_2 \leq y_{1.} - y_{2.} + c_0 s_p\sqrt{\frac{N_1 + N_2}{N_1 N_2}} \tag{3.7.9}$$

derived from the complement of the rejection region for testing H_0, where

$$c_0 = t^{\alpha/2}(N_1 + N_2 - 2)$$

An alternative approach to the two-sample problem, commonly used in analysis-of-variance situations for more than two groups, is to assume that the model for the jth observation within the ith group is

$$y_{ij} = \mu + \alpha_i + \varepsilon_{ij} \qquad (3.7.10)$$
$$E(\varepsilon_{ij}) = 0 \quad \text{and} \quad V(\varepsilon_{ij}) = \sigma^2$$

Thus the $E(Y_{ij}) = \mu + \alpha_i$ and the $V(Y_{ij}) = \sigma^2$, where μ is a mean common to all samples and α_i is the parameter that depicts the group effect. The general linear model representation of (3.7.10) is

$$\underset{(N\times 1)}{\mathbf{y}} = \underset{(N\times 3)}{\mathbf{X}} \underset{(\beta_3\times 1)}{\boldsymbol{\beta}} + \underset{(N\times 1)}{\boldsymbol{\varepsilon}}$$

$$\begin{bmatrix} y_{11} \\ y_{12} \\ \vdots \\ y_{1N_1} \\ y_{21} \\ y_{22} \\ \vdots \\ y_{2N_2} \end{bmatrix} = \begin{bmatrix} 1 & 1 & 0 \\ 1 & 1 & 0 \\ \vdots & \vdots & \vdots \\ 1 & 1 & 0 \\ 1 & 0 & 1 \\ 1 & 0 & 1 \\ \vdots & \vdots & \vdots \\ 1 & 0 & 1 \end{bmatrix} \begin{bmatrix} \mu \\ \alpha_1 \\ \alpha_2 \end{bmatrix} + \begin{bmatrix} \varepsilon_{11} \\ \varepsilon_{12} \\ \vdots \\ \varepsilon_{1N_1} \\ \varepsilon_{21} \\ \varepsilon_{22} \\ \vdots \\ \varepsilon_{2N_2} \end{bmatrix} \qquad (3.7.11)$$

With the introduction of the overall parameter μ, $\mathbf{X'X}$ is no longer of full rank, so that $(\mathbf{X'X})^{-1}$ does not exist. The rank of $\mathbf{X'X}$ is $2 < q = 3$.

Before considering the general case, attention is again given the example discussed in Section 1.6. Equation (3.7.11) becomes

$$\begin{bmatrix} y_{11} \\ y_{12} \\ y_{21} \\ y_{22} \end{bmatrix} = \begin{bmatrix} 1 & 1 & 0 \\ 1 & 1 & 0 \\ 1 & 0 & 1 \\ 1 & 0 & 1 \end{bmatrix} \begin{bmatrix} \mu \\ \alpha_1 \\ \alpha_2 \end{bmatrix} + \begin{bmatrix} \varepsilon_{11} \\ \varepsilon_{12} \\ \varepsilon_{21} \\ \varepsilon_{22} \end{bmatrix} \qquad (3.7.12)$$

The normal equations are

$$(\mathbf{X'X})\hat{\boldsymbol{\beta}} = \mathbf{X'y}$$

$$\begin{aligned} 4\hat{\mu} + 2\hat{\alpha}_1 + 2\hat{\alpha}_2 &= 4y_{..} \\ 2\hat{\mu} + 2\hat{\alpha}_1 \qquad\quad &= 2y_{1.} \\ 2\hat{\mu} \qquad\quad + 2\hat{\alpha}_2 &= 2y_{2.} \end{aligned} \qquad (3.7.13)$$

At this point, most authors impose the condition that $\hat{\alpha}_1 + \hat{\alpha}_2 = 0$ to solve the normal equations. Then a solution to (3.7.13) becomes

$$\begin{bmatrix} \hat{\mu} \\ \hat{\alpha}_1 \\ \hat{\alpha}_2 \end{bmatrix} = \begin{bmatrix} y_{..} \\ y_{1.} - y_{..} \\ y_{2.} - y_{..} \end{bmatrix} \qquad (3.7.14)$$

From Example 1.5.6, a generalized inverse of $\mathbf{X}'\mathbf{X}$ that yields this solution is

$$(3.7.15) \qquad (\mathbf{X}'\mathbf{X})^{-} = \begin{bmatrix} 1/2 & -1/4 & -1/4 \\ -1/4 & 1/2 & 0 \\ -1/4 & 0 & 1/2 \end{bmatrix}$$

where $(\mathbf{X}'\mathbf{X})^{-} = (\mathbf{X}'\mathbf{X} + \mathbf{R}'\mathbf{R})^{-1}$. The matrix $\mathbf{R}$ associated with the condition that $\hat{\alpha}_1 + \hat{\alpha}_2 = 0$ is $\mathbf{R} = [0, 1, 1]$. John (1964) shows that any matrix of the form $(\mathbf{X}'\mathbf{X} + \mathbf{R}'\mathbf{R})^{-1}$ is a g-inverse of $\mathbf{X}'\mathbf{X}$ for appropriately chosen restrictions.

By employing the general theory of g-inverses and estimable functions, the matrix $\mathbf{H}$ is formed;

$$(3.7.16) \qquad \mathbf{H} = (\mathbf{X}'\mathbf{X})^{-}\mathbf{X}'\mathbf{X} = \begin{bmatrix} 1 & 1/2 & 1/2 \\ 0 & 1/2 & -1/2 \\ 0 & -1/2 & 1/2 \end{bmatrix}$$

and $\mathbf{c}'\boldsymbol{\beta}$ has a unique linear unbiased estimator if and only if $\mathbf{c}'\mathbf{H} = \mathbf{c}'$. By selecting $\mathbf{c}' = [0, 1, -1]$, $\mathbf{c}'\mathbf{H} = \mathbf{c}'$ so that $\mathbf{c}'\boldsymbol{\beta} = \alpha_1 - \alpha_2$ is estimable. For $\mathbf{c}' = [1, 0, 0]$ or $\mathbf{c}' = [0, 1, 0]$, $\mathbf{c}'\mathbf{H} \neq \mathbf{c}'$, so that μ and α_1 are not estimable. However, for $\mathbf{c}' = [1, 1, 0]$, $\mathbf{c}' = \mathbf{c}'\mathbf{H}$, or $\mu + \alpha_1$ is estimable. For arbitrary vectors $\mathbf{t}' = [t_0, t_1, t_2]$,

$$\mathbf{c}'\boldsymbol{\beta} = \mathbf{t}'\mathbf{H}\boldsymbol{\beta} = t_0\mu + \frac{t_0 + t_1 - t_2}{2}\alpha_1 + \frac{t_0 - t_1 + t_2}{2}\alpha_2$$

and

$$\mathbf{c}'\hat{\boldsymbol{\beta}} = \mathbf{t}'\mathbf{H}\hat{\boldsymbol{\beta}} = t_0 y_{..} + \frac{t_0 + t_1 - t_2}{2}(y_{1.} - y_{..}) + \frac{t_0 - t_1 + t_2}{2}(y_{2.} - y_{..})$$

By setting $\mathbf{t}' = [0, 1, -1]$, $\mathbf{c}'\boldsymbol{\beta} = \alpha_1 - \alpha_2$ is estimated by $\mathbf{c}'\hat{\boldsymbol{\beta}} = y_{1.} - y_{2.}$. Similarly, with $\mathbf{t}' = [1, 1, 0]$, $\mathbf{c}'\boldsymbol{\beta} = \mu + \alpha_1$ is estimated by $\mathbf{c}'\hat{\boldsymbol{\beta}} = y_{1.}$. Choosing $\mathbf{t}' = [1, 1, 1]$,

$$\mathbf{c}'\boldsymbol{\beta} = \mu + \frac{\alpha_1 + \alpha_2}{2} \quad \text{and} \quad \mathbf{c}'\hat{\boldsymbol{\beta}} = y_{..}$$

No value of $\mathbf{t}$ can be found to estimate μ or α_i. Thus using the condition $\hat{\alpha}_1 + \hat{\alpha}_2 = 0$ to solve the normal equations does not ensure estimability of μ or α_i. In general, conditions imposed to find a solution to the normal equations do not necessarily have to be applied to the model parameters. However, adding the side condition $\alpha_1 + \alpha_2 = 0$ to the model parameters, in this case, allows an estimate of μ and α_i. This now gives $\hat{\mu} = y_{..}$ and $\hat{\alpha}_i = y_{i.} - y_{..}$, which are unique for the model with the side conditions $\alpha_1 + \alpha_2 = 0$ added; such a model is referred to as a *restricted model* since the restrictions become part of the model.

Instead of using the condition that $\hat{\alpha}_1 + \hat{\alpha}_2 = 0$ to solve the normal equations, consider setting $\hat{\alpha}_2 = 0$. Then a solution to (3.7.12) would be

$$(3.7.17) \qquad \begin{bmatrix} \hat{\mu} \\ \hat{\alpha}_1 \\ \hat{\alpha}_2 \end{bmatrix} \begin{matrix} = \\ = \\ = \end{matrix} \begin{bmatrix} y_{2.} \\ y_{1.} - y_{2.} \\ 0 \end{bmatrix}$$

From Example 1.5.10, a g-inverse that yields this solution is

$$(\mathbf{X'X})^- = \begin{bmatrix} 1/2 & -1/2 & 0 \\ -1/2 & 1 & 0 \\ 0 & 0 & 0 \end{bmatrix} \tag{3.7.18}$$

so that

$$\mathbf{H} = \begin{bmatrix} 1 & 0 & 1 \\ 0 & 1 & -1 \\ 0 & 0 & 0 \end{bmatrix} = (\mathbf{X'X})^-\mathbf{X'X} \tag{3.7.19}$$

In this case,

$$\mathbf{c'\beta} = \mathbf{t'H\beta} = t_0(\mu + \alpha_2) + t_1(\alpha_1 - \alpha_2) \quad \text{and} \quad \mathbf{c'\hat{\beta}} = t\mathbf{H\hat{\beta}} = t_0 y_{2.} + t_{1.}(y_{1.} - y_{2.})$$

TABLE 3.7.1. Estimable Functions when $\hat{\alpha}_2 = 0$.

t_0	t_1	t_2	Parameter	Estimate
0	1	0	$\alpha_1 - \alpha_2$	$y_{1.} - y_{2.}$
1	1	0	$\mu + \alpha_1$	$y_{1.}$
1	0	0	$\mu + \alpha_2$	$y_{2.}$
1	$\frac{1}{2}$	0	$\mu + \frac{\alpha_1 + \alpha_2}{2}$	$y_{..}$

Table 3.7.1 gives estimable functions for various values of t_0, t_1, and t_2. No values for $\mathbf{t}$ exist that provide estimates for μ or α_i. These results, as expected, are consistent with those discussed using (3.7.16). In this case, however, there is no reason for including as part of the model the side condition $\alpha_2 = 0$ since then $\mathbf{c'\beta} = \mu + \alpha_1/2$ is estimated by $\mathbf{c'\hat{\beta}} = y_{..}$, while $\hat{\mu} = y_{2.}$ and $\hat{\alpha}_1 = y_{1.} - y_{2.}$.

Many other conditions besides $\hat{\alpha}_1 + \hat{\alpha}_2 = 0$ and $\hat{\alpha}_2 = 0$ may be placed on the estimates of the population parameters for the model to obtain a solution to (3.7.13) or to any system of normal equations and to determine which linear combinations of the population parameters are estimable. However, there may be no reason for imposing the same conditions or side conditions on population parameters and including these as part of the model.

Returning to the general problem, by setting $\hat{\mu} = 0$ in the normal equations, a convenient g-inverse for $\mathbf{X'X}$ is

$$(\mathbf{X'X})^- = \begin{bmatrix} 0 & 0 & 1 \\ 0 & 1/N_1 & 0 \\ 0 & 0 & 1/N_2 \end{bmatrix} \tag{3.7.20}$$

A solution for $\hat{\boldsymbol{\beta}}$ is then

$$\hat{\boldsymbol{\beta}} = (\mathbf{X'X})^-\mathbf{X'y} = \begin{bmatrix} 0 \\ y_{1.} \\ y_{2.} \end{bmatrix} \tag{3.7.21}$$

To find unique unbiased estimates for linear combinations of the elements of $\boldsymbol{\beta}$, (3.5.9) is applied. Here

$$\mathbf{H} = (\mathbf{X}'\mathbf{X})^{-}\mathbf{X}'\mathbf{X} = \begin{bmatrix} 0 & 0 & 0 \\ 1 & 1 & 0 \\ 1 & 0 & 1 \end{bmatrix} \tag{3.7.22}$$

and $\mathbf{c}'\boldsymbol{\beta}$ has a unique estimator if and only if $\mathbf{c}'\mathbf{H} = \mathbf{c}'$. For arbitrary $\mathbf{t}' = [t_0, t_1, t_2]$,

$$\begin{aligned} \mathbf{c}'\boldsymbol{\beta} &= \mathbf{t}'\mathbf{H}\boldsymbol{\beta} = (t_1 + t_2)\mu + t_1\alpha_1 + t_2\alpha_2 \\ \mathbf{c}'\hat{\boldsymbol{\beta}} &= \mathbf{t}'\mathbf{H}\hat{\boldsymbol{\beta}} = t_1 y_{1.} + t_2 y_{2.} \end{aligned} \tag{3.7.23}$$

The unique minimum-variance unbiased estimate of $\mathbf{c}'\boldsymbol{\beta} = \alpha_1 - \alpha_2$ is obtained by setting $\mathbf{t}' = [0, 1, -1]$, so that $\mathbf{c}'\hat{\boldsymbol{\beta}} = y_{1.} - y_{2.}$, as seen above.

No value of $\mathbf{t}$ in (3.7.23) can be chosen to estimate μ since to estimate μ $\mathbf{c}' = [1, 0, 0]$ and $\mathbf{c}'\mathbf{H} \neq \mathbf{c}'$. However, by letting $t_1 = N_1/(N_1 + N_2)$ and $t_2 = N_2/(N_1 + N_2)$, $\mathbf{c}'\boldsymbol{\beta} = \mu + (N_1\alpha_1 + N_2\alpha_2)/(N_1 + N_2)$. Adding the side condition that $N_1\alpha_1 + N_2\alpha_2 = 0$ now allows an estimate of μ:

$$\hat{\mu} = (N_1 y_{1.} + N_2 y_{2.})/(N_1 + N_2) = \bar{y}_{..}$$

Alternatively, setting $t_1 = t_2 = 1/2$, $\mathbf{c}'\boldsymbol{\beta} = \mu + (\alpha_1 + \alpha_2)/2$ and choosing $\alpha_1 + \alpha_2 = 0$ implies $\hat{\mu} = (y_{1.} + y_{2.})/2 = y_{..}$.

In using conditions like $\hat{\mu} = 0$, $\hat{\alpha}_2 = 0$, and $(\hat{\alpha}_1 + \hat{\alpha}_2)/2$ to solve the normal equations, arbitrary conditions cannot be used. All conditions illustrated involved combinations of estimates of parameters in the original model that were not estimable. To increase a model that is less than full rank to a full-rank model (see Section 1.6), the rows of the matrix $\mathbf{R}$ in $\mathbf{R}\boldsymbol{\beta} = \boldsymbol{\theta}$ must be linearly independent of the rows of the design matrix $\mathbf{X}$. That is, $\mathbf{R} \neq \mathbf{AX}$, so conditions are applied such that $\mathbf{R}\boldsymbol{\beta}$ is not estimable. Furthermore, the conditions do not have to be invoked to impose side conditions on the model parameters but can be simply employed to solve the normal equations.

When using the method of reparameterization to solve the normal equations (see Section 1.6), the matrix

$$\mathbf{C} = \begin{bmatrix} 1 & 1/2 & 1/2 \\ 0 & 1 & -1 \end{bmatrix}$$

is employed to find linear combinations of the parameters of interest to the experimenter. Each row of $\mathbf{C}$ corresponds to linear combinations of parameters that are estimable in model (3.7.10). Since the

$$R\begin{bmatrix} \mathbf{X} \\ \mathbf{C} \end{bmatrix} = R(\mathbf{X}) = R(\mathbf{C}) = r$$

$\mathbf{C} = \mathbf{AX}$, the condition for estimability.

Using model (3.7.10), the hypothesis of interest that corresponds to H_0: $\mu_1 = \mu_2$ in the less-than-full-rank case is H_0: $\mu + \alpha_1 = \mu + \alpha_2$. This can be tested since both $\mu + \alpha_1$ and $\mu + \alpha_2$ are estimable. However, in terms of the α_i's, H_0 is equivalent to H_0: $\alpha_1 = \alpha_2$, the most frequently employed form of the hypothesis. Using matrix notation, the hypothesis is represented by $\mathbf{C}\boldsymbol{\beta} = \boldsymbol{\xi}$. With $\mathbf{C} = [0, 1, -1]$

and $\boldsymbol{\xi} = \mathbf{0}$, H_0 is equivalently written as

$$H_0: \mathbf{C}\boldsymbol{\beta} = \mathbf{0} \qquad \alpha_1 - \alpha_2 = 0$$
(3.7.24) $$\text{or}$$
$$H_1: \mathbf{C}\boldsymbol{\beta} \neq \mathbf{0} \qquad \alpha_1 - \alpha_2 \neq 0$$

Employing (3.5.18),

$$\begin{aligned} Q_h = Q_t - Q_e &= (\mathbf{C}\hat{\boldsymbol{\beta}})'[\mathbf{C}(\mathbf{X}'\mathbf{X})^{-}\mathbf{C}']^{-1}(\mathbf{C}\hat{\boldsymbol{\beta}}) \\ &= (y_{1.} - y_{2.})\left(\frac{1}{N_1} + \frac{1}{N_2}\right)^{-1}(y_{1.} - y_{2.}) \\ &= \frac{N_1 N_2}{N_1 + N_2}(y_{1.} - y_{2.})^2 \end{aligned}$$

(3.7.25)

$$\begin{aligned} Q_e &= \mathbf{y}'[\mathbf{I} - \mathbf{X}(\mathbf{X}'\mathbf{X})^{-}\mathbf{X}']\mathbf{y} \\ &= \sum_{i=1}^{N_1} (y_{1i} - y_{1.})^2 + \sum_{i=1}^{N_2} (y_{2i} - y_{2.})^2 \end{aligned}$$

When the null hypothesis is true,

$$F = \frac{Q_h/1}{Q_e/(N_1 + N_2 - 1)} = \frac{N_1 N_2}{N_1 + N_2} \frac{(y_{1.} - y_{2.})^2}{s_p^2}$$

or

$$t = \frac{y_{1.} - y_{2.}}{s_p\sqrt{(1/N_1) + (1/N_2)}}$$

as previously obtained.

Hypothesis testing in the full-rank model (3.7.1) and the less-than-full-rank model (3.7.10) is facilitated by considering the test that the "overall mean" is zero. The correspondence between the parameters in the two models for the two-sample

TABLE 3.7.2

	Group 1	Group 2
Full-Rank (FR) Model	μ_1	μ_2
Less-Than-Full-Rank (LFR) Model	$\mu + \alpha_1$	$\mu + \alpha_2$

case is shown in Table 3.7.2. Test the hypothesis that the overall mean is zero in the FR model

(3.7.26) $$H_0: \frac{N_1\mu_1 + N_2\mu_2}{N_1 + N_2} = 0$$

The parameter vector $\boldsymbol{\beta}' = [\mu_1, \mu_2]$, $\mathbf{c}' = [N_1/(N_1 + N_2),\ N_2/(N_1 + N_2)]$, $\mathbf{c}'\hat{\boldsymbol{\beta}} = (N_1 y_{1.} + N_2 y_{2.})/(N_1 + N_2) = y_{..}$, and $[\mathbf{c}'(\mathbf{X}'\mathbf{X})^{-1}\mathbf{c}]^{-1} = N$. The test statistic for testing (3.7.26) becomes

$$F = \frac{(\mathbf{c}'\hat{\boldsymbol{\beta}})'[\mathbf{c}'(\mathbf{X}'\mathbf{X})^{-1}\mathbf{c}]^{-1}\mathbf{c}'\hat{\boldsymbol{\beta}}}{s_p^2}$$

$$= \frac{Ny_{..}^2}{s_p^2}$$

where s_p^2 is the unbiased estimate of σ^2.

Using the LFR model, estimable functions have the form

$$\mathbf{c}'\boldsymbol{\beta} = (t_1 + t_2)\mu + t_1\alpha_1 + t_2\alpha_2$$

By associating the parameter μ_i in the FR model with the combination of parameters $\mu + \alpha_i$ in the LFR model, a test of the hypothesis

(3.7.27) $$H_0 : \mu + \frac{N_1\alpha_1 + N_2\alpha_2}{N_1 + N_2} = 0$$

is equivalent to testing (3.7.26). The test statistic is determined by choosing

$$\mathbf{c}' = [1, N_1/(N_1 + N_2), N_2/(N_1 + N_2)]$$

With $\hat{\boldsymbol{\beta}}' = [0, y_{1.}, y_{2.}]$ and $(\mathbf{c}'\hat{\boldsymbol{\beta}})'[\mathbf{c}(\mathbf{X}'\mathbf{X})^{-}\mathbf{c}]^{-1}\mathbf{c}'\hat{\boldsymbol{\beta}} = Ny_{..}^2$, the F statistic for testing (3.7.27) becomes $F = Ny_{..}^2/s_p^2$ as expected. Thus the parameter μ in the FR model is identified with the parameter $\mu + (N_1\alpha_1 + N_2\alpha_2)/(N_1 + N_2)$ in the LFR model. The parameter μ in the LFR model has no interpretation in the FR model. Furthermore, the hypothesis $H_0 : \mu = 0$ is not testable in the LFR model since μ is not estimable; that is, $\mathbf{c}'\mathbf{H} \neq \mathbf{c}'$ for $\mathbf{c}' = [1, 0, 0]$ in the LFR model.

To test the hypothesis

(3.7.28) $$H_0 : \frac{\mu_1 + \mu_2}{2} = 0$$

in the FR model, we set $\mathbf{c}' = [1/2, 1/2]$; then $\mathbf{c}'\hat{\boldsymbol{\beta}} = (y_{1.} + y_{2.})/2$, $[\mathbf{c}'(\mathbf{X}'\mathbf{X})^{-1}\mathbf{c}]^{-1} = (N_1/4 + N_2/4)^{-1}$, and

$$F = \frac{[(y_{1.} + y_{2.})/2]^2}{s_p^2(N_1/4 + N_2/4)} = \frac{4[(N_1N_2)/(N_1 + N_2)][(y_{1.} + y_{2.})/2]^2}{s_p^2}$$

To test (3.7.28) by using the LFR model, the function

$$\mathbf{c}'\boldsymbol{\beta} = (t_1 + t_2)\mu + t_1\alpha_1 + t_2\alpha_2$$

is considered. Letting $t_1 = t_2 = 1/2$, $\mathbf{c}'\boldsymbol{\beta} = \mu + (\alpha_1 + \alpha_2)/2$. The parameter $\mu + (\alpha_1 + \alpha_2)/2$ in the LFR model is the mean associated with $(\mu_1 + \mu_2)/2$ in the FR case. However, the parameter μ in the LFR model does not have the same interpretation as μ in the FR model. In order to test the hypothesis

(3.7.29) $$H_0 : \mu + \frac{\alpha_1 + \alpha_2}{2} = 0$$

for the LFR model, we set $\mathbf{c}' = [1, 1/2, 1/2]$; then $\mathbf{c}'\hat{\boldsymbol{\beta}} = y_{1.} + y_{2.}$ and

$$F = \frac{4[(N_1 N_2)/(N_1 + N_2)][(y_{1.} + y_{2.})/2]^2}{s_p^2}$$

Thus testing (3.7.28) in the FR model is equivalent to testing (3.7.29) in the LFR model. This test is different than the test specified in (3.7.27).

Finally, suppose that the hypothesis

$$H_0\colon \mu_2 = 0 \tag{3.7.30}$$

is tested in the FR case. Then, with $\mathbf{c}' = [0, 1]$, $\mathbf{c}'\hat{\boldsymbol{\beta}} = y_{2.}$ and $[\mathbf{c}'(\mathbf{X}'\mathbf{X})^{-1}\mathbf{c}]^{-1} = 1/N_2$. The F statistic reduces to $F = N_2 y_{2.}^2/s_p^2$. To test the hypothesis $H_0\colon \mu_2 = 0$ by using the LFR model, the estimable function $\mu + \alpha_2$ is of interest. By setting $\mathbf{c}' = [1, 0, 1]$, $\mathbf{c}'\hat{\boldsymbol{\beta}} = y_{2.}$. Thus the test of (3.7.30) becomes

$$H_0\colon \mu + \alpha_2 = 0 \tag{3.7.31}$$

and the F statistic is shown to be $N_2 y_{2.}^2/s_p^2$.

The previous examples illustrate that hypotheses that can be tested by using the FR model can also be tested by using the LFR model. However, the form of the hypothesis changes because the interpretation given the parameters in the two cases differ. Some hypotheses in the LFR model can not be tested because the parameters involved are not estimable or have no meaning in terms of the FR model. To avoid this confusion, some authors choose suitable side conditions that use nonestimable functions so that the interpretation given parameters in the new model (which includes the side conditions) is the same as in the FR model. Such a model is called a reparameterized model by Graybill (1961), an FR model with side conditions by Scheffé (1959), and a restricted model by Searle (1971).

For models with side conditions, parameters not estimable in the LFR model become estimable (have meaning) in terms of the FR model and hypotheses not testable using the LFR model therefore become testable in terms of the restricted FR model. The correspondence among the hypotheses for the three models is illustrated in Table 3.7.3. The F statistics to test each hypothesis, across models, are the same.

TABLE 3.7.3. Comparing Hypotheses Across Models.

LFR Model	FR Model with Side Conditions	FR Model
$H_0\colon \mu + \frac{N_1\alpha_1 + N_2\alpha_2}{2} = 0$	$H_0\colon \mu = 0$ where $N_1\alpha_1 + N_2\alpha_2 = 0$	$H_0\colon \frac{N_1\mu_1 + N_2\mu_2}{N_1 + N_2} = 0$
$H_0\colon \mu + \frac{\alpha_1 + \alpha_2}{2} = 0$	$H_0\colon \mu = 0$ where $\alpha_1 + \alpha_2 = 0$	$H_0\colon \frac{\mu_1 + \mu_2}{2} = 0$
$H_0\colon \mu + \alpha_2 = 0$	$H_0\colon \mu = 0$ where $\alpha_2 = 0$	$H_0\colon \mu_2 = 0$

By using conditions like $\hat{\alpha}_2 = 0$, $\hat{\alpha}_1 + \hat{\alpha}_2 = 0$, and $\hat{\mu} = 0$, among others, we are able to find a solution to the normal equations, provided linear combinations of the estimates chosen to obtain a solution are not estimable. Furthermore, the estimable parametric functions $\psi = \mathbf{c}'\boldsymbol{\beta}$ do not depend on the conditions selected. However, by imposing the same conditions (side conditions) on the population parameters, we change not only the form of the estimable functions but also the form of the hypotheses tested. This has caused considerable confusion for researchers. To avoid this confusion, we can either choose to discuss the LFR model or the FR model, since for either approach we always know which of the parametric functions are estimable and hence which hypotheses may be tested. However, this is not necessarily the case for FR models with side conditions.

As a final example, consider the FR model with side conditions (restrictions)

$$\begin{aligned} y_{ij} &= \mu + \alpha_i + \varepsilon_{ij} \\ \alpha_2 &= 0 \\ \varepsilon_{ij} &\sim IN(0, \sigma^2) \end{aligned} \tag{3.7.32}$$

Beginning with the FR model and making the correspondence $\mu_1 = \mu + \alpha_1$ and $\mu_2 = \mu$, when the side condition $\alpha_2 = 0$ is part of the model, we see that $(\mu_1 + \mu_2)/2$ becomes $\mu + \alpha_1/2$ by using (3.7.32). Since the hypothesis $H_0: (\mu_1 + \mu_2)/2 = 0$ can be tested by using the FR model, the hypothesis $H_0: \mu + \alpha_1/2 = 0$ would be testable by using (3.7.32). However, by using the LFR model, the parametric function $\mu + \alpha_1/2$ is not estimable and hence not testable since, for $\mathbf{c}' = [1, 1/2, 0]$, $\mathbf{c}'\mathbf{H} \neq \mathbf{c}'$. That is, the function $\mu + \alpha_1/2$ has no meaning in terms of the FR model with the correspondence $\mu_1 = \mu + \alpha_1$ and $\mu_2 = \mu + \alpha_2$.

The equivalent hypothesis using the LFR model is $H_0: \mu + (\alpha_1 + \alpha_2)/2 = 0$, which is testable. By using either the LFR model or the FR model, we always know exactly the hypotheses being tested and whether a hypothesis is testable. When using the FR model with side conditions, this is not always the case. Furthermore, by using the LFR model, the hypothesis being tested employing the FR model with side conditions is always evident. For the example, we merely set $\alpha_2 = 0$.

Although the FR model can also be employed to test hypotheses in analysis-of-variance situations, we will not pursue this approach in this book since it is not in keeping with the traditional concept of estimability to test hypotheses. For FR models, all elements in the parameter vector are estimable. For further details, see Speed (1969), Urquhart et al. (1970), and Timm and Carlson (1973) where full-rank, analysis-of-variance models are discussed. In this case, side conditions on estimable parameters are essential to the development of hypothesis test theory. For a discussion on this topic, see Rao (1954), Chipman and Rao (1964), and Milliken (1971).

The advantage of employing a general approach to the analysis of data, assuming an LFR model, is with the flexibility it allows. For example, suppose that, in a two-sample problem, samples of sizes $N_1 = 10$ and $N_2 = 5$ are obtained and a test of the overall mean being zero is desired. Furthermore, suppose it is known that the proportion of subjects exposed to a new treatment is 3:2 in the population. Instead of using the standard side condition $(N_1\alpha_1 + N_2\alpha_2)/(N_1 + N_2) = 0$ to test the overall mean hypothesis, it would be more meaningful to test $H_0: 5\mu + 3\alpha_1 + 2\alpha_2 = 0$ or to assume that $3\alpha_1 + 2\alpha_2 = 0$.

The implications of estimating parameters in a restricted model, which were not estimable in a LFR model, is that careful interpretation should be given the quantities estimated. Furthermore, tests of hypotheses must be given the same consideration.

EXERCISES 3.7

1. Use the LR test criterion to derive the LRT of size α for testing $H_0: \mu_1 = \mu_2$ against $H_1: \mu_1 \neq \mu_2$ under normality and known variance $\sigma^2 = \sigma_0^2$. Also, determine a $100(1 - \alpha)\%$ confidence interval for $\psi = \mu_1 - \mu_2$.

2. For the LFR model specified by

$$y_{ij} = \mu + \alpha_i + \varepsilon_{ij} \qquad i = 1, 2;\ j = 1, \ldots, N_i$$
$$\varepsilon_{ij} \sim IN(0, \sigma^2)$$

   a. Derive the F statistic to test the hypothesis

$$H_0: \mu + \frac{t_1\alpha_1}{t_1 + t_2} + \frac{t_2\alpha_2}{t_1 + t_2} = 0$$

   b. With the side condition $t_1\alpha_1 + t_2\alpha_2 = 0$ added to the LFR model, what is the form of the hypothesis given by H_0?

   c. What hypothesis would the F statistic derived in part a test if the FR model were used?

3. a. For the linear model $\mathbf{y} = \mathbf{X}\boldsymbol{\beta} + \boldsymbol{\varepsilon}$, with the restrictions $\mathbf{R}\boldsymbol{\beta} = \boldsymbol{\theta}$, determine an estimate of $\boldsymbol{\beta}$ that minimizes

$$Q = (\mathbf{y} - \mathbf{X}\boldsymbol{\beta})'(\mathbf{y} - \mathbf{X}\boldsymbol{\beta})$$

   subject to the condition that $\mathbf{R}\boldsymbol{\beta} = \boldsymbol{\theta}$.

   b. Show that the estimate obtained in part a satisfies the normal equations obtained with restrictions if the $\mathbf{R}\boldsymbol{\beta}$ involve estimable functions.

   c. Assuming normality, derive the test statistic for testing the hypothesis $H_0: \mathbf{C}\boldsymbol{\beta} = \boldsymbol{\xi}$ with nonestimable restrictions and with estimable restrictions.

3.8 TWO-SAMPLE PROBLEM—MULTIVARIATE

Instead of observing a one-response random variable, suppose that two independent random samples of sizes N_1 and N_2 on p-vector variables are obtained:

$$\begin{aligned}
\mathbf{Y}'_1 &= [Y_{111}, Y_{112}, \ldots, Y_{11p}] \\
&\vdots \\
\mathbf{Y}'_{N_1} &= [Y_{1N_11}, Y_{1N_12}, \ldots, Y_{1N_1p}] \\
\mathbf{Y}'_{N_1+1} &= [Y_{2(N_1+1)1}, Y_{2(N_1+1)2}, \ldots, Y_{2(N_1+1)p}] \\
&\vdots \\
\mathbf{Y}'_{N_1+N_2} &= [Y_{2(N_1+N_2)1}, Y_{2(N_1+N_2)2}, \ldots, Y_{2(N_1+N_2)p}]
\end{aligned} \tag{3.8.1}$$

where each row $\mathbf{Y}_i \sim IN_p(\boldsymbol{\mu}_1, \boldsymbol{\Sigma})$, $i = 1, \ldots, N_1$, and $\mathbf{Y}_i \sim IN_p(\boldsymbol{\mu}_2, \boldsymbol{\Sigma})$, $i = N_1 + 1, \ldots, N_1 + N_2$. The subscripts of Y_{ijk} denote the ith group, the jth individual, and the kth variable, respectively.

Using the general linear model, the natural representation for the observations is as follows:

$$\underset{[(N_1+N_2)\times p]}{\mathbf{Y}} = \underset{[(N_1+N_2)\times 2]}{\mathbf{X}} \underset{(2\times p)}{\mathbf{B}} + \underset{[(N_1+N_2)\times p]}{\mathbf{E}_0}$$

$$\begin{bmatrix} y_{111} & \cdots & y_{11p} \\ \vdots & \vdots & \vdots \\ y_{1N_1} & \cdots & y_{1N_1p} \\ y_{2(N_1+1)1} & \cdots & y_{2(N_1+1)p} \\ \vdots & \vdots & \vdots \\ y_{2(N_1+N_2)1} & \cdots & y_{2(N_1+N_2)p} \end{bmatrix} = \begin{bmatrix} 1 & 0 \\ \vdots & \vdots \\ 1 & 0 \\ 0 & 1 \\ \vdots & \vdots \\ 0 & 1 \end{bmatrix} \begin{bmatrix} \mu_{11} & \mu_{12} & \cdots & \mu_{1p} \\ \mu_{21} & \mu_{22} & \cdots & \mu_{2p} \end{bmatrix} + \begin{bmatrix} \varepsilon_{111} & \cdots & \varepsilon_{11p} \\ \vdots & \vdots & \vdots \\ \varepsilon_{1N_11} & \cdots & \varepsilon_{1N_1p} \\ \varepsilon_{2(N_1+N_1)1} & \cdots & \varepsilon_{2(N_1+1)p} \\ \vdots & \vdots & \vdots \\ \varepsilon_{2(N_1+N_2)1} & \cdots & \varepsilon_{2(N_1+N_2)p} \end{bmatrix}$$

where

(3.8.2) $$E(\mathbf{Y}) = \mathbf{XB} \quad \text{and} \quad V(\mathbf{Y}) = \mathbf{I}_{(N_1+N_2)} \otimes \mathbf{\Sigma}$$

To estimate **B** and **Σ**, the multivariate Gauss-Markoff theorem is applied, so that

(3.8.3) $$\hat{\mathbf{B}} = (\mathbf{X}'\mathbf{X})^{-}\mathbf{X}'\mathbf{Y} = \begin{bmatrix} \mathbf{y}'_{1.} \\ \mathbf{y}'_{2.} \end{bmatrix}$$

where

$$\mathbf{y}'_{1.} = [y_{1.1}, \ldots, y_{1.p}] \qquad y_{1.k} = \frac{\sum_{j=1}^{N_1} y_{1jk}}{N_1}$$

$$\mathbf{y}'_{2.} = [y_{2.1}, \ldots, y_{2.p}] \qquad y_{2.k} = \frac{\sum_{j=N+1}^{N_1+N_2} y_{2jk}}{N_2}$$

and

(3.8.4) $$\mathbf{S} = \frac{\mathbf{Y}'[\mathbf{I} - \mathbf{X}(\mathbf{X}'\mathbf{X})^{-1}\mathbf{X}']\mathbf{Y}}{N_1 + N_2 - 2} = \frac{\mathbf{A}_1 + \mathbf{A}_2}{N_1 + N_2 - 2}$$

where

$$\mathbf{A}_1 = \sum_{i=1}^{N_1} (\mathbf{y}_i - \mathbf{y}_{1.})(\mathbf{y}_i - \mathbf{y}_{1.})'$$

$$\mathbf{A}_2 = \sum_{i=N_1+1}^{N_1+N_2} (\mathbf{y}_i - \mathbf{y}_{2.})(\mathbf{y}_i - \mathbf{y}_{2.})'$$

Both estimates are again in complete agreement with the univariate results.

To test the null hypothesis that the means for all variables are the same

(3.8.5)
$$H_0: \begin{bmatrix} \mu_{11} \\ \mu_{12} \\ \vdots \\ \mu_{1p} \end{bmatrix} = \begin{bmatrix} \mu_{21} \\ \mu_{22} \\ \vdots \\ \mu_{2p} \end{bmatrix}$$

with the alternative hypothesis

(3.8.6)
$$H_1: \boldsymbol{\mu}_1 \neq \boldsymbol{\mu}_2$$

consider the univariate t statistic. First write

$$t^2 = \frac{(y_{1.} - y_{2.})^2}{s_p^2[(N_1 + N_2)/(N_1 N_2)]}$$

(3.8.7)
$$= \frac{N_1 N_2}{N_1 + N_2}(y_{1.} - y_{2.})' S^{-1}(y_{1.} - y_{2.})$$

By replacing the means and s_p^2 by their multivariate analogues, t^2 becomes

(3.8.8)
$$T^2 = \frac{N_1 N_2}{N_1 + N_2}(\mathbf{y}_{1.} - \mathbf{y}_{2.})' \mathbf{S}^{-1}(\mathbf{y}_{1.} - \mathbf{y}_{2.})$$

called *Hotelling's two-sample T^2 statistic.* As with the univariate case,

(3.8.9)
$$T^2 = \frac{N_1 N_2}{N_1 + N_2} D^2$$

where now $D^2 = (\mathbf{y}_{1.} - \mathbf{y}_{2.})' \mathbf{S}^{-1}(\mathbf{y}_{1.} - \mathbf{y}_{2.})$, called *Mahalanobis' two-sample D^2 statistic.* For T^2 to have an F distribution, the expression for T^2 is now weighted by the factor $(N_1 + N_2 - p - 1)/p(N_1 + N_2 - 2)$, so that

(3.8.10)
$$\frac{N_1 + N_2 - p - 1}{p(N_1 + N_2 - 2)} T^2 = \frac{N_1 N_2 (N_1 + N_2 - p - 1)}{p(N_1 + N_2 - 2)} D^2 \sim F(p, N_1 + N_2 - p - 1)$$

Using the F distribution, H_0 is rejected at the significance level α if

(3.8.11)
$$\frac{N_1 + N_2 - p - 1}{p(N_1 + N_2 - 2)} T^2 > F^\alpha(p, N_1 + N_2 - p - 1)$$

where F^α denotes the upper α percentage point of the F distribution.

Employing the T^2 distribution, H_0 is rejected if

(3.8.12)
$$T^2 > T^\alpha(p, N_1 + N_2 - 2)$$

where T^α denotes the upper α percentage point of Hotelling's T^2 distribution.

As in the one-sample problem, the test of H_0 is an LRT. With

$$T^2 = (N_1 + N_2 - 2)\left\{\frac{|\mathbf{Q}_e + [N_1N_2/(N_1 + N_2)](\mathbf{y}_{1.} - \mathbf{y}_{2.})(\mathbf{y}_{1.} - \mathbf{y}_{2.})'|}{|\mathbf{Q}_e|} - 1\right\}$$

(3.8.13)

$$\Lambda = \frac{|\mathbf{Q}_e|}{|\mathbf{Q}_e + [N_1N_2/(N_1 + N_2)](\mathbf{y}_{1.} - \mathbf{y}_{2.})(\mathbf{y}_{1.} - \mathbf{y}_{2.})'|} = \frac{1}{1 + [T^2/(N_1 + N_2 - 2)]}$$

or

$$\Lambda = \frac{1}{1 + [(pF)/(N_1 + N_2 - p - 1)]}$$

Furthermore, this implies that

$$\frac{N_1 + N_2 - p - 1}{p}\,\frac{1 - \Lambda}{\Lambda} \sim F(p, N_1 + N_2 - p - 1) \tag{3.8.14}$$

where $\Lambda \sim U(p, 1, N_1 + N_2 - 2)$. Using Wilks' Λ criterion, H_0 is rejected if

$$\frac{N_1 + N_2 - p - 1}{p}\,\frac{1 - \Lambda}{\Lambda} > F^{\alpha}(p, N_1 + N_2 - p - 1) \tag{3.8.15}$$

or

$$\Lambda < U(p, 1, N_1 + N_2 - 2)$$

Instead of proceeding as above, suppose that the general fundamental least-squares theorem is applied to test H_0. For this approach, the matrices

$$\underset{(1\times 2)}{\mathbf{C}} = [1, -1], \quad \underset{(2\times p)}{\mathbf{B}} = \begin{bmatrix} \mu_{11} & \cdots & \mu_{1p} \\ \mu_{21} & \cdots & \mu_{2p} \end{bmatrix}, \quad \mathbf{A} = \mathbf{I}_p, \quad \text{and} \quad \underset{(1\times p)}{\mathbf{\Gamma}} = [0] \tag{3.8.16}$$

so that the null hypothesis $\mathbf{CBA} = \mathbf{\Gamma}$ becomes

$$\begin{gathered} \mathbf{C} \qquad\qquad \mathbf{B} \qquad\qquad \mathbf{A} \quad = \quad \mathbf{\Gamma} \\ [1, -1]\begin{bmatrix} \mu_{11} & \cdots & \mu_{1p} \\ \mu_{21} & \cdots & \mu_{2p} \end{bmatrix}\begin{bmatrix} 1 & 0 \\ 0 & 1 \end{bmatrix} = [0, \ldots, 0] \\ [\mu_{11} - \mu_{21}, \ldots, \mu_{1p} - \mu_{2p}] = [0, \ldots, 0] \\ \begin{bmatrix} \mu_{11} \\ \vdots \\ \mu_{1p} \end{bmatrix} = \begin{bmatrix} \mu_{21} \\ \vdots \\ \mu_{2p} \end{bmatrix} \end{gathered} \tag{3.8.17}$$

which is H_0 as stated in (3.8.5).

Using the matrices in (3.8.17),

$$\mathbf{Q}_h = (\mathbf{C}\hat{\mathbf{B}}\mathbf{A} - \mathbf{\Gamma})'[\mathbf{C}(\mathbf{X}'\mathbf{X})^{-1}\mathbf{C}']^{-1}(\mathbf{C}\hat{\mathbf{B}}\mathbf{A} - \mathbf{\Gamma})$$
$$= (\mathbf{y}'_{1.} - \mathbf{y}'_{2.})'\frac{N_1 + N_2}{N_1 N_2}(\mathbf{y}_{1.} - \mathbf{y}_{2.})$$
$$= \frac{N_1 + N_2}{N_1 N_2}(\mathbf{y}_{1.} - \mathbf{y}_{2.})(\mathbf{y}_{1.} - \mathbf{y}_{2.})'$$

to verify (3.8.13),

$$\Lambda = \frac{|\mathbf{Q}_e|}{|\mathbf{Q}_e + \mathbf{Q}_h|} = \frac{1}{1 + [T^2/(N_1 + N_2 - 2)]} \sim U(p, 1, N_1 + N_2 - 2)$$

It will be recalled that the Roy criterion was obtained by considering the largest root λ_1 of the equation $|\mathbf{Q}_h - \lambda \mathbf{Q}_e| = 0$. Again, there is only one root; thus Wilks' criterion and Roy's criterion yield consistent results.

(3.8.18) In summary, $H_0 : \boldsymbol{\mu}_1 = \boldsymbol{\mu}_2$ is rejected at the significance level α if

(a) Hotelling:

$$T^2 > T^{\alpha}(p, N_1 + N_2 - 2) \quad \text{or}$$
$$\frac{N_1 + N_2 - p - 1}{p(N_1 + N_2 - 2)} T^2 > F^{\alpha}(p, N_1 + N_2 - p - 1)$$

(b) Wilks:

$$\Lambda < U^{\alpha}(p, 1, N_1 + N_2 - 2) \quad \text{or}$$
$$\frac{N_1 + N_2 - p - 1}{p} \frac{1 - \Lambda}{\Lambda} > F^{\alpha}(p, N_1 + N_2 - p - 1)$$

(c) Roy:

$$\theta_s = \frac{\lambda_1}{1 + \lambda_1} > \theta^{\alpha}(s, m, n) \quad \text{or}$$
$$\frac{N_1 + N_2 - p - 1}{p} \frac{\theta_s}{1 - \theta_s} > F^{\alpha}(p, N_1 + N_2 - p - 1)$$

(d) Lawley-Hotelling:

$$U^{(s)} = \frac{T_0^2}{N_1 + N_2 - 2} > U_0^{\alpha}(s, m, n) \quad \text{or}$$
$$\frac{N_1 + N_2 - p - 1}{p} U^{(s)} > F^{\alpha}(p, N_1 + N_2 - p - 1)$$

where $s = 1, m = (|p - 1| - 1)/2$, and $n = (N_1 + N_2 - p - 3)/2$.

Whenever the degrees of freedom for the hypothesis are $v_h = 1$ or $s = \min(v_h, u) = 1$, where $u = R(\mathbf{A})$ in the hypothesis matrix product $\mathbf{CBA} = \mathbf{\Gamma}$, Wilks', Roy's, and the Lawley-Hotelling trace criteria are equivalent and may be referred to the F

distribution. In particular,

$$\frac{\nu_e - u + 1}{|u - \nu_h| + 1}\frac{1 - \Lambda}{\Lambda} \sim F(|u - \nu_h| + 1, \nu_e - u + 1)$$

(3.8.19) $$\frac{\nu_e - u + 1}{|u - \nu_h| + 1}\frac{\theta_s}{1 - \theta_s} \sim F(|u - \nu_h| + 1, \nu_e - u + 1)$$

$$\frac{(\nu_e - u + 1)T_0^2}{(|u - \nu_h| + 1)\nu_e} = \frac{\nu_e - u + 1}{|u - \nu_h| + 1}U^{(s)} \sim F(|u - \nu_h| + 1, \nu_e - u + 1)$$

or, by letting $m = (|u - \nu_h| - 1)/2$ and $n = (\nu_e - u - 1)/2$, (3.8.19) becomes

$$\frac{n + 1}{m + 1}\frac{1 - \Lambda}{\Lambda} \sim F(2m + 2, 2n + 2)$$

$$\frac{n + 1}{m + 1}\frac{\theta_s}{1 - \theta_s} \sim F(2m + 2, 2n + 2)$$

$$\frac{(n + 1)T_0^2}{(m + 1)\nu_e} = \frac{n + 1}{m + 1}U^{(s)} \sim F(2m + 2, 2n + 2)$$

Although the above test statistics used to test $H_0 : \boldsymbol{\mu}_1 = \boldsymbol{\mu}_2$ are equivalent for the two-sample problems, by tradition, Hotelling's T^2 statistic is usually employed to test the hypothesis. Confidence intervals will be constructed, using only critical values of the T^2 probability distribution.

In the univariate case, confidence intervals were obtained by considering the expression

(3.8.20) $$t^2 = \frac{[(y_{1.} - y_{2.}) - (\mu_1 - \mu_2)]^2}{s_p^2} \leq \frac{N_1 + N_2}{N_1 N_2} F^{\alpha}(N_1 + N_2 - 2)$$

In the multivariate situation, an ellipsoidal confidence region for the difference $\boldsymbol{\Delta} = \boldsymbol{\mu}_1 - \boldsymbol{\mu}_2$ is constructed by analogy. The $100(1 - \alpha)\%$ joint confidence set are those vectors $\boldsymbol{\Delta}$ satisfying the inequality

(3.8.21) $$T^2 = (\mathbf{y}_{1.} - \mathbf{y}_{2.} - \boldsymbol{\Delta})'\mathbf{S}^{-1}(\mathbf{y}_{1.} - \mathbf{y}_{2.} - \boldsymbol{\Delta}) \leq \frac{N_1 + N_2}{N_1 N_2} T^{\alpha}(p, N_1 + N_2 - 2)$$

By an argument similar to that given for (3.3.27),

(3.8.22) $$T^2 = \max_{\mathbf{a}} \frac{[N_1 N_2/(N_1 + N_2)][\mathbf{a}'(\mathbf{y}_{1.} - \mathbf{y}_{2.}) - (\boldsymbol{\mu}_1 - \boldsymbol{\mu}_2)]^2}{\mathbf{a}'\mathbf{S}\mathbf{a}}$$

so that $100(1 - \alpha)\%$ simultaneous confidence intervals for all linear combinations

(3.8.23) $$\psi = \mathbf{a}'\boldsymbol{\Delta}$$

of the mean differences are defined by

(3.8.24) $$\mathbf{a}'(\mathbf{y}_{1.} - \mathbf{y}_{2.}) - c_0\sqrt{\mathbf{a}'\mathbf{S}\mathbf{a}} \leq \mathbf{c}'\boldsymbol{\Delta} \leq \mathbf{a}'(\mathbf{y}_{1.} - \mathbf{y}_{2.}) + c_0\sqrt{\mathbf{a}'\mathbf{S}\mathbf{a}}$$

whcre

$$c_0^2 = \frac{N_1 + N_2}{N_1 N_2} T^{\alpha}(p, N_1 + N_2 = 2)$$

Alternatively, Bonferroni intervals may be constructed, as explained for (3.3.31) and the one-sample problem.

As in the univariate case, there is an alternative approach to the multivariate problem from an analysis-of-variance viewpoint. That is, the p-variate observation vector is represented by the linear model

(3.8.25)
$$\mathbf{y}_i = \boldsymbol{\mu} + \boldsymbol{\alpha}_i + \boldsymbol{\varepsilon}_i$$
$$E(\boldsymbol{\varepsilon}_i) = \mathbf{0} \quad \text{and} \quad V(\boldsymbol{\varepsilon}_i) = \boldsymbol{\Sigma}$$

The general linear model representation of (3.8.25) is

(3.8.26)
$$\underset{[(N_1+N_2)\times p]}{\mathbf{Y}} = \underset{[(N_1+N_2)\times 3]}{\mathbf{X}} \underset{[(3\times p)]}{\mathbf{B}} + \underset{[(N_1+N_2)\times p]}{\mathbf{E}_0}$$
$$\begin{bmatrix} \mathbf{y}'_1 \\ \vdots \\ \mathbf{y}'_{N_1} \\ \mathbf{y}'_{N_1+1} \\ \vdots \\ \mathbf{y}'_{N_1+N_2} \end{bmatrix} = \begin{bmatrix} 1 & 1 & 0 \\ \vdots & \vdots & \vdots \\ 1 & 1 & 0 \\ 1 & 0 & 1 \\ \vdots & \vdots & \vdots \\ 1 & 0 & 1 \end{bmatrix} \begin{bmatrix} \mu_{01} & \mu_{02} & \cdots & \mu_{0p} \\ \alpha_{11} & \alpha_{12} & \cdots & \alpha_{1p} \\ \alpha_{21} & \alpha_{22} & \cdots & \alpha_{2p} \end{bmatrix} + \begin{bmatrix} \varepsilon_{111} & \cdots & \varepsilon_{11p} \\ \vdots & \vdots & \vdots \\ \varepsilon_{1N_1} & \cdots & \varepsilon_{1N_1p} \\ \varepsilon_{2(N_1+1)} & \cdots & \varepsilon_{2(N_1+1)p} \\ \vdots & \vdots & \vdots \\ \varepsilon_{2(N_1+N_2)1} & \cdots & \varepsilon_{2(N_1+N_2)} \end{bmatrix}$$

By setting $\boldsymbol{\mu}' = [\mu_{01}, \mu_{02}, \ldots, \mu_{0p}]$, $\boldsymbol{\alpha}'_i = [\alpha_{i1}, \alpha_{i2}, \ldots, \alpha_{ip}]$, and the observation vector $\mathbf{y}'_i$ as the ith row of $\mathbf{Y}$ in (3.8.26), representation (3.8.25) is evidenced. The extension from the univariate model to the multivariate model is more easily seen in (3.8.25) than in (3.8.26).

The presentation of the overall mean vector $\boldsymbol{\mu}$ in the linear model implies the nonexistence of $(\mathbf{X}'\mathbf{X})^{-1}$ so that $(\mathbf{X}'\mathbf{X})^{-}$ must be constructed. With $(\mathbf{X}'\mathbf{X})^{-}$ defined as in the univariate case (3.7.20), an estimate of $\mathbf{B}$ is

(3.8.27)
$$\hat{\mathbf{B}} = (\mathbf{X}'\mathbf{X})^{-}\mathbf{X}'\mathbf{Y} = \begin{bmatrix} \mathbf{0}' \\ \mathbf{y}'_{1.} \\ \mathbf{y}'_{2.} \end{bmatrix}$$

which is a direct extension of (3.7.21). The unbiased estimate of $\boldsymbol{\Sigma}$ is as before,

$$\mathbf{S} = \frac{\mathbf{Y}'[\mathbf{I} - \mathbf{X}(\mathbf{X}'\mathbf{X})^{-}\mathbf{X}']\mathbf{Y}}{N_1 + N_2 - 2}$$

To find the unique linear unbiased estimates for linear combinations of the elements of $\mathbf{B}$, (3.6.17) is applied with the condition that $\mathbf{c}'\mathbf{H} = \mathbf{c}'$, where $\mathbf{H}$ is given by (3.7.22). The estimable function ψ and the estimates $\hat{\psi}$ have the following general form:

(3.8.28)
$$\begin{aligned} \psi = \mathbf{c}'\mathbf{B}\mathbf{a} &= \sum_{k=1}^{p} a_k(\mathbf{c}'\boldsymbol{\beta}_k) = \sum_{k=1}^{p} a_k(\mathbf{t}'\mathbf{H}\boldsymbol{\beta}_k) \\ &= \mathbf{a}'[(t_1 + t_2)\mu + t_1\alpha_1 + t_2\alpha_2] \\ \hat{\psi} = \mathbf{c}'\hat{\mathbf{B}}\mathbf{a} &= \sum_{k=1}^{p} a_k(\mathbf{c}'\hat{\boldsymbol{\beta}}_k) = \sum_{k=1}^{p} a_k(\mathbf{t}'\mathbf{H}\boldsymbol{\beta}_k) \\ &= \mathbf{a}'(t_1\mathbf{y}_{1.} + t_2\mathbf{y}_{2.}) \end{aligned}$$

where $\boldsymbol{\beta}_k$ denotes the kth column of $\mathbf{B}$ in (3.8.26) and $\mathbf{t}' = [t_0, t_1, t_2]$ is an arbitrary vector, thus completing the analogy between univariate and multivariate estimation theory. Imposing side conditions or reparameterizing also follow the univariate case.

To use model (3.8.26), the hypothesis of interest is $H_0: \boldsymbol{\alpha}_1 = \boldsymbol{\alpha}_2$ or, in the form $\mathbf{CBA} = \boldsymbol{\Gamma}$, with $\mathbf{C}$ defined as in the univariate case by $\mathbf{C} = [0, 1, -1]$, $\boldsymbol{\Gamma} = \mathbf{0}$, $\mathbf{A} = \mathbf{I}$, and H_0 becomes

$$
\begin{aligned}
&H_0: \mathbf{CB} = \mathbf{0} \quad \text{or} \quad \boldsymbol{\alpha}_1 - \boldsymbol{\alpha}_2 = \mathbf{0} \\
&H_1: \mathbf{CB} \neq \mathbf{0} \quad \text{or} \quad \boldsymbol{\alpha}_1 - \boldsymbol{\alpha}_2 \neq \mathbf{0}
\end{aligned}
\tag{3.8.29}
$$

Employing (3.6.19),

$$
\begin{aligned}
\mathbf{Q}_e &= \mathbf{A}'\mathbf{Y}'[\mathbf{I} - \mathbf{X}(\mathbf{X}'\mathbf{X})^{-}\mathbf{X}']\mathbf{YA} \\
&= \mathbf{A}_1 + \mathbf{A}_2 \\
\mathbf{Q}_h &= (\mathbf{C}\hat{\mathbf{B}}\mathbf{A})'[\mathbf{C}(\mathbf{X}'\mathbf{X})^{-}\mathbf{C}']^{-1}(\mathbf{C}\hat{\mathbf{B}}\mathbf{A}) \\
&= (\mathbf{y}'_{1.} - \mathbf{y}'_{2.})'\left(\frac{N_1 + N_2}{N_1 N_2}\right)^{-1}(\mathbf{y}_{1.} - \mathbf{y}_{2.})' \\
&= \frac{N_1 N_2}{N_1 + N_2}(\mathbf{y}_{1.} - \mathbf{y}_{2.})(\mathbf{y}_{1.} - \mathbf{y}_{2.})' \\
\Lambda &= \frac{|\mathbf{Q}_e|}{|\mathbf{Q}_e + \mathbf{Q}_h|} \sim U(p, 1, N_1 + N_2 - 2)
\end{aligned}
\tag{3.8.30}
$$

Having defined $\mathbf{Q}_e$ and $\mathbf{Q}_h$ for the g-inverse approach, the other test criteria follow the presentation given for the full-rank model.

EXERCISES 3.8

1. Prove (3.8.22) by following the argument in Section 3.3 for (3.3.27).
2. Use (3.8.25) to determine the test statistic for testing $H_0: \boldsymbol{\mu} + (\boldsymbol{\alpha}_1 + \boldsymbol{\alpha}_2)/2 = \mathbf{0}$.

3.9 EXAMPLE—TWO-SAMPLE MULTIVARIATE CASE

In this section, the theory of the two-sample multivariate problem is illustrated with matrix computational formulas. In addition, the two-sample discrimination problem is introduced.

TABLE 3.9.1. Sample Data: Hotelling's Two-Sample T^2.

Method I BM, WP	Method II BM, WP
190, 90	160, 120
160, 80	190, 150
180, 80	150, 90
200, 120	160, 130
150, 60	140, 110
180, 70	145, 130

As in Section 3.4, suppose another mathematics teacher gave a final exam to his two "equal ability" mathematics classes, in which individuals were randomly assigned to two different teaching methods. The scores that represent the basic mathematics and word problem skills for the two classes are summarized in Table 3.9.1. To determine if the two classes are the same with respect to the two variables, a test of the hypotheses

$$H_0: \boldsymbol{\alpha}_1 = \boldsymbol{\alpha}_2 \quad \text{or} \quad \begin{bmatrix} \alpha_{11} \\ \alpha_{12} \end{bmatrix} = \begin{bmatrix} \alpha_{21} \\ \alpha_{22} \end{bmatrix}$$

(3.9.1)

$$H_1: \boldsymbol{\alpha}_1 \neq \boldsymbol{\alpha}_2 \quad \text{or} \quad \begin{bmatrix} \alpha_{11} \\ \alpha_{12} \end{bmatrix} \neq \begin{bmatrix} \alpha_{21} \\ \alpha_{22} \end{bmatrix}$$

is made at a significance level of $\alpha = .10$, where $N_1 = N_2 = 6$ and $p = 2$. Depending on the test statistic chosen to test H_0, the null hypothesis is rejected at the .10 level if

(3.9.2) (a) Hotelling:

$$T^2 > T^\alpha(p, N_1 + N_2 - 2) = T^{.10}_{(2,10)} = 6.681$$

or

$$\frac{(N_1 + N_2 - p - 1)T^2}{(N_1 + N_2 - 2)p} = \frac{9T^2}{20} > F^\alpha(p, N_1 + N_2 - p - 1)$$

$$= F^{.10}_{(2,9)} = 3.006$$

(b) Wilks:

$$\Lambda < U^\alpha(p, 1, N_1 + N_2 - 2) = U^{.10}_{(2,1,10)} = .59952$$

or

$$\frac{N_1 + N_2 - p - 1}{p} \frac{1 - \Lambda}{\Lambda} = \frac{9}{2} \frac{1 - \Lambda}{\Lambda} > F^{.05}_{(2,9)} = 3.006$$

(c) Roy:

$$\theta_s = \frac{\lambda_1}{1 + \lambda_1} > \theta^\alpha(s, m, n) = \theta^{.10}_{(1,0,3,5)} = .40048$$

or

$$\frac{9}{2} \frac{\theta_s}{1 - \theta_s} > F^{.10}_{(2,9)} = 3.006$$

(d) Pillai:

$$U^{(s)} = \frac{T_0^2}{N_1 + N_2 - 2} = \lambda_1 > U^{.10}_{0(s,m,n)} = U^{.10}_{0(1,0,3.5)} = .6681$$

or

$$\frac{9}{2} U^{(s)} > F^{.10}_{(2,9)} = 3.006$$

By employing the above critical values, the following relations may be verified for the two-sample hypothesis:

$$\frac{(N_1 + N_2 - 2)p}{N_1 + N_2 - p - 1} F^{\alpha}(p, N_1 + N_2 - p - 1) = T^{\alpha}(p, N_1 + N_2 - 2)$$

$$U^{\alpha}(p, 1, N_1 + N_2 - 2) = \frac{1}{1 + \dfrac{T^{\alpha}(p, N_1 + N_2 - 2)}{(N_1 + N_2 - 2)}} = \frac{1}{1 + \dfrac{pF^{\alpha}(p, N_1 + N_2 - p - 1)}{(N_1 + N_2 - p - 1)}}$$

(3.9.3)

$$\theta^{\alpha}(s, m, n) = \frac{pF^{\alpha}(p, N_1 + N_2 - p - 1)}{N_1 + N_2 - p - 1 + pF^{\alpha}(p, N_1 + N_2 - p - 1)}$$

$$= 1 - U^{\alpha}(p, 1, N_1 + N_2 - 1)$$

$$U_0^{\alpha}(s, m, n) = \frac{T^{\alpha}(p, N_1 + N_2 - 1)}{N_1 + N_2 - 2} = \frac{pF^{\alpha}(p, N_1 + N_2 - p - 1)}{N_1 + N_2 - p - 1}$$

again indicating that all statistics are equivalent.

Although the general linear model representation given in (3.8.26) is not natural for the sample problem, it will be used to illustrate the theory of Section 3.8 to help us familiarize ourselves with the more general approach. For the model $\mathbf{Y} = \mathbf{XB} + \mathbf{E}_0$, using the example data,

$$\mathbf{Y} = \begin{bmatrix} 190 & 90 \\ 170 & 80 \\ 180 & 80 \\ 200 & 120 \\ 150 & 60 \\ 180 & 70 \\ 160 & 120 \\ 190 & 150 \\ 150 & 90 \\ 160 & 130 \\ 140 & 110 \\ 145 & 130 \end{bmatrix}, \quad \mathbf{X} = \begin{bmatrix} 1 & 1 & 0 \\ 1 & 1 & 0 \\ 1 & 1 & 0 \\ 1 & 1 & 0 \\ 1 & 1 & 0 \\ 1 & 1 & 0 \\ 1 & 0 & 1 \\ 1 & 0 & 1 \\ 1 & 0 & 1 \\ 1 & 0 & 1 \\ 1 & 0 & 1 \\ 1 & 0 & 1 \end{bmatrix}, \quad \text{and} \quad \mathbf{B} = \begin{bmatrix} \mu_{01} & \mu_{02} \\ \alpha_{11} & \alpha_{12} \\ \alpha_{21} & \alpha_{22} \end{bmatrix} \tag{3.9.4}$$

To test hypotheses about the elements of **B** and to estimate **B**, a g-inverse of $\mathbf{X'X}$ is needed; for the example,

$$\mathbf{X'X} = \begin{bmatrix} 12 & 6 & 6 \\ 6 & 6 & 0 \\ 6 & 0 & 6 \end{bmatrix} \quad \text{and} \quad (\mathbf{X'X})^{-} = \begin{bmatrix} 0 & 0 & 0 \\ 0 & 1/6 & 0 \\ 0 & 0 & 1/6 \end{bmatrix}$$

To estimate **B**,

$$\hat{\mathbf{B}} = (\mathbf{X}'\mathbf{X})^{-}\mathbf{X}'\mathbf{Y} = \begin{bmatrix} 0 & 0 & 0 \\ 0 & .16667 & 0 \\ 0 & 0 & .16667 \end{bmatrix} \begin{bmatrix} 2015 & 1230 \\ 1070 & 500 \\ 945 & 730 \end{bmatrix}$$

$$= \begin{bmatrix} 0 & 0 \\ 178.33 & 83.33 \\ 157.50 & 121.67 \end{bmatrix} = \begin{bmatrix} \mathbf{0} \\ \mathbf{y}'_{1.} \\ \mathbf{y}'_{2.} \end{bmatrix} \tag{3.9.5}$$

The estimate of $\mathbf{\Sigma}$ is **S**, obtained from

$$\mathbf{Q}_e = \mathbf{Y}'[\mathbf{I} - \mathbf{X}(\mathbf{X}'\mathbf{X})^{-}\mathbf{X}']\mathbf{Y}$$

$$= \begin{bmatrix} 3070.83 & \\ 2808.33 & 4216.67 \end{bmatrix}$$

$$\mathbf{S} = \frac{\mathbf{Q}_e}{N_1 + N_2 - 2} \tag{3.9.6}$$

$$= \begin{bmatrix} 307.083 & \\ 280.833 & 421.667 \end{bmatrix}$$

To test the hypothesis

$$H_0 : \boldsymbol{\alpha}_1 = \boldsymbol{\alpha}_2 \tag{3.9.7}$$

which is equivalent to the hypothesis

$$H_0 : \begin{bmatrix} \mu_{11} \\ \mu_{12} \end{bmatrix} = \begin{bmatrix} \mu_{21} \\ \mu_{22} \end{bmatrix}$$

for the full-rank model, we set

$$\mathbf{C} = [0, 1, -1], \quad \mathbf{A} = \mathbf{I}_2, \quad \text{and} \quad \mathbf{\Gamma} = \begin{bmatrix} 0 \\ 0 \end{bmatrix}$$

Then the hypothesis becomes

$$\begin{matrix} \mathbf{C} & \mathbf{B} & \mathbf{A} & = \mathbf{\Gamma} \end{matrix}$$

$$[0, 1, -1] \begin{bmatrix} \mu_{01} & \mu_{02} \\ \alpha_{11} & \alpha_{12} \\ \alpha_{21} & \alpha_{22} \end{bmatrix} \begin{bmatrix} 1 & 0 \\ 0 & 1 \end{bmatrix} = \begin{bmatrix} 0 \\ 0 \end{bmatrix} \tag{3.9.8}$$

or

$$\begin{bmatrix} \alpha_{11} - \alpha_{21} \\ \alpha_{12} - \alpha_{22} \end{bmatrix} = \boldsymbol{\alpha}_1 - \boldsymbol{\alpha}_2 = \mathbf{0}$$

To test the estimability of the hypothesis, the condition $\mathbf{c}'\mathbf{H} = \mathbf{c}'$ must be satisfied for each row of $\mathbf{C}$. Since $\mathbf{H} = (\mathbf{X}'\mathbf{X})^{-}\mathbf{X}'\mathbf{X}$, we have, for the example, $\mathbf{c}'\mathbf{H} = \mathbf{c}'$:

$$[0, 1, -1]\begin{bmatrix} 0 & 0 & 0 \\ 0 & 1/6 & 0 \\ 0 & 0 & 1/6 \end{bmatrix}\begin{bmatrix} 12 & 6 & 6 \\ 6 & 6 & 0 \\ 6 & 0 & 6 \end{bmatrix} = [0, 1, -1]$$

To test H_0, the form of T^2,

$$T^2 = (N_1 + N_2 - 2)\left(\frac{|\mathbf{Q}_e + \mathbf{Q}_h|}{|\mathbf{Q}_e|} - 1\right) \tag{3.9.9}$$

is used to illustrate the linear model approach. Alternatively, T^2 could be calculated from (3.8.8). Calculating $\mathbf{Q}_h$ in (3.9.9), we observe that

$$\mathbf{C}\hat{\mathbf{B}}\mathbf{A} = (\mathbf{y}_{1.} - \mathbf{y}_{2.})' \tag{3.9.10}$$

so that

$$\begin{aligned} \mathbf{Q}_h &= (\mathbf{C}\hat{\mathbf{B}}\mathbf{A})'[\mathbf{C}(\mathbf{X}'\mathbf{X})^{-}\mathbf{C}']^{-1}(\mathbf{C}\hat{\mathbf{B}}\mathbf{A}) \\ &= \left[\frac{1}{N_1} + \frac{1}{N_2}\right]^{-1}(\mathbf{y}_{1.} - \mathbf{y}_{2.})(\mathbf{y}_{1.} - \mathbf{y}_{2.})' \\ &= 3\begin{bmatrix} 20.83 \\ -38.33 \end{bmatrix}[20.83, -38.33] \\ &= \begin{bmatrix} 1302.08 & \\ -2395.83 & 4408.33 \end{bmatrix} \end{aligned} \tag{3.9.11}$$

With $\mathbf{Q}_e$ given in (3.9.6),

$$T^2 = 10\left(\frac{37{,}546{,}250}{5{,}061{,}944.4} - 1\right) = 64.17 \tag{3.9.12}$$

Furthermore, $T^2 = 64.17 > T^{.10}_{(2,10)} = 6.681$, so the hypothesis is rejected. Other criteria are easily obtained from (3.9.12) by using (3.9.3); that is,

$$\begin{aligned} \Lambda &= .1348 \\ \theta_s &= .8652 \\ U^{(s)} &= 6.417 \\ D^2 &= \frac{N_1 + N_2}{N_1 N_2}T^2 = 21.39 \end{aligned} \tag{3.9.13}$$

Following the overall test with Fisher's LSD procedure, as suggested by some authors, try to determine which of the differences are significantly different from zero.

To test $\alpha_{11} = \alpha_{21}$,

$$|t| = \left|\frac{178.33 - 157.50}{\sqrt{[(1/6) + (1/6)](307.08)}}\right| = 2.06 \tag{3.9.14}$$

and to test $\alpha_{12} = \alpha_{22}$,

$$|t| = \left| \frac{83.33 - 121.67}{\sqrt{[(1/6) + (1/6)](421.67)}} \right| = 3.23 \tag{3.9.15}$$

Comparing each t with the critical value $t_{(10)}^{.05} = 1.812$, both hypotheses are rejected, giving the appearance that the students differ with respect to both variables. This analysis does not take into account the correlation between the variables, which is $r = .78043$. In addition, α may be larger than the nominal level of .10.

More clearly, consider plotting the entire sample in the (y_1, y_2) plane, where y_1 denotes basic mathematical ability and y_2 denotes the word-problem skills. Notice that the projection of these points on each axis shows little overlap (see Figure 3.9.1), thus producing significant univariate t's. However, by considering all the points

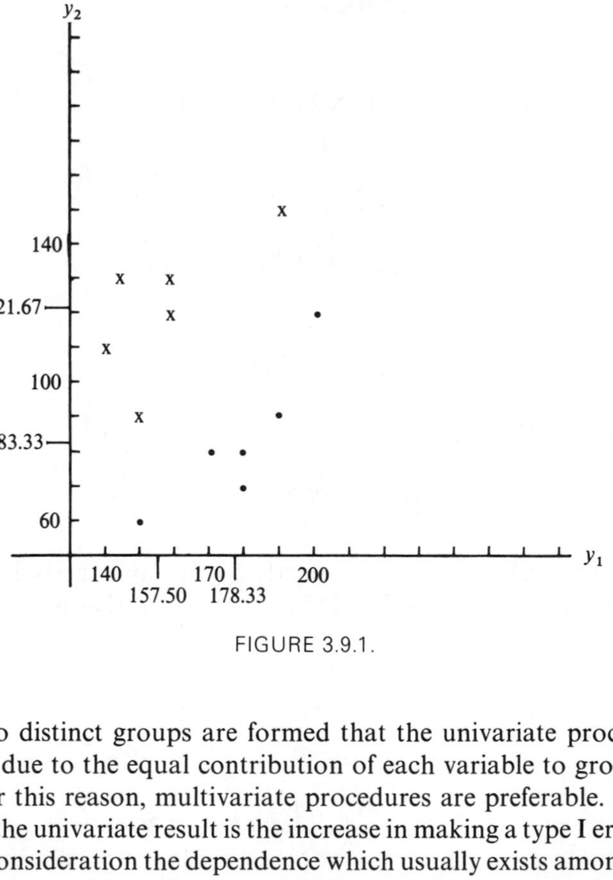

FIGURE 3.9.1.

in the plane, two distinct groups are formed that the univariate procedure fails to uncover. This is due to the equal contribution of each variable to group separation in the plane. For this reason, multivariate procedures are preferable. Another clear disadvantage of the univariate result is the increase in making a type I error because of not taking into consideration the dependence which usually exists among educational and psychological variables. For a few variables, this can be controlled by the Bonferroni method. However, when p is large, this procedure is not always the best.

Returning to the plot of the two variables in the two-sample example, observe that the groups may be separated by a plane passing through the origin. The name given this plane

$$L = a_1 y_1 + a_2 y_2 \tag{3.9.16}$$

is the *linear discriminant function*, originally investigated by Fisher (1936), and discussed in detail by Kendall (1961, p. 144).

From Figure 3.9.2, we see that a plane needs to be determined such that the values of between groups are maximally distant, showing little overlap and at the

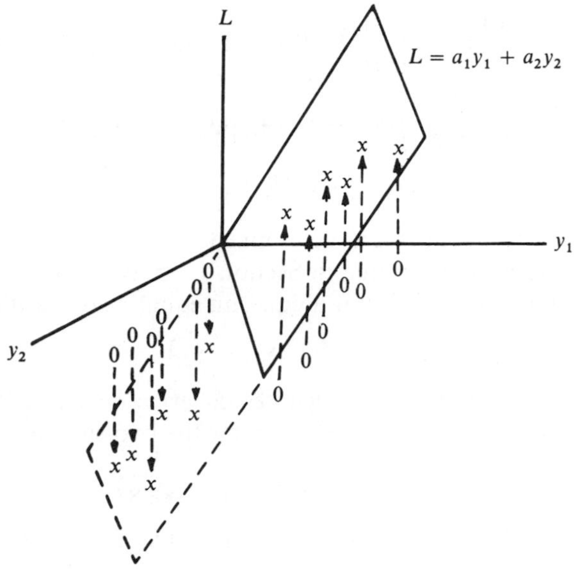

FIGURE 3.9.2. Discriminant function.

same time minimizing the variability of values of L within each group. That is, the variability between groups is to be maximum relative to the variability within groups. For p-variates, the coefficient vector is represented by the $(p \times 1)$ vector $\mathbf{a}$. Then, from (1.8.48),

(3.9.17) $$\lambda_1 = \max_{\mathbf{a}} \frac{\mathbf{a}'\mathbf{Q}_h\mathbf{a}}{\mathbf{a}'\mathbf{Q}_e\mathbf{a}}$$

where λ_1 is the largest root of determinantal equation $|\mathbf{Q}_h - \lambda\mathbf{Q}_e| = 0$, $\mathbf{Q}_h$ is the sum of squares between groups, $\mathbf{Q}_e$ is the sum of squares within groups, and $\mathbf{a}$ is the eigenvector associated with λ_1.

The elements of $\mathbf{a}$ yield the coefficients of the linear discriminant function that satisfy the relationship

(3.9.18) $$(\mathbf{Q}_h - \lambda\mathbf{Q}_e)\mathbf{a} = \mathbf{0}$$

A solution to (3.9.17) and (3.9.18) is seen to be

(3.9.19) $$\mathbf{a}_w = \mathbf{Q}_e^{-1}(\mathbf{y}_{1.} - \mathbf{y}_{2.})$$

in the two-group case. To derive the expression for $\mathbf{a}_w$ in (3.9.18), we merely have to substitute into the equation $(\mathbf{Q}_h - \lambda\mathbf{Q}_e)\mathbf{a} = \mathbf{0}$ the expression for $\mathbf{Q}_h$ in (3.9.11) and simplify. That is,

$$(\mathbf{Q}_h - \lambda\mathbf{Q}_e)\mathbf{a}_w = \mathbf{0}$$

$$\left[\frac{N_1N_2}{N_1 + N_2}(\mathbf{y}_{1.} - \mathbf{y}_{2.})(\mathbf{y}_{1.} - \mathbf{y}_{2.})' - \lambda\mathbf{Q}_e\right]\mathbf{a}_w = \mathbf{0}$$

$$\frac{1}{\lambda}\left(\frac{N_1N_2}{N_1 + N_2}\right)\mathbf{Q}_e^{-1}(\mathbf{y}_{1.} - \mathbf{y}_{2.})(\mathbf{y}_{1.} - \mathbf{y}_{2.})'\mathbf{a}_w = \mathbf{a}_w$$

$$\left[\frac{1}{\lambda}\left(\frac{N_1N_2}{N_1 + N_2}\right)(\mathbf{y}_{1.} - \mathbf{y}_{2.})'\mathbf{a}_w\right]\mathbf{Q}_e^{-1}(\mathbf{y}_{1.} - \mathbf{y}_{2.}) = \mathbf{a}_w$$

$$(\text{Constant})\mathbf{Q}_e^{-1}(\mathbf{y}_{1.} - \mathbf{y}_{2.}) = \mathbf{a}_w$$

as required, since eigenvectors are determined only up to an arbitrary constant. In general, the procedure illustrated in Section 1.7 may be employed to find an $\mathbf{a}_w$. Because the solution to (3.9.18) is not unique, an equally good value of $\mathbf{a}$ would be

$$\mathbf{a}_s = \mathbf{S}^{-1}(\mathbf{y}_{1.} - \mathbf{y}_{2.}) \tag{3.9.20}$$

so that $\mathbf{a}_w(N - r) = \mathbf{a}_s$, where $N - r$ denotes the error degrees of freedom.

To determine $\mathbf{a}_w$ and the values of L for the two-sample example, compute

$$\mathbf{Q}_e^{-1} = \begin{bmatrix} 3070.83 & \\ 2808.33 & 4216.67 \end{bmatrix}^{-1} = \begin{bmatrix} .0008330 & \\ -.0005548 & .0006166 \end{bmatrix}$$

and

$$\mathbf{y}_{1.} - \mathbf{y}_{2.} = \begin{bmatrix} 20.83 \\ -38.33 \end{bmatrix}$$

so that

$$\mathbf{a}_w = \mathbf{Q}_e^{-1}(\mathbf{y}_{1.} - \mathbf{y}_{2.}) = \begin{bmatrix} .03862 \\ -.03481 \end{bmatrix} \tag{3.9.21}$$

These results could have been obtained from the BMD 04M statistical routine (see Dixon, 1967).

Applying these coefficients to the original variables, the linear discriminant function shows perfect discrimination:

(3.9.22)

	Method I	Method II
1	4.2049	2.0019
2	3.7806	2.1161
3	4.1668	2.6600
4	3.5467	1.6537
5	3.7044	1.5776
6	4.5150	1.0744
Mean	3.9864	1.8473
Variance	.13497	.29286

By using (3.9.22) and applying a univariate t test to the composite discriminant scores,

$$t = \frac{\bar{L}_1 - \bar{L}_2}{s_p\sqrt{[(1/N_1) + (1/N_2)]}} = \frac{3.9864 - 1.8473}{.21392\sqrt{[(1/6) + (1/6)]}} = 8.0108$$

where $s_p^2 = [5(.13497) + 5(.29286)]/10$, so that $t^2 = 64.17 = T^2$. The relationship between discriminant analysis and Hotelling's T^2 statistic is evident.

Alternatively, by using

$$\mathbf{a}_s = (N - r)\mathbf{a}_w = 10\mathbf{a}_w = \begin{bmatrix} .3862 \\ -.3481 \end{bmatrix} \tag{3.9.23}$$

the values in (3.9.22) are multiplied by $N - r = 10$ and perfect discrimination is maintained. With (3.9.23), the relation between D^2 and T^2 is made clear. Evaluating the linear discriminant function, $L = a_1y_1 + a_2y_2$, where a_1 and a_2 are coefficients from $\mathbf{a}_s$ and y_1 and y_2 are replaced by the variable means for each group,

$$\bar{L}_1 = (.3862)(178.33) + (-.3481)(83.33) = 39.864$$

and

$$\bar{L}_2 = (.3862)(157.50) + (-.3481)(121.67) = 18.473$$

and it is observed that D^2 is the difference between the means in the discriminant space when $\mathbf{a}_s$ is employed; that is, $D^2 = \bar{L}_1 - \bar{L}_2 = 21.39$. Thus a significant $T^2 = [N_1N_2/(N_1 + N_2)]D^2$ implies good separation of the group centroids, which completes the relationship between hypothesis testing and discrimination.

For uniformity, many authors choose to have the leading coefficient of the first variable set to 1, then

$$\mathbf{a}_c = \begin{bmatrix} 1.0000 \\ -.9014 \end{bmatrix} \tag{3.9.24}$$

However, since the unit of scale is arbitrary, it is advantageous to scale the coefficient vector $\mathbf{a}_w$ so that the within-group variance of the discriminant scores is unity. This new vector is called the *standardized discriminant vector*. To accomplish this, the coefficient vector becomes

$$\mathbf{a}_{ws} = \frac{\mathbf{a}_w}{\sqrt{\mathbf{a}_w'\mathbf{S}\mathbf{a}_w}} \tag{3.9.25}$$

For the example, $\sqrt{\mathbf{a}_w'\mathbf{S}\mathbf{a}_w} = .46248$, so

$$\mathbf{a}_{ws} = \begin{bmatrix} .08351 \\ -.07527 \end{bmatrix} \quad \text{or} \quad \mathbf{a}_{ws} = \begin{bmatrix} -.08351 \\ .07527 \end{bmatrix}$$

Using these coefficients, the sum of the mean distances weighted by $\mathbf{a}_{ws}$ directly yields v, the number of within-group standard deviations separating the mean discriminant scores of the two groups. Thus our example shows extremely good discrimination since

$$v = \mathbf{a}_{ws}'(\mathbf{y}_{1.} - \mathbf{y}_{2.}) = .08531(20.83) + (-.07527)(-38.33) = 4.625 \tag{3.9.26}$$

To determine v from the T^2, observe that $v = \sqrt{D^2} = D$. To prove this, observe that

$$\begin{aligned} v &= \mathbf{a}'_{ws}(\mathbf{y}_{1.} - \mathbf{y}_{2.}) \\ &= \frac{\mathbf{a}'_w}{\sqrt{\mathbf{a}'_w \mathbf{S} \mathbf{a}_w}}(\mathbf{y}_{1.} - \mathbf{y}_{2.}) \\ &= \frac{(\mathbf{y}_{1.} - \mathbf{y}_{2.})'\mathbf{Q}_e^{-1}(\mathbf{y}_{1.} - \mathbf{y}_{2.})}{\sqrt{(\mathbf{y}_{1.} - \mathbf{y}_{2.})'\mathbf{Q}_e^{-1}\mathbf{S}\mathbf{Q}_e^{-1}(\mathbf{y}_{1.} - \mathbf{y}_{2.})}} \\ &= \frac{(N_1 + N_2 - 2)D^2}{\sqrt{(N_1 + N_2 - 2)^2 D^2}} \\ &= D \end{aligned}$$

The discriminant function is often used for classification purposes. For any new value of L (under equal probability of error and equal variance), an individual would be assigned to groups in the following manner:

$$\text{Group I} \quad \text{if } L \geq \frac{\bar{L}_1 + \bar{L}_2}{2}$$

$$\text{Group II} \quad \text{if } L \leq \frac{\bar{L}_1 + \bar{L}_2}{2}$$

where any coefficient vector is employed for the evaluation of L and, for convenience, it is assumed that $\bar{L}_1 \geq \bar{L}_2$. (See also, Rulon et al., 1967, and Cooley and Lohnes, 1971, who discuss applications, and Anderson, 1958, Chap. 6, and Porebski 1966b, who present theoretical foundations.)

The elements of the vector $\mathbf{a}_{ws}$ are often used in determining the relative importance of the contribution of each variable to the statistic T^2, if the variances of each variable are nearly equal. Should this not occur, the coefficients are adjusted by multiplying each by their within-group standard deviations. For the example, the adjusted coefficient vector becomes

$$\mathbf{a}_{wsa} = \begin{bmatrix} .08351 \times 17.523 \\ -.07527 \times 20.534 \end{bmatrix} = \begin{bmatrix} 1.4633 \\ -1.5456 \end{bmatrix}$$

This type of analysis must be used with extreme caution since the degree of correlation among the variables affects the coefficients' magnitudes.

Alternatively, Bargmann (1970) and Porebski (1966a) suggest looking at the correlations between the discriminant function and each variable. Letting $\mathbf{a}$ denote a coefficient vector and $\mathbf{b}$ a vector of 0_s except for a 1 in the ith position, the correlation between the ith variable and $L = \mathbf{a}'\mathbf{y}$ is

$$\begin{aligned} \rho_{y_i L} &= \text{cor}(\mathbf{b}'\mathbf{y}, \mathbf{a}'\mathbf{y}) \\ &= \frac{\text{cov}(\mathbf{b}'\mathbf{y}, \mathbf{a}'\mathbf{y})}{\sqrt{V(\mathbf{b}'\mathbf{y})}\sqrt{V(\mathbf{a}'\mathbf{y})}} \end{aligned}$$

$$= \frac{\mathbf{b}'\boldsymbol{\Sigma}\mathbf{a}}{\sqrt{\mathbf{b}'\boldsymbol{\Sigma}\mathbf{b}}\sqrt{\mathbf{a}'\boldsymbol{\Sigma}\mathbf{a}}}$$

$$= \frac{(\sigma_i)^{-1}(\boldsymbol{\Sigma}\mathbf{a})_i}{\sqrt{\mathbf{a}'\boldsymbol{\Sigma}\mathbf{a}}}$$

where $(\boldsymbol{\Sigma}\mathbf{a})_i$ denotes the ith row of $\boldsymbol{\Sigma}\mathbf{a}$. For all variables, the vector of correlations is written as

$$\boldsymbol{\rho} = \frac{(\text{Dia } \boldsymbol{\Sigma})^{-1/2}\boldsymbol{\Sigma}\mathbf{a}}{\sqrt{\mathbf{a}'\boldsymbol{\Sigma}\mathbf{a}}} \tag{3.9.27}$$

where

$$(\text{Dia } \boldsymbol{\Sigma})^{-1/2} = \begin{bmatrix} 1/\sigma_1 & 0 & \cdots & 0 \\ 0 & 1/\sigma_2 & \cdots & 0 \\ \vdots & \vdots & \vdots & \vdots \\ 0 & 0 & \cdots & 1/\sigma_p \end{bmatrix}$$

When estimating $\boldsymbol{\rho}$ from the data, $\boldsymbol{\Sigma}$ is replaced by $\mathbf{S}$, so that

$$\hat{\boldsymbol{\rho}} = \frac{(\text{Dia } \mathbf{S})^{-1/2}\mathbf{S}\mathbf{a}}{\sqrt{\mathbf{a}'\mathbf{S}\mathbf{a}}} \tag{3.9.28}$$

By letting $\mathbf{a} = \mathbf{a}_w$ in (3.9.28),

$$\hat{\boldsymbol{\rho}} = (\text{Dia } \mathbf{S})^{-1/2}\mathbf{S}\mathbf{a}_{ws} = (\text{Dia } \mathbf{S})^{-1/2}\mathbf{S}(\text{Dia } \mathbf{S})^{-1/2}\mathbf{a}_{wsa} \tag{3.9.29}$$

or

$$\hat{\boldsymbol{\rho}} = \mathbf{R}_e\mathbf{a}_{wsa}$$

where $\mathbf{R}_e$ denotes the error correlation matrix. For the two-sample example,

$$\mathbf{R}_e = (\text{Dia } \mathbf{S})^{-1/2}\mathbf{S}(\text{Dia } \mathbf{S})^{-1/2} = \begin{bmatrix} 1.0000 & \\ .78043 & 1.0000 \end{bmatrix}$$

so

$$\hat{\boldsymbol{\rho}} = \begin{bmatrix} 1.0000 & .78043 \\ .78043 & 1.0000 \end{bmatrix} \begin{bmatrix} 1.4633 \\ -1.5456 \end{bmatrix}$$

$$= \begin{bmatrix} .26 \\ -.40 \end{bmatrix}$$

indicating that the second variable (word problem skills rather than basic mathematics ability) is contributing most to discrimination and thus to the rejection of the hypothesis. Just looking at the elements of $\mathbf{a}_w$, $\mathbf{a}_{ws}$, or $\mathbf{a}_{wsa}$ might have indicated equal contribution of the variables to discrimination.

Step-down analysis may also be employed to help determine the importance of each variable or set of variables in a discriminant function (see Rao, 1952, p. 252). Although Finn's program provides the researcher with a means of testing the contribution of a variable, given that the others are already included in the function by means of a step-down F statistic, the BMD 07M routine is more conveniently utilized for this purpose.

Having rejected H_0, confidence-interval procedures are applied to determine whether differences in the first, second, or both variables contributed significantly to rejection. To obtain simultaneous confidence intervals, (3.8.24) is evaluated:

$$\mathbf{a}'(\mathbf{y}_{1.} - \mathbf{y}_{2.}) - c_0\sqrt{\mathbf{a}'\mathbf{S}\mathbf{a}} \leq \mathbf{a}'\boldsymbol{\Delta} \leq \mathbf{a}'(\mathbf{y}_{1.} - \mathbf{y}_{2.}) + c_0\sqrt{\mathbf{a}'\mathbf{S}\mathbf{a}}$$

where

$$c_0^2 = \frac{N_1 + N_2}{N_1 N_2} T^{\alpha}(p, N_1 + N_2 - 2) \quad \text{and} \quad \boldsymbol{\Delta} = \boldsymbol{\alpha}_1 - \boldsymbol{\alpha}_2$$

For the example, $N_1 = N_2 = 6$, $p = 2$, and $T^{\alpha}(p, N_1 + N_2 - 2) = 9.459$, so that $c_0^2 = (12/36)(6.681) = 2.24$, or $c_0 = 1.5$. Since

$$\mathbf{S} = \begin{bmatrix} 307.08 & \\ 280.83 & 421.67 \end{bmatrix} \quad \text{and} \quad \mathbf{y}_{1.} - \mathbf{y}_{2.} = \begin{bmatrix} 20.83 \\ -38.33 \end{bmatrix}$$

the confidence intervals for each variable become, for $\mathbf{a}' = [1, 0]$,

$$20.83 - 1.5\sqrt{307.08} \leq \alpha_{11} - \alpha_{21} \leq 20.83 + 1.5\sqrt{307.08}$$
$$-5.46 \leq \alpha_{11} - \alpha_{21} \leq 47.12$$

and for $\mathbf{a}' = [0, 1]$, the interval is

$$-38.33 - 1.5\sqrt{421.67} \leq \alpha_{12} - \alpha_{22} \leq -38.33 + 1.5\sqrt{421.67}$$
$$-69.13 \leq \alpha_{12} - \alpha_{22} \leq -7.53$$

From these results, zero is contained in the confidence interval for only the mathematics skills subtest mean difference. Accordingly, it may be concluded that the two classes differ with respect to the mean subtest word problem scores, but the classes do not differ in basic mathematics skills. This agrees with the discriminant analysis results, which also indicated that the second variable, word problem ability, contributes to rejection more than the first variable, basic mathematics skills. Note, however, that our univariate analysis approach indicated that both variables were significant.

When using the Bonferroni inequality to compute mean differences for one variable at a time, intervals would be obtained from the expression

$$\hat{\psi}_i - \hat{\sigma}_{\hat{\psi}_i} t_{(v_e)}^{\alpha/2p} \leq \psi_i \leq \hat{\psi}_i + \hat{\sigma}_{\hat{\psi}_i} t_{(v_e)}^{\alpha/2p} \tag{3.9.30}$$

where

$$\hat{\psi}_i = y_{1 \cdot i} - y_{2 \cdot i} \qquad \hat{\sigma}_{\hat{\psi}} = \left(\frac{N_1 + N_2}{N_1 N_2}\right)^{1/2} s_{ii} \qquad t_{(v_e)}^{\alpha/2p} = t_{(10)}^{.10/2p} = t_{(10)}^{.025} = 2.228$$

and $p = 2$. For variable 1, (3.9.30) becomes

$$20.83 - 2.228\sqrt{\frac{12}{36}(307.08)} \leq \alpha_{11} - \alpha_{21} \leq 20.83 + 2.228\sqrt{\frac{12}{36}(307.08)}$$
$$20.83 - (2.228)(10.12) \leq \alpha_{11} - \alpha_{21} \leq 20.83 + (2.228)(10.12)$$
$$-1.72 \leq \alpha_{11} - \alpha_{21} \leq 43.38$$

and, for variable 2, we have

$$-38.33 - 2.228\sqrt{\frac{12}{36}(421.67)} \leq \alpha_{12} - \alpha_{22} \leq -38.33 + 2.228\sqrt{\frac{12}{36}(421.67)}$$

$$-38.33 - (2.228)(11.86) \leq \alpha_{12} - \alpha_{22} \leq -38.33 + (2.228)(11.86)$$

$$-64.75 \leq \alpha_{12} - \alpha_{22} \leq 11.90$$

These results again agree with the multivariate results; however, the intervals are shorter. This analysis is performed without calculating the multivariate test statistic and should be used only if $t_{(\nu_e)}^{\alpha/2k} < T_{(p,\nu_e)}^{\alpha}$, for k specific comparisons.

EXERCISES 3.9

1. Use $\alpha = .05$ to verify (3.9.3).
2. Rework the example using $\alpha = .05$ and $\alpha = .01$. What can you conclude about the choice of α for multivariate problems?
3. From (3.8.22), derive (3.10.20). What does this tell you about simultaneous inference procedures limited to single variables?
4. The following data resulted from an independent sample of students for the experiment described in Exercise 1 (Section 3.4).

Subject	G	R	S
1	31	50	20
2	60	40	15
3	65	36	12
4	70	29	18
5	78	48	24
6	90	47	26
7	98	18	40
8	95	10	10

a. Using $\alpha = .05$, test the hypothesis that the mean performance on the three dependent variables is the same for both groups; that is, $H_0: \boldsymbol{\mu}_1 = \boldsymbol{\mu}_2$.

b. Use discriminant analysis and confidence intervals to determine which variables are most critical to group separation.

c. Use the Bonferroni procedure to obtain confidence intervals for group differences for one variable at a time.

3.10 PAIRED OBSERVATIONS—UNIVARIATE

In comparing two treatments in univariate analysis, attempts are made to control for extraneous factors by matching individuals, making sure that the two members of any pair are alike in other respects. Instead of two random samples, one random sample of pairs exists to control for effects that are not of interest and that can cause differences in means. A common example of this type occurs when data is gathered before and after some treatment for well-defined populations to determine if the change in behavior was significant over some time period. The analysis of the data is accomplished by the matched-pair t test.

Let y_{1i} be the first element of the ith pair and y_{2i} be the second element. Suppose that the N independent paired observations

$$\begin{bmatrix} y_{11} \\ y_{21} \end{bmatrix}, \begin{bmatrix} y_{12} \\ y_{22} \end{bmatrix}, \ldots, \begin{bmatrix} y_{1N} \\ y_{2N} \end{bmatrix}$$

follow a joint bivariate normal distribution

$$\mathbf{Y}_i = \begin{bmatrix} y_{1i} \\ y_{2i} \end{bmatrix} \sim IN_2\left\{\begin{bmatrix} \mu_1 \\ \mu_2 \end{bmatrix}, \quad \mathbf{\Sigma} = \begin{bmatrix} \sigma_1^2 & \sigma_{12} \\ \sigma_{12} & \sigma_2^2 \end{bmatrix}\right\}$$

For each pair, a difference is computed. Denote these by $d_1, d_2, \ldots, d_N$, where $d_i = y_{1i} - y_{2i}, i = 1, \ldots, N$. The d_i form a random sample from a normal population with mean $\tau = \mu_1 - \mu_2$ and variance $\sigma_d^2 = \sigma_1^2 + \sigma_2^2 - 2\rho\sigma_1\sigma_2$. If the notations $\bar{d}$ and s_d^2 are used to indicate the sample mean and variance, respectively, of the differences d_i

$$\bar{d} = \frac{\sum_{i=1}^{N} d_i}{N} \tag{3.10.1}$$

$$s_d^2 = \frac{\sum_{i=1}^{N} (d_i - \bar{d})^2}{N - 1}$$

then, by the arguments of Section 3.5, the LRT statistic for testing

$$H_0: \tau = 0 \qquad H_1: \tau \neq 0 \tag{3.10.2}$$

is

$$t = \frac{\bar{d}\sqrt{N}}{s_d} = \frac{(y_{1.} - y_{2.})\sqrt{N}}{s_d} \tag{3.10.3}$$

which is distributed as t with $N - 1$ degrees of freedom when H_0 is true.

Alternatively, the distance concept could be employed;

$$t = D\sqrt{N} \tag{3.10.4}$$

so that $t^2(N - 1) = F(1, N - 1)$.

If the test of (3.10.2) is performed at the significance level α, H_0 is rejected if

$$|t| > t^{\alpha/2}(N - 1) \tag{3.10.5}$$

A $100(1 - \alpha)\%$ confidence interval for the parameter τ is obtained by evaluating the expression

$$\frac{d - t_{(N-1)}^{\alpha/2} s_d}{\sqrt{N}} \leq \tau \leq \frac{\bar{d} + t_{(N-1)}^{\alpha/2} s_d}{\sqrt{N}} \tag{3.10.6}$$

The analogy between this problem and the one-sample unpaired t test should be clear; hence linear model details will not be given.

EXERCISES 3.10

1. Derive the LRT for testing (3.10.2).

3.11 PAIRED OBSERVATIONS—MULTIVARIATE

Instead of observing paired observations of one response variable, let us assume the existence of N p-vector paired observations. Represent the observations as follows:

$$\begin{matrix} y_{111} & y_{112} & \cdots & y_{11p} \\ y_{121} & y_{122} & \cdots & y_{12p} \\ \vdots & \vdots & \vdots & \vdots \\ y_{1N1} & y_{1N2} & \cdots & y_{1Np} \end{matrix} \qquad \begin{matrix} y_{211} & y_{212} & \cdots & y_{21p} \\ y_{221} & y_{222} & \cdots & y_{22p} \\ \vdots & \vdots & \vdots & \vdots \\ y_{2N1} & y_{2N2} & \cdots & y_{2Np} \end{matrix}$$

where the subscripts ijk denote the kth characteristic for the jth subject and the ith treatment.

As in the univariate case, differences are formed where the matrix of differences is denoted by

$$\Delta = \begin{bmatrix} d_{11} & d_{12} & \cdots & d_{1p} \\ d_{21} & d_{22} & \cdots & d_{2p} \\ \vdots & \vdots & \vdots & \vdots \\ d_{N1} & d_{N2} & \cdots & d_{Np} \end{bmatrix} \tag{3.11.1}$$

where $d_{ij} = y_{1jk} - y_{2jk}$, for $j = 1, \dots, N$ and $k = 1, \dots, p$. Similarly,

$$\mathbf{d}_i' = [d_{i1}, d_{i2}, \dots, d_{ip}] \sim N(\boldsymbol{\mu}_1 - \boldsymbol{\mu}_2, \boldsymbol{\Sigma}_d) \tag{3.11.2}$$

where

$$\boldsymbol{\tau} = \boldsymbol{\mu}_1 - \boldsymbol{\mu}_2 = \begin{bmatrix} \mu_{11} - \mu_{21} \\ \mu_{12} - \mu_{22} \\ \vdots \\ \mu_{1p} - \mu_{2p} \end{bmatrix}$$

and $\boldsymbol{\Sigma}_d$ is the variance-covariance matrix of the differences.

To test the hypotheses

$$\begin{aligned} H_0 &: \boldsymbol{\tau} = \mathbf{0} \\ H_1 &: \boldsymbol{\tau} \neq \mathbf{0} \end{aligned} \tag{3.11.4}$$

where $\boldsymbol{\tau}' = [\tau_1, \tau_2, \dots, \tau_p]$ and $\tau_i = \mu_{1i} - \mu_{2i}$, we shall consider the univariate statistic

$$t^2 = \frac{\bar{d}^2 N}{s_d^2} = N\bar{d}(s_d^2)^{-1}\bar{d} \tag{3.11.5}$$

Replacing scalars by vectors,

(3.11.6) $$T^2 = N\bar{\mathbf{d}}'\mathbf{S}_d^{-1}\bar{\mathbf{d}} = ND^2$$

where $\bar{\mathbf{d}} = [\bar{d}_1, \ldots, \bar{d}_p]$,

$$\bar{d}_j = \sum_{i=1}^{N} \frac{d_{ij}}{N}$$

and $\mathbf{S}_d$ is the sample variance-covariance matrix of the differences.

For T^2 to have an F distribution, we must weight the expression for T^2 by $(N - p)/(N - 1)p$, so that

(3.11.7) $$\frac{(N-p)T^2}{(N-1)p} = \frac{(N-p)ND^2}{(N-1)p} \sim F(p, N-p)$$

The test of H_0 is rejected at the significance level α if

(3.11.8) $$\frac{N-p}{p(N-1)} T^2 > F^\alpha(p, N-p)$$

or if $T^2 > T^\alpha(p, N-1)$. Results for the other test statistics follow if the form of T^2 given in (3.11.6) is substituted for T^2 as discussed in the one-sample problem.

If $T^\alpha(p, N - 1)$ is the upper α percentage point of the T^2 distribution, by analogy with the one-sample problem, $100(1 - \alpha)\%$ confidence intervals may be obtained for $\psi = \mathbf{a}'\boldsymbol{\tau}$ by employing the expression

(3.11.9) $$\mathbf{a}'\bar{\mathbf{d}} - c_0\sqrt{\frac{\mathbf{a}'\mathbf{S}_d\mathbf{a}}{N}} \le \mathbf{a}'\boldsymbol{\tau} \le \mathbf{a}'\bar{\mathbf{d}} + c_0\sqrt{\frac{\mathbf{a}'\mathbf{S}_d\mathbf{a}}{N}}$$

where $c_0^2 = T^\alpha(p, N - 1)$ and $\mathbf{a}$ is an arbitrary vector.

Again, the Bonferroni procedure could be applied to test these hypotheses.

3.12 EXAMPLE—MULTIVARIATE "MATCHED-PAIR" *T* TEST

In the study of motor skill performance under two conditions of body cooling with two types of hand protection, the data in Table 3.12.1, which reflect the accuracy

TABLE 3.12.1. Sample Data: Hotelling's Matched-Pair T^2.

Subject	Condition 1		Condition 2	
	y_{1j1}	y_{1j2}	y_{2j1}	y_{2j2}
1	33.5	6.0	19.8	5.0
2	52.8	65.0	30.0	5.5
3	52.0	37.0	21.9	17.9
4	56.9	14.0	28.9	2.0
5	40.9	6.2	43.8	8.1
6	72.0	78.4	64.3	49.0
7	55.6	31.0	47.2	20.0
8	64.4	10.0	36.4	3.0
9	72.8	68.0	50.5	37.9
10	62.7	19.0	47.4	39.0

of a complex training task, were obtained. To test the hypothesis

$$H_0: \boldsymbol{\tau} = \mathbf{0} \qquad H_1: \boldsymbol{\tau} \neq \mathbf{0} \tag{3.12.1}$$

or

$$H_0: \begin{bmatrix} \mu_{11} - \mu_{21} \\ \mu_{12} - \mu_{22} \end{bmatrix} = \begin{bmatrix} 0 \\ 0 \end{bmatrix}$$

the array of differences $\boldsymbol{\Delta}$ is formed:

$$\boldsymbol{\Delta} = \begin{bmatrix} 13.7 & 1.0 \\ 22.8 & 59.5 \\ 30.1 & 19.1 \\ 28.0 & 12.0 \\ -2.9 & -1.9 \\ 7.7 & 29.4 \\ 8.4 & 11.0 \\ 28.0 & 7.0 \\ 22.3 & 30.1 \\ 15.3 & -20.0 \end{bmatrix} \tag{3.12.2}$$

The mean vector $\bar{\mathbf{d}}$ and variance-covariance matrix $\mathbf{S}_d$ from (3.12.2) are

$$\bar{\mathbf{d}}' = [17.34, 14.72]$$

$$\mathbf{S}_d = \boldsymbol{\Delta}'[\mathbf{I} - \mathbf{X}(\mathbf{X}'\mathbf{X})^{-1}\mathbf{X}']\boldsymbol{\Delta} = \begin{bmatrix} 116.047 & \\ 68.254 & 469.695 \end{bmatrix}$$

and $\mathbf{X}$ is a matrix of N 1s. Hence

$$T^2 = N\bar{\mathbf{d}}'\mathbf{S}_d^{-1}\bar{\mathbf{d}} = 10[17.34, 14.72]\begin{bmatrix} .0094225 & -.0013692 \\ -.0013692 & .0023280 \end{bmatrix}\begin{bmatrix} 17.34 \\ 14.72 \end{bmatrix}$$

$$= 26.32 \tag{3.12.3}$$

Using a significance level $\alpha = .05$, the hypothesis is rejected if

$$T^2 > T^{\alpha}(p, N-1) = T^{.05}_{(2,9)} = 10.033 \tag{3.12.4}$$

Since $T^2 = 26.32 > 10.033$, the hypothesis is rejected. There is a difference between the two conditions of body cooling with two types of hand protection for the complex training task.

To determine which variables led to the rejection of the hypothesis, the expression

$$\mathbf{a}'\bar{\mathbf{d}} - c_0\sqrt{\frac{\mathbf{a}'\mathbf{S}_d\mathbf{a}}{N}} \leq \mathbf{a}'\boldsymbol{\tau} \leq \mathbf{a}'\bar{\mathbf{d}} + c_0\sqrt{\frac{\mathbf{a}'\mathbf{S}_d\mathbf{a}}{N}} \tag{3.12.5}$$

is evaluated, where $c_0^2 = 10.033$ or $c_0 = 3.1675$. Selecting $\mathbf{a}' = [1, 0]$, the confidence interval for $\tau_1 = \mu_{11} - \mu_{21}$ is obtained:

$$17.34 - 3.1675\sqrt{\frac{116.047}{10}} \le \mu_{11} - \mu_{21} \le 17.34 + 3.1675\sqrt{\frac{116.047}{10}}$$

$$17.34 - (3.1675)(3.4066) \le \mu_{11} - \mu_{21} \le 17.34 + (3.1675)(3.4066)$$

$$6.55 \le \mu_{11} - \mu_{21} \le 28.13 \tag{3.12.6}$$

Selecting $\mathbf{a}' = [0, 1]$ yields the confidence interval for $\tau_2 = \mu_{12} - \mu_{22}$. Therefore

$$14.72 - 3.1675\sqrt{\frac{469.695}{10}} \le \mu_{12} - \mu_{22} \le 14.72 + 3.1675\sqrt{\frac{469.695}{10}}$$

$$-6.99 \le \mu_{12} - \mu_{22} \le 36.42 \tag{3.12.7}$$

These intervals indicate that hand protection one, but not two, contributed significantly to the rejection of the hypothesis.

Observe that the form of (3.12.6) and (3.12.7) is

$$\hat{\psi}_i - \sqrt{T^\alpha(p, N-1)}\hat{\sigma}_{\hat{\psi}_i} \le \psi_i \le \hat{\psi}_i + \sqrt{T^\alpha(p, N-1)}\hat{\sigma}_{\hat{\psi}_i} \tag{3.12.8}$$

where $\psi = \mu_{1i} - \mu_{2i}, \hat{\sigma}_{\hat{\psi}_1} = 3.415$, and $\hat{\sigma}_{\hat{\psi}_2} = 6.8486$.

To determine Bonferroni intervals, the constant multiplier $c_0 = \sqrt{T^\alpha(p, N-1)} = 3.1675$ in (3.12.8) becomes $c_0 = t_{(N-1)}^{\alpha/2p} = 2.57$, for $p = 2$. Since these intervals are shorter, this again illustrates that when interest is restricted to individual variable comparisons, the Bonferroni method should be employed.

EXERCISES 3.12

1. After matching subjects according to age, education, former language training, intelligence, and language aptitude, Postovsky (1970) investigated the effects of delay in oral practice at the beginning of second-language learning. Using an experimental condition with a 4-week delay in oral practice and a control condition with no delay, evaluation was carried out for language skills: listening (L), speaking (S), reading (R), and writing (W). The data for a comprehensive examination given at the end of the first 6 weeks follows.

	Experimental Group				Control Group			
Subject	L	S	R	W	L	S	R	W
1	34	66	39	97	33	56	36	81
2	35	60	39	95	21	39	33	74
3	32	57	39	94	29	47	35	89
4	29	53	39	97	22	42	34	85
5	37	58	40	96	39	61	40	97
6	35	57	34	90	34	58	38	94
7	34	51	37	84	29	38	34	76
8	25	42	37	80	31	42	38	83
9	29	52	37	85	18	35	28	58
10	25	47	37	94	36	51	36	83
11	34	55	35	88	25	45	36	67
12	24	42	35	88	33	43	36	86
13	25	59	32	82	29	50	37	94

Subject	Experimental Group L	S	R	W	Control Group L	S	R	W
14	34	57	35	89	30	50	34	84
15	35	57	39	97	34	49	38	94
16	29	41	36	82	30	42	34	77
17	25	44	30	65	25	47	36	66
18	28	51	39	96	32	37	38	88
19	25	42	38	86	22	44	22	85
20	30	43	38	91	30	35	35	77
21	27	50	39	96	34	45	38	95
22	25	46	38	85	31	50	37	96
23	22	33	27	72	21	36	19	43
24	19	30	35	77	26	42	33	73
25	26	45	37	90	30	49	36	88
26	27	38	33	77	23	37	36	82
27	30	36	22	62	21	43	30	85
28	36	50	39	92	30	45	34	70

a. Test the hypothesis of no difference between experimental and control conditions using $\alpha = .05$.

b. Summarize your findings.

3.13 ONE-SAMPLE PROFILE ANALYSIS—MULTIVARIATE

An experimenter frequently finds that instead of observing two responses for a given subject, the same subject is exposed to p response situations at successive times. This type of experiment is common in learning theory research; the response observations might be trials to criteria for p exposures to a learning device. The measurements taken at the p response conditions are assumed to be in the same units.

Let the observations be represented as follows:

Subject	Conditions 1	2	$\cdots$	p
1	y_{11}	y_{12}	$\cdots$	y_{1p}
2	y_{21}	y_{22}	$\cdots$	y_{2p}
$\vdots$	$\vdots$	$\vdots$	$\vdots$	$\vdots$
N	y_{N1}	y_{N2}	$\cdots$	y_{Np}

where $\mathbf{Y}_i' = [y_{i1}, y_{i2}, \ldots, y_{ip}] \sim IN_p(\boldsymbol{\mu}, \boldsymbol{\Sigma})$, and $\boldsymbol{\mu}' = [\mu_1, \ldots, \mu_p]$.

Employing the general linear model, the observations are represented as

$$\underset{(N \times p)}{\mathbf{Y}} = \underset{(N \times 1)}{\mathbf{X}} \underset{(1 \times p)}{\boldsymbol{\mu}'} + \underset{(N \times p)}{\mathbf{E}_0} \tag{3.13.1}$$

$$\begin{bmatrix} y_{11} & y_{12} & \cdots & y_{1p} \\ y_{21} & y_{22} & \cdots & y_{2p} \\ \vdots & \vdots & \vdots & \vdots \\ y_{N1} & y_{N2} & \cdots & y_{Np} \end{bmatrix} = \begin{bmatrix} 1 \\ 1 \\ \vdots \\ 1 \end{bmatrix} [\mu_1, \mu_2, \ldots, \mu_p] + \begin{bmatrix} \varepsilon_{11} & \varepsilon_{12} & \cdots & \varepsilon_{1p} \\ \varepsilon_{21} & \varepsilon_{22} & \cdots & \varepsilon_{2p} \\ \vdots & \vdots & \vdots & \vdots \\ \varepsilon_{N1} & \varepsilon_{N2} & \cdots & \varepsilon_{Np} \end{bmatrix}$$

where

$$E(\mathbf{Y}) = \mathbf{X}\boldsymbol{\mu}' \qquad (3.13.2)$$
$$V(\mathbf{Y}) = \mathbf{I} \otimes \boldsymbol{\Sigma}$$

The null hypothesis of equal-response effects is

$$H_0: \mu_1 = \mu_2 = \cdots = \mu_p \qquad (3.13.3)$$
$$H_1: H_0 \text{ is false}$$

Using (3.6.19), where $H_0: \mathbf{CBA} = \boldsymbol{\Gamma}$, (3.13.3) is transformed to

$$\underset{(1\times 1)}{\mathbf{C}} = [1] \qquad \underset{(1\times p)}{\mathbf{B}} = [\mu_1, \ldots, \mu_p] \qquad \underset{1\times(p-1)}{\boldsymbol{\Gamma}} = [0, \ldots, 0]$$

where

$$\underset{p\times(p-1)}{\mathbf{A}} = \begin{bmatrix} 1 & 0 & \cdots & 0 \\ 0 & 1 & \cdots & 0 \\ \vdots & \vdots & \vdots\vdots\vdots & \vdots \\ 0 & 0 & \cdots & 1 \\ -1 & -1 & \cdots & -1 \end{bmatrix}$$

Simplifying,

$$\begin{matrix} \mathbf{C} & \mathbf{B} & \mathbf{A} & = & \boldsymbol{\Gamma} \end{matrix} \qquad (3.13.4)$$

$$[1]\,[\mu_1, \ldots, \mu_p] \begin{bmatrix} 1 & 0 & \cdots & 0 \\ 0 & 1 & \cdots & 0 \\ \vdots & \vdots & \vdots\vdots\vdots & \vdots \\ 0 & 0 & \cdots & 1 \\ -1 & -1 & \cdots & -1 \end{bmatrix} = [0, \ldots, 0]$$

reduces to

$$H_0: \begin{bmatrix} \mu_1 & -\mu_p \\ \mu_2 & -\mu_p \\ \vdots & \vdots \\ \mu_{p-1} & -\mu_p \end{bmatrix} = \begin{bmatrix} 0 \\ 0 \\ \vdots \\ 0 \end{bmatrix} \qquad (3.13.5)$$

That each element of the hypothesis is estimable follows from the fact that $\mathbf{X}$ is of full rank.

To test (3.13.5), the general multivariate fundamental least-squares theorem (3.6.19) is employed. Thus

$$\underset{[(p-1)\times(p-1)]}{\mathbf{Q}_e} = \mathbf{A}'\mathbf{Y}'[\mathbf{I} - \mathbf{X}(\mathbf{X}'\mathbf{X})^{-1}\mathbf{X}']\mathbf{Y}\mathbf{A}$$
$$\underset{[(p-1)\times(p-1)]}{\mathbf{Q}_h} = (\mathbf{C}\hat{\mathbf{B}}\mathbf{A} - \boldsymbol{\Gamma})'[\mathbf{C}(\mathbf{X}'\mathbf{X})^{-1}\mathbf{C}']^{-1}(\mathbf{C}\hat{\mathbf{B}}\mathbf{A} - \boldsymbol{\Gamma}) \qquad (3.13.6)$$
$$= \mathbf{A}'\mathbf{Y}'\mathbf{X}(\mathbf{X}'\mathbf{X})^{-1}\mathbf{X}'\mathbf{X}(\mathbf{X}'\mathbf{X})^{-1}\mathbf{X}'\mathbf{Y}\mathbf{A}$$

where $\hat{\mathbf{B}} = (\mathbf{X}'\mathbf{X})^{-1}\mathbf{X}'\mathbf{Y}\mathbf{A}$ and, if H_0 is true,

$$\Lambda = \frac{|\mathbf{Q}_e|}{|\mathbf{Q}_e + \mathbf{Q}_h|} \sim U(p-1, 1, N-1) \qquad (3.13.7)$$

H_0 is rejected if

(3.13.8) $$\Lambda < U^{\alpha}(p-1, 1, N-1)$$

or if

$$\frac{n+1}{m+1}\frac{1-\Lambda}{\Lambda} = \frac{1-\Lambda}{\Lambda}\frac{N-p+1}{p-1} > F^{\alpha}(p-1, N-p+1)$$

since $s = \min(g = v_h, u) = \min(1, p-1) = 1$.

The product defined by $\mathbf{Q}_h$ may be simplified by noticing that

(3.13.9) $$\mathbf{A}'\mathbf{Y}'\mathbf{X}(\mathbf{X}'\mathbf{X})^{-1} = \bar{\mathbf{y}}_d$$

where the ith element of $\bar{\mathbf{y}}_d$ is $\bar{y}_i^* = y_{.i} - y_{.p}$. The matrix $\mathbf{Q}_h$ has the form

(3.13.10) $$\mathbf{Q}_h = N\bar{\mathbf{y}}_d\bar{\mathbf{y}}_d'$$

The matrix $\mathbf{Q}_e$ is the sum of squares and cross-products matrix of the differences $y_{ij}^* = y_{ij} - y_{ip}$, for $i = 1, \ldots, N$ and $j = 1, \ldots, p-1$.

From these relationships,

(3.13.11) $$(N-1)\left(\frac{|\mathbf{Q}_e + \mathbf{Q}_h|}{|\mathbf{Q}_e|} - 1\right) = N\bar{\mathbf{y}}_d'\mathbf{S}_d^{-1}\bar{\mathbf{y}}_d = T^2$$

where $\mathbf{S}_d^{-1}$ is the $(p-1) \times (p-1)$ variance-covariance matrix of the transformed differences y_{ij}^*.

With this as motivation, and from the fact that $s = \min(g, u) = 1$, the relationships that exist among the various test statistics to test H_0 are noted. In summary, $H_0\colon \mu_1 = \mu_2 = \cdots = \mu_p$ is rejected at the significance level α if

(3.13.12) (a) Hotelling:

$$T^2 = N\bar{\mathbf{y}}_d'\mathbf{S}_d^{-1}\bar{\mathbf{y}}_d > T^{\alpha}(p-1, N-1)$$

(b) Wilks:

$$\Lambda = \frac{|\mathbf{Q}_e|}{|\mathbf{Q}_e + \mathbf{Q}_h|} < U^{\alpha}(p-1, 1, N-1)$$

(c) Roy:

$$\theta_s = \frac{\lambda_1}{1+\lambda_1} > \theta^{\alpha}(s, m, n)$$

where

$$m = \frac{|p-2| - 1}{2} \quad \text{and} \quad n = \frac{N-p-1}{2}$$

(d) Lawley-Hotelling:

$$U^{(s)} = \lambda_1 > U_0^{\alpha}(s, m, n)$$

To obtain confidence intervals, the statistic T^2, where

(3.13.13) $$T^2 = N\bar{\mathbf{y}}_d'\mathbf{S}_d^{-1}\bar{\mathbf{y}}_d = N\bar{\mathbf{y}}'\mathbf{A}(\mathbf{A}'\mathbf{S}\mathbf{A})^{-1}\mathbf{A}'\bar{\mathbf{y}}$$

is considered, where $\bar{\mathbf{y}}$ is the mean vector for the original variables and $\mathbf{S}$ is the variance-covariance matrix of the original variables. It may be shown, by using (3.3.27) for arbitrary vectors $\mathbf{a}$, that

$$T^2 = N \max_{\mathbf{a}} \frac{[\mathbf{a}'(\mathbf{A}'\bar{\mathbf{y}} - \mathbf{A}'\boldsymbol{\mu})]^2}{\mathbf{a}'\mathbf{A}'\mathbf{S}\mathbf{A}\mathbf{a}}$$

$$= N \max_{\mathbf{a}} \frac{[\mathbf{a}'\mathbf{A}'(\bar{\mathbf{y}} - \boldsymbol{\mu})]^2}{\mathbf{a}'\mathbf{A}'\mathbf{S}\mathbf{A}\mathbf{a}} \tag{3.13.14}$$

By setting $\mathbf{c}' = \mathbf{a}'\mathbf{A}'$, $\psi = \mathbf{c}'\boldsymbol{\mu} = \mathbf{a}'\mathbf{A}'\boldsymbol{\mu}$ is a contrast in the elements of $\boldsymbol{\mu}$, and $100(1 - \alpha)\%$ simultaneous confidence intervals are immediately obtained:

$$\mathbf{a}'\mathbf{A}'\bar{\mathbf{y}} - c_0\sqrt{\frac{\mathbf{a}'\mathbf{A}'\mathbf{S}\mathbf{A}\mathbf{a}}{N}} \le \mathbf{a}'\mathbf{A}'\boldsymbol{\mu} \le \mathbf{a}'\mathbf{A}'\bar{\mathbf{y}} + c_0\sqrt{\frac{\mathbf{a}'\mathbf{A}'\mathbf{S}\mathbf{A}\mathbf{a}}{N}} \tag{3.13.15}$$

or, if $\mathbf{c}' = \mathbf{a}'\mathbf{A}'$ and $\psi = \mathbf{c}'\boldsymbol{\mu}$,

$$\hat{\psi} - c_0\sqrt{\frac{\mathbf{c}'\mathbf{S}\mathbf{c}}{N}} \le \psi \le \hat{\psi} + c_0\sqrt{\frac{\mathbf{c}'\mathbf{S}\mathbf{c}}{N}} \tag{3.13.16}$$

where $c_0^2 = T^\alpha(p - 1, N - 1)$ and $\hat{\psi} = \mathbf{c}'\bar{\mathbf{y}}$.

The transformation matrix $\mathbf{A}$ used to test (3.13.3) is, of course, not unique; other matrices of rank $p - 1$ can be employed. For example,

$$\mathbf{A}_1 = \begin{bmatrix} 1 & 0 & \cdots & 0 \\ -1 & 1 & \cdots & 0 \\ 0 & -1 & \cdots & 0 \\ \vdots & \vdots & \vdots & \vdots \\ 0 & 0 & \cdots & 1 \\ 0 & 0 & \cdots & -1 \end{bmatrix} \quad \text{or} \quad \mathbf{A}_2 = \begin{bmatrix} -1 & 0 & \cdots & 0 \\ 0 & -1 & \cdots & 0 \\ \vdots & \vdots & \vdots & \vdots \\ 0 & 0 & \cdots & -1 \\ 1 & 1 & \cdots & 1 \end{bmatrix}$$

However, the definitions given by $\bar{y}_i^*$ and y_{ij}^* would have to be modified appropriately. To test for trends in the means, the matrix $\mathbf{A}$ would consist of coefficients from orthogonal polynomials. Trend analysis is discussed in Chapter 5. For this situation, the p response conditions must have a natural order.

3.14 EXAMPLE—ONE-SAMPLE PROFILE ANALYSIS

As an example illustrating the techniques of Section 3.13, partial data—provided by Dr. Paul Ammon of the University of California at Berkeley—are employed to show whether a subject's immediate memory of a sentence is organized according to the sentence's phrase structure. These data are given in Table 3.14.1. The experiment was set up as follows.

> *Procedure:* Subjects listened to tape-recorded sentences. Each sentence was followed by a "probe word" taken from one of five positions in the sentence. The subject was to respond with the word that came immediately after the probe word in the sentence.

Example Statement: The tall man met the young girl who got the new hat.
1 2 3 4 5
Dependent Variable: Speed of response (transformed reaction time).

TABLE 3.14.1. Sample Data: One-Sample Profile Analysis.

	Probe-Word Positions				
Subject	1	2	3	4	5
1	51	36	50	35	42
2	27	20	26	17	27
3	37	22	41	37	30
4	42	36	32	34	27
5	27	18	33	14	29
6	43	32	43	35	40
7	41	22	36	25	38
8	38	21	31	20	16
9	36	23	27	25	28
10	26	31	31	32	36
11	29	20	25	26	25

The hypothesis is that the mean reaction time is the same for the five probe positions. To test this we use

$$\begin{aligned} &H_0\colon \mu_1 = \mu_2 = \mu_3 = \mu_4 = \mu_5 \\ &H_1\colon H_0 \text{ is false} \end{aligned} \tag{3.14.1}$$

The transformation matrix **A**, where

$$\mathbf{A} = \begin{bmatrix} 1 & 0 & 0 & 0 \\ 0 & 1 & 0 & 0 \\ 0 & 0 & 1 & 0 \\ 0 & 0 & 0 & 1 \\ -1 & -1 & -1 & -1 \end{bmatrix} \tag{3.14.2}$$

is employed. The equivalent form of H_0 is

$$H_0\colon \begin{bmatrix} \mu_1 - \mu_5 \\ \mu_2 - \mu_5 \\ \mu_3 - \mu_5 \\ \mu_4 - \mu_5 \end{bmatrix} = \begin{bmatrix} 0 \\ 0 \\ 0 \\ 0 \end{bmatrix}$$

To estimate **B**, employ the formula

$$\hat{\mathbf{B}} = (\mathbf{X}'\mathbf{X})^{-1}\mathbf{X}'\mathbf{Y}\mathbf{A} = \begin{bmatrix} \bar{y}_1^* \\ \bar{y}_2^* \\ \bar{y}_3^* \\ \bar{y}_4^* \end{bmatrix} = \begin{bmatrix} y_{.1} - y_{.5} \\ y_{.2} - y_{.5} \\ y_{.3} - y_{.5} \\ y_{.4} - y_{.5} \end{bmatrix}$$

$$= \begin{bmatrix} 36.09091 - 30.72727 \\ 25.54545 - 30.72727 \\ 34.09091 - 30.72727 \\ 27.27273 - 30.72727 \end{bmatrix}$$

(3.14.3)
$$= \begin{bmatrix} 5.36364 \\ -5.18182 \\ 3.36364 \\ -3.45455 \end{bmatrix}$$

so that $\bar{\mathbf{y}}_d = \hat{\mathbf{B}}$. Thus, to compute $\mathbf{Q}_h$,

(3.14.4)
$$\mathbf{Q}_h = N\bar{\mathbf{y}}_d\bar{\mathbf{y}}_d' = \begin{bmatrix} 316.4545 & & & \\ -305.7273 & 295.3636 & & \\ 198.4546 & -191.7273 & 124.4546 & \\ -203.8182 & 196.9091 & -127.8182 & 131.2727 \end{bmatrix}$$

From (3.13.6),

$$\mathbf{Q}_e = \mathbf{A}'\mathbf{Y}[\mathbf{I} - \mathbf{X}(\mathbf{X}'\mathbf{X})^{-1}\mathbf{X}']\mathbf{Y}\mathbf{A}$$

(3.14.5)
$$= \begin{bmatrix} 724.5455 & & & \\ 380.7273 & 475.6364 & & \\ 392.5455 & 176.7273 & 366.5455 & \\ 378.8182 & 385.0909 & 227.8182 & 576.7273 \end{bmatrix}$$

so that

$$\begin{aligned} T^2 &= (N-1)\left(\frac{|\mathbf{Q}_e + \mathbf{Q}_h|}{|\mathbf{Q}_e|} - 1\right) \\ &= 10\left(\frac{29{,}224{,}701{,}600.0000}{7{,}254{,}315{,}351.2727} - 1\right) \\ &= 30.286 \end{aligned}$$

Using a significance level of $\alpha = .05$, the hypothesis H_0 is rejected if

(3.14.6)
$$T^2 > T^{\alpha}(p-1, N-1) = T^{.05}_{(4,10)} = 23.545$$

Since $T^2 = 30.286$, H_0 is rejected.

To locate differences that might have led to the rejection of the null hypothesis, the expression

(3.14.7)
$$\mathbf{c}'\bar{\mathbf{y}} - c_0\sqrt{\frac{\mathbf{c}'\mathbf{S}\mathbf{c}}{N}} \leq \mathbf{c}'\boldsymbol{\mu} \leq \mathbf{c}'\bar{\mathbf{y}} + c_0\sqrt{\frac{\mathbf{c}'\mathbf{S}\mathbf{c}}{N}}$$

or

$$\hat{\psi} - c_0\hat{\sigma}_{\hat{\psi}} \leq \psi \leq \hat{\psi} + c_0\hat{\sigma}_{\hat{\psi}}$$

where $\mathbf{c}$ is a contrast vector such that the sum of the elements of $\mathbf{c}$ is 0, is employed with $c_0^2 = 23.545$ or $c_0 = 4.8523$. However, $\mathbf{c}'\mathbf{S}\mathbf{c} = \mathbf{a}'\mathbf{A}'\mathbf{S}\mathbf{A}\mathbf{a} = \mathbf{a}'\mathbf{S}_d\mathbf{a}$, where $\mathbf{a}$ is an arbitrary vector. Letting $\mathbf{A}$ be defined by (3.14.2) and $\mathbf{a}$ be a vector of 0s except for a 1 in the ith position, the standard errors of the contrasts $\psi_i = \mu_i - \mu_p$ are immediately obtained from the ith diagonal elements of $\mathbf{S}_d$. Thus, from $\mathbf{B}$ and $\hat{\mathbf{B}}$,

$$\begin{aligned}
\psi_1 &= \mu_1 - \mu_5 & \hat{\psi}_1 &= 5.3636 \\
\psi_2 &= \mu_2 - \mu_5 & \hat{\psi}_2 &= -5.1818 \\
\psi_3 &= \mu_3 - \mu_5 & \hat{\psi}_3 &= 3.3636 \\
\psi_4 &= \mu_4 - \mu_5 & \hat{\psi}_4 &= -3.4546
\end{aligned}$$

and from $\mathbf{Q}_e$, $\mathbf{S}_d = \mathbf{Q}_e/(N - 1)$, so that

$$\hat{\sigma}_{\hat{\psi}_1} = 2.56647 = \sqrt{\frac{72.4546}{11}}$$

$$\hat{\sigma}_{\hat{\psi}_2} = 2.07942 = \sqrt{\frac{47.5636}{11}}$$

$$\hat{\sigma}_{\hat{\psi}_3} = 1.82544 = \sqrt{\frac{36.6546}{11}}$$

$$\hat{\sigma}_{\hat{\psi}_4} = 2.28975 = \sqrt{\frac{57.6727}{11}}$$

By substituting these into (3.14.7), the following simultaneous intervals are obtained:

$$\begin{aligned}
-\ 7.09 &\leq \mu_1 - \mu_5 \leq 17.82 & \text{(N.S.)} \\
-15.27 &\leq \mu_2 - \mu_5 \leq 4.91 & \text{(N.S.)} \\
-\ 5.63 &\leq \mu_3 - \mu_5 \leq 12.36 & \text{(N.S.)} \\
-14.57 &\leq \mu_4 - \mu_5 \leq 7.66 & \text{(N.S.)}
\end{aligned}$$

Thus contrasts obtained immediately from $\mathbf{S}_d$ are not significant (N.S.). Investigating some others as examples, consider

$$\begin{aligned}
\psi_5 &= \psi_1 - \psi_3 & \hat{\psi}_5 &= y_{.1} - y_{.3} = 2.0000 \\
\psi_6 &= \psi_2 - \psi_4 & \hat{\psi}_6 &= y_{.2} - y_{.4} = -1.7272 \\
\psi_7 &= \psi_1 - \psi_2 & \hat{\psi}_7 &= y_{.1} - y_{.2} = 10.5454 \\
\psi_8 &= \psi_3 - \psi_4 & \hat{\psi}_8 &= y_{.3} - y_{.4} = 6.8182 \\
\psi_9 &= \frac{1}{2}(\mu_1 - \mu_2) + \frac{1}{2}(\mu_3 - \mu_4) & \hat{\psi}_9 &= \frac{\hat{\psi}_7 + \hat{\psi}_8}{2} = 8.6818
\end{aligned}$$

Finding $\hat{\sigma}_{\hat{\psi}}$ for the contrasts defined by ψ_i, for $i = 5, \ldots, 9$, the following procedure is employed. As an example, to find $\hat{\sigma}_{\hat{\psi}_5}$, $\mathbf{a}' = [1, 0, -1, 0]$, so that

$$\hat{\sigma}_{\hat{\psi}_5} = \sqrt{\frac{\mathbf{a}'\mathbf{S}_d\mathbf{a}}{N}}$$

$$= \sqrt{\frac{72.4546 + 36.6564 - 2(39.2546)}{11}}$$

$$= \sqrt{2.78182}$$

$$= 1.6679$$

In a similar manner,

$$\hat{\sigma}_{\hat{\psi}_6} = 1.6017 \qquad \hat{\sigma}_{\hat{\psi}_8} = 2.1055$$

$$\hat{\sigma}_{\hat{\psi}_7} = 1.9971 \qquad \hat{\sigma}_{\hat{\psi}_9} = 1.9997$$

By substituting $\hat{\psi}$'s and $\hat{\sigma}_{\hat{\psi}}$'s into (3.14.7), the confidence intervals are obtained:

$$-6.09 \leq \mu_1 - \mu_3 \leq 10.09 \qquad \text{(N.S.)}$$

$$-9.50 \leq \mu_2 - \mu_4 \leq 6.04 \qquad \text{(N.S.)}$$

$$.86 \leq \mu_1 - \mu_2 \leq 20.24 \qquad \text{(Sig.)}$$

$$-3.40 \leq \mu_3 - \mu_4 \leq 17.03 \qquad \text{(N.S.)}$$

and

$$-1.02 \leq \frac{1}{2}(\mu_1 - \mu_2) + \frac{1}{2}(\mu_3 - \mu_4) \leq 18.39 \qquad \text{(N.S.)}$$

It may be concluded that the mean rate of response differs with respect to probe positions in the sentences. However, among the contrasts investigated, only the difference between positions 1 and 2 was significant.

EXERCISES 3.14

1. Kirk (1968, p. 133), in an experiment involving eight helicopter pilots with 500 to 3000 flying hours, investigated the pilots' ability in reading an altimeter at low altitudes. Using the amount of reading error as the dependent variable, each subject made 100 readings with four experimental altimeters. The average of each pilot's reading errors is recorded below.

	Altimeters			
Subject	1	2	3	4
1	3	7	4	7
2	6	8	5	8
3	3	7	4	9
4	3	6	3	8
5	1	5	2	10
6	2	6	3	10
7	2	5	4	9
8	2	6	3	11

a. Use Hotelling's statistic to determine whether the mean performance on the four altimeters is significantly different for $\alpha = .05$.

b. Altimeters 2 and 4 are more expensive than 1 and 3. Compare the pilot's performances on 2 and 4 with their performances on 1 and 3 following the overall test and using the Bonferroni method.

c. Summarize your findings.

3.15 TWO-SAMPLE PROFILE ANALYSIS—MULTIVARIATE

One of the most frequently encountered designs in education and psychology is the two-sample, repeated-measures design. Instead of observing p responses for one group of subjects, two independent samples of individuals are compared.

Letting $\mathbf{y}'_{ij} = [y_{ij1}, y_{ij2}, \ldots, y_{ijp}]$ denote the observation vector of the ith group and the jth subject within that group, represent the observations as in Table 3.15.1.

The observation vectors $\mathbf{Y}_{ij}$ are independent and normally distributed with mean $\boldsymbol{\mu}_i$ and common variance-covariance matrix $\boldsymbol{\Sigma}$, so that

$$\mathbf{Y}_{ij} \sim IN_p(\boldsymbol{\mu}_i, \boldsymbol{\Sigma}) \tag{3.15.1}$$

By employing the general linear model, the natural representation of the observation vectors is

$$\underset{(N \times p)}{\mathbf{Y}} = \underset{(N \times 2)}{\mathbf{X}} \quad \underset{(2 \times p)}{\mathbf{B}} + \underset{(N \times p)}{\mathbf{E}_0} \tag{3.15.2}$$

$$\begin{bmatrix} \mathbf{y}'_{11} \\ \mathbf{y}'_{12} \\ \vdots \\ \mathbf{y}'_{1N_1} \\ \mathbf{y}'_{21} \\ \mathbf{y}'_{22} \\ \vdots \\ \mathbf{y}'_{2N_2} \end{bmatrix} = \begin{bmatrix} 1 & 0 \\ 1 & 0 \\ \vdots & \vdots \\ 1 & 0 \\ 0 & 1 \\ 0 & 1 \\ \vdots & \vdots \\ 0 & 1 \end{bmatrix} \begin{bmatrix} \mu_{11} & \mu_{12} & \cdots & \mu_{1p} \\ \mu_{21} & \mu_{22} & \cdots & \mu_{2p} \end{bmatrix} + \begin{bmatrix} \boldsymbol{\varepsilon}'_{11} \\ \boldsymbol{\varepsilon}'_{12} \\ \vdots \\ \boldsymbol{\varepsilon}'_{1N_1} \\ \boldsymbol{\varepsilon}'_{21} \\ \boldsymbol{\varepsilon}'_{22} \\ \vdots \\ \boldsymbol{\varepsilon}'_{2N_2} \end{bmatrix}$$

Although many authors (for example, Greenhouse and Geisser, 1959; Kirk, 1968, Chap. 8; Myers, 1966, Chap. 7; Scheffé, 1959, Chap. 8; and Winer, 1971, Chap. 7) treat this design as specified by (3.15.1) and (3.15.2) from a univariate viewpoint, by adding the oftentimes unreasonable assumption of compound symmetry to $\boldsymbol{\Sigma}$ (see Huynh and Feldt, 1970), it is more natural to analyze the design by multivariate methods (see Bock, 1963, and Koch, 1969). This section will be addressed to multivariate considerations; in Chapter 5 the analogy between the univariate and multivariate approach will be made more explicit.

The repeated-measures design is applied in two important situations in psychology and education. The first is in *profile analysis* where the two independent groups receive the same set of p psychological tests, and the second is in growth-curve experiments. In the first case, it is important that the tests are scores in comparable

TABLE 3.15.1. Data Representation for Two-Sample, Repeated-Measures Design.

	Conditions			
Group 1	$\mathbf{y}'_{11} = y_{111}$	y_{112}	$\cdots$	y_{11p}
	$\mathbf{y}'_{12} = y_{121}$	y_{122}	$\cdots$	y_{12p}
	$\vdots$	$\vdots$		$\vdots$
	$\mathbf{y}'_{1N_1} = y_{1N_11}$	y_{1N_12}	$\cdots$	y_{1N_1p}
Means	$y_{1.1}$	$y_{1.2}$	$\cdots$	$y_{1.p}$
Group 2	$\mathbf{y}'_{21} = y_{211}$	y_{212}	$\cdots$	y_{21p}
	$\mathbf{y}'_{22} = y_{221}$	y_{222}	$\cdots$	y_{22p}
	$\vdots$	$\vdots$		$\vdots$
	$\mathbf{y}'_{2N_2} = y_{2N_21}$	y_{2N_22}	$\cdots$	y_{2N_2p}
Means	$y_{2.1}$	$y_{2.2}$	$\cdots$	$y_{2.p}$

units, whereas, in the second case, subjects are measured repeatedly at p distant *ordered time points.*

The primary hypotheses of interest in profile analysis, where the "repeated" measures usually have no natural order, are as follows:

H_{01}: Are the profiles for the two groups parallel?
H_{02}: Are there differences among conditions?
H_{03}: Are there significant differences between groups?

The first hypothesis tested in this design is that of *parallelism* or the group-by-condition interaction hypothesis H_{01}. The acceptance or rejection of this hypothesis may affect how H_{02} is tested. To aid in determining whether the parallelism hypothesis is satisfied, plots of the mean vectors for each group should be made. From this it is seen that parallelism follows if the slopes of each segment are the same for each group. That is, the test of parallelism of profiles in terms of the parameters is

$$H_{01}: \begin{bmatrix} \mu_{11} - \mu_{12} \\ \mu_{12} - \mu_{13} \\ \vdots \\ \mu_{1(p-1)} - \mu_{1p} \end{bmatrix} = \begin{bmatrix} \mu_{21} - \mu_{22} \\ \mu_{22} - \mu_{23} \\ \vdots \\ \mu_{2(p-1)} - \mu_{2p} \end{bmatrix} \tag{3.15.3}$$

Using the notation $\mathbf{CBA} = \mathbf{\Gamma}$ to state H_{01}, (3.15.3) becomes

$$\underset{(1\times 2)}{\mathbf{C}} \quad \underset{(2\times p)}{\mathbf{B}} \quad \underset{[(p\times(p-1)]}{\mathbf{A}} = \underset{[1\times(p-1)]}{\mathbf{\Gamma}} \tag{3.15.4}$$

$$[1, -1]\begin{bmatrix} \mu_{11} & \cdots & \mu_{1p} \\ \mu_{21} & \cdots & \mu_{2p} \end{bmatrix}\begin{bmatrix} 1 & 0 & \cdots & 0 & 0 \\ -1 & 1 & \cdots & 0 & 0 \\ 0 & -1 & \cdots & 0 & 0 \\ \vdots & \vdots & & \vdots & \vdots \\ 0 & 0 & \cdots & -1 & 1 \\ 0 & 0 & \cdots & 0 & -1 \end{bmatrix} = [0, \ldots, 0]$$

Since the design matrix $\mathbf{X}$ is of full rank, $(\mathbf{X}'\mathbf{X})^{-1}$ exists, so that

(3.15.5) $$\hat{\mathbf{B}} = (\mathbf{X}'\mathbf{X})^{-1}\mathbf{X}'\mathbf{Y} = \begin{bmatrix} y_{1.1} & y_{1.2} & \cdots & y_{1.p} \\ y_{2.1} & y_{2.2} & \cdots & y_{2.p} \end{bmatrix} = \begin{bmatrix} \mathbf{y}'_{1.} \\ \mathbf{y}'_{2.} \end{bmatrix}$$

and

(3.15.6) $$(\mathbf{C}\hat{\mathbf{B}}\mathbf{A})' = \mathbf{y}^*_{1.} - \mathbf{y}^*_{2.})$$

where

$$\mathbf{y}^{*\prime}_{i.} = [y_{i.1} - y_{i.2}, \cdots, y_{i.(p-1)} - y_{i.p}]$$

Thus the matrix products $\mathbf{Q}_e$ and $\mathbf{Q}_h$ used to test H_{01} are:

$$\mathbf{Q}_e = \mathbf{A}'\mathbf{Y}'[\mathbf{I} - \mathbf{X}(\mathbf{X}'\mathbf{X})^{-1}\mathbf{X}']\mathbf{Y}\mathbf{A}$$

$$\mathbf{Q}_h = (\mathbf{C}\hat{\mathbf{B}}\mathbf{A})'[\mathbf{C}(\mathbf{X}'\mathbf{X})^{-1}\mathbf{C}']^{-1}(\mathbf{C}\hat{\mathbf{B}}\mathbf{A})$$

(3.15.7) $$= \frac{N_1N_2}{N_1 + N_2}(\mathbf{y}^*_{1.} - \mathbf{y}^*_{2.})(\mathbf{y}^*_{1.} - \mathbf{y}^*_{2.})'$$

$$= \frac{N_1N_2}{N_1 + N_2}\bar{\mathbf{y}}_d\bar{\mathbf{y}}'_d$$

For Wilks' Λ criterion,

$$\Lambda = \frac{|\mathbf{Q}_e|}{|\mathbf{Q}_e + \mathbf{Q}_h|} \sim U(p-1, 1, N_1 + N_2 - 2)$$

The hypothesis H_{01} is rejected at the significance level α if

(3.15.8) $$\Lambda < U^\alpha(p-1, 1, N_1 + N_2 - 2)$$

Alternatively,

$$T^2 = (N_1 + N_2 - 2)\left\{\frac{|\mathbf{Q}_e + [N_1N_2/(N_1 + N_2)](\mathbf{y}^*_{1.} - \mathbf{y}^*_{2.})(\mathbf{y}^*_{1.} - \mathbf{y}^*_{2.})'|}{|\mathbf{Q}_e|} - 1\right\}$$

(3.15.9)

or

$$T^2 = \frac{N_1N_2}{N_1 + N_2}\bar{\mathbf{y}}'_d\mathbf{S}_d^{-1}\bar{\mathbf{y}}_d \sim T^2(p-1, N_1 + N_2 - 2)$$

where $\mathbf{S}_d = \mathbf{Q}_e/(N_1 + N_2 - 2)$. Hence H_{01} is rejected if

(3.15.10) $$T^2 > T^\alpha(p-1, N_1 + N_2 - 2)$$

at the significance level α.

In terms of the original observations and sample variance-covariance matrix $\mathbf{S}$, defined by

(3.15.11) $$\mathbf{S} = \frac{\mathbf{Y}'[\mathbf{I} - \mathbf{X}(\mathbf{X}'\mathbf{X})^{-1}\mathbf{X}']\mathbf{Y}}{N_1 + N_2 - 2}$$

the expression given for T^2 in (3.15.9) reduces to

$$T^2 = \frac{N_1 N_2}{N_1 + N_2}(\mathbf{y}_{1.} - \mathbf{y}_{2.})'\mathbf{A}(\mathbf{A}'\mathbf{S}\mathbf{A})^{-1}\mathbf{A}'(\mathbf{y}_{1.} - \mathbf{y}_{2.}) \tag{3.15.12}$$

Alternative approaches are available to test for no condition effects. If H_{01} is accepted so that there is no group-by-condition interaction, then H_{02} may be tested in two ways (Koch, 1969). H_{02} becomes, in terms of the model parameters,

$$H_{02.}: \frac{\mu_{11} + \mu_{21}}{2} = \frac{\mu_{12} + \mu_{22}}{2} = \cdots = \frac{\mu_{1p} + \mu_{2p}}{2} \tag{3.15.13}$$

(3.15.13) can be represented in terms of $\mathbf{CBA} = \mathbf{\Gamma}$, in which case it becomes

(3.15.14)

$$\underset{(1\times 2)}{\mathbf{C}} \quad \underset{(2\times p)}{\mathbf{B}} \quad \underset{[p\times(p-1)]}{\mathbf{A}} = \underset{[1\times(p-1)]}{\mathbf{\Gamma}}$$

$$[1/2, 1/2]\begin{bmatrix} \mu_{11} & \mu_{12} & \cdots & \mu_{1p} \\ \mu_{21} & \mu_{22} & \cdots & \mu_{2p} \end{bmatrix}\begin{bmatrix} 1 & 0 & \cdots & 0 \\ 0 & 1 & \cdots & 0 \\ \vdots & \vdots & \vdots & \vdots \\ 0 & 0 & \cdots & 1 \\ -1 & -1 & \cdots & -1 \end{bmatrix} = [0, \ldots, 0]$$

With $\hat{\mathbf{B}}$ as defined in (3.15.5), $\mathbf{Q}_e$ as defined in (3.15.7), and $\mathbf{Q}_h$ as defined by

$$\begin{aligned} \mathbf{Q}_h &= (\mathbf{C}\hat{\mathbf{B}}\mathbf{A})'[\mathbf{C}(\mathbf{X}'\mathbf{X})^{-1}\mathbf{C}']^{-1}(\mathbf{C}\hat{\mathbf{B}}\mathbf{A}) \\ &= \frac{4N_1N_2}{N_1 + N_2}\mathbf{A}'\left(\frac{\mathbf{y}_{1.} + \mathbf{y}_{2.}}{2}\right)\left(\frac{\mathbf{y}_{1.} + \mathbf{y}_{2.}}{2}\right)'\mathbf{A} \\ &= \frac{4N_1N_2}{N_1 + N_2}(\mathbf{A}'\mathbf{y}_{..})(\mathbf{y}_{..}\mathbf{A}) \end{aligned} \tag{3.15.15}$$

Wilks' Λ criterion for testing H_{02} is given by

$$\Lambda = \frac{|\mathbf{Q}_e|}{|\mathbf{Q}_e + \mathbf{Q}_h|} \sim U(p - 1, 1, N_1 + N_2 - 2) \tag{3.15.16}$$

Alternatively, using (3.15.9), with $\mathbf{S}$ as given in (3.15.11),

$$T^2 = 4\left(\frac{N_1N_2}{N_1 + N_2}\right)\mathbf{y}'_{..}\mathbf{A}(\mathbf{A}'\mathbf{S}\mathbf{A})^{-1}\mathbf{A}'\mathbf{y}_{..} \tag{3.15.17}$$

The hypothesis H_{02} is rejected at the significance level α if

$$T^2 > T^\alpha(p - 1, N_1 + N_2 - 2) \tag{3.15.18}$$

or

$$\Lambda < U^\alpha(p - 1, 1, N_1 + N_2 - 2)$$

Using an overall average, rather than a simple average, $\mathbf{C}$ is chosen such that

$$\mathbf{C} = \left[\frac{N_1}{N_1 + N_2}, \frac{N_2}{N_1 + N_2}\right] \tag{3.15.19}$$

Then $\mathbf{Q}_h$ becomes

$$\mathbf{Q}_h = (N_1 + N_2)\mathbf{A}'\left(\frac{N_1\mathbf{y}_{1.} + N_2\mathbf{y}_{2.}}{N_1 + N_2}\right)\left(\frac{N_1\mathbf{y}_{1.} + N_2\mathbf{y}_{2.}}{N_1 + N_2}\right)'\mathbf{A}$$

$$= (N_1 + N_2)(\mathbf{A}'\bar{\mathbf{y}}_{..})(\bar{\mathbf{y}}'_{..}\mathbf{A}) \tag{3.15.20}$$

so that T^2 is identified by the expression

$$T^2 = (N_1 + N_2)\bar{\mathbf{y}}'_{..}\mathbf{A}(\mathbf{A}'\mathbf{S}\mathbf{A})^{-1}\mathbf{A}'\bar{\mathbf{y}}_{..} \tag{3.15.21}$$

where $\mathbf{S}$ is defined by (3.15.11) and the rules for rejection are the same as those given in (3.15.18). Expression (3.15.21) reduces to (3.15.17) under equal N_i's.

In using (3.15.21) to test for differences in conditions, the elements in the matrix $\mathbf{C}$ are data dependent since they are chosen proportional to the number of subjects in each treatment condition. This is not the case for the elements of $\mathbf{C}$ in (3.15.17). If an unequal number of subjects results from the nature of the experimental treatments, an unequal weighting procedure is appropriate; otherwise, equal weights are usually appropriate.

When the parallelism hypothesis is rejected, it may be best to analyze each group separately as a one-sample, repeated-measures design and then to follow the procedure developed in Section 3.13. Alternatively, a multivariate procedure is available to test for differences in conditions. This procedure does not depend on the acceptance of the parallelism test H_{01}. In this case, the test for differences in conditions is stated as in (3.15.22).

$$H_{02}^* = \begin{bmatrix}\mu_{11}\\ \mu_{21}\end{bmatrix} = \begin{bmatrix}\mu_{12}\\ \mu_{22}\end{bmatrix} = \cdots = \begin{bmatrix}\mu_{1p}\\ \mu_{2p}\end{bmatrix} \tag{3.15.22}$$

H_{02} is equivalent to H_{02}^* under parallelism. Representing (3.15.22) in terms of $\mathbf{CBA} = \mathbf{\Gamma}$, $\mathbf{C}$ and $\mathbf{\Gamma}$ in expression (3.15.14) are replaced with

$$\underset{(2\times 2)}{\mathbf{C}} = \begin{bmatrix}1 & 0\\ 0 & 1\end{bmatrix} \quad \text{and} \quad \underset{[2\times(p-1)]}{\mathbf{\Gamma}} = [0] \tag{3.15.23}$$

With $\mathbf{B}$ as defined in (3.15.5), $\mathbf{Q}_e$ as defined in (3.15.7), and $\mathbf{Q}_h$ as defined by

$$\mathbf{Q}_h = (\mathbf{C}\hat{\mathbf{B}}\mathbf{A})'[\mathbf{C}(\mathbf{X}'\mathbf{X})^{-1}(\mathbf{C}\hat{\mathbf{B}}\mathbf{A})$$

$$= \mathbf{A}'[\hat{\mathbf{B}}'(\mathbf{X}'\mathbf{X})\hat{\mathbf{B}}]\mathbf{A} \tag{3.15.24}$$

where the degrees of freedom for the hypothesis are $v_h = 2$, the rank of $\mathbf{C}$, Wilks' Λ criterion is

$$\Lambda = \frac{|\mathbf{Q}_e|}{|\mathbf{Q}_e + \mathbf{Q}_h|} \sim U(p - 1, 2, N_1 + N_2 - 2) \tag{3.15.25}$$

and H_{02}^* is rejected if $\Lambda < U^\alpha(p - 1, 2, N_1 + N_2 - 2)$ at the significance level α. However, all criteria are not equivalent for testing H_{02}^*. Using the Roy criterion, H_{02}^* is rejected if

$$\theta_s = \frac{\lambda_1}{1 + \lambda_1} > \theta^\alpha(s, m, n) \tag{3.15.26}$$

where $s = \min(\nu_h, p - 1)$, $m = (|p - 3| - 1)/2$, $n = (N_1 + N_2 - p - 2)/2$, and λ_1 is the largest root of $|\mathbf{Q}_h - \lambda \mathbf{Q}_e| = 0$. More will be presented about this test procedure in Chapter 5, where confidence intervals and tests are discussed in some detail for the many-group problem.

By employing a multivariate procedure, the test for group differences does not depend on the hypothesis H_{01}. Hotelling's two-sample T^2 statistic, discussed in Section 3.8, is employed. That is, the hypothesis H_{03} becomes, in terms of the parameters,

$$H_{03}: \begin{bmatrix} \mu_{11} \\ \mu_{12} \\ \vdots \\ \mu_{1p} \end{bmatrix} = \begin{bmatrix} \mu_{21} \\ \mu_{22} \\ \vdots \\ \mu_{2p} \end{bmatrix} \tag{3.15.27}$$

This hypothesis is equivalent to the univariate hypothesis

$$H_{03}^{*}: \frac{\sum_{j=1}^{p} \mu_{1j}}{p} = \frac{\sum_{i=1}^{p} \mu_{2i}}{p} \tag{3.15.28}$$

only if H_{01} is true. To test (3.15.28), the means computed over the p conditions are taken as the observations and a two-sample univariate t statistic is used for testing.

The Roy criterion or the Lawley-Hotelling trace criterion can also be employed to test H_{01}, H_{02}, or H_{03}; however, since $\nu_h = 1$ and $s = \min(\nu_h, u) = 1$, all procedures are again equivalent.

To obtain simultaneous confidence intervals for the three multivariate hypotheses, the method employed for (3.3.31) and (3.8.24) is used with the T^2 statistic. For H_{01}, (parallelism), the intervals are, for arbitrary $[1 \times (p - 1)]$ vectors $\mathbf{a}'$,

$$\mathbf{a}'\mathbf{A}'(\mathbf{y}_{1.} - \mathbf{y}_{2.}) - c_0\sqrt{\frac{N_1 + N_2}{N_1 N_2}\mathbf{a}'\mathbf{A}'\mathbf{S}\mathbf{A}\mathbf{a}} < \mathbf{a}'\mathbf{A}'(\boldsymbol{\mu}_1 - \boldsymbol{\mu}_2)$$

$$\leq \mathbf{a}'\mathbf{A}'(\mathbf{y}_{1.} - \mathbf{y}_{2.}) + c_0\sqrt{\frac{N_1 + N_2}{N_1 N_2}\mathbf{a}'\mathbf{A}'\mathbf{S}\mathbf{A}\mathbf{a}}$$

or, letting $\mathbf{c}' = \mathbf{a}'\mathbf{A}'$, where $\mathbf{c}$ is a contrast vector,

$$\mathbf{c}'(\mathbf{y}_{1.} - \mathbf{y}_{2.}) - c_0\sqrt{\frac{N_1 + N_2}{N_1 N_2}\mathbf{c}'\mathbf{S}\mathbf{c}} \leq \mathbf{c}'(\boldsymbol{\mu}_1 - \boldsymbol{\mu}_2) \leq \mathbf{c}'(\mathbf{y}_{1.} - \mathbf{y}_{2.}) + c_0\sqrt{\frac{N_1 + N_2}{N_1 N_2}\mathbf{c}'\mathbf{S}\mathbf{c}} \tag{3.15.29}$$

where $c_0^2 = T^{\alpha}(p - 1, N_1 + N_2 - 2)$.

Simultaneous confidence intervals for H_{02} depend on which average is being employed. For the weighted average, the intervals become, for $\mathbf{c}' = \mathbf{a}'\mathbf{A}$,

$$\mathbf{c}'\bar{\mathbf{y}}_{..} - c_0\sqrt{\frac{\mathbf{c}'\mathbf{S}\mathbf{c}}{N_1 + N_2}} \leq \mathbf{c}'\bar{\boldsymbol{\mu}}_{..} \leq \mathbf{c}'\bar{\mathbf{y}}_{..} + c_0\sqrt{\frac{\mathbf{c}'\mathbf{S}\mathbf{c}}{N_1 + N_2}} \tag{3.15.30}$$

where $c_0^2 = T^{\alpha}(p - 1, N_1 + N_2 - 2)$ and $\bar{\boldsymbol{\mu}}_{..}$ is used to denote the weighted average of the population mean vectors $\bar{\boldsymbol{\mu}}_{..} = (N_1\boldsymbol{\mu}_1 + N_2\boldsymbol{\mu}_2)/(N_1 + N_2)$.

Confidence intervals for H_{03} are identical with those discussed in Section 3.8 for the two-sample case. Confidence intervals for H_{02}^* are discussed in Chapter 5.

In profile analysis, three null hypotheses are of primary interest.

(1) Parallelism (interaction between groups and conditions):

$$H_{01}: \begin{bmatrix} \mu_{11} - \mu_{12} \\ \mu_{12} - \mu_{13} \\ \vdots \\ \mu_{1(p-1)} - \mu_{1p} \end{bmatrix} = \begin{bmatrix} \mu_{21} - \mu_{22} \\ \mu_{22} - \mu_{23} \\ \vdots \\ \mu_{2(p-1)} - \mu_{2p} \end{bmatrix}$$

(2) Differences among conditions:

$$H_{02}: \frac{\mu_{11} + \mu_{21}}{2} = \frac{\mu_{12} + \mu_{22}}{2} = \cdots = \frac{\mu_{1p} + \mu_{2p}}{2}$$

$$H_{02}^*: \begin{bmatrix} \mu_{11} \\ \mu_{21} \end{bmatrix} = \begin{bmatrix} \mu_{12} \\ \mu_{22} \end{bmatrix} = \cdots = \begin{bmatrix} \mu_{1p} \\ \mu_{2p} \end{bmatrix}$$

(3) Differences between groups:

$$H_{03}: \begin{bmatrix} \mu_{11} \\ \mu_{12} \\ \vdots \\ \mu_{1p} \end{bmatrix} = \begin{bmatrix} \mu_{21} \\ \mu_{22} \\ \vdots \\ \mu_{2p} \end{bmatrix}$$

$$H_{03}^*: \frac{\sum_{j=1}^{p} \mu_{1j}}{p} = \frac{\sum_{j=1}^{p} \mu_{2j}}{p}$$

Given that H_{01} is tenable, H_{02} and H_{03}^* are equivalent to the usual univariate tests for differences among conditions and differences between groups. However, when H_{01} is not known to hold, H_{02} and H_{03}^* are not equivalent to the univariate tests. We might still test H_{03}^*, but the test for differences among conditions is better analyzed for each group separately. Alternatively, tests H_{03} and H_{02}^* do not depend on the parallelism assumption and may always be tested by using multivariate procedures.

You should, however, be aware that in (3.15.1) $\mathbf{\Sigma}$ is of full rank p, so that, in each group of subjects $N_i \geq p$ for multivariate tests. If this is not the case, univariate methods must be employed in profile studies. That is, restrictions on $\mathbf{\Sigma}$ are imposed to have exact F ratios. The usual restriction is compound symmetry,

$$\mathbf{\Sigma} = \rho\sigma^2\mathbf{J} + (1 - \rho)\sigma^2\mathbf{I}$$

$$= \sigma^2 \begin{bmatrix} 1 & \rho & \rho & \cdots & \rho \\ \rho & 1 & \rho & \cdots & \rho \\ \rho & \rho & 1 & \cdots & \rho \\ \vdots & \vdots & \vdots & \vdots & \vdots \\ \rho & \rho & \rho & \cdots & 1 \end{bmatrix}$$

where **J** is a matrix of 1s. The variance-covariance matrix may have other patterns that would yield exact F ratios since compound symmetry is a sufficient condition but not a necessary condition (see Huynh and Feldt, 1970).

EXERCISE 3.15

1. Derive the confidence intervals given in (3.15.29) and (3.15.30).

3.16 EXAMPLE—TWO-SAMPLE PROFILE ANALYSIS

The data in Table 3.16.1 were provided by Dr. Paul Ammon. They were collected as in the one-sample, profile analysis example, except that group I data were obtained from subjects with low short-term memory capacity and group II data were obtained from subjects with high short-term memory capacity.

TABLE 3.16.1. Sample Data: Two-Sample Profile Analysis.

		Probe-Word Position				
		1	2	3	4	5
Group I	S_1	20	21	42	32	32
	S_2	67	29	56	39	41
	S_3	37	25	28	31	34
	S_4	42	38	36	19	35
	S_5	57	32	21	30	29
	S_6	39	38	54	31	28
	S_7	43	20	46	42	31
	S_8	35	34	43	35	42
	S_9	41	23	51	27	30
	S_{10}	39	24	35	26	32
	Mean	42.0	28.4	41.2	31.2	33.4
Group II	S_1	47	25	36	21	27
	S_2	53	32	48	46	54
	S_3	38	33	42	48	49
	S_4	60	41	67	53	50
	S_5	37	35	45	34	46
	S_6	59	37	52	36	52
	S_7	67	33	61	31	50
	S_8	43	27	36	33	32
	S_9	64	53	62	40	43
	S_{10}	41	34	47	37	46
	Mean	50.9	35.0	49.6	37.9	44.9

Before testing any hypotheses, the profile curves for each group are plotted by using the means at each profile, as in Figure 3.16.1. Although the profile plot clearly indicates no group-by-condition interaction, the test is performed for illustration purposes.

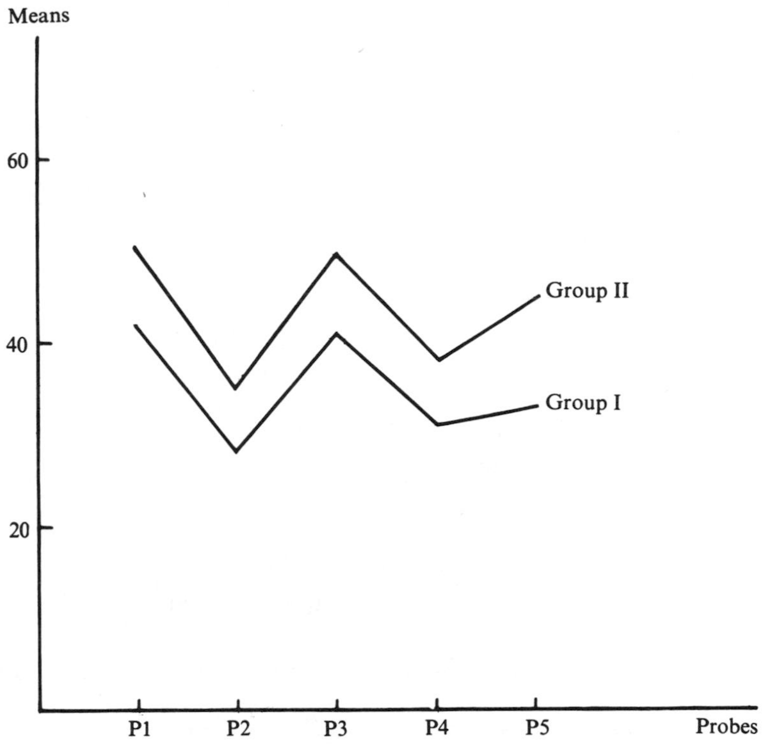

FIGURE 3.16.1. Two-sample, profile analysis plot.

To begin the analysis of this example problem, the least-squares estimator of **B** is computed.

$$\hat{\mathbf{B}} = (\mathbf{X}'\mathbf{X})^{-1}\mathbf{X}'\mathbf{Y} = \begin{bmatrix} y_{1.1} & y_{1.2} & y_{1.3} & y_{1.4} & y_{1.5} \\ y_{2.1} & y_{2.2} & y_{2.3} & y_{2.4} & y_{2.5} \end{bmatrix}$$

$$= \begin{bmatrix} .10 & 0 \\ 0 & .10 \end{bmatrix} \begin{bmatrix} 420 & 284 & 412 & 312 & 334 \\ 509 & 350 & 496 & 379 & 449 \end{bmatrix}$$

$$(3.16.1) \qquad = \begin{bmatrix} 42.0 & 28.4 & 41.2 & 31.2 & 33.4 \\ 50.9 & 35.0 & 49.6 & 37.9 & 44.9 \end{bmatrix}$$

The transformation matrices **C** and **A** to test H_{01} (parallelism) are

$$(3.16.2) \qquad \mathbf{C} = [1, -1] \quad \text{and} \quad \mathbf{A} = \begin{bmatrix} 1 & 0 & 0 & 0 \\ -1 & 1 & 0 & 0 \\ 0 & -1 & 1 & 0 \\ 0 & 0 & -1 & 1 \\ 0 & 0 & 0 & -1 \end{bmatrix}$$

so that the contrast matrix $\mathbf{C\hat{B}A}$ becomes

(3.16.3) $$\mathbf{C\hat{B}A} = [-2.3, 1.8, -1.7, 4.8] = \bar{\mathbf{y}}'_d$$

The matrix products $\mathbf{Q}_e$ and $\mathbf{Q}_h$ are

$$\mathbf{Q}_e = \mathbf{A'Y'[I - X(X'X)^{-1}X']YA}$$
$$= \begin{bmatrix} 2313.30 & & & \\ -740.80 & 2026.00 & & \\ 247.70 & -1103.80 & 1996.10 & \\ 16.20 & -230.60 & -505.00 & 917.60 \end{bmatrix}$$

(3.16.4) $$\mathbf{Q}_h = (\mathbf{C\hat{B}A})'[\mathbf{C(X'X)^{-1}C'}]^{-1}(\mathbf{C\hat{B}A})$$
$$= \begin{bmatrix} 26.45 & & & \\ -20.70 & 16.20 & & \\ 19.55 & -15.30 & 14.45 & \\ -55.20 & 43.20 & -40.80 & 115.20 \end{bmatrix}$$

Furthermore, the variance-covariance matrix $\mathbf{S}$ based on the original data is

$$\mathbf{S} = \frac{\mathbf{Y'[I - X(X'X)^{-1}X']Y}}{N_1 + N_2 - 2}$$
$$= \frac{1}{18}\begin{bmatrix} 2546.9 & & & & \\ 597.0 & 960.4 & & & \\ 946.6 & 569.2 & 2204.0 & & \\ 257.9 & 155.2 & 686.2 & 1164.5 & \\ 385.9 & 299.4 & 599.8 & 573.1 & 899.3 \end{bmatrix}$$
$$= \begin{bmatrix} 141.494 & & & & \\ 33.167 & 53.356 & & & \\ 52.589 & 31.622 & 122.444 & & \\ 14.328 & 8.622 & 38.122 & 64.694 & \\ 21.439 & 16.633 & 33.222 & 31.834 & 49.961 \end{bmatrix}$$

Testing the hypothesis of parallelism,

(3.16.5) $$H_{01}: \begin{bmatrix} \mu_{11} - \mu_{12} \\ \mu_{12} - \mu_{13} \\ \mu_{13} - \mu_{14} \\ \mu_{14} - \mu_{15} \end{bmatrix} = \begin{bmatrix} \mu_{21} - \mu_{22} \\ \mu_{22} - \mu_{23} \\ \mu_{23} - \mu_{24} \\ \mu_{24} - \mu_{25} \end{bmatrix}$$

the statistic T^2 is formed by using (3.16.4), so that

$$\begin{aligned} T^2 &= (N_1 + N_2 - 2)\left(\frac{|\mathbf{Q}_e + \mathbf{Q}_h|}{|\mathbf{Q}_e|} - 1\right) \\ &= 18\left(\frac{4{,}091{,}585{,}397{,}759}{3{,}433{,}079{,}600{,}323} - 1\right) \\ &= 3.453 \end{aligned} \tag{3.16.6}$$

Since T^2 is less than the critical value of $T^{.05}_{(4,18)} = 14.667$, for $\alpha = .05$, H_{01}, the parallel hypothesis, is not rejected.

To test H_{02}, the matrices

$$\mathbf{C} = [1/2, 1/2] \quad \text{and} \quad \mathbf{A} = \begin{bmatrix} 1 & 0 & 0 & 0 \\ 0 & 1 & 0 & 0 \\ 0 & 0 & 1 & 0 \\ 0 & 0 & 0 & 1 \\ -1 & -1 & -1 & -1 \end{bmatrix}$$

are chosen so that

$$\mathbf{C\hat{B}A} = [7.30, -7.45, 6.25, -4.60] \tag{3.16.7}$$

where $\hat{\mathbf{B}}$ is defined by (3.16.1).

Using formulas (3.15.20) and (3.15.7),

$$\mathbf{Q}_h = \begin{bmatrix} 1065.80 & & & \\ -1087.70 & 1110.05 & & \\ 912.50 & -931.25 & 781.25 & \\ -671.60 & 685.40 & -575.00 & 423.20 \end{bmatrix}$$

$$\mathbf{Q}_e = \begin{bmatrix} 2674.00 & & & \\ 811.00 & 1260.90 & & \\ 860.20 & 569.30 & 1903.70 & \\ 198.20 & 182.00 & 412.60 & 917.60 \end{bmatrix} \tag{3.16.8}$$

The hypothesis H_{02}, difference in conditions,

$$H_{02}: \frac{\mu_{11} + \mu_{21}}{2} = \cdots = \frac{\mu_{15} + \mu_{25}}{2} \tag{3.16.9}$$

is rejected at the significance level $\alpha = .05$ if $T^2 > T^{.05}_{(4,18)} = 14.667$. Following (3.16.6), $T^2 = 63.868$. Thus the hypothesis of no difference in conditions is rejected.

Testing the hypothesis of no difference between groups, H_{03},

$$\mathbf{C} = [1, -1], \quad \mathbf{A} = \mathbf{I}, \quad \text{and} \quad \mathbf{C\hat{B}A} = [-8.9, -6.6, -8.4, -6.7, -11.5] \tag{3.16.10}$$

Furthermore, the error and hypothesis sum of squares and product matrices are

$$\mathbf{Q}_e = \begin{bmatrix} 2546.9 & & & & \\ 597.0 & 960.4 & & & \\ 946.6 & 569.2 & 2204.0 & & \\ 257.9 & 155.2 & 686.2 & 1164.5 & \\ 385.9 & 299.4 & 599.8 & 573.1 & 899.3 \end{bmatrix}$$

(3.16.11)

$$\mathbf{Q}_h = \begin{bmatrix} 396.05 & & & & \\ 293.70 & 217.80 & & & \\ 373.80 & 277.20 & 352.80 & & \\ 298.15 & 221.10 & 281.40 & 224.45 & \\ 511.75 & 379.50 & 483.00 & 385.25 & 661.25 \end{bmatrix}$$

and $T^2 = 14.370$. This hypothesis is not rejected at the significance level $\alpha = .05$ since $T^2 < T^{;.05}_{(5,18)} = 19.017$.

The two-sample t statistic computed from the 10 subject means in each group is significant when employed to test for group differences. The value of the test statistic is $t = 2.984$. The univariate test for group differences leads to significance since t is compared with critical value $t^{;.025}_{(18)} = 2.101$.

The reason for this apparent discrepancy between the univariate and multivariate results is readily explained. The univariate test H^*_{03} given by (3.15.28) examines only one contrast among many in the multivariate hypothesis H_{03}. If the single comparison tested by the univariate procedure were the only comparison of interest under the multivariate hypothesis, Bonferroni intervals would be computed to test the comparison. In this case, it is reduced to the ordinary t test since there is only one comparison.

The multivariate approach using Bonferroni intervals may be generally preferred since at most $p + 1$ intervals would be desired in many multivariate situations; that is, one interval for each variable and one for the overall difference in group averages. The critical value in this example problem would be for $k = 6$, $t^{;.05/2k}_{(18)} = 2.98$, leading to rejection of H_{03}.

To determine simultaneous confidence intervals, we shall need the following least-squares estimates:

$$\hat{\mathbf{B}} = \begin{bmatrix} 42.0 & 28.4 & 41.2 & 31.2 & 33.4 \\ 50.9 & 35.0 & 49.6 & 37.9 & 44.9 \end{bmatrix}$$

$$\bar{\mathbf{y}}'_{..} = [46.45, 31.7, 45.4, 34.55, 39.15]$$

To find differences in conditions for this example, the expression

$$\mathbf{c}'\bar{\mathbf{y}}_{..} - c_0\sqrt{\frac{\mathbf{c}'\mathbf{S}\mathbf{c}}{N_1 + N_2}} \le \mathbf{c}'\bar{\boldsymbol{\mu}}_{..} \le \mathbf{c}'\bar{\mathbf{y}}_{..} + c_0\sqrt{\frac{\mathbf{c}'\mathbf{S}\mathbf{c}}{N_1 + N_2}} \tag{3.16.12}$$

is evaluated with $c_0 = \sqrt{14.667} = 3.830$. For differences in conditions, the following contrasts are examined.

$$\begin{aligned}
\psi_1 &= \mu_1 - \mu_5 & \hat{\psi}_1 &= 7.30\\
\psi_2 &= \mu_2 - \mu_5 & \hat{\psi}_2 &= -7.45\\
\psi_3 &= \mu_1 - \mu_5 & \hat{\psi}_3 &= 6.25\\
\psi_4 &= \mu_4 - \mu_5 & \hat{\psi}_4 &= -4.60\\
\psi_5 &= \mu_1 - \mu_2 & \hat{\psi}_5 &= 14.75
\end{aligned}$$

The standard errors for these contrasts are obtained from $\sqrt{\mathbf{c}'\mathbf{S}\mathbf{c}/(N_1 + N_2)}$, where $\mathbf{c}$ is chosen such that $\hat{\psi}_i = \mathbf{c}'\mathbf{y}_{..}$. Thus, for each ψ considered,

$$\begin{aligned}
\hat{\sigma}_{\hat{\psi}_1} &= \sqrt{\frac{\mathbf{c}'\mathbf{S}\mathbf{c}}{N_1 + N_2}} = \sqrt{\frac{141.494 + 49.961 - 2(21.439)}{20}}\\
&= \sqrt{7.429} = 2.7256\\
\hat{\sigma}_{\hat{\psi}_2} &= \sqrt{3.503} = 1.8715\\
\hat{\sigma}_{\hat{\psi}_3} &= \sqrt{5.288} = 2.2996\\
\hat{\sigma}_{\hat{\psi}_4} &= \sqrt{2.549} = 1.5965\\
\hat{\sigma}_{\hat{\psi}_5} &= \sqrt{\frac{141.494 + 53.356 - 2(33.167)}{20}} = \sqrt{6.426} = 2.5349
\end{aligned}$$

By evaluating expression (3.16.12), which, in general, is of the form

$$\hat{\psi}_i - c_0\hat{\sigma}_{\hat{\psi}_i} \leq \psi_i \leq \hat{\psi}_i + c_0\hat{\sigma}_{\hat{\psi}_i} \tag{3.16.13}$$

an interval for each contrast is obtained;

$$\begin{aligned}
-3.14 &\leq \mu_1 - \mu_5 \leq 17.74 & &\text{(N.S.)}\\
-14.62 &\leq \mu_2 - \mu_5 \leq -.28 & &\text{(Sig.)}\\
-2.56 &\leq \mu_3 - \mu_5 \leq 15.07 & &\text{(N.S.)}\\
-10.71 &\leq \mu_4 - \mu_5 \leq 1.51 & &\text{(N.S.)}\\
5.04 &\leq \mu_1 - \mu_2 \leq 24.46 & &\text{(Sig.)}
\end{aligned}$$

for the population elements in the grand mean vector.

To obtain a confidence interval for the mean difference in groups using a simple t test, compute for $\psi = \mu_1 - \mu_2$:

$$\hat{\psi} = \mathbf{C}\hat{\mathbf{B}}\mathbf{A} = [1, -1] \quad \hat{\mathbf{B}} \begin{bmatrix} 1/5 \\ 1/5 \\ 1/5 \\ 1/5 \\ 1/5 \end{bmatrix} = -8.42 \tag{3.16.14}$$

To establish $\hat{\sigma}_{\hat{\psi}}$, the formula $\hat{\sigma}_{\hat{\psi}} = \sqrt{\mathbf{a}'\mathbf{S}\mathbf{a}\mathbf{c}'(\mathbf{X}'\mathbf{X})^{-1}\mathbf{c}}$ from Section 3.6 is employed, so that $\hat{\sigma}_{\hat{\psi}} = [(39.804)(.20)]^{1/2} = 2.822$. Thus, substituting into (3.16.14), with $c_0 = 2.101$, the interval becomes $-14.35 \leq \mu_1 - \mu_2 \leq -2.49$, which indicates significance since zero is not covered by the interval.

EXERCISES 3.16

1. In an experiment designed to investigate problem-solving ability for two groups of subjects, experimental (E) and control (C), subjects were required to solve four different mathematics problems presented in a random order for each subject. The time required to solve each problem was recorded. All problems were thought to be of the same level of difficulty. The data for the experiment are summarized below.

		Problems			
	Subject	1	2	3	4
C	1	43	90	51	67
	2	87	36	12	14
	3	18	56	22	68
	4	34	73	34	87
	5	81	55	29	54
	6	45	58	62	44
	7	16	35	71	37
	8	43	47	87	27
	9	22	91	37	78
E	1	10	81	43	33
	2	58	84	35	43
	3	26	49	55	84
	4	18	30	49	44
	5	13	14	25	45
	6	12	8	40	48
	7	9	55	10	30
	8	31	45	9	66

a. Can you conclude that the profiles for the two groups are similar?

b. Depending on your answer to a, are there any differences between the groups or among the four conditions?

c. Using simultaneous inference procedures where are the differences, if any, most pronounced?

3.17 EQUALITY OF VARIANCE-COVARIANCE MATRICES

In multivariate analysis, as in univariate analysis, when testing statistical hypotheses about means, three assumptions are fundamental to the statistical theory:

(3.17.1)
(a) independence,
(b) multivariate normality, and
(c) equality of variance-covariance matrices.

A statistical test procedure is said to be *robust or insensitive* if departures from these assumptions do not greatly affect the significance level or the power of the test.

Ito and Schull (1964) and Ito (1969) have investigated the robustness of Hotelling's T_0^2 statistic when violating either (3.17.1a) or (3.17.1b) for large sample sizes. No extensive work has been conducted in this area for small samples; thus the results stated here are asymptotic. Their work represents a multivariate extension of the univariate considerations summarized by Scheffé (1959, Chap. 10) and Petrinovich and Hardyck (1969).

Ito and Schull found that unequal variance-covariance matrices have little affect on the significance level α or power if the sample sizes are equal when testing for differences in mean vectors. However, as the ratio of the sample sizes increases, serious changes in both α and power occur. Ito further demonstrates that nonnormality has little effect on the hypothesis about means, but serious consequences when testing for equality of variance-covariance matrices.

In this section, some commonly employed procedures for evaluating the equality of variance-covariance matrices under normality and nonnormality are reviewed. In addition, tests for studying the form of the variance-covariance matrix are also considered.

In the two-sample univariate case, the hypothesis

$$\begin{aligned} H_0 &: \sigma_1^2 = \sigma_2^2 \\ H_1 &: \sigma_1^2 \neq \sigma_2^2 \end{aligned} \tag{3.17.2}$$

is usually tested by calculating the statistic

$$f = \frac{s_1^2}{s_2^2} \sim F(v_1, v_2) \tag{3.17.3}$$

when H_0 is true. The quantities s_i^2 are the sample variances obtained from the two independent random normal samples represented by $y_{11}, y_{12}, \ldots, y_{1N_1}$ and $y_{21}, y_{22}, \ldots, y_{2N_2}$, where $Y_{ij} \sim IN(\mu_i, \sigma_i^2)$. The degrees of freedom v_1 and v_2 are related to the sample sizes; $v_1 = N_1 - 1$ and $v_2 = N_2 - 1$.

In applying f to test (3.17.2), f is compared to an F critical value and (3.17.2) is rejected if $f > F^{\alpha/2}(v_1, v_2)$ or $f < F^{1-\alpha/2}(v_1, v_2)$ with significance level α.

In using (3.17.3) to test (3.17.1), the assumption of normality is critical (see Scheffé, 1959, p. 337). Numerous test procedures have been proposed that are robust and reasonably powerful for testing equality of variances under nonnormality. Miller (1968), through a Monte Carlo study, compares five commonly used robust procedures; he finds that Box's test, the Box-Anderson test, and the jackknife test are the most powerful procedures. However, Box's test, though not as powerful as the other two, is easier to compute. For an example of Box's procedure, see Scheffé (1959, pp. 83–87).

In testing the hypothesis of equal variance-covariance matrices in the multivariate case,

$$\begin{aligned} H_0 &: \mathbf{\Sigma}_1 = \cdots = \mathbf{\Sigma}_g \\ H_1 &: H_0 \text{ is false} \end{aligned} \tag{3.17.4}$$

under multivariate normality, a procedure suggested by Box (1950) is discussed. This is a generalization of Bartlett's test in the k-sample univariate situation (see Bartlett, 1937a, and Brownlee, 1965, p. 290).

To test (3.17.4), let $\mathbf{S}_i$ denote the unbiased estimate of $\mathbf{\Sigma}_i$ for the ith group, with N_i independent p-vector-valued observations from a multivariate normal distribution with mean $\boldsymbol{\mu}_i$ and variance-covariance matrix $\mathbf{\Sigma}_i$. Setting

$$N = \sum_{i=1}^{g} N_i, \qquad v_i = N_i - 1$$

and

$$\mathbf{S} = \frac{1}{N - g} \sum_{i=1}^{g} v_i \mathbf{S}_i \tag{3.17.5}$$

where $\mathbf{S}$ is the pooled-within estimate of the variance-covariance matrix and g denotes the number of groups, the statistic

$$M = (N - g) \log |\mathbf{S}| - \sum_{i=1}^{g} v_i \log |\mathbf{S}_i| \tag{3.17.6}$$

is formed. Anderson (1958, p. 248) and Kullback (1959, p. 318) use this statistic to test (3.17.4). However, multiplying M by $1 - C$, where

$$C = \frac{2p^2 + 3p - 1}{6(p + 1)(g - 1)} \left(\sum_{i=1}^{g} \frac{1}{v_i} - \frac{1}{N - g} \right) \tag{3.17.7}$$

the quantity

$$X_B^2 = (1 - C)M \tag{3.17.8}$$

more rapidly approximates a chi-square distribution with degrees of freedom

$$v = \frac{p(p + 1)(g - 1)}{2} \tag{3.17.9}$$

when H_0 is true and $N \to \infty$.

Hypothesis (3.17.4) is rejected at the significance level α if

$$X_B^2 > \chi_\alpha^2(v) \tag{3.17.10}$$

and reasonably approximates a χ^2 distribution when each $N_i > 20$ and N_i is large relative to $p < 6$ and $g < 6$.

A more precise approximation, when the values of v_i are small and/or p and g are larger than 6, is obtained by using the F distribution (see Box, 1949, 1950). Employing the F approximation to test (3.17.4), calculate

$$\begin{aligned} C_0 &= \frac{(p - 1)(p + 2)}{6(g - 1)} \left[\sum_{i=1}^{g} \frac{1}{v_i^2} - \frac{1}{(N - g)^2} \right] \\ v_0 &= \frac{v + 2}{C_0 - C^2} \\ F &= \frac{1 - C - v/v_0}{v} M \sim F(v, v_0) \end{aligned} \tag{3.17.11}$$

where M, C, v, and v_i are defined in the evaluation of X_B^2.

The hypothesis of homogeneity is rejected at the significance level α if $F > F^\alpha(v, v_0)$, where F^α denotes the upper α percentage point of the F distribution.

As in the univariate case, the tests of (3.17.4) are very sensitive to nonnormality. However, Layard (1969) proposes several asymptotically robust procedures. One procedure is a generalization of Box's univariate test (Box, 1953). For the two-sample case, each sample of size N_1 and N_2 is divided at random into k_1 and $k_2 > 2$ subsamples of approximately equal size such that $N_1/k_1 = n_1$ and $N_2/k_2 = n_2$. Let $\mathbf{S}_1^{(1)}, \mathbf{S}_2^{(1)}, \ldots, \mathbf{S}_{k_1}^{(1)}$, and $\mathbf{S}_1^{(2)}, \mathbf{S}_2^{(2)}, \ldots, \mathbf{S}_{k_2}^{(2)}$ denote variance-covariance matrices obtained from the two groups and subsamples. Considering only the diagonal elements of each of these matrices and the correlations, we compute, for each sample $i = 1, 2$, the vectors

$$\mathbf{L}_{ij} = \begin{bmatrix} \log s_{11}^{(j)} \\ \log s_{22}^{(j)} \\ \vdots \\ \log s_{pp}^{(j)} \\ \tanh^{-1} r_{12}^{(j)} \\ \vdots \\ \tanh^{-1} r_{p-1,p}^{(i)} \end{bmatrix} \qquad j = 1, \ldots, k_i$$

where $\tanh^{-1}$ denotes the inverse of the hyperbolic tangent transformation, and variance-covariance matrices $\mathbf{\Gamma}_1$ and $\mathbf{\Gamma}_2$ from the vectors $\mathbf{L}_{11}, \ldots, \mathbf{L}_{1k_1}$ and $\mathbf{L}_{21}, \ldots, \mathbf{L}_{2k_2}$, respectively. Averaging the vectors $\mathbf{L}_{ij}$, where

(3.17.12)
$$\bar{\mathbf{L}}_i = \frac{\sum_{j=1}^{k_i} \mathbf{L}_{ij}}{k_i} \qquad \text{for } i = 1, 2$$

the statistic

(3.17.13)
$$T^2 = \frac{k_1 k_2}{k_1 + k_2}(\bar{\mathbf{L}}_1 - \bar{\mathbf{L}}_2)'\mathbf{\Gamma}^{-1}(\bar{\mathbf{L}}_1 - \bar{\mathbf{L}}_2)$$

where $\mathbf{\Gamma} = [(k - 1)\mathbf{\Gamma}_1 + (k_2 - 1)\mathbf{\Gamma}_2]/(k_1 + k_2 - 2)$, is formed. T^2 has approximately (when H_0 is true) a Hotelling's T^2 distribution with $p(p + 1)/2$ and $k_1 + k_2 - 2$ degrees of freedom or, as k_1 and k_2 tend to infinity, $T^2 \mathrel{\dot\sim} \chi^2[p(p + 1)/2]$. The hypothesis specified by (3.17.4) is rejected if

(3.17.14)
$$T^2 > T^{\alpha}\left[\frac{p(p + 1)}{2}, k_1 + k_2 - 2\right]$$

To extend this procedure to the g-sample case, Wilks' MANOVA criterion (discussed in Chapter 4) would be employed by using the vectors $\bar{\mathbf{L}}_i$, $i = 1, \ldots, g$.

In addition to testing for the equality of variance-covariance matrices, a common problem in multivariate analysis is testing that a single variance-covariance matrix has a specified form. Some examples are

(3.17.15) (a) The hypothesis of compound symmetry (equal variances and correlations):

$$H_0 : \mathbf{\Sigma} = \sigma^2 \begin{bmatrix} 1 & \rho & \cdots & \rho \\ \rho & 1 & \cdots & \rho \\ \cdot & \cdot & \cdots & \cdot \\ \rho & \rho & \cdots & 1 \end{bmatrix}$$

(b) The "independence" hypothesis:

$$H_0 : \mathbf{\Sigma} = \begin{bmatrix} \sigma_1^2 & \cdots & \mathbf{0} \\ & \ddots & \\ \mathbf{0} & \cdots & \sigma_p^2 \end{bmatrix}$$

(c) The sphericity hypothesis:

$$H_0: \mathbf{\Sigma} = \sigma^2\mathbf{I}$$

(d) The two-set independence hypothesis:

$$H_0: \mathbf{\Sigma} = \begin{bmatrix} \mathbf{\Sigma}_1 & \mathbf{0} \\ \mathbf{0} & \mathbf{\Sigma}_2 \end{bmatrix}$$

Again, see Layard's (1969) work for a discussion of asymptotically robust techniques. Methods that assume multivariate normality will receive attention here.

Anderson (1958, Chaps. 9, 10) and Kullback (1959, Chap. 12) present the general theory necessary to test the hypotheses summarized in (3.17.15). Although exact distributions for the proposed statistics are unknown, the statistics presented are based on the LRT criterion and the fact that $-2 \log \lambda$ can be approximated asymptotically with a chi-square distribution. With $\lambda = L(\hat{\omega})/L(\hat{\Omega})$, the degrees of freedom for the χ^2 distribution are $v_{\hat{\Omega}} - v_{\hat{\omega}}$, where $v_{\hat{\Omega}}$ denotes the number of independent parameters estimated under Ω and $v_{\hat{\omega}}$ the number of independent parameters estimated under ω. However, numerous authors have suggested modified approximations of the general theory to facilitate convergence; these techniques are discussed here to evaluate the hypotheses stated in (3.17.15) under moderately large sample sizes.

The test for compound symmetry is based on Box's (1949, 1950) work. The assumption of compound symmetry of the variance-covariance matrix given by (3.17.15a) is often suggested when employing univariate procedures to analyze factorial repeated-measures designs (see Winer, 1971, Chap. 7, and Gaito and Wiley, 1963). However, as Huynh and Feldt (1970) show, this condition is only sufficient and not necessary for exact F ratios.

To develop the test of (3.17.15a) in this context, let g denote the number of groups and p the number of variables. To test for compound symmetry under normality, use the notation introduced to test (3.17.4). Letting

(3.17.16)
$$M_x = -(N - g) \log \frac{|\mathbf{S}|}{|\bar{\mathbf{S}}|}$$
$$C_x = \frac{p(p + 1)^2(2p - 3)}{6(p - 1)(N - g)(p^2 + p - 4)}$$

where

$$\bar{\mathbf{S}} = \begin{bmatrix} \bar{s}^2 & \bar{s} & \cdots & \bar{s} \\ \bar{s} & \bar{s}^2 & \cdots & \bar{s} \\ \vdots & \vdots & & \vdots \\ \bar{s} & \bar{s} & \cdots & \bar{s}^2 \end{bmatrix}$$

$\bar{s}^2$ denotes the average of the diagonal elements of $\mathbf{S}$, and $\bar{s}$ is the mean of the off-diagonal elements of $\mathbf{S}$, with $\mathbf{S}$ as defined in (3.17.5).

By forming the statistic

(3.17.17)
$$X^2_{BX} = (1 - C_x)M_x$$

X_{BX}^2 is distributed approximately as a χ^2 distribution with degrees of freedom $v = (p^2 + p - 4)/2$ when H_0 is true. The hypothesis of compound symmetry is rejected at the significance level α if $X_{BX}^2 > X_\alpha^2(v)$.

A more precise approximation, when p and g are large (greater than 6) and/or $N - g$ is small ($N - g < 20$), is obtained by calculating

$$C_{0x} = \frac{(p-1)p(p+1)(p+2)}{6(N-g)^2(p^2+p-4)}$$

(3.17.18)
$$v_{0x} = \frac{v+2}{C_{0x} - C_x^2}$$

$$F = \frac{(1 - C_x - v)/v_{0x}}{v} M_x$$

where M_x, v, and C_x are defined in the calculation of X_{BX}^2 (Box, 1950, p. 375). The hypothesis of compound symmetry is rejected in this case if $F > F^\alpha(v, v_{0x})$, where $F^\alpha(v, v_{0x})$ denotes the upper α percentage point of the F distribution.

The test of independence, that $\mathbf{\Sigma}$ is a diagonal matrix, is equivalent to testing that the population correlation matrix $\mathbf{P}$ is the identity matrix. Thus (3.17.15b) is equivalent to testing

(3.17.19)
$$H_0 : \mathbf{P} = \mathbf{I}$$

Given independence, application of univariate procedures to analyze a set of p variables, a variable at a time, is indicated, rather than proceeding with a multivariate analysis.

Bartlett (1950, 1954) suggests using the statistic

(3.17.20)
$$X_I^2 = -\left[(N-1) - \frac{2p+5}{6}\right] \log |\mathbf{R}|$$

to test for independence. The matrix $\mathbf{R}$ denotes the sample correlation matrix obtained from the random sample of p-variate vectors $\mathbf{Y}_i \sim IN(\boldsymbol{\mu}, \mathbf{\Sigma})$, $i = 1, \ldots, N$.

The hypothesis of independence is rejected at the significance level α if

(3.17.21)
$$X_I^2 > \chi_\alpha^2 \frac{p(p-1)}{2}$$

since $\chi_I^2 \mathrel{\dot\sim} \chi^2[p(p-1)/2]$ when H_0 is true.

Again, when $p \geq 6$, an F approximation should be employed to test (3.17.15b) (see Box, 1949).

Extending the hypothesis of independence, it might be of interest to see if the variances are equal. This is achieved with the test of sphericity, $\mathbf{\Sigma} = \sigma^2\mathbf{I}$, which is equivalent to testing the equality of the eigenvalues of the variance-covariance matrix $\mathbf{\Sigma}$. Bartlett (1954) and Lawley (1956) consider this test. More recently, James (1969) investigates their approximations, while Box (1949) discusses an F approximation for large p.

Assuming $\mathbf{Y}_i \sim IN(\boldsymbol{\mu}, \mathbf{\Sigma})$ and employing Bartlett's approximation to test the equality of the p roots of $\mathbf{\Sigma}$, the test statistic

(3.17.22)
$$X_s^2 = -\left[(N-1) - \frac{2p^2+p+2}{6p}\right]\left[\log |\mathbf{S}| - p \log \frac{\operatorname{Tr}(\mathbf{S})}{p}\right]$$

is computed where X_s^2 is distributed approximately as a χ^2 random variable with degrees of freedom $v = (p-1)(p+2)/2$ when H_0 is true. Thus the test of sphericity is rejected if $X_s^2 > \chi^2[(p-1)(p+2)/2]$.

The test of independence for two groups, (3.17.15d), is related to canonical correlation analysis to be discussed in Chapter 4. A test proposed by Box (1949) is presented that enables us to test for independence of g groups with p_i variables per group.

Let $\mathbf{Y}_i \sim IN(\boldsymbol{\mu}, \boldsymbol{\Sigma})$, for $i = 1, \ldots, N$, where

$$\boldsymbol{\mu} = \begin{bmatrix} \boldsymbol{\mu}_1 \\ \boldsymbol{\mu}_2 \\ \vdots \\ \boldsymbol{\mu}_g \end{bmatrix} \quad \text{and} \quad \boldsymbol{\Sigma} \begin{bmatrix} \boldsymbol{\Sigma}_{11} & \boldsymbol{\Sigma}_{12} & \cdots & \boldsymbol{\Sigma}_{1g} \\ & \boldsymbol{\Sigma}_{22} & \cdots & \boldsymbol{\Sigma}_{2g} \\ & & \ddots & \vdots \\ & & & \boldsymbol{\Sigma}_{gg} \end{bmatrix}$$

then the test of independence becomes

$$H_0: \boldsymbol{\Sigma}_{ij} = \mathbf{0} \qquad \text{for } i = j \tag{3.17.23}$$

Letting

$$V = \frac{|\mathbf{S}|}{|\mathbf{S}_{11}| \cdots |\mathbf{S}_{gg}|} = \frac{|\mathbf{R}|}{|\mathbf{R}_{11}| \cdots |\mathbf{R}_{gg}|} \tag{3.17.24}$$

where $\mathbf{S}$ is the unbiased estimate of $\boldsymbol{\Sigma}$ and $\mathbf{R}$ is the sample correlation matrix, compute the statistic

$$X_{I_g}^2 = -c^{-1}(N-1)\log V \tag{3.17.25}$$

where

$$c^{-1} = 1 - \frac{2\boldsymbol{\Sigma}_3 + 3\boldsymbol{\Sigma}_2}{12v(N-1)} \tag{3.17.26}$$

$$v = \frac{\boldsymbol{\Sigma}_2}{2}$$

$$\boldsymbol{\Sigma}_s = \left(\sum_{i=1}^{g} p_i\right)^s - \sum_{i=1}^{g} p_i^s \qquad s = 2, 3$$

to test (3.17.23).

The hypothesis of independence for g sets is rejected at the significance level α if

$$X_{I_g}^2 > \chi^2(v) \tag{3.17.27}$$

since $X_{I_g}^2 \mathrel{\dot\sim} \chi^2(v)$ when H_0 is true. For more precise approximations for large values of p, see Box (1949).

To illustrate some of the test statistics developed in this section, we will consider some of the examples already discussed in the preceding sections of this chapter.

In Section 3.9, Hotelling's T^2 statistic was employed to test for the equality of two mean vectors. Implicit in employing T^2 was the assumption that $\boldsymbol{\Sigma}_1 = \boldsymbol{\Sigma}_2$. Although each N_i is too small to use equation (3.17.8) to test the equality of variance-covariance matrices, the example is used to illustrate the procedure. Using the data given for the two methods in Section 3.9, the sample variance-covariance matrices for the methods are

$$\mathbf{S}_1 = \begin{bmatrix} 296.66 & \\ 306.66 & 426.67 \end{bmatrix} \quad \text{and} \quad \mathbf{S}_2 = \begin{bmatrix} 317.50 & \\ 255.00 & 416.67 \end{bmatrix}$$

where $N_1 = 6$, $N_2 = 6$, and $p = 2$. Using equation (3.17.5), the estimate of the common (pooled) within-variance-covariance matrix is

$$\begin{aligned} \mathbf{S} &= \frac{1}{N - g} \sum_{i=1}^{g} v_i \mathbf{S}_i \\ &= \frac{1}{12 - 2}[(6 - 1)\mathbf{S}_1 + (6 - 1)\mathbf{S}_2] \\ &= \begin{bmatrix} 307.08 & \\ 280.83 & 421.67 \end{bmatrix} \end{aligned}$$

which of course agrees with the result given by (3.9.6). Evaluating C, M, and v,

$$\begin{aligned} C &= \frac{2p^2 + 3p - 1}{6(p + 1)(g - 1)} \left(\sum_{i=1}^{g} \frac{1}{v_i} - \frac{1}{N - g} \right) \\ &= \frac{2(2)^2 + 3(2) - 1}{6(2 + 1)(2 - 1)} \left(\frac{1}{5} + \frac{1}{5} - \frac{1}{10} \right) \\ &= .2167 \end{aligned}$$

$$\begin{aligned} M &= (N - g) \log |\mathbf{S}| - \sum_{i=1}^{g} v_i \log |\mathbf{S}_i| \\ &= (N - g) \log 50{,}620.93 - 5(\log 32{,}535.57 + \log 67{,}267.73) \\ &= 10(10.8321) - 5(10.3901 + 11.1164) \\ &= .6485 \end{aligned}$$

and

$$\begin{aligned} v &= \frac{p(p + 1)(g - 1)}{2} \\ &= \frac{2(2 + 1)(2 - 1)}{2} \\ &= 3 \end{aligned}$$

The statistic X_B^2 for testing $H_0: \boldsymbol{\Sigma}_1 = \boldsymbol{\Sigma}_2$ becomes $X_B^2 = (1 - C)M = .2279$. For $\alpha = .05$, the hypothesis is not rejected since $X_B^2 < \chi^2_{.05}(3) = 7.81$. This analysis shows that the equality of variance assumption, required when using Hotelling's T^2 statistic, is tenable.

To test the hypotheses discussed in the two-sample, repeated-measures example (Section 3.16), the variance-covariance matrices must first satisfy the condition that $\boldsymbol{\Sigma}_1 = \boldsymbol{\Sigma}_2$. However, in addition to this assumption, the hypothesis of compound symmetry must be tested if univariate methods are employed to analyze the data. Huynh and Feldt (1970) review a test for more general patterns of $\mathbf{\Sigma}$ for which the one considered here is a special case. For the data in Section 3.16, the hypothesis of compound symmetry is

$$H_0: \mathbf{\Sigma} = \sigma^2 \begin{bmatrix} 1 & \rho & \rho & \cdots & \rho \\ \rho & 1 & \rho & \cdots & \rho \\ \rho & \rho & 1 & \cdots & \rho \\ \vdots & \vdots & \vdots & \vdots & \vdots \\ \rho & \rho & \rho & \cdots & 1 \end{bmatrix}$$

The value of the within-variance-covariance matrix $\mathbf{S}$ is given in Section 3.16, where the sample size $N = 20$, the number of variables, $p = 5$, and the number of groups $g = 2$. From $\mathbf{S}$, $\bar{\mathbf{S}}$ is computed so that

$$\bar{\mathbf{S}} = \begin{bmatrix} 86.3898 & & & & \\ 28.1578 & 86.3898 & & & \\ 28.1578 & 28.1578 & 86.3898 & & \\ 28.1578 & 28.1578 & 28.1578 & 86.3898 & \\ 28.1578 & 28.1578 & 28.1578 & 28.1578 & 86.3898 \end{bmatrix}$$

and the $|\mathbf{S}| = 9.8069504 \times 10^8$ and the $|\bar{\mathbf{S}}| = 2.288469 \times 10^9$. To find the $|\bar{\mathbf{S}}|$, it is convenient to apply the following general formula, which applies to matrices of the form given by $\bar{\mathbf{S}}$.

$$|\bar{\mathbf{S}}| = (\bar{s}^2 - \bar{s})^{p-1}[\bar{s}^2 + (p-1)\bar{s}]$$

In order to obtain the value of the test statistic X^2_{BX} defined by (3.17.17), the following need to be evaluated:

$$\begin{aligned} M_X &= -(N-g)\log\frac{|\mathbf{S}|}{|\bar{\mathbf{S}}|} \\ &= 15.2528 \\ C_x &= \frac{p(p+1)^2(2p-3)}{6(p-1)(N-g)(p^2+p-4)} \\ &= \frac{5(6)^2[2(5)-3]}{6(4)(18)(26)} \\ &= .1122 \\ v &= \frac{p^2+p-4}{2} = 13 \end{aligned}$$

so that $X^2_{BX} = (1 - C_x)M_x = 13.541$. The critical value for $\alpha = .10$ is $\chi^2_{.10}(13) = 19.8$, indicating that the hypothesis of compound symmetry is not rejected, so that univariate procedures may be applied to analyze the data.

Because of the dependence that existed between the two mathematical ability measures, Hotelling's one-sample T^2 statistic was used to analyze the data of

Section 3.4. Given independence, however, univariate techniques could be applied to study the variables individually. To test for independence, $H_0: \mathbf{P} = \mathbf{I}$, equation (3.17.20) is employed (although there are better procedures for $p = 2$). For the data summarized in Section 3.4, where $N = 28$ and $p = 2$, the sample correlation matrix $\mathbf{R}$ is

$$\mathbf{R} = \begin{bmatrix} 1.0 & \\ .8854 & 1.0 \end{bmatrix}$$

Evaluating the criterion,

$$\begin{aligned} X_I^2 &= -\left[(N-1) - \frac{2p+5}{6}\right] \log|\mathbf{R}| \\ &= \left[(28-1) - \frac{2(2)+5}{6}\right] \log .2161 \\ &= \left(27 - \frac{9}{6}\right) 1.53244 \\ &= 39.08 \end{aligned}$$

and comparing X_I^2 with the chi-square value $\chi_\alpha^2[p(p-1)/2] = \chi_{.05}^2(1) = 3.84$, for $\alpha = .05$, the hypothesis of independence is rejected. We shall use the same data to test for sphericity, $H_0: \boldsymbol{\Sigma} = \sigma^2 I$, which implies independence and equal variances. From $\mathbf{R}$, the variance-covariance matrix $\mathbf{S}$ is

$$\mathbf{S} = \begin{bmatrix} 290.4074 & \\ 223.7519 & 219.9296 \end{bmatrix}$$

and the $|\mathbf{S}| = 13{,}804.27$ and $\text{Tr}(\mathbf{S}) = 510.337$. Evaluating the test statistic X_s^2,

$$\begin{aligned} X_s^2 &= -\left(N - 1 - \frac{2p^2 + p + 2}{6p}\right)\left(\log|\mathbf{S}| - p \log \frac{\text{Tr}(\mathbf{S})}{p}\right) \\ &= -(28 - 1 - 12/12)(9.4327 - 11.08385) \\ &= 40.33 \end{aligned}$$

By comparing X_s^2 with the chi-square critical value of

$$\chi_\alpha^2\left[\frac{(p-1)(p+2)}{2}\right] = \chi_{.05}^2(2) = 5.99$$

for $\alpha = .05$, the test of sphericity is rejected.

The primary purpose of the test procedures discussed in this section is to provide the researcher with a formal set of techniques to check the model assumption of equal variance-covariance matrices. Less formal graphical procedures have been developed by Gnanadesikan and Lee (1970).

Although multivariate normality is fundamental to the tests discussed in this section, nothing has been said about ways in which this assumption can be checked. Wilk and Gnanadesikan (1968) discuss informal gamma-plotting procedures for this purpose. Examples of the techniques may be found in Roy, Gnanadesikan, and Strivastava (1971). For an introduction to gamma plotting, see Wilk, Gnanadesikan, and Huyett (1962). If evidence of nonnormality is detected by using gamma plots, transformation of the data to multivariate normality may be warranted. Andrews, Gnanadesikan, and Warner (1971) propose a class of useful marginal and joint transformations to obtain normality for various kinds of nonnormal data.

EXERCISES 3.17

1. Employ the data in Exercise 1, Section 3.4 and Exercise 2, Section 3.9 to test the hypothesis of equal variance-covariance matrices by using Bartlett's, Box's, and Layard's test statistics (using $\alpha = .05$).
2. For the sample correlation matrix given by

$$\mathbf{R} = \begin{bmatrix} 1.000 & .347 & .170 & .114 \\ & 1.000 & .478 & .433 \\ & & 1.000 & .512 \\ & & & 1.000 \end{bmatrix}$$

   based on 500 observations, test the hypothesis of independence.
3. Use both Bartlett's X^2_{BX} statistic and Box's F statistic to test the hypothesis of compound symmetry for Exercise 1, Section 3.4.
4. For the data in Exercise 1, Section 3.4, test the sphericity hypothesis by using (3.17.22). How would the acceptance of the hypothesis affect your analysis of the data?

3.18 THE TWO-SAMPLE, BEHRENS-FISHER PROBLEM

Suppose there are two independent samples from normal populations and unknown variances σ_1^2 and σ_2^2. To test the hypothesis of equal means, we assume or test the assumption of equal variances to apply the t statistic for determining mean differences. When $\sigma_1^2 \neq \sigma_2^2$, the t statistic may no longer be applied. Testing for equality of means with unknown and unequal variances is referred to as the *Behrens-Fisher problem.* Although numerous procedures have been provided to solve this problem, Welch's (1937, 1947) solution is presented since it is a satisfactory practical solution (see also Scheffé, 1970, and Mehta and Srinivasan, 1970).

Applying Welch's solution, which is an approximate t solution, to test $H_0: \mu_1 = \mu_2$ in the two-sample univariate situation, the statistic

$$t_v = \frac{y_{1.} - y_{2.}}{\sqrt{s_1^2/N_1 + s_2^2/N_2}} \tag{3.18.1}$$

is computed, which has an approximate t distribution with degrees of freedom

$$v = \frac{(s_1^2/N_1 + s_2^2/N_2)^2}{\dfrac{(s_1^2/N_1)^2}{(N_1 + 1)} + \dfrac{(s_2^2/N_2)^2}{(N_2 + 1)}} - 2 \tag{3.18.2}$$

to the nearest integer. The hypothesis of equal means is rejected at the significance level α if

$$|t_v| > t^{\alpha/2}(v) \tag{3.18.3}$$

By using (3.18.1), an approximate $100(1 - \alpha)\%$ confidence interval for the difference $\mu_1 - \mu_2$ is immediately obtained:

$$\begin{aligned} y_{1.} - y_{2.} - t^{\alpha/2}(v)\sqrt{s_1^2/N_1 + s_2^2/N_2} &\le \mu_1 - \mu_2 \\ &\le y_{1.} - y_{2.} + t^{\alpha/2}(v)\sqrt{s_1^2/N_1 + s_2^2/N_2} \end{aligned} \tag{3.18.4}$$

Multivariate extensions of the Behrens–Fisher problem have been treated by Anderson (1958, p. 118), Bennett (1951), Eaton (1969), and Ito (1969). Most of their procedures extend Scheffé's (1943) technique of randomization to solve the problem. However, as Scheffé (1970) and Eaton (1969) mention, any Scheffé procedure has two very serious shortcomings: (1) the procedures necessitate randomized pairing of the samples, and (2) when the sample sizes are unequal, data are discarded and not used in the test procedure. Scheffé states that "the effect of this in practice would be deplorable," and therefore refers to the procedures as impractical solutions.

A test lacking the aforementioned shortcomings, proposed by James (1954) and treated by Ito (1969), will now be discussed. However, the test procedure is asymptotic and should be used only if each sample size is near 50.

For the two-sample case, let p variate vectors $\mathbf{Y}_{ij} \sim IN(\boldsymbol{\mu}_i, \boldsymbol{\Sigma}_i)$, for $i = 1, 2$ and $j = 1, \ldots, N_i$. To test the hypothesis

$$H_0: \boldsymbol{\mu}_1 = \boldsymbol{\mu}_2 \qquad \text{when } \boldsymbol{\Sigma}_1 \ne \boldsymbol{\Sigma}_2 \tag{3.18.5}$$

the test statistic

$$T_v^2 = (\mathbf{y}_{1.} - \mathbf{y}_{2.})'\left(\frac{\mathbf{S}_1}{N_1} + \frac{\mathbf{S}_2}{N_2}\right)^{-1}(\mathbf{y}_{1.} - \mathbf{y}_{2.}) \tag{3.18.6}$$

is formed where the $\mathbf{y}_{i.}$ denote sample mean vectors and the $\mathbf{S}_i$ denote sample variance-covariance matrices.

When H_0 is true, the test statistic T_v^2 is distributed approximately, as the sample sizes tend to infinity, as a chi-squared distribution with degrees of freedom $v = p$.

The hypothesis of equal mean vectors would be rejected at the significance level α if

$$T_v^2 > \chi_\alpha^2(p) = T^\alpha(p, \infty) \tag{3.18.7}$$

By using (3.18.7), approximate $100(1 - \alpha)\%$ confidence intervals may be obtained by evaluating

$$\begin{aligned} \mathbf{a}'(\mathbf{y}_{1.} - \mathbf{y}_{2.}) - c_0\sqrt{\mathbf{a}'\left(\frac{\mathbf{S}_1}{N_1} + \frac{\mathbf{S}_2}{N_2}\right)\mathbf{a}} &\le \mathbf{a}'(\boldsymbol{\mu}_1 - \boldsymbol{\mu}_2) \\ &\le \mathbf{a}'(\mathbf{y}_{1.} - \mathbf{y}_{2.}) + c_0\sqrt{\mathbf{a}'\left(\frac{\mathbf{S}_1}{N_1} + \frac{\mathbf{S}_2}{N_2}\right)\mathbf{a}} \end{aligned} \tag{3.18.8}$$

where $c_0^2 = T^\alpha(p, \infty)$.

Employing James' (1954) first-order approximation for the statistic T_v^2, an approximation for the distribution of T_v^2 is obtained and may be used for small samples. To test (3.18.7), the hypothesis is rejected if

$$T_v^2 > \chi_\alpha^2(p)\left\{1 + \frac{1}{2}\left[\frac{k_1}{p} + \frac{k_2\chi_\alpha^2(p)}{p(p+2)}\right]\right\} \tag{3.18.9}$$

where

$$k_1 = \frac{\sum_{i=1}^{2} \mathrm{Tr}(\mathbf{W}^{-1}\mathbf{W}_i)^2}{N_i - 1}$$

$$k_2 = \frac{\sum_{i=1}^{2} \mathrm{Tr}(\mathbf{W}^{-1}\mathbf{W}_i)^2 + 2\,\mathrm{Tr}(\mathbf{W}^{-1}\mathbf{W}_i\mathbf{W}^{-1}\mathbf{W}_i)}{N_i - 1}$$

$$\mathbf{W}_i = \frac{\mathbf{S}_i}{N_i} \quad \text{and} \quad \mathbf{W} = \sum_i \mathbf{W}_i$$

The important contribution of Yao's (1965) work to the multivariate Behrens–Fisher problem is in the extension of Welch's approximate t solution to the multivariate case. The test statistic is again defined by T_v^2; however, the critical region for the test is to reject at the significance level α if

$$T_v^2 > T^\alpha(p, v) \tag{3.18.10}$$

where

$$\frac{1}{v} = \sum_{i=1}^{2}\left(\frac{1}{N_i - 1}\right)\left[\frac{(\mathbf{y}_{1.} - \mathbf{y}_{2.})'\mathbf{W}^{-1}\mathbf{W}_i\mathbf{W}^{-1}(\mathbf{y}_{1.} - \mathbf{y}_{2.})}{T_v^2}\right]^2 \tag{3.18.11}$$

Furthermore, with the approximate T^2 solution, $100(1-\alpha)\%$ simultaneous confidence intervals for $\psi = \mathbf{a}'(\boldsymbol{\mu}_1 - \boldsymbol{\mu}_2)$ are also evidenced.

Comparing Yao's solution with James' first-order approximate solution, Yao (1965) shows that the level of significance α for his procedure is always smaller than James'. Both are near the nominal level α when the sample sizes are equal; however, as the ratio of the sample sizes increases, Yao's procedure seems to be superior.

EXAMPLE 3.18.1 To test the equality of means when $\boldsymbol{\Sigma}_1 \neq \boldsymbol{\Sigma}_2$, values of the mean vectors and variance-covariance matrices given by

$$\mathbf{y}_{1.} = \begin{bmatrix} 45.0 \\ 90.0 \end{bmatrix} \qquad \mathbf{y}_{2.} = \begin{bmatrix} 40.0 \\ 80.0 \end{bmatrix}$$

$$\mathbf{S}_1 = \begin{bmatrix} 80.0 & \\ 30.0 & 20.0 \end{bmatrix} \qquad \mathbf{S}_2 = \begin{bmatrix} 120.0 & \\ -100.0 & 200.0 \end{bmatrix}$$

respectively, where $N_1 = 10$ and $N_2 = 20$, are considered for illustration. The techniques of this section are applied to test $H_0: \boldsymbol{\mu}_1 = \boldsymbol{\mu}_2$ since $\boldsymbol{\Sigma}_1 \neq \boldsymbol{\Sigma}_2$.

The test statistic is

$$T_v^2 = (\mathbf{y}_{1.} - \mathbf{y}_{2.})'\left(\frac{\mathbf{S}_1}{N_1} + \frac{\mathbf{S}_2}{N_2}\right)^{-1}(\mathbf{y}_{1.} - \mathbf{y}_{2.})$$

$$= [5.0, 10.0]'\begin{bmatrix} .0731707 & \\ .0121951 & .0853659 \end{bmatrix}\begin{bmatrix} 5.0 \\ 10.0 \end{bmatrix}$$

$$= 11.5854$$

For $\alpha = .01$, $\chi^2_{.01}(2) = 9.210$. The values for N_i are not large enough to use asymptotic results; however, by using James' first-order approximation, the critical value for the test of H_0 is

$$J_\alpha = \chi^2_\alpha(p)\left\{1 + \frac{1}{2}\left[\frac{k_1}{p} + \frac{k_2\chi^2_\alpha(p)}{p(p+2)}\right]\right\}$$

$$= 9.210\left[1 + \frac{1}{2}\left(\frac{.1485}{2} + \frac{.3229(9.210)}{2(4)}\right)\right]$$

$$= 11.264$$

Following Yao's method,

$$\frac{1}{v} = \sum_{i=1}^{2}\left(\frac{1}{N_i - 1}\right)\frac{[(\mathbf{y}_{1.} - \mathbf{y}_{2.})'\mathbf{W}^{-1}\mathbf{W}_i\mathbf{W}^{-1}(\mathbf{y}_{1.} - \mathbf{y}_{2.})]^2}{(T_v^2)^2}$$

$$= \left(\frac{1}{11.5854}\right)^2\left[\frac{(6.2537)^2}{9} + \frac{(5.3317)^2}{19}\right]$$

$$= .04352$$

so that $v = 22.976$. The T^2 critical value for the test, for $\alpha = .01$, is $T^{.01}_{(2,23)} = 11.958$. In summary, the critical values for the test of equal mean vectors, for $\alpha = .01$, are

Asymptotic	James	Yao
9.210	11.264	11.958

With Yao's procedure, the null hypothesis is not rejected; however, if either James' asymptotic critical value or the adjusted asymptotic value had been used, H_0 would have been rejected.

From Yao's Monte Carlo results, the α level for his procedure and James' method are near the nominal level α. The large-sample theory is certainly in error. The best procedure for testing (3.18.5), when $\mathbf{\Sigma}_1 \neq \mathbf{\Sigma}_2$, seems to be Yao's method, especially when $N_1 > N_2$.

EXERCISES 3.18

1. In Exercise 4a, Section 3.9, test the hypothesis of equal mean performance by using the statistic T_v^2 in (3.18.6) with Yao's critical value.

3.19 POWER CALCULATIONS

In Section 2.7, power and sample-size calculations using the F distribution were illustrated. Since Hotelling's T^2 distribution is related to the F distribution, these methods may also be applied to determine sample size and power when employing the T^2 statistic.

From (2.7.8), if $\mathbf{Y} \sim N_p(\boldsymbol{\mu}, \boldsymbol{\Sigma})$, $n\mathbf{S} \sim W_p(n, \boldsymbol{\Sigma})$, and $\mathbf{Y}$ and $\mathbf{S}$ are independent, then

$$\frac{n-p-1}{p}\frac{T^2}{n} \sim F(p, n-p-1, \delta) \tag{3.19.1}$$

where $T^2 = \mathbf{Y}'\mathbf{S}^{-1}\mathbf{Y}$ and the noncentrality parameter $\delta = \boldsymbol{\mu}'\boldsymbol{\Sigma}^{-1}\boldsymbol{\mu}$. Applying (3.19.1) for testing $H_0: \boldsymbol{\mu} = \boldsymbol{\mu}_0$ in the one-sample problem,

$$T^2 = N(\overline{\mathbf{Y}} - \mu_0)'\mathbf{S}^{-1}(\overline{\mathbf{Y}} - \mu_0) \tag{3.19.2}$$

and

$$\delta = N(\boldsymbol{\mu} - \boldsymbol{\mu}_0)'\boldsymbol{\Sigma}^{-1}(\boldsymbol{\mu} - \boldsymbol{\mu}_0) \tag{3.19.3}$$

For the two-sample problem,

$$T^2 = \frac{N_1 N_2}{N_1 + N_2}(\mathbf{y}_{1.} - \mathbf{y}_{2.})'\mathbf{S}^{-1}(\mathbf{y}_{1.} - \mathbf{y}_{2.}) \tag{3.19.4}$$

and

$$\delta = \frac{N_1 N_2}{N_1 + N_2}(\boldsymbol{\mu}_1 - \boldsymbol{\mu}_2)'\boldsymbol{\Sigma}^{-1}(\boldsymbol{\mu}_1 - \boldsymbol{\mu}_2) \tag{3.19.5}$$

In particular, the noncentrality parameter is obtained from the test statistic by replacing sample estimates by their expectations.

Finding power and sample sizes by using T^2 is similar to the procedure outlined in Section 2.7 for the F distribution since the noncentral T^2 distribution is related to the noncentral F distribution.

For example, to determine power for the two-sample multivariate problem, the type I error α is first specified. Then, given an estimate of $\boldsymbol{\Sigma}$ (based on past experience) and the magnitude of the difference $\boldsymbol{\mu}_1 - \boldsymbol{\mu}_2$ desired for detection, ϕ is computed. Entering the charts with the parameters associated with the F distribution, since $T^2 = [(N_1 + N_2 - 2)p/(N_1 + N_2 - p - 1)]F(p, N_1 + N_2 - p - 1)$, where $v_h = p$, $v_e = N_1 + N_2 - p - 1, \alpha$, and ϕ, power $= 1 - \beta$ is obtained by using Table V in the Appendix.

Using the data from the two-sample problem (see Section 3.9), suppose location of a mean difference of

$$\boldsymbol{\mu}_1 - \boldsymbol{\mu}_2 = \begin{bmatrix} 20.38 \\ -38.33 \end{bmatrix} \qquad \text{where } \boldsymbol{\Sigma} = \begin{bmatrix} 307.08 & 421{\cdot}67 \\ 280.83 & 421.67 \end{bmatrix}$$

was sought. Then

$$\delta = \frac{N_1 N_2}{N_1 + N_2}(\boldsymbol{\mu}_1 - \boldsymbol{\mu}_2)'\boldsymbol{\Sigma}^{-1}(\boldsymbol{\mu}_1 - \boldsymbol{\mu}_2) = 64.17$$

and

$$\phi = \left(\frac{\delta}{p+1}\right)^{1/2} = \left(\frac{64.17}{3}\right)^{1/2} = 4.62$$

If $\alpha = .01$, $\nu_h = 2$, $\nu_e = 9$, and $\phi = 4.62$, then power $= 1 - \beta = .99$.

In the determination of power for a given difference $\boldsymbol{\mu}_1 - \boldsymbol{\mu}_2$, an estimate of $\boldsymbol{\Sigma}$ obtained from the data was employed. However, in determining sample size, given a mean difference of interest to the experimenter, an accurate estimate of $\boldsymbol{\Sigma}$ must be available. This is usually obtained from previous research or from a pilot study. For $N_1 = N_2 = N_0$, a guess for N_0 is used in the formula for δ to compute ϕ and power, with α and ν_h specified. The value of N_0 is incremented until the value of $1 - \beta$ specified by the experimenter is obtained. Thus, to determine sample size for a given experimental situation, power calculations are performed iteratively until the value of the desired power is attained. For example, with

$$\boldsymbol{\mu}_1 - \boldsymbol{\mu}_2 = \begin{bmatrix} 20.83 \\ -38.33 \end{bmatrix} \quad \text{and} \quad \boldsymbol{\Sigma} = \begin{bmatrix} 307.08 & \\ 280.83 & 421.67 \end{bmatrix}$$

$$\delta = \frac{N_0}{2}(\boldsymbol{\mu}_1 - \boldsymbol{\mu}_2)'\boldsymbol{\Sigma}^{-1}(\boldsymbol{\mu}_1 - \boldsymbol{\mu}_2)$$

By setting $N_0 = 5$, then $\nu_e = 7$, $\phi = 4.23$, and $1 - \beta = 0.979$, for $\alpha = .01$. With $N_0 = 6$, $\nu_e = 9$, $\phi = 4.62$, and $1 - \beta = .99$, if $\alpha = .01$.

In this chapter, assuming multivariate normality and homogeneity of variance-covariance matrices, we have summarized a set of frequently used formal statistical procedures for testing hypotheses about means in the one- and two-sample case through the general linear model. All the models considered are multivariate functionally related models or, in particular, multivariate experimental design models. Further consideration is given these models in Chapter 5. In the next chapter multivariate conditional regression models are investigated.

EXERCISES 3.19

1. A researcher desires to detect differences of 1, 3, and 5 units on three dependent variables in an experiment comparing two treatments. Randomly assigning an equal number of subjects to the two conditions, with

$$\boldsymbol{\Sigma} = \begin{bmatrix} 10 & & \\ 5 & 10 & \\ 5 & 5 & 10 \end{bmatrix}$$

and $\alpha = .05$, how large a sample size is required if the experimenter wants the power of the test to be at least .80?

2. Estimate the power of the tests for testing the hypotheses in Exercise 1, Section 3.4, in Exercise 2, Section 3.9, and in Exercise 1, Section 3.12.

CHAPTER 4

MULTIVARIATE REGRESSION AND CANONICAL CORRELATION ANALYSIS

4.1 INTRODUCTION

In Chapter 2, we found that the mean of the conditional distribution of a random variable Y, given $X = x$, is a linear function of x if $\boldsymbol{\mu}' = [Y, X]$ is assumed to have a bivariate normal distribution. The means of the conditional distribution followed a theoretical regression line for Y, given $X = x$, $\mu_{y \cdot x} = \mu_y + \rho(\sigma_y/\sigma_x)(x - \mu_x)$. Since both variables were random, the regression line of X, given $Y = y$, was also obtained and found to be a linear function of y. We used the population correlation coefficient ρ to determine the extent to which Y and X were linearly related. A model of this type, which assumes that both Y and X are random variables sampled from a single normal population, is a special case of the *random regression model* and is often referred to as a *normal correlation model*. In terms of the linear model, $Y = \beta_0 + \beta_1 X + \varepsilon$, Y and X are random observable variables measured without error, $\boldsymbol{\beta}' = [\beta_0, \beta_1]$ is a nonrandom vector, ε is an error component, and ε and X are assumed to be independent. The experimenter is interested in the degree of the joint variation between the two random variables. The interpretation, given ρ or r in the sample, is dependent on the form of the joint distribution of Y and X. In the nonnormal case, the meaning of a correlation coefficient is not well defined (see Carroll, 1961). In the multivariate case, the researcher is investigating the mutual dependencies that exist between two random vectors **Y** and **X**.

Although the normal correlation model may be used for prediction, it is more often the case that the model assumptions are not tenable; that is, knowledge about the joint distribution of Y and X does not exist. In behavioral research, experimenters are more likely to select fixed values of X to predict Y. In this case, X is not a random variable, and the interpretation of the sample correlation coefficient is no longer meaningful when used as a measure of the linear relationship between Y and x. The value of r will depend on the choices of the values of x. In these cases, a conditional regression model is most appropriate.

A conditional regression model is not a one-sample problem, but a k-sample problem where values of the dependent variable Y are obtained at each of k fixed x values. The relationship between Y and x is of the form $Y = f(x) + \varepsilon$, where $E(Y|x) = f(x)$ and $V(Y|x) = \sigma^2$, so that the variance of Y, given $X = x$, is constant at every fixed x value and no distributional assumption is made about the conditional distribution of Y. A special case of interest is when $f(x)$ is linear in the parameters, such as $f(x) = \alpha + \beta x + \gamma x^2 + \cdots$. Such a model is called the *classical linear regression model*. If we can assume that Y has a normal distribution at each $X = x$, the model is termed the *normal linear regression model*. Use of such models enables a researcher to predict Y from fixed values of X measured without error.

When $f(x)$ is of the form $f(x) = \alpha + \beta x$, the correlation model and regression model lead to similar mathematical results; however, keep in mind that different assumptions are required when using the models. The correlation model requires more stringent distributional assumptions. This chapter provides some of the theory and differences in the models and extends univariate procedures to their multivariate analogues.

4.2 MULTIPLE LINEAR REGRESSION—UNIVARIATE

Univariate multiple regression analysis is a statistical technique practiced in experimental work to examine the manner in which several variables considered simultaneously can be used to understand or predict the behavior of one response or dependent variable.

Defining **y** to be the vector of observations, **X** to be the matrix of fixed known independent variables, **β** to be the vector of parameters estimated, and **ε** to be the vector of errors, it is assumed that there is a linear relationship (linear in the parameters) of the form

$$y = \beta_0 + \beta_1 x_1 + \cdots + \beta_k x_k + \varepsilon \tag{4.2.1}$$

between a dependent variable y and k independent variables $x_1, x_2, \ldots, x_k$ in multiple linear regression analysis. Employing matrix notation, we represent the observations as

$$\begin{bmatrix} y_1 \\ y_2 \\ \vdots \\ y_N \end{bmatrix} = \begin{bmatrix} 1 & x_{11} & \cdots & x_{1k} \\ 1 & x_{21} & \cdots & x_{2k} \\ \vdots & \vdots & \vdots & \vdots \\ 1 & x_{N1} & \cdots & x_{Nk} \end{bmatrix} \begin{bmatrix} \beta_0 \\ \beta_1 \\ \vdots \\ \beta_k \end{bmatrix} + \begin{bmatrix} \varepsilon_1 \\ \varepsilon_2 \\ \vdots \\ \varepsilon_N \end{bmatrix} \tag{4.2.2}$$

or

$$\underset{(N\times 1)}{\mathbf{y}} = \underset{(N\times q)}{\mathbf{X}}\ \underset{(q\times 1)}{\boldsymbol{\beta}} + \underset{(N\times 1)}{\boldsymbol{\varepsilon}} \tag{4.2.3}$$

where

$$E(\mathbf{Y}) = \mathbf{X}\boldsymbol{\beta} \qquad V(\mathbf{Y}) = \sigma^2\mathbf{I} \tag{4.2.4}$$

$q = k + 1$, and $N > q$. Result (4.2.3) is known as the *classical linear regression model* in the above context. In order to apply statistical tests, we make the further assumption that

$$\mathbf{Y} \sim N_N(\mathbf{X}\boldsymbol{\beta}, \sigma^2\mathbf{I}) \tag{4.2.5}$$

Expression (4.2.3) with assumption (4.2.5) is known as the *normal linear regression model.*

By applying the Gauss-Markoff theorem, (3.5.14), to (4.2.3), the elements β_i in (4.2.1) that minimize the error sum of squares $Q_e = (\mathbf{y} - \mathbf{X}\boldsymbol{\beta})'(\mathbf{y} - \mathbf{X}\boldsymbol{\beta})$ are estimated by

$$\hat{\boldsymbol{\beta}} = (\mathbf{X}'\mathbf{X})^{-1}\mathbf{X}'\mathbf{y} \tag{4.2.6}$$

where

$$E(\hat{\boldsymbol{\beta}}) = \boldsymbol{\beta} \quad \text{and} \quad V(\hat{\boldsymbol{\beta}}) = \sigma^2(\mathbf{X}'\mathbf{X})^{-1} \tag{4.2.7}$$

If the $|\mathbf{X}'\mathbf{X}| = 0$ in the conditional regression model, the inverse of $\mathbf{X}'\mathbf{X}$ does not exist and the estimate $\hat{\boldsymbol{\beta}}$ is no longer unique. The term *multicollinearity* is associated with the condition $|\mathbf{X}'\mathbf{X}| = 0$ and occurs if some independent variables are dependent on other independent variables. Writing $\hat{\boldsymbol{\beta}}$ as in (4.2.6), we are assuming no multicollinearity in the data. Often the $|\mathbf{X}'\mathbf{X}| \neq 0$, but $\mathbf{X}'\mathbf{X}$ is nearly singular. When this occurs, the variances of the elements of the vector $\hat{\boldsymbol{\beta}}$ will be large. This frequently happens in multiple curvilinear regression models (see Section 4.4). In order to obtain an unbiased estimate of σ^2, (3.5.14c) is applied, and σ^2 is estimated by

$$s^2 = \frac{\mathbf{y}'[\mathbf{I} - \mathbf{X}(\mathbf{X}'\mathbf{X})^{-1}\mathbf{X}']\mathbf{y}}{N - g} \tag{4.2.8}$$

For the simple linear regression model in Section 1.8, where $y = \alpha + \beta x$, estimates α_0 and β_0 for α and β were found by using (4.2.6). Letting $\hat{\alpha} = \alpha_0$ and $\hat{\beta} = \beta_0$, (4.2.6) may be used to find the variance of the parameters in the case of simple linear regression; then the

$$V(\hat{\boldsymbol{\beta}}) = \sigma^2(\mathbf{X}'\mathbf{X})^{-1} = \sigma^2 \begin{bmatrix} \dfrac{\sum_i x_i^2}{N\sum_i (x_i - \bar{x})^2} & \\ \dfrac{-\sum_i x_i}{N\sum_i (x_i - \bar{x})^2} & \dfrac{1}{\sum_i (x_i - \bar{x})^2} \end{bmatrix} \tag{4.2.9}$$

so that

$$V(\hat{\alpha}) = \sigma^2 \frac{\sum_i x_i^2}{N \sum_i (x_i - \bar{x})^2}$$

(4.2.10)
$$V(\hat{\beta}) = \frac{\sigma^2}{\sum_i (x_i - \bar{x})^2}$$

$$\text{cov}(\hat{\alpha}, \hat{\beta}) = \frac{-\sigma^2 \sum_i x_i}{N \sum_i (x_i - \bar{x})^2}$$

Alternatively, since $\hat{\alpha} = \bar{y} - \hat{\beta}\bar{x}$,

$$V(\hat{\alpha}) = V(\bar{y}) + V(\hat{\beta}\bar{x}) - 2\,\text{cov}(\bar{y}, \hat{\beta}\bar{x})$$

$$= \frac{\sigma^2}{N} + \frac{\bar{x}^2\sigma^2}{\sum_i (x_i - \bar{x})^2} - 0$$

(4.2.11)
$$= \sigma^2\left[\frac{1}{N} + \frac{\bar{x}^2}{\sum_i (x_i - \bar{x})^2}\right]$$

which is the familiar form in elementary texts for the variance of the constant term. To prove that the covariance term in (4.2.11) is indeed 0, recall that

$$\hat{\beta} = \frac{\sum_i (x_i - \bar{x})(y_i - \bar{y})}{\sum_i (x_i - \bar{x})^2} = \frac{\sum_i (x_i - \bar{x})y_i}{\sum_i (x_i - \bar{x})^2}$$

so that

$$\text{cov}(\bar{y}, \bar{x}\hat{\beta}) = \text{cov}(\bar{y}, \hat{\beta}) = \sigma^2 \sum_i \frac{1}{N} \frac{x_i - \bar{x}}{\sum_i (x_i - \bar{x})^2} = 0$$

Having obtained $\hat{\boldsymbol{\beta}}$, the fitted or predicted value of $\mathbf{y}$ is given by $\hat{\mathbf{y}} = \mathbf{X}\hat{\boldsymbol{\beta}}$. The error made in using $\hat{\mathbf{y}}$ for $\mathbf{y}$ is given by the residual vector $\mathbf{e} = \mathbf{y} - \hat{\mathbf{y}}$. The mean and variance of $\mathbf{e}$ are

(4.2.12)
$$E(\mathbf{e}) = E(\mathbf{Y} - \hat{\mathbf{Y}}) = \mathbf{X}\boldsymbol{\beta} - \mathbf{X}\boldsymbol{\beta} = \mathbf{0}$$
$$V(\mathbf{e}) = V(\mathbf{Y} - \hat{\mathbf{Y}}) = [\mathbf{I} - \mathbf{X}(\mathbf{X}'\mathbf{X})^{-1}\mathbf{X}']V(\mathbf{Y})$$
$$= \sigma^2[\mathbf{I} - \mathbf{X}(\mathbf{X}'\mathbf{X})^{-1}\mathbf{X}']$$

Critical analysis of the residual vector $\mathbf{e}$ aids us in determining outliers. Plots of residuals against the fitted values, against the independent variables, and sometimes against variables not included in the model help to determine (1) whether model assumptions are reasonable, (2) the linearity of the regression function, and (3) whether important variables have been left out of the model. These procedures are reviewed in detail by Draper and Smith (1966, Chap. 3) and Daniel and Wood (1971).

The ith element of the vector $\mathbf{y}$ specified by (4.2.2) has the form

$$y_i = \beta_0 + \beta_1 x_{i1} + \cdots + \beta_k x_{ik} + \varepsilon_i \tag{4.2.13}$$

and is called the *raw form* of the linear representation of $\mathbf{y}$ on $\mathbf{x}$. Although representation (4.2.2) of the multiple regression model is a straightforward application of the linear model representation and theory, it is often convenient to use an equivalent alternative representation of (4.2.2) called the *deviation* or *reparameterized model.* Observing that (4.2.13) is equivalent to

$$y_i = \beta_0 + \sum_{j=1}^{k} \beta_j \bar{x}_j + \sum_{j=1}^{k} \beta_j (x_{ij} - \bar{x}_j) + \varepsilon_i \tag{4.2.14}$$

where $\bar{x}_j$ denotes the means of the independent variables, and letting

$$\alpha = \beta_0 + \sum_{j=1}^{k} \beta_j \bar{x}_j$$

$$\boldsymbol{\eta} = \begin{bmatrix} \alpha \\ \boldsymbol{\gamma} \end{bmatrix} \qquad \boldsymbol{\gamma}' = [\beta_1, \beta_2, \ldots, \beta_k]$$

$$\mathbf{X}_d = \begin{bmatrix} 1 & d_{11} & d_{12} & \cdots & d_{1k} \\ 1 & d_{21} & d_{22} & \cdots & d_{2k} \\ \vdots & \vdots & \vdots & & \vdots \\ 1 & d_{N1} & d_{N2} & \cdots & d_{Nk} \end{bmatrix} = [\mathbf{1} \quad \mathbf{D}] \tag{4.2.15}$$

where $d_{ij} = x_{ij} - \bar{x}_j$, (4.2.2) becomes

$$\underset{(N \times 1)}{\mathbf{y}} = \underset{(N \times q)}{\mathbf{X}_d} \, \underset{(q \times 1)}{\boldsymbol{\eta}} + \underset{(N \times 1)}{\boldsymbol{\varepsilon}} \tag{4.2.16}$$

which has been constructed such that each column of $\mathbf{D}$ is orthogonal to $\mathbf{1}$. Applying the Gauss-Markoff theorem to (4.2.16),

$$\begin{aligned} \hat{\boldsymbol{\eta}} = \begin{bmatrix} \hat{\alpha} \\ \hat{\boldsymbol{\gamma}} \end{bmatrix} &= (\mathbf{X}_d'\mathbf{X}_d)^{-1}\mathbf{X}_d'\mathbf{y} \\ &= \begin{bmatrix} N & \mathbf{0}' \\ \mathbf{0} & \mathbf{D}'\mathbf{D} \end{bmatrix}^{-1} \begin{bmatrix} \sum_i y_i \\ \mathbf{D}'\mathbf{y} \end{bmatrix} \\ &= \begin{bmatrix} \bar{y} \\ (\mathbf{D}'\mathbf{D})^{-1}\mathbf{D}'\mathbf{y} \end{bmatrix} \\ &= \begin{bmatrix} \bar{y} \\ (\mathbf{D}'\mathbf{D})^{-1}\mathbf{D}'\mathbf{y}_d \end{bmatrix} \end{aligned} \tag{4.2.17}$$

where the ith element of $\mathbf{y}_d$ is $y_i - \bar{y}$. That $(\mathbf{D}'\mathbf{D})^{-1}\mathbf{D}'\mathbf{y} = (\mathbf{D}'\mathbf{D})^{-1}\mathbf{D}'\mathbf{y}_d$ in (4.2.17) follows from the fact that $\Sigma_j d_{ij} = 0$, for $i = 1, \ldots, N$.

A distinct advantage of (4.2.16) over (4.2.2) is that the variance-covariance matrix of the x's, S_{22}, and the covariances of the x's with Y, s_{21}, are directly obtained from $\hat{\boldsymbol{\gamma}}$. By multiplying $\hat{\boldsymbol{\gamma}}$ by $(N-1)/(N-1)$,

$$\hat{\boldsymbol{\gamma}} = \begin{bmatrix} \hat{\beta}_1 \\ \cdot \\ \cdot \\ \hat{\beta}_k \end{bmatrix} = (\mathbf{D}'\mathbf{D})^{-1}\mathbf{D}'\mathbf{y}_d = \left[\frac{\mathbf{D}'\mathbf{D}}{N-1}\right]^{-1}\left[\frac{\mathbf{D}'\mathbf{y}_d}{N-1}\right] = \mathbf{S}_{22}^{-1}\mathbf{s}_{21} \tag{4.2.18}$$

Using model (4.2.16), the variance-covariance matrix of $\hat{\boldsymbol{\eta}}$ becomes

$$\begin{aligned} V(\hat{\boldsymbol{\eta}}) &= \sigma^2(\mathbf{X}_d'\mathbf{X}_d)^{-1} \\ &= \sigma^2\begin{bmatrix} 1/N & \mathbf{0}' \\ \mathbf{0} & (\mathbf{D}'\mathbf{D})^{-1} \end{bmatrix} \end{aligned} \tag{4.2.19}$$

so that $V(\hat{\boldsymbol{\gamma}}) = \sigma^2(\mathbf{D}'\mathbf{D})^{-1}$, $V(\hat{\alpha}) = \sigma^2/N$, and $\hat{\alpha}$ and $\hat{\boldsymbol{\gamma}}$ are uncorrelated. In model (4.2.2), $\hat{\beta}_0$ was correlated with $\hat{\beta}_1, \ldots, \hat{\beta}_k$.

To find the variance of $\hat{\beta}_0$ by using model (4.2.16),

$$\hat{\beta}_0 = \hat{\alpha} - \sum_{j=1}^{k} \hat{\beta}_j \bar{x}_j = \bar{y} - \hat{\boldsymbol{\gamma}}'\bar{\mathbf{x}} \tag{4.2.20}$$

where $\bar{\mathbf{x}}' = [\bar{x}_1, \ldots, \bar{x}_k]$ is the vector of means for the k independent variables. The variance of $\hat{\beta}_0$ is

$$\begin{aligned} V(\hat{\beta}_0) &= V(\bar{y}) + V(\bar{\mathbf{x}}'\hat{\boldsymbol{\gamma}}) \\ &= \sigma^2\left[\frac{1}{N} + \bar{\mathbf{x}}'(\mathbf{D}'\mathbf{D})^{-1}\bar{\mathbf{x}}\right] \end{aligned} \tag{4.2.21}$$

which of course agrees with model (4.2.2). For an example, review the simple regression situation (4.2.11). Furthermore, the

$$\begin{aligned} \operatorname{cov}(\hat{\beta}_0, \hat{\boldsymbol{\gamma}}) &= \operatorname{cov}(\bar{y} - \hat{\boldsymbol{\gamma}}'\bar{\mathbf{x}}, \hat{\boldsymbol{\gamma}}) \\ &= \operatorname{cov}(\bar{y}, \hat{\boldsymbol{\gamma}}) - \operatorname{cov}(\hat{\boldsymbol{\gamma}}, \bar{\mathbf{x}}'\hat{\boldsymbol{\gamma}}) \\ &= -\sigma^2(\mathbf{D}'\mathbf{D})^{-1}\bar{\mathbf{x}} \end{aligned} \tag{4.2.22}$$

Extending (4.2.16) one step further, suppose the elements of $\mathbf{y}_d$ and $\mathbf{X}_d$ are standardized. The model then becomes

$$\begin{bmatrix} y_{1z} \\ y_{2z} \\ \vdots \\ y_{Nz} \end{bmatrix} = \begin{bmatrix} z_{11} & z_{12} & \cdots & z_{1k} \\ z_{21} & z_{22} & \cdots & z_{2k} \\ \vdots & \vdots & \vdots & \vdots \\ z_{N1} & z_{N2} & \cdots & z_{Nk} \end{bmatrix} \begin{bmatrix} \gamma_{1z} \\ \gamma_{2z} \\ \vdots \\ \gamma_{kz} \end{bmatrix} + \begin{bmatrix} \varepsilon_1 \\ \varepsilon_2 \\ \vdots \\ \varepsilon_N \end{bmatrix} \tag{4.2.23}$$

or

$$\underset{(N\times 1)}{\mathbf{y}_z} = \underset{(N\times k)}{\mathbf{Z}} \quad \underset{(k\times 1)}{\boldsymbol{\gamma}_z} + \underset{(N\times 1)}{\boldsymbol{\varepsilon}}$$

Then

$$\hat{\boldsymbol{\gamma}}_z = (\mathbf{Z}'\mathbf{Z})^{-1}\mathbf{Z}'\mathbf{y}_z = \mathbf{R}_{22}^{-1}\mathbf{r}_{21} \tag{4.2.24}$$

represents the vector of standardized coefficients. $\mathbf{R}_{22}$ is the correlation matrix among the x's, and $\mathbf{r}_{21}$ is the correlation between the x's, and Y.

Obtaining the vector of deviation coefficients $\hat{\boldsymbol{\gamma}}$ from the standardized coefficients $\hat{\boldsymbol{\gamma}}_z$, let

$$\mathbf{S}_{22} = \text{Dia}\,[s_{x_i}]\mathbf{R}_{22}\,\text{Dia}\,[s_{x_i}]$$
$$\mathbf{s}_{21} = \text{Dia}\,[s_{x_i}]\mathbf{r}_{21}s_y$$

where Dia $[s_{x_i}]$ is a diagonal matrix with standard deviations of the k independent variables on the diagonal and s_y is the standard deviation of the dependent variable; then,

$$\hat{\boldsymbol{\gamma}} = \mathbf{S}_{22}^{-1}\mathbf{s}_{21} = \text{Dia}\,[s_{x_i}^{-1}]\mathbf{R}_{22}^{-1}\,\text{Dia}\,[s_{x_i}^{-1}]\,\text{Dia}\,[s_{x_i}]\mathbf{r}_{21}s_y$$
$$= \text{Dia}\left[\frac{s_y}{s_{x_i}}\right]\mathbf{R}_{22}^{-1}\mathbf{r}_{21}$$
$$(4.2.25) \qquad = \text{Dia}\left[\frac{s_y}{s_{x_i}}\right]\hat{\boldsymbol{\gamma}}_z$$

The raw, deviation, and standardized forms of the multiple linear regression model yield three equivalent approaches to regression analysis. We will be primarily concerned with the development of models (4.2.2) and (4.2.16). See Bock and Haggard (1968) for a detailed discussion of the regression model using the standardized form, model (4.2.25). Using representation (4.2.2) and assuming normality, so that $\mathbf{y} \sim N_N(\mathbf{X}\boldsymbol{\beta}, \sigma^2\mathbf{I})$, suppose that the joint test

$$(4.2.26) \qquad H_0\colon \boldsymbol{\beta} = \mathbf{0}$$

is of interest and tests whether the entire linear relationship is significant. To test (4.2.26), the univariate least-squares theorem is used with the hypothesis test matrix $\mathbf{C} = \mathbf{I}_q$ so that the rank of $\mathbf{C}$ is $q = k + 1$. The sum of squares due to error and the hypothesis are then easily obtained:

$$(4.2.27) \qquad \begin{aligned} Q_e &= \mathbf{y}'[\mathbf{I} - \mathbf{X}(\mathbf{X}'\mathbf{X})^{-1}\mathbf{X}']\mathbf{y} \\ &= \mathbf{y}'\mathbf{y} - \hat{\boldsymbol{\beta}}'\mathbf{X}'\mathbf{X}\hat{\boldsymbol{\beta}} \\ Q_h &= (\mathbf{C}\hat{\boldsymbol{\beta}})'[\mathbf{C}(\mathbf{X}'\mathbf{X})^{-1}\mathbf{C}']^{-1}(\mathbf{C}\hat{\boldsymbol{\beta}}) \\ &= \hat{\boldsymbol{\beta}}'\mathbf{X}'\mathbf{X}\hat{\boldsymbol{\beta}} \end{aligned}$$

so that

$$F = \frac{\hat{\boldsymbol{\beta}}'\mathbf{X}'\mathbf{X}\hat{\boldsymbol{\beta}}/(k+1)}{[\mathbf{y}'\mathbf{y} - \hat{\boldsymbol{\beta}}'\mathbf{X}'\mathbf{X}\hat{\boldsymbol{\beta}}]/(N-k-1)}$$

Hypothesis (4.2.26) is rejected if $F > F^{\alpha}(q, N - q)$ with q and $N - q$ degrees of freedom (df) for the significance level α. The quantity $\hat{\boldsymbol{\beta}}'\mathbf{X}'\mathbf{X}\hat{\boldsymbol{\beta}}$ is usually referred to as the sum of squares due to total regression, rather than as the hypothesis sum of squares, and is denoted by SSR in many texts.

By forming an ANOVA table (Table 4.2.1) to test (4.2.26), a summary of the analysis is obtained. The expression for the E(MS) in the table is an application of equation (3.5.24).

The test of (4.2.26) is usually not the test of primary interest to researchers. Concern is usually associated with the hypothesis involving only $\beta_1, \beta_2, \ldots, \beta_k$

TABLE 4.2.1. ANOVA Table for Testing $\boldsymbol{\beta} = \mathbf{0}$.

Source	df	SS	E(MS)
Total regression	$k + 1$	$Q_h = \hat{\boldsymbol{\beta}}'\mathbf{X}'\mathbf{X}\hat{\boldsymbol{\beta}}$	$\sigma^2 + \dfrac{\boldsymbol{\beta}'\mathbf{X}'\mathbf{X}\boldsymbol{\beta}}{k+1}$
Residual	$N - k - 1$	$Q_e = \mathbf{y}'\mathbf{y} - \hat{\boldsymbol{\beta}}'\mathbf{X}'\mathbf{X}\hat{\boldsymbol{\beta}}$	σ^2
Total	N	$Q_t = \mathbf{y}'\mathbf{y}$	

or some subset of the elements of β with no hypothesis on the other elements. Such tests are often referred to as *partial joint tests*.

Using model (4.2.2), consider testing

$$H_0: \gamma = \begin{bmatrix} \beta_1 \\ \beta_2 \\ \vdots \\ \beta_k \end{bmatrix} = \begin{bmatrix} 0 \\ 0 \\ \vdots \\ 0 \end{bmatrix} \tag{4.2.28}$$

where model (4.2.2) is written as

$$y = [\mathbf{1} \quad \mathbf{X}_2] \begin{bmatrix} \beta_0 \\ \gamma \end{bmatrix} + \boldsymbol{\varepsilon} \tag{4.2.29}$$

and $\boldsymbol{\gamma}' = [\beta_1, \beta_2, \ldots, \beta_k]$. Then

$$\begin{bmatrix} \hat{\beta}_0 \\ \hat{\gamma} \end{bmatrix} = \begin{bmatrix} \mathbf{1}'\mathbf{1} & \mathbf{1}'\mathbf{X}_2 \\ \mathbf{X}_2'\mathbf{1} & \mathbf{X}_2'\mathbf{X}_2 \end{bmatrix}^{-1} \begin{bmatrix} \mathbf{1}'\mathbf{y} \\ \mathbf{X}_2'\mathbf{y} \end{bmatrix} \tag{4.2.30}$$

Applying (1.5.28) with $\mathbf{A}_{11} = \mathbf{1}'\mathbf{1}$, $\mathbf{A}_{12} = \mathbf{1}'\mathbf{X}_2$, $\mathbf{A}_{21} = \mathbf{X}_2'\mathbf{1}$, and $\mathbf{A}_{22} = \mathbf{X}_2'\mathbf{X}_2$,

$$\begin{bmatrix} \mathbf{A}_{11} & \mathbf{A}_{12} \\ \mathbf{A}_{21} & \mathbf{A}_{22} \end{bmatrix}^{-1} = \begin{bmatrix} (\mathbf{1}'\mathbf{1})^{-1} + (\mathbf{1}'\mathbf{1})^{-1}\mathbf{1}'\mathbf{X}_2(\mathbf{X}_2'\mathbf{Q}\mathbf{X}_2)^{-1}\mathbf{X}_2'\mathbf{1}(\mathbf{1}'\mathbf{1})^{-1} & -(\mathbf{1}'\mathbf{1})^{-1}\mathbf{1}'\mathbf{X}_2(\mathbf{X}_2'\mathbf{Q}\mathbf{X}_2)^{-1} \\ -(\mathbf{X}_2'\mathbf{Q}\mathbf{X}_2)^{-1}\mathbf{X}_2'\mathbf{1}(\mathbf{1}'\mathbf{1})^{-1} & (\mathbf{X}_2'\mathbf{Q}\mathbf{X}_2)^{-1} \end{bmatrix}$$

where $\mathbf{Q} = \mathbf{I} - \mathbf{1}(\mathbf{1}'\mathbf{1})^{-1}\mathbf{1}'$, so that

$$\begin{bmatrix} \hat{\beta}_0 \\ \hat{\gamma} \end{bmatrix} = \begin{bmatrix} (\mathbf{1}'\mathbf{1})^{-1}\mathbf{1}'\mathbf{y} - (\mathbf{1}'\mathbf{1})^{-1}\mathbf{1}'\mathbf{X}_2(\mathbf{X}_2'\mathbf{Q}\mathbf{X}_2)^{-1}\mathbf{X}_2'\mathbf{Q}\mathbf{y} \\ (\mathbf{X}_2'\mathbf{Q}\mathbf{X}_2)^{-1}\mathbf{X}_2'\mathbf{Q}\mathbf{y} \end{bmatrix} \tag{4.2.31}$$

From expression (4.2.31), we see that the elements of $\boldsymbol{\gamma}$ estimated by model (4.2.16) are indeed identical to the corresponding elements of $\boldsymbol{\beta}$ when model (4.2.2) was used. This was previously implied but not shown explicitly:

$$\begin{aligned} \hat{\boldsymbol{\gamma}} &= (\mathbf{X}_2'\mathbf{Q}\mathbf{X}_2)^{-1}\mathbf{X}'\mathbf{Q}\mathbf{y} \\ &= \{\mathbf{X}_2'[\mathbf{I} - \mathbf{1}(\mathbf{1}'\mathbf{1})^{-1}\mathbf{1}']\mathbf{X}_2\}^{-1}\mathbf{X}_2'[\mathbf{I} - \mathbf{1}(\mathbf{1}'\mathbf{1})^{-1}\mathbf{1}']\mathbf{y} \\ &= (\mathbf{D}'\mathbf{D})^{-1}\mathbf{D}'\mathbf{y} \\ &= (\mathbf{D}'\mathbf{D})^{-1}\mathbf{D}'\mathbf{y}_d \end{aligned} \tag{4.2.32}$$

since $\mathbf{Q}$ is an idempotent matrix.

To test hypothesis (4.2.28) that $\boldsymbol{\gamma} = \mathbf{0}$, the fundamental least-squares theorem is applied by selecting the $k \times q$ matrix $\mathbf{C} = [\mathbf{0} \quad \mathbf{I}_k]$. Employing model (4.2.2) and the notation of (4.2.16), the hypothesis test matrix and error matrix, respectively, are derived:

$$
\begin{aligned}
Q_h &= (\mathbf{C}\hat{\boldsymbol\beta})'[\mathbf{C}(\mathbf{X}'\mathbf{X})^{-1}\mathbf{C}']^{-1}(\mathbf{C}\hat{\boldsymbol\beta}) \\
&= \hat{\boldsymbol\gamma}'[\mathbf{X}_2'\mathbf{X}_2 - \mathbf{X}_2'\mathbf{1}(\mathbf{1}'\mathbf{1})^{-1}\mathbf{1}'\mathbf{X}_2]\hat{\boldsymbol\gamma} \\
&= \hat{\boldsymbol\gamma}'(\mathbf{D}'\mathbf{D})\hat{\boldsymbol\gamma} \\
Q_e &= \mathbf{y}'[\mathbf{I} - \mathbf{X}(\mathbf{X}'\mathbf{X})^{-1}\mathbf{X}']\mathbf{y} \\
&= \mathbf{y}'\mathbf{y} - \mathbf{y}'\mathbf{X}(\mathbf{X}'\mathbf{X})^{-1}\mathbf{X}'\mathbf{y} \\
&= \mathbf{y}'\mathbf{y} - \mathbf{y}'[\mathbf{1}(\mathbf{1}'\mathbf{1})^{-1}\mathbf{1}' - \mathbf{1}(\mathbf{1}'\mathbf{1})^{-1}\mathbf{1}'\mathbf{X}_2(\mathbf{X}_2'\mathbf{Q}\mathbf{X}_2)^{-1}\mathbf{X}_2'\mathbf{Q} \\
&\qquad + \mathbf{X}_2(\mathbf{X}_2'\mathbf{Q}\mathbf{X}_2)^{-1}\mathbf{X}_2'\mathbf{Q}]\mathbf{y} \qquad (4.2.33)\\
&= \mathbf{y}'\mathbf{y} - N\bar{y}^2 - \mathbf{y}'[\mathbf{X}_2(\mathbf{X}_2\mathbf{Q}\mathbf{X}_2)^{-1}\mathbf{X}'\mathbf{Q} - \mathbf{1}(\mathbf{1}'\mathbf{1})^{-1}\mathbf{1}'\mathbf{X}_2(\mathbf{X}_2'\mathbf{Q}\mathbf{X}_2)^{-1}\mathbf{X}_2'\mathbf{Q}]\mathbf{y} \\
&= \mathbf{y}'\mathbf{y} - N\bar{y}^2 - \mathbf{y}'[\mathbf{I} - \mathbf{1}(\mathbf{1}'\mathbf{1})^{-1}\mathbf{1}'][\mathbf{X}_2(\mathbf{X}_2'\mathbf{Q}\mathbf{X}_2)^{-1}\mathbf{X}_2'\mathbf{Q}]\mathbf{y} \\
&= \mathbf{y}'\mathbf{y} - N\bar{y}^2 - \mathbf{y}'\mathbf{Q}\mathbf{X}_2(\mathbf{X}_2'\mathbf{Q}\mathbf{X}_2)^{-1}\mathbf{X}_2'\mathbf{Q}\mathbf{y} \\
&= \mathbf{y}'\mathbf{y} - N\bar{y}^2 - \hat{\boldsymbol\gamma}'(\mathbf{D}'\mathbf{D})\hat{\boldsymbol\gamma} \\
&= \mathbf{y}_d'\mathbf{y}_d - \hat{\boldsymbol\gamma}'(\mathbf{D}'\mathbf{D})\hat{\boldsymbol\gamma}
\end{aligned}
$$

To test the hypothesis of no linear regression in (4.2.28), the F ratio

$$F = \frac{\hat{\boldsymbol\gamma}'(\mathbf{D}'\mathbf{D})\hat{\boldsymbol\gamma}/k}{[\mathbf{y}_d'\mathbf{y}_d - \hat{\boldsymbol\gamma}'\mathbf{D}'\mathbf{D}\hat{\boldsymbol\gamma}]/(N-k-1)} = \frac{[\hat{\boldsymbol\beta}'\mathbf{X}'\mathbf{X}\hat{\boldsymbol\beta} - N\bar{y}^2]/k}{[\mathbf{y}'\mathbf{y} - \hat{\boldsymbol\beta}'\mathbf{X}'\mathbf{X}\hat{\boldsymbol\beta}]/(N-k-1)} \qquad (4.2.34)$$

or

$$F = \frac{\mathbf{s}_{12}'\mathbf{S}_{22}^{-1}\mathbf{s}_{21}/k}{(\mathbf{s}_y^2 - \mathbf{s}_{12}'\mathbf{S}_{22}^{-1}\mathbf{s}_{21})/(N-k-1)}$$

is formed, and rejection occurs if $F > F(k, N-k-1)$. Again, an ANOVA table (Table 4.2.2) is obtained to summarize the test of (4.2.28).

From equations (2.5.9), the maximum likelihood estimate of the square of the population multiple correlation coefficient of the variable Y, with the best linear combination of the random X's, is given by

$$\bar{R}_Y^2(X_1, \ldots, X_k) = \frac{\mathbf{s}_{12}'\mathbf{S}_{22}^{-1}\mathbf{s}_{21}}{s_y^2}$$

In the derivation of the population multiple correlation coefficient, the assumption is made that the vector $[Y, X_1, X_2, \ldots, X_k]$ has a multivariate normal distribution.

TABLE 4.2.2. ANOVA Table for Testing $\boldsymbol\gamma = \mathbf{0}$.

Source	df	SS	E(MS)
α	1	$N\bar{y}^2$	$\sigma^2 + N\alpha^2$
$\beta_1, \ldots, \beta_2 \mid \alpha$	k	$\hat{\boldsymbol\gamma}'\mathbf{D}'\mathbf{D}\hat{\boldsymbol\gamma}$	$\sigma^2 + \dfrac{\boldsymbol\gamma'\mathbf{D}'\mathbf{D}\boldsymbol\gamma}{k}$
Residual	$N-k-1$	$\mathbf{y}_d'\mathbf{y}_d - \hat{\boldsymbol\gamma}'\mathbf{D}'\mathbf{D}\hat{\boldsymbol\gamma}$	σ^2
Total	N	$y'y$	

No such assumption is made for the normal linear regression model. In this case, the x's are not random but fixed variates so that, strictly speaking, a correlation coefficient is not defined.

A measure of the accuracy of prediction or goodness of fit of the regression, which has the same distribution as $\bar{R}_Y^2(X_1, \ldots, X_k)$ when the null hypothesis given in (4.2.28) is true, is called the *coefficient of determination* in the fixed variate case (see Kendall and Stuart, 1961, p. 339). Using representation (4.2.14), the population coefficient of determination squared is defined by

$$P_Y^2(X_1, \ldots, X_k) = \frac{\boldsymbol{\beta}'\mathbf{X}'\mathbf{Q}\mathbf{X}\boldsymbol{\beta}}{\boldsymbol{\beta}'\mathbf{X}'\mathbf{Q}\mathbf{X}\boldsymbol{\beta} + N\sigma^2} \tag{4.2.35}$$

where $\mathbf{Q} = \mathbf{I} - \mathbf{1}(\mathbf{1}'\mathbf{1})^{-1}\mathbf{1}'$. To estimate $P_Y^2(X_1, \ldots, X_k) \equiv P^2$,

$$R_Y^2(X_1, \ldots, X_k) = \frac{\hat{\boldsymbol{\beta}}'\mathbf{X}'\mathbf{X}\hat{\boldsymbol{\beta}} - N\bar{y}^2}{\mathbf{y}'\mathbf{y} - N\bar{y}^2} \equiv R^2 \tag{4.2.36}$$

is employed, which is the square of the product moment correlation between the Y_i's and the $\hat{Y}_i$'s. R^2 defined in (4.2.36) is a measure of the total variance in the dependent variable accounted for by the set of independent variables. By utilizing R^2, (4.2.28) is rejected if

$$F = \frac{R^2/k}{(1 - R^2)/(N - k - 1)} > F^\alpha(k, N - k - 1) \tag{4.2.37}$$

Solving (4.2.37) for R^2, it is noted that the computed F statistic is related to R^2 by

$$R^2 = \frac{kF}{N - k - 1 + kF} \tag{4.2.38}$$

R^2, when used to help determine whether the variables included in the model adequately account for the variability in the dependent variable, is often misleading and should be used with caution (see Barten, 1962). When H_0 in (4.2.28) is true, $E(R^2) \neq 0$. Then, $E(R^2) = (q - 1)/(N - 1)$ so that, as $q \to N$, the expected value of $R^2 \to 1$. Thus the sheer magnitude of R^2 is not the best indication of a "good" fit. To help eliminate this, an alternate estimate of P^2 has been proposed by Barten to correct for the number of variables:

$$\hat{R}^2 = 1 - (1 - R^2)\frac{N - 1}{N - q} \tag{4.2.39}$$

so that $E(\hat{R}^2) = 0$, when $\boldsymbol{\gamma} = \mathbf{0}$.

Testing that all the coefficients $\beta_1, \ldots, \beta_k$ are equal to 0 is a special case of the more general problem of testing that some subset of the coefficients is 0, with no restrictions on the other elements. In general, let

$$\boldsymbol{\beta} = \begin{bmatrix} \boldsymbol{\beta}_1 \\ \boldsymbol{\beta}_2 \end{bmatrix} \tag{4.2.40}$$

and write (4.2.2) as

$$\mathbf{y} = [\mathbf{X}_1 \quad \mathbf{X}_2]\begin{bmatrix} \boldsymbol{\beta}_1 \\ \boldsymbol{\beta}_2 \end{bmatrix} + \boldsymbol{\varepsilon} \tag{4.2.41}$$

Suppose it is desirable to test the hypothesis that $\boldsymbol{\beta}_2 = \mathbf{0}$, where $\boldsymbol{\beta}_2$ consists of the elements $\beta_{m+1}, \ldots, \beta_k$; the procedure is the same as that outlined to test (4.2.28), except that the vector $\mathbf{1}$ is replaced by the matrix $\mathbf{X}_1$. Thus to test

$$H_0: \boldsymbol{\beta}_2 = \mathbf{0} \tag{4.2.42}$$

the hypothesis sum of squares Q_h becomes

$$\begin{aligned} Q_h &= \hat{\boldsymbol{\beta}}_2'[\mathbf{X}_2'\mathbf{X}_2 - \mathbf{X}_2'\mathbf{X}_1(\mathbf{X}_1'\mathbf{X}_1)^{-1}\mathbf{X}_1'\mathbf{X}_2]\hat{\boldsymbol{\beta}}_2 \\ &= \mathbf{y}'\mathbf{Q}\mathbf{X}_2(\mathbf{X}_2'\mathbf{Q}\mathbf{X}_2)^{-1}\mathbf{X}_2'\mathbf{Q}\mathbf{y} \end{aligned} \tag{4.2.43}$$

where

$$\hat{\boldsymbol{\beta}}_2 = [\mathbf{X}_2'\mathbf{X}_2 - \mathbf{X}_2'\mathbf{X}_1(\mathbf{X}_1'\mathbf{X}_1)^{-1}\mathbf{X}_1\mathbf{X}_2]^{-1}\mathbf{X}_2'[\mathbf{I} - \mathbf{X}_1(\mathbf{X}_1'\mathbf{X}_1)^{-1}]\mathbf{y} = (\mathbf{X}_2'\mathbf{Q}\mathbf{X}_2)^{-1}\mathbf{X}_2'\mathbf{Q}\mathbf{y},$$

$Q = \mathbf{I} - \mathbf{X}_1(\mathbf{X}_1'\mathbf{X}_1)^{-1}\mathbf{X}_1$, and Q_e is given by (4.2.27). To test (4.2.42),

$$F = \frac{Q_h/(k-m)}{Q_e/(N-k-1)} \sim F(k-m, N-k-1) \tag{4.2.44}$$

when (4.2.42) is true. Again, an ANOVA table could be formed.

An alternative procedure for testing (4.2.42), in which the last $k - m$ variables do not significantly contribute to accounting for the variability in the dependent variable, may be approached through the use of R^2 and two separate regressions. Let Q_{h_m} denote the regression sum of squares for fitting $x_1, x_2, \ldots, x_m$, and let Q_{h_k} denote the regression sum of squares for fitting $x_1, x_2, \ldots, x_k$, where Q_e, Q_{h_m}, and Q_{h_k} are defined by (4.2.27) for testing that all the coefficients are 0. Let $\mathbf{X}_1$ denote the design matrix associated with the set $1, x_1, \ldots, x_m$, and let $[\mathbf{X}_1 \quad \mathbf{X}_2]$ be the matrix associated with the whole set. Then

$$F = \frac{(Q_{h_k} - Q_{h_m})/(k-m)}{Q_e/(N-k-1)} \sim F(k-m, N-k-1) \tag{4.2.45}$$

is exactly equal to (4.2.44). Expressions for Q_{h_k} and Q_{h_m} are seen to be

$$\begin{aligned} Q_{h_k} &= \hat{\boldsymbol{\beta}}'\mathbf{X}'\mathbf{X}\hat{\boldsymbol{\beta}} = \mathbf{y}'\mathbf{X}\hat{\boldsymbol{\beta}} = \mathbf{y}'\mathbf{X}_1\hat{\boldsymbol{\beta}}_1 + \mathbf{y}'\mathbf{X}_2\hat{\boldsymbol{\beta}}_2 \\ &= \mathbf{y}'\mathbf{X}_1(\mathbf{X}_1'\mathbf{X}_1)^{-1}\mathbf{X}_1'\mathbf{y} + \mathbf{y}'\mathbf{Q}\mathbf{X}_2(\mathbf{X}_2'\mathbf{Q}\mathbf{X}_2)^{-1}\mathbf{X}_2'\mathbf{Q}\mathbf{y} \\ Q_{h_m} &= \mathbf{y}'\mathbf{X}_1(\mathbf{X}_1'\mathbf{X}_1)^{-1}\mathbf{X}_1'\mathbf{y} \end{aligned}$$

F is now transformed to an expression involving R^2's,

$$F = \frac{N-k-1}{k-m}\,\frac{R_k^2 - R_m^2}{1 - R_k^2} \sim F(k-m, N-k-1) \tag{4.2.46}$$

so that either (4.2.44) or (4.2.46) may be used to test (4.2.42) (see Goldberger, 1964, p. 177), where $R_k^2 = R^2_{Y(x_1,\ldots,x_k)}$, $R_m^2 = R^2_{Y(x_1,\cdots,x_m)}$, and $k > m$.

Equation 4.2.45 is used to examine the extent to which prediction of the dependent variable can be improved by including more variables in the model. $1 - R_m^2$ is the amount of variability in y not accounted for by $x_1, x_2, \ldots, x_m$, and $1 - R_k^2$ is a measure of the variability not accounted for by $x_1, \ldots, x_k$. Thus the difference

$$(1 - R_m^2) - (1 - R_k^2) = R_k^2 - R_m^2 \tag{4.2.47}$$

is a measure of the reduction of the variability due to the use of the extra variables. The proportional reduction of the variation in y remaining after all other variables $x_1, \ldots, x_m$ have been used (accounted for by $x_{m+1}, \ldots, x_k$) is given by

(4.2.48)
$$R^2_{Y(m+1,\ldots,k)\cdot(1,\ldots,m)} = \frac{R_k^2 - R_m^2}{1 - R_m^2}$$

which is called the *square of the partial multiple coefficient of determination* of y, with $x_{m+1}, \ldots, x_k$ removing the influences of $x_1, \ldots, x_m$.

Relating (4.2.46) to (4.2.48),

(4.2.49)
$$F = \frac{R^2_{Y(m+1,\ldots,k)\cdot(1,\ldots,m)}/(k-m)}{[1 - R^2_{Y(m+1,\ldots,k)\cdot(1,\ldots,m)}]/(N-k-1)}$$

or

$$R^2_{Y(m+1,\ldots,k)\cdot(1,\ldots,m)} = \frac{(k-m)F}{(N-k-1)+(k-m)F}$$

When $m = k - 1$, equation (4.2.46) is of special interest; it enables us to determine the importance of an individual variable added to a set of existing variables. In this case equation (4.2.48) reduces to a *partial correlation coefficient* for fixed variates; thus

(4.2.50)
$$r^2_{Yk\cdot(1,\ldots,k-1)} = \frac{R_k^2 - R_{k-1}^2}{1 - R_{k-1}^2}$$

and (4.2.46) is used to test that an individual regression weight is 0. Result (4.2.49) now relates the square of the partial correlation coefficient to the $F(1, N-k-1)$ distribution or to $t^2(N-k-1)$.

Definition (4.2.50) is not the usual definition of a partial correlation (which is the simple correlation between two residuals after removing the linear dependence on the set of variables $x_1, x_2, \ldots, x_{k-1}$); however, the two definitions are equivalent (see Goldberger, 1964, p. 199).

Tests of significance given by (4.2.26), (4.2.28), and (4.2.40) help to determine the number of variables for inclusion in a regression function. However, on replication of a study, the tests do not help researchers to judge whether the individual weights in their studies are within the limits of previous studies. In addition, tests (4.2.26) and (4.2.28) do not indicate which variable or linear combination of variables contributed "most" significantly to the rejection of the hypothesis. These omissions are partially answered by obtaining confidence intervals for individual or linear combinations of the regression coefficients.

Following any regression analysis, the first thing a researcher should report is confidence intervals for individual coefficients. Since $\hat{\boldsymbol{\beta}} \sim N[\boldsymbol{\beta}, \sigma^2(\mathbf{X}'\mathbf{X})^{-1}]$, intervals for each element β_i of $\boldsymbol{\beta}$ are obtained by use of the t distribution. That is, a $100(1-\alpha)\%$ confidence interval for each β_i is acquired by evaluating

(4.2.51)
$$\hat{\beta}_i - t^{\alpha/2}(v_e)\hat{\sigma}_{\hat{\beta}_i} \leq \beta_i \leq \hat{\beta}_i + t^{\alpha/2}(v_e)\hat{\sigma}_{\hat{\beta}_i}$$

where the v_e are the degrees of freedom for error in estimating σ^2, and where $\hat{\sigma}_{\hat{\beta}_i}$ is the estimated standard error of $\hat{\beta}_i$.

Since the construction of individual confidence intervals fails to take into account the other variables in the regression function, the primary purpose of the intervals is to allow for comparisons with studies involving the same variable. The intervals are not to be used to judge the influence of a variable after an overall test. Because of the dependencies that exist among a set of independent variables, it might happen that all intervals include 0 while the fit is still reasonable.

An alternative to individual intervals are joint or simultaneous intervals, which take the dependencies into account. However, these intervals are ellipsoids and in general are not very helpful for more than two variables (see Huang, 1970, p. 92). In order that we may approximate the joint-interval procedure, the S method of multiple comparisons is employed. Following the test of the hypothesis

$$H_0\colon \boldsymbol{\beta}_2 = \mathbf{0} \tag{4.2.52}$$

where $\boldsymbol{\beta}' = [\boldsymbol{\beta}_1' \quad \boldsymbol{\beta}_2']$, $100(1-\alpha)\%$ simultaneous confidence intervals for expressions of the form $\psi = \mathbf{a}'\boldsymbol{\beta}_2$ are acquired for arbitrary weight vectors $\mathbf{a}$, where all possible intervals generate the ellipsoid. The confidence intervals are obtained by evaluating the expression

$$\mathbf{a}'\hat{\boldsymbol{\beta}}_2 - c_0\sqrt{\frac{Q_e}{\nu_e}\mathbf{a}'(\mathbf{X}_2'\mathbf{Q}\mathbf{X}_2)^{-1}\mathbf{a}} \le \psi \le \mathbf{a}'\hat{\boldsymbol{\beta}}_2 + c_0\sqrt{\frac{Q_e}{\nu_e}\mathbf{a}'(\mathbf{X}_2'\mathbf{Q}\mathbf{X}_2)^{-1}\mathbf{a}} \tag{4.2.53}$$

where $c_0^2 = \nu_h F^\alpha(\nu_h, \nu_e)$, ν_h and ν_e are the degrees of freedom for the hypothesis and error, Q_e is the residual sum of squares for estimating σ^2, and F^α is the upper α percentage point of the F distribution. Rejection of the hypothesis implies the existence of at least one interval that does not contain 0. For the simple linear regression case, where the hypothesis is that the regression weight associated with the independent variable is 0, (4.2.53) reduces to

$$\hat{\beta}_1 - t^{\alpha/2}(N-2)\sqrt{\frac{Q_e}{N-2}(\mathbf{D}'\mathbf{D})^{-1}} \le \beta_1 \le \hat{\beta}_1 + t^{\alpha/2}(N-2)\sqrt{\frac{Q_e}{N-2}(\mathbf{D}'\mathbf{D})^{-1}} \tag{4.2.54}$$

where $\mathbf{D}$ is defined by (4.2.15). By using (4.2.53) to test the hypothesis that $\boldsymbol{\gamma} = \mathbf{0}$, where $\boldsymbol{\gamma}' = [\beta_1, \beta_2, \ldots, \beta_k]$, the following interval results:

$$\mathbf{a}'\boldsymbol{\gamma} - c_0\sqrt{\frac{Q_e}{\nu_e}\mathbf{a}'(\mathbf{D}'\mathbf{D})^{-1}\mathbf{a}} \le \psi \le \mathbf{a}'\boldsymbol{\gamma} + c_0\sqrt{\frac{Q_e}{\nu_e}\mathbf{a}'(\mathbf{D}'\mathbf{D})^{-1}\mathbf{a}} \tag{4.2.55}$$

where $c_0^2 = kF^\alpha(k, N-k-1)$ and $\psi = \mathbf{a}'\boldsymbol{\gamma}$.

The intervals specified by (4.2.53) and (4.2.54) are very conservative; however, they are very useful for studies involving the same set of highly correlated variables.

To prove (4.2.53), Scheffé-type simultaneous intervals are derived when testing the hypothesis $H_0\colon \boldsymbol{\beta} = \mathbf{0}$ by using the definition of a confidence interval. From the test of the hypothesis, it follows that the

$$P\left[\frac{(\hat{\boldsymbol{\beta}} - \boldsymbol{\beta})'\mathbf{X}'\mathbf{X}(\hat{\boldsymbol{\beta}} - \boldsymbol{\beta})}{\nu_h(Q_e/\nu_e)} \le F^\alpha(\nu_h, \nu_e)\right] = 1 - \alpha \tag{4.2.56}$$

where $F^{\alpha}(v_h, v_e)$ is the upper α percentage point of the F distribution. However,

$$\max_{\mathbf{a}} \frac{[\mathbf{a}'(\hat{\boldsymbol{\beta}} - \boldsymbol{\beta})]^2}{\mathbf{a}'(\mathbf{X}'\mathbf{X})^{-1}\mathbf{a}} \leq (\hat{\boldsymbol{\beta}} - \boldsymbol{\beta})'\mathbf{X}'\mathbf{X}(\hat{\boldsymbol{\beta}} - \boldsymbol{\beta}) \tag{4.2.57}$$

To prove (4.2.57), the Cauchy-Schwarz inequality is employed (see Exercise 12, Section 1.3) for two column vectors $\mathbf{x}$ and $\mathbf{y}$:

$$(\mathbf{x}'\mathbf{y})^2 \leq (\mathbf{x}'\mathbf{x})(\mathbf{y}'\mathbf{y}) \tag{4.2.58}$$

Letting $\mathbf{G}$ denote a p.d. matrix so that $\mathbf{G} = \mathbf{F}\mathbf{F}'$, $\mathbf{x} = \mathbf{F}'\mathbf{a}$, and $\mathbf{y} = \mathbf{F}^{-1}\mathbf{b}$, (4.2.58) becomes

$$\begin{aligned} (\mathbf{a}'\mathbf{F}\mathbf{F}^{-1}\mathbf{b})^2 &\leq (\mathbf{a}'\mathbf{F}\mathbf{F}'\mathbf{a})\mathbf{b}'(\mathbf{F}^{-1})'\mathbf{F}^{-1}\mathbf{b} \\ (\mathbf{a}'\mathbf{b})^2 &\leq (\mathbf{a}'\mathbf{G}\mathbf{a})(\mathbf{b}'\mathbf{G}^{-1}\mathbf{b}) \end{aligned} \tag{4.2.59}$$

From (4.2.59), it follows that the

$$\max_{\mathbf{a}} \frac{(\mathbf{a}'\mathbf{b})^2}{\mathbf{a}'\mathbf{G}\mathbf{a}} \leq \mathbf{b}'\mathbf{G}^{-1}\mathbf{b}$$

so that (4.2.57) is proven. Hence (4.2.56) becomes

$$P\left[\max_{\mathbf{a}} \frac{|\mathbf{a}'(\hat{\boldsymbol{\beta}} - \boldsymbol{\beta})|}{\sqrt{\mathbf{a}'(\mathbf{X}'\mathbf{X})^{-1}\mathbf{a}}} \leq \sqrt{v_h F^{\alpha}(v_h, v_e)\frac{Q_e}{v_e}}\right] \geq 1 - \alpha \tag{4.2.60}$$

or

$$P\left[\mathbf{a}'(\hat{\boldsymbol{\beta}} - \boldsymbol{\beta}) \leq \sqrt{v_h F^{\alpha}(v_h, v_e)\frac{Q_e}{v_e}\mathbf{a}'(\mathbf{X}'\mathbf{X})^{-1}\mathbf{a}} \quad \text{for all } \mathbf{a}\right] = 1 - \alpha$$

Thus the confidence interval for linear combinations of the parameters is

$$\mathbf{a}'\hat{\boldsymbol{\beta}} - c_0\sqrt{\frac{Q_e}{v_e}\mathbf{a}'(\mathbf{X}'\mathbf{X})^{-1}\mathbf{a}} \leq \mathbf{a}'\boldsymbol{\beta} \leq \mathbf{a}'\hat{\boldsymbol{\beta}} + c_0\sqrt{\frac{Q_e}{v_e}\mathbf{a}'(\mathbf{X}'\mathbf{X})^{-1}\mathbf{a}} \tag{4.2.61}$$

where $c_0^2 = v_h F^{\alpha}(v_h, v_e)$ has a confidence interval of at least $1 - \alpha$ for any linear combination of the elements of $\boldsymbol{\beta}$. Equation (4.2.61) is in agreement with (4.2.53), given the appropriate modification.

Having determined a regression function in which all the weights are "significantly" different from 0, it is desirable to learn the most important variables in accounting for the variability in the dependent variable. One procedure is to consider the magnitude of the standardized beta weights. This should be used only if the average correlation among the independent variables is not large (for example, $\bar{r} \leq .6$), since multicollinearity in the independent variables results in inaccurate estimation of the regression coefficients (see Kendall, 1961, p. 74).

An alternative procedure, which may be helpful in determining which variables in a regression function are most important, is to compute the correlations between the independent variables and the fitted values. Let $\mathbf{a}$ be a vector with 1 in the ith position and 0s elsewhere. Then

$$\begin{aligned}
\operatorname{cor}(\mathbf{a}'\mathbf{x}, \hat{\boldsymbol{\gamma}}'\mathbf{x}) &= \frac{\operatorname{cov}(\mathbf{a}'\mathbf{x}, \hat{\boldsymbol{\gamma}}'\mathbf{x})}{\sqrt{V(\mathbf{a}'\mathbf{x})}\sqrt{V(\hat{\boldsymbol{\gamma}}'\mathbf{x})}} \\
&= \frac{\mathbf{a}'\mathbf{S}_{22}\hat{\boldsymbol{\gamma}}}{\sqrt{\mathbf{a}'\mathbf{S}_{22}\mathbf{a}}\sqrt{\hat{\boldsymbol{\gamma}}'\mathbf{S}_{22}\hat{\boldsymbol{\gamma}}}} \\
&= \frac{\mathbf{a}'\mathbf{S}_{22}\hat{\boldsymbol{\gamma}}}{s_y\sqrt{\mathbf{a}'\mathbf{S}_{22}\mathbf{a}}\sqrt{\dfrac{\hat{\boldsymbol{\gamma}}'\mathbf{S}_{22}\hat{\boldsymbol{\gamma}}}{s_y^2}}} \\
&= \frac{\mathbf{a}'\mathbf{s}_{21}}{\mathbf{s}_y\mathbf{s}_{x_i}R} \\
&= \frac{\mathbf{a}'\mathbf{r}_{21}}{R} \qquad (4.2.62)
\end{aligned}$$

This measure is not overly influenced by the size of the coefficients, which only affects R.

Given a model of the form

$$E(Y) = \beta_0 + \beta_1 x_1 + \beta_2 x_2 + \cdots + \beta_k x_k$$

we have shown, using Table 4.2.1, how we would test the hypothesis

$$H_0 : \beta_0 = \hat{\boldsymbol{\beta}}_1 = \beta_2 = \cdots = \beta_k = 0$$

that all elements in the vector $\beta' = [\beta_0, \beta_1, \ldots, \beta_k]$ are 0. Partitioning $\boldsymbol{\beta}'$ as $\boldsymbol{\beta}' = [\beta_0, \boldsymbol{\gamma}']$, and using Table 4.2.2, we can test that only the coefficients associated with $x_1, \ldots, x_k$ are 0 with no stipulation on β_0 other than that it is in the model. This test is

$$H_0 : \beta_1 = \beta_2 = \cdots = \beta_k = 0 \quad \text{or} \quad \boldsymbol{\gamma} = \mathbf{0}$$

Finally, partitioning $\boldsymbol{\beta}$ as $\boldsymbol{\beta}' = [\boldsymbol{\beta}_1', \boldsymbol{\beta}_2']$, the contribution of the variables associated with the vector $\boldsymbol{\beta}_2$ is determined through the test

$$H_0 : \boldsymbol{\beta}_2 = 0$$

with $\boldsymbol{\beta}_1$ included in the model. In the special case that $\boldsymbol{\beta}_2 = \beta_k$, the last element of $\boldsymbol{\beta}$, the contribution of a single variable above those already in the model is easily assessed.

EXERCISES 4.2

1. The deviation form of the regression model was used to determine the hypothesis sum of squares for testing (4.2.28). Use (4.2.2) to determine the projection matrix $\mathbf{P}$ such that $Q_h = \mathbf{y}'\mathbf{P}\mathbf{y}$ and test (4.2.28). What is the rank of the matrix $\mathbf{P}$?

2. Employ the simple linear regression model to find simple expressions for the SS given in Table 4.2.2. Also, construct the ANOVA table for the simple linear regression model without an intercept.

3. Use (4.2.2) to derive the ANOVA table, including the simple formula for the SS, for testing that the last beta weight, β_k, equals 0. Furthermore, relate the F statistic obtained to the partial correlation coefficient.

4.3 EXAMPLE—MULTIPLE LINEAR REGRESSION

The following example, provided by Dr. William D. Rohwer of the University of California at Berkeley, is considered as an illustration of the procedures discussed in Section 4.2.

Selecting 32 students at random from an upper-class, white, residential school, the author was interested in determining how well data from a set of paired-associated (PA), learning-proficiency tests may be used to predict children's

TABLE 4.3.1. Sample Data: Multiple Linear Regression.

	PPVT	N	S	NS	NA	SS
	68	0	10	8	21	22
	82	7	3	21	28	21
	82	7	9	17	31	30
	91	6	11	16	27	25
	82	20	7	21	28	16
	100	4	11	18	32	29
	100	6	7	17	26	23
	96	5	2	11	22	23
	63	3	5	14	24	20
	91	16	12	16	27	30
	87	5	3	17	25	24
	105	2	11	10	26	22
	87	1	4	14	25	19
	76	11	5	18	27	22
	66	0	0	3	16	11
	74	5	8	11	12	15
	68	1	6	10	28	23
	98	1	9	12	30	18
	63	0	13	13	19	16
	94	4	6	14	27	19
	82	4	5	16	21	24
	89	1	6	15	23	28
	80	5	8	14	25	24
	61	4	5	11	16	22
	102	5	7	17	26	15
	71	0	4	8	16	14
	102	4	17	21	27	31
	96	5	8	20	28	26
	55	4	7	19	20	13
	96	4	7	10	23	19
	74	2	6	14	25	17
	78	5	10	18	27	26
Means	83.0938	4.5938	7.2500	14.5000	24.3125	21.4688
Standard Deviations	13.9501	4.3244	3.5012	4.2880	4.6865	5.2363

performances on the Peabody Picture Vocabulary Test (PPVT). The independent variables were the sum of the number of items correct out of 20 (on two exposures) to five types of PA tasks. The basic tasks were *named* (N), *still* (S), *named still* (NS), *named action* (NA), and *sentence still* (SS). The values of the PPVT were obtained for fixed values of the students' performance on the PA tasks. The data for this study are summarized in Table 4.3.1.

Employing representation (4.2.2), the least-squares estimate of $\boldsymbol{\beta}$ for the data in Table 4.3.1 is

$$\hat{\boldsymbol{\beta}} = (\mathbf{X}'\mathbf{X})^{-1}\mathbf{X}'\mathbf{y} = \begin{bmatrix} 39.69710 \\ .06728 \\ .36998 \\ -.37438 \\ 1.52301 \\ .41016 \end{bmatrix} \begin{matrix} \\ \text{(N)} \\ \text{(S)} \\ \text{(NS)} \\ \text{(NA)} \\ \text{(SS)} \end{matrix} \tag{4.3.1}$$

where

$$\mathbf{X}'\mathbf{X} = \begin{bmatrix} 32 & & & & & \\ 147 & 1255 & & & & \\ 232 & 1115 & 2062 & & & \\ 464 & 2455 & 3508 & 7298 & & \\ 778 & 3789 & 5800 & 11{,}647 & 19{,}596 & \\ 687 & 3306 & 5240 & 10{,}260 & 17{,}126 & 15{,}599 \end{bmatrix} \tag{4.3.2}$$

Alternatively, using (4.2.16),

$$\hat{\boldsymbol{\eta}} = \begin{bmatrix} \hat{\alpha} \\ \hat{\boldsymbol{\beta}} \end{bmatrix} = \begin{bmatrix} \bar{y} \\ (\mathbf{D}'\mathbf{D})^{-1}\mathbf{D}'\mathbf{y}_d \end{bmatrix} = \begin{bmatrix} 83.09375 \\ .06728 \\ .36998 \\ -.37438 \\ 1.52301 \\ .41016 \end{bmatrix} \tag{4.3.3}$$

where

$$\mathbf{D}'\mathbf{D} = \begin{bmatrix} 579.72 & & & & \\ 49.25 & 380.00 & & & \\ 323.50 & 144.00 & 570.00 & & \\ 215.06 & 159.50 & 366.00 & 680.88 & \\ 150.09 & 259.25 & 298.50 & 423.31 & 849.97 \end{bmatrix} \tag{4.3.4}$$

By transforming $\mathbf{D'D}$ to $\mathbf{S}_{22}$, the variance-covariance matrix for the independent variables is

$$\mathbf{S}_{22} = \begin{bmatrix} 18.701 & & & & \\ 1.589 & 12.258 & & & \\ 10.435 & 4.645 & 18.387 & & \\ 6.937 & 5.145 & 11.806 & 21.964 & \\ 4.842 & 8.363 & 9.629 & 13.655 & 27.418 \end{bmatrix} \tag{4.3.5}$$

so that $\mathbf{R}_{22}$ is easily obtained:

$$\mathbf{R}_{22} = \begin{bmatrix} 1.000 & & & & \\ .105 & 1.000 & & & \\ .563 & .309 & 1.000 & & \\ .342 & .314 & .588 & 1.000 & \\ .214 & .456 & .429 & .556 & 1.000 \end{bmatrix} \tag{4.3.6}$$

Since the average of the off-diagonal elements of $\mathbf{R}_{22}$ is $\bar{r} = .39$, near multicollinearity in the data is not a problem for the set of independent variables selected. By using model (4.2.23) and the fact that

$$\mathbf{r}_{21} = \begin{bmatrix} .174 \\ .290 \\ .292 \\ .566 \\ .436 \end{bmatrix} \tag{4.3.7}$$

where $\mathbf{r}_{21}$ denotes the correlations between the five independent variables and y, the standardized coefficients are

$$\boldsymbol{\gamma}_z = \mathbf{R}_{22}^{-1}\mathbf{r}_{21} = \begin{bmatrix} .020857 \\ .092858 \\ -.115079 \\ .511657 \\ .153956 \end{bmatrix} \tag{4.3.8}$$

Testing the hypothesis of no linear regression, that the coefficients associated with the independent variables are 0,

$$H_0 : \boldsymbol{\gamma} = \mathbf{0} \tag{4.3.9}$$

where $\boldsymbol{\gamma}' = [\beta_1, \beta_2, \beta_3, \beta_4, \beta_5]$, ANOVA Table 4.3.2 is obtained. Testing (4.3.9) at the significance level $\alpha = .05$, the hypothesis is rejected if $F = 2.846$ is greater than $F^{.05}_{(5,26)} = 2.592$. Since $2.846 > 2.592$, the hypothesis is rejected. The p value for the test is $\alpha_p = .0353$.

To determine individual confidence intervals for the regression weights, equation (4.2.52) is employed. To locate the standard errors for the coefficients, use either (4.2.7) or (4.2.19) and (4.2.20). In either case, the standard error of each regression

TABLE 4.3.2. ANOVA Table for Testing $\gamma = \mathbf{0}$ for Multiple Linear Regression.

Source	df	SS	MS	F
α	1	220,946.281		
$\beta_1, \ldots, \beta_5 \mid \alpha$	5	2,133.725	426.745	2.846
Residual	26	3,898.994	149.916	
Total	32	226,979.000		

coefficient is the square root of the diagonal elements of the variance-covariance matrix of the estimates:

$$\hat{\boldsymbol{\sigma}}_{\hat{\boldsymbol{\beta}}} = (\text{Dia}\,[V(\hat{\boldsymbol{\beta}})])^{1/2} = \begin{bmatrix} 12.26872 \\ .61813 \\ .71555 \\ .73699 \\ .63847 \\ .54409 \end{bmatrix} \tag{4.3.10}$$

Using $t^{\alpha/2}(26) = 2.056$, where $\alpha = .05$, the confidence interval for each coefficient is obtained by evaluating

$$\hat{\beta}_i - t^{\alpha/2}(26)\hat{\sigma}_{\hat{\beta}_i} \le \beta_i \le \hat{\beta}_i + t^{\alpha/2}(26)\hat{\sigma}_{\hat{\beta}_i} \tag{4.3.11}$$

For our example, the intervals are

$$39.69710 - 2.056(12.26872) \le \beta_0 \le 39.69710 + 2.056(12.26872)$$

$$\begin{aligned} 14.473 &\le \beta_0 \le 64.922 \\ -1.204 &\le \beta_1 \le 1.388 \\ -1.101 &\le \beta_2 \le 1.841 \\ -1.890 &\le \beta_3 \le 1.139 \\ .210 &\le \beta_4 \le 2.836 \\ -.709 &\le \beta_5 \le 1.529 \end{aligned} \tag{4.3.12}$$

Computing R^2 and R for the hypothesis,

$$\begin{aligned} R^2 &= \frac{\hat{\boldsymbol{\gamma}}'\mathbf{D}'\mathbf{D}\hat{\boldsymbol{\gamma}}}{\mathbf{y}_d'\mathbf{y}_d} = \frac{2133.725}{6032.719} = .3537 \\ R &= .5947 \end{aligned} \tag{4.3.13}$$

so that approximately 35% of the variability in y is accounted for by the regression function

$$\begin{aligned} y = 39.69710 &+ .067283(\text{N}) + .369984(\text{S}) - .374381(\text{NS}) \\ &+ 1.523008(\text{NA}) + .410158(\text{SS}) \end{aligned} \tag{4.3.14}$$

However, by using $\hat{R}^2$ to estimate the population coefficient, only 23% of the variability is accounted for since

$$\hat{R}^2 = 1 - (1 - R^2)\left(\frac{N - 1}{N - q}\right) = .2294 \tag{4.3.15}$$

To determine the predictability of the values of y by using (4.3.1), Table 4.3.3 of observed and fitted values is included, which indicates that some of the independent variables are more important to prediction than others. This becomes obvious by noting that residuals for the same values of y are larger for certain students.

To establish which variables among the five are "most" important to prediction, the correlations between each independent variable and the fitted values

TABLE 4.3.3. Observations, Fitted Values, and Residuals.

y	$\hat{y}$	Residual $\hat{y} - y$
68.000	81.409	−13.409
82.000	84.674	− 2.674
82.000	96.651	−14.651
91.000	89.556	1.444
82.000	84.977	− 2.977
100.000	97.928	2.072
100.000	85.358	14.642
96.000	79.595	16.405
63.000	81.263	−18.263
91.000	92.649	− 1.649
87.000	82.698	4.302
105.000	88.779	16.221
87.000	81.871	5.129
76.000	85.693	− 9.693
66.000	67.454	− 1.454
74.000	63.304	10.696
68.000	90.318	−22.318
98.000	91.675	6.325
63.000	75.140	−12.140
94.000	85.859	8.141
82.000	77.653	4.347
89.000	82.882	6.118
80.000	85.671	− 5.671
61.000	71.090	−10.090
102.000	82.010	19.990
71.000	68.292	2.708
102.000	92.230	9.770
96.000	88.814	7.186
55.000	71.235	−16.235
96.000	81.634	14.366
74.000	81.858	− 7.858
78.000	88.780	−10.780

are examined for guidance:

$$\mathbf{r}_{x_i\hat{y}} = \frac{\mathbf{r}_{21}}{R} = \frac{1}{.5947}\begin{bmatrix} .174 \\ .290 \\ .292 \\ .566 \\ .436 \end{bmatrix} = \begin{bmatrix} .293 \\ .488 \\ .491 \\ .952 \\ .733 \end{bmatrix} \begin{matrix} \text{(N)} \\ \text{(S)} \\ \text{(NS)} \\ \text{(NA)} \\ \text{(SS)} \end{matrix} \tag{4.3.16}$$

From (4.3.16), it appears that the variables may be separated into two groups with NA and SS in one and N, S, and NS in the other.

To determine the contribution of N, S, and NS to the prediction of y, we employ equation (4.2.46). Fitting the five variables N, S, NS, NA, and SS, $R_5^2 = .3537$. Eliminating variables N, S, and NS from the model and recomputing, $R_2^2 = .3416$. Hence

$$F = \frac{N - k - 1}{k - m} \frac{R_k^2 - R_m^2}{1 - R_k^2}$$

$$= \frac{26}{3}\left(\frac{.0121}{.6463}\right) = .1622 \tag{4.3.17}$$

so that variables N, S, and NS do not significantly contribute to the regression. The partial multiple correlation squared of y, with N, S, and NS removing the influences of SS and NA, is

$$R^2_{y(123)\cdot(45)} = \frac{R_5^2 - R_2^2}{1 - R_2^2}$$

$$= \frac{.0121}{.6584}$$

$$= .0184 \tag{4.3.18}$$

which indicates that the reduction in variability due to variables N, S, and NS is less than 2%. Recomputing the regression function by using the variables SS and NA,

$$y = 39.16331 + 1.39391(\text{NA}) + .46771(\text{SS}) \tag{4.3.19}$$

Applying (4.2.46) again to (4.3.19), R^2 calculated by fitting only variable NA is $R^2 = .3203$. The F value for significance is again less than 1. The partial correlation squared between y and SS, removing NA, is only

$$r^2_{y(\text{SS})\cdot(\text{NA})} = \frac{.3416 - .3203}{.6797} = .0313 \tag{4.3.20}$$

so that the partial correlation is $r_{y(\text{SS})\cdot(\text{NA})} = .177$. Thus the final regression equation for predicting the dependent variable y (PPVT) becomes

$$y = 42.1347 + 1.6847(\text{NA}) \tag{4.3.21}$$

where $R = .5660$, $R^2 = .3203$, and $\hat{R}^2 = .2976$. Notice that, although R^2 is smaller for the one-variable model than for the five-variable model, $\hat{R}^2$ has increased; $R_5^2 = .3537$, $\hat{R}_5^2 = .2294$, $R_1^2 = .3203$, and $\hat{R}_1^2 = .2976$. The ANOVA table for equation

TABLE 4.3.4. ANOVA Table for Testing $\beta_1 = 0$, Simple Linear Regression.

Source	df	SS	MS	F
$\beta_1 \mid \beta_0$	1	1932.445	1932.445	14.139
Residual	30	4100.275	136.675	
"Total" about mean	31	6032.719		

(4.3.21), without separating the constant term, is given in Table 4.3.4. Confidence intervals for the individual coefficients in (4.3.21) are

$$42.1347 - 2.042(11.08717) \leq \beta_0 \leq 42.1347 + 2.042(11.08717)$$

$$19.50 \leq \beta_0 \leq 64.77$$

(4.3.22)

$$1.6847 - 2.042(.448035) \leq \beta_1 \leq 1.6847 + 2.042(.448035)$$

$$.7698 \leq \beta_1 \leq 2.600$$

TABLE 4.3.5. Observations, Fitted Values, and Residuals.

y	$\hat{y}_5$	$\hat{y}_1$	Residual_5	Residual_1
68.000	81.409	77.513	−13.409	− 9.513
82.000	84.674	89.306	− 2.674	− 7.306
82.000	96.651	94.360	−14.651	−21.360
91.000	89.556	87.621	1.444	3.379
82.000	84.977	89.306	− 2.977	− 7.306
100.000	97.928	96.045	2.072	3.955
100.000	85.358	85.937	14.642	14.063
96.000	79.595	79.198	16.405	16.802
63.000	81.263	82.567	−18.263	−19.567
91.000	92.649	87.621	− 1.649	3.379
87.000	82.698	84.252	4.302	2.748
105.000	88.779	85.937	16.221	19.063
87.000	81.871	84.252	5.129	2.748
76.000	85.693	87.621	− 9.693	−11.621
66.000	67.454	69.090	− 1.454	− 3.090
74.000	63.304	62.351	10.696	11.649
68.000	90.318	89.306	−22.318	−21.306
98.000	91.675	92.675	6.325	5.325
63.000	75.140	74.144	−12.140	−11.144
94.000	85.859	87.621	8.141	6.379
82.000	77.653	77.513	4.347	4.487
89.000	82.882	80.883	6.118	8.117
80.000	85.671	84.252	− 5.671	− 4.252
61.000	71.090	69.090	−10.090	− 8.090
102.000	82.010	85.937	19.990	16.063
71.000	68.292	69.090	2.708	1.910
102.000	92.230	87.621	9.770	14.379
96.000	88.814	89.306	7.186	6.694
55.000	71.235	75.829	−16.235	−20.829
96.000	81.634	80.883	14.366	15.117
74.000	81.858	84.252	− 7.858	−10.252
78.000	88.780	87.621	−10.780	− 9.621

Comparing the observed and fitted values for the one-variable equation and the five-variable equation (Table 4.3.5), the differences are not overly large. Originally it was thought that five variables would be necessary to predict scores on the PPVT; however, analysis shows that the variability can be reasonably accounted for by one variable. The process of eliminating variables one at a time in a given regression equation is known as the *backward elimination technique.*

The stepwise regression and forward selection procedures for selecting an optimal regression equation are discussed by Draper and Smith (1966, Chap. 6). Researchers using these procedures tend to employ them without examining the regression equation with all variables included in the model. They often accept, without question, the subset of variables determined by these methods with no consideration directed at investigating the variables excluded by the techniques. It is for these reasons that the backward elimination procedure was chosen for discussion in this text.

EXERCISES 4.3

1. The data below were obtained on the percent of expenditure in a state on public higher education compared to the national total (y_i) for the years 1955 to 1965 ($x = 0$ for 1960).

x:	−5	−3	−1	1	3	5
y_i:	12.6	13.7	16.9	17.4	17.7	17.0

   a. Use the model $y_i = \beta_0 + \beta_1 x + \varepsilon_i$ to calculate the normal equations.

   b. Estimate β_0 and β_1.

   c. Construct the ANOVA table to test the hypothesis $H_0: \beta_1 = 0$.

   d. Using $\alpha = .05$, what are the confidence limits for β_1?

   e. Determine the residuals for the model and plot the residuals against time. Are there any indications that a different model should be tried? Why?

2. A researcher in higher education obtained data on the instructional cost (C) for 35 institutions. In addition, data on the number of students (S), number of faculty (F), and the average wages for the faculty were obtained. The data for the study follow.

	C(000)	S	F	W
1	1,071	1,564	79	7,156
2	5,640	3,790	289	11,941
3	15,579	11,383	464	11,299
4	6,902	5,340	430	10,506
5	10,308	8,028	472	11,243
6	4,862	2,841	221	7,839
7	15,148	13,744	929	11,099
8	15,831	12,421	808	11,853
9	5,359	4,348	320	11,656
10	9,278	7,128	482	10,776
11	12,088	9,578	653	11,879
12	15,823	13,489	857	10,508
13	19,245	16,545	1,051	11,838
14	1,763	1,149	92	11,063
15	2,505	4,045	211	8,742

	C(000)	S	F	W
16	3,139	5,105	275	11,089
17	3,634	2,102	163	10,107
18	1,669	2,660	136	7,773
19	2,946	2,754	163	12,277
20	2,167	2,475	201	10,675
21	2,721	3,432	148	8,657
22	2,987	3,259	174	8,957
23	1,893	3,036	162	10,520
24	1,259	1,733	89	10,271
25	2,965	4,088	223	9,797
26	6,122	6,013	321	11,204
27	1,118	1,094	65	8,757
28	4,077	4,387	219	9,970
29	3,231	3,157	180	11,012
30	3,918	4,611	235	11,306
31	24,615	18,908	1,020	12,710
32	14,758	15,745	874	11,872
33	1,749	3,602	133	9,495
34	950	905	58	7,884
35	1,253	1,913	81	9,016

a. Use the data above to determine the "best" linear regression equation to predict instructional cost (C) for the sample of 35 institutions under study.

b. Summarize your findings.

3. Using instructional costs divided by the number of students (C/S) in Exercise 2 as a dependent variable and the number of students (S) as an independent variable, determine an "appropriate" model for predicting C/S based on the number of students.

4.4 MULTIPLE CURVILINEAR REGRESSION—UNIVARIATE

The procedure for fitting the curvilinear (polynomial) model

(4.4.1) $$y = \beta_0 + \beta_1 x + \beta_2 x^2 + \cdots + \beta_k x^k + \varepsilon$$

is identical in principle to fitting a model of the form

(4.4.2) $$y = \beta_0 + \beta_1 x_1 + \cdots + \beta_k x_k + \varepsilon$$

since models of any order can be represented by (4.4.2).

For each set of independent variables $\{x_1, \ldots, x_k\}$ in (4.4.2), there is usually only one value of y. In polynomial regression we notice the frequent appearance of several values of y for each x. For example, if $x_1 = x$ and $x_2 = x^2$, a model of the form $y = \beta_0 + \beta_1 x + \beta_2 x^2$ is being fitted. If $x = 1, 2, 3, 4, 5$ (say), a plot of the data might be given by Figure 4.4.1. Note that numerous values of y are obtained at each x. The theory applicable to (4.4.1), with only one value of y at each x, is not the same when several observations are gathered at each point x, but reduces to the single-observation regression problem developed in Section 4.2.

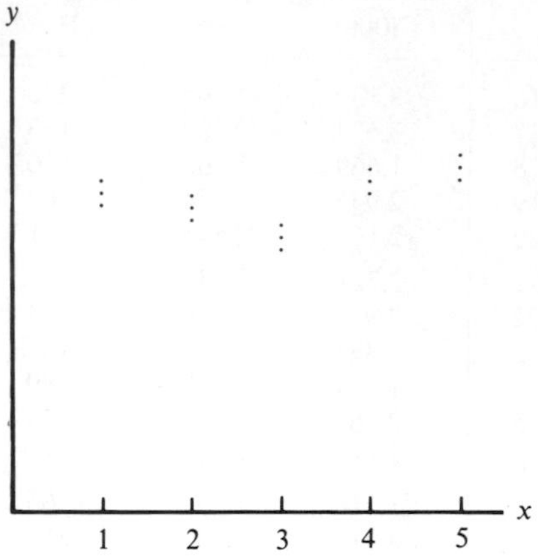

FIGURE 4.4.1. Curvilinear data plot.

For several observations, represent the data by

$$\begin{array}{ccccc} & & & & \text{Means} \\ y_{11} & y_{12} & \cdots & y_{1N_1} & y_{1.} \\ y_{21} & y_{22} & \cdots & y_{2N_2} & y_{2.} \\ \vdots & \vdots & \vdots & \vdots & \vdots \\ y_{n1} & y_{n2} & \cdots & y_{nN_n} & y_{n.} \end{array} \tag{4.4.3}$$

where N_i data points are acquired at each independent value x_i. For this case, instead of fitting a model to the original observations, we shall fit a model to the mean vector $\mathbf{y}'_{.} = [y_{1.}, y_{2.}, \ldots, y_{n.}]$. The polynomial model for these data in matrix form is

$$\begin{bmatrix} y_{1.} \\ y_{2.} \\ \vdots \\ y_{n.} \end{bmatrix} = \overset{\begin{array}{ccccc} 1 & x & x^2 & \cdots & x^k \end{array}}{\begin{bmatrix} 1 & x_{11} & x_{12} & \cdots & x_{1k} \\ 1 & x_{21} & x_{22} & \cdots & x_{2k} \\ \vdots & \vdots & \vdots & \vdots & \vdots \\ 1 & x_{n1} & x_{n2} & \cdots & x_{nk} \end{bmatrix}} \begin{bmatrix} \beta_0 \\ \beta_1 \\ \vdots \\ \beta_k \end{bmatrix} + \begin{bmatrix} \varepsilon_{1.} \\ \varepsilon_{2.} \\ \vdots \\ \varepsilon_{n.} \end{bmatrix} \tag{4.4.4}$$

or

$$\underset{(n \times 1)}{\mathbf{y}_{.}} = \underset{(n \times q)}{\mathbf{X}} \quad \underset{(q \times 1)}{\boldsymbol{\beta}} + \underset{(n \times 1)}{\boldsymbol{\varepsilon}} \tag{4.4.5}$$

so that

$$E(\mathbf{Y}_{.}) = \mathbf{X}\boldsymbol{\beta} \quad \text{and} \quad V(\mathbf{Y}) = \sigma^2\mathbf{V} \tag{4.4.6}$$

where

$$\mathbf{V} = \begin{bmatrix} 1/N_1 & 0 & \cdots & 0 \\ 0 & 1/N_2 & \cdots & 0 \\ \vdots & \vdots & \vdots & \vdots \\ 0 & 0 & \cdots & 1/N_n \end{bmatrix} \tag{4.4.7}$$

$q = k + 1$, and $n > q$.

Employing the theory of (3.5.28), let $\mathbf{y}_0 = \mathbf{V}^{-1/2}\mathbf{y}_.$, $\mathbf{Q} = \mathbf{V}^{-1/2}\mathbf{X}$, and $\boldsymbol{\varepsilon} = \mathbf{V}^{-1/2}\boldsymbol{\varepsilon}_.$, then (4.4.5) reduces to the usual model

$$\mathbf{y}_0 = \mathbf{Q}\boldsymbol{\beta} + \boldsymbol{\varepsilon} \tag{4.4.8}$$

where $E(\boldsymbol{\varepsilon}) = 0$ and $V(\boldsymbol{\varepsilon}) = \sigma^2\mathbf{I}$.

Applying (3.5.14) to (4.4.8), where now

$$Q_e^* = \boldsymbol{\varepsilon}'\boldsymbol{\varepsilon} = \boldsymbol{\varepsilon}_.'\mathbf{V}^{-1}\boldsymbol{\varepsilon}_. = (\mathbf{y}_. - \mathbf{X}\boldsymbol{\beta})'\mathbf{V}^{-1}(\mathbf{y}_. - \mathbf{X}\boldsymbol{\beta}) \tag{4.4.9}$$

is to be minimized, the estimate of $\boldsymbol{\beta}$ is

$$\hat{\boldsymbol{\beta}} = (\mathbf{Q}'\mathbf{Q})^{-1}\mathbf{Q}'\mathbf{y}_0 = (\mathbf{X}'\mathbf{V}^{-1}\mathbf{X})^{-1}\mathbf{X}'\mathbf{V}^{-1}\mathbf{y}_. \tag{4.4.10}$$

where the

$$V(\hat{\boldsymbol{\beta}}) = \sigma^2(\mathbf{X}'\mathbf{V}^{-1}\mathbf{X})^{-1}$$

An unbiased estimate of σ^2, given the correctness of model (4.4.5), is

$$\begin{aligned} s_r^2 &= \frac{Q_e^*}{n-q} = \frac{\mathbf{y}_0'[\mathbf{I} - \mathbf{Q}(\mathbf{Q}'\mathbf{Q})^{-1}\mathbf{Q}']\mathbf{y}_0}{n-q} \\ &= \frac{\mathbf{y}_.'[\mathbf{V}^{-1} - \mathbf{V}^{-1}\mathbf{X}(\mathbf{X}'\mathbf{V}^{-1}\mathbf{X})^{-1}\mathbf{X}'\mathbf{V}^{-1}]\mathbf{y}_.}{n-k-1} \\ &= \frac{\mathbf{y}_.'\mathbf{V}^{-1}\mathbf{y}_. - \hat{\boldsymbol{\beta}}'(\mathbf{X}'\mathbf{V}^{-1}\mathbf{X})\hat{\boldsymbol{\beta}}}{n-k-1} \end{aligned} \tag{4.4.11}$$

If it is now assumed that

$$\mathbf{y}_. \sim N_n(\mathbf{X}\boldsymbol{\beta}, \sigma^2\mathbf{V}) \tag{4.4.12}$$

then, to test the hypothesis

$$H_0: \boldsymbol{\beta} = \mathbf{0} \tag{4.4.13}$$

for model (4.4.5), the hypothesis sum of squares, using (4.2.27) and (4.4.11), is

$$\begin{aligned} Q_h &= (\mathbf{C}\hat{\boldsymbol{\beta}})'[\mathbf{C}(\mathbf{X}'\mathbf{V}^{-1}\mathbf{X})^{-1}\mathbf{C}']^{-1}(\mathbf{C}\hat{\boldsymbol{\beta}}) \\ &= \mathbf{y}_.'\mathbf{V}^{-1}\mathbf{X}(\mathbf{X}'\mathbf{V}^{-1}\mathbf{X})^{-1}\mathbf{X}'\mathbf{V}^{-1}\mathbf{y}_. \\ &= \hat{\boldsymbol{\beta}}'\mathbf{X}'\mathbf{V}^{-1}\mathbf{X}\hat{\boldsymbol{\beta}} \end{aligned} \tag{4.4.14}$$

An ANOVA table (Table 4.4.1) is formed to test (4.4.13). It might be tempting at this point to use the entries in Table 4.4.1 to test the hypothesis $H_0: \boldsymbol{\beta} = \mathbf{0}$ for the

TABLE 4.4.1

Source	df	SS	E(MS)
Total regression	$k + 1$	$Q_h = \hat{\boldsymbol{\beta}}'\mathbf{X}'\mathbf{V}^{-1}\mathbf{X}\hat{\boldsymbol{\beta}}$	$\sigma^2 + \dfrac{\boldsymbol{\beta}'\mathbf{X}'\mathbf{V}^{-1}\mathbf{X}\boldsymbol{\beta}}{k+1}$
Residual	$n - k - 1$	$Q_e^* = \mathbf{y}_.'\mathbf{V}^{-1}\mathbf{y}_. - \hat{\boldsymbol{\beta}}'\mathbf{X}'\mathbf{V}^{-1}\mathbf{X}\hat{\boldsymbol{\beta}}$	σ^2
Total (between)	n	$Q_{tb}^* = \mathbf{y}_.'\mathbf{V}^{-1}\mathbf{y}_.$	

model given by (4.4.5), so that

(4.4.15) $$F = \frac{Q_h/(k+1)}{Q_e^*/(n-k-1)} \sim F(k+1, n-k-1)$$

and (4.4.13) would be rejected if $F > F^\alpha(k+1, n-k-1)$. This is the procedure to follow if there is only one y observation at each value of x. However, with repeated observations, another estimate of σ^2 is obtained and is independent of model (4.4.5). That is, for each point $x_i, i = 1, \ldots, n$, an estimate of the sample variance for y is acquired. Let s_i^2 denote the sample variance of the y_{ij}'s at each x_i, so that

(4.4.16) $$s_i^2 = \frac{\sum_{j=1}^{N_i} (y_{ij} - y_{i.})^2}{N_i - 1}$$

Pooling these estimates for the n groups,

$$s_e^2 = \frac{\sum_{i=1}^{n} (N_i - 1)s_i^2}{N - n}$$

$$= \frac{\sum_{i=1}^{n}\sum_{j=1}^{N_j} (y_{ij} - y_{i.})^2}{N - n}$$

(4.4.17) $$= \frac{\mathbf{y}'\mathbf{y} - \mathbf{y}_.'\mathbf{V}^{-1}\mathbf{y}_.}{N - n}$$

is an unbiased estimate of σ^2 and is independent of whether the model fits the data, since it is only based on the within-variability of the repeated observations. The vector $\mathbf{y}$ in (4.4.17) is the $N \times 1$ vector of all the observations and $N = \Sigma_i N_i$.

Given this second estimate of variance, Table 4.4.1 is completed to include this source of variation (see Table 4.4.2). From Table 4.4.1, there are two possible estimators of σ^2 that may be used to test (4.4.13) when repeated observations are present: (1) the estimator s_r^2 when the model is valid and (2) the estimator s_e^2 independent of the validity of the model. Comparison of these two variances would indicate, on the basis of the data, that the proposed model was plausible (see Draper and Smith, 1966, p. 26). This test is known as the test for *lack of fit* and is obtained by forming the F ratio

(4.4.18) $$F = \frac{Q_e^*/(n-k-1)}{Q_e/(N-n)} \sim F^\alpha(n-k-1, N-n)$$

TABLE 4.4.2. ANOVA Table for Testing $\boldsymbol{\beta} = \mathbf{0}$ (Lack-Of-Fit Test).

Source	df	SS	E(MS)
Total regression	$k+1$	$Q_h = \hat{\boldsymbol{\beta}}'\mathbf{X}'\mathbf{V}^{-1}\mathbf{X}\hat{\boldsymbol{\beta}}$	$\sigma^2 + \dfrac{\boldsymbol{\beta}'\mathbf{X}'\mathbf{V}^{-1}\mathbf{X}\boldsymbol{\beta}}{k+1}$
Residual	$n-k-1$	$Q_e^* = \mathbf{y}_.'\mathbf{V}^{-1}\mathbf{y}_. - \hat{\boldsymbol{\beta}}\mathbf{X}'\mathbf{V}^{-1}\mathbf{X}\hat{\boldsymbol{\beta}}$	σ^2
Total (between)	n	$Q_{tb}^* = \mathbf{y}_.'\mathbf{V}^{-1}\mathbf{y}_.$	
Within error	$N-n$	$Q_e = \mathbf{y}'\mathbf{y} - \mathbf{y}_.'\mathbf{V}^{-1}\mathbf{y}_.$	σ^2
Total	N	$Q_t = \mathbf{y}'\mathbf{y}$	

Let us be more explicit. Suppose an experimenter hypothesized that the model for a set of data is of degree k, but the degree of the polynomial that best fits the data is of degree s, where $n > s > k$. To test the hypothesis that the degree of the polynomial is in fact k, let

(4.4.19)
$$\mathbf{y}_. = [\mathbf{X}_1 \quad \mathbf{X}_2]\begin{bmatrix}\boldsymbol{\beta}_1 \\ \boldsymbol{\beta}_2\end{bmatrix} + \boldsymbol{\varepsilon}$$

represent the model that best fits the data, where $\mathbf{X}_1$ contains the terms $1, x_1, \ldots, x_k$ and $\mathbf{X}_2$ contains the terms $x_{k+1}, \ldots, x_s$. To test the hypothesis that the degree of the polynomial is of order k, the hypothesis that

(4.4.20)
$$H_0 : \boldsymbol{\beta}_2 = \mathbf{0}$$

is tested where $\boldsymbol{\beta}_2' = [\beta_{k+1}, \ldots, \beta_s]$. To test (4.4.20), (4.2.43) is employed where now

(4.4.21)
$$\hat{\boldsymbol{\beta}}_2 = (\mathbf{X}_2'\mathbf{Q}\mathbf{X}_2)^{-1}\mathbf{X}_2'\mathbf{Q}\mathbf{y}$$
$$\mathbf{Q} = \mathbf{V}^{-1} - \mathbf{V}^{-1}\mathbf{X}_1(\mathbf{X}_1'\mathbf{V}^{-1}\mathbf{X}_1)^{-1}\mathbf{X}_1'\mathbf{V}^{-1}$$

and the hypothesis sum of squares is

(4.4.22)
$$Q_{h_2} = \mathbf{y}_.'\mathbf{Q}\mathbf{X}_2(\mathbf{X}_2'\mathbf{Q}\mathbf{X}_2)^{-1}\mathbf{X}_2'\mathbf{Q}\mathbf{y}_. = \hat{\boldsymbol{\beta}}_2'\mathbf{X}_2'\mathbf{Q}\mathbf{X}_2\hat{\boldsymbol{\beta}}_2$$

with degrees of freedom $s - k$. To test the hypothesis that $\boldsymbol{\beta}_1 = \mathbf{0}$, the hypothesis sum of squares is

(4.4.23)
$$Q_{h_1} = \mathbf{y}_.'\mathbf{V}^{-1}\mathbf{X}_1(\mathbf{X}_1'\mathbf{V}^{-1}\mathbf{X}_1)^{-1}\mathbf{X}_1'\mathbf{V}^{-1}\mathbf{y}_. = \hat{\boldsymbol{\beta}}_1'\mathbf{X}_1'\mathbf{V}^{-1}\mathbf{X}_1\hat{\boldsymbol{\beta}}_1$$

with degrees of freedom $k + 1$. Under the "true" model, the residual sum of squares is

(4.4.24)
$$Q_e^* = \mathbf{y}_.'\mathbf{V}^{-1}\mathbf{y}_. - \hat{\boldsymbol{\beta}}_1'\mathbf{X}_1'\mathbf{V}^{-1}\mathbf{X}_1\hat{\boldsymbol{\beta}}_1$$

with degrees of freedom $n - k - 1$. Forming an ANOVA table, Table 4.4.3, the following observations are made. Q_{h_1} in Table 4.4.3 and Q_h in Table 4.4.2 are identical. Hence $Q_e' = Q_e^* - Q_{h_2}$ or $Q_e^* = Q_e' + Q_{h_2}$ with degrees of freedom $n - k - 1 = (n - s - 1) + (s - k)$. Now the expected mean square of Q_e^* is

(4.4.25)
$$E\left(\frac{Q_e^*}{n-k-1}\right) = \sigma^2 + \frac{\boldsymbol{\beta}_2'\mathbf{X}_2'\mathbf{Q}\mathbf{X}_2\boldsymbol{\beta}_2}{n-k-1}$$

TABLE 4.4.3

Source	df	SS	E(MS)
Total regression 1	$k + 1$	$Q_{h_1} = \hat{\boldsymbol{\beta}}_1'\mathbf{X}_1'\mathbf{V}^{-1}\mathbf{X}_1\hat{\boldsymbol{\beta}}_1$	$\sigma^2 + \dfrac{\boldsymbol{\beta}_1'\mathbf{X}_1'\mathbf{V}^{-1}\boldsymbol{\beta}_1}{k+1}$
Regression 2	$s - k$	$Q_{h_2} = \hat{\boldsymbol{\beta}}_2'\mathbf{X}_2'\mathbf{Q}\mathbf{X}_2\hat{\boldsymbol{\beta}}_2$	$\sigma_2 + \dfrac{\boldsymbol{\beta}_2'\mathbf{X}_2'\mathbf{Q}\mathbf{X}_2\boldsymbol{\beta}_2}{s-k}$
Residual	$n - s - 1$	$Q_e' = \mathbf{y}_.'\mathbf{V}^{-1}\mathbf{y}_. - \mathbf{Q}_{h_1} - \mathbf{Q}_{h_2}$	σ^2
Total (between)	n	$Q_{tb}^* = \mathbf{y}_.'\mathbf{V}^{-1}\mathbf{y}_.$	
Within error	$N - n$	$Q_e = \mathbf{y}'\mathbf{y} - \mathbf{y}_.'\mathbf{V}^{-1}\mathbf{y}_.$	σ^2
Total	N	$Q_t = \mathbf{y}'\mathbf{y}$	

so that Q_e^* is the appropriate estimate for σ^2 only if $\boldsymbol{\beta}_2 = \mathbf{0}$. However, s_e^2 is an unbiased estimate of σ^2, independent of the correctness of the model, so that (4.4.18) may be used to test the appropriateness of the model.

Rejection after using (4.4.18) indicates that the model is not adequate. To find the "correct" model, residual plots may be employed. Given that (4.4.18) is nonsignificant indicates only that there is no reason to question the proposed model; it does not mean it is "absolutely" correct. We may go ahead and test (4.4.13) or test that some subset of the elements of $\boldsymbol{\beta}$ is 0. To do this, we must select the appropriate error term. Acceptance of (4.4.18) means there is no evidence that s_r^2 and s_e^2 are not estimating the same variance. A better estimate may be obtained by pooling the two; then the common variance would be estimated by

(4.4.26) $$s_p^2 = \frac{Q_e^* + Q_e}{N - k - 1}$$

However, when the number of subjects at each of the x points is large, little precision is gained by pooling. In this case, s_e^2 is usually used. In summary then, to test (4.4.13), where $\boldsymbol{\beta}$ contains $k + 1$ elements, the test statistic is

(4.4.27) $$F = \frac{Q_h/(k+1)}{Q_e/(N-n)} \sim F(k + 1, N - n)$$

when $\boldsymbol{\beta} = 0$. It is rejected if $F > F^{\alpha}(k + 1, N - n)$, given that the lack-of-fit test is tenable.

As an alternative to the lack-of-fit test, when no specific order of a polynomial model is known a priori, a step-down procedure is employed to help determine the "best" model for a set of data. The step-down procedure advances sequentially from the higher-order terms to the lower-order terms to evaluate the contribution of each to a hypothesized regression function known not to exceed some degree. This analysis is best achieved by using orthogonal polynomials; if the columns of the design matrix are orthogonal, the contribution of each term is immediately evaluated. Rather than testing that all the weights of the regression equation are 0 after a lack-of-fit test, the sequential step-down procedure may be applied to evaluate the contribution of each term in the polynomial.

Before discussing orthogonal polynomials, we will investigate how an orthogonal design matrix simplifies regression analysis. Suppose the columns $\mathbf{p}_i$ of the design matrix $\mathbf{P}$ are orthogonal, so that the model

(4.4.28)
$$\underset{(N\times 1)}{\mathbf{y}} = \underset{(N\times q)}{\mathbf{P}}\underset{(q\times 1)}{\boldsymbol{\xi}} + \underset{(N\times 1)}{\boldsymbol{\varepsilon}}$$

where $E(\boldsymbol{\varepsilon}) = \mathbf{0}$ and $V(\boldsymbol{\varepsilon}) = \sigma^2\mathbf{I}$ may be represented as

(4.4.29)
$$\mathbf{y} = [\mathbf{p}_0 \quad \mathbf{p}_1 \quad \cdots \quad \mathbf{p}_k]\boldsymbol{\xi} + \boldsymbol{\varepsilon}$$

where $\mathbf{p}_i'\mathbf{p}_j = 0$, for $i \neq j$ and $i = 0, 1, \ldots, k$. $\boldsymbol{\xi}' = [\xi_0, \xi_1, \ldots, \xi_k]$ and $q = k + 1$.

Finding the least-squares estimator $\hat{\boldsymbol{\xi}}$ of $\boldsymbol{\xi}$, when the columns of the design matrix are orthogonal, is simplified since $\mathbf{P}'\mathbf{P}$ is now a diagonal matrix:

$$\hat{\boldsymbol{\xi}} = (\mathbf{P}'\mathbf{P})^{-1}\mathbf{P}'\mathbf{y}$$

$$= \begin{bmatrix} \mathbf{p}_0'\mathbf{p}_0 & 0 & \cdots & 0 \\ 0 & \mathbf{p}_1'\mathbf{p}_1 & \cdots & 0 \\ \vdots & \vdots & \vdots\vdots\vdots & \vdots \\ 0 & 0 & \cdots & \mathbf{p}_k'\mathbf{p}_k \end{bmatrix}^{-1} \begin{bmatrix} \mathbf{p}_0'\mathbf{y} \\ \mathbf{p}_1'\mathbf{y} \\ \vdots \\ \mathbf{p}_k'\mathbf{y} \end{bmatrix}$$

(4.4.30)
$$= \begin{bmatrix} (\mathbf{p}_0'\mathbf{p}_0)^{-1}\mathbf{p}_0'\mathbf{y} \\ \vdots \\ (\mathbf{p}_k'\mathbf{p}_k)^{-1}\mathbf{p}_k'\mathbf{y} \end{bmatrix}$$

Thus the elements of $\hat{\boldsymbol{\xi}}$ are directly obtained from the columns of $\mathbf{P}$ and $\mathbf{y}$: $\hat{\boldsymbol{\xi}}_i = (\mathbf{p}_i'\mathbf{y})/\|\mathbf{p}_i\|^2$, for $i = 0, \ldots, k$, and are the coefficients of the projection of $\mathbf{y}$ onto the space spanned by the columns of $\mathbf{P}$. Furthermore, the elements of $\hat{\boldsymbol{\xi}}$ are uncorrelated and the $V(\hat{\xi}_i) = \sigma^2/\|\mathbf{p}_i\|^2$.

Representation (4.4.29) also simplifies the hypothesis sum of squares Q_h for testing

(4.4.31)
$$H_0\colon \boldsymbol{\xi} = \mathbf{0}$$

into q independent sums of squares:

$$Q_h = \hat{\boldsymbol{\xi}}'\mathbf{P}'\mathbf{P}\hat{\boldsymbol{\xi}} = \sum_i \hat{\xi}_i'\mathbf{p}_i'\mathbf{p}_i\hat{\xi}_i$$

$$= \sum_i \hat{\xi}_i^2 \|\mathbf{p}_i\|^2$$

(4.4.32)
$$= \frac{\sum_i (\mathbf{p}_i'\mathbf{y})^2}{\|\mathbf{p}_i\|^2}$$

where each sum of squares is testing $H_{0i}\colon \xi_i = 0$ and eliminating the previous terms. Hence the coefficients do not have to be recomputed in the regression function when terms are added in an orthogonal model, and further tests of their significance are immediate.

Due to the simplicity of an orthogonal design matrix, it is often desirable to transform a nonorthogonal matrix to an orthogonal matrix for computational purposes. This is particularly helpful in polynomial regression or trend analysis when the time points are equally spaced and the number of observations at each time point is the same.

Given a nonorthogonal design matrix $\mathbf{X}$, an orthogonal design matrix may be established by applying the Gram-Schmidt orthogonalization process to the columns of $\mathbf{X}$. For example, if

$$\mathbf{X} = \begin{bmatrix} 1 & 1 & 1 \\ 1 & 2 & 4 \\ 1 & 3 & 9 \end{bmatrix} = [\mathbf{x}_1 \quad \mathbf{x}_2 \quad \mathbf{x}_3]$$

then

$$\mathbf{p}_i = \mathbf{x}_i - \sum_{j=1}^{i-1} c_{ij}\mathbf{p}_j$$

where $c_{ij} = \mathbf{x}_i'\mathbf{p}_j/\|\mathbf{p}_j\|^2$, so that

$$\mathbf{P} = \begin{bmatrix} 1 & -1 & 1 \\ 1 & 0 & -2 \\ 1 & 1 & 1 \end{bmatrix} \quad \text{and} \quad \mathbf{P'P} = \begin{bmatrix} 3 & 0 & 0 \\ 0 & 2 & 0 \\ 0 & 0 & 6 \end{bmatrix}$$

and the matrix $\mathbf{P'P}$ is a diagonal matrix with the length squared of each column of $\mathbf{P}$ as diagonal elements. By finding an orthonormal basis for $\mathbf{X}$ by replacing the columns of $\mathbf{P}$ by vectors of unit length, $\mathbf{P'P}$ simplifies to $\mathbf{P'P} = \mathbf{I}$.

The orthogonalization of $\mathbf{X}$ through the Gram-Schmidt process has only been defined for a Euclidean space where the metric is the identity matrix $\mathbf{I}$. To orthogonalize $\mathbf{X}$ in any metric $\mathbf{A}$ for an affine space, two vectors are considered orthogonal in the metric of $\mathbf{A}$ if $\mathbf{x'Ay} = 0$. Generalization of the orthogonalization of $\mathbf{X}$ in any metric $\mathbf{A}$ is realized by finding

(4.4.33)
$$\mathbf{p}_i = \mathbf{x}_i - \sum_{j=1}^{i-1} c_{ij}\mathbf{p}_j$$

where $c_{ij} = \mathbf{x}_i'\mathbf{A}\mathbf{p}_j/\mathbf{p}_j'\mathbf{A}\mathbf{p}_j$, so that $\mathbf{p}_i'\mathbf{A}\mathbf{p}_j = 0$, for $i \neq j$.

To apply the orthogonalization procedure to a nonorthogonal model

(4.4.34)
$$\mathbf{y} = \mathbf{X}\boldsymbol{\beta} + \boldsymbol{\varepsilon}$$

where $E(\boldsymbol{\varepsilon}) = \mathbf{0}$ and $V(\boldsymbol{\varepsilon}) = \sigma^2\mathbf{I}$, let $\mathbf{P} = \mathbf{XG}^{-1}$ and $\boldsymbol{\xi} = \mathbf{G}\boldsymbol{\beta}$, so that (4.4.34) becomes

(4.4.35)
$$\mathbf{y} = \mathbf{P}\boldsymbol{\xi} + \boldsymbol{\varepsilon}$$

where the matrix $\mathbf{G}^{-1}$ is operating on the columns of $\mathbf{X}$. Given $\mathbf{P}$, notice that $\mathbf{G}^{-1} = (\mathbf{X'X})^{-1}\mathbf{X'P}$ and that $\mathbf{P'P} = \mathbf{D}$ is a diagonal matrix. Furthermore, $\boldsymbol{\beta} = \mathbf{G}^{-1}\boldsymbol{\xi}$. Applying the process to the model

(4.4.36)
$$\mathbf{y}_{.} = \mathbf{X}\boldsymbol{\beta} + \boldsymbol{\varepsilon}_{.}$$

where $E(\boldsymbol{\varepsilon}_{.}) = \mathbf{0}$ and $V(\boldsymbol{\varepsilon}_{.}) = \sigma^2\mathbf{V}$, $\mathbf{X}$ must be orthogonalized in the metric of $\mathbf{V}^{-1}$ to make $\mathbf{P'V}^{-1}\mathbf{P} = \mathbf{D}$, where $\mathbf{D}$ is a diagonal matrix. Under orthogonalization, (4.4.36) becomes

(4.4.37)
$$\mathbf{y}_{.} = \mathbf{P}\boldsymbol{\xi} + \boldsymbol{\varepsilon}_{.}$$

where, again for some $\mathbf{G}, \mathbf{P} = \mathbf{X}\mathbf{G}^{-1}$ and $\boldsymbol{\xi} = \mathbf{G}\boldsymbol{\beta}$. The least-squares estimator of $\boldsymbol{\xi}$ is now

$$\hat{\boldsymbol{\xi}} = (\mathbf{P}'\mathbf{V}^{-1}\mathbf{P})^{-1}\mathbf{P}'\mathbf{V}^{-1}\mathbf{y}_.$$

$$= \mathbf{D}^{-1}\hat{\boldsymbol{\eta}} \tag{4.4.38}$$

where $\hat{\boldsymbol{\eta}} = \mathbf{P}\mathbf{V}^{-1}\mathbf{y}_{.}$. Furthermore, the hypothesis sum of squares for testing $\boldsymbol{\xi} = \mathbf{0}$ again simplifies to

$$Q_h = \hat{\boldsymbol{\xi}}'\mathbf{P}'\mathbf{V}^{-1}\mathbf{P}\hat{\boldsymbol{\xi}}$$

$$= \hat{\boldsymbol{\eta}}\mathbf{D}^{-1}\hat{\boldsymbol{\eta}}$$

$$= \frac{\sum_i \hat{\boldsymbol{\eta}}_i^2}{d_{ii}} \tag{4.4.39}$$

where d_{ii} is the ith diagonal element of $\mathbf{D}$. When all the N_i's in $\mathbf{V}^{-1}$ are equal and the x_i's are at equal intervals, $\mathbf{P}$ can be obtained in the metric of $\mathbf{I}$ and (4.4.39) becomes

$$Q_h = \frac{\sum_i (\mathbf{p}_i'\mathbf{y}_.)^2 n_0}{\|\mathbf{p}_i\|^2}$$

$$= \frac{\sum_i \hat{\psi}_i^2 n_0}{\|\mathbf{p}_i\|^2} \tag{4.4.40}$$

where $N_i = n_0$, for all i, and $\hat{\psi}_i = \mathbf{p}_i'\mathbf{y}_{.}$.

Although in theory the Gram-Schmidt process can be employed to orthogonalize a design matrix $\mathbf{X}$ in any metric, the computations involved are prohibitive when q is large. With equally spaced x_i's and an equal number of observations n_0 at each x_i, tables of matrices $\mathbf{P}$ exist and facilitate transforming from a nonorthogonal to an orthogonal model. That is, given a model of the form

$$y_. = \beta_0 + \beta_1 x + \cdots + \beta_q x_q + \varepsilon_. \tag{4.4.41}$$

to be fitted, orthogonal polynomials

$$\begin{aligned} f_0(x_i) &= 1 \\ f_1(x_i) &= a_1 x_i + c_1 \\ f_2(x_i) &= a_x x_i^2 + b_2 x_i + c_2 \\ &\vdots \\ f_q(x_i) &= a_q x_i^q + b_q x_i^{q-1} + \cdots + c_q \end{aligned} \tag{4.4.42}$$

exist such that

$$\sum_{i=1}^{n} f_j(x_i) f_k(x_i) = 0$$

for $j \neq k$ and $k < n - 1$. By using (4.4.42), (4.4.41) becomes

$$y_. = \xi_0 f_0(x) + \xi_1 f_1(x) + \cdots + \xi_q f_q(x) + \varepsilon_. \tag{4.4.43}$$

where $f_0(x), f_1(x), f_2(x), \ldots$ represent the constant, linear, quadratic, and so on trends.

For example, the design matrix

$$\mathbf{X} = \begin{bmatrix} 1 & 1 & 1 \\ 1 & 2 & 4 \\ 1 & 3 & 9 \end{bmatrix}$$

represents the model $y_{.} = \beta_0 + \beta_1 x + \beta_2 x^2$. Transforming to $\mathbf{P}$, where

$$\mathbf{P} = \begin{bmatrix} 1 & -1 & 1 \\ 1 & 0 & -2 \\ 1 & 1 & 1 \end{bmatrix}$$

the model becomes $y_{.} = \xi_0 + \xi_1(x - 2) + \xi_2(3x^2 - 12x + 10)$, where evaluation of the polynomials at $x_i = 1, 2, 3$ yields the columns of $\mathbf{P}$. Tables for $\mathbf{P}$ have been constructed so that the Gram-Schmidt process does not have to be employed to find $\mathbf{P}$ when the values of N_i are equal and the x's are equally spaced (see Pearson and Hartley, 1966, p. 236). A simple procedure for the construction of $\mathbf{P}$ when the x's are not equally spaced but the N_i's are equal has been given by Robson (1959). When the x's are not equally spaced and the N_i's are unequal, (4.4.33) must be employed.

With the somewhat restricted conditions of equal $N_i = n_0$ and equally spaced x's, utilization of the step-down procedure for the determination of the order of a polynomial that "best" fits a set of data is illustrated. Suppose the model of best fit is hypothesized to be of order $p = s + 1$. Then the model is

$$\underset{(n \times 1)}{\mathbf{y}_{.}} = \underset{(n \times p)}{\mathbf{X}} \underset{(p \times 1)}{\boldsymbol{\beta}} + \underset{(n \times 1)}{\boldsymbol{\varepsilon}_{.}} \tag{4.4.44}$$

Letting $\boldsymbol{\beta} = \mathbf{G}^{-1}\boldsymbol{\xi}$ and $\mathbf{P} = \mathbf{X}\mathbf{G}^{-1}$, the orthogonal model is

$$\mathbf{y}_{.} = \mathbf{P}\boldsymbol{\xi} + \boldsymbol{\varepsilon}_{.} \tag{4.4.45}$$

where $\mathbf{P}$ is obtained from Table XIII in the Appendix. From (4.4.38),

$$\begin{aligned} \hat{\boldsymbol{\xi}} &= (\mathbf{P}'\mathbf{V}^{-1}\mathbf{P})^{-1}\mathbf{P}'\mathbf{V}^{-1}\mathbf{y}_{.} \\ &= (\mathbf{P}'\mathbf{P})^{-1}\mathbf{P}'\mathbf{y}_{.} \\ &= \mathbf{D}^{-1}\hat{\boldsymbol{\psi}} \\ &= \frac{\hat{\boldsymbol{\psi}}}{\|\mathbf{p}_i\|^2} \end{aligned} \tag{4.4.46}$$

and $\hat{\boldsymbol{\beta}} = \mathbf{G}^{-1}\hat{\boldsymbol{\xi}} = (\mathbf{X}'\mathbf{V}^{-1}\mathbf{X})^{-1}\mathbf{X}'\mathbf{V}^{-1}\mathbf{y}_{.}$.

To find Q_h for testing

$$\begin{aligned} H_0 &: \boldsymbol{\xi} = \mathbf{0} \\ H_1 &: \boldsymbol{\xi} \neq \mathbf{0} \end{aligned} \tag{4.4.47}$$

which is equivalent to testing

$$\begin{aligned} H_0 &: \boldsymbol{\beta} = \mathbf{0} \\ H_1 &: \boldsymbol{\beta} \neq \mathbf{0} \end{aligned} \quad \text{or} \quad \begin{aligned} H_0 &: \boldsymbol{\psi} = \mathbf{0} \\ H_1 &: \boldsymbol{\psi} \neq \mathbf{0} \end{aligned} \tag{4.4.48}$$

the hypothesis sum of squares is partitioned into orthogonal components:

$$Q_h = \hat{\boldsymbol{\xi}}'\mathbf{P}'\mathbf{V}^{-1}\mathbf{P}\hat{\boldsymbol{\xi}} = \sum_i (\mathbf{p}_i'\mathbf{y}_.)^2 \frac{n_0}{\|\mathbf{p}_i\|^2}$$

(4.4.49)
$$= \sum_i \hat{\psi}_i^2 \frac{n_0}{\|\mathbf{p}_i\|^2}$$

By forming ANOVA Table 4.4.4, each test

(4.4.50)
$$\begin{aligned} H_0 &: \psi_i = 0 \qquad H_0 : \xi_i = 0 \\ H_1 &: \psi_i \neq 0 \qquad H_1 : \xi_i \neq 0 \end{aligned}$$

TABLE 4.4.4. ANOVA Table for Testing $\boldsymbol{\beta} = \mathbf{0}$ (Orthogonal Polynomials).

Source	df	SS	E(MS)
Total regression	$s + 1$	$Q_h = \hat{\boldsymbol{\beta}}'(\mathbf{X}'\mathbf{V}^{-1}\mathbf{X})^{-1}\hat{\boldsymbol{\beta}}$	$\sigma^2 + \frac{\boldsymbol{\beta}'\mathbf{X}'\mathbf{V}^{-1}\mathbf{X}\boldsymbol{\beta}}{s+1}$
ξ_0	1	$Q_{h_0} = \hat{\psi}_0^2 \frac{n_0}{\|\mathbf{p}_0\|^2}$	$\sigma_2 + \hat{\psi}_0^2 \frac{n_0}{\|\mathbf{p}_0\|^2}$
$\xi_1 \mid \xi_0$	1	$Q_{h_1} = \hat{\psi}_1^2 \frac{n_0}{\|\mathbf{p}_1\|^2}$	$\sigma^2 + \psi_1^2 \frac{n_0}{\|\mathbf{p}_1\|^2}$
$\vdots$	$\vdots$	$\vdots$	
$\xi_s \mid \xi_0, \ldots, \xi_{s-1}$	1	$Q_{h_s} = \hat{\psi}_s^2 \frac{n_0}{\|\mathbf{p}_s\|^2}$	$\sigma^2 + \psi_s^2 \frac{n_0}{\|\mathbf{p}_s\|^2}$
Residual	$n - s - 1$	$Q_e^* = Q_{tb}^* - Q_h$	σ^2
Total (between)	n	$Q_{tb}^* = \mathbf{y}_.'\mathbf{V}^{-1}\mathbf{y}_.$	
Within error	$N - n$	$Q_e = \mathbf{y}'\mathbf{y} - \mathbf{y}_.'\mathbf{V}^{-1}\mathbf{y}_.$	σ^2
Total	N	$Q_t = \mathbf{y}.\mathbf{y}$	

is tested by forming the F ratio

(4.4.51)
$$F_i = \frac{Q_{h_i}}{Q_e/(N - n)} \sim F(1, N - n)$$

when (4.4.50) is true. The tests are performed in a step-down manner from the highest-order term to the lowest-order term and compared to critical values $F_i^{\alpha}(1, N - n)$. The model is said to be of degree $m' \leq s + 1$ if the m'th F_i is significant, then testing stops. That we should test the ξ_i from highest to lowest terms is demonstrated by considering a simple quadratic equation

$$\begin{aligned} y_. &= \beta_0 + \beta_1 x + \beta_2 x^2 \\ &= \xi_0 + \xi_1(x - 2) + \xi_2(3x^2 - 12x + 10) \end{aligned}$$

for $n = 3$. Testing $\xi_2 = 0$ is equivalent to testing $\beta_2 = 0$; however, testing $\xi_1 = 0$ is equivalent to testing $\beta_1 = 0$ only if $\xi_2 = 0$. This is due to the fact that β_1 is associated with $\xi_1 - 12\xi_2$ in the orthogonal polynomial model, which is confounded with ξ_2 if $\xi_2 \neq 0$.

The value of α is usually chosen for the whole experiment. The probability of at least one type I error is then

$$\alpha = 1 - \prod_{i=1}^{m} (1 - \alpha_i) \tag{4.4.52}$$

If the values of α_i are chosen such that the $\Sigma_i \alpha_i = \alpha$, then the error for the whole experiment is approximately α. For each α_i, $\alpha_i = \alpha/(s + 1)$ is chosen.

Having fitted the model $\mathbf{y}_. = \mathbf{P}\boldsymbol{\xi} + \boldsymbol{\varepsilon}_.$, where $\mathbf{P}$ is of rank m', the relationship $\hat{\boldsymbol{\beta}} = \mathbf{G}^{-1}\hat{\boldsymbol{\xi}} = (\mathbf{X}'\mathbf{V}^{-1}\mathbf{X})^{-1}\mathbf{X}'\mathbf{V}^{-1}\mathbf{y}_.$ may be employed to determine the form of the model in terms of the nonorthogonal design matrix, where $\mathbf{X}$ is of rank m'.

EXERCISES 4.4

1. a. (Equally spaced data, unequal N_i.) Orthogonalize the matrix

$$\mathbf{X} = \begin{bmatrix} 1 & 1 & 1 \\ 1 & 2 & 4 \\ 1 & 3 & 9 \end{bmatrix}$$

in the metric of $\mathbf{V}^{-1}$ where

$$\mathbf{V}^{-1} = \begin{bmatrix} 2 & 0 & 0 \\ 0 & 4 & 0 \\ 0 & 0 & 9 \end{bmatrix}$$

b. (Unequal N_i, data not equally spaced.) Repeat part a for

$$\mathbf{X} = \begin{bmatrix} 1 & 1 & 1 \\ 1 & 2 & 4 \\ 1 & 4 & 16 \end{bmatrix}$$

4.5 EXAMPLE—MULTIPLE CURVILINEAR REGRESSION

To determine the effect of practice on a manual learning task, 24 students were assigned at random to six experimental conditions. Each group of four students was given a number of practice trials and then the time required to solve the problem on a test was recorded. The data for the experiment are summarized in Table 4.5.1. Rather than performing a one-way analysis of variance, which merely helps to locate differences in means, polynomial regression analysis is used to analyze the data to determine whether the means follow a simple trend as the number of practice trials increase.

Plotting the observations (Figure 4.5.1), where × denotes the mean value of the data points at each value of the independent variable, either a linear or quadratic model may be fitted to the means. Hypothesizing a quadratic model, a lack-of-fit test to check whether the model is "correct" is performed. The mean data represented in

TABLE 4.5.1. Sample Data: Multiple Curvilinear Regression.

	Number of Practice Trials					
	1	2	3	4	5	6
	103	60	58	57	55	52
	99	63	82	68	49	48
	85	71	41	50	60	55
	104	64	60	39	44	42
Means	97.75	64.50	60.25	53.50	52.00	49.25

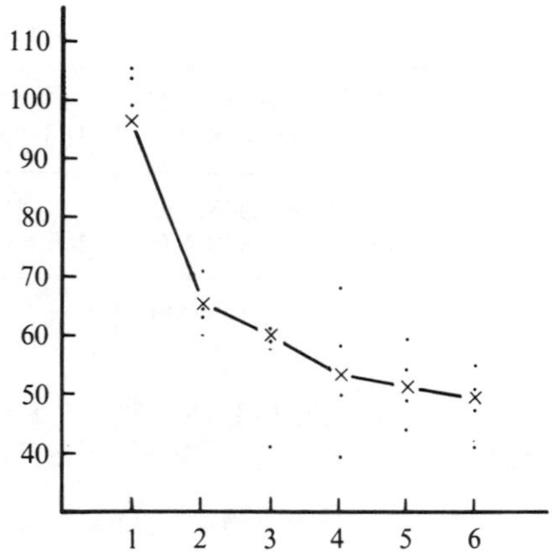

FIGURE 4.5.1. Plot of means.

matrix form, using (4.4.4), are illustrated:

$$\begin{bmatrix} 97.75 \\ 64.50 \\ 60.25 \\ 53.50 \\ 52.00 \\ 49.25 \end{bmatrix} = \begin{bmatrix} 1 & 1 & 1 \\ 1 & 2 & 4 \\ 1 & 3 & 9 \\ 1 & 4 & 16 \\ 1 & 5 & 25 \\ 1 & 6 & 36 \end{bmatrix} \begin{bmatrix} \beta_0 \\ \beta_1 \\ \beta_2 \end{bmatrix} + \begin{bmatrix} \varepsilon_{1.} \\ \varepsilon_{2.} \\ \varepsilon_{3.} \\ \varepsilon_{4.} \\ \varepsilon_{5.} \\ \varepsilon_{6.} \end{bmatrix} \tag{4.5.1}$$

or

$$\mathbf{y}_{.} = \mathbf{X} \quad \boldsymbol{\beta} + \boldsymbol{\varepsilon}_{.}$$

Employing (4.4.10), the regression coefficients of the quadratic function are

$$\hat{\boldsymbol{\beta}} = (\mathbf{X}'\mathbf{V}^{-1}\mathbf{X})^{-1}\mathbf{X}'\mathbf{V}^{-1}\mathbf{y}_{.} = \begin{bmatrix} 118.80000 \\ -28.63036 \\ 2.91964 \end{bmatrix} \tag{4.5.2}$$

where

$$\underset{(6\times 6)}{\mathbf{V}} = \begin{bmatrix} 1/4 & 0 & \cdots & 0 \\ 0 & 1/4 & \cdots & 0 \\ \vdots & \vdots & \vdots & \vdots \\ 0 & 0 & \cdots & 1/4 \end{bmatrix}$$

Table 4.5.2, a table similar to Table 4.4.2, summarizes the lack-of-fit test (4.4.18) for the example. The model is rejected at the significance level $\alpha = .05$ if the computed F

TABLE 4.5.2. ANOVA Table for Lack-Of-Fit Test.

Source	df	SS	MS	F
Total regression	3	100,849.943		
Residual	3	498.807	166.269	1.64
Total (between)	6	101,348.750		
Within error	18	1,830.250	101.681	
Total	24	103,179.000		

value for the lack-of-fit test is larger than $F(3, 18) = 3.16$. Since $F = 1.64 < 3.16$, the model as hypothesized is tenable and, from (4.5.2), has the form given by

$$y = 118.8 - 28.36036x + 2.91964x^2 \tag{4.5.3}$$

The lack-of-fit test indicates that the residual and within-mean squares are not estimating different variances. Thus, from the within variation, the estimation of σ^2 is $s_e^2 = 101.681$ and by pooling it is, with $s_r^2 = 166.269$, $s_p^2 = (498.807 + 1830.250)/21 = 110.907$. Because the number of subjects within each group for this study is small, the difference between the two estimates is large. Computing the estimated standard errors for the regression coefficients from the formula $V(\hat{\boldsymbol{\beta}}) = \hat{\sigma}^2(\mathbf{X}'\mathbf{V}^{-1}\mathbf{X})^{-1}$, using both estimates for $\hat{\sigma}^2$,

(4.5.4)

	$\hat{\sigma}_{\hat{\beta}_0}$	$\hat{\sigma}_{\hat{\beta}_1}$	$\hat{\sigma}_{\hat{\beta}_2}$
No pool	9.0191	5.9006	.82538
Pool	9.4194	6.1624	.86202

where $\hat{\sigma}^2$ in the formula for the variance of $\hat{\boldsymbol{\beta}}$ is first replaced by s_e^2 (no pool) and then by s_p^2 (pool). For this example, it is best to use the pooled estimate for σ^2; however, if there is uncertainty about the tenability of the model, do not pool. Individual confidence intervals may be constructed for the regression coefficients by using (4.2.51). Here $t_{(21)}^{\alpha/2} = 2.080$, for $\alpha = .05$ and $v_e = 21$; the intervals become

$$118.8000 - 2.080(9.4194) \leq \beta_0 \leq 118.8000 + 2.080(9.4194)$$

$$\begin{aligned} 99.21 &\leq \beta_0 \leq 138.39 \\ -41.45 &\leq \beta_1 \leq -15.81 \\ 1.13 &\leq \beta_2 \leq 4.71 \end{aligned} \tag{4.5.5}$$

Notice that, although the standard errors seem to be decreasing relative to the value of the coefficients, the errors are actually increasing for the higher-order coefficients. This illustrates the danger of overfitting; good estimates for higher-degree terms are harder to obtain. Furthermore, as in multiple regression, tests on the individual weights are not independent. The correlations among the estimates are

$$\mathbf{R} = \begin{bmatrix} 1.00 & -.93 & .86 \\ & 1.00 & -.98 \\ & & 1.00 \end{bmatrix} \tag{4.5.6}$$

which indicates high multicollinearity.

Hypothesizing a linear model, the lack-of-fit test is rejected. As witnessed from Table 4.5.4, the residual sum of squares is 498.807 + 1272.964 = 1771.771, and an estimate of error becomes 1771.771/4 = 442.942. Computing the F ratio, $F = 442.942/101.681 = 4.38$, so the computed F value is larger than the critical F^{α} value of 2.93 for $\alpha = .05$ with $v_h = 4$ and $v_e = 18$. A plot of the residuals against the fitted values for the linear model produces a simple curved pattern (Figure 4.5.2), indicating that a quadratic model is more appropriate. The fit for the quadratic and linear models is illustrated in Table 4.5.3.

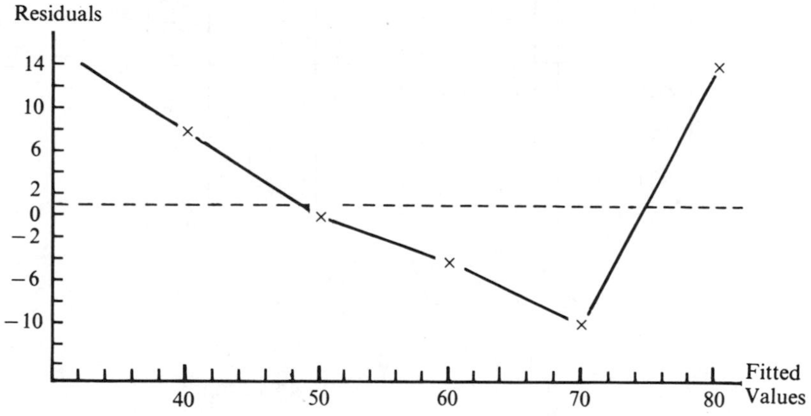

FIGURE 4.5.2. Plot of residuals versus fitted values for linear model.

TABLE 4.5.3. Observations, Fitted Values, and Residuals for Linear and Quadratic Models.

Linear $y_.$	$\hat{y}_.$	Residual	Quadratic $\hat{y}_.$	Residual
97.75	83.357	14.393	93.089	4.661
64.50	75.164	−10.664	73.218	−8.718
60.25	66.971	− 6.721	59.186	1.064
53.50	58.779	− 5.279	50.993	2.507
52.00	50.586	1.414	48.639	3.361
49.25	42.393	6.857	52.125	−2.875

For any model of order two or greater, we would not have rejected the lack-of-fit test. The fit, evaluated by examining the residuals, would be better for higher-order models. However, overfitting a set of data can occur using this test. Aside from the fact that estimates in polynomial regression are correlated and that higher-order estimates are difficult to estimate, overfitted data are difficult to reproduce in subsequent studies.

In attempting to avoid overfitting, a second lack-of-fit test on a lower-order model should be rejected; however, due to the dependencies of the tests, this is not always the best way to proceed. Alternatively, the orthogonalization procedure may be employed to transform the nonorthogonal model to an orthogonal model, so that terms in a regression function may be independently evaluated. This technique does not require hypothesizing the order of the model prior to the study and avoids any lack-of-fit tests. The procedure locates the "best" model in a sequential manner.

For illustration, let us suppose that the degree of the polynomial to be fitted to our example is not greater than three. Using the step-down orthogonal procedure, we try to determine the appropriate polynomial that best represents the data. Since the data for the example are equally spaced with an equal number of observations at each value of the independent variable, the matrix **P** for a model of order three is obtained from Table XIII in the Appendix.

Equation (4.4.36) becomes

$$\begin{bmatrix} 97.75 \\ 64.50 \\ 60.25 \\ 53.50 \\ 52.00 \\ 49.25 \end{bmatrix} = \begin{bmatrix} 1 & -5 & 5 & -5 \\ 1 & -3 & -1 & 7 \\ 1 & -1 & -4 & 4 \\ 1 & 1 & -4 & -4 \\ 1 & 3 & -1 & -7 \\ 1 & 5 & 5 & 5 \end{bmatrix} \begin{bmatrix} \xi_0 \\ \xi_1 \\ \xi_2 \\ \xi_3 \end{bmatrix} + \begin{bmatrix} \varepsilon_{1.} \\ \varepsilon_{2.} \\ \varepsilon_{3.} \\ \varepsilon_{4.} \\ \varepsilon_{5.} \\ \varepsilon_{6.} \end{bmatrix} \tag{4.5.7}$$

or

$$\mathbf{y}_{.} = \mathbf{P} \qquad \boldsymbol{\xi} + \boldsymbol{\varepsilon}_{.}$$

so that

$$\hat{\boldsymbol{\xi}} = (\mathbf{P}'\mathbf{V}^{-1}\mathbf{P})^{-1}\mathbf{P}'\mathbf{V}^{-1}\mathbf{y}_{.} = \begin{bmatrix} 62.87500 \\ -4.09643 \\ 1.94643 \\ -.71111 \end{bmatrix} \tag{4.5.8}$$

where **V** is defined as in the nonorthogonal case. Using (4.4.48), Table 4.5.4, an ANOVA table for the example, is obtained.

Choosing $\alpha = .05$ and dividing it equally over the four regression coefficients, each F test is performed at $\alpha = .05/4 = .0125$, where the critical F value is $F(1, 18) = 7.762$. Alternatively, Kimball's improvement of the Bonferroni inequality can be employed here (see Miller, 1966, p. 101). Starting with the highest hypothesized order, each computed F is compared to the critical value of 7.762, and testing stops with the first significant result. The model is then concluded to be of the order where significance occurs. With this procedure, the "best" model for the example is again a quadratic function.

TABLE 4.5.4. ANOVA Table for Orthogonal Polynomials.

Source	df	SS	MS	F
Total regression	4	101,214.032	25,303.508	
ξ_0	1	94,878.375	94,878.375	933.10
$\xi_1 \mid \xi_0$	1	4,698.604	4,698.604	46.21
$\xi_2 \mid \xi_1 \xi_0$	1	1,272.964	1,272.964	12.52
$\xi_3 \mid \xi_2 \xi_1 \xi_0$	1	364.089	364.089	3.58
Residual	2	134.718	67.359	
Total (between)	6	101,348.750		
Within error	18	1,830.250	101.681	
Total	24	103,179.000		

In terms of the ε_i's, the fitted model is

$$y_{.} = 62.8750 - 4.09643(2x - 7) + 1.94643\left(\frac{3x^2}{2} - \frac{21x}{2 + 14}\right) \tag{4.5.9}$$

or

$$y_{.} = 118.8 - 28.63036x + 2.91964x^2 \tag{4.5.10}$$

as was obtained with the lack-of-fit procedure. In general, the degree of the polynomial arrived at by using both procedures need not agree.

To procure confidence intervals for the regression coefficients in (4.5.10), standard errors of the weights are usually determined by using the no-pool estimate of σ^2—that is, using s_e^2 rather than s_p^2. From (4.5.4), the confidence intervals are

$$118.8 - 2.101(9.01916) \leq \beta_0 \leq 118.8 + 2.101(9.01916)$$

$$\begin{aligned} 99.85 &\leq \beta_0 \leq 137.75 \\ -41.03 &\leq \beta_1 \leq -16.23 \\ 1.19 &\leq \beta_2 \leq 4.65 \end{aligned} \tag{4.5.11}$$

where $t^{\alpha/2}(18) = 2.101$, for $\alpha = .05$, is the critical constant showing only 18 degrees of freedom for error.

Instead of representing the final equation in terms of (4.5.10), often just the ξ_i's are reported with confidence intervals; then equation (4.5.9) is implied. Alternatively, since the columns of the matrix **P** represent constant, linear, and quadratic components, they are readily used to summarize the trend information of the data. The trend components are represented by

$$\begin{aligned} \theta_{\text{constant}} &= \frac{1}{6}(\mu_1 + \mu_2 + \mu_3 + \mu_4 + \mu_5 + \mu_6) \\ \theta_{\text{linear}} &= \frac{1}{\sqrt{70}}(-5\mu_1 - 3\mu_2 - \mu_3 + \mu_4 + 3\mu_5 + 5\mu_6) \\ \theta_{\text{quadratic}} &= \frac{1}{\sqrt{84}}(5\mu_1 - \mu_2 - 4\mu_3 - 4\mu_4 - \mu_5 + 5\mu_6) \end{aligned} \tag{4.5.12}$$

where the μ_i's are the population means of the observations for the ith group, $i = 1, \ldots, n$. Then

$$\hat{\theta}_C = \frac{1}{6}(y_{1.} + y_{2.} + y_{3.} + y_{4.} + y_{5.} + y_{6.}) = \frac{377.250}{6} = 62.875$$

$$\hat{\theta}_L = \frac{1}{\sqrt{70}}(-5y_{1.} - 3y_{2.} - y_{3.} + y_{4.} + 3y_{5.} + 5y_{6.}) = \frac{-286.75}{\sqrt{70}} = -34.273$$

$$\hat{\theta}_Q = \frac{1}{84}(5y_{1.} - y_{2.} - 4y_{3.} - 4y_{4.} - y_{5.} + 5y_{6.}) = \frac{163.50}{\sqrt{84}} = 17.839$$

so that the evaluation of

$$\hat{\theta} - c_0\hat{\sigma}_{\hat{\theta}} \leq \theta \leq \hat{\theta} + c_0\hat{\sigma}_{\hat{\theta}} \tag{4.5.14}$$

where, if $\hat{\theta} = \Sigma_i a_i y_{i.}$ generally, $\hat{\sigma}_{\hat{\theta}} = s(\Sigma_i a_i^2/n_0)^{1/2}$ is the standard error of the trend component and $c_0^2 = F^{\alpha_i}(1, v_e)$ summarizes the trend analysis. In the example, the standard errors are

$$\hat{\sigma}_{\hat{\theta}_C} = (10.0837)\left[\frac{6}{(36)(4)}\right]^{1/2} = 2.05832$$

$$\hat{\sigma}_{\hat{\theta}_L} = \frac{10.0837}{2} = 5.04185 \tag{4.5.15}$$

$$\hat{\sigma}_{\hat{\theta}_Q} = \frac{10.0837}{2} = 5.04185$$

and simultaneous confidence intervals for the θ_i's become

$$62.875 - (7.762)^{1/2}(2.05832) \leq \theta_C \leq 62.875 + (7.762)^{1/2}(2.05832)$$

$$57.14 \leq \theta_C \leq 68.61$$

$$-48.32 \leq \theta_L \leq -20.23 \tag{4.5.16}$$

$$3.79 \leq \theta_Q \leq 31.89$$

In summary, the analysis of the data indicates that the means are reasonably fitted by a quadratic model, with significant lower-order trend components.

EXERCISES 4.5

1. To investigate the effect of several drug doses on maze learning, 40 rates were randomly assigned to five drug levels. The mean time in seconds for two runs of a maze is reported:

Drug Levels

0	1	2	3	4
6	20	12	28	38
14	18	28	34	46
12	26	16	38	39
10	22	30	26	47
18	16	26	22	52
10	28	24	36	54
4	20	28	36	49
8	22	10	30	56

a. A researcher hypothesized that the appropriate model for the data is quadratic. Use the lack-of-fit test to evaluate the proposed model.

b. A second researcher was not able to hypothesize a given model; instead, he indicated that the model should be not more than cubic. Use orthogonal polynomials to determine the most appropriate model.

c. For part b, carry out a lack-of-fit test for a cubic model.

d. For the models obtained in parts b and c, determine the 95% confidence intervals for the regression weights.

4.6 MULTIVARIATE LINEAR REGRESSION ANALYSIS

In multiple linear regression analysis, an experimenter is concerned with predicting the behavior of one dependent variable from a set of independent variables; however, in the behavioral sciences it is often the case that several dependent variables are amassed simultaneously with a given independent set. Relationships among the sets of variables are desired. This latter analysis is accomplished through multivariate regression analysis.

Defining $\mathbf{Y}$ to be the $N \times p$ data matrix of N independent observations on p responses, $\mathbf{X}$ to be the $N \times q$ design matrix of fixed known independent variables, $\mathbf{B}$ to be the $q \times p$ matrix of parameters to be estimated, and $\mathbf{E}_0$ to be the matrix of random errors, the multivariate linear model for multivariate regression analysis is

(4.6.1)
$$\underset{(N \times p)}{\mathbf{Y}} = \underset{(N \times q)}{\mathbf{X}} \underset{(q \times p)}{\mathbf{B}} + \underset{(N \times p)}{\mathbf{E}_0}$$

or

(4.6.2)
$$\begin{bmatrix} y_{11} & y_{12} & \cdots & y_{1p} \\ y_{21} & y_{22} & \cdots & y_{2p} \\ \vdots & \vdots & & \vdots \\ y_{N1} & y_{N2} & \cdots & y_{Np} \end{bmatrix} = \begin{bmatrix} 1 & x_{11} & x_{12} & \cdots & x_{1k} \\ 1 & x_{21} & x_{22} & \cdots & x_{2k} \\ \vdots & \vdots & \vdots & & \vdots \\ 1 & x_{N1} & x_{N2} & \cdots & x_{Nk} \end{bmatrix} \begin{bmatrix} \beta_{01} & \beta_{02} & \cdots & \beta_{0p} \\ \beta_{11} & \beta_{12} & \cdots & \beta_{1p} \\ \vdots & \vdots & & \vdots \\ \beta_{k1} & \beta_{k2} & \cdots & \beta_{kp} \end{bmatrix} + \begin{bmatrix} \varepsilon_{11} & \varepsilon_{12} & \cdots & \varepsilon_{1p} \\ \varepsilon_{21} & \varepsilon_{22} & \cdots & \varepsilon_{2p} \\ \vdots & \vdots & & \vdots \\ \varepsilon_{N1} & \varepsilon_{N2} & \cdots & \varepsilon_{Np} \end{bmatrix}$$

where

(4.6.3)
$$\mathcal{E}(\mathbf{Y}) = \mathbf{XB} \quad \text{and} \quad V(\mathbf{Y}) = \mathbf{I} \otimes \mathbf{\Sigma}$$

$q = k + 1$ and $N > q$. From (4.6.1) to (4.6.3), which is the *multivariate classical linear regression model*, each dependent variable follows a univariate model where the ijth element of the matrix $\mathbf{Y}$ has the representation

(4.6.4)
$$y_{ij} = \beta_{0j} + \beta_{1j}x_{i1} + \cdots + \beta_{kj}x_{ik} + \varepsilon_{ij}$$

called the *raw form* of the regression equation.

By applying the multivariate Gauss-Markoff theorem to (4.6.3), the elements β_{ij} in $\mathbf{B}$, which minimize the $\mathrm{Tr}[(\mathbf{Y} - \mathbf{XB})'(\mathbf{Y} - \mathbf{XB})]$, are acquired by evaluation:

(4.6.5)
$$\underset{(q \times p)}{\hat{\mathbf{B}}} = (\mathbf{X}'\mathbf{X})^{-1}\mathbf{X}'\mathbf{Y} \quad \text{or} \quad \underset{(p \times q)}{\hat{\mathbf{B}}'} = \mathbf{Y}'\mathbf{X}(\mathbf{X}'\mathbf{X})^{-1}$$

when using the notation of (3.6.12),

(4.6.6)
$$\mathcal{E}(\hat{\mathbf{B}}) = \mathbf{B} \quad \text{and} \quad V(\boldsymbol{\beta}^*) = \mathbf{\Sigma} \otimes (\mathbf{X}'\mathbf{X})^{-1}$$

In order to test hypotheses about the elements of **B**, it is usually assumed that each row of **Y** follows a multivariate normal distribution. This assumption and (4.6.3) constitute the *multivariate normal linear regression model.*

To obtain the deviation and standardization regression forms of (4.6.2), note that (4.6.4) is identical to

$$y_{ij} = \beta_{0j} + \sum_{h=1}^{k} \beta_{hj}\bar{x}_h + \sum_{h=1}^{k} \beta_{hj}(x_{ih} - \bar{x}_h) + \varepsilon_{ij} \tag{4.6.7}$$

where $\bar{x}_1, \bar{x}_2, \ldots, \bar{x}_k$ are the means of the k fixed independent variables. Setting

$$\alpha_{0j} = \beta_{0j} + \sum_{h=1}^{k} \beta_{hj}\bar{x}_h \qquad j = 1, \ldots, p$$

$$\alpha_0' = [\alpha_{01}, \alpha_{02}, \ldots, \alpha_{0p}]$$

$$\mathbf{H} = \begin{bmatrix} \alpha_0' \\ \mathbf{\Gamma} \end{bmatrix}$$

$$\mathbf{\Gamma} = \begin{bmatrix} \beta_{11} & \beta_{12} & \cdots & \beta_{1p} \\ \beta_{21} & \beta_{22} & \cdots & \beta_{2p} \\ \vdots & \vdots & \vdots & \vdots \\ \beta_{k1} & \beta_{k2} & \cdots & \beta_{kp} \end{bmatrix} \tag{4.6.8}$$

$$\mathbf{X}_d = \begin{bmatrix} 1 & d_{11} & \cdots & d_{1k} \\ 1 & d_{21} & \cdots & d_{2k} \\ \vdots & \vdots & \vdots & \vdots \\ 1 & d_{N1} & \cdots & d_{Nk} \end{bmatrix} = [\mathbf{1} \quad \mathbf{D}]$$

where $d_{ij} = x_{ij} - \bar{x}_j$, the equivalent reparameterized representation for (4.6.1) is

$$\underset{(N \times p)}{\mathbf{Y}} = \underset{(N \times q)}{\mathbf{X}_d} \; \underset{(q \times p)}{\mathbf{H}} + \underset{(N \times p)}{\mathbf{E}_0} \tag{4.6.9}$$

so that

$$\hat{H} = \begin{bmatrix} \alpha_0' \\ \mathbf{\Gamma} \end{bmatrix} = (\mathbf{X}_d'\mathbf{X}_d)^{-1}\mathbf{X}_d'\mathbf{Y}_d = \begin{bmatrix} \bar{\mathbf{y}}' \\ (\mathbf{D}'\mathbf{D})^{-1}\mathbf{D}'\mathbf{Y}_d \end{bmatrix} \tag{4.6.10}$$

where $y_{ij} - \bar{y}_j$ is the ijth element of $\mathbf{Y}_d$, $\bar{y}_j$ is the mean for the jth dependent variable, and $\bar{y}'$ is a row vector of the means for the p dependent variables. Proceeding as in the univariate case,

$$\underset{(k \times p)}{\hat{\mathbf{\Gamma}}} = \mathbf{S}_{22}^{-1}\mathbf{S}_{21} \quad \text{or} \quad \underset{(p \times k)}{\hat{\mathbf{\Gamma}}'} = \mathbf{S}_{12}\mathbf{S}_{22}^{-1} \tag{4.6.11}$$

where $\mathbf{S}_{22}$ is the $k \times k$ variance-covariance matrix of the x's and $\mathbf{S}_{21}$ is the $k \times p$ variance-covariance matrix of the x's with the y's.

By employing standardized variables, the model becomes

$$\underset{(N \times p)}{\mathbf{Y}_z} = \underset{(N \times k)}{\mathbf{Z}} \; \underset{(k \times p)}{\mathbf{\Gamma}_z} + \underset{(N \times p)}{\mathbf{E}_0} \tag{4.6.12}$$

where

$$\underset{(k \times p)}{\hat{\boldsymbol{\Gamma}}_z} = \mathbf{R}_{22}^{-1}\mathbf{R}_{21} \quad \text{or} \quad \underset{(p \times k)}{\hat{\boldsymbol{\Gamma}}'_z} = \mathbf{R}_{12}\mathbf{R}_{22}^{-1} \tag{4.6.13}$$

where $\mathbf{R}_{22}$ is the $k \times k$ correlation matrix of the x's and $\mathbf{R}_{21}$ is the $k \times p$ intercorrelation matrix of the x's with the y's.

Estimation theory, using the multivariate linear model, is no different from employing p univariate models. It is not until hypothesis test theory is employed that the models really differ. Univariate analysis does not address itself to the dependency that exists among a set of p response variables.

Using model (4.6.1) and the multivariate normal regression model, suppose that the joint test

$$H_0 : \mathbf{B} = \mathbf{0} \tag{4.6.14}$$

is of interest. Applying the fundamental multivariate least-squares theorem (3.6.18), with $\mathbf{C} = \mathbf{I}_q$ and the $R(\mathbf{C}) = k + 1$, the sum of squares and products (SSP) matrices due to error and the hypothesis are:

$$\begin{aligned} \mathbf{Q}_e &= \mathbf{Y}'[\mathbf{I} - \mathbf{X}(\mathbf{X}'\mathbf{X})]^{-1}\mathbf{X}'\mathbf{Y} \\ &= \mathbf{Y}'\mathbf{Y} - \hat{\mathbf{B}}'\mathbf{X}'\mathbf{X}\hat{\mathbf{B}} \\ \mathbf{Q}_h &= \hat{\mathbf{B}}'\mathbf{X}'\mathbf{X}\hat{\mathbf{B}} \end{aligned} \tag{4.6.15}$$

These SSP matrices are represented by using a multivariate analysis-of-variance (MANOVA) table, Table 4.6.1.

TABLE 4.6.1. MANOVA Table for Testing $\mathbf{B} = \mathbf{0}$.

Source	df	SS	E(MS)
Total regression	$k + 1$	$\mathbf{Q}_h = \hat{\mathbf{B}}'\mathbf{X}'\mathbf{X}\hat{\mathbf{B}}$	$\boldsymbol{\Sigma} + \dfrac{\mathbf{B}'\mathbf{X}'\mathbf{X}\mathbf{B}}{k+1}$
Residual	$N - k - 1$	$\mathbf{Q}_e = \mathbf{Y}'\mathbf{Y} - \hat{\mathbf{B}}'\mathbf{X}'\mathbf{X}\hat{\mathbf{B}}$	$\boldsymbol{\Sigma}$
Total	N	$\mathbf{Q}_t = \mathbf{Y}'\mathbf{Y}$	

To test (4.6.14), numerous criteria are available where each of the statistics are functions of the roots $\lambda_1, \lambda_2, \ldots, \lambda_s$ of the characteristic equation

$$|\mathbf{Q}_h - \lambda\mathbf{Q}_e| = 0 \tag{4.6.16}$$

where $s = \min(v_h, p)$ as in (2.7.15). Applying the general theory of (2.7.12) and (3.6.18) to test (4.6.14), where $\mathbf{Q}_e$ and $\mathbf{Q}_h$ are defined as in Table 4.6.1,

$$\begin{aligned} s &= \min(v_h, p) = \min(k + 1, p) \\ m &= \frac{|u - v_h| - 1}{2} = \frac{|p - k - 1| - 1}{2} \\ n &= \frac{v_e - u - 1}{2} = \frac{N - k - p - 2}{2} \end{aligned} \tag{4.6.17}$$

$u = R(\mathbf{A}) = p$, $v_h = k + 1$, and $v_e = N - k - 1$, the hypothesis $H_0\colon \mathbf{B} = \mathbf{0}$ is rejected at the significance level α if

(4.6.18) (a) Wilks:

$$\Lambda = \frac{|\mathbf{Q}_e|}{|\mathbf{Q}_e + \mathbf{Q}_h|} = \sum_{i=1}^{s} (1 + \lambda_i)^{-1} < U^{\alpha}(p, k + 1, N - k - 1)$$

(b) Roy:

$$\theta_s = \frac{\lambda_1}{1 + \lambda} > \theta^{\alpha}(s, m, n)$$

(c) Lawley-Hotelling:

$$U^{(s)} = \frac{T_0^2}{N - k - 1} = \sum_{i=1}^{s} \lambda_i > U_0^{\alpha}(s, m, n)$$

(d) Pillai:

$$V^{(s)} = \sum_{i=1}^{s} \frac{\lambda_i}{1 + \lambda_i} > V^{\alpha}(s, m, n)$$

where s, m, and n are defined as in (4.6.17). Tables for (4.6.18a) and (4.6.18b) may be found in Tables IX and VIII in the Appendix, while tables for (4.6.18c) and (4.6.18d) are given in Tables X and XI.

An application of these criteria is deferred until Section 4.7. All statistics were equivalent in Chapter 2: since this is not generally the case, the different criteria may yield contradictory results concerning the rejection of (4.6.14) (see Jones, 1966). The choice of a statistic is left to the researcher since no criteria is uniformly best (most powerful); however, certain criteria possess merit over others.

The test of no linear relationship between the two sets of variables is most often used in multivariate analysis:

(4.6.19) $$H_0\colon \boldsymbol{\Gamma} = \mathbf{0}$$

Following the derivation of the univariate test, equation (4.2.27) with $\mathbf{y}$ replaced by the data matrix $\mathbf{Y}$, the MANOVA table necessary to test (4.6.19) is immediately obtained and given in Table 4.6.2. Using variance-covariance matrices to evaluate

TABLE 4.6.2. MANOVA Table for Testing $\boldsymbol{\Gamma} = \mathbf{0}$.

Source	df	SSP	E(MS)
$\boldsymbol{\alpha}_0$	1	$N\overline{\mathbf{y}\mathbf{y}}'$	$\boldsymbol{\Sigma} + N\boldsymbol{\alpha}_0\boldsymbol{\alpha}_0'$
$\boldsymbol{\Gamma}\|\boldsymbol{\alpha}_0$	k	$\mathbf{Q}_h = \hat{\boldsymbol{\Gamma}}'\mathbf{D}'\mathbf{D}\hat{\boldsymbol{\Gamma}}$	$\boldsymbol{\Sigma} + \frac{\boldsymbol{\Gamma}'\mathbf{D}'\mathbf{D}\boldsymbol{\Gamma}}{k}$
Residual	$N - k - 1$	$\mathbf{Q}_e = \mathbf{Y}_d'\mathbf{Y}_d - \hat{\boldsymbol{\Gamma}}'\mathbf{D}'\mathbf{D}\hat{\boldsymbol{\Gamma}}$	$\boldsymbol{\Sigma}$
Total	N	$\mathbf{Y}'\mathbf{Y}$	

$\mathbf{Q}_h$ and $\mathbf{Q}_e$ in Table 4.6.2, $\mathbf{Q}_h$ and $\mathbf{Q}_e$ become

(4.6.20)
$$\mathbf{Q}_h = \mathbf{S}_{12}\mathbf{S}_{22}^{-1}\mathbf{S}_{21}$$
$$\mathbf{Q}_e = \mathbf{S}_{11} - \mathbf{S}_{12}\mathbf{S}_{22}^{-1}\mathbf{S}_{21}$$

Again, in testing (4.6.19), the test statistics are functions of the roots of the determinantal equation

(4.6.21)
$$|\mathbf{Q}_h - \lambda\mathbf{Q}_e| = 0$$

or

$$|\mathbf{S}_{12}\mathbf{S}_{22}^{-1}\mathbf{S}_{21} - \lambda(\mathbf{S}_{11} - \mathbf{S}_{22}^{-1}\mathbf{S}_{21})| = 0$$

From the relationship between the roots of (4.6.21) and

(4.6.22)
$$|\mathbf{Q}_h - \theta(\mathbf{Q}_h + \mathbf{Q}_e)| = 0 \quad \text{or} \quad |\mathbf{S}_{12}\mathbf{S}_{22}^{-1}\mathbf{S}_{21} - \theta\mathbf{S}_{11}| = 0$$

Wilks' Λ criterion to test (4.6.19) is

(4.6.23)
$$\Lambda = \prod_{i=1}^{s} (1 + \lambda_i)^{-1} = \prod_{i=1}^{s} (1 - \theta_i)$$

where $s = \min(k, p)$.

With $m = (|p - k| - 1)/2$ and $n = (N - k - p - 2)/2$, the various criteria used to test (4.6.19) at the significance level α become

(4.6.24) (a) Wilks:

$$\Lambda = \prod_{i=1}^{s} (1 - \theta_i) < U^{\alpha}(p, k, N - k - 1)$$

(b) Roy:

$$\theta_s = \frac{\lambda_1}{1 + \lambda_1} = \theta_1 > \theta^{\alpha}(s, m, n)$$

(c) Lawley-Hotelling:

$$U^{(s)} = \frac{T_0^2}{N - k - 1} = \sum_{i=1}^{s} \lambda_i = \sum_{i=1}^{s} \frac{\theta_i}{1 - \theta_i} > U_0^{\alpha}(s, m, n)$$

(d) Pillai:

$$V^{(s)} = \sum_{i=1}^{s} \theta_i > V^{\alpha}(s, m, n)$$

(e) Bartlett:

$$X_B^2 = -[(N - 1) - \frac{1}{2}(p + k + 1)] \log \Lambda > \chi_{\alpha}^2(pk)$$

The roots θ_i in (4.6.22) are important in statistical analysis as they are the squares of the canonical correlations of Hotelling (1936a) when the multivariate correlation model is being employed and $\mathbf{X}$ and $\mathbf{Y}$ are both random. More will be presented on this topic in Section 4.14.

To test whether some subset of the independent set is contributing significantly to the predictability of the dependent set, we shall compute two separate regressions. Writing the multivariate model as

$$\mathbf{Y} = [\mathbf{X}_1 \quad \mathbf{X}_2]\begin{bmatrix}\mathbf{B}_1 \\ \mathbf{B}_2\end{bmatrix} + \mathbf{E}_0 \tag{4.6.25}$$

where

$$\mathbf{B}_1 = \begin{bmatrix} \beta_{01} & \cdots & \beta_{0p} \\ \beta_{11} & \cdots & \beta_{1p} \\ \vdots & \vdots & \vdots \\ \beta_{m1} & \cdots & \beta_{mp} \end{bmatrix} \quad \text{and} \quad \mathbf{B}_2 = \begin{bmatrix} \beta_{m+1,1} & \cdots & \beta_{m+1,p} \\ \vdots & \vdots & \vdots \\ \beta_{k1} & \cdots & \beta_{kp} \end{bmatrix}$$

and $q = k + 1$, suppose the test of the hypothesis

$$H_0 : \mathbf{B}_2 = \mathbf{0} \tag{4.6.26}$$

is of interest. Following the arguments of the univariate case (4.2.43),

$$\mathbf{Q}_h = \mathbf{Y}'\mathbf{Q}\mathbf{X}_2(\mathbf{X}_2'\mathbf{Q}\mathbf{X}_2)^{-1}\mathbf{X}_2'\mathbf{Q}\mathbf{Y} = \mathbf{Q}_{h_k} - \mathbf{Q}_{h_m} \tag{4.6.27}$$

and $\mathbf{Q}_e$ is defined as in the tests of (4.6.19) or (4.6.14). Given $\mathbf{Q}_h$ and $\mathbf{Q}_e$, a test of (4.6.26) using any of the criteria can be employed.

Individual confidence intervals for each of the qp regression weights are obtained by evaluating

$$\hat{\beta}_{ij} - t^{\alpha/2}(\nu_e)\hat{\sigma}_{\hat{\beta}_{ij}} \le \beta_{ij} \le \hat{\beta}_{ij} + t^{\alpha/2}(\nu_e)\hat{\sigma}_{\hat{\beta}_{ij}} \tag{4.6.28}$$

for $i = 0, \ldots, q$ and $j = 1, \ldots, p$.

Acquiring joint simultaneous confidence intervals for the parametric estimable functions $\psi = \mathbf{c}'\mathbf{B}\mathbf{a}$, for arbitrary weight vectors $\mathbf{c}$ and $\mathbf{a}$, Roy (1957, Chap. 14) found intervals when the largest root was employed to test an overall multivariate hypothesis. When the number of variables is one, the Scheffé intervals of Section 4.2 are a special case (see Scheffé, 1953, and Roy and Bose, 1953).

Roy and Bose (1953) extend Scheffe's results to the multivariate case. Letting $\mathbf{V}$ denote the variance-covariance matrix of $\hat{\boldsymbol{\beta}}^*$, where $\hat{\boldsymbol{\beta}}^*$ is the columnwise, rolled-out vector for the $\mathbf{B}$ matrix, they show that the

$$P\left[(\hat{\boldsymbol{\beta}}^* - \boldsymbol{\beta}^*)'\mathbf{V}^{-1}(\hat{\boldsymbol{\beta}}^* - \boldsymbol{\beta}) \le \frac{\nu_e\theta^\alpha}{1 - \theta^\alpha}\right] = 1 - \alpha \tag{4.6.29}$$

which reduces to (4.2.56) when $p = 1$. Employing the Cauchy-Schwarz inequality, the

$$\max_{\mathbf{r}^*} \frac{[\mathbf{r}^{*\prime}(\hat{\boldsymbol{\beta}}^* - \boldsymbol{\beta}^*)]^2}{\mathbf{r}^{*\prime}\mathbf{V}\mathbf{r}^*} \le (\hat{\boldsymbol{\beta}}^* - \boldsymbol{\beta})'\mathbf{V}^{-1}(\hat{\boldsymbol{\beta}}^* - \boldsymbol{\beta})$$

and (4.6.29) reduces to

$$P\left[\max_{\mathbf{r}^*} \frac{|\mathbf{r}^{*\prime}(\hat{\boldsymbol{\beta}}^* - \boldsymbol{\beta})|}{\mathbf{r}^{*\prime}\mathbf{V}\mathbf{r}^*} \le \sqrt{\nu_e\left(\frac{\theta^\alpha}{1 - \theta^\alpha}\right)} \quad \text{for all } \mathbf{r}^*\right] = 1 - \alpha \tag{4.6.30}$$

Thus $100(1 - \alpha)\%$ simultaneous confidence intervals for $\psi = \mathbf{r}^{*\prime}\boldsymbol{\beta}^* = \mathbf{c}'\mathbf{B}\mathbf{a}$ become

$$\mathbf{c}'\hat{\mathbf{B}}\mathbf{a} - c_0\sqrt{\mathbf{a}'\left(\frac{\mathbf{Q}_e}{\nu_e}\right)\mathbf{a}\mathbf{c}'(\mathbf{X}'\mathbf{X})^{-1}\mathbf{c}} \le \mathbf{c}'\mathbf{B}\mathbf{a} \le \mathbf{c}'\hat{\mathbf{B}}\mathbf{a} + c_0\sqrt{\mathbf{a}'\left(\frac{\mathbf{Q}_e}{\nu_e}\right)\mathbf{a}\mathbf{c}'(\mathbf{X}'\mathbf{X})^{-1}\mathbf{c}} \tag{4.6.31}$$

where $c_0^2 = \nu_e\theta^\alpha/(1 - \theta^\alpha)$ and $\mathbf{Q}_e$ is the residual error matrix.

The intervals derived in (4.6.31) are to be used when testing $H_0: \mathbf{B} = \mathbf{0}$ while using the Roy criterion. Modifications for testing $H_0: \mathbf{B}_2 = \mathbf{0}$ are immediate. However, the intervals should not be employed when other criteria are used to test multivariate hypotheses. Gabriel (1968, 1969) indicates that, for the other test procedures, the critical constant c_0^2 in (4.6.31) changes. In summary, they are

(4.6.32) (a) Wilks:

$$c_0^2 = v_e\left(\frac{1 - U^\alpha}{U^\alpha}\right)$$

(b) Roy:

$$c_0^2 = v_e\left(\frac{\theta^\alpha}{1 - \theta^\alpha}\right)$$

(c) Lawley-Hotelling:

$$c_0^2 = v_e U_0^\alpha$$

(d) Pillai:

$$c_0^2 = v_e\left(\frac{V^\alpha}{1 - V^\alpha}\right)$$

where the critical values U^α, θ^α, U_0^α, and V^α were procured in testing the multivariate hypothesis at the significance level α. As in the univariate case, no matter which test procedure is employed, the confidence intervals obtained are very conservative.

EXERCISES 4.6

1. Use the method of (4.6.31) to determine the appropriate expression for obtaining $100(1 - \alpha)\%$ simultaneous confidence intervals for the elements of $\mathbf{B}_2$ after testing the hypothesis $H_0: \mathbf{B}_2 = \mathbf{0}$.

4.7 EXAMPLE—MULTIVARIATE REGRESSION

Rohwer's data in Section 4.3 were applied to kindergarten students in a high-socioeconomic-status area. Similar data were collected for subjects in a low-socioeconomic-status area. The dependent variables for the study were scores on a student achievement test (SAT), the Peabody Picture Vocabulary Test (PPVT), and the Ravin Progressive Matrices Test (RPMT). The data for the study are given in Table 4.7.1, where the sample size $N = 37$.

The general linear model for this example is

(4.7.1)
$$\underset{(37 \times 3)}{\mathbf{Y}} = \underset{(37 \times 6)}{\mathbf{X}} \; \underset{(6 \times 3)}{\mathbf{B}} + \underset{(37 \times 3)}{\mathbf{E}_0}$$

TABLE 4.7.1. Sample Data: Multivariate Regression Analysis.

	SAT	PPVT	RPMT	N	S	NS	NA	SS
	49	48	8	1	2	6	12	16
	47	76	13	5	14	14	30	27
	11	40	13	0	10	21	16	16
	9	52	9	0	2	5	17	8
	69	63	15	2	7	11	26	17
	35	82	14	2	15	21	34	25
	6	71	21	0	1	20	23	18
	8	68	8	0	0	10	19	14
	49	74	11	0	0	7	16	13
	8	70	15	3	2	21	26	25
	47	70	15	8	16	15	35	24
	6	61	11	5	4	7	15	14
	14	54	12	1	12	13	27	21
	30	55	13	2	1	12	20	17
	4	54	10	3	12	20	26	22
	24	40	14	0	2	5	14	8
	19	66	13	7	12	21	35	27
	45	54	10	0	6	6	14	16
	22	64	14	12	8	19	27	26
	16	47	16	3	9	15	18	10
	32	48	16	0	7	9	14	18
	37	52	14	4	6	20	26	26
	47	74	19	4	9	14	23	23
	5	57	12	0	2	4	11	8
	6	57	10	0	1	16	15	17
	60	80	11	3	8	18	28	21
	58	78	13	1	18	19	34	23
	6	70	16	2	11	9	23	11
	16	47	14	0	10	7	12	8
	45	94	19	8	10	28	32	32
	9	63	11	2	12	5	25	14
	69	76	16	7	11	18	29	21
	35	59	11	2	5	10	23	24
	19	55	8	0	1	14	19	12
	58	74	14	1	0	10	18	18
	58	71	17	6	4	23	31	26
	79	54	14	0	6	6	15	14
Means	31.2703	62.6486	13.2432	2.5405	6.9189	13.4865	22.3784	18.3784
Variances	488.4258	156.9044	9.5783	8.6997	25.6877	40.3681	52.4649	40.5759

Employing (4.6.5), the estimate of $\mathbf{B}$ that minimizes the $\mathrm{Tr}[(\mathbf{Y} - \mathbf{XB})'(\mathbf{Y} - \mathbf{XB})]$ is

$$\hat{\mathbf{B}} = (\mathbf{X}'\mathbf{X})^{-1}\mathbf{X}'\mathbf{Y}$$

$$\begin{matrix} & \text{(SAT)} & \text{(PPVT)} & \text{(RPMT)} & \\ \end{matrix}$$

$$(4.7.2) \qquad = \begin{bmatrix} 4.1511 & 33.0058 & 11.1734 \\ -.6089 & -.0806 & .2110 \\ -.0502 & -.7211 & .0646 \\ -1.7324 & -.2983 & .2136 \\ .4946 & 1.4704 & -.0373 \\ 2.2477 & .3240 & -.0521 \end{bmatrix} \begin{matrix} \\ \text{(N)} \\ \text{(S)} \\ \text{(NS)} \\ \text{(NA)} \\ \text{(SS)} \end{matrix}$$

so that the error sums of squares and products matrix $\mathbf{Q}_e$ is

$$\mathbf{Q}_e = \mathbf{Y}'[\mathbf{I} - \mathbf{X}(\mathbf{X}'\mathbf{X})^{-1}\mathbf{X}']\mathbf{Y}$$

(4.7.3)
$$= \begin{bmatrix} 13{,}929.5241 & & \\ 1530.5169 & 2764.7565 & \\ 457.1995 & 213.6780 & 268.2822 \end{bmatrix}$$

Using $\hat{\mathbf{B}}$ and the vector of means denoted by $\hat{\boldsymbol{\alpha}}_0$, the matrix $\hat{\boldsymbol{H}}$ for the deviation model is immediately constructed. Forming the matrix

(4.7.4)
$$\mathbf{R} = \begin{bmatrix} \mathbf{R}_{11} & \mathbf{R}_{12} \\ \mathbf{R}_{21} & \mathbf{R}_{22} \end{bmatrix}$$

where $\mathbf{R}_{11}$ is the correlation among the dependent variables, $\mathbf{R}_{22}$ the correlations among the independent variables, and $\mathbf{R}_{21}$ the correlations between the two sets, $\mathbf{R}$ becomes

(4.7.5)
$$\mathbf{R} = \left[\begin{array}{ccc|ccccc} 1.000 & & & & & & & \\ .3703 & 1.000 & & & & & & \\ .2114 & .3548 & 1.000 & & & & & \\ \hline .1860 & .4444 & .3504 & 1.0000 & & & & \\ .1609 & .2682 & .2386 & .4007 & 1.0000 & & & \\ .0685 & .4692 & .4388 & .5370 & .3523 & 1.0000 & & \\ .2617 & .6720 & .3390 & .6481 & .6478 & .7136 & 1.0000 & \\ .3341 & .5876 & .3404 & .6704 & .4252 & .7695 & .7951 & 1.0000 \end{array}\right]$$

With (4.7.5), the standardized estimates of $\boldsymbol{\Gamma}_z$ are easily obtained,

$$\hat{\boldsymbol{\Gamma}}_z = \mathbf{R}_{22}^{-1}\mathbf{R}_{21}$$

(4.7.6)
$$\hat{\boldsymbol{\Gamma}}_z = \begin{bmatrix} -.0813 & -.0190 & .2011 \\ -.0115 & -.2918 & .1057 \\ -.4980 & -.1513 & .4385 \\ .1621 & .8503 & -.0873 \\ .6479 & .1648 & -.1073 \end{bmatrix}$$

Having estimated the coefficient matrices for regression models commonly used in the literature, the data is first employed to test the hypothesis of no linear relationship:

(4.7.7)
$$H_0\colon \boldsymbol{\Gamma} = \mathbf{0}$$

To test (4.7.7), we construct Table 4.7.2, using Table 4.6.2. By (4.6.24), the test of the hypothesis $H_0\colon \boldsymbol{\Gamma} = \mathbf{0}$ is rejected at the significance level α, where $\alpha = .05$, say, if

(4.7.8) (a) Wilks:

$$\Lambda < U^{\alpha}(p, k, N - k - 1) = U^{.05}_{(3,5,31)} = .449$$

TABLE 4.7.2. MANOVA Table for Testing $\boldsymbol{\Gamma} = \mathbf{0}$ for Multivariate Regression Model

Source	df	SSP
$\boldsymbol{\alpha}_0$	1	$\begin{bmatrix} 36{,}179.7027 & & \\ 72{,}484.4865 & 145{,}219.5676 & \\ 15{,}322.4325 & 30{,}697.8378 & 6489.1892 \end{bmatrix}$
$\boldsymbol{\Gamma}\vert\boldsymbol{\alpha}_0$	5	$\begin{bmatrix} 3{,}653.7732 & & \\ 2{,}159.9966 & 2{,}883.6759 & \\ 63.3680 & 281.4842 & 76.5286 \end{bmatrix} = \mathbf{Q}_h$
Residual	31	Given by (4.7.3)
Total	37	$\begin{bmatrix} 57{,}363.0 & 0 & \\ 76{,}175.0 & 150{,}868.0 & \\ 15{,}843.0 & 31{,}193.0 & 6{,}834.0 \end{bmatrix} = \mathbf{Y'Y}$

(b) Roy:

$$\theta_s > \theta^{\alpha}(s, m, n) = \theta^{.05}_{(3,1/2,13.5)} = .420$$

(c) Lawley-Hotelling:

$$U^{(s)} > U^{\alpha}(s, m, n) = U_0^{.05}(3, .5, 13.5) = .999$$

(d) Pillai:

$$V^{(s)} > V^{\alpha}(s, m, n) = V^{.05}_{(3,1/2,13.5)} = .664$$

(e) Bartlett:

$$X_B^2 > \chi_{\alpha}^2(pk) = \chi_{.05}^2(15) = 25.0$$

To solve the characteristic equation

$$|\mathbf{Q}_h - \lambda \mathbf{Q}_e| = 0 \tag{4.7.9}$$

the roots of (4.7.9) are $\lambda_1 = 1.0551$, $\lambda_2 = .3170$, and $\lambda_3 = .0767$. By (4.6.24) and the fact that $\theta_i = \lambda_i/(1 + \lambda_i)$, the statistical criteria given in (4.7.8) become

$$\begin{aligned} \Lambda &= .3432 & &\text{(Wilks)} \\ \theta_s &= .4866 & &\text{(Roy)} \\ U^{(s)} &= 1.4588 & &\text{(Lawley-Hotelling)} \\ V^{(s)} &= 2.1747 & &\text{(Pillai)} \\ X_B^2 &= 33.6907 & &\text{(Bartlett)} \end{aligned} \tag{4.7.10}$$

From (4.7.10) and the decision rules given in (4.7.8), all criteria are consistent in leading to the rejection of (4.7.7). A similar procedure would be used to test $H_0: \mathbf{B} = \mathbf{0}$ as outlined in Section 4.6.

To find individual confidence intervals for the kp coefficients of $\mathbf{\Gamma}$, formula (4.6.28) is utilized. Representing the standard deviations $\hat{\sigma}_{\hat{\beta}_{ij}}$ in matrix (4.7.11), where $\hat{\sigma}_{\hat{\beta}_{ij}}$ is an estimate of an element in the diagonal of $\mathbf{\Sigma} \otimes (\mathbf{X}'\mathbf{X})^{-1}$,

(4.7.11)
$$\begin{array}{c} \\ (N) \\ (S) \\ (NS) \\ (NA) \\ (SS) \end{array} \begin{array}{c} \begin{array}{ccc} (SAT) & (PPVT) & (RPMT) \end{array} \\ \begin{bmatrix} 1.6711 & .7445 & .2319 \\ .9415 & .4195 & .1307 \\ .9105 & .4056 & .1264 \\ 1.0369 & .4619 & .1439 \\ 1.101 & .4905 & .1528 \end{bmatrix} \end{array} = [\hat{\sigma}_{\hat{\beta}ij}]$$

individual confidence intervals are immediately obtained. For $\alpha = .05, t^{\alpha/2}(31) = 2.040$, so that the intervals become

(4.7.12)
$$\begin{array}{lll}
-4.018 \leq \beta_{11} \leq 2.800 & -1.599 \leq \beta_{12} \leq 1.438 & -.262 \leq \beta_{13} \leq .684 \\
-1.971 \leq \beta_{21} \leq 1.870 & -1.577 \leq \beta_{22} \leq .135 & -.202 \leq \beta_{23} \leq .331 \\
-3.590 \leq \beta_{31} \leq .125 & -1.126 \leq \beta_{32} \leq .529 & -.044 \leq \beta_{33} \leq .471 \\
-1.621 \leq \beta_{41} \leq 2.610 & .528 \leq \beta_{42} \leq 2.413 & -.331 \leq \beta_{43} \leq .256 \\
.002 \leq \beta_{51} \leq 4.494 & -.677 \leq \beta_{52} \leq 1.325 & -.364 \leq \beta_{53} \leq .260
\end{array}$$

Simultaneous intervals for each coefficient are gained by evaluating (4.6.31), with c_0^2 chosen to match the criterion used for the overall test. Since $\hat{\sigma}_{\hat{\beta}_{ij}}$ is the same for all confidence procedures, the only difference in the procedures is due to c_0, which affects the width of the intervals. The value of c_0 for each criterion, by (4.6.32), follows.

(4.7.13)
$$\begin{array}{ll}
c_0 = 4.7380 & \text{(Roy)} \\
c_0 = 5.5650 & \text{(Lawley-Hotelling)} \\
c_0 = 6.1680 & \text{(Wilks)} \\
c_0 = 7.8305 & \text{(Pillai)}
\end{array}$$

From (4.7.13), it is seen that Roy's criterion leads to the shortest intervals; however, these are much too conservative to be of any practical value for this study. In general, if the number of variables in an experiment is larger than four, simultaneous intervals are usually very wide. Comparing (4.7.13) with simultaneous univariate intervals a variable at a time, $c_0 = 3.55$, where $c_0^2 = v_h F^{\alpha}(v_h, v_e)$.

From (4.7.12), we see that not all independent variables are necessarily important to the prediction of three dependent variables. Using (4.2.62) a dependent variable at a time, as a set, different independent variables are important to the prediction of the dependent set:

$$\mathbf{r}_{x_i\hat{y}}\text{SAT} = \frac{\mathbf{r}_{21}}{R_1} = \frac{1}{.4559}\begin{bmatrix} .1860 \\ .1609 \\ .0685 \\ .2617 \\ .3341 \end{bmatrix} = \begin{bmatrix} .408 \\ .353 \\ .150 \\ .574 \\ .733 \end{bmatrix} \begin{matrix} \text{(N)} \\ \text{(S)} \\ \text{(NS)} \\ \text{(NA)} \\ \text{(SS)} \end{matrix}$$

(4.7.14)
$$\mathbf{r}_{x_i\hat{y}}\text{PPVT} = \frac{1}{.7145}\begin{bmatrix} .4444 \\ .2682 \\ .4692 \\ .6720 \\ .5876 \end{bmatrix} = \begin{bmatrix} .622 \\ .375 \\ .657 \\ .941 \\ .822 \end{bmatrix} \begin{matrix} \text{(NA)} \\ \text{(S)} \\ \text{(NS)} \\ \text{(NA)} \\ \text{(SS)} \end{matrix}$$

$$\mathbf{r}_{x_i\hat{y}}\text{RMPT} = \frac{1}{.4711}\begin{bmatrix} .3504 \\ .2386 \\ .4388 \\ .3390 \\ .3404 \end{bmatrix} = \begin{bmatrix} .744 \\ .506 \\ .931 \\ .720 \\ .723 \end{bmatrix} \begin{matrix} \text{(N)} \\ \text{(S)} \\ \text{(NS)} \\ \text{(NA)} \\ \text{(SS)} \end{matrix}$$

To determine each coefficient of determination in (4.7.14), merely compute $\mathbf{R}_{12}\mathbf{R}_{22}^{-1}\mathbf{R}_{21}$. The square roots of the diagonal elements of this matrix yield the values of $\mathbf{R}$ for each dependent variable:

$$[\text{Dia}\,\mathbf{R}_{12}\mathbf{R}_{22}^{-1}\mathbf{R}_{21}]^{1/2} = \begin{bmatrix} R_1 & 0 & 0 \\ 0 & R_2 & 0 \\ 0 & 0 & R_3 \end{bmatrix} = \begin{bmatrix} .4559 & 0 & 0 \\ 0 & .7145 & 0 \\ 0 & 0 & .4711 \end{bmatrix}$$

From (4.7.12) and (4.7.14), it appears that SS is most important for SAT, NA for PPVT, and NS for RPMT. Averaging the squares of the correlations for the three vectors a variable at a time, some indication is attained of how the independent set simultaneously predicts the dependent set:

(4.7.15)
$$\bar{\mathbf{r}}^2 = \begin{bmatrix} .369 \\ .174 \\ .440 \\ .578 \\ .578 \end{bmatrix} \begin{matrix} \text{(N)} \\ \text{(S)} \\ \text{(NS)} \\ \text{(NA)} \\ \text{(SS)} \end{matrix}$$

In addition, the square of the correlation of each of the independent variables should be examined with the linear function of the dependent variables. This allows the maximum correlation in the sample. These values are acquired from the matrix given in (4.7.5). The elements on the

$$\text{Dia}\,[\mathbf{R}_{21}\mathbf{R}_{11}^{-1}\mathbf{R}_{12}] = \begin{bmatrix} .240 \\ .098 \\ .325 \\ .463 \\ .378 \end{bmatrix} \begin{matrix} \text{(N)} \\ \text{(S)} \\ \text{(NS)} \\ \text{(NA)} \\ \text{(SS)} \end{matrix} \tag{4.7.16}$$

yield the desired squared correlations. The unions of the independent variables that have maximum correlations in (4.7.15) and (4.7.16) tend to separate the dependent variables important to the simultaneous prediction of **Y**. Inspection of (4.7.15) and (4.7.16) indicates that the variables NA and SS are "more important" than the variables N, S, and NS.

To determine whether N, S, and NS significantly contribute to the prediction of **Y**, employ equation (4.6.27). By forming MANOVA Table 4.7.3, a test of the contribution of N, S, and NS is obtained.

To test the hypothesis that the variables N, S, and NS are not significantly important to the prediction of **Y**, the characteristic equation

$$|\mathbf{Q}_h - \lambda \mathbf{Q}_e| = 0 \tag{4.7.17}$$

is evaluated where $\mathbf{Q}_h$ is the SSP matrix in Table 4.7.3 with three degrees of freedom and $\mathbf{Q}_e$ is given by (4.7.3). Solving (4.7.17), the eigenvalues are $\lambda_1 = .3354$, $\lambda_2 = .0985$,

TABLE 4.7.3. MANOVA Table to Test $\mathbf{B}_2 = \mathbf{0}$.

Source	df	SSP
Regression on NA and SS	2	$\begin{bmatrix} 1963.4839 & & \\ 1934.7295 & 2594.4108 & \\ 278.2477 & 332.2817 & 44.3386 \end{bmatrix}$
Regression on N, S, and NS above NA and SS	3	$\begin{bmatrix} 1690.2893 & & \\ 225.2671 & 289.2651 & \\ -214.8797 & -50.7975 & 32.1900 \end{bmatrix} = \mathbf{Q}_h$
$\boldsymbol{\Gamma}\vert\boldsymbol{\alpha}_0$	5	Given in Table 4.7.2
Residual	31	Given in (4.7.3)

and $\lambda_3 = .0045$. Thus the values of the criteria become

$$\begin{aligned} \Lambda &= .6786 \\ \theta_s &= .2512 \\ U^{(s)} &= .4884 \\ V^{(s)} &= .3454 \\ X_B^2 &= 11.8248 \end{aligned} \tag{4.7.18}$$

Letting $\alpha = .05$, the decision rules for the criteria are to reject if

$$
\begin{aligned}
\Lambda &< U^{.05}(3, 3, 31) &&= .5719 \\
\theta_s &> U^{.05}(3, -.5, 13.5) &&= .3520 \\
U^{(s)} &= U^{.05}(3, -.5, 13.5) &&= .6698 \\
V^{(s)} &= V^{.05}(3, -.5, 13.5) &&= .479 \\
X_B^2 &> U_{.05}^2(9) &&= 19.0
\end{aligned}
\tag{4.7.19}
$$

From (4.7.18) and (4.7.19), the variables N, S, and NS are found not to contribute significantly to the prediction of **Y**. Utilizing the variables NA and SS, the regression equation becomes

$$
\hat{\mathbf{Y}} = \mathbf{X}\hat{\mathbf{B}} \qquad \text{where } \hat{\mathbf{B}} = \begin{bmatrix} 10.1547 & 35.8637 & 9.7459 \\ -.0327 & .9629 & .0794 \\ 1.1888 & .2850 & .0936 \end{bmatrix}
\tag{4.7.20}
$$

By using (4.2.62) a variable at a time, the relation between each independent variable and the dependent variables is evaluated:

$$
\begin{aligned}
\mathbf{r}_{x_i\hat{y}}\text{SAT} &= \frac{1}{.3342}\begin{bmatrix} .2617 \\ .3341 \end{bmatrix} = \begin{bmatrix} .7831 \\ .9997 \end{bmatrix} \begin{matrix} (\text{NA}) \\ (\text{SS}) \end{matrix} \\
\mathbf{r}_{x_i\hat{y}}\text{PPVT} &= \frac{1}{.6777}\begin{bmatrix} .6720 \\ .5876 \end{bmatrix} = \begin{bmatrix} .9916 \\ .8671 \end{bmatrix} \begin{matrix} (\text{NA}) \\ (\text{SS}) \end{matrix} \\
\mathbf{r}_{x_i\hat{y}}\text{RPMT} &= \frac{1}{.3586}\begin{bmatrix} .3390 \\ .3404 \end{bmatrix} = \begin{bmatrix} .9453 \\ .9492 \end{bmatrix} \begin{matrix} (\text{NA}) \\ (\text{SS}) \end{matrix}
\end{aligned}
\tag{4.7.21}
$$

By averaging the squares of the elements for the three vectors,

$$
\bar{\mathbf{r}}^2 = \begin{bmatrix} .830 \\ .884 \end{bmatrix} \begin{matrix} (\text{NA}) \\ (\text{SS}) \end{matrix}
\tag{4.7.22}
$$

an indication of the simultaneous prediction of NA and SS to the dependent set is obtained. This suggests that NA and SS adequately predict the three dependent variables SAT, PPVT, and RPMT. We would now obtain confidence intervals or simultaneous confidence intervals for the coefficients in **B** as estimated by $\hat{\mathbf{B}}$ in (4.7.20).

EXERCISES 4.7

1. Verify the calculations in Table 4.7.3 and obtain 95% simultaneous confidence intervals for the coefficients in **B** as estimated by $\hat{\mathbf{B}}$ in (4.7.20) by using all multivariate criteria.
2. Summary data on the number of MA and BA degrees awarded at the end of one academic year by 163 institutions of higher learning, along with data on instructional cost (C), number of students (S), number of full-time faculty (F), and average wages (W) are summarized below.

	BA	MA	C(000)	S	F	W
Means:	586.4	108.9	3978.8	4115.2	230.5	9781.8

Correlations:	BA	MA	C	S	F	W
BA	1.000					
MA	.830	1.000				
C	.945	.812	1.000			
S	.941	.817	.959	1.000		
F	.944	.796	.966	.972	1.000	
W	.528	.443	.589	.552	.593	1.000

	BA	MA	C(000)	S	F	W
Standard Deviations:	575.2	157.6	4015.4	3411.5	198.7	1236.8

With the data provided, determine the simplest model for predicting the number of MA and BA degrees conferred, and obtain simultaneous confidence intervals for the regression weights.

4.8 MULTIVARIATE POLYNOMIAL REGRESSION

In Section 4.6, the extension of multiple regression to multivariate regression was discussed. The results of Section 4.4 are now extended to the multivariate situation.

In multivariate polynomial regression, N_i p-vector valued observations are acquired at each independent point x_i. The data at each independent point i, $i = 1, \ldots, n$ are represented as follows:

$$\begin{array}{ccccc} & & & & \text{means} \\ \mathbf{y}_{11} & \mathbf{y}_{12} & \cdots & \mathbf{y}_{1N_1} & \mathbf{y}_{1.} \\ \mathbf{y}_{21} & \mathbf{y}_{22} & \cdots & \mathbf{y}_{2N_2} & \mathbf{y}_{2.} \\ \vdots & \vdots & \cdots & \vdots & \vdots \\ \mathbf{y}_{n1} & \mathbf{y}_{n2} & \cdots & \mathbf{y}_{nN_n} & \mathbf{y}_{n.} \end{array} \tag{4.8.1}$$

where the $\mathbf{y}'_{ij} = [y_{ij1}, y_{ij2}, \ldots, y_{ijp}]$ denote the independent p-variate response vectors.

Instead of fitting a model to the original observations, a model is fitted to the mean vectors. Let the p-variate mean vector for the ith group be denoted by

$$\mathbf{y}_{i.} = \frac{1}{N_i} \sum_{j=1}^{N_i} \mathbf{y}_{ij} \tag{4.8.2}$$

Then the polynomial model for these means becomes, using matrix notation,

$$\underset{(n \times p)}{\mathbf{Y}_.} = \underset{(n \times q)}{\mathbf{X}} \underset{(q \times p)}{\mathbf{B}} + \underset{(n \times p)}{\mathbf{E}_.} \tag{4.8.3}$$

or

$$\begin{bmatrix} \mathbf{y}'_{1.} \\ \mathbf{y}'_{2.} \\ \vdots \\ \mathbf{y}'_{n.} \end{bmatrix} = \overset{\begin{matrix} 1 & x & x^2 & \cdots & x^k \end{matrix}}{\begin{bmatrix} 1 & x_{11} & x_{12} & \cdots & x_{1k} \\ 1 & x_{21} & x_{22} & \cdots & x_{2k} \\ \vdots & \vdots & \vdots & \cdots & \vdots \\ 1 & x_{n1} & x_{n2} & \cdots & x_{nk} \end{bmatrix}} \begin{bmatrix} \beta_{01} & \beta_{02} & \cdots & \beta_{0p} \\ \beta_{11} & \beta_{12} & \cdots & \beta_{1p} \\ \vdots & \vdots & \cdots & \vdots \\ \beta_{k1} & \beta_{k2} & \cdots & \beta_{kp} \end{bmatrix} + \begin{bmatrix} \boldsymbol{\varepsilon}'_{1.} \\ \boldsymbol{\varepsilon}'_{2.} \\ \vdots \\ \boldsymbol{\varepsilon}'_{n.} \end{bmatrix} \tag{4.8.4}$$

so that

$$E(\mathbf{Y}_.) = \mathbf{XB} \quad \text{and} \quad V(\mathbf{Y}_.) = \mathbf{V} \otimes \mathbf{\Sigma} \tag{4.8.5}$$

where

$$\mathbf{V} = \begin{bmatrix} 1/N_1 & 0 & \cdots & 0 \\ 0 & 1/N_2 & \cdots & \cdot \\ \vdots & \vdots & \vdots & \vdots \\ 0 & 0 & \cdots & 1/N_n \end{bmatrix} \tag{4.8.6}$$

$q = k + 1$, and $n > q > p$.

The same design matrix $\mathbf{X}$ is applied to each dependent variable. This implies that all of the dependent variables in the study are described by polynomials of the same order. However, in some instances, this may lead to overfitting some of the variables. By employing the notation of Section 3.6, we roll out the columns of (4.8.3) so that it becomes

$$\mathbf{y}_.^* = (\mathbf{I} \otimes \mathbf{X})\boldsymbol{\beta}^* + \boldsymbol{\varepsilon}_.^* \tag{4.8.7}$$

where

$$E(\boldsymbol{\varepsilon}_.^*) = \mathbf{0} \quad \text{and} \quad V(\boldsymbol{\varepsilon}_.^*) = \mathbf{\Sigma} \otimes \mathbf{V} \tag{4.8.8}$$

By transforming each column of $\mathbf{Y}_.$ and $\mathbf{E}_.$ by $\mathbf{V}^{-1/2}$ and letting $\mathbf{Q} = \mathbf{V}^{-1/2}\mathbf{X}$, (4.8.7) becomes

$$\mathbf{y}_0^* = (\mathbf{I} \otimes \mathbf{Q})\boldsymbol{\beta}^* + \boldsymbol{\varepsilon}^* \tag{4.8.9}$$

where

$$E(\boldsymbol{\varepsilon}^*) = \mathbf{0} \quad \text{and} \quad V(\boldsymbol{\varepsilon}^*) = \mathbf{\Sigma} \otimes \mathbf{I} \tag{4.8.10}$$

Transforming back to matrix notation, (4.8.3) reduces to

$$\mathbf{Y}_0 = \mathbf{QB} + \mathbf{E}_0 \tag{4.8.11}$$

where

$$E(\mathbf{E}_0) = \mathbf{0} \quad \text{and} \quad V(\mathbf{E}_0) = \mathbf{I} \otimes \mathbf{\Sigma} \tag{4.8.12}$$

Applying the multivariate Gauss-Markoff theorem to (4.8.12), where $\mathbf{B}$ is found to minimize the $\mathrm{Tr}(\mathbf{E}_0'\mathbf{E}_0)$,

$$\hat{\mathbf{B}} = (\mathbf{Q}'\mathbf{Q})^{-1}\mathbf{Q}'\mathbf{Y}_0 = (\mathbf{X}'\mathbf{V}^{-1}\mathbf{X})^{-1}\mathbf{X}'\mathbf{V}^{-1}\mathbf{Y}_. \tag{4.8.13}$$

Rolling out the elements of $\hat{\mathbf{B}}$, the

$$V(\boldsymbol{\beta}^*) = \mathbf{\Sigma} \otimes (\mathbf{X}'\mathbf{V}^{-1}\mathbf{X})^{-1} \tag{4.8.14}$$

Furthermore, an unbiased estimate of $\mathbf{\Sigma}$, given that model (4.8.3) is correct, is

$$\mathbf{S}_r = \frac{\mathbf{Y}_0'[\mathbf{I} - \mathbf{Q}(\mathbf{Q}'\mathbf{Q})^{-1}\mathbf{Q}']\mathbf{Y}_0}{n - q} = \frac{\mathbf{Y}_.'\mathbf{V}^{-1}\mathbf{Y}_. - \hat{\mathbf{B}}'\mathbf{X}'\mathbf{V}^{-1}\mathbf{X}\hat{\mathbf{B}}}{n - k - 1} \tag{4.8.15}$$

The similarity between the univariate and multivariate results should be evident; each $\mathbf{y}_.$ in the univariate case is replaced by $\mathbf{Y}_.$ in the multivariate situation.

TABLE 4.8.1. MANOVA Table for the Lack-Of-Fit Test.

Source	df	SSP	E(MS)
Total regression	$k+1$	$\mathbf{Q}_h = \hat{\mathbf{B}}'\mathbf{X}'\mathbf{V}^{-1}\mathbf{X}\hat{\mathbf{B}}$	$\mathbf{\Sigma} + \dfrac{\mathbf{B}'\mathbf{X}'\mathbf{V}^{-1}\mathbf{X}\mathbf{B}}{k+1}$
Residual	$n-k-1$	$\mathbf{Q}_e^* = \mathbf{Y}_.'\mathbf{V}^{-1}\mathbf{Y}_. - \hat{\mathbf{B}}'\mathbf{X}'\mathbf{V}^{-1}\mathbf{X}\hat{\mathbf{B}}$	$\mathbf{\Sigma}$
Total (between)	n	$\mathbf{Q}_{tb}^* = \mathbf{Y}_.'\mathbf{V}^{-1}\mathbf{Y}_.$	
Within error	$N-n$	$\mathbf{Q}_e = \mathbf{Y}'\mathbf{Y} - \mathbf{Y}_.'\mathbf{V}^{-1}\mathbf{Y}_.$	$\mathbf{\Sigma}$
Total	N	$\mathbf{Q}_t = \mathbf{Y}'\mathbf{Y}$	

Proceeding as in the univariate case, suppose an experimenter hypothesized that the degree of the polynomial that best fits all variables is of order k. To test this hypothesis using the *multivariate* lack-of-fit test, Table 4.8.1 is constructed. Note the similarity between Table 4.8.1 and Table 4.4.2. The multivariate lack-of-fit test is testing the hypothesis that $\mathbf{B}_2 = \mathbf{0}$, or, equivalently, that $\mathbf{S}_r = \mathbf{Q}_e^*/(n-k-1)$ and $\mathbf{S} = \mathbf{Q}_e/(N-n)$ are estimating the same variance-covariance matrix $\mathbf{\Sigma}$.

The multivariate criteria available to test

(4.8.16) $$H_0 : \mathbf{B}_2 = \mathbf{0}$$

depend on the roots of the equation

(4.8.17) $$|\mathbf{Q}_e^* - \lambda \mathbf{Q}_e| = 0$$

With parameters to enter the tables to test (4.8.16), employing various criteria defined by

(4.8.18) $$s = \min(v_h, p) \qquad m = \frac{|v_h - p| - 1}{2} \qquad n = \frac{v_e - p - 1}{2}$$

where $v_h = n - k - 1$ and $v_e = N - n$, (4.8.16) is rejected at the significance level α if

(4.8.19) (a) Wilks:

$$\Lambda = \frac{|\mathbf{Q}_e|}{|\mathbf{Q}_e^* + \mathbf{Q}_e|} = \prod_{i=1}^{s} (1 + \lambda_i)^{-1} < U^\alpha(p, v_h, v_e)$$

(b) Roy:

$$\theta_s = \frac{\lambda_1}{1 + \lambda_1} > \theta^\alpha(s, m, n)$$

(c) Lawley-Hotelling:

$$U^{(s)} = \frac{T_0^2}{v_e} = \sum_{i=1}^{s} \lambda_i > U_0^\alpha(s, m, n)$$

(d) Pillai:

$$V^{(s)} = \sum_{i=1}^{s} \frac{\lambda_i}{1 + \lambda_i} > V^{\alpha}(s, m, n)$$

Given that (4.8.16) is tenable, we may pool the residual sums of the products matrices to obtain a better estimate of $\mathbf{\Sigma}$. The pooled estimate of $\mathbf{\Sigma}$ becomes

$$\mathbf{S}_p = \frac{\mathbf{Q}_e^* + \mathbf{Q}_e}{N - k - 1} \tag{4.8.20}$$

However, if the number of observations in each group is large, $\mathbf{S} = \mathbf{Q}_e/(N - n)$ is usually employed as an estimate of $\mathbf{\Sigma}$. If (4.8.16) is rejected, residual plots, a variable at a time, might be used to determine which variables did not fit the proposed model.

When no specific order of a polynomial is known a priori, we employ a step-down procedure to obtain the "best" model as an alternative to the multivariate lack-of-fit test. Suppose that the number of points at each value of the independent variable is the same and that the intervals between points are equally spaced, as in the univariate case. The advantage of the step-down procedure over the lack-of-fit test, when considering multivariate data, is that all multivariate criteria are equivalent for testing the contribution of each "term" in the model.

Proceeding as in Section 4.4, suppose the model of best fit is hypothesized to be of order $q = s + 1$. We then transform the nonorthogonal model given by (4.8.3) to be the orthogonal model

$$\underset{(n \times p)}{\mathbf{Y}} = \underset{(n \times q)}{\mathbf{P}} \underset{(q \times p)}{\mathbf{\Xi}} + \underset{(n \times p)}{\mathbf{E}} \tag{4.8.21}$$

where $\mathbf{P}$ is obtained from Table XIII in the Appendix. Given equal intervals and an equal number of observations at each x,

$$\begin{aligned} \hat{\mathbf{\Xi}} &= (\mathbf{P}'\mathbf{V}^{-1}\mathbf{P})^{-1}\mathbf{P}'\mathbf{V}^{-1}\mathbf{Y} \\ &= (\mathbf{P}'\mathbf{P})^{-1}\mathbf{P}'\mathbf{Y} \\ &= \mathbf{D}^{-1}\hat{\mathbf{\Psi}} \\ &= \frac{\hat{\mathbf{\Psi}}}{\|\mathbf{p}_i\|^2} \end{aligned} \tag{4.8.22}$$

where $\mathbf{p}_i$ denotes the ith column of the design matrix $\mathbf{P}$.

To find $\mathbf{Q}_h$ for testing

$$\begin{aligned} H_0&: \mathbf{\Xi} = \mathbf{0} \\ H_1&: \mathbf{\Xi} \neq \mathbf{0} \end{aligned} \tag{4.8.23}$$

the hypothesis sum of squares and products matrix is partitioned into q separate components:

$$\begin{aligned} \mathbf{Q}_h &= \hat{\mathbf{\Psi}}'\mathbf{P}'\mathbf{V}^{-1}\mathbf{P}\hat{\mathbf{\Psi}} \\ &= \sum_i \left(\frac{n_0}{\|\mathbf{p}_i\|^2} \right) \hat{\mathbf{\Psi}}_i \hat{\mathbf{\Psi}}_i' \\ &= \sum_i \mathbf{Q}_{h_i} \end{aligned} \tag{4.8.24}$$

where $\hat{\mathbf{\Psi}}_i$ is the ith *row* of $\hat{\mathbf{\Psi}}$.

Each of the matrices $\mathbf{Q}_{h_i}$ in $\mathbf{Q}_h$ have degrees of freedom $v_h = 1$. Thus any criteria may be employed to test the significance of each term because they are all equivalent when $s = \min(v_h, p) = 1$. Using the lambda criterion,

$$\Lambda = \frac{|\mathbf{Q}_e|}{|\mathbf{Q}_{h_i} + \mathbf{Q}_e|} \sim U^{\alpha}(p, 1, v_e)$$

$$= (1 + \lambda)^{-1} \tag{4.8.25}$$

where λ is the root of

$$|\mathbf{Q}_{h_i} - \lambda \mathbf{Q}_e| = 0 \tag{4.8.26}$$

and $\mathbf{Q}_e$ is the residual sum of products matrix with $v_e = N - n$ degrees of freedom. Alternatively,

$$T^2 = v_e\left(\frac{|\mathbf{Q}_e + \mathbf{Q}_{h_i}|}{|\mathbf{Q}_e|} - 1\right) = v_e\lambda \sim T^2(p, v_e) \tag{4.8.27}$$

An example of the procedures discussed in this section, including methods for acquiring confidence intervals, is given in the next section.

4.9 EXAMPLE—MULTIVARIATE POLYNOMIAL REGRESSION

In Section 4.5 an example of multiple curvilinear regression was given to determine the effect of practice on a manual learning task. The 24 students in that study were ambidextrous so that the time to solve the problem on the test was recorded for both their right- and left-hand performance, given a number of practice trials. The data are shown in Table 4.9.1.

TABLE 4.9.1. Sample Data: Multivariate Curvilinear Regression.

		Number of Practice Trials											
		1		2		3		4		5		6	
		R	L	R	L	R	L	R	L	R	L	R	L
		103	93	60	65	58	63	57	53	55	39	52	54
		99	84	63	59	82	61	68	59	49	53	48	63
		85	96	71	55	41	54	50	52	60	59	55	50
		104	77	64	70	60	69	39	68	44	70	42	49
Means	R	97.75		64.50		60.25		53.50		52.00		49.25	
	L	87.50		62.25		61.75		58.00		55.25		54.00	

To find the functional form of the right- and left-hand mean performance on an equal number of practice trials, a multivariate polynomial regression procedure is applied to the means where each pair of data points has a joint bivariate normal distribution. Hypothesizing a quadratic trend, a multivariate lack-of-fit test is first considered.

Plotting the observations, where the crosses denote the mean values for variable **R**, and the dots denote the mean values for variable L, we observe from Figure 4.9.1 that the mean trend seems to be quadratic for both variables. Using matrices, the mean data for the study is represented by

$$\mathbf{Y} \quad = \quad \mathbf{X} \quad \mathbf{B} \quad + \quad \mathbf{E}.$$

(4.9.1)
$$\begin{bmatrix} 97.75 & 87.50 \\ 64.50 & 62.25 \\ 60.25 & 61.75 \\ 53.50 & 58.00 \\ 52.00 & 55.25 \\ 49.25 & 54.00 \end{bmatrix} = \begin{bmatrix} 1 & 1 & 1 \\ 1 & 2 & 4 \\ 1 & 3 & 9 \\ 1 & 4 & 16 \\ 1 & 5 & 25 \\ 1 & 6 & 36 \end{bmatrix} \begin{bmatrix} \beta_{01} & \beta_{02} \\ \beta_{11} & \beta_{12} \\ \beta_{21} & \beta_{22} \end{bmatrix} + \begin{bmatrix} \boldsymbol{\varepsilon}'_{1.} \\ \boldsymbol{\varepsilon}'_{2.} \\ \boldsymbol{\varepsilon}'_{3.} \\ \boldsymbol{\varepsilon}'_{4.} \\ \boldsymbol{\varepsilon}'_{5.} \\ \boldsymbol{\varepsilon}'_{6.} \end{bmatrix}$$

Employing (4.8.13),

(4.9.2)
$$\hat{\mathbf{B}} = (\mathbf{X}'\mathbf{V}^{-1}\mathbf{X})^{-1}\mathbf{X}'\mathbf{V}^{-1}\mathbf{Y}_{.} = \begin{bmatrix} 118.80000 & 100.85000 \\ -29.63036 & -19.36786 \\ 2.91964 & 1.98214 \end{bmatrix}$$

where

$$\underset{(6\times 6)}{\mathbf{V}} = \begin{bmatrix} 1/4 & 0 & \cdots & 0 \\ 0 & 1/4 & \cdots & 0 \\ \vdots & \vdots & \vdots\vdots\vdots & \vdots \\ 0 & 0 & \cdots & 1/4 \end{bmatrix}$$

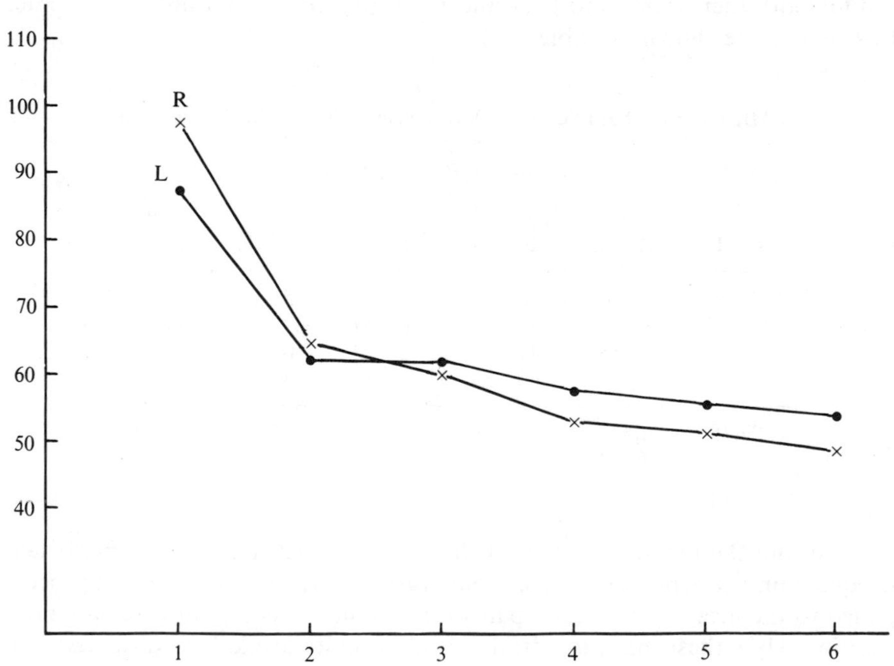

FIGURE 4.9.1. Plot of means.

TABLE 4.9.2. MANOVA Table for Multivariate Lack-Of-Fit Test.

Source	df	SSP
Total regression	3	$\mathbf{Q}_h = \begin{bmatrix} 100{,}849.943 & \\ 99{,}269.993 & 98{,}333.093 \end{bmatrix}$
Residual	3	$\mathbf{Q}_e^* = \begin{bmatrix} 498.807 & \\ 426.757 & 374.657 \end{bmatrix}$
Total	6	$\mathbf{Q}_{tb}^* = \begin{bmatrix} 101{,}348.750 & \\ 99{,}696.750 & 98{,}707.750 \end{bmatrix}$
Within error	18	$\mathbf{Q}_e = \begin{bmatrix} 1{,}830.250 & \\ -334.750 & 12{,}555.250 \end{bmatrix}$
Total	24	$\mathbf{Q}_t = \begin{bmatrix} 103{,}179.000 & \\ 99{,}362.000 & 99{,}963.000 \end{bmatrix}$

By constructing Table 4.9.2, a MANOVA table for the lack-of-fit test, entries are computed from those expressions given in Table 4.8.1.

Employing equation (4.8.17),

$$|\mathbf{Q}_e^* - \lambda \mathbf{Q}_e| = 0 \tag{4.9.3}$$

where $\mathbf{Q}_e^*$ and $\mathbf{Q}_e$ are given in Table 4.9.2, the hypothesized degree of the model, which for the example is two, is tested by using (4.8.19). Since the roots of (4.9.3) are $\lambda_1 = .7342$ and $\lambda_2 = .0014$, the value of Wilks' and Roy's criteria are

$$\begin{aligned} \Lambda &= .5758 \\ \theta_s &= .4233 \end{aligned} \tag{4.9.4}$$

The lack-of-fit test is rejected at the significance level $\alpha = .05$ if

$$\begin{aligned} \Lambda &< U^{\alpha}(p, v_h, v_e) = U^{.05}_{(2,3,18)} = .495888 \\ \theta_s &> \theta^{\alpha}(s, m, n) = \theta^{.05}_{(2,0,7.5)} = .450 \end{aligned} \tag{4.9.5}$$

Thus the lack-of-fit test is not rejected.

Because the number of observations in each group is small, it is best to pool the estimates obtained for estimating $\mathbf{\Sigma}$:

$$\mathbf{S}_p = \frac{1}{3 + 18}(\mathbf{Q}_e^* + \mathbf{Q}_e) = \begin{bmatrix} 110.907 & \\ 3.905 & 77.615 \end{bmatrix} \tag{4.9.6}$$

By using the pooled estimate of $\mathbf{\Sigma}$, confidence intervals for the elements of $\mathbf{B}$ are obtained by the formula

(4.9.7)

$$V(\hat{\boldsymbol{\beta}}^*) = \mathbf{S}_p \otimes (\mathbf{X}'\mathbf{V}^{-1}\mathbf{X})^{-1} = \begin{bmatrix} 110.907 & \\ 3.905 & 77.615 \end{bmatrix} \otimes \begin{bmatrix} .80000 & & \\ -.48750 & .34241 & \\ .06250 & -.04687 & .00670 \end{bmatrix}$$

where $\mathbf{S}_p$ is the pooled estimate of $\boldsymbol{\Sigma}$ in this example. To evaluate (4.9.7), the diagonal elements are arranged in a matrix,

$$\begin{array}{c} \\ \beta_0 \\ \beta_1 \\ \beta_2 \end{array}\begin{array}{c} \begin{array}{cc} \text{R} & \text{L} \end{array} \\ \begin{bmatrix} 88.7256 & 62.0920 \\ 39.9757 & 26.5762 \\ .7431 & .5200 \end{bmatrix} \end{array} \tag{4.9.8}$$

where the square roots of the elements in (4.9.8) are the quantities $\hat{\sigma}_{\hat{\beta}ij}$ needed for finding confidence intervals for individual regression coefficients. The formula discussed for the multivariate regression case in Section 4.6 is now applicable with $(\mathbf{X}'\mathbf{X})^{-1}$ replaced by $(\mathbf{X}'\mathbf{V}^{-1}\mathbf{X})^{-1}$. For example, individual intervals using Roy's criterion become

$$\begin{array}{ll} 83.55 \le \beta_{01} \le 154.05 & 71.36 \le \beta_{02} \le 130.34 \\ -52.69 \le \beta_{11} \le \quad 6.57 & 38.66 \le \beta_{12} \le \quad .08 \\ -.31 \le \beta_{21} \le \quad 6.15 & -.72 \le \beta_{22} \le \quad 4.68 \end{array} \tag{4.9.9}$$

where $c_0 = \sqrt{v_e[\theta^\alpha/(1-\theta\alpha)]} = \sqrt{21(.4/.6)} = 3.472$.

Alternatively, orthogonal polynomials may be employed in a step-down manner to find the model of best fit. Hypothesizing at most a cubic polynomial for this example, (4.9.1) becomes by using (4.8.21),

(4.9.10)

$$\begin{array}{ccccc} \mathbf{Y}_. & = & \mathbf{P} & \boldsymbol{\Xi} & + \quad \mathbf{E}_. \end{array}$$

$$\begin{bmatrix} 97.75 & 87.50 \\ 64.50 & 62.25 \\ 60.25 & 61.75 \\ 53.50 & 58.00 \\ 52.00 & 55.25 \\ 49.25 & 54.00 \end{bmatrix} = \begin{bmatrix} 1 & -5 & 5 & -5 \\ 1 & -3 & -1 & 7 \\ 1 & -1 & -4 & 4 \\ 1 & 1 & -4 & -4 \\ 1 & 3 & -1 & -7 \\ 1 & 5 & 5 & 5 \end{bmatrix} \begin{bmatrix} \xi_{01} & \xi_{02} \\ \xi_{11} & \xi_{12} \\ \xi_{21} & \xi_{22} \\ \xi_{31} & \xi_{32} \end{bmatrix} + \begin{bmatrix} \boldsymbol{\varepsilon}'_{1.} \\ \boldsymbol{\varepsilon}'_{2.} \\ \boldsymbol{\varepsilon}'_{3.} \\ \boldsymbol{\varepsilon}'_{4.} \\ \boldsymbol{\varepsilon}'_{5.} \\ \boldsymbol{\varepsilon}'_{6.} \end{bmatrix}$$

so that

$$\hat{\boldsymbol{\Xi}} = (\mathbf{P}'\mathbf{V}^{-1}\mathbf{P})^{-1}\mathbf{P}'\mathbf{V}^{-1}\mathbf{Y}_. = \begin{bmatrix} 62.87500 & 63.12500 \\ -4.09643 & -2.74643 \\ 1.94643 & 1.32143 \\ -.71111 & -.57500 \end{bmatrix} \tag{4.9.11}$$

From (4.8.24),

$$\hat{\boldsymbol{\Psi}} = \begin{bmatrix} 377.25 & 378.75 \\ -286.75 & -192.25 \\ 163.50 & 111.00 \\ -128.00 & -103.50 \end{bmatrix} = \begin{bmatrix} \hat{\boldsymbol{\Psi}}'_1 \\ \hat{\boldsymbol{\Psi}}'_2 \\ \hat{\boldsymbol{\Psi}}'_3 \\ \hat{\boldsymbol{\Psi}}'_4 \end{bmatrix} \tag{4.9.12}$$

TABLE 4.9.3. MANOVA Table for Orthogonal Polynomials.

Source	df	SSP	T^2
ξ_0	1	$\hat{\psi}_1\hat{\psi}_1' = \begin{bmatrix} 94{,}878.375 & \\ 95{,}255.625 & 95{,}634.375 \end{bmatrix}$	2947.9
$\xi_1 \mid \xi_0$	1	$\hat{\psi}_2\hat{\psi}_2' = \begin{bmatrix} 1272.964 & \\ 864.214 & 586.714 \end{bmatrix}$	97.8
$\xi_2 \mid \xi_1, \xi_0$	1	$\hat{\psi}_3\hat{\psi}_3' = \begin{bmatrix} 4698.604 & \\ 3150.154 & 2112.004 \end{bmatrix}$	26.8
$\xi_3 \mid \xi_2, \xi_1, \xi_0$	1	$\hat{\psi}_4\hat{\psi}_4' = \begin{bmatrix} 364.089 & \\ 294.400 & 238.050 \end{bmatrix}$	
Residual	2	$\mathbf{Q}_e^* = \begin{bmatrix} 134.718 & \\ 132.357 & 136.607 \end{bmatrix}$	
Total (between)	6	$\mathbf{Q}_{tb}^* = \begin{bmatrix} 101{,}348.750 & \\ 99{,}696.750 & 98{,}707.750 \end{bmatrix}$	
Within error	18	$\mathbf{Q}_e = \begin{bmatrix} 1830.250 & \\ -334.750 & 1255.250 \end{bmatrix}$	
Total	24	$\mathbf{Q}_t = \begin{bmatrix} 103{,}179.000 & \\ 99{,}362.000 & 99{,}963.000 \end{bmatrix}$	

MANOVA Table 4.9.3 is formed to determine the degree of the polynomial that "best" fits the data.

Following the univariate procedure, the degree of the polynomial that "best" fits the data is of that order where the first multivariate test is significant. Setting $\alpha = .05$, each test is performed at $\alpha_i = \alpha/4 = .0125$, so that the critical value is $T_i^\alpha = 21.335$ when using Hotelling's T^2 statistic. Comparing the computed T^2 values in Table 4.9.3 with the critical constant, the degree of the polynomial is two. Thus

$$\hat{\boldsymbol{\Xi}} = \begin{bmatrix} 62.87500 & 63.12500 \\ -4.09643 & -2.74643 \\ 1.94643 & 1.32143 \end{bmatrix} \tag{4.9.13}$$

or, in terms of the original nonorthogonal model,

$$\hat{\mathbf{B}} = \begin{bmatrix} 118.8000 & 100.8500 \\ -29.63036 & -19.36786 \\ 2.91964 & 1.98214 \end{bmatrix} \tag{4.9.14}$$

To obtain confidence intervals for the elements of $\mathbf{B}$, the matrix associated with the within-error term in Table 4.9.3 is used for $\boldsymbol{\Sigma}$ in equation (4.8.14), where the degrees of freedom for error are 18. The interval becomes, when using Roy's criterion,

$$\begin{array}{ll} 84.19 \le \beta_{01} \le 153.42 & 72.18 \le \beta_{02} \le 129.52 \\ -52.28 \le \beta_{11} \le 6.98 & -38.12 \le \beta_{12} \le -.61 \\ -97 \le \beta_{21} \le 6.09 & -.64 \le \beta_{22} \le 4.61 \end{array} \tag{4.9.15}$$

where $c_0 = \sqrt{v_e[\theta^\alpha/(1 - \theta^\alpha)]} = \sqrt{18(.450/.550)} = 3.838$. Using either (4.9.13) or (4.9.14) produces the fit of the quadratic model, which is illustrated in Table 4.9.4.

TABLE 4.9.4. Observations, Fitted Values, and Residuals.

$\mathbf{y}_{.}$	$\hat{\mathbf{y}}_{.}$	Residual
[95.75, 87.50]	[93.0893, 87.5001]	[4.6607, −.0001]
[64.50, 62.25]	[73.2178, 62.2701]	[−8.7178, −.0001]
[60.25, 61.75]	[59.1857, 61.7504]	[1.0643, −.0004]
[53.50, 58.00]	[50.9928, 58.0007]	[2.5072, −.0007]
[52.00, 55.25]	[48.6393, 55.2510]	[3.3607, −.0010]
[49.25, 54.00]	[52.1250, 53.9973]	[−2.8750, −.0027]

Instead of using either (4.9.13) or (4.9.14) to summarize the model for the data, trend components for each variable are often utilized. The trends are represented by

$$\theta^{(i)}_{\text{constant}} = \frac{1}{6}(\mu_1 + \mu_2 + \mu_3 + \mu_4 + \mu_5 + \mu_6)$$

$$\theta^{(i)}_{\text{linear}} = \frac{1}{\sqrt{70}}(-5\mu_1 - 3\mu_2 - \mu_3 + \mu_4 + 3\mu_5 + 5\mu_6) \tag{4.9.16}$$

$$\theta^{(i)}_{\text{quadratic}} = \frac{1}{\sqrt{84}}(5\mu_1 - \mu_2 - 4\mu_3 - 4\mu_4 - \mu_5 + 5\mu_6)$$

for $i = 1, 2$, where i denotes the variable and $\mu_j\,(j = 1, \ldots, n)$ the means. Thus

$$\begin{aligned} \hat{\theta}^{(1)}_{\text{C}} &= 62.875 & \hat{\theta}^{(2)}_{\text{C}} &= 63.125 \\ \hat{\theta}^{(1)}_{\text{L}} &= -34.273 & \hat{\theta}^{(2)}_{\text{L}} &= -22.978 \\ \hat{\theta}^{(1)}_{\text{Q}} &= 17.839 & \hat{\theta}^{(2)}_{\text{Q}} &= 12.111 \end{aligned} \tag{4.9.17}$$

To find confidence intervals for each $\theta^{(i)}$, we evaluate the equation

$$\hat{\theta}^{(i)} - c_0\hat{\sigma}_{\hat{\theta}_{(i)}} \leq \theta^{(i)} \leq \hat{\theta}^{(i)} + c_0\hat{\sigma}_{\hat{\theta}_{(i)}} \tag{4.9.18}$$

where $\hat{\theta}^{(i)} = \Sigma_j a_j y_{j.}$, $\hat{\sigma}_{\hat{\theta}_{(i)}} = s_i(\Sigma_j a_j^2/n_0)^{1/2}$, s_i^2 is the estimate of σ^2 for the ith variable, and $c_0^2 = T^{\alpha_i}(p, v_e)$. In our example,

$$\begin{aligned} \hat{\sigma}_{\hat{\theta}^{(1)}_{\text{C}}} &= 2.05832 & \hat{\sigma}_{\hat{\theta}^{(2)}_{\text{C}}} &= 1.70460 \\ \hat{\sigma}_{\hat{\theta}^{(1)}_{\text{L}}} &= 5.04185 & \hat{\sigma}_{\hat{\theta}^{(2)}_{\text{L}}} &= 4.17541 \\ \hat{\sigma}_{\hat{\theta}^{(1)}_{\text{Q}}} &= 5.04185 & \hat{\sigma}_{\hat{\theta}^{(2)}_{\text{Q}}} &= 4.18541 \end{aligned} \tag{4.9.19}$$

so that simultaneous confidence intervals for the $\theta^{(i)}$'s are

$$62.875 - (21.335)^{1/2}(2.05832) \leq \theta_{C}^{(1)} \leq 62.875 + (21.335)^{1/2}(2.05832)$$

$$\begin{aligned} 53.37 &\leq \theta_{C}^{(1)} \leq 72.38 \\ 55.25 &\leq \theta_{C}^{(2)} \leq 71.00 \\ -57.56 &\leq \theta_{L}^{(1)} \leq -10.99 \\ -42.46 &\leq \theta_{L}^{(2)} \leq -3.69 \\ -5.45 &\leq \theta_{Q}^{(1)} \leq 41.13 \\ -7.18 &\leq \theta_{Q}^{(2)} \leq 31.40 \end{aligned} \tag{4.9.20}$$

EXERCISE 4.9

1. In a psychological experiment, 150 subjects were randomly assigned to five treatment conditions to investigate reaction time (in seconds) to two stimuli. For the treatment conditions, the subjects responded to the stimuli after 0, 2, 4, 6, and 8 hours of sleep. Where the variance-covariance matrix for the variables S_1 and S_2 is

$$\mathbf{S} = \begin{bmatrix} 100 & \\ 50 & 80 \end{bmatrix}$$

the mean data based on 30 subjects per condition was as follows.

	Treatment Conditions (Hours of Sleep)									
	8		6		4		2		0	
Stimuli	S_1	S_2	S_1	S_2	S_1	S_2	S_1	S_2	S_1	S_2
Reaction time	64	84	70	80	72	86	69	98	82	100

a. Use orthogonal polynomials to determine the lowest-order polynomial that may be used to describe the data for $\alpha = .05$.

b. Obtain trend components and construct 95% confidence intervals for the regression coefficients.

4.10 HETEROGENEOUS DATA—UNIVARIATE

The preceding section emphasized fitting a regression function to a single population. In behavioral research, several measurements on a number of independent groups are often obtained, and interest is focused on whether the same regression function may be applied to each data set. Typical situations arise when data are gathered from two groups such as boys and girls, blacks and whites, and high- and low-socioeconomic-status children.

A dependent variable existing with several independent variables for several groups is typical of the univariate case. Some hypotheses of interest are (1) complete equality of regressions and (2) parallelism of regressions. The test of (2) is the familiar test made before performing an analysis of covariance (ANCOVA). The test of (1), often termed the *test of coincidence*, allows us to apply the same regression function to all groups under study. Williams (1959, Chap. 8) and Rao (1965a, p. 237) treat the topic of heterogeneous data in some detail. In this chapter, our discussion is limited to testing (1) and (2).

In multivariate analysis, several dependent variables occur with the independent set and the same hypotheses are of interest in this situation. The derivation of the tests for the multivariate case depends on understanding the univariate procedure; thus a review of univariate analysis, employing the notation of this chapter, is helpful (the review is based on Williams, 1959).

Suppose data are collected on I independent groups and k variables. Represent the I separate regression equations as

$$E(y_i) = \beta_0^{(i)} + \beta_1^{(i)}x_1 + \cdots + \beta_k^{(i)}x_k \qquad i = 1, \ldots, I \tag{4.10.1}$$

The general linear model for each group is

$$\underset{(N_i \times 1)}{\mathbf{y}_i} = \underset{(N_i \times q)}{\mathbf{X}_i} \underset{(q \times 1)}{\boldsymbol{\beta}_i} + \underset{(N_i \times 1)}{\boldsymbol{\varepsilon}_i} \tag{4.10.2}$$

where

$$E(\mathbf{y}_i) = \mathbf{X}_i\boldsymbol{\beta}_i$$
$$V(\mathbf{y}_i) = \sigma^2\mathbf{I}$$
$$\boldsymbol{\beta}_i' = [\beta_0^{(i)}, \beta_1^{(i)}, \ldots, \beta_k^{(i)}]$$

and

$$\mathbf{X}_i = \begin{bmatrix} 1 & x_{11}^{(i)} & \cdots & x_{1k}^{(i)} \\ \vdots & \vdots & & \vdots \\ 1 & x_{N_i 1}^{(i)} & \cdots & x_{N_i k}^{(i)} \end{bmatrix}$$

N_i is the number of observations for each group, $q = k + 1$, and $\mathbf{Y}_i \sim IN(\mathbf{X}_i\boldsymbol{\beta}_i, \sigma^2\mathbf{I})$. Representing all of the observations in a single-model equation, the general linear model becomes

$$\underset{(N \times 1)}{\mathbf{y}} = \underset{(N \times Iq)}{\mathbf{X}} \quad \underset{(Iq \times 1)}{\boldsymbol{\beta}} + \underset{(N \times 1)}{\boldsymbol{\varepsilon}}$$

$$\begin{bmatrix} \mathbf{y}_1 \\ \mathbf{y}_2 \\ \vdots \\ \mathbf{y}_I \end{bmatrix} = \begin{bmatrix} \mathbf{X}_1 & \mathbf{0} & \cdots & \mathbf{0} \\ \mathbf{0} & \mathbf{X}_2 & \cdots & \mathbf{0} \\ \vdots & \vdots & & \vdots \\ \mathbf{0} & \mathbf{0} & \cdots & \mathbf{X}_I \end{bmatrix} \begin{bmatrix} \boldsymbol{\beta}_1 \\ \boldsymbol{\beta}_2 \\ \vdots \\ \boldsymbol{\beta}_I \end{bmatrix} + \begin{bmatrix} \boldsymbol{\varepsilon}_1 \\ \boldsymbol{\varepsilon}_2 \\ \vdots \\ \boldsymbol{\varepsilon}_I \end{bmatrix} \tag{4.10.3}$$

By reparameterization of each of the models given in (4.10.2), as in univariate multiple regression, (4.10.3) becomes

$$\underbrace{\begin{bmatrix} \mathbf{y}_1 \\ \mathbf{y}_2 \\ \vdots \\ \mathbf{y}_I \end{bmatrix}}_{(N \times 1)} = \underbrace{\begin{bmatrix} \mathbf{X}_{d1} & \mathbf{0} & \cdots & \mathbf{0} \\ \mathbf{0} & \mathbf{X}_{d2} & \cdots & \mathbf{0} \\ \vdots & \vdots & & \vdots \\ \mathbf{0} & \mathbf{0} & \cdots & \mathbf{X}_{dI} \end{bmatrix}}_{(N \times Iq)} \underbrace{\begin{bmatrix} \boldsymbol{\eta}_1 \\ \boldsymbol{\eta}_2 \\ \vdots \\ \boldsymbol{\eta}_I \end{bmatrix}}_{(Iq \times 1)} + \underbrace{\begin{bmatrix} \boldsymbol{\varepsilon}_1 \\ \boldsymbol{\varepsilon}_2 \\ \vdots \\ \boldsymbol{\varepsilon}_I \end{bmatrix}}_{(N \times I)} \tag{4.10.4}$$

where

$$\mathbf{X}_{di} = [\mathbf{1} \quad \mathbf{D}_i]$$

$$\boldsymbol{\eta}_i = \begin{bmatrix} \alpha_{0i} \\ \boldsymbol{\gamma}_i \end{bmatrix} \quad \text{and} \quad \boldsymbol{\gamma}_i = \begin{bmatrix} \beta_1^{(i)} \\ \beta_2^{(i)} \\ \vdots \\ \beta_k^{(i)} \end{bmatrix}$$

Thus

$$\begin{aligned} \hat{\boldsymbol{\beta}}_i &= (\mathbf{X}_i'\mathbf{X}_i)^{-1}\mathbf{X}_i'\mathbf{y}_i \\ \hat{\boldsymbol{\gamma}}_i &= (\mathbf{D}_i'\mathbf{D}_i)^{-1}\mathbf{D}_i'\mathbf{y}_{di} \\ \hat{\alpha}_{0i} &= \bar{y}_i \end{aligned} \tag{4.10.5}$$

Combining all of the observations into a single sample, the model for the total sample is

$$\underset{(N \times 1)}{\mathbf{y}_t} = \underset{(N \times q)}{\mathbf{X}_t} \underset{(q \times 1)}{\boldsymbol{\beta}_t} + \underset{(N \times 1)}{\boldsymbol{\varepsilon}_t} \tag{4.10.6}$$

where the total regression equation is

$$E(y_t) = \beta_0 + \beta_1 x_1 + \cdots + \beta_k x_k \tag{4.10.7}$$

and $\Sigma_i N_i = N$.

By reparameterization of (4.10.6), the model is

$$\mathbf{y}_t = \mathbf{X}_{dt}\boldsymbol{\eta}_t + \boldsymbol{\varepsilon}_t \tag{4.10.8}$$

where

$$\mathbf{X}_{dt} = \begin{bmatrix} X_{dt1} \\ X_{dt2} \\ \vdots \\ X_{dtI} \end{bmatrix} = [\mathbf{1} \quad \mathbf{D}]$$

and

$$\boldsymbol{\eta}_t = \begin{bmatrix} \alpha_{0t} \\ \boldsymbol{\gamma}_t \end{bmatrix}$$

Incorporating the preceding notation, the first hypothesis of interest is the test of coincidence of regressions:

$$\begin{aligned} H_0: \beta_0^{(1)} &= \beta_0^{(2)} = \cdots = \beta_0^{(I)} \\ \beta_1^{(1)} &= \beta_1^{(2)} = \cdots = \beta_1^{(I)} \\ &\vdots \\ \beta_k^{(1)} &= \beta_k^{(2)} = \cdots = \beta_k^{(I)} \end{aligned} \quad \text{or} \quad \boldsymbol{\beta}_1 = \boldsymbol{\beta}_2 = \cdots = \boldsymbol{\beta}_I \tag{4.10.9}$$

From the univariate fundamental least-squares theorem, the sum of squares due to error to test (4.10.9) is

$$\begin{aligned} Q_e &= \mathbf{y}'[\mathbf{I} - \mathbf{X}(\mathbf{X}'\mathbf{X})^{-1}\mathbf{X}']\mathbf{y} \\ &= \sum_i \{\mathbf{y}_i'[\mathbf{I} - \mathbf{X}_i(\mathbf{X}_i'\mathbf{X}_i)^{-1}\mathbf{X}_i']\mathbf{y}_i\} \\ &= \sum_i (\mathbf{y}_i'\mathbf{y}_i - \hat{\boldsymbol{\beta}}_i'\mathbf{X}_i'\mathbf{X}_i\hat{\boldsymbol{\beta}}_i) \\ &= \sum_i (\mathbf{y}_{id}'\mathbf{y}_{id} - \boldsymbol{\gamma}_i'\mathbf{D}_i'\mathbf{D}_i\hat{\boldsymbol{\gamma}}_i) \end{aligned} \tag{4.10.10}$$

with degrees of freedom $N - r$; but the rank of the design matrix is $r = qI$, so that the degrees of freedom for error are $N - qI = N - Ik - I$.

The matrix $\mathbf{C}$ to test (4.10.9) is

$$\underset{[q(I-1)\times qI)}{\mathbf{C}} = \begin{bmatrix} \mathbf{I}_q & \mathbf{0} & \cdots & \mathbf{0} & -\mathbf{I}_q \\ \mathbf{0} & \mathbf{I}_q & \cdots & \mathbf{0} & -\mathbf{I}_q \\ \vdots & \vdots & & \vdots & \vdots \\ \mathbf{0} & \mathbf{0} & \cdots & \mathbf{I}_q & -\mathbf{I}_q \end{bmatrix} \tag{4.10.11}$$

the rank of $\mathbf{C}$ is $R(\mathbf{C}) = q(I - 1) = (k + 1)(I - 1)$. Although $Q_h = Q_t - Q_e$ may be found by using the general theory,

$$Q_h = (\mathbf{C}\hat{\mathbf{B}})'[\mathbf{C}(\mathbf{X}'\mathbf{X})^{-1}\mathbf{C}']^{-1}(\mathbf{C}\hat{\boldsymbol{\beta}}) \tag{4.10.12}$$

where $\mathbf{C}$ is defined as in (4.10.11) and $\hat{\boldsymbol{\beta}} = (\mathbf{X}'\mathbf{X})^{-1}\mathbf{X}'\mathbf{y}$, this is not the only way to proceed. That is, (4.10.9) can be tested by doing several multiple regression runs. Recall that $Q_h = Q_t - Q_e$, where $Q_t = \min(\mathbf{y} - \mathbf{X}\boldsymbol{\beta})'(\mathbf{y} - \mathbf{X}\boldsymbol{\beta})$ subject to $\mathbf{C}\boldsymbol{\beta} = \boldsymbol{\xi}$ (specified). In testing for coincidence, when the hypothesis is true, model (4.10.3) reduces to model (4.10.6), so that

$$Q_t = \mathbf{y}_t'\mathbf{y}_t - \hat{\boldsymbol{\beta}}_t'\mathbf{X}_t'\mathbf{X}_t\hat{\boldsymbol{\beta}}_t \tag{4.10.13}$$

Thus

$$\begin{aligned} Q_h &= (\mathbf{y}_t'\mathbf{y}_t - \hat{\boldsymbol{\beta}}_t'\mathbf{X}_t'\mathbf{X}_t\hat{\boldsymbol{\beta}}_t) - \sum_i (\mathbf{y}_i'\mathbf{y}_i - \hat{\boldsymbol{\beta}}_i'\mathbf{X}_i'\mathbf{X}_i\hat{\boldsymbol{\beta}}_i) \\ &= \sum_i \hat{\boldsymbol{\beta}}_i'\mathbf{X}_i'\mathbf{X}_i\hat{\boldsymbol{\beta}}_i - \hat{\boldsymbol{\beta}}_t'\mathbf{X}_t'\mathbf{X}_t\hat{\boldsymbol{\beta}}_t \\ &= (\mathbf{y}_{td}'\mathbf{y}_{td} - \hat{\boldsymbol{\gamma}}_t'\mathbf{D}'\mathbf{D}\hat{\boldsymbol{\gamma}}_t) - \sum_i (\mathbf{y}_{id}'\mathbf{y}_{id} - \hat{\boldsymbol{\gamma}}_i'\mathbf{D}_i'\mathbf{D}_i\hat{\boldsymbol{\gamma}}_i) \end{aligned} \tag{4.10.14}$$

so (4.10.9) may be tested by running separate regressions for each group and one for the total combined sample. Q_e is obtained by adding the three separate errors, Q_t is the error term for the total model, and $Q_h = Q_t - Q_e$.

The hypothesis of coincidence of regression planes is rejected at the significance level α if

$$F = \frac{Q_h/(k + 1)(I - 1)}{Q_e/(N - Ik - I)} > F^{\alpha}[(k + 1)(I - 1), (N - Ik - I)] \tag{4.10.15}$$

TABLE 4.10.1. ANOVA Table for the Test of Coincidence.

Source	df	SS	E(MS)
Coincidence	$(k+1)(I-1)$	$Q_h = Q_t - Q_e$	$\sigma^2 + \dfrac{(\mathbf{C}\boldsymbol{\beta})'[\mathbf{C}(\mathbf{X}'\mathbf{X})^{-1}\mathbf{C}']^{-1}(\mathbf{C}\boldsymbol{\beta})}{(k+1)(I-1)}$
Residual (separate regressions)	$N - Ik - I$	$Q_e = \sum_i (\mathbf{y}_i'\mathbf{y}_i - \hat{\boldsymbol{\beta}}_i'\mathbf{X}_i'\mathbf{X}_i\hat{\boldsymbol{\beta}}_i)$	σ^2
Residual (total regression)	$N - k - 1$	$Q_t = \mathbf{y}_t'\mathbf{y}_t - \boldsymbol{\beta}_t'\mathbf{X}_t'\mathbf{X}_t\boldsymbol{\beta}_t$	σ_t^2

The test of coincidence is conveniently represented in an ANOVA table, Table 4.10.1. Acceptance of the coincidence hypothesis implies that the separate regression functions are the same for the I groups. The best estimate of the weights of the common regression function is $\hat{\boldsymbol{\beta}}_t$.

Alternatively, the regression functions may be the same with the exception of the constant terms. Testing whether the coefficients on the independent variables are the same irrespective of the β_0's is the *test of parallelism*. The test for parallelism of regressions, in terms of the parameters, is

$$
\begin{aligned}
H_0\colon \beta_1^{(1)} &= \beta_1^{(2)} = \cdots = \beta_1^{(I)} \\
\beta_2^{(1)} &= \beta_2^{(2)} = \cdots = \beta_2^{(I)} \qquad \text{or} \quad \boldsymbol{\gamma}_1 = \boldsymbol{\gamma}_2 = \cdots = \boldsymbol{\gamma}_I \\
&\ \ \vdots \\
\beta_k^{(1)} &= \beta_k^{(2)} = \cdots = \beta_k^{(I)}
\end{aligned}
\tag{4.10.16}
$$

The error sum of squares to test (4.10.16) is the same as Q_e given in (4.10.10). To determine the hypothesis sum of squares Q_{h_p}, the hypothesis test matrix $\mathbf{C}$ is defined by

$$
\underset{[k(I-1)\times qI]}{\mathbf{C}} = \begin{bmatrix}
\mathbf{0} \,\vdots\, \mathbf{I}_k & \mathbf{0} & \cdots & \mathbf{0} & \mathbf{0} \,\vdots\, -\mathbf{I}_k \\
\mathbf{0} & \mathbf{0} \,\vdots\, \mathbf{I}_k & \cdots & \mathbf{0} & \mathbf{0} \,\vdots\, -\mathbf{I}_k \\
\vdots & \vdots & \ddots & \vdots & \vdots \\
\mathbf{0} & \mathbf{0} & \cdots & \mathbf{0} \,\vdots\, \mathbf{I}_k & \mathbf{0} \,\vdots\, -\mathbf{I}_k
\end{bmatrix}
\tag{4.10.17}
$$

and the $R(\mathbf{C}) = k(I-1)$. Again, the general theory may be used to test (4.10.16); however, computationally it may be easier to find the total sum of squares Q_{tp} to test for parallelism. Using the deviation form of the model,

$$
Q_{tp} = \sum_i (\mathbf{y}_{id}'\mathbf{y}_{id} - \hat{\boldsymbol{\gamma}}_p'\mathbf{D}_i'\mathbf{D}_i\hat{\boldsymbol{\gamma}}_p)
\tag{4.10.18}
$$

where

$$
\hat{\boldsymbol{\gamma}}_p = \left(\sum_i \mathbf{D}_i'\mathbf{D}_i\right)^{-1} \sum_i \mathbf{D}_i'\mathbf{y}_{id}
\tag{4.10.19}
$$

so that

$$
\begin{aligned}
Q_{hp} &= \sum_i (\mathbf{y}_{id}'\mathbf{y}_{id} - \hat{\boldsymbol{\gamma}}_p'\mathbf{D}_i'\mathbf{D}_i\hat{\boldsymbol{\gamma}}_p) - \sum_i (\mathbf{y}_{id}'\mathbf{y}_{id} - \hat{\boldsymbol{\gamma}}_i'\mathbf{D}_i'\mathbf{D}_i\hat{\boldsymbol{\gamma}}_i) \\
&= \sum_i (\hat{\boldsymbol{\gamma}}_i - \hat{\boldsymbol{\gamma}}_p)'\mathbf{D}_i'\mathbf{y}_{id}
\end{aligned}
\tag{4.10.20}
$$

The test of the hypothesis of parallelism is rejected at the significance level α if

$$F = \frac{Q_{hp}/k(I-1)}{Q_e/(N-Ik-I)} > F^\alpha[k(I-1),(N-Ik-I)] \tag{4.10.21}$$

where Q_{hp} is defined as in (4.10.20), or $Q_{hp} = (\mathbf{C}\hat{\boldsymbol{\beta}})'[\mathbf{C}(\mathbf{X}'\mathbf{X})\mathbf{C}']^{-1}(\mathbf{C}\hat{\boldsymbol{\beta}})$, with $\mathbf{C}$ defined as in (4.10.17). Acceptance of (4.10.16) would imply that the pooled regression weights $\hat{\gamma}_p$ would be used for each regression plane, with, however, different intercepts.

The tests of coincidence and parallelism in elementary statistics books are usually introduced by examining the variation about various regression lines. To make the transition to the derivations presented here, the test of coincidence compares two estimates for the residual sum of squares; Q_e is the sum of squares of the separate residual sum of squares, and Q_t is the total residual sum of squares. The test of parallelism compares Q_e to Q_{tp}, where Q_{tp} now is the pooled or common residual sum of squares. If the variation about the pooled (total) regression plane is not significantly greater than the residual variation when separate planes are fitted, the lines are considered parallel (coincident). Again the hypothesis of parallelism may be represented in an ANOVA table, Table 4.10.2.

TABLE 4.10.2. ANOVA Table for the Test of Parallelism.

Source	df	SS	E(MS)
Parallelism	$k(I-1)$	$Q_{hp} = Q_{tp} - Q_e$	$\sigma^2 + \dfrac{(\mathbf{C}\boldsymbol{\beta})'[\mathbf{C}(\mathbf{X}'\mathbf{X})^{-1}\mathbf{C}]^{-1}(\mathbf{C}\boldsymbol{\beta})}{k(I-1)}$
Residual (separate regression)	$N-Ik-I$	$Q_e = \sum_i \left(\mathbf{y}'_{id}\mathbf{y}_{id} - \hat{\gamma}'_i\mathbf{D}'_i\mathbf{D}_i\hat{\gamma}_i\right)$	σ^2
Residual (pooled regression)	$N-I-k$	$Q_{tp} = \sum_i \left(\mathbf{y}'_{id}\mathbf{y}_{id} - \hat{\gamma}'_p\mathbf{D}'_i\mathbf{D}_i\hat{\gamma}_p\right)$	σ_p^2

A detailed example of these procedures is given in Williams (1959, Chap. 8). Brownlee (1965, Sect. 11.14) also considers the problem of several regressions with geometric illustrations.

Given that the planes are parallel, a test for significant differences of intercepts for the regression functions is usually pursued. This is the topic of univariate analysis of covariance. This procedure will be discussed in considerable detail in Chapter 5; for completeness of this section, observation of the manner in which the test would proceed is made, as the necessary calculations have already been completed.

The test of coincidence can be thought of as two tests, one for parallelism and one for intercepts given parallelism. The sum of squares Q_h for coincidence was

$$Q_h = Q_t - Q_e = (\mathbf{y}'_{td}\mathbf{y}_{td} - \hat{\gamma}'_t\mathbf{D}'\mathbf{D}\hat{\gamma}_t) - Q_e \tag{4.10.22}$$

where Q_e is given in (4.10.11). The sum of squares Q_{hp} for parallelism was

$$Q_{hp} = \sum_i (\mathbf{y}'_{id}\mathbf{y}_{id} - \gamma'_p\mathbf{D}'_i\mathbf{D}_i\gamma_p) - Q_e = Q_{tp} - Q_e \tag{4.10.23}$$

Subtraction of the two sums of squares yields the sum of squares for intercepts

$$Q'_h = Q_t - Q_{tp} \tag{4.10.24}$$

TABLE 4.10.3. ANOVA Table for Coincidence, Intercepts, and Parallelism.

Source	df	SS
Coincidence	$(k+1)(I-1)$	$Q_h = Q_t - Q_e$
Intercepts	$I-1$	$Q_h' = Q_t - Q_{tp}$
Parallelism	$k(I-1)$	$Q_{hp} = Q_{tp} - Q_e$
Residual (separate regressions)	$N - Ik - I$	$Q_e = \sum_i (\mathbf{y}_{id}'\mathbf{y}_{id} - \hat{\boldsymbol{\gamma}}_i'\mathbf{D}_i'\mathbf{D}_i\hat{\boldsymbol{\gamma}}_i)$
Residual (total regression)	$N - k - 1$	$Q_t = \mathbf{y}_{td}'\mathbf{y}_{td} - \hat{\boldsymbol{\gamma}}_t'\mathbf{D}'\mathbf{D}\hat{\boldsymbol{\gamma}}_t$

The calculations immediately yield ANOVA Table 4.10.3, which may be used to test for coincidence, parallelism, and differences in intercepts. The tests for coincidence and parallelism have already been discussed. To test for differences in intercepts, given parallelism, the parallelism sum of squares is pooled with Q_e to obtain

$$Q_{tp} = \sum_i (\mathbf{y}_{id}'\mathbf{y}_{id} - \hat{\boldsymbol{\gamma}}_p'\mathbf{D}_i'\mathbf{D}_i\hat{\boldsymbol{\gamma}}_p) \tag{4.10.25}$$

a new pooled residual sum of squares. Using Q_h' for intercepts and the *new* estimate of error, differences in intercepts, given parallelism, are tested by using the statistic

$$F = \frac{(Q_t - Q_{tp})/(I-1)}{Q_{tp}/(N-k-I)} \sim F(I-1, N-k-I) \tag{4.10.26}$$

This is the usual analysis-of-covariance test for differences in the adjusted means, with k covariates and I groups.

When the hypothesis of parallelism is not tenable, differences in intercepts using a no-pool rule may still be tested from the entries in Table 4.10.3.

4.11 EXAMPLE—HETEROGENEOUS DATA—UNIVARIATE

Using Rohwer's data in Section 4.3 obtained for high-socioeconomic-status kindergarten children and his data in Section 4.7 for low-socioeconomic-status kindergarten children, the theory developed in Section 4.10 is illustrated. Considering PPVT as the dependent variable and N, S, NS, NA, and SS as the independent variables, the primary question of interest for the study is coincidence of regression planes for the two groups.

The model for this problem is

$$\underset{(69\times 1)}{\mathbf{y}} = \underset{(69\times 12)}{\mathbf{X}} \; \underset{(12\times 1)}{\boldsymbol{\beta}} + \underset{(69\times 1)}{\boldsymbol{\varepsilon}} \tag{4.11.1}$$

$$\begin{bmatrix} \mathbf{y}_1 \\ \mathbf{y}_2 \end{bmatrix} = \begin{bmatrix} \mathbf{X}_1 & \mathbf{0} \\ \mathbf{0} & \mathbf{X}_2 \end{bmatrix} \begin{bmatrix} \boldsymbol{\beta}_1 \\ \boldsymbol{\beta}_2 \end{bmatrix} + \begin{bmatrix} \boldsymbol{\varepsilon}_1 \\ \boldsymbol{\varepsilon}_2 \end{bmatrix}$$

where $\mathbf{y}_1$ and $\mathbf{y}_2$ are the observation vectors of the dependent variable PPVT for the high- and low-socioeconomic-status groups, respectively,

$$\boldsymbol{\beta}_i = \begin{bmatrix} \beta_1^{(i)} \\ \beta_2^{(i)} \\ \vdots \\ \beta_5^{(i)} \end{bmatrix} \qquad i = 1, 2$$

is the parameter vector of regression coefficients to be estimated, $N_1 = 32$ and $N_2 = 37$ are the sample sizes, and $N = 69$ is the total number of observations for the study.

Before testing either the hypothesis of coincidence or parallelism, the least-squares estimates of effects are acquired. Using equations (4.10.5),

$$\hat{\boldsymbol{\beta}} = \begin{bmatrix} \hat{\boldsymbol{\beta}}_1 \\ \hat{\boldsymbol{\beta}}_2 \end{bmatrix} = (\mathbf{X}'\mathbf{X})^{-1}\mathbf{X}'\mathbf{y} = \begin{bmatrix} 39.69709 \\ .06728 \\ .36998 \\ -.37438 \\ 1.52301 \\ .41016 \\ \hline 33.00577 \\ -.08057 \\ -.72105 \\ -.29830 \\ 1.47042 \\ .32396 \end{bmatrix} \tag{4.11.2}$$

Thus the error sum of squares is, by (4.10.10),

$$Q_e = \mathbf{y}'[\mathbf{I} - \mathbf{X}(\mathbf{X}'\mathbf{X})^{-1}\mathbf{X}']\mathbf{y} = 6663.75032 \tag{4.11.3}$$

Employing the general linear model approach to test for coincidence

$$H_0\colon \begin{matrix} \beta_0^{(1)} = \beta_0^{(2)} \\ \beta_1^{(1)} = \beta_1^{(2)} \\ \vdots \\ \beta_5^{(1)} = \beta_5^{(2)} \end{matrix} \quad \text{or} \quad \boldsymbol{\beta}_1 = \boldsymbol{\beta}_2 \tag{4.11.4}$$

the matrix $\mathbf{C}$ from (4.10.11) becomes

$$\mathbf{C} = \begin{bmatrix} 1 & 0 & 0 & 0 & 0 & 0 & -1 & 0 & 0 & 0 & 0 & 0 \\ 0 & 1 & 0 & 0 & 0 & 0 & 0 & -1 & 0 & 0 & 0 & 0 \\ 0 & 0 & 1 & 0 & 0 & 0 & 0 & 0 & -1 & 0 & 0 & 0 \\ 0 & 0 & 0 & 1 & 0 & 0 & 0 & 0 & 0 & -1 & 0 & 0 \\ 0 & 0 & 0 & 0 & 1 & 0 & 0 & 0 & 0 & 0 & -1 & 0 \\ 0 & 0 & 0 & 0 & 0 & 1 & 0 & 0 & 0 & 0 & 0 & -1 \end{bmatrix} \tag{4.11.5}$$

TABLE 4.11.1. ANOVA Table for the Test of Coincidence.

Source	df	SS	MS	F
Coincidence	6	4595.66929	765.9449	6.55
Residual (separate regressions)	57	6663.75032	116.9079	
Residual (total regression)	63	11,259.41961		

With Q_e defined as in (4.11.3) and Q_h defined by $Q_h = (\mathbf{C}\hat{\boldsymbol{\beta}})'[\mathbf{C}(\mathbf{X}'\mathbf{X})^{-1}\mathbf{C}'](\mathbf{C}\hat{\boldsymbol{\beta}})$, Table 4.11.1, an ANOVA table corresponding to Table 4.10.1, is immediately obtained.

With $\alpha = .01$, the hypothesis of coincidence is rejected if the computed F ratio is larger than $F^{\alpha}[(k+1)(I-1), (N-Ik-I)] = F^{.01}(6, 57) = 3.14$. The hypothesis of coincidence is rejected.

To test the hypothesis of parallelism when using these data,

$$H_0: \beta_1^{(1)} = \beta_1^{(2)}$$
$$\beta_2^{(1)} = \beta_2^{(2)} \tag{4.11.6}$$
$$\vdots$$
$$\beta_5^{(1)} = \beta_5^{(2)}$$

the matrix $\mathbf{C}$ becomes

$$\mathbf{C} = \begin{bmatrix} 0 & 1 & 0 & 0 & 0 & 0 & 0 & -1 & 0 & 0 & 0 & 0 \\ 0 & 0 & 1 & 0 & 0 & 0 & 0 & 0 & -1 & 0 & 0 & 0 \\ 0 & 0 & 0 & 1 & 0 & 0 & 0 & 0 & 0 & -1 & 0 & 0 \\ 0 & 0 & 0 & 0 & 1 & 0 & 0 & 0 & 0 & 0 & -1 & 0 \\ 0 & 0 & 0 & 0 & 0 & 1 & 0 & 0 & 0 & 0 & 0 & -1 \end{bmatrix} \tag{4.11.7}$$

With $Q_{hp} = (\mathbf{C}\hat{\boldsymbol{\beta}})'[\mathbf{C}(\mathbf{X}'\mathbf{X})^{-1}\mathbf{C}']^{-1}(\mathbf{C}\hat{\boldsymbol{\beta}})$, the ANOVA table for the parallelism hypothesis is tenable, as Table 4.11.2 illustrates.

The regression equation under parallelism is acquired from the estimate of the pooled regression weights,

$$\hat{\boldsymbol{\gamma}}_p = \begin{bmatrix} .00232 \\ -.35110 \\ -.29887 \\ 1.29375 \\ .47892 \end{bmatrix} \tag{4.11.8}$$

TABLE 4.11.2. ANOVA Table for the Test of Parallelism.

Source	df	SS	MS	F
Parallelism	5	343.29906	68.6598	.587
Residual (separate regression)	57	6663.75032	116.9079	
Residual (pooled regression)	62	7007.04938		

where the intercepts are given by

$$\hat{\beta}_0^{(1)} = \hat{\alpha}_0^{(1)} - \hat{\gamma}_p'\bar{x}_1 = 83.0938 - 34.8677 = 48.2261$$
$$\hat{\beta}_0^{(2)} = \hat{\alpha}_0^{(2)} - \hat{\gamma}_p'\bar{x}_2 = 62.6484 - 31.2998 = 31.3488 \tag{4.11.9}$$

The coefficients $\hat{\beta}_0^{(i)}$ are more commonly known as the *adjusted means* in an analysis of covariance.

EXERCISE 4.11

1. Kirk (1968, p. 473) gives data for one dependent variable Y (arithmetic achievement) and two independent variables X and Z (intelligence and arithmetic achievement, respectively) measured prior to the beginning of the experiment, for an analysis-of-covariance design used to evaluate $I = 4$ methods of teaching arithmetic.

Method 1			Method 2			Method 3			Method 4		
Y	X	Z	Y	X	Z	Y	X	Z	Y	X	Z
3	42	3	4	47	4	7	61	5	7	65	2
6	57	5	5	49	6	8	65	7	8	74	4
3	33	4	4	42	5	7	64	5	9	80	5
3	47	4	3	41	2	6	56	4	8	73	5
1	32	0	2	38	1	5	52	2	10	85	6
2	35	1	3	43	2	6	58	3	10	82	6
2	33	0	4	48	5	5	53	3	9	78	5
2	39	2	3	45	3	6	54	4	11	89	7

Using the linear model,

$$\begin{bmatrix} \mathbf{y}_1 \\ \mathbf{y}_2 \\ \mathbf{y}_3 \end{bmatrix} = \begin{bmatrix} \mathbf{X}_1 & \mathbf{0} & \mathbf{0} \\ \mathbf{0} & \mathbf{X}_1 & \mathbf{0} \\ \mathbf{0} & \mathbf{0} & \mathbf{X}_1 \end{bmatrix} \begin{bmatrix} \boldsymbol{\beta}_1 \\ \boldsymbol{\beta}_2 \\ \boldsymbol{\beta}_3 \end{bmatrix} + \begin{bmatrix} \boldsymbol{\varepsilon}_1 \\ \boldsymbol{\varepsilon}_2 \\ \boldsymbol{\varepsilon}_3 \end{bmatrix}$$

and the data provided, test for (a) coincidence, (b) parallelism of regression functions, and (c) differences in intercepts.

4.12 HETEROGENEOUS DATA—MULTIVARIATE

Suppose that several dependent variables are observed for individuals in I independent groups when analyzing multivariate heterogeneous data. As with multivariate multiple regression, tests will follow by analogy with the univariate case when the fundamental multivariate least-squares theorem is applied. Of course, the other multivariate test criteria may also be utilized.

Suppose N_i independent p-vector random variables are collected for the ith independent group, where $i = 1, \ldots, I$. Represent the I separate regression equations as

$$\underset{(N \times p)}{\mathbf{Y}} = \underset{(N \times Iq)}{\mathbf{X}} \underset{(Iq \times p)}{\mathbf{B}} + \underset{(N \times p)}{\mathbf{E}_0}$$

(4.12.1)
$$\begin{bmatrix} \mathbf{Y}_1 \\ \mathbf{Y}_2 \\ \vdots \\ \mathbf{Y}_I \end{bmatrix} = \begin{bmatrix} \mathbf{X}_1 & \mathbf{0} & \cdots & \mathbf{0} \\ \mathbf{0} & \mathbf{X}_2 & \cdots & \mathbf{0} \\ \vdots & \vdots & \vdots & \vdots \\ \mathbf{0} & \mathbf{0} & \cdots & \mathbf{X}_I \end{bmatrix} \begin{bmatrix} \mathbf{B}_1 \\ \mathbf{B}_2 \\ \vdots \\ \mathbf{B}_I \end{bmatrix} + \begin{bmatrix} \mathbf{E}_1 \\ \mathbf{E}_2 \\ \vdots \\ \mathbf{E}_I \end{bmatrix}$$

where

(4.12.2)
$$E(\mathbf{Y}_i) = \mathbf{X}_i\mathbf{B}_i$$
$$V(\mathbf{Y}_i) = \mathbf{I} \otimes \mathbf{\Sigma}$$

and

$$\mathbf{B}_i = \begin{bmatrix} \beta_{01}^{(i)} & \beta_{02}^{(i)} & \cdots & \beta_{0p}^{(i)} \\ \beta_{11}^{(i)} & \beta_{12}^{(i)} & \cdots & \beta_{1p}^{(i)} \\ \vdots & \vdots & \vdots & \vdots \\ \beta_{k1}^{(i)} & \beta_{k2}^{(i)} & \cdots & \beta_{kp}^{(i)} \end{bmatrix}$$

$k + 1 = q$, $\mathbf{X}_i$ is defined by (4.10.2), and each row of $\mathbf{Y}_i$ has a multivariate normal distribution.

By reparameterization of each group, (4.12.1) becomes

(4.12.3)
$$\underbrace{\begin{bmatrix} \mathbf{Y}_1 \\ \mathbf{Y}_2 \\ \vdots \\ \mathbf{Y}_I \end{bmatrix}}_{(N \times p)} = \underbrace{\begin{bmatrix} \mathbf{X}_{d1} & \mathbf{0} & \cdots & \mathbf{0} \\ \mathbf{0} & \mathbf{0} & \cdots & \mathbf{0} \\ \vdots & \vdots & \vdots & \vdots \\ \mathbf{0} & \mathbf{0} & \cdots & \mathbf{X}_{dI} \end{bmatrix}}_{(N \times Iq)} \underbrace{\begin{bmatrix} \mathbf{H}_1 \\ \mathbf{H}_2 \\ \vdots \\ \mathbf{H}_I \end{bmatrix}}_{(Iq \times p)} + \underbrace{\begin{bmatrix} \mathbf{E}_1 \\ \mathbf{E}_2 \\ \vdots \\ \mathbf{E}_I \end{bmatrix}}_{(N \times I)}$$

where

$$\mathbf{H}_i = \begin{bmatrix} \boldsymbol{\alpha}_{0i}' \\ \mathbf{\Gamma}_i \end{bmatrix} \quad \text{and} \quad \mathbf{X}_{di} = [\mathbf{1} \quad \mathbf{D}_i]$$

Combining all observations into a single sample, the model equation is

(4.12.4)
$$\underset{(N \times p)}{\mathbf{Y}_t} = \underset{(N \times q)}{\mathbf{X}_t} \underset{(q \times p)}{\mathbf{B}_t} + \underset{(N \times p)}{\mathbf{E}_t}$$

or

$$\underset{(N \times p)}{\mathbf{Y}_t} = \underset{(N \times q)}{\mathbf{X}_{dt}} \underset{(q \times p)}{\mathbf{H}_t} + \underset{(N \times p)}{\mathbf{E}_t}$$

by reparameterization where

(4.12.5)
$$\mathbf{H}_t = \begin{bmatrix} \boldsymbol{\alpha}_{0t}' \\ \mathbf{\Gamma}_t \end{bmatrix} \quad \text{and} \quad \mathbf{X}_{dt} = [\mathbf{1} \quad \mathbf{D}]$$

The least-squares estimates for these models are

$$\begin{aligned}\hat{\mathbf{B}}_i &= (\mathbf{X}_i'\mathbf{X}_i)^{-1}\mathbf{X}_i'\mathbf{Y}_i \\ \hat{\mathbf{\Gamma}}_i &= (\mathbf{D}_i'\mathbf{D}_i)^{-1}\mathbf{D}_i'\mathbf{Y}_{id} \\ \boldsymbol{\alpha}_{0i} &= \bar{\mathbf{y}}_i \\ \hat{\mathbf{\Gamma}}_t &= (\mathbf{D}'\mathbf{D})^{-1}\mathbf{D}'\mathbf{Y}_{td} \\ \boldsymbol{\alpha}_{0t} &= \bar{\mathbf{y}}_t\end{aligned} \tag{4.12.6}$$

Using the notation as defined above, the multivariate test of coincidence is

$$H_0: \mathbf{B}_1 = \mathbf{B}_2 = \cdots = \mathbf{B}_I \tag{4.12.7}$$

Applying the fundamental multivariate least-squares theorem,

$$\begin{aligned}\mathbf{Q}_e &= \mathbf{Y}[\mathbf{I} - \mathbf{X}(\mathbf{X}'\mathbf{X})^{-1}\mathbf{X}']\mathbf{Y} \\ &= \sum_i \{\mathbf{Y}_i'[\mathbf{I} - \mathbf{X}_i(\mathbf{X}_i'\mathbf{X}_i)^{-1}\mathbf{X}_i']\mathbf{Y}_i\} \\ &= \sum_i (\mathbf{Y}_i'\mathbf{Y}_i - \mathbf{B}_i'\mathbf{X}_i'\mathbf{X}_i\mathbf{B}_i) \\ &= \sum_i (\mathbf{Y}_{id}'\mathbf{Y}_{id} - \hat{\mathbf{\Gamma}}_i'\mathbf{D}_i'\mathbf{D}_i\hat{\mathbf{\Gamma}}_i)\end{aligned} \tag{4.12.8}$$

where the degrees of freedom for error are $\nu_e = N - Ik - I$. Using $\mathbf{C}$ as defined in (4.10.10),

$$\mathbf{Q}_h = (\mathbf{C}\hat{\mathbf{B}})'[\mathbf{C}(\mathbf{X}'\mathbf{X})^{-1}\mathbf{C}']^{-1}(\mathbf{C}\hat{\mathbf{B}}) \tag{4.12.9}$$

or, alternatively,

$$\mathbf{Q}_t = \mathbf{Y}_t'\mathbf{Y}_t - \hat{\mathbf{B}}_t'\mathbf{X}_t'\mathbf{X}_t\hat{\mathbf{B}}_t \tag{4.12.10}$$

and $\mathbf{Q}_h = \mathbf{Q}_t - \mathbf{Q}_e$. By forming the ratio

$$\Lambda = \frac{|\mathbf{Q}_e|}{|\mathbf{Q}_h + \mathbf{Q}_e|} = \frac{|\mathbf{Q}_e|}{|\mathbf{Q}_t|} \sim U[p, (k+1)(I-1), N - Ik - I] \tag{4.12.11}$$

the hypothesis of coincidence is rejected at the significance level α if

$$\Lambda = \frac{|\mathbf{Q}_e|}{|\mathbf{Q}_t|} = \prod_{i=1}^{s} v_i < U^\alpha[p, (k+1)(I-1), N - Ik - I] \tag{4.12.12}$$

where the v_i are the roots of the determinantal equation

$$|\mathbf{Q}_e - v\mathbf{Q}_t| = 0 \tag{4.12.13}$$

Acceptance of the coincidence hypothesis implies that the whole regression matrix is the same for all I groups. The appropriate regression equation would be the equation obtained from the total sample.

The test for parallelism in the multivariate case

$$H_0: \mathbf{\Gamma}_1 = \mathbf{\Gamma}_2 = \cdots = \mathbf{\Gamma}_I \tag{4.12.14}$$

is again similar to the univariate result. Following the arguments of the univariate case, a comparison of two sums of squares and products matrices is employed to test the hypothesis. Using the multivariate least-squares theorem,

$$\mathbf{Q}_e = \sum_i (\mathbf{Y}'_{id}\mathbf{Y}_{id} - \hat{\mathbf{\Gamma}}'_i\mathbf{D}'_i\mathbf{D}\hat{\mathbf{\Gamma}}_i)$$

$$\mathbf{Q}_{tp} = \sum_i (\mathbf{Y}'_{id}\mathbf{Y}_{id} - \hat{\mathbf{\Gamma}}'_p\mathbf{D}'_i\mathbf{D}_i\hat{\mathbf{\Gamma}}_p) \tag{4.12.15}$$

where

$$\hat{\mathbf{\Gamma}}_p = \left(\sum_i \mathbf{D}'_i\mathbf{D}_i\right)^{-1} \sum_i \mathbf{D}'_i\mathbf{Y}_{id} \tag{4.12.16}$$

denotes the estimated pooled regression weights matrix, the test of parallelism is rejected at the significance level α if

$$\Lambda = \frac{|\mathbf{Q}_e|}{|\mathbf{Q}_{tp}|} < U^\alpha[p, k(I-1), N - Ik - I] \tag{4.12.17}$$

Alternatively, the matrix **C** defined by (4.10.16) could be employed; then,

$$\Lambda = \frac{|\mathbf{Q}_e|}{|\mathbf{Q}_h + \mathbf{Q}_e|}$$

would be used.

Various other test criteria may be employed for testing the hypotheses of coincidence and parallelism. In using any of the criteria, the assumption of homogeneity of covariance matrices is implicitly made for the I groups. This should be tested before performing tests of coincidence or parallelism. The test procedure is similar to the method discussed for the equality-of-covariance matrices for two or more groups (Section 3.16), except that the variation due to the independent variables is first removed.

By analogy with the univariate results, a test for differences in intercepts, given parallelism, in the multivariate situation should be apparent. The difference of intercepts is rejected at the significance level α if

$$\Lambda = \frac{|\mathbf{Q}_{tp}|}{|\mathbf{Q}_t|} = \frac{\left|\sum_i \mathbf{Y}'_{id}\mathbf{Y}_{id} - \hat{\mathbf{\Gamma}}'_p\mathbf{D}'_i\mathbf{D}_i\hat{\mathbf{\Gamma}}_p\right|}{\mathbf{Y}'_{td}\mathbf{Y}_{td} - \hat{\mathbf{\Gamma}}'_t\mathbf{D}'\mathbf{D}\hat{\mathbf{\Gamma}}_t} < U^\alpha(p, I-1, N-I-k)$$

which is the ratio of the Λ value obtained for coincidence divided by the value obtained for testing parallelism. This and other criteria will be discussed in Chapter 5 when the topic of multivariate analysis of covariance is considered.

Prior to investigating an example, a table is constructed of the required matrix products to perform the discussed tests on heterogeneous data when using Wilks' lambda criterion in the multivariate case. (See Table 4.12.1.) For convenience, lowercase letters are used to signify that all observations have been centered and are in deviation-score form. The calculations in Table 4.12.1 are conveniently performed through a matrix-operator Fortran package.

Since the relationship between the Λ statistic and the F statistic is

$$F(v_h, v_e) = \frac{v_e}{v_h}\frac{1-\Lambda}{\Lambda} \tag{4.12.18}$$

TABLE 4.12.1. Computing Formulas for Tests of Coincidence, Parallelism, and Intercepts.

(1) For each data group $i = 1, \ldots, I$, for the following table:

	$\mathbf{y'y}$		$\hat{\mathbf{\Gamma}} = (\mathbf{x'x})^{-1}\mathbf{x'y}$		$\hat{\mathbf{\Gamma}}'\mathbf{x'x}\hat{\mathbf{\Gamma}}$
1	$\mathbf{y}_1'\mathbf{y}_1$	$\mathbf{x}_1'\mathbf{x}_1$	$\mathbf{x}_1'\mathbf{y}_1$	$\hat{\mathbf{\Gamma}}_1 = (\mathbf{x}_1'\mathbf{x}_1)^{-1}\mathbf{x}_1'\mathbf{y}_1$	$\hat{\mathbf{\Gamma}}_1'\mathbf{x}_1'\mathbf{x}_1\hat{\mathbf{\Gamma}}_1$
⋮	⋮	⋮	⋮	⋮	⋮
I	$\mathbf{y}_I'\mathbf{y}_I$	$\mathbf{x}_I'\mathbf{x}_I$	$\mathbf{x}_I'\mathbf{y}_I$	$\hat{\mathbf{\Gamma}}_1 = (\mathbf{x}_I'\mathbf{x}_I)^{-1}\mathbf{x}_I'\mathbf{y}_I$	$\hat{\mathbf{\Gamma}}_I'\mathbf{x}_I'\mathbf{x}_I\hat{\mathbf{\Gamma}}_I$
Sum	$\mathbf{S}_1$	$\mathbf{S}_2$	$\mathbf{S}_3$	$\hat{\mathbf{\Gamma}}_p = \mathbf{S}_2^{-1}\mathbf{S}_3$	$\mathbf{S}_4$

(2) Putting all observations into one group, form

$$\mathbf{y}_t'\mathbf{y}_t \quad \mathbf{x}_t'\mathbf{x}_t \quad \mathbf{x}_t'\mathbf{y}_t \quad \hat{\mathbf{\Gamma}}_t = (\mathbf{x}_t'\mathbf{x}_t)^{-1}\mathbf{x}_t'\mathbf{y}_t \quad \hat{\mathbf{\Gamma}}_t'\mathbf{x}_t'\mathbf{x}_t\hat{\mathbf{\Gamma}}_t$$

(3) Form the determinants:

$$|\mathbf{Q}_e| = |\mathbf{S}_1 - \mathbf{S}_4|$$
$$|\mathbf{Q}_{tp}| = |\mathbf{S}_1 - \hat{\mathbf{\Gamma}}_p'\mathbf{S}_3|$$
$$|\mathbf{Q}_t| = |\mathbf{y}_t'\mathbf{y}_t - \hat{\mathbf{\Gamma}}_t'\mathbf{x}_t'\mathbf{x}_t\hat{\mathbf{\Gamma}}_t|$$

(4) Evaluate:

$$\Lambda_c = \frac{|\mathbf{Q}_e|}{|\mathbf{Q}_t|} \qquad \text{coincidence}$$

$$\Lambda_p = \frac{|\mathbf{Q}_e|}{|\mathbf{Q}_{tp}|} \qquad \text{parallelism}$$

$$\Lambda_I = \frac{|\mathbf{Q}_{tp}|}{|\mathbf{Q}_t|} \qquad \text{intercept (given parallelism)}$$

when the number of variables $p = 1$, the univariate case is seen to be a special application of the multivariate procedure.

4.13 EXAMPLE—HETEROGENEOUS DATA—MULTIVARIATE

Rohwer's data in Section 4.7 were for kindergarten students in a low-socioeconomic-status area. Adding the dependent variables RPMT and SAT to the data set used in Section 4.3, where the data were for a high-socioeconomic-status population, we want to find whether the multivariate regressions for the two groups, using three dependent variables and five independent variables, are coincident or parallel. Given that they are parallel, it is of interest to determine whether the intercepts are significantly different. The data for the two variables RPMT and SAT in the high-socioeconomic-status group are summarized in Table 4.13.1; the data for the low-socioeconomic-status group are given in Section 4.7.

TABLE 4.13.1. Sample Data: Multivariate Heterogeneous Data.

$N_1 = 32$

RPMT	SAT	RPMT	SAT
15	24	13	64
11	8	16	88
13	88	15	14
18	82	16	99
13	90	18	50
15	77	15	36
13	58	19	88
12	14	11	14
10	1	20	24
18	98	12	24
10	8	16	24
21	88	13	50
14	4	16	8
16	14	18	98
14	38	15	98
15	4	19	50

The model for the stated problem is

$$\begin{array}{cccccc} \mathbf{Y} & = & \mathbf{X} & \mathbf{B} & + & \mathbf{E} \end{array}$$

$$\begin{bmatrix} \mathbf{Y}_1 \\ \mathbf{Y}_1 \end{bmatrix} = \begin{bmatrix} \mathbf{X}_1 & \mathbf{0} \\ \mathbf{0} & \mathbf{X}_2 \end{bmatrix} \begin{bmatrix} \mathbf{B}_1 \\ \mathbf{B}_2 \end{bmatrix} + \begin{bmatrix} \mathbf{E}_1 \\ \mathbf{E}_2 \end{bmatrix} \tag{4.13.1}$$

where $\mathbf{Y}_1$ and $\mathbf{Y}_2$ denote the data matrices for the high- and low-socioeconomic-status groups, respectively,

$$\mathbf{B}_i = \begin{bmatrix} \beta_{01}^{(i)} & \beta_{02}^{(i)} & \beta_{03}^{(i)} \\ \beta_{11}^{(i)} & \beta_{12}^{(i)} & \beta_{13}^{(i)} \\ \vdots & \vdots & \vdots \\ \beta_{51}^{(i)} & \beta_{52}^{(i)} & \beta_{53}^{(i)} \end{bmatrix} \qquad i = 1, 2$$

$N_1 = 32$, $N_2 = 37$, $N = N_1 + N_2 = 69$, and each row of $\mathbf{Y}_i$ has a multivariate normal distribution.

When testing for coincidence or parallelism, the matrices $\mathbf{A}$ and $\mathbf{\Gamma}$, for the general linear hypothesis $\mathbf{CBA} = \mathbf{\Gamma}$, are $\mathbf{A} = \mathbf{I}$ and $\mathbf{\Gamma} = \mathbf{0}$. The hypothesis test matrix $\mathbf{C}$ for the coincidence hypothesis

$$H_{01}: \mathbf{B}_1 = \mathbf{B}_2 \tag{4.13.2}$$

is seen to be

$$\mathbf{C}_{01} = \begin{bmatrix} 1 & 0 & 0 & 0 & 0 & 0 & -1 & 0 & 0 & 0 & 0 & 0 \\ 0 & 1 & 0 & 0 & 0 & 0 & 0 & -1 & 0 & 0 & 0 & 0 \\ 0 & 0 & 1 & 0 & 0 & 0 & 0 & 0 & -1 & 0 & 0 & 0 \\ 0 & 0 & 0 & 1 & 0 & 0 & 0 & 0 & 0 & -1 & 0 & 0 \\ 0 & 0 & 0 & 0 & 1 & 0 & 0 & 0 & 0 & 0 & -1 & 0 \\ 0 & 0 & 0 & 0 & 0 & 1 & 0 & 0 & 0 & 0 & 0 & -1 \end{bmatrix} \tag{4.13.3}$$

and the **C** matrix for the parallelism test

$$H_{02}: \mathbf{\Gamma}_1 = \mathbf{\Gamma}_2 \tag{4.13.4}$$

is

$$\mathbf{C}_{02} = \begin{bmatrix} 0 & 1 & 0 & 0 & 0 & 0 & -1 & 0 & 0 & 0 & 0 \\ 0 & 0 & 1 & 0 & 0 & 0 & 0 & -1 & 0 & 0 & 0 \\ 0 & 0 & 0 & 1 & 0 & 0 & 0 & 0 & -1 & 0 & 0 \\ 0 & 0 & 0 & 0 & 1 & 0 & 0 & 0 & 0 & -1 & 0 \\ 0 & 0 & 0 & 0 & 0 & 1 & 0 & 0 & 0 & 0 & -1 \end{bmatrix} \tag{4.13.5}$$

By employing equations (4.12.8), (4.12.9), and (4.12.11), the test of coincidence is obtained with

$$\begin{aligned} \mathbf{Q}_e &= \mathbf{Y}'[\mathbf{I} - \mathbf{X}(\mathbf{X}'\mathbf{X})^{-1}\mathbf{X}']\mathbf{Y} \\ &= \begin{bmatrix} 31{,}063.64377 & & \\ 2810.11792 & 6663.75032 & \\ 1080.97254 & 495.06422 & 445.31307 \end{bmatrix} \end{aligned} \tag{4.13.6}$$

and

$$\begin{aligned} \mathbf{Q}_h &= \mathbf{Y}'\mathbf{X}(\mathbf{X}'\mathbf{X})^{-1}\mathbf{C}'[\mathbf{C}(\mathbf{X}'\mathbf{X})^{-1}\mathbf{C}']^{-1}\mathbf{C}(\mathbf{X}'\mathbf{X})^{-1}\mathbf{X}'\mathbf{Y} \\ &= \begin{bmatrix} 12{,}997.43415 & & \\ 3395.43164 & 4596.66929 & \\ 727.57924 & 474.40502 & 108.21002 \end{bmatrix} \end{aligned} \tag{4.13.7}$$

forming Λ:

$$\Lambda = \frac{|\mathbf{Q}_e|}{|\mathbf{Q}_e + \mathbf{Q}_h|} = .387538 \tag{4.13.8}$$

Since $p = 3$, $k = 5$, and $I = 2$, the hypothesis of coincidence is rejected at the significance level $\alpha = .05$ if

$$\Lambda < U^{\alpha}[p, (k+1)(I-1), N - Ik - I] = U^{.05}_{(3,6,57)} = .603297 \tag{4.13.9}$$

Thus the hypothesis of coincidence is rejected.

To test the hypothesis of parallelism, following the general linear model approach with **C** as defined in (4.13.5),

$$\mathbf{Q}_h = \begin{bmatrix} 11{,}841.90878 & & \\ 1178.74004 & 343.29906 & \\ 519.30017 & 74.85476 & 70.66851 \end{bmatrix} \tag{4.13.10}$$

so that, by (4.12.17),

$$\Lambda = \frac{|\mathbf{Q}_e|}{|\mathbf{Q}_e + \mathbf{Q}_h|} = .623582 \tag{4.13.11}$$

Since $\Lambda = .623582 < U^{.05}_{(3,5,57)} = .642961$ at the significance level $\alpha = .05$, the hypothesis of parallelism is rejected. Thus $\mathbf{B}_1$ and $\mathbf{B}_2$ are estimated by

(5.13.12)
$$\hat{\mathbf{B}}_1 = \begin{bmatrix} -29.46747 & 39.69709 & 13.24384 \\ 3.25713 & .06728 & .05935 \\ 2.99658 & .36998 & .49244 \\ -5.85906 & -.37438 & -.16402 \\ 5.66622 & 1.52301 & .11898 \\ -.62265 & .41016 & -.12116 \end{bmatrix}$$

$$\hat{\mathbf{B}}_2 = \begin{bmatrix} 4.15106 & 33.00577 & 11.17338 \\ -.60887 & -.08057 & .21100 \\ -.05016 & -.72105 & .06457 \\ -1.73240 & -.29830 & .21358 \\ .49456 & 1.47042 & -.03732 \\ 2.24772 & .32396 & -.05214 \end{bmatrix}$$

Confidence intervals for the elements of $\mathbf{B}_i$ follow the procedure outlined in the multivariate regression example, (4.6.31).

The analysis of the data for this section illustrates that the same function is not employed simultaneously to predict PPVT, RPMT, and SAT scores for high- and low-socioeconomic-status kindergarten children. Furthermore, the functions are not even parallel. To overcome nonparallelism, a subset of the independent variables may be sought that would provide parallel regression planes. A test of differences could then be pursued.

In doing multivariate analysis of covariance, researchers often fail to check the parallelism assumption when testing for differences in intercepts. The primary purpose of the inclusion of the analysis of heterogeneous data in this chapter was to provide a means with which to test this assumption; further, the techniques developed also allow experimenters to determine whether regression functions are coincident. Multivariate analysis of covariance (MANCOVA) is discussed in Chapter 5.

EXERCISE 4.13

1. For the multivariate data in Exercise 1, Section 5.15, test the hypothesis of parallelism for the multivariate regression functions obtained for the three groups.

4.14 CANONICAL CORRELATION ANALYSIS

In Section 2.5, the multiple correlation coefficient was defined as the maximum correlation between the random variable Y_i and the linear combination $\boldsymbol{\beta}'\mathbf{X}$, where

(4.14.1)
$$\mathbf{U} = \begin{bmatrix} Y \\ X \end{bmatrix} \sim N_{p+q}\left\{\begin{bmatrix} \boldsymbol{\mu}_1 \\ \boldsymbol{\mu}_2 \end{bmatrix}, \quad \boldsymbol{\Sigma} = \begin{bmatrix} \boldsymbol{\Sigma}_{11} & \boldsymbol{\Sigma}_{12} \\ \boldsymbol{\Sigma}_{21} & \boldsymbol{\Sigma}_{22} \end{bmatrix}\right\}$$

$\mathbf{Y}$ is a p-vector and $\mathbf{X}$ is a q-vector. For convenience, we shall assume that $p \leq q$. In this section, we will consider maximizing the correlations between two sets of variables where the joint distribution of the variables is given by (4.14.1).

As developed by Hotelling (1936a), the process of maximizing the correlation between two linear functions of two sets of random variables is known as *canonical correlation analysis*. The linear functions that yield the maximum correlations are termed *canonical variates*. The goal of canonical analysis is to find two sets of weights $\mathbf{a}$ and $\mathbf{b}$ such that each canonical variate is maximally correlated, subject to the restriction that each variate be orthogonal or uncorrelated with the previous linear combinations.

Canonical correlation analysis is employed in testing for independence of two sets of variables. More importantly, it is used as a data-reduction method. Given a large number of variables, an investigator may be able to find a few linear combinations of the variables in each set to study the intercorrelations of the canonical variates and thus simplify the analysis in a coordinate system that will clarify the interrelationships.

As an example of the usefulness of canonical correlation analysis, suppose a researcher had a set of ability variables and a set of personality variables. The researcher may want to determine what sort of personality traits can be associated with various ability measures. Canonical correlation analysis aids in determining linear combinations of personality measures that are highly correlated with linear combinations of abilities. These linear combinations are the canonical variates.

Letting one set of variables exist with p elements and another set with q elements were $p \leq q$ and where $\mathbf{Y}' = [Y_1, Y_2, \ldots, Y_p]$ and $\mathbf{X}' = [X_1, X_2, \ldots, X_q]$, two linear functions are needed, $U = \mathbf{a}'\mathbf{Y}$ and $V = \mathbf{b}'\mathbf{X}$, of unit variance such that the correlation between U and V is maximum. Thus we need to maximize

$$F_{UV} = \max_{\mathbf{a},\mathbf{b}} \mathbf{a}'\mathbf{\Sigma}_{12}\mathbf{b} \tag{4.14.2}$$

subject to the constraints that $\mathbf{a}'\mathbf{\Sigma}_{11}\mathbf{a} = \mathbf{b}'\mathbf{\Sigma}_{22}\mathbf{b} = 1$. Using Lagrange multipliers, the function

$$F = \mathbf{a}'\mathbf{\Sigma}_{12}\mathbf{b} - \frac{\rho_1}{2}(\mathbf{a}'\mathbf{\Sigma}_{11}\mathbf{a} - 1) - \frac{\rho_2}{2}(\mathbf{b}'\mathbf{\Sigma}_{22}\mathbf{b} - 1) \tag{4.14.3}$$

is to be maximized. Taking partial derivatives with respect to $\mathbf{a}$, $\mathbf{b}$, ρ_1, and ρ_2 and equating them to 0, we have

$$\begin{aligned}
\frac{\partial F}{\partial \mathbf{a}} &= \mathbf{\Sigma}_{12}\mathbf{b} - \rho_1\mathbf{\Sigma}_{11}\mathbf{a} = \mathbf{0} \\
\frac{\partial F}{\partial \mathbf{b}} &= \mathbf{\Sigma}'_{12}\mathbf{a} - \rho_2\mathbf{\Sigma}_{22}\mathbf{b} = \mathbf{0} \\
\frac{\partial F}{\partial \rho_1} &= \mathbf{a}'\mathbf{\Sigma}_{11}\mathbf{a} - 1 = 0 \\
\frac{\partial F}{\partial \rho_2} &= \mathbf{b}'\mathbf{\Sigma}_{22}\mathbf{b} - 1 = 0
\end{aligned} \tag{4.14.4}$$

Multiplying the first equation in (4.14.4) by $\mathbf{a}'$, the second by $\mathbf{b}'$, and employing the constraints, the system of equations to solve becomes

(4.14.5)
$$\begin{aligned} -\rho\mathbf{\Sigma}_{11}\mathbf{a} + \mathbf{\Sigma}_{12}\mathbf{b} &= 0 \\ \mathbf{\Sigma}_{21}\mathbf{a} - \rho\mathbf{\Sigma}_{22}\mathbf{b} &= 0 \end{aligned}$$

where $\rho = \rho_1 = \rho_2 = \mathbf{a}'\mathbf{\Sigma}_{12}\mathbf{b}$. Multiplying the first equation in (4.14.5) by ρ and the second by $\mathbf{\Sigma}_{12}\mathbf{\Sigma}_{22}^{-1}$, (4.14.5) becomes

(4.14.6)
$$\begin{aligned} -\rho^2\mathbf{\Sigma}_{11}\mathbf{a} + \rho\mathbf{\Sigma}_{12}\mathbf{b} &= \mathbf{0} \\ \mathbf{\Sigma}_{12}\mathbf{\Sigma}_{22}^{-1}\mathbf{\Sigma}_{21}\mathbf{a} - \rho\mathbf{\Sigma}_{12}\mathbf{\Sigma}_{22}^{-1}\mathbf{\Sigma}_{22}\mathbf{b} &= \mathbf{0} \end{aligned}$$

Adding the two equations in (4.14.6) yields the equation

(4.14.7)
$$(\mathbf{\Sigma}_{12}\mathbf{\Sigma}_{22}^{-1}\mathbf{\Sigma}_{21} - \rho^2\mathbf{\Sigma}_{11})\mathbf{a} = \mathbf{0}$$

Hence $\rho_1^2, \rho_2^2, \ldots, \rho_p^2$ and $\mathbf{a}_1, \mathbf{a}_2, \ldots, \mathbf{a}_p$ are the roots and vectors, respectively, of the characteristic equation

(4.14.8)
$$|\mathbf{\Sigma}_{12}\mathbf{\Sigma}_{22}^{-1}\mathbf{\Sigma}_{21} - \rho^2\mathbf{\Sigma}_{11}| = 0$$

From (1.7.25), let $\mathbf{A} = [\mathbf{a}_1, \mathbf{a}_2, \ldots, \mathbf{a}_p]$, then $\mathbf{A}'\mathbf{\Sigma}_{11}\mathbf{A} = \mathbf{I}_p$ and $\mathbf{A}'\mathbf{\Sigma}_{12}\mathbf{\Sigma}_{22}^{-1}\mathbf{\Sigma}_{21}\mathbf{A} = \mathbf{\Lambda}_1$, where $\mathbf{\Lambda}_1$ is a diagonal matrix with roots $\rho_1^2, \rho_2^2, \ldots, \rho_p^2$.

Similarly, by multiplying the second equation in (4.14.5) by ρ and the first by $\mathbf{\Sigma}_{21}\mathbf{\Sigma}_{11}^{-1}$, (4.14.5) becomes

(4.14.9)
$$\begin{aligned} -\rho\mathbf{\Sigma}_{21}\mathbf{a} + \mathbf{\Sigma}_{21}\mathbf{\Sigma}_{11}^{-1}\mathbf{\Sigma}_{12}\mathbf{b} &= \mathbf{0} \\ \rho\mathbf{\Sigma}_{21}\mathbf{a} - \rho^2\mathbf{\Sigma}_{22}\mathbf{b} &= \mathbf{0} \end{aligned}$$

Adding these two equations leads to the characteristic equation

(4.14.10)
$$|\mathbf{\Sigma}_{21}\mathbf{\Sigma}_{11}^{-1}\mathbf{\Sigma}_{12} - \rho^2\mathbf{\Sigma}_{22}| = 0$$

The roots of this equation are $\rho_1^2, \rho_2^2, \ldots, \rho_q^2$, with corresponding vectors $\mathbf{b}_1, \mathbf{b}_2, \ldots, \mathbf{b}_q$. By (1.7.25), let $\mathbf{B} = [\mathbf{b}_1, \mathbf{b}_2, \ldots, \mathbf{b}_q]$, then $\mathbf{B}'\mathbf{\Sigma}_{22}\mathbf{B} = \mathbf{I}_q$ and $\mathbf{B}'\mathbf{\Sigma}_{21}\mathbf{\Sigma}_{11}^{-1}\mathbf{\Sigma}_{12}\mathbf{B} = \mathbf{\Lambda}_2$, where $\mathbf{\Lambda}_2$ is a diagonal matrix with roots $\rho_1^2, \rho_2^2, \ldots, \rho_q^2$.

The nonzero positive square roots ρ_i of the roots ρ_i^2 are called the *canonical correlations* between the *canonical variates* $U_i = \mathbf{a}_i'\mathbf{Y}$ and $V_i = \mathbf{b}_i'\mathbf{X}$, for $i = 1, \ldots, p \leq q$. From the second equation in (4.14.6), the relationship between $\mathbf{a}_i$ and $\mathbf{b}_i$ is given by

(4.14.11)
$$\mathbf{b}_i = \frac{\mathbf{\Sigma}_{22}^{-1}\mathbf{\Sigma}_{21}\mathbf{a}_i}{\rho_i}$$

The set of canonical variates U_i and V_i are clearly uncorrelated and have unit variance,

$$\text{cov}(U_i, U_j) = \text{cov}(V_i, V_j) = \begin{cases} 1 & i = j \\ 0 & i \neq j \end{cases}$$

Furthermore, the covariance between V_i and U_i is ρ_i, for $i = 1, \ldots, p$, and 0, otherwise:

$$\begin{aligned} \text{cov}(U_i, V_i) &= \rho_i \qquad i = 1, \ldots, p \\ \text{cov}(U_i, V_j) &= 0 \qquad i \neq j \end{aligned}$$

Thus if U_1, U_2, and V_1, V_2, and V_3 are canonical variates, the correlation matrix for U_1, U_2, V_1, V_2, and V_3 has the form

$$
\begin{array}{c} \\ U_1 \\ U_2 \\ V_1 \\ V_2 \\ V_3 \end{array}
\begin{array}{c}
\begin{array}{ccccc} U_1 & U_2 & V_1 & V_2 & V_3 \end{array} \\
\left[\begin{array}{cc|ccc}
1 & 0 & \rho_1 & 0 & 0 \\
0 & 1 & 0 & \rho_2 & 0 \\
\hline
\rho_1 & 0 & 1 & 0 & 0 \\
0 & \rho_2 & 0 & 1 & 0 \\
0 & 0 & 0 & 0 & 1
\end{array}\right]
\end{array}
$$

More generally, these results imply that the original observation vector's covariance matrix given in (4.14.1) has been transformed to the correlation matrix

$$\text{(4.14.12)} \qquad \mathbf{R}_{UV} = \begin{bmatrix} \mathbf{I}_p & \mathbf{\Delta} \\ \mathbf{\Delta}' & \mathbf{I}_q \end{bmatrix}$$

where $\mathbf{\Delta}$ is a $p \times q$ matrix containing the first p canonical correlations between U_i and V_i. Hence canonical correlation affects two sets of original variables by representing all the correlations existing between the sets through the canonical correlation. The quantity ρ_i^2 can be interpreted as a squared multiple correlation coefficient relating a canonical variate with the set of complete variables. Thus a significant ρ_i^2 with t variables is t times a significant squared multiple correlation coefficient.

To apply the theory developed above for a sample where $\mathbf{U}_1, \mathbf{U}_2, \ldots, \mathbf{U}_N \sim IN(\boldsymbol{\mu}, \mathbf{\Sigma})$, replace the population variance-covariance matrices by their sample counterparts $\mathbf{S}_{ij}$. Alternatively, the sample correlation matrices $\mathbf{R}_{ij}$ may also be used since the roots of $\mathbf{S}_{11}^{-1/2}\mathbf{S}_{12}\mathbf{S}_{22}^{-1}\mathbf{S}_{21}\mathbf{S}_{11}^{-1/2}$ and $\mathbf{R}_{11}^{-1/2}\mathbf{R}_{12}\mathbf{R}_{22}^{-1}\mathbf{R}_{21}\mathbf{R}_{11}^{-1/2}$ are equal. Although the vectors $\mathbf{a}_i$ and $\mathbf{b}_i$ associated with the characteristic equations (4.14.8) and (4.14.10), respectively, with the $\mathbf{S}_{ij}$'s replacing the $\mathbf{\Sigma}_{ij}$'s, will have units of measurement proportional to the original variables, the units of U_i and V_i may not be meaningful. Canonical variates obtained by using correlation matrices have no units of measurement and should be evaluated in terms of standardized variables when the analysis is to be employed in a data-reduction capacity.

Bartlett (1938a) outlines a procedure for testing the significance of the canonical correlations when the sample size is large. He defines

$$\text{(4.14.13)} \qquad \Lambda = \prod_{i=1}^{s} (1 - r_i^2) = \frac{|\mathbf{S}_{11} - \mathbf{S}_{12}\mathbf{S}_{22}^{-1}\mathbf{S}_{21}|}{|\mathbf{S}_{11}|}$$

where r_i^2 is the sample estimate of ρ_i^2, and employs his chi-square approximation for the distribution of Λ to test the null hypothesis that the p-variates are unrelated to the q-variates:

$$\text{(4.14.14)} \qquad \begin{aligned} H_0 &: \mathbf{\Sigma}_{12} = 0 \\ H_1 &: \mathbf{\Sigma}_{12} \neq 0 \end{aligned}$$

Hypothesis (4.14.14) is rejected if

$$\text{(4.14.15)} \qquad X_B^2 = -\left[(N-1) - \frac{1}{2}(p+q+1)\right] \log \Lambda > \chi_\alpha^2(pq)$$

From (4.6.23), Λ defined by (4.14.13) is Wilks' lambda criterion used in testing $\mathbf{\Gamma} = \mathbf{0}$ in the multivariate regression model. Thus, as in univariate analysis, the close relationship between correlation and regression studies extends to the multivariate case.

To test (4.14.14) by using the largest-root criterion, the test statistic would be the largest canonical correlation coefficient squared. The other criteria follow immediately from (4.6.24) by replacing k by q.

If the null hypothesis of no relationship or independence can be rejected, the contribution of the first root of Λ may be removed and the significance of the remaining roots evaluated (see Bartlett, 1951, or Rao, 1952, p. 370). In general, with the $s' < s = \min(p, q)$ roots removed,

$$\Lambda^* = \prod_{i=s'+1}^{s} (1 - r_i^2) \tag{4.14.16}$$

Partitioning Bartlett's chi-square statistic

$$X_B^2 = -\left[(N-1) - \frac{1}{2}(p + q + 1)\right] \log \Lambda^* \tag{4.14.17}$$

we find a chi-square distribution with $(p - s')(q - s')$ degrees of freedom, which is used to test the significance of the roots $s' + 1$ to s. The tests for significant canonical correlations, other than the first, are very conservative unless the correlations removed are very close to 1 (see Williams, 1967). Alternatively, the U distribution may be employed directly.

Testing the importance of the roots in (4.14.13) aids researchers to determine how many canonical variates can be retained for further analysis. Also, considering correlations between canonical variates of the two sets and correlations between variables with canonical variates within a set helps in the interpretation of a canonical correlation analysis. This will be considered in Section 4.15 through an example.

Hypothesis (4.14.14) in the multivariate situation reduces promptly to some familiar univariate results. If the number of variables in the p set is one, (4.14.14) becomes

$$\begin{aligned} H_0 &: \sigma_{12} = 0 \qquad \text{or} \qquad \rho_{0(1,\ldots,q)} = 0 \\ H_1 &: \sigma_{12} \neq 0 \qquad \qquad \rho_{0(1,\ldots,q)} \neq 0 \end{aligned} \tag{4.14.18}$$

and is tested by using $F = [R^2/q]/[(1 - R^2)/(N - q - 1)]$. In the bivariate case, (4.14.14) reduces to

$$\begin{aligned} H_0 &: \rho = 0 \\ H_1 &: \rho \neq 0 \end{aligned} \tag{4.14.19}$$

where the test statistic is $t = r\sqrt{N-2}/\sqrt{1 - r^2}$.

Other procedures commonly used in univariate analysis have multivariate analogues. For example, to test that a partial correlation coefficient is significantly different from 0 in the simplest case

$$\begin{aligned} H_0 &: \rho_{12\cdot3} = 0 \\ H_1 &: \rho_{12\cdot3} \neq 0 \end{aligned} \tag{4.14.20}$$

(see Anderson, 1958, p. 85), a t statistic similar to the simple correlation case is formed where $t = r_{12\cdot 3}\sqrt{N-3}/\sqrt{1-r_{12\cdot 3}^2}$.

To generalize (4.14.20) to the multivariate case, let

$$\mathbf{U} = \begin{bmatrix} \mathbf{Y} \\ \mathbf{X} \\ \mathbf{Z} \end{bmatrix} \sim N_{p+q+r}\left\{\begin{bmatrix} \boldsymbol{\mu}_1 \\ \boldsymbol{\mu}_2 \\ \boldsymbol{\mu}_3 \end{bmatrix}, \boldsymbol{\Sigma} = \begin{bmatrix} \boldsymbol{\Sigma}_{11} & \boldsymbol{\Sigma}_{12} & \boldsymbol{\Sigma}_{13} \\ \boldsymbol{\Sigma}_{21} & \boldsymbol{\Sigma}_{22} & \boldsymbol{\Sigma}_{23} \\ \boldsymbol{\Sigma}_{31} & \boldsymbol{\Sigma}_{32} & \boldsymbol{\Sigma}_{33} \end{bmatrix}\right\} \tag{4.14.21}$$

where $\mathbf{Y}, \mathbf{X}$, and $\mathbf{Z}$ are p, q, and r random vectors, respectively. From Anderson (1958, p. 33) or Rao (1969), the variance-covariance matrix of the conditional distribution of $\mathbf{Y}$ and $\mathbf{X}$, given $\mathbf{Z}$, is

$$\begin{aligned} \boldsymbol{\Sigma}_{.3} &= \begin{bmatrix} \boldsymbol{\Sigma}_{11\cdot 3} & \boldsymbol{\Sigma}_{12\cdot 3} \\ \boldsymbol{\Sigma}_{21\cdot 3} & \boldsymbol{\Sigma}_{22\cdot 3} \end{bmatrix} \\ &= \begin{bmatrix} \boldsymbol{\Sigma}_{11} - \boldsymbol{\Sigma}_{13}\boldsymbol{\Sigma}_{33}^{-1}\boldsymbol{\Sigma}_{31} & \boldsymbol{\Sigma}_{12} - \boldsymbol{\Sigma}_{13}\boldsymbol{\Sigma}_{33}^{-1}\boldsymbol{\Sigma}_{32} \\ \boldsymbol{\Sigma}_{21} - \boldsymbol{\Sigma}_{23}\boldsymbol{\Sigma}_{33}^{-1}\boldsymbol{\Sigma}_{31} & \boldsymbol{\Sigma}_{22} - \boldsymbol{\Sigma}_{23}\boldsymbol{\Sigma}_{33}^{-1}\boldsymbol{\Sigma}_{32} \end{bmatrix} \end{aligned} \tag{4.14.22}$$

so that a test of the partial independence of the two sets of variables $\mathbf{Y}$ and $\mathbf{X}$, after partialing out $\mathbf{Z}$ from both $\mathbf{Y}$ and $\mathbf{X}$, becomes

$$\begin{aligned} H_0 &: \boldsymbol{\Sigma}_{12\cdot 3} = 0 \\ H_1 &: \boldsymbol{\Sigma}_{12\cdot 3} \neq 0 \end{aligned} \tag{4.14.23}$$

To test (4.14.23), apply the procedure described to test (4.14.13). With $\boldsymbol{\Sigma}_{11}$, $\boldsymbol{\Sigma}_{12}$, $\boldsymbol{\Sigma}_{21}$, and $\boldsymbol{\Sigma}_{22}$ replaced by $\boldsymbol{\Sigma}_{11\cdot 3}$, $\boldsymbol{\Sigma}_{12\cdot 3}$, $\boldsymbol{\Sigma}_{21\cdot 3}$, and $\boldsymbol{\Sigma}_{22\cdot 3}$, respectively, the determinantal equation

$$|\boldsymbol{\Sigma}_{12\cdot 3}\boldsymbol{\Sigma}_{22\cdot 3}^{-1}\boldsymbol{\Sigma}_{21\cdot 3} - \rho^2\boldsymbol{\Sigma}_{11\cdot 3}| = 0 \tag{4.14.24}$$

is evaluated in the sample, where $\boldsymbol{\Sigma}_{ij\cdot 3}$ is replaced by $\mathbf{S}_{ij\cdot 3}$ or $\mathbf{R}_{ij\cdot 3}$. The roots obtained from (4.14.24), are to be represented by $\rho_{i\cdot 3}^2$ and are called the *squares* of the *partial canonical correlations.* The positive square roots $\rho_{i\cdot 3}$ represent the maximal correlation between the *partial canonical variates* $U_i' = \mathbf{a}_i'\mathbf{e}_y$ and $V_i' = \mathbf{b}_i'\mathbf{e}_x$, where $\mathbf{e}_y$ and $\mathbf{e}_x$ denote residual vectors after regressing $\mathbf{Y}$ on $\mathbf{Z}$ and $\mathbf{X}$ on $\mathbf{Z}$. The partial canonical variates acquired from (4.14.24) may also be employed for data-reduction purposes. Use the "partial" roots of (4.14.24) to test (4.14.23) by also using

$$\Lambda = \prod_{i=1}^{s} (1 - r_{i\cdot 3}^2) = \frac{|\boldsymbol{\Sigma}_{11\cdot 3} - \boldsymbol{\Sigma}_{12\cdot 3}\boldsymbol{\Sigma}_{22\cdot 3}^{-1}\boldsymbol{\Sigma}_{21\cdot 3}|}{|\boldsymbol{\Sigma}_{11\cdot 3}|} \tag{4.14.25}$$

where $\Lambda \sim U(p, q, N - r - q - 1)$ and r is the number of elements in the partialed-out $\mathbf{Z}$ set. Other criteria may also be used to test (4.14.23). For example, by using the Roy criterion with

$$\begin{aligned} s &= \min(p, q) \\ m &= \frac{|p - q| - 1}{2} \\ n &= \frac{N - r - p - q - 2}{2} \end{aligned}$$

(4.14.23) is rejected if $r_{i\cdot 3}^2 > \theta^\alpha(s, m, n)$ at the significance level α. In general, following the rules of univariate analysis, every v_e in the criteria to test (4.14.14) is replaced by $v_e - r$. If $\mathbf{\Sigma}_{11}$ is a scalar in (4.14.20), then $r_{1\cdot 3}^2$ is just the square of the partial multiple correlation coefficient.

By modifying (4.14.22) to

$$\mathbf{\Sigma}_{1(2\cdot 3)} = \begin{bmatrix} \mathbf{\Sigma}_{11} & \mathbf{\Sigma}_{12\cdot 3} \\ \mathbf{\Sigma}_{21\cdot 3} & \mathbf{\Sigma}_{22\cdot 3} \end{bmatrix} \tag{4.14.26}$$

part canonical correlations and *variates* may be examined where the variation in the **Z** set is removed from the **X** set but not the **Y** set. This is accomplished by considering the determinantal equations

$$\begin{aligned} |\mathbf{\Sigma}_{12\cdot 3}\mathbf{\Sigma}_{22\cdot 3}^{-1}\mathbf{\Sigma}_{21\cdot 3} - \rho^2\mathbf{\Sigma}_{11}| &= 0 \\ |\mathbf{\Sigma}_{21\cdot 3}\mathbf{\Sigma}_{11}^{-1}\mathbf{\Sigma}_{12\cdot 3} - \rho^2\mathbf{\Sigma}_{22\cdot 3}| &= 0 \end{aligned} \tag{4.14.27}$$

These represent a generalization of a *part correlation analysis* in univariate analysis. In the simplest case, a part correlation coefficient is computed by using the formula

$$r_{1(2\cdot 3)} = \frac{r_{12} - r_{13}r_{32}}{\sqrt{1 - r_{32}^2}} \tag{4.14.28}$$

(see, for example, Glass and Stanley, 1970, p. 184, or Nunnally, 1967, p. 154).

Extending the part correlation in the simplest case one step further, suppose that the influence of variable 4 is to be removed from variable 1 and not 2, and variable 3 is to be removed from 2 but not 1. Then a *bipartial correlation coefficient* is obtained:

$$r_{(1\cdot 4)(2\cdot 3)} = \frac{r_{12} - r_{14}r_{42} - r_{13}r_{32} + r_{14}r_{43}r_{32}}{\sqrt{1 - r_{41}^2}\sqrt{1 - r_{32}^2}} \tag{4.14.29}$$

Equation (4.14.29) is useful when variables 4 and 3 are highly correlated. To generalize to the multivariate case, let

$$\mathbf{U} = \begin{bmatrix} \mathbf{Y} \\ \mathbf{X} \\ \mathbf{W} \\ \mathbf{Z} \end{bmatrix} \sim N_{p+q+r+s}\left\{\begin{bmatrix} \boldsymbol{\mu}_1 \\ \boldsymbol{\mu}_2 \\ \boldsymbol{\mu}_3 \\ \boldsymbol{\mu}_4 \end{bmatrix}, \quad \mathbf{\Sigma} = \begin{bmatrix} \mathbf{\Sigma}_{11} & \mathbf{\Sigma}_{12} & \mathbf{\Sigma}_{13} & \mathbf{\Sigma}_{14} \\ \mathbf{\Sigma}_{21} & \mathbf{\Sigma}_{22} & \mathbf{\Sigma}_{23} & \mathbf{\Sigma}_{24} \\ \mathbf{\Sigma}_{31} & \mathbf{\Sigma}_{32} & \mathbf{\Sigma}_{33} & \mathbf{\Sigma}_{34} \\ \mathbf{\Sigma}_{41} & \mathbf{\Sigma}_{42} & \mathbf{\Sigma}_{43} & \mathbf{\Sigma}_{44} \end{bmatrix}\right\} \tag{4.14.30}$$

Forming the matrix

$$\mathbf{\Sigma}_{(1\cdot 4)(2\cdot 3)} = \begin{bmatrix} \mathbf{\Sigma}_{11} - \mathbf{\Sigma}_{14}\mathbf{\Sigma}_{44}^{-1}\mathbf{\Sigma}_{41} & \mathbf{\Sigma}_{12} - \mathbf{\Sigma}_{14}\mathbf{\Sigma}_{44}^{-1}\mathbf{\Sigma}_{42} - \mathbf{\Sigma}_{13}\mathbf{\Sigma}_{33}^{-1}\mathbf{\Sigma}_{32} + \mathbf{\Sigma}_{14}\mathbf{\Sigma}_{44}^{-1}\mathbf{\Sigma}_{43}\mathbf{\Sigma}_{33}^{-1}\mathbf{\Sigma}_{32} \\ & \mathbf{\Sigma}_{22} - \mathbf{\Sigma}_{23}\mathbf{\Sigma}_{33}^{-1}\mathbf{\Sigma}_{32} \end{bmatrix}$$

and substituting into the determinantal equations allows us to obtain *bipartial canonical variates* and *bipartial canonical correlations.*

To use canonical analysis as a data-analysis tool, it is only necessary to assume that the conditional mean of **Y**, given **X**, is linear in **X**; however, to test the hypothesis, we make normality assumptions.

EXERCISE 4.14

1. a. Derive the result given by equation (4.14.11).
   b. Let U_i and V_i denote canonical variates for two sets of variables, and show that

$$\operatorname{cov}(U_i, U_j) = \operatorname{cov}(V_i, V_j) = \begin{cases} 1 & i = j \\ 0 & i \neq j \end{cases}$$

and that

$$\begin{aligned} \operatorname{cov}(U_i, V_i) &= \rho_i \qquad i = 1, \ldots, p \\ \operatorname{cov}(U_i, V_j) &= 0 \qquad i \neq j \end{aligned}$$

4.15 EXAMPLE—CANONICAL CORRELATION ANALYSIS

Employing Rohwer's data in Section 4.7 for the low-socioeconomic-status group where the sample size $N = 37$, let NA and SS be the variables in one set and SAT, PEA, and RAV be the variables in the other set, where

$$U_i = \left\{ \begin{bmatrix} \mathbf{Y} = \begin{bmatrix} \text{NA} \\ \text{SS} \end{bmatrix} \\ \mathbf{X} = \begin{bmatrix} \text{SAT} \\ \text{PEA} \\ \text{RAV} \end{bmatrix} \end{bmatrix} \right\} \sim N_5 \left\{ \begin{bmatrix} \boldsymbol{\mu}_1 \\ \boldsymbol{\mu}_2 \end{bmatrix}, \quad \boldsymbol{\Sigma} = \begin{bmatrix} \boldsymbol{\Sigma}_{11} & \boldsymbol{\Sigma}_{12} \\ \boldsymbol{\Sigma}_{21} & \boldsymbol{\Sigma}_{22} \end{bmatrix} \right\} \tag{4.15.1}$$

Applying the theory developed in Section 4.14, we employ canonical correlation analysis to test for independence. Linear combinations of the original variables are also acquired to study the structure of the relationships between the sets.

By forming the sample correlation matrix

$$\mathbf{R} = \begin{bmatrix} \mathbf{R}_{11} & \mathbf{R}_{12} \\ \mathbf{R}_{21} & \mathbf{R}_{22} \end{bmatrix} = \left[\begin{array}{cc|ccc} 1.0000 & & & & \\ .7951 & 1.0000 & & & \\ \hline .2617 & .3341 & 1.0000 & & \\ .6720 & .5876 & .3703 & 1.0000 & \\ .3390 & .3404 & .2114 & .3548 & 1.0000 \end{array}\right] \tag{4.15.2}$$

for the example, and solving the determinantal equation

$$|\mathbf{R}_{12}\mathbf{R}_{22}^{-1}\mathbf{R}_{21} - \rho^2\mathbf{R}_{11}| = 0 \tag{4.15.3}$$

the sample canonical correlations r_1 and r_2 are obtained:

$$\begin{aligned} r_1^2 &= .4746 \qquad & r_2^2 &= .0375 \\ r_1 &= .6889 \qquad & r_2 &= .1936 \end{aligned} \tag{4.15.4}$$

Thus, by employing Bartlett's criterion to test for independence, $H_0: \boldsymbol{\Sigma}_{12} = \mathbf{0}$,

$$\Lambda = \prod_{i=1}^{2} (1 - r_i^2) = .5057 \tag{4.15.5}$$

so that $X_B^2 = 22.4982$, where $N = 37$, $p = 2$, and $k = 3$. Since $X_B^2 > \chi_\alpha^2(6) = 11.07$, for $\alpha = .05$, the test of independence is rejected.

Using (4.14.7) and (4.14.9), with the $\boldsymbol{\Sigma}_{ij}$'s replaced by $\mathbf{R}_{ij}$'s, the weights associated with the canonical variates are as follows:

(4.15.6) (a) First canonical variate:

	a		**b**
(NA)	−.7752	(SAT)	−.0520
(SS)	—.2662	(PEA)	−.8991
		(RAV)	−.1831

(b) Second canonical variate:

(NA)	−1.4554	(SAT)	.9939
(SS)	1.6274	(PEA)	−.5912
		(RAV)	.3125

From (4.14.16),

(4.15.7)
$$\Lambda^* = (1 - r_2^2) = .9625$$

so that $X_B^2 = 1.2605 < \chi_\alpha^2(2) = 5.99$, for $\alpha = .05$, which indicates that only the first canonical correlation is significantly different from 0. In summary, the significant canonical variates become, using standardized variables,

(4.15.8)
$$\begin{aligned} U_1 &= -.7752\,(\text{NA}) - .2662\,(\text{SS}) \\ V_1 &= -.0520\,(\text{SAT}) - .8991\,(\text{PEA}) - .1831\,(\text{RAV}) \end{aligned}$$

where the correlation between U_1 and V_1 is $r_1 = .6889$, so that the proportion of variance common to the two canonical variates is $r_1^2 = .4746$.

Investigating the canonical variates further, it is of interest to determine the correlation of each canonical variate for a set with the individual variables within the set. These correlations indicate the contribution of each variable to the composite canonical variate and aid in the interpretation of the canonical variates. To find these correlations generally:

(4.15.9)
$$\begin{aligned} \text{cor}(\mathbf{Y}, U_i) &= \text{cor}(\mathbf{Y}, \mathbf{a}_i'\mathbf{Y}) \\ &= \frac{\boldsymbol{\Sigma}_{11}\mathbf{a}_i}{\sigma_{y_i}} \\ \text{cor}(\mathbf{X}, V_i) &= \frac{\boldsymbol{\Sigma}_{22}\mathbf{b}_i}{\sigma_{x_i}} \end{aligned}$$

Using standardized variables, (4.15.9) becomes

(4.15.10)
$$\begin{aligned} \text{cor}(\mathbf{Z}_y, U_i) &= \mathbf{R}_{11}\mathbf{a}_i \\ \text{cor}(\mathbf{Z}_x, V_i) &= \mathbf{R}_{22}\mathbf{b}_i \end{aligned}$$

For the example, (4.15.10) reduces to

$$\text{(4.15.11)}\qquad \begin{aligned} \text{cor}(\mathbf{Z}_y, U_i) &= \mathbf{R}_{11}\mathbf{a}_1 = \begin{bmatrix} -.987 \\ -.883 \end{bmatrix} \begin{matrix} \text{(NA)} \\ \text{(SS)} \end{matrix} \\ \text{cor}(\mathbf{Z}_x, V_i) &= \mathbf{R}_{22}\mathbf{b}_1 = \begin{bmatrix} -.424 \\ -.983 \\ -.513 \end{bmatrix} \begin{matrix} \text{(SAT)} \\ \text{(PEA)} \\ \text{(RAV)} \end{matrix} \end{aligned}$$

Expression (4.15.11) indicates that both variables NA and SS are equally important to the first canonical variate U_1 and that the variable PEA is about twice as important to the second variate V_1 than either SAT or RAV, which appear to be contributing equally. For a large number of variables in a given domain of definition, "meaning" is usually associated with the canonical variates by the magnitude of the correlations obtained from (4.15.11):

$$\text{(4.15.12)}\qquad \begin{aligned} U_Y^2 &= \frac{(-.987)^2 + (-.883)^2}{2} = .877 \\ V_X^2 &= \frac{(.424)^2 + (-.983)^2 + (-.513)^2}{2} = .470 \end{aligned}$$

Thus 88% of the variance of the first set is accounted for by the first canonical variate U_1 and only 47% of the variance in the second set is accounted for by V_1.

Besides using correlations within a set to better understand canonical variates, we should also examine those relationships between canonical variates in one set and individual variables in the other. Thus

$$\text{(4.15.13)}\qquad \begin{aligned} \text{cor}(\mathbf{X}, U_i) &= \text{cor}(\mathbf{X}, \mathbf{a}_i'\mathbf{Y}) \\ &= \frac{\mathbf{\Sigma}_{21}\mathbf{a}_i}{\sigma_{x_i}} \\ &= \frac{\rho_i \mathbf{\Sigma}_{22}\mathbf{b}_i}{\sigma_{x_i}} \\ &= \text{cor}(\mathbf{X}, V_i)\rho_i \\ \text{cor}(\mathbf{Y}, V_i) &= \text{cor}(\mathbf{Y}, U_i)\rho_i \end{aligned}$$

For the example, (4.15.13) reduces to

$$\text{(4.15.14)}\qquad \begin{aligned} \text{cor}(\mathbf{X}, U_i) &= \begin{bmatrix} -.424 \\ -.983 \\ -.513 \end{bmatrix} (.6889) = \begin{bmatrix} -.292 \\ -.677 \\ -.353 \end{bmatrix} \\ \text{cor}(\mathbf{Y}, V_i) &= \begin{bmatrix} -.987 \\ -.883 \end{bmatrix} (.6889) = \begin{bmatrix} -.680 \\ -.608 \end{bmatrix} \end{aligned}$$

so that variable 2 (PEA) in the second set is most influenced by the first canonical variable of the first set; whereas the first canonical variate of the second set influences both variables in the first set equally. More specifically, the proportion of variance in the set SAT, PEA, and RAV accounted for by the first canonical variate $U_1 = \mathbf{a}'\mathbf{Y}$

of the other set is only

$$U^2_{X|U_1} = \frac{(-.292)^2 + (-.677)^2 + (-.355)^2}{3} = .223 \tag{4.15.15}$$

whereas the proportion of variance in the set of variables NA and SS accounted for by $V_1 = \mathbf{b}'\mathbf{X}$ is

$$V^2_{Y|V_1} = \frac{(-.680)^2 + (-.608)^2}{2} = .416 \tag{4.15.16}$$

That is, 22% of the variance common to the variables SAT, PEA, and RAV can be accounted for by a linear combination of the variables NA and SS, and 42% of the variability in NA and SS is accounted for by a linear combination of the variables SAT, PEA, and RAV.

In summary, given the two sets of variables

$$\mathbf{Y} = \{\text{NA, SS}\} \quad \text{and} \quad \mathbf{X} = \{\text{SAT, PEA, RAV}\}$$

where

$$U_1 = -.7753\,(\text{NA}) - .2662\,(\text{SS})$$

and

$$V_1 = .0520\,(\text{SAT}) - .8991\,(\text{PEA}) - .1831\,(\text{RAV})$$

it appears that the proportion of variance "in common" to the two canonical variates is about 47% since $r_1^2 = .4746$. However, 88% of the variance in the set **Y** is accounted for by U_1, and only 42% of the variance in **Y** is accounted for by the canonical variate V_1. Similarly, 47% of the variance in the set **X** is accounted for by V_1, but only 23% of the variance in **X** is accounted for by the canonical variate U_1.

Stewart and Love (1968) observed that

$$\begin{aligned} V^2_{Y|V_1} &= U^2_Y r_1^2 = (.877)(.4746) = .416 \\ U^2_{X|U_1} &= V^2_X r_1^2 = (.470)(.4746) = .226 \end{aligned} \tag{4.15.17}$$

and termed $V^2_{Y|V_1}$ and $U^2_{X|U_1}$ *redundancy indexes* since they better summarize the overlap between two sets of variables than the squares of canonical correlations. Thus the redundancy in the set **Y**, given the set **X**, is .416, and the redundancy in the set **X**, given the set **Y**, is only .226. The larger the redundancy indexes, the larger the overlap of the variables in each domain.

For more than one significant canonical correlation, we would sum the redundancy indexes to obtain an overall index. Hence

$$V^2_{Y|V_1,\ldots,V_k} = \sum_{i=1}^{k} V^2_{Y|V_i}$$

is the total redundancy index for the set **Y**, given $V_1, \ldots, V_k$, and

$$U^2_{X|U_1,\ldots,U_k} = \sum_{i=1}^{k} U^2_{Y|U_i}$$

is the total redundancy index for the set **X**, given $U_1, \ldots, U_k$. Since $V^2_{Y|V_1,\ldots,V_k} \neq U^2_{X|U_1,\ldots,U_k}$, this index also allows us to determine from two test domains which is more redundant given the other. Pure canonical correlations do not allow this flexibility and tend to be high measures of overlap.

The example discussed in this section can be immediately extended to partial canonical correlation analysis, part canonical correlation analysis, and bipartial canonical correlation analysis. Cooley and Lohnes (1971, p. 207) illustrate some of these procedures using Project Talent data.

EXERCISES 4.15

1. Shin (1971) collected data on intelligence using six creativity measures and six achievement measures for 116 subjects in the eleventh grade in suburban Pittsburgh. The correlation matrix for the study is given in Table 6.3.1 and an explanation of the variables is summarized in Example 6.3.1. Inspection of Table 6.3.1 indicates that four out of the six creativity measures (tests 4, 5, 6, and 7) are highly correlated with synthesis and evaluation (tests 12 and 13), as is IQ (test 1). What else do you notice?
   a. Use canonical correlation analysis to investigate the relationship between the six achievement variables and IQ and the six creativity variables.
   b. Use canonical correlation analysis to investigate the relationship between the six achievement variables and IQ and the six creativity variables.
   c. By partialing out the IQ variable from both sets of variables, use partial canonical correlation analysis to analyze the data and compare your results with parts a and b.
   d. Would a part canonical correlation analysis of the data be meaningful? If so, why? If not, why not?
   e. Summarize your findings.
2. Using a random sample of 502 twelfth-grade students from the Project Talent survey (supplied by William W. Cooley at the University of Pittsburgh), data were collected on 11 tests: (1) general information test, part I, (2) general information test, part II, (3) English, (4) reading comprehension, (5) creativity, (6) mechanical reasoning, (7) abstract reasoning, (8) mathematics, (9) sociability inventory, (10) physical science interest inventory, and (11) office work interest inventory. Tests (1) through (5) were verbal ability tests, tests (6) through (8) were nonverbal ability tests, and tests (9) through (11) were interest measures (for a description of the variables, see Cooley and Lohnes, 1971). The correlation matrix for the 11 variables and 502 subjects follows. Use various canonical correlation analysis procedures to analyze the correlation matrix. Summarize your findings.

	1	2	3	4	5	6	7	8	9	10	11
1	1.000										
2	.861	1.000									
3	.492	.550	1.000								
4	.698	.765	.613	1.000							
5	.644	.621	.418	.595	1.000						
6	.661	.519	.160	.413	.522	1.000					
7	.487	.469	.456	.530	.433	.451	1.000				
8	.761	.649	.566	.641	.556	.547	.517	1.000			
9	−.011	.062	.083	.021	.001	−.075	.007	.030	1.000		
10	.573	.397	.094	.275	.340	.531	.202	.500	.055	1.000	
11	−.349	−.234	.109	−.087	−.119	−.364	−.079	−.191	.084	−.246	1.000

CHAPTER 5

MULTIVARIATE ANALYSIS OF VARIANCE AND COVARIANCE ANALYSIS

5.1 INTRODUCTION

Univariate analysis of variance and covariance analysis are commonly used statistical techniques employed in the behavioral sciences. Multivariate generalizations of some of the most often used procedures are discussed and illustrated in this chapter. These techniques are again special cases of the general linear model. The dependence of the methods on the theory developed in Sections 3.5 and 3.6 will be evident. Of particular importance are the fundamental univariate and multivariate least-squares theorems.

5.2 ONE-WAY ANALYSIS OF VARIANCE—UNIVARIATE

The one-way layout refers to the comparison of means of several univariate treatment populations. Denote the means by $\mu_1, \mu_2, \ldots, \mu_k$. Assigning subjects to the k treatment groups at random, assume that the k populations are normal with equal variance σ^2, and that independent random samples of sizes $N_1, N_2, \ldots, N_k$ have been obtained. Letting y_{ij} denote the jth subject in the ith treatment population, assume that the observed sample observations have the form

$$y_{ij} = \mu_i + \varepsilon_{ij} \qquad i = 1, \ldots, k;\ j = 1, \ldots, N_i \tag{5.2.1}$$

where

$$\varepsilon_{ij} \sim IN(0, \sigma^2)$$

and the ε_{ij}'s are experimental errors that include measurement errors and other uncontrolled variations affecting the observations.

Many writers refer to the one-way layout design as a completely randomized design. Kirk (1968, Chap. 4) uses the abbreviation CR-k for such designs since subjects are assigned completely at random to the k treatment populations. For this procedure to be effective, it is desirable to have a relatively homogeneous group of subjects. The observations are conveniently represented in the following tabular form.

	Treatments			
	1	2	$\cdots$	k
	y_{11}	y_{21}	$\cdots$	y_{k1}
	y_{12}	y_{22}	$\cdots$	y_{k2}
	$\vdots$	$\vdots$	$\vdots$	$\vdots$
	y_{1N_1}	y_{2N_2}	$\cdots$	y_{kN_k}
Means	$y_{1.}$	$y_{2.}$	$\cdots$	$y_{k.}$

By employing the general linear model $\mathbf{y} = \mathbf{X}\boldsymbol{\beta} + \boldsymbol{\varepsilon}$, (5.2.1) becomes

$$\underset{(N \times 1)}{\begin{bmatrix} y_{11} \\ y_{12} \\ \vdots \\ y_{1N_1} \\ \hline \vdots \\ \hline y_{k1} \\ y_{k2} \\ \vdots \\ y_{kN_k} \end{bmatrix}} = \underset{(N \times k)}{\begin{bmatrix} 1 & 0 & \cdots & 0 \\ 1 & 0 & \cdots & 0 \\ \vdots & \vdots & \vdots & \vdots \\ 1 & 0 & \cdots & 0 \\ \hline \vdots & \vdots & \vdots & \vdots \\ \hline 0 & 0 & \cdots & 1 \\ 0 & 0 & \cdots & 1 \\ \vdots & \vdots & \vdots & \vdots \\ 0 & 0 & \cdots & 1 \end{bmatrix}} \underset{(k \times 1)}{\begin{bmatrix} \mu_1 \\ \mu_2 \\ \vdots \\ \mu_k \end{bmatrix}} + \underset{(N \times 1)}{\begin{bmatrix} \varepsilon_{11} \\ \varepsilon_{12} \\ \vdots \\ \varepsilon_{1N_1} \\ \hline \vdots \\ \hline \varepsilon_{k1} \\ \varepsilon_{k2} \\ \vdots \\ \varepsilon_{kN_k} \end{bmatrix}} \tag{5.2.2}$$

where $N = \Sigma_i N_i$ and

$$\begin{aligned} E(\mathbf{Y}) &= \mathbf{X}\boldsymbol{\beta} \\ V(\mathbf{Y}) &= \sigma^2 \mathbf{I} \end{aligned} \tag{5.2.3}$$

Although model (5.2.1) is the most natural representation for the one-way layout, the more familiar representation is given by

$$\text{(5.2.4)} \qquad \begin{aligned} y_{ij} &= \mu + \alpha_i + \varepsilon_{ij} \qquad i = 1, \ldots, k;\ j = 1, \ldots, N_i \\ \varepsilon_{ij} &\sim IN(0, \sigma^2) \end{aligned}$$

where μ denotes an overall constant and α_i the ith treatment effect. The design matrix $\mathbf{X}$ and parameter vector $\boldsymbol{\beta}$ for representation (5.2.4) become

$$\text{(5.2.5)} \qquad \mathbf{X} = \underbrace{\begin{bmatrix} 1 & 1 & \cdots & 0 \\ 1 & 1 & \cdots & 0 \\ \vdots & \vdots & \vdots & \vdots \\ 1 & 1 & \cdots & 0 \\ \hline 1 & 0 & \cdots & 0 \\ 1 & 0 & \cdots & 0 \\ \vdots & \vdots & \vdots & \vdots \\ 1 & 0 & \cdots & 0 \\ \hline \vdots & \vdots & \vdots & \vdots \\ \hline 1 & 0 & \cdots & 1 \\ 1 & 0 & \cdots & 1 \\ \vdots & \vdots & \vdots & \vdots \\ 1 & 0 & \cdots & 1 \end{bmatrix}}_{(N \times q)} \qquad \boldsymbol{\beta} = \underbrace{\begin{bmatrix} \mu \\ \alpha_1 \\ \alpha_2 \\ \vdots \\ \alpha_k \end{bmatrix}}_{(q \times 1)}$$

where $q = k + 1$. In (5.2.2) the design matrix $\mathbf{X}$ is of full rank; however, in (5.2.5), $\mathbf{X}$ is less than full rank so that all parameters in $\boldsymbol{\beta}$ are not individually estimable since $(\mathbf{X}'\mathbf{X})^{-1}$ does not exist. Using model (5.2.4), the general theory of Chapter 3 is applied.

By employing the univariate Gauss-Markoff theorem, with

$$\text{(5.2.6)} \qquad (\mathbf{X}'\mathbf{X})^{-} = \begin{bmatrix} 0 & 0 & \cdots & 0 \\ 0 & 1/N_1 & \cdots & 0 \\ \vdots & \vdots & \vdots & \vdots \\ 0 & 0 & \cdots & 1/N_k \end{bmatrix}$$

the estimate of $\boldsymbol{\beta}$ is

$$\text{(5.2.7)} \qquad \hat{\boldsymbol{\beta}} = (\mathbf{X}'\mathbf{X})^{-}\mathbf{X}'\mathbf{y} = \begin{bmatrix} 0 \\ y_{1.} \\ y_{2.} \\ \vdots \\ y_{k.} \end{bmatrix}$$

where

$$y_{i.} = \sum_{j=1}^{N_i} y_{ij}/N_i$$

is the sample mean of the observation in the ith treatment population.

To locate unique linear unbiased estimates for linear combinations of the elements of $\boldsymbol{\beta}$ in (5.2.5), theorem (3.5.8) is employed. Here

$$\mathbf{H} = (\mathbf{X}'\mathbf{X})^{-}\mathbf{X}'\mathbf{X} = \begin{bmatrix} 0 & 0 & 0 & \cdots & 0 \\ 1 & 1 & 0 & \cdots & 0 \\ 1 & 0 & 1 & \cdots & 0 \\ \vdots & \vdots & \vdots & \cdots & \vdots \\ 1 & 0 & 0 & \cdots & 1 \end{bmatrix} \tag{5.2.8}$$

and $\psi = \mathbf{c}'\boldsymbol{\beta}$ has a unique estimate if and only if $\mathbf{c}'\mathbf{H} = \mathbf{c}'$. For arbitrary vectors $\mathbf{t}' = [t_0, t_1, \ldots, t_k]$,

$$\begin{aligned} \psi &= \mathbf{c}'\boldsymbol{\beta} = \mathbf{t}'\mathbf{H}\boldsymbol{\beta} = \sum_{i=1}^{k} t_i\mu + \sum_{i=1}^{k} t_i\alpha_i \\ \hat{\psi} &= \mathbf{c}'\hat{\boldsymbol{\beta}} = \sum_{i=1}^{k} t_i y_{i.} \end{aligned} \tag{5.2.9}$$

when ψ is estimable or $\mathbf{c}'\mathbf{H} = \mathbf{c}'$. As in the two-sample case, $\psi = \mu$ and $\psi = \alpha_i$ are not estimable. To estimate μ or α_i, side conditions of the form $\Sigma_i\, t_i\alpha_i = 0$ are added to the model specification given in (5.2.4).

To test the hypothesis of equal treatment effects

$$H_0\colon \alpha_1 = \alpha_2 = \cdots = \alpha_k \tag{5.2.10}$$

an estimate of the common variance σ^2 is required; an unbiased estimator of σ^2 is

$$\begin{aligned} s^2 &= \frac{\mathbf{y}'[\mathbf{I} - \mathbf{X}(\mathbf{X}'\mathbf{X})^{-}\mathbf{X}']\mathbf{y}}{N - k} \\ &= \frac{\mathbf{y}'\mathbf{y} - \mathbf{y}'[\mathbf{X}(\mathbf{X}'\mathbf{X})^{-}\mathbf{X}']\mathbf{y}}{N - k} \end{aligned} \tag{5.2.11}$$

However,

$$\mathbf{X}(\mathbf{X}'\mathbf{X})^{-}\mathbf{X}' = \begin{bmatrix} N_1^{-1}\mathbf{J}_{11} & 0 & \cdots & 0 \\ 0 & N_2^{-1}\mathbf{J}_{22} & \cdots & 0 \\ \vdots & \vdots & \cdots & \vdots \\ 0 & 0 & \cdots & N_k^{-1}\mathbf{J}_{kk} \end{bmatrix} \tag{5.2.12}$$

where $\mathbf{J}_{ij}$ is an $N_i \times N_j$ matrix of unities. Thus (5.2.11) reduces to

$$s^2 = \frac{\sum_i \sum_j y_{ij}^2 - \sum_i N_i y_{i.}^2}{N - k}$$

$$= \frac{\sum_{i=1}^{k} \sum_{j=1}^{N_i} (y_{ij} - y_{i.})^2}{N - k} \tag{5.2.13}$$

where the error sum of squares Q_e is

$$Q_e = \sum_{i=1}^{k} \sum_{j=1}^{N_i} (y_{ij} - y_{i.})^2 \tag{5.2.14}$$

A matrix **C** to test (5.2.10) is of the form

$$\underset{[(k-1)\times q]}{\mathbf{C}} = \begin{bmatrix} 0 & 1 & 0 & \cdots & 0 & -1 \\ 0 & 0 & 1 & \cdots & 0 & -1 \\ \vdots & \vdots & \vdots & \vdots & \vdots & \vdots \\ 0 & 0 & 0 & \cdots & 1 & -1 \end{bmatrix} \tag{5.2.15}$$

where the $R(\mathbf{C}) = k - 1$. By using (5.2.15), the null hypothesis specified in (5.2.10) becomes

$$H_0: \mathbf{C\beta} = \mathbf{0} \quad \text{or} \quad \begin{bmatrix} \alpha_1 - \alpha_k \\ \alpha_2 - \alpha_k \\ \vdots \\ \alpha_{k-1} - \alpha_k \end{bmatrix} = \begin{bmatrix} 0 \\ 0 \\ \vdots \\ 0 \end{bmatrix} \tag{5.2.16}$$

The hypothesis in (5.2.16) can be tested since for each row vector $\mathbf{c}'$ of **C**, $\mathbf{c}'\mathbf{H} = \mathbf{c}'$. The matrix **C** used to test (5.2.10) is not unique; another equivalent choice for **C** is

$$\underset{[(k-1)\times q]}{\mathbf{C}} = \begin{bmatrix} 0 & 1 & -1 & 0 & \cdots & 0 & 0 \\ 0 & 0 & 1 & -1 & \cdots & 0 & 0 \\ \vdots & \vdots & \vdots & \vdots & \vdots & \vdots & \vdots \\ 0 & 0 & 0 & 0 & \cdots & 1 & -1 \end{bmatrix}$$

Applying the univariate least-squares theory, we test (5.2.10) by using the hypothesis sum of squares defined by

$$Q_h = (\mathbf{C\hat{\beta}})'[\mathbf{C}(\mathbf{X}'\mathbf{X})^{-}\mathbf{C}']^{-1}(\mathbf{C\hat{\beta}}) \tag{5.2.17}$$

To simplify (5.2.17), by using **C** as defined in (5.2.15), we observe that the matrix

$$\mathbf{C}(\mathbf{X}'\mathbf{X})^{-}\mathbf{C}' = \text{Dia}\,[N_i^{-1}] + N_k^{-1}\mathbf{J}_{k-1}{}'_{k-1} \tag{5.2.18}$$

of order $(k - 1) \times (k - 1)$ is of the form $\mathbf{A} + \mathbf{uw}'$, where **u** and **w** are column vectors. Applying Householder's (1964, p. 3) definition of an elementary matrix, it is seen that the inverse of $(\mathbf{A} + \mathbf{uw}')$ is

$$(\mathbf{A} + \mathbf{uw}')^{-1} = \mathbf{A}^{-1} - \frac{(\mathbf{A}^{-1}\mathbf{u})(\mathbf{w}'\mathbf{A}^{-1})}{1 + \mathbf{w}'\mathbf{A}^{-1}\mathbf{u}} \tag{5.2.19}$$

which gives a method for inverting the matrix in (5.2.18). Thus

$$(5.2.20)\qquad [\mathbf{C}(\mathbf{X}'\mathbf{X})^{-}\mathbf{C}']^{-1} = [c_{ij}] \qquad \text{where } c_{ij} = \begin{cases} N^{-1}N_i(N - N_i) & i = j \\ -N^{-1}N_iN_j & i \neq j \end{cases}$$

By using (5.2.20) and the fact that symmetric matrix

$$N(\mathbf{C}(\mathbf{X}'\mathbf{X})^{-}\mathbf{X}']'[\mathbf{C}(\mathbf{X}'\mathbf{X})^{-}\mathbf{C}']^{-1}[\mathbf{C}(\mathbf{X}'\mathbf{X})^{-}\mathbf{X}'] = N\mathbf{Q}$$

$$= \begin{bmatrix} (N - N_1)\mathbf{J}_{11}/N_1 & & & \\ -\mathbf{J}_{21} & (N - N_2)\mathbf{J}_{22}/N_2 & & \\ \vdots & \vdots & \ddots & \\ -\mathbf{J}_{k1} & -\mathbf{J}_{k2} & \cdots & (N - N_k)\mathbf{J}_{kk}/N_k \end{bmatrix}$$

Q_h is seen to reduce to

$$(5.2.21)\qquad Q_h = \mathbf{y}'\mathbf{Q}\mathbf{y} = \sum_{i=1}^{k} N_i(y_{i.} - y_{..})^2$$

where

$$y_{..} = \frac{\sum_{i=1}^{k}\sum_{j=1}^{N_i} y_{ij}}{N}$$

Table 5.2.1, the ANOVA table for testing (5.2.10), can be constructed from (5.2.21) and (5.2.14), which were derived by applying the general linear model least-squares theory. To determine the expectation of the sum of squares in Table 5.2.1,

TABLE 5.2.1. ANOVA Table for a CR-k Design.

Source	df	SS
Treatment	$k - 1$	$Q_h = (\mathbf{C}\hat{\boldsymbol{\beta}})'[\mathbf{C}(\mathbf{X}'\mathbf{X})^{-}\mathbf{C}']^{-1}(\mathbf{C}\hat{\boldsymbol{\beta}}) = \sum_{i=1}^{k} N_i(y_{i.} - y_{..})^2$
Within error	$N - k$	$Q_e = \mathbf{y}'[\mathbf{I} - \mathbf{X}(\mathbf{X}'\mathbf{X})^{-}\mathbf{X}']\mathbf{y} = \sum_{i=1}^{k}\sum_{j=1}^{N_i} (y_{ij} - y_{i.})^2$
"Total"	$N - 1$	$Q_t = \sum_{i=1}^{k}\sum_{j=1}^{N_i} (y_{ij} - y_{..})^2$

recall from (2.6.8) that if $\mathbf{Y} \sim N_p(\boldsymbol{\mu}, \boldsymbol{\Sigma})$, then the $E(\mathbf{Y}'\mathbf{A}\mathbf{Y}) = \text{Tr}(\mathbf{A}\boldsymbol{\Sigma}) + \boldsymbol{\mu}'\mathbf{A}\boldsymbol{\mu}$. For $Q_e = \mathbf{Y}'[\mathbf{I} - \mathbf{X}(\mathbf{X}'\mathbf{X})^{-}\mathbf{X}']\mathbf{Y}$, with the observed observation vector $\mathbf{y}$ replaced by the random vector $\mathbf{Y}$, the

$$\begin{aligned} E(Q_e) &= \text{Tr}\{[\mathbf{I} - \mathbf{X}(\mathbf{X}'\mathbf{X})^{-}\mathbf{X}']\sigma^2\mathbf{I}\} + (\mathbf{X}\boldsymbol{\beta})'[\mathbf{I} - \mathbf{X}(\mathbf{X}'\mathbf{X})^{-}\mathbf{X}']\mathbf{X}\boldsymbol{\beta} \\ &= (N - k)\sigma^2 + \boldsymbol{\beta}'\mathbf{X}'\mathbf{X}\boldsymbol{\beta} - \boldsymbol{\beta}'\mathbf{X}'\mathbf{X}(\mathbf{X}'\mathbf{X})^{-}\mathbf{X}'\mathbf{X}\boldsymbol{\beta} \\ &= (N - k)\sigma^2 + \boldsymbol{\beta}'\mathbf{X}'\mathbf{X}\boldsymbol{\beta} - \boldsymbol{\beta}'\mathbf{X}'\mathbf{X}\boldsymbol{\beta} \\ &= (N - k)\sigma^2 \end{aligned}$$

where $k = R(\mathbf{X})$ for the one-way design.

Using (2.6.8) in (3.5.24), it was shown that

$$E(Q_h) = g\sigma^2 + \delta\sigma^2$$

in general, where $g = R(\mathbf{C})$ and the noncentrality parameter

$$\delta = \frac{(\mathbf{C}\boldsymbol{\beta} - \boldsymbol{\xi})'[\mathbf{C}(\mathbf{X}'\mathbf{X})^{-}\mathbf{C}']^{-1}(\mathbf{C}\boldsymbol{\beta} - \boldsymbol{\xi})}{\sigma^2}$$

However, under the hypothesis for the one-way layout, $\boldsymbol{\xi} = \mathbf{0}$, so that

$$\sigma^2\delta = (\mathbf{C}\boldsymbol{\beta})'[\mathbf{C}(\mathbf{X}'\mathbf{X})^{-}\mathbf{C}']^{-1}(\mathbf{C}\boldsymbol{\beta})$$

For any $\mathbf{C}$ and $(\mathbf{X}'\mathbf{X})^{-}$ constructed to test (5.2.10), where $\boldsymbol{\beta}$ is defined as in (5.2.5), it can be shown that

$$\sigma^2\delta = \sum_{i=1}^{k} N_i(\alpha_i - \alpha_.)^2 \tag{5.2.22}$$

for

$$\alpha_. = \frac{\sum_{i=1}^{k} N_i\alpha_i}{N}$$

That is, each observation in (5.2.21) is replaced by its expected value and equated to $\sigma^2\delta$. The expectation of Q_h in Table 5.2.1 becomes

$$E(Q_h) = (k - 1)\sigma^2 + \sum_{i=1}^{k} N_i(\alpha_i - \alpha_.)^2$$

which reduces to the usual expression found in most beginning statistical texts if we add the side condition that $\alpha_. = 0$ to the model. To find the expected mean squares, merely divide by the corresponding degrees of freedom. For Table 5.2.1, the

$$E(MS_h) = \sigma^2 + \frac{\sum_{i=1}^{k} N_i(\alpha_i - \alpha_.)^2}{k - 1} \tag{5.2.23}$$

and

$$E(MS_e) = \sigma^2$$

The hypothesis of equal treatment effects is rejected at the significance level α if

$$F = \frac{Q_h/(k - 1)}{Q_e/(N - k)} > F^{\alpha}(k - 1, N - k)$$

The preceding discussion illustrates how the theory of the general linear model of Chapter 3 is employed to test the equality of treatment effects. Graybill (1961) and other authors prefer the *reduction in sum of squares method* to analyze experimental designs. This procedure follows the discussion associated with testing the significance of a subset of beta weights $\boldsymbol{\beta}_2$ and of a complete set of weights $\boldsymbol{\beta}$ in regression analysis (see Chapter 4). However, it was not specifically termed the reduction method because I have chosen to stress the determination of the appropriate matrix

C for testing any testable hypothesis employing linear model theory. To facilitate your ability to transfer the theory and procedures learned in this text when reading other books, the reduction method terminology is reviewed for the univariate regression model and applied to the one-way layout.

Testing the hypothesis that a subset of beta weights is equal to 0 in a regression analysis, the vector $\boldsymbol{\beta}$ is partitioned as $\boldsymbol{\beta}' = [\boldsymbol{\beta}_1' \quad \boldsymbol{\beta}_2']$ and the hypothesis $\boldsymbol{\beta}_2 = \mathbf{0}$ is tested by selecting the appropriate matrix **C** and using equation (4.2.43). Alternatively, it was shown in (4.2.45) that the same hypothesis sum of squares $(Q_{h_k} - Q_{h_m})$ led to the correct sum of squares for testing $\boldsymbol{\beta}_2 = \mathbf{0}$. Transforming the regression sum of squares Q_{h_k} and Q_{h_m} (defined in Section 4.2) into the reduction method terminology, the regression sum of squares Q_{h_m} is the reduction in the total sum of squares for fitting the reduced model $\mathbf{y} = \mathbf{X}_1\boldsymbol{\beta}_1 + \boldsymbol{\varepsilon}$. That is, since $Q_t = Q_e + Q_h$, the quantity Q_h reduces the total sum of squares by an amount Q_h. We represent Q_{h_m} by $R(\boldsymbol{\beta}_1)$ since it is obtained by assuming a reduced model that includes only the parameter vector $\boldsymbol{\beta}_1$. Similarly, let $R(\boldsymbol{\beta}_1, \boldsymbol{\beta}_2) = Q_{h_k}$ denote the reduction in the total sum of squares for fitting the full model $\mathbf{y} = \mathbf{X}_1\boldsymbol{\beta}_1 + \mathbf{X}_2\boldsymbol{\beta}_2 + \boldsymbol{\varepsilon}$ or $\mathbf{y} = \mathbf{X}\boldsymbol{\beta} + \boldsymbol{\varepsilon}$. Then $R(\boldsymbol{\beta}_1, \boldsymbol{\beta}_2) - R(\boldsymbol{\beta}_1) = Q_{h_k} - Q_{h_m}$ represents the reduction in the sum of squares resulting from fitting $\boldsymbol{\beta}_2$, already fitting $\boldsymbol{\beta}_1$. This is the hypothesis sum of squares and is often described as the reduction of fitting $\boldsymbol{\beta}_2$, adjusting for $\boldsymbol{\beta}_1$. Therefore it is convenient to write $R(\boldsymbol{\beta}_1, \boldsymbol{\beta}_2) - R(\boldsymbol{\beta}_1)$ as

$$R(\boldsymbol{\beta}_2|\boldsymbol{\beta}_1) = R(\boldsymbol{\beta}_1, \boldsymbol{\beta}_2) - R(\boldsymbol{\beta}_1) \tag{5.2.24}$$

The reduction $R(\boldsymbol{\beta}_1)$ is also called the reduction due to fitting $\boldsymbol{\beta}_1$, ignoring $\boldsymbol{\beta}_2$. For a nonorthogonal design matrix, it is clear that $R(\boldsymbol{\beta}_2|\boldsymbol{\beta}_1) \neq R(\boldsymbol{\beta}_2)$, by (4.3.43). However, if $\mathbf{X}_1'\mathbf{X}_2 = \mathbf{0}$, then $R(\boldsymbol{\beta}_2) = R(\boldsymbol{\beta}_2|\boldsymbol{\beta}_1)$, and $\boldsymbol{\beta}_1$ is said to be orthogonal to $\boldsymbol{\beta}_2$.

Extending the reduction notation to a partitioned vector $\boldsymbol{\beta}' = [\boldsymbol{\beta}_1', \boldsymbol{\beta}_2', \boldsymbol{\beta}_3']$, it is further witnessed that the order chosen for fitting affects the sum of squares. That is, $R(\boldsymbol{\beta}_2|\boldsymbol{\beta}_1) = R(\boldsymbol{\beta}_1, \boldsymbol{\beta}_2) - R(\boldsymbol{\beta}_1)$ is not equal to $R(\boldsymbol{\beta}_2|\boldsymbol{\beta}_1, \boldsymbol{\beta}_3) = R(\boldsymbol{\beta}_1, \boldsymbol{\beta}_2, \boldsymbol{\beta}_3) - R(\boldsymbol{\beta}_1, \boldsymbol{\beta}_3)$ unless the design is orthogonal.

Translating the regression notation to the one-way ANOVA design, let

$$\boldsymbol{\beta} = \begin{bmatrix} \mu \\ \hline \alpha_1 \\ \alpha_2 \\ \vdots \\ \alpha_k \end{bmatrix} = \begin{bmatrix} \boldsymbol{\beta}_1 \\ \boldsymbol{\beta}_2 \end{bmatrix} \tag{5.2.25}$$

The linear model $\mathbf{y} = \mathbf{X}\boldsymbol{\beta} + \boldsymbol{\varepsilon}$, where $y_{ij} = \mu + \alpha_i + \varepsilon_{ij}$, is represented as

$$\mathbf{y} = \mathbf{X}_1\boldsymbol{\beta}_1 + \mathbf{X}_2\boldsymbol{\beta}_2 + \boldsymbol{\varepsilon} \tag{5.2.26}$$

where $\mathbf{X}_1$ is a vector of N 1s and $\mathbf{X}_2$ is the design matrix **X** given in (5.2.2).

To test the equality of the α_i's by using the reduction method, set each α_i in the model equal to α_0 (say) so that $y_{ij} = \mu + \alpha_0 + \varepsilon_{ij} = \mu_0 + \varepsilon_{ij}$ is the reduced model. The sum of squares for fitting the model $y_{ij} = \mu_0 + \varepsilon_{ij}$ is clearly the same as for fitting the model $y_{ij} = \mu + \varepsilon_{ij}$, which is obtained from the original model by setting all the α_i's = 0. Thus, for the reduced model, testing all α_i's equal to zero or all α_i's equal is indistinguishable. However, because each α_i is not estimable, the hypothesis all $\alpha_i = 0$ cannot be formally tested. Correctly stated, the reduction method tests that all the treatment effects are equal.

Proceeding, we set $\boldsymbol{\beta}_2 = \mathbf{0}$ to obtain the reduced model

$$\mathbf{y} = \mathbf{X}_1\boldsymbol{\beta}_1 + \boldsymbol{\varepsilon} \tag{5.2.27}$$

The reduction in the total sum of squares for fitting (5.2.27) is

$$R(\boldsymbol{\beta}_1) = \mathbf{y}'\mathbf{X}_1(\mathbf{X}_1'\mathbf{X}_1)^{-}\mathbf{X}_1'\mathbf{y} = \hat{\boldsymbol{\beta}}_1'\mathbf{X}_1'\mathbf{X}_1\hat{\boldsymbol{\beta}}_1 \tag{5.2.28}$$

For fitting the full model (5.2.26),

$$R(\boldsymbol{\beta}_1, \boldsymbol{\beta}_2) = \mathbf{y}'\mathbf{X}(\mathbf{X}'\mathbf{X})^{-}\mathbf{X}'\mathbf{y} = \hat{\boldsymbol{\beta}}'\mathbf{X}'\mathbf{X}\hat{\boldsymbol{\beta}} \tag{5.2.29}$$

The hypothesis sum of squares is, using (5.2.12),

$$\begin{aligned} R(\boldsymbol{\beta}_2|\boldsymbol{\beta}_1) &= R(\boldsymbol{\beta}_1, \boldsymbol{\beta}_2) - R(\boldsymbol{\beta}_1) \\ &= \mathbf{y}'\mathbf{X}(\mathbf{X}'\mathbf{X})^{-}\mathbf{X}'\mathbf{y} - \mathbf{y}'\mathbf{X}_1(\mathbf{X}_1'\mathbf{X}_1)^{-}\mathbf{X}_1'\mathbf{y} \\ &= \sum_{i=1}^{k} N_i y_{i.}^2 - N y_{..}^2 \\ &= \sum_{i=1}^{k} N_i(y_{i.} - y_{..})^2 \end{aligned} \tag{5.2.30}$$

where $y_{..} = \Sigma_i N_i y_{i.}/N$, which is in agreement with (5.2.21).

The verification of (5.2.23), using the reduction method, is immediate since the quadratic forms in the expression for the $R(\boldsymbol{\beta}_2|\boldsymbol{\beta}_1)$ are in the appropriate form to apply (2.6.8) directly. Using the full model $\mathbf{y} = \mathbf{X}\boldsymbol{\beta} + \boldsymbol{\varepsilon}$, which is written as $\mathbf{y} = \mathbf{X}_1\boldsymbol{\beta}_1 + \mathbf{X}_2\boldsymbol{\beta}_2 + \boldsymbol{\varepsilon}$, where the $E(\mathbf{Y}) = \mathbf{X}\boldsymbol{\beta}$ and the $V(\mathbf{Y}) = \sigma^2\mathbf{I}$, the

$$\begin{aligned} E[R(\boldsymbol{\beta}_1, \boldsymbol{\beta}_2)] &= \mathrm{Tr}[\mathbf{X}(\mathbf{X}'\mathbf{X})^{-}\mathbf{X}'\sigma^2\mathbf{I}] + \boldsymbol{\beta}'\mathbf{X}'\mathbf{X}(\mathbf{X}'\mathbf{X})^{-}\mathbf{X}'\mathbf{X}\boldsymbol{\beta} \\ &= R(\mathbf{X})\sigma^2 + \boldsymbol{\beta}'\mathbf{X}'\mathbf{X}\boldsymbol{\beta} \\ &= R(\mathbf{X})\sigma^2 + [\boldsymbol{\beta}_1'\boldsymbol{\beta}_2']\begin{bmatrix} \mathbf{X}_1'\mathbf{X}_1 & \mathbf{X}_1'\mathbf{X}_2 \\ \mathbf{X}_2'\mathbf{X}_1 & \mathbf{X}_2'\mathbf{X}_2 \end{bmatrix}\begin{bmatrix} \boldsymbol{\beta}_1 \\ \boldsymbol{\beta}_2 \end{bmatrix} \end{aligned} \tag{5.2.31}$$

and the

$$\begin{aligned} E[R(\boldsymbol{\beta}_1)] &= \mathrm{Tr}[\mathbf{X}_1(\mathbf{X}_1'\mathbf{X}_1)^{-}\mathbf{X}_1'\sigma^2\mathbf{I}] + \boldsymbol{\beta}'\mathbf{X}_1'\mathbf{X}_1(\mathbf{X}_1'\mathbf{X}_1)^{-}\mathbf{X}_1'\mathbf{X}_1\boldsymbol{\beta} \\ &= R(\mathbf{X}_1)\sigma^2 + \boldsymbol{\beta}'\left\{\begin{bmatrix} \mathbf{X}_1'\mathbf{X}_1 \\ \mathbf{X}_2'\mathbf{X}_2 \end{bmatrix}(\mathbf{X}_1'\mathbf{X}_1)^{-}[\mathbf{X}_1'\mathbf{X}_1 \quad \mathbf{X}_1'\mathbf{X}_2]\right\}\boldsymbol{\beta} \\ &= R(\mathbf{X}_1)\sigma^2 + [\boldsymbol{\beta}_1'\boldsymbol{\beta}_2']\begin{bmatrix} \mathbf{X}_1'\mathbf{X}_1 & \mathbf{X}_1'\mathbf{X}_2 \\ \mathbf{X}_2'\mathbf{X}_1 & \mathbf{X}_2'\mathbf{X}_1(\mathbf{X}_1'\mathbf{X}_1)^{-}\mathbf{X}_1'\mathbf{X}_2 \end{bmatrix}\begin{bmatrix} \boldsymbol{\beta}_1 \\ \boldsymbol{\beta}_2 \end{bmatrix} \end{aligned} \tag{5.2.32}$$

Hence the expectation of the sum of squares for the hypothesis becomes, by the reduction method,

$$\begin{aligned} E[R(\boldsymbol{\beta}_2|\boldsymbol{\beta}_1)] &= E[R(\boldsymbol{\beta}_1, \boldsymbol{\beta}_2)] - E[R(\boldsymbol{\beta}_1)] \\ &= [R(\mathbf{X}) - R(\mathbf{X}_1)]\sigma^2 + \boldsymbol{\beta}_2'[\mathbf{X}_2'\mathbf{X}_2 - \mathbf{X}_2'\mathbf{X}_1(\mathbf{X}_1'\mathbf{X}_1)^{-}\mathbf{X}_1'\mathbf{X}_2]\boldsymbol{\beta}_2 \end{aligned} \tag{5.2.33}$$

The general formula derived in (5.2.33) is only applicable if the elements in the vectors $\boldsymbol{\beta}_1$ and $\boldsymbol{\beta}_2$ represent all the elements in the parameter vector $\boldsymbol{\beta}$. If $\boldsymbol{\beta}$ was partitioned as $\boldsymbol{\beta}' = [\boldsymbol{\beta}_1' \quad \boldsymbol{\beta}_2' \quad \boldsymbol{\beta}_3']$, it could not be employed to find the $E[R(\boldsymbol{\beta}_1|\boldsymbol{\beta}_2)]$ since the elements in $\boldsymbol{\beta}_1$ and $\boldsymbol{\beta}_2$ do not represent all the elements in $\boldsymbol{\beta}$; however, it could be used to find the $E[R(\boldsymbol{\beta}_3|\boldsymbol{\beta}_2, \boldsymbol{\beta}_1)]$. Since the partition is such for the one-way design that all the elements in $\boldsymbol{\beta}$ are specified by the reduction procedure, equation (5.2.33) is applied to find the expected mean squares for the hypothesis. In this case, the

$$
\begin{aligned}
E[R(\boldsymbol{\beta}_2|\boldsymbol{\beta}_1)] &= (k-1)\sigma^2 + \sum_{i=1}^{k} N_i\alpha_i^2 - N\alpha_{.}^2 \\
&= (k-1)\sigma^2 + \sum_{i=1}^{k} N_i(\alpha_i - \alpha_{.})^2
\end{aligned}
$$

which, on dividing by $k - 1$, is in agreement with (5.2.23).

The advantage of the general linear model theory over the reduction method is that the experimenter knows exactly what is being tested. This is not as obvious when using the reduction method. However, a real advantage of the reduction method over the general linear model approach is the ease in which we may compute some expected sums of squares; this is especially true for mixed models and unbalanced data, as illustrated by Searle (1968).

On rejection of (5.2.10), a general multiple comparison procedure (see Scheffé, 1959, p. 68) is available to determine $100(1-\alpha)\%$ simultaneous confidence intervals for all estimable functions in the row space of the hypothesis test matrix $\mathbf{C}$. To establish these bounds, (4.2.61) is applied—that is, with confidence $1 - \alpha$,

$$
\mathbf{c}'\hat{\boldsymbol{\beta}} - c_0(Q_e/v_e)(\mathbf{c}'(X'X)^{-}\mathbf{c}) \le \mathbf{c}'\boldsymbol{\beta} \le \mathbf{c}'\hat{\boldsymbol{\beta}} + c_0(Q_e/v_e)(\mathbf{c}'(X'X)^{-}\mathbf{c})
$$

$$
\hat{\psi} - c_0\hat{\sigma}_{\hat{\psi}} \le \psi \le \hat{\psi} + c_0\hat{\sigma}_{\hat{\psi}} \tag{5.2.34}
$$

for all $\mathbf{c}'$ in the row space of $\mathbf{C}$ and $c_0^2 = v_h F^{\alpha}(v_h, v_e)$. For the one-way layout, (5.2.34) reduces to

$$
\sum_i c_i y_{i.} - c_0 s\left(\frac{\sum_i c_i^2}{N_i}\right)^{1/2} \le \sum_i c_i\alpha_i \le \sum_i c_i y_{i.} + c_0 s\left(\frac{\sum_i c_i^2}{N_i}\right)^{1/2} \tag{5.2.35}
$$

where $s = (Q_e/v_e)^{1/2}$ and the $\Sigma_i c_i = 0$; the estimable function $\psi = \Sigma_i c_i\alpha_i$ is called a *contrast in the effects.*

Alternatively, when interest is focused on q estimable functions, Bonferroni t statistics are employed to test the overall hypothesis. To construct confidence intervals for q comparisons, the expression

$$
\hat{\psi}_i - t^{\alpha/2q}(N-k)\hat{\sigma}_{\hat{\psi}_i} \le \psi_i \le \hat{\psi}_i + t^{\alpha/2q}(N-k)\hat{\sigma}_{\hat{\psi}i} \tag{5.2.36}
$$

is evaluated. Other commonly employed techniques are attributed to Tukey, Duncan, Newman-Keuls, Fisher, and others. These techniques are summarized in Miller (1966) and Kirk (1968). For a comparison of type I error rates for these procedures, see Boardman (1971).

EXERCISES 5.2

1. Show that the expression for s^2 in (5.2.11) reduces to the equation for s^2 given in (5.2.13).

2. Find two matrices $\mathbf{C}$ to test the null hypothesis $H_0: \alpha_1 = \alpha_2 = \alpha_3$ for the one-way layout, and verify that the F statistic is invariant for the matrices chosen.

3. Verify (5.2.18), (5.2.20), (5.2.21) and, using (4.2.61), prove (5.2.34).

5.3 ONE-WAY ANALYSIS OF VARIANCE—MULTIVARIATE

Observations were obtained for one dependent variable in the univariate one-way layout, CR-k design. For the multivariate one-way layout, p response measures are simultaneously obtained for each subject. In this situation the experimenter is interested in differences among k treatments on several measures. As with the univariate design, the subjects are randomly assigned to the k treatments. Hence we can abbreviate this design as an MCR-k design to denote a multivariate, completely randomized design with k treatment levels.

Denote the vector-valued observation from the ith population on the jth subject by

(5.3.1)
$$\mathbf{y}'_{ij} = [y_{ij1}, y_{ij2}, \ldots, y_{ijp}]$$

Employing the general linear model $\mathbf{Y} = \mathbf{XB} + \mathbf{E}_0$, we write the MCR-$k$ design as follows:

(5.3.2)
$$\underbrace{\begin{bmatrix} \mathbf{y}'_{11} \\ \mathbf{y}'_{12} \\ \vdots \\ \mathbf{y}'_{1N_1} \\ \hline \mathbf{y}'_{21} \\ \mathbf{y}'_{22} \\ \vdots \\ \mathbf{y}'_{2N_2} \\ \hline \vdots \\ \hline \mathbf{y}'_{k1} \\ \mathbf{y}'_{k2} \\ \vdots \\ \mathbf{y}'_{kNk} \end{bmatrix}}_{(N \times p)} = \underbrace{\begin{bmatrix} 1 & 1 & 0 & \cdots & 0 \\ 1 & 1 & 0 & \cdots & 0 \\ \vdots & \vdots & \vdots & \vdots & \vdots \\ 1 & 1 & 0 & \cdots & 0 \\ \hline 1 & 0 & 1 & \cdots & 0 \\ 1 & 0 & 1 & \cdots & 0 \\ \vdots & \vdots & \vdots & \vdots & \vdots \\ 1 & 0 & 1 & \cdots & 0 \\ \hline \vdots & \vdots & \vdots & \vdots & \vdots \\ \hline 1 & 0 & \cdot & \cdots & 1 \\ 1 & 0 & \cdot & \cdots & 1 \\ \vdots & \vdots & \vdots & \vdots & \vdots \\ 1 & 0 & \cdot & \cdots & 1 \end{bmatrix}}_{(N \times q)} \underbrace{\begin{bmatrix} \mu_{11} & \mu_{12} & \cdots & \mu_{1p} \\ \alpha_{11} & \alpha_{12} & \cdots & \alpha_{1p} \\ \vdots & \vdots & \vdots & \vdots \\ \alpha_{k1} & \alpha_{k2} & \cdots & \alpha_{kp} \end{bmatrix}}_{(q \times p)} + \underset{(N \times p)}{\underset{\sim}{E_0}}$$

where $N = \Sigma_i N_i$,

$$\begin{aligned} E(\mathbf{Y}) &= \mathbf{XB} \\ V(\mathbf{Y}) &= \mathbf{I} \otimes \mathbf{\Sigma} \end{aligned} \tag{5.3.3}$$

and each row of $\mathbf{Y}$ is drawn from a multivariate normal distribution. Alternatively, letting

$$\begin{aligned} \boldsymbol{\mu}' &= [\mu_{11}, \mu_{12}, \ldots, \mu_{1p}] \\ \boldsymbol{\alpha}'_1 &= [\alpha_{11}, \alpha_{12}, \ldots, \alpha_{1p}] \\ &\vdots \\ \boldsymbol{\alpha}'_k &= [\alpha_{k1}, \alpha_{k2}, \ldots, \alpha_{kp}] \end{aligned} \tag{5.3.4}$$

(5.3.2) becomes

$$\mathbf{y}_{ij} = \boldsymbol{\mu} + \boldsymbol{\alpha}_i + \boldsymbol{\varepsilon}_{ij} \tag{5.3.5}$$

where $\mathbf{y}_{ij}$ is a p-valued vector observation.

By using (5.2.6) and the multivariate Gauss-Markoff theorem, an estimate of the least-squares estimate of effects is

$$\hat{\mathbf{B}} = (\mathbf{X}'\mathbf{X})^{-}\mathbf{X}'\mathbf{Y} = \begin{bmatrix} 0 & 0 & \cdots & 0 \\ y_{1.1} & y_{1.2} & \cdots & y_{1.p} \\ y_{2.1} & y_{2.2} & \cdots & y_{2.p} \\ \vdots & \vdots & \vdots & \vdots \\ y_{k.1} & y_{k.2} & \cdots & y_{k.p} \end{bmatrix} = \begin{bmatrix} \mathbf{0}' \\ \mathbf{y}'_{1.} \\ \mathbf{y}'_{2.} \\ \vdots \\ \mathbf{y}'_{k.} \end{bmatrix} \tag{5.3.6}$$

where

$$y_{i.r} = \frac{\sum_{j=1}^{N_i} y_{ijr}}{N_i} \qquad (r = 1, \ldots, p)$$

or

$$\mathbf{y}_{i.} = \frac{\sum_{j=1}^{N_i} \mathbf{y}_{ij}}{N_i}$$

From (5.2.11), the unbiased estimator of $\mathbf{\Sigma}$ is

$$\begin{aligned} \mathbf{S} &= \frac{\mathbf{Y}'[\mathbf{I} - \mathbf{X}(\mathbf{X}'\mathbf{X})^{-}\mathbf{X}']\mathbf{Y}}{N-k} \\ &= \frac{\mathbf{Y}'\mathbf{Y} - \mathbf{Y}'\mathbf{X}(\mathbf{X}'\mathbf{X})^{-}\mathbf{X}'\mathbf{Y}}{N-k} \\ &\quad \frac{\mathbf{Y}'\mathbf{Y} - \sum_{i=1}^{k} N_i\mathbf{y}_{i.}\mathbf{y}'_{i.}}{N-k} \\ &= \frac{\sum_{i=1}^{k}\sum_{j=1}^{N_j} (\mathbf{y}_{ij} - \mathbf{y}_{i.})(\mathbf{y}_{ij} - \mathbf{y}_{i.})'}{N-k} \end{aligned} \tag{5.3.7}$$

so that

$$\mathbf{Q}_e = \sum_{i=1}^{k} \sum_{j=1}^{N_i} (\mathbf{y}_{ij} - \mathbf{y}_{i.})(\mathbf{y}_{ij} - \mathbf{y}_{i.})' \tag{5.3.8}$$

To find unique linear unbiased estimates for linear combinations of the elements of **B**, (3.6.17) is applied with the condition that $\mathbf{c}'\mathbf{H} = \mathbf{c}'$, where **H** is given by (5.2.8). Estimable functions ψ are of the form

$$\begin{aligned} \psi = \mathbf{c}'\mathbf{B}\mathbf{a} &= \sum_{r=1}^{p} a_r(\mathbf{c}'\boldsymbol{\beta}_r) = \sum_{r=1}^{p} a_r(\mathbf{t}'\mathbf{H}\boldsymbol{\beta}_r) \\ &= \mathbf{a}'\left[\sum_{i=1}^{k} t_i\boldsymbol{\mu} + \sum_{i=1}^{k} t_i\boldsymbol{\alpha}_i\right] \\ \hat{\psi} = \mathbf{c}'\hat{\mathbf{B}}\mathbf{a} &= \sum_{r=1}^{p} a_r(\mathbf{c}'\hat{\boldsymbol{\beta}}_r) = \mathbf{a}'\left(\sum_{i=1}^{k} t_i\mathbf{y}_{i.}\right) \end{aligned} \tag{5.3.9}$$

where $\boldsymbol{\beta}_r$ denotes the rth column of **B** and $\mathbf{t}' = [t_0, t_1, \ldots, t_k]$ is an arbitrary vector.

To test the hypothesis of equal treatment effects, the hypothesis

$$H_0: \boldsymbol{\alpha}_1 = \boldsymbol{\alpha}_2 = \cdots = \boldsymbol{\alpha}_k \tag{5.3.10}$$

is represented in the form

$$H_0: \mathbf{CBA} = \boldsymbol{\Gamma} \tag{5.3.11}$$

where **C** is identical to (5.2.15) in the univariate case, **A** is a $p \times p$ identity matrix, and $\boldsymbol{\Gamma}$ is a $(k-1) \times p$ matrix of 0s.

Applying the multivariate fundamental least-squares theorem,

$$\mathbf{Q}_h = (\mathbf{C}\hat{\mathbf{B}}\mathbf{A})'[\mathbf{C}(\mathbf{X}'\mathbf{X})^{-}\mathbf{C}']^{-1}(\mathbf{C}\hat{\mathbf{B}}\mathbf{A}) \tag{5.3.12}$$

Using (5.2.21), $\mathbf{Q}_h$ becomes

$$\mathbf{Q}_h = \sum_{i=1}^{k} N_i(\mathbf{y}_{i.} - \mathbf{y}_{..})(\mathbf{y}_{i.} - \mathbf{y}_{..})' \tag{5.3.13}$$

where

$$\mathbf{y}_{..} = \frac{\sum_{i=1}^{k} \sum_{j=1}^{N_i} \mathbf{y}_{ij}}{N}$$

TABLE 5.3.1. MANOVA Table for a MCR-k Design.

Source	df	SSP
Treatment	$k-1$	$\mathbf{Q}_h = \sum_{i=1}^{k} N_i(\mathbf{y}_{i.} - \mathbf{y}_{..})(\mathbf{y}_{i.} - \mathbf{y}_{..})'$
Within error	$N-k$	$\mathbf{Q}_e = \sum_{i=1}^{k} \sum_{j=1}^{N_i} (\mathbf{y}_{ij} - \mathbf{y}_{i.})(\mathbf{y}_{ij} - \mathbf{y}_{i.})'$
"Total"	$N-1$	$\mathbf{Q}_t = \sum_{i=1}^{k} \sum_{j=1}^{N_i} (\mathbf{y}_{ij} - \mathbf{y}_{..})(\mathbf{y}_{ij} - \mathbf{y}_{..})'$

Thus Table 5.3.1, the MANOVA table for testing (5.3.10) is obtained, so that the analogy between univariate and multivariate analysis is complete. Individual univariate observations are replaced with vector-valued observations to perform a multivariate one-way analysis of variance. Following the univariate case exactly, the expected values of the sum of squares and products (SSP) matrices in Table 5.3.1 are immediately seen to be as follows:

$$\begin{aligned} E(\mathbf{Q}_h) &= (k-1)\boldsymbol{\Sigma} + \sum_{i=1}^{k} N_i(\boldsymbol{\alpha}_i - \boldsymbol{\alpha}_.)(\boldsymbol{\alpha}_i - \boldsymbol{\alpha}_.)' \\ E(\mathbf{Q}_e) &= (N-k)\boldsymbol{\Sigma} \end{aligned} \tag{5.3.14}$$

for $\boldsymbol{\alpha}_i$ as defined in (5.3.4) and

$$\boldsymbol{\alpha}_. = \frac{\sum_{i=1}^{k} N_i \boldsymbol{\alpha}_i}{N}$$

Numerous criteria are available to test (5.3.10). All of them are functions of the roots $\lambda_1, \lambda_2, \ldots, \lambda_s$ of the determinantal equation $|\mathbf{Q}_h - \lambda \mathbf{Q}_e| = 0$, where $s = \min(v_h, u)$. Applying the general theory of (2.7.12) and (3.6.19) to test (5.3.10), s, m, and n are evaluated, and $\mathbf{Q}_e$ and $\mathbf{Q}_h$ in Table 5.3.1 are substituted into the determinantal equation. For the MCR-k design,

$$\begin{aligned} s &= \min(v_h, u) = \min(k-1, p) \\ m &= \frac{|u - v_h| - 1}{2} = \frac{|k-p-1| - 1}{2} \\ n &= \frac{v_e - u - 1}{2} = \frac{N-k-p-1}{2} \end{aligned} \tag{5.3.15}$$

where $u = R(\mathbf{A}) = p$, $v_h = k - 1$, and $v_e = N - k$. Hypothesis (5.3.10) is rejected at the significance level α if

(5.3.16) (a) Wilks:

$$\Lambda = \frac{|\mathbf{Q}_e|}{|\mathbf{Q}_e + \mathbf{Q}_h|} = \sum_{i=1}^{s} (1 + \lambda_i)^{-1} < U^{\alpha}(p, k-1, N-k)$$

(b) Roy:

$$\theta_s = \frac{\lambda_1}{1 + \lambda_1} > \theta^{\alpha}(s, m, n)$$

(c) Lawley-Hotelling:

$$U^{(s)} = \frac{T_0^2}{N-k} = \sum_{i=1}^{s} \lambda_i > U_0^{\alpha}(s, m, n)$$

(d) Pillai:

$$V^{(s)} = \sum_{i=1}^{s} \frac{\lambda_i}{1 + \lambda_i} > V^{\alpha}(s, m, n)$$

Given that the hypothesis of equal treatment effects is rejected, the researcher hopes to determine those linear combinations of parameters that led to the rejection of the hypothesis. The generalization of Scheffé's method by Roy and Bose (1953) allows the researcher to accomplish this.

Applying (4.6.31), using Roy's largest-root criterion,

$$\mathbf{c}'\hat{\mathbf{B}}\mathbf{a} - c_0\sqrt{\mathbf{a}'\left(\frac{\mathbf{Q}_e}{v_e}\right)\mathbf{a}\mathbf{c}'(\mathbf{X}'\mathbf{X})^{-}\mathbf{c}} \le \mathbf{c}'\mathbf{B}\mathbf{a} \le \mathbf{c}'\hat{\mathbf{B}}\mathbf{a} + c_0\sqrt{\mathbf{a}'\left(\frac{\mathbf{Q}_e}{v_e}\right)\mathbf{a}\mathbf{c}'(\mathbf{X}'\mathbf{X})^{-}\mathbf{c}} \tag{5.3.17}$$

or

$$\hat{\psi} - c_0\hat{\sigma}_{\hat{\psi}} \le \psi \le \hat{\psi} + c_0\hat{\sigma}_{\hat{\psi}}$$

where $c_0^2 = v_e\theta^\alpha/(1 - \theta^\alpha)$, $\mathbf{c}'$ satisfies the condition that $\mathbf{c}'\mathbf{H} = \mathbf{c}'$, $\mathbf{a}'$ is an arbitrary vector, and $\mathbf{Q}_e$ is the residual sum of squares and products matrix, $100(1 - \alpha)\%$ simultaneous confidence intervals for estimable functions are obtained.

As an example, pairwise comparisons are constructed by selecting $\mathbf{c}' = [0, \ldots, 0, 1_i, 0, \ldots, 0, -1_{i'}, 0, \ldots, 0]$; then

$$\hat{\psi} = \mathbf{c}'\hat{\mathbf{B}}\mathbf{a} = \sum_{r=1}^{p} a_r(y_{i.r} - y_{i'.r})$$

$$\mathbf{c}'(\mathbf{X}'\mathbf{X})^{-}\mathbf{c} = \frac{1}{N_i} + \frac{1}{N_{i'}}$$

so that (5.3.17) reduces to

$$\sum_{r=1}^{k} a_r(y_{i.r} - y_{i'.r}) - \sqrt{\frac{\theta^\alpha}{1-\theta^\alpha}\mathbf{a}'\mathbf{Q}_e\mathbf{a}\left(\frac{1}{N_i} + \frac{1}{N_{i'}}\right)} \le \sum_{r=1}^{k} a_r(\alpha_{ir} - \alpha_{i'r})$$

$$\le \sum_{r=1}^{k} a_r(y_{i.r} - y_{i'.r}) + \sqrt{\frac{\theta^\alpha}{1-\theta^\alpha}\mathbf{a}'\mathbf{Q}_e\mathbf{a}\left(\frac{1}{N_i} + \frac{1}{N_{i'}}\right)} \tag{5.3.18}$$

If an interval does not contain a 0, it may be concluded that the particular contrast on the response variables is significant for the treatments under study.

Simultaneous test procedures may be similarly employed for the other criteria; however, Gabriel (1968, 1969) indicates that the largest-root criterion is most resolute and yields the narrowest intervals for criteria based on the roots of (5.3.14). The critical constants c_0 are given in (4.6.32), where they were evaluated for the regression example. For $p = 1$, the intervals reduce to the familiar Scheffé intervals.

As shown by Hummel and Sligo (1971), Roy's criterion is extremely conservative when the main focus of the researcher is in restricting the analysis of an experiment to comparisons within variables across groups. When this is the case, the combination of a significant multivariate test result with univariate F tests is often recommended; however, the disadvantages of this procedure (discussed for the two-sample problem) hold also for the k-sample case.

Again, the Bonferroni procedure, using formula (5.2.36), may be employed to test (5.3.10). With this procedure, comparisons need not be restricted within variables. However, it is best used when a specific number of planned comparisons are sought. To illustrate the fact that the Bonferroni procedure is not too restrictive, suppose the Bonferroni procedure was employed to locate differences among individual means for the estimable functions $\psi = \mathbf{c}'\mathbf{B}\mathbf{a}$, with Gabriel's (1968) data on skull measures of $N = 48$ anteaters for $k = 6$ groups and $p = 3$ skull measurements. Comparing the critical constant $c_0 = 4.572$ for the Roy criterion at the significant level $\alpha = .05$, the intervals obtained would be shorter for more than 250 planned comparisons since the critical constant for the Bonferroni technique is only $c_0 \doteq 4.1$ for 250 comparisons. In addition, for $p > 3$ the critical value c_0 for the Roy criterion increases while the Bonferroni constant for any number of comparisons remains fixed. However, reducing the number of groups does affect the number of possible shorter intervals since c_0 for the Roy criterion decreases. Reducing k from 6 to 4 restricts the experimenter to about 100 specific comparisons. For $p = 2$ and $k = 3$, the restriction is about 15. Similar comparisons can be made for other values of N, p, and k; in general, the Bonferroni intervals are usually shorter than those acquired by using Roy's criterion when a few comparisons are of interest.

Discriminant analysis to determine standardized discriminant vectors, as used in Section 3.8, may also be applied in the k-group problem or, more generally, to any MANOVA analysis; examples appear in Rao (1952, p. 370), Seal (1964, p. 136), and Bock and Haggard (1968). By solving equations (5.3.14), linear combinations of the variables that maximize the variation in the hypothesis sum of squares $\mathbf{Q}_h$ relative to the error sum of squares $\mathbf{Q}_e$ are procured through the eigenvectors. These mathematical constructs are called *discriminants* or *canonical variates* by these authors, and indicate those combinations of the variables that are most sensitive to deviations from the null hypothesis. When the coefficients are applied to the group means, a graphical representation of the contrasts or group means for significant roots yields a pictorial representation of group differences. Furthermore, correlations between the discriminants and the individual variables help to locate variables in the set that are contributing to group differences in orthogonal directions of variation relative to the error metric (see Bargmann, 1970). An illustration of this technique is discussed with an example in Section 5.4.

When testing for the equality of k means, the assumptions of independence, multivariate normality, and equal variance-covariance matrices are fundamental to the analysis. Under multivariate normality, a procedure for testing the equality of k variance-covariance matrices was outlined in Section 3.16. Under nonnormality, a nonparametric MANOVA technique has been developed by Gabriel and Sen (1968). For further details of multivariate nonparametric methods, see Puri and Sen (1971).

From a simultaneous testing point of view, we have indicated that the largest-root criterion is the preferred procedure for minimal hypotheses of the form $\psi = \mathbf{c}'\mathbf{B}\mathbf{a}$. However, for the overall test, the other criteria may be more powerful (see, for example, Schatzoff, 1966; Roy, 1966; Pillai and Jayachandran, 1967, 1968; and Roy, Gnanadesikan, and Srivastava, 1971, p. 72), especially for small deviations in means, at least when the number of variables $p = 2$. Pillai's trace seems to have greater power for small differences and Roy's largest root is best for large differences. However, no test is uniformly most powerful (Potthoff and Roy, 1964), and all tests have monotonically increasing power functions (Das Gupta, Anderson, and Mudholkar, 1964, and Srivastava, 1964).

EXERCISES 5.3

1. Follow the univariate case to determine the expected value for the hypothesis and error sum of products matrix for the one-way MANOVA design, equations (5.3.14).

2. For $k = 4$, 6, and 8, $p = 2$, 3, and 4, and $N = 25$, 50, and 100, compare the Bonferroni and Roy procedures.

5.4 EXAMPLE—MULTIVARIATE ONE-WAY ANALYSIS OF VARIANCE

The data used in this study were taken from a larger study by Dr. Stanley Jacobs and Mr. Ronald Hritz at the University of Pittsburgh to investigate risk-taking behavior. Students were randomly assigned to three different direction treatments known as Arnold and Arnold (AA), Coombs (C), and Coombs with no penalty in the directions (NC). Using the three direction conditions, students were administered two parallel forms of a test given under high and low penalty. The data for the study are summarized in Table 5.4.1. The sample sizes for the three treatments are, respectively, $N_1 = 30$, $N_2 = 28$, and $N_3 = 29$. The total sample size is $N = 87$, the number of treatments is $k = 3$, and the number of variables is $p = 2$ for the study.

TABLE 5.4.1. Sample Data: MCR-3 Design.

AA				C				NC			
Low	High	Low	High	Low	High	Low	High	Low	High	Low	High
8	28	31	24	46	13	25	9	50	55	55	43
18	28	11	20	26	10	39	2	57	51	52	49
8	23	17	23	47	22	34	7	62	52	67	62
12	20	14	32	44	14	44	15	56	52	68	61
15	30	15	23	34	4	36	3	59	40	65	58
12	32	8	20	34	4	40	5	61	68	46	53
12	20	17	31	44	7	49	21	66	49	46	49
18	31	7	20	39	5	42	7	57	49	47	40
29	25	12	23	20	0	35	1	62	58	64	22
6	28	15	20	43	11	30	2	47	58	64	54
7	28	12	20	43	25	31	13	53	40	63	64
6	24	21	20	34	2	53	12	60	54	63	56
14	30	27	27	25	10	40	4	55	48	64	44
11	23	18	20	50	9	26	4	56	65	63	40
12	20	25	27					67	56		

The null hypothesis of interest is whether the mean vectors for the two variables are the same across the three treatments. In terms of the effects, the hypothesis may be written as

$$H_0: \begin{bmatrix} \alpha_{11} \\ \alpha_{12} \end{bmatrix} = \begin{bmatrix} \alpha_{21} \\ \alpha_{22} \end{bmatrix} = \begin{bmatrix} \alpha_{31} \\ \alpha_{32} \end{bmatrix} \tag{5.4.1}$$

by using hypothesis (5.3.10).

Depending on the test statistic employed to test (5.4.1), H_0 is rejected at the significance level $\alpha = .05$ if

(5.4.2)
(a) $\Lambda < U^{\alpha}_{(p,k-1,N-k)} = U^{.05}_{(2,2,84)} = .892$ (Wilks)
(b) $\theta_s > \theta^{\alpha}(s, m, n) = \theta^{.05}_{(2,-.5,40.5)} = .101$ (Roy)
(c) $U^{(s)} > U^{\alpha}_0(s, m, n) = U^{.05}_{(2,-.5,40.5)} \doteq .120$ (Lawley-Hotelling)
(d) $V^{(s)} > V(s, m, n) = V^{.05}_{(2,-.5,40.5)} = .110$ (Pillai)

With $(\mathbf{X}'\mathbf{X})^-$ defined by (5.2.6), an estimator of $\mathbf{B}$, using (5.3.6), becomes

$$\hat{\mathbf{B}} = (\mathbf{X}'\mathbf{X})^-\mathbf{X}'\mathbf{Y} = \begin{bmatrix} 0 & 0 \\ 14.60 & 24.67 \\ 37.61 & 8.61 \\ 58.45 & 51.38 \end{bmatrix} = \begin{bmatrix} \mathbf{0}' \\ \mathbf{y}'_{1.} \\ \mathbf{y}'_{2.} \\ \mathbf{y}'_{3.} \end{bmatrix} \tag{5.4.3}$$

where $\mathbf{Y}$ denotes the 87×2 data matrix and $\mathbf{X}$ the 87×4 design matrix of rank 3:

$$\mathbf{X} = \begin{bmatrix} 1 & 1 & 0 & 0 \\ 1 & 1 & 0 & 0 \\ \vdots & \vdots & \vdots & \vdots \\ 1 & 1 & 0 & 0 \\ \hline 1 & 0 & 1 & 0 \\ 1 & 0 & 1 & 0 \\ \vdots & \vdots & \vdots & \vdots \\ 1 & 0 & 1 & 0 \\ \hline 1 & 0 & 0 & 1 \\ 1 & 0 & 0 & 1 \\ \vdots & \vdots & \vdots & \vdots \\ 1 & 0 & 0 & 1 \end{bmatrix} \begin{matrix} 30 = N_1 \\ \\ 28 = N_2 \\ \\ 29 = N_3 \end{matrix} \tag{5.4.4}$$

From (5.3.7), the estimator for $\mathbf{\Sigma}$ is

$$\mathbf{S} = \frac{\mathbf{Y}'[\mathbf{I} - \mathbf{X}(\mathbf{X}'\mathbf{X})^-\mathbf{X}']\mathbf{Y}}{N - k}$$

$$= \frac{1}{84}\begin{bmatrix} 4523.0510 & \\ 1137.7475 & 4248.1728 \end{bmatrix}$$

$$= \begin{bmatrix} 53.8458 & \\ 13.5446 & 50.5735 \end{bmatrix} \tag{5.4.5}$$

Using equation (5.3.12), with the matrix $\mathbf{C}$ defined by

$$\mathbf{C} = \begin{bmatrix} 0 & 1 & 0 & -1 \\ 0 & 0 & 1 & -1 \end{bmatrix} \tag{5.4.6}$$

$\mathbf{A} = \mathbf{I}$, and $\mathbf{\Gamma} = \mathbf{0}$, the hypothesis sum of squares $\mathbf{Q}_h$ for testing (5.4.1) is

$$\mathbf{Q}_h = (\mathbf{C}\hat{\mathbf{B}}\mathbf{A})'[\mathbf{C}(\mathbf{X}'\mathbf{X})^{-}\mathbf{C}']^{-1}(\mathbf{C}\hat{\mathbf{B}}\mathbf{A})$$

$$= \sum_{i=1}^{k} N_i(\mathbf{y}_{i.} - \mathbf{y}_{..})(\mathbf{y}_{i.} - \mathbf{y}_{..})'$$

(5.4.7)
$$= \begin{bmatrix} 28{,}391.4318 & \\ 16{,}465.5283 & 26{,}700.7467 \end{bmatrix}$$

Thus we form MANOVA Table 5.4.2 for testing the hypothesis.

TABLE 5.4.2. MANOVA Table for the MCR-3 Design Example.

Source	df	SSP
Treatment	2	$\mathbf{Q}_h = \begin{bmatrix} 28{,}391.4318 & \\ 16{,}465.5283 & 26{,}700.7467 \end{bmatrix}$
Within error	84	$\mathbf{Q}_e = \begin{bmatrix} 4523.0510 & \\ 1137.7475 & 4248.1728 \end{bmatrix}$
"Total"	86	$\mathbf{Q}_t = \begin{bmatrix} 32{,}914.4828 & \\ 17{,}603.2758 & 30{,}948.9195 \end{bmatrix}$

Solving the characteristic equation $|\mathbf{Q}_h - \lambda\mathbf{Q}_e| = 0$, where $\mathbf{Q}_h$ and $\mathbf{Q}_e$ are defined in Table 5.4.2, the eigenvalues are $\lambda_1 = 7.9684$ and $\lambda_2 = 3.4099$. Employing (5.3.16), the test criteria are evaluated:

(5.4.8)
(a) $\Lambda = .02528$ (Wilks)
(b) $\theta_s = .8885$ (Roy)
(c) $U^{(s)} = 11.3789$ (Lawley-Hotelling)
(d) $V^{(s)} = 1.6617$ (Pillai)

By comparing the values of the criteria in (5.4.8) with the critical values in (5.4.2), the hypothesis of equal treatment effects is rejected for all criteria. By using formula (5.3.17), the critical constants c_0, for contrasts in the parameters, are

(5.4.9)
(a) $c_0 = \left[\nu_e \left(\frac{1 - U^\alpha}{U^\alpha} \right) \right]^{1/2} = 3.1924$ (Wilks)

(b) $c_0 = \left[\nu_e \left(\frac{\theta^\alpha}{1 - \theta^\alpha} \right) \right]^{1/2} = 3.072$ (Roy)

(c) $c_0 = (\nu_e U_0)^{1/2} \doteq 3.1749$ (Lawley-Hotelling)

(d) $c_0 = \left[\nu_e \left(\frac{V^\alpha}{1 - V^\alpha} \right) \right]^{1/2} = 3.2221$ (Pillai)

for the various criteria. From (5.4.9), we see that the Roy procedure yields the shortest intervals.

To determine some linear combinations of the parameters significantly different from 0, (5.3.17) is evaluated. For the example,

$$\mathbf{c}'(\mathbf{X}'\mathbf{X})^{-}\mathbf{c} = \mathbf{c}'\begin{bmatrix} 0 & 0 & 0 & 0 \\ 0 & .03333 & 0 & 0 \\ 0 & 0 & .03571 & 0 \\ 0 & 0 & 0 & .03448 \end{bmatrix}\mathbf{c}' \tag{5.4.10}$$

Letting $\mathbf{a}' = \mathbf{a}_1' = [1, 0]$ and $\mathbf{a}' = \mathbf{a}_2' = [0, 1]$

$$\begin{aligned} \mathbf{a}_1'\left(\frac{\mathbf{Q}_e}{\nu_e}\right)\mathbf{a}_1 &= 53.8458 \\ \mathbf{a}_2'\left(\frac{\mathbf{Q}_e}{\nu_e}\right)\mathbf{a}_2 &= 50.5735 \end{aligned} \tag{5.4.11}$$

For $\mathbf{c}' = \mathbf{c}_1' = [0, 1, 0, -1]$ and $\mathbf{c}' = \mathbf{c}_2' = [0, 0, 1, -1]$, the following contrasts are formed by employing the general expression $\psi = \mathbf{c}'\mathbf{B}\mathbf{a}$:

$$\begin{aligned} \psi_{11} &= \mathbf{c}_1'\mathbf{B}\mathbf{a}_1 = \alpha_{11} - \alpha_{31} & \psi_{12} &= \mathbf{c}_2'\mathbf{B}\mathbf{a}_1 = \alpha_{21} - \alpha_{31} \\ \psi_{21} &= \mathbf{c}_1'\mathbf{B}\mathbf{a}_2 = \alpha_{12} - \alpha_{32} & \psi_{22} &= \mathbf{c}_2'\mathbf{B}\mathbf{a}_2 = \alpha_{22} - \alpha_{32} \end{aligned} \tag{5.4.12}$$

Thus the best linear unbiased estimators for the contrasts in (5.4.12), using $\hat{\psi} = \mathbf{c}'\hat{\mathbf{B}}\mathbf{a}$ as defined in (5.3.9), become

$$\begin{aligned} \hat{\psi}_{11} &= 14.60 - 58.45 = -43.85 & \hat{\psi}_{12} &= 37.61 - 58.45 = -20.84 \\ \hat{\psi}_{21} &= 24.67 - 51.38 = -26.71 & \hat{\psi}_{22} &= 8.61 - 51.38 = -42.77 \end{aligned} \tag{5.4.13}$$

By (5.4.10),

$$\hat{\sigma}_{\hat{\psi}_{ij}} = \sqrt{[\mathbf{a}_i'(\mathbf{Q}_e/\nu_e)\mathbf{a}_i][\mathbf{c}_j'(\mathbf{X}'\mathbf{X})^{-}\mathbf{c}_j]} \tag{5.4.14}$$

so that

$$\begin{aligned} \hat{\sigma}_{\hat{\psi}_{11}} &= 1.9109 & \hat{\sigma}_{\hat{\psi}_{12}} &= 1.9442 \\ \hat{\sigma}_{\hat{\psi}_{21}} &= 1.8520 & \hat{\sigma}_{\hat{\psi}_{22}} &= 1.8842 \end{aligned}$$

By evaluating the expression

$$\hat{\psi}_{ij} - c_0\hat{\sigma}_{\hat{\psi}_{ij}} \le \psi_{ij} \le \hat{\psi}_{ij} + c_0\hat{\sigma}_{\hat{\psi}_{ij}}$$

where $c_0 = 3.0720$ for the Roy criterion, simultaneous confidence intervals for the contrasts ψ_{ij} are

$$\begin{aligned} -43.85 - (3.0720)(1.9109) \le \psi_{11} &\le -43.85 + (3.0720)(1.9109) \\ -49.72 \le \psi_{11} &\le -37.98 \\ -26.81 \le \psi_{12} &\le -14.87 \\ -32.40 \le \psi_{21} &\le -21.02 \\ -48.56 \le \psi_{22} &\le -36.98 \end{aligned} \tag{5.4.15}$$

In a similar manner, confidence intervals for $\psi_1 = \psi_{11} - \psi_{12} = \alpha_{11} - \alpha_{21}$ and $\psi_2 = \psi_{21} - \psi_{22} = \alpha_{12} - \alpha_{22}$ may be determined by choosing $\mathbf{c}' = [0, 1, -1, 0]$ and

$\mathbf{a} = \mathbf{a}_1' = [1, 0]$ and $\mathbf{a}_2' = [0, 1]$. The intervals now become

$$(5.4.16)\quad \begin{aligned} -23.01 - (3.0720)(1.9282) \leq \psi_1 &\leq -23.01 + (3.0720)(1.9282) \\ -28.93 \leq \psi_1 &\leq -17.09 \\ 16.02 - (3.0720)(1.8687) < \psi_2 &\leq 16.02 + (3.0720)(1.8687) \\ 10.28 < \psi_2 &\leq 21.76 \end{aligned}$$

In summary, there is a significant difference in the direction treatments under study. In particular, all pairwise comparisons under both low and high penalty are significantly different from 0.

As an alternative to the Roy procedure, the Bonferroni method for a few comparisons may be used to an experimenter's advantage in this study. If a total of $q = 6$ comparisons were made, the critical constant would be $c_0 = t_{(84)}^{.05/2q} = 2.71$. This technique leads to shorter intervals than the Roy procedure. In fact, for 20 specific comparisons, the Bonferroni method leads to shorter intervals.

In Section 5.3, it was indicated that discriminant function analysis could be employed in a k-group problem to determine a linear combination of variables most sensitive to deviations from the null hypothesis. That is, by solving the characteristic equation

$$(5.4.17)\quad |\mathbf{Q}_h - \lambda\mathbf{Q}_e| = 0$$

there are $s = \min(\nu_h, p)$ eigenvalues; associated with each eigenvalue is an eigenvector whose elements are the coefficients of the linear discriminant functions.

Using (1.7.27) to locate the eigenvectors of (5.4.17) and standardizing them according to (3.9.25), so that the variance of the discriminant scores in the sample is unity, standardized discriminant vectors for the example are

$$(5.4.18)\quad \mathbf{a}_1 = \begin{bmatrix} .0858 \\ .0888 \end{bmatrix} \quad \text{and} \quad \mathbf{a}_2 = \begin{bmatrix} .1121 \\ -.1155 \end{bmatrix}$$

The elements of $\mathbf{a}_1$ are the weights to be applied to the dependent variables low and high penalty, respectively. The linear combination

$$L_1 = (.0857)\,\text{low} + (.0888)\,\text{high}$$

is a linear combination of the dependent variables with the property that its distribution in the three treatments shows the largest separation among all possible linear combinations of variables. Similarly, the second discriminant function

$$L_2 = (.1121)\,\text{low} + (-.1155)\,\text{high}$$

is the linear combination of the dependent variables that, among those orthogonal to L_1 in the metric of $\mathbf{Q}_e$, presents the least overlap of the three groups.

After computing the discriminant functions, it is necessary to determine how many are needed to illustrate the least overlap among the groups and to establish which set of dependent variables is most important for this purpose. To evaluate this, determination must first be made of the number of discriminant functions statistically significant for maximal separation. Several procedures are available; however, the simplest is Bartlett's (1951). This technique was used in (4.14.15) to test for the significance of canonical variates. Alternatively, Roy's criterion or Wilks' Λ criterion may be employed. For Roy,

$$s = \min(\nu_h - i + 1, p - i + 1)$$

$$m = \frac{|\nu_h - i - p + 1| - 1}{2} \qquad i = 1, \ldots, \min(\nu_h, p)$$

$$n = \frac{\nu_e - p - i}{2}$$

By using Bartlett's procedure,

$$X_B^2 = -\left[\nu_e - \frac{1}{2}(p - \nu_h + 1)\right] \log \Lambda \mathrel{\dot\sim} \chi^2(\nu_h p)$$

By partitioning Bartlett's chi-square statistic to test the significance of succession roots, the test of overall significance is first achieved by computing

$$X_B^2 = -\left[84 - \frac{1}{2}(2 - 2 + 1)\right] \sum_{i=1}^{s} \log(1 + \lambda_i)^{-1}$$
$$= 307.08 \mathrel{\dot\sim} \chi^2(4)$$

which is significant. Thus the two functions taken together are significant. To determine whether the second function is significant, the first root is removed. Then

$$X_B^2 = -\left[84 - \frac{1}{2}(2 - 2 + 1)\right] \sum_{i=2}^{s} \log(1 + \lambda_i)^{-1}$$
$$= 123.90 \mathrel{\dot\sim} \chi^2(1)$$

which is also significant; the degrees of freedom are $1 = (p - 1)(\nu_h - 1)$. It is concluded for this example that the axes corresponding to both discriminant functions denote the variation along which significant differences among the treatments exist. Figure 5.4.1 of the group means evaluated for L_1 and L_2 yields a visual verification of this fact:

$$\bar{L}_1^{(1)} = .0857(14.60) + .0888(24.67) = 3.44$$
$$\bar{L}_2^{(1)} = .1121(14.60) - .1155(24.67) = -1.21$$
$$\bar{L}_1^{(2)} = .0857(37.61) + .0888(8.61) = 3.99$$
$$\bar{L}_2^{(2)} = .1121(37.61) - .1155(8.61) = 3.22$$
$$\bar{L}_1^{(3)} = .0857(58.45) + .0888(51.38) = 9.57$$
$$\bar{L}_2^{(3)} = .1121(58.45) - .1155(51.38) = .62$$

As a rough indication of the separation and uncertainty of each of the centroids, a circle with a radius of $1.96/\sqrt{N_i}$ may be placed at each point. Although the axes in Figure 5.4.1 are drawn orthogonal, the vectors $\mathbf{a}_1$ and $\mathbf{a}_2$ are not orthogonal. This is common practice when the original variables are not included in the system.

From Figure 5.4.1, L_1 serves primarily to separate groups 2 and 1 from group 3, while L_2 is used to separate groups 2 and 3 from group 1. To see which of the variables are most important to the discrimination along the two dimensions,

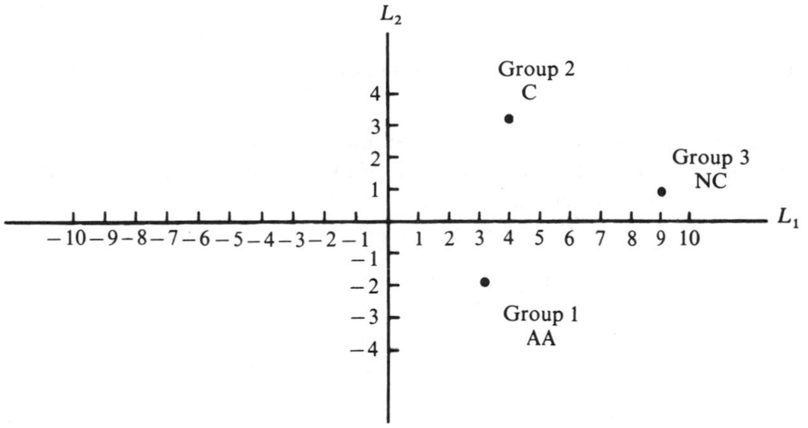

FIGURE 5.4.1. Discriminant function analysis for the MCR-3 design example.

correlations between the discriminant functions and the variables are used rather than the magnitudes of the coefficients.

By using formula (3.9.29), the correlations are found from the formula $\hat{\boldsymbol{\rho}} = \mathbf{R}_e\mathbf{a}_{wsa}$, where $\mathbf{R}_e$ is the within-error correlation matrix obtained from $\mathbf{Q}_e$, and $\mathbf{a}_{wsa}$ is the vector of the standardized adjusted coefficients. From (5.4.18), the standardized adjusted coefficients are computed by multiplying the elements of the standardized vectors by the standard deviation of the dependent variable to adjust for unequal variances; thus the standardized adjusted coefficients become

$$\mathbf{a}_1^{(A)} = \begin{bmatrix} .0858 \times 7.3380 \\ .0888 \times 7.1115 \end{bmatrix} = \begin{bmatrix} .6296 \\ .6318 \end{bmatrix}$$

$$\mathbf{a}_2^{(A)} = \begin{bmatrix} .1121 \times 7.3380 \\ -.1155 \times 7.1115 \end{bmatrix} = \begin{bmatrix} .8226 \\ -.8214 \end{bmatrix}$$

Transforming (5.4.5) to a correlation matrix,

$$\mathbf{R}_e = \begin{bmatrix} 1.000 & \\ .260 & 1.000 \end{bmatrix}$$

so that the correlations between the variables and the discriminant functions are

$$\hat{\boldsymbol{\rho}}_{y_i}L_1 = \mathbf{R}_e \begin{bmatrix} .6296 \\ .6308 \end{bmatrix} = \begin{bmatrix} .79 \\ .79 \end{bmatrix}$$

$$\hat{\boldsymbol{\rho}}_{y_i}L_2 = \mathbf{R}_e \begin{bmatrix} .8226 \\ -.8214 \end{bmatrix} = \begin{bmatrix} .61 \\ -.61 \end{bmatrix}$$

This indicates that both variables in both directions seem to be contributing equally to discrimination (group differences); furthermore, the separation along both directions appears to be nearly equal.

EXERCISES 5.4

1. Determine the number of within-group standard deviations separating the centroids in Figure 5.4.1.

2. An experiment was performed to investigate four different methods for teaching schoolchildren multiplication (M) and addition (A) of two four-digit numbers. The data for four independent groups of students are summarized below.

Group 1		Group 2		Group 3		Group 4	
A	M	A	M	A	M	A	M
97	66	76	29	66	34	100	79
94	61	60	22	60	32	96	64
96	52	84	18	58	27	90	80
84	55	86	32	52	33	90	90
90	50	70	33	56	34	87	82
88	43	70	32	42	28	83	72
82	46	73	17	55	32	85	67
65	41	85	29	41	28	85	77
95	58	58	21	56	32	78	68
90	56	65	25	55	29	86	70
95	55	80	20	40	33	67	67
84	40	75	16	50	30	57	57
71	46	74	21	42	29	83	79
76	32	84	30	46	33	60	50
90	44	62	32	32	34	89	77
77	39	71	23	30	31	92	81
61	37	71	19	47	27	86	86
91	50	75	18	50	28	47	45
93	64	92	23	35	28	90	85
88	68	70	27	47	27	86	65

a. On the basis of these data, is there any reason to believe that any one method or set of methods is superior or inferior for teaching skills for multiplication and addition of four-digit numbers?

b. What assumptions must you make to answer part a? Are they satisfied?

c. Are there any significant differences between addition and multiplication skills within the various groups?

d. Construct the linear discriminant function(s) for part a.

5.5 RANDOMIZED BLOCK DESIGN—UNIVARIATE AND MULTIVARIATE

The experimenter is interested in the effects of one factor (k treatments) in a one-way layout of a CR-k design. In this type of design, the subjects should be relatively homogeneous since large variability among the subjects may obscure treatment effects unless the number of subjects available for the experiment is large.

To try to minimize individual differences such as previous experience, attitudes, and intelligence levels among subjects, blocks of an equal number of homogeneous subjects are formed such that the variability among the subjects within each block is less than the variability among the blocks. Then subjects are randomly assigned to the various treatment levels. For example, suppose achievement is used as a blocking variable. Given r blocks and c treatments, each block would consist of c subjects selected at random for each block (for example, low, middle, and high achievement). Then the c subjects in each block would be randomly assigned to the c treatment levels. When the variability among blocks is large, this design is more efficient than a completely randomized design.

For randomized block designs, the primary interest of the experimenter is in differences among the c treatments. The block differences are usually checked, but are not of real interest. However, nonsignificance of the block effect would indicate that the blocking variable was not useful and thus that a completely randomized design would be more appropriate.

Some authors consider a one-sample, repeated-measures design as a randomized block design. In this case, each subject is a block and receives all treatment levels. This type of design is best analyzed by using the multivariate methods discussed in Chapter 3.

Another type of two-way design used in the behavioral sciences is the completely randomized factorial design with one observation per cell. For these designs, the experimenter is interested in two factors where each factor consists of two or more treatments. If one treatment has r levels and the other c levels, rc subjects are randomly assigned to the rc treatment combinations.

In this chapter we use the terminology of the completely randomized block design, denoted sometimes by RB-c, to study the statistical analysis of the design. However, since the statistical models for the RB-c and the completely randomized factorial design with one observation per cell are the same, except for the randomization procedure, the analysis-of-variance tables in this section can be readily modified for an additive two-way completely randomized factorial design with only one observation in each cell.

Univariate Procedure. For the randomized block design, let y_{ij} denote the observation for the ith block and the jth column treatment classification where the data are represented in tabular form as follows:

		Treatments			
		1	2		c
	1	y_{11}	y_{12}	$\cdots$	y_{1c}
	2	y_{21}	y_{22}	$\cdots$	y_{2c}
Blocks	3	y_{31}	y_{32}	$\cdots$	y_{3c}
	$\vdots$	$\vdots$	$\vdots$	$\vdots$	$\vdots$
	r	y_{r1}	y_{r2}	$\cdots$	y_{rc}

The general linear model for the observation y_{ij} is additive:

$$y_{ij} = \mu + \alpha_i + \beta_j + \varepsilon_{ij} \qquad i = 1, \ldots, r; \quad j = 1, \ldots, c$$
$$\varepsilon_{ij} \sim IN(0, \sigma^2) \tag{5.5.1}$$

That is, $E(Y_{ij}) = \mu + \alpha_i + \beta_j$, so that the effect of treatment j on $E(Y_{ij})$ is the same no matter what block is employed. Using matrix notation, (5.5.1) is written as

(5.5.2)

$$
\underset{(N\times 1)}{\left[\begin{array}{c} y_{11}\\ y_{12}\\ \vdots\\ y_{1c}\\ \hline y_{21}\\ y_{22}\\ \vdots\\ y_{2c}\\ \hline \vdots\\ \hline y_{r1}\\ y_{r2}\\ \vdots\\ y_{rc} \end{array}\right]}
=
\underset{(N\times q)}{\left[\begin{array}{cccccccccc}
1 & 1 & 0 & \cdots & 0 & 1 & 0 & \cdots & 0\\
1 & 1 & 0 & \cdots & 0 & 0 & 1 & \cdots & 0\\
\vdots & \vdots & \vdots & \vdots & \vdots & \vdots & \vdots & \vdots & \vdots\\
1 & 1 & 0 & \cdots & 0 & 0 & 0 & \cdots & 1\\ \hline
1 & 0 & 1 & \cdots & 0 & 1 & 0 & \cdots & 0\\
1 & 0 & 1 & \cdots & 0 & 0 & 1 & \cdots & 0\\
\vdots & \vdots & \vdots & \vdots & \vdots & \vdots & \vdots & \vdots & \vdots\\
1 & 0 & 1 & \cdots & 0 & 0 & 0 & \cdots & 1\\ \hline
\vdots & \vdots & \vdots & \vdots & \vdots & \vdots & \vdots & \vdots & \vdots\\ \hline
1 & 0 & \cdot & \cdots & 1 & 1 & 0 & \cdots & 0\\
1 & 0 & \cdot & \cdots & 1 & 0 & 1 & \cdots & 0\\
\vdots & \vdots & \vdots & \vdots & \vdots & \vdots & \vdots & \vdots & \vdots\\
1 & 0 & \cdot & \cdots & 1 & 0 & 0 & \cdots & 1
\end{array}\right]}
\underset{(q\times 1)}{\begin{bmatrix} \mu\\ \alpha_1\\ \alpha_2\\ \vdots\\ \alpha_r\\ \beta_1\\ \beta_2\\ \vdots\\ \beta_c \end{bmatrix}}
+
\underset{(N\times 1)}{\left[\begin{array}{c} \varepsilon_{11}\\ \varepsilon_{12}\\ \vdots\\ \varepsilon_{1c}\\ \hline \varepsilon_{21}\\ \varepsilon_{22}\\ \vdots\\ \varepsilon_{2c}\\ \hline \vdots\\ \hline \varepsilon_{r1}\\ \varepsilon_{r2}\\ \vdots\\ \varepsilon_{rc} \end{array}\right]}
$$

$$
\underset{(N\times 1)}{\mathbf{y}} = \underset{(N\times q)}{\mathbf{X}} \quad \underset{(q\times 1)}{\boldsymbol{\beta}} + \underset{(N\times 1)}{\boldsymbol{\varepsilon}}
$$

where $N = rc$ and $q = r + c + 1$.

As in the CR-k design, the model as specified by (5.5.2) is not of full rank. The rank of $\mathbf{X}$ and $\mathbf{X}'\mathbf{X}$ is $r + c - 1$, where $\mathbf{X}'\mathbf{X}$ is defined by

$$
\mathbf{X}'\mathbf{X} = \left[\begin{array}{c|c|c}
N & c \; c \; \cdots \; c & r \; r \; \cdots \; r\\ \hline
\begin{array}{c} c\\ c\\ \vdots\\ c \end{array} & c\mathbf{I}_r & \mathbf{J}_{rc}\\ \hline
\begin{array}{c} r\\ r\\ \vdots\\ r \end{array} & \mathbf{J}_{cr} & r\mathbf{I}_c
\end{array}\right] \tag{5.5.3}
$$

As an example of (5.5.2), consider the simple 3 × 2 design represented by

$$(5.5.4)\qquad \underbrace{\begin{bmatrix} y_{11} \\ y_{12} \\ y_{21} \\ y_{22} \\ y_{31} \\ y_{32} \end{bmatrix}}_{(6\times 1)} = \underbrace{\begin{bmatrix} 1 & 1 & 0 & 0 & 1 & 0 \\ 1 & 1 & 0 & 0 & 0 & 1 \\ 1 & 0 & 1 & 0 & 1 & 0 \\ 1 & 0 & 1 & 0 & 0 & 1 \\ 1 & 0 & 0 & 1 & 1 & 0 \\ 1 & 0 & 0 & 1 & 0 & 1 \end{bmatrix}}_{(6\times 6)} \underbrace{\begin{bmatrix} \mu \\ \alpha_1 \\ \alpha_2 \\ \alpha_3 \\ \beta_1 \\ \beta_2 \end{bmatrix}}_{(6\times 1)} + \underbrace{\begin{bmatrix} \varepsilon_{11} \\ \varepsilon_{12} \\ \varepsilon_{21} \\ \varepsilon_{22} \\ \varepsilon_{31} \\ \varepsilon_{32} \end{bmatrix}}_{(6\times 1)}$$

Then

$$(5.5.5)\qquad \mathbf{X}'\mathbf{X} = \left[\begin{array}{c|ccc|cc} 6 & 2 & 2 & 2 & 3 & 3 \\ \hline 2 & 2 & 0 & 0 & 1 & 1 \\ 2 & 0 & 2 & 0 & 1 & 1 \\ 2 & 0 & 0 & 2 & 1 & 1 \\ \hline 3 & 1 & 1 & 1 & 3 & 0 \\ 3 & 1 & 1 & 1 & 0 & 3 \end{array}\right]$$

and a g-inverse of $\mathbf{X}'\mathbf{X}$ is given by

$$(5.5.6)\qquad (\mathbf{X}'\mathbf{X})^{-} = \begin{bmatrix} -1/6 & 0 & 0 & 0 & 0 & 0 \\ 0 & 1/2 & 0 & 0 & 0 & 0 \\ 0 & 0 & 1/2 & 0 & 0 & 0 \\ 0 & 0 & 0 & 1/2 & 0 & 0 \\ 0 & 0 & 0 & 0 & 1/3 & 0 \\ 0 & 0 & 0 & 0 & 0 & 1/3 \end{bmatrix}$$

so that

$$(5.5.7)\qquad \mathbf{H} = (\mathbf{X}'\mathbf{X})^{-}\mathbf{X}'\mathbf{X} = \begin{bmatrix} -1 & -1/3 & -1/3 & -1/3 & -1/2 & -1/2 \\ 1 & 1 & 0 & 0 & 1/2 & 1/2 \\ 1 & 0 & 1 & 0 & 1/2 & 1/2 \\ 1 & 0 & 0 & 1 & 1/2 & 1/2 \\ 1 & 1/3 & 1/3 & 1/3 & 1 & 0 \\ 1 & 1/3 & 1/3 & 1/3 & 0 & 1 \end{bmatrix}$$

More generally,

$$(5.5.8)\qquad (\mathbf{X}'\mathbf{X})^{-} = \begin{bmatrix} -1/N & \mathbf{0}' & \mathbf{0}' \\ \mathbf{0} & c^{-1}\mathbf{I}_r & \mathbf{0} \\ \mathbf{0} & \mathbf{0} & r^{-1}\mathbf{I}_c \end{bmatrix}$$

and

(5.5.9)
$$\mathbf{H} = \begin{bmatrix} -1 & -r^{-1}\mathbf{1}_r' & -c^{-1}\mathbf{1}_c' \\ \mathbf{1}_r & \mathbf{I}_r & c^{-1}\mathbf{J}_{rc} \\ \mathbf{1}_c & r^{-1}\mathbf{J}_{cr} & \mathbf{I}_c \end{bmatrix}$$

By applying the Gauss-Markoff theorem to (5.5.4), we have

(5.5.10)
$$\hat{\boldsymbol{\beta}} = (\mathbf{X}'\mathbf{X})^{-}\mathbf{X}'\mathbf{y} = \begin{bmatrix} -y_{..} \\ y_{1.} \\ y_{2.} \\ y_{3.} \\ y_{.1} \\ y_{.2} \end{bmatrix}$$

or, for $(\mathbf{X}'\mathbf{X})^{-}$ as defined in (5.5.8),

(5.5.11)
$$\hat{\boldsymbol{\beta}} = \begin{bmatrix} -y_{..} \\ \hline y_{1.} \\ y_{2.} \\ \vdots \\ y_{r.} \\ \hline y_{.1} \\ y_{.2} \\ \vdots \\ y_{.c} \end{bmatrix}$$

where

$$y_{i.} = \frac{\sum_{j=1}^{c} y_{ij}}{c}, \quad y_{.j} = \frac{\sum_{i=1}^{r} y_{ij}}{r}, \quad \text{and} \quad y_{..} = \frac{\sum_{i=1}^{r}\sum_{j=1}^{c} y_{ij}}{N}$$

Unique linear unbiased estimates for the parametric functions $\psi = \mathbf{c}'\boldsymbol{\beta}$ in (5.5.2), where $\boldsymbol{\beta}$ is defined as in (5.5.11), are obtained by using (3.5.8) with $\mathbf{c}'\mathbf{H} = \mathbf{c}'$. The form of ψ and $\hat{\psi}$ is given by

(5.5.12)

$$\psi = \mathbf{c}'\boldsymbol{\beta} = \mathbf{t}'\mathbf{H}\boldsymbol{\beta} = -t_0(\mu + \alpha_. + \beta_.) + \sum_{i=1}^{r} t_i(\mu + \alpha_i + \beta_.) + \sum_{j=1}^{c} t_j'(\mu + \alpha_. + \beta_j)$$

$$\hat{\psi} = \mathbf{c}'\hat{\boldsymbol{\beta}} = \mathbf{t}'\mathbf{H}\hat{\boldsymbol{\beta}} = -t_0 y_{..} + \sum_{i=1}^{r} t_i y_{i.} + \sum_{j=1}^{c} t_j' y_{.j}$$

for arbitrary vectors $\mathbf{t}' = [t_0, t_1, \ldots, t_r, t_1', \ldots, t_c']$ with

$$\alpha_. = \frac{\sum_{i=1}^{r} \alpha_i}{r} \quad \text{and} \quad \beta_. = \frac{\sum_{j=1}^{c} \beta_j}{c}$$

For the example given in (5.5.4), (5.5.12) reduces to

$$\begin{aligned} \psi &= -t_0(\mu + \alpha_. + \beta_.) + (\mu + \alpha_1 + \beta_.)t_1 + (\mu + \alpha_2 + \beta_.)t_2 + (\mu + \alpha_3 + \beta_.)t_3 \\ &\quad + (\mu + \alpha_. + \beta_1)t_1' + (\mu + \alpha_. + \beta_2)t_2' \\ \hat{\psi} &= -t_0 y_{..} + t_1 y_{1.} + t_2 y_{2.} + t_3 y_{3.} + t_1' y_{.} + t_2' y_{.2} \end{aligned} \tag{5.5.13}$$

From (5.5.13), estimable parametric functions are determined. For example, $\psi = \beta_1 - \beta_2$ and $\psi = \mu + (\alpha_1 + \alpha_2 + \alpha_3)/3 + (\beta_1 + \beta_2)/2$ are estimable, and are estimated by $\hat{\psi} = y_{.1} - y_{.2}$ and $\hat{\psi} = y_{..}$, respectively. However, μ and individual effects are not estimable since, for $\mathbf{c}' = [0, 1_i, 0, \ldots, 0]$, $\mathbf{c}'\mathbf{H} \neq \mathbf{c}'$, for any vector $\mathbf{c}$ with a 1 in the ith position. By choosing the side conditions $\alpha_. = 0$ and $\beta_. = 0$, μ becomes estimable and is estimated by $\hat{\mu} = y_{..}$. This then allows the individual effects to be estimated. For example, $\psi = \mu + \beta_1 + \beta_.$ and $\hat{\psi} = y_{.1}$, so that $\hat{\beta}_1 = y_{.1} - y_{..}$.

Although the hypothesis of differences among blocks is not of primary interest, it is usually tested to evaluate the effect of the blocking variable. To test the hypothesis of equal block effects,

$$H_0 : \alpha_1 = \alpha_2 = \cdots = \alpha_r \tag{5.5.14}$$

we need a matrix $\mathbf{C}$ to represent (5.5.14) in the form $\mathbf{C\beta} = \mathbf{0}$. Let $\mathbf{C}$ be defined by

$$\underset{[(r-1)\times q]}{\mathbf{C}} = \left[\begin{array}{c|cccc|c|ccc} 0 & 1 & 0 & \cdots & 0 & -1 & 0 & \cdots & 0 \\ 0 & 0 & 1 & \cdots & 0 & -1 & 0 & \cdots & 0 \\ \vdots & \vdots & \vdots & & \vdots & \vdots & \vdots & & \vdots \\ 0 & 0 & \cdot & \cdots & 1 & -1 & 0 & \cdots & 0 \end{array}\right] = [\mathbf{0} \quad \mathbf{I}_{r-1} \quad -\mathbf{1} \quad \mathbf{0}_c] \tag{5.5.15}$$

where the $R(\mathbf{C}) = r - 1$. To find Q_h for testing (5.5.11), where

$$Q_h = (\mathbf{C}\hat{\boldsymbol{\beta}})'[\mathbf{C}(\mathbf{X}'\mathbf{X})^-\mathbf{C}']^{-1}(\mathbf{C}\hat{\boldsymbol{\beta}}) \tag{5.5.16}$$

observe that $c[\mathbf{C}(\mathbf{X}'\mathbf{X})^-\mathbf{C}'] = \mathbf{I}_{r-1} + \mathbf{J}_{r-1,r-1}$; by (5.2.19), $[\mathbf{C}(\mathbf{X}'\mathbf{X})^-\mathbf{C}']^{-1} = c[\mathbf{I}_{r-1} - (1/r)\mathbf{J}_{r-1,r-1}]$. Furthermore, the symmetric matrix

$$N[\mathbf{C}(\mathbf{X}'\mathbf{X})^-\mathbf{X}']'[\mathbf{C}(\mathbf{X}'\mathbf{X})^-\mathbf{C}']^{-1}\mathbf{C}(\mathbf{X}'\mathbf{X})^-\mathbf{X}' = N\mathbf{Q}$$

$$= \begin{bmatrix} (r-1)\mathbf{J}_{cc} & & & \\ -\mathbf{J}_{cc} & (r-1)\mathbf{J}_{cc} & & \\ \vdots & \vdots & \ddots & \\ -\mathbf{J}_{cc} & -\mathbf{J}_{cc} & \cdots & (r-1)\mathbf{J}_{cc} \end{bmatrix} \tag{5.5.17}$$

so that

$$Q_h = \mathbf{y}'\mathbf{Q}\mathbf{y} = c \sum_{i=1}^{r} (y_{i.} - y_{..})^2 \tag{5.5.18}$$

where the degrees of freedom for Q_h are $r - 1$. In a similar manner, the hypothesis sum of squares for testing for treatment effects,

$$H_0: \beta_1 = \beta_2 = \cdots = \beta_c \tag{5.5.19}$$

is

$$Q_h = r \sum_{j=1}^{c} (y_{.j} - y_{..})^2 \tag{5.5.20}$$

when $\mathbf{C}$ is defined by

$$\underset{[(c-1)\times q]}{\mathbf{C}} = [\mathbf{0} \quad \mathbf{0}_r \quad \mathbf{I}_{c-1} \quad -\mathbf{1}] \tag{5.5.21}$$

and the $R(\mathbf{C}) = c - 1$. To evaluate the sum of squares Q_e for error, it may be shown that

$$\begin{aligned} Q_e &= \mathbf{y}'[\mathbf{I} - \mathbf{X}(\mathbf{X}'\mathbf{X})^-\mathbf{X}']\mathbf{y} \\ &= \sum_{i=1}^{r} \sum_{j=1}^{c} (y_{ij} - y_{i.} - y_{.j} + y_{..})^2 \end{aligned} \tag{5.5.22}$$

where the degrees of freedom for Q_e are $N - (r + c - 1) = rc - r - c + 1 = (r - 1)(c - 1)$. ANOVA Table 5.5.1 is now constructed to test (5.5.14) and (5.5.19).

TABLE 5.5.1. ANOVA Table for an RB-c Design.

Source	df	SS
Blocks A	$r - 1$	$Q_{h_r} = c \sum_{i=1}^{r} (y_{i.} - y_{..})^2$
Treatments B	$c - 1$	$Q_{h_c} = r \sum_{j=1}^{c} (y_{.j} - y_{..})^2$
Within error	$(r - 1)(c - 1)$	$Q_e = \sum_{i=1}^{r} \sum_{j=1}^{c} (y_{ij} - y_{i.} - y_{.j} + y_{..})^2$
"Total"	$N - 1$	$Q_t = \sum_{i=1}^{r} \sum_{j=1}^{c} (y_{ij} - y_{..})^2$

The hypotheses of equal block and treatment effects are rejected at the significance level α if

$$\begin{aligned} &\text{Blocks:} \quad F = \frac{Q_{h_r}/(r - 1)}{Q_e/(r - 1)(c - 1)} > F^{\alpha}[r - 1, (r - 1)(c - 1)] \\ &\text{Treatments:} \quad F = \frac{Q_{h_c}/(c - 1)}{Q_e/(r - 1)(c - 1)} > F^{\alpha}[c - 1, (r - 1)(c - 1)] \end{aligned} \tag{5.5.23}$$

On rejection of the treatment hypothesis in (5.5.23), Scheffé-type simultaneous intervals are constructed to help determine which contrasts in the effects are significant. Simultaneous intervals for contrasts in the row space of $\mathbf{C}$ are

$$\text{(5.5.24)} \quad \sum_{j=1}^{c} c_j y_{.j} - c_0 s \left(\frac{\sum_{j=1}^{c} c_j^2}{r} \right)^{1/2} \leq \sum_{j=1}^{c} c_j \beta_{j}. \leq \sum_{j=1}^{c} c_j y_{.j} + c_0 s \left(\frac{\sum_{j=1}^{c} c_j^2}{r} \right)^{1/2}$$

which are of the form $\hat{\psi} - c_0\hat{\sigma}_{\hat{\psi}} \leq \psi \leq \hat{\psi} + c_0\hat{\sigma}_{\hat{\psi}}$, where $c_0^2 = (c-1)F^{\alpha}[(c-1), (r-1)(c-1)]$ and $s^2 = [Q_e/(r-1)(c-1)]$.

Alternatively, when q comparisons are of interest, the Bonferroni t procedure may be employed. For q comparisons, intervals are obtained by evaluating

$$\text{(5.5.25)} \qquad \hat{\psi}_i - t^{\alpha/2q}_{(r-1)(c-1)}\hat{\sigma}_{\hat{\psi}_i} \leq \psi_i \leq \hat{\psi}_i + t^{\alpha/2q}_{(r-1)(c-1)}\hat{\sigma}_{\hat{\psi}_i} \qquad i = 1, \ldots, q$$

Multivariate Procedure. To extend the univariate model to the multivariate situation, individual observations y_{ij} are replaced by vector-valued observations $\mathbf{y}_{ij}$. To be more explicit, let $\mathbf{y}_{ij}$ denote the p-variate vector observation in the ith block and jth treatment classification, where $\mathbf{y}'_{ij} = [y_{ij1}, y_{ij2}, \ldots, y_{ijp}]$. Then (5.5.2) becomes

(5.5.26)

$$\begin{bmatrix} \mathbf{y}'_{11} \\ \mathbf{y}'_{12} \\ \vdots \\ \mathbf{y}'_{1c} \\ \hline \mathbf{y}'_{21} \\ \mathbf{y}'_{22} \\ \vdots \\ \mathbf{y}'_{2c} \\ \hline \vdots \\ \hline \mathbf{y}'_{r1} \\ \mathbf{y}'_{r2} \\ \vdots \\ \mathbf{y}'_{rc} \end{bmatrix} = \begin{bmatrix} 1 & 1 & 0 & \cdots & 0 & 1 & 0 & \cdots & 0 \\ 1 & 1 & 0 & \cdots & 0 & 0 & 1 & \cdots & 0 \\ \vdots & \vdots & \vdots & \vdots & \vdots & \vdots & \vdots & \vdots & \vdots \\ 1 & 1 & 0 & \cdots & 0 & 0 & 0 & \cdots & 1 \\ \hline 1 & 0 & 1 & \cdots & 0 & 1 & 0 & \cdots & 0 \\ 1 & 0 & 1 & \cdots & 0 & 0 & 1 & \cdots & 0 \\ \vdots & \vdots & \vdots & \vdots & \vdots & \vdots & \vdots & \vdots & \vdots \\ 1 & 0 & 1 & \cdots & 0 & 0 & 0 & \cdots & 1 \\ \hline \vdots & \vdots & \vdots & \vdots & \vdots & \vdots & \vdots & \vdots & \vdots \\ \hline 1 & 0 & 0 & \cdots & 1 & 1 & 0 & \cdots & 0 \\ 1 & 0 & 0 & \cdots & 1 & 0 & 1 & \cdots & 0 \\ \vdots & \vdots & \vdots & \vdots & \vdots & \vdots & \vdots & \vdots & \vdots \\ 1 & 0 & 0 & \cdots & 1 & 0 & 0 & \cdots & 1 \end{bmatrix} \begin{bmatrix} \mu_{11} & \mu_{12} & \cdots & \mu_{1p} \\ \alpha_{11} & \alpha_{12} & \cdots & \alpha_{1p} \\ \alpha_{21} & \alpha_{22} & \cdots & \alpha_{2p} \\ \vdots & \vdots & \vdots & \vdots \\ \alpha_{r1} & \alpha_{r2} & \cdots & \alpha_{rp} \\ \beta_{11} & \beta_{12} & \cdots & \beta_{1p} \\ \beta_{21} & \beta_{22} & \cdots & \beta_{2p} \\ \vdots & \vdots & \vdots & \vdots \\ \beta_{c1} & \beta_{c2} & \cdots & \beta_{cp} \end{bmatrix} + \mathbf{E}_0$$

$$\underset{(N \times p)}{\mathbf{Y}} = \underset{(N \times q)}{\mathbf{X}} \quad \underset{(q \times p)}{\mathbf{B}} + \underset{(N \times p)}{\mathbf{E}_0}$$

where

$$E(\mathbf{Y}) = \mathbf{XB} \qquad \text{(5.5.27)}$$
$$V(\mathbf{Y}) = \mathbf{I} \otimes \boldsymbol{\Sigma}$$

and each row of $\mathbf{Y}$ has a multivariate normal distribution, with $N = rc$ and $q = r + c + 1$.

Alternatively, let

$$\mathbf{B}' = [\boldsymbol{\mu} \quad \boldsymbol{\alpha}_1 \quad \boldsymbol{\alpha}_2 \quad \cdots \quad \boldsymbol{\alpha}_r \quad \boldsymbol{\beta}_1 \quad \boldsymbol{\beta}_2 \quad \cdots \quad \boldsymbol{\beta}_c] \qquad \text{(5.5.28)}$$

then each row of $\mathbf{Y}$ has the form

$$\mathbf{y}_{ij} = \boldsymbol{\mu} + \boldsymbol{\alpha}_i + \boldsymbol{\beta}_j + \boldsymbol{\varepsilon}_{ij} \qquad \boldsymbol{\varepsilon}_{ij} \sim IN_p(\mathbf{0}, \boldsymbol{\Sigma}) \qquad \text{(5.5.29)}$$

where $\boldsymbol{\varepsilon}_{ij}$ is a row from $\mathbf{E}_0$, which is a direct extension of (5.5.1). Following the abbreviated notation of the univariate randomized block design, the multivariate design could be denoted by MRB-c.

In the multivariate model, $(\mathbf{X}'\mathbf{X})^-$ and $\mathbf{H}$ are identical to (5.5.8) and (5.5.9) given in the univariate case; however,

$$\mathbf{B} = \begin{bmatrix} -\mathbf{y}'_{..} \\ \mathbf{y}'_{1.} \\ \vdots \\ \mathbf{y}'_{r.} \\ \mathbf{y}'_{.1} \\ \mathbf{y}'_{.2} \\ \vdots \\ \mathbf{y}'_{.c} \end{bmatrix} \qquad \text{(5.5.30)}$$

where now

$$\mathbf{y}_{i.} = \frac{\sum_{j=1}^{c} \mathbf{y}_{ij}}{c}$$
$$\mathbf{y}_{.j} = \frac{\sum_{i=1}^{r} \mathbf{y}_{ij}}{r} \qquad \text{(5.5.31)}$$
$$\mathbf{y}_{..} = \frac{\sum_{i=1}^{r} \sum_{j=1}^{c} \mathbf{y}_{ij}}{N}$$

represent vector averages.

With $\mathbf{c}'\mathbf{H} = \mathbf{c}'$, unique linear unbiased estimators for the estimable parametric functions $\psi = \mathbf{c}'\mathbf{Ba}$, a generalization of (5.5.12), are acquired by using (3.6.17)

$$\psi = \mathbf{c}'\mathbf{Ba} = \mathbf{a}'\left[-t_0(\boldsymbol{\mu} + \boldsymbol{\alpha}_. + \boldsymbol{\beta}_.) + \sum_{i=1}^{r} t_i(\boldsymbol{\mu} + \boldsymbol{\alpha}_i + \boldsymbol{\beta}_.) + \sum_{j=1}^{c} t'_j(\boldsymbol{\mu} + \boldsymbol{\alpha}_. + \boldsymbol{\beta}_j)\right] \qquad \text{(5.5.32)}$$
$$\hat{\psi} = \mathbf{c}'\hat{\mathbf{B}}\mathbf{a} = \mathbf{a}'\left(-t_0\mathbf{y}_{..} + \sum_{i=1}^{r} t_i\mathbf{y}_{i.} + \sum_{j=1}^{c} t'_j\mathbf{y}_{.j}\right)$$

where $\mathbf{t}' = [t_0, t_1, \ldots, t_r, t'_1, \ldots, t'_c]$,

$$\boldsymbol{\alpha}_. = \frac{\sum_{i=1}^{r} \boldsymbol{\alpha}_i}{r} \quad \text{and} \quad \boldsymbol{\beta}_. = \frac{\sum_{j=1}^{c} \boldsymbol{\beta}_j}{c}$$

To test the null hypotheses of equal block or treatment effects

$$H_0: \boldsymbol{\alpha}_1 = \boldsymbol{\alpha}_2 = \cdots = \boldsymbol{\alpha}_r \quad \text{(blocks)} \tag{5.5.33}$$

$$H_0: \boldsymbol{\beta}_1 = \boldsymbol{\beta}_2 = \cdots = \boldsymbol{\beta}_c \quad \text{(treatments)} \tag{5.5.34}$$

matrices $\mathbf{C}$, $\mathbf{A}$, and $\boldsymbol{\Gamma}$ are defined to represent (5.5.33) and (5.5.34) by $H_0: \mathbf{CBA} = \boldsymbol{\Gamma}$. Setting $\mathbf{A} = \mathbf{I}, \boldsymbol{\Gamma} = \mathbf{0}$, and $\mathbf{C}$ as in the univariate case, the hypothesis sums of products matrices to test for block and treatment effects are:

$$\begin{aligned}
\mathbf{Q}_{h_r} &= (\mathbf{C\hat{B}A})'[\mathbf{C}(\mathbf{X}'\mathbf{X})^-\mathbf{C}']^{-1}(\mathbf{C\hat{B}A}) \\
&= c \sum_{i=1}^{r} (\mathbf{y}_{i.} - \mathbf{y}_{..})(\mathbf{y}_{i.} - \mathbf{y}_{..})' \\
\mathbf{Q}_{h_c} &= (\mathbf{C\hat{B}A})'[\mathbf{C}(\mathbf{X}'\mathbf{X})^-\mathbf{C}']^{-1}(\mathbf{C\hat{B}A}) \\
&= r \sum_{j=1}^{c} (\mathbf{y}_{.j} - \mathbf{y}_{..})(\mathbf{y}_{.j} - \mathbf{y}_{..})'
\end{aligned} \tag{5.5.35}$$

The MANOVA table to test (5.5.33) and (5.5.34) is given in Table 5.5.2.

In Table 5.5.2, the characteristic equations

$$|\mathbf{Q}_{h_r} - \lambda \mathbf{Q}_e| = 0 \quad \text{and} \quad |\mathbf{Q}_{h_c} - \lambda \mathbf{Q}_e| = 0 \tag{5.5.36}$$

must be solved. Using the Roy criterion, the parameters s, m, and n become

$$\begin{array}{ll}
r \text{ blocks} & c \text{ treatments} \\
s = \min(r - 1, p) & s = \min(c - 1, p) \\
m = \dfrac{|r - p - 1| - 1}{2} & m = \dfrac{|c - p - 1| - 1}{2} \\
n = \dfrac{N - r - c - p}{2} & n = \dfrac{N - r - c - p}{2}
\end{array} \tag{5.5.37}$$

TABLE 5.5.2. MANOVA Table for an MRB-c Design.

Source	df	SSP
Blocks A	$r - 1$	$\mathbf{Q}_{h_r} = c \sum_{i=1}^{r} (\mathbf{y}_{i.} - \mathbf{y}_{..})(\mathbf{y}_{i.} - \mathbf{y}_{..})'$
Treatments B	$c - 1$	$\mathbf{Q}_{h_c} = r \sum_{j=1}^{c} (\mathbf{y}_{.j} - \mathbf{y}_{..})(\mathbf{y}_{.j} - \mathbf{y}_{..})'$
Within error	$(r - 1)(c - 1)$	$\mathbf{Q}_e = \sum_{i=1}^{r} \sum_{j=1}^{c} (\mathbf{y}_{ij} - \mathbf{y}_{i.} - \mathbf{y}_{.j} + \mathbf{y}_{..})(\mathbf{y}_{ij} - \mathbf{y}_{i.} - \mathbf{y}_{ij} + \mathbf{y}_{..})'$
"Total"	$N - 1$	$\mathbf{Q}_t = \sum_{i=1}^{r} \sum_{j=1}^{c} (\mathbf{y}_{ij} - \mathbf{y}_{..})(\mathbf{y}_{ij} - \mathbf{y}_{..})'$

where the hypotheses are rejected at the significance level α if

$$\theta_s > \theta^\alpha(s, m, n) \tag{5.5.38}$$

The other criteria follow immediately from (5.3.16).

Following (4.6.31), simultaneous intervals may be constructed for the estimable functions $\psi = \mathbf{c}'\mathbf{B}\mathbf{a}$. From (5.3.17), intervals for simple comparisons are established for treatment effects:

$$\begin{aligned}
\psi &= \sum_{s=1}^{p} a_s(\beta_{js} - \beta_{j's}) \\
\hat{\psi} &= \sum_{s=1}^{p} a_s(y_{.js} - y_{.j's}) \\
\hat{\sigma}_{\hat{\psi}} &= \sqrt{\left(\frac{2}{r}\right)\left(\frac{\mathbf{a}'\mathbf{Q}_e\mathbf{a}}{v_e}\right)} \\
c_0^2 &= \frac{v_e\theta^\alpha}{1 - \theta^\alpha}
\end{aligned} \tag{5.5.39}$$

$$\hat{\psi} - c_0\hat{\sigma}_{\hat{\psi}} \leq \psi \leq \hat{\psi} + c_0\hat{\sigma}_{\hat{\psi}}$$

Using the Bonferroni procedure, confidence intervals are procured by evaluating

$$\hat{\psi} - t^{\alpha/2q}_{(r-1)(c-1)}\hat{\sigma}_{\hat{\psi}} \leq \psi \leq \hat{\psi} + t^{\alpha/2q}_{(r-1)(c-1)}\hat{\sigma}_{\hat{\psi}} \tag{5.5.40}$$

for q comparisons.

As in the one-way layout, discriminants using $\mathbf{Q}_{h_c}$ with $\mathbf{Q}_e$ may be evaluated to help locate variables that are significant to the rejection of the overall treatment hypothesis. These procedures are illustrated through the use of an example in Section 5.7.*

EXERCISES 5.5

1. Verify (5.5.17) and, by using (5.5.16), prove (5.5.18).
2. Determine the expected value of the SSP matrices in Table 5.5.2.
3. In a learning experiment with four males and four females, 150 trials were given to each subject in the following manner.

(1)	Morning	With training	Nonsense words
(2)	Afternoon	With training	Letter words
(3)	Afternoon	No training	Nonsense words
(4)	Morning	No training	Letter words

Using the number of trials to criterion for five-letter (F) and seven-letter (S) "words," the data for the randomized block design follows.

* The results discussed in sections 5.3 and 5.5 have also been given by John (1970).

Treatment Conditions

		1	2	3	4
Blocks	M	F 120 S 130	F 90 S 100	F 140 S 150	F 70 S 85
	F	F 70 S 80	F 30 S 60	F 100 S 110	F 20 S 35

Using $\alpha = .05$, analyze the data by using the most appropriate multivariate methods.

4. In a pilot study designed to investigate the mental abilities of four ethnic groups and four socioeconomic-status (SES) classifications, high SES males (HM), high SES females (HF), low SES males (LM), and low SES females (LF), the following table of data on three different measures, mathematics (MAT), English (ENG), and general knowledge (GK), was obtained.

Ethnic Groups

		I			II			III			IV		
		MAT	ENG	GK	MAT	ENG	GK	MAT	ENG	GK	MAT	ENG	GK
SES	HM	80	60	70	85	65	75	90	70	80	95	75	85
	HF	85	65	75	89	69	80	94	76	85	96	80	90
	LM	89	78	81	91	73	82	99	81	90	100	85	97
	LF	92	82	85	100	80	90	105	84	93	110	90	101

a. Carry out a multivariate, analysis-of-variance procedure to investigate differences in SES and ethnic groups using $\alpha = .05$ for each test.

b. Discuss the effectiveness of the blocking variable, socioeconomic status.

5.6 TWO-WAY LAYOUT WITH INTERACTION—UNIVARIATE AND MULTIVARIATE

In Section 5.5, the RB-c and the MRB-c designs were discussed. Depending on the assignment of the subjects in these designs, it was indicated that the designs may be viewed as completely randomized, two-way factorial designs with one observation per cell. This design is reviewed in this section; however, the observations per cell n_0 are greater than one. For this design, the experimenter is interested in two factors or treatment combinations for which a random sample of rcn_0 subjects are randomly divided into rc samples of size n_0 that are randomly assigned to rc treatments.

The notation sometimes used for this type of design is CRF-rc for the univariate case and MCRF-rc for the multivariate case. This indicates that we have a two-way layout with two factors, one with r levels and one with c levels with complete randomization of subjects. For $n_0 = 1$, the model for this design is identical to a randomized block design; however, in this case the experimenter is interested in both factors. For the additive case of $n_0 = 1$, it is assumed that differences between row (column) treatments are constant from column (row) to column (row). This is typically evidenced by the parallelism of cell means.

Univariate Procedure. A CRF-*rc* design is used to analyze two treatments (factor *A* and factor *B*) with $n_0 > 1$ observations per cell. This allows us to investigate the interaction between factors, so that the effect differences are not necessarily constant from treatment to treatment. To take this into account, the linear model with interaction becomes

$$\begin{aligned} &y_{ijk} = \mu + \alpha_i + \beta_j + \gamma_{ij} + \varepsilon_{ijk} \\ &\varepsilon_{ijk} \sim IN(0, \sigma^2) \qquad i = 1, \ldots, r; j = 1, \ldots, c; k = 1, \ldots, n_0 \end{aligned} \tag{5.6.1}$$

The data for the design are represented as follows:

		Treatment (factor *B*)			
		1	2	$\cdots$	3
		y_{111}	y_{121}	$\cdots$	y_{1c1}
		y_{112}	y_{122}	$\cdots$	y_{1c2}
	1	$\cdot$	$\cdot$	$\cdots$	$\cdot$
		y_{11n_0}	y_{12n_0}	$\cdots$	y_{1cn_0}
		y_{211}	y_{221}	$\cdots$	y_{2c1}
		y_{212}	y_{222}	$\cdots$	y_{2c2}
Treatment	2	$\cdot$	$\cdot$	$\cdots$	$\cdot$
(factor A)		y_{21n_0}	y_{22n_0}	$\cdots$	y_{2cn_0}
		$\vdots$	$\vdots$	$\vdots$	$\vdots$
		y_{r11}	y_{r21}	$\cdots$	y_{rc1}
		y_{r12}	y_{r22}	$\cdots$	y_{rc2}
	r	$\cdot$	$\cdot$	$\cdots$	$\cdot$
		y_{rin_0}	y_{r2n_0}	$\cdots$	y_{rcn_0}

Considering a simple example, with $i = 1, 2, 3$; $j = 1, 2$; and $k = 1, 2$, the general linear model for (5.6.1) is written as

$$\begin{bmatrix} y_{111} \\ y_{112} \\ y_{121} \\ y_{122} \\ y_{211} \\ y_{212} \\ y_{221} \\ y_{222} \\ y_{311} \\ y_{312} \\ y_{321} \\ y_{322} \end{bmatrix} = \begin{bmatrix} 1 & 1 & 0 & 0 & 1 & 0 & 1 & 0 & 0 & 0 & 0 & 0 \\ 1 & 1 & 0 & 0 & 1 & 0 & 1 & 0 & 0 & 0 & 0 & 0 \\ 1 & 1 & 0 & 0 & 0 & 1 & 0 & 1 & 0 & 0 & 0 & 0 \\ 1 & 1 & 0 & 0 & 0 & 1 & 0 & 1 & 0 & 0 & 0 & 0 \\ 1 & 0 & 1 & 0 & 1 & 0 & 0 & 0 & 1 & 0 & 0 & 0 \\ 1 & 0 & 1 & 0 & 1 & 0 & 0 & 0 & 1 & 0 & 0 & 0 \\ 1 & 0 & 1 & 0 & 0 & 1 & 0 & 0 & 0 & 1 & 0 & 0 \\ 1 & 0 & 1 & 0 & 0 & 1 & 0 & 0 & 0 & 1 & 0 & 0 \\ 1 & 0 & 0 & 1 & 1 & 0 & 0 & 0 & 0 & 0 & 1 & 0 \\ 1 & 0 & 0 & 1 & 1 & 0 & 0 & 0 & 0 & 0 & 1 & 0 \\ 1 & 0 & 0 & 1 & 0 & 1 & 0 & 0 & 0 & 0 & 0 & 1 \\ 1 & 0 & 0 & 1 & 0 & 1 & 0 & 0 & 0 & 0 & 0 & 1 \end{bmatrix} \begin{bmatrix} \mu \\ \alpha_1 \\ \alpha_2 \\ \alpha_3 \\ \beta_1 \\ \beta_2 \\ \gamma_{11} \\ \gamma_{12} \\ \gamma_{21} \\ \gamma_{22} \\ \gamma_{31} \\ \gamma_{32} \end{bmatrix} + \begin{bmatrix} \varepsilon_{111} \\ \varepsilon_{112} \\ \varepsilon_{121} \\ \varepsilon_{122} \\ \varepsilon_{211} \\ \varepsilon_{212} \\ \varepsilon_{221} \\ \varepsilon_{222} \\ \varepsilon_{311} \\ \gamma_{312} \\ \varepsilon_{321} \\ \varepsilon_{322} \end{bmatrix} \tag{5.6.2}$$

which is of the form

(5.6.3)
$$\underset{(N\times 1)}{\mathbf{y}} = \underset{(N\times q)}{\mathbf{X}}\underset{(q\times 1)}{\boldsymbol{\beta}} + \underset{(N\times 1)}{\boldsymbol{\varepsilon}}$$

where $N = rcn_0$ and $q = 1 + r + c + rc$. From (5.6.2), we see that the $R(\mathbf{X}) = rc$ since the first six columns are linear combinations of the last six. Thus $(\mathbf{X}'\mathbf{X})^{-1}$ does not exist. By partitioning $\mathbf{X}$ into

(5.6.4)
$$\mathbf{X} = [\mathbf{X}_1 \quad \mathbf{X}_2]$$

where the order of $\mathbf{X}_1$ is $N \times q_1$ $(q_1 = 1 + r + c)$ and the order of $\mathbf{X}_2$ is $N \times q_2$ $(q_2 = rc)$,

$$\mathbf{X}'\mathbf{X} = \begin{bmatrix} \mathbf{X}_1'\mathbf{X}_1 & \mathbf{X}_1'\mathbf{X}_2 \\ \mathbf{X}_2'\mathbf{X}_1 & \mathbf{X}_2'\mathbf{X}_2 \end{bmatrix}$$

and a g-inverse of $\mathbf{X}'\mathbf{X}$ is seen to be

(5.6.5)
$$(\mathbf{X}'\mathbf{X})^{-} = \begin{bmatrix} \mathbf{0} & \mathbf{0} \\ \mathbf{0} & \text{Dia}\,[1/n_0] \end{bmatrix}$$

where Dia $[1/n_0]$ is a $rc \times rc$ diagonal matrix.

Applying the Gauss-Markoff theorem to (5.6.2), with $(\mathbf{X}'\mathbf{X})^{-}$ as defined in (5.6.5), an estimate of $\boldsymbol{\beta}$ is given by

(5.6.6)
$$\hat{\boldsymbol{\beta}}' = [0, 0, 0, 0, 0, 0, y_{11.}, y_{12.}, y_{21.}, y_{22.}, y_{31.}, y_{32.}]$$

where

(5.6.7)
$$y_{ij.} = \frac{\sum_{k=1}^{n_0} y_{ijk}}{n_0}$$
$$y_{i..} = \frac{\sum_{j=1}^{c}\sum_{k=1}^{n_0} y_{ijk}}{cn_0}$$
$$y_{.j.} = \frac{\sum_{i=1}^{r}\sum_{k=1}^{n_0} y_{ijk}}{rn_0}$$
$$y_{...} = \frac{\sum_{i=1}^{r}\sum_{j=1}^{c}\sum_{k=1}^{n_0} y_{ijk}}{N}$$

From the form of $\hat{\boldsymbol{\beta}}$ in (5.6.6), $\hat{\boldsymbol{\beta}}$ in the general case is written by analogy.

Unique linear unbiased estimates for the parametric functions $\psi = \mathbf{c}'\boldsymbol{\beta}$ are found by using (3.5.8):

(5.6.8)
$$\psi = \mathbf{c}'\boldsymbol{\beta} = \mathbf{t}'\mathbf{H}\boldsymbol{\beta} = \sum_{i=1}^{r}\sum_{j=1}^{c} t_{ij}(\mu + \alpha_i + \beta_j + \gamma_{ij})$$
$$= \mu\sum_i\sum_j t_{ij} + \sum_i\sum_j t_{ij}\alpha_i + \sum_i\sum_j t_{ij}\beta_j + \sum_i\sum_j y_{ij}\gamma_{ij}$$
$$\hat{\psi} = \mathbf{c}'\hat{\boldsymbol{\beta}} = \mathbf{t}'\mathbf{H}\hat{\boldsymbol{\beta}} = \sum_{i=1}^{r}\sum_{j=1}^{c} t_{ij}y_{ij.}$$

where $\mathbf{t}' = [t_0, t_1, \ldots, t_r, t'_1, \ldots, t'_c, t_{11}, t_{12}, \ldots, t_{rc}]$, $\mathbf{c}' = \mathbf{t}'\mathbf{H}$, and $\mathbf{H} = (\mathbf{X}'\mathbf{X})^{-}\mathbf{X}'\mathbf{X}$ for $(\mathbf{X}'\mathbf{X})^{-}$ defined by (5.6.5). From (5.6.8), $\psi = \mu + \alpha_i + \beta_j + \gamma_{ij} = \mu_{ij}$ (say) is estimable and can be estimated by $\hat{\psi} = \hat{\mu}_{ij} = y_{ij.}$. However, for this design, as with the randomized block design, the individual parameters are not estimable. Furthermore, without side conditions, ψ's involving only α's and β's are not obtainable with (5.6.8). For example, if

$$\sum_j t_{ij} = \sum_j t_{i'j} = 1 \qquad \text{for } i \neq i'$$

$$\psi = \alpha_i - \alpha_{i'} + \sum_{j=1}^{c} t_{ij}(\beta_j + \gamma_{ij}) - \sum_{j=1}^{c} t_{i'j}(\beta_j + \gamma_{i'j})$$

is estimable and can be estimated by

$$\hat{\psi} = \sum_j t_{ij} y_{i.j} - \sum_j t_{i'j} y_{i'j.}$$

Similarly,

$$\psi = \beta_j - \beta_{j'} + \sum_{i=1}^{r} t_{ij}(\alpha_i + \gamma_{ij}) - \sum_{i=1}^{r} t_{ij'}(\alpha_i + \gamma_{ij'})$$

is estimable if

$$\sum_i t_{ij} = \sum_i t_{ij'} = 1 \qquad \text{for } j \neq j'$$

By choosing $t_{ij} = 1/c$ for differences in the α's, $t_{ij} = 1/r$ for differences in the β's, and the side conditions $\beta = \Sigma_j \beta_j/c = 0$, $\alpha = \Sigma_i \alpha_i/r = 0$, $\gamma_{i.} = \Sigma_j \gamma_{ij}/c = 0$ and $\gamma_{.j} = \Sigma_i \gamma_{ij}/r = 0$ for all i and j, differences in α's and β's are acquired. The contrasts, $\psi = \alpha_i - \alpha_{i'}$ are then estimated by $\hat{\psi} = \Sigma_j y_{ij.}/c - \Sigma_j y_{i'j.}/c = y_{i..} - y_{i'..}$. Similarly, for $\psi = \beta_j - \beta_{j'}$, $\hat{\psi} = y_{.j.} - y_{.j'.}$.

Another parameter of interest in (5.6.1) is the interaction term γ_{ij}. Even though these terms are not individually estimable, *contrasts* in the γ_{ij}'s *are estimable.* To see this, 2×2 cells in our simple example are considered in Table 5.6.1 (solid

TABLE 5.6.1.

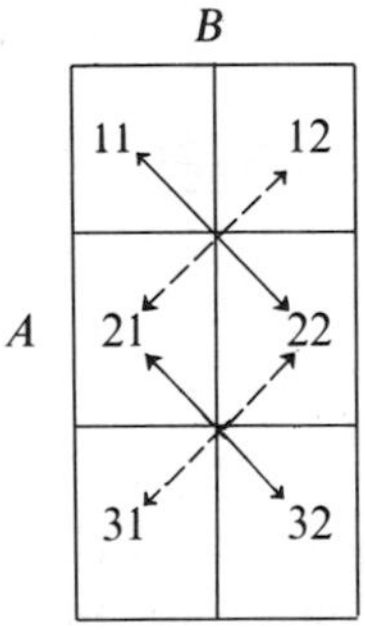

arrows indicate addition and broken arrows indicate subtraction). By letting $\psi = (\mu_{11} + \mu_{22}) - (\mu_{21} + \mu_{12})$, we see that $\psi = \gamma_{11} - \gamma_{21} - \gamma_{12} + \gamma_{22}$ is an estimable function. More generally, $\psi = \gamma_{ij} - \gamma_{i'j} - \gamma_{ij'} + \gamma_{i'j'}$ is estimable and can be estimated by $\hat{\psi} = y_{ij.} - y_{i'j.} + y_{ij'.} + y_{i'j'.}$ and *no side conditions are needed* to determine the estimable functions ψ involving only the γ_{ij}'s. To see more clearly what the contrasts in the γ_{ij}'s represent, consider plotting the means for A at each level of B. For example, in Figure 5.6.1, the lines are parallel if the slopes for each segment are equal. That is, $21 - 11 = 22 - 12$ or $\psi = \mu_{11} - \mu_{21} - \mu_{12} + \mu_{22} = 0$, and $32 - 22 = 31 - 21$ or

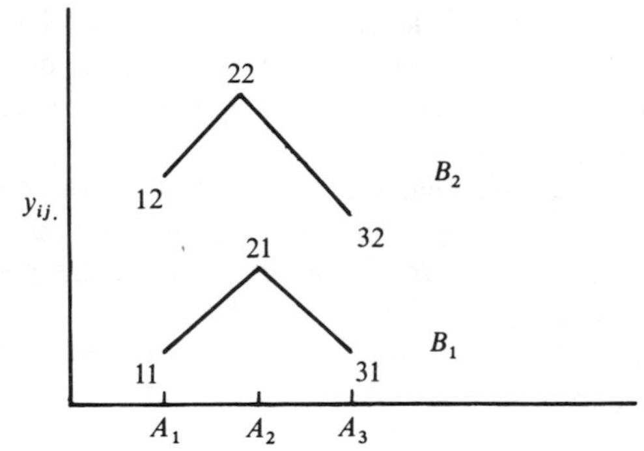

FIGURE 5.6.1. Plots of means of A at each level of B.

$\psi = \mu_{21} - \mu_{31} - \mu_{22} + \mu_{32} = 0$. Alternatively, by plotting the means of B at each level of A, Figure 5.6.2 results, where the lines are parallel if the slope for each segment is the same. That is, $11 - 12 = 21 - 22 = 31 - 32$ or $\psi = \mu_{11} - \mu_{21} - \mu_{12} + \mu_{22} = 0$ and $\psi = \mu_{21} - \mu_{31} - \mu_{22} + \mu_{32} = 0$. One other way to interpret the contrasts is to consider a general layout as represented in Figure 5.6.3. The contrast $\psi = \mu_{ij} - \mu_{i'j} - \mu_{ij'} + \mu_{i'j'} = (\mu_{ij} - \mu_{i'j}) - (\mu_{ij'} - \mu_{i'j'})$ represents the difference between the

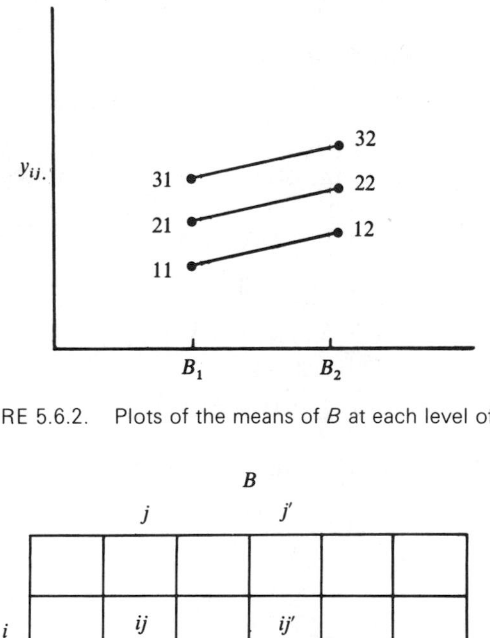

FIGURE 5.6.2. Plots of the means of B at each level of A.

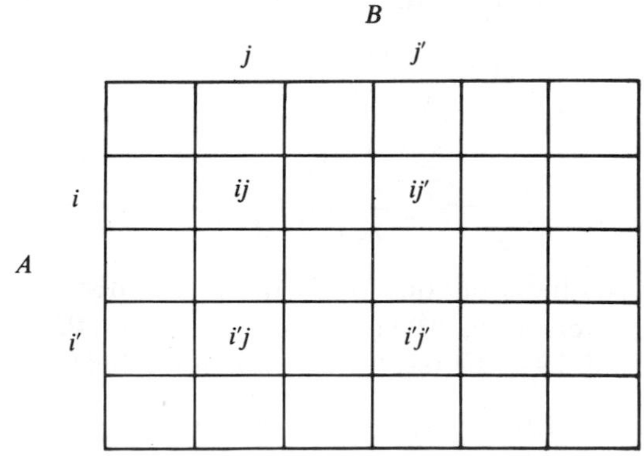

FIGURE 5.6.3.

differences of treatments i and i' for level j and i and i' at level j'. When these are not constant, an interaction is said to exist. Thus, *not* the γ_{ij}'s, but the *functions* of the γ_{ij}'s are of primary interest to the researcher. They indicate at which levels a crossing of segments or interaction exists.

It should be emphasized that not all contrasts in the cell means $y_{ij.}$ lead to estimable functions involving only γ_{ij}'s. For example, consider $\hat{\psi} = y_{11.} - y_{12.}$, a difference in cell means for B_1 and B_2 at the first level of the A effect. The estimator $\hat{\psi}$ is the estimate for $\psi = \mu_{11} - \mu_{12} = (\mu + \alpha_1 + \beta_1 + \gamma_{11}) - (\mu + \alpha_1 + \beta_2 + \gamma_{12}) = (\beta_1 - \beta_2) + (\gamma_{11} - \gamma_{12})$, which involves γ_{ij}'s confounded by the β_j's.

With the side conditions,

(5.6.9)
$$\begin{aligned} \alpha_. &= 0 \\ \beta_. &= 0 \\ \gamma_{i.} &= 0 \qquad \text{for all } i \\ \gamma_{.j} &= 0 \qquad \text{for all } j \end{aligned}$$

added to the model specification given in (5.6.1) the following familiar estimators for the model parameters are

(5.6.10)
$$\begin{aligned} \hat{\mu} &= y_{...} \\ \hat{\alpha}_i &= y_{i..} - y_{...} \\ \hat{\beta}_j &= y_{.j.} - y_{...} \\ \hat{\gamma}_{ij} &= y_{ij.} - y_{i..} - y_{.j.} + y_{...} \end{aligned}$$

for the CRF-rc design.

To test any hypothesis, an estimate of σ^2 is required. From the Gauss-Markoff theorem,

(5.6.11)
$$\begin{aligned} Q_e &= \mathbf{y}'[\mathbf{I} - \mathbf{X}(\mathbf{X}'\mathbf{X})^{-}\mathbf{X}']\mathbf{y} \\ &= \mathbf{y}'\mathbf{y} - \mathbf{y}'\mathbf{X}\hat{\boldsymbol{\beta}} \\ &= \sum_i \sum_j \sum_k y_{ijk}^2 - \sum_i \sum_j n_0 y_{ij.}^2 \\ & \sum_i \sum_j \sum_k (y_{ijk} - y_{ij.})^2 \end{aligned}$$

where $R(\mathbf{X}) = rc$ and $N = rcn_0$. Thus an unbiased estimator of σ^2 is

(5.6.12)
$$\begin{aligned} s^2 &= \frac{Q_e}{N - R(\mathbf{X})} \\ &= \frac{\sum_i \sum_j \sum_k (y_{ijk} - y_{ij.})^2}{rc(n_0 - 1)} \end{aligned}$$

The primary hypothesis of concern to the researcher in a CRF-rc design is the test for significant interactions or parallelism. Following the discussion associated with Figures 5.9.1 and 5.9.2, the test for no interaction becomes

(5.6.13)
$$\begin{aligned} H_0\colon \gamma_{11} - \gamma_{21} - \gamma_{12} + \gamma_{22} &= 0 \\ \gamma_{21} - \gamma_{31} - \gamma_{22} + \gamma_{32} &= 0 \end{aligned}$$

for our simple example. Representing (5.6.11) by $\mathbf{C}\boldsymbol{\beta} = \mathbf{0}$,

$$\underset{(2\times 12)}{\mathbf{C}} = \left[\begin{array}{c|cccccc} \underset{(2\times 6)}{\mathbf{0}} & 1 & -1 & -1 & 1 & 0 & 0 \\ & 0 & 0 & 1 & -1 & -1 & 1 \\ & \multicolumn{6}{c}{(2\times 6)} \end{array}\right] \tag{5.6.14}$$

is used to test (5.6.13). The rank of $\mathbf{C}$ is 2. More generally, let

$$\underset{[(c-1)\times c]}{\mathbf{G}} = \begin{bmatrix} 1 & -1 & 0 & 0 & 0 & \cdots & 0 & 0 \\ 0 & 1 & -1 & 0 & 0 & \cdots & 0 & 0 \\ \vdots & \vdots & \vdots & \vdots & \vdots & \vdots & \vdots & \vdots \\ 0 & 0 & 0 & 0 & 0 & \cdots & 1 & -1 \end{bmatrix} \tag{5.6.15}$$

Then a matrix $\mathbf{C}$ to test for parallelism becomes

$$\underset{[(r-1)(c-1)\times q]}{\mathbf{C}} = \left[\begin{array}{c|cccccc} & \mathbf{G} & -\mathbf{G} & \mathbf{0} & \cdots & \mathbf{0} & \mathbf{0} \\ \underset{[(r-1)(c-1)\times(1+r+c)]}{\mathbf{0}} & \mathbf{0} & \mathbf{G} & -\mathbf{G} & \cdots & \mathbf{0} & \mathbf{0} \\ & \vdots & \vdots & \vdots & \vdots & \vdots & \vdots \\ & \mathbf{0} & \mathbf{0} & \mathbf{0} & \cdots & \mathbf{G} & -\mathbf{G} \\ & \multicolumn{6}{c}{[(c-1)\times rc]} \end{array}\right] \tag{5.6.16}$$

where $q = 1 + r + c + rc$ and the $R(\mathbf{C}) = (r-1)(c-1)$. The matrix $\mathbf{C}$ in (5.6.16) is directed at testing whether the linearly independent functions, $\psi = \gamma_{ij} - \gamma_{i'j} - \gamma_{ij'} + \gamma_{i'j'} = 0$. When the side conditions in (5.6.9) are added to the model equation (5.6.1), this is equivalent to testing whether all γ_{ij}'s are 0.

An alternative form of $\mathbf{C}$ often used to test for interactions is

$$\underset{[(r-1)(c-1)\times q]}{\mathbf{C}} = \left[\begin{array}{c|cccccc} & \mathbf{G}^* & \mathbf{0} & \mathbf{0} & \cdots & \cdot & -\mathbf{G}^* \\ \mathbf{0} & \mathbf{0} & \mathbf{G}^* & \mathbf{0} & \cdots & \cdot & -\mathbf{G}^* \\ & \vdots & \vdots & \vdots & \vdots & \vdots & \vdots \\ & \mathbf{0} & \mathbf{0} & \mathbf{0} & \cdots & \mathbf{G}^* & -\mathbf{G}^* \end{array}\right] \tag{5.6.17}$$

where

$$\underset{[(c-1)\times c]}{\mathbf{G}^*} = \begin{bmatrix} 1 & 0 & \cdots & 0 & -1 \\ 0 & 1 & \cdots & 0 & -1 \\ \vdots & \vdots & \vdots & \vdots & \vdots \\ 0 & 0 & \cdots & 1 & -1 \end{bmatrix} \tag{5.6.18}$$

By using (5.6.17) in (5.6.2), the interaction hypothesis becomes

$$\begin{aligned} H_0\colon \gamma_{11} - \gamma_{31} - \gamma_{12} + \gamma_{32} &= 0 \\ \gamma_{21} - \gamma_{31} - \gamma_{22} + \gamma_{32} &= 0 \end{aligned} \tag{5.6.19}$$

By forming a table similar to Table 5.6.1, (5.6.18) is used to evaluate different 2×2 tables. This is illustrated in Table 5.6.2.

TABLE 5.6.2.

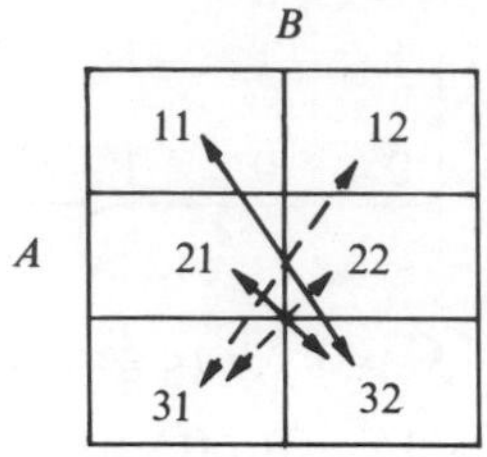

With **C** defined by (5.6.17) or (5.6.18), Q_h for the test of significant interactions can be shown to be

$$
\begin{aligned}
Q_h &= (\mathbf{C}\hat{\boldsymbol{\beta}})'[\mathbf{C}(\mathbf{X}'\mathbf{X})^{-}\mathbf{C}']^{-1}(\mathbf{C}\hat{\boldsymbol{\beta}}) \\
&= n_0 \sum_i \sum_j \sum_k (y_{ij.} - y_{i..} - y_{.j.} + y_{...})^2 \equiv Q_{AB} \qquad (5.6.20)
\end{aligned}
$$

where the degrees of freedom for Q_h are $(r - 1)(c - 1)$.

Following the test for interactions, tests for significant main effects are of concern. The hypotheses of interest, using (5.6.8), are

$$H_0: \alpha_1 + \sum_j t_{1j}(\beta_j + \gamma_{1j}) = \cdots = \alpha_r + \sum_j t_{rj}(\beta_j + \gamma_{rj}) \qquad (5.6.21)$$

$$H_0: \beta_1 + \sum_i t_{i1}(\alpha_i + \gamma_{i1}) = \cdots = \beta_c + \sum_i t_{ic}(\alpha_i + \gamma_{ic}) \qquad (5.6.22)$$

where

$$
\begin{aligned}
\sum_j t_{ij} &= \sum_j t_{i'j} = 1 \qquad i \neq i' \\
\sum_i t_{ij} &= \sum_i t_{ij'} = 1 \qquad j \neq j'
\end{aligned} \qquad (5.6.23)
$$

Selecting in (5.6.21) all weights equal to $1/c$ and in (5.6.22) all weights equal to $1/r$, (5.6.21) and (5.6.22) reduce to the null hypotheses

$$
\begin{aligned}
H_A: \alpha_i + \frac{\sum_j \gamma_{ij}}{c} &\text{ are equal for all } i \\
H_B: \beta_j + \frac{\sum_i \gamma_{ij}}{r} &\text{ are equal for all } j
\end{aligned} \qquad (5.6.24)
$$

respectively. The tests specified in (5.6.24) are often referred to as *main-effect tests.* They are used to help determine if there are mean differences in the levels of one factor when averaged over the levels of the other factor. If the model specified in (5.6.1) includes the side conditions $\Sigma_j \gamma_{ij} = 0$, for all j, and $\Sigma_i \gamma_{ij} = 0$, for all i, then the main-effect tests given in (5.6.24) have the following familiar form:

$$H_A: \alpha_1 = \alpha_2 = \cdots = \alpha_r \quad \text{and} \quad H_B: \beta_1 = \beta_2 = \cdots = \beta_c \qquad (5.6.25)$$

The sum of squares for testing (5.6.24) and (5.6.25) are, of course, identical; however, by employing (5.6.24), we know exactly what is being tested. This is especially important in unbalanced designs (see Section 5.17).

Using equal weights, the matrices $\mathbf{C}$ to test H_A and H_B in (5.6.24) become, for H_A,

(5.6.26)

$$\underset{[(r-1)\times q]}{\mathbf{C}} = \left[\begin{array}{c|ccccc|c|cccc|cccc|c|cccc|cccc} & 1 & 0 & \cdots & 0 & -1 & & 1/c & 1/c & \cdots & 1/c & 0 & 0 & \cdots & 0 & \cdot & 0 & 0 & \cdots & 0 & -1/c & -1/c & \cdots & -1/c \\ & 0 & 1 & \cdots & 0 & -1 & & 0 & 0 & \cdots & 0 & 1/c & 1/c & \cdots & 1/c & \cdot & 0 & 0 & \cdots & 0 & -1/c & -1/c & \cdots & -1/c \\ \mathbf{0} & \vdots & \vdots & & \vdots & \vdots & \mathbf{0}_c & \vdots & \vdots & & \vdots & \vdots & \vdots & & \vdots & \vdots & \vdots & \vdots & & \vdots & \vdots & \vdots & & \vdots \\ & 0 & 0 & \cdots & 1 & -1 & & 0 & 0 & \cdots & 0 & 0 & 0 & \cdots & 0 & \cdot & 1/c & 1/c & \cdots & 1/c & -1/c & -1/c & \cdots & -1/c \end{array}\right]$$

and, for H_B,

(5.6.27)

$$\underset{[(c-1)\times q]}{\mathbf{C}} = \left[\begin{array}{c|c|cccccc|ccccc|ccccc|c|ccccc} & & 1 & 0 & 0 & \cdots & 0 & -1 & 1/r & 0 & \cdots & 0 & -1/r & 1/r & 0 & \cdots & 0 & -1/r & \cdot & 1/r & 0 & \cdots & 0 & -1/r \\ & & 0 & 1 & 0 & \cdots & 0 & -1 & 0 & 1/r & \cdots & 0 & -1/r & 0 & 1/r & \cdots & 0 & -1/r & \cdot & 0 & 1/r & \cdots & 0 & -1/r \\ \mathbf{0} & \mathbf{0}_c & \vdots & \vdots & \vdots & & \vdots & \vdots & \vdots & \vdots & & \vdots & \vdots & \vdots & \vdots & & \vdots & \vdots & \vdots & \vdots & \vdots & & \vdots & \vdots \\ & & 0 & 0 & 0 & \cdots & 1 & -1 & 0 & 0 & \cdots & 1/r & -1/r & 0 & 0 & \cdots & 1/r & -1/r & \cdot & 0 & 0 & \cdots & 1/r & -1/r \end{array}\right]$$

where $q = 1 + r + c + rc$.

To test for mean differences for factor A, first observe that

$$n_0 c[\mathbf{C}(\mathbf{X}'\mathbf{X})^{-}\mathbf{C}'] = \mathbf{I}_{r-1} + \mathbf{J}_{r-1,r-1} \tag{5.6.28}$$

and, by (5.2.19)

$$[\mathbf{C}(\mathbf{X}'\mathbf{X})^{-}\mathbf{C}']^{-1} = n_0 c\left[\mathbf{I}_{r-1} - \frac{1}{r}(\mathbf{J}_{r-1,r-1})\right]$$

Furthermore, the symmetric matrix

$$N[\mathbf{C}(\mathbf{X}'\mathbf{X})^{-}\mathbf{X}']'[\mathbf{C}(\mathbf{X}'\mathbf{X})^{-}\mathbf{C}']^{-1}[\mathbf{C}(\mathbf{X}'\mathbf{X})^{-}\mathbf{X}'] = N\mathbf{Q}$$

$$= \begin{bmatrix} (r-1)\mathbf{J}_{nc,nc} & & & \\ -\mathbf{J}_{nc,nc} & (r-1)\mathbf{J}_{nc,nc} & & \\ \vdots & \vdots & \ddots & \\ -\mathbf{J}_{nc,nc} & -\mathbf{J}_{nc,nc} & \cdots & (r-1)\mathbf{J}_{nc,nc} \end{bmatrix} \tag{5.6.29}$$

so that Q_h becomes, for testing factor A,

$$Q_h = \mathbf{y}'\mathbf{Q}\mathbf{y} = n_0 c \sum_{i=1}^{r} (y_{i..} - y_{...})^2 \equiv Q_A \tag{5.6.30}$$

In a similar manner, Q_h for testing factor B can be shown to be

$$\begin{aligned} Q_h &= (\mathbf{C}\hat{\boldsymbol{\beta}})'[\mathbf{C}(\mathbf{X}'\mathbf{X})^{-}\mathbf{X}']^{-1}(\mathbf{C}\hat{\boldsymbol{\beta}}) \\ &= n_0 r \sum_{j=1}^{c} (y_{.j.} - y_{...})^2 \equiv Q_B \end{aligned} \tag{5.6.31}$$

TABLE 5.6.3. ANOVA Table for a CRF-rc Design.

Source	df	SS
Rows A	$v_A = r - 1$	$Q_A = n_0 c \sum_i (y_{i..} - y_{...})^2$
Columns B	$v_B = c - 1$	$Q_B = n_0 r \sum_j (y_{.j.} - y_{...})^2$
Interaction AB	$v_{AB} = (r-1)(c-1)$	$Q_{AB} = n_0 \sum_i \sum_j (y_{ij.} - y_{i..} - y_{.j.} + y_{...})^2$
Within error	$v_e = rc(n_0 - 1)$	$Q_e = \sum_i \sum_j \sum_k (y_{ijk} - y_{ij.})^2$
"Total"	$N - 1$	$Q_t = \sum_i \sum_j \sum_k (y_{ijk} - y_{...})^2$

The ANOVA table for the tests is summarized in Table 5.6.3. Each hypothesis is rejected at the significance level α if

(5.6.32)
$$H_A\colon \frac{Q_A/v_A}{Q_e/v_e} > F^\alpha(v_A, v_e)$$
$$H_B\colon \frac{Q_B/v_B}{Q_e/v_e} > F^\alpha(v_B, v_e)$$
$$H_{AB}\colon \frac{Q_{AB}/v_{AB}}{Q_e/v_e} > F^\alpha(v_{AB}, v_e)$$

Scheffé-type simultaneous confidence intervals may be employed to determine which contrasts among the effects are significantly different from zero. $100(1 - \alpha)\%$ confidence intervals following the tests summarized in (5.6.32) (with side conditions) become

(5.6.33)
$$\text{Rows:}\quad \sum_{i=1}^{r} c_i y_{i..} - c_A s\sqrt{\left(\frac{1}{cn_0}\sum_{i=1}^{r} c_i^2\right)} \le \sum_{i=1}^{r} c_i \alpha_i \le \sum_{i=1}^{r} c_i y_{i..} + c_A s\sqrt{\left(\frac{1}{cn_0}\sum_{i=1}^{r} c_i^2\right)}$$
$$\text{Columns:}\quad \sum_{j=1}^{c} c_j y_{.j.} - c_B s\sqrt{\left(\frac{1}{rn_0}\sum_{j=1}^{c} c_j^2\right)} \le \sum_{j=1}^{c} c_j \beta_j \le \sum_{j=1}^{c} c_j y_{.j.} + c_B s\sqrt{\left(\frac{1}{rn_0}\sum_{j=1}^{c} c_j^2\right)}$$
$$\text{Interaction:}\quad \hat{\psi}_m - c_{AB} s\sqrt{\frac{4}{n_0}} \le \psi_m \le \hat{\psi}_m - c_{AB}\sqrt{\frac{4}{n_0}}.$$

where $\hat{\psi}_m = y_{ij.} - y_{i'j.} - y_{ij'.}$ and $\psi_m = \gamma_{ij} - \gamma_{i'j} - \gamma_{ij'} + \gamma_{i'j'}$ and

(5.6.34)
$$c_A^2 = v_A F^\alpha(v_A, v_e)$$
$$c_B^2 = v_B F^\alpha(v_B, v_e)$$
$$c_{AB}^2 = v_{AB} F^\alpha(v_{AB}, v_e)$$

Alternatively, Bonferroni t statistics may also be employed to analyze the two-way design. For example, the first expression in (5.6.33), for q comparisons, is

(5.6.35)
$$\sum_{i=1}^{r} c_{ki} y_{i..} - s t_{rc(n_0-1)}^{\alpha/2q}\sqrt{\frac{1}{cn_0}\sum_{i=1}^{r} c_{ki}^2} \le \sum_{i=1}^{r} c_{ki}\alpha_i \le \sum_{i=1}^{r} c_{ki} y_{i..} + s\sqrt{\frac{1}{cn_0}\sum_{i=1}^{r} c_{ki}^2 t_{rc(n_0-1)}^{\alpha/2q}}$$

for $k = 1, \ldots, q$.

Multivariate Procedure. The results of the univariate case extend immediately. Following the analogy of extending the univariate results of the preceding sections to their multivariate counterparts, we replace the observations y_{ijk} by p-variate vector observations $\mathbf{y}_{ijk}$. Using this principle, model (5.6.1) is written as

$$\begin{aligned} \mathbf{y}_{ijk} &= \boldsymbol{\mu} + \boldsymbol{\alpha}_i + \boldsymbol{\beta}_j + \boldsymbol{\gamma}_{ij} + \boldsymbol{\varepsilon}_{ijk} \\ \boldsymbol{\varepsilon}_{ijk} &\sim IN(\mathbf{0}, \boldsymbol{\Sigma}) \end{aligned} \tag{5.6.36}$$

where $\mathbf{y}_{ijk}$ is a p-variate random vector.

By using (5.6.36), the linear model representation for (5.6.2) becomes

(5.6.37)

$$\begin{bmatrix} \mathbf{y}'_{111} \\ \mathbf{y}'_{112} \\ \mathbf{y}'_{121} \\ \mathbf{y}'_{122} \\ \mathbf{y}'_{211} \\ \mathbf{y}'_{212} \\ \mathbf{y}'_{221} \\ \mathbf{y}'_{222} \\ \mathbf{y}'_{311} \\ \mathbf{y}'_{312} \\ \mathbf{y}'_{321} \\ \mathbf{y}'_{322} \end{bmatrix} = \begin{bmatrix} 1 & 1 & 0 & 0 & 1 & 0 & 1 & 0 & 0 & 0 & 0 & 0 \\ 1 & 1 & 0 & 0 & 1 & 0 & 1 & 0 & 0 & 0 & 0 & 0 \\ 1 & 1 & 0 & 0 & 0 & 1 & 0 & 1 & 0 & 0 & 0 & 0 \\ 1 & 1 & 0 & 0 & 0 & 1 & 0 & 1 & 0 & 0 & 0 & 0 \\ 1 & 0 & 1 & 0 & 1 & 0 & 0 & 0 & 1 & 0 & 0 & 0 \\ 1 & 0 & 1 & 0 & 1 & 0 & 0 & 0 & 1 & 0 & 0 & 0 \\ 1 & 0 & 1 & 0 & 0 & 1 & 0 & 0 & 0 & 1 & 0 & 0 \\ 1 & 0 & 1 & 0 & 0 & 1 & 0 & 0 & 0 & 1 & 0 & 0 \\ 1 & 0 & 0 & 1 & 1 & 0 & 0 & 0 & 0 & 0 & 1 & 0 \\ 1 & 0 & 0 & 1 & 1 & 0 & 0 & 0 & 0 & 0 & 1 & 0 \\ 1 & 0 & 0 & 1 & 0 & 1 & 0 & 0 & 0 & 0 & 0 & 1 \\ 1 & 0 & 0 & 1 & 0 & 1 & 0 & 0 & 0 & 0 & 0 & 1 \end{bmatrix} \begin{bmatrix} \mu_{11} & \mu_{12} & \cdots & \mu_{1p} \\ \alpha_{11} & \alpha_{12} & \cdots & \alpha_{1p} \\ \alpha_{21} & \alpha_{22} & \cdots & \alpha_{2p} \\ \alpha_{31} & \alpha_{32} & \cdots & \alpha_{3p} \\ \beta_{11} & \beta_{12} & \cdots & \beta_{1p} \\ \beta_{21} & \beta_{22} & \cdots & \beta_{2p} \\ \gamma_{111} & \gamma_{112} & \cdots & \gamma_{11p} \\ \gamma_{121} & \gamma_{122} & \cdots & \gamma_{12p} \\ \gamma_{211} & \gamma_{212} & \cdots & \gamma_{21p} \\ \gamma_{221} & \gamma_{222} & \cdots & \gamma_{22p} \\ \gamma_{311} & \gamma_{312} & \cdots & \gamma_{31p} \\ \gamma_{321} & \gamma_{322} & \cdots & \gamma_{32p} \end{bmatrix} + \mathbf{E}_0$$

which is a special case of the general multivariate linear model

$$\underset{(N \times p)}{\mathbf{Y}} = \underset{(N \times q)}{\mathbf{X}} \; \underset{(q \times p)}{\mathbf{B}} + \underset{(N \times p)}{\mathbf{E}_0} \tag{5.6.38}$$

Each row of $\mathbf{Y}$ is the observation vector $\mathbf{y}'_{ijk} = [y^{(1)}_{ijk}, \ldots, y^{(p)}_{ijk}]$, where $N = rcn_0$ and $q = 1 + r + c + rc$.

As in the univariate case, the rank of $\mathbf{X}$ is rc, so that $\mathbf{X}$ is less than full rank. Using (5.6.5), a least-squares estimator for $\mathbf{B}$ is

$$\hat{\mathbf{B}} = (\mathbf{X}'\mathbf{X})^{-}\mathbf{X}'\mathbf{Y} = \begin{bmatrix} \mathbf{0}' \\ \mathbf{0}' \\ \vdots \\ \mathbf{0}' \\ \mathbf{y}'_{11.} \\ \mathbf{y}'_{12.} \\ \vdots \\ \mathbf{y}'_{rc.} \end{bmatrix} \tag{5.6.39}$$

where

$$\mathbf{y}_{ij.} = \frac{\sum_{k=1}^{n_0} \mathbf{y}_{ijk}}{n_0}$$

(5.6.40)
$$\mathbf{y}_{i..} = \frac{\sum_{j=1}^{c} \sum_{k=1}^{n_0} \mathbf{y}_{ijk}}{cn_0}$$

$$\mathbf{y}_{.j.} = \frac{\sum_{i=1}^{r} \sum_{k=1}^{n_0} \mathbf{y}_{ijk}}{rn_0}$$

$$\mathbf{y}_{...} = \frac{\sum_i \sum_j \sum_k \mathbf{y}_{ijk}}{N_.}$$

Following (5.6.8), unique linear unbiased estimators for the parametric functions $\psi = \mathbf{c}'\mathbf{Ba}$ are

(5.6.41)
$$\psi = \mathbf{c}'\mathbf{Ba} = \mathbf{a}'\left[\sum_{i=1}^{r} \sum_{j=1}^{c} t_{ij}(\boldsymbol{\mu} + \boldsymbol{\alpha}_i + \boldsymbol{\beta}_j + \boldsymbol{\gamma}_{ij})\right]$$

$$\hat{\psi} = \mathbf{c}'\hat{\mathbf{B}}\mathbf{a} = \mathbf{a}'\left(\sum_{i=1}^{r} \sum_{j=1}^{c} t_{ij}\mathbf{y}_{ij.}\right)$$

where $\mathbf{t}' = [t_0, t_1, \ldots, t_r, t'_1, \ldots, t'_c, t_{11}, \ldots, t_{rc}]$. As in the univariate case, without side conditions, ψ's involving only $\boldsymbol{\alpha}$'s and $\boldsymbol{\beta}$'s are not directly estimable by using (5.6.41) under the model assumptions given in (5.6.36). However, if $\Sigma_j t_{ij} = \Sigma_j t_{i'j} = 1$, then

$$\psi = \alpha_{is} - \alpha_{i's} + \sum_{j=1}^{c} t_{ij}(\beta_{js} + \gamma_{ijs}) - \sum_{j=1}^{c} t_{i'j}(\beta_{js} + \gamma_{i'js})$$

for $s = 1, \ldots, p$ is estimated by $\hat{\psi} = \Sigma_j t_{ij} y_{ij.}^{(s)} - \Sigma_j t_{i'j} y_{i'j.}^{(s)}$. A similar expression is available for the $\boldsymbol{\beta}$'s. Letting $t_{ij} = 1/c$ for differences in $\boldsymbol{\alpha}$'s and $t_{ij} = 1/r$ for differences in $\boldsymbol{\beta}$'s, estimable functions involving only $\boldsymbol{\alpha}$'s and $\boldsymbol{\beta}$'s are gained with the side conditions $\boldsymbol{\beta} = \Sigma_j \boldsymbol{\beta}_j/c = \mathbf{0}$, $\boldsymbol{\alpha} = \Sigma_j \boldsymbol{\alpha}_i/r = \mathbf{0}$, $\boldsymbol{\gamma}_{i.} = \Sigma_j \boldsymbol{\gamma}_{ij}/c = \mathbf{0}$ for all i, and $\boldsymbol{\gamma}_{.j} = \Sigma_i \boldsymbol{\gamma}_{ij}/r = \mathbf{0}$ for all j. Following the univariate case by analogy, $\psi = \gamma_{ijs} - \gamma_{i'js} - \gamma_{ij's} + \gamma_{i'j's}$ is estimable for $s = 1, \ldots, p$, and can be estimated by $\hat{\psi} = y_{ij.}^{(s)} - y_{i'j.}^{(s)} - y_{ij'.}^{(s)} + y_{i'j'.}^{(s)}$.

With the side conditions

(5.6.42)
$$\begin{aligned} \boldsymbol{\alpha}_. &= \mathbf{0} \\ \boldsymbol{\beta}_. &= \mathbf{0} \\ \boldsymbol{\gamma}_{i.} &= \mathbf{0} \quad \text{for all } i \\ \boldsymbol{\gamma}_{.j} &= \mathbf{0} \quad \text{for all } j \end{aligned}$$

added to (5.6.36), individual parameters become estimable for the MCRF-*rc* design:

$$
\begin{aligned}
\hat{\boldsymbol{\mu}} &= \mathbf{y}_{...} \\
\hat{\boldsymbol{\alpha}}_i &= \mathbf{y}_{i..} - \mathbf{y}_{...} \\
\hat{\boldsymbol{\beta}}_j &= \mathbf{y}_{.j.} - \mathbf{y}_{...} \\
\hat{\boldsymbol{\gamma}}_{ij} &= \mathbf{y}_{ij.} - \mathbf{y}_{i..} - \mathbf{y}_{.j.} + \mathbf{y}_{...}
\end{aligned}
\tag{5.6.43}
$$

To test hypotheses in the multivariate design, the matrices $\mathbf{C}$ used in the univariate case are substituted into $\mathbf{CBA} = \boldsymbol{\Gamma}$, with $\mathbf{A} = \mathbf{I}$, $\boldsymbol{\Gamma} = \mathbf{0}$, and $\mathbf{B}$ equal to the matrix of parameters. By replacing scalar observations in the univariate results by vector observations, Table 5.6.3 is immediately extended to Table 5.6.4.

TABLE 5.6.4. MANOVA Table for an MCRF-*rc* Design.

Source	df	SSP
H_A, row A	$v_A = r - 1$	$\mathbf{Q}_A = n_0 c \sum_i (\mathbf{y}_{i..} - \mathbf{y}_{...})(\mathbf{y}_{i..} - \mathbf{y}_{...})'$
H_B, column B	$v_B = c - 1$	$\mathbf{Q}_B = n_0 r \sum_j (\mathbf{y}_{.j.} - \mathbf{y}_{...})(\mathbf{y}_{.j.} - \mathbf{y}_{...})'$
H_{AB}, interaction AB	$v_{AB} = (r-1)(c-1)$	$\mathbf{Q}_{AB} = n_0 \sum_i \sum_j (\mathbf{y}_{ij.} - \mathbf{y}_{i..} - \mathbf{y}_{.j.} + \mathbf{y}_{...})(\mathbf{y}_{ij.} - \mathbf{y}_{i..} - \mathbf{y}_{.j.} + \mathbf{y}_{...})'$
Within error	$v_e = rc(n_0 - 1)$	$\mathbf{Q}_e = \sum_i \sum_j \sum_k (\mathbf{y}_{ijk} - \mathbf{y}_{ij.})(\mathbf{y}_{ijk} - \mathbf{y}_{ij.})'$
"Total"	$N - 1$	$\mathbf{Q}_t = \sum_i \sum_j \sum_k (\mathbf{y}_{ijk} - \mathbf{y}_{...})(\mathbf{y}_{ijk} - \mathbf{y}_{...})'$

The sum of squares and products matrix $\mathbf{Q}_{AB}$ is used to test for parallelism or significant interactions. Using a procedure similar to that given for Figures 5.6.1 and 5.6.2 in the univariate case, the test of no interactions for the simple example given in (5.6.37) becomes

$$
\begin{aligned}
H_0 : \boldsymbol{\gamma}_{11} - \boldsymbol{\gamma}_{21} - \boldsymbol{\gamma}_{12} + \boldsymbol{\gamma}_{22} &= \mathbf{0} \\
\boldsymbol{\gamma}_{21} - \boldsymbol{\gamma}_{31} - \boldsymbol{\gamma}_{22} + \boldsymbol{\gamma}_{32} &= \mathbf{0}
\end{aligned}
\tag{5.6.44}
$$

When the side conditions given in (5.6.42) are added to the model, the hypothesis is equivalent to testing that all $\boldsymbol{\gamma}_{ij}$'s $= 0$.

The hypotheses for the main effects are written as

$$
\begin{aligned}
H_A^* &: \boldsymbol{\alpha}_i + \sum_j t_{ij}(\boldsymbol{\beta}_j + \boldsymbol{\gamma}_{ij}) \text{ are equal for all } i \\
H_B^* &: \boldsymbol{\beta}_j + \sum_i t_{ij}(\boldsymbol{\alpha}_i + \boldsymbol{\gamma}_{ij}) \text{ are equal for all } j
\end{aligned}
\tag{5.6.45}
$$

in a multivariate design. Employing equal weights $t_{ij} = 1/c$ for H_A^* and $t_{ij} = 1/r$ for H_B^*, (5.6.45) reduces to

$$
\begin{aligned}
H_A &: \boldsymbol{\alpha}_i + \frac{\sum_j \boldsymbol{\gamma}_{ij}}{c} \text{ are equal for all } i \\
H_B &: \boldsymbol{\beta}_j + \frac{\sum_i \boldsymbol{\gamma}_{ij}}{r} \text{ are equal for all } j
\end{aligned}
\tag{5.6.46}
$$

or

$$H_A : \boldsymbol{\alpha}_1 = \boldsymbol{\alpha}_2 = \cdots = \boldsymbol{\alpha}_r$$
$$H_B : \boldsymbol{\beta}_1 = \boldsymbol{\beta}_2 = \cdots = \boldsymbol{\beta}_c \tag{5.6.47}$$

when side conditions of the form given in (5.6.42) are imposed on the general model. The matrices $\mathbf{Q}_A$ and $\mathbf{Q}_B$ are then employed to test H_A and H_B.

The parameters s, m, and n, required to test the multivariate hypotheses, are summarized by

$$s = \min(\nu_h, p)$$
$$m = \frac{|\nu_h - p| - 1}{2} \tag{5.6.48}$$
$$n = \frac{\nu_e - p - 1}{2}$$

where ν_h equals ν_A, ν_B, or ν_{AB}, depending on the hypothesis of interest, and $\nu_e = rc(n_0 - 1)$. The testing procedure is apparent by using (5.3.16). The assumptions of no interactions in the CRF-rc design, with one observation per cell, and in the MCRF-rc design are of concern to researchers. A procedure developed by Tukey (1949) is discussed and extended to the MANOVA designs in Section 5.14.

Insight into a rejected hypothesis is aided by considering Roy-Bose-type confidence intervals of contrasts in the model parameters if the Roy criterion is employed for testing purposes. By letting $\psi = \mathbf{c}'\mathbf{B}\mathbf{a}$ denote an estimable function, a $100(1 - \alpha)\%$ confidence interval for ψ is

$$\hat{\psi} - c_0\hat{\sigma}_{\hat{\psi}} \leq \psi \leq \hat{\psi} + c_0\hat{\sigma}_{\hat{\psi}} \tag{5.6.49}$$

where

$$c_0^2 = \nu_e\left(\frac{\theta^\alpha}{1 - \theta^\alpha}\right)$$
$$\hat{\sigma}_{\hat{\psi}} = \sqrt{\left[\mathbf{a}'\left(\frac{\mathbf{Q}_e}{\nu_e}\right)\mathbf{a}\right][\mathbf{c}'(\mathbf{X}'\mathbf{X})^-\mathbf{c}]} \tag{5.6.50}$$

For pairwise comparisons using equal weights we have for row effects

$$\psi = \sum_{s=1}^{p} a_s\left[(\alpha_{is} - \alpha_{i's}) + \frac{\sum_j (\gamma_{ijs} - \gamma_{i'js})}{c}\right]$$
$$\hat{\psi} = \sum_{s=1}^{p} a_s(y_{i..}^{(s)} - y_{i'..}^{(s)}) \tag{5.6.51}$$
$$\hat{\sigma}_{\hat{\psi}} = \sqrt{\frac{2}{cn_0}\left[\mathbf{a}'\left(\frac{\mathbf{Q}_e}{\nu_e}\right)\mathbf{a}\right]}$$

For column effects

$$\psi = \sum_{s=1}^{p} a_s \left[(\beta_{js} - \beta_{j's}) + \frac{\sum_i (\gamma_{ijs} - \gamma_{ij's})}{r} \right]$$

(5.6.52)
$$\hat{\psi} = \sum_{s=1}^{p} a_s (y_{.j.}^{(s)} - y_{.j'.}^{(s)})$$

$$\hat{\sigma}_{\hat{\psi}} = \sqrt{\frac{2}{rn_0} \left[\mathbf{a}' \left(\frac{\mathbf{Q}_e}{v_e} \right) \mathbf{a} \right]}$$

For the interaction hypothesis, let $\theta_s = \gamma_{ijs} - \gamma_{i'js} - \gamma_{ij's} + \gamma_{i'j's}$ and

$$\hat{\theta}_s = y_{ij.}^{(s)} - y_{i'j.}^{(s)} - y_{ij'.}^{(s)} + y_{i'j'.}^{(s)}$$

Confidence intervals for linear functions of the θ's become

$$\hat{\psi} = \sum_{s=1}^{p} a_s \hat{\theta}_s$$

(5.6.53)
$$\psi = \sum_{s=1}^{p} a_s \theta_s$$

$$\hat{\sigma}_{\hat{\psi}} = \sqrt{\frac{4}{n_0} \left[\mathbf{a}' \left(\frac{\mathbf{Q}_e}{v_e} \right) \mathbf{a} \right]}$$

Again, the Bonferroni procedure is applicable in this design, as is discriminant analysis.

The procedures outlined to analyze an MCRF-*rc* design can be immediately extended to higher-order factorial designs by using linear model theory. For a three-way, completely randomized factorial design, abbreviated as MCRF-*rcv*, with $n_0 > 1$ observations per cell and factors *A*, *B*, and *C*, the multivariate model is

(5.6.54)
$$\mathbf{y}_{ijkm} = \boldsymbol{\mu} + \boldsymbol{\alpha}_i + \boldsymbol{\beta}_j + \boldsymbol{\tau}_k + \boldsymbol{\theta}_{ij} + \boldsymbol{\delta}_{jk} + \boldsymbol{\omega}_{ik} + \boldsymbol{\gamma}_{ijk} + \boldsymbol{\varepsilon}_{ijkm}$$
$$\boldsymbol{\varepsilon}_{ijkm} \sim IN_p(\mathbf{0}, \boldsymbol{\Sigma})$$

where $i = 1, \ldots, r$; $j = 1, \ldots, c$; $k = 1, \ldots, v$; and $m = 1, \ldots, n_0$. The subscripts i, j, and k in the model equation are associated with the levels of factors *A*, *B*, and *C*, respectively.

As with the MCRF-*rc* design, none of the individual parameter vectors in (5.6.1) are estimable; however, the function

$$\boldsymbol{\mu}_{ijk} = \boldsymbol{\mu} + \boldsymbol{\alpha}_i + \boldsymbol{\beta}_j + \boldsymbol{\tau}_k + \boldsymbol{\theta}_{ij} + \boldsymbol{\delta}_{jk} + \boldsymbol{\omega}_{ik} + \boldsymbol{\gamma}_{ijk}$$

is estimable and can be estimated by $\hat{\boldsymbol{\mu}}_{ijk} = \mathbf{y}_{ijk.}$. Furthermore, using only $\boldsymbol{\mu}_{ijk}$'s, estimable functions involving functions of only $\boldsymbol{\alpha}_i$'s, or $\boldsymbol{\beta}_j$'s, or $\boldsymbol{\theta}_{ij}$'s, or $\boldsymbol{\delta}_{jk}$'s, or $\boldsymbol{\omega}_{jk}$'s are not possible, unless side conditions are added to the model, because of the $\boldsymbol{\gamma}_{ijk}$ term in the model. This was true for the parameters $\boldsymbol{\alpha}_i$ and $\boldsymbol{\beta}_j$ in the two-way design because of the term $\boldsymbol{\gamma}_{ij}$. It is possible, however, to obtain functions of the $\boldsymbol{\mu}_{ijk}$'s that involve only $\boldsymbol{\gamma}_{ijk}$'s. These functions have the form

(5.6.55)
$$\boldsymbol{\mu}_{ijk} - \boldsymbol{\mu}_{i'jk} - \boldsymbol{\mu}_{ij'k} + \boldsymbol{\mu}_{i'j'k} - \boldsymbol{\mu}_{ijk'} + \boldsymbol{\mu}_{i'jk'} + \boldsymbol{\mu}_{ij'k'} - \boldsymbol{\mu}_{i'j'k'}$$

for all i, i', j, j', k, and k'. They are estimated by

$$\mathbf{y}_{ijk.} - \mathbf{y}_{i'jk.} - \mathbf{y}_{ij'k.} + \mathbf{y}_{i'j'k.} - \mathbf{y}_{ijk'.} + \mathbf{y}_{i'jk'.} + \mathbf{y}_{ij'k'.} - \mathbf{y}_{i'j'k'.} \tag{5.6.56}$$

and are estimates for

$$\gamma_{ijk} - \gamma_{i'jk} - \gamma_{ij'k} + \gamma_{i'j'k} - \gamma_{ijk'} + \gamma_{i'jk'} + \gamma_{ij'k'} - \gamma_{i'j'k'} \tag{5.6.57}$$

Hence the three-way interaction hypothesis becomes

$$H_{ABC}: \gamma_{ijk} - \gamma_{i'jk} - \gamma_{ij'k} + \gamma_{i'j'k} - \gamma_{ijk'} + \gamma_{i'jk'} + \gamma_{ij'k'} - \gamma_{i'j'k'} = \mathbf{0} \tag{5.6.58}$$

for all i, i', j, j', k, and k' is testable.

To test the three-way-interaction hypothesis, a matrix $\mathbf{C}$ is chosen to generate a linearly independent set satisfying (5.6.58). The rank of the matrix $\mathbf{C}$ is $(r - 1)(c - 1)(v - 1)$, which is the degrees of freedom of the hypothesis. For example, suppose $i = 1, 2, 3$; $j = 1, 2, 3$; and $k = 1, 2, 3$. The set used to test for no three-way interaction might be

$$\begin{aligned}
\gamma_{111} - \gamma_{211} - \gamma_{121} + \gamma_{221} - \gamma_{112} + \gamma_{212} + \gamma_{122} - \gamma_{222} &= \mathbf{0} \\
\gamma_{111} - \gamma_{211} - \gamma_{121} + \gamma_{221} - \gamma_{113} + \gamma_{213} + \gamma_{123} - \gamma_{223} &= \mathbf{0} \\
\gamma_{121} - \gamma_{131} - \gamma_{221} + \gamma_{231} - \gamma_{122} + \gamma_{132} + \gamma_{222} - \gamma_{232} &= \mathbf{0} \\
\gamma_{121} - \gamma_{131} - \gamma_{221} + \gamma_{231} - \gamma_{123} + \gamma_{133} + \gamma_{223} - \gamma_{233} &= \mathbf{0} \\
\gamma_{211} - \gamma_{221} - \gamma_{311} + \gamma_{321} - \gamma_{212} + \gamma_{222} + \gamma_{312} - \gamma_{322} &= \mathbf{0} \\
\gamma_{211} - \gamma_{221} - \gamma_{311} + \gamma_{321} - \gamma_{213} + \gamma_{223} + \gamma_{313} - \gamma_{323} &= \mathbf{0} \\
\gamma_{221} - \gamma_{231} - \gamma_{321} + \gamma_{331} - \gamma_{222} + \gamma_{232} + \gamma_{322} - \gamma_{332} &= \mathbf{0} \\
\gamma_{221} - \gamma_{231} - \gamma_{321} + \gamma_{331} - \gamma_{223} + \gamma_{233} + \gamma_{323} - \gamma_{333} &= \mathbf{0}
\end{aligned}$$

Associated with these eight expressions is a matrix $\mathbf{C}$. Hence the extension to higher-way designs is complete.

Having obtained a matrix $\mathbf{C}$ to test for significant three-way interactions, we are in a position to interpret these in terms of cell means, as was the case for the MCR-rc design and two-way interactions. This would not be obvious from the standard interaction statement that all γ_{ijk}'s are equal to 0. This hypothesis is not directly testable, but is equivalent to H_{ABC} in (5.6.58) for equal observations in all cells when the standard side conditions are added to the model. This is not, however, the case for unbalanced data.

Another common variation of the MCRF-rc design is the multivariate randomized block factorial design, abbreviated as MRBF-rc. For a MCRF-rc design, rc randomly formed subsamples of size n_0 are randomly assigned to the treatment combinations. Alternatively, a blocking variable may be employed to form n_0 blocks of size rc. Then the subjects within each block are randomly assigned to the rc treatment combinations. By forming blocks, it is hoped that the variation within any block is less than the variation among the blocks. In univariate experiments, the blocking variable is a nuisance variable that is correlated with the single dependent variable. For multivariate experiments, the blocking variable is usually a variable such as location or time. That is, all rc subjects are exposed to the treatment combinations at different times during the day or days of the week or at different schools within a system. This variation is not usually of interest to the experimenter; however, we would want to remove this variation from the within-error term of an MCRF-*rc* design. In this situation, an MRBF-*rc* design may be most appropriate. The blocks would be the time of day or days of the week.

The multivariate model for an MRBF-*rc* design is

$$\mathbf{y}_{ijk} = \boldsymbol{\mu} + \boldsymbol{\alpha}_i + \boldsymbol{\beta}_j + \boldsymbol{\gamma}_{ij} + \boldsymbol{\tau}_k + \boldsymbol{\varepsilon}_{ijk} \tag{5.6.59}$$
$$\boldsymbol{\varepsilon}_{ijk} \sim IN_p(\mathbf{0}, \boldsymbol{\Sigma})$$

where $\boldsymbol{\alpha}_i$ and $\boldsymbol{\beta}_j$ denote treatment effects, $\boldsymbol{\gamma}_{ij}$ is the interaction term, $\boldsymbol{\tau}_k$ is the block effect, and $\boldsymbol{\varepsilon}_{ijk}$ is the error. The analysis of this design is similar to the MCRF-*rc* design, except that the within-error SSP matrix is now partitioned into two matrices, an SSP matrix for blocks and a residual SSP matrix.

EXERCISES 5.6

1. Find two hypothesis test matrices **C** to test the three-way-interaction hypothesis specified in (5.6.58) for $i = 1, 2, 3$; $j = 1, 2, 3$; and $k = 1, 2, 3$. Use (5.6.20) to derive a general expression for the hypothesis test matrix.
2. Given the MRBF-*rc* design in (5.6.59), for $i = 1, 2, 3$; $j = 1, 2, 3$; and $k = 1, 2, 3$, determine the form of the estimable functions. What are some interesting hypotheses for the model, and how would they be tested?

5.7 EXAMPLE—MULTIVARIATE TWO-WAY LAYOUT

The data for this section were obtained from a larger study, by Mr. Joseph Raffaele at the University of Pittsburgh, to analyze reading comprehension (C) and reading rate (R), using subtest scores of the Iowa Test of Basic Skills. After

TABLE 5.7.1. Sample Data: MCRF-32 Design.

		Factor *B*			
		Contract Classes		Noncontract Classes	
		R	C	R	C
	Teacher 1	10	21	9	14
		12	22	8	15
		9	19	11	16
		10	21	9	17
		14	23	9	17
Factor *A*	Teacher 2	11	23	11	15
		14	27	12	18
		13	24	10	16
		15	26	9	17
		14	24	9	18
	Teacher 3	8	17	9	22
		7	15	8	18
		10	18	10	17
		8	17	9	19
		7	19	8	19

randomly selecting $N = 30$ students for the study and randomly dividing them into six subsamples of size 5, the groups were randomly assigned to two treatment conditions—contract classes and noncontract classes—and three teachers; a total of $n_0 = 5$ observations are in each cell. The achievement data for the experiment are conveniently represented by cells in Table 5.7.1. Calculating the means for each cell

TABLE 5.7.2. Cell Means for Example Data.

	B_1	B_2	Means
A_1	$\mathbf{y}'_{11.} = [11.00, 21.20]$	$\mathbf{y}'_{12.} = [\ 9.20, 15.80]$	$\mathbf{y}'_{1..} = [10.10, 18.50]$
A_2	$\mathbf{y}'_{21.} = [13.40, 24.80]$	$\mathbf{y}'_{22.} = [10.20, 16.80]$	$\mathbf{y}'_{2..} = [11.80, 20.80]$
A_3	$\mathbf{y}'_{31.} = [\ 8.00, 17.20]$	$\mathbf{y}'_{32.} = [\ 8.80, 19.00]$	$\mathbf{y}'_{3..} = [\ 8.40, 18.10]$
Means	$\mathbf{y}'_{.1.} = [10.80, 21.07]$	$\mathbf{y}'_{.2.} = [\ 9.40, 17.20]$	$\mathbf{y}'_{...} = [10.10, 19.13]$

of the study, Table 5.7.2 is obtained. The mathematical model for the example is

$$\begin{aligned} &\mathbf{y}_{ijk} = \boldsymbol{\mu} + \boldsymbol{\alpha}_i + \boldsymbol{\beta}_j + \boldsymbol{\gamma}_{ij} + \boldsymbol{\varepsilon}_{ijk} \\ &\boldsymbol{\varepsilon}_{ijk} \sim IN(\mathbf{0}, \boldsymbol{\Sigma}) \qquad i = 1, 2, 3; j = 1, 2; k = 1, 2, \ldots, 5 \end{aligned} \tag{5.7.1}$$

The 12×2 matrix $\mathbf{B}$ of unknown parameters, where the number of dependent variables $p = 2$, is given in (5.6.37). By using (5.6.39) and the information in Table

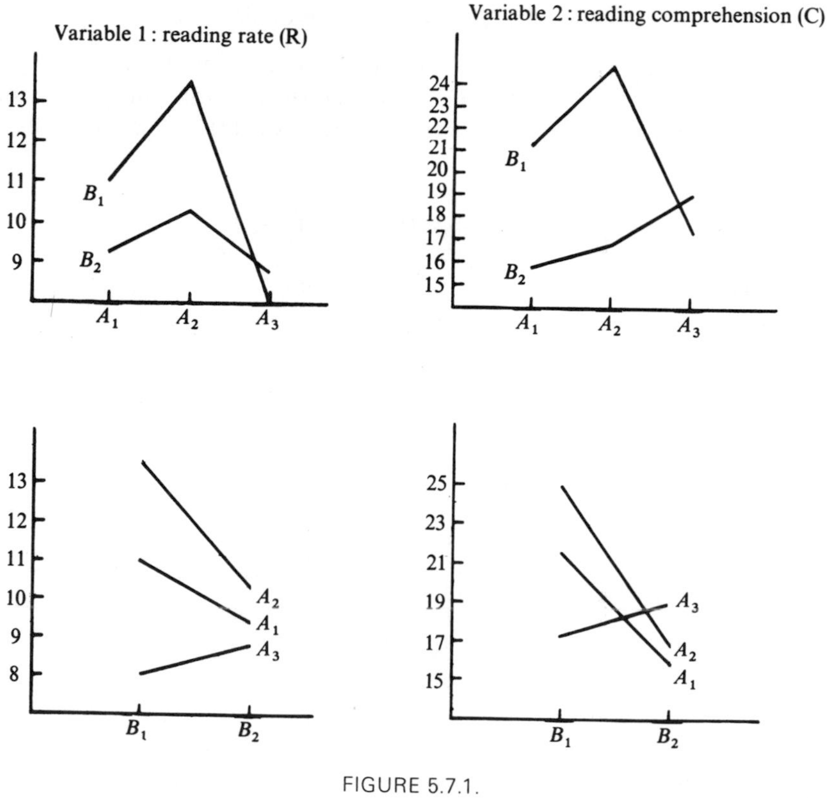

FIGURE 5.7.1.

5.7.2, an estimator for **B** is

$$\underset{(12\times 2)}{\hat{\mathbf{B}}} = \begin{bmatrix} \multicolumn{2}{c}{\underset{(6\times 2)}{\mathbf{0}}} \\ 11.00 & 21.20 \\ 9.20 & 15.80 \\ 13.40 & 24.80 \\ 10.20 & 16.80 \\ 8.00 & 17.20 \\ 8.80 & 19.00 \end{bmatrix} \tag{5.7.2}$$

where

$$\underset{(12\times 12)}{(\mathbf{X}'\mathbf{X})^-} = \begin{bmatrix} \mathbf{0} & \mathbf{0} \\ \mathbf{0} & \text{Dia}\,[1/5] \end{bmatrix} \tag{5.7.3}$$

Before testing the three main hypotheses of interest for the two-way layout, plots of the cell means, a variable at a time, are given in Figure 5.7.1. From the figure, it is apparent that a significant interaction exists in the data. The hypotheses of interest become

$$\begin{aligned} H_A&: \boldsymbol{\alpha}_1 + \frac{\sum_j \boldsymbol{\gamma}_{1j}}{c} = \boldsymbol{\alpha}_2 + \frac{\sum_j \boldsymbol{\gamma}_{2j}}{c} = \boldsymbol{\alpha}_3 + \frac{\sum_j \boldsymbol{\gamma}_{3j}}{c} \\ H_B&: \boldsymbol{\beta}_1 + \frac{\sum_i \boldsymbol{\gamma}_{i1}}{r} = \boldsymbol{\beta}_2 + \frac{\sum_i \boldsymbol{\gamma}_{i2}}{r} \\ H_{AB}&: \boldsymbol{\gamma}_{11} - \boldsymbol{\gamma}_{31} - \boldsymbol{\gamma}_{12} - \boldsymbol{\gamma}_{32} = \mathbf{0} \\ &\quad\ \boldsymbol{\gamma}_{21} - \boldsymbol{\gamma}_{31} - \boldsymbol{\gamma}_{22} + \boldsymbol{\gamma}_{32} = \mathbf{0} \end{aligned} \tag{5.7.4}$$

To test any of the hypotheses in (5.7.4), the estimate of $\mathbf{Q}_e$ is needed. The formula for $\mathbf{Q}_e$ is

$$\begin{aligned} \mathbf{Q}_e &= \mathbf{Y}'[\mathbf{I} - \mathbf{X}(\mathbf{X}'\mathbf{X})^-\mathbf{X}']\mathbf{Y} \\ &= \sum_i \sum_j \sum_k (\mathbf{y}_{ijk} - \mathbf{y}_{ij.})(\mathbf{y}_{ijk} - \mathbf{y}_{ij.})' \\ &= \begin{bmatrix} 45.6000 & \\ 19.8000 & 56.0000 \end{bmatrix} \end{aligned} \tag{5.7.5}$$

Thus

$$\mathbf{S} = \frac{\mathbf{Q}_e}{v_e} = \begin{bmatrix} 1.9000 & \\ .8250 & 2.3333 \end{bmatrix} \tag{5.7.6}$$

where $v_e = N - R(\mathbf{X}) = 30 - 6 = 24$.

By using the expressions for $\mathbf{Q}_A$, $\mathbf{Q}_B$, and $\mathbf{Q}_{AB}$ in Table 5.6.4, the MANOVA Table 5.7.3 for the example is acquired. Depending on the test statistic employed

TABLE 5.7.3. Sample Data: MCR-32-Design Example.

Source	df	SSP
Row A (teachers)	2	$\mathbf{Q}_A = \begin{bmatrix} 57.8000 & \\ 45.9000 & 42.4664 \end{bmatrix}$
Column B (classes)	1	$\mathbf{Q}_B = \begin{bmatrix} 14.7000 & \\ 40.6000 & 112.1332 \end{bmatrix}$
Interaction AB	2	$\mathbf{Q}_{AB} = \begin{bmatrix} 20.6000 & \\ 51.3000 & 129.8667 \end{bmatrix}$
Within error	24	$\mathbf{Q}_e$ = [given in (5.7.5)]
"Total"	29	$\mathbf{Q}_t = \begin{bmatrix} 138.7000 & \\ 157.6000 & 339.4666 \end{bmatrix}$

to test H_A, H_B, H_{AB}, each hypothesis is rejected at the significance level $\alpha = .05$ if

$$
\begin{array}{lcccl}
 & H_A & H_B & H_{AB} & \\
\Lambda < & .668 & .771 & .668 & \text{(Wilks)} \\
\theta_s > & .310 & .229 & .310 & \text{(Roy)} \\
U^{(s)} > & .475 & .297 & .475 & \text{(Lawley-Hotelling)} \\
V^{(s)} > & .347 & .229 & .347 & \text{(Pillai)}
\end{array} \tag{5.7.7}
$$

where $p = 2$, ν_h is defined in Table 5.7.3, $\nu_e = 24$, $s = \min(\nu_h, p)$, $m = (|\nu_h - p| - 1)/2$, and $n = (\nu_e - p - 1)/2$. For H_A and H_{AB}, $s = 2$, $m = -1/2$, and $n = 10.5$. For H_B, $s = 1$, $m = 0$, and $n = 10.5$. Solving the determinantal equations

$$
\begin{aligned}
|\mathbf{Q}_A - \lambda \mathbf{Q}_e| &= 0 \\
|\mathbf{Q}_B - \lambda \mathbf{Q}_e| &= 0 \\
|\mathbf{Q}_{AB} - \lambda \mathbf{Q}_e| &= 0
\end{aligned} \tag{5.7.8}
$$

the eigenvalues to test the three hypotheses are

$$
\begin{array}{ll}
H_A: \lambda_1 = 1.441 & \lambda_2 = .112 \\
H_B: \lambda_1 = 2.003 & \\
H_{AB}: \lambda_1 = 2.307 & \lambda_2 = .005
\end{array} \tag{5.7.9}
$$

By employing (5.3.16), the test statistics are

$$
\begin{array}{lcccl}
 & H_A & H_B & H_{AB} & \\
\Lambda & .369 & .331 & .301 & \text{(Wilks)} \\
\theta_s & .590 & .667 & .698 & \text{(Roy)} \\
U^{(s)} & 1.552 & 2.003 & 2.312 & \text{(Lawley-Hotelling)} \\
V^{(s)} & .691 & .667 & .702 & \text{(Pillai)}
\end{array} \tag{5.7.10}
$$

By comparing the test statistics in (5.7.10) with the critical values in (5.7.7), all hypotheses are rejected for all criteria. Now using (5.7.7) and (5.4.9), the critical constants for the criteria become

$$
\begin{array}{llllll}
 & & H_A & H_B & H_{AB} & \\
 & c_0 = & 3.454 & 2.669 & 3.454 & \text{(Wilks)} \\
(5.7.11) & c_0 = & 3.283 & 2.669 & 3.283 & \text{(Roy)} \\
 & c_0 = & 3.376 & 2.669 & 3.376 & \text{(Lawley-Hotelling)} \\
 & c_0 = & 3.571 & 2.669 & 3.571 & \text{(Pillai)}
\end{array}
$$

Since $s = \min(v_h, p) = 1$ for the hypothesis H_B, all criteria are equivalent, and c_0 is the same for every procedure. For H_A and H_{AB}, Roy's criterion again yields the shortest intervals.

For illustration purposes, contrasts for the hypotheses H_A, H_B, and H_{AB} are examined. From Table 5.7.2,

$$
(5.7.12) \qquad
\begin{aligned}
\hat{\boldsymbol{\psi}}_1 &= \mathbf{y}_{1..} - \mathbf{y}_{3..} = \begin{bmatrix} 10.10 \\ 18.50 \end{bmatrix} - \begin{bmatrix} 8.40 \\ 18.10 \end{bmatrix} = \begin{bmatrix} 1.70 \\ .40 \end{bmatrix} \\
\hat{\boldsymbol{\psi}}_2 &= \mathbf{y}_{2..} - \mathbf{y}_{3..} = \begin{bmatrix} 11.80 \\ 20.80 \end{bmatrix} - \begin{bmatrix} 8.40 \\ 18.10 \end{bmatrix} = \begin{bmatrix} 3.40 \\ 2.70 \end{bmatrix}
\end{aligned}
$$

are the best estimates of $\boldsymbol{\psi}_1 = \boldsymbol{\alpha}_1 - \boldsymbol{\alpha}_3$ and $\boldsymbol{\psi}_2 = \boldsymbol{\alpha}_2 - \boldsymbol{\alpha}_3$ with side conditions added to the model. Without side conditions, $\hat{\boldsymbol{\psi}}_1$ and $\hat{\boldsymbol{\psi}}_2$ are estimating $\boldsymbol{\psi}_1 = (\boldsymbol{\alpha}_1 - \boldsymbol{\alpha}_3) + \Sigma_j(\boldsymbol{\gamma}_{ij} - \boldsymbol{\gamma}_{3j})/2$ and $\boldsymbol{\psi}_2 = (\boldsymbol{\alpha}_2 - \boldsymbol{\alpha}_3) + \Sigma_j(\boldsymbol{\gamma}_{2j} - \boldsymbol{\gamma}_{3j})/2$. For H_B,

$$
(5.7.13) \qquad \hat{\boldsymbol{\psi}}_3 = \mathbf{y}_{.1.} - \mathbf{y}_{.2.} = \begin{bmatrix} 10.80 \\ 21.07 \end{bmatrix} - \begin{bmatrix} 9.40 \\ 17.20 \end{bmatrix} = \begin{bmatrix} 1.40 \\ 3.87 \end{bmatrix}
$$

is the best estimate for $\boldsymbol{\psi}_3 = (\boldsymbol{\beta}_1 - \boldsymbol{\beta}_2) + \Sigma_i(\boldsymbol{\gamma}_{i1} - \boldsymbol{\psi}_{i2})/3$. For H_{AB},

$$
(5.7.14) \qquad
\begin{aligned}
\hat{\boldsymbol{\psi}}_4 &= \mathbf{y}_{11.} - \mathbf{y}_{31.} - \mathbf{y}_{12.} + \mathbf{y}_{32.} = \begin{bmatrix} 2.60 \\ 7.20 \end{bmatrix} \\
\hat{\boldsymbol{\psi}}_5 &= \mathbf{y}_{21.} - \mathbf{y}_{31.} - \mathbf{y}_{22.} + \mathbf{y}_{32.} = \begin{bmatrix} 4.00 \\ 9.80 \end{bmatrix}
\end{aligned}
$$

are estimating $\boldsymbol{\psi}_4 = \boldsymbol{\gamma}_{11} - \boldsymbol{\gamma}_{31} - \boldsymbol{\gamma}_{12} + \boldsymbol{\gamma}_{32}$ and $\boldsymbol{\psi}_5 = \boldsymbol{\gamma}_{21} - \boldsymbol{\gamma}_{31} - \boldsymbol{\gamma}_{22} + \boldsymbol{\gamma}_{32}$.

To find the variances of the contrasts given in (5.7.12), (5.7.13), and (5.7.14), a variable at a time, (5.6.50) is employed. Arranging the standard errors in vectors to correspond to the contrasts, the $\hat{\sigma}_{\hat{\psi}_i}$ become

$$
(5.7.15) \qquad
\begin{aligned}
&\hat{\sigma}_{\hat{\psi}_1} = \begin{bmatrix} .6164 \\ .6832 \end{bmatrix} \quad \hat{\sigma}_{\hat{\psi}_2} = \begin{bmatrix} .6164 \\ .6832 \end{bmatrix} \quad \hat{\sigma}_{\hat{\psi}_3} = \begin{bmatrix} .5033 \\ .5578 \end{bmatrix} \\
&\hat{\sigma}_{\hat{\psi}_4} = \begin{bmatrix} 1.2329 \\ 1.3664 \end{bmatrix} \quad \hat{\sigma}_{\hat{\psi}_5} = \begin{bmatrix} 1.2329 \\ 1.3664 \end{bmatrix}
\end{aligned}
$$

Employing the Roy criterion and the formula

$$\hat{\psi}_i - c_0\hat{\sigma}_{\hat{\psi}_i} \leq \psi_i \hat{\psi}_i + c_0\hat{\sigma}_{\hat{\psi}_i}$$

where $c_0 = 3.283$, $c_0 = 2.669$, and $c_0 = 3.283$, for H_A, H_B, and H_{AB}, respectively, confidence intervals, a variable at a time, are determined by using (5.7.12) through (5.7.15). The intervals for the simple contrasts are as follows:

$$\begin{aligned}
1.70 - (3.283)(.6143) \le (\alpha_{11} - \alpha_{31}) + (\gamma_{1.1} - \gamma_{3.1}) &\le 1.70 + (3.283)(.6143) \\
-.323 \le (\alpha_{11} - \alpha_{31}) + (\gamma_{1.1} - \gamma_{3.1}) &\le 3.723 \qquad \text{(N.S.)} \\
.40 - (3.283)(.6832) \le (\alpha_{12} - \alpha_{32}) + (\gamma_{1.2} - \gamma_{3.2}) &\le .40 + (3.283)(.6832) \\
-1.843 \le (\alpha_{12} - \alpha_{32}) + (\gamma_{1.2} - \gamma_{3.2}) &\le 2.643 \qquad \text{(N.S.)} \\
1.377 \le (\alpha_{21} - \alpha_{31}) + (\gamma_{2.1} - \gamma_{3.1}) &\le 5.243 \qquad \text{(Sig.)} \\
.457 \le (\alpha_{22} - \alpha_{32}) + (\gamma_{2.2} - \gamma_{3.2}) &\le 4.943 \qquad \text{(Sig.)} \\
.057 \le (\beta_{11} - \beta_{21}) + (\gamma_{.11} - \gamma_{.21}) &\le 2.743 \qquad \text{(Sig.)} \\
2.381 \le (\beta_{12} - \beta_{22}) + (\gamma_{.12} - \gamma_{.22}) &\le 5.359 \qquad \text{(Sig.)} \\
-1.477 \le \theta_{11} &\le 6.647 \qquad \text{(N.S.)} \\
2.714 \le \theta_{12} &\le 11.686 \qquad \text{(Sig.)} \\
-.047 \le \theta_{21} &\le 8.047 \qquad \text{(N.S.)} \\
5.314 \le \theta_{22} &\le 14.286 \qquad \text{(Sig.)}
\end{aligned} \tag{5.7.16}$$

where $\theta_{11} = \gamma_{111} - \gamma_{311} - \gamma_{121} + \gamma_{321}$, $\theta_{12} = \gamma_{112} - \gamma_{312} + \gamma_{122} + \gamma_{322}$, $\theta_{21} = \gamma_{211} - \gamma_{311} - \gamma_{221} + \gamma_{321}$, and $\theta_{22} = \gamma_{212} - \gamma_{312} - \gamma_{222} + \gamma_{321}$.

From (5.7.16), there appear to be differences between classes and teachers when the averages over each classification for each variable are calculated. Furthermore, there is a significant interaction for variable 2, comprehension, but not for reading rate for the second and third teachers and classes.

Employing discriminant analysis to help in the investigation of the experiment for the hypotheses H_A and H_{AB}, standardized discriminant functions, using (3.9.25), are obtained for the example by finding the eigenvectors of the equations $|\mathbf{Q}_A - \lambda\mathbf{Q}_e| = 0$ and $|\mathbf{Q}_{AB} - \lambda\mathbf{Q}_e| = 0$. The standardized vectors for H_A and H_{AB} are

$$\begin{array}{cc} H_A & H_{AB} \\ \mathbf{a}_1 = \begin{bmatrix} .5650 \\ .2568 \end{bmatrix} \quad \mathbf{a}_2 = \begin{bmatrix} .5500 \\ -.6635 \end{bmatrix} & \mathbf{a}_1 = \begin{bmatrix} .0423 \\ .6388 \end{bmatrix} \quad \mathbf{a}_2 = \begin{bmatrix} .7874 \\ -.3135 \end{bmatrix} \end{array} \tag{5.7.17}$$

Adjusting for unequal variances for each dependent variable, the vectors in (5.7.17) become

$$\begin{array}{cc} H_A & H_{AB} \\ \begin{array}{cc} \mathbf{a}_1^{(A)} & \mathbf{a}_2^{(A)} \\ \begin{bmatrix} .7788 \\ .3924 \end{bmatrix} & \begin{bmatrix} .7581 \\ -1.0137 \end{bmatrix} \end{array} & \begin{array}{cc} \mathbf{a}_1^{(A)} & \mathbf{a}_2^{(A)} \\ \begin{bmatrix} .0583 \\ .9757 \end{bmatrix} & \begin{bmatrix} 1.0854 \\ -.4790 \end{bmatrix} \end{array} \end{array} \tag{5.7.18}$$

Employing Roy's criterion or Bartlett's procedure, only the first vector is significant for both hypotheses. Illustrating with H_{AB},

$$
\begin{aligned}
X_B^2 &= -[v_e - \tfrac{1}{2}(n - v_h + 1)] \log \Lambda \sim X^2(v_h p) \\
&= -\left[24 - \frac{1}{2}(2 - 2 + 1)\right] \sum_{i=1}^{s} \log (1 + \lambda_i)^{-1} \\
&= 28.22 \mathrel{\dot{\sim}} \chi^2(4) \qquad (5.7.19)
\end{aligned}
$$

is significant when both roots are employed. Using only the second root,

$$
\begin{aligned}
X_B^2 &= -[24 - \tfrac{1}{2}(2 + 2 - 1)] \log \lambda_2 \\
&= .1079 \mathrel{\dot{\sim}} \chi^2(1)
\end{aligned} \qquad (5.7.20)
$$

is not significant. For Roy's criterion,

$$
\theta_s = .697 \qquad \text{where } m = -1/2, n = 10.5, \text{ and } s = 2 \qquad (5.7.21)
$$

for the first root, and

$$
\theta_s = .005 \qquad \text{where } m = 0, n = 10.0, \text{ and } s = 1 \qquad (5.7.22)
$$

for the second, with only the first being significant.

By transforming the variance-covariance matrix $\mathbf{S}$ in (5.7.6) to a correlation matrix

$$
\mathbf{R}_e = \begin{bmatrix} 1.0000 & \\ .3918 & 1.0000 \end{bmatrix} \qquad (5.7.23)
$$

the correlations between the discriminant functions and each variable are obtained:

$$
\begin{aligned}
H_A: \hat{\boldsymbol{\rho}} &= \mathbf{R}_e \begin{bmatrix} .7789 \\ .3924 \end{bmatrix} = \begin{bmatrix} .933 \\ .698 \end{bmatrix} \\
H_{AB}: \hat{\boldsymbol{\rho}} &= \mathbf{R}_e \begin{bmatrix} .0583 \\ .9757 \end{bmatrix} = \begin{bmatrix} .441 \\ .979 \end{bmatrix}
\end{aligned} \qquad (5.7.24)
$$

From (5.7.24), we see that the dependent variable reading rate seems to be more important for differences among teachers while reading comprehension is the important variable for the interaction hypothesis; this is due to the fact that, for the hypothesis H_B, the reading-comprehension variable is more important. These conclusions are confirmed through (5.7.16).

For a graphical representation of these results, a figure similar to Figure 5.4.1 may be constructed. When the units of measurement for the dependent variables are not the same, it is best to divide the means by the error standard deviation of the corresponding variable and then apply the adjusted coefficients to these means. The procedure may also be used with contrasts.

EXERCISES 5.7

1. For the data given in Table 5.7.1, verify the results summarized in Table 5.7.3.
2. An experiment is conducted to compare two different methods of teaching physics during the morning, afternoon, and evening using the traditional lecture approach and the discovery method. The following table summarizes the test scores obtained in the areas of mechanics (M), heat (H), and sound (S) for the 24 students in the study.

	Traditional			Discovery		
	M	H	S	M	H	S
Morning 8 A.M.	30	131	34	51	140	36
	26	126	28	44	145	37
	32	134	33	52	141	30
	31	137	31	50	142	33
Afternoon 2 P.M.	41	104	36	57	120	31
	44	105	31	68	130	35
	40	102	33	58	125	34
	42	102	27	62	150	39
Evening 8 P.M.	30	74	35	52	91	33
	32	71	30	50	89	28
	29	69	27	50	90	28
	28	67	29	53	95	41

a. Analyze the data, testing for (1) effects of treatments, (2) effects of time of day, and (3) interaction effects. Include in the analysis a test of the equality of the variance-covariance matrices.

b. In this study does trend analysis make any sense? If so, incorporate it into your analysis.

c. Summarize the results of this experiment in one paragraph.

5.8 TWO-WAY NESTED DESIGN—UNIVARIATE AND MULTIVARIATE

In Section 5.6, the CRF-*rc* design and the MCRF-*rc* design with $n_0 > 1$ observations per cell were discussed. The first factor, *A*, was termed the row effect and the second factor, *B*, the column effect, where every level of factor *A* occurred with every level of factor *B*. When this is evidenced, the two factors are said to be *completely crossed*. However, in many research situations it may not be possible to design an experiment so that factors (treatments) are crossed. A *nested* or *hierarchal design* may then be appropriate.

A factor *B* is said to be nested within the levels of factor *A* if every level of *B* occurs with only a single level of factor *A*. The difference between crossed and nested factors is illustrated in Table 5.8.1. In the crossed design, every level of *B* appears with every level of *A*. In the nested design, levels B_1 and B_2 appear only with level A_1, whereas B_3, B_4, and B_5 appear with A_2. As a result of scheduling or other administrative procedures, it is not always possible for all classrooms to be evaluated under all teachers. In such a situation, a nested design is more appropriately employed for testing.

TABLE 5.8.1. Example of (a) Crossed and (b) Nested Designs.

(a)

A \ *B*	B_1	B_2	B_3	B_4	B_5
A_1	×	×	×	×	×
A_2	×	×	×	×	×

(b)

A \ *B*	B_1	B_2	B_3	B_4	B_5
A_1	×	×			
A_2			×	×	×

Univariate Procedure. In the RB-c design of Section 5.5, the effect β_j must be modified to take into account the nested factor. The model becomes

$$y_{ijk} = \mu + \alpha_i + \beta_{(i)j} + \varepsilon_{ijk} \qquad \begin{cases} i = 1, \ldots, r \\ j = 1, \ldots, c_i \\ k = 1, \ldots, N_{ij} \end{cases}$$

(5.8.1)

$$\varepsilon_{ijk} \sim IN(0, \sigma^2)$$

such that

$$N_i = \sum_{j=1}^{c_i} N_{ij} \quad \text{and} \quad N = \sum_{i=1}^{r} \sum_{j=1}^{c_i} N_{ij}$$

and the notation $\beta_{(i)j}$ is employed to denote the effects of the jth level of B nested in the ith level of A. Following Kirk, the abbreviation for this design would be CRH-$r(c)$, indicating that treatment B is a nested treatment where the levels of B are randomly assigned to levels of A.

The observations in (5.8.1) are represented as follows:

Factor A

A_1			$\cdots$	A_r		
B_1	$\cdots$	B_{c_1}	$\cdots$	B'_1	$\cdots$	B'_{c_r}
y_{111}	$\cdots$	y_{1c_11}	$\cdots$	y_{r11}	$\cdots$	y_{rc_r1}
y_{112}	$\cdots$	y_{1c_12}	$\cdots$	y_{r12}	$\cdots$	y_{rc_r2}
$\vdots$	$\vdots$	$\vdots$	$\vdots$	$\vdots$	$\vdots$	$\vdots$
$y_{11N_{11}}$	$\cdots$	$y_{1c_1N_{1c_1}}$	$\cdots$	$y_{r1N_{r1}}$	$\cdots$	$y_{rc_rN_{rc_r}}$
$y_{11.}$	$\cdots$	$y_{1c_1.}$	$\cdots$	$y_{r1.}$	$\cdots$	$y_{rc_r.}$

where

$$N = \sum_{i=1}^{r} N_i = \sum_i \sum_j N_{ij}$$

$$y_{ij.} = \frac{\sum_{k=1}^{N_{ij}} y_{ijk}}{N_{ij}}$$

(5.8.2)

$$y_{i..} = \frac{\sum_{j=1}^{c_i} \sum_{k=1}^{N_{ij}} y_{ijk}}{N_i}$$

$$= \frac{\sum_j N_{ij} y_{ij.}}{N_i} \qquad \text{(a weighted average of } y_{ij.}\text{)}$$

$$y_{...} = \frac{\sum_{i=1}^{r} \sum_{j=1}^{c_i} \sum_{k=1}^{N_{ij}} y_{ijk}}{N}$$

$$= \frac{\sum_i N_i y_{i..}}{N} \qquad \text{(a weighted average of } y_{i..}\text{)}$$

By using matrix notation, the general linear model for (5.8.1) becomes

$$
\text{(5.8.3)}\quad
\underbrace{\begin{bmatrix} y_{111} \\ y_{112} \\ \vdots \\ y_{11N_{11}} \\ \hdashline \vdots \\ \hdashline y_{1c_11} \\ y_{1c_12} \\ \vdots \\ y_{1c_1N_{1c_1}} \\ \hdashline \vdots \\ \hdashline y_{r11} \\ y_{r12} \\ \vdots \\ y_{r1N_{r1}} \\ \hdashline \vdots \\ \hdashline y_{rc_r1} \\ y_{rc_r2} \\ \vdots \\ y_{rc_rN_{rc_r}} \end{bmatrix}}_{(N \times 1)}
=
\underbrace{\left[\begin{array}{c:ccc:ccccccc}
\mathbf{1}_N & \mathbf{1}_{N_1} & & \mathbf{0} & \mathbf{1}_{N_{11}} & & & & & & \mathbf{0} \\
 & & \ddots & & & \ddots & & & & & \\
 & \mathbf{0} & & \mathbf{1}_{N_r} & & & \mathbf{1}_{N_{1c_1}} & & & & \\
 & & & & & & & \ddots & & & \\
 & & & & & & & & \mathbf{1}_{N_{r1}} & & \\
 & & & & & & & & & \ddots & \\
 & & & & \mathbf{0} & & & & & & \mathbf{1}_{N_{rc_r}}
\end{array}\right]}_{[N \times (\sum_i c_i + r + 1)]}
\underbrace{\begin{bmatrix} \mu \\ \alpha_1 \\ \vdots \\ \alpha_r \\ \beta_{(1)1} \\ \beta_{(2)1} \\ \vdots \\ \beta_{(r)c_r} \end{bmatrix}}_{[(\sum_i c_i + r + 1) \times 1]}
+
\underbrace{\begin{bmatrix} \varepsilon_{111} \\ \varepsilon_{112} \\ \vdots \\ \varepsilon_{11N_{11}} \\ \hdashline \vdots \\ \hdashline \varepsilon_{1c_11} \\ \varepsilon_{1c_12} \\ \vdots \\ \varepsilon_{1c_1N_{1c_1}} \\ \hdashline \vdots \\ \hdashline \varepsilon_{r11} \\ \varepsilon_{r12} \\ \vdots \\ \varepsilon_{r1N_{r1}} \\ \hdashline \vdots \\ \hdashline \varepsilon_{rc_r1} \\ \varepsilon_{rc_r2} \\ \vdots \\ \varepsilon_{rc_rN_{rc_r}} \end{bmatrix}}_{(N \times 1)}
$$

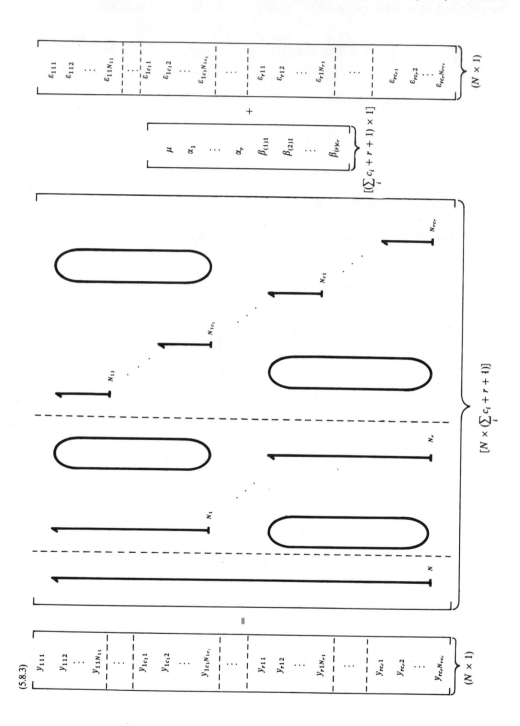

Since this specification is very involved, the theory of nested design will be illustrated by a simple example. Then extending the results and forms of standard expressions to the general case should be apparent. The data for the simple example is summarized in Table 5.8.2, where the levels B_1 and B_2 under A_1 are not the same as B'_1 and B'_2 under A_2, so that B is nested in A. Following (5.8.3), the linear model for the data in Table 5.8.2 is easily specified and given in (5.8.4).

TABLE 5.8.2.

A_1		A_2		
B_1	B_2	B'_1	B'_2	B'_3
y_{111}	y_{121}	y_{211}	y_{221}	y_{231}
y_{112}	y_{122}	y_{212}	y_{222}	y_{232}
	y_{123}			y_{233}

(5.8.4)

$$\underset{(12\times 1)}{\begin{bmatrix} y_{111}\\ y_{112}\\ y_{121}\\ y_{122}\\ y_{123}\\ y_{211}\\ y_{212}\\ y_{221}\\ y_{222}\\ y_{231}\\ y_{232}\\ y_{233}\end{bmatrix}} = \underset{(12\times 8)}{\begin{bmatrix} 1&1&0&1&0&0&0&0\\ 1&1&0&1&0&0&0&0\\ 1&1&0&0&1&0&0&0\\ 1&1&0&0&1&0&0&0\\ 1&1&0&0&1&0&0&0\\ 1&0&1&0&0&1&0&0\\ 1&0&1&0&0&1&0&0\\ 1&0&1&0&0&0&1&0\\ 1&0&1&0&0&0&1&0\\ 1&0&1&0&0&0&0&1\\ 1&0&1&0&0&0&0&1\\ 1&0&1&0&0&0&0&1\end{bmatrix}} \underset{(8\times 1)}{\begin{bmatrix} \mu\\ \alpha_1\\ \alpha_2\\ \beta_{(1)1}\\ \beta_{(1)2}\\ \beta_{(2)1}\\ \beta_{(2)2}\\ \beta_{(2)3}\end{bmatrix}} + \underset{(12\times 1)}{\begin{bmatrix} \varepsilon_{111}\\ \varepsilon_{112}\\ \varepsilon_{121}\\ \varepsilon_{122}\\ \varepsilon_{123}\\ \varepsilon_{211}\\ \varepsilon_{212}\\ \varepsilon_{221}\\ \varepsilon_{222}\\ \varepsilon_{231}\\ \varepsilon_{232}\\ \varepsilon_{233}\end{bmatrix}}$$

The design matrix in (5.8.4) is not of full rank; the first three columns are linear combinations of the last five columns. To procure an estimate of $\boldsymbol{\beta}$ in (5.8.4), $\mathbf{X}'\mathbf{X}$ is calculated:

(5.8.5)

$$\mathbf{X}'\mathbf{X} = \begin{bmatrix} N & N_1 & N_2 & N_{11} & N_{12} & N_{21} & N_{22} & N_{23}\\ N_1 & N_1 & 0 & N_{11} & N_{12} & 0 & 0 & 0\\ N_2 & 0 & N_2 & 0 & 0 & N_{21} & N_{22} & N_{23}\\ N_{11} & N_{11} & 0 & N_{11} & 0 & 0 & 0 & 0\\ N_{12} & N_{12} & 0 & 0 & N_{12} & 0 & 0 & 0\\ N_{21} & 0 & N_{21} & 0 & 0 & N_{21} & 0 & 0\\ N_{22} & 0 & N_{22} & 0 & 0 & 0 & N_{22} & 0\\ N_{23} & 0 & N_{23} & 0 & 0 & 0 & 0 & N_{23}\end{bmatrix} = \begin{bmatrix} 12&5&7&2&3&2&2&3\\ 5&5&0&2&3&0&0&0\\ 7&0&7&0&0&2&2&3\\ 2&2&0&2&0&0&0&0\\ 3&3&0&0&3&0&0&0\\ 2&0&2&0&0&2&0&0\\ 2&0&2&0&0&0&2&0\\ 3&0&3&0&0&0&0&3\end{bmatrix}$$

We obtain a g-inverse for $\mathbf{X'X}$ by deleting the first three rows and columns from $\mathbf{X'X}$. Thus

$$(\mathbf{X'X})^- = \left[\begin{array}{ccc|ccccc} 0 & 0 & 0 & 0 & 0 & 0 & 0 & 0 \\ 0 & 0 & 0 & 0 & 0 & 0 & 0 & 0 \\ 0 & 0 & 0 & 0 & 0 & 0 & 0 & 0 \\ \hline 0 & 0 & 0 & 1/2 & 0 & 0 & 0 & 0 \\ 0 & 0 & 0 & 0 & 1/3 & 0 & 0 & 0 \\ 0 & 0 & 0 & 0 & 0 & 1/2 & 0 & 0 \\ 0 & 0 & 0 & 0 & 0 & 0 & 1/2 & 0 \\ 0 & 0 & 0 & 0 & 0 & 0 & 0 & 1/3 \end{array}\right] = \begin{bmatrix} \mathbf{0} & \mathbf{0} \\ \mathbf{0} & \text{Dia}\,[N_{ij}^{-1}] \end{bmatrix} \tag{5.8.6}$$

so that

$$\hat{\boldsymbol{\beta}} = (\mathbf{X'X})^-\mathbf{X'y} = \begin{bmatrix} 0 \\ 0 \\ 0 \\ y_{11.} \\ y_{12.} \\ y_{21.} \\ y_{22.} \\ y_{23.} \end{bmatrix} \tag{5.8.7}$$

and

$$\mathbf{H} = (\mathbf{X'X})^-\mathbf{X'X} = \begin{bmatrix} 0 & 0 & 0 & 0 & 0 & 0 & 0 & 0 \\ 0 & 0 & 0 & 0 & 0 & 0 & 0 & 0 \\ 0 & 0 & 0 & 0 & 0 & 0 & 0 & 0 \\ 1 & 1 & 0 & 1 & 0 & 0 & 0 & 0 \\ 1 & 1 & 0 & 0 & 1 & 0 & 0 & 0 \\ 1 & 0 & 1 & 0 & 0 & 1 & 0 & 0 \\ 1 & 0 & 1 & 0 & 0 & 0 & 1 & 0 \\ 1 & 0 & 1 & 0 & 0 & 0 & 0 & 1 \end{bmatrix} \tag{5.8.8}$$

The general form of $\hat{\boldsymbol{\beta}}$ is obvious from (5.8.7).

By applying (3.5.8), unique linear unbiased estimators for linear combinations of the elements of $\boldsymbol{\beta}$ are easily acquired. Letting $\mathbf{t}' = [t_0, t_1, t_2, t_{11}, t_{12}, t_{21}, t_{22}, t_{23}]$,

$$\begin{aligned} \psi &= \mathbf{c}'\boldsymbol{\beta} = \mathbf{t}'\mathbf{H}\boldsymbol{\beta} = \sum_i \sum_j t_{ij}\mu + \sum_i \sum_j t_{ij}\alpha_i + \sum_i \sum_j t_{ij}\beta_{(i)j} \\ \hat{\psi} &= \mathbf{c}'\hat{\boldsymbol{\beta}} = \sum_i \sum_j t_{ij}y_{ij.} \end{aligned} \tag{5.8.9}$$

denote the form of the estimable functions and their estimates. Before being too general, those functions of the parameters in (5.8.4) that are estimable and therefore might be of interest for hypothesis testing are determined. Clearly, $\psi = \mu + \alpha_i + \beta_{(i)j}$ is estimable and can be estimated by $\hat{\psi} = y_{ij.}$; furthermore, since linear combinations of estimable functions are estimable, $\psi = \beta_{(i)j} - \beta_{(i)j'}$, for $j \neq j'$, is estimable and can be estimated by $\hat{\psi} = y_{ij.} - y_{ij'.}$. Thus differences among the levels of B within the levels of A are testable. To test for differences among the levels of A, an estimate of $\psi = \alpha_i - \alpha_{i'}$ is needed. Using (5.8.8), $\psi = \alpha_i - \alpha_{i'} + \Sigma_j t_{ij}\beta_{(i)j} - \Sigma_j t_{i'j}\beta_{(i')j}$ is estimated with $\hat{\psi} = \Sigma_j t_{ij} y_{ij.} - \Sigma_j t_{i'j} y_{i'j.}$, if $\Sigma_j t_{ij} = \Sigma_j t_{i'j} = 1$; however, $\psi = \alpha_i - \alpha_{i'}$ is not estimable. For $\psi = \alpha_i - \alpha_{i'}$ to be estimable, the condition $\Sigma_j t_{ij}\beta_{(i)j}$, for all i, must be 0. Thus the test for factor A depends on the weights t_{ij}. A common choice of weights is $t_{ij} = N_{ij}/N_i$ (see Scheffé, 1959, p. 184). Without the side conditions, differences in α_i's are confounded by the $\beta_{(i)j}$'s. Thus, with the side condition $\Sigma_j N_{ij}\beta_{(i)j}/N_i = 0$, the functions $\psi = \alpha_i - \alpha_{i'}$ become estimable and can be estimated by

$$\hat{\psi} = \sum_j \frac{N_{ij} y_{ij.}}{N_i} - \sum_j \frac{N_{i'j} y_{i'j.}}{N_{i'}}$$

Furthermore, $\psi = \mu + \alpha_i$ is now estimable and can be estimated by $\hat{\psi} = y_{i..}$. By combining the functions $\psi = \mu + \alpha_i$ and adding the condition that $\Sigma_i N_i \alpha_i = 0$, an estimate of μ is obtained, $\hat{\mu} = y_{...}$. This second side condition allows μ, α_i, and $\beta_{(i)j}$ to be individually estimated. This is the rationale behind the addition of the "convenient" side conditions

$$\sum_i N_i \alpha_i = \frac{\sum_j N_{ij}\beta_{(i)j}}{N_i} = 0 \tag{5.8.10}$$

in many texts to the model equations represented by (5.8.1). However, it must be remembered that the weights t_{ij} were chosen *proportional* to the number of observations in the sample. A nested design with an equal number of levels of B within each level of A and equal N_{ij} is said to be *balanced*. This type of design does not depend on the weights.

With (5.8.10) added to (5.8.2), the familiar estimators for the parameters become

$$\begin{aligned} \hat{\mu} &= y_{...} \\ \hat{\alpha}_i &= y_{i..} - y_{...} \\ \hat{\beta}_{(i)j} &= y_{ij.} - y_{i..} \end{aligned} \tag{5.8.11}$$

Prior to testing any hypothesis, an estimate of σ^2 is required; an unbiased estimator of σ^2 is

$$\begin{aligned} Q_e &= \mathbf{y}'[\mathbf{I} - \mathbf{X}(\mathbf{X}'\mathbf{X})^{-}\mathbf{X}']\mathbf{y} \\ &= \mathbf{y}'\mathbf{y} - \mathbf{y}'\mathbf{X}\hat{\boldsymbol{\beta}} \\ &= \sum_{i=1}^{r} \sum_{j=1}^{c_i} \sum_{k=1}^{N_{ij}} y_{ijk} - \sum_{i=1}^{r} \sum_{j=1}^{c_i} N_{ij} y_{ij.}^2 \\ &= \sum_i \sum_j \sum_k (y_{ijk} - y_{ij.})^2 \end{aligned} \tag{5.8.12}$$

Since the

$$R(\mathbf{X}) = \sum_{i=1}^{r} c_i \quad \text{and} \quad v_e = N - R(\mathbf{X}) = \sum_{i=1}^{r} \sum_{j=1}^{c_i} (N_{ij} - 1)$$

an unbiased estimate of σ^2 is given by

$$s^2 = \frac{\sum_i \sum_j \sum_k (y_{ijk} - y_{ij.})^2}{\sum_i \sum_j (N_{ij} - 1)} \tag{5.8.13}$$

The primary hypothesis of interest in a nested design is the test for differences in the levels of B within the levels of A:

$$\begin{aligned} H_0\colon \beta_{(1)1} &= \beta_{(1)2} = \cdots = \beta_{(1)c_1} \\ \beta_{(2)1} &= \beta_{(2)2} = \cdots = \beta_{(2)c_2} \\ &\vdots \\ \beta_{(r)1} &= \beta_{(r)2} = \cdots = \beta_{(r)c_r} \end{aligned} \tag{5.8.14}$$

By representing (5.8.14) as $H_0\colon \mathbf{C}\boldsymbol{\beta} = \mathbf{0}$, the matrix $\mathbf{C}$ to test (5.8.14) has the form

$$\underset{[\Sigma_i (c_i - 1) \times q]}{\mathbf{C}} = \left[\begin{array}{c|cccc} & \mathbf{C}_1 & \mathbf{0} & \cdots & \mathbf{0} \\ \underset{[\Sigma_i (c_i - 1) \times (r+1)]}{\mathbf{0}} & \mathbf{0} & \mathbf{C}_2 & \cdots & \mathbf{0} \\ & \vdots & \vdots & & \vdots \\ & \mathbf{0} & \mathbf{0} & \cdots & \mathbf{C}_r \end{array}\right] \tag{5.8.15}$$

$$\underset{[(c_i - 1) \times c_i]}{\mathbf{C}_i} = \begin{bmatrix} 1 & 0 & 0 & \cdots & -1 \\ 0 & 1 & 0 & \cdots & -1 \\ \vdots & \vdots & \vdots & & \vdots \\ 0 & 0 & 0 & \cdots & -1 \end{bmatrix} \tag{5.8.16}$$

and $q = r + 1 + \Sigma_i c_i$. Thus the hypothesis sum of squares to test (5.8.14) is

$$Q_h = (\mathbf{C}\hat{\boldsymbol{\beta}})'[\mathbf{C}(\mathbf{X}'\mathbf{X})^{-}\mathbf{C}']^{-1}(\mathbf{C}\hat{\boldsymbol{\beta}}) \tag{5.8.17}$$

where the degrees of freedom for the hypothesis are $v_h = \Sigma_i (c_i - 1)$. To find a simple form for Q_h, note that $\hat{\boldsymbol{\beta}}$ in the nested case is of the same form as $\hat{\boldsymbol{\beta}}$ for the one-way ANOVA design with $y_{i.}$ replaced by $y_{ij.}$ [see (5.2.7)]. Further, $\mathbf{C}_i$ in (5.8.15) is like $\mathbf{C}$ in (5.2.15). Thus

$$\mathbf{C}(\mathbf{X}'\mathbf{X})^{-}\mathbf{C}' = \text{Dia} \begin{bmatrix} \text{Dia}\,[1/N_{11}, \ldots, 1/N_{1(c_1-1)}] + N_{1c_1}^{-1}\mathbf{J}_{c_1-1, c_1-1} \\ \vdots \\ \text{Dia}\,[1/N_{r1}, \ldots, 1/N_{r(c_r-1)}] + N_{rc_r}^{-1}\mathbf{J}_{c_r-1, c_r-1} \end{bmatrix} \tag{5.8.18}$$

And, by analogy, it may be shown that

$$Q_h = \sum_{i=1}^{r} \sum_{j=1}^{c} N_{ij}(y_{ij.} - y_{i..})^2 \tag{5.8.19}$$

From the form of $\mathbf{C}$, Q_h is partitioned into r separate orthogonal sums of squares. That is, corresponding to each $\mathbf{C}_i$, there is a Q_{h_i} with degrees of freedom $v_{h_i} = c_i - 1$, where

$$Q_{h_i} = \sum_{j=1}^{c_i} N_{ij}(y_{ij.} - y_{i..})^2 \qquad i = 1, \ldots, r \tag{5.8.20}$$

The hypothesis sum of squares in (5.8.20) may be used to test

$$H_{0i}: \beta_{(i)1} = \cdots = \beta_{(i)c_i} \qquad i = 1, \ldots, r \tag{5.8.21}$$

The other test of interest, using model (5.8.1), is the test of equal levels of A. That is,

$$H_0: \alpha_1 = \alpha_2 = \cdots = \alpha_r \tag{5.8.22}$$

To test (5.8.22), simple differences in the levels of A, $\psi = \alpha_i - \alpha_{i'}$, must be estimable. Following (5.8.9), ψ is not estimable. However, by selecting weights such that $\Sigma_j t_{ij} = 1$, the hypothesis

$$H_0: \alpha_1 + \sum_j t_{1j}\beta_{(1)j} = \cdots = \alpha_r + \sum_j t_{rj}\beta_{(r)j} \tag{5.8.23}$$

is testable. Choosing weights $t_{ij} = N_{ij}/N_i$ implies a test of whether the *weighted average* of the means for each level B within A are the same across the levels of A. The selection of weights $t_{ij} = 1/c_i$ assumes the use of an *unweighted* comparison of those averages. Without knowing the weights chosen to test (5.8.22), the specification of the hypotheses tested is misleading where there are an unequal number of observations for each level of the nested factor. By having a researcher construct the hypothesis test matrix $\mathbf{C}$ for testing, this ambiguity is avoided. Imposing the side condition

$$\sum_j t_{ij}\beta_{(i)j} = 0 \qquad \text{where } t_{ij} = \frac{N_{ij}}{N_i}$$

(5.8.23) reduces to (5.8.22), and $\psi = \alpha_i - \alpha_{i'}$ is estimable and can be tested by using the same F statistic.

The matrix $\mathbf{C}$ to test (5.8.23) with $t_{ij} = N_{ij}/N_i$ becomes

(5.8.24)

$$\underset{[(r-1)\times q]}{\mathbf{C}} = \left[\begin{array}{c|ccccc|ccc|ccc|c|ccc} 0 & 1 & 0 & \cdots & 0 & -1 & N_{11}/N_1 & \cdots & N_{1c_1}/N_{c_1} & 0 & \cdots & 0 & \cdot & -N_{r_1}/N_{c_r} & \cdots & -N_{rc_r}/N_{c_r} \\ 0 & 0 & 1 & \cdots & 0 & -1 & 0 & \cdots & 0 & N_{21}/N_2 & \cdots & N_{2c_2}/N_2 & \cdot & \cdot & \cdots & \cdot \\ \vdots & \vdots & \vdots & \vdots & \vdots & \vdots & \vdots & \vdots & \vdots & \vdots & \vdots & \vdots & \vdots & \vdots & \vdots & \vdots \\ 0 & 0 & 0 & \cdots & 1 & -1 & 0 & \cdots & 0 & 0 & \cdots & 0 & \cdot & -N_{r_1}/N_{c_r} & \cdots & -N_{rc_r}/N_{c_r} \end{array}\right]$$

where $q = r + 1 + \Sigma_i c_i$. The hypothesis sum of squares Q_h reduces to

$$\begin{aligned} Q_h &= (\mathbf{C}\hat{\boldsymbol{\beta}})'[\mathbf{C}(\mathbf{X}'\mathbf{X})^{-}\mathbf{C}']^{-1}(\mathbf{C}\hat{\boldsymbol{\beta}}) \\ &= \sum_{i=1}^{r} N_i(y_{i..} - y_{...})^2 \end{aligned} \tag{5.8.25}$$

TABLE 5.8.3. ANOVA Table for a Two-Way Nested Design.

Source	df	SS
$H_A: A$	$r - 1 = v_A$	$Q_A = \sum_i N_i(y_{i..} - y_{...})^2$
$H_{(A)B}: (A)B$	$\sum_i (c_i - 1) = v_{(A)B}$	$Q_{(A)B} = \sum_i \sum_j N_{ij}(y_{ij.} - y_{i..})^2$
$H_{(A_1)B}: (A_1)B$	$c_1 - 1 = v_1$	$Q_{(A_1)B} = \sum_j N_{1j}(y_{1j.} - y_{1..})^2$
$\vdots$	$\vdots$	$\vdots$
$H_{(A_r)B}: (A_r)B$	$c_r - 1 = v_r$	$Q_{(A_r)B} = \sum_j N_{rj}(y_{rj.} - y_{r..})^2$
Within error	$\sum_i \sum_j (N_{ij} - 1) = v_e$	$Q_e = \sum_i \sum_j \sum_k (y_{ijk} - y_{ij.})$
"Total"	$N - 1$	$Q_t = \sum_i \sum_j \sum_k (y_{ijk} - y_{...})^2$

The ANOVA table for the tests is now formed and summarized in Table 5.8.3. Each hypothesis is rejected at the significance level α if

$$H_A: \frac{Q_A/(r-1)}{Q_e/v_e} > F^{\alpha}(r-1, v_e)$$

(5.8.26)

$$H_{(A)B}: \frac{Q_{(A)B}/v_{(A)B}}{Q_e/v_e} > F^{\alpha}(v_{(A)B}, v_e)$$

or, alternatively testing each hypothesis within levels of A,

(5.8.27)
$$H_{(A_i)B}: \frac{Q_{(A_i)B}/v_i}{Q_e/v_e} > F^{\alpha_i}(v_i, v_e)$$

such that $\Sigma_i \alpha_i = \alpha$. On the rejection of a hypothesis, Scheffé-type intervals are again available.

For the test H_A, define

(5.8.28)
$$\psi = \mathbf{c}'\boldsymbol{\beta} = \sum_i c_i \alpha_i \text{ such that } \sum_i c_i = 0$$

then, under the general linear model with the side conditions given in (5.8.10), the probability is $1 - \alpha$ that, simultaneously for all ψ,

(5.8.29)
$$\hat{\psi} - c_0 \hat{\sigma}_{\hat{\psi}} \le \psi \le \hat{\psi} + c_0 \hat{\sigma}_{\hat{\psi}}$$

where $c_0^2 = (r-1)F^{\alpha}(r-1, v_e)$ and $\hat{\sigma}_{\hat{\psi}} = \sqrt{(Q_e/v_e)[\mathbf{c}'(X'X)^{-}\mathbf{c}]}$. For a simple comparison, (5.8.29) reduces to

$$y_{i..} - y_{i'..} - c_0 \sqrt{\frac{Q_e}{v_e}\left(\sum_{j=1}^{c_i} N_{ij}/N_i^2 + \sum_{j=1}^{c_{i'}} N_{i'j}/N_{i'}^2\right)} \le \alpha_i - \alpha_{i'}$$

(5.8.30)
$$\le y_{i..} - y_{i'..} + c_0 \sqrt{\frac{Q_e}{v_e}\left(\sum_{j=1}^{c_i} N_{ij}/N_i^2 + \sum_{j=1}^{c_{i'}} N_{i'j}/N_{i'}^2\right)}$$

Implicit in (5.8.30) is the assumption that $t_{ij} = N_{ij}/N_i$.

For the tests $H_{(A_i)B}$, let

(5.8.31)
$$\psi_i = \mathbf{c}'\boldsymbol{\beta} = \sum_j c_j \beta_{(i)j} \quad \text{and} \quad \sum_j c_j = 0$$

Again, under the general model, the probability is $1 - \alpha_i$ that simultaneous confidence intervals exist for a given value of i,

$$\hat{\psi}_i - c_{0i}\hat{\sigma}_{\hat{\psi}_i} \leq \psi \leq \hat{\psi}_i + c_{0i}\hat{\sigma}_{\hat{\psi}_i} \tag{5.8.32}$$

where $c_{0i}^2 = v_i F^{\alpha_i}(v_i, v_e)$, $i = 1, \ldots, r$. Employing a simple comparison, (5.8.32) reduces to

$$\begin{aligned} y_{ij.} - y_{ij'.} - c_{0i}\sqrt{\frac{Q_e}{v_e}\left(\frac{1}{N_{ij}} + \frac{1}{N_{ij'}}\right)} &\leq \beta_{(i)j} - \beta_{(i)j'} \\ &\leq y_{ij.} - y_{ij'.} + c_{0i}\sqrt{\frac{Q_e}{v_e}\left(\frac{1}{N_{ij}} + \frac{1}{N_{ij'}}\right)} \end{aligned} \tag{5.8.33}$$

As in previous discussions, Bonferroni t's may be applied in any case.

Multivariate Procedure. The multivariate generalization of the univariate results is immediate. Instead of scalar-valued observations, vector-valued observations are available for each individual. For the multivariate two-way nested design N_{ij}, p vector-valued observations are in each cell. Let $\mathbf{y}'_{ijk} = [y_{111}^{(1)}, y_{111}^{(2)}, \ldots, y_{111}^{(p)}]$ denote the $1 \times p$ sample vector of the kth observation under treatment j within treatment i, where the data are represented as follows:

Factor A

A_1				$\cdots$		A_r	
B_1	$\cdots$	B_{c_1}	$\cdots$	B'_1	$\cdots$	B'_{c_r}	
$\mathbf{y}'_{111}$	$\cdots$	$\mathbf{y}_{1c_11}$	$\cdots$	$\mathbf{y}'_{r11}$	$\cdots$	$\mathbf{y}'_{rc_r1}$	
$\mathbf{y}'_{112}$	$\cdots$	$\mathbf{y}_{1c_12}$	$\cdots$	$\mathbf{y}'_{r12}$	$\cdots$	$\mathbf{y}'_{rc_r2}$	
$\vdots$	$\vdots$	$\vdots$	$\vdots$	$\vdots$	$\vdots$	$\vdots$	
$\mathbf{y}'_{11N_{11}}$	$\cdots$	$\mathbf{y}_{1c_1N_{1c_1}}$	$\cdots$	$\mathbf{y}'_{r1N_{r1}}$	$\cdots$	$\mathbf{y}'_{rc_r N_{rc_r}}$	
$y_{11.}$	$\cdots$	$y_{1c_1.}$	$\cdots$	$y'_{r1.}$	$\cdots$	$y'_{rc_r.}$	

where

$$\begin{aligned} \mathbf{y}_{ij.} &= \frac{\sum_{k=1}^{N_{ij}} \mathbf{y}_{ijk}}{N_{ij}} \\ \mathbf{y}_{i..} &= \frac{\sum_{j=1}^{c_i}\sum_{k=1}^{N_{ij}} \mathbf{y}_{ijk}}{N_i} = \frac{\sum_{j=1}^{c_i} N_{ij}\mathbf{y}_{ij.}}{N_i} \\ \mathbf{y}_{...} &= \frac{\sum_{i=1}^{r}\sum_{j=1}^{c_i}\sum_{k=1}^{N_{ij}} \mathbf{y}_{ijk}}{N} = \frac{\sum_{i=1}^{c_i} N_i\mathbf{y}_{i..}}{N} \\ N &= \sum_{i=1}^{r} N_i = \sum_i\sum_j N_{ij} \end{aligned} \tag{5.8.34}$$

The general linear model for the observation $\mathbf{y}'_{ijk}$ is

$$\begin{aligned} \mathbf{y}_{ijk} &= \boldsymbol{\mu} + \boldsymbol{\alpha}_i + \boldsymbol{\beta}_{(i)j} + \boldsymbol{\varepsilon}_{ijk} \qquad i = 1, \ldots, r;\, j = 1, \ldots, c_i;\, k = 1, \ldots, N_{ij} \\ \boldsymbol{\varepsilon}_{ijk} &\sim IN_p(\mathbf{0}, \boldsymbol{\Sigma}) \end{aligned} \tag{5.8.35}$$

Since the design matrix of the univariate and multivariate models are identical, an unbiased estimate of $\mathbf{B}$, where

$$\mathbf{B} = \begin{bmatrix} \boldsymbol{\mu}' \\ \boldsymbol{\alpha}'_1 \\ \vdots \\ \boldsymbol{\alpha}'_r \\ \boldsymbol{\beta}'_{(i)1} \\ \vdots \\ \boldsymbol{\beta}'_{(r)c_r} \end{bmatrix} = \begin{bmatrix} \mu_{11} & \mu_{12} & \cdots & \mu_{1p} \\ \alpha_{11} & \alpha_{12} & \cdots & \alpha_{1p} \\ \vdots & \vdots & \vdots & \vdots \\ \alpha_{r1} & \alpha_{r2} & \cdots & \alpha_{rp} \\ \beta_{(1)11} & \beta_{(1)12} & \cdots & \beta_{(1)1p} \\ \vdots & \vdots & \vdots & \vdots \\ \beta_{(r)c_r1} & \beta_{(r)c_r2} & \cdots & \beta_{(r)c_rp} \end{bmatrix} \tag{5.8.36}$$

is given by

$$\hat{\mathbf{B}} = \begin{bmatrix} 0 \\ \mathbf{y}'_{11.} \\ \mathbf{y}'_{12.} \\ \vdots \\ \mathbf{y}'_{rc_r.} \end{bmatrix} \tag{5.8.37}$$

using the g-inverse defined by (5.8.6).

Generalizing (5.8.9), unique linear unbiased estimators for the estimable functions $\psi = \mathbf{c}'\mathbf{B}\mathbf{a}$ are obtained by using (3.6.17). Letting $\mathbf{t}' = [t_0, t_1, \ldots, t_r, t_{11}, \ldots, t_{rc_r}]$,

$$\begin{aligned} \psi &= \mathbf{c}'\mathbf{B}\mathbf{a} = \mathbf{a}'\left(\sum_i \sum_j t_{ij}\boldsymbol{\mu} + \sum_i \sum_j t_{ij}\boldsymbol{\alpha}_i + \sum_i \sum_j t_{ij}\boldsymbol{\beta}_{(i)j}\right) \\ \hat{\psi} &= \mathbf{c}'\hat{\mathbf{B}}\mathbf{a} = \mathbf{a}'\left(\sum_i \sum_j t_{ij}\mathbf{y}_{ij.}\right) \end{aligned} \tag{5.8.38}$$

denote the form of the estimable functions. The restrictions imposed in the univariate case extend to the multivariate case

$$\sum_i N_i \boldsymbol{\alpha}_i = \frac{\sum_j N_{ij}\boldsymbol{\beta}_{(i)j}}{N_i} = \mathbf{0} \tag{5.8.39}$$

where the weights are chosen proportional to the number of observations, so that $t_{ij} = N_{ij}/N_i$.

With $\mathbf{C}$ defined as in the univariate case, $\mathbf{B}$ defined by (5.8.36), and $\mathbf{A} = \mathbf{I}$, tests of the hypotheses, with side conditions given in (5.8.39) on the model, become

$$H_{(A_i)B}\colon \boldsymbol{\beta}_{(i)1} = \boldsymbol{\beta}_{(i)2} = \cdots = \boldsymbol{\beta}_{(i)c_i} \qquad (i = 1, \ldots, r) \tag{5.8.40}$$

$$H_A\colon \boldsymbol{\alpha}_1 = \boldsymbol{\alpha}_2 = \cdots = \boldsymbol{\alpha}_r \tag{5.8.41}$$

TABLE 5.8.4. MANOVA Table for a Two-Way Nested Design.

Source	df	SSP
A	$r - 1 = v_A$	$\mathbf{Q}_A = \sum_i N_i(\mathbf{y}_{i..} - \mathbf{y}_{...})(\mathbf{y}_{i..} - \mathbf{y}_{...})'$
$(A)B$	$\sum_i (c_i - 1) = v_{(A)B}$	$\mathbf{Q}_{(A)B} = \sum_i \sum_j N_{ij}(\mathbf{y}_{ij.} - \mathbf{y}_{i..})(\mathbf{y}_{ij.} - \mathbf{y}_{...})'$
$(A_1)B$	$c_1 - 1 = v_1$	$\mathbf{Q}_{(A_1)B} = \sum_j N_{ij}(\mathbf{y}_{1j.} - \mathbf{y}_{i..})(\mathbf{y}_{1j.} - \mathbf{y}_{...})'$
$\vdots$	$\vdots$	$\vdots$
$(A_r)B$	$c_r - 1 = v_r$	$\mathbf{Q}_{(A_r)B} = \sum_j N_{rj}(\mathbf{y}_{rj.} - \mathbf{y}_{r..})(\mathbf{y}_{rj.} - \mathbf{y}_{r..})'$
Within error	$\sum_i \sum_j (N_{ij} - 1) = v_e$	$\mathbf{Q}_e = \sum_i \sum_j \sum_k (\mathbf{y}_{ijk} - \mathbf{y}_{ij.})(\mathbf{y}_{ijk} - \mathbf{y}_{ij.})'$
"Total"	$N - 1$	$\mathbf{Q}_t = \sum_i \sum_j \sum_k (\mathbf{y}_{ijk} - \mathbf{y}_{...})(\mathbf{y}_{ijk} - \mathbf{y}_{...})'$

Table 5.8.4 summarizes the sum of squared and product matrices for the tests. A determinantal equation of the form $|\mathbf{Q}_h - \lambda\mathbf{Q}_e| = 0$ must be solved to test the hypotheses summarized in Table 5.8.4. Employing the numerous criteria, the parameters s, m, and n are necessary and given by

(5.8.42)

H_A	$H_{(A)B}$	$H_{(A_i)B}$
$s = \min(r - 1, p)$	$s = \min(v_{(A)B}, p)$	$s = \min(v_i, p)$
$m = \dfrac{|(r - p - 1)| - 1}{2}$	$m = \dfrac{|v_{(A)B} - p| - 1}{2}$	$m = \dfrac{|c_i - p - 1| - 1}{2}$
$n = \dfrac{v_e - p - 1}{2}$	$n = \dfrac{v_e - p - 1}{2}$	$n = \dfrac{v_e - p - 1}{2}$
$v_h: v_A = r - 1$	$v_{(A)B} = \sum_i (c_i - 1)$	$v_i = c_i - 1$

where the hypotheses are rejected if

(5.8.43) (a) $$\Lambda = \frac{|\mathbf{Q}_e|}{|\mathbf{Q}_e + \mathbf{Q}_h|} = \sum_{i=1}^{s} (1 + \lambda_i)^{-1} < U^\alpha(p, v_h, v_e) \qquad \text{(Wilks)}$$

(b) $$\theta_s = \frac{1}{1 + \lambda_1} > \theta^\alpha(s, m, n) \qquad \text{(Roy)}$$

(c) $$U^{(s)} = \sum_{i=1}^{s} \lambda_i > U_0^\alpha(s, m, n) \qquad \text{(Lawley-Hotelling)}$$

(d) $$V^{(s)} = \sum_{i=1}^{s} \frac{\lambda_i}{1 + \lambda_i} > V^\alpha(s, m, n) \qquad \text{(Pillai)}$$

The choice of the overall test criterion determines the confidence-interval procedure for parametric functions. The Roy-Bose method will be illustrated. Let $\psi = \mathbf{c}'\mathbf{B}\mathbf{a}$ be any estimable function of the parameters. The Roy-Bose theorem states that with probability $1 - \alpha$, for all ψ,

$$\hat{\psi} - c_0\hat{\sigma}_{\hat{\psi}} \leq \psi \leq \hat{\psi} + c_0\hat{\sigma}_{\hat{\psi}} \tag{5.8.44}$$

where

$$c_0^2 = \frac{\theta^\alpha}{1 - \theta^\alpha}\nu_e$$

$$\hat{\sigma}_{\hat{\psi}}^2 = \left[\mathbf{a}'\left(\frac{\mathbf{Q}_e}{\nu_e}\right)\mathbf{a}\right][\mathbf{c}'(\mathbf{X}'\mathbf{X})^-\mathbf{c}]$$

For pairwise comparisons for factor A, the symbols in (5.8.44), with the side conditions given in (5.8.39), simplify to

$$\begin{aligned} \hat{\psi} &= \sum_{s=1}^{p} a_s(\mathbf{y}_{i..}^{(s)} - \mathbf{y}_{i..}^{\prime(s)}) \\ \psi &= \sum_{s=1}^{p} a_s(\alpha_{is} - \alpha_{i's}) \\ \hat{\sigma}_{\hat{\psi}} &= \sqrt{\mathbf{a}'\left(\frac{\mathbf{Q}_e}{\nu_e}\right)\mathbf{a}\left[\sum_j\left(\frac{N_{ij}}{N_i^2}\right) + \sum_j\left(\frac{N_{i'j}}{N_{i'}^2}\right)\right]} \end{aligned} \tag{5.8.45}$$

If there are c levels of B in each A, and n_0 observations per cell, $\hat{\sigma}_{\hat{\psi}}$ in (5.8.45) reduces to

$$\hat{\sigma}_{\hat{\psi}} = \sqrt{\frac{2}{cn_0}\left[\mathbf{a}'\left(\frac{\mathbf{Q}_e}{\nu_e}\right)\mathbf{a}\right]} \tag{5.8.46}$$

Performing only tests $H_{(A_i)B}$, formation of intervals for pairwise comparisons is obtained by using (5.8.44) with

$$\begin{aligned} \hat{\psi} &= \sum_{s=1}^{p} a_s(\mathbf{y}_{ij.}^{(s)} - \mathbf{y}_{ij'.}^{(s)}) \\ \psi &= \sum_{s=1}^{p} a_s(\beta_{(i)js} - \beta_{(i)j's}) \\ \hat{\sigma}_{\hat{\psi}} &= \sqrt{\left(\frac{1}{N_{ij}} + \frac{1}{N_{ij'}}\right)\mathbf{a}'\left(\frac{\mathbf{Q}_e}{\nu_e}\right)\mathbf{a}} \\ c_0^2 &= \frac{\theta^{\alpha_i}}{1 - \theta^{\alpha_i}}\nu_e \end{aligned} \tag{5.8.47}$$

As in other analyses, Bonferroni and discriminant analysis procedures may also be applied with this design.

As with completely randomized factorial designs, higher-order nesting is also possible. That is, with three factors A, B, and C, we might have a nested design in which B is nested in A and C is nested in B. The multivariate model for this design would be as follows:

(5.8.48)
$$\mathbf{y}_{ijkm} = \boldsymbol{\mu} + \boldsymbol{\alpha}_i + \boldsymbol{\beta}_{(i)j} + \boldsymbol{\tau}_{(ij)k} + \boldsymbol{\varepsilon}_{ijkm}$$
$$\boldsymbol{\varepsilon}_{ijkm} \sim IN_p(\mathbf{0}, \boldsymbol{\Sigma})$$

where $i = 1, \ldots, r$; $j = 1, \ldots, c_i$; $k = 1, \ldots, N_{ij}$; $m = 1, \ldots, M_{ijkm}$. The expressions for the sum of squares necessary for testing, as given by Scheffé (1959, p. 186) for the univariate model, are immediately extended to the multivariate design with appropriate side conditions.

Another common variation of a nested design is to have both nesting and crossing factors. This kind of design is sometimes called a *partial hierarchal design.* For example, B could be nested in A, but C might be crossed with A and B. For this design the model would be

(5.8.49)
$$\mathbf{y}_{ijkm} = \boldsymbol{\mu} + \boldsymbol{\alpha}_i + \boldsymbol{\beta}_{(i)j} + \boldsymbol{\tau}_k + \boldsymbol{\gamma}_{ik} + \boldsymbol{\delta}_{(i)jk} + \boldsymbol{\varepsilon}_{ijkm}$$
$$\boldsymbol{\varepsilon}_{ijkm} \sim IN_p(\mathbf{0}, \boldsymbol{\Sigma})$$

These models, as well as numerous other variations, may be analyzed by using the general linear model theory; they are simple extensions of common univariate designs.

EXERCISES 5.8

1. Prove the result given in equation (5.8.19).
2. For an experimental design with three factors A, B, and C, and several observations per cell, specify the multivariate linear model if A and B are crossed but C is nested in B.
3. Marascuilo (1969) gives data from an experiment designed to investigate the effects of variety on the learning of the mathematical concepts of Boolean set unit and intersection. For the study, students were selected from fifth and seventh grades from four schools representing different socioeconomic areas. The students were then randomly assigned to four experimental conditions that varied in two ways. Subjects in the small variety (S) conditions were given eight problems to solve with each problem repeated six times, while the students in the large-variety (L) conditions solved 48 different problems. Half the students were given familiar geometric forms (G), and the remaining students were given nonsense forms (N) generated from random numbers. The data for the experiment follow.

5th Grade								7th Grade							
School 1				School 2				School 1′				School 2′			
S		L		S		L		S		L		S		L	
N	G	N	G	N	G	N	G	N	G	N	G	N	G	N	G
18	9	11	20	38	33	35	21	3	25	14	16	19	39	44	41
38	32	6	13	19	22	31	36	10	27	8	6	41	41	39	40
24	6	6	10	44	13	26	34	14	2	7	15	28	36	38	44
17	4	2	0	40	21	27	30	25	21	39	9	40	45	36	46

a. What are the mathematical model and the statistical assumptions for the design? Test to see that the assumptions are satisfied.

b. Construct an analysis-of-variance table for the design and test the appropriate hypotheses.

c. Summarize your findings.

5.9 EXAMPLE—MULTIVARIATE TWO-WAY NESTED DESIGN

In the investigation of the data given in Section 5.7, suppose that the teachers were nested within the classes. In addition, suppose that the third teacher under noncontract classes was unavailable for the study. The design for the analysis would then be a nested design represented diagrammatically as follows:

	T_1	T_2	T_3	T_4	T_5
Noncontract classes	×	×			
Contract classes			×	×	×

where × denotes data available for the study. The layout for the data is summarized in Table 5.9.1, where factor A, classes, has two levels, A_1 for the noncontract class and A_2 for the contract class. Factor B has two levels within A_1 and three levels within A_2; the factor B denotes teachers. Computing means for the data, the observed cell means are illustrated:

TABLE 5.9.1. Summary Data: Multivariate Nested Design.

A_1				A_2					
B_1		B_2		B'_1		B'_2		B'_3	
R	C	R	C	R	C	R	C	R	C
9	14	11	15	10	21	11	23	8	17
8	15	12	18	12	22	14	27	7	15
11	16	10	16	9	19	13	24	10	18
9	17	9	17	10	21	15	26	8	17
9	17	9	18	14	23	14	24	7	19

	A_1		A_2		
	B_1	B_2	B'_1	B'_2	B'_3
$\bar{\mathbf{y}}'_{ij.}$	[9.20, 15.80]	[10.20, 16.80]	[11.00, 21.20]	[13.40, 24.80]	[8.00, 17.20]
$\bar{\mathbf{y}}'_{i..}$	[9.70, 16.30]		[10.80, 21.07]		

The general linear model for the observation vector $\mathbf{y}_{ijk}$ is

$$\begin{aligned} \mathbf{y}_{ijk} &= \boldsymbol{\mu} + \boldsymbol{\alpha}_i + \boldsymbol{\beta}_{(i)j} + \boldsymbol{\varepsilon}_{ijk} \\ \boldsymbol{\varepsilon}_{ijk} &\sim IN_2(\mathbf{0}, \boldsymbol{\Sigma}) \end{aligned} \tag{5.9.1}$$

where the matrix **B** of parameters is

$$\underset{(8\times 2)}{\mathbf{B}} = \begin{bmatrix} \mu_{11} & \mu_{12} \\ \alpha_{11} & \alpha_{12} \\ \alpha_{21} & \alpha_{22} \\ \beta_{(1)11} & \beta_{(1)12} \\ \beta_{(1)21} & \beta_{(1)22} \\ \beta_{(2)11} & \beta_{(2)12} \\ \beta_{(2)21} & \beta_{(2)22} \\ \beta_{(2)31} & \beta_{(2)32} \end{bmatrix} = \begin{bmatrix} \boldsymbol{\mu}' \\ \boldsymbol{\alpha}_1' \\ \boldsymbol{\alpha}_2' \\ \boldsymbol{\beta}_{(1)1}' \\ \boldsymbol{\beta}_{(1)2}' \\ \boldsymbol{\beta}_{(2)1}' \\ \boldsymbol{\beta}_{(2)2}' \\ \boldsymbol{\beta}_{(2)3}' \end{bmatrix} \tag{5.9.2}$$

The total number of observations for the investigation is $N = 25$. The number of observations N_i for each level of A is $N_1 = 10$ and $N_2 = 15$. The number of observations in each cell is $N_{ij} = 5$. By employing (5.8.5) and (5.8.6),

$$(\mathbf{X}'\mathbf{X})^- = \begin{bmatrix} \underset{(3\times 3)}{\mathbf{0}} & \underset{(3\times 5)}{\mathbf{0}} \\ \underset{(5\times 3)}{\mathbf{0}} & \underset{(5\times 5)}{\text{Dia}\,[1/5]} \end{bmatrix} \tag{5.9.3}$$

so that

$$\hat{\mathbf{B}} = (\mathbf{X}'\mathbf{X})^-\mathbf{X}'\mathbf{Y}$$

$$= \begin{bmatrix} \mathbf{0}' \\ \mathbf{0}' \\ \mathbf{0}' \\ \mathbf{y}_{11.}' \\ \mathbf{y}_{12.}' \\ \mathbf{y}_{21.}' \\ \mathbf{y}_{22.}' \\ \mathbf{y}_{23.}' \end{bmatrix} = \begin{bmatrix} 0 & 0 \\ 0 & 0 \\ 0 & 0 \\ 9.20 & 15.80 \\ 10.20 & 16.80 \\ 11.00 & 21.20 \\ 13.40 & 24.80 \\ 8.00 & 17.20 \end{bmatrix} \tag{5.9.4}$$

By using the formula

$$\mathbf{S} = \frac{\mathbf{Y}'[\mathbf{I} - \mathbf{X}(\mathbf{X}'\mathbf{X})^-\mathbf{X}']\mathbf{Y}}{N - R(\mathbf{X})}$$

$$= \frac{\sum_i \sum_j \sum_k (\mathbf{y}_{ijk} - \mathbf{y}_{...})(\mathbf{y}_{ijk} - \mathbf{y}_{...})'}{\sum_i \sum_j (N_{ij} - 1)} \tag{5.9.5}$$

the unbiased estimator for $\boldsymbol{\Sigma}$ is

$$\mathbf{S} = \begin{bmatrix} 2.1400 & \\ 1.0403 & 2.1007 \end{bmatrix} \tag{5.9.6}$$

To test the hypotheses

$$H_{(A_1)B}: \boldsymbol{\beta}_{(1)1} = \boldsymbol{\beta}_{(1)2}$$

(5.9.7)
$$H_{(A_2)B}: \boldsymbol{\beta}_{(2)1} = \boldsymbol{\beta}_{(2)2} = \boldsymbol{\beta}_{(2)3}$$

$$H_A: \boldsymbol{\alpha}_i + \frac{\sum_{j=1}^{c_1} N_{1j}\boldsymbol{\beta}_{(1)j}}{N_1} = \boldsymbol{\alpha}_2 + \frac{\sum_{j=1}^{c_1} N_{2j}\boldsymbol{\beta}_{(2)j}}{N_2}$$

Table 5.9.2, a MANOVA table for the form given by Table 5.8.4, is constructed.

TABLE 5.9.2. MANOVA Table for Nested Design.

Source	df	SSP
A (classes)	1	$\mathbf{Q}_A = \begin{bmatrix} 7.2600 & \\ 31.4600 & 136.3267 \end{bmatrix}$
$(A)B$ (teachers within classes)	3	$\mathbf{Q}_{(A)B} = \begin{bmatrix} 75.7000 & \\ 105.3000 & 147.0330 \end{bmatrix}$
$(A_1)B$ (teachers within A_1)	1	$\mathbf{Q}_{(A_1)B} = \begin{bmatrix} 2.5000 & \\ 2.5000 & 2.5000 \end{bmatrix}$
$(A_2)B$ (teachers within A_2)	2	$\mathbf{Q}_{(A_2)B} = \begin{bmatrix} 73.2000 & \\ 102.8000 & 144.5333 \end{bmatrix}$
Within error	20	$\mathbf{Q}_e$ is obtained from (5.9.6)
"Total"	24	

By solving the characteristic equations of the form $|\mathbf{Q}_h - \lambda\mathbf{Q}_e| = 0$, the eigenvalues for each hypothesis are determined. In summary, they are

$$H_A: \lambda_1 = 3.538$$

(5.9.8)
$$H_{(A_1)B}: \lambda_1 = .0791$$

$$H_{(A_2)B}: \lambda_1 = 3.648 \qquad \lambda_2 = .0024$$

From (5.8.42), the parameters for the hypotheses in Table 5.9.2 are

(5.9.9)

	A	$(A)B$	$(A_1)B$	$(A_2)B$
s	1	2	1	2
m	0	0	0	$-.5$
n	8.5	8.5	8.5	8.5
ν_h	1	3	1	2

where each hypothesis is rejected according to the criteria given in (5.8.43).

Employing Roy's criterion for (5.9.7), significant differences exist between classes and among teachers within contract classes. Using $\alpha = .01$, Roy's critical values for H_A and $H_{(A_2)B}$ are $\theta^\alpha = .384$ and $\theta^\alpha = .470$, respectively. Using (5.9.8), the sample statistics are $\theta_s = .780$ and $\theta_s = .785$.

Looking at the hypothesis $H_{(A_2)B}$ in more detail, there is only one significant standardized vector. The standardized and adjusted vectors are

$$\mathbf{a}_1 = \begin{bmatrix} .1873 \\ .5773 \end{bmatrix} \qquad \mathbf{a}_1^{(A)} = \begin{bmatrix} .2739 \\ .8367 \end{bmatrix} \tag{5.9.10}$$

Thus the correlation between the significant discriminant function and each dependent variable is

$$\begin{aligned} \hat{\boldsymbol{\rho}} = \mathbf{R}_e \mathbf{a}_1^{(A)} &= \begin{bmatrix} 1.0000 & \\ .4906 & 1.0000 \end{bmatrix} \begin{bmatrix} .2739 \\ .8367 \end{bmatrix} \\ &= \begin{bmatrix} .684 \\ .971 \end{bmatrix} \end{aligned} \tag{5.9.11}$$

indicating that the comprehension variable contributes more to difference than the reading rate variable.

By using (5.8.47), confidence intervals are acquired for comparisons within or between levels of A.

EXERCISES 5.9

1. In an experiment designed to investigate two driver-training programs, students in the eleventh grade in schools I, II, and III were trained with one program and eleventh-grade students in schools IV, V, and VI were trained with another program. After the completion of a 6-week training period, a test measuring knowledge (K) of traffic laws and driving ability (A) was administered to the 138 students in the study.

Program 1						Program 2					
S1(I)		S2(II)		S3(III)		S1(IV)		S2(V)		S3(VI)	
K	A	K	A	K	A	K	A	K	A	K	A
48	66	36	43	82	51	54	79	46	46	21	5
68	16	24	24	79	55	30	79	13	13	51	16
28	22	24	24	82	46	9	36	59	76	52	43
42	21	37	30	65	33	23	79	26	42	53	54
73	10	78	21	33	67	18	66	38	84	11	9
46	13	82	60	79	12	15	82	29	65	34	52
46	17	56	24	33	48	16	89	12	47	40	16
76	15	24	12	75	48	14	82	6	56	43	48
52	11	82	63	33	35	48	83	15	46	11	32
44	64	78	34	67	28	31	65	18	34	39	69
33	14	44	39	67	52	11	74	41	23	32	25
43	25	92	34	67	37	41	88	26	33	33	26
76	60	68	85	33	35	56	67	15	29	45	12
76	18	43	50	33	30	51	93	54	50	27	12
36	49	68	28	67	33	23	83	36	83	49	21
39	75	53	90	67	54	16	76	27	82	44	56
76	16	76	41	75	63	40	69	64	79	64	18
73	11	35	10	83	93	21	37	55	67	27	11
34	12	24	11	94	92	88	82	10	70	79	57
68	63	23	21	79	61	7	32	12	65	47	64
52	69	66	65	89	53	38	75	34	93	17	29
46	22	76	78	94	92	20	77	21	83	39	42
26	18	34	36	93	91	13	70	34	81	21	11

a. Analyze the data for this experiment by using the procedures outlined in Section 5.9. Include in your analysis the construction of linear discriminant function(s). Use $\alpha = .05$ for all tests.

b. On the basis of your analysis in part a, what would you conclude about the different training programs and the differences among the schools?

5.10 LATIN-SQUARE DESIGN—UNIVARIATE AND MULTIVARIATE

The randomized block design discussed in Section 5.5 was employed to minimize heterogeneity on one variable. A more complex device called a *Latin square* allows an experimenter to produce a double blocking that eliminates nuisance variables in two directions from experimental error. The randomization procedure is to assign treatment levels to the blocks such that each treatment level appears in only one row block and one column block according to the Latin-square design. For example, if an investigator was interested in examining a concept learning task for five experimental treatments and used as blocking variables days of the week and hours of the day, the following Latin-square design might be employed

(5.10.1)

	Hours of day 1	2	3	4	5
Mon.	T_2	T_5	T_4	T_3	T_1
Tues.	T_3	T_1	T_2	T_5	T_4
Wed.	T_4	T_2	T_3	T_1	T_5
Thurs.	T_5	T_3	T_1	T_4	T_2
Fri.	T_1	T_4	T_5	T_2	T_3

where each treatment condition appears only once in each row and column. Many authors compare a Latin-square design with a completely randomized three-way factorial design with one observation per cell. For a three-way factorial design with d levels per factor and one observation per cell, d^3 observations are required for the analysis of main-effect hypotheses. A Latin-square design requires only d^2 observations and may be thought of as an incomplete balanced three-way factorial design with one observation per cell. However, one would seldom use the design in this way since, for three treatment factors, it is more likely that an interaction would exist among the treatments. This is less likely for two nuisance variables and several treatment levels.

For a discussion of the construction of Latin-square designs with tables of designs, see Fisher and Yates (1963). An elementary introduction is offered by either Kirk (1968) or Scheffé (1959).

Given that a Latin square D has been selected for an experiment, let y_{ijk} denote the observation in the ith row block, jth column block, and kth cell (treatment) of the Latin square where the tuples i, j, and k take on only the d^2 values for the design, which has d levels per factor.

Univariate Procedure. The general linear model for the observation y_{ijk} for the Latin square is

$$\begin{aligned} y_{ijk} &= \mu + \alpha_i + \beta_j + \gamma_k + \varepsilon_{ijk} \qquad (i, j, k) \in D \\ \varepsilon_{ijk} &\sim IN(0, \sigma^2) \end{aligned} \tag{5.10.2}$$

No interactions are assumed from the model.

To obtain the Gauss-Markoff estimators of a Latin-square design, a simple example is considered. Suppose that $d = 3$ and that the design is given by

$$\begin{matrix} A & B & C \\ B & C & A \\ C & A & B \end{matrix} \tag{5.10.3}$$

where $A = 1$, $B = 2$, and $C = 3$; then (5.10.2) becomes

$$\begin{bmatrix} y_{111} \\ y_{122} \\ y_{133} \\ y_{212} \\ y_{223} \\ y_{231} \\ y_{313} \\ y_{321} \\ y_{332} \end{bmatrix} = \begin{bmatrix} 1 & 1 & 0 & 0 & 1 & 0 & 0 & 1 & 0 & 0 \\ 1 & 1 & 0 & 0 & 0 & 1 & 0 & 0 & 1 & 0 \\ 1 & 1 & 0 & 0 & 0 & 0 & 1 & 0 & 0 & 1 \\ 1 & 0 & 1 & 0 & 1 & 0 & 0 & 0 & 1 & 0 \\ 1 & 0 & 1 & 0 & 0 & 1 & 0 & 0 & 0 & 1 \\ 1 & 0 & 1 & 0 & 0 & 0 & 1 & 1 & 0 & 0 \\ 1 & 0 & 0 & 1 & 1 & 0 & 0 & 0 & 0 & 1 \\ 1 & 0 & 0 & 1 & 0 & 1 & 0 & 1 & 0 & 0 \\ 1 & 0 & 0 & 1 & 0 & 0 & 1 & 0 & 1 & 0 \end{bmatrix} \begin{bmatrix} \mu \\ \alpha_1 \\ \alpha_2 \\ \alpha_3 \\ \beta_1 \\ \beta_2 \\ \beta_3 \\ \gamma_1 \\ \gamma_2 \\ \gamma_3 \end{bmatrix} + \begin{bmatrix} \varepsilon_{111} \\ \varepsilon_{122} \\ \varepsilon_{133} \\ \varepsilon_{212} \\ \varepsilon_{223} \\ \varepsilon_{231} \\ \varepsilon_{313} \\ \varepsilon_{321} \\ \varepsilon_{332} \end{bmatrix} \tag{5.10.4}$$

$$\underset{(9\times 1)}{\mathbf{y}} = \underset{(9\times 10)}{\mathbf{X}} \; \underset{(10\times 1)}{\boldsymbol{\beta}} + \underset{(9\times 1)}{\boldsymbol{\varepsilon}}$$

The $R(\mathbf{X}) = 3(d-1) + 1 = 3d - 2$ or 7 for the design given in (5.10.3). Thus $N - R(\mathbf{X}) = d^2 - 3d + 2 = (d-1)(d-2)$ are the degrees of freedom associated with Q_e, the error sum of squares.

From (5.10.4),

$$\mathbf{X}'\mathbf{X} = \begin{bmatrix} 9 & 3 & 3 & 3 & 3 & 3 & 3 & 3 & 3 & 3 \\ 3 & 3 & 0 & 0 & 1 & 1 & 1 & 1 & 1 & 1 \\ 3 & 0 & 3 & 0 & 1 & 1 & 1 & 1 & 1 & 1 \\ 3 & 0 & 0 & 3 & 1 & 1 & 1 & 1 & 1 & 1 \\ 3 & 1 & 1 & 1 & 3 & 0 & 0 & 1 & 1 & 1 \\ 3 & 1 & 1 & 1 & 0 & 3 & 0 & 1 & 1 & 1 \\ 3 & 1 & 1 & 1 & 0 & 0 & 3 & 1 & 1 & 1 \\ 3 & 1 & 1 & 1 & 1 & 1 & 1 & 3 & 0 & 0 \\ 3 & 1 & 1 & 1 & 1 & 1 & 1 & 0 & 3 & 0 \\ 3 & 1 & 1 & 1 & 1 & 1 & 1 & 0 & 0 & 3 \end{bmatrix} \tag{5.10.5}$$

By setting $\hat{\mu}$, $\hat{\beta}_3$, and $\hat{\gamma}_3$ equal to zero, $\mathbf{X}'\mathbf{X}$ becomes

$$(5.10.6)\qquad \mathbf{A}'\mathbf{A} = \begin{bmatrix} 3 & 0 & 0 & 1 & 1 & 1 & 1 \\ 0 & 3 & 0 & 1 & 1 & 1 & 1 \\ 0 & 0 & 3 & 1 & 1 & 1 & 1 \\ 1 & 1 & 1 & 3 & 0 & 1 & 1 \\ 1 & 1 & 1 & 0 & 3 & 1 & 1 \\ 1 & 1 & 1 & 1 & 1 & 3 & 0 \\ 1 & 1 & 1 & 1 & 1 & 0 & 3 \end{bmatrix}$$

which is of the form

$$(5.10.7)\qquad \mathbf{A}'\mathbf{A} = \begin{bmatrix} \text{Dia}\,[k] & \mathbf{J} & \mathbf{J} \\ \mathbf{J} & \text{Dia}\,[k] & \mathbf{J} \\ \mathbf{J} & \mathbf{J} & \text{Dia}\,[k] \end{bmatrix} \begin{matrix} (k) \\ (k-1) \\ (k-1) \end{matrix}$$
$$\qquad\qquad (k-1) \quad (k-1) \quad (k-1)$$

where $k = 3$.

From Roy and Sarhan (1956), the inverse of $\mathbf{A}'\mathbf{A}$ is of the form

$$(5.10.8)\qquad (\mathbf{A}'\mathbf{A})^{-1} = \begin{bmatrix} \text{Dia}\,[1/k] + \alpha\mathbf{J} & \beta\mathbf{J} & \gamma\mathbf{J} \\ \beta\mathbf{J} & \text{Dia}\,[1/k] + \delta\mathbf{J} & \phi\mathbf{J} \\ \gamma\mathbf{J} & \phi\mathbf{J} & \text{Dia}\,[1/k] + \xi\mathbf{J} \end{bmatrix} \begin{matrix} (k) \\ (k-1) \\ (k-1) \end{matrix}$$

where $\alpha = 2(k-1)/k^2$, $\beta = \gamma = -1/k$, $\delta = \xi = 1/k$, and $\phi = 0$.

By using (5.10.8),

$$(\mathbf{A}'\mathbf{A})^{-1} = \begin{bmatrix} 7/9 & 4/9 & 4/9 & -1/3 & -1/3 & -1/3 & -1/3 \\ 4/9 & 7/9 & 4/9 & -1/3 & -1/3 & -1/3 & -1/3 \\ 4/9 & 4/9 & 7/9 & -1/3 & -1/3 & -1/3 & -1/3 \\ -1/3 & -1/3 & -1/3 & 2/3 & 1/3 & 0 & 0 \\ -1/3 & -1/3 & -1/3 & 1/3 & 2/3 & 0 & 0 \\ -1/3 & -1/3 & -1/3 & 0 & 0 & 2/3 & 1/3 \\ -1/3 & -1/3 & -1/3 & 0 & 0 & 1/3 & 2/3 \end{bmatrix}$$

so that a g-inverse of $\mathbf{X}'\mathbf{X}$ is

$$(5.10.9)\qquad (\mathbf{X}'\mathbf{X})^{-} = \begin{bmatrix} 0 & 0 & 0 & 0 & 0 & 0 & 0 & 0 & 0 & 0 \\ 0 & 7/9 & 4/9 & 4/9 & -1/3 & -1/3 & 0 & -1/3 & -1/3 & 0 \\ 0 & 4/9 & 7/9 & 4/9 & -1/3 & -1/3 & 0 & -1/3 & -1/3 & 0 \\ 0 & 4/9 & 4/9 & 7/9 & -1/3 & -1/3 & 0 & -1/3 & -1/3 & 0 \\ 0 & -1/3 & -1/3 & -1/3 & 2/3 & 1/3 & 0 & 0 & 0 & 0 \\ 0 & -1/3 & -1/3 & -1/3 & 1/3 & 2/3 & 0 & 0 & 0 & 0 \\ 0 & 0 & 0 & 0 & 0 & 0 & 0 & 0 & 0 & 0 \\ 0 & -1/3 & -1/3 & -1/3 & 0 & 0 & 0 & 2/3 & 1/3 & 0 \\ 0 & -1/3 & -1/3 & -1/3 & 0 & 0 & 0 & 1/3 & 2/3 & 0 \\ 0 & 0 & 0 & 0 & 0 & 0 & 0 & 0 & 0 & 0 \end{bmatrix}$$

and

$$\mathbf{H} = (\mathbf{X}'\mathbf{X})^{-}\mathbf{X}'\mathbf{X} = \begin{bmatrix} 0 & 0 & 0 & 0 & 0 & 0 & 0 & 0 & 0 & 0 \\ 1 & 1 & 0 & 0 & 0 & 0 & 1 & 0 & 0 & 0 \\ 1 & 0 & 1 & 0 & 0 & 0 & 1 & 0 & 0 & 0 \\ 1 & 0 & 0 & 1 & 0 & 0 & 1 & 0 & 0 & 0 \\ 0 & 0 & 0 & 0 & 1 & 0 & -1 & 0 & 0 & 0 \\ 0 & 0 & 0 & 0 & 0 & 1 & -1 & 0 & 0 & 0 \\ 0 & 0 & 0 & 0 & 0 & 0 & 0 & 0 & 0 & 0 \\ 0 & 0 & 0 & 0 & 0 & 0 & 0 & 1 & 0 & -1 \\ 0 & 0 & 0 & 0 & 0 & 0 & 0 & 1 & 1 & -1 \\ 0 & 0 & 0 & 0 & 0 & 0 & 0 & 1 & 0 & 0 \end{bmatrix} \tag{5.10.10}$$

By checking that $\mathbf{c}' = \mathbf{c}'\mathbf{H}$, differences between pairs of α_i's, β_j's, and γ_k's can be estimated. However, the individual parameters in the model are not estimable.

The estimable functions for the example are

$$\begin{aligned} \psi = \mathbf{c}'\boldsymbol{\beta} = \mathbf{t}'\mathbf{H}\boldsymbol{\beta} = \sum_i t_i(\mu + \alpha_i + \beta_3 + \gamma_3) + t_4(\beta_1 - \beta_3) \\ + t_5(\beta_2 - \beta_3) + t_6(\gamma_1 - \gamma_3) + t_7(\gamma_2 - \gamma_3) \end{aligned} \tag{5.10.11}$$

By setting $t_1 = 1$ and $t_2 = -1$, $\psi = \alpha_1 - \alpha_3$ is estimable. To estimate $\psi = \alpha_1 - \alpha_3$, $\hat{\psi} = \mathbf{c}'\hat{\boldsymbol{\beta}} = \mathbf{t}'\mathbf{H}\hat{\boldsymbol{\beta}}$ is evaluated for $\mathbf{c}' = [0, 1, 0, -1, 0, 0, 0, 0, 0, 0]$. It can be shown that $\hat{\psi} = y_{1..} - y_{3..}$, where the dot notation for the Latin-square design indicates the average of the d observations for the first factor. That is,

$$\begin{aligned} y_{i..} &= \sum_{(j,k)\in D} \frac{y_{ijk}}{d} \\ y_{.j.} &= \sum_{(i,k)\in D} \frac{y_{ijk}}{d} \\ y_{..k} &= \sum_{(i,j)\in D} \frac{y_{ijk}}{d} \\ y_{...} &= \sum_{(i,j,k)\in D} \frac{y_{ijk}}{d^2} \end{aligned} \tag{5.10.12}$$

Since differences in the effects are estimable, the main-effect hypotheses

$$\begin{aligned} &H_\alpha: \text{all } \alpha_i\text{'s are equal} \\ &H_\beta: \text{all } \beta_j\text{'s are equal} \\ &H_\gamma: \text{all } \gamma_k\text{'s are equal} \end{aligned} \tag{5.10.13}$$

can be tested. The matrices **C** to test the hypotheses in (5.10.13) are, respectively,

$$\underset{[(d-1)\times(3d+1)]}{\mathbf{C}_\alpha} = \begin{bmatrix} 0 & 1 & 0 & \cdots & 0 & -1 & & \\ 0 & 0 & 1 & \cdots & 0 & -1 & & \\ \vdots & \vdots & \vdots & \vdots & \vdots & \vdots & \mathbf{0} & \mathbf{0} \\ \vdots & \vdots & \vdots & \vdots & \vdots & \vdots & & \\ 0 & 0 & 0 & \cdots & 1 & -1 & & \end{bmatrix}$$

$$(5.10.14) \qquad \underset{[(d-1)\times(3d+1)]}{\mathbf{C}_\beta} = \begin{bmatrix} 0 & & 1 & 0 & \cdots & 0 & -1 & \\ 0 & & 0 & 1 & \cdots & 0 & -1 & \\ \vdots & \mathbf{0} & \vdots & \vdots & \vdots & \vdots & \vdots & \mathbf{0} \\ \vdots & & \vdots & \vdots & \vdots & \vdots & \vdots & \\ 0 & & 0 & 0 & \cdots & 1 & -1 & \end{bmatrix}$$

$$\underset{[(d-1)\times(3d+1)]}{\mathbf{C}_\gamma} = \begin{bmatrix} 0 & & & 1 & 0 & \cdots & 0 & -1 \\ 0 & & & 0 & 1 & \cdots & 0 & -1 \\ \vdots & \mathbf{0} & \mathbf{0} & \vdots & \vdots & \vdots & \vdots & \vdots \\ \vdots & & & \vdots & \vdots & \vdots & \vdots & \vdots \\ 0 & & & 1 & 0 & \cdots & 1 & -1 \end{bmatrix}$$

where the rank of each **C** is $d - 1$ and the order of **0** is $(d - 1) \times d$.

TABLE 5.10.1. ANOVA Table for a $d \times d$ Latin-Square Design.

Source	df	SS
Rows (α)	$d - 1$	$Q_{h_\alpha} = d \sum_i (y_{i..} - y_{...})^2$
Columns (β)	$d - 1$	$Q_{h_\beta} = d \sum_j (y_{.j.} - y_{...})^2$
Treatments (γ)	$d - 1$	$Q_{h_\gamma} = d \sum_k (y_{..k} - y_{...})^2$
Within error	$(d - 1)(d - 2)$	$Q_e = \sum_{i,j,k} (y_{ijk} - y_{i..} - y_{.j.} - y_{..k} + 2y_{...})^2$
"Total"	$d^2 - 1$	$Q_t = \sum_{i,j,k} (y_{ijk} - y_{...})^2$

By utilizing the Gauss-Markoff theorem and the fundamental least-squares theorem, the ANOVA table for a $d \times d$ Latin-square design is displayed in Table 5.10.1.

The hypotheses of equal main effects are rejected at the significance level α if

$$(5.10.15) \qquad F = \frac{Q_{h_i}/(d-1)}{Q_e/(d-1)(d-2)} > F^\alpha[d-1, (d-1)(d-2)] \qquad i = \alpha, \beta, \gamma$$

Again employing Scheffé's theorem, multiple comparisons in the parametric estimable functions are immediately gained. For any of the three main-effect tests, we define $\psi = \Sigma_i c_i \theta_i$ such that $\Sigma_i c_i = 0$, where θ_i is either α_i, β_i, or γ_i. Then the probability is $1 - \alpha$ that, simultaneously for all ψ,

$$\hat{\psi} - c_0\hat{\sigma}_{\hat{\psi}} \leq \psi \leq \hat{\psi} + c_0\hat{\sigma}_{\hat{\psi}}$$

where

$$c_0^2 = (d-1)F^{\alpha}[d-1, (d-1)(d-2)]$$

For pairwise comparisons in any of the main effects,

$$\hat{\sigma}_{\hat{\psi}} = s\sqrt{\frac{2}{d}}$$

The Latin-square design used in this section is a restrictive application of its use in the behavioral sciences where it is often employed as a building block for more complex designs. The general linear model theory, although still applicable, is more complex. For further uses and illustrations of Latin squares, see Kirk (1968).

Multivariate Procedure. For the multivariate Latin-square design, the d^2 observations are vector valued. Let $\mathbf{y}_{ijk}$ denote the p-vector-valued observation for $(i, j, k) \in D$.

The general linear model for the observations $\mathbf{y}_{ijk}$ is

$$\begin{aligned} \mathbf{y}_{ijk} &= \boldsymbol{\mu} + \boldsymbol{\alpha}_i + \boldsymbol{\beta}_j + \boldsymbol{\gamma}_k + \boldsymbol{\varepsilon}_{ijk} \\ \boldsymbol{\varepsilon}_{ijk} &\sim IN_p(\mathbf{0}, \boldsymbol{\Sigma}) \end{aligned} \tag{5.10.16}$$

The hypotheses of interest are

$$\begin{aligned} &H_\alpha: \text{all } \boldsymbol{\alpha}_i \text{ are equal} \\ &H_\beta: \text{all } \boldsymbol{\beta}_j \text{ are equal} \\ &H_\gamma: \text{all } \boldsymbol{\gamma}_k \text{ are equal} \end{aligned} \tag{5.10.17}$$

Extending the results of the univariate case, MANOVA Table 5.10.2 is constructed.

TABLE 5.10.2. MANOVA Table for a $d \times d$ Latin-Square Design.

Source	df	SSP
Rows (α)	$d-1$	$\mathbf{Q}_{h_\alpha} = d\sum_i (\mathbf{y}_{i..} - \mathbf{y}_{...})(\mathbf{y}_{i..} - \mathbf{y}_{...})'$
Columns (β)	$d-1$	$\mathbf{Q}_{h_\beta} = d\sum_j (\mathbf{y}_{.j.} - \mathbf{y}_{...})(\mathbf{y}_{.j.} - \mathbf{y}_{...})'$
Treatments (γ)	$d-1$	$\mathbf{Q}_{h_\gamma} = d\sum_k (\mathbf{y}_{..k} - \mathbf{y}_{...})(\mathbf{y}_{..k} - \mathbf{y}_{...})'$
Within error	$(d-1)(d-2)$	$\mathbf{Q}_e = \sum_{i,j,k} (\mathbf{y}_{ijk} - \mathbf{y}_{i..} - \mathbf{y}_{ij.} + 2\mathbf{y}_{...})(\mathbf{y}_{ijk} - \mathbf{y}_{i..} - \mathbf{y}_{ij.} + 2\mathbf{y}_{...})'$
"Total"	$d^2 - 1$	$\mathbf{Q}_t = \sum_{i,j,k} (\mathbf{y}_{ijk} - \mathbf{y}_{...})(\mathbf{y}_{ijk} - \mathbf{y}_{...})'$

All main-effect null hypotheses H_i $(i = \alpha, \beta, \gamma)$ are rejected at the significance level α if

(5.10.18) (a) Wilks:

$$\Lambda = \frac{|\mathbf{Q}_e|}{|\mathbf{Q}_{h_i} + \mathbf{Q}_e|} < U^{\alpha}(p, d-1, \nu_e)$$

(b) Roy:

$$\theta_s = \frac{\lambda_1}{\lambda_1 + 1} > \theta^{\alpha}(s, m, n) \qquad \text{where } s = \min(d - 1, p)$$

$$m = \frac{|(d - 1) - p| - 1}{2}$$

$$n = \frac{\nu_e - p - 1}{2}$$

(c) Lawley-Hotelling:

$$U^{(s)} = \sum_{i=1}^{s} \lambda_i > U_0^{\alpha}(s, m, n)$$

(d) Pillai:

$$V^{(s)} = \sum_{i=1}^{s} \frac{\lambda_i}{1 + \lambda_i} > V(s, m, n)$$

Confidence intervals must correspond to the criteria selected to test for main effects since the procedures are not equivalent. The intervals for this design are closely related to the previously discussed design; therefore the multiple comparison procedure is postponed until the example is discussed in the next section.

EXERCISES 5.10

1. Derive the expressions for the SS given in Table 5.10.1 and determine their expected values. By analogy, what are the expected values of the SSP matrices in Table 5.10.2?
2. For the Latin square given in (5.10.3), suppose that two observations are obtained for each treatment condition A, B, and C and that the multivariate linear model for the design is

$$\mathbf{y}_{ijkm} = \boldsymbol{\mu} + \boldsymbol{\alpha}_i + \boldsymbol{\beta}_j + \boldsymbol{\gamma}_k + \boldsymbol{\theta}_{ijk} + \boldsymbol{\varepsilon}_{ijkm}$$

where $\boldsymbol{\varepsilon}_{ijkm} \sim N(\mathbf{0}, \boldsymbol{\Sigma})$. Utilizing the general theory developed in Section 3.6, construct a MANOVA table for the design.

5.11 EXAMPLE—MULTIVARIATE LATIN-SQUARE DESIGN

An investigator is interested in examining a concept learning task for five experimental treatments. Primary to the investigation is deciding whether treatments T_1, T_2, T_3, and T_4 differ from T_5. The Latin-square design for the five experimental treatments, using as blocking variables days of the week and hours of the day, is given in Section 5.10. The dependent variables for the experiment are the number of trials to criterion used to measure learning (V_1) and the number of errors in the test set on one presentation 10 minutes later (V_2) used to measure retention.

This represents a hypothetical example to illustrate a 5×5 multivariate Latin-square design. The data for the example are best represented by using the cell indexes where the first index denotes days of the week, the second index hours of the day, and the third index treatment conditions. The data are given in Table 5.11.1.

TABLE 5.11.1.

Cell	V_1	V_2	Cell	V_1	V_2
112	8	4	333	4	17
125	18	8	341	8	8
134	5	3	355	14	8
143	8	16	415	11	9
151	6	12	423	4	15
213	1	6	431	14	17
221	6	19	444	1	5
232	5	7	452	7	8
245	18	9	511	9	14
254	9	23	524	9	13
314	5	11	535	16	23
322	4	5	542	3	7
			553	2	10

The observed combined means for the data, using (5.10.12) and replacing the individual observations by vectors, are

Factor (days) $\mathbf{y}'_{i..}$		Factor (hours) $\mathbf{y}'_{.j.}$		Factor (treatments) $\mathbf{y}'_{..k}$	
9.00	8.60	6.80	8.80	8.60	14.00
7.80	12.80	8.20	12.00	5.40	6.20
7.00	9.80	8.80	13.40	3.80	12.80
7.40	10.80	7.60	9.00	5.80	11.00
7.80	13.40	7.60	12.20	15.40	11.40

with the overall mean $\mathbf{y}'_{...} = [7.80, 11.08]$. MANOVA Table 5.11.2 is formed to test the hypotheses

(5.11.1)
$$\begin{aligned} H_\alpha &: \text{all } \boldsymbol{\alpha}_i \text{ are equal} \\ H_\beta &: \text{all } \boldsymbol{\beta}_j \text{ are equal} \\ H_\gamma &: \text{all } \boldsymbol{\gamma}_k \text{ are equal} \end{aligned}$$

by utilizing the formula in Table 5.10.2 with $d = 5$. The model for the data is

(5.11.2)
$$\begin{aligned} \mathbf{y}_{ijk} &= \boldsymbol{\mu} + \boldsymbol{\alpha}_i + \boldsymbol{\beta}_j + \boldsymbol{\gamma}_k + \boldsymbol{\varepsilon}_{ijk} \\ \boldsymbol{\varepsilon}_{ijk} &\sim IN(\mathbf{0}, \boldsymbol{\Sigma}) \qquad \text{for } (i, j, k) \in D \end{aligned}$$

From Table 5.11.2, the sample variance-covariance matrix $\mathbf{S}$ is

(5.11.3)
$$\mathbf{S} = \begin{bmatrix} 12.2333 & \\ 9.8333 & 35.2267 \end{bmatrix}$$

Using Roy's criterion for illustration purposes, the three hypotheses in (5.11.1) are tested. The critical constant for all three tests, with $\alpha = .05$, is $\theta^\alpha = .590$ for $s = 2$, $m = .5$, and $n = 4.5$. The value of θ_s for each hypothesis is computed to be $\theta_s = .2540$,

TABLE 5.11.2. MANOVA Table for the Latin-Square Design.

Source	df	SSP
Days (α)	4	$\mathbf{Q}_\alpha = \begin{bmatrix} 11.20 & \\ -9.20 & 81.04 \end{bmatrix}$
Hours (β)	4	$\mathbf{Q}_\beta = \begin{bmatrix} 11.20 & \\ 25.80 & 85.04 \end{bmatrix}$
Treatments (γ)	4	$\mathbf{Q}_\gamma = \begin{bmatrix} 420.80 & \\ 48.80 & 177.04 \end{bmatrix}$
Within error	12	$\mathbf{Q}_e = \begin{bmatrix} 146.80 & \\ 118.00 & 422.72 \end{bmatrix}$
"Total"	24	$\mathbf{Q}_t = \begin{bmatrix} 590.00 & \\ 183.40 & 1355.84 \end{bmatrix}$

$\theta_s = .1678$, and $\theta_s = .7815$, respectively. Thus the only significant difference in the experiment is among the five treatments. Since the other two tests proved not to be significant, the value of these as blocking variables is questionable.

By solving the determinantal equation

$$|\mathbf{Q}_\gamma - \lambda \mathbf{Q}_e| = 0 \tag{5.11.4}$$

the eigenvalues and corresponding standardized vectors are

$$\begin{aligned} \lambda_1 &= 3.5776 & \lambda_2 &= .4188 \\ \mathbf{a}_1 &= \begin{bmatrix} .3246 \\ -.0908 \end{bmatrix} & \mathbf{a}_2 &= \begin{bmatrix} .0003 \\ -.1684 \end{bmatrix} \end{aligned} \tag{5.11.5}$$

By utilizing Roy's procedure or Bartlett's chi-square test, it is seen that only the first root is significant—$X_B^2 = 23.288 \mathrel{\dot{\sim}} \chi^2(8)$. By calculating $\mathbf{R}_e$, where

$$\mathbf{R}_e = \begin{bmatrix} 1.0000 & \\ .4737 & 1.0000 \end{bmatrix} \tag{5.11.6}$$

the correlation between the discriminant functions and each variable is

$$\hat{\boldsymbol{\rho}} = \mathbf{R}_e \begin{bmatrix} 1.1355 \\ .5388 \end{bmatrix} = \begin{bmatrix} .880 \\ -.001 \end{bmatrix} \tag{5.11.7}$$

Thus only the first dependent variable seems to be contributing equally to the differences among treatments.

The Roy-Bose theorem states that, with probability $1 - \alpha$,

$$\hat{\psi} - c_0 \hat{\sigma}_{\hat{\psi}} \leq \psi \leq \hat{\psi} + c_0 \hat{\sigma}_{\hat{\psi}} \tag{5.11.8}$$

where

$$\hat{\psi} = \mathbf{c}'\hat{\mathbf{B}}\mathbf{a}$$

$$\hat{\sigma}_{\hat{\psi}} = \sqrt{\left[\mathbf{a}'\left(\frac{\mathbf{Q}_e}{v_e}\right)\mathbf{a}\right][\mathbf{c}'(\mathbf{X}'\mathbf{X})^- \mathbf{c}]}$$

$$c_0^2 = \frac{\theta^\alpha}{1 - \theta^\alpha} v_e$$

For contrasts comparing each treatment condition with treatment 5, for each dependent variable, the parametric functions

$$\psi_{ij} = \gamma_{ij} - \gamma_{5j} \qquad j = 1, 2; i = 1, 2, 3, 4 \tag{5.11.9}$$

are investigated. The values of $\hat{\psi}_{ij}$, using the combined means, are

$$\begin{aligned} \hat{\psi}_{11} &= -6.80 & \hat{\psi}_{12} &= 2.60 \\ \hat{\psi}_{21} &= -10.00 & \hat{\psi}_{22} &= -5.20 \\ \hat{\psi}_{31} &= -11.60 & \hat{\psi}_{32} &= 1.40 \\ \hat{\psi}_{41} &= -9.60 & \hat{\psi}_{42} &= -.40 \end{aligned} \tag{5.11.10}$$

For the contrasts in (5.11.10),

$$\hat{\sigma}_{\hat{\psi}_{i1}} = 2.21209 \quad \text{and} \quad \hat{\sigma}_{\hat{\psi}_{i2}} = 3.75375 \qquad i = 1, 2, 3, 4 \tag{5.11.11}$$

and $c_0 = 4.16$. Multiplying the values in (5.11.11) by c_0,

$$\begin{aligned} c_0\hat{\sigma}_{\hat{\psi}_{i1}} &= 9.202 \\ c_0\hat{\sigma}_{\hat{\psi}_{i2}} &= 15.616 \end{aligned} \qquad i = 1, 2, 3, 4$$

for each variable; the only significant differences exist between groups 2 and 5, 3 and 5, and 4 and 5, for variable 1, where

$$\begin{aligned} -19.202 &\le \gamma_{21} - \gamma_{51} \le -.798 \\ -11.802 &\le \gamma_{31} - \gamma_{51} \le -2.398 \\ -18.802 &\le \gamma_{41} - \gamma_{51} \le -.398 \end{aligned}$$

Other contrasts may be found by following the procedure outlined in the preceding sections. The above contrasts indicate that only the learning variable contributed to the significant overall hypothesis.

EXERCISE 5.11

1. Using as blocking variables ability tracks and teaching machines, an investigator interested in the evaluation of four teaching units in science employed the following Latin-square design.

		Teaching Machines 1	2	3	4
Ability Groups	1	T_2	T_1	T_3	T_4
	2	T_4	T_3	T_1	T_2
	3	T_1	T_4	T_2	T_3
	4	T_3	T_2	T_4	T_1

The treatments T_1, T_2, T_3, and T_4 are four versions of measuring astronomical distances in the solar system and beyond the solar system. The dependent variables for the study are subtest scores on one test designed to measure the students' ability in determining solar system distances within (W) and beyond (B) the solar system. The data for the study follow.

Cell	W	B	Cell	W	B
112	33	15	311	10	5
121	40	4	324	20	16
133	31	16	332	17	16
144	37	10	343	12	4
214	25	20	413	24	15
223	30	18	422	20	13
231	22	6	434	19	14
242	25	18	441	29	20

a. Use the MANOVA model for the Latin-square design to test the main-effect hypothesis and to form discriminant functions for each effect. What are your conclusions?

b. Can you suggest an alternative analysis for the data that uses univariate methods? What information would be gained or lost?

5.12 PROFILE ANALYSIS

The repeated-measures design, restricted to a profile analysis of p commensurable responses and two independent groups, was introduced in Section 3.15. In this section, the technique for analyzing this type of design for many groups is presented by a simple extension of the tests of the two-sample case. Utilizing the previous notation, let

(5.12.1)
$$\mathbf{y}'_{ij} = [y_{ij1}, y_{ij2}, \ldots, y_{ijp}] \sim IN(\boldsymbol{\mu}_i, \boldsymbol{\Sigma})$$

denote the vector observation of the jth subject within the ith group for $i = 1, 2, \ldots, I$; $j = 1, 2, \ldots, N_i$; and $N = \Sigma_i N_i$.

The general linear model representation for the observation vectors is conveniently taken to be

(5.12.2)
$$\underbrace{\begin{bmatrix} \mathbf{y}'_{11} \\ \vdots \\ \mathbf{y}'_{1N_1} \\ \hline \mathbf{y}'_{21} \\ \vdots \\ \mathbf{y}'_{2N_2} \\ \hline \vdots \\ \hline \mathbf{y}'_{I1} \\ \vdots \\ \mathbf{y}'_{IN_I} \end{bmatrix}}_{(N \times p)} = \underbrace{\begin{bmatrix} 1 & 0 & \cdots & 0 \\ 1 & 0 & \cdots & 0 \\ \vdots & \vdots & \vdots & \vdots \\ 1 & 0 & \cdots & 0 \\ \hline 0 & 1 & \cdots & 0 \\ 0 & 1 & \cdots & 0 \\ \vdots & \vdots & \vdots & \vdots \\ 0 & 1 & \cdots & 0 \\ \hline \vdots & \vdots & \vdots & \vdots \\ \hline 0 & \cdot & \cdots & 1 \\ \vdots & \vdots & \vdots & \vdots \\ 0 & \cdot & \cdots & 1 \end{bmatrix}}_{(N \times I)} \underbrace{\begin{bmatrix} \mu_{11} & \mu_{12} & \cdots & \mu_{1p} \\ \mu_{21} & \mu_{22} & \cdots & \mu_{2p} \\ \vdots & \vdots & \vdots & \vdots \\ \mu_{I1} & \mu_{I2} & \cdots & \mu_{Ip} \end{bmatrix}}_{(I \times p)} + \underbrace{\mathbf{E}_0}_{(N \times p)}$$

where

$$E(\mathbf{Y}) = \mathbf{XB}$$
$$V(\mathbf{Y}) = \mathbf{I} \otimes \mathbf{\Sigma} \tag{5.12.3}$$

and the rank of the matrix **X** is I. The data for the I-group profile analysis is arranged as in Table 5.12.1.

TABLE 5.12.1. Data Arrangement for I-Group Profile Analysis.

	Conditions			
	1	2	$\cdots$	p
Group 1	y_{111}	y_{112}	$\cdots$	y_{11p}
	y_{121}	y_{122}	$\cdots$	y_{12p}
	$\vdots$	$\vdots$	$\vdots$	$\vdots$
	y_{1N_11}	y_{1N_12}	$\cdots$	y_{1N_1p}
Means	$[y_{1.1}$	$y_{1.2}$	$\cdots$	$y_{1.p}] = \mathbf{y}_{1.}$
Group i	y_{i11}	y_{i12}	$\cdots$	y_{i1p}
	y_{i21}	y_{i22}	$\cdots$	y_{i2p}
	$\vdots$	$\vdots$	$\vdots$	$\vdots$
	y_{iN_i1}	y_{iN_i2}	$\cdots$	y_{iN_ip}
Means	$[y_{i.1}$	$y_{i.2}$	$\cdots$	$y_{i.p}] = \mathbf{y}'_{i.}$
Group I	y_{I11}	y_{I12}	$\cdots$	y_{I1p}
	y_{I21}	y_{I22}	$\cdots$	y_{I2p}
	$\vdots$	$\vdots$	$\vdots$	$\vdots$
	y_{IN_I1}	y_{IN_I2}	$\cdots$	y_{IN_Ip}
Means	$[y_{I.1}$	$y_{I.2}$	$\cdots$	$y_{I.p}] = \mathbf{y}'_{I.}$
Grand mean	$[y_{..1}$	$y_{..2}$	$\cdots$	$y_{..p}] = \mathbf{y}'_{..}$

As in the two-sample case, the primary hypotheses of interest with I groups are

H_{01}: are the profiles for the I groups parallel?

H_{02}: are there differences among conditions?

H_{03}: are there significant differences between groups?

To test for differences in groups, the hypothesis is stated as

$$H_{03}: \begin{bmatrix} \mu_{11} \\ \mu_{12} \\ \vdots \\ \mu_{1p} \end{bmatrix} = \begin{bmatrix} \mu_{21} \\ \mu_{22} \\ \vdots \\ \mu_{2p} \end{bmatrix} = \cdots = \begin{bmatrix} \mu_{I1} \\ \mu_{I2} \\ \vdots \\ \mu_{Ip} \end{bmatrix} \tag{5.12.4}$$

To test (5.12.4) the theory developed in Section 5.3 may be applied for the one-way multivariate analysis-of-variance model. That is, the matrices **C**, **B**, **A**, and **Γ** in the expression **CBA** = **Γ** have the following form:

(5.12.5)
$$\underset{[(I-1)\times I]}{\mathbf{C}} = \begin{bmatrix} 1 & 0 & \cdots & 0 & -1 \\ 0 & 1 & \cdots & 0 & -1 \\ \vdots & \vdots & \vdots & \vdots & \vdots \\ 0 & 0 & \cdots & 1 & -1 \end{bmatrix}$$
$$\underset{(I\times p)}{\mathbf{B}} = \begin{bmatrix} \mu_{11} & \mu_{12} & \cdots & \mu_{1p} \\ \mu_{21} & \mu_{22} & \cdots & \mu_{2p} \\ \vdots & \vdots & \vdots & \vdots \\ \mu_{I1} & \mu_{I2} & \cdots & \mu_{Ip} \end{bmatrix}$$
$$\mathbf{A} = \mathbf{I}_p$$
$$\mathbf{\Gamma} = \mathbf{0}$$

so that $\mathbf{Q}_h$ becomes

(5.12.6)
$$\begin{aligned}\mathbf{Q}_h &= (\mathbf{C\hat{B}A})'[\mathbf{C}(\mathbf{X}'\mathbf{X})^{-1}\mathbf{C}']^{-1}(\mathbf{C\hat{B}A}) \\ &= \sum_{i=1}^{I} N_i(\mathbf{y}_{i.} - \mathbf{y}_{..})(\mathbf{y}_{i.} - \mathbf{y}_{..})'\end{aligned}$$

where

(5.12.7)
$$\mathbf{\hat{B}} = (\mathbf{X}'\mathbf{X})^{-1}\mathbf{X}'\mathbf{Y} = \begin{bmatrix} \mathbf{y}'_{1.} \\ \mathbf{y}'_{2.} \\ \vdots \\ \mathbf{y}'_{I.} \end{bmatrix}$$

The error matrix $\mathbf{Q}_e$ is

(5.12.8)
$$\mathbf{Q}_e = \mathbf{Y}'[\mathbf{I} - \mathbf{X}(\mathbf{X}'\mathbf{X})^{-1}\mathbf{X}']\mathbf{Y}$$

with degrees of freedom $v_e = N - I$. The MANOVA table for testing H_{03} is summarized in Table 5.12.2. After solving the characteristic equation

(5.12.9)
$$|\mathbf{Q}_h - \lambda\mathbf{Q}_e| = 0$$

TABLE 5.12.2. MANOVA Table for Differences among Groups

Source	df	SSP
Groups	$I - 1$	$\mathbf{Q}_h = \sum_{i=1}^{I} N_i(\mathbf{y}_{i.} - \mathbf{y}_{..})(\mathbf{y}_{i.} - \mathbf{y}_{..})'$
Within error	$N - I$	$\mathbf{Q}_e = \sum_{i=1}^{I}\sum_{j=1}^{N_i} (\mathbf{y}_{ij} - \mathbf{y}_{i.})(\mathbf{y}_{ij} - \mathbf{y}_{i.})'$
"Total"	$N - 1$	$\mathbf{Q}_t = \sum_{i=1}^{I}\sum_{j=1}^{N_i} (\mathbf{y}_{ij} - \mathbf{y}_{..})(\mathbf{y}_{ij} - \mathbf{y}_{..})'$

with $\mathbf{Q}_h$ defined by (5.12.6) and $\mathbf{Q}_e$ defined by (5.12.8), the results of Section 5.3 are employed to test for group differences.

The parameters for the multivariate test criteria are

$$s = \min(\nu_h, p) = \min(I - 1, p)$$

(5.12.10) $$m = \frac{|\nu_h - p| - 1}{2} = \frac{|I - p - 1| - 1}{2}$$

$$n = \frac{\nu_e - p - 1}{2} = \frac{N - I - p - 1}{2}$$

To test for differences in conditions

(5.12.11) $$H_{02}: \begin{bmatrix} \mu_{11} \\ \mu_{21} \\ \vdots \\ \mu_{I1} \end{bmatrix} = \begin{bmatrix} \mu_{12} \\ \mu_{22} \\ \vdots \\ \mu_{I2} \end{bmatrix} = \cdots = \begin{bmatrix} \mu_{1p} \\ \mu_{2p} \\ \vdots \\ \mu_{Ip} \end{bmatrix}$$

the matrices $\mathbf{C}$, $\mathbf{A}$, and $\mathbf{\Gamma}$ are of the form

(5.12.12) $$\underset{[p \times (p-1)]}{\mathbf{A}} = \begin{bmatrix} 1 & 1 & \cdots & 0 \\ 0 & 1 & \cdots & 0 \\ \vdots & \vdots & \vdots & \vdots \\ 0 & 0 & \cdots & 1 \\ -1 & -1 & \cdots & -1 \end{bmatrix}$$

$$\mathbf{C} = \mathbf{I}_I$$

$$\underset{[I \times (p-1)]}{\mathbf{\Gamma}} = \mathbf{0}$$

where the $R(\mathbf{C}) = \nu_h = I$ and $R(\mathbf{A}) = u = p - 1$. Thus

$$\mathbf{Q}_h = (\mathbf{C}\hat{\mathbf{B}}\mathbf{A})'[\mathbf{C}(\mathbf{X}'\mathbf{X})^{-1}\mathbf{C}']^{-1}(\mathbf{C}\hat{\mathbf{B}}\mathbf{A})$$

$$= \mathbf{A}'(\hat{\mathbf{B}}'\mathbf{X}'\mathbf{X}\hat{\mathbf{B}})\mathbf{A}$$

(5.12.13) $$= \sum_{i=1}^{I} N_i(\mathbf{y}_{i.}^{*}\mathbf{y}_{i.}^{*\prime})$$

where $\mathbf{y}_{i.}^{*\prime} = [\mathbf{y}_{i.1} - y_{i.p}, \ldots, y_{i.(p-1)} - y_{i.p}]$ for $\mathbf{A}$ as defined in (5.12.12).

With $\mathbf{Q}_e$ defined by

(5.12.14) $$\mathbf{Q}_e = \mathbf{A}'\mathbf{Y}[\mathbf{I} - \mathbf{X}(\mathbf{X}'\mathbf{X})^{-1}\mathbf{X}']\mathbf{Y}\mathbf{A}$$

and degrees of freedom $\nu_e = N - 1$, the test for difference in conditions is obtained by evaluating (5.12.9) with expressions (5.12.13) and (5.12.14) defining $\mathbf{Q}_h$ and $\mathbf{Q}_e$, respectively. The parameters for the multivariate criteria in this case are

$$s = \min(v_h, u) = \min(I, p - 1)$$

$$m = \frac{|v_h - u| - 1}{2} = \frac{|I - p + 1| - 1}{2}$$

$$n = \frac{v_e - u - 1}{2} = \frac{N - I - p}{2}$$

To test for parallelism of profiles,

$$H_{01}: \begin{bmatrix} \mu_{11} - \mu_{12} \\ \mu_{12} - \mu_{13} \\ \vdots \\ \mu_{1(p-1)} - \mu_{1p} \end{bmatrix} = \cdots = \begin{bmatrix} \mu_{I1} - \mu_{I2} \\ \mu_{I2} - \mu_{I3} \\ \vdots \\ \mu_{I(p-1)} - \mu_{Ip} \end{bmatrix} \tag{5.12.15}$$

the matrices **C**, **A**, and **Γ** are

$$\underset{[(I-1)\times I]}{\mathbf{C}} = \begin{bmatrix} 1 & 0 & \cdots & 0 & -1 \\ 0 & 1 & \cdots & 0 & -1 \\ \vdots & \vdots & \vdots & \vdots & \vdots \\ 0 & 0 & \cdots & 1 & -1 \end{bmatrix}$$

$$\underset{[p\times(p-1)]}{\mathbf{A}} = \begin{bmatrix} 1 & 0 & \cdots & 0 \\ -1 & 1 & \cdots & 0 \\ 0 & -1 & \cdots & 0 \\ \vdots & \vdots & \vdots & \vdots \\ 0 & 0 & \cdots & 1 \\ 0 & 0 & \cdots & -1 \end{bmatrix} \tag{5.12.16}$$

$$\underset{[(I-1)\times(p-1)]}{\mathbf{\Gamma}} = \mathbf{0}$$

where the $R(\mathbf{C}) = v_h = I - 1$ and the $R(\mathbf{A}) = u = p - 1$.

By using **A** as defined in (5.12.16), $\mathbf{Q}_h$ and $\mathbf{Q}_e$ become

$$\mathbf{Q}_h = (\mathbf{C\hat{B}A})'[\mathbf{C}(\mathbf{X}'\mathbf{X})^{-1}\mathbf{C}']^{-1}(\mathbf{C\hat{B}A}) \tag{5.12.17}$$

$$\mathbf{Q}_e = \mathbf{A}'\mathbf{Y}'[\mathbf{I} - \mathbf{X}(\mathbf{X}'\mathbf{X})^{-1}\mathbf{X}']\mathbf{YA} \tag{5.12.18}$$

The parameters for the test are

$$s = \min(v_h, u) = \min(I - 1, p - 1)$$

$$m = \frac{|v_h - u| - 1}{2} = \frac{|I - p| - 1}{2}$$

$$n = \frac{v_e - u - 1}{2} = \frac{N - I - p}{2}$$

TABLE 5.12.3. Summary of Primary Profile Hypotheses

	H_{01} (parallelism)	H_{02} (conditions)	H_{03} (groups)
$\mathbf{C}$	$\underbrace{\begin{bmatrix}1&0&\cdots&0&-1\\0&1&\cdots&0&-1\\\vdots&\vdots&&\vdots&\vdots\\0&0&\cdots&1&-1\end{bmatrix}}_{[(I-1)\times I]}$	$\underset{(I\times I)}{\mathbf{I}_I}$	$\underbrace{\begin{bmatrix}1&0&\cdots&0&-1\\0&1&\cdots&0&-1\\\vdots&\vdots&&\vdots&\vdots\\0&0&\cdots&1&-1\end{bmatrix}}_{[(I-1)\times I]}$
$\mathbf{A}$	$\underbrace{\begin{bmatrix}1&0&\cdots&0\\-1&1&\cdots&0\\0&-1&\cdots&0\\\vdots&\vdots&&\vdots\\0&0&\cdots&1\\0&0&\cdots&-1\end{bmatrix}}_{[p\times(p-1)]}$	$\underbrace{\begin{bmatrix}1&0&\cdots&0\\0&1&\cdots&0\\\vdots&\vdots&&\vdots\\-1&-1&\cdots&-1\end{bmatrix}}_{[p\times(p-1)]}$	$\underset{(p\times p)}{\mathbf{I}_p}$
$\mathbf{\Gamma}$	$\underset{[(I-1)\times(p-1)]}{\mathbf{0}}$	$\underset{[I\times(p-1)]}{\mathbf{0}}$	$\underset{[(I-1)\times p]}{\mathbf{0}}$
$\mathbf{Q}_h$	$(\mathbf{C\hat{B}A})'[\mathbf{C(X'X)^{-1}C'}]^{-1}(\mathbf{C\hat{B}A})$	$(\mathbf{C\hat{B}A})'[\mathbf{C(X'X)^{-1}C'}]^{-1}(\mathbf{C\hat{B}A})$	$(\mathbf{C\hat{B}A})'[\mathbf{C(X'X)^{-1}C'}]^{-1}(\mathbf{C\hat{B}A})$
$\mathbf{Q}_e$	$\mathbf{A'Y'[I-X(X'X)^{-1}X']YA}$	$\mathbf{A'Y'[I-X(X'X)^{-1}X']YA}$	$\mathbf{A'Y'[I-X(X'X)^{-1}X']YA}$
ν_e	$N-I$	$N-I$	$N-I$
ν_h	$I-1$	I	$I-1$
u	$p-1$	$p-1$	p
s	$\min(I-1, p-1)$	$\min(I, p-1)$	$\min(I-1, p)$
m	$\dfrac{\lvert I-p\rvert-1}{2}$	$\dfrac{\lvert I-p+1\rvert-1}{2}$	$\dfrac{\lvert I-p-1\rvert-1}{2}$
n	$\dfrac{N-I-p}{2}$	$\dfrac{N-I-p}{2}$	$\dfrac{N-I-p-1}{2}$

To summarize, the multivariate tests for group differences, differences in conditions, and parallelism may be tested by using the matrices and parameters given in Table 5.12.3. Each test H_{0i} $(i = 1, 2, 3)$ is rejected at the significance level α if

(5.12.19) (a) Wilks:

$$\Lambda = \frac{|\mathbf{Q}_e|}{|\mathbf{Q}_e + \mathbf{Q}_h|} < U^{\alpha}(u, \nu_h, \nu_e)$$

(b) Roy:

$$\theta_s = \frac{\lambda_1}{\lambda_1 + 1} > \theta^{\alpha}(s, m, n)$$

(c) Lawley-Hotelling:

$$U^{(s)} = \sum_{i=1}^{s} \lambda_i > U_0^{\alpha}(s, m, n)$$

(d) Pillai:

$$V^{(s)} = \sum_{i=1}^{s} \frac{\lambda_i}{1 + \lambda_i} > V^{\alpha}(s, m, n)$$

where the λ_i are the roots of the equation $|\mathbf{Q}_h - \lambda \mathbf{Q}_e| = 0$.

On rejection of any hypothesis, simultaneous confidence intervals for the parametric functions $\psi = \mathbf{c}'\mathbf{B}\mathbf{a}$ may be directly acquired by using the formula

$$\mathbf{c}'\hat{\mathbf{B}}\mathbf{a} - c_0 \sqrt{\left[\mathbf{a}'\left(\frac{\mathbf{Q}_e}{\nu_e}\right)\mathbf{a}\right][\mathbf{c}'(\mathbf{X}'\mathbf{X})^{-1}\mathbf{c}]} \leq \mathbf{c}'\mathbf{B}\mathbf{a}$$

(5.12.20)
$$\leq \mathbf{c}'\hat{\mathbf{B}}\mathbf{a} + c_0 \sqrt{\left[\mathbf{a}'\left(\frac{\mathbf{Q}_e}{\nu_e}\right)\mathbf{a}\right][\mathbf{c}'(\mathbf{X}'\mathbf{X})^{-1}\mathbf{c}]}$$

where c_0 is chosen to agree with the overall test criterion.

The multivariate tests for differences in groups, (5.12.4), and differences in conditions, (5.12.11), do not assume that the parallelism hypothesis is tenable for valid tests of these hypotheses. However, if there is no interaction between groups and conditions, so that μ_{ij} has the following additive form

(5.12.21)
$$\mu_{ij} = \mu + \alpha_i + \beta_j \qquad \begin{cases} i = 1, \ldots, I \\ j = 1, \ldots, p \end{cases}$$

where α_i denotes the effect due to the ith group and β_j is the effect due to the jth condition, then the test of differences between groups, H_{03}, is equivalent to testing the hypothesis

(5.12.22)
$$H_{03}^{*}: \alpha_1 = \alpha_2 = \cdots = \alpha_I$$

Furthermore, the test for differences in conditions, H_{02}, is equivalent to

(5.12.23)
$$H_{02}^{*}: \beta_1 = \beta_2 = \cdots = \beta_p$$

The tests for differences in groups and conditions reduce, respectively, to

(5.12.24)
$$H_{03}^{*}: \sum_{j=1}^{p} \mu_{1j} = \cdots = \sum_{j=1}^{p} \mu_{Ij}$$

(5.12.25)
$$H_{02}^{*}: \frac{\sum_{i=1}^{I} \mu_{i1}}{I} = \cdots = \frac{\sum_{i=1}^{I} \mu_{ip}}{I}$$

The hypothesis H_{03}^{*} of equal group means can be expressed as $\mathbf{CBA} = \mathbf{0}$ if

(5.12.26)
$$\underset{[(I-1)\times p]}{\mathbf{C}} = \begin{bmatrix} 1 & 0 & \cdots & 0 & -1 \\ 0 & 1 & \cdots & 0 & -1 \\ \vdots & \vdots & \vdots & \vdots & \vdots \\ 0 & 0 & \cdots & 1 & -1 \end{bmatrix} \quad \text{and} \quad \underset{(p\times 1)}{\mathbf{A}} = \begin{bmatrix} 1 \\ 1 \\ \vdots \\ 1 \end{bmatrix}$$

Applying the transformation $\mathbf{A}$ to the data matrix $\mathbf{Y}$ yields a univariate analysis-of-variance analysis on the subject totals. A univariate analysis of variance on means is obtained by selecting

$$\underset{(p \times 1)}{\mathbf{A}} = \begin{bmatrix} 1/p \\ 1/p \\ \vdots \\ 1/p \end{bmatrix} \tag{5.12.27}$$

The equivalent hypothesis, using (5.12.27) is

$$H_{03}^{*}: \text{all means } \frac{\sum_{j=1}^{p} \mu_{ij}}{p} \text{ are equal}$$

By using the formulas developed in Section 5.2, the analysis of H_{03}^{*} is immediately obtained.

To test the hypothesis H_{02}^{*} of equal response conditions, the results of Section 3.15 are extended. That is, by setting

$$\underset{(1 \times I)}{\mathbf{C}} = [1/I, 1/I, \ldots, 1/I] \quad \text{and} \quad \underset{[p \times (p-1)]}{\mathbf{A}} = \begin{bmatrix} 1 & 0 & \cdots & 0 \\ 0 & 1 & \cdots & 0 \\ \vdots & \vdots & \ddots & \vdots \\ 0 & 0 & \cdots & 1 \\ -1 & -1 & \cdots & -1 \end{bmatrix} \tag{5.12.28}$$

the statistic for testing H_{02}^{*}, by (3.15.17), becomes

$$T^2 = I^2\left(\sum_{i=1}^{I} \frac{1}{N_i}\right)^{-1} \mathbf{y}_{..}'\mathbf{A}(\mathbf{A}'\mathbf{S}\mathbf{A})^{-1}\mathbf{A}'\mathbf{y}_{..} \sim T^2(p-1, N-I) \tag{5.12.29}$$

where

$$\mathbf{y}_{..} = \frac{\sum_{i=1}^{I} \mathbf{y}_{i.}}{I} \quad \text{and} \quad \mathbf{S} = \frac{\mathbf{Y}'[\mathbf{I} - \mathbf{X}(\mathbf{X}'\mathbf{X})^{-1}\mathbf{X}']\mathbf{Y}}{N - I}$$

Alternatively, by using an overall average

$$\underset{(1 \times I)}{\mathbf{C}} = \left[\frac{N_1}{N}, \frac{N_2}{N}, \ldots, \frac{N_I}{N}\right] \tag{5.12.30}$$

and, by (3.15.20),

$$T^2 = N\bar{\mathbf{y}}_{..}'\mathbf{A}(\mathbf{A}'\mathbf{S}\mathbf{A})^{-1}\mathbf{A}'\bar{\mathbf{y}}_{..} \sim T^2(p-1, N-I)$$

where

$$\bar{\mathbf{y}}_{..} = \frac{\sum_i N_i \mathbf{y}_{i.}}{N}$$

Confidence intervals are obtained from expression (3.14.30).

Given homogeneity of the variance-covariance matrices across groups where the number of subjects N_i within each group is larger than p, the analysis-of-profile data using multivariate techniques allow the variance-covariance matrix $\mathbf{\Sigma}$ to have any structure. The traditional analysis-of-profile data have been through the use of the two-way, mixed-model, analysis-of-variance procedure, (see, for example, Greenhouse and Geisser, 1959, and Danford, Hughes, and McNee, 1960). The univariate design for the mixed-model analysis is often called a split-plot design because it was first used in agricultural experiments. Kirk (1968, Chap. 8) uses the notation SPF-$r.c$ for this design. The model equation for the univariate procedure may be written as

(5.12.31)

$$y_{ijk} = \mu + \alpha_i + \beta_k + \gamma_{ik} + s_{(i)j} + \varepsilon_{(i)jk} \qquad i = 1, \ldots, I; j = 1, \ldots, N_i; k = 1, \ldots, p$$

where the $s_{(i)j} \sim IN(0, \rho\sigma^2)$, $\varepsilon_{(i)jk} \sim IN[0, (1 - \rho)\sigma^2]$, and $s_{(i)j}$ and $\varepsilon_{(i)jk}$ are jointly independent. Thus the variance-covariance matrix $\mathbf{\Sigma}$ is of the form

$$\mathbf{\Sigma} = \rho\sigma^2\mathbf{J} + (1 - \rho)\sigma^2\mathbf{I} = \sigma_s^2\mathbf{J} + \sigma_e^2\mathbf{I} \tag{5.12.32}$$

which satisfies the assumption of compound symmetry (that is, equal variances and equal covariances). The compound-symmetry assumption ensures that the variance is constant between condition means. There are, however, other patterns of $\mathbf{\Sigma}$ besides (5.12.32) that satisfy this requirement (see, for example, Huynh and Feldt, 1970, and Winer, 1971, p. 282). The symbols in (5.12.31) are defined as follows.

(5.12.33)

μ = overall constant
α_i = ith-group effect
$s_{(i)j}$ = effect of jth subject measured at ith group
β_k = kth condition effect
γ_{ik} = group by condition interaction
$\varepsilon_{(i)jk}$ = subject by condition interaction plus usual random error component

The ANOVA table for the model given by (5.12.31) has been given by several authors, including Geisser and Greenhouse (1958), Kirk (1968, p. 250), Myers (1966, p. 223), and Winer (1971, p. 520); however, the reports by Hughes and Danford (1958) and Danford and Hughes (1957) include details often omitted by other authors. The analysis-of-variance procedure is employed to derive the results when using the general linear model. The procedure is to assume that all factors are fixed to derive the sum of squares and then to compute the expected mean squares to determine the appropriate ratios for testing. Alternatively, rules of thumb have been developed to write down the ANOVA tables for crossed and/or balanced nested classifications when the number of observations in the cells is the same. See, for example, Henderson (1969), Millman and Glass (1967), or Kirk (1968, p. 208). The ANOVA table for the model specified by (5.12.31) is summarized in Table 5.12.4. The expected mean squares of the sums of squares provided in Table 5.12.4 are given in Table 5.12.5.

By using the information in Tables 5.12.4 and 5.12.5, tests for group differences in conditions, and group $\times$ conditions, are found by evaluating the following F

TABLE 5.12.4. ANOVA Table for Simple Repeated-Measures Design.

Source	df	SS
Between groups	$I - 1$	$Q_1 = p \sum_i N_i(y_{i..} - y_{...})^2$
Subjects within groups (within error 1)	$N - 1$	$Q_{e_1} = p \sum_i \sum_j (y_{ij.} - y_{i..})^2$
Between conditions	$p - 1$	$Q_2 = N \sum_k (y_{..k} - y_{...})^2$
Groups × conditions	$(I - 1)(p - 1)$	$Q_3 = \sum_i \sum_k N_i(y_{i.k} - y_{..k} - y_{i..} + y_{...})^2$
Subjects × conditions within groups (within error 2)	$(N - I)(p - 1)$	$Q_{e_2} = \sum_i \sum_j \sum_k (y_{ijk} - y_{i.k} - y_{ij.} + y_{i..})^2$
"Total"	$Np - 1$	$Q_t = \sum_{i,j,k} (y_{ijk} - y_{...})^2$

TABLE 5.12.5. Expected Mean Squares for Simple Repeated-Measures Design.

Between groups	$\sigma_e^2 + p\sigma_s^2 + \sigma_\alpha^2$
Subjects within groups	$\sigma_e^2 + p\sigma_s^2$
Between conditions	$\sigma_e^2 + \sigma_\beta^2$
Groups × conditions	$\sigma_e^2 + \sigma_\gamma^2$
Subjects × conditions within groups	σ_e^2

statistics for the univariate model:

$$\text{Groups}: F_g = \frac{Q_1/(I - 1)}{Q_{e_1}/(N - I)} \sim F(I - 1, N - I)$$

$$\text{(5.12.34)} \quad \text{Conditions}: F_c = \frac{Q_2/(p - 1)}{Q_{e_2}/(N - I)(p - 1)} \sim F[(p - 1), (N - I)(p - 1)]$$

$$\text{Group} \times \text{conditions}: F_{gc} = \frac{Q_3/(I - 1)(p - 1)}{Q_{e_2}/(N - I)(p - 1)} \sim F[(I - 1)(p - 1), (N - I)(p - 1)]$$

If we find that the statistic F_{gc} is not significant, then F_g and F_c may be used to test for differences in groups and conditions, respectively.

When the variances are unequal and the covariances heterogeneous in Σ, F_c and F_{gc} will usually no longer have central F distributions under the null hypothesis. Greenhouse and Geisser (1959) propose conservative tests for this case. However, multivariate procedures should be employed since the univariate model is no longer appropriate. A major disadvantage of the multivariate model is that test statistics are not defined when the number of response conditions is larger than the number of subjects in each group.

In analyzing profile data, multivariate procedures should be employed whenever the compound-symmetry assumption is not satisfied; if this assumption is tenable, univariate methods should be used. The univariate results can be easily obtained from a multivariate analysis by orthogonalizing the matrix **A** in the expression **CBA** = **Γ**. The correspondence is most clearly witnessed through the example given in Section 5.13.

EXERCISES 5.12

1. When the matrix **A** in (5.12.16) is substituted into the expression for $\mathbf{Q}_h$ in (5.12.13), what hypothesis is being tested? How would you now write $\mathbf{Q}_h$?
2. Derive the results summarized in Tables 5.12.4 and 5.12.5.

5.13 EXAMPLE—PROFILE ANALYSIS

The data used in this section are selected from Edwards (1968, p. 276) to illustrate the multivariate methods presented in Section 5.12. Edwards analyzed the data by using univariate procedures. The investigation concerns the influence of three drugs, each at a standard dosage, on learning. Fifteen subjects are assigned at random to the three drug levels so that five subjects are tested with each drug on three different trials. The data for the study are shown in Table 5.13.1. The sample sizes for the three

TABLE 5.13.1. Edwards' Repeated-Measure Data.

		Trials		
	Subjects	1	2	3
Drug group 1	1	2	4	7
	2	2	6	10
	3	3	7	10
	4	7	9	11
	5	6	9	12
Means		4	7	10
Drug group 2	1	5	6	10
	2	4	5	10
	3	7	8	11
	4	8	9	11
	5	11	12	13
Means		7	8	11
Drug group 3	1	3	4	7
	2	3	6	9
	3	4	7	9
	4	8	8	10
	5	7	10	10
Means		5	7	9
Grand mean		5.333	7.333	10

groups are $N_1 = N_2 = N_3 = 5$. The total sample size is $N = 15$, the number of groups is $I = 3$, and the number of dependent variables is $p = 3$ for the study.

The linear model for the experiment is

$$\underset{(15\times 3)}{\mathbf{Y}} = \underset{(15\times 3)}{\mathbf{X}}\ \underset{(3\times 3)}{\mathbf{B}} + \underset{(15\times 3)}{\mathbf{E}_0} \tag{5.13.1}$$

where

(5.13.2)

$$\mathbf{Y} = \begin{bmatrix} 2 & 4 & 7 \\ 2 & 6 & 10 \\ 3 & 7 & 10 \\ 7 & 9 & 11 \\ 6 & 9 & 12 \\ 5 & 6 & 10 \\ 4 & 5 & 10 \\ 7 & 8 & 11 \\ 8 & 9 & 11 \\ 11 & 12 & 13 \\ 3 & 4 & 7 \\ 3 & 6 & 9 \\ 4 & 7 & 9 \\ 8 & 8 & 10 \\ 7 & 10 & 10 \end{bmatrix} \qquad \mathbf{X} = \begin{bmatrix} 1 & 0 & 0 \\ 1 & 0 & 0 \\ 1 & 0 & 0 \\ 1 & 0 & 0 \\ 1 & 0 & 0 \\ 0 & 1 & 0 \\ 0 & 1 & 0 \\ 0 & 1 & 0 \\ 0 & 1 & 0 \\ 0 & 1 & 0 \\ 0 & 0 & 1 \\ 0 & 0 & 1 \\ 0 & 0 & 1 \\ 0 & 0 & 1 \\ 0 & 0 & 1 \end{bmatrix} \qquad \mathbf{B} = \begin{bmatrix} \mu_{11} & \mu_{12} & \mu_{13} \\ \mu_{21} & \mu_{22} & \mu_{23} \\ \mu_{31} & \mu_{32} & \mu_{33} \end{bmatrix} = \begin{bmatrix} \boldsymbol{\mu}'_1 \\ \boldsymbol{\mu}'_2 \\ \boldsymbol{\mu}'_3 \end{bmatrix}$$

and

$$\begin{aligned} E(\mathbf{Y}) &= \mathbf{XB} \\ V(\mathbf{Y}) &= \mathbf{I} \otimes \boldsymbol{\Sigma} \end{aligned} \tag{5.13.3}$$

Each row of $\mathbf{Y}$ is distributed $IN_3(\boldsymbol{\mu}_i, \boldsymbol{\Sigma})$ and $\boldsymbol{\Sigma}$ is allowed any pattern.

By employing the multivariate Gauss-Markoff theorem, the least-squares estimator for $\mathbf{B}$ is

$$\hat{\mathbf{B}} = (\mathbf{X}'\mathbf{X})^{-}\mathbf{X}'\mathbf{Y} = \begin{bmatrix} \mathbf{y}'_{1.} \\ \mathbf{y}'_{2.} \\ \mathbf{y}'_{3.} \end{bmatrix}$$

$$= \begin{bmatrix} 4 & 7 & 10 \\ 7 & 8 & 11 \\ 5 & 7 & 9 \end{bmatrix} \tag{5.13.4}$$

To test the hypothesis of significant differences among drug groups,

(5.13.5)
$$H_{03}: \begin{bmatrix} \mu_{11} \\ \mu_{12} \\ \mu_{13} \end{bmatrix} = \begin{bmatrix} \mu_{21} \\ \mu_{22} \\ \mu_{23} \end{bmatrix} = \begin{bmatrix} \mu_{31} \\ \mu_{32} \\ \mu_{33} \end{bmatrix}$$

the theory developed in Section 5.3 is applied for a one-way MANOVA design. Setting

$$\mathbf{C} = \begin{bmatrix} 1 & 0 & -1 \\ 0 & 1 & -1 \end{bmatrix} \qquad \mathbf{B} = \begin{bmatrix} \mu_{11} & \mu_{12} & \mu_{13} \\ \mu_{21} & \mu_{22} & \mu_{23} \\ \mu_{31} & \mu_{32} & \mu_{33} \end{bmatrix}$$

$$\mathbf{A} = \begin{bmatrix} 1 & 0 & 0 \\ 0 & 1 & 0 \\ 0 & 0 & 1 \end{bmatrix} \quad \text{and} \quad \mathbf{\Gamma} = \begin{bmatrix} 0 & 0 & 0 \\ 0 & 0 & 0 \end{bmatrix}$$

H_{03} is placed in the form

$$H_{03}: \mathbf{CBA} = \mathbf{\Gamma}$$

The hypothesis sum of products matrix is

(5.13.6)
$$\begin{aligned} \mathbf{Q}_h &= (\mathbf{C\hat{B}A})'[\mathbf{C}(\mathbf{X'X})^{-1}\mathbf{C'}]^{-1}(\mathbf{C\hat{B}A}) \\ &= \begin{bmatrix} 23.3333 & & \\ 8.3333 & 3.3333 & \\ 10.0000 & 5.0000 & 10.0000 \end{bmatrix} \end{aligned}$$

while the error matrix $\mathbf{Q}_e$ is

(5.13.7)
$$\begin{aligned} \mathbf{Q}_e &= \mathbf{Y'}[\mathbf{I} - \mathbf{X}(\mathbf{X'X})^{-1}\mathbf{X'}]\mathbf{Y} \\ &= \begin{bmatrix} 74 & & \\ 65 & 68 & \\ 35 & 38 & 26 \end{bmatrix} \end{aligned}$$

By employing Wilks' Λ criterion to test H_{03}:

$$\Lambda = \frac{|\mathbf{Q}_e|}{|\mathbf{Q}_e + \mathbf{Q}_h|} = .2368$$

Setting $\alpha = .05$, H_{03} is rejected if $\Lambda < U^{\alpha}(p, I-1, N-I) = U^{.05}(3, 2, 12) = .3157$. Thus, by employing the multivariate test for differences among groups, equality of group means is not tenable. To locate significant differences, the procedures outlined in Section 5.3 are employed.

Testing for differences in conditions,

(5.13.8)
$$H_{02}: \begin{bmatrix} \mu_{11} \\ \mu_{21} \\ \mu_{31} \end{bmatrix} = \begin{bmatrix} \mu_{12} \\ \mu_{22} \\ \mu_{32} \end{bmatrix} = \begin{bmatrix} \mu_{13} \\ \mu_{23} \\ \mu_{33} \end{bmatrix}$$

the matrices **C**, **A**, and **Γ** are of the form

$$\mathbf{C} = \begin{bmatrix} 1 & 0 & 0 \\ 0 & 1 & 0 \\ 0 & 0 & 1 \end{bmatrix} \qquad \mathbf{A} = \begin{bmatrix} 1 & 0 \\ 0 & 1 \\ -1 & -1 \end{bmatrix} \qquad \mathbf{\Gamma} = \begin{bmatrix} 0 & 0 \\ 0 & 0 \\ 0 & 0 \end{bmatrix} \tag{5.13.9}$$

where the $R(\mathbf{C}) = 3$ and $R(\mathbf{A}) = 2$. The hypothesis sum of squares matrix is

$$\begin{aligned} \mathbf{Q}_h &= (\mathbf{C\hat{B}A})'[\mathbf{C}(\mathbf{X'X})^{-1}\mathbf{C'}]^{-1}(\mathbf{C\hat{B}A}) \\ &= \mathbf{A'\hat{B}'X'X\hat{B}A} \\ &= \sum_{i=1}^{I} N_i(\mathbf{y}_{i.}^{*}\mathbf{y}_{i.}^{*\prime}) \end{aligned} \tag{5.13.10}$$

where

$$\mathbf{y}_{1.}^{*\prime} = [-6, -3] \qquad \mathbf{y}_{2.}^{*\prime} = [-4, -3] \qquad \mathbf{y}_{3.}^{*\prime} = [-4, -2]$$

Thus

$$\mathbf{Q}_h = \begin{bmatrix} 340.00 & \\ 190.00 & 110.00 \end{bmatrix}$$

while the matrix $\mathbf{Q}_e$ is

$$\begin{aligned} \mathbf{Q}_e &= \mathbf{A'Y'}[\mathbf{I} - \mathbf{X}(\mathbf{X'X})^{-1}\mathbf{X'}]\mathbf{YA} \\ &= \begin{bmatrix} 1 & 0 & -1 \\ 0 & 1 & -1 \end{bmatrix} \begin{bmatrix} 74 & 65 & 35 \\ 65 & 68 & 38 \\ 35 & 38 & 26 \end{bmatrix} \begin{bmatrix} 1 & 0 \\ 0 & 1 \\ -1 & -1 \end{bmatrix} \\ &= \begin{bmatrix} 30 & \\ 18 & 18 \end{bmatrix} \end{aligned} \tag{5.13.11}$$

Again employing Wilks' Λ criterion,

$$\Lambda = \frac{|\mathbf{Q}_e|}{|\mathbf{Q}_e + \mathbf{Q}_h|} = .0527 \tag{5.13.12}$$

The decision rule for $\alpha = .05$ is to reject H_{02} if $\Lambda < U^{\alpha}(p - 1, I, N - I) = U^{.05}_{(2,3,12)} = .3480$. Thus H_{02} is rejected. Alternatively, Roy's criterion is $\theta_s = \lambda_1/(1 + \lambda_1) = .9195$, and H_{02} is rejected at the significance level $\alpha = .05$ if $\theta_s > \theta^{\alpha}(2, 0, 4.5) = .450$. The two criteria are consistent.

By employing formula (5.12.20), simultaneous confidence intervals for the parametric functions $\psi = \mathbf{c'Ba}$ are immediately obtained. Letting $\mathbf{a}' = [1, 0, -1]$ and $\mathbf{c}' = [1, 0, 0]$, $\mathbf{c}' = [0, 1, 0]$, and $\mathbf{c}' = [0, 0, 1]$,

$$\begin{aligned} \psi_1 &= \mu_{11} - \mu_{12} & \hat{\psi}_1 &= 4 - 10 = -6 \\ \psi_2 &= \mu_{21} - \mu_{23} & \hat{\psi}_2 &= 7 - 11 = -4 \\ \psi_3 &= \mu_{31} - \mu_{33} & \hat{\psi}_3 &= 5 - 9 = -4 \end{aligned} \tag{5.13.13}$$

and

$$c_0\sqrt{\left[\mathbf{a}'\left(\frac{\mathbf{Q}_e}{\nu_e}\right)\mathbf{a}\right][\mathbf{c}'(\mathbf{X}'\mathbf{X})^{-1}\mathbf{c}]} = \sqrt{\nu_e\left(\frac{\theta^\alpha}{1-\theta^\alpha}\right)}\sqrt{\frac{30}{12}\frac{1}{5}}$$
$$= 3.1333(.7071)$$
$$= 2.215$$

Confidence intervals for the parametric functions given in (5.13.13) become

$$\hat{\psi}_1 - c_0\hat{\sigma}_{\hat{\psi}_1} \le \psi_1 \le \hat{\psi}_1 + c_0\hat{\sigma}_{\hat{\psi}_1}$$

(5.13.14)
$$-8.22 \le \psi_1 \le -3.79 \qquad \text{(Sig.)}$$
$$-6.22 \le \psi_2 \le -1.79 \qquad \text{(Sig.)}$$
$$-6.22 \le \psi_3 \le -1.79 \qquad \text{(Sig.)}$$

Other confidence intervals are similarly formed.

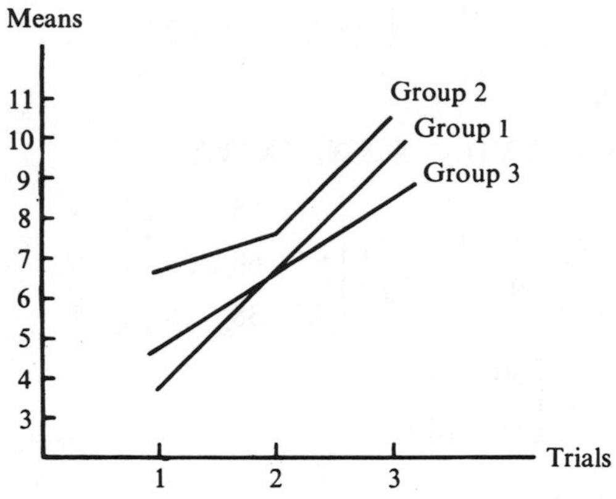

FIGURE 5.13.1. Plot of means for Edwards' data.

Before testing the hypothesis of parallelism, a plot of the means is made (Figure 5.13.1). Figure 5.13.1 would seem to indicate that the parallelism hypothesis is not tenable. To test the hypothesis of parallelism

(5.13.15)
$$H_{01}: \begin{bmatrix}\mu_{11}-\mu_{12}\\ \mu_{12}-\mu_{13}\end{bmatrix} = \begin{bmatrix}\mu_{21}-\mu_{22}\\ \mu_{22}-\mu_{23}\end{bmatrix} = \begin{bmatrix}\mu_{31}-\mu_{32}\\ \mu_{32}-\mu_{33}\end{bmatrix}$$

matrices **C**, **A**, and **Γ** are chosen such that

(5.13.16)
$$\mathbf{C} = \begin{bmatrix}1 & 0 & -1\\ 0 & 1 & -1\end{bmatrix} \qquad \mathbf{A} = \begin{bmatrix}1 & 0\\ -1 & 1\\ 0 & -1\end{bmatrix} \qquad \mathbf{\Gamma} = \begin{bmatrix}0 & 0\\ 0 & 0\end{bmatrix}$$

By using **C** and **A** as defined by (5.13.16),

(5.13.17)
$$\mathbf{Q}_h = (\mathbf{C}\hat{\mathbf{B}}\mathbf{A})'[\mathbf{C}(\mathbf{X}'\mathbf{X})^{-1}\mathbf{C}']^{-1}(\mathbf{C}\hat{\mathbf{B}}\mathbf{A}) = \begin{bmatrix} 10.000 & \\ 0 & 3.33 \end{bmatrix}$$
$$\mathbf{Q}_e = \mathbf{A}'\mathbf{Y}'[\mathbf{I} - \mathbf{X}(\mathbf{X}'\mathbf{X})^{-1}\mathbf{X}']\mathbf{Y}\mathbf{A} = \begin{bmatrix} 12 & \\ 0 & 18 \end{bmatrix}$$

Now employing Wilks' Λ criterion, $\Lambda = .4602$. By comparing Λ with the critical value at the significance level $\alpha = .05$, $U^{\alpha}(2, 2, 12) = .4373$, the hypothesis of parallelism is tenable. The p value for the test is $\alpha = .064$. The other criteria give consistent results.

As an alternative, the values of $\mathbf{Q}_h$ and $\mathbf{Q}_e$ in (5.13.17) may be obtained from (5.13.6) and (5.13.7). That is,

(5.13.18)
$$\mathbf{Q}_h = \mathbf{A}'\left[\sum_i N_i(\mathbf{y}_{i.} - \mathbf{y}_{..})(\mathbf{y}_{i.} - \mathbf{y}_{..})'\right]\mathbf{A}$$
$$\mathbf{Q}_e = \mathbf{A}'\left[\sum_{i=1}^{I}\sum_{j=1}^{N_i}(\mathbf{y}_{ij} - \mathbf{y}_{i.})(\mathbf{y}_{i.} - \mathbf{y}_{i.})'\right]\mathbf{A}$$

The matrix **A** associated with the test of parallelism is not unique. For example, by setting

(5.13.19)
$$\mathbf{A} = \begin{bmatrix} 1 & 0 \\ 0 & 1 \\ -1 & -1 \end{bmatrix}$$

an equivalent test of parallelism becomes

(5.13.20)
$$H'_{01}: \begin{bmatrix} \mu_{11} - \mu_{13} \\ \mu_{12} - \mu_{13} \end{bmatrix} = \begin{bmatrix} \mu_{21} - \mu_{23} \\ \mu_{22} - \mu_{23} \end{bmatrix} = \begin{bmatrix} \mu_{31} - \mu_{33} \\ \mu_{32} - \mu_{33} \end{bmatrix}$$

Alternatively, suppose

(5.13.21)
$$\mathbf{A} = \begin{bmatrix} 1 & 0 \\ -1/2 & 1 \\ -1/2 & -1 \end{bmatrix}$$

then an equivalent test of parallelism becomes

(5.13.22)
$$H''_{01}: \begin{bmatrix} \mu_{11} - \dfrac{\mu_{12} + \mu_{13}}{2} \\ \mu_{12} - \mu_{13} \end{bmatrix} = \begin{bmatrix} \mu_{21} - \dfrac{\mu_{22} + \mu_{23}}{2} \\ \mu_{22} - \mu_{23} \end{bmatrix} = \begin{bmatrix} \mu_{31} - \dfrac{\mu_{32} + \mu_{33}}{2} \\ \mu_{32} - \mu_{33} \end{bmatrix}$$

By employing Table 5.12.4, where the data for this example are represented in Table 5.13.2, the ANOVA Table 5.13.3 for the study is obtained. The test of parallelism (interaction), using Table 5.13.3, becomes

(5.13.23)
$$F_{gt} = \frac{8.89/4}{20.00/24} = \frac{2.22}{.83} = 2.67$$

TABLE 5.13.2. ANOVA Representation for Edwards' Data.

	Trials			Means
Drug group 1	y_{111}	y_{112}	y_{113}	$y_{11.}$
	y_{121}	y_{122}	y_{123}	$y_{12.}$
	y_{131}	y_{132}	y_{133}	$y_{13.}$
	y_{141}	y_{142}	y_{143}	$y_{14.}$
	y_{151}	y_{152}	y_{153}	$y_{15.}$
Means	$y_{1.1}$	$y_{1.2}$	$y_{1.3}$	$y_{1..}$
Drug group 2	y_{211}	y_{212}	y_{213}	$y_{21.}$
	y_{221}	y_{222}	y_{223}	$y_{22.}$
	y_{231}	y_{232}	y_{233}	$y_{23.}$
	y_{241}	y_{242}	y_{243}	$y_{24.}$
	y_{251}	y_{252}	y_{253}	$y_{25.}$
Means	$y_{2.1}$	$y_{2.2}$	$y_{2.3}$	$y_{2..}$
Drug group 3	y_{311}	y_{312}	y_{313}	$y_{31.}$
	y_{321}	y_{322}	y_{323}	$y_{32.}$
	y_{331}	y_{332}	y_{333}	$y_{33.}$
	y_{341}	y_{342}	y_{343}	$y_{34.}$
	y_{351}	y_{352}	y_{353}	$y_{35.}$
Means	$y_{3.1}$	$y_{3.2}$	$y_{3.3}$	$y_{3..}$
Grand means	$y_{..1}$	$y_{..2}$	$y_{..3}$	$y_{...}$

TABLE 5.13.3. ANOVA Analysis for Edwards' Data.

Source	df	SS	MS
Between drugs	2	27.78	13.89
Subjects within drugs (within error 1)	12	148.00	12.33
Between trials	2	164.44	82.22
Drugs × trials	4	8.89	2.22
Subjects × trials within drugs (within error 2)	24	20.00	.83
"Total"	44	389.11	

which is not significant for $\alpha = .05$ since the critical F value is $F^{.05}_{(4,24)} = 2.78$. The univariate analysis is only valid for restrictive patterns of the variance-covariance matrix $\mathbf{\Sigma}$. One common pattern is that of compound symmetry. Procedures for testing for compound symmetry were discussed in Section 3.16.

Given a multivariate analysis of the profile data and a variance-covariance matrix that satisfies the compound-symmetry assumption, it was mentioned in Section 5.12 that the univariate ANOVA results could be located from the multivariate analysis if the transformation $\mathbf{A}$ is orthogonalized. That is, if $\mathbf{A'A} = \mathbf{I}$. For example, with $\mathbf{A}$ defined by (5.13.16), the orthogonal matrix $\mathbf{A}$ is

$$\text{(5.13.24)} \qquad \mathbf{A} = \begin{bmatrix} \sqrt{2}/2 & \sqrt{6}/6 \\ -\sqrt{2}/2 & \sqrt{6}/6 \\ 0 & \sqrt{6}/3 \end{bmatrix} = \begin{bmatrix} .707107 & .408248 \\ -.707107 & .408248 \\ 0 & -.816497 \end{bmatrix}$$

For $\mathbf{A}$ defined in (5.13.21), the matrix $\mathbf{A}$ becomes

$$\text{(5.13.25)} \qquad \mathbf{A} = \begin{bmatrix} \sqrt{6}/3 & 0 \\ -\sqrt{6}/6 & \sqrt{2}/2 \\ -\sqrt{6}/6 & -\sqrt{2}/2 \end{bmatrix} = \begin{bmatrix} .816497 & 0 \\ -.408248 & .707107 \\ -.408248 & -.707107 \end{bmatrix}$$

By using $\mathbf{A}$ as defined in (5.13.24), the matrices $\mathbf{Q}_h$ and $\mathbf{Q}_e$ for the parallelism test are

$$\text{(5.13.26)} \qquad \mathbf{Q}_h = \begin{bmatrix} 4.0000 & \\ 2.8868 & 4.8889 \end{bmatrix} \qquad \mathbf{Q}_e = \begin{bmatrix} 6.0000 & \\ 3.4641 & 14.0000 \end{bmatrix}$$

Dividing the diagonal entries of $\mathbf{Q}_h$ and $\mathbf{Q}_e$ in (5.13.26) by $\nu_h = 2$ and $\nu_e = 12$, respectively, and averaging,

$$\text{(5.13.27)} \qquad \text{MS}_h = \text{ave. of Dia}\left[\frac{\mathbf{Q}_h}{\nu_h}\right] = 2.2222 \qquad \text{MS}_e = \text{ave. of Dia}\left[\frac{\mathbf{Q}_e}{\nu_e}\right] = .8333$$

the values of the mean squares due to parallelism and the appropriate error term found in (5.13.23) are evidenced. To calculate the degrees of freedom for the parallelism hypothesis in the univariate case, the rank of $\mathbf{A}$ is used. For the error degrees of freedom, the rank of $\mathbf{A}$ is multiplied by ν_e, $R(\mathbf{A})\nu_e = 2(12) = 24$. For the hypothesis degrees of freedom, the rank of $\mathbf{A}$ is multiplied by ν_h, $R(\mathbf{A})\nu_h = 2(2) = 4$. Although a matrix $\mathbf{A}$ chosen such that $\mathbf{A}'\mathbf{A} = \mathbf{I}$ helps in determining univariate results, the transformed variables are less meaningful for a multivariate analysis.

The univariate tests for differences in drugs and differences in trials may not be gained from the multivariate tests as given in (5.13.5) and (5.13.8) since the univariate tests are only valid if the parallelism hypothesis is tenable and those multivariate tests do not assume parallelism. However, from the multivariate tests (5.12.22) and (5.12.23), which do assume parallelism, the univariate results found in Table 5.13.3 are immediate.

To illustrate the procedure, the multivariate test for differences in drugs under parallelism is first discussed. From (5.12.24), the hypothesis is written as

$$\text{(5.13.28)} \qquad H_{03}^{*}: \sum_{j=1}^{3} \mu_{1j} = \sum_{j=1}^{3} \mu_{2j} = \sum_{j=1}^{3} \mu_{3j}$$

The hypothesis as stated in (5.13.28) would use

$$\underset{(3\times 1)}{\mathbf{A}} = \begin{bmatrix} 1 \\ 1 \\ 1 \end{bmatrix} \quad \text{and} \quad \underset{(2\times 3)}{\mathbf{C}} = \begin{bmatrix} 1 & 0 & -1 \\ 0 & 1 & -1 \end{bmatrix} \tag{5.13.29}$$

This is equivalent to performing a one-way analysis of variance on the subject totals. The mean performance on each drug is investigated with this test. Alternatively, a matrix **A** equal to

$$\underset{(3\times 1)}{\mathbf{A}} = \begin{bmatrix} 1/3 \\ 1/3 \\ 1/3 \end{bmatrix} \quad \text{or} \quad \mathbf{A} = \begin{bmatrix} \sqrt{3}/3 \\ \sqrt{3}/3 \\ \sqrt{3}/3 \end{bmatrix} = \begin{bmatrix} .57735 \\ .57735 \\ .57735 \end{bmatrix} \tag{5.13.30}$$

could be employed. Since the univariate and multivariate tests are equivalent, assuming parallelism, the choice of **A** depends not on normalization but on whether totals or averages are utilized. In the two-sample case, averages were used; so here, totals are employed. Choosing **A** and **C** as in (5.13.29),

$$\begin{aligned} \mathbf{Q}_h &= (\mathbf{C}\hat{\mathbf{B}}\mathbf{A})'[\mathbf{C}(\mathbf{X}'\mathbf{X})^{-1}\mathbf{C}']^{-1}(\mathbf{C}\hat{\mathbf{B}}\mathbf{A}) \\ &= [1, 1, 1]' \begin{bmatrix} 23.3333 & & \\ 8.3333 & 3.3333 & \\ 10.0000 & 5.0000 & 10.0000 \end{bmatrix} \begin{bmatrix} 1 \\ 1 \\ 1 \end{bmatrix} \\ &= 83.3333 \\ \mathbf{Q}_e &= 444.0000 \end{aligned} \tag{5.13.31}$$

Thus

$$\Lambda = \frac{|\mathbf{Q}_e|}{|\mathbf{Q}_e + \mathbf{Q}_h|} = \frac{444.0000}{527.3333} = .84197$$

where $\Lambda \sim U(1, 2, 12)$. Alternatively, since $v_h = 1$,

$$F_g = v_e\left(\frac{1 - \Lambda}{\Lambda}\right) = 1.13 \sim F(2, 12)$$

which is not significant at the $\alpha = .05$ level of significance. From Table 5.13.3,

$$F_g = \frac{27.28/2}{148.00/12} = \frac{13.89}{12.33} = 1.13 \sim F(2, 12)$$

as in the multivariate test of H_{03}^*. If **A** had been chosen such that $\mathbf{A}'\mathbf{A} = \mathbf{I}$, then $\mathbf{Q}_e =$ 148.00 and $\mathbf{Q}_h = 27.78$. By applying the rule of dividing by degrees of freedom and averaging, the univariate F statistic results. To determine the appropriate degrees of freedom, the $R(\mathbf{A})$ and the degrees of freedom for the multivariate test are utilized. That is, $R(\mathbf{A})v_h = 1(2) = 2$ and $R(\mathbf{A})v_e = 1(12) = 12$, which are the degrees of freedom for the statistic F_g.

To test the hypothesis of differences in trials, under parallelism,

$$H_{02}^*: \frac{\sum_{i=1}^{3} \mu_{i1}}{3} = \frac{\sum_{i=1}^{3} \mu_{21}}{3} = \frac{\sum_{i=1}^{3} \mu_{31}}{3} \tag{5.13.32}$$

the matrices **A** and **C** are chosen such that

$$\mathbf{A} = \begin{bmatrix} 1 & 0 \\ 0 & 1 \\ -1 & -1 \end{bmatrix} \quad \text{and} \quad \mathbf{C} = [1/3, 1/3, 1/3] \tag{5.13.33}$$

The test statistic is

$$T^2 = 3^2 \left(\sum_{i=1}^{3} \frac{1}{N_i} \right)^{-1} \mathbf{y}'_{..} \mathbf{A}(\mathbf{A}'\mathbf{S}\mathbf{A})^{-1}\mathbf{A}'\mathbf{y}_{..} \tag{5.13.34}$$

where

$$\mathbf{y}_{..} = \frac{\sum_{i=1}^{3} \mathbf{y}_{i.}}{3}$$

Evaluating (5.13.34),

$$T^2 = 9\left(\frac{5}{3}\right) [-14/3, -8/3] \begin{bmatrix} 30/12 & 18/12 \\ 18/12 & 18/12 \end{bmatrix}^{-1} [-14/3, -8/3] = 131.111$$

and $T^2 \sim T(2, 12)$. The critical value of the T^2 statistic for $\alpha = .05$ is 8.689; hence there seems to be a significant difference among trials. The criterion used in (5.13.34) is employed when an unweighted test is of interest. For a weighted test, the criterion

$$T^2 = N\bar{\mathbf{y}}'_{..}\mathbf{A}(\mathbf{A}'\mathbf{S}\mathbf{A})^{-1}\mathbf{A}'\bar{\mathbf{y}}'_{..} \tag{5.13.35}$$

would be appropriate.

The univariate result assuming compound symmetry and parallelism is easily obtained by using (5.13.34) or (5.13.35). Alternatively, recall that when $v_h = 1 = R(\mathbf{C})$,

$$\Lambda = \frac{1}{1 + T^2/v_e} \tag{5.13.36}$$

By substituting the value of T^2 into (5.13.36), $\Lambda = .0839 \sim U(2, 1, 12)$.

To compute Λ, the matrices **C** and **A** are chosen such that

(5.13.37)

$$\mathbf{C} = [1/3, 1/3, 1/3] \quad \text{and} \quad \mathbf{A} = \begin{bmatrix} \sqrt{2}/2 & -\sqrt{6}/6 \\ 0 & \sqrt{6}/3 \\ -\sqrt{2}/2 & -\sqrt{6}/6 \end{bmatrix} = \begin{bmatrix} .707107 & -.408258 \\ 0 & .816497 \\ -.707107 & -.408248 \end{bmatrix}$$

so that **A** given in (5.13.33) satisfies $\mathbf{A}'\mathbf{A} = I$. Then

$$\mathbf{Q}_h = (\mathbf{C}\hat{\mathbf{B}}\mathbf{A})'[\mathbf{C}(\mathbf{X}'\mathbf{X})^{-1}\mathbf{C}']^{-1}(\mathbf{C}\hat{\mathbf{B}}\mathbf{A}) = \begin{bmatrix} 163.3334 & \\ 13.4711 & 1.1111 \end{bmatrix} \tag{5.13.38}$$

$$\mathbf{Q}_e = \mathbf{A}'\mathbf{Y}'[\mathbf{I} - \mathbf{X}(\mathbf{X}'\mathbf{X})^{-1}\mathbf{X}']\mathbf{Y}\mathbf{A} = \begin{bmatrix} 15.0000 & \\ 1.7321 & 5.0000 \end{bmatrix}$$

and

$$\Lambda = \frac{|\mathbf{Q}_e|}{|\mathbf{Q}_e + \mathbf{Q}_h|} = .0839$$

as indicated by (5.13.36). However, dividing the diagonal elements of $\mathbf{Q}_h$ by $v_h = 1$ and averaging, $\text{MS}_h = 82.22$. Similarly, by dividing the diagonal element of $\mathbf{Q}_e$ by $v_e = 12$ and averaging, $\text{MS}_e = .8333$. The ratio

$$F_t = \frac{82.22}{.83} = 99.06 \tag{5.13.39}$$

again agrees with the univariate result from Table 5.13.3. To determine the degrees of freedom for the univariate test, the $R(\mathbf{A})$ is employed. That is, $R(\mathbf{A})v_h = 2(1) = 2$ and $R(\mathbf{A})v_e = 2(12) = 24$, so that the correct degrees of freedom for F_t are therefore (2,12) as found in Table 5.13.3.

Alternatively, the matrices **A** and **C** given in (5.13.33) could have been employed to determine Λ. However, the univariate results are not easily determined for that choice of the matrix **A** since $\mathbf{A}'\mathbf{A} \neq \mathbf{I}$.

To construct simultaneous confidence intervals for the hypothesis H_{02}^* in the multivariate case, either T^2 or Λ may be employed, for when $v_h = 1$ the tests are equivalent. Using T^2, let $\mu = \Sigma_i \mu_i/3$, then the simultaneous confidence intervals for $\mu = [\mu_1, \mu_2, \ldots, \mu_p]$ become

$$\mathbf{c}'\frac{\left(\sum_i \mathbf{y}_{i.}\right)}{3} - c_0\sqrt{\frac{\mathbf{c}'\mathbf{S}\mathbf{c}}{I^2\left(\sum_i 1/N_i\right)^{-1}}} \leq \mathbf{c}'\boldsymbol{\mu} \leq \mathbf{c}'\frac{\left(\sum_i \mathbf{y}_{i.}\right)}{3} + c_0\sqrt{\frac{\mathbf{c}'\mathbf{S}\mathbf{c}}{I^2\left(\sum_i 1/N_i\right)^{-1}}} \tag{5.13.40}$$

where $\mathbf{c}' = \mathbf{a}'\mathbf{A}$, $c_0^2 = T^\alpha(p-1, N-I)$, and **a** is an arbitrary vector.

For differences in trials, the following contrasts are examined

$$\begin{array}{lll} \psi_1 = \mu_1 - \mu_3 & \psi_2 = \mu_2 - \mu_3 & \psi_3 = \mu_1 - \mu_2 \\ \hat{\psi}_1 = -14/3 & \hat{\psi}_2 = -8/3 & \hat{\psi}_3 = -6/3 \\ \quad = -4.66 & \quad = -2.66 & \quad = -2 \end{array} \tag{5.13.41}$$

and

(5.13.42)
$$\hat{\sigma}_{\hat{\psi}_1} = 2.5/15 = .4082$$
$$\hat{\sigma}_{\hat{\psi}_2} = 1.5/15 = .3162$$
$$\hat{\sigma}_{\hat{\psi}_3} = 1.0/15 = .2582$$

Evaluating (5.13.40) with $c_0 = [T^{.05}_{(2,12)}]^{1/2} = 2.948$, simultaneous confidence intervals for the contrasts ψ_i defined in (5.13.42) are

(5.13.43)
$$-5.86 \leq \psi_1 \leq -3.46 \quad \text{(Sig.)}$$
$$-3.59 \leq \psi_2 \leq -1.73 \quad \text{(Sig.)}$$
$$-2.76 \leq \psi_3 \leq -1.24 \quad \text{(Sig.)}$$

These contrasts help to locate differences in conditions while ignoring the group effect, while (5.13.8) allows for examination of the condition effect within groups. Furthermore, the test of H^*_{02} is only valid given that the parallelism hypothesis is tenable.

The tests for differences in conditions and differences among groups given parallelism are easily applied through most general MANOVA programs (see Finn, 1972). Employing Finn's program, the matrix **A** is usually chosen to be an orthogonal matrix of normalized Helmert contrasts for a profile analysis. For the Edwards' example, the matrix **A** is

$$\mathbf{A} = \begin{bmatrix} \sqrt{3}/3 & \sqrt{6}/6 & 0 \\ \sqrt{3}/3 & -\sqrt{6}/3 & \sqrt{2}/2 \\ \sqrt{3}/3 & -\sqrt{6}/3 & -\sqrt{2}/2 \end{bmatrix}$$

Alternatively, the BMD-X63 routine could be used, among others (see Dixon, 1969).

When employing profile analysis to study p commensurable ordered or non-ordered responses, the pattern of $\mathbf{\Sigma}$ should be checked. If the compound-symmetry assumption is tenable, the univariate mixed model with two fixed effects and one random effect may be used. Otherwise, multivariate methods as illustrated in this section are easily employed. The first hypothesis tested when using multivariate analysis is the test of parallelism. Given parallelism, either hypothesis H^*_{02} or H^*_{03} may be tested. Alternatively, whether the parallelism hypothesis is satisfied or not, the experimenter may always test hypothesis H_{02} or H_{03}.

The methods illustrated in this section may be extended to more complicated models. For example, consider the mixed model discussed by Kirk (1968, p. 299), Myers (1966, p. 198), and Winer (1971, p. 546) which is a split-split-plot design with three fixed effects and one random effect where subjects are random and nested in one fixed effect and repeated measures are obtained over two other fixed factors. The univariate model for the design is

(5.13.44)
$$y_{ijkm} = \mu + \alpha_i + \beta_k + \gamma_m + \eta_{ik} + \omega_{im} + \theta_{km} + \pi_{ikm} + s_{(i)j} + b_{(i)jk} + c_{(i)jm} + \varepsilon_{(i)jkm}$$

where $i = 1, \ldots, I; j = 1, \ldots, n; k = 1, \ldots, K$; and $m = 1, \ldots, M$, and where $s_{(i)j} \sim IN(0, \rho_1\sigma^2)$, $b_{(i)jk} \sim IN(0, \rho_2\sigma^2)$, $c_{(i)jm} \sim IN(0, \rho_3)\sigma^2$, $\varepsilon_{(i)jk} \sim IN[0, (1 - \rho_i)\sigma^2]$, and $s_{(i)j}, b_{(i)jk}, c_{(i)jm}$, and $\varepsilon_{(i)jkm}$ are jointly independent. The symbols in (5.13.44) are defined as follows:

$$\begin{aligned}
\mu &= \text{overall constant} \\
\alpha_i &= i\text{th group effect (factor } A) \\
s_{(i)j} &= \text{effect of } j\text{th subject at } i\text{th group} \\
\beta_k &= k\text{th treatment effect (factor } B) \\
\eta_{ik} &= AB \text{ interaction} \\
b_{(i)jk} &= \text{subject by } B \text{ interaction within } i\text{th group} \\
\gamma_m &= m\text{th treatment effect (factor } C) \\
\omega_{im} &= AC \text{ interaction} \\
c_{(i)jm} &= \text{subject by } C \text{ interaction within } i\text{th group} \\
\theta_{km} &= BC \text{ interaction} \\
\pi_{ikm} &= ABC \text{ interaction} \\
\varepsilon_{(i)jkm} &= \text{subject by } BC \text{ interaction plus usual} \\
&\quad\ \text{random error component}
\end{aligned}$$

With $I = 2$, $K = 3$, and $M = 3$, the data for model (5.13.44) are represented in Table 5.13.4. To test hypotheses under (5.13.44), compound-symmetry assumptions are usually imposed for the $K \times K$, the $M \times M$, and the $KM \times KM$ variance-covariance matrices.

Alternatively, if $n \geq KM$, to test hypotheses of interest that do not require any pattern on $\mathbf{\Sigma}$, multivariate methods may be employed. That is, let $\mathbf{Y}$ of order $N = nI$ by $p = KM$ denote the data matrix for a set of data represented in Table 5.13.4. The general linear model is conveniently given as

$$\underbrace{\begin{bmatrix} \mathbf{y}'_{11} \\ \mathbf{y}'_{12} \\ \vdots \\ \mathbf{y}'_{1n} \\ \hline \mathbf{y}'_{21} \\ \mathbf{y}'_{22} \\ \vdots \\ \mathbf{y}'_{2n} \\ \hline \vdots \\ \hline \mathbf{y}'_{I1} \\ \mathbf{y}'_{I2} \\ \vdots \\ \mathbf{y}'_{In} \end{bmatrix}}_{(N \times p)} = \underbrace{\begin{bmatrix} 1 & 0 & \cdots & 0 \\ 1 & 0 & \cdots & 0 \\ \vdots & \vdots & \vdots & \vdots \\ 1 & 0 & \cdots & 0 \\ \hline 0 & 1 & \cdots & 0 \\ 0 & 1 & \cdots & 0 \\ \vdots & \vdots & \vdots & \vdots \\ 0 & 1 & \cdots & 0 \\ \hline \vdots & \vdots & \vdots & \vdots \\ \hline 0 & 0 & \cdots & 1 \\ 0 & 0 & \cdots & 1 \\ \vdots & \vdots & \vdots & \vdots \\ 0 & 0 & \cdots & 1 \end{bmatrix}}_{(N \times I)} \underbrace{\begin{bmatrix} \mu_{11} & \mu_{12} & \cdots & \mu_{1p} \\ \mu_{21} & \mu_{22} & \cdots & \mu_{2p} \\ \vdots & \vdots & \vdots & \vdots \\ \mu_{I1} & \mu_{I2} & \cdots & \mu_{Ip} \end{bmatrix}}_{(I \times p)} + \underbrace{\mathbf{E}_0}_{(N \times p)} \tag{5.13.45}$$

where

$$\begin{aligned} E(\mathbf{Y}) &= \mathbf{XB} \\ V(\mathbf{Y}) &= \mathbf{I} \otimes \mathbf{\Sigma} \end{aligned} \tag{5.13.46}$$

TABLE 5.13.4. Data for Split-Split-Plot Design.

	B:	b_1			b_2			b_3		
	C:	c_1	c_2	c_3	c_1	c_2	c_3	c_1	c_2	c_3
A:	a_1	y_{1111}	y_{1112}	y_{1113}	y_{1121}	y_{1122}	y_{1123}	y_{1131}	y_{1132}	y_{1133}
		y_{1121}	y_{1212}	y_{1213}	y_{1221}	y_{1222}	y_{1223}	y_{1231}	y_{1232}	y_{1233}
		$\vdots$	$\vdots$	$\vdots$	$\vdots$	$\vdots$	$\vdots$	$\vdots$	$\vdots$	$\vdots$
		y_{1n11}	y_{1n12}	y_{1n13}	y_{1n21}	y_{1n22}	y_{1n23}	y_{1n31}	y_{1n32}	y_{1n33}
	a_2	y_{2111}	y_{2112}	y_{2113}	y_{2121}	y_{2122}	y_{2123}	y_{2131}	y_{2132}	y_{2113}
		y_{2211}	y_{2212}	y_{2213}	y_{2221}	y_{2222}	y_{2223}	y_{2231}	y_{2232}	y_{2233}
		$\vdots$	$\vdots$	$\vdots$	$\vdots$	$\vdots$	$\vdots$	$\vdots$	$\vdots$	$\vdots$
		y_{2n11}	y_{2n12}	y_{2n13}	y_{2n21}	y_{2n22}	y_{2n23}	y_{2n31}	y_{2n32}	y_{2n33}

Letting $I = 2$, $n = 9$, $K = 3$, and $M = 3$, suppose we wanted to test the ABC, BC, AC, and AB hypotheses employing the multivariate model. Putting these hypotheses in the form $\mathbf{CBA} = \boldsymbol{\Gamma}$, where $\mathbf{B}$ is equal to

$$\mathbf{B} = \left[\begin{array}{ccc|ccc|ccc} \mu_{11} & \mu_{12} & \mu_{13} & \mu_{14} & \mu_{15} & \mu_{16} & \mu_{17} & \mu_{18} & \mu_{19} \\ \mu_{21} & \mu_{22} & \mu_{23} & \mu_{24} & \mu_{25} & \mu_{26} & \mu_{27} & \mu_{28} & \mu_{29} \end{array}\right]$$

and $\boldsymbol{\Gamma} = 0$, the matrices $\mathbf{A}$ and $\mathbf{C}$ are defined for the hypotheses:

$$ABC: \quad \mathbf{C} = [1, -1] \qquad \mathbf{A} = \left[\begin{array}{cccc} 1 & 0 & 0 & 0 \\ 0 & 1 & 0 & 0 \\ -1 & -1 & 0 & 0 \\ \hdashline 0 & 0 & 1 & 0 \\ 0 & 0 & 0 & 1 \\ 0 & 0 & -1 & -1 \\ \hdashline -1 & 0 & -1 & 0 \\ 0 & -1 & 0 & -1 \\ 1 & 1 & 1 & 1 \end{array}\right]$$

$$BC: \quad \mathbf{C} = [1, 1] \qquad \mathbf{A} = \left[\begin{array}{cccc} 1 & 0 & 0 & 0 \\ 0 & 1 & 0 & 0 \\ -1 & -1 & 0 & 0 \\ \hdashline 0 & 0 & 1 & 0 \\ 0 & 0 & 0 & 1 \\ 0 & 0 & -1 & -1 \\ \hdashline -1 & 0 & -1 & 0 \\ 0 & -1 & 0 & -1 \\ 1 & 1 & 1 & 1 \end{array}\right] \tag{5.13.47}$$

$$AC: \quad \mathbf{C} = [1, -1] \qquad \mathbf{A} = \begin{bmatrix} 1 & 0 \\ 1 & 1 \\ -1 & -1 \\ \hline 1 & 0 \\ 0 & 1 \\ -1 & -1 \\ \hline 1 & 0 \\ 0 & 1 \\ -1 & -1 \end{bmatrix}$$

(5.13.47)

$$AB: \quad \mathbf{C} = [1, -1] \qquad \mathbf{A} = \begin{bmatrix} 1 & 0 \\ 1 & 0 \\ 1 & 0 \\ \hline 0 & 1 \\ 0 & 1 \\ 0 & 1 \\ \hline -1 & -1 \\ -1 & -1 \\ -1 & -1 \end{bmatrix}$$

To test for main effects under no significant interactions, the following compounding matrices are needed.

$$C: \quad \mathbf{C} = [1, 1] \qquad \mathbf{A} = \begin{bmatrix} 1 & 0 \\ 0 & 1 \\ -1 & -1 \\ \hline 1 & 0 \\ 0 & 1 \\ -1 & -1 \\ \hline 1 & 0 \\ 0 & 1 \\ -1 & -1 \end{bmatrix}$$

$$B: \quad \mathbf{C} = [1, 1] \qquad \mathbf{A} = \begin{bmatrix} 1 & 0 \\ 1 & 0 \\ 1 & 0 \\ \hline 0 & 1 \\ 0 & 1 \\ 0 & 1 \\ \hline -1 & -1 \\ -1 & -1 \\ -1 & -1 \end{bmatrix}$$

(5.13.48)

$$A: \quad \mathbf{C} = [1, -1] \qquad \mathbf{A} = \begin{bmatrix} 1 \\ 1 \\ 1 \\ \hline 1 \\ 1 \\ 1 \\ \hline 1 \\ 1 \\ 1 \end{bmatrix}$$

By employing matrices $\mathbf{A}$ such that $\mathbf{A}'\mathbf{A} = I$, univariate results are obtained from the multivariate analysis by the averaging procedure discussed with the previous design in this section. However, due to the amount of calculation involved in more complex designs, a computer program is essential.

It is important to note that the tests of ABC, AB, and AC in (5.13.47) are subtests of the multivariate test of parallelism of profiles and, when combined into one overall test, are used to test for parallel profiles. Furthermore, BC should only be tested if the test of ABC is not significant.

A multivariate test of A which does not depend on nonsignificant tests of interaction is obtained by forming the matrices

(5.13.49) $$\mathbf{C} = [1, -1] \quad \text{and} \quad \mathbf{A} = \mathbf{I}_9$$

for our simple example and is used to test $H_0: \boldsymbol{\mu}_1 = \boldsymbol{\mu}_2$. Given that the test of BC is not significant, multivariate tests of B and C are defined by the following matrices.

$$B:\quad \mathbf{C} = \begin{bmatrix} 1 & 0 \\ 0 & 1 \end{bmatrix} \qquad \mathbf{A} = \begin{bmatrix} 1 & 0 \\ 1 & 0 \\ 1 & 0 \\ 0 & 1 \\ 0 & 1 \\ 0 & 1 \\ -1 & -1 \\ -1 & -1 \\ -1 & -1 \end{bmatrix}$$

(5.13.50)

$$C:\quad \mathbf{C} = \begin{bmatrix} 1 & 0 \\ 0 & 1 \end{bmatrix} \qquad \mathbf{A} = \begin{bmatrix} 1 & 0 \\ 0 & 1 \\ -1 & -1 \\ 1 & 0 \\ 0 & 1 \\ -1 & -1 \\ 1 & 0 \\ 0 & 1 \\ -1 & -1 \end{bmatrix}$$

So far our entire discussion of repeated-measures designs in this text has been restricted to profile analysis; for data with one underlying attribute, these designs have been measured in the same units. In many situations, a number of distinct variables are measured at several conditions. This is called a *multivariate-response, repeated-measures design* and is discussed in Section 5.16 along with trend analysis of repeated-measures designs.

EXERCISE 5.13

1. In a study of maze learning, 15 rats were randomly assigned to three different reinforcement schedules and then given a maze to run under four experimental conditions. The sequence in which the four conditions were presented in the experiment was randomized independently for each animal. The dependent variable for the study was the number of seconds taken to run the maze. The data for the experiment follow.

Reinforcement Schedule	Rat	Conditions 1	2	3	4
1	1	29	20	21	18
	2	24	15	10	8
	3	31	19	10	31
	4	41	11	15	42
	5	30	20	20	53
2	1	25	17	19	17
	2	20	12	8	8
	3	35	16	9	28
	4	35	8	14	40
	5	26	18	18	51
3	1	10	18	16	14
	2	9	10	18	11
	3	7	18	19	12
	4	8	19	20	5
	5	11	20	17	6

Analyze the data by using multivariate methods and summarize your results. How does your analysis compare with a univariate analysis of the data?

5.14 ANALYSIS OF COVARIANCE—UNIVARIATE AND MULTIVARIATE

In employing the linear model $\mathbf{y} = \mathbf{X}\boldsymbol{\beta} + \boldsymbol{\varepsilon}$ in the study of various experimental design models in this chapter, the design matrix $\mathbf{X}$ had entries that were only 0 and 1. In many cases, the design matrix was often not of full rank. We will now investigate linear models where some of the elements of the design matrix are continuous mathematical variables or fixed values of a random variable and where other elements in the matrix are assigned the values 0 or 1. These types of models are still referred to as *covariance models*. In particular, such models are used in analysis of covariance (ANCOVA).

General Theory. Before considering a specific model, the general case is discussed. By assuming that the relationship between a dependent random variable Y on a set of fixed independent variables $z_1, z_2, \ldots, z_k$ is linear (in the parameters), the general covariance linear model is written as

(5.14.1)
$$\underset{(N \times 1)}{\mathbf{y}} = \underset{(N \times q)}{\mathbf{X}} \underset{(q \times 1)}{\boldsymbol{\beta}} + \underset{(N \times h)}{\mathbf{Z}} \underset{(h \times 1)}{\boldsymbol{\gamma}} + \underset{(N \times 1)}{\boldsymbol{\varepsilon}}$$

or

$$\underset{(N \times 1)}{\mathbf{y}} = \underset{[N \times (q+h)]}{[\mathbf{X} \quad \mathbf{Z}]} \underset{[(q+h) \times 1]}{\begin{bmatrix} \boldsymbol{\beta} \\ \boldsymbol{\gamma} \end{bmatrix}} + \underset{(N \times 1)}{\boldsymbol{\varepsilon}}$$

where $E(\boldsymbol{\varepsilon}) = 0$ and $V(\boldsymbol{\varepsilon}) = \sigma^2 I$. This implies that the appropriate form of the functional relationship between the dependent and independent variables is used by the experimenter.

When using (5.14.1), the design matrix $\mathbf{X}$ is not necessarily of full rank, and the elements of $\boldsymbol{\beta}$ correspond to the effects in an experimental design model so that $\mathbf{X}$ contains only 0s or 1s. The matrix $\mathbf{Z}$ is assumed to be of full rank, and the columns of $\mathbf{Z}$ are linearly independent of those of $\mathbf{X}$. $\mathbf{Z}$ contains continuous fixed observed variables called *concomitant variables* or *covariates.*

Applying the general theory of Chapter 3 to (5.14.1) does not result in any new problems since (5.14.1) is of the form $\mathbf{y} = \mathbf{X}^*\boldsymbol{\beta}^* + \boldsymbol{\varepsilon}$ with $\mathbf{X}^* = [\mathbf{X} \quad \mathbf{Z}]$ and $\boldsymbol{\theta}' = [\boldsymbol{\beta}' \quad \boldsymbol{\gamma}']$. The normal equations for (5.14.1) become

$$\begin{bmatrix} \mathbf{X}'\mathbf{X} & \mathbf{X}'\mathbf{Z} \\ \mathbf{Z}'\mathbf{X} & \mathbf{Z}'\mathbf{Z} \end{bmatrix} + \begin{bmatrix} \boldsymbol{\beta} \\ \boldsymbol{\gamma} \end{bmatrix} = \begin{bmatrix} \mathbf{X}'\mathbf{y} \\ \mathbf{Z}'\mathbf{y} \end{bmatrix} \tag{5.14.2}$$

or

$$\begin{aligned} (\mathbf{X}'\mathbf{X})\boldsymbol{\beta} + (\mathbf{X}'\mathbf{Z})\boldsymbol{\gamma} &= \mathbf{X}'\boldsymbol{\gamma} \\ (\mathbf{Z}'\mathbf{X})\boldsymbol{\beta} + (\mathbf{Z}'\mathbf{Z})\boldsymbol{\gamma} &= \mathbf{Z}'\mathbf{y} \end{aligned} \tag{5.14.3}$$

By using the first equation in (5.14.3), an estimate of $\boldsymbol{\beta}$ is obtained,

$$\begin{aligned} \hat{\boldsymbol{\beta}} &= (\mathbf{X}'\mathbf{X})^-(\mathbf{X}'\mathbf{y} - \mathbf{X}'\mathbf{Z}\hat{\boldsymbol{\gamma}}) \\ &= (\mathbf{X}'\mathbf{X})^-\mathbf{X}'\mathbf{y} - (\mathbf{X}'\mathbf{X})^-\mathbf{X}'\mathbf{Z}\hat{\boldsymbol{\gamma}} \\ &= \hat{\boldsymbol{\beta}}_0 - (\mathbf{X}'\mathbf{X})^-\mathbf{X}'\mathbf{Z}\hat{\boldsymbol{\gamma}} \end{aligned} \tag{5.14.4}$$

where $\hat{\boldsymbol{\beta}}_0$ is a solution of the normal equations without the covariates. By substituting $\hat{\boldsymbol{\beta}}$ for $\boldsymbol{\beta}$ in the second equation of (5.14.3),

$$\hat{\boldsymbol{\gamma}} = \{\mathbf{Z}'[\mathbf{I} - \mathbf{X}(\mathbf{X}'\mathbf{X})^-\mathbf{X}']\mathbf{Z}\}^-\mathbf{Z}'[\mathbf{I} - \mathbf{X}(\mathbf{X}'\mathbf{X})^-\mathbf{X}']\mathbf{y} \tag{5.14.5}$$

Allowing $\mathbf{Q} = \mathbf{I} - \mathbf{X}(\mathbf{X}'\mathbf{X})^-\mathbf{X}'$, so that $\mathbf{Q}$ is a projection operator, (5.14.5) becomes

$$\hat{\boldsymbol{\gamma}} = (\mathbf{Z}'\mathbf{Q}\mathbf{Z})^{-1}\mathbf{Z}'\mathbf{Q}\mathbf{y} \tag{5.14.6}$$

Since $\mathbf{Z}$ is of full rank and the columns of $\mathbf{Z}$ are linearly independent of $\mathbf{X}$, an ordinary inverse exists. From (5.14.6) we also notice that $\hat{\boldsymbol{\gamma}}$ is the best linear unbiased estimator for $\boldsymbol{\gamma}$ in the model $\mathbf{y} = \mathbf{Q}\mathbf{Z}\boldsymbol{\gamma} + \boldsymbol{\varepsilon}$.

From the theory of Chapter 3, an estimate for the parameters of model (5.14.1) is given by

$$\hat{\boldsymbol{\theta}} = \begin{bmatrix} \hat{\boldsymbol{\beta}} \\ \hat{\boldsymbol{\gamma}} \end{bmatrix} = \begin{bmatrix} \hat{\boldsymbol{\beta}}_0 - (\mathbf{X}'\mathbf{X})^-\mathbf{X}'\mathbf{Z}\hat{\boldsymbol{\gamma}} \\ (\mathbf{Z}'\mathbf{Q}\mathbf{Z})^{-1}\mathbf{Z}'\mathbf{Q}\mathbf{y} \end{bmatrix} \tag{5.14.7}$$

Furthermore, $\boldsymbol{\gamma}$ is always estimable, and linear combinations $\psi = \mathbf{c}'\boldsymbol{\beta}$ have a unique linear unbiased estimator whenever the parametric function ψ is estimable in the linear model without concomitant variables.

An alternative procedure for obtaining (5.14.7) is to use a result from Rohode (1965) for finding a g-inverse for a partitioned symmetric matrix. That is, if $\mathbf{M}$ is a nonsingular partitioned symmetric matrix of the form

$$\mathbf{M} = \begin{bmatrix} \mathbf{A} & \mathbf{B} \\ \mathbf{B}' & \mathbf{C} \end{bmatrix}$$

$$\mathbf{M}^- = \begin{bmatrix} \mathbf{A}^- + \mathbf{A}^-\mathbf{B}\mathbf{F}^-\mathbf{B}'\mathbf{A}' & -\mathbf{A}^-\mathbf{B}\mathbf{F}^- \\ -\mathbf{F}^-\mathbf{B}'\mathbf{A}' & \mathbf{F}^- \end{bmatrix}$$

$$= \begin{bmatrix} \mathbf{A}^- & \mathbf{0} \\ \mathbf{0} & \mathbf{0} \end{bmatrix} + \begin{bmatrix} -\mathbf{A}^-\mathbf{B} \\ \mathbf{I} \end{bmatrix} \mathbf{F}^-[-\mathbf{B}'\mathbf{A}^- \quad \mathbf{I}]$$

where $\mathbf{F} = \mathbf{C} - \mathbf{B}'\mathbf{A}^-\mathbf{B}$. With $\mathbf{M}$ defined by

$$\mathbf{M} = \begin{bmatrix} \mathbf{X}'\mathbf{X} & \mathbf{X}'\mathbf{Z} \\ \mathbf{Z}'\mathbf{X} & \mathbf{Z}'\mathbf{Z} \end{bmatrix}$$

a g-inverse of $\mathbf{M}$ satisfying the condition $\mathbf{M}\mathbf{M}^-\mathbf{M} = \mathbf{M}$ is

$$\text{(5.14.8)} \quad \mathbf{M}^- = \begin{bmatrix} (\mathbf{X}'\mathbf{X})^- & \mathbf{0} \\ \mathbf{0} & \mathbf{0} \end{bmatrix} + \begin{bmatrix} -(\mathbf{X}'\mathbf{X})^-\mathbf{X}'\mathbf{Z} \\ \mathbf{I} \end{bmatrix} (\mathbf{Z}'\mathbf{Q}\mathbf{Z})^{-1}[-\mathbf{Z}'\mathbf{X}(\mathbf{X}'\mathbf{X})^- \quad \mathbf{I}]$$

where $\mathbf{Q} = \mathbf{I} - \mathbf{X}(\mathbf{X}'\mathbf{X})^-\mathbf{X}'$ and result (5.14.7) follows (Searle, 1971, p. 340).

By using (5.14.8), the variance-covariance matrix for estimable functions is also determined:

$$\begin{aligned} V(\hat{\boldsymbol{\gamma}}) &= (\mathbf{Z}'\mathbf{Q}\mathbf{Z})^{-1}\sigma^2 \\ V(\mathbf{c}'\hat{\boldsymbol{\beta}}) &= \sigma^2\mathbf{c}'[(\mathbf{X}'\mathbf{X})^- + (\mathbf{X}'\mathbf{X})^-\mathbf{X}'\mathbf{Z}(\mathbf{Z}'\mathbf{Q}\mathbf{Z})^{-1}\mathbf{Z}'\mathbf{X}(\mathbf{X}'\mathbf{X})^-]\mathbf{c} \qquad \text{(5.14.9)} \\ \operatorname{cov}(\mathbf{c}'\hat{\boldsymbol{\beta}}, \hat{\boldsymbol{\gamma}}) &= -\sigma^2\mathbf{c}'[(\mathbf{X}'\mathbf{X})^-\mathbf{X}'\mathbf{Z}(\mathbf{Z}'\mathbf{Q}\mathbf{Z})^{-1}] \end{aligned}$$

To evaluate the expressions given in (5.14.9), an unbiased estimate for σ^2 is required. Employing the general theory of Chapter 3, the sum of squares resulting from error is

$$\begin{aligned} Q_e^* &= \mathbf{y}'\left\{\mathbf{I} - [\mathbf{X} \quad \mathbf{Z}] \begin{bmatrix} \mathbf{X}'\mathbf{X} & \mathbf{X}'\mathbf{Z} \\ \mathbf{Z}'\mathbf{X} & \mathbf{Z}'\mathbf{Z} \end{bmatrix}^- [\mathbf{X}' \quad \mathbf{Z}']\right\}\mathbf{y} \\ &= \mathbf{y}'\mathbf{y} - \mathbf{y}'[\mathbf{X} \quad \mathbf{Z}] \begin{bmatrix} \hat{\boldsymbol{\beta}} \\ \hat{\boldsymbol{\gamma}} \end{bmatrix} \\ &= \mathbf{y}'\mathbf{y} - \mathbf{y}'\mathbf{X}[\hat{\boldsymbol{\beta}}_0 - (\mathbf{X}'\mathbf{X})^-\mathbf{X}'\mathbf{Z}\hat{\boldsymbol{\gamma}}] - \mathbf{y}'\mathbf{Z}\hat{\boldsymbol{\gamma}} \\ &= \mathbf{y}'\mathbf{y} - \mathbf{y}'\mathbf{X}\hat{\boldsymbol{\beta}}_0 - \mathbf{y}'\mathbf{Q}\mathbf{Z}\hat{\boldsymbol{\gamma}} \\ &= \mathbf{y}'\mathbf{y} - \mathbf{y}'\mathbf{X}(\mathbf{X}'\mathbf{X})^-\mathbf{X}'\mathbf{y} - \mathbf{y}'\mathbf{Q}\mathbf{Z}(\mathbf{Z}'\mathbf{Q}\mathbf{Z})^{-1}\mathbf{Z}'\mathbf{Q}\mathbf{y} \end{aligned}$$

Thus the residual sum of squares may be written as

$$\text{(5.14.10)} \qquad Q_e^* = Q_e - \mathbf{y}'\mathbf{Q}\mathbf{Z}\hat{\boldsymbol{\gamma}} = Q_e - \hat{\boldsymbol{\gamma}}'\mathbf{Z}'\mathbf{Q}\mathbf{Z}\hat{\boldsymbol{\gamma}}$$

where Q_e is the usual ANOVA residual sum of squares and $\hat{\boldsymbol{\gamma}}'\mathbf{Z}'\mathbf{Q}\mathbf{Z}\hat{\boldsymbol{\gamma}}$ is the sum of squares due to regression for the conditional regression model $\mathbf{y} = \mathbf{Q}\mathbf{Z}\boldsymbol{\gamma} + \boldsymbol{\varepsilon}$. The degrees of freedom associated with Q_e^* are $v_e = N - r - h$, where the $R(\mathbf{X}) = r$

(assuming $|\mathbf{Z}'\mathbf{Z}| \neq 0$), and thus

$$s_{y|z}^2 = \frac{Q_e^*}{N - r - h}$$

is an unbiased estimator for the residual variance σ^2.

To test hypotheses by using model (5.14.1), it is assumed that $\boldsymbol{\varepsilon} \sim N(\mathbf{0}, \sigma^2\mathbf{I})$. The first hypothesis of interest in covariance analysis might be that the vector of parameters $\boldsymbol{\gamma}$ is equal to some specified value. Using the general theory of Chapter 3, a hypothesis of the form

$$H_0: \mathbf{C}_2\boldsymbol{\gamma} = \boldsymbol{\xi} \tag{5.14.11}$$

may be tested by using (5.14.1). Letting $\mathbf{C} = [\mathbf{0} \quad \mathbf{C}_2]$, where $\mathbf{C}_2 = \mathbf{I}_h$, the hypothesis sum of squares becomes

$$Q_h = (\hat{\boldsymbol{\gamma}} - \boldsymbol{\xi})'\mathbf{Z}'\mathbf{Q}\mathbf{Z}(\hat{\boldsymbol{\gamma}} - \boldsymbol{\xi}) \tag{5.14.12}$$

Since the $R(\mathbf{C}) = h$, the F statistic to test (5.14.11) is

$$F = \frac{(\hat{\boldsymbol{\gamma}} - \boldsymbol{\xi})'\mathbf{Z}'\mathbf{Q}\mathbf{Z}(\hat{\boldsymbol{\gamma}} - \boldsymbol{\xi})/h}{Q_e^*/(N - r - h)} \sim F(h, N - r - h) \tag{5.14.13}$$

To test hypotheses of the form

$$H_0: \mathbf{C}_1\boldsymbol{\beta} = \boldsymbol{\xi} \tag{5.14.14}$$

where the linear combinations of the elements of $\boldsymbol{\beta}$ are individually estimable, the matrix $\mathbf{C}$ is chosen such that $\mathbf{C} = [\mathbf{C}_1 \quad \mathbf{0}]$.

The sum of squares for the hypothesis is

(5.14.15)

$$Q_h^* = (\mathbf{C}_1\hat{\boldsymbol{\beta}} - \boldsymbol{\xi})'[\mathbf{C}_1(\mathbf{X}'\mathbf{X})^-\mathbf{C}_1' + \mathbf{C}_1(\mathbf{X}'\mathbf{X})^-\mathbf{X}'\mathbf{Z}(\mathbf{Z}'\mathbf{Q}\mathbf{Z})^{-1}\mathbf{Z}'\mathbf{X}(\mathbf{X}'\mathbf{X})^-\mathbf{C}_1']^{-1}(\mathbf{C}_1\hat{\boldsymbol{\beta}} - \boldsymbol{\xi})$$

If the $R(\mathbf{C}) = q$, the F statistic is

$$F = \frac{Q_h^*/q}{Q_e^*/(N - r - h)} \sim F(q, N - r - h) \tag{5.14.16}$$

From (5.14.9), simultaneous confidence intervals for estimable functions $\psi = \mathbf{c}'\boldsymbol{\beta}$ are immediately obtained by using Scheffé's method. That is,

$$\hat{\psi} - c_0\sqrt{\left(\frac{Q_e^*}{v_e}\right)\mathbf{c}'[(\mathbf{X}'\mathbf{X})^- + (\mathbf{X}'\mathbf{X})^-\mathbf{X}'\mathbf{Z}(\mathbf{Z}'\mathbf{Q}\mathbf{Z})^{-1}\mathbf{Z}'\mathbf{X}(\mathbf{X}'\mathbf{X})^-]\mathbf{c}'} \leq \psi$$

(5.14.17)

$$\leq \hat{\psi} + c_0\sqrt{\left(\frac{Q_e^*}{v_e}\right)\mathbf{c}'[(\mathbf{X}'\mathbf{X})^- + (\mathbf{X}'\mathbf{X})^-\mathbf{X}'\mathbf{Z}(\mathbf{Z}'\mathbf{Q}\mathbf{Z})^{-1}\mathbf{Z}'\mathbf{X}(\mathbf{X}'\mathbf{X})^-]\mathbf{c}'}$$

where $c_0^2 = v_h F(v_h, v_e)$. This completes the formal theory necessary for covariance analysis. The general procedure is now illustrated by using the one-way ANCOVA model.

One-way ANCOVA Model. The ANCOVA model for a one-way design with one covariate, which is often called a *completely randomized, analysis-of-covariance design* and abbreviated as CRAC-k, is

(5.14.18)
$$\begin{aligned} y_{ij} &= \mu + \alpha_i + \gamma z_{ij} + \varepsilon_{ij} & i &= 1, \ldots, k \\ \varepsilon_{ij} &\sim IN(0, \sigma^2) & j &= 1, \ldots, N_i \end{aligned}$$

Employing (5.14.1), the model in (5.14.18) is written as

(5.14.19)

$$\underbrace{\begin{bmatrix} y_{11} \\ y_{12} \\ \vdots \\ y_{1N_1} \\ \hline y_{21} \\ y_{22} \\ \vdots \\ y_{2N_2} \\ \hline \vdots \\ \hline y_{k1} \\ y_{k2} \\ \vdots \\ y_{kN_k} \end{bmatrix}}_{(N \times 1)} = \underbrace{\begin{bmatrix} 1 & 1 & \cdots & 0 & z_{11} \\ 1 & 1 & \cdots & 0 & z_{12} \\ \vdots & \vdots & \vdots & \vdots & \vdots \\ 1 & 1 & \cdots & 0 & z_{1N_1} \\ \hline 1 & 0 & \cdots & 0 & z_{21} \\ 1 & 0 & \cdots & 0 & z_{22} \\ \vdots & \vdots & \vdots & \vdots & \vdots \\ 1 & 0 & \cdots & 0 & z_{2N_2} \\ \hline \vdots & \vdots & \vdots & \vdots & \vdots \\ \hline 1 & 0 & \cdots & 1 & z_{k1} \\ 1 & 0 & \cdots & 1 & z_{k2} \\ \vdots & \vdots & \vdots & \vdots & \vdots \\ 1 & 0 & \cdots & 1 & z_{kN_k} \end{bmatrix}}_{[N \times (k+2)]} \underbrace{\begin{bmatrix} \mu \\ \alpha_1 \\ \alpha_2 \\ \vdots \\ \alpha_k \\ \gamma \end{bmatrix}}_{[(k+2) \times 1]} + \underbrace{\begin{bmatrix} \varepsilon_{11} \\ \varepsilon_{12} \\ \vdots \\ \varepsilon_{1N_1} \\ \hline \varepsilon_{21} \\ \varepsilon_{22} \\ \vdots \\ \varepsilon_{2N_2} \\ \hline \vdots \\ \hline \varepsilon_{k1} \\ \varepsilon_{k2} \\ \vdots \\ \varepsilon_{kN_k} \end{bmatrix}}_{(N \times 1)}$$

From the univariate Gauss-Markoff theorem, with $(\mathbf{X}'\mathbf{X})^-$ as defined in (5.2.6),

(5.14.20)
$$\begin{aligned} \hat{\boldsymbol{\beta}} &= \hat{\boldsymbol{\beta}}_0 - (\mathbf{X}'\mathbf{X})^-\mathbf{X}'\mathbf{Z}\hat{\gamma} \\ &= (\mathbf{X}'\mathbf{X})^-\mathbf{X}'\mathbf{y} - (\mathbf{X}'\mathbf{X})^-\mathbf{X}'\mathbf{Z}\hat{\gamma} \\ &= \begin{bmatrix} 0 \\ y_{1.} \\ y_{2.} \\ \vdots \\ y_{k.} \end{bmatrix} - \begin{bmatrix} 0 \\ \hat{\gamma} z_{1.} \\ \hat{\gamma} z_{2.} \\ \vdots \\ \hat{\gamma} z_{k.} \end{bmatrix} \end{aligned}$$

Thus the ith element in $\hat{\boldsymbol{\beta}}$ has the form $\hat{\boldsymbol{\beta}}_i = y_{i.} - \hat{\gamma} z_{i.}$. Since $\hat{\gamma}$ is always estimable, functions of the form

$$\psi = [\mathbf{c}' \quad \mathbf{a}'] \begin{bmatrix} \boldsymbol{\beta} \\ \boldsymbol{\gamma} \end{bmatrix}$$

are estimable if and only if $\psi = \mathbf{c}'\boldsymbol{\beta}$ is estimable. From (5.2.9),

$$\begin{aligned} \psi &= \mathbf{c}'\boldsymbol{\beta} = \sum_i t_i\mu + \sum_i t_i\alpha_i \\ \hat{\psi} &= \mathbf{c}'\hat{\boldsymbol{\beta}} = \sum_i t_i(y_{i.} - \hat{\gamma}z_{i.}) \end{aligned} \tag{5.14.21}$$

yield the forms of the estimable functions.

Before testing hypotheses, we have to estimate σ^2. By using $(\mathbf{X}'\mathbf{X})^-$ as defined by (5.2.6), an expression for Q_e^* is immediate. From (5.2.13),

$$Q_e = \sum_{i,j} y_{ij}^2 - \sum_i N_i y_{i.}^2$$

and, from (5.2.12),

$$\begin{aligned} \mathbf{y}'\mathbf{QZ} &= \mathbf{y}'[\mathbf{I} - \mathbf{X}(\mathbf{X}'\mathbf{X})^-\mathbf{X}']\mathbf{z} \\ &= \mathbf{y}'\mathbf{z} - \mathbf{y}'\mathbf{X}(\mathbf{X}'\mathbf{X})^-\mathbf{X}'\mathbf{z} \\ &= \sum_{i,j} y_{ij}z_{ij} - \sum_i N_i y_{i.}z_{i.} \end{aligned}$$

Thus

$$Q_e^* = \left(\sum_{i,j} y_{ij}^2 - \sum_i N_i y_{i.}^2\right) - \left(\sum_{i,j} y_{ij}z_{ij} - \sum_i N_i y_{i.}z_{i.}\right)\hat{\gamma}$$

However, $\hat{\gamma}$ reduces to

$$\begin{aligned} \hat{\gamma} &= (\mathbf{Z}'\mathbf{QZ})^{-1}\mathbf{Z}'\mathbf{Qy} = (\mathbf{z}'\mathbf{Qz})^{-1}\mathbf{z}'\mathbf{Qy} \\ &= \left(\sum_{i,j} z_{ij}^2 - \sum_i N_i z_{i.}\right)^{-1}\left(\sum_{i,j} z_{ij}y_{ij} - \sum_i N_i y_{i.}z_{i.}\right) \end{aligned} \tag{5.14.22}$$

For convenience, let

$$\begin{aligned} E_{yy} &= \sum_{i,j} y_{ij}^2 - \sum_i N_i y_{i.}^2 \\ E_{zz} &= \sum_{i,j} z_{ij}^2 - \sum_i N_i z_{i.}^2 \\ E_{zy} &= \sum_{i,j} z_{ij}y_{ij} - \sum_i N_i z_{i.}y_{i.} \end{aligned} \tag{5.14.23}$$

Then

$$Q_e^* = E_{yy} - E_{yz}E_{zz}^{-1}E_{zy}$$

and

$$\hat{\gamma} = E_{zz}^{-1}E_{zy}$$

where the degrees of freedom for Q_e^* are $v_e = N - k - 1$. Hence the estimate for σ^2 becomes

$$s_{y|z}^2 = \frac{E_{yy} - E_{yz}E_{zz}^{-1}E_{zy}}{N - k - 1} \tag{5.14.24}$$

To test the hypothesis

$$H_0: \gamma = 0 \tag{5.14.25}$$

(5.14.12) reduces to

$$Q_h = E_{yz}E_{zz}^{-1}E_{zy} \tag{5.14.26}$$

so that

$$F = \frac{E_{yz}E_{zz}^{-1}E_{zy}}{(E_{yy} - E_{yz}(E_{zz})^{-1}E_{zy})/(N-k-1)} \sim F(1, N-k-1)$$

is used to test (5.14.25).

To test the hypothesis

$$H_0: \alpha_1 = \alpha_2 = \cdots = \alpha_k \tag{5.14.27}$$

(5.14.15) may be employed directly with $\mathbf{C}_1$ equal to

$$\underset{[(k-1)\times(k+1)]}{\mathbf{C}_1} = \begin{bmatrix} 0 & 1 & 0 & \cdots & 0 & -1 \\ 0 & 0 & 1 & \cdots & 0 & -1 \\ \vdots & \vdots & \vdots & & \vdots & \vdots \\ 0 & 0 & 0 & \cdots & 1 & -1 \end{bmatrix} \tag{5.14.28}$$

where the $R(\mathbf{C}_1) = k - 1$ and $\boldsymbol{\xi} = \mathbf{0}$.

Alternatively, in the derivation of Q_h^*, $Q_h^* = Q_t^* - Q_e^*$, where $Q_t^* = \min (\mathbf{y} - \mathbf{X}\boldsymbol{\beta} - \mathbf{Z}\boldsymbol{\gamma})'(\mathbf{y} - \mathbf{X}\boldsymbol{\beta} - \mathbf{Z}\boldsymbol{\gamma})$ subject to the hypothesis $\mathbf{C}\boldsymbol{\theta} = \boldsymbol{\xi}$, where $\boldsymbol{\theta}' = [\boldsymbol{\beta}' \quad \boldsymbol{\gamma}']$. Using this procedure, which is the reduction method reviewed in Section 5.2, the reduced model under (5.14.27) becomes

$$y_{ij} = \mu + \gamma_r z_{ij} + \varepsilon_{ij} \tag{5.14.29}$$

which is the simple regression model. From simple linear regression, (1.8.22),

$$\hat{\gamma}_r = \frac{\sum_{i,j} (y_{ij} - y_{..})(z_{ij} - z_{..})}{\sum_{i,j} (z_{ij} - z_{..})^2} \tag{5.14.30}$$

and Q_e under (5.14.29) is Q_t^*; thus

$$\begin{aligned} Q_t^* &= \mathbf{y}'\mathbf{y} - Ny_{..}^2 - \hat{\gamma}_r\mathbf{D}'\mathbf{D}\hat{\gamma}_r \\ &= \sum_{i,j} y_{ij}^2 - Ny_{..}^2 - \frac{\left[\sum_{i,j} (y_{ij} - y_{..})(z_{ij} - z_{..})\right]^2}{\sum_{i,j} (z_{ij} - z_{..})^2} \\ &= \sum_{i,j} (y_{ij} - y_{..})^2 - \left[\sum_{i,j} (y_{ij} - y_{..})(z_{ij} - z_{..})\right]\hat{\gamma}_r \end{aligned} \tag{5.14.31}$$

Allowing

$$T_{yy} = \sum_{i,j} (y_{ij} - y_{..})^2$$

(5.14.32)
$$T_{zz} = \sum_{i,j} (z_{ij} - z_{..})^2$$

$$T_{yz} = \sum_{i,j} (y_{ij} - y_{..})(z_{ij} - z_{..})$$

expressions for $\hat{\gamma}_r$ and Q_t^* become

(5.14.33)
$$\hat{\gamma}_r = T_{zz}^{-1} T_{zy}$$
$$Q_t^* = T_{yy} - T_{yz} T_{zz}^{-1} T_{zy}$$

Since $Q_h^* = Q_t^* - Q_e^*$, a simple expression for computing the hypothesis sum of squares is

(5.14.34)
$$Q_h^* = (T_{yy} - T_{yz}T_{zz}^{-1}T_{zy}) - (E_{yy} - E_{yz}E_{zz}^{-1}E_{zy})$$
$$= \frac{T_{yy} - T_{zy}^2}{T_{zz}} - \frac{E_{yy} - E_{zy}^2}{E_{zz}}$$

for testing (5.14.27). The expressions Q_t^* and Q_e^* are commonly referred to as the adjusted total and the adjusted error sum of squares, respectively. The ANCOVA table, Table 5.14.1, for testing (5.14.27) by the one-way ANCOVA model may now be constructed.

TABLE 5.14.1. ANCOVA Summary Table.

Source	df	SS (Adj)
Between	$k - 1$	$Q_h^* =$ (by subtraction)
Within error	$N - k - 1$	$Q_e^* = E_{yy} - E_{yz}E_{zz}^{-1}E_{zy}$
"Total"	$N - 2$	$Q_t^* = T_{yy} - T_{yz}T_{zz}^{-1}T_{zy}$

The hypothesis of equal treatment effects, adjusting for concomitant variables, is rejected at the significance level α if

(5.14.35)
$$F = \frac{Q_h^*/(k-1)}{Q_e^*/(N-k-1)} > F^{\alpha}(k-1, N-k-1)$$

Confidence intervals for contrasts of the form $\psi = \mathbf{c}'\boldsymbol{\beta}$ are acquired by using formula (5.14.17),

$$V(\mathbf{c}'\hat{\boldsymbol{\beta}}) = \frac{Q_e^*}{v_e}\mathbf{c}'[(\mathbf{X}'\mathbf{X})^- + (\mathbf{X}'\mathbf{X})^-\mathbf{X}'\mathbf{Z}(\mathbf{Z}'\mathbf{Q}\mathbf{Z})^{-1}\mathbf{Z}'\mathbf{X}(\mathbf{X}'\mathbf{X})^-]\mathbf{c}$$

which, for our simple case, becomes

(5.14.36)
$$V(\mathbf{c}'\hat{\boldsymbol{\beta}}) = \frac{Q_e^*}{v_e} \frac{\sum_i c_i^2 \Big/ N_i + \left(\sum_i c_i z_{i.}\right)^2}{\sum_{i,j} (z_{ij} - z_{i.})^2}$$

By employing the general covariance analysis procedure with the model $y_{ij} = \mu + \alpha_i + \gamma z_{ij} + \varepsilon_{ij}$, we have seen how the method is applied when there is only one concomitant variable. The method can be extended to experiments having two or more concomitant variables and ways of classification. For example, the ANCOVA model for a randomized block design with two concomitant variables is of the form

$$y_{ij} = \mu + \alpha_i + \beta_j + \gamma z_{ij} + \delta w_{ij} + \varepsilon_{ij} \tag{5.14.37}$$

For this model, the matrix **Z** has two columns while the design matrix **X** is the same as that obtained from Section 5.5 for analysis of variance. Another frequently used design is

$$y_{ijk} = \mu + \alpha_i + \beta_j + \gamma_{ij} + \delta z_{ij} + \varepsilon_{ij} \tag{5.14.38}$$

which is the model for a completely randomized factorial design with one covariate.

More generally, analysis of covariance models are applicable for Latin-square designs, repeated-measures designs, and many others. Federer (1955, Chap. 16) considers a number of different designs. *Biometrics* (Vol. 13, Sept. 1957) reviews a number of complex applications of analysis of covariance designs.

Analysis of covariance designs, as given in (5.14.37) and (5.14.38), assume that the within-group regression coefficients with each concomitant variable for the k populations are identical. Such designs are special cases of the general covariance-analysis model given in (5.14.1). Designs that do not have the same "slopes," although difficult to interpret, may also be analyzed by using (5.14.1). Searle (1971, p. 356) provides an example.

A useful by-product of the analysis of covariance model is that it may be employed to estimate missing observations, in a least-squares sense, in an analysis-of-variance analysis (see Bartlett, 1937b). Estimates of the missing observations are obtained by using covariance analysis to transform an unbalanced design to a balanced-data analysis (see Seber, 1966, p. 72).

General Multivariate Theory. The model for multivariate covariance analysis is

$$\underset{(N\times p)}{\mathbf{Y}} = \underset{(N\times q)}{\mathbf{X}}\ \underset{(q\times p)}{\mathbf{B}} + \underset{(N\times h)}{\mathbf{Z}}\ \underset{(h\times p)}{\mathbf{\Gamma}}\ \underset{(N\times p)}{\mathbf{E}_0} \tag{5.14.39}$$

or

$$\underset{(N\times p)}{\mathbf{Y}} = \underset{[N\times(q+h)]}{[\mathbf{X} \,\vdots\, \mathbf{Z}]} \underset{[(q+h)\times p]}{\begin{bmatrix}\boldsymbol{\beta}\\ \mathbf{\Gamma}\end{bmatrix}} + \underset{(N\times p)}{\mathbf{E}_0}$$

where $E(\mathbf{E}_0) = \mathbf{0}$ and $V(\mathbf{E}_0) = \mathbf{I} \otimes \mathbf{\Sigma}$. The number of dependent variables in (5.14.39) is p, and the number of concomitant or independent variables is h.

By replacing vectors in the univariate model by matrices in the multivariate model, the following results are gained:

$$\hat{\mathbf{B}} = (\mathbf{X}'\mathbf{X})^{-}\mathbf{X}'\mathbf{Y} - (\mathbf{X}'\mathbf{X})^{-}\mathbf{X}'\mathbf{Z}\hat{\mathbf{\Gamma}} = \hat{\mathbf{B}}_0 - (\mathbf{X}'\mathbf{X})^{-}\mathbf{X}'\mathbf{Z}\hat{\mathbf{\Gamma}}$$

$$\hat{\mathbf{\Gamma}} = (\mathbf{Z}'\mathbf{Q}\mathbf{Z})^{-1}\mathbf{Z}'\mathbf{Q}\mathbf{Y}$$

(5.14.40) $$\mathbf{Q}_e^* = \mathbf{Y}'\mathbf{Y} - \mathbf{Y}'\mathbf{X}(\mathbf{X}'\mathbf{X})^{-}\mathbf{X}'\mathbf{Y} - \mathbf{Y}'\mathbf{Z}\mathbf{Q}(\mathbf{Z}'\mathbf{Q}\mathbf{Z})^{-1}\mathbf{Z}'\mathbf{Q}\mathbf{Y}$$

$$= \mathbf{Q}_e - \hat{\mathbf{\Gamma}}'(\mathbf{Z}'\mathbf{Q}\mathbf{Z})\hat{\mathbf{\Gamma}}$$

$$\mathbf{S}_{Y|Z} = \frac{\mathbf{Q}_e^*}{N - r - h}$$

Using (5.14.8), the variance-covariance matrix for the estimable functions is

$$V(\hat{\mathbf{\Gamma}}) = \mathbf{\Sigma}(\mathbf{Z}'\mathbf{Q}\mathbf{Z})^{-1}$$

(5.14.41) $$V(\mathbf{c}'\hat{\mathbf{\Gamma}}\mathbf{a}) = \mathbf{a}'\mathbf{\Sigma}\mathbf{a}[\mathbf{c}'(\mathbf{Z}'\mathbf{Q}\mathbf{Z})^{-1}\mathbf{c}]$$

$$V(\mathbf{c}'\hat{\mathbf{B}}\mathbf{a}) = \mathbf{a}'\mathbf{\Sigma}\mathbf{a}[\mathbf{c}'(\mathbf{X}'\mathbf{X})^{-} + (\mathbf{X}'\mathbf{X})^{-}\mathbf{X}'\mathbf{Z}(\mathbf{Z}'\mathbf{Q}\mathbf{Z})^{-1}\mathbf{Z}'\mathbf{X}(\mathbf{X}'\mathbf{X})^{-}\mathbf{c}]$$

$$\operatorname{cov}(\mathbf{c}'\hat{\mathbf{B}}\mathbf{a}, \mathbf{c}'\hat{\mathbf{\Gamma}}\mathbf{a}) = -(\mathbf{a}'\mathbf{\Sigma}\mathbf{a})[\mathbf{c}'(\mathbf{X}'\mathbf{X})^{-}\mathbf{X}'\mathbf{Z}(\mathbf{Z}'\mathbf{Q}\mathbf{Z})^{-1}\mathbf{c}]$$

where $\mathbf{\Sigma}$ is estimated by $S_{Y|Z} = \mathbf{Q}_e^*/v_e$.

Testing a hypothesis, employing the multivariate model, proceeds in a straightforward manner using the general theory of Chapter 3, where it is assumed that each row of $\mathbf{Y}$ has a multivariate normal distribution. To test the hypothesis

(5.14.42) $$H_0 : \mathbf{\Gamma} = \mathbf{0}$$

the hypothesis sum of squares becomes

(5.14.43) $$\mathbf{Q}_h = \hat{\mathbf{\Gamma}}'(\mathbf{Z}'\mathbf{Q}\mathbf{Z})\hat{\mathbf{\Gamma}}$$

To obtain $\mathbf{Q}_h$, (5.14.42) is put in the form

$$\mathbf{C}\begin{bmatrix}\mathbf{B}\\ \mathbf{\Gamma}\end{bmatrix}\mathbf{A} = \mathbf{0}$$

by selecting $\mathbf{C}$ as in the univariate case, where

$$\mathbf{A} = \begin{bmatrix}\mathbf{0}\\ \mathbf{I}\end{bmatrix}, \quad v_h = h, \quad \text{and} \quad v_e = N - r - h$$

Solving the characteristic equation

(5.14.44) $$|\mathbf{Q}_h - \mathbf{Q}_e^*| = 0$$

with $\mathbf{Q}_e^*$ defined as in (5.14.40), (2.7.12) and (3.6.18) are invoked to test (5.14.42). The parameters for the criteria are

$$s = \min(v_h, p) = \min(h, p)$$

(5.14.45) $$m = \frac{|u - v_h| - 1}{2} = \frac{|p - h| - 1}{2}$$

$$n = \frac{v_e - u - 1}{2} = \frac{N - r - h - p - 1}{2}$$

where $u = R(\mathbf{A}) = p$, $v_h = h$, and $v_e = N - r - h$.

Testing hypotheses for elements of **B**, usually in the form

$$H_0: \mathbf{C}_1\mathbf{B}\mathbf{A} = 0 \tag{5.14.46}$$

the hypothesis sum of squares becomes

(5.14.47)

$$\mathbf{Q}_h^* = (\mathbf{C}_1\hat{\mathbf{B}}\mathbf{A})'[\mathbf{C}_1(\mathbf{X}'\mathbf{X})^-\mathbf{C}_1' + \mathbf{C}_1(\mathbf{X}'\mathbf{X})^-\mathbf{X}'\mathbf{Z}(\mathbf{Z}'\mathbf{Q}\mathbf{Z})^{-1}\mathbf{Z}'\mathbf{X}(\mathbf{X}'\mathbf{X})^-\mathbf{C}_1']^{-1}(\mathbf{C}_1\hat{\mathbf{B}}\mathbf{A})$$

for $\mathbf{C} = [\mathbf{C}_1 \quad \mathbf{0}]$, where the $R(\mathbf{C}) = g$ and the $R(\mathbf{A}) = u$. Using $\mathbf{Q}_h^*$ and $\mathbf{Q}_e^*$ in (5.14.44), the parameters for the criteria are

$$\begin{aligned} s &= \min(g, u) \\ m &= \frac{|u - g| - 1}{2} \\ n &= \frac{N - r - h - u - 1}{2} \end{aligned} \tag{5.14.48}$$

Simultaneous confidence intervals for estimable functions $\psi = \mathbf{c}'\mathbf{B}\mathbf{a}$ are of the form

$$\hat{\psi} - c_0\hat{\sigma}_{\hat{\psi}} \le \psi \le \hat{\psi} + c_0\hat{\sigma}_{\hat{\psi}} \tag{5.14.49}$$

where

$$\hat{\sigma}_{\hat{\psi}} = \sqrt{\mathbf{a}'\left(\frac{\mathbf{Q}_e^*}{\nu_e}\right)\mathbf{a}[\mathbf{c}'(\mathbf{X}'\mathbf{X})^-\mathbf{c}' + \mathbf{c}'(\mathbf{X}'\mathbf{X})^-\mathbf{X}'\mathbf{Z}(\mathbf{Z}'\mathbf{Q}\mathbf{Z})^{-1}\mathbf{Z}'\mathbf{X}(\mathbf{X}'\mathbf{X})^-\mathbf{c}]}$$

and c_0 is chosen to correspond to the overall criteria used in testing (5.14.46).

As with the univariate case, the general theory is applied most successfully to complex designs; however, for simple designs where formulas are explicitly given, the multivariate analogies are easily determined. For example, by defining

$$\begin{aligned} \mathbf{E}_{YY} &= \sum_{i,j}(\mathbf{y}_{ij} - \mathbf{y}_{i.})(\mathbf{y}_{ij} - \mathbf{y}_{i.})' \\ \mathbf{E}_{ZZ} &= \sum_{i,j}(\mathbf{z}_{ij} - \mathbf{z}_{i.})(\mathbf{z}_{ij} - \mathbf{z}_{i.})' \\ \mathbf{E}_{YZ} &= \sum_{i,j}(\mathbf{y}_{ij} - \mathbf{y}_{i.})(\mathbf{z}_{ij} - \mathbf{z}_{i.})' \end{aligned}$$

which are the multivariate extensions of (5.14.23), $\mathbf{Q}_e^*$ and $\hat{\mathbf{\Gamma}}$ become

$$\begin{aligned} \mathbf{Q}_e^* &= \mathbf{E}_{YY} - \mathbf{E}_{YZ}\mathbf{E}_{ZZ}^{-1}\mathbf{E}_{ZY} \\ \hat{\mathbf{\Gamma}} &= \mathbf{E}_{ZZ}^{-1}\mathbf{E}_{ZY} \end{aligned} \tag{5.14.50}$$

An example of the one-way MANCOVA design with one covariate is given in the next section.

Tests of Nonadditivity. An additive model for the randomized block design was assumed in Section 5.5. A test for nonadditivity or no interaction was first proposed by Tukey (1949) and then generalized by Scheffé (1959, p. 144, prob. 4.19) for the univariate model. Milliken and Graybill (1970, 1971) examine the test of nonadditivity using the expanded linear model (ELM) that includes nonlinear terms with conventional linear model theory. Their results lead to numerous tests for noninteraction models commonly employed in data analysis. A summary of the general theory for obtaining a test of nonadditivity for the RB-c design or the CRF-rc design with one observation per cell is presented.

Using multivariate linear model theory, McDonald and Milliken (1971) extend the results provided by Milliken and Graybill, employing what they call the expanded multiple design multivariate (EMDM) model, the multivariate analogue of the ELM.

For the univariate case, the ELM is an extension of the usual linear model $\mathbf{y} = \mathbf{X}\boldsymbol{\beta} + \boldsymbol{\varepsilon}$ to

(5.14.51)
$$\underset{(N \times 1)}{\mathbf{y}} = \underset{(N \times q)}{\mathbf{X}} \underset{(q \times 1)}{\boldsymbol{\beta}} + \underset{(N \times h)}{\mathbf{F}} \underset{(h \times 1)}{\boldsymbol{\alpha}} + \underset{(N \times 1)}{\boldsymbol{\varepsilon}}$$
$$\boldsymbol{\varepsilon} \sim IN(0, \sigma^2\mathbf{I})$$

where $\mathbf{F} = [f_{ij}(\cdot)]$ is a matrix of known functions of the unknown elements of $\mathbf{X}\boldsymbol{\beta}$, the $R(\mathbf{X}) = r$, the $R[\mathbf{X} \quad \mathbf{F}] = r + h < N$ where $[\mathbf{X} \quad \mathbf{F}]$ is spanned by the rows of $\mathbf{X}\boldsymbol{\beta}$, and $\boldsymbol{\alpha}$ is an unknown vector of parameters.

For Tukey's test of nonadditivity, the design considered in (5.5.4), using (5.14.51), becomes

$$\underset{(6 \times 1)}{\begin{bmatrix} y_{11} \\ y_{12} \\ y_{21} \\ y_{22} \\ y_{31} \\ y_{33} \end{bmatrix}} = \underset{(6 \times 6)}{\begin{bmatrix} 1 & 1 & 0 & 0 & 1 & 0 \\ 1 & 1 & 0 & 0 & 0 & 1 \\ 1 & 0 & 1 & 0 & 1 & 0 \\ 1 & 0 & 1 & 0 & 0 & 1 \\ 1 & 0 & 0 & 1 & 1 & 0 \\ 1 & 0 & 0 & 1 & 0 & 1 \end{bmatrix}} \underset{(6 \times 1)}{\begin{bmatrix} \mu \\ \alpha_1 \\ \alpha_2 \\ \alpha_3 \\ \beta_1 \\ \beta_2 \end{bmatrix}} + \underset{(6 \times 1)}{\begin{bmatrix} \alpha_1\beta_1 \\ \alpha_1\beta_2 \\ \alpha_2\beta_1 \\ \alpha_2\beta_2 \\ \alpha_3\beta_1 \\ \alpha_3\beta_2 \end{bmatrix}} \underset{(1 \times 1)}{\alpha} + \underset{(6 \times 1)}{\begin{bmatrix} \varepsilon_{11} \\ \varepsilon_{12} \\ \varepsilon_{21} \\ \varepsilon_{22} \\ \varepsilon_{31} \\ \varepsilon_{32} \end{bmatrix}}$$

which represents the model $y_{ij} = \mu + \alpha_i + \beta_j + \alpha_i\beta_j + \varepsilon_{ij}$, where $\varepsilon_{ij} \sim IN(0, \sigma^2)$. A test of the hypothesis $H_0: \alpha = 0$ against the alternative hypothesis $H_1: \alpha \neq 0$ assesses the significance of the interaction term $\alpha_i\beta_j$. If $\alpha = 0$, the additive model would be taken as satisfactory.

Since the elements of $\mathbf{F}$ can be chosen to be any known function of the elements of $\mathbf{X}\boldsymbol{\beta}$, a test of the interaction hypothesis

(5.14.52)
$$H_0: \boldsymbol{\alpha} = \mathbf{0}$$

is of interest when employing the ELM given in (5.14.51).

To test (5.14.52), the vector $\boldsymbol{\alpha}$ must be estimable. From the general covariance model, $\boldsymbol{\alpha}$ is estimable if and only if the $R\{[\mathbf{I} - \mathbf{X}(\mathbf{X}'\mathbf{X})^-\mathbf{X}']\mathbf{F}\} = h$ since

(5.14.53)
$$\hat{\boldsymbol{\alpha}} = (\mathbf{F}'\mathbf{Q}\mathbf{F})^{-1}\mathbf{F}'\mathbf{Q}\mathbf{y}$$

where $\mathbf{Q} = \mathbf{I} - \mathbf{X}(\mathbf{X}'\mathbf{X})^-\mathbf{X}'$, when using the model $\mathbf{y} = \mathbf{Q}\mathbf{F}\boldsymbol{\alpha} + \boldsymbol{\varepsilon}$.

Assuming $\mathbf{F}$ is known, a test of $H_0 : \boldsymbol{\alpha} = \mathbf{0}$ is found to be similar to the theory developed for the general covariance model. That is, the statistic for testing (5.14.52) is

$$F = \frac{\hat{\boldsymbol{\alpha}}'\mathbf{F}'\mathbf{QF}\hat{\boldsymbol{\alpha}}/h}{Q_e^*/(N-r-h)} \tag{5.14.54}$$

where $Q_e^* = Q_e - \hat{\boldsymbol{\alpha}}'\mathbf{F}'\mathbf{QF}\hat{\boldsymbol{\alpha}}$. By substituting for $\hat{\boldsymbol{\alpha}}$ in (5.14.54), the numerator and denominator sum of squares become, respectively,

$$\begin{aligned} Q_h &= \mathbf{y}'\mathbf{Q}'\mathbf{F}(\mathbf{F}'\mathbf{QF})^{-1}\mathbf{F}'\mathbf{Qy} \\ Q_e^* &= \mathbf{y}'[\mathbf{I} - \mathbf{X}(\mathbf{X}'\mathbf{X})^{-}\mathbf{X}' - \mathbf{Q}'\mathbf{F}(\mathbf{F}'\mathbf{QF})^{-1}\mathbf{F}'\mathbf{Q}]\mathbf{y} \\ &= Q_e - Q_h \end{aligned} \tag{5.14.55}$$

where $Q_e = \mathbf{y}'[\mathbf{I} - \mathbf{X}(\mathbf{X}'\mathbf{X})^{-}\mathbf{X}']\mathbf{y}$.

The matrix $\mathbf{F}$ in (5.14.54) is unknown and may not be used directly to test (5.14.52); however, the elements of $\mathbf{F}$ are known functions of the unknown quantities of $\mathbf{X}\boldsymbol{\beta}$, thus $\mathbf{F}$ is replaced by $\hat{\mathbf{F}}$, where $\hat{\mathbf{F}}$ is procured by substituting $\mathbf{X}(\mathbf{X}'\mathbf{X})^{-}\mathbf{X}'\mathbf{y}$ for $\mathbf{X}\hat{\boldsymbol{\beta}}$ in $f_{ij}(\mathbf{X}\hat{\boldsymbol{\beta}})$. That is, $\mathbf{X}\hat{\boldsymbol{\beta}}$ is substituted for $\mathbf{X}\boldsymbol{\beta}$ where $\hat{\boldsymbol{\beta}}$ is a least-squares estimator of $\boldsymbol{\beta}$ when H_0 is true. Furthermore, the matrix $[\mathbf{X} \quad \hat{\mathbf{F}}]$ is of rank $r + h$ with probability 1. The test statistic becomes

$$\hat{F} = \frac{\mathbf{y}'\mathbf{Q}'\hat{\mathbf{F}}(\hat{\mathbf{F}}'\mathbf{Q}\hat{\mathbf{F}})^{-1}\hat{\mathbf{F}}'\mathbf{Qy}/h}{[\mathbf{y}'\mathbf{Qy} - \mathbf{y}'\mathbf{Q}'\hat{\mathbf{F}}(\hat{\mathbf{F}}'\mathbf{Q}\hat{\mathbf{F}})^{-1}\hat{\mathbf{F}}'\mathbf{Qy}]/(N-r-h)} = \frac{\hat{Q}_h/h}{\hat{Q}_e^*/(N-r-h)} \tag{5.14.56}$$

The hypothesis given in (5.14.52) is rejected at the significance level α if

$$\hat{F} > F^{\alpha}(h, N-r-h)$$

To illustrate the above procedure, Tukey's test for nonadditivity is derived for the randomized block design. Since $\hat{\mathbf{F}}$ is a vector $\hat{\mathbf{f}}$ (say) and not a matrix in this case, (5.14.56) is simplified. The hypothesis sum of squares $\hat{Q}_h$ becomes

$$\begin{aligned} \hat{Q}_h &= \mathbf{y}'\mathbf{Q}'\hat{\mathbf{f}}(\hat{\mathbf{f}}'\mathbf{Q}\hat{\mathbf{f}})^{-1}\hat{\mathbf{f}}'\mathbf{Qy} \\ &= \frac{\mathbf{y}'\mathbf{Q}'\hat{\mathbf{f}}\hat{\mathbf{f}}'\mathbf{Qy}}{\hat{\mathbf{f}}'\mathbf{Q}\mathbf{f}} \\ &= \frac{\{\mathbf{y}'[\mathbf{I} - \mathbf{X}(\mathbf{X}'\mathbf{X})^{-}\mathbf{X}']\hat{\mathbf{f}}\}^2}{\hat{\mathbf{f}}'[\mathbf{I} - \mathbf{X}(\mathbf{X}'\mathbf{X})^{-}\mathbf{X}']\hat{\mathbf{f}}} \end{aligned}$$

where the error sum of squares is

$$\hat{Q}_e^* = \mathbf{y}'[\mathbf{I} - \mathbf{X}(\mathbf{X}'\mathbf{X})^{-}\mathbf{X}']\mathbf{y} - \hat{Q}_h$$

From Section 5.5,

$$\mathbf{y}'[\mathbf{I} - \mathbf{X}(\mathbf{X}'\mathbf{X})^{-}\mathbf{X}']\mathbf{y} = \sum_i \sum_j (y_{ij} - y_{i.} - y_{.j} + y_{..})^2$$

The general element of $\mathbf{X}\hat{\boldsymbol{\beta}}$, which is not dependent on the choice of the g inverse $(\mathbf{X}'\mathbf{X})^{-}$ because $\mathbf{X}(\mathbf{X}'\mathbf{X})^{-}\mathbf{X}'$ is unique, is from (5.5.11): $-y_{..} + y_{i.} + y_{.j}$. Hence the general element in the vector $\hat{\mathbf{f}}$ is $\hat{\alpha}_i\hat{\beta}_j = y_{i.}y_{.j}$. Alternatively, by implementing the relationship $-y_{..} - y_{i.} + y_{.j} = y_{..} + (y_{i.} - y_{..}) + (y_{.j} - y_{..})$, the general element in the vector $\hat{\mathbf{f}}$ would be $\hat{\alpha}_i\hat{\beta}_j = (y_{i.} - y_{..})(y_{.j} - y_{..})$. In either case, the equations for $\hat{Q}_h$ and $\hat{Q}_e^*$ can be reduced to

$$\hat{Q}_h = \frac{\left[\sum_i \sum_j y_{ij}(y_{i.} - y_{..})(y_{.j} - y_{..})\right]^2}{\sum_i (y_{i.} - y_{..})^2 \sum_j (y_{.j} - y_{..})^2}$$

$$\hat{Q}_e^* = \sum_i \sum_j (y_{ij} - y_{i.} - y_{.j} + y_{..})^2 - \hat{Q}_h$$

and

$$\hat{F} = \frac{\hat{Q}_h/h}{\hat{Q}_e^*/[N - R(X) - h]} = \frac{\hat{Q}_h}{\hat{Q}_e^*/[(r-1)(c-1)-1]} \sim F(1, [(r-1)(c-1)-1])$$

where r denotes the number of rows (blocks) and c the number of columns (treatments) in a CRF-rc (RB-k) design with one observation per cell.

The EMDM model is a simple extension of the ELM. Following the univariate case, the multivariate extension of (5.14.51) is

(5.14.57)
$$\underset{(N\times p)}{\mathbf{Y}} = \underset{(N\times q)}{\mathbf{X}}\ \underset{(q\times p)}{\mathbf{B}} + \underbrace{[\mathbf{F}_1\ \ \mathbf{F}_2\ \ \cdots\ \ \mathbf{F}_p]}_{(N\times h)} \underbrace{\begin{bmatrix} \boldsymbol{\alpha}_1 & \mathbf{0} & \cdots & \mathbf{0} \\ \mathbf{0} & \boldsymbol{\alpha}_2 & \cdots & \mathbf{0} \\ \vdots & \vdots & \ddots & \vdots \\ \mathbf{0} & \mathbf{0} & \cdots & \boldsymbol{\alpha}_p \end{bmatrix}}_{(h\times p)} + \underset{(N\times p)}{\mathbf{E}_0}$$

where $\boldsymbol{\alpha}_j$ is a vector of unknown parameters, $\mathbf{F}_j$ is a $N \times k_j$ matrix of known functions of the unknown elements of $\mathbf{X}\boldsymbol{\beta}_j$, and $\boldsymbol{\beta}_j$ is the jth column of the matrix $\mathbf{B}$. Letting $h = \Sigma_j k_j$ and r denote the rank of the matrix $\mathbf{X}$, the rank of the matrix $[\mathbf{X}\ \ \mathbf{F}]$ is $r + h < N$ in the vector space spanned by the rows of $\mathbf{XB}$ and $\mathbf{F} = [\mathbf{F}_1\ \ \mathbf{F}_2\ \ \cdots\ \ \mathbf{F}_p]$.

By assuming that each row of the matrix $\mathbf{Y}$ is sampled from a p-variate normal distribution with common variance-covariance matrix $\boldsymbol{\Sigma}$, the interaction hypothesis becomes

(5.14.58)
$$H_0 : \boldsymbol{\alpha}_j = \mathbf{0} \qquad j = 1, \ldots, p$$

In the univariate case, the ELM had the form of a general linear univariate model

$$\mathbf{y} = [\mathbf{X}\ \ \mathbf{F}]\begin{bmatrix}\boldsymbol{\beta} \\ \boldsymbol{\alpha}\end{bmatrix} + \boldsymbol{\varepsilon}$$

This does not hold in the multivariate situation. The standard linear model, using the notation of Section 3.8, is

(5.14.59)
$$\underset{\mathbf{y}^*}{\begin{bmatrix}\mathbf{y}_1 \\ \mathbf{y}_2 \\ \vdots \\ \mathbf{y}_p\end{bmatrix}} = \underset{(\mathbf{I}\otimes\mathbf{X})}{\begin{bmatrix}\mathbf{X} & \mathbf{0} & \cdots & \mathbf{0} \\ \mathbf{0} & \mathbf{X} & \cdots & \mathbf{0} \\ \vdots & \vdots & \ddots & \vdots \\ \mathbf{0} & \mathbf{0} & \cdots & \mathbf{X}\end{bmatrix}} \underset{\boldsymbol{\beta}^*}{\begin{bmatrix}\boldsymbol{\beta}_1 \\ \boldsymbol{\beta}_2 \\ \vdots \\ \boldsymbol{\beta}_p\end{bmatrix}} + \underset{\boldsymbol{\varepsilon}^*}{\begin{bmatrix}\boldsymbol{\varepsilon}_1 \\ \boldsymbol{\varepsilon}_2 \\ \vdots \\ \boldsymbol{\varepsilon}_p\end{bmatrix}}$$

where

$$\begin{aligned} E(\mathbf{Y}_j) &= \mathbf{X}\boldsymbol{\beta}_j \\ \text{cov}(\mathbf{Y}_j, \mathbf{Y}_{j'}) &= \sigma_{jj'}\mathbf{I}_N \end{aligned} \qquad j = 1, 2, \ldots, p$$

By employing the standard multivariate linear model, the design matrix $\mathbf{X}$ is common to all p responses. Alternatively, using (5.14.59), (5.14.57) is

$$\text{(5.14.60)} \qquad \begin{bmatrix} \mathbf{y}_1 \\ \mathbf{y}_2 \\ \vdots \\ \mathbf{y}_p \end{bmatrix} = \begin{bmatrix} \mathbf{XF}_1 & \mathbf{0} & \cdots & \mathbf{0} \\ \mathbf{0} & \mathbf{XF}_2 & \cdots & \mathbf{0} \\ \vdots & \vdots & & \vdots \\ \mathbf{0} & \mathbf{0} & \cdots & \mathbf{XF}_p \end{bmatrix} \begin{bmatrix} \boldsymbol{\beta}_1 \\ \boldsymbol{\alpha}_1 \\ \boldsymbol{\beta}_2 \\ \boldsymbol{\alpha}_2 \\ \vdots \\ \boldsymbol{\beta}_p \\ \boldsymbol{\alpha}_p \end{bmatrix} + \begin{bmatrix} \boldsymbol{\varepsilon}_1 \\ \boldsymbol{\varepsilon}_2 \\ \vdots \\ \boldsymbol{\varepsilon}_p \end{bmatrix}$$

where

$$\begin{aligned} E(\mathbf{Y}_j) &= [\mathbf{X} \quad \mathbf{F}_j]\begin{bmatrix} \boldsymbol{\beta}_j \\ \boldsymbol{\alpha}_j \end{bmatrix} \\ \text{cov}(\mathbf{Y}_j, \mathbf{Y}_j') &= \sigma_{jj'}\mathbf{I}_N \end{aligned} \qquad j = 1, \ldots, p$$

so that the design matrix $[\mathbf{X} \quad \mathbf{F}_j]$ is *different* for each dependent variable. However, each response for $j = 1, \ldots, p$ does follow the ELM.

When the elements of $\mathbf{F}_j$, $j = 1, \ldots, p$, are known and the design matrix for each variable is $[\mathbf{X}_j \quad \mathbf{F}_j]$, the design is called a *multiple-design multivariate model*, an MDM model. In this text we have only considered the general theory for a multivariate model that has the same design matrix. Srivastava (1966, 1967) considers MDM models and some of their applications.

For the multivariate test of nonadditivity, the matrix $\mathbf{X}$ is the same for each vector $\mathbf{y}_j, j = 1, \ldots, p$, and the multivariate extension follows easily by analogy. For further details, see McDonald and Milliken (1971).

To test the multivariate interaction hypothesis specified in (5.14.58), the following hypothesis and error sum of squares and cross-products matrices are formed

$$\text{(5.14.61)} \qquad \begin{aligned} \hat{\mathbf{Q}}_h &= \mathbf{Y}'\mathbf{Q}\hat{\mathbf{F}}(\hat{\mathbf{F}}'\hat{\mathbf{Q}}\hat{\mathbf{F}})^{-1}\hat{\mathbf{F}}'\mathbf{QY} \\ \hat{\mathbf{Q}}_e^* &= \mathbf{Y}'\mathbf{QY} - \hat{\mathbf{Q}}_h = \mathbf{Q}_e - \hat{\mathbf{Q}}_h \end{aligned}$$

where the jth matrix of the matrix $\mathbf{F}$ is obtained by substituting $\mathbf{X}(\mathbf{X}'\mathbf{X})^-\mathbf{X}'\mathbf{y}_j$ for $\mathbf{X}\hat{\boldsymbol{\beta}}_j$ in the matrix $\hat{\mathbf{F}}_j = [f_{ikj}(\mathbf{X}\hat{\boldsymbol{\beta}}_j)]$, for $j = 1, \ldots, p$. That is, the product $\hat{\mathbf{B}} = (\mathbf{X}'\mathbf{X})^-\mathbf{X}'\mathbf{Y}$ is formed and each matrix $\hat{\mathbf{F}}_j$ is constructed as in the univariate case by using the jth column $\mathbf{X}\hat{\boldsymbol{\beta}}_j$ of the matrix $\mathbf{X}\hat{\mathbf{B}}$.

When the null hypothesis $H_0 : \boldsymbol{\alpha}_j = 0$ for all j is true, then

$$\begin{aligned} \hat{\mathbf{Q}}_h &\sim W_p(h, \boldsymbol{\Sigma}) \\ \hat{\mathbf{Q}}_e &\sim W_p(N - r - h, \boldsymbol{\Sigma}) \end{aligned}$$

and $\hat{\mathbf{Q}}_h$ and $\hat{\mathbf{Q}}_e^*$ are independent. If λ_i is a root of the characteristic equation $|\hat{\mathbf{Q}}_h - \lambda\hat{\mathbf{Q}}_e^*| = 0$ and the parameters s, m, and n are defined by

(5.14.62)
$$s = \min(h, p)$$
$$m = \frac{|p - h| - 1}{2}$$
$$n = \frac{N - r - h - p - 1}{2}$$

then the interaction hypothesis is rejected at the significance level α if

(5.14.63)
$$\Lambda = \prod_{i=1}^{s} (1 + \lambda_i)^{-1} < U^{\alpha}(p, h, N - r - h) \qquad \text{(Wilks)}$$
$$\theta_s = \frac{\lambda_1}{1 + \lambda_1} > \theta^{\alpha}(s, m, n) \qquad \text{(Roy)}$$

The other criteria follow similarly.

For the MRB-k design considered in Section 5.5, (5.14.57) is represented by

(5.14.64)

$$\begin{bmatrix} y_{111} & y_{112} \\ y_{121} & y_{122} \\ y_{211} & y_{212} \\ y_{221} & y_{222} \\ y_{311} & y_{312} \\ y_{321} & y_{322} \end{bmatrix} = \begin{bmatrix} 1 & 1 & 0 & 0 & 1 & 0 \\ 1 & 1 & 0 & 0 & 0 & 1 \\ 1 & 0 & 1 & 0 & 1 & 0 \\ 1 & 0 & 1 & 0 & 0 & 1 \\ 1 & 0 & 0 & 1 & 1 & 0 \\ 1 & 0 & 0 & 1 & 0 & 1 \end{bmatrix} \begin{bmatrix} \mu_{11} & \mu_{12} \\ \alpha_{11} & \alpha_{12} \\ \alpha_{21} & \alpha_{22} \\ \alpha_{31} & \alpha_{32} \\ \beta_{11} & \beta_{12} \\ \beta_{21} & \beta_{23} \end{bmatrix} + \begin{bmatrix} \alpha_{11}\beta_{11} & \alpha_{12}\beta_{12} \\ \alpha_{11}\beta_{21} & \alpha_{12}\beta_{22} \\ \alpha_{21}\beta_{11} & \alpha_{22}\beta_{12} \\ \alpha_{21}\beta_{21} & \alpha_{22}\beta_{22} \\ \alpha_{31}\beta_{11} & \alpha_{32}\beta_{12} \\ \alpha_{31}\beta_{21} & \alpha_{32}\beta_{22} \end{bmatrix} \begin{bmatrix} \alpha_1 & 0 \\ 0 & \alpha_2 \end{bmatrix} + \mathbf{E}_0$$

$$\underset{(6\times 2)}{\mathbf{Y}} = \underset{(6\times 6)}{\mathbf{X}} \; \underset{(6\times 2)}{\mathbf{B}} + \underbrace{[\mathbf{F}_1 \;\; \mathbf{F}_2]}_{(6\times 2)} \underbrace{\begin{bmatrix} \alpha_1 & \mathbf{0} \\ \mathbf{0} & \alpha_2 \end{bmatrix}}_{(2\times 2)} + \underset{(6\times 2)}{\mathbf{E}_0}$$

Since $\mathbf{F} = [\mathbf{F}_1 \;\; \mathbf{F}_2]$ is a matrix in this case and not a vector, Tukey's multivariate test of nonadditivity is tested by employing (5.14.61).

EXERCISES 5.14

1. Derive the results given in equations (5.14.9).
2. Modify the model given in (5.14.18) to

$$y_{ij} = \mu + \alpha_i + \gamma_i z_{ij} + \varepsilon_{ij} \qquad i = 1, \ldots, k$$
$$\varepsilon_{ij} \sim IN(0, \sigma^2) \qquad j = 1, \ldots, n_0$$

and derive an expression to test the hypothesis H_0: all γ_i's are equal. Use the data given by Kirk (1968, p. 466) and summarized below to illustrate your derived formula with the data provided.

T_1		T_2		T_3	
Y	Z	Y	Z	Y	Z
3	42	4	47	7	61
6	57	5	49	8	65
3	33	4	42	7	64
3	47	3	41	6	56
1	32	2	38	5	52
2	35	3	43	6	58
2	33	4	48	5	53
2	39	3	45	6	54

3. Derive the analysis-of-variance table for the ANCOVA model given in (5.14.37), where the indices i and j are defined by $i = 1, \dots, k$ and $j = 1, \dots, n_0$. Include in your table tests of the hypotheses H_γ: all γ's equal zero, H_δ: all δ's equal zero, and $H_{\gamma,\delta}$: all γ's and δ's equal zero.

4. Use the univariate model as a guide to derive the results given in (5.14.40), (5.14.41), and (5.14.50). How would you express the SSP matrix for the one-way MANCOVA model $y_{ij} = \mu + \alpha_i + \gamma z_{ij} + \varepsilon_{ij}$ to test the hypothesis H_0: all α_i's are equal?

5. Use only the dependent variable labeled F for the data in Exercise 3, Section 5.9 to illustrate Tukey's test for nonadditivity. For the same data, use both dependent variables to carry out a multivariate test for nonadditivity.

5.15 EXAMPLE—ONE-WAY MULTIVARIATE ANALYSIS OF COVARIANCE

To illustrate the one-way MANCOVA design with one covariate, Rohwer's data for high- and low-socioeconomic-state data will be used. The experimental setting and data for his study were given in Sections 4.3, 4.7, and 4.13. The dependent variables were PEA, RAV, and SAT. Using as a covariate the independent variable SS, analysis employing the simple one-way MANCOVA model is presented. Table 5.15.1 includes the means for the data. The model for this example is

$$\underset{(69\times 3)}{\mathbf{Y}} \quad \underset{(69\times 2)}{\mathbf{X}}\ \underset{(2\times 3)}{\mathbf{B}} + \underset{(69\times 1)}{\mathbf{Z}}\ \underset{(1\times 3)}{\mathbf{\Gamma}} + \underset{(69\times 3)}{\mathbf{E}_0} \tag{5.15.1}$$

so that the number of dependent variables is $p = 3$, and the number of covariates is $h = 1$.

TABLE 5.15.1. Summary Data: MANCOVA Example.

Socioeconomic Status	SAT	PEA	RAV	SS
High ($N_1 = 32$)	47.6563	83.0938	15.0000	21.4688
Low ($N_2 = 37$)	31.2703	62.6486	13.2432	18.3784

In Chapter 3, it was indicated that, for the one-way MANOVA model, the design matrix **X** could be of full rank or less than full rank, depending on whether the overall mean was excluded or included in the parameter matrix **B**. Since the case of less than full rank has been thoroughly discussed, this example will select **X** such that **X** is of full rank. That is,

(5.15.2)
$$\underset{(69\times 2)}{\mathbf{X}} = \begin{bmatrix} \underset{(32\times 1)}{\mathbf{1}} & \mathbf{0} \\ \mathbf{0} & \underset{(37\times 1)}{\mathbf{1}} \end{bmatrix}$$

which is the natural representation for the one-way model. The parameter matrix **B** is defined as

(5.15.3)
$$\underset{(2\times 3)}{\mathbf{B}} = \begin{bmatrix} u_{11} & u_{12} & u_{13} \\ u_{21} & u_{22} & u_{23} \end{bmatrix}$$

Z is a vector of covariates for the variable SS, and **Γ** is a vector of regression coefficients

(5.15.4)
$$\underset{(1\times 3)}{\mathbf{\Gamma}} = [\gamma_{11} \quad \gamma_{12} \quad \gamma_{13}]$$

By employing equations (5.14.40),

(5.15.5)
$$\hat{\mathbf{\Gamma}} = [1.3327 \quad 1.1580 \quad .1197]$$

$$\hat{\mathbf{B}} = \begin{bmatrix} 19.0454 & 58.2352 & 12.4301 \\ 6.7779 & 41.3684 & 11.0433 \end{bmatrix}$$

$$\mathbf{Q}_e^* = \begin{bmatrix} 52{,}158.7155 & & \\ 6491.9578 & 8583.1911 & \\ 1579.9571 & 619.8949 & 567.7016 \end{bmatrix}$$

$\hat{\mathbf{\Gamma}}$ is the pooled estimator of **Γ**, given that the hypothesis of parallelism, $H_0: \mathbf{\Gamma}_1 = \mathbf{\Gamma}_2$, is tenable. Methods for checking this assumption were discussed in Chapter 4. The matrix $\hat{\mathbf{B}}$ is often referred to as the matrix of adjusted means, and $\mathbf{Q}_e^*$ is the sum of squares and cross-products matrix after eliminating the set of covariates. The degrees of freedom associated with $\mathbf{Q}_e^*$ are $v_e = N - r - h = 69 - 2 - 1 = 66$.

The first hypothesis of interest in multivariate analysis of covariance is determining whether some or all covariates may be eliminated from the model representation. For the one-way design, this is stated as

(5.15.6)
$$H_0: \mathbf{\Gamma} = \mathbf{0}$$

To test (5.15.6), the matrices **C** and **A** become

(5.15.7)
$$\mathbf{C} = [0, 0, 1] \quad \text{and} \quad \mathbf{A} = \begin{bmatrix} 1 & 0 & 0 \\ 0 & 1 & 0 \\ 0 & 0 & 1 \end{bmatrix}$$

where $v_h = h = 1$ and $v_e = 66$. With **C** and **A** defined by (5.15.7),

$$\mathbf{Q}_h = \hat{\mathbf{\Gamma}}\mathbf{Z}'\mathbf{Q}\mathbf{Z}\hat{\mathbf{\Gamma}}'$$

$$= \begin{bmatrix} 4103.8006 & & \\ 3565.5870 & 8583.1911 & \\ 1579.9571 & 619.8949 & 567.7016 \end{bmatrix}$$

Having $\mathbf{Q}_e^*$ defined in (5.15.5), the Λ criterion to test (5.15.6) is

$$\Lambda = \frac{|\mathbf{Q}_e^*|}{|\mathbf{Q}_e^* + \mathbf{Q}_h|} = .7277 < U^{.05}_{(3,1,66)} = .881$$

for $\alpha = .05$. Alternatively, using Roy's criterion,

$$\theta_s = .2723 > \theta^{.05}_{(1,1/2,31)}$$

Thus the hypothesis is rejected.

Proceeding, a test for differences between groups, eliminating the linear association of $\mathbf{Y}$ on $\mathbf{Z}$, is made. That is,

$$H_0: \begin{bmatrix} \mu_{11} \\ \mu_{12} \\ \mu_{13} \end{bmatrix} = \begin{bmatrix} \mu_{21} \\ \mu_{22} \\ \mu_{23} \end{bmatrix} \tag{5.15.8}$$

To test (5.15.8), the matrices $\mathbf{C}$ and $\mathbf{A}$ are

$$\mathbf{C} = [1, -1, 0] \quad \text{and} \quad \mathbf{A} = \begin{bmatrix} 1 & 0 & 0 \\ 0 & 1 & 0 \\ 0 & 0 & 1 \end{bmatrix} \tag{5.15.9}$$

Evaluating (5.14.47),

$$\mathbf{Q}_h^* = \begin{bmatrix} 2411.3378 & & \\ 3315.3807 & 4588.3612 & \\ 272.5991 & 374.8001 & 30.8170 \end{bmatrix} \tag{5.15.10}$$

Using Roy's criterion, the determinantal equation $|\mathbf{Q}_h^* - \lambda \mathbf{Q}_e^*| = 0$ is solved. The eigenvalue is $\lambda_1 = .5322$, so that

$$\theta_s = \frac{\lambda_1}{1 + \lambda_1} = .3473 > \theta^{.05}_{(1,1/2,31)} = .119$$

which is significant for $\alpha = .05$.

To determine confidence intervals for contrasts in the parameters, which are contrasts in the adjusted means, expression (5.14.49) must be evaluated. For this study, the primary contrasts of interest were

$$\begin{aligned} \psi_1 &= \mu_{11} - \mu_{21} & \hat{\psi}_1 &= 12.27 \\ \psi_2 &= \mu_{12} - \mu_{22} & \hat{\psi}_2 &= 16.87 \\ \psi_3 &= \mu_{13} - \mu_{23} & \hat{\psi}_3 &= 1.38 \end{aligned}$$

By evaluating (5.14.49), the standard errors for the contrasts may be shown to be

$$\begin{aligned} \hat{\sigma}_{\hat{\psi}_1} &= 7.0229 \\ \hat{\sigma}_{\hat{\psi}_2} &= 2.8489 \\ \hat{\sigma}_{\hat{\psi}_3} &= .7327 \end{aligned}$$

With c_0 defined by $c_0^2 = \nu_e[\theta^\alpha/(1 - \theta^\alpha)] = 8.915$, the following intervals for the contrasts are obtained:

$$-8.70 \leq \psi_1 \leq 33.24$$

$$8.36 \leq \psi_2 \leq 25.38$$

$$-.81 \leq \psi_3 \leq 3.57$$

Thus the main differences between groups seem to be due to the dependent variable PEA. As in the case of MANOVA analysis, discriminant analysis in MANCOVA designs may also be pursued by solving the characteristic equation $|\mathbf{Q}_h^* - \lambda \mathbf{Q}_e^*| = 0$ to provide further insight into an analysis.

EXERCISES 5.15

1. In an experiment designed to study two new reading and mathematics programs in the fourth grade, subjects in the school were randomly assigned to three treatment conditions, one being the old program and two being experimental programs. Before beginning the experiment, a test was administered to obtain grade-equivalent reading and mathematics levels for the subjects, labeled R_1 and M_1, respectively, in the table below. At the end of the study, 6 months later, similar data (R_2 and M_2) were obtained for each subject.

Control				Experimental				Experimental			
Y		*Z*		*Y*		*Z*		*Y*		*Z*	
R_2	M_2	R_1	M_1	R_2	M_2	R_1	M_1	R_2	M_2	R_1	M_1
4.1	5.3	3.2	4.7	5.5	6.2	5.1	5.1	6.1	7.1	5.0	5.1
4.6	5.0	4.2	4.5	5.0	7.1	5.3	5.3	6.3	7.0	5.2	5.2
4.8	6.0	4.5	4.6	6.0	7.0	5.4	5.6	6.5	6.2	5.3	5.6
5.4	6.2	4.6	4.8	6.2	6.1	5.6	5.7	6.7	6.8	5.4	5.7
5.2	6.1	4.9	4.9	5.9	6.5	5.7	5.7	7.0	7.1	5.8	5.9
5.7	5.9	4.8	5.0	5.2	6.8	5.0	5.8	6.5	6.9	4.8	5.1
6.0	6.0	4.9	5.1	6.4	7.1	6.0	5.9	7.1	6.7	5.9	6.1
5.9	6.1	5.0	6.0	5.4	6.1	5.0	4.9	6.9	7.0	5.0	4.8
4.6	5.0	4.2	4.5	6.1	6.0	5.5	5.6	6.7	6.9	5.6	5.1
4.2	5.2	3.3	4.8	5.8	6.4	5.6	5.5	7.2	7.4	5.7	6.0

   a. Is there any reason to believe that the programs differ?

   b. Summarize your findings.

5.16 GROWTH-CURVE ANALYSIS

The analysis of multivariate data using profile-analysis procedures, discussed in Sections 3.15 and 5.12, does not require that the commensurable responses for a subject be ordered. However, in looking at the example used in Section 5.13 (Figure 5.13.1), it is apparent that the means follow a specific trend. Analysis of data containing a natural order for the responses is often termed *trend analysis* or *growth-curve analysis* since a polynomial is fitted to the data.

Although growth-curve models have been studied by many authors, including Wishart (1938), Elston and Grizzle (1962), and Bock (1963), among others, the work of Potthoff and Roy (1964), Rao (1959, 1965b, 1966a, 1967), and Khatri

(1966) has been essential to the development of growth-curve models using a multivariate approach. Traditionally, the mixed-model approach has been employed (see Gaito and Wiley, 1963).

The repeated-measures design for p commensurable responses was presented in Section 5.12 as

$$
\begin{aligned}
E(\underset{(N\times p)}{\mathbf{Y}}) &= \underset{(N\times I)}{\mathbf{X}}\ \underset{(I\times p)}{\mathbf{B}} \\
V(\mathbf{Y}) &= \mathbf{I}\otimes\mathbf{\Sigma}
\end{aligned}
\tag{5.16.1}
$$

where each row of $\mathbf{Y}$ has a multivariate normal distribution with mean vector $\boldsymbol{\mu}_i$ and variance-covariance matrix $\mathbf{\Sigma}$.

The standard MANOVA model in (5.16.1) has been generalized by Potthoff and Roy (1964) as follows

$$
\begin{aligned}
E(\underset{(N\times q)}{\mathbf{Y}_0}) &= \underset{(N\times I)}{\mathbf{X}}\ \underset{(I\times p)}{\mathbf{B}}\ \underset{(p\times q)}{\mathbf{Q}} \\
V(\mathbf{Y}_0) &= \mathbf{I}\otimes\mathbf{\Sigma}
\end{aligned}
\tag{5.16.2}
$$

by appending a postmatrix $\mathbf{Q}$ of order $p \times q$, where the $R(\mathbf{Q}) = p$ and $p \leq q$, to the original model. Again, it is assumed that each row of the $N \times q$ data matrix $\mathbf{Y}_0$ is distributed independently and follows a multivariate normal distribution with variance-covariance matrix $\mathbf{\Sigma}_0$ of order $q \times q$. Model (5.16.2) is especially applicable to growth-curve or trend analysis in experiments with ordered, commensurable repeated measures. It also provides procedures for testing hypotheses of the form

$$
H_0\colon \underset{(g\times I)}{\mathbf{C}}\ \underset{(I\times p)}{\mathbf{B}}\ \underset{(p\times u)}{\mathbf{A}} = \underset{(g\times u)}{\mathbf{\Gamma}}
\tag{5.16.3}
$$

where the $R(\mathbf{C}) = g$ and the $R(\mathbf{A}) = u$, and in obtaining confidence bounds for growth curves.

To reduce model (5.16.2) to model (5.16.1), Potthoff and Roy suggested the following transformation of $\mathbf{Y}_0$ to $\mathbf{Y}$:

$$
\mathbf{Y} = \mathbf{Y}_0\mathbf{G}^{-1}\mathbf{Q}'(\mathbf{Q}\mathbf{G}^{-1}\mathbf{Q}')^{-1}
\tag{5.16.4}
$$

where $\mathbf{G}$, of order $q \times q$, is any symmetric, positive, definite matrix. The matrix $\mathbf{Y}$ defined in (5.16.4) is of order $N \times p$ such that each row of $\mathbf{Y}$ has a multivariate normal distribution with variance-covariance matrix

$$
\underset{(p\times p)}{\Sigma} = (\mathbf{Q}(\mathbf{G}')^{-1}\mathbf{Q}')^{-1}\mathbf{Q}(\mathbf{G}')^{-1}\mathbf{\Sigma}_0\mathbf{G}^{-1}\mathbf{Q}'(\mathbf{Q}\mathbf{G}^{-1}\mathbf{Q}')^{-1}
\tag{5.16.5}
$$

where $E(\mathbf{Y}) = \mathbf{XB}$ and $V(\mathbf{Y}) = \mathbf{I}\otimes\mathbf{\Sigma}$. $\mathbf{Y}$ in (5.16.4) satisfies the conditions of the usual MANOVA model defined in (5.16.1) for any choice of $\mathbf{G}$, subject to the limitations mentioned. Hence all the theory developed under the usual model is now applicable by utilizing the transformed data matrix. Hypothesis testing and the determination of simultaneous confidence intervals are included.

As an illustration of the applicability of (5.16.2), the I-group case with N_i subjects per group, discussed in Section 5.12, is used. The growth curve associated with the ith group is

$$
\beta_{i0} + \beta_{i1}t + \beta_{i2}t^2 + \cdots + \beta_{i,p-1}t^{p-1}
\tag{5.16.6}
$$

Letting $\mathbf{Y}_0$ denote the data matrix of order $N \times q$, where q commensurable responses are obtained on each subject, the matrices **X**, **B**, and **Q** are defined as follows:

(5.16.7)
$$\underset{(N \times I)}{\mathbf{X}} = \begin{bmatrix} \mathbf{1}_{N_1} & \mathbf{0} & \cdots & \mathbf{0} \\ \mathbf{0} & \mathbf{1}_{N_2} & \cdots & \mathbf{0} \\ \mathbf{0} & \mathbf{0} & \cdots & \mathbf{0} \\ \vdots & \vdots & \ddots & \vdots \\ \mathbf{0} & \mathbf{0} & \cdots & \mathbf{1}_{N_I} \end{bmatrix} \qquad \underset{(I \times p)}{\mathbf{B}} = \begin{bmatrix} \beta_{10} & \beta_{11} & \cdots & \beta_{1,p-1} \\ \beta_{20} & \beta_{21} & \cdots & \beta_{2,p-1} \\ \vdots & \vdots & \ddots & \vdots \\ \beta_{I0} & \beta_{I1} & \cdots & \beta_{I,p-1} \end{bmatrix}$$

$$\underset{(p \times q)}{\mathbf{Q}} = \begin{bmatrix} 1 & 1 & \cdots & 1 \\ t_1 & t_2 & \cdots & t_q \\ t_1^2 & t_2^2 & \cdots & t_q^2 \\ \vdots & \vdots & \ddots & \vdots \\ t_1^{p-1} & t_2^{p-1} & \cdots & t_q^{p-1} \end{bmatrix}$$

where $\mathbf{1}_{N_i}$ denotes a $N_i \times 1$ vector of unities in the design matrix **X**.

As in the case of heterogeneous data (Section 4.10),

(a) complete equality of regressions and
(b) parallelism of regressions

may be tested by using model (5.16.2). To test (a), the matrices **C**, **A**, and **Γ** are defined as

(5.16.8)
$$\underset{[(I-1) \times I]}{\mathbf{C}} = \begin{bmatrix} 1 & 0 & \cdots & -1 \\ 0 & 1 & \cdots & -1 \\ \vdots & \vdots & \ddots & \vdots \\ 0 & 0 & \cdots & -1 \end{bmatrix}, \quad \mathbf{A} = \mathbf{I}_p, \quad \text{and} \quad \mathbf{\Gamma} = \mathbf{0}$$

To test (b), the matrices **C**, **A**, and **Γ** are defined by

(5.16.9)
$$\underset{[(I-1) \times I]}{\mathbf{C}} = \begin{bmatrix} 1 & 0 & \cdots & 0 & -1 \\ 0 & 1 & \cdots & 0 & -1 \\ \vdots & \vdots & \ddots & \vdots & \vdots \\ 0 & 0 & \cdots & 1 & -1 \end{bmatrix}$$

$$\underset{[p \times (p-1)]}{\mathbf{A}} = \begin{bmatrix} 0 & 0 & \cdots & 0 \\ 1 & 0 & \cdots & 0 \\ 0 & 1 & \cdots & 0 \\ \vdots & \vdots & \ddots & \vdots \\ 0 & 0 & \cdots & 1 \end{bmatrix} \quad \text{and} \quad \mathbf{\Gamma} = \mathbf{0}$$

Many other hypotheses may also be tested; for example, to test that the I growth curves are of degree $p - 2$ or less, we set

$$\mathbf{C} = \mathbf{I}_I, \quad \underset{(p \times 1)}{\mathbf{A}} = \begin{bmatrix} 0 \\ \vdots \\ 0 \\ 1 \end{bmatrix}, \quad \text{and} \quad \mathbf{\Gamma} = \mathbf{0} \tag{5.16.10}$$

Model (5.16.2) may also be applied in simpler cases; for example, in the one-group case,

$$\underset{(N \times 1)}{\mathbf{X}} = \begin{bmatrix} 1 \\ 1 \\ \vdots \\ 1 \end{bmatrix} \quad \text{and} \quad \underset{(1 \times p)}{\mathbf{B}} = [\beta_{10}, \beta_{11}, \ldots, \beta_{1,p-1}] \tag{5.16.11}$$

where $\mathbf{Q}$ is defined as in (5.16.7). Or, with $N = 1$ and $\mathbf{Y}_0$ denoting a vector of means for the q responses, model (5.16.2) is conveniently written as

$$\underset{(q \times 1)}{\mathbf{y}_{\cdot}} = \underset{(q \times p)}{\mathbf{Q}'} \underset{(p \times 1)}{\boldsymbol{\beta}} + \underset{(q \times 1)}{\boldsymbol{\varepsilon}_{\cdot}} \tag{5.16.12}$$

In Section 4.4, model (5.16.12) was examined by assuming independent time points, so that $E(\boldsymbol{\varepsilon}) = \sigma^2\mathbf{V}$ where $\mathbf{V}$ was known and diagonal. Now, by using (5.16.2), the model can be investigated for correlated time points where $E(\boldsymbol{\varepsilon}_{\cdot}) = \mathbf{\Sigma}/N$. More importantly, the Potthoff-Roy model is also applicable to more complex designs such as factorial designs and multivariate-response, repeated-measures designs.

As an example of an $r \times c$ factorial design, consider a two-way layout with interaction and an equal number of observations per cell. For illustration, suppose there are three levels for the first factor and two levels for the second; then the growth curve for the subjects in the ijth cell is of the form

$$\begin{aligned}(\alpha_{i0} + \alpha_{i1}t + \cdots + \alpha_{i,p-1}t^{p-1}) + (\beta_{j0} + \beta_{ji}t + \cdots + \beta_{j,p-1}t^{p-1}) \\ + (\gamma_{ij0} + \gamma_{ij1} + \cdots + \gamma_{ij,p-1}t^{p-1})\end{aligned} \tag{5.16.13}$$

The matrices $\mathbf{X}$ and $\mathbf{B}$ are defined as

$$\underset{[N \times (r+c+rc)]}{\mathbf{X}} = \left[\begin{array}{ccc|cc|c} \mathbf{1}_r & \mathbf{0} & \mathbf{0} & \mathbf{1}_c & \mathbf{0} & \\ & & & \mathbf{0} & \mathbf{1}_c & \\ \mathbf{0} & \mathbf{1}_r & \mathbf{0} & \mathbf{1}_c & \mathbf{0} & \mathbf{I}_{rc} \\ & & & \mathbf{0} & \mathbf{1}_c & \\ \mathbf{0} & \mathbf{0} & \mathbf{1}_r & \mathbf{1}_c & \mathbf{0} & \\ & & & \mathbf{0} & \mathbf{1}_c & \end{array}\right] \tag{5.16.14}$$

(5.16.14) *continued*

$$\underset{[(r+c+rc)\times p]}{\mathbf{B}} = \begin{bmatrix} \alpha_{10} & \alpha_{11} & \cdots & \alpha_{1,p-1} \\ \alpha_{20} & \alpha_{21} & \cdots & \alpha_{2,p-1} \\ \alpha_{30} & \alpha_{31} & \cdots & \alpha_{3,p-1} \\ \beta_{10} & \beta_{11} & \cdots & \beta_{1,p-1} \\ \beta_{20} & \beta_{21} & \cdots & \beta_{2,p-1} \\ \gamma_{110} & \gamma_{111} & \cdots & \gamma_{11,p-1} \\ \vdots & \vdots & \vdots & \vdots \\ \gamma_{320} & \gamma_{321} & \cdots & \gamma_{32,p-1} \end{bmatrix}$$

where $r = 3$, $c = 2$, and $\mathbf{Q}$ is defined as in (5.16.7).

To apply model (5.16.2) to a multivariate-response, repeated-measures design, where vector-valued observations are obtained at each time point, the one-way layout with I groups is selected for illustration. For simplicity, suppose that three measures are procured at each of the q time points, so that each subject will have $3q$ measurements, all correlated with the unknown variance-covariance matrix $\mathbf{\Sigma}_0$. Letting the ith growth curve for each of the three multivariate responses be represented by

(5.16.15)
$$\begin{aligned} &\beta_{i0} + \beta_{i1}t + \cdots + \beta_{1,p_1-1}t^{p_1-1} \\ &\theta_{i0} + \theta_{i1}t + \cdots + \theta_{i,p_2-1}t^{p_1-1} \\ &\xi_{i0} + \xi_{i1}t + \cdots + \xi_{i,p_3-1}t^{p_3-1} \end{aligned}$$

where the data matrix $\mathbf{Y}_0$ is arranged so that the first q columns contain the measurements of the first variable, and so on. Then the matrices $\mathbf{B}$ and $\mathbf{Q}$ are defined as follows:

(5.16.16)

$$\underset{[I\times(p_1+p_2+p_3)]}{\mathbf{B}} = \begin{bmatrix} \beta_{10} & \beta_{11} & \cdots & \beta_{1,p_1-1} & \theta_{10} & \theta_{11} & \cdots & \theta_{1,p_2-1} & \xi_{10} & \xi_{11} & \cdots & \xi_{1,p_3-1} \\ \beta_{20} & \beta_{21} & \cdots & \beta_{2,p_1-1} & \theta_{20} & \theta_{21} & \cdots & \theta_{2,p_2-1} & \xi_{20} & \xi_{21} & \cdots & \xi_{2,p_3-1} \\ \vdots & \vdots & \vdots & \vdots & \vdots & \vdots & \vdots & \vdots & \vdots & \vdots & \vdots & \vdots \\ \beta_{I0} & \beta_{I1} & \cdots & \beta_{I,p_1-1} & \theta_{I0} & \theta_{I,p_1-1} & \cdots & \theta_{I,p_2-1} & \xi_{I0} & \xi_{I1} & \cdots & \xi_{I,p_3-1} \end{bmatrix}$$

$$\underset{[(p_1+p_2+p_3)\times 3q]}{\mathbf{Q}} = \begin{bmatrix} 1 & 1 & \cdots & 1 & 0 & 0 & \cdots & 0 & 0 & 0 & \cdots & 0 \\ t_1 & t_2 & \cdots & t_q & 0 & 0 & \cdots & 0 & 0 & 0 & \cdots & 0 \\ \vdots & \vdots & \vdots & \vdots & \vdots & \vdots & \vdots & \vdots & \vdots & \vdots & \vdots & \vdots \\ t_1^{p_1-1} & t_2^{p_1-1} & \cdots & t_q^{p_1-1} & 0 & 0 & \cdots & 0 & 0 & 0 & \cdots & 0 \\ 0 & 0 & \cdots & 0 & 1 & 1 & \cdots & 1 & 0 & 0 & \cdots & 0 \\ 0 & 0 & \cdots & 0 & t_1 & t_2 & \cdots & t_q & 0 & 0 & \cdots & 0 \\ \vdots & \vdots & \vdots & \vdots & \vdots & \vdots & \vdots & \vdots & \vdots & \vdots & \vdots & \vdots \\ 0 & 0 & \cdots & 0 & t_1^{p_2-1} & t_2^{p_2-1} & \cdots & t_q^{p_2-1} & 0 & 0 & \cdots & 0 \\ 0 & 0 & \cdots & 0 & 0 & 0 & \cdots & 0 & 1 & 1 & \cdots & 1 \\ 0 & 0 & \cdots & 0 & 0 & 0 & \cdots & 0 & t_1 & t_2 & \cdots & t_q \\ \vdots & \vdots & \vdots & \vdots & \vdots & \vdots & \vdots & \vdots & \vdots & \vdots & \vdots & \vdots \\ 0 & 0 & \cdots & 0 & 0 & 0 & \cdots & 0 & t_1^{p_3-1} & t_2^{p_3-1} & \cdots & t_q^{p_3-1} \end{bmatrix}$$

The design matrix $\mathbf{X}$ is the same as the I-group case with only q responses [given in (5.16.7)].

In the use of the transformation defined by Potthoff and Roy, a problem in the selection of the matrix $\mathbf{G}$ arises. The linear unbiased estimator for $\mathbf{B}$ when using (5.16.4) is

$$\hat{\mathbf{B}} = (\mathbf{X}'\mathbf{X})^{-1}\mathbf{X}'\mathbf{Y}_0\mathbf{G}^{-1}\mathbf{Q}'(\mathbf{Q}\mathbf{G}^{-1}\mathbf{Q}')^{-1} \tag{5.16.17}$$

from the theory of model (5.16.1). Furthermore,

$$\begin{aligned} E(\hat{\mathbf{B}}) &= (\mathbf{X}'\mathbf{X})^{-1}\mathbf{X}'(\mathbf{XBQ})\mathbf{G}^{-1}\mathbf{Q}'(\mathbf{Q}\mathbf{G}^{-1}\mathbf{Q}')^{-1} \\ &= \mathbf{B} \end{aligned} \tag{5.16.18}$$

so that $\hat{\mathbf{B}}$ is independent of the choice of the matrix $\mathbf{G}$. However, the minimum-variance unbiased estimator of $\hat{\mathbf{B}}$ under model (5.16.2) is

$$\hat{\mathbf{B}} = (\mathbf{X}'\mathbf{X})^{-1}\mathbf{X}'\mathbf{Y}_0\mathbf{\Sigma}_0^{-1}\mathbf{Q}'(\mathbf{Q}\mathbf{\Sigma}_0^{-1}\mathbf{Q}')^{-1} \tag{5.16.19}$$

as shown by Potthoff and Roy (1964). Thus, based on the minimum-variance, unbiased-estimation criterion, the optimal choice of $\mathbf{G}$ is $\mathbf{\Sigma}_0$. But, in practice, $\mathbf{\Sigma}_0$ is usually unknown.

When $p = q$, the transformation defined by (5.16.4) becomes

$$\mathbf{Y} = \mathbf{Y}_0\mathbf{Q}^{-1} \tag{5.16.20}$$

so that there is no need to choose $\mathbf{G}$. Bock (1963) developed a procedure just for this case by using step-down F tests and orthogonal polynomials.

If $p < q$, however, the choice of $\mathbf{G}$ is important since it affects the power of tests, width of confidence bands, and the variance of estimators. The variance of the estimator $\hat{\mathbf{B}}$ increases as $\mathbf{G}^{-1}$ departs from $\mathbf{\Sigma}_0^{-1}$. A simple choice of $\mathbf{G}$ is to set $\mathbf{G} = \mathbf{I}$. Then

$$\mathbf{Y} = \mathbf{Y}_0\mathbf{Q}'(\mathbf{Q}\mathbf{Q}')^{-1} \tag{5.16.21}$$

Such a choice of $\mathbf{G}$ will certainly simplify the calculations and allow the experimenter to substitute orthogonal polynomials in the matrix $\mathbf{Q}$. However, it may not be the best choice in terms of power since information is lost by reducing $\mathbf{Y}_0$ to $\mathbf{Y}$ unless $\mathbf{G}$ is set equal to $\mathbf{\Sigma}_0$. Potthoff and Roy suggest using an estimate of $\mathbf{\Sigma}_0$ obtained from an independent experiment. They did not, however, develop the theory for allowing $\mathbf{G} = \mathbf{S}$, where $\mathbf{S}$ is proportional to the estimate of $\mathbf{\Sigma}_0$ calculated from the data used to estimate $\mathbf{B}$. Note that the analysis under model (5.16.2) is valid regardless of the selection of $\mathbf{G}$.

To try to avoid the arbitrary choice of the matrix $\mathbf{G}$ under the Potthoff-Roy development, Rao (1965b, 1966a, 1967) and Khatri (1966) independently developed an alternative reduction of model (5.16.2) to a conditional model of the form

$$E(\mathbf{Y}) = \mathbf{XB} + \mathbf{Z\Gamma} \tag{5.16.22}$$

where

> $\mathbf{Y}$ is an $N \times p$ data matrix,
> $\mathbf{X}$ is an $N \times I$ design matrix,
> $\mathbf{B}$ is an $I \times p$ matrix of unknown parameters,
> $\mathbf{Z}$ is an $N \times h$ matrix of covariates,
> $\mathbf{\Gamma}$ is an $h \times p$ matrix of unknown regression coefficients,

and each row of $\mathbf{Y}$ is independently and normally distributed with variance-covariance matrix $\mathbf{\Sigma}$. Model (5.16.22) is the multivariate covariance-analysis model, discussed in some detail in Section 5.14.

To reduce (5.16.2) to (5.16.22), a $q \times q$ nonsingular matrix $\mathbf{H} = [\mathbf{H}_1 \quad \mathbf{H}_2]$ is constructed so that the columns of $\mathbf{H}_1$ form a basis for the vector space generated by the rows of $\mathbf{Q}$, $\mathbf{QH}_1 = \mathbf{I}$, and $\mathbf{QH}_2 = \mathbf{0}$. When the rank of the matrix $\mathbf{Q}$ is p, so that $\mathbf{Q}$ is of full row rank, $\mathbf{H}_1$ and $\mathbf{H}_2$ can be selected as

(5.16.23)
$$\underset{(q \times p)}{\mathbf{H}_1} = \mathbf{G}^{-1}\mathbf{Q}'(\mathbf{QG}^{-1}\mathbf{Q}')^{-1} \qquad \underset{[q \times (q-p)]}{\mathbf{H}_2} = \mathbf{I} - \mathbf{H}_1\mathbf{Q}$$

where $\mathbf{G}$ is an arbitrary positive definite matrix. Such a matrix $\mathbf{H}$ is not unique; however, estimates and tests are invariate for all choices of $\mathbf{H}$, satisfying the specified conditions (see Khatri, 1966). By letting $\mathbf{Y} = \mathbf{Y}_0\mathbf{H}_1$ and $\mathbf{Z} = \mathbf{Y}_0\mathbf{H}_2$, $E(\mathbf{Y}) = \mathbf{XB}$ and $E(\mathbf{Z}) = \mathbf{0}$; thus the expected value of $\mathbf{Y}$, given $\mathbf{Z}$, can be shown to be of the form specified by (5.16.22) (see Khatri, 1966). In this case the information contained in the covariates $\mathbf{Z} = \mathbf{Y}_0\mathbf{H}_2$, which is ignored in the Potthoff-Roy reduction, is utilized.

To better understand the Rao-Khatri reduction, when $p < q$, and the Potthoff-Roy reduction, suppose $R(\mathbf{Q}) = p$, where $\mathbf{Q}$ is a matrix of orthogonal polynomials:

$$\mathbf{Q} = \begin{bmatrix} t_{01} & t_{02} & \cdots & t_{0q} \\ t_{11} & t_{12} & \cdots & t_{1q} \\ \vdots & \vdots & \vdots & \vdots \\ t_{p-1,1} & t_{p-1,2} & \cdots & t_{p-1,q} \end{bmatrix}$$

$$\mathbf{H}_1 = \mathbf{Q}'$$

and

$$\mathbf{H}_2 = \begin{bmatrix} t_{p,1} & \cdots & t_{q-1,1} \\ t_{p,2} & \cdots & t_{q-1,2} \\ \vdots & \vdots & \vdots \\ t_{p,q} & \cdots & t_{q-1,q} \end{bmatrix}$$

Then

$$E(\mathbf{Y}_0\mathbf{Q}') = \mathbf{XBQQ}' \quad \text{and} \quad E(\mathbf{Y}_0\mathbf{H}_2) = \mathbf{0}$$

or

$$E[\mathbf{Y}_0\mathbf{Q}'(\mathbf{QQ}')^{-1}] = \mathbf{XB}$$

Hence the conditional model employing $\mathbf{Y}_0\mathbf{H}_2$ as covariates is

$$E[\mathbf{Y}_0\mathbf{Q}'(\mathbf{QQ}')^{-1}|\mathbf{Y}_0\mathbf{H}_2] = \mathbf{XB} + \mathbf{Y}_0\mathbf{H}_2\mathbf{\Gamma}$$

However, when the covariates are not included in the Rao-Khatri model, the Rao-Khatri reduction is identical to the Potthoff-Roy reduction since

$$\mathbf{Y} = \mathbf{Y}_0\mathbf{G}^{-1}\mathbf{Q}'(\mathbf{Q}\mathbf{G}^{-1}\mathbf{Q}')^{-1} = \mathbf{Y}_0\mathbf{Q}'(\mathbf{Q}\mathbf{Q}')^{-1}$$

for $\mathbf{G} = \mathbf{I}$. Thus, by setting $\mathbf{G} = \mathbf{I}$ in the Potthoff-Roy transformation, information that may improve the estimator $\mathbf{B}$ is not incorporated.

Khatri (1966) shows that

$$\hat{\mathbf{B}} = (\mathbf{X}'\mathbf{X})^{-1}\mathbf{X}'\mathbf{Y}_0\mathbf{S}^{-1}\mathbf{Q}'(\mathbf{Q}\mathbf{S}^{-1}\mathbf{Q}')^{-1} \tag{5.16.24}$$

is the maximum-likelihood estimator for $\mathbf{B}$ when employing model (5.16.2) where the $q \times q$ matrix $\mathbf{S}$ is the sample estimate of $\boldsymbol{\Sigma}_0$. However, when using the reductions

$$\mathbf{Y} = \mathbf{Y}_0\mathbf{S}^{-1}\mathbf{Q}'(\mathbf{Q}\mathbf{S}^{-1}\mathbf{Q}')^{-1} = \mathbf{Y}_0\mathbf{H}_1$$
$$\mathbf{Z} = \mathbf{Y}_0[\mathbf{I} - \mathbf{S}^{-1}\mathbf{Q}'(\mathbf{Q}\mathbf{S}^{-1}\mathbf{Q}')^{-1}\mathbf{Q}] = \mathbf{Y}_0\mathbf{H}_2$$

the least-squares estimator for model (5.16.22) and the maximum-likelihood estimator for $\mathbf{B}$ under model (5.16.2) are equal (Khatri, 1966, and Rao, 1967). From (5.14.40),

$$\begin{aligned}\hat{\mathbf{B}} &= (\mathbf{X}'\mathbf{X})^{-1}\mathbf{X}'\mathbf{Y} - (\mathbf{X}'\mathbf{X})^{-1}\mathbf{X}'\mathbf{Z}\{[\mathbf{Z}'(\mathbf{I} - \mathbf{X}(\mathbf{X}'\mathbf{X})^{-1}\mathbf{X}'\mathbf{Z}]^{-1}\mathbf{Z}'[\mathbf{I} - \mathbf{X}(\mathbf{X}'\mathbf{X})^{-1}\mathbf{X}']\mathbf{Y}\} \\
&= (\mathbf{X}'\mathbf{X})^{-1}\mathbf{X}'\mathbf{Y} - (\mathbf{X}'\mathbf{X})^{-1}\mathbf{X}'\mathbf{Y}_0[\mathbf{H}_2(\mathbf{H}_2'\mathbf{S}\mathbf{H}_2)^{-1}\mathbf{H}_2']\mathbf{S}\mathbf{H}_1 \\
&= (\mathbf{X}'\mathbf{X})^{-1}\mathbf{X}'\mathbf{Y}_0\mathbf{H}_1 - (\mathbf{X}'\mathbf{X})^{-1}\mathbf{X}'\mathbf{Y}_0[\mathbf{S}^{-1} - \mathbf{S}^{-1}\mathbf{Q}'(\mathbf{Q}\mathbf{S}^{-1}\mathbf{Q}')^{-1}\mathbf{Q}\mathbf{S}^{-1}]\mathbf{S}\mathbf{H}_1 \\
&= (\mathbf{X}'\mathbf{X})^{-1}\mathbf{X}'\mathbf{Y}_0\mathbf{S}^{-1}\mathbf{Q}'(\mathbf{Q}\mathbf{S}^{-1}\mathbf{Q}')^{-1}\mathbf{Q}\mathbf{H}_1 \\
&= (\mathbf{X}'\mathbf{X})^{-1}\mathbf{X}'\mathbf{Y}_0\mathbf{S}^{-1}\mathbf{Q}'(\mathbf{Q}\mathbf{S}^{-1}\mathbf{Q}')^{-1}\end{aligned}$$

Thus the estimates obtained by Rao using $q - p$ covariates, by Khatri using the maximum-likelihood procedure, and by Potthoff and Roy weighting $\mathbf{G}^{-1} = \mathbf{S}^{-1}$ are identical.

In summary, when $p < q$, the Potthoff-Roy reduction using $\mathbf{G} = \mathbf{I}$ is equivalent to not using covariates in the Rao-Khatri reduction. Setting the compounding matrix $\mathbf{G}$ equal to the $q \times q$ matrix $\mathbf{S}$ in the Potthoff-Roy transformation is the same as employing $q - p$ covariates in the Rao-Khatri reduction. When $p = q$, $\mathbf{H}_2$ does not exist; thus the Rao-Khatri model is not applicable.

There are, however, as indicated by Rao (1966a), Grizzle and Allan (1969), and Lee (1970), a few serious problems in using the Rao-Khatri model. First, the estimate of $\mathbf{B}$ is more efficient only if the covariances between $\mathbf{Y}_0\mathbf{H}_1$ and $\mathbf{Y}_0\mathbf{H}_2$ are larger than zero. It is necessary to determine whether all or a best subset of variables should be included. Second, while inspecting the correlations between the variables included and the covariates may be helpful in the decision for inclusion or exclusion of variables, preliminary tests based on the sample data are also required. Furthermore, the selected set of covariates may vary from sample to sample, while the best subset often depends on a linear structure for $\boldsymbol{\Sigma}$. Third, when the number of groups or treatments is larger than one, since the covariates are not determined prior to the experiment, it is difficult to determine whether the covariates affect the treatment conditions. In addition, there is no reason to believe that the same set of covariates is "best" for all groups. Finally, the model does not lend itself easily to the analysis of multivariate-response, repeated-measures designs in the selection of a best set of covariates.

In attempting to implement the Potthoff-Roy model or the Rao-Khatri reduction in practice when $p < q$, the degree of the polynomial that best fits the growth curves is either assumed to be known or determined through a sequence of preliminary tests. For the purposes of our discussion in this section, a polynomial will be fitted to the data for $p = q$. Hence it will not be necessary to choose the appropriate compounding matrix **G** in the Potthoff-Roy transformation or to select a best subset of covariates in the Rao-Khatri reduction.

For $p = q$, the Potthoff-Roy transformation is defined by

$$\mathbf{Y} = \mathbf{Y}_0\mathbf{Q}^{-1} \tag{5.16.25}$$

Furthermore, by letting **Q** be a set of orthogonal polynomials for equally spaced data so that $\mathbf{Q}'\mathbf{Q} = \mathbf{Q}\mathbf{Q}' = \mathbf{I}$, (5.16.25) reduces to

$$\mathbf{Y} = \mathbf{Y}_0\mathbf{Q}' \tag{5.16.26}$$

Using (5.16.26) amounts to virtually the same analysis as that performed under the usual model (5.16.1), since (5.16.2) reduces to

$$\begin{aligned} E(\mathbf{Y}) &= \mathbf{XB} \\ V(\mathbf{Y}) &= \mathbf{I} \otimes \mathbf{Q\Sigma Q}' \end{aligned} \tag{5.16.27}$$

The least-squares estimator for **B** is

$$\hat{\mathbf{B}} = (\mathbf{X}'\mathbf{X})^{-1}\mathbf{X}'\mathbf{Y}_0\mathbf{Q}' \tag{5.16.28}$$

The estimator acquired in (5.16.28) is the same as $\hat{\mathbf{B}}$ obtained by using model (5.16.1) except that **Y** is replaced by $\mathbf{Y}_0\mathbf{Q}'$. In general, the theory developed by using (5.16.1) may be directly applied to (5.16.2) by using (5.16.27) if the data matrix **Y** is replaced by $\mathbf{Y}_0\mathbf{Q}'$. In testing a hypothesis of the form

$$H_0\colon \underset{(q \times l)}{\mathbf{C}}\ \underset{(l \times p)}{\mathbf{B}}\ \underset{(p \times u)}{\mathbf{A}} = \underset{(q \times u)}{\mathbf{\Gamma}} \tag{5.16.29}$$

where the $R(\mathbf{C}) = g$ and the $R(\mathbf{A}) = u$, the hypothesis sum of squares and the error sum of squares matrices are defined as follows:

$$\begin{aligned} \mathbf{Q}_h &= (\mathbf{C}\hat{\mathbf{B}}\mathbf{A})'[\mathbf{C}(\mathbf{X}'\mathbf{X})^{-1}\mathbf{C}']^{-1}(\mathbf{C}\hat{\mathbf{B}}\mathbf{A}) \\ &= \mathbf{A}'\mathbf{Q}\mathbf{Y}_0'\mathbf{X}(\mathbf{X}'\mathbf{X})^{-1}[\mathbf{C}(\mathbf{X}'\mathbf{X})^{-1}\mathbf{C}']^{-1}(\mathbf{X}'\mathbf{X})^{-1}\mathbf{X}'\mathbf{Y}_0\mathbf{Q}'\mathbf{A} \\ \mathbf{Q}_e &= \mathbf{A}'\mathbf{Q}\mathbf{Y}_0'[\mathbf{I} - \mathbf{X}(\mathbf{X}'\mathbf{X})^{-1}\mathbf{X}']\mathbf{Y}_0\mathbf{Q}'\mathbf{A} \end{aligned} \tag{5.16.30}$$

Confidence intervals for the parameters are also immediate. To illustrate the Potthoff-Roy model using the transformation defined by (5.16.26), the example discussed in Section 5.13 will be used. The generalized linear model for the experiment is

$$\underset{(15 \times 3)}{\mathbf{Y}_0} = \underset{(15 \times 3)}{\mathbf{X}}\ \underset{(3 \times 3)}{\mathbf{B}}\ \underset{(3 \times 3)}{\mathbf{Q}} + \underset{(15 \times 3)}{\mathbf{E}_0} \tag{5.16.31}$$

where $\mathbf{Y}_0$, $\mathbf{X}$, and $\mathbf{B}$ are defined as in (5.13.2):

(5.16.32)

$$\mathbf{Y}_0 = \begin{bmatrix} 2 & 4 & 7 \\ 2 & 6 & 10 \\ 3 & 7 & 10 \\ 7 & 9 & 11 \\ 6 & 9 & 12 \\ 5 & 6 & 10 \\ 4 & 5 & 10 \\ 7 & 8 & 11 \\ 8 & 9 & 11 \\ 11 & 12 & 13 \\ 3 & 4 & 7 \\ 3 & 6 & 9 \\ 4 & 7 & 9 \\ 8 & 8 & 10 \\ 7 & 10 & 10 \end{bmatrix} \qquad \mathbf{X} = \begin{bmatrix} 1 & 0 & 0 \\ 1 & 0 & 0 \\ 1 & 0 & 0 \\ 1 & 0 & 0 \\ 1 & 0 & 0 \\ 0 & 1 & 0 \\ 0 & 1 & 0 \\ 0 & 1 & 0 \\ 0 & 1 & 0 \\ 0 & 1 & 0 \\ 0 & 0 & 1 \\ 0 & 0 & 1 \\ 0 & 0 & 1 \\ 0 & 0 & 1 \\ 0 & 0 & 1 \end{bmatrix} \qquad \mathbf{B} = \begin{bmatrix} \beta_{10} & \beta_{11} & \beta_{12} \\ \beta_{20} & \beta_{21} & \beta_{22} \\ \beta_{30} & \beta_{31} & \beta_{32} \end{bmatrix}$$

Furthermore,

(5.16.33)
$$E(\mathbf{Y}_0) = \mathbf{XBQ}$$
$$V(\mathbf{Y}_0) = \mathbf{I} \otimes \mathbf{\Sigma}_0$$

and each row of $\mathbf{Y}_0$ is independently and normally distributed.

By using a matrix of normalized orthogonal polynomials for the matrix $\mathbf{Q}$,

(5.16.34)
$$\mathbf{Q} = \begin{bmatrix} .577350 & .577350 & .577350 \\ -.707107 & 0 & .707107 \\ .408248 & -.816497 & .408248 \end{bmatrix}$$

the functional form of the polynomial employing (5.16.34) is

(5.16.35)
$$E(Y_{0ij}) = \hat{\beta}_{i0}\left(\frac{1}{\sqrt{3}}\right) + \hat{\beta}_{i1}\left(\frac{x-2}{\sqrt{2}}\right) + \hat{\beta}_{i2}\left(\frac{3x^2 - 12x + 10}{\sqrt{6}}\right) \qquad x = 1, 2, 3$$

Alternatively, a nonnormalized matrix of orthogonal polynomials could have been used. Then

(5.16.36)
$$\mathbf{Q} = \begin{bmatrix} 1 & 1 & 1 \\ -1 & 0 & 1 \\ 1 & -2 & 1 \end{bmatrix}$$

and

(5.16.37)
$$E(Y_{0ij}) = \hat{\xi}_{i0} + \hat{\xi}_{i}(x - 2) + \hat{\xi}_{i2}(3x^2 - 12x + 10) \qquad x = 1, 2, 3$$

Finally, if a Vandermode matrix is used,

(5.16.38)
$$\mathbf{Q} = \begin{bmatrix} 1 & 1 & 1 \\ 1 & 2 & 3 \\ 1 & 4 & 9 \end{bmatrix}$$

and

(5.16.39)
$$E(Y_{0ij}) = \hat{\theta}_{i0} + \hat{\theta}_{i1} + \hat{\theta}_{i2}x^2 \qquad x = 1, 2, 3$$

However, if (5.16.36) or (5.16.38) is used, the inverse of **Q** would have to be determined.

Utilizing **Q** as defined in (5.16.34),

$$\hat{\mathbf{B}} = (\mathbf{X}'\mathbf{X})^{-1}\mathbf{X}'\mathbf{Y}_0\mathbf{Q}' = \mathbf{Y}_{.}\mathbf{Q}'$$

$$= \begin{bmatrix} 4 & 7 & 10 \\ 7 & 8 & 11 \\ 5 & 7 & 9 \end{bmatrix} \begin{bmatrix} .577350 & -.707107 & .408248 \\ .577350 & 0 & -.816497 \\ .577350 & .707107 & .408248 \end{bmatrix}$$

(5.16.40)
$$= \begin{bmatrix} 12.1244 & 4.2426 & .0000 \\ 15.0111 & 2.8284 & .8165 \\ 12.1244 & 2.8284 & .0000 \end{bmatrix}$$

From (5.16.37), the elements $\hat{\xi}_{i0}$, $\hat{\xi}_{i1}$, and $\hat{\xi}_{i2}$ are obtained from

$$\hat{\Xi} = \hat{\mathbf{B}}\mathbf{D} = \hat{\mathbf{B}} \begin{bmatrix} 1/\sqrt{3} & 0 & 0 \\ 0 & 1/\sqrt{2} & 0 \\ 0 & 0 & 1/\sqrt{6} \end{bmatrix} = \begin{bmatrix} \hat{\xi}_{10} & \hat{\xi}_{11} & \hat{\xi}_{12} \\ \hat{\xi}_{20} & \hat{\xi}_{21} & \hat{\xi}_{22} \\ \hat{\xi}_{30} & \hat{\xi}_{31} & \hat{\xi}_{32} \end{bmatrix}$$

(5.16.41)
$$= \begin{bmatrix} 7.0000 & 3.0000 & .0000 \\ 8.6667 & 2.0000 & .3333 \\ 7.0000 & 2.0000 & .0000 \end{bmatrix}$$

and from (5.16.4) and (5.16.37), (5.16.39) follows. Alternatively, we may invert the matrix **Q** as defined in (5.16.38) to gain

$$\hat{\boldsymbol{\theta}} = [(\mathbf{X}'\mathbf{X})^{-1}\mathbf{X}'\mathbf{Y}'\mathbf{Y}_0]\mathbf{Q}^{-1}$$

$$= \mathbf{Y}_{.} \begin{bmatrix} 3 & -2.5 & .5 \\ -3 & 4 & -1 \\ 1 & -1.5 & .5 \end{bmatrix}$$

(5.16.42)
$$= \begin{bmatrix} 1.0 & 3.0 & .0 \\ 8.0 & -2.0 & 1.0 \\ 3.0 & 2.0 & .0 \end{bmatrix}$$

the coefficients for the model defined in (5.16.39).

The first hypothesis of interest in growth-curve-analysis experiments is determining whether the curves for the groups are parallel. From Figure 5.13.1, it would appear that this is not the case. To test the parallelism hypothesis

(5.16.43)
$$H_{01}: \begin{bmatrix} \beta_{11} \\ \beta_{12} \end{bmatrix} = \begin{bmatrix} \beta_{21} \\ \beta_{22} \end{bmatrix} = \begin{bmatrix} \beta_{31} \\ \beta_{32} \end{bmatrix}$$

the matrices **C** and **A** in (5.16.29) are defined by

$$\text{(5.16.44)} \qquad \mathbf{C} = \begin{bmatrix} 1 & 0 & -1 \\ 0 & 1 & -1 \end{bmatrix} \quad \text{and} \quad \mathbf{A} = \begin{bmatrix} 0 & 0 \\ 1 & 0 \\ 0 & 1 \end{bmatrix}$$

With $\hat{\mathbf{B}}$ as defined in (5.16.40), the hypothesis sum of products matrix $\mathbf{Q}_h$ becomes

$$\begin{aligned} \mathbf{Q}_h &= (\mathbf{C}\hat{\mathbf{B}}\mathbf{A})'[\mathbf{C}(\mathbf{X}'\mathbf{X})\mathbf{C}']^{-1}(\mathbf{C}\hat{\mathbf{B}}\mathbf{A}) \\ &= [\mathbf{C}(\mathbf{X}'\mathbf{X})^{-1}\mathbf{X}'\mathbf{Y}_0\mathbf{Q}'\mathbf{A}]'[\mathbf{C}(\mathbf{X}'\mathbf{X})^{-1}\mathbf{C}']^{-1}[\mathbf{C}(\mathbf{X}'\mathbf{X})^{-1}\mathbf{X}'\mathbf{Y}_0\mathbf{Q}'\mathbf{A}] \\ &= (\mathbf{C}\mathbf{Y}_.\mathbf{Q}'\mathbf{A})'[\mathbf{C}(\mathbf{X}'\mathbf{X})^{-1}\mathbf{C}']^{-1}(\mathbf{C}\mathbf{Y}_.\mathbf{Q}'\mathbf{A}) \\ \text{(5.16.45)} \qquad &= (\mathbf{C}\mathbf{Y}_.\mathbf{A}^*)'[\mathbf{C}(\mathbf{X}'\mathbf{X})^{-1}\mathbf{C}']^{-1}(\mathbf{C}\mathbf{Y}_.\mathbf{A}^*) \end{aligned}$$

and

$$\mathbf{A}^* = \mathbf{Q}'\mathbf{A} = \begin{bmatrix} -.707107 & .408248 \\ 0 & -.816497 \\ .707107 & .408248 \end{bmatrix}$$

which satisfies the condition $\mathbf{A}^{*\prime}\mathbf{A}^* = \mathbf{I}$ and is in a form similar to **A** as defined in (5.13.24). Calculating $\mathbf{Q}_h$,

$$\text{(5.16.46)} \qquad \mathbf{Q}_h = \begin{bmatrix} 6.6667 & \\ -1.9245 & 2.2222 \end{bmatrix}$$

The sum of squares and products matrix due to error is

$$\begin{aligned} \mathbf{Q}_e &= \mathbf{A}'\mathbf{Q}\mathbf{Y}_0'[\mathbf{I} - \mathbf{X}(\mathbf{X}'\mathbf{X})^{-1}\mathbf{X}']\mathbf{Y}_0\mathbf{Q}'\mathbf{A} \\ &= \mathbf{A}^{*\prime}\mathbf{Y}_0'[\mathbf{I} - \mathbf{X}(\mathbf{X}'\mathbf{X})^{-1}\mathbf{X}']\mathbf{Y}_0\mathbf{A}^* \\ \text{(5.16.47)} \qquad &= \begin{bmatrix} 15.0000 & \\ 1.7321 & 5.0000 \end{bmatrix} \end{aligned}$$

Employing Wilks' Λ criterion

$$\Lambda = \frac{|\mathbf{Q}_e|}{|\mathbf{Q}_e + \mathbf{Q}_h|} = \frac{72.0003}{156.4448} = .4602$$

By comparing Λ with the critical value at the significance level $\alpha = .05$, $U^{.05}_{(2,2,12)} = .4374$, the parallelism hypothesis is tenable. The p value for the test is $\alpha = .064$. As expected, this result is identical to the test of parallelism for profile analysis for the transformation given in (5.16.26).

Following a significant multivariate test of parallelism under the assumption of a general variance-covariance matrix, tests for linear and quadratic trends are no longer independent. Independence follows only if $\mathbf{Q\Sigma Q}'$ is a diagonal matrix. This is unlikely in educational and psychological studies. In this example, the parallelism hypothesis was not rejected; however, for illustration purposes, tests of linear and quadratic interaction trends are examined. For a linear trend, the ratio

$$F = \frac{6.6667/2}{15.0000/12} = 2.67 \sim F(2, 12)$$

is used. While, for a quadratic trend, the ratio

$$F = \frac{2.2222/2}{5.0000/12} = 2.67 \sim F(2, 12)$$

is employed. Since these tests are not independent, an α level equal to $\alpha^* = \alpha/2$ is recommended. The critical value for the test is $F^{.025}_{(2,12)} = 3.096$. Following the procedure for curvilinear regression, we would proceed by testing higher-order trends first and stop when the first significant F ratio is reached.

Employing the mixed model to analyze trends, the drugs × trials sum of squares is obtained from the multivariate results. Again, however, the compound-symmetry assumption is usually assumed. Proceeding as in profile analysis,

$$\text{MS}_h = \text{ave. of Dia}\left[\frac{\mathbf{Q}_h}{v_h}\right] = \frac{6.667 + 2.222}{4} = 2.2222$$

$$\text{MS}_e = \text{ave. of Dia}\left[\frac{\mathbf{Q}_e}{v_e}\right] = \frac{15.000 + 5.000}{4} = .8333$$

Furthermore, under the mixed model, the drugs × trials sum of squares can be partitioned into a linear and quadratic component as in Table 5.16.1 (see Edwards, 1968, p. 284). These results can also be directly obtained from multivariate analysis, since $\mathbf{A}^{*\prime}\mathbf{A}^* = \mathbf{I}$, by comparing the diagonal elements of

$$\text{MS}_h = \text{ave. of Dia}\left[\frac{\mathbf{Q}_h}{v_h}\right]$$

with

$$\text{MS}_e = \text{ave. of Dia}\left[\frac{\mathbf{Q}_e}{v_e}\right]$$

For example, using the upper-left diagonal element in $\mathbf{Q}_h$ and dividing by 2, $\text{MS}_h = 3.33$. The $\text{MS}_e = (15.00 + 5.00)/24 = .83$, so $F = 4.01$ with 2 and 24 degrees of freedom as indicated in Table 5.16.1. In a similar manner, the test for a quadratic interaction for the univariate case is acquired. Edwards pooled the linear and quadratic estimates for error to test for linear and quadratic trends. This is not common

TABLE 5.16.1. ANOVA Trend Analysis for Edwards' Data.

Source	df	SS	MS	F
Drugs × trials	4	8.89	2.22	2.67
Linear	2	6.67	3.33	4.01
Quadratic	2	2.22	1.11	1.34
Within error 2	24	20.00	.83	
Linear	12	15.00	1.25	
Quadratic	12	5.00	.42	

practice in most statistical work since usually the linear and quadratic estimates of error differ, as they do in Edwards' example. In this case, separate estimates of error variance are used to construct the F tests, which are then equivalent with the multivariate procedure for detecting linear and quadratic interaction trends.

Given parallelism, coincidence of regressions is next tested:

(5.16.48)
$$H_{03}: \begin{bmatrix} \beta_{10} \\ \beta_{11} \\ \beta_{12} \end{bmatrix} = \begin{bmatrix} \beta_{20} \\ \beta_{21} \\ \beta_{22} \end{bmatrix} = \begin{bmatrix} \beta_{30} \\ \beta_{31} \\ \beta_{32} \end{bmatrix}$$

The matrices $\mathbf{C}$ and $\mathbf{A}$ to test for coincidence are defined by

$$\mathbf{C} = \begin{bmatrix} 1 & 0 & -1 \\ 0 & 1 & -1 \end{bmatrix} \quad \text{and} \quad \mathbf{A} = \begin{bmatrix} 1 & 0 & 0 \\ 0 & 1 & 0 \\ 0 & 0 & 1 \end{bmatrix}$$

With $\mathbf{B}$ as given in (5.16.40),

(5.16.49)
$$\mathbf{Q}_h = (\mathbf{C\hat{B}A})'[\mathbf{C}(\mathbf{X}'\mathbf{X})^{-1}\mathbf{C}']^{-1}(\mathbf{C\hat{B}A})$$
$$= \begin{bmatrix} 27.7778 & & \\ -6.0841 & 6.0667 & \\ 7.8567 & -1.9245 & 2.2222 \end{bmatrix}$$
$$\mathbf{Q}_e = \mathbf{QY}_0'[\mathbf{I} - \mathbf{X}(\mathbf{X}'\mathbf{X})^{-1}\mathbf{X}']\mathbf{YQ}'$$
$$= \begin{bmatrix} 148.0000 & & \\ -30.6186 & 15.0000 & \\ -16.2635 & 1.7321 & 5.0000 \end{bmatrix}$$

Employing Wilks' Λ criterion to test for coincidence,

(5.16.50)
$$\Lambda = \frac{|\mathbf{Q}_e|}{|\mathbf{Q}_e + \mathbf{Q}_h|} = .2368$$

With $\alpha = .05$, H_{03} is rejected if $\Lambda < U^{\alpha}(u, v_h, v_e) = U^{.05}_{(3,2,12)} = .3157$. Thus, by employing the multivariate test for coincidence, the coincidence hypothesis is rejected. As usual, any other criterion discussed for multivariate hypotheses could have been employed to test (5.16.48). For example, using Roy's criterion, $\theta_s = .6256$. For $\alpha = .05$, $s = \min(v_h, u) = \min(2, 3) = 2$, $m = (|u - v_h| - 1)/2 = 0$, and $n = (v_e - u - 1)/2 = (12 - 3 - 1)/2 = 4.0$, the hypothesis is rejected if $\theta_s > \theta^{.05}_{(2,0,4)} = .560$, which is consistent with Wilks' criterion.

Confidence intervals for the elements of $\mathbf{B}$, using the transformation $\mathbf{Y} = \mathbf{Y}_0\mathbf{Q}'$ with $\mathbf{G} = \mathbf{I}$, are obtained by evaluating the following expression:

(5.16.51)
$$\mathbf{c}'\mathbf{\hat{B}a} - c_0\hat{\sigma} \leq \mathbf{c}'\mathbf{Ba} \leq \mathbf{c}'\mathbf{\hat{B}a} + c_0\hat{\sigma}$$

where

$$c_0^2 = v_e \frac{\theta^{\alpha}}{1 - \theta^{\alpha}}$$

and

$$\hat{\sigma} = \sqrt{\mathbf{a}'\left\{\frac{\mathbf{QY}_0'[\mathbf{I} - \mathbf{X}(\mathbf{X}'\mathbf{X})^{-1}\mathbf{X}']\mathbf{YQ}'}{\nu_e}\right\}\mathbf{a}[\mathbf{c}'(\mathbf{X}'\mathbf{X})^{-1}\mathbf{c}]}$$

for Roy's criterion.

Given that the coincidence hypothesis is rejected and the parallelism hypothesis is tenable, the overall polynomial trend on trials may be tested even though there exist significant group differences. The trend hypothesis is

$$H_{02}^{*}: \sum_{i=1}^{3} \beta_{i1} = \sum_{i=1}^{3} \beta_{i2} = 0 \tag{5.16.52}$$

To test (5.16.52), the matrices **C** and **A** are defined by

$$\mathbf{C} = [1, 1, 1] \quad \text{and} \quad \mathbf{A} = \begin{bmatrix} 0 & 0 \\ 1 & 0 \\ 0 & 1 \end{bmatrix}$$

Then, with $\hat{\mathbf{B}}$ as defined in (5.16.40),

$$\begin{aligned} \mathbf{Q}_h &= (\mathbf{C}\hat{\mathbf{B}}\mathbf{A})'[\mathbf{C}(\mathbf{X}'\mathbf{X})^{-1}\mathbf{C}']^{-1}(\mathbf{C}\hat{\mathbf{B}}\mathbf{A}) \\ &= \begin{bmatrix} 163.3334 & \\ 13.4711 & 1.1111 \end{bmatrix} \\ \mathbf{Q}_e &= \mathbf{QY}_0'[\mathbf{I} - \mathbf{X}(\mathbf{X}'\mathbf{X})^{-1}\mathbf{X}']\mathbf{YQ}' \\ &= \begin{bmatrix} 15.0000 & \\ 1.7321 & 5.0000 \end{bmatrix} \end{aligned} \tag{5.16.53}$$

and

$$\Lambda = \frac{|\mathbf{Q}_e|}{|\mathbf{Q}_e + \mathbf{Q}_h|} = .0839$$

Comparing $\Lambda = .0839$ with the critical value $U^{.05}_{(2,1,12)} = .5800$, for $\alpha = .05$, the hypothesis of no trend is rejected. Employing Roy's criterion, $\theta_s = .9161$. For $\alpha = .05$, H_{02}^{*} is rejected if $\theta_s > \theta^{.05}_{(1,0,4.5)} = .4200$, which is consistent with Wilks' criterion.

To determine if a linear or quadratic model best represents the data following an overall significant multivariate test, dependent F tests are used. By selecting the compounding matrices

$$\mathbf{C} = [1, 1, 1] \quad \text{and} \quad \mathbf{A} = \begin{bmatrix} 0 \\ 0 \\ 1 \end{bmatrix}$$

$$\mathbf{C} = [1, 1, 1] \quad \text{and} \quad \mathbf{A} = \begin{bmatrix} 0 \\ 1 \\ 0 \end{bmatrix}$$

or by using the diagonal elements of $\mathbf{Q}_h$ and $\mathbf{Q}_e$ as computed in (5.16.53), tests for significant quadratic and linear trends are obtained. The F ratios become, from (5.16.53),

$$\text{Quadratic:} \quad F = \frac{1.1111/1}{5.0000/12} = 2.667 \sim F(1, 12)$$

(5.16.54)

$$\text{Linear:} \quad F = \frac{163.3332/1}{15.0000/12} = 130.6644 \sim F(1, 12)$$

The p values for the tests in (5.16.54) are .129 and .001, respectively. Thus a linear model best represents the relationship for trials.

The model for the trend on trials is, from (5.16.41),

$$\begin{aligned} E(Y_{0ij}) &= 7.5556 + 2.3333(x - 2) \\ &= 2.889 + 2.333x \end{aligned} \tag{5.16.55}$$

Following (5.16.51), confidence intervals for the coefficients in (5.16.55) can also be established.

The tests summarized in (5.16.54) are not the same as the tests for trend using the mixed model, as given by Edwards, since again he pooled the linear and quadratic estimates for error. For Edwards, the overall test for trials is immediately procured from the overall multivariate test, as stated in H_{02}^*, as are the tests for linear and quadratic trend assuming parallelism and compound symmetry. Then, the sum of squares due to trials (Edwards, 1968, p. 281) is partitioned according to Table 5.16.2. To obtain the results given in Table 5.16.2 from the multivariate test (5.16.52), the diagonal elements of $\mathbf{Q}_h$ and the sum of the diagonal elements in $\mathbf{Q}_e$ are used.

TABLE 5.16.2. ANOVA Trend Analysis for Edwards' Data.

Source	df	SS	F
Trials	2	164.44	99.06
Linear	1	163.33	196.78
Quadratic	1	1.11	1.34
Within error 2	24	20.00	

Given that both the coincidence and parallelism hypotheses are tenable, a test for trend on trials may proceed by applying the Potthoff-Roy model for a one-group, growth-curve analysis. The advantage to this approach is that the degrees of freedom for error increase since the group × trials variation is pooled with error.

The model is

$$\underset{(15\times 3)}{E(\mathbf{Y}_0)} = \underset{(15\times 1)}{\mathbf{X}}\;\underset{(1\times 3)}{\mathbf{B}}\;\underset{(3\times 3)}{\mathbf{Q}} \tag{5.16.56}$$

where the design matrix **X** and parameter matrix **B** become

(5.16.57) $$\mathbf{X} = \begin{bmatrix} 1 \\ 1 \\ \vdots \\ 1 \end{bmatrix} \quad \text{and} \quad \mathbf{B} = [\beta_{10}, \beta_{11}, \beta_{12}]$$

Assuming tenable coincidence and parallelism for illustration purposes, with **Q** defined as in (5.16.34),

(5.16.58) $$\begin{aligned} \hat{\mathbf{B}} &= (\mathbf{X}'\mathbf{X})^{-1}\mathbf{X}'\mathbf{Y}_0\mathbf{Q}' \\ &= [13.0866 \quad 3.29983 \quad .27216] \end{aligned}$$

The first hypothesis of interest is that there is no mean trend:

(5.16.59) $$H_0: \beta_{11} = \beta_{12} = 0$$

To test (5.16.59), use matrices **C** and **A**, where

(5.16.60) $$\underset{(1\times 1)}{\mathbf{C}} = 1 \quad \text{and} \quad \underset{(3\times 1)}{\mathbf{A}} = \begin{bmatrix} 0 & 0 \\ 1 & 0 \\ 0 & 1 \end{bmatrix}$$

With **C** and **A** as defined in (5.16.60), the hypothesis and error sum of products matrices become

(5.16.61) $$\begin{aligned} \mathbf{Q}_h &= (\mathbf{C}\hat{\mathbf{B}}\mathbf{A})'[\mathbf{C}(\mathbf{X}'\mathbf{X})^{-1}\mathbf{C}']^{-1}(\mathbf{C}\hat{\mathbf{B}}\mathbf{A}) \\ &= \begin{bmatrix} 163.3334 & \\ 13.4711 & 1.1111 \end{bmatrix} \\ \mathbf{Q}_e &= \mathbf{Q}\mathbf{Y}_0'[\mathbf{I} - \mathbf{X}(\mathbf{X}'\mathbf{X})^{-1}\mathbf{X}']\mathbf{Y}_0\mathbf{Q}' \\ &= \begin{bmatrix} 21.6667 & \\ -.1924 & 7.2222 \end{bmatrix} \end{aligned}$$

Then Wilks' Λ criterion becomes

$$\Lambda = .1146 < U^{.05}_{(2,1,14)} = .6307$$

Alternatively, for Roy's criterion,

$$\theta_s = .8854 > \theta^{.05}_{(1,0,5.5)} = .3693$$

Hence the hypothesis of no trend is rejected. To determine if the trend is linear or quadratic, the following F ratios are examined:

(5.16.62) $$\begin{aligned} \text{Quadratic:} \quad F &= \frac{1.1111/1}{7.2222/14} = 2.15 \\ \text{Linear:} \quad F &= \frac{163.3332/1}{21.6667/14} = 105.54 \end{aligned}$$

The p values for these tests are .1644 and .0001, respectively. Hence the linear model again is adequate to describe the trend on trials.

In this section we have shown how the Potthoff-Roy model is used to analyze trends when p is set equal to q. Although this procedure is always permissible, it may not always be the best procedure when a set of data contains several time points. Then we would usually choose $p < q$. Even in our example, where only a few points were available, we found that p should be less than q. However, then the choice of the compounding matrix $\mathbf{G}$ is critical since its choice affects the regression weights, the width of confidence intervals, and the power of statistical tests.

Now assume that $2 = p < q$, so that the formula for $\hat{\mathbf{B}}$, given in (5.16.17), is

$$\hat{\mathbf{B}} = (\mathbf{X}'\mathbf{X})^{-1}\mathbf{X}'\mathbf{Y}_0\mathbf{G}^{-1}\mathbf{Q}'(\mathbf{Q}\mathbf{G}^{-1}\mathbf{Q}')^{-1}$$

Using $\mathbf{Q}$ and $\mathbf{G}$ defined by

$$\mathbf{Q} = \begin{bmatrix} 1 & 1 & 1 \\ -1 & 0 & 1 \end{bmatrix} \quad \text{and} \quad \mathbf{G} = \mathbf{S} \tag{5.16.63}$$

we have

$$\hat{\mathbf{B}} = \begin{bmatrix} 7.000 & 3.000 \\ 10.200 & 1.800 \\ 7.000 & 2.000 \end{bmatrix}$$

Selecting $\mathbf{G} = \mathbf{I}$ and $\mathbf{Q}$ as in (5.16.63), we have, by (5.16.41), that

$$\hat{\mathbf{B}} = \begin{bmatrix} 7.0000 & 3.0000 \\ 8.6667 & 2.0000 \\ 7.0000 & 2.0000 \end{bmatrix}$$

The model for trend on trials becomes, for $\mathbf{Q}$ and $\mathbf{G}$ in (5.16.63),

$$\begin{aligned} E(Y_{0ij}) &= 8.067 + 2.267(x - 2) \\ &= 3.533 + 2.267x \end{aligned} \tag{5.16.64}$$

Thus the selection of the compounding matrix $\mathbf{G}$ affects the magnitude of the regression coefficients. In addition, it affects the size of the standard errors and also the tests of hypotheses.

Preliminary results by Tan (1972) indicate that one should choose $\mathbf{G}$ equal to the sample covariance matrix $\mathbf{S}$, rather than $\mathbf{G} = \mathbf{I}$, when $p < q$ and $q \gg p$. This choice tends to increase the power of statistical tests and to reduce the size of confidence intervals. However, this occurs with a slight increase in the type I error. Until the Potthoff-Roy model is experimented with by more behavioral science researchers, firm guidelines for the choice of the matrix $\mathbf{G}$ cannot be given.

As indicated in (5.16.16), the Potthoff-Roy model is readily extended to the multivariate-response, growth-curve design where vector-valued observations are obtained at each time point. By employing transformation (5.16.26), $\mathbf{Y} = \mathbf{Y}_0\mathbf{Q}'$, with $\mathbf{Q}$ defined by

$$\mathbf{Q} = \begin{bmatrix} \mathbf{Q}_1 & \mathbf{0} & \mathbf{0} \\ \mathbf{0} & \mathbf{Q}_1 & \mathbf{0} \\ \mathbf{0} & \mathbf{0} & \mathbf{Q}_1 \end{bmatrix}$$

When three measures are obtained at q time points and $\mathbf{Q}_i$ is normalized so that $\mathbf{QQ}' = \mathbf{Q}'\mathbf{Q} = \mathbf{I}$, (5.16.2) reduces to (5.16.1). With $\mathbf{B}$ defined by

$$\mathbf{B} = \begin{bmatrix} \beta_{10} & \beta_{11} & \cdots & \beta_{1,q-1} & \theta_{10} & \theta_{11} & \cdots & \theta_{1,q-1} & \xi_{10} & \xi_{11} & \cdots & \xi_{1,q-1} \\ \beta_{20} & \beta_{21} & \cdots & \beta_{2,q-1} & \theta_{20} & \theta_{21} & \cdots & \theta_{2,q-1} & \xi_{20} & \xi_{21} & \cdots & \xi_{2,q-1} \\ \vdots & \vdots & \vdots & \vdots & \vdots & \vdots & \vdots & \vdots & \vdots & \vdots & \vdots & \vdots \\ \beta_{I0} & \beta_{I1} & \cdots & \beta_{I,q-1} & \theta_{I0} & \theta_{I1} & \cdots & \theta_{I,q-1} & \xi_{I0} & \xi_{J1} & \cdots & \xi_{I,q-1} \end{bmatrix} \tag{5.16.65}$$

for an I-group, one-way layout.

EXERCISES 5.16

1. Marascuilo (1969) gives data for a concept formation study in which 12 subjects were randomly assigned to three different experimental groups and then given four trials in the solution of a problem. For each trial, the number of minutes to solve the problem was recorded. The data for the experiment follow.

Group	Subject	Trial 1	2	3	4
	1	9	8	8	5
1	2	12	11	11	4
	3	15	18	13	10
	4	14	15	12	7
	1	20	11	12	9
2	2	15	15	14	10
	3	12	19	18	7
	4	13	10	15	9
	1	8	7	6	6
3	2	10	5	5	9
	3	8	7	7	8
	4	9	13	12	6

Using univariate methods, the analysis-of-variance table is

Source	SS	df	MS	F
G (Groups)	216.13	2	108.06	7.92
Within error 1	133.44	9	14.83	
T (Trials)	156.063	3	52.02	8.39
Linear	121.838	1	121.84	
Quadratic	28.521	1	28.52	
Cubic	5.704	1	5.70	
G × T (Group × trials)	56.875	6	9.48	1.53
Within error 2	167.310	27	6.20	
Total	729.810	47		

a. Analyze the data by using the Potthoff-Roy, growth-curve model with $p = q = 4$. Following the analysis of Edwards' data in Section 5.16, test for parallelism, coincidence, and differences among groups. What are your conclusions?

b. Suppose we assume in this experiment that no more than a quadratic growth curve should be employed with the data. Test whether the coefficient associated with the second-order term is 0. Use the compounding matrices $\mathbf{G} = \mathbf{I}$ using the Potthoff-Roy model and $\mathbf{G} = \mathbf{S}$ using the Rao-Khatri model.

c. For $\mathbf{G} = \mathbf{I}$ and $\mathbf{G} = \mathbf{S}$ in part b, can we conclude that the growth curves for the separate groups are identical? Compare your conclusions in this case with those obtained in part a.

d. Construct confidence bands for each growth curve.

2. Lee (1970) gives data for two dependent variables (time on target in seconds and the number of hits on target) and five trials to investigate bilateral transfer of reminiscence of teaching performance under four treatment conditions: (1) distributed practice on a linear circular-tracking task, (2) distributed practice on a nonlinear hexagonal-tracking task, (3) massed practice on a linear circular-tracking task, and (4) massed practice on a nonlinear hexagonal-tracking task. Subjects, randomly assigned to each group, performed on the given task under each condition with one hand for ten trials and then transferred to the other hand for the same number of trials after a prescribed interval. The two sets of measurements taken for the ten trials were blocked into five blocks of two trials and averaged, yielding five repeated measures for each dependent variable. The data obtained for groups 1 and 2 are given below.

Group 1	Time on Target/Sec.				
	1	2	3	4	5
1	13.95	12.00	14.20	14.40	13.00
2	18.15	22.60	19.30	18.25	20.45
3	19.65	21.60	19.70	19.55	21.00
4	20.80	21.35	21.25	21.25	20.90
5	17.80	20.05	20.35	19.80	18.30
6	17.35	20.85	20.95	20.30	20.70
7	16.15	16.70	19.25	16.50	18.55
8	19.10	18.35	22.95	22.70	22.65
9	12.05	15.40	14.75	13.45	11.60
10	8.55	9.00	9.10	10.50	9.55
11	7.35	5.85	6.20	7.05	9.15
12	17.85	17.95	19.05	18.40	16.85
13	14.50	17.70	16.00	17.40	17.10
14	22.30	22.30	21.90	21.65	21.45
15	19.70	19.25	19.85	18.00	17.80
16	13.25	17.40	18.75	18.40	18.80

Group 1	Hits on Target				
	1	2	3	4	5
1	31.50	37.50	36.50	35.50	34.00
2	22.50	12.00	17.50	19.00	16.50
3	18.50	18.00	21.50	18.50	14.50
4	20.50	18.50	17.00	16.50	16.50
5	29.00	21.00	19.00	23.00	21.00
6	22.00	15.50	18.00	18.00	22.50
7	36.00	29.50	22.00	26.00	25.50
8	18.00	9.50	10.50	10.50	14.50
9	28.00	30.50	37.50	31.50	28.00
10	36.00	37.00	36.00	36.00	33.00
11	33.50	32.00	33.00	32.50	36.50
12	23.00	26.00	20.00	21.50	30.00
13	31.00	31.50	33.00	26.00	29.50
14	16.00	14.00	16.00	19.50	18.00
15	32.00	25.50	24.00	30.00	26.50
16	23.50	24.00	22.00	20.50	21.50

Group 2	Time on Target/Sec. 1	2	3	4	5
1	11.30	13.25	11.90	11.30	9.40
2	6.70	6.50	4.95	4.00	6.65
3	13.70	18.70	16.10	16.20	17.55
4	14.90	15.95	15.40	15.60	15.45
5	10.90	12.10	12.10	13.15	13.35
6	7.55	11.40	12.15	13.00	11.75
7	12.40	14.30	15.80	15.70	15.85
8	12.85	14.45	15.00	14.80	13.35
9	7.50	10.10	12.40	12.40	14.95
10	8.85	9.15	10.70	10.05	9.50
11	12.95	12.25	12.00	12.05	11.35
12	3.35	6.70	6.60	6.70	6.60
13	7.75	8.25	10.40	9.20	10.40
14	14.25	16.20	15.25	17.60	16.25
15	11.40	14.85	17.20	17.15	16.05
16	11.60	13.75	13.25	12.80	10.90

Group 2	Hits on Target 1	2	3	4	5
1	49.00	46.50	44.00	45.50	50.00
2	32.50	42.50	46.00	43.00	47.00
3	47.00	49.50	50.50	45.50	48.50
4	42.50	46.50	46.00	44.00	41.50
5	24.00	44.00	43.00	44.50	41.00
6	42.50	42.50	53.50	46.00	54.50
7	48.00	46.00	44.50	42.00	45.00
8	39.00	42.50	47.50	37.50	34.50
9	36.00	43.00	30.00	29.50	39.50
10	40.00	34.50	35.50	35.00	38.00
11	44.50	56.00	53.50	52.50	56.00
12	41.50	36.00	44.50	45.00	43.50
13	33.50	51.50	49.00	43.00	47.00
14	50.50	51.50	51.00	47.50	46.50
15	54.50	54.00	52.00	49.00	49.50
16	43.00	52.50	45.50	47.00	43.00

a. Arrange the data matrix such that the first five columns represent the measurements on the first variable and the last five columns the second dependent variable, so that the observation matrix is a 32×10 matrix. Letting

$$\mathbf{B} = \begin{bmatrix} \beta_{10} & \beta_{11} & \beta_{12} & \beta_{13} & \beta_{14} & \theta_{10} & \theta_{11} & \theta_{12} & \theta_{13} & \theta_{14} \\ \beta_{20} & \beta_{21} & \beta_{22} & \beta_{23} & \beta_{24} & \theta_{20} & \theta_{21} & \theta_{22} & \theta_{23} & \theta_{24} \end{bmatrix}$$

so that $p_1 = p_2 = q = 5$ and $p_3 = 0$ in (5.16.16), test the overall hypothesis of parallelism

$$H_0: \begin{bmatrix} \beta_{11} \\ \beta_{12} \\ \beta_{13} \\ \beta_{14} \\ --- \\ \theta_{11} \\ \theta_{12} \\ \theta_{13} \\ \theta_{14} \end{bmatrix} \begin{bmatrix} \beta_{21} \\ \beta_{22} \\ \beta_{23} \\ \beta_{24} \\ --- \\ \theta_{21} \\ \theta_{22} \\ \theta_{23} \\ \theta_{24} \end{bmatrix} \quad \text{or} \quad \begin{bmatrix} \boldsymbol{\beta}_1 \\ \boldsymbol{\theta}_1 \end{bmatrix} = \begin{bmatrix} \boldsymbol{\beta}_2 \\ \boldsymbol{\theta}_2 \end{bmatrix}$$

and tests of parallelism separately, a variable at a time,

$$H_0: \boldsymbol{\beta}_1 = \boldsymbol{\beta}_2 \quad \text{and} \quad H_0: \boldsymbol{\theta}_1 = \boldsymbol{\theta}_2$$

b. For the same data, test that the third- and fourth-order terms are all 0.

c. Are the growth curves for both variables coincident, for either variable? Determine confidence intervals for the regression weights.

d. Assume that no more than a quadratic function should be fitted to the data, so that $p = 6 < q$, and use $\mathbf{G} = \mathbf{S}$ to test the hypothesis that the quadratic term is 0 using the Rao-Khatri model.

e. From your analysis in part d, summarize your findings.

5.17 NONORTHOGONAL MULTIVARIATE DESIGNS

In multivariate factorial designs, it is not uncommon in practice for an experimenter to find that vector-valued observations for some subjects are not obtained. The result is unequal and disproportionate numbers of vector observations in each cell and empty cells, or a *nonorthogonal design.* The analysis of designs with this unbalance requires careful consideration since the subspaces associated with the effects are no longer orthogonal. In addition to nonorthogonality, an experimenter may find that observations within a vector are missing. The results are *incomplete multivariate data* and nonorthogonality.

Several authors (for example, Afifi and Elashoff, 1966, 1967, 1969; Timm, 1970; and Hartley and Hocking, 1971) have studied the problem of estimating $\boldsymbol{\mu}$ and $\boldsymbol{\Sigma}$ from incomplete multivariate data. Srivastava (1967) has considered the problems associated with tests of significance using incomplete nonorthogonal data, and Rao (1956), Trawinski and Bargmann (1964), and McDonald (1971), among others, have considered tests of significance in MANOVA designs with incomplete or nonorthogonal multivariate data.

Assuming complete vector-valued observations, an introduction to the analysis of a nonorthogonal MCR-*rc* design is discussed in this section. For a thorough discussion of the univariate case, which extends immediately to multivariate analysis, see Bancroft (1968), Searle (1971), or Timm and Carlson (in press). Before considering the multivariate two-way factorial design with an unequal number of observations per cell with interaction, the no-interaction model is presented.

The statistical model for the additive two-way layout with $N_{ij} > 0$ observations per cell is

$$\begin{aligned} &\mathbf{y}_{ijk} = \boldsymbol{\mu} + \boldsymbol{\alpha}_i + \boldsymbol{\beta}_j + \boldsymbol{\varepsilon}_{ijk} \\ &\boldsymbol{\varepsilon}_{ijk} \sim IN(\mathbf{0}, \boldsymbol{\Sigma}) \qquad i = 1, \ldots, r; j = 1, \ldots, c; k = 1, \ldots, N_{ij} \end{aligned} \tag{5.17.1}$$

Consideration of a specific example will suffice to illustrate the general theory and provide the similarity between univariate and multivariate analysis for non-orthogonal designs. With $i = 1, 2, 3, j = 1, 2; N_{11} = 2,\ N_{12} = 2,\ N_{21} = 2,\ N_{22} = 1,$

$N_{31} = 1$, and $N_{32} = 1$, the model using (5.17.1) becomes

$$
\begin{bmatrix} \mathbf{y}'_{111} \\ \mathbf{y}'_{112} \\ \mathbf{y}'_{121} \\ \mathbf{y}'_{122} \\ \mathbf{y}'_{211} \\ \mathbf{y}'_{212} \\ \mathbf{y}'_{221} \\ \mathbf{y}'_{311} \\ \mathbf{y}'_{321} \end{bmatrix}
=
\begin{bmatrix} 1 & 1 & 0 & 0 & 1 & 0 \\ 1 & 1 & 0 & 0 & 1 & 0 \\ 1 & 1 & 0 & 0 & 0 & 1 \\ 1 & 1 & 0 & 0 & 0 & 1 \\ 1 & 0 & 1 & 0 & 1 & 0 \\ 1 & 0 & 1 & 0 & 1 & 0 \\ 1 & 0 & 1 & 0 & 0 & 1 \\ 1 & 0 & 0 & 1 & 1 & 0 \\ 1 & 0 & 0 & 1 & 0 & 1 \end{bmatrix}
\begin{bmatrix} \mu_{11} & \mu_{12} \\ \alpha_{11} & \alpha_{12} \\ \alpha_{21} & \alpha_{22} \\ \alpha_{31} & \alpha_{32} \\ \beta_{11} & \beta_{12} \\ \beta_{21} & \beta_{22} \end{bmatrix}
+
\begin{bmatrix} \varepsilon'_{111} \\ \varepsilon'_{112} \\ \varepsilon'_{121} \\ \varepsilon'_{122} \\ \varepsilon'_{211} \\ \varepsilon'_{212} \\ \varepsilon'_{221} \\ \varepsilon'_{311} \\ \varepsilon'_{321} \end{bmatrix}
\tag{5.17.2}
$$

$$
\underset{(9 \times 2)}{\mathbf{Y}} = \underset{(9 \times 6)}{\mathbf{X}} \quad \underset{(6 \times 2)}{\mathbf{B}} + \underset{(9 \times 2)}{\mathbf{E}_0}
$$

where the $R(\mathbf{X}) = 4$.

The data for the example is represented as in Table 5.17.1. The notation employed in the table follows.

$$
\begin{aligned}
N_{ij} &= \text{number of observations in cell } AB_{ij} \\
N_{i+} &= \text{number of observation in } A_i \\
N_{+j} &= \text{number of observations in } B_j \\
N &= \text{total number of observations} \\
N &= \sum_i \sum_j N_{ij} = \sum_i N_{i+} = \sum_j N_{+j}
\end{aligned}
$$

Associated with the cell, row, and column totals for the observations in Table 5.17.1 are means. Letting

$$
\begin{aligned}
\mathbf{y}_{ij.} &= \frac{\sum_k \mathbf{y}_{ijk}}{N_{ij}} \\
\mathbf{y}_{i..} &= \frac{\sum_j \sum_k \mathbf{y}_{ijk}}{N_{i+}} = \frac{\sum_j N_{ij}\mathbf{y}_{ij.}}{N_{i+}} \\
\mathbf{y}_{.j.} &= \frac{\sum_i \sum_k \mathbf{y}_{ijk}}{N_{+j}} = \frac{\sum_i N_{ij}\mathbf{y}_{ij.}}{N_{+j}}
\end{aligned}
\tag{5.17.3}
$$

TABLE 5.17.1. Nonorthogonal-Design Example.

		Factor B: B_1	Factor B: B_2	Totals
Factor A	A_1	$\mathbf{y}'_{111}$, $\mathbf{y}'_{112}$ (N_{11})	$\mathbf{y}'_{121}$, $\mathbf{y}'_{122}$ (N_{12})	N_{1+}
	A_2	$\mathbf{y}'_{211}$, $\mathbf{y}'_{212}$ (N_{21})	$\mathbf{y}'_{221}$ (N_{22})	N_{2+}
	A_3	$\mathbf{y}'_{311}$ (N_{31})	$\mathbf{y}'_{321}$ (N_{32})	N_{3+}
Totals		N_{+1}	N_{+2}	$N_{++} = N$

TABLE 5.17.2. Table of Means for Nonorthogonal Design.

		Factor B B_1	B_2	Means
	A_1	$\mathbf{y}_{11.}$	$\mathbf{y}_{12.}$	$\mathbf{y}_{1..}$
Factor A	A_2	$\mathbf{y}_{21.}$	$\mathbf{y}_{22.}$	$\mathbf{y}_{2..}$
	A_3	$\mathbf{y}_{31.}$	$\mathbf{y}_{32.}$	$\mathbf{y}_{3..}$
Means		$\mathbf{y}_{.1.}$	$\mathbf{y}_{.2.}$	$\mathbf{y}_{...}$

a table of means corresponding to the data in Table 5.17.1 is shown in Table 5.17.2.

To estimate the parameters in a nonorthogonal design, the normal equations

$$(\mathbf{X}'\mathbf{X})\mathbf{B} = \mathbf{X}'\mathbf{Y} \tag{5.17.4}$$

are solved. For the example,

$$\mathbf{X}'\mathbf{X} = \begin{bmatrix} 9 & 4 & 3 & 2 & 5 & 4 \\ 4 & 4 & 0 & 0 & 2 & 2 \\ 3 & 0 & 3 & 0 & 2 & 1 \\ 2 & 0 & 0 & 2 & 1 & 1 \\ 5 & 2 & 2 & 1 & 5 & 0 \\ 4 & 2 & 1 & 1 & 0 & 4 \end{bmatrix} \tag{5.17.5}$$

For all designs discussed in this text, a simple expression for the generalized inverse of $\mathbf{X}'\mathbf{X}$ was always reported; however, in this case, no explicit expression can be given. Instead, since the $R(\mathbf{X}'\mathbf{X}) = 4$, two vector parameters in $\hat{\mathbf{B}}$ are set to $\mathbf{0}$ to obtain a submatrix of $\mathbf{X}'\mathbf{X}$ of full rank. For this example, we set $\hat{\boldsymbol{\mu}} = \mathbf{0}$ and $\hat{\boldsymbol{\beta}}_2 = \mathbf{0}$. The resulting matrix that has a regular inverse is the matrix

$$(\mathbf{X}'\mathbf{X})_s = \left[\begin{array}{ccc|c} 4 & 0 & 0 & 2 \\ 0 & 3 & 0 & 2 \\ 0 & 0 & 2 & 1 \\ \hline 2 & 2 & 1 & 5 \end{array}\right]$$

whose inverse is established by applying (1.5.28), the formula for a partitioned matrix. The inverse is computed to be

$$(\mathbf{X}'\mathbf{X})_s^{-1} = \frac{1}{(12)(13)} \begin{bmatrix} 57 & 24 & 18 & -36 \\ 24 & 84 & 24 & -48 \\ 18 & 24 & 96 & -36 \\ -36 & -48 & -36 & 72 \end{bmatrix}$$

so that

$$(5.17.6) \qquad (\mathbf{X}'\mathbf{X})^{-} = \frac{1}{(12)(13)}\begin{bmatrix} 0 & 0 & 0 & 0 & 0 & 0 \\ 0 & 57 & 24 & 18 & -36 & 0 \\ 0 & 24 & 84 & 24 & -48 & 0 \\ 0 & 18 & 24 & 96 & -36 & 0 \\ 0 & -36 & -48 & -36 & 72 & 0 \\ 0 & 0 & 0 & 0 & 0 & 0 \end{bmatrix}$$

Using the notation in (5.17.3),

$$\hat{\mathbf{B}} = (\mathbf{X}'\mathbf{X})^{-}\mathbf{X}'\mathbf{Y} = (\mathbf{X}'\mathbf{X})^{-}\begin{bmatrix} 9\mathbf{y}'_{...} \\ 4\mathbf{y}'_{1..} \\ 3\mathbf{y}'_{2..} \\ 2\mathbf{y}'_{3..} \\ 5\mathbf{y}'_{.1.} \\ 4\mathbf{y}'_{.2.} \end{bmatrix}$$

$$(5.17.7) \qquad = \begin{bmatrix} \mathbf{0}' \\ (19\mathbf{y}'_{1..} + 6\mathbf{y}'_{2..} + 3\mathbf{y}'_{3..} - 15\mathbf{y}'_{.1.})/13 \\ (8\mathbf{y}'_{1..} + 21\mathbf{y}'_{2..} + 4\mathbf{y}'_{3..} - 20\mathbf{y}'_{.1.})/13 \\ (6\mathbf{y}'_{1..} + 6\mathbf{y}'_{2..} + 16\mathbf{y}'_{3..} - 15\mathbf{y}'_{.1.})/13 \\ (-12\mathbf{y}'_{1..} - 12\mathbf{y}'_{1..} - 6\mathbf{y}'_{3..} + 30\mathbf{y}'_{.1.})/13 \\ \mathbf{0}' \end{bmatrix} = \begin{bmatrix} \hat{\boldsymbol{\mu}}' \\ \hat{\boldsymbol{\alpha}}'_1 \\ \hat{\boldsymbol{\alpha}}'_2 \\ \hat{\boldsymbol{\alpha}}'_3 \\ \hat{\boldsymbol{\beta}}'_1 \\ \hat{\boldsymbol{\beta}}'_2 \end{bmatrix}$$

Unique linear unbiased estimators for the parametric functions $\psi = \mathbf{c}'\mathbf{B}\mathbf{a}$ are determined by using

$$(5.17.8) \qquad \mathbf{H} = (\mathbf{X}'\mathbf{X})^{-}\mathbf{X}'\mathbf{X} = \begin{bmatrix} 0 & 0 & 0 & 0 & 0 & 0 \\ 1 & 1 & 0 & 0 & 0 & 1 \\ 1 & 0 & 1 & 0 & 0 & 1 \\ 1 & 0 & 0 & 1 & 0 & 1 \\ 0 & 0 & 0 & 0 & 1 & -1 \\ 0 & 0 & 0 & 0 & 0 & 0 \end{bmatrix}$$

and an arbitrary vector $\mathbf{t}' = [t_0, t_1, t_2, t_3, t'_1, t'_2]$. The estimable functions are

$$\psi = \mathbf{c}'\mathbf{B}\mathbf{a} = \mathbf{t}'\mathbf{H}\mathbf{B}\mathbf{a}$$

$$(5.17.9) \qquad = \mathbf{a}'\left[\sum_i t_i\boldsymbol{\mu} + \sum_i t_i\boldsymbol{\alpha}_i + t'_1\boldsymbol{\beta}_1 + \sum_i (t_i - t'_1)\boldsymbol{\beta}_2\right]$$

with estimators

$$\hat{\psi} = \mathbf{c}'\hat{\mathbf{B}}\mathbf{a} = \mathbf{t}'\mathbf{H}\hat{\mathbf{B}}\mathbf{a}$$

$$(5.17.10) \qquad = \mathbf{a}'\left(\sum_i t_i\boldsymbol{\alpha}_i + t'_1\hat{\boldsymbol{\beta}}_1\right)$$

When analyzing nonorthogonal designs, expressions of the form given in (5.17.9) and (5.17.10) are always calculated to determine which functions of the parameters are estimable for testing purposes and also to establish the value of the estimates for obtaining simultaneous confidence intervals. For nonorthogonal designs, the form of the estimators is in general different from those found in the orthogonal case. For the nonorthogonal design under investigation, it is seen that, for $\mathbf{c}' = [0, 1, -1, 0, 0, 0]$, $\mathbf{c}'\mathbf{H} = \mathbf{c}'$, so that $\psi = \alpha_1 - \alpha_2$ is estimable. By setting $t_1 = 1$ and $t_2 = -1$, $\hat{\psi} = (11\mathbf{y}_{1..} - 15\mathbf{y}_{2..} - \mathbf{y}_{3..} + 5\mathbf{y}_{.1.})/13$. Other functions and their estimates are summarized in Table 5.17.3.

TABLE 5.17.3.

Parametric Function	Estimator
$\psi_1 = \alpha_1 - \alpha_3$	$\hat{\psi}_1 = \mathbf{y}_{1..} - \mathbf{y}_{3..}$
$\psi_2 = \alpha_2 - \alpha_3$	$\hat{\psi}_2 = (2\mathbf{y}_{1..} + 15\mathbf{y}_{2..} - 12\mathbf{y}_{3..} - 5\mathbf{y}_{.1.})/13$
$\psi_3 = \alpha_1 - \alpha_2$	$\hat{\psi}_3 = (11\mathbf{y}_{1..} - 15\mathbf{y}_{2..} - \mathbf{y}_{3..} + 5\mathbf{y}_{.1.})/13$
$\psi_4 = \beta_1 - \beta_2$	$\hat{\psi}_4 = (-12\mathbf{y}_{1..} - 12\mathbf{y}_{2..} - 6\mathbf{y}_{3..} + 30\mathbf{y}_{.1.})/13$

TABLE 5.17.4. Sample Data for a Nonorthogonal Design.

		Factor B		
		B_1	B_2	Totals
	A_1	[10, 21] [12, 22] $N_{11} = 2$	[9, 17] [8, 15] $N_{12} = 2$	$N_{1+} = 4$
Factor A	A_2	[14, 27] [11, 23] $N_{21} = 2$	[12, 18] $N_{22} = 1$	$N_{2+} = 3$
	A_3	[7, 15] $N_{31} = 1$	[8, 18] $N_{32} = 1$	$N_{3+} = 2$
Totals		$N_{+1} = 5$	$N_{+2} = 4$	$N = 9$

TABLE 5.17.5. Means for Sample Data.

		Factor B		
		B_1	B_2	Means
	A_1	$\mathbf{y}'_{11.} = [11.0, 21.5]$	$\mathbf{y}'_{12.} = [8.5, 16.0]$	$\mathbf{y}'_{1..} = [9.75, 18.75]$
Factor A	A_2	$\mathbf{y}'_{21.} = [12.5, 25.0]$	$\mathbf{y}'_{22.} = [12.0, 18.0]$	$\mathbf{y}'_{2..} = [12.33, 22.67]$
	A_3	$\mathbf{y}'_{31.} = [7.0, 15.0]$	$\mathbf{y}'_{32.} = [8.0, 18.0]$	$\mathbf{y}'_{3..} = [7.50, 16.50]$
Means		$\mathbf{y}'_{.1.} = [10.80, 21.60]$	$\mathbf{y}'_{.2.} = [9.25, 17.00]$	$\mathbf{y}'_{...} = [10.11, 19.56]$

From Table 5.17.3, the comparison of the first level of A to the third level of A leads to the usual estimator since the values of N_{ij} for the cells are equal. However, for the other comparisons, the estimators are quite different. To evaluate the estimates in Table 5.17.3, consideration is given the sample data in Table 5.17.4 and the summary data in Table 5.17.5.

The values of the estimates, assuming no interaction or an additive model, become

$$\psi_1 = \alpha_1 - \alpha_3 \qquad \hat{\psi}_1 = \begin{bmatrix} 2.25 \\ 2.25 \end{bmatrix}$$

$$\psi_2 = \alpha_2 - \alpha_3 \qquad \hat{\psi}_2 = \begin{bmatrix} 4.65 \\ 5.50 \end{bmatrix}$$

$$\hat{\psi}_3 = \alpha_1 - \alpha_2 \qquad \hat{\psi}_3 = \begin{bmatrix} -2.40 \\ -3.25 \end{bmatrix}$$

$$\hat{\psi}_4 = \beta_1 - \beta_2 \qquad \hat{\psi}_4 = \begin{bmatrix} 1.08 \\ 4.00 \end{bmatrix}$$

Since differences in α's and β's are estimable under an additive model, the hypotheses

$$H_A: \text{all } \alpha_i\text{'s are equal}$$

$$H_B: \text{all } \beta_j\text{'s are equal}$$

are testable. The matrix $\mathbf{C}$ to test H_A is

$$\mathbf{C} = \begin{bmatrix} 0 & 1 & 0 & -1 & 0 & 0 \\ 0 & 0 & 1 & -1 & 0 & 0 \end{bmatrix} \tag{5.17.11}$$

and the matrix $\mathbf{C}$ to test H_B is

$$\mathbf{C} = [0, 0, 0, 0, 1, -1] \tag{5.17.12}$$

so that the degrees of freedom associated with the hypotheses H_A and H_B are, respectively, $v_A = 2$ and $v_B = 1$, the rank of the corresponding $\mathbf{C}$ matrix.

From the matrix $\hat{\mathbf{B}}$ defined in (5.17.7),

$$\hat{\mathbf{B}} = \begin{bmatrix} 0 & 0 \\ 9.21 & 16.75 \\ 11.61 & 20.00 \\ 6.96 & 14.50 \\ 1.08 & 4.00 \\ 0 & 0 \end{bmatrix}$$

the matrices **C** given in (5.17.11) and (5.17.12), and $\mathbf{A} = \mathbf{I}$, the hypotheses SSP matrices to test H_A and H_B are calculated:

$$\mathbf{Q}_{h_A} = (\mathbf{C}\hat{\mathbf{B}})'[\mathbf{C}(\mathbf{X}'\mathbf{X})^{-}\mathbf{C}']^{-1}(\mathbf{C}\hat{\mathbf{B}}) = \begin{bmatrix} 26.1461 & \\ 31.3500 & 37.9500 \end{bmatrix}$$

(5.17.13)

$$\mathbf{Q}_{h_B} = (\mathbf{C}\hat{\mathbf{B}})'[\mathbf{C}(\mathbf{X}'\mathbf{X})^{-}\mathbf{X}']^{-1}(\mathbf{C}\hat{\mathbf{B}}) = \begin{bmatrix} 2.5128 & \\ 9.3334 & 34.6667 \end{bmatrix}$$

With the SSP matrix for error defined as

$$\mathbf{Q}_e = \mathbf{Y}'[\mathbf{I} - \mathbf{X}(\mathbf{X}'\mathbf{X})^{-}\mathbf{X}']\mathbf{Y} = \begin{bmatrix} 11.4041 & \\ 16.2506 & 43.2512 \end{bmatrix} \tag{5.17.14}$$

the degrees of freedom associated with error are $v_e = N - R(\mathbf{X}) = 9 - 4 = 5$, or $N - r - c + 1$. By solving the determinantal equations $|\mathbf{Q}_{h_A} - \lambda \mathbf{Q}_e| = 0$ and $|\mathbf{Q}_{h_B} - \lambda \mathbf{Q}_e| = 0$, the eigenvalues for the hypotheses H_A and H_B are determined:

H_A	H_B
$\lambda_1 = 2.3596$	$\lambda_1 = .8757$
$\lambda_2 = .0174$	

Employing Roy's criterion, H_A and H_B are rejected at the significance level $\alpha = .05$ if

(5.17.15)

$$H_A: \theta_s = \frac{\lambda_1}{1 + \lambda_1} = .702 > \theta^{\alpha}_{(s,m,n)} = \theta^{.05}_{(2,-1/2,1.0)} \doteq .725$$

$$H_B: \theta_s = \frac{\lambda_1}{1 + \lambda_1} = .467 > \theta^{\alpha}_{(s,m,n)} = \theta^{.05}_{(1,0,1.0)} = .776$$

where

$$s = \min(v_h, p)$$

$$m = \frac{|v_h - p| - 1}{2}$$

$$n = \frac{N - r - c - p}{2}$$

Because of the nonorthogonality of the design, the sums of squares for the main-effect tests are not independent. Thus, if it is essential to control the type I error for the analysis.

Simultaneous confidence intervals for the estimable functions $\psi = \mathbf{c}'\mathbf{B}\mathbf{a}$ are evaluated by using the general expression

$$\hat{\psi} - c_0\hat{\sigma}_{\hat{\psi}} \leq \psi \leq \hat{\psi} + c_0\hat{\sigma}_{\hat{\psi}} \tag{5.17.16}$$

where

$$\hat{\sigma}_{\hat{\psi}} = \sqrt{\left[\mathbf{a}'\left(\frac{\mathbf{Q}_e}{v_e}\right)\mathbf{a}\right][\mathbf{c}'(\mathbf{X}'\mathbf{X})^-\mathbf{c}]} \tag{5.17.17}$$

and, for Roy's criterion,

$$c_0^2 = v_e\left(\frac{\theta^\alpha}{1-\theta^\alpha}\right)$$

For the comparison in column effects, a variable at a time, the estimates of

$$\psi_1 = \beta_{11} - \beta_{21} \quad \text{and} \quad \psi_2 = \beta_{12} - \beta_{22}$$

are

$$\hat{\psi}_1 = 1.08 \quad \text{and} \quad \hat{\psi}_2 = 4.00$$

while

$$\hat{\sigma}_{\hat{\psi}1} = 1.026 \quad \text{and} \quad \hat{\sigma}_{\hat{\psi}2} = 1.998$$

With $c_0^2 = v_e[\theta^\alpha/(1-\theta^\alpha)] = 4(.776/.224) = 13.86$, or $c_0 = 3.723$, simultaneous confidence intervals for ψ_1 and ψ_2 become

$$-2.74 \leq \beta_{11} - \beta_{21} \leq 4.90$$
$$-3.44 \leq \beta_{12} - \beta_{22} \leq 11.44$$

In deriving tests for main effects for the additive two-way multivariate layout, no empty cells were present in the design. However, the general theory may also be applied when some empty cells are present in the design. The only condition necessary for the analysis is that all differences $\boldsymbol{\alpha}_i - \boldsymbol{\alpha}_{i'}$ and $\boldsymbol{\beta}_j - \boldsymbol{\beta}_{j'}$, for $i \neq i'$ and $j \neq j'$, are estimable. This condition is readily checked from the general expression for estimable functions.

Most authors (for example, Graybill, 1961, Bancroft, 1968, and Searle, 1971) prefer to use the reduction in sum of squares method when analyzing nonorthogonal designs. However, when using this method, the experimenter is never really clear about what is being tested and if a significant test is obtained, he will usually not know how to find the correct estimates for the effects. This is not the case when one must determine estimable functions by using the general linear model theory and specifying the matrix **C** to test a hypothesis of interest.

To indicate the confusion that might arise when using the reduction method, we will now analyze the two-way factorial design with one observation per cell by the reduction in sum of squares procedure. Translating the notation used in Section 5.2 to the MANOVA model specified by (5.17.1) two sequences of elimination of effects are needed to test for main effects:

A	B
$R(\boldsymbol{\mu})$	$R(\boldsymbol{\mu})$
$R(\boldsymbol{\beta}\mid\boldsymbol{\mu})$	$R(\boldsymbol{\alpha}\mid\boldsymbol{\mu})$
$R(\boldsymbol{\alpha}\mid\boldsymbol{\mu}, \boldsymbol{\beta})$	$R(\boldsymbol{\beta}\mid\boldsymbol{\mu}, \boldsymbol{\alpha})$

$R(\boldsymbol{\alpha}\mid\boldsymbol{\mu}, \boldsymbol{\beta})$ is the reduction in sum of squares for fitting $\boldsymbol{\alpha}$, adjusting for $\boldsymbol{\mu}$ and $\boldsymbol{\beta}$, and is the appropriate SSP matrix for testing the null hypothesis H_0: all $\boldsymbol{\alpha}_i$'s are equal; $R(\boldsymbol{\beta}\mid\boldsymbol{\mu}, \boldsymbol{\alpha})$ is used to test H_0: all $\boldsymbol{\beta}_j$'s are equal. For an orthogonal design, $R(\boldsymbol{\beta}\mid\boldsymbol{\mu}, \boldsymbol{\alpha}) = R(\boldsymbol{\beta}\mid\boldsymbol{\mu})$, and $R(\boldsymbol{\alpha}|\boldsymbol{\mu}, \boldsymbol{\beta}) = R(\boldsymbol{\alpha}|\boldsymbol{\mu})$; however, this is not true for a nonorthogonal design. Furthermore, the sums of squares obtained from $R(\boldsymbol{\alpha}|\boldsymbol{\mu}, \boldsymbol{\beta})$ and $R(\boldsymbol{\beta}|\boldsymbol{\mu}, \boldsymbol{\alpha})$ are not independent.

The reduction sum of squares $R(\boldsymbol{\beta}|\boldsymbol{\mu})$ is obtained by fitting a one-way MANOVA model with an unequal number of observations to the levels of B, ignoring the A way of classification. For the one-way MANOVA model, functions of the form $\boldsymbol{\beta}_j^* - \boldsymbol{\beta}_{j'}^*$ are estimable where the model is $\mathbf{y}_{ijk} = \boldsymbol{\mu} + \boldsymbol{\beta}_j^* + \boldsymbol{\varepsilon}$. In terms of the cell means $\boldsymbol{\mu}_{ij}$ for the two-way additive model,

$$\boldsymbol{\beta}_j^* = \frac{\sum_i N_{ij}\boldsymbol{\mu}_{ij}}{N_{+j}}$$

Since $\boldsymbol{\mu}_{ij} = \boldsymbol{\mu} + \boldsymbol{\alpha}_i + \boldsymbol{\beta}_j$,

$$\boldsymbol{\beta}_j^* = \frac{\sum_i N_{ij}(\boldsymbol{\mu} + \boldsymbol{\alpha}_i + \boldsymbol{\beta}_j)}{N_{+j}} = \frac{\boldsymbol{\mu} + \boldsymbol{\beta}_j + \sum_i N_{ij}\boldsymbol{\alpha}_i}{N_{+j}}$$

For the additive model, functions of the form

$$\boldsymbol{\psi} = \boldsymbol{\mu} + \boldsymbol{\beta}_j + \frac{\sum_i N_{ij}\boldsymbol{\alpha}_i}{N_{+j}}$$

are estimable. To illustrate for our simple example, set $t_1' = 1$ and $t_i = N_{i1}/N_{+1}$, then $\boldsymbol{\psi}$ becomes

$$\boldsymbol{\psi} = \boldsymbol{\mu} + \frac{\sum_i N_{i1}\boldsymbol{\alpha}_i}{N_{+1} + \boldsymbol{\beta}_1}$$

Similarly, for $t_1' = 0$ and $t_i = N_{i2}/N_{+2}$,

$$\boldsymbol{\psi} = \boldsymbol{\mu} + \frac{\sum_i N_{i2}\boldsymbol{\alpha}_i}{N_{+2} + \boldsymbol{\beta}_2}$$

so that

$$\boldsymbol{\psi} = \boldsymbol{\beta}_1 - \boldsymbol{\beta}_2 + \frac{N_{11}\boldsymbol{\alpha}_1 + N_{21}\boldsymbol{\alpha}_1 + N_{31}\boldsymbol{\alpha}_3}{N_{+1}} - \frac{N_{12}\boldsymbol{\alpha}_1 + N_{22}\boldsymbol{\alpha}_2 + N_{32}\boldsymbol{\alpha}_3}{N_{+2}}$$

is estimable and hence testable. That is, $R(\boldsymbol{\beta}|\boldsymbol{\mu})$ is testing the null hypothesis

$$H_0: \boldsymbol{\beta}_j + \frac{\sum_i N_{ij}\boldsymbol{\alpha}_i}{N_{+j}} \text{ are all equal}$$

Similarly, $R(\boldsymbol{\alpha}|\boldsymbol{\mu})$ is testing

$$H_0: \boldsymbol{\alpha}_i + \frac{\sum_j N_{ij}\boldsymbol{\beta}_j}{N_{i+}} \text{ are all equal}$$

If the numbers of observations in the cells are unequal but proportionate for any two columns or rows, so that $N_{ij} = N_{i+}N_{+j}/N$, for all i and j, then $R(\boldsymbol{\beta}|\boldsymbol{\mu}) = R(\boldsymbol{\beta}|\boldsymbol{\mu}, \boldsymbol{\alpha})$ and $R(\boldsymbol{\alpha}|\boldsymbol{\mu}, \boldsymbol{\beta}) = R(\boldsymbol{\alpha}|\boldsymbol{\mu})$. Substituting for N_{ij} in the test obtained by evaluating $R(\boldsymbol{\beta}|\boldsymbol{\mu})$, the null hypothesis becomes

$$H_0: \boldsymbol{\beta}_j + \frac{\sum_i N_{i+}\boldsymbol{\alpha}_i}{N} \text{ are all equal}$$

which is equivalent to the equality of the $\boldsymbol{\beta}_j$'s since the coefficients N_{i+}/N do not depend on the subscript j. Two examples of designs with proportional cell numbers are given in Table 5.17.6.

TABLE 5.17.6. Proportional-Cell, Two-Way Designs.

		B		
		B_1	B_2	B_3
	A_1	$N_{11} = 8$	$N_{12} = 4$	$N_{13} = 12$
A	A_2	$N_{21} = 6$	$N_{22} = 2$	$N_{23} = 9$
	A_3	$N_{31} = 10$	$N_{32} = 4$	$N_{33} = 15$

		B	
		B_1	B_2
	A_1	$N_{11} = 7$	$N_{12} = 15$
A	A_2	$N_{21} = 7$	$N_{22} = 15$
	A_3	$N_{31} = 7$	$N_{32} = 15$

The reduction method does not clearly indicate what hypothesis is being tested or what to estimate if significance is observed. Furthermore, several orders of elimination must be determined to obtain the appropriate sum of squares and products matrices for testing. These shortcomings are not evidenced when the general linear model theory is used.

An important consideration in applying the general theory to test main-effect hypotheses in an additive model is the appropriateness of the specification of the model. The additive model assumes no interaction term. If an interaction is present in the population and an additive model is employed to analyze the data, the tests for main effects given above are not correct.

In practice, it is more common for an experimenter to hypothesize a non-additive model. The two-way multivariate factorial design with an unequal and disproportionate number of observations per cell, $N_{ij} > 0$, with interaction is

$$
\text{(5.17.18)} \qquad
\begin{aligned}
&\mathbf{y}_{ijk} = \boldsymbol{\mu} + \boldsymbol{\alpha}_i + \boldsymbol{\beta}_j + \boldsymbol{\gamma}_{ij} + \boldsymbol{\varepsilon}_{ijk} \\
&\boldsymbol{\varepsilon}_{ijk} \sim IN(\mathbf{0}, \boldsymbol{\Sigma}) \qquad i = 1, \ldots, r;\ j = 1, \ldots, c;\ k = 1, \ldots, N_{ij}
\end{aligned}
$$

In the simple example of this section, (5.17.18) becomes

(5.17.19)

$$
\begin{bmatrix} \mathbf{y}'_{111} \\ \mathbf{y}'_{112} \\ \mathbf{y}'_{121} \\ \mathbf{y}'_{122} \\ \mathbf{y}'_{211} \\ \mathbf{y}'_{212} \\ \mathbf{y}'_{221} \\ \mathbf{y}'_{311} \\ \mathbf{y}'_{321} \end{bmatrix}
=
\begin{bmatrix}
1 & 1 & 0 & 0 & 1 & 0 & 1 & 0 & 0 & 0 & 0 & 0 \\
1 & 1 & 0 & 0 & 1 & 0 & 1 & 0 & 0 & 0 & 0 & 0 \\
1 & 1 & 0 & 0 & 0 & 1 & 0 & 1 & 0 & 0 & 0 & 0 \\
1 & 1 & 0 & 0 & 0 & 1 & 0 & 1 & 0 & 0 & 0 & 0 \\
1 & 0 & 1 & 0 & 1 & 0 & 0 & 0 & 1 & 0 & 0 & 0 \\
1 & 0 & 1 & 0 & 1 & 0 & 0 & 0 & 1 & 0 & 0 & 0 \\
1 & 0 & 1 & 0 & 0 & 1 & 0 & 0 & 0 & 1 & 0 & 0 \\
1 & 0 & 0 & 1 & 1 & 0 & 0 & 0 & 0 & 0 & 1 & 0 \\
1 & 0 & 0 & 1 & 0 & 1 & 0 & 0 & 0 & 0 & 0 & 1
\end{bmatrix}
\begin{bmatrix}
\mu_{11} & \mu_{12} \\
\alpha_{11} & \alpha_{12} \\
\alpha_{21} & \alpha_{22} \\
\alpha_{31} & \alpha_{32} \\
\beta_{11} & \beta_{12} \\
\beta_{21} & \beta_{22} \\
\gamma_{111} & \gamma_{112} \\
\gamma_{121} & \gamma_{122} \\
\gamma_{211} & \gamma_{212} \\
\gamma_{221} & \gamma_{222} \\
\gamma_{311} & \gamma_{312} \\
\gamma_{321} & \gamma_{322}
\end{bmatrix}
+
\begin{bmatrix} \boldsymbol{\varepsilon}'_{111} \\ \vdots \\ \boldsymbol{\varepsilon}'_{321} \end{bmatrix}
$$

$$
\underset{(9 \times 2)}{\mathbf{Y}} = \underset{(9 \times 12)}{\mathbf{X}} \quad \underset{(12 \times 2)}{\mathbf{B}} + \underset{(12 \times 2)}{\mathbf{E}_0}
$$

From (5.17.19), the $R(\mathbf{X}) = 6$, and, by (5.6.6), a solution is

$$\hat{\mathbf{B}} = (\mathbf{X}'\mathbf{X})^{-}\mathbf{X}'\mathbf{Y} = \begin{bmatrix} \mathbf{0} \\ \mathbf{y}'_{11.} \\ \mathbf{y}'_{12.} \\ \mathbf{y}'_{21.} \\ \mathbf{y}'_{22.} \\ \mathbf{y}'_{31.} \\ \mathbf{y}'_{32.} \end{bmatrix} \tag{5.17.20}$$

using, for a g inverse of $\mathbf{X}'\mathbf{X}$,

$$(\mathbf{X}'\mathbf{X})^{-} = \left[\begin{array}{c|cccccc} \mathbf{0} & & & & \mathbf{0} & & \\ \hline & 1/2 & & & & \mathbf{0} & \\ & & 1/2 & & & & \\ & & & 1/2 & & & \\ \mathbf{0} & & & & 1 & & \\ & & & & & 1 & \\ & & \mathbf{0} & & & & 1 \end{array}\right]$$

where

$$\mathbf{X}'\mathbf{X} = \left[\begin{array}{cccccc|cccccc} 9 & 4 & 3 & 2 & 5 & 4 & 2 & 2 & 2 & 1 & 1 & 1 \\ 4 & 4 & 0 & 0 & 2 & 2 & 2 & 2 & 0 & 0 & 0 & 0 \\ 3 & 0 & 3 & 0 & 2 & 1 & 0 & 0 & 2 & 1 & 0 & 0 \\ 2 & 0 & 0 & 2 & 1 & 1 & 0 & 0 & 0 & 0 & 1 & 1 \\ 5 & 2 & 2 & 1 & 5 & 0 & 2 & 0 & 2 & 0 & 1 & 0 \\ 4 & 2 & 1 & 1 & 0 & 4 & 0 & 2 & 0 & 1 & 0 & 1 \\ \hline 2 & 2 & 0 & 0 & 2 & 0 & 2 & 0 & 0 & 0 & 0 & 0 \\ 2 & 2 & 0 & 0 & 0 & 2 & 0 & 2 & 0 & 0 & 0 & 0 \\ 2 & 0 & 2 & 0 & 2 & 0 & 0 & 0 & 2 & 0 & 0 & 0 \\ 1 & 0 & 1 & 0 & 0 & 1 & 0 & 0 & 0 & 1 & 0 & 0 \\ 1 & 0 & 0 & 1 & 1 & 0 & 0 & 0 & 0 & 0 & 1 & 0 \\ 1 & 0 & 0 & 1 & 0 & 1 & 0 & 0 & 0 & 0 & 0 & 1 \end{array}\right]$$

Furthermore,

$$\mathbf{H} = (\mathbf{X}'\mathbf{X})^{-}\mathbf{X}'\mathbf{X} = \begin{bmatrix} \mathbf{0} & \mathbf{0} \\ \mathbf{0} & \mathbf{I}_6 \end{bmatrix} \tag{5.17.21}$$

By utilizing (5.17.21) with an arbitrary vector $\mathbf{t}' = [t_0, t_1, t_2, t_3, t_1', t_2', t_{11}, t_{12}, \ldots, t_{32}]$,

$$\psi = \mathbf{c}'\mathbf{B}\mathbf{a} = \mathbf{t}'\mathbf{H}\mathbf{B}\mathbf{a} = \mathbf{a}'\left[\sum_i \sum_i t_{ij}(\boldsymbol{\mu} + \boldsymbol{\alpha}_i + \boldsymbol{\beta}_j + \boldsymbol{\gamma}_{ij})\right]$$

$$= \mathbf{a}'\left(\sum_i \sum_j t_{ij}\boldsymbol{\mu}_{ij}\right) \tag{5.17.22}$$

is estimated by

$$\hat{\psi} = \mathbf{c}'\hat{\mathbf{B}}\mathbf{a} = \mathbf{t}'\mathbf{H}\hat{\mathbf{B}}\mathbf{a} = \mathbf{a}'\left(\sum_i \sum_j t_{ij}\mathbf{y}_{ij.}\right)$$

$$= \mathbf{a}'\left(\sum_{ij} t_{ij}\hat{\boldsymbol{\mu}}_{ij}\right) \tag{5.17.23}$$

For factorial designs with empty cells, t_{ij} is set equal to zero when $N_{ij} = 0$. In this case, all interactions of the form

$$\boldsymbol{\psi} = \boldsymbol{\mu}_{ij} - \boldsymbol{\mu}_{i'j} - \boldsymbol{\mu}_{ij'} + \boldsymbol{\mu}_{i'j'}$$

$$= \boldsymbol{\gamma}_{ij} - \boldsymbol{\gamma}_{i'j} - \boldsymbol{\gamma}_{ij'} + \boldsymbol{\gamma}_{i'j'} \tag{5.17.24}$$

are not estimable. However, by considering sums and differences of the functions given in (5.17.24), which eliminate the $\boldsymbol{\mu}_{ij}$'s associated with the empty cells, $v_{AB} = f - r - c + 1$, where f is used to denote the number of cells filled, linearly independent functions are estimable. For example, suppose cell number 32 is empty, so that $N_{32} = 0$, then

$$\boldsymbol{\psi}_1 = \boldsymbol{\gamma}_{11} - \boldsymbol{\gamma}_{31} - \boldsymbol{\gamma}_{12} + \boldsymbol{\gamma}_{32} \quad \text{and} \quad \boldsymbol{\psi}_2 = \boldsymbol{\gamma}_{21} - \boldsymbol{\gamma}_{31} - \boldsymbol{\gamma}_{22} + \boldsymbol{\gamma}_{32}$$

would not be estimable since they involve $\boldsymbol{\mu}_{32}$, which is not estimable. However, the function $\boldsymbol{\psi} = \boldsymbol{\psi}_1 - \boldsymbol{\psi}_2 = \boldsymbol{\gamma}_{11} - \boldsymbol{\gamma}_{21} - \boldsymbol{\gamma}_{12} + \boldsymbol{\gamma}_{22}$ is estimable. There is only one independent estimable function involving the $\boldsymbol{\gamma}_{ij}$'s for the design considered in (5.17.19) if $N_{32} = 0$ since $v_{AB} = 5 - 3 - 2 + 1 = 1$. When all the N_{ij}'s > 0, all functions of the form given in (5.17.24) are estimable.

By writing the interaction hypothesis as $H_0 : \mathbf{CBA} = \mathbf{0}$, where the rows of $\mathbf{C}$ involve v_{AB} linearly independent functions of the form given in (5.17.24), a variable at a time, the hypothesis is evaluated by using

$$\mathbf{Q}_{AB} = (\mathbf{C}\hat{\mathbf{B}}\mathbf{A})'[\mathbf{C}(\mathbf{X}'\mathbf{X})^{-}\mathbf{C}']^{-1}(\mathbf{C}\hat{\mathbf{B}}\mathbf{A})$$

and

$$\mathbf{Q}_e = \mathbf{Y}'[\mathbf{I} - \mathbf{X}(\mathbf{X}'\mathbf{X})^{-}\mathbf{X}']\mathbf{Y}$$

The degrees of freedom associated with the hypothesis SSP matrix are $v_{AB} = f - r - c + 1$. The degrees of freedom for error are $v_e = N - f$.

For the data given in Table 5.17.3, using model (5.17.8), where $v_{AB} = 2$, linearly independent interactions are estimable. With the matrix $\mathbf{C}$ defined by

$$\mathbf{C} = \begin{bmatrix} 0 & 0 & 0 & 0 & 0 & 0 & 1 & -1 & 0 & 0 & -1 & 1 \\ 0 & 0 & 0 & 0 & 0 & 0 & 0 & 0 & 1 & -1 & -1 & 1 \end{bmatrix}$$

$$\hat{\mathbf{B}} = \begin{bmatrix} \multicolumn{2}{c}{\underset{(6 \times 2)}{\mathbf{0}}} \\ 11.0 & 21.5 \\ 8.5 & 16.0 \\ 12.5 & 25.0 \\ 12.0 & 18.0 \\ 7.0 & 15.0 \\ 8.0 & 18.0 \end{bmatrix}$$

and $\mathbf{A} = \mathbf{I}_2$, the interaction hypothesis

$$\begin{aligned} H_0 : \gamma_{11} - \gamma_{31} - \gamma_{12} + \gamma_{32} &= \mathbf{0} \\ \gamma_{21} - \gamma_{31} - \gamma_{22} + \gamma_{32} &= \mathbf{0} \end{aligned} \tag{5.17.25}$$

is tested.

For Wilks' Λ criterion or Roy's largest-root statistic, the interaction hypothesis is rejected at the significance level α if

$$\begin{aligned} \Lambda &= \frac{|\mathbf{Q}_e|}{|\mathbf{Q}_e + \mathbf{Q}_{AB}|} < U^{\alpha}(p, v_{AB}, N - f) \qquad \text{(Wilks)} \\ \theta_s &= \frac{\lambda_1}{1 + \lambda_1} > \theta^{\alpha}(s, m, n) \qquad \text{(Roy)} \end{aligned} \tag{5.17.26}$$

where s, m, and n are defined by

$$s = \min(v_{AB}, p)$$

$$m = \frac{|p - v_{AB}| - 1}{2}$$

$$n = \frac{N - f - p - 1}{2}$$

and λ_1 is the largest root of the characteristic equation $|\mathbf{Q}_{AB} - \lambda \mathbf{Q}_e| = 0$.

For the data in Table 5.17.3,

$$\mathbf{Q}_{AB} = \begin{bmatrix} 4.4039 & \\ 8.2500 & 32.7500 \end{bmatrix}$$

and

$$\begin{aligned} \mathbf{Q}_e &= \mathbf{Y}'[\mathbf{I} - \mathbf{X}(\mathbf{X}'\mathbf{X})^{-}\mathbf{X}']\mathbf{Y} \\ &= \sum_i \sum_j \sum_k \mathbf{y}_{ijk}\mathbf{y}'_{ijk} - \sum_i \sum_j N_{ij}\mathbf{y}_{ij.}\mathbf{y}'_{ij.} \\ &= \begin{bmatrix} 7.000 & \\ 8.000 & 10.500 \end{bmatrix} \end{aligned}$$

The roots of the characteristic equation $|\mathbf{Q}_{AB} - \lambda \mathbf{Q}_e| = 0$ are $\lambda_1 = 14.5534$ and $\lambda_2 = .5509$.

Employing Wilks' lambda criterion, the interaction hypothesis is rejected at the significance level $\alpha = .05$ if $\Lambda < U^{\alpha}(p, \nu_h, \nu_e) = U^{.05}_{(2,2,3)} = .018318$. Since $\Lambda = .0415$, the interaction hypothesis is not rejected. The estimators of $\boldsymbol{\psi}_1 = \boldsymbol{\gamma}_{11} - \boldsymbol{\gamma}_{31} - \boldsymbol{\gamma}_{12} + \boldsymbol{\gamma}_{32}$ and $\boldsymbol{\psi}_2 = \boldsymbol{\gamma}_{21} - \boldsymbol{\gamma}_{31} - \boldsymbol{\gamma}_{22} + \boldsymbol{\gamma}_{32}$ are

$$\hat{\boldsymbol{\psi}}_1 = \mathbf{y}_{11.} - \mathbf{y}_{31.} - \mathbf{y}_{12.} + \mathbf{y}_{32.} = \begin{bmatrix} 3.5 \\ 8.5 \end{bmatrix}$$

and

$$\hat{\boldsymbol{\psi}}_2 = \mathbf{y}_{21.} - \mathbf{y}_{31.} - \mathbf{y}_{22.} + \mathbf{y}_{32.} = \begin{bmatrix} 1.5 \\ 10.0 \end{bmatrix}$$

respectively.

In a two-way multivariate design with interaction, the usual tests of main effects that use the general theory, as in the familiar equal-cell-frequency case, do not exist. This is because, with unequal N_{ij} and empty cells, estimates involving only $\boldsymbol{\alpha}$'s or only $\boldsymbol{\beta}$'s are confounded by the $\boldsymbol{\gamma}_{ij}$'s as well as the other main-effect parameters and are dependent on the system of weights t_{ij}. For example, when using (5.17.22) to obtain simple differences in the $\boldsymbol{\beta}$'s, the parametric functions

$$\boldsymbol{\psi} = \boldsymbol{\beta}_j - \boldsymbol{\beta}_{j'} + \sum_i t_{ij}(\boldsymbol{\alpha}_i + \boldsymbol{\gamma}_{ij}) - \sum_i t_{ij'}(\boldsymbol{\alpha}_i + \boldsymbol{\gamma}_{ij'}) \tag{5.17.27}$$

for $j \neq j'$, are estimable provided the $\Sigma_i t_{ij} = \Sigma_i t_{ij'} = 1$, with $t_{ij} = 0$ if $N_{ij} = 0$ and $t_{ij'} = 0$ if $N_{ij'} = 0$. The estimator for ψ is

$$\hat{\boldsymbol{\psi}} = \sum_i t_{ij}\mathbf{y}_{ij.} - \sum_i t_{ij'}\mathbf{y}_{ij'.} \tag{5.17.28}$$

The parametric function $\boldsymbol{\psi}$ in (5.17.27) is confounded by differences in the $\boldsymbol{\alpha}$'s as well as the $\boldsymbol{\gamma}_{ij}$'s. For designs with proportional cell numbers, we can select $t_{ij} = N_{ij}/N_{+j}$ to eliminate the $\boldsymbol{\alpha}$'s. This is accomplished by selecting weights according to the pattern of the N_{ij}'s determined by the data. For the nonproportional-cell case, such weights are usually not appropriate. Similar results are evident for finding simple differences in the $\boldsymbol{\alpha}$'s; that is,

$$\boldsymbol{\psi} = \boldsymbol{\alpha}_i - \boldsymbol{\alpha}_{i'} + \sum_j t_{ij}(\boldsymbol{\beta}_j + \boldsymbol{\gamma}_{ij}) - \sum_j t_{i'j}(\boldsymbol{\beta}_j + \boldsymbol{\gamma}_{i'j}) \tag{5.17.29}$$

for $i \neq i'$, are estimable provided the $\Sigma_j t_{ij} = \Sigma_j t_{i'j} = 1$, and they can be estimated by

$$\hat{\boldsymbol{\psi}} = \sum_j t_{ij}\mathbf{y}_{ij.} - \sum_j t_{i'j}\mathbf{y}_{ij.} \tag{5.17.30}$$

where $t_{ij} = 0$ if $N_{ij} = 0$ and $t_{i'j} = 0$ if $N_{i'j} = 0$.

In summary, when cell frequencies are unequal and empty cells exist in a two-way design with interaction, main-effect hypotheses in the usual sense are not testable. By choosing weights that are data dependent, main-effect tests

$$\begin{aligned} H^A&: \text{all } \boldsymbol{\alpha}_i + \frac{\sum_j N_{ij}(\boldsymbol{\beta}_j + \boldsymbol{\gamma}_{ij})}{N_{i+}} \text{ are equal} \\ H^B&: \text{all } \boldsymbol{\beta}_j + \frac{\sum_i N_{ij}(\boldsymbol{\alpha}_i + \boldsymbol{\gamma}_{ij})}{N_{+j}} \text{ are equal} \end{aligned} \tag{5.17.31}$$

are testable.

If the interaction hypotheses tested by (5.17.26) are nonsignificant, many authors employing the reduction method suggest using the hypothesis SSP matrix [obtained in the additive model $R(\boldsymbol{\beta}|\boldsymbol{\mu}, \boldsymbol{\alpha})$ and $R(\boldsymbol{\alpha}|\boldsymbol{\mu}, \boldsymbol{\beta})$] and the error SSP matrix (obtained in the interaction model) to test for main-effect significant difference. This is incorrect. If one really had an additive model, the interaction SSP matrix would be included in the error SSP matrix. Furthermore, estimates of parametric functions $\boldsymbol{\psi}$ change since the design is nonorthogonal. This difficulty is avoided by constructing the matrix $\mathbf{C}$ to test hypotheses when using the general linear model theory.

When all cells are filled in a nonorthogonal two-way design with interaction, an appropriate set of weights might be $t_{ij} = 1/c$ when testing for an A-factor main effect. Then (5.17.29) becomes

$$\boldsymbol{\psi} = \boldsymbol{\alpha}_i - \boldsymbol{\alpha}_{i'} + \frac{\sum_j \boldsymbol{\gamma}_{ij} - \sum_j \boldsymbol{\gamma}_{i'j}}{c} \tag{5.17.32}$$

where $\boldsymbol{\psi}$ is estimated by

$$\hat{\boldsymbol{\psi}} = \frac{\sum_j \mathbf{y}_{ij.}}{c} - \frac{\sum_j \mathbf{y}_{i'j.}}{c} \tag{5.17.33}$$

By using these weights, the null hypothesis

$$H_A: \text{all } \boldsymbol{\alpha}_i + \frac{\sum_j \boldsymbol{\gamma}_{ij}}{c} \text{ are equal} \tag{5.17.34}$$

is testable. Similarly, by selecting weights $t_{ij} = 1/r$, the hypothesis

$$H_B: \boldsymbol{\beta}_j + \frac{\sum_i \boldsymbol{\gamma}_{ij}}{r} \tag{5.17.35}$$

is testable. However, the sums of squares for the tests are not independent.

To test (5.17.34), the matrix $\mathbf{C}$ for the example becomes

$$\mathbf{C} = \begin{bmatrix} 0 & 1 & 0 & -1 & 0 & 0 & 1/2 & 1/2 & 0 & 0 & -1/2 & -1/2 \\ 0 & 0 & 1 & -1 & 0 & 0 & 0 & 0 & 1/2 & 1/2 & -1/2 & -1/2 \end{bmatrix}$$

With $\hat{\mathbf{B}}$ defined as in the test for a significant interaction and $\mathbf{A} = \mathbf{I}_2$, the hypothesis SSP matrix becomes

$$\begin{aligned} \mathbf{Q}_A &= (\mathbf{C}\hat{\mathbf{B}})'[\mathbf{C}(\mathbf{X}'\mathbf{X})^-\mathbf{C}']^{-1}(\mathbf{C}\hat{\mathbf{B}}) \\ &= \begin{bmatrix} 26.2500 & \\ 27.7500 & 29.3650 \end{bmatrix} \end{aligned}$$

The error SSP matrix is

$$\begin{aligned} \mathbf{Q}_e &= \mathbf{Y}'[\mathbf{I} - \mathbf{X}(\mathbf{X}'\mathbf{X})^-\mathbf{X}']\mathbf{Y} \\ &= \begin{bmatrix} 7.000 & \\ 8.000 & 10.500 \end{bmatrix} \end{aligned}$$

Wilks' criterion is $\Lambda = .2051$. By comparing this with the critical constant $U^{.05}_{(2,2,3)} = .018318$, there do not appear to be significant differences among the levels of A when averaged over the levels of the B factor. Alternatively, by using the largest root of $|\mathbf{Q}_A - \lambda \mathbf{Q}_e| = 0$, $\lambda_1 = 3.868$, the Roy criterion could have been employed to test the hypothesis.

By adding the usual side conditions $\Sigma_i \gamma_{ij} = \Sigma_j \gamma_{ij} = \mathbf{0}$, (5.17.34) becomes

(5.17.36) $$H_A : \boldsymbol{\alpha}_1 = \boldsymbol{\alpha}_2 = \cdots = \boldsymbol{\alpha}_r$$

Then

$$\boldsymbol{\psi}_1 = \boldsymbol{\alpha}_1 - \boldsymbol{\alpha}_3 \qquad \hat{\boldsymbol{\psi}}_1 = \begin{bmatrix} 2.25 \\ 2.25 \end{bmatrix}$$

$$\boldsymbol{\psi}_2 = \boldsymbol{\alpha}_2 - \boldsymbol{\alpha}_3 \qquad \hat{\boldsymbol{\psi}}_2 = \begin{bmatrix} 4.75 \\ 5.00 \end{bmatrix}$$

$$\boldsymbol{\psi}_3 = \boldsymbol{\alpha}_1 - \boldsymbol{\alpha}_2 \qquad \hat{\boldsymbol{\psi}}_3 = \begin{bmatrix} -2.50 \\ -2.75 \end{bmatrix}$$

The standard errors of these contrasts, a variable at a time, are obtained in the usual manner by finding $\hat{\sigma}_{\hat{\psi}_i}$.

An approximate solution to the analysis of a two-way layout with interaction and all cells filled is to use an *unweighted-means analysis.* Then the cell means are used to compute the hypothesis SSP matrices $\mathbf{Q}_A$, $\mathbf{Q}_B$, and $\mathbf{Q}_{AB}$ as if a one-observation-per-cell, two-way design were being investigated. The matrix $\mathbf{Q}_e$ is computed as usual, using the observations and not the means. However, it is multiplied by the inverse of the harmonic mean of the cell frequencies where the *harmonic mean* is the reciprocal of the average of the reciprocals of the cell frequencies:

$$\bar{h} = \frac{1}{\sum_{i=1}^{r} \sum_{j=1}^{c} (1/N_{ij})/rc}$$

Thus the modified SSP matrix for error becomes $\mathbf{Q}_e^* = (1/\bar{h})\mathbf{Q}_e$ with degrees of freedom $v_e = N - rc$. The analysis now proceeds as in the two-way layout for the nonorthogonal case with interaction and all cells filled. The matrices $\mathbf{Q}_A$, $\mathbf{Q}_B$, and $\mathbf{Q}_{AB}$ are the hypothesis SSP matrices, with degrees of freedom $r - 1$, $c - 1$, and $(r - 1)(c - 1)$, respectively. $\mathbf{Q}_e^*$ defines the error SSP matrix, and $v_e = N - rc$ degrees of freedom. For further details on this procedure and others, see Yates (1934) and Searle (1971, p. 365).

Due to the complexities in analyzing nonorthogonal designs generally, the best solution to the problem is not to have unequal N_{ij}'s and empty cells in experimental designs.

EXERCISES 5.17

1. Given a two-way classification with two rows and two columns and the following univariate data; analyze the data.

		B: B_1	B_2	Totals
A	A_1	1 3	3	$N_{1+} = 3$
	A_2	4	5	$N_{2+} = 2$
Totals		$N_{+1} = 3$	$N_{+2} = 2$	$N_{++} = 5$

a. Use the model $y_{ij} = \mu + \alpha_i + \beta_j + \gamma_{ij}, \quad \varepsilon_{ij} \sim IN(0, \sigma^2)$, to determine point estimates for $\psi = \alpha_1 - \alpha_2$ and $\psi = \beta_1 - \beta_2$. Test the hypotheses $H_0: \alpha_1 = \alpha_2$ and $H_0: \beta_1 = \beta_2$.

b. Use the model $y_{ij} = \mu + \alpha_i + \beta_j + \gamma_{ij} + \varepsilon_{ij}, \quad \varepsilon_{ij} \sim IN(0, \sigma^2)$, to test the following hypotheses

$$H_0: \gamma_{11} - \gamma_{21} - \gamma_{12} + \gamma_{22} = 0$$

$$H_0: \alpha_i + \sum_{j=1}^{2} \frac{\gamma_{ij}}{2}$$

$$H_0: \alpha_i + \frac{1}{N_{i+}} \sum_{j=1}^{2} N_{ij}(\beta_j + \gamma_{ij})$$

Also, determine point estimates for appropriate contrasts.

c. Repeat parts a and b but switch the values in cells 12 and 22.

d. Summarize your findings.

2. John Levine and Leonard Saxe at the University of Pittsburgh obtained data to investigate the effects of social-support characteristics (allies and assessors) on conformity reduction under normative social pressure. The subjects were placed in a situation where three persons gave incorrect answers and a fourth person gave the correct answer. The dependent variables for the study are mean opinion (O) scores and mean visual-perception (V) scores for a nine-item test. High scores indicate more conformity. Analyze the following data from the unpublished study and summarize your findings.

		Assessor: Good		Assessor: Poor	
		O	V	O	V
Ally	Good	2.67	.67	1.44	.11
		1.33	.22	2.78	1.00
		.44	.33	1.00	.11
		.89	.11	1.44	.22
		.44	.22	2.22	.11
		1.44	−.22	.89	.11
		.33	.11	2.89	.22
		.78	−.11	.67	.11
				1.00	.67
	Poor	1.89	.78	2.22	.11
		1.44	.00	1.89	.33
		1.67	.56	1.67	.33
		1.78	−.11	1.89	.78
		1.00	1.11	.78	.22
		.78	.44	.67	.00
		.44	.00	2.89	.67
		.78	.33	2.67	.67
		2.00	.22	2.78	.44
		1.89	.56		
		2.00	.56		
		.67	.56		
		1.33	.22		

CHAPTER 6

PRINCIPAL-COMPONENT ANALYSIS

6.1 INTRODUCTION

In multivariate data analysis, an experimenter is often confronted with a large set of correlated variables. In the exploration of the interdependencies among the variates, he may be satisfied with a few linear functions of the variables containing most of the information of the original set, provided "meaning" is evident for the hypothetical linear functions.

Principal-component analysis is a procedure developed by Hotelling (1933) that offers reduction of a large set of correlated variables to a smaller number of uncorrelated hypothetical components. The technique is useful in understanding the dependencies existing among variables of a set and also in determining whether subsets of variables cluster, or go with one another.

6.2 PRINCIPAL COMPONENTS

Given a random p vector, $\mathbf{Y}$ with mean $\boldsymbol{\mu} = \mathbf{0}$ and variance-covariance matrix $\boldsymbol{\Sigma}$ of rank p, the object of principal-component analysis becomes one of obtaining a set of new variates called *principal components* or *principal variates*, which are linear combinations of the elements of $\mathbf{Y}$ such that the components are uncorrelated and the variance of the jth principal component is maximal. If the assumption is made that $\boldsymbol{\mu} \neq \mathbf{0}$, the mathematical development of the theory of principal-component analysis is not affected, since variances are being maximized. Since principal components are linear combinations of the elements of the random vector $\mathbf{Y}$, it is essential to the analysis of principal components that the units of measurement of the responses making up $\mathbf{Y}$ are commensurable. If this were not the case, the new principal variates would lack meaning; in addition, the criterion of maximum variance is questionable with unequal units of measurement.

The first principal component of the random vector $\mathbf{Y}$ is the linear combination

$$X_1 = \mathbf{p}_1'\mathbf{Y} \tag{6.2.1}$$

of the elements of $\mathbf{Y}$ such that the variance of X_1 is maximal. To determine the unique linear function $\mathbf{p}_1'\mathbf{Y}$ that maximizes the variance of X_1, a vector $\mathbf{p}_1$ is sought such that the variance

$$V(X_1) = V(\mathbf{p}_1'\mathbf{Y}) = \mathbf{p}_1'\boldsymbol{\Sigma}\mathbf{p}_1 \tag{6.2.2}$$

is maximal, subject to the constraint that $\mathbf{p}_1'\mathbf{p}_1 = 1$. The condition that $\mathbf{p}_1'\mathbf{p}_1 = 1$ is imposed to ensure the uniqueness of the coefficients for the principal variates (except for sign). From (1.8.48), the vector that maximizes (6.2.2), subject to the constraint that $\mathbf{p}_1'\mathbf{p}_1 = 1$, is the characteristic vector associated with the largest root of the equation

$$|\boldsymbol{\Sigma} - \lambda\mathbf{I}| = 0 \tag{6.2.3}$$

The largest variance of X_1 is the largest root λ_1 of (6.2.3).

To determine the second principal component, the linear combination

$$X_2 = \mathbf{p}_2'\mathbf{Y} \tag{6.2.4}$$

must be formed to be uncorrelated with X_1 and to have maximal variance. For X_2 to be uncorrelated with X_1, the covariance between X_2 and X_1 must be zero. That the

$$\operatorname{cov}(X_2, X_1) = \mathbf{p}_2'\boldsymbol{\Sigma}\mathbf{p}_1 = \mathbf{p}_2'\mathbf{p}_1\lambda_1 = 0 \tag{6.2.5}$$

implies that $\mathbf{p}_2'\mathbf{p}_1 = 0$. Using the method of Lagrange multipliers, the function $F = \mathbf{p}_2'\boldsymbol{\Sigma}\mathbf{p}_2 - \lambda(\mathbf{p}_2'\mathbf{p}_2 - 1) - \theta(\mathbf{p}_2'\mathbf{p}_1)$ is maximized. By taking the partial derivative of F with respect to $\mathbf{p}_2$, the equation

$$\frac{\partial F}{\partial \mathbf{p}_2} = 2\boldsymbol{\Sigma}\mathbf{p}_2 - 2\lambda\mathbf{p}_2 - \theta\mathbf{p}_1 \tag{6.2.6}$$

is set equal to zero. Premultiplication of (6.2.6) by $\mathbf{p}_1'$ and using (6.2.5) indicates that $\theta = 0$. Hence the vector $\mathbf{p}_2$ must satisfy the equation $(\boldsymbol{\Sigma} - \lambda\mathbf{I})\mathbf{p}_2 = 0$. That is, the vector associated with the second largest root of $|\boldsymbol{\Sigma} - \lambda\mathbf{I}| = 0$ yields the second principal component, which has maximal variance λ_2. Continuing, the remaining principal components may be determined in a similar manner.

Alternatively, since $\mathbf{\Sigma}$ is a symmetric and positive-semidefinite matrix, there exists an orthogonal matrix $\mathbf{P}$ such that

$$\mathbf{P}'\mathbf{\Sigma}\mathbf{P} = \mathbf{\Lambda} = \text{Dia}\,[\lambda_j] \tag{6.2.7}$$

where $\lambda_1 \geq \lambda_2 \geq \cdots \geq \lambda_p \geq 0$ by the spectral decomposition theorem. Setting

$$\mathbf{X} = \mathbf{P}'\mathbf{Y} \tag{6.2.8}$$

the jth element X_j of $\mathbf{X}$ is the jth principal component of $\mathbf{Y}$. The variance of the jth principal component is λ_j and the components are uncorrelated since the

$$V(\mathbf{X}) = \mathbf{P}'V(\mathbf{Y})\mathbf{P} = \mathbf{P}'\mathbf{\Sigma}\mathbf{P} = \mathbf{\Lambda} \tag{6.2.9}$$

The geometry of principal-component analysis is observable from the discussion associated with maxima and minima, the geometry of quadratic forms, and the multivariate normal distribution. Given that $\mathbf{Y}$ has a multivariate normal distribution with mean $\boldsymbol{\mu} = \mathbf{0}$ and variance-covariance matrix $\mathbf{\Sigma}$, the density function of $\mathbf{Y}$ is

$$f(\mathbf{y}) = (2\pi)^{-p/2}|\mathbf{\Sigma}|^{-1/2}\,e^{-\mathbf{y}'\mathbf{\Sigma}^{-1}\mathbf{y}/2}$$

Associated with the quadratic form $\mathbf{y}'\mathbf{\Sigma}^{-1}\mathbf{y} = Q$ are ellipsoids of constant density centered at $\boldsymbol{\mu} = \mathbf{0}$. The maximal distance from the origin to the surface of the ellipsoid, for a given value of Q, is obtained by maximizing $f(y_1, y_2, \ldots, y_p) = \mathbf{y}'\mathbf{y} = y_1^2 + y_2^2 + \cdots + y_p^2$, subject to the constraint that $\mathbf{y}'\mathbf{\Sigma}^{-1}\mathbf{y} = Q$. From (1.8.44), this problem is reduced to finding the eigenvalues of $\mathbf{\Sigma}$. That is, the length of the longest semimajor axis of the family of ellipsoids defined by $\mathbf{y}'\mathbf{\Sigma}^{-1}\mathbf{y} = Q$ is $\sqrt{\lambda_1 Q}$, where λ_1 is the largest root of $\mathbf{\Sigma}$. By solving the characteristic equation $|\mathbf{\Sigma} - \lambda\mathbf{I}| = 0$, the lengths of the principal axes of the ellipsoid $\mathbf{y}'\mathbf{\Sigma}^{-1}\mathbf{y} = Q$ are acquired from the eigenvalues. The orientation of the principal axes is located by employing the eigenvectors that contain the direction cosines of the ith old axis with the jth new axis for a rigid rotation:

$$\mathbf{P} = \begin{bmatrix} \cos\theta_{11} & \cos\theta_{12} & \cdots & \cos\theta_{1p} \\ \cos\theta_{21} & \cos\theta_{22} & \cdots & \cos\theta_{2p} \\ \vdots & \vdots & \ddots & \vdots \\ \cos\theta_{p1} & \cos\theta_{p2} & \cdots & \cos\theta_{pp} \end{bmatrix} = [\cos\theta_{ij}] \tag{6.2.10}$$

EXAMPLE 6.2.1. Let

$$\mathbf{\Sigma} = \begin{bmatrix} 1 & 1/2 \\ 1/2 & 1 \end{bmatrix}$$

Then $\lambda_1 = 3/2$ and $\lambda_2 = 1/2$, so that $\sqrt{\lambda_1 Q}$ and $\sqrt{\lambda_2 Q}$ are the lengths of the semimajor principal axes of the ellipse $\mathbf{y}'\mathbf{\Sigma}^{-1}\mathbf{y} = Q$. Since

$$\mathbf{\Sigma}^{-1} = \begin{bmatrix} 4/3 & -2/3 \\ -2/3 & 4/3 \end{bmatrix}$$

$$Q = \mathbf{y}'\mathbf{\Sigma}^{-1}\mathbf{y} = \frac{4y_1^2}{3} - \frac{4y_1y_2}{3} + \frac{4y_2^2}{3}$$

or $y_1^2 - y_1y_2 + y_2^2 = 3/4$ if $Q = 1$. The ellipse for $Q = 1$ is represented in Figure 6.2.1, where $a = \sqrt{3/2}$ and $b = \sqrt{1/2}$.

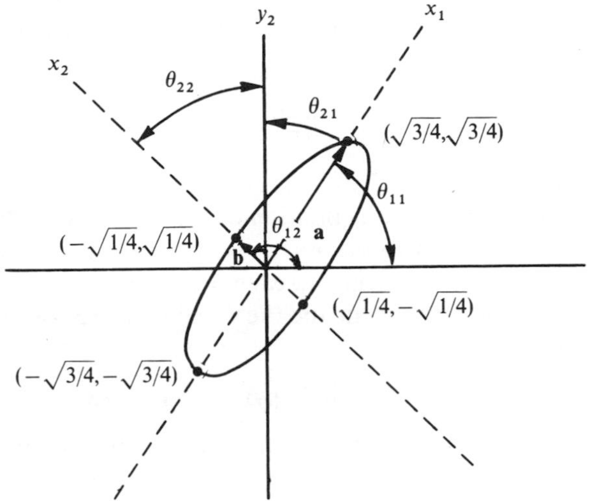

FIGURE 6.2.1.

The matrix of characteristic vectors associated with the roots is

$$\mathbf{P} = \begin{bmatrix} \cos\theta_{11} & \cos\theta_{12} \\ \cos\theta_{21} & \cos\theta_{22} \end{bmatrix} = \begin{bmatrix} \cos 45^\circ & \cos 135^\circ \\ \cos 45^\circ & \cos 45^\circ \end{bmatrix} = \begin{bmatrix} 1/\sqrt{2} & -1/\sqrt{2} \\ 1/\sqrt{2} & 1/\sqrt{2} \end{bmatrix}$$

The principal components are found by evaluating the expression $\mathbf{X} = \mathbf{P}'\mathbf{Y}$. The matrix $\mathbf{P}'$ rotates the old system through a 45° angle to the new principal-axes system that maximizes variances.

The theory of principal-component analysis has been developed by using a known variance-covariance matrix $\mathbf{\Sigma}$. In applying the method to sample data, the unbiased estimate of the variance-covariance matrix $\mathbf{S}$ is used in place of $\mathbf{\Sigma}$ and is computed by using the $N \times p$ data matrix $\mathbf{Y}$,

$$\mathbf{Y} = \begin{bmatrix} y_{11} & y_{12} & \cdots & y_{1p} \\ y_{21} & y_{22} & \cdots & y_{2p} \\ \vdots & \vdots & \vdots & \vdots \\ y_{N1} & y_{N2} & \cdots & y_{Np} \end{bmatrix} \tag{6.2.11}$$

where the rows of $\mathbf{Y}$ are independent random vectors sampled from a multivariate distribution with mean $\boldsymbol{\mu} = \mathbf{0}$ and variance-covariance matrix $\mathbf{\Sigma}$. Provided the units of measurement of the p variables are conformable, the characteristic equation

$$|\mathbf{S} - \lambda\mathbf{I}| = 0 \tag{6.2.12}$$

is solved for its roots and vectors such that $\mathbf{P}'\mathbf{S}\mathbf{P} = \mathbf{\Lambda}$ or $\mathbf{S} = \mathbf{P}\mathbf{\Lambda}\mathbf{P}' = \Sigma_j \lambda_j \mathbf{p}_j \mathbf{p}'_j$ to determine principal components in a sample. The jth component has the form

$$X_j = \mathbf{p}'_j \mathbf{Y} = p_{1j}Y_1 + p_{2j}Y_2 + \cdots + p_{pj}Y_p \tag{6.2.13}$$

where p_j is the jth column of the matrix $\mathbf{P}$ so that $\mathbf{p}'_j$ is the jth row of $\mathbf{P}'$.

Since the sum of the roots of the equation given in (6.2.12) equals the trace of $\mathbf{S}$, the ratio

$$\frac{\lambda_j}{\mathrm{Tr}(\mathbf{S})} \tag{6.2.14}$$

is a measure of the importance of the jth component in accounting for the total variability in the system. Therefore evaluation of the proportion of the total variance each component "explains" is possible. As in discriminant analysis, which establishes the importance of each principal variate to each of the variables in the sample, we use the correlation between the component and the variable. The covariance between each variable and the jth component is

$$\mathrm{cov}(\mathbf{Y}, X_j) = \mathrm{cov}(\mathbf{Y}, \mathbf{p}'_j\mathbf{Y}) = \mathbf{S}\mathbf{p}_j = \lambda_j \mathbf{p}_j \tag{6.2.15}$$

since $(\mathbf{S} - \lambda_j \mathbf{I})\mathbf{p}_j = 0$. Hence the correlation between the ith variable and the jth component is

$$\mathrm{cor}(Y_i, X_j) = \frac{\mathrm{cov}(Y_j, X_j)}{\sqrt{V(Y_i)}\sqrt{V(X_j)}} = \frac{\lambda_j p_{ij}}{s_i\sqrt{\lambda_j}} = \frac{p_{ij}\sqrt{\lambda_j}}{s_i}$$

The matrix of correlations is denoted by $[\mathrm{Dia}\ \mathbf{S}]^{-1/2}\mathbf{P}\mathbf{\Lambda}^{1/2}$.

The relationship between the components and the variables is represented by

$$\begin{aligned} x_1 &= p_{11}y_1 + p_{21}y_2 + \cdots + p_{p1}y_p \\ x_2 &= p_{12}y_1 + p_{22}y_2 + \cdots + p_{p2}y_p \\ &\ \ \vdots \\ x_p &= p_{1p}y_1 + p_{2p}y_2 + \cdots + p_{pp}y_p \end{aligned} \quad \text{or} \quad \mathbf{x} = \mathbf{P}'\mathbf{y} \tag{6.2.16}$$

and

$$\begin{aligned} y_1 &= p_{11}x_1 + p_{12}x_2 + \cdots + p_{1p}x_p \\ y_2 &= p_{21}x_1 + p_{22}x_2 + \cdots + p_{2p}x_p \\ &\ \ \vdots \\ y_p &= p_{p1}x_1 + p_{p2}x_2 + \cdots + p_{pp}x_p \end{aligned} \quad \text{or} \quad \mathbf{y} = \mathbf{P}\mathbf{x} \tag{6.2.17}$$

The values of x_j in (6.2.16) are called *component scores* and the scores y_i based on the components are termed *variate scores*. The scores for y_i will agree with the original observations only if all p components are in use when the $R(\mathbf{S}) = p$. In practice, an experimenter employs only a subset of the p components to represent the variability in the y's, realizing that some information in the original set of p variables is sacrificed for fewer hypothetical variables that account for most of the variance.

To acquire component scores by using the centered data matrix $\mathbf{Y}$, the transformation

$$\underset{(N \times k)}{\mathbf{X}} = \underset{(N \times p)}{\mathbf{Y}} \underset{(p \times k)}{\mathbf{P}} \tag{6.2.18}$$

is utilized where the columns of **P** are the first k normalized characteristic vectors of **S**. The matrix **X** contains uncorrelated principal variates that have maximal variance λ_j. Fortunately, in practice, λ_j is not usually equal to $\lambda_{j'}$. If $\lambda_{j'} = \lambda_j$, the principal components would have little interpretive value since any set of rigid rotations of axes with equal lengths is as good as any other (see Anderson, 1958, p. 275).

Throughout the development of the principal-component technique, the assumption has been that $\boldsymbol{\mu} = \mathbf{0}$ since a scale factor does not affect maximizing a variance. However, if $E(\mathbf{Y}) \neq \mathbf{0}$ for the data matrix **Y**, the following formula is employed to compute component scores,

$$\underset{(N \times k)}{\mathbf{X}} = \underset{(N \times N)}{[\mathbf{I} - \mathbf{1}(\mathbf{1}'\mathbf{1})^{-1}\mathbf{1}']} \underset{(N \times p)}{\mathbf{Y}} \underset{(p \times k)}{\mathbf{P}} \tag{6.2.19}$$

as the ellipsoid that is being rotated is not centered at the origin; thus **Y** in (6.2.19) is a raw data matrix rather than a deviation matrix as in (6.2.18).

By employing correlations to depict the relationship between the components and the variables, Table 6.2.1 is constructed to summarize the correlations, the standard deviations of the variables under investigation, the eigenvalues, and the percentage of the total variance each component accounts for in the sample. From Table 6.2.1, representations (6.2.16) and (6.2.17) are immediately determinable.

TABLE 6.2.1. Summary of Principal-Component Analysis Using a Variance-Covariance Matrix.

Variables	Components X_1	X_2	$\cdots$	X_p	Standard deviations
Y_1	$p_{11}\sqrt{\lambda_1}/s_1$	$p_{12}\sqrt{\lambda_2}/s_1$	$\cdots$	$p_{1p}\sqrt{\lambda_p}/s_1$	s_1
Y_2	$p_{21}\sqrt{\lambda_1}/s_2$	$p_{22}\sqrt{\lambda_2}/s_2$	$\cdots$	$p_{2p}\sqrt{\lambda_p}/s_2$	s_2
$\vdots$	$\vdots$	$\vdots$	$\vdots$	$\vdots$	$\vdots$
Y_p	$p_{p1}\sqrt{\lambda_1}/s_p$	$p_{p2}\sqrt{\lambda_2}/s_p$	$\cdots$	$p_{pp}\sqrt{\lambda_p}/s_p$	s_p
Eigenvalues	λ_1	λ_2	$\cdots$	λ_p	
Percentage of total variance	$\lambda_1/\mathrm{Tr}(\mathbf{S})$	$\lambda_2/\mathrm{Tr}(\mathbf{S})$	$\cdots$	$\lambda_p/\mathrm{Tr}(\mathbf{S})$	
Cumulative percentage of total variance	$\lambda_1/\mathrm{Tr}(\mathbf{S})$	$(\lambda_1 + \lambda_2)/\mathrm{Tr}(\mathbf{S})$	$\cdots$	1	

In education and psychology, the measures on the variables Y_i are not often commensurable. This situation can be avoided by standardizing all p variables so that **Y** is replaced by the standard-score-data matrix **Z**. The matrix used in the analysis of principal components is the sample-correlation matrix **R**. Generally, the components obtained from **S** are not the same as those gained from **R**. That is, principal variates are not invariant to changes in scale. This is due to the fact that the roots of $|\mathbf{D\Sigma D} - \delta\mathbf{I}| = 0$ are not generally the same as the roots of $|\mathbf{\Sigma} - \lambda\mathbf{I}| = 0$ for a diagonal matrix **D**. Furthermore, the distribution theory of components acquired from a correlation matrix is more complex than for a variance-covariance matrix (see Girshick, 1939b, and Anderson, 1963). If the original observations have a multivariate normal distribution, the standardized observations are only asymptotically normal. Another disadvantage in using **R** is that the original measurements may be distorted when all variables are treated "equally."

When extracting principal variates from a correlation matrix, the characteristic equation

$$|\mathbf{R} - \delta \mathbf{I}| = 0 \qquad (6.2.20)$$

is solved for its roots and vectors. For a $p \times p$ correlation matrix, let $\delta_1, \delta_2, \ldots, \delta_p$ and $\mathbf{q}_1, \mathbf{q}_2, \ldots, \mathbf{q}_p$ denote the roots and vectors of **R**. Then

$$\mathbf{\Delta} = \mathbf{Q}'\mathbf{R}\mathbf{Q}$$

$$\mathbf{R} = \mathbf{Q}'\mathbf{\Delta}\mathbf{Q}' = \sum_j \delta_j \mathbf{q}_j \mathbf{q}_j'$$

$$\mathbf{Q}\mathbf{Q}' = \mathbf{I}$$

where $\mathbf{\Delta}$ is a diagonal matrix with δ_i as diagonal elements and **Q** is the matrix of eigenvectors.

The representation of components and variates when using a correlation matrix is

$$\begin{aligned} x_1 &= q_{11}z_1 + q_{21}z_1 + \cdots + q_{p1}z_p \\ x_2 &= q_{12}z_1 + q_{22}z_2 + \cdots + q_{p2}z_p \\ &\vdots \\ x_p &= q_{1p}z_1 + q_{2p}z_2 + \cdots + q_{pp}z_p \end{aligned} \quad \text{or} \quad \mathbf{x} = \mathbf{Q}'\mathbf{z} \qquad (6.2.21)$$

and

$$\begin{aligned} z_1 &= q_{11}x_1 + q_{12}x_2 + \cdots + q_{1p}x_p \\ z_2 &= q_{21}x_1 + q_{22}x_2 + \cdots + q_{2p}x_p \\ &\vdots \\ z_p &= q_{p1}x_1 + q_{p2}x_2 + \cdots + q_{pp}x_p \end{aligned} \quad \text{or} \quad \mathbf{z} = \mathbf{Q}\mathbf{x} \qquad (6.2.22)$$

Since the $\text{Tr}(\mathbf{R}) = p$, the percentage of total variance accounted for by the jth component is

$$\frac{\delta_j}{\text{Tr}(\mathbf{R})} = \frac{\delta_j}{p} \qquad (6.2.23)$$

Component scores using the data matrix **Z** are obtained from the transformation

$$\underset{(N \times k)}{\mathbf{X}} = \underset{(N \times p)}{\mathbf{Z}} \; \underset{(p \times k)}{\mathbf{Q}} \qquad (6.2.24)$$

where the columns of **Q** are the first k characteristic vectors of **R**.

The correlation between the ith variable and the jth component is $q_{ij}\sqrt{\delta_j}$. A summary of a principal-component analysis using correlations is presented in Table 6.2.2. Letting $\mathbf{F} = \mathbf{Q}\mathbf{\Delta}^{1/2} = [q_{ij}\sqrt{\delta_j}]$, the matrix product $\mathbf{F}'\mathbf{F} = \mathbf{\Delta}$ yields the matrix of eigenvalues. This demonstrates that the jth column sum of squares of the elements in Table 6.2.2 equals the eigenvalue δ_j. The matrix product

$$\mathbf{F}\mathbf{F}' = \mathbf{R} = \sum_{j=1} \delta_j \mathbf{p}_j \mathbf{p}_j'$$

TABLE 6.2.2. Summary of Principal-Component Analysis Using a Correlation Matrix.

Variables	Components			
	X_1	X_2	$\cdots$	X_p
Z_1	$q_{11}\sqrt{\delta_1}$	$q_{12}\sqrt{\delta_2}$	$\cdots$	$q_{1p}\sqrt{\delta_p}$
Z_2	$q_{21}\sqrt{\delta_1}$	$q_{22}\sqrt{\delta_2}$	$\cdots$	$q_{2p}\sqrt{\delta_p}$
$\vdots$	$\vdots$	$\vdots$	$\vdots$	$\vdots$
Z_p	$q_{p1}\sqrt{\delta_1}$	$q_{p2}\sqrt{\delta_2}$	$\cdots$	$q_{pp}\sqrt{\delta_p}$
Eigenvalues	δ_1	δ_2	$\cdots$	δ_p
Percentage of total variance	δ_1/p	δ_2/p	$\cdots$	δ_p/p
Cumulative percentage of total variance	δ_1/p	$(\delta_1 + \delta_2)/p$	$\cdots$	1

allows the unique factorization of $\mathbf{R}$ in terms of principal components. The row sum of squares in Table 6.2.2 is 1. If the rank of $\mathbf{R} = r \leq p$, fewer than p principal variates may generate $\mathbf{R}$. However, if the number of components employed in the representation is less than the rank of the matrix factored, the correlations are never reproduced exactly. In addition, some of the variance in the system will be unaccounted for by the components. In practice, an experimenter is usually satisfied if 70% to 80% of the total variance can be accounted for by no more than five or six components. If the variance-covariance matrix $\mathbf{S}$ is analyzed, the matrix products $\mathbf{WW}' = \mathbf{S}$ and $\mathbf{W}'\mathbf{W} = \mathbf{\Lambda}$, where $\mathbf{W} = \mathbf{P}\mathbf{\Lambda}^{1/2}$, are formed to generate $\mathbf{S}$ and $\mathbf{\Lambda}$.

When employing a correlation matrix in component analysis, many authors in the social sciences standardize the components to achieve unit variance by dividing the jth component by the $\sqrt{\delta_j}$. Expressions (6.2.21) and (6.2.22) are then represented by

(6.2.25)

$$
\begin{aligned}
x_1^* &= q_{11}z_1/\sqrt{\delta_1} + q_{21}z_2/\sqrt{\delta_1} + \cdots + q_{p1}z_p/\sqrt{\delta_1} \\
x_2^* &= q_{12}z_1/\sqrt{\delta_2} + q_{22}z_2/\sqrt{\delta_2} + \cdots + q_{p2}z_p/\sqrt{\delta_2} \\
&\vdots \\
x_p^* &= q_{1p}z_1/\sqrt{\delta_p} + q_{p2}z_2/\sqrt{\delta_p} + \cdots + q_{pp}z_p/\sqrt{\delta_p}
\end{aligned}
\qquad \text{or} \qquad \mathbf{x}^* = \mathbf{\Delta}^{-1/2}\mathbf{Q}'\mathbf{z}
$$

and

(6.2.26)

$$
\begin{aligned}
z_1 &= q_{11}\sqrt{\delta_1}x_1^* + q_{12}\sqrt{\delta_2}x_2^* + \cdots + q_{1p}\sqrt{\delta_p}x_p^* \\
z_2 &= q_{21}\sqrt{\delta_1}x_1^* + q_{22}\sqrt{\delta_2}x_2^* + \cdots + q_{2p}\sqrt{\delta_p}x_p^* \\
&\vdots \\
z_p &= z_{p1}\sqrt{\delta_1}x_1^* + q_{p2}\sqrt{\delta_2}x_2^* + \cdots + q_{pp}\sqrt{\delta_p}x_p^*
\end{aligned}
\qquad \text{or} \qquad \mathbf{z} = \mathbf{Q}\mathbf{\Delta}^{1/2}\mathbf{x}^*
$$

respectively. From (6.2.15), the coefficients in (6.2.26) are the correlations between the ith variable and the jth component. Component scores determined by using (6.2.25) have mean 0 and variance 1, the same as the original standardized variables Z_i. Thus the matrix of component scores is acquired by using the expression $\mathbf{X}^* = \mathbf{ZQ\Lambda}^{-1/2} = \mathbf{ZF(F'F)}^{-1}$.

EXERCISES 6.2

1. For the sample variance-covariance matrix

$$\mathbf{S} = \begin{bmatrix} 26.64 & & \\ 8.52 & 9.85 & \\ 18.29 & 8.27 & 22.08 \end{bmatrix}$$

based on 151 observations, construct Table 6.2.1.

   a. What is the correlation between the second component and the first variable?

   b. Express each component as a linear combination of the variables. What is the sample variance of the first component? What proportion of the total sample variance is accounted for by the first component?

   c. By writing $\mathbf{S}$ as $\mathbf{S} = \mathbf{P\Lambda P'}$, where $\mathbf{W} = \mathbf{P\Lambda}^{1/2}$, verify that $\mathbf{W'W} = \mathbf{\Lambda}$.

2. For the sample correlation matrix

$$\mathbf{R} = \begin{bmatrix} 1.00 & & & & \\ .086 & 1.000 & & & \\ -.031 & .187 & 1.000 & & \\ -.034 & .242 & .197 & 1.000 & \\ .085 & .129 & .080 & .327 & 1.000 \end{bmatrix}$$

based on 151 observations, construct Table 6.2.2 for the two sample eigenvalues larger than 1.

   a. What is the correlation between the second component and the first variable?

   b. What proportion of the total sample variance is accounted for by the first and second components?

   c. Express the first component as a linear combination of the five variables such that the mean of the component scores would be 0 and the variance 1.

6.3 EXAMPLES—PRINCIPAL-COMPONENT ANALYSIS

Principal-component analysis is a descriptive procedure for analyzing relationships that may exist in a set of quantitative variables. Usually the technique is not utilized as an end in itself, but as a method for illustrating, modeling, and combining variables for further analysis.

Whether insight into data is best achieved by using the variance-covariance matrix or a correlation matrix is not determinable by statistical procedures. Psychological and educational differences in the scales of different subtests are often arbitrary. In this case, it has been common practice to use the correlation matrix.

EXAMPLE 6.3.1. Shin (1971) collected data on intelligence, creativity, and achievement for 116 subjects in the eleventh grade in suburban Pittsburgh. The Otis Quick Scoring Mental Ability Test, Guilford's Divergent Productivity Battery, and Kropp and Stoker's Lisbon Earthquake Achievement Test were used to gather one IQ score, six creativity measures, and six achievement measures for each subject. In addition to the IQ variable (1), the variables included in the creativity test were ideational fluency (2), spontaneous flexibility (3), associational fluency (4), expressional fluency (5), originality (6), and elaboration (7). The achievement measures were knowledge (8), comprehensive (9), application (10), analysis (11), synthesis (12), and evaluation (13). The correlation matrix for the study is presented in Table 6.3.1.

TABLE 6.3.1. Matrix of Intercorrelations among IQ, Creativity, and Achievement Variables.

Tests	1	2	3	4	5	6	7	8	9	10	11	12	13
1	1.00												
2	.16	1.00											
3	.32	.71	1.00										
4	.24	.12	.12	1.00									
5	.43	.34	.45	.43	1.00								
6	.30	.27	.33	.24	.33	1.00							
7	.43	.21	.11	.42	.46	.32	1.00						
8	.67	.13	.27	.21	.39	.27	.38	1.00					
9	.63	.18	.24	.15	.36	.33	.26	.62	1.00				
10	.57	.08	.14	.09	.25	.13	.23	.44	.66	1.00			
11	.59	.10	.16	.09	.25	.12	.28	.58	.66	.64	1.00		
12	.45	.13	.23	.42	.50	.41	.47	.46	.47	.37	.53	1.00	
13	.24	.08	.15	.36	.28	.21	.26	.30	.24	.19	.29	.58	1.00

By subjecting the correlation matrix in Table 6.3.1 to a principal-component analysis, Shin's results are summarized in Table 6.3.2, using the form specified in Table 6.2.2.

TABLE 6.3.2. Summary of Principal-Component Analysis Using 13 × 13 Correlation Matrix.

Tests	Components 1	2	3
1 (IQ)	.79	−.22	−.14
2 (Ideational fluency)	.36	.62	−.52
3 (Spontaneous flexibility)	.47	.56	−.54
4 (Associational fluency)	.45	.34	.55
5 (Expressional fluency)	.67	.38	.07
6 (Originality)	.50	.36	.02
7 (Elaboration)	.59	.18	.34
8 (Knowledge)	.75	−.23	−.08
9 (Comprehension)	.76	−.34	−.23
10 (Application)	.64	−.48	−.21
11 (Analysis)	.71	−.48	−.13
12 (Synthesis)	.76	.04	.35
13 (Evaluation)	.50	.09	.46
Eigenvalues	5.11	1.81	1.45
Percentage of total variance	39.3	13.9	11.2
Cumulative percentage of total variance	39.3	53.2	64.4

Examination of the results in Table 6.3.2 indicates that the first component is a general measure of mental ability where only one measure, ideational fluency, has a correlation considerably lower than .50. The second component is bipolar, comparing low levels of creativity with low levels of achievement. The correlations of synthesis and evaluation with the component are low. This second component may be termed verbal creativity. The third component compares higher levels of creativity and achievement with lower levels of creativity and achievement. This last component is a higher-level cognitive component consisting of higher levels of creativity and achievement.

Since the total variance acquired by the three components is only 64.4%, using the components for data reduction is not too meaningful. The analysis does lead to a better understanding of creativity, achievement, and intelligence variables in an experimental setting. That is, perhaps creativity may be considered a learning aptitude that might affect the relationship between achievement and general intelligence.

Shin employed the rule proposed by Kaiser (1960) to retain only those components with eigenvalues larger than 1; however, this applies to population values and not sample estimates. Occasions can arise wherein components with roots less than 1 may be included and even where components with eigenvalues greater than 1 may be excluded. A researcher should not be bound by a strict rule but allow himself some flexibility in the analysis of data. For example, extract a fourth component for this example by using the power method summarized in Section 1.7 and interpret the results. With four components, over 70% of the total variance is accounted for in the sample and a natural interpretation can be given the fourth component.

EXAMPLE 6.3.2. Di Vesta and Walls (1970) studied mean semantic differential ratings given by fifth-grade children for 487 words. The semantic differential ratings were obtained on the following eight scales: friendly/unfriendly (1), good/bad (2), nice/awful (3), brave/not brave (4), big/little (5), strong/weak (6), moving/still (7), and fast/slow (8). Table 6.3.3 shows the intercorrelations among mean semantic differential ratings for one list of 292 words.

TABLE 6.3.3. Intercorrelations of Ratings on the Following Semantic Differential Scales: (1) Friendly/Unfriendly, (2) Good/Bad, (3) Nice/Awful, (4) Brave/Not Brave, (5) Big/Little, (6) Strong/Weak, (7) Moving/Still, and (8) Fast/Slow.

Scale	1	2	3	4	5	6	7	8
1	1.00							
2	.95	1.00						
3	.96	.98	1.00					
4	.68	.70	.68	1.00				
5	.33	.35	.31	.52	1.00			
6	.60	.63	.61	.79	.61	1.00		
7	.21	.19	.19	.43	.31	.42	1.00	
8	.30	.31	.31	.57	.29	.57	.68	1.00

Table 6.3.4. results from submitting the correlation matrix in Table 6.3.3 to a principal-component analysis, following the format of Table 6.2.2. The first component in Table 6.3.4 presents an overall response-set component of the subjects to the list of

TABLE 6.3.4. Summary of Principal-Component Analysis Using 8 × 8 Correlation Matrix.

	Components		
	1	2	3
1 (Friendly/Unfriendly)	.87	−.42	−.13
2 (Good/Bad)	.88	−.42	−.10
3 (Nice/Awful)	.87	−.44	−.15
4 (Brave/Not Brave)	.89	.10	.06
5 (Big/Little)	.58	.26	.72
6 (Strong/Weak)	.85	.19	.22
7 (Moving/Still)	.49	.70	−.29
8 (Fast/Slow)	.61	.61	−.32
Eigenvalues	4.77	1.53	.81
Percentage of total variance	59.6	19.1	10.1
Cumulative percentage of total variance	59.6	78.7	88.8

words. The second component is bipolar, comparing evaluative behavior (friendly/unfriendly, good/bad, and nice/awful) with activity judgments (moving/still and fast/slow). The third component is dominated by the weight .72 and may be termed a size component.

Since Di Vesta and Walls also report the standard deviations of the ratings on each scale, an analysis of the variance-covariance matrix was submitted to a principal-component analysis. The variance-covariance matrix is included in Table 6.3.5. Following the format of Table 6.2.1, a summary of the principal-component analysis is found in Table 6.3.6.

Comparing the principal-component analysis, using S, with the analysis using **R**, only two components are required to account for the same amount of variance in the sample. The first component is again a general response-set component, but in this case it is dominated by evaluation behavior. The second component is an activity component. Employing S, a potency component is not evidenced.

TABLE 6.3.5. Variance-Covariance Matrix of Ratings on Semantic Differential Scales.

Scale	1	2	3	4	5	6	7	8
1	1.44							
2	1.58	1.93						
3	1.55	1.83	1.82					
4	.61	.73	.69	.56				
5	.23	.28	.24	.23	.34			
6	.56	.68	.64	.46	.28	.61		
7	.19	.20	.17	.24	.13	.25	.56	
8	.21	.25	.24	.24	.10	.25	.29	.33

TABLE 6.3.6 Summary of Principal-Component Analysis Using 8 × 8 Variance-Covariance Matrix.

Variables	Components 1	Components 2	Standard Deviations
1 (Friendly/Unfriendly)	.97	−.13	1.20
2 (Good/Bad)	.98	−.12	1.39
3 (Nice/Awful)	.98	−.16	1.35
4 (Brave/Not Brave)	.78	.43	.75
5 (Big/Little)	.42	.47	.58
6 (Strong/Weak)	.72	.50	.78
7 (Moving/Still)	.28	.80	.75
8 (Fast/Slow)	.40	.71	.57
Eigenvalues	5.77	.95	
Percentage of total variance	76.1	12.5	
Cumulative percentage of total variance	76.1	88.6	

By using the first two eigenvectors of $|\mathbf{S} - \lambda\mathbf{I}| = 0$, the first two components are represented by

$$x_1 = .484(y_1 - \bar{y}_1) + .569(y_2 - \bar{y}_2) + .551(y_3 - \bar{y}_3) + .243(y_4 - \bar{y}_4) + .100(y_5 - \bar{y}_5) + .232(y_6 - \bar{y}_6) + .087(y_7 - \bar{y}_7) + .096(y_8 - \bar{y}_8)$$

$$x_2 = -.156(y_1 - \bar{y}_1) - .178(y_2 - \bar{y}_2) - .215(y_3 - \bar{y}_3) + .328(y_4 - \bar{y}_4) + .278(y_5 - \bar{y}_5) + .404(y_6 - \bar{y}_6) + .612(y_7 - \bar{y}_7) + .417(y_8 - \bar{y}_8)$$

for the raw data matrix given by Di Vesta and Walls.

After a principal-component analysis, it is often helpful to plot the correlations between the variables and the components in the component space. This allows evaluation of those variables that tend to be associated with one another. For example, using the matrix **S** in the Di Vesta and Walls study, Figure 6.3.1 shows such a plot. Observation of the plot indicates that variables 1, 2, and 3 form a cluster while the rest of the variables form a less distinct cluster. Following such a plot, researchers sometimes rotate the axes in Figure 6.3.1 for "meaning" by postmultiplying the matrix of weights by

$$\mathbf{T} = \begin{bmatrix} \cos\theta & -\sin\theta \\ \sin\theta & \cos\theta \end{bmatrix} \quad \text{or} \quad \mathbf{T}' = \begin{bmatrix} \cos\theta & \sin\theta \\ -\sin\theta & \cos\theta \end{bmatrix}$$

for counterclockwise and clockwise rigid-orthogonal rotations of axes. However, this practice should be restricted to factor analysis. In component analysis, the criterion of maximum variance is destroyed by such transformations. In principal-component analysis, a "factorization" of **R** or **S** is unique (provided the λ_i are not equal) except for sign.

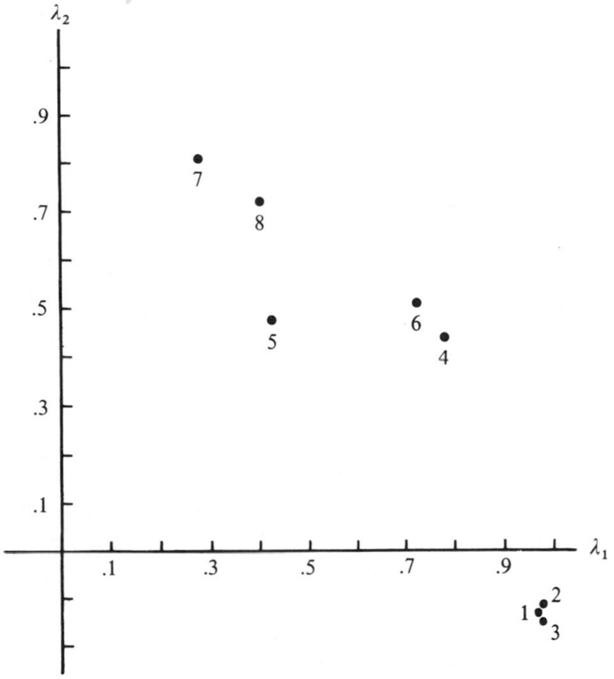

FIGURE 6.3.1. Plot of first two components using **S**.

EXERCISES 6.3

1. Use Table 6.3.1, the correlation matrix reported by Shin (1971), to extract and interpret the fourth principal component.

2. Based on scores obtained on two quizzes (Q), one midterm (M), and a final examination (F) for 30 students, a teacher obtained the following correlation matrix.

$$\mathbf{R} = \begin{bmatrix} 1.00 & & & \\ .80 & 1.00 & & \\ .85 & .79 & 1.00 & \\ .81 & .86 & .83 & 1.00 \end{bmatrix}$$

(columns: Q_1, Q_2, M, F)

How might one combine the scores to obtain an index of each student's performance?

3. Postovsky (1970) collected data on listening (L), speaking (S), reading (R), and writing (W) Russian for 28 subjects in a language-learning experiment. The correlation matrix and standard deviations of the variables (measured on a common scale) were reported as follows.

$$\mathbf{R} = \begin{bmatrix} 1.000 & & & \\ .727 & 1.000 & & \\ .593 & .571 & 1.000 & \\ .596 & .589 & .781 & 1.000 \end{bmatrix}$$

(columns: L, S, R, W)

Standard Deviations

L: 5.228 S: 8.534 R: 3.432 W: 8.480

a. Following the analysis of Walls' data, analyze both **R** and **S** for Postovsky's data.

b. Summarize your findings.

6.4 SOME STATISTICAL TESTS IN PRINCIPAL-COMPONENT ANALYSIS

Distribution assumptions are not necessary when using principal-component analysis as an exploratory procedure to describe a set of variables. However, to use the method as a quasi-confirmatory procedure, tests of hypotheses and confidence intervals for population roots and vectors may be of interest. To employ principal-component analysis in this manner, it is assumed that each of the N p-vector observations is sampled from a multivariate normal distribution.

Tests Using a Variance-Covariance Matrix. In this text a root λ_i, extracted from a variance-covariance matrix, denotes both a population and a sample eigenvalue. In this section it is essential that a clear distinction be made between the population value and the estimated value of λ_i. Thus let $\lambda_1, \lambda_2, \ldots, \lambda_p > 0$ denote the distinct roots, in the population, of the determinantal equation $|\mathbf{\Sigma} - \lambda\mathbf{I}| = 0$, and let $\mathbf{p}_1, \mathbf{p}_2, \ldots, \mathbf{p}_p$ denote the corresponding vectors for the $p \times p$ variance-covariance matrix $\mathbf{\Sigma}$ such that $\mathbf{p}_i'\mathbf{p}_j = 0$ and $\mathbf{p}_i'\mathbf{p}_i = 1$. For the sample, let $\hat{\lambda}_1, \hat{\lambda}_2, \ldots, \hat{\lambda}_p > 0$ and $\hat{\mathbf{p}}_1, \hat{\mathbf{p}}_2, \ldots, \hat{\mathbf{p}}_p$ represent the roots and vectors, respectively, of the equation $|\mathbf{S} - \hat{\lambda}\mathbf{I}| = 0$, where $\mathbf{S}$ is the usual unbiased estimator of $\mathbf{\Sigma}$.

To construct approximate confidence intervals for the population parameters λ_j when using large-sample theory, the following result is utilized (see Girshick, 1939b, and Anderson, 1963). As the sample size N tends to infinity ($N \to \infty$), $\hat{\lambda}_j - \lambda_j$ divided by its standard error $\sigma_{\hat{\lambda}_j} = \lambda_j\sqrt{2/(N-1)}$ approaches a normal distribution with mean 0 and variance 1 in the limit such that each is independently distributed; that is,

$$L_j = \frac{\hat{\lambda}_j - \lambda_j}{\lambda_j\sqrt{2/(N-1)}} \to IN(0, 1) \qquad \text{for } j = 1, \ldots, p \tag{6.4.1}$$

Result (6.4.1) is used to construct an approximate $100(1-\alpha)\%$ confidence interval for λ_j. That is, letting $Z^{\alpha/2}$ denote the upper $\alpha/2$ percentage point of the standard normal distribution, the

$$P\left[-Z^{\alpha/2} \le \frac{\hat{\lambda}_j - \lambda_j}{\lambda_j\sqrt{2/(N-1)}} \le Z^{\alpha/2}\right] \doteq 1 - \alpha \tag{6.4.2}$$

From (6.4.2), a $100(1-\alpha)\%$ confidence interval for λ_j is

$$\frac{\hat{\lambda}_j}{1 + Z^{\alpha/2}\sqrt{2/(N-1)}} \le \lambda_j \le \frac{\hat{\lambda}_j}{1 - Z^{\alpha/2}\sqrt{2/(N-1)}} \tag{6.4.3}$$

EXAMPLE 6.4.1. For Di Vesta and Walls' data, an approximate 95% confidence interval for the first eigenvalue is

$$\frac{5.774}{1 + 1.96(\sqrt{2/291})} \leq \lambda_1 \leq \frac{5.774}{1 - 1.96(\sqrt{2/291})}$$

$$4.967 \leq \lambda_1 \leq 6.894$$

A test for the null hypothesis that the vector of coefficients $\mathbf{p}_j$ of the jth principal component is equal to a specified value $\mathbf{p}_{0j}$—that is,

(6.4.4) $$H_0 : \mathbf{p}_j = \mathbf{p}_{0j}$$

for the jth distinct root λ_j, has been derived by Anderson (1963). Employing the fact that $\sqrt{N-1}(\hat{p}_j - p_j)$ has a limiting multivariate normal distribution with mean $\mathbf{0}$ and variance-covariance matrix,

(6.4.5) $$\lambda_{j'} \sum_{\substack{j=1 \\ j \neq j'}}^{p} \left[\frac{\lambda_j}{(\lambda_{j'} - \lambda_j)^2}\right] \mathbf{p}_j \mathbf{p}_{j'}$$

Anderson demonstrates that

(6.4.6) $$X^2 = (N-1)(\hat{\lambda}_j \mathbf{p}'_{0j} \mathbf{S}^{-1} \mathbf{p}_{0j} + \hat{\lambda}_j^{-1} \mathbf{p}'_{0j} \mathbf{S} \mathbf{p}_{0j} - 2)$$

is distributed asymptotically as a χ^2 distribution when H_0 is true with $\nu = p - 1$ degrees of freedom. The null hypothesis given in (6.4.6) is rejected at the significance level α if

(6.4.7) $$X^2 > \chi^2_\alpha(p-1)$$

where $\chi^2_\alpha(p-1)$ denotes the upper α percentage point of a χ^2 distribution with $p - 1$ degrees of freedom.

The confidence intervals for the roots λ_j given in (6.4.3) are appropriate only if they are distinct. As a result and with the realization that the interpretive value of principal components with equal roots is meaningless due to the lack of uniqueness of position in the component space, it is important to be able to test whether a subset of the roots or the last m roots are equal.

Bartlett (1954) proposed a test procedure for determining whether the smallest m roots of $\mathbf{\Sigma}$ are equal. Lawley (1956) tried to introduce more precision to Bartlett's procedure; however, James (1969) indicates that Lawley's correction is a little conservative.

Anderson (1963) considered the problem of testing the equality of any subset of the roots that includes as a special case the equality of the m smallest roots and is asymptotically equivalent to Bartlett's test when $m = p - k$. Modifying Anderson's procedure according to Bartlett, the hypothesis

(6.4.8) $$H_0 : \lambda_{k+1} = \lambda_{k+2} = \cdots = \lambda_{k+m}$$

that any m roots of $\mathbf{\Sigma}$ are equal is tested by using the statistic

(6.4.9) $$X^2 = \left(N - k - 1 - \frac{2m^2 + m + 2}{6m}\right)\left(m \log \bar{\lambda} - \sum_{j=k+1}^{k+m} \log \hat{\lambda}_j\right)$$

where

$$\bar{\lambda} = \frac{\sum_{j=k+1}^{k+m} \hat{\lambda}_j}{m}$$

When the null hypothesis (6.4.8) is true, X^2 has a χ^2 distribution with $v = (m-1)(m+2)/2$ degrees of freedom as $N \to \infty$. If $k + m = p$, so that $m = p - k$, (6.4.9) reduces to the criterion proposed by Bartlett (1954) and Lawley (1956) for testing the equality of the last m roots of $\mathbf{\Sigma}$; if $k = 0$, (6.4.9) reduces to Bartlett's test for the equality of all roots, equation (3.17.22).

EXAMPLE 6.4.2. Employing Di Vesta and Walls' data, the roots of the variance-covariance matrix $\mathbf{S}$ are

$$\begin{array}{ll} \hat{\lambda}_1 = 5.7735 & \hat{\lambda}_5 = .1167 \\ \hat{\lambda}_2 = .9481 & \hat{\lambda}_6 = .0961 \\ \hat{\lambda}_3 = .3564 & \hat{\lambda}_7 = .0803 \\ \hat{\lambda}_4 = .1869 & \hat{\lambda}_8 = .0314 \end{array}$$

To test the hypothesis $H_0: \lambda_2 = \lambda_3$ against the alternative $H_1: \lambda_2 \neq \lambda_3$, $k = 1$ and $m = 2$. The formula for X^2 becomes

$$\begin{aligned} X^2 &= \left[292 - 1 - 1 - \frac{2(2)^2 + 2 + 2}{6(2)}\right](2 \log .6523 - \log .9481 - \log .3564) \\ &= 289(-.8547 + .05330 + 1.0317) \\ &= 66.56 \end{aligned}$$

Since $v = (m-1)(m+2)/2 = 2$, the critical value for the test is, with $\alpha = .05$, $\chi^2_\alpha(2) = 5.991$. Thus it is concluded that $\lambda_2 \neq \lambda_3$, so that the directions represented by λ_2 and λ_3 are distinct.

If the multiplicity of λ_j is m, Anderson (1963) showed that an approximate $100(1-\alpha)\%$ confidence interval for λ_j is

$$\text{(6.4.10)} \qquad \frac{\bar{\lambda}}{1 + Z^{\alpha/2}\sqrt{2/(N-1)m}} \leq \lambda_j \leq \frac{\bar{\lambda}}{1 - Z^{\alpha/2}\sqrt{2/(N-1)m}}$$

In addition to testing hypotheses for roots and vectors of a variance-covariance matrix $\mathbf{\Sigma}$, confidence intervals for the elements of $\mathbf{\Sigma}$ are also of some interest in affording the variability of the entries. To test the null hypothesis that $\mathbf{\Sigma}$ has a specified value $\mathbf{\Sigma}_0$,

$$\text{(6.4.11)} \qquad H_0: \mathbf{\Sigma} = \mathbf{\Sigma}_0 \quad \text{(specified)}$$

Bartlett (1954), Kullback (1959, p. 302), and Anderson (1958, p. 264) propose asymptotic chi-squared criteria. The criterion proposed by Bartlett is

$$\text{(6.4.12)} \qquad X^2 = \left[(N-1) - \frac{1}{6}\left(2p + 1 - \frac{2}{p-1}\right)\right]\left[\log \frac{|\mathbf{\Sigma}_0|}{|\mathbf{S}|} - \text{Tr}(\mathbf{\Sigma}_0^{-1}\mathbf{S}) - p\right]$$

As $N \to \infty$, X^2 has a χ^2 distribution with $v = p(p-1)/2$ degrees of freedom under the null hypothesis. Korin (1968) investigated the asymptotic behavior of (6.4.12) for various values of N and p and illustrated that the approximation is very good for any N and p so long as $N > p$.

Constructing simultaneous confidence intervals for the elements of $\mathbf{\Sigma}$ is difficult when using (6.4.12); however, S. N. Roy (1953; 1957, p. 106) demonstrates that, for arbitrary nonzero vectors, $100(1-\alpha)\%$ confidence intervals for functions of the form $\psi = \mathbf{a}'\mathbf{\Sigma}\mathbf{a}$ are obtained by evaluating

$$\frac{(N-1)\mathbf{a}'\mathbf{S}\mathbf{a}}{u} \leq \mathbf{a}'\mathbf{\Sigma}\mathbf{a} \leq \frac{(N-1)\mathbf{a}'\mathbf{S}\mathbf{a}}{\ell} \tag{6.4.13}$$

if values u and ℓ are found such that the $P(u > v_1 > \cdots > v_p > \ell) = 1 - \alpha$, where $v_1, v_2, \ldots, v_p$ are the roots of the determinantal equation $|(N-1)\mathbf{S} - v\mathbf{\Sigma}| = 0$ or $|(N-1)\mathbf{S}\mathbf{\Sigma}^{-1} - v\mathbf{I}| = 0$. Hanumara and Thompson (1968) have generated the percentage points of the extreme roots of a Wishart matrix $\mathbf{W} \sim W_p(n = N-1, \mathbf{\Sigma})$ such that the $P(u > v_1 > v_p > \ell) = 1 - \alpha$ for the case $\mathbf{\Sigma} = \mathbf{I}$. Since $(N-1)\mathbf{S} \sim W_p(N-1, \mathbf{\Sigma})$ or $(N-1)\mathbf{S}\mathbf{\Sigma}^{-1} \sim W_p(N-1, \mathbf{I})$, the tabled values for u and ℓ given by Hanumara and Thompson are utilized to find simultaneous $100(1-\alpha)\%$ confidence intervals for $\psi = \mathbf{a}'\mathbf{\Sigma}\mathbf{a}$. Their table is included in the Appendix (Table XIV). To secure a confidence interval for the variance of the ith variate, the vector $\mathbf{a}$ to be substituted into (6.4.13) is of the form $\mathbf{a}' = [0, \ldots, 0, 1_i, 0, \ldots, 0]$, so that (6.4.13) reduces to

$$\frac{(N-1)s_i^2}{u(\alpha/2)} \leq \sigma_i^2 \leq \frac{(N-1)s_i^2}{\ell(\alpha/2)}$$

where $u\alpha/2$ and $\ell\alpha/2$ denote the upper and lower percentage points, respectively, of the extreme roots in Hanumara and Thompson's table.

When using (6.4.13), intervals for σ_{ij}, for $i \neq j$, are not immediate. However, Thompson (1962) shows that, with probability at least $1 - \alpha$, the expressions

$$\begin{aligned} &\frac{(N-1)s_i^2}{\mu(\alpha/2)} \leq \sigma_i^2 \leq \frac{(N-1)s_i^2}{\ell(\alpha/2)} \\ &\left|\sigma_{ij} - \frac{N-1}{2}\left(\frac{1}{u(\alpha/2)} + \frac{1}{\ell(\alpha/2)}\right)s_{ij}\right| \leq \frac{N-1}{2}\left(\frac{1}{\ell(\alpha/2)} - \frac{1}{u(\alpha/2)}\right)s_i s_j \end{aligned} \tag{6.4.14}$$

hold simultaneously if inference is restricted to the elements of $\mathbf{\Sigma}$. The intervals for the individual variances agree with Roy's findings.

Tests Using a Correlation Matrix. The tests and confidence intervals presented in the preceding paragraphs involve a variance-covariance matrix. For a correlation matrix, the problems become more complex. A general procedure is unavailable for testing whether a vector is equal to a specified vector or for finding confidence intervals for roots, except in very special cases (see Anderson, 1963). However, some tests involving correlation matrices are available and of interest.

Assuming multivariate normality and using the same convention as for a variance-covariance matrix, let $\delta_1, \delta_2, \ldots, \delta_p > 0$ denote the roots and $\mathbf{q}_1, \ldots, \mathbf{q}_p$ the corresponding vectors of a population-correlation matrix $\mathbf{P}$. In the sample, let $\hat{\delta}_1, \hat{\delta}_2, \ldots, \hat{\delta}_p$ and $\hat{\mathbf{q}}_1, \hat{\mathbf{q}}_2, \ldots, \hat{\mathbf{q}}_p$ represents the roots and vectors of the determinantal equation $|\mathbf{R} - \delta\mathbf{I}| = 0$, where $\mathbf{R}$ is the maximum-likelihood estimator for $\mathbf{P}$.

Bartlett (1950, 1954), Anderson (1963), and Lawley (1963) proposed tests for the equality of all correlations in a population-correlation matrix. This is equivalent to saying that $\mathbf{P} = (1 - \rho)\mathbf{I}_p + \rho\mathbf{J}_p$ or that the last $p - 1$ roots of a population-correlation matrix are equal. Aitkin et al. (1968) also suggest a test statistic for testing

$$H_0: \delta_2 = \delta_3 = \cdots = \delta_p \quad \text{or equivalently} \qquad (6.4.15)$$
$$H_0: \rho_{ij} = \rho \text{ for all } i \neq j$$

However, as Gleser (1968) has shown, the asymptotic null distribution of their statistic, as well as Bartlett's and Anderson's, depends on the unknown value of ρ. Lawley's proposed statistic is asymptotically independent of ρ. Furthermore, for $p \leq 6$, Lawley's statistic converges to a χ^2 distribution for N as small as 25 (Aitkin et al., 1968).

Lawley proposes that the statistic

$$X^2 = \frac{N-1}{\lambda^2}\left[\sum_{i<j}\sum (r_{ij} - \bar{r})^2 - \mu \sum_k (\bar{r}_k - \bar{r})^2\right] \qquad (6.4.16)$$

where

$$\lambda = 1 - \rho \qquad \mu = \frac{(p-1)^2(1-\lambda^2)}{p - (p-2)^2} \qquad \bar{r}_k = \frac{\sum_{i<k} r_{ik}}{p-1} \quad \text{and} \quad \bar{r} = \frac{2\sum\sum_{i \neq j} r_{ij}}{p(p-1)}$$

is substituted for the unknown value of ρ used to test (6.4.15). Under the null hypothesis, χ^2 approaches a χ^2 distribution with $v = (p+1)(p-2)/2$ degrees of freedom as N becomes large.

Aside from using (6.4.16) to test for compound symmetry, in education, the largest root of a correlation matrix may account for 70% or more of the total variance in a principal-component analysis. If the analysis is being used to determine a single number expressing the general relationship among variables in a set, the researcher hopes that the entire meaningful variation in the sample can be expressed with only one principal variate. All the other roots should then be approximately equal. This is tested by using the statistic given in (6.4.16).

An exact asymptotic procedure was given to test the equality of the last m roots of a variance-covariance matrix. When $\mathbf{R}$ is used in place of $\mathbf{S}$, no proposed criterion has a χ^2 distribution even as $N \to \infty$. However, Bartlett (1954) and Lawley (1956) suggest a very approximate procedure when the roots removed are twice the magnitude of the remaining roots. To test the hypothesis

$$H_0: \delta_{k+1} = \delta_{k+2} = \cdots = \delta_p \qquad (6.4.17)$$

that the last $m = p - k$ roots are equal, the criterion, ignoring the multiplying factor, is

$$T = (N-1)\left(m \log \bar{\delta} - \sum_{j=k+1}^{p} \log \hat{\delta}_j\right) \qquad (6.4.18)$$

where

$$\bar{\delta} = \frac{\sum_{j=k+i}^{p} \hat{\delta}_j}{m}$$

The degrees of freedom associated with T are approximately $v = (m - 1)(m + 2)/2$ when all correlations tend to 1 and $\bar{\delta} \to 0$ in the limit. The criterion T is compared to a χ^2 distribution with v degrees of freedom to obtain a very approximate test of (6.4.17) when using a correlation matrix.

When working with correlation matrices, another test of some interest is whether the population-correlation matrix $\mathbf{P}$ has a specific value $\mathbf{P}_0$. The test statistic experimenters usually use to test the null hypothesis

(6.4.19) $$H_0 : \mathbf{P} = \mathbf{P}_0 \qquad \text{(specified)}$$

is the criterion

(6.4.20) $$X = \left[(N - 1) - \frac{2p + 11}{6}\right]\left[\log\frac{|\mathbf{P}_0|}{|\mathbf{R}|} + \operatorname{Tr}(\mathbf{P}_0^{-1}\mathbf{R}) - p\right]$$

first proposed by Bartlett and Rajalakshman (1953). Although Bartlett (1954) expressed some reservations about using the statistic in (6.4.20) to test (6.4.19), Kullback (1959, p. 302) claims that, when (6.4.19) is true, X is distributed approximately as a χ^2 random variable on $v = p(p - 1)/2$ degrees of freedom as $N \to \infty$. He employs it and similar types of expressions to test hypotheses about correlation matrices (see Kullback, 1967). Aitken et al. (1968) investigated the behavior of X and indicated that the approximation is not adequate for samples of sizes $N \geq 200$. This is due to the fact that, as Aitkin (1969) demonstrates, X is not distributed as a χ^2 random variable on $v = p(p - 1)/2$ degrees of freedom under the null hypothesis, even asymptotically, unless $\mathbf{P}_0 = \mathbf{I}$. The statistic X is distributed as a linear function of $p(p - 1)/2$ independent χ^2 random variables on one degree of freedom. The coefficients of the linear function required to find the critical value to test (6.14.19) also involve finding the roots of a matrix product. The matrix product also necessitates considerable calculation. In Aitkin's notation, let

(6.4.21) $$\begin{aligned} &\mathbf{P}_0 = [\rho_{ij}] \qquad \mathbf{P}_0^{-1} = [\rho^{ij}] \\ &\mathbf{A} = [a_{jk,\ell m}] = [\rho^{j\ell}\rho^{km} + \rho^{jm}\rho^{k\ell}] \qquad \text{for } j < k \text{ and } \ell < m \\ &\mathbf{\Psi} = [\psi_{jk,\ell m}] \end{aligned}$$

where the order of $\mathbf{P}_0$ and $\mathbf{P}_0^{-1}$ is p, the order of the symmetric matrices $\mathbf{A}$ and $\mathbf{\Psi}$ is $v = p(p - 1)/2$, and

(6.4.22) $$\begin{aligned} \psi_{jk,jk} &= (1 - \rho_{jk}^2)^2 \\ \psi_{jk,j\ell} &= -\frac{1}{2}\rho_{jk}\rho_{j\ell}(1 - \rho_{jk}^2 - \rho_{j\ell}^2 - \rho_{k\ell}^2) + \rho_{k\ell}(1 - \rho_{jk}^2 - \rho_{j\ell}^2) \\ \psi_{jk,\ell m} &= \frac{1}{2}\rho_{jk}\rho_{\ell m}(\rho_{j\ell}^2 + \rho_{jm}^2 + \rho_{k\ell}^2 + \rho_{km}^2) - \rho_{jk}\rho_{j\ell}\rho_{k\ell} - \rho_{jk}\rho_{jm}\rho_{km} \\ &\quad - \rho_{j\ell}\rho_{jm}\rho_{\ell m} - \rho_{k\ell}\rho_{km}\rho_{\ell m} \end{aligned}$$

For example, if $p = 4$,

$$\mathbf{P}_0 = \begin{bmatrix} \rho_{11} & \rho_{12} & \rho_{13} & \rho_{14} \\ & \rho_{22} & \rho_{23} & \rho_{24} \\ & & \rho_{33} & \rho_{34} \\ & & & \rho_{44} \end{bmatrix} \qquad \mathbf{P}_0^{-1} = \begin{bmatrix} \rho^{11} & \rho^{12} & \rho^{13} & \rho^{14} \\ & \rho^{22} & \rho^{23} & \rho^{24} \\ & & \rho^{33} & \rho^{34} \\ & & & \rho^{44} \end{bmatrix}$$

$$\mathbf{A} = \begin{bmatrix} a_{12,12} & a_{12,13} & a_{12,14} & a_{12,23} & a_{12,24} & a_{12,34} \\ & a_{12,13} & a_{13,14} & a_{13,23} & a_{13,24} & a_{13,34} \\ & & a_{14,14} & a_{14,23} & a_{14,24} & a_{14,34} \\ & & & a_{23,23} & a_{23,24} & a_{23,34} \\ & & & & a_{24,24} & a_{24,34} \\ & & & & & a_{34,34} \end{bmatrix}$$

and the matrix $\mathbf{\Psi}$ is formed like $\mathbf{A}$ using (6.4.22). By forming the matrix product $\mathbf{A\Psi}$, the roots of the equation $|\mathbf{A\Psi} - \lambda\mathbf{I}| = 0$ are acquired and used as weights for the linear function

$$X_0 = \sum_{i=1}^{v} \lambda_i \chi_\alpha^2(1) \tag{6.4.23}$$

The matrix $\mathbf{A\Psi}$ is not, in general, symmetric. Solving for the roots of $|\mathbf{A\Psi} - \lambda\mathbf{I}| = 0$ is equivalent to solving for the roots of $|\mathbf{A} - \lambda\mathbf{\Psi}^{-1}| = 0$. Furthermore, there exists an orthogonal matrix $\mathbf{P}$ such that

$$\mathbf{\Psi}^{-1} = \mathbf{PDP}' = (\mathbf{PD}^{1/2}\mathbf{P}')(\mathbf{PD}^{1/2}\mathbf{P}') = \mathbf{\Psi}^{-1/2}\mathbf{\Psi}^{-1/2}$$

$$\mathbf{\Psi} = \mathbf{PD}^{-1}\mathbf{P}' = (\mathbf{PD}^{-1/2}\mathbf{P}')(\mathbf{PD}^{-1/2}\mathbf{P}') = \mathbf{\Psi}^{1/2}\mathbf{\Psi}^{1/2}$$

where $\mathbf{PP}' = \mathbf{P}'\mathbf{P} = \mathbf{I}$ and $\mathbf{D}$ is a diagonal matrix. By forming the symmetric matrix $\mathbf{\Psi}^{1/2}\mathbf{A}\mathbf{\Psi}^{1/2}$, the eigenvalues of $|\mathbf{A} - \lambda\mathbf{\Psi}^{-1}| = 0$ are the same as the roots of $|\mathbf{\Psi}^{1/2}\mathbf{A}\mathbf{\Psi}^{1/2} - \lambda\mathbf{I}| = 0$ and are more easily utilized to locate the coefficients in (6.4.23). The hypothesis in (6.4.19) is rejected at the significance level α if $X > X_0$, where each $\chi_\alpha^2(1)$ is computed at the upper α percentage point. To test (6.4.19) without finding the roots of $|\mathbf{A\Psi} - \lambda\mathbf{I}| = 0$, Aitkins suggests replacing X_0 with the approximate value $X_0 = \chi_\alpha^2(v)\mathrm{Tr}(\mathbf{A\Psi})/v$, where $v = p(p-1)/2$.

Before using any of Kullback's (1967; 1959, p. 320) proposed test statistics for evaluating the equality of correlation matrices, the remarks made by Aitkin (1969) should be utilized when asymptotic chi-squared procedures show borderline significance (see also Bartlett, 1954).

By employing (6.4.13) and the percentage points in Table XIV of the Appendix, Aitkin demonstrates that simultaneous confidence intervals for the elements of $\mathbf{P}$ are given by

$$\max\left(\frac{u(\alpha/2)}{\ell(\alpha/2)}r_{ij} + \frac{u(\alpha/2)}{\ell(\alpha/2)} - 1, \frac{\ell(\alpha/2)}{u(\alpha/2)}r_{ij} - \frac{\ell(\alpha/2)}{u(\alpha/2)} + 1\right) \le \rho_{ij}$$

$$\le \min\left(\frac{\ell(\alpha/2)}{u(\alpha/2)}r_{ij} + \frac{\ell(\alpha/2)}{u(\alpha/2)} - 1, \frac{u(\alpha/2)}{\ell(\alpha/2)}r_{ij} - \frac{u(\alpha/2)}{\ell(\alpha/2)} + 1\right) \tag{6.4.24}$$

with at least $100(1-\alpha)\%$ confidence intervals where $\mathbf{R} = [r_{ij}]$, and $u(\alpha/2)$ and $\ell(\alpha/2)$ are the extreme-root percentage points found in Table XIV. He further warns that for small values of p, the intervals will be especially large.

EXERCISES 6.4

1. For the variance-covariance matrix given in Exercise 1, Section 6.2, do the following:

   a. Find an appropriate 95% confidence interval for the first eigenvalue. How would you use this to test the hypothesis that the first eigenvalue has a specified variance?

   b. Test the hypothesis $H_0: \lambda_2 = \lambda_3$ that the last two eigenvalues are equal.

   c. Determine approximate 95% confidence intervals for the variances $\sigma_i^2 (i = 1, 2, 3)$.

   d. Using $\alpha = .05$, test the hypothesis $H_0: \mathbf{\Sigma} = \mathbf{\Sigma}_0$ that $\mathbf{\Sigma}$ has a specified value, if $\mathbf{\Sigma}_0$ is defined by

$$\mathbf{\Sigma}_0 = \begin{bmatrix} 40.43 & & \\ 9.38 & 10.40 & \\ 30.33 & 7.14 & 30.46 \end{bmatrix}$$

2. For the sample correlation matrix given by

$$\mathbf{R} = \begin{bmatrix} 1.000 & & \\ .586 & 1.000 & \\ .347 & .611 & 1.000 \end{bmatrix}$$

   constructed from $N = 101$ observations, do each of the following:

   a. Test the hypothesis $H_0: \rho_{ij} = \rho$, for all $i \neq j$, using $\alpha = .05$.

   b. If the test in part a is accepted, test the equality of the last two roots, $H_0: \delta_2 = \delta_3$.

   c. The matrix $\mathbf{R}$ appears to have simplex form

$$\begin{bmatrix} 1 & & \\ \rho & 1 & \\ \rho^2 & \rho & 1 \end{bmatrix}$$

   with $\rho = .6$. Using $\alpha = .05$, test the hypothesis that

$$\mathbf{P} = \begin{bmatrix} 1.00 & & \\ .60 & 1.00 & \\ .36 & .60 & 1.00 \end{bmatrix}$$

   and determine approximate 95% confidence intervals for each ρ_{ij}.

3. For a sample correlation matrix of rank 10 computed from 500 observations, the following sample eigenvalues were obtained: $\delta_i = 2.969, 1.485, 1.230, .825, .799, .670, .569, .555, .480, .439$. Test the hypothesis that the last six roots are equal, $H_0: \delta_5 = \delta_6 = \cdots = \delta_{10}$.

6.5 COMMON-FACTOR ANALYSIS

Common-factor analysis (or factor analysis), a technique originating with Spearman (1904) and developed extensively by Thurstone (1931, 1947), is the most familiar multivariate procedure in use by behavioral science researchers. However, since common-factor analysis and principal-component analysis are both invoked to analyze the structure of variance-covariance and correlation matrices, they are at times confused.

In using principal-component analysis to improve understanding of the interdependencies existing among a set of random variables $\mathbf{Y}$, the researcher transforms a large number of uncorrelated variables to a smaller number of components that account for most of the total variance in the system. That is, a matrix $\mathbf{P}'$ is determined such that the components $\mathbf{X} = \mathbf{P}'\mathbf{Y}$. In terms of the original variables $\mathbf{Y}$, the mathematical model for the observations is

$$\mathbf{Y} = \mathbf{P}\mathbf{X} \tag{6.5.1}$$

By letting the $V(\mathbf{Y}) = \mathbf{\Sigma}$, (6.5.1) implies that the matrix $\mathbf{\Sigma}$ has the form

$$\begin{aligned} \mathbf{\Sigma} &= \mathbf{P}V(\mathbf{X})\mathbf{P}' = \mathbf{P}\mathbf{\Lambda}\mathbf{P}' \\ &= \mathbf{P}\mathbf{\Lambda}^{1/2}/\mathbf{\Lambda}^{1/2}\mathbf{P}' \\ &= \mathbf{W}\mathbf{W}' \end{aligned} \tag{6.5.2}$$

yielding a unique decomposition for $\mathbf{\Sigma}$. However, the procedure is not invariant to changes in scales of the original variables, so the decision to analyze the correlation matrix or the variance-covariance matrix is critical.

Common-factor analysis, on the other hand, although similar to principal-component analysis, is considerably more complex. The difference between the two methods is best explained by Lawley and Maxwell (1963). Common-factor analysis is covariance (or correlation) oriented, whereas principal-component analysis is variance oriented. In principal-component analysis, all components are necessary in reproducing an intercorrelation (covariance) exactly, whereas, in common-factor analysis, a few factors (less than the number of variables) will reproduce intercorrelations (covariances) exactly. However, these same factors will not account for or "explain" as much of the total variance as the same number of principal components. Another important difference between the two procedures lies in the direction of the method of analysis. Principal-component analysis originates with observations that describe and reduce the dimensions of variation in the data to establish a hypothetical mathematical model. Common-factor analysis works the other way around. First a linear statistical model is hypothesized, and then the data are used to try to determine whether the model is in agreement with the data (Kendall, 1961, p. 37). That is, in common-factor analysis a model is fitted to data, whereas in principal-component analysis data are fitted to a model.

The statistical model of common-factor analysis is

$$\underset{(p\times 1)}{\mathbf{Y}} = \underset{(p\times k)}{\mathbf{\Lambda}}\,\underset{(k\times 1)}{\mathbf{X}} + \underset{(p\times 1)}{\boldsymbol{\varepsilon}} \tag{6.5.3}$$

where $\mathbf{Y}$ is a random vector of observations, $\mathbf{X}$ is a vector of unobserved factors, $\boldsymbol{\varepsilon}$ is a random error vector, and $\mathbf{\Lambda}$ is a matrix of regression weights with the $R(\mathbf{\Lambda}) = k \ll p$. It is further assumed that $\mathbf{X}$ and $\boldsymbol{\varepsilon}$ are statistically independent with $E(\mathbf{Y}) = E(\mathbf{X}) = E(\boldsymbol{\varepsilon}) = \mathbf{0}$, $V(\mathbf{Y}) = \mathbf{\Sigma}$, $V(\mathbf{X}) = \mathbf{I}$, and $V(\boldsymbol{\varepsilon}) = \mathbf{\Psi}$, a diagonal matrix with elements greater than 0. Then (6.5.3) implies that the matrix $\mathbf{\Sigma}$ has the structure

$$
\begin{aligned}
\mathbf{\Sigma} &= V(\mathbf{Y}) = V(\mathbf{\Lambda X} + \boldsymbol{\varepsilon}) \\
&= \mathbf{\Lambda} V(\mathbf{X})\mathbf{\Lambda}' + V(\boldsymbol{\varepsilon}) \\
&= \mathbf{\Lambda\Lambda}' + \mathbf{\Psi} \qquad (6.5.4)
\end{aligned}
$$

Comparing (6.5.4) with (6.5.2) illustrates that k factors replace the p principal components in $\mathbf{X}$ and that the error vector $\boldsymbol{\varepsilon}$ does not exist in the principal-component formulization. The vector $\boldsymbol{\varepsilon}$ in (6.5.4) removes $\mathbf{\Psi}$ from the diagonal of $\mathbf{\Sigma}$.

By expanding (6.5.3), the linear common-factor-analysis model equations become

$$
\begin{aligned}
y_1 &= \lambda_{11}x_1 + \lambda_{12}x_2 + \cdots + \lambda_{1k}x_k + \varepsilon_1 \\
&\vdots \\
y_i &= \lambda_{i1}x_1 + \lambda_{i2}x_2 + \cdots + \lambda_{ik}x_k + \varepsilon_i \qquad (6.5.5) \\
&\vdots \\
y_p &= \lambda_{p1}x_1 + \lambda_{p2}x_2 + \cdots + \lambda_{pk}x_k + \varepsilon_p
\end{aligned}
$$

where the correlation among all pairs of errors is 0, for $i \neq j$, and the λ_{ij} are regression weights. For psychologists, the y_i usually represent test scores, the λ_{ij} are termed *factor loadings*, and the residuals ε_i are *unique factors*. Thus each test is divided into a *common part* and a *unique part*:

$$y_i = c_i + \varepsilon_i \qquad (6.5.6)$$

Common-factor analysis is used to investigate the unobservable c_i's, whereas the principal-component-analysis model is used to study the observable y_i's. The unique part of a test is often thought of as a specific factor and a random error of measurement; however, since these do not contribute to any correlation or covariance among tests, they are omitted in our discussion (see Harmon, 1967, p. 19).

Corresponding to (6.5.3), a specific structure is given for $\mathbf{\Sigma}$ that partitions the variance of a random observation Y_i as

$$
\begin{aligned}
\sigma_i^2 &= \lambda_{i1}^2 + \lambda_{i2}^2 + \cdots + \lambda_{ik}^2 + \psi_i \\
&= V(c_i) + V(\varepsilon_i) \qquad (6.5.7)
\end{aligned}
$$

where the variance of the common part of Y_i is called the *common variance* or *communality* of the response and $V(\varepsilon_i) = \psi_i$, the ith diagonal element of $\mathbf{\Psi}$, is termed the *unique variance* or the *uniqueness* of Y_i. The uniqueness is that part of the total variance not accounted for by the common factors, while the communality is that portion of the variance attributed to the common factors.

The covariance between Y_i and Y_j, for $i \neq j$, is

$$\sigma_{ij} = \lambda_{i1}\lambda_{j1} + \lambda_{i2}\lambda_{j2} + \cdots + \lambda_{ik}\lambda_{jk} \qquad (6.5.8)$$

where λ_{ij} is the covariance between the Y_i and the jth common factor X_j. Using matrix notation, the

$$\begin{aligned} \operatorname{cov}(\mathbf{Y}, \mathbf{X}) &= \operatorname{cov}(\mathbf{\Lambda X} + \boldsymbol{\varepsilon}, \mathbf{X}) \\ &= \mathbf{\Lambda} \operatorname{cov}(\mathbf{X}, \mathbf{X}) + \operatorname{cov}(\boldsymbol{\varepsilon}, \mathbf{X}) \\ &= \mathbf{\Lambda} \end{aligned} \tag{6.5.9}$$

since $\operatorname{cov}(\boldsymbol{\varepsilon}, \mathbf{X}) = \mathbf{0}$. If $\mathbf{\Sigma}$ is a population-correlation matrix, (6.5.8) becomes

$$\rho_{ij} = \lambda_{i1}\lambda_{j1} + \cdots + \lambda_{ik}\lambda_{jk} \tag{6.5.10}$$

where λ_{ij} denotes the correlation between Y_i and X_j. The principal aim of common-factor analysis is in estimating $\mathbf{\Lambda}$ and $\mathbf{\Psi}$.

In comparing the statistical common-factor-analysis model with the mathematical principal-component-analysis model, it has been assumed that the factors are orthogonal or uncorrelated since the $V(\mathbf{X}) = \mathbf{I}$. If the $V(\mathbf{X}) = \mathbf{\Phi}$ and $\mathbf{\Phi} \neq \mathbf{I}$, $\mathbf{\Sigma}$ is written as

$$\mathbf{\Sigma} = \mathbf{\Lambda \Phi \Lambda}' + \mathbf{\Psi} \tag{6.5.11}$$

and the resulting factors are said to be *oblique*. For the treatment of factor analysis in this text, $\mathbf{\Phi}$ is always taken to be equal to the identity matrix.

Given that a structure $\mathbf{\Sigma}$, as defined by (6.5.4), exists for a given value of k, where a necessary and sufficient condition for the existence of a structure is that there is a matrix $\mathbf{\Psi}$ such that $\mathbf{\Sigma} - \mathbf{\Psi}$ is p.s.d. and of rank k (see, for example, Anderson and Rubin, 1956, and Jöreskog, 1963), an indeterminacy exists in common-factor analysis that did not occur in principal-component analysis. In principal-component analysis, the "factorization" of $\mathbf{\Sigma}$ is unique (provided the eigenvalues are not equal); in common-factor analysis the matrix $\mathbf{\Lambda}$ may be postmultiplied by any orthogonal matrix $\mathbf{T}$, and $\mathbf{\Sigma}$ is unaltered. Thus, if $\mathbf{\Lambda}$ is postmultiplied by $\mathbf{T}$, $\mathbf{\Gamma} = \mathbf{\Lambda T}$,

$$\begin{aligned} \mathbf{\Gamma\Gamma}' + \mathbf{\Psi} &= \mathbf{\Lambda TT}'\mathbf{\Lambda} + \mathbf{\Psi} \\ &= \mathbf{\Lambda\Lambda}' + \mathbf{\Psi} \\ &= \mathbf{\Sigma} \end{aligned}$$

and the model equation in (6.5.3) becomes $\mathbf{Y} = \mathbf{\Gamma TX} + \boldsymbol{\varepsilon} = \mathbf{\Gamma f} + \boldsymbol{\varepsilon}$, which is mathematically equivalent to (6.5.3).

The problem of selecting the "best" matrix of loadings to reproduce the covariances or correlations is known as the *rotation* or *transformation* problem. This allows researchers using common-factor analysis to transform the axes representing the common factors for "meaning." Thurstone (1947, p. 335) proposed the idea of *simple structure* as a means for finding interpretable factors. The simple-structure criterion states:

1. Each row of $\mathbf{\Lambda}$ should have at least one zero.
2. Each column of $\mathbf{\Lambda}$ should have at least k zeros.
3. For all pairs of columns in $\mathbf{\Lambda}$, there should be several rows of zero and nonzero loadings.
4. If $k \geq 4$, several pairs of columns in $\mathbf{\Lambda}$ should have two zero loadings and a small number of nonzero loadings.

Various graphical and analytic techniques have been proposed to provide simple structure. The most widely used procedure in the orthogonal case is Kaiser's varimax (see Kaiser, 1958, and Harris and Kaiser, 1964). For a comprehensive discussion of the transformation problem in the orthogonal and oblique case, see Harmon (1967) and Rummel (1970). Transforming for meaning is called *exploratory* (*unrestricted*) *factor analysis*.

Howe (1955), Lawley and Maxwell (1963), Jöreskog (1966, 1969b), and others suggest that a pattern of loadings be specified a priori as a target matrix for $\boldsymbol{\Lambda}$. Having obtained loadings as close to the predetermined specification as possible, the researcher tests to see if the target matrix is acquired. This procedure is called *confirmatory* (*restricted*) *factor analysis*. One advantage of a confirmatory approach is the elimination of the transformation problem. However, to test that the target matrix is attained, normality assumptions are added to the statistical model.

Aside from the transformation problem, there are several statistical problems associated with the common-factor-analysis model (see Anderson and Rubin, 1956): (1) estimation of the model parameters, (2) testing the hypothesis that the model fits the data, (3) determining the number of common factors, and (4) estimating factor scores. In this section, primary consideration is given to problems 1 and 2.

The primary equation of the common-factor-analysis model is

$$\underset{(p \times p)}{\boldsymbol{\Sigma}} = \underset{(p \times k)}{\boldsymbol{\Lambda}} \underset{(k \times p)}{\boldsymbol{\Lambda}'} + \underset{(p \times p)}{\boldsymbol{\Psi}} \tag{6.5.12}$$

where $\boldsymbol{\Sigma}$ is the variance-covariance matrix of the observations; $\boldsymbol{\Psi}$ is the variance-covariance matrix of $\boldsymbol{\varepsilon}$, which is diagonal and nonsingular; and $\boldsymbol{\Lambda\Lambda}'$ is the variance-covariance matrix of the common parts of the observations and of rank k such that $\boldsymbol{\Sigma} - \boldsymbol{\Psi}$ is p.s.d. and of minimal rank. The matrix $\boldsymbol{\Lambda\Lambda}'$ has the same off-diagonal elements as $\boldsymbol{\Sigma}$. The aim of common-factor analysis is to obtain a few factors that adequately reproduce these covariances or, if $\boldsymbol{\Sigma}$ is a correlation matrix, the intercorrelations.

In (6.5.12), the $q = p(p + 1)/2$ elements of $\boldsymbol{\Sigma}$ that are not equal are represented in terms of the $p(1 + k)$ unknown parameters of $\boldsymbol{\Psi}$ and $\boldsymbol{\Lambda}$. Because of the transformation problem, $\boldsymbol{\Lambda}$ may be made to satisfy $k(k - 1)/2$ independent conditions. Thus the effective number of unknown parameters is not $p + pk$ but $f = [p + pk - k(k - 1)/2$. The degrees of freedom for the model are $v = [(p - k)^2 - (p + k)]/2$. the number of variances and covariances less the effective number of unknown parameters. For $v > 0$, k must satisfy the condition $p + k < (p - k)^2$ (Anderson and Rubin, 1956).

To estimate $\boldsymbol{\Lambda}$ and $\boldsymbol{\Psi}$ in (6.5.12), based on a sample of N p-variate observations, $\boldsymbol{\Sigma}$ is replaced by the sample variance-covariance matrix $\mathbf{S}$ or the sample correlation matrix $\mathbf{R}$ if the variables are standardized. Suppose that k is known and that a reasonable estimate $\hat{\boldsymbol{\Psi}}_0$ of $\boldsymbol{\Psi}$ is available. Then, for the sample, (6.5.12) becomes

$$\mathbf{S} - \hat{\boldsymbol{\Psi}}_0 = \hat{\boldsymbol{\Lambda}}_0\hat{\boldsymbol{\Lambda}}_0' \tag{6.5.13}$$

By associating $\mathbf{S}_r$, the reduced variance-covariance matrix, with $\mathbf{S} - \hat{\boldsymbol{\Psi}}_0$, equation (6.5.13) suggests that a matrix $\hat{\boldsymbol{\Lambda}}_0$ be found, by using principal-component analysis, such that $\mathbf{S}_r = \hat{\boldsymbol{\Lambda}}_0\hat{\boldsymbol{\Lambda}}_0'$. By letting $\mathbf{S}_r$ replace $\mathbf{S}$ in a principal-component analysis, a matrix $\hat{\mathbf{\Lambda}}_0$ is found by solving the determinantal equation $|\mathbf{S}_r - \delta\mathbf{I}| = 0$. Then

$$\underset{(p \times k)}{\hat{\mathbf{\Lambda}}_0} = \underset{(p \times k)}{\mathbf{P}} \underset{(k \times k)}{\mathbf{\Delta}^{1/2}} \tag{6.5.14}$$

where $\mathbf{P}$ contains the first k eigenvectors such that $\mathbf{p}_i\mathbf{p}'_j = 1$ (for $i = j$) and 0 (for $i \neq j$) and $\mathbf{\Delta}$ is a diagonal matrix of the eigenvalues of $\mathbf{S}_r$. By investigating the degree to which $\hat{\mathbf{\Lambda}}_0\hat{\mathbf{\Lambda}}'_0$ reproduces the off-diagonal elements of $\mathbf{S}_r$, the residual matrix $\mathbf{S}_r - \hat{\mathbf{\Lambda}}_0\hat{\mathbf{\Lambda}}'_0$ is utilized (see Harmon, 1967). This procedure of estimating $\mathbf{\Lambda}$ and $\mathbf{\Psi}$ is referred to as *principal-factor analysis* and was suggested by Thompson (1934).

In the discussion of the principal-factor-analysis solution, two assumptions are made: (1) a reasonable estimate of $\mathbf{\Psi}$ is available, and (2) k is known. Given $\mathbf{S}$ and treating it as a population matrix $\mathbf{\Sigma}$, ignoring sampling variation, the number of common factors k is equal to the $R(\mathbf{S} - \mathbf{\Psi}) = k$, if $\mathbf{\Psi}$ is known. Since $\mathbf{\Psi}$ is not usually known, the rank of $\mathbf{S} - \mathbf{\Psi}$ is affected by the elements used in $\mathbf{\Psi}$ or by the values of the communalities. Furthermore, the estimates of the communalities are influenced by the number of common factors, which are only determinable if k is known. But k is not obtained until factoring is complete, and even if k is known, it is not sufficient for $\mathbf{\Psi}$ to be uniquely determined. This unfortunate difficulty has led many statisticians to avoid the common-factor-analysis model.

The problem of selecting k and a reasonable estimate for $\mathbf{\Psi}$ is referred to as the *communality problem* (Guttman, 1954, 1956). For details, see Harmon (1967, Chap. 5). In practice, the estimate for $\mathbf{\Psi}_0$,

$$\hat{\mathbf{\Psi}}_0 = [\text{Dia}\,\mathbf{S}^{-1}]^{-1} = \text{Dia}\left[\frac{1}{s^{ii}}\right] \tag{6.5.15}$$

is usually selected when using a variance-covariance matrix, and

$$\hat{\mathbf{\Psi}}_0 = [\text{Dia}\,\mathbf{R}^{-1}]^{-1} \tag{6.5.16}$$

is chosen when employing a correlation matrix. For a correlation matrix, the diagonal elements of $\mathbf{I} - \text{Dia}\,[\mathbf{R}^{-1}]^{-1} = \text{Dia}\,[1 - 1/r^{ii}]$ yield the estimates for the squared multiple-correlation coefficients between Y_i and the remaining $p - 1$ variables. Furthermore, the matrix $\text{Dia}\,[1/s^{ii}]$ has diagonal element estimates of the residual variances between Y_i and the remaining variables. To prove these observations, recall that the square of the multiple correlation between Y_1 and the remaining $p - 1$ variables is

$$\rho^2_{1\cdot 2,\ldots,p-1} = \frac{\boldsymbol{\sigma}'_{12}\mathbf{\Sigma}^{-1}_{22}\boldsymbol{\sigma}_{21}}{\sigma_{11}}$$

for a random regression model. Hence

$$\sigma_{11} - \boldsymbol{\sigma}'_{12}\mathbf{\Sigma}^{-1}_{22}\boldsymbol{\sigma}_{21} = \sigma_{11}(1 - \rho^2_{1\cdot 2,\ldots,p-1}) \quad \text{or} \quad \frac{1}{\sigma^{11}} = \sigma_{11}(1 - \rho^2_{1\cdot 2,\ldots,p-1})$$

where σ^{11} denotes the inverse of σ_{11} in $\mathbf{\Sigma}$ partitioned as

$$\mathbf{\Sigma} = \begin{bmatrix} \sigma_{11} & \boldsymbol{\sigma}'_{12} \\ \boldsymbol{\sigma}_{21} & \mathbf{\Sigma}_{22} \end{bmatrix}$$

Thus, for any variable i, it follows that

$$\rho^2_{i\cdot\text{remaining}} = 1 - \frac{1}{\sigma_{ii}\sigma^{ii}}$$

which is estimated by $1 - 1/r^{ii}$ for standardized variables as claimed. To show that the matrix Dia $[1/\sigma^{ii}]$ has diagonal elements, which are the residual variances between Y_i and the remaining variables, we merely observe that $\sigma_{11} - \boldsymbol{\sigma}'_{12}\boldsymbol{\Sigma}_{22}^{-1}\boldsymbol{\sigma}_{21} = |\boldsymbol{\Sigma}|/|\boldsymbol{\Sigma}_{22}| = 1/\sigma^{11}$. The generalization is now immediate.

Given an estimate of $\boldsymbol{\Psi}$, a sequential progression for various values of k is made until the fit of the off-diagonal elements of $\mathbf{S}$ or $\mathbf{R}$ is adequate, using $\hat{\boldsymbol{\Lambda}}_0\hat{\boldsymbol{\Lambda}}'_0$. For guidance in the choice of k, Guttman's strongest lower-bound k_G is often used, where k_G = the number of nonnegative roots of $\mathbf{S} - \hat{\boldsymbol{\Psi}}_0$ (see Guttman, 1954, and Jöreskog, 1963, p. 30).

As an improvement to principal-factor analysis, an iterative process is employed by repeated principal factoring. Having used $\hat{\boldsymbol{\Psi}}_0$ to calculate $\hat{\boldsymbol{\Lambda}}_0$, we compute $\hat{\boldsymbol{\Psi}}_1 = \text{Dia}(\mathbf{S} - \hat{\boldsymbol{\Lambda}}_0\hat{\boldsymbol{\Lambda}}'_0)$, which is the new calculated estimate for $\boldsymbol{\Psi}$. Equation (6.5.13) becomes $\mathbf{S} - \hat{\boldsymbol{\Psi}}_1 = \hat{\boldsymbol{\Lambda}}_1\hat{\boldsymbol{\Lambda}}'_1$, where the principal-factor-analysis procedure used to compute $\hat{\boldsymbol{\Lambda}}_0$ is employed to compute $\hat{\boldsymbol{\Lambda}}_1$. Principal-factor analysis is applied repeatedly until the estimates for $\boldsymbol{\Psi}$ and $\boldsymbol{\Lambda}$ hopefully converge. In this way, communalities are determined by iteration for a *specified value* of k. Convergence is said to be satisfactory if the estimates $\hat{\boldsymbol{\Lambda}}_i$ and $\hat{\boldsymbol{\Psi}}_{i+1}$ agree with $\hat{\boldsymbol{\Lambda}}_{i-1}$ and $\hat{\boldsymbol{\Psi}}_i$ to a predetermined number of significant figures. This technique is termed the *iterative principal-factor-analysis method.*

A method that yields solution mathematically equivalent to the iterative principal-factor-analysis technique and Harmon and Jones' (1966) minres method, is realized in least-squares theory. The first use of the least-squares method in the psychometric literature was illustrated by Eckart and Young (1936), who interpreted the location of principal components as a least-squares problem. By the method of least squares, a means for estimating $\boldsymbol{\Lambda}$ and $\boldsymbol{\Psi}$ is apparent by making $\boldsymbol{\Sigma}$ near $\mathbf{S}$, using the definition of a matrix norm. That is, the function

$$\begin{aligned} F(\boldsymbol{\Lambda}, \boldsymbol{\Psi}) &= \text{Tr}[(\boldsymbol{\Sigma} - \mathbf{S})'(\boldsymbol{\Sigma} - \mathbf{S})] \\ &= \text{Tr}[(\boldsymbol{\Sigma} - \mathbf{S})^2] \end{aligned} \tag{6.5.17}$$

is minimized for a *known value of* k where $\boldsymbol{\Sigma} = \boldsymbol{\Lambda}\boldsymbol{\Lambda}' + \boldsymbol{\Psi}$.

By taking the partial derivatives of $F(\boldsymbol{\Lambda}, \boldsymbol{\Psi})$ with respect to the loadings λ_{ij}, the

$$\begin{aligned} \frac{\partial F(\boldsymbol{\Lambda}, \boldsymbol{\Psi})}{\partial \lambda_{ij}} &= 2\,\text{Tr}\left[(\boldsymbol{\Sigma} - \mathbf{S})\left(\frac{\partial \boldsymbol{\Sigma}}{\partial \lambda_{ij}}\right)\right] \\ &= 2\,\text{Tr}\left\{(\boldsymbol{\Sigma} - \mathbf{S})\left[\boldsymbol{\Lambda}\left(\frac{\partial \boldsymbol{\Lambda}'}{\partial \lambda_{ij}}\right) + \left(\frac{\partial \boldsymbol{\Lambda}}{\partial \lambda_{ij}}\right)\boldsymbol{\Lambda}'\right]\right\} \\ &= 4\,\text{Tr}\left[(\boldsymbol{\Sigma} - \mathbf{S})\boldsymbol{\Lambda}\left(\frac{\partial \boldsymbol{\Lambda}'}{\partial \lambda_{ij}}\right)\right] \\ &= 4s^*_{ij} \end{aligned} \tag{6.5.18}$$

where s_{ij}^* is the ij element in the matrix $(\mathbf{\Sigma} - \mathbf{S})\mathbf{\Lambda}$. By representing the pk derivatives in matrix form, the

$$\frac{\partial F(\mathbf{\Lambda}, \mathbf{\Psi})}{\partial \mathbf{\Lambda}} = 4(\mathbf{\Sigma} - \mathbf{S})\mathbf{\Lambda} \tag{6.5.19}$$

By equating (6.5.19) to $\mathbf{0}$ and setting $\hat{\mathbf{\Sigma}} = \hat{\mathbf{\Lambda}}\hat{\mathbf{\Lambda}} + \hat{\mathbf{\Psi}}$, the first equation, using least-squares theory, becomes

$$(\mathbf{S} - \hat{\mathbf{\Lambda}}\hat{\mathbf{\Lambda}}' - \hat{\mathbf{\Psi}})\hat{\mathbf{\Lambda}} = \mathbf{0} \tag{6.5.20}$$

By now taking the partial derivative of $F(\mathbf{\Lambda}, \mathbf{\Psi})$ with respect to the ith diagonal element of $\mathbf{\Psi}$, the

$$\frac{\partial F(\mathbf{\Lambda}, \mathbf{\Psi})}{\partial \psi_i} = 2\,\mathrm{Tr}\left[(\mathbf{\Sigma} - \mathbf{S})\left(\frac{\partial \mathbf{\Psi}}{\partial \psi_i}\right)\right] \tag{6.5.21}$$

But the $\partial \mathbf{\Psi}/\partial \psi_i$ is a diagonal matrix with the value 1 in the iith position. Thus, setting all derivatives to $\mathbf{0}$, the

$$\frac{\partial F(\mathbf{\Lambda}, \mathbf{\Psi})}{\partial \mathbf{\Psi}} = 2\,\mathrm{Dia}\,[\hat{\mathbf{\Sigma}} - \mathbf{S}] = \mathbf{0} \tag{6.5.22}$$

and

$$\hat{\mathbf{\Psi}} = \mathrm{Dia}\,[\mathbf{S} - \hat{\mathbf{\Lambda}}\hat{\mathbf{\Lambda}}'] \tag{6.5.23}$$

which states that the sample variance minus the estimated unique variance is equal to the estimated communality, which is also true for the minres and iterative principal-factor-analysis procedures.

By simplifying (6.5.20), the two equations for determining $\hat{\mathbf{\Lambda}}$ and $\hat{\mathbf{\Psi}}$ become

$$\begin{aligned} (\mathbf{S} - \hat{\mathbf{\Psi}})\hat{\mathbf{\Lambda}} &= \hat{\mathbf{\Lambda}}(\hat{\mathbf{\Lambda}}'\hat{\mathbf{\Lambda}}) \\ \hat{\mathbf{\Psi}} &= \mathrm{Dia}\,[\mathbf{S} - \hat{\mathbf{\Lambda}}\hat{\mathbf{\Lambda}}'] \end{aligned} \tag{6.5.24}$$

These equations cannot be solved algebraically; however, an iterative solution is available through principal factoring. Since $\mathbf{\Lambda}$ is only determinable up to a post-multiplication by an orthogonal matrix $\mathbf{T}$, it is convenient for the least-squares solution to impose the $k(k-1)/2$ restrictions on $\mathbf{\Lambda}$ by requiring that $\mathbf{\Lambda}'\mathbf{\Lambda}$ be a diagonal matrix. Then, for a given value of k and an estimate $\mathbf{\Psi}_0$ of $\mathbf{\Psi}$, principal-component analysis provides the existence of a matrix $\mathbf{P}$ such that

$$\mathbf{S}_r = \mathbf{P\Delta P}' = (\mathbf{P\Delta}^{1/2})(\mathbf{P\Delta}^{1/2})' = \mathbf{\Lambda\Lambda}' \tag{6.5.25}$$

where $\mathbf{PP}' = \mathbf{P}'\mathbf{P} = \mathbf{I}$ and $\mathbf{\Delta}$ is a diagonal matrix containing the roots of $\mathbf{S}_r$ with at least k roots larger than 0. Furthermore,

$$\begin{aligned} \mathbf{S}_r\mathbf{P} &= \mathbf{P\Delta} \\ \mathbf{S}_r\mathbf{P\Delta}^{1/2} &= \mathbf{P\Delta}^{1/2}(\mathbf{\Delta}^{1/2}\mathbf{P}'\mathbf{P\Delta}^{1/2}) \\ \mathbf{S}_r\hat{\mathbf{\Lambda}} &= \hat{\mathbf{\Lambda}}(\hat{\mathbf{\Lambda}}'\hat{\mathbf{\Lambda}}) \end{aligned} \tag{6.5.26}$$

where $\hat{\mathbf{\Lambda}}'\hat{\mathbf{\Lambda}}$ is diagonal. By associating $\mathbf{S}_r$ with $\mathbf{S} - \hat{\mathbf{\Psi}}_0$, where an initial estimate for $\mathbf{\Psi}$ is obtained by using (6.5.15),

$$\hat{\mathbf{\Lambda}} = \mathbf{P\Delta}^{1/2} \quad \text{and} \quad \hat{\mathbf{\Lambda}}\hat{\mathbf{\Lambda}}' = \mathbf{P\Delta P}' \tag{6.5.27}$$

Continuing iteratively, as in iterative principal-factor analysis, estimates for $\mathbf{\Lambda}$ and $\mathbf{\Psi}$ can be determined.

Although the procedure outlined for solving (6.5.24) is mathematically correct, it is not feasible for use on a computer since convergence would be slow and would not work if $\mathbf{S}$ were non-Gramian. A new rapid algorithm has been developed by Jöreskog and van Thillo (1971) that utilizes the Newton-Raphson minimization procedure. The least-squares technique is called the unweighted least-squares method by these authors. The function $F(\mathbf{\Lambda}, \mathbf{\Psi})$ defined in (6.5.17) is minimized in two steps.

For a given value of $\mathbf{\Psi}$ and k, equation (6.5.20) is employed to find a conditional solution for $\mathbf{\Lambda}$, $\hat{\mathbf{\Lambda}} = \mathbf{P}\mathbf{\Delta}^{1/2}$, where $\mathbf{P} = [\mathbf{p}_1, \mathbf{p}_2, \ldots, \mathbf{p}_k]$ and $\mathbf{\Delta}$ is a diagonal matrix with roots $\delta_1, \delta_2, \ldots, \delta_k$. For $\hat{\mathbf{\Lambda}} = \mathbf{P}\mathbf{\Delta}^{1/2}$, the conditional minimum for $F[\mathbf{\Lambda}, \mathbf{\Psi}$ (given)] is equal to the sum of squares of the smallest $p - k$ roots of $\mathbf{S} - \mathbf{\Psi}$. That is, the $\min_{\Lambda} F[\mathbf{\Lambda}, \mathbf{\Psi}$ (given)] is

$$
\begin{aligned}
F(\mathbf{\Psi}) &= \operatorname{Tr}[\mathbf{S} - \mathbf{\Psi} - \hat{\mathbf{\Lambda}}\hat{\mathbf{\Lambda}}']^2 \\
&= \operatorname{Tr}[(\mathbf{S} - \mathbf{\Psi})^2] - 2\operatorname{Tr}[\hat{\mathbf{\Lambda}}'(\mathbf{S} - \mathbf{\Psi})\hat{\mathbf{\Lambda}}] + \operatorname{Tr}[(\hat{\mathbf{\Lambda}}'\hat{\mathbf{\Lambda}})^2] \\
&= \operatorname{Tr}[(\mathbf{S} - \mathbf{\Psi})^2] - \operatorname{Tr}[(\hat{\mathbf{\Lambda}}'\hat{\mathbf{\Lambda}})^2] \\
&= \operatorname{Tr}[(\mathbf{S} - \mathbf{\Psi})^2] - \operatorname{Tr}[\mathbf{\Delta}^2] \\
&= \sum_{j=1}^{p} \delta_j^2 - \sum_{j=1}^{k} \delta_j^2 \\
&= \sum_{j=k+1}^{p} \delta_j^2 \qquad (6.5.28)
\end{aligned}
$$

The expression in (6.5.28) is a function of the elements of $\mathbf{\Psi}$ for a given $\hat{\mathbf{\Lambda}}$ and is minimized numerically employing the Newton-Raphson procedure to obtain an absolute minimum for $F(\mathbf{\Lambda}, \mathbf{\Psi})$. See, for example, Jennrich and Robinson (1969), Jöreskog and Goldberger (1971), and Jöreskog and van Thillo (1971). By using (6.5.23), another estimate of $\mathbf{\Psi}$ is located by evaluating

$$
\psi_{ii} = s_{ii} - \sum_{j=1}^{k} \delta_j p_{ij}^2 \qquad \text{for } i = 1, \ldots, p \tag{6.5.29}
$$

and the process is repeated. This method must converge since $\mathbf{F}(\mathbf{\Psi})$ is bounded; it works when $\mathbf{S}$ is non-Gramian and successfully handles Heywood cases, a term defined for situations in which $\psi_{ii} \to 0$ (Jöreskog and van Thillo, 1971).

Although the unweighted least-squares method effectively estimates $\mathbf{\Psi}$ and $\mathbf{\Lambda}$ of the common-factor-analysis model, for a given value of k, it, like the other procedures discussed, has a serious shortcoming. The method is not scale free. The technique depends on the unit of measurement used for the observations, so that no simple relationship exists for solutions obtained when using $\mathbf{S}$ and $\mathbf{R}$. A method for analyzing the common-factor-analysis model is said to be *scale free* or *scale invariant* if, when an observation y_i is multiplied by a constant, the factor loadings for the observation are multiplied by the same constant and the estimated unique variance is multiplied by the square of the constant (see Anderson and Rubin, 1956). To illustrate that the unweighted least-squares technique is not scale free, let $\hat{\mathbf{\Lambda}}$ and $\hat{\mathbf{\Psi}}$ satisfy the equations given in (6.5.24), where $\mathbf{K} = \hat{\mathbf{\Lambda}}'\hat{\mathbf{\Lambda}}$ is diagonal and the common-factor-analysis model is $\mathbf{\Sigma} = \mathbf{\Lambda\Lambda}' + \mathbf{\Psi}$. By letting

$$\mathbf{Y}^* = \mathbf{DY} \tag{6.5.30}$$

where $\mathbf{D}$ is a nonsingular diagonal matrix, the new covariance matrix for the analysis is

$$\begin{aligned} V(\mathbf{Y}^*) &= \mathbf{D\Sigma D} = \mathbf{D\Lambda\Lambda' D} + \mathbf{D\Psi D} \\ &= \mathbf{D\Lambda}(\mathbf{D\Lambda})' + \mathbf{D\Psi D} \\ &= \mathbf{\Lambda}^*\mathbf{\Lambda}^{*\prime} + \mathbf{\Psi}^* \\ &= \mathbf{\Sigma}^* \end{aligned} \tag{6.5.31}$$

so that $\hat{\mathbf{\Psi}}^* = \mathbf{D}\hat{\mathbf{\Psi}}\mathbf{D}$ and $\hat{\mathbf{\Lambda}}^*\hat{\mathbf{\Lambda}}^{*\prime} = \mathbf{D}\hat{\mathbf{\Lambda}}(\mathbf{D}\hat{\mathbf{\Lambda}})'$. For a scale-free procedure, $\hat{\mathbf{\Lambda}}^* = \mathbf{D}\hat{\mathbf{\Lambda}}$ is necessary; however, in the unweighted least-squares procedure, $\hat{\mathbf{\Lambda}}'\hat{\mathbf{\Lambda}}$ was required to be a diagonal matrix. With this condition, if $\hat{\mathbf{\Lambda}}^* = \mathbf{D}\hat{\mathbf{\Lambda}}$, $\hat{\mathbf{\Lambda}}^{*\prime}\hat{\mathbf{\Lambda}}^* = \hat{\mathbf{\Lambda}}'\mathbf{D}^2\hat{\mathbf{\Lambda}}$ would not be diagonal. Furthermore, no orthogonal transformation exists such that the equations associated with the original model and the transformed model can be made to agree in accordance with the scale-free criterion (Anderson and Rubin, 1956, p. 131).

Scale-free, common-factor-analysis procedures have been developed, and Harris (1964) states, "I believe that in time, only such solutions (scale-free solutions) will be regarded as desirable." Attention is now directed toward a discussion of some of the proposed scale-free methods.

Scale-Free Methods. The first scale-free procedure was developed by Lawley (1940) utilizing the maximum-likelihood method. For this procedure it is assumed that the vector observations are sampled from a multivariate normal distribution with mean $\boldsymbol{\mu} = \mathbf{0}$ and variance-covariance matrix $\mathbf{\Sigma} = \mathbf{\Lambda\Lambda}' + \mathbf{\Psi}$. A sequence of contributions relevant to this method includes Lawley and Maxwell (1963), Fuller and Hemmerle (1966), Jöreskog (1967), Jöreskog and Lawley (1968), and Jöreskog and van Thillo (1971).

Given the common-factor-analysis model as specified in (6.5.3), where k is known, it is assumed that $\mathbf{Y} \sim N_p(\mathbf{0}, \mathbf{\Sigma} = \mathbf{\Lambda\Lambda}' + \mathbf{\Psi})$, so that the density function of $\mathbf{Y}$ is

$$f(\mathbf{y}) = (2\pi)^{-p/2}|\mathbf{\Sigma}|^{-1/2} \exp -\frac{1}{2}\mathbf{y}'\mathbf{\Sigma}^{-1}\mathbf{y} \tag{6.5.32}$$

The log of the likelihood function is represented as

$$L^*(\mathbf{\Lambda}, \mathbf{\Psi}) = -\frac{1}{2}Np \log 2\pi - \frac{1}{2}N \log |\mathbf{\Sigma}| - \frac{1}{2}N \operatorname{Tr}(\mathbf{\Sigma}^{-1}\mathbf{S}) \tag{6.5.33}$$

for a random sample of size N. To determine the likelihood equations, the partial derivatives of $L^*(\mathbf{\Lambda}, \mathbf{\Psi})$ with respect to $\mathbf{\Lambda}$ and $\mathbf{\Psi}$ are established and equated to $\mathbf{0}$.

Taking the partial derivative of $L^*(\mathbf{\Lambda}, \mathbf{\Psi})$ with respect to the loadings λ_{ij}, the

$$\frac{\partial L^*(\mathbf{\Lambda}, \mathbf{\Psi})}{\partial \lambda_{ij}} = -\frac{N}{2}\frac{\partial}{\partial \lambda_{ij}} \log |\mathbf{\Sigma}| - \frac{N}{2}\frac{\partial}{\partial \lambda_{ij}} \operatorname{Tr}(\mathbf{\Sigma}^{-1}\mathbf{S}) \tag{6.5.34}$$

However,

$$-\frac{N}{2}\frac{\partial}{\partial\lambda_{ij}}\log|\boldsymbol{\Sigma}| = -\frac{N}{2}\frac{1}{|\boldsymbol{\Sigma}|}\frac{\partial|\boldsymbol{\Sigma}|}{\partial\lambda_{ij}}$$

$$= -\frac{N}{2}\frac{1}{|\boldsymbol{\Sigma}|}\sum_{t=1}^{p}\sum_{u=1}^{p} C_{ij}\left(\frac{\partial\sigma_{tu}}{\partial\lambda_{ij}}\right)$$

(6.5.35)
$$= -\frac{N}{2}\operatorname{Tr}\left[\boldsymbol{\Sigma}^{-1}\left(\frac{\partial\boldsymbol{\Sigma}}{\partial\lambda_{ij}}\right)\right]$$

where C_{ij} is the cofactor of σ_{ij} in $\boldsymbol{\Sigma}$ [see equation (1.8.38)]. By employing (6.5.18),

(6.5.36)
$$-\frac{N}{2}\frac{\partial}{\partial\boldsymbol{\Lambda}}\log|\boldsymbol{\Sigma}| = -N\boldsymbol{\Sigma}^{-1}\boldsymbol{\Lambda}$$

Consider the second term in (6.5.34),

$$-\frac{N}{2}\frac{\partial}{\partial\lambda_{ij}}\operatorname{Tr}(\boldsymbol{\Sigma}^{-1}\mathbf{S}) = \frac{N}{2}\operatorname{Tr}\left[\boldsymbol{\Sigma}^{-1}\left(\frac{\partial\boldsymbol{\Sigma}}{\partial\lambda_{ij}}\right)\boldsymbol{\Sigma}^{-1}\mathbf{S}\right]$$

so that

$$\frac{\partial}{\partial\boldsymbol{\Lambda}}\left[-\frac{N}{2}\operatorname{Tr}(\boldsymbol{\Sigma}^{-1}\mathbf{S})\right] = N\boldsymbol{\Sigma}^{-1}\mathbf{S}\boldsymbol{\Sigma}^{-1}\boldsymbol{\Lambda}$$

by using (6.5.18) and (1.8.36). Hence the

(6.5.37)
$$\frac{\partial L^*(\boldsymbol{\Lambda},\boldsymbol{\Psi})}{\partial\boldsymbol{\Lambda}} = -N\boldsymbol{\Sigma}^{-1}\boldsymbol{\Lambda} + N\boldsymbol{\Sigma}^{-1}\mathbf{S}\boldsymbol{\Sigma}^{-1}\boldsymbol{\Lambda}$$

where $\boldsymbol{\Sigma} = \boldsymbol{\Lambda}\boldsymbol{\Lambda}' + \boldsymbol{\Psi}$. By equating (6.5.37) to $\mathbf{0}$, the first set of equations employing the maximum-likelihood method becomes

(6.5.38)
$$\mathbf{S}\hat{\boldsymbol{\Sigma}}^{-1}\hat{\boldsymbol{\Lambda}} = \hat{\boldsymbol{\Lambda}}$$

Using the identity

(6.5.39)
$$\hat{\boldsymbol{\Sigma}}^{-1} = \hat{\boldsymbol{\Psi}}^{-1} - \hat{\boldsymbol{\Psi}}^{-1}\hat{\boldsymbol{\Lambda}}(\mathbf{I} + \hat{\boldsymbol{\Lambda}}'\hat{\boldsymbol{\Psi}}^{-1}\hat{\boldsymbol{\Lambda}})^{-1}\hat{\boldsymbol{\Lambda}}'\hat{\boldsymbol{\Psi}}^{-1}$$

(see Lawley and Maxwell, 1963, p. 13) and following Jöreskog (1967),

$$\hat{\boldsymbol{\Sigma}}^{-1}\hat{\boldsymbol{\Lambda}} = \hat{\boldsymbol{\Psi}}^{-1}\hat{\boldsymbol{\Lambda}} - \hat{\boldsymbol{\Psi}}^{-1}\hat{\boldsymbol{\Lambda}}(\mathbf{I} + \hat{\boldsymbol{\Lambda}}'\hat{\boldsymbol{\Psi}}^{-1}\hat{\boldsymbol{\Lambda}})^{-1}\hat{\boldsymbol{\Lambda}}'\hat{\boldsymbol{\Psi}}^{-1}\hat{\boldsymbol{\Lambda}}$$

$$= \hat{\boldsymbol{\Psi}}^{-1}\hat{\boldsymbol{\Lambda}}[\mathbf{I} - (\mathbf{I} + \hat{\boldsymbol{\Lambda}}'\hat{\boldsymbol{\Psi}}^{-1}\hat{\boldsymbol{\Lambda}})^{-1}\hat{\boldsymbol{\Lambda}}'\hat{\boldsymbol{\Psi}}^{-1}\hat{\boldsymbol{\Lambda}}]$$

$$= \hat{\boldsymbol{\Psi}}^{-1}\hat{\boldsymbol{\Lambda}}(\mathbf{I} + \hat{\boldsymbol{\Lambda}}'\hat{\boldsymbol{\Psi}}^{-1}\hat{\boldsymbol{\Lambda}})^{-1}[(\mathbf{I} + \hat{\boldsymbol{\Lambda}}'\hat{\boldsymbol{\Psi}}^{-1}\hat{\boldsymbol{\Lambda}}) - \hat{\boldsymbol{\Lambda}}'\hat{\boldsymbol{\Psi}}^{-1}\hat{\boldsymbol{\Lambda}}]$$

(6.5.40)
$$= \hat{\boldsymbol{\Psi}}^{-1}\hat{\boldsymbol{\Lambda}}(\mathbf{I} + \hat{\boldsymbol{\Lambda}}'\hat{\boldsymbol{\Psi}}^{-1}\hat{\boldsymbol{\Lambda}})^{-1}$$

(6.5.38) reduces to

(6.5.41)
$$\mathbf{S}\hat{\boldsymbol{\psi}}^{-1}\hat{\boldsymbol{\Lambda}} = \hat{\boldsymbol{\Lambda}}(\mathbf{I} + \hat{\boldsymbol{\Lambda}}'\hat{\boldsymbol{\Psi}}^{-1}\hat{\boldsymbol{\Lambda}})$$

by substituting (6.5.40) into (6.5.38). Finally, premultiplying (6.5.41) by $\boldsymbol{\Psi}^{-1/2}$, the last equation is written as

(6.5.42)
$$\hat{\boldsymbol{\Psi}}^{-1/2}\mathbf{S}\hat{\boldsymbol{\Psi}}^{-1/2}(\hat{\boldsymbol{\Psi}}^{-1/2}\hat{\boldsymbol{\Lambda}}) = \hat{\boldsymbol{\Psi}}^{-1/2}\hat{\boldsymbol{\Lambda}}(\mathbf{I} + \hat{\boldsymbol{\Lambda}}'\hat{\boldsymbol{\Psi}}^{-1}\hat{\boldsymbol{\Lambda}})$$

By taking the partial derivative of $L^*(\mathbf{\Lambda}, \mathbf{\Psi})$ with respect to the elements of $\mathbf{\Psi}$,

$$\frac{\partial L^*(\mathbf{\Lambda}, \mathbf{\Psi})}{\partial \psi_i} = -\frac{N}{2}\frac{1}{|\mathbf{\Sigma}|}\frac{\partial |\mathbf{\Sigma}|}{\partial \psi_i} + \frac{N}{2}\text{Tr}\left[\mathbf{\Sigma}^{-1}\left(\frac{\partial \mathbf{\Psi}}{\partial \psi_i}\right)\mathbf{\Sigma}^{-1}\mathbf{S}\right]$$

$$= -\frac{N}{2}\text{Tr}\left[\mathbf{\Sigma}^{-1}\left(\frac{\partial \mathbf{\Psi}}{\partial \psi_i}\right)\right] + \frac{N}{2}\text{Tr}\left[\mathbf{\Sigma}^{-1}\mathbf{S}\mathbf{\Sigma}^{-1}\left(\frac{\partial \mathbf{\Psi}}{\partial \psi_i}\right)\right] \tag{6.5.43}$$

Since the $\partial \mathbf{\Psi}/\psi_i$ is a matrix of all 0s except for a 1 in the iith position, the

$$\frac{\partial L^*(\mathbf{\Lambda}, \mathbf{\Psi})}{\partial \mathbf{\Psi}} = -\text{Dia}\,[N\mathbf{\Sigma}^{-1}] + \text{Dia}\,[N\mathbf{\Sigma}^{-1}\mathbf{S}\mathbf{\Sigma}^{-1}] \tag{6.5.44}$$

By equating (6.5.44) to 0, the second set of equations becomes

$$\text{Dia}\,[\hat{\mathbf{\Sigma}}^{-1} - \hat{\mathbf{\Sigma}}^{-1}\mathbf{S}\hat{\mathbf{\Sigma}}^{-1}] = \mathbf{0} \quad \text{or} \quad \text{Dia}\,[\mathbf{I} - \mathbf{S}\hat{\mathbf{\Sigma}}^{-1}] = \mathbf{0} \tag{6.5.45}$$

Following Lawley and Maxwell (1963, p. 13),

$$(\mathbf{I} - \mathbf{S}\hat{\mathbf{\Sigma}}^{-1})(\hat{\mathbf{\Sigma}} - \hat{\mathbf{\Lambda}}\hat{\mathbf{\Lambda}}') = \hat{\mathbf{\Sigma}} - \hat{\mathbf{\Lambda}}\hat{\mathbf{\Lambda}}' - \mathbf{S} + \mathbf{S}\hat{\mathbf{\Sigma}}^{-1}\hat{\mathbf{\Lambda}}\hat{\mathbf{\Lambda}}'$$
$$= \hat{\mathbf{\Sigma}} - \mathbf{S}$$

since, by (6.5.38), $\mathbf{S}\hat{\mathbf{\Sigma}}^{-1}\hat{\mathbf{\Lambda}}\hat{\mathbf{\Lambda}}' = \hat{\mathbf{\Lambda}}\hat{\mathbf{\Lambda}}'$. By substituting for $\hat{\mathbf{\Sigma}} = \hat{\mathbf{\Lambda}}\hat{\mathbf{\Lambda}}' + \hat{\mathbf{\Psi}}$, (6.5.45) becomes

$$\hat{\mathbf{\Psi}} = \text{Dia}\,[\mathbf{S} - \hat{\mathbf{\Lambda}}\hat{\mathbf{\Lambda}}'] \tag{6.5.46}$$

The system of equations for the determination of $\hat{\mathbf{\Lambda}}$ and $\hat{\mathbf{\Psi}}$ is therefore

$$\hat{\mathbf{\Psi}}^{-1/2}\mathbf{S}^{-1/2}\hat{\mathbf{\Psi}}^{-1/2}(\hat{\mathbf{\Psi}}^{-1/2}\hat{\mathbf{\Lambda}}) = \hat{\mathbf{\Psi}}^{-1/2}\hat{\mathbf{\Lambda}}(\mathbf{I} + \hat{\mathbf{\Lambda}}'\hat{\mathbf{\Psi}}^{-1}\hat{\mathbf{\Lambda}})$$
$$\hat{\mathbf{\Psi}} = \text{Dia}\,[\mathbf{S} - \hat{\mathbf{\Lambda}}\hat{\mathbf{\Lambda}}'] \tag{6.5.47}$$

To solve (6.5.47), the procedure is as in the unweighted least-squares case with the exception that the matrix $\hat{\mathbf{\Lambda}}'\mathbf{\Psi}^{-1}\hat{\mathbf{\Lambda}}$ is assumed to be diagonal. This allows a unique solution to (6.5.47) by avoiding the rotation problem (Anderson and Rubin, 1956). Given k and $\hat{\mathbf{\Psi}}_0$, equation (6.5.42) implies that the matrix $\mathbf{Q} = \hat{\mathbf{\Psi}}_0^{-1/2}\hat{\mathbf{\Lambda}}$ contains the eigenvectors of the $p \times p$ matrix $\hat{\mathbf{\Psi}}_0^{-1/2}\mathbf{S}\hat{\mathbf{\Psi}}_0^{-1/2}$ and that the diagonal matrix $\mathbf{\Delta} = \mathbf{I} + \hat{\mathbf{\Lambda}}'\hat{\mathbf{\Psi}}_0^{-1}\hat{\mathbf{\Lambda}}$ contains the corresponding k nonnegative roots. However, the matrix product $\mathbf{Q}'\mathbf{Q} = \hat{\mathbf{\Lambda}}'\hat{\mathbf{\Psi}}_0^{-1}\mathbf{\Lambda} = \mathbf{\Delta} - \mathbf{I} \neq \mathbf{I}$. To have $\mathbf{Q}'\mathbf{Q} = \mathbf{I}$, set $\mathbf{P} = \mathbf{Q}(\mathbf{\Delta} - \mathbf{I})^{-1/2}$, so that $\mathbf{P}'\mathbf{P} = \mathbf{I}$. Then $\mathbf{Q}$ becomes

$$\mathbf{Q} = \hat{\mathbf{\Psi}}_0^{-1/2}\hat{\mathbf{\Lambda}} = \mathbf{P}(\mathbf{\Delta} - \mathbf{I})^{1/2} \tag{6.5.48}$$

or $\mathbf{Q}'\hat{\mathbf{\Psi}}_0^{-1/2}\mathbf{S}\hat{\mathbf{\Psi}}_0^{-1/2}\mathbf{Q} = \mathbf{\Delta}$ as required. Hence a solution for the loadings is, from (6.5.48),

$$\hat{\mathbf{\Lambda}} = \hat{\mathbf{\Psi}}_0^{1/2}\mathbf{P}(\mathbf{\Delta} - \mathbf{I})^{1/2} \tag{6.5.49}$$

By forming the matrix product $\hat{\mathbf{\Lambda}}\hat{\mathbf{\Lambda}}'$ and using the equation $\hat{\mathbf{\Psi}} = \text{Dia}\,[\mathbf{S} - \hat{\mathbf{\Lambda}}\hat{\mathbf{\Lambda}}']$ for a calculated $\hat{\mathbf{\Lambda}}$, an iterative procedure is established to determine $\hat{\mathbf{\Psi}}$ and $\hat{\mathbf{\Lambda}}$. However, as in the unweighted least-squares case, this method of computation, although straightforward, lacks computational efficiency. Instead, a two-step algorithm is employed to initially establish a conditional minimum and then locate an absolute minimum. See Jöreskog (1967), Jennrich and Robinson (1969), and Jöreskog and van Thillo (1971). Continuing iteratively, as is characteristic of the unweighted least-squares case, estimates for $\hat{\mathbf{\Lambda}}$ and $\hat{\mathbf{\Psi}}$ are obtained.

Although the computations involved in determining a maximum-likelihood solution are considerable, the procedure is scale free. For example, let $\mathbf{DY} = \mathbf{Y}^*$ for a nonsingular diagonal matrix $\mathbf{D}$; then, for the transformed variables, the maximum-likelihood equations become

$$\begin{aligned}
\mathbf{S}^*\hat{\boldsymbol{\Sigma}}^{*-1}\hat{\boldsymbol{\Lambda}}^* &= \hat{\boldsymbol{\Lambda}}^* \\
(\mathbf{DSD})\mathbf{D}^{-1}\hat{\boldsymbol{\Sigma}}^{-1}\mathbf{D}^{-1}\mathbf{D}\hat{\boldsymbol{\Lambda}} &= \mathbf{D}\hat{\boldsymbol{\Lambda}} \\
\mathbf{DS}\hat{\boldsymbol{\Sigma}}^{-1}\hat{\boldsymbol{\Lambda}} &= \mathbf{D}\hat{\boldsymbol{\Lambda}} \\
\mathbf{S}\hat{\boldsymbol{\Sigma}}^{-1}\hat{\boldsymbol{\Lambda}} &= \hat{\boldsymbol{\Lambda}} \\
\hat{\boldsymbol{\Psi}}^* &= \text{Dia}\,[\mathbf{S}^* - \hat{\boldsymbol{\Lambda}}^*\hat{\boldsymbol{\Lambda}}^{*\prime}] \\
\mathbf{D}\hat{\boldsymbol{\Psi}}\mathbf{D} &= \text{Dia}\,[\mathbf{DSD} - \mathbf{D}\hat{\boldsymbol{\Lambda}}\hat{\boldsymbol{\Lambda}}'\mathbf{D}] \\
\hat{\boldsymbol{\Psi}} &= \text{Dia}\,[\mathbf{S} - \hat{\boldsymbol{\Lambda}}\hat{\boldsymbol{\Lambda}}'] \\
\hat{\mathbf{K}}^* &= \hat{\boldsymbol{\Lambda}}^{*\prime}\hat{\boldsymbol{\Sigma}}^{*-1}\hat{\boldsymbol{\Lambda}}^* \\
&= (\mathbf{D}\hat{\boldsymbol{\Lambda}})'\mathbf{D}^{-1}\hat{\boldsymbol{\Sigma}}^{-1}\mathbf{D}^{-1}(\mathbf{D}\hat{\boldsymbol{\Lambda}}) \\
&= \hat{\boldsymbol{\Lambda}}'\hat{\boldsymbol{\Sigma}}^{-1}\hat{\boldsymbol{\Lambda}} \\
&= \mathbf{K}
\end{aligned}$$

for $\hat{\boldsymbol{\Lambda}}^* = \mathbf{D}\hat{\boldsymbol{\Lambda}}$ and $\hat{\boldsymbol{\Psi}}^* = \mathbf{D}\hat{\boldsymbol{\Psi}}\mathbf{D}$, where $\hat{\boldsymbol{\Lambda}}$ and $\hat{\boldsymbol{\Psi}}$ are solutions to (6.5.47) with diagonality condition and are equal to the equations for the original variables. Thus, if $\boldsymbol{\Lambda}$ is located by using $\mathbf{S}$, $\mathbf{D}\boldsymbol{\Lambda}$ is the set of loadings employing $\mathbf{R}$.

Another scale-free procedure, *canonical-factor analysis*, was developed by Rao (1955b). Rather than maximizing the total variance of the variables, as in principal-component analysis, or requiring that a few common factors account for as much of the common variance as possible, Rao utilized canonical-correlational analysis in determining linear combinations of the unobserved factors $\mathbf{X}$ that are maximally correlated with linear combinations of the observation vector $\mathbf{Y}$. In this way, the predictability of $\mathbf{Y}$ from $\mathbf{X}$ through a set of uncorrelated common factors is possible.

By assuming that

(6.5.50) $$\begin{bmatrix}\mathbf{Y}\\ \mathbf{X}\end{bmatrix} \sim N_{p+k}\left\{\begin{bmatrix}\mathbf{0}\\ \mathbf{0}\end{bmatrix}, \begin{bmatrix}\boldsymbol{\Sigma} & \boldsymbol{\Lambda}\\ \boldsymbol{\Lambda}' & \mathbf{I}\end{bmatrix}\right\}$$

linear combinations of the variables $\mathbf{a}_j'\mathbf{Y} = U_j$ and $\mathbf{b}_j'\mathbf{X} = V_j$ with maximal correlation are located by solving the determinantal equation

(6.5.51) $$|\hat{\boldsymbol{\Lambda}}\hat{\boldsymbol{\Lambda}}' - \theta\mathbf{S}| = 0$$

where θ_j is the square of the canonical correlation between U_j and V_j. By letting $\hat{\boldsymbol{\Lambda}}\hat{\boldsymbol{\Lambda}}' = \hat{\boldsymbol{\Sigma}} - \hat{\boldsymbol{\Psi}}$, equation (6.5.51) becomes

(6.5.52) $$|\mathbf{S} - \delta\hat{\boldsymbol{\Psi}}| = 0$$

for $\delta_j = 1/(1 - \theta_j)$. Now by letting $\boldsymbol{\Delta}$ denote the roots of $\hat{\boldsymbol{\Psi}}^{-1/2}\mathbf{S}\hat{\boldsymbol{\Psi}}^{-1/2}$ and $\mathbf{P}$ be the matrix of eigenvectors, canonical-factor analysis is mathematically equivalent to the maximum-likelihood procedure. That is, the common-factor loadings are $\hat{\boldsymbol{\Lambda}} = \hat{\boldsymbol{\Psi}}^{1/2}\mathbf{P}(\boldsymbol{\Delta} - \mathbf{I})^{1/2}$ for given values of k and $\hat{\boldsymbol{\Psi}}$. Furthermore, the condition $\hat{\boldsymbol{\Psi}} =$

Dia $[\mathbf{S} - \hat{\mathbf{\Lambda}}\hat{\mathbf{\Lambda}}']$ is also satisfied. However, different iterative procedures have been proposed for the two methods; see, for example, Harris (1962, 1963) and Pingel (1966). McDonald (1968) provides a unified treatment of canonical-factor analysis.

Although algorithms for the maximum-likelihood method are the best developed for a scale-free procedure, a researcher may also appreciate having a "best" procedure without assuming normality, especially if N is small. A new technique that does not depend on normality was first developed by Jöreskog and Goldberger (1971). It is known as the generalized least-squares method. They used it to provide a rapidly convergent algorithm for a scale-free technique independent of normality and yielding asymptotic maximum-likelihood estimates, utilizing the Newton-Raphson minimization technique. This procedure outperforms the scale-free, maximum-determinant method developed by Howe (1955) and Bargmann (1957), which employs the Gauss-Siedel method. However, it needs to be investigated more fully.

Additional topics related to factor analysis include alpha-factor analysis and image analysis. A discussion of the former technique is provided by Kaiser and Caffrey (1965) and Glass (1966). Alpha-factor analysis is a scale-free technique; however, instead of scaling in the metric of the unique part of the observations, it uses the common portion. An introduction to image analysis and some of its extensions can be found in Guttman (1953), Harris (1962, 1971), Kaiser (1963), and Jöreskog (1969a).

When employing the various iteration procedures to estimate the model parameters, it has always been assumed that the number of factors $k = k^*$ was "known." By assuming multivariate normality, Lawley (1940) derived a likelihood ratio test to test the null hypothesis

$$H_0: \mathbf{\Sigma} = \hat{\mathbf{\Lambda}}\hat{\mathbf{\Lambda}}' + \mathbf{\Psi} \quad \text{or} \quad k = k^* \tag{6.5.53}$$

for a specified value of k^* against the alternative that $\mathbf{\Sigma}$ has any structure not specified by (6.5.53). Bartlett (1954) suggested multiplying the likelihood ratio criterion by a multiplication factor for improved convergence. Using Bartlett's modification, the test statistic

$$X_B^2 = \left(N - 1 - \frac{2p + 5}{6} - \frac{2k}{3}\right)[\log|\hat{\mathbf{\Sigma}}| - \log|\mathbf{S}| + \mathrm{Tr}(\mathbf{S}\hat{\mathbf{\Sigma}}^{-1}) - p] \tag{6.5.54}$$

is used to test (6.5.53). As $N \to \infty$, X_B^2 tends towards a χ^2 distribution with $\nu = [(p - k)^2 - (p + k)]/2$ degrees of freedom when the null hypothesis is true. The null hypothesis is rejected at the significance level α if $X_B^2 > \chi_\alpha^2(\nu)$.

In using (6.5.54) to test (6.5.53), it is still assumed that k can be specified; however, the object of a factor-analytic study is to find as few factors as possible such that the model fits the data. After accomplishing this, several values of k are examined and tested using the same data. That is, $X_{B,k_0}^2, X_{B,k_0+1,\ldots}^2$ are computed where the starting point for k_0 is usually Guttman's strongest lower-bound criterion. Given that each test is calculated at a significance level α, the process is continued until the value of k that yields nonsignificance is reached. This value is then selected for the appropriate value of k. When employing such a procedure, the tests are clearly not independent; however, if k^* is the true value of k, $P(k > k^*) \leq \alpha$ for the test of $k = k^*$.

By locating the matrix $\mathbf{P}$ after a principal-component decomposition of a data set, component scores for each individual are obtained for further data analysis.

The component scores are uncorrelated, fewer in number than the number of variables, and uniquely determined by a linear combination of the observation vectors. In factor-analytic studies, it might also be of interest to represent the data in terms of a few factors after having established $\hat{\mathbf{\Lambda}}$, $\hat{\mathbf{\Psi}}$, and k. However, in this case, factor scores cannot be exactly obtained but merely estimated since the number of common- and unique-factor scores exceed the number of observed variables. For this reason most researchers tend to avoid computation of factor scores. Assuming that $\mathbf{\Lambda}$, $\mathbf{\Psi}$, and k are known, linear functions of the observations may be located and, in some sense, yield reasonable estimates of the common-factor scores. The usual approach to the problem is to apply a least-squares or regression procedure to estimate $\mathbf{x}$.

To obtain factor scores when using least-squares theory, it is understood that $\mathbf{x}$ is a fixed vector and that $\mathbf{\Lambda}$, $\mathbf{\Psi}$, and k are known; the model for the ith vector observation $\mathbf{y}_i$ is

$$\underset{(p\times 1)}{\mathbf{y}_i} = \underset{(p\times k)}{\mathbf{\Lambda}}\underset{(k\times 1)}{\mathbf{x}_i} + \underset{(p\times 1)}{\boldsymbol{\varepsilon}_i} \qquad i = 1, 2, \ldots, N$$

$$E(\boldsymbol{\varepsilon}_i) = \mathbf{0} \tag{6.5.55}$$

$$V(\boldsymbol{\varepsilon}_i) = \mathbf{\Psi}$$

By (4.8.13), the least-squares estimate of $\mathbf{x}_i$, which is the best linear unbiased estimate of $\mathbf{x}_i$, is given by

$$\hat{\mathbf{x}}_i = (\hat{\mathbf{\Lambda}}'\hat{\mathbf{\Psi}}^{-1}\hat{\mathbf{\Lambda}})^{-1}\hat{\mathbf{\Lambda}}'\hat{\mathbf{\Psi}}^{-1}\mathbf{y}_i \tag{6.5.56}$$

so that the matrix of factor scores is written as

$$\hat{\mathbf{X}} = \mathbf{Y}\hat{\mathbf{\Psi}}^{-1}\hat{\mathbf{\Lambda}}(\hat{\mathbf{\Lambda}}'\hat{\mathbf{\Psi}}^{-1}\hat{\mathbf{\Lambda}})^{-1} \tag{6.5.57}$$

This procedure for obtaining factor scores was proposed by Bartlett (1938b). If $\mathbf{x}$ is assumed to be a random vector so that

$$E\begin{bmatrix}\mathbf{Y}\\ \mathbf{X}\end{bmatrix} = \begin{bmatrix}\mathbf{0}\\ \mathbf{0}\end{bmatrix} = \begin{bmatrix}\boldsymbol{\mu}_1\\ \boldsymbol{\mu}_2\end{bmatrix}$$

$$V\begin{bmatrix}\mathbf{Y}\\ \mathbf{X}\end{bmatrix} = \begin{bmatrix}\mathbf{\Sigma} & \mathbf{\Lambda}\\ \mathbf{\Lambda}' & \mathbf{I}\end{bmatrix} = \begin{bmatrix}\mathbf{\Sigma}_{11} & \mathbf{\Sigma}_{12}\\ \mathbf{\Sigma}_{21} & \mathbf{\Sigma}_{22}\end{bmatrix} \tag{6.5.58}$$

then

$$E(\mathbf{Y}|\mathbf{x}) = \boldsymbol{\mu}_1 + \mathbf{\Sigma}_{12}\mathbf{\Sigma}_{22}^{-1}(\mathbf{x} - \boldsymbol{\mu}_2) = \mathbf{\Lambda}\mathbf{x}$$

$$V(\mathbf{Y}|\mathbf{x}) = \mathbf{\Sigma}_{11} - \mathbf{\Sigma}_{12}\mathbf{\Sigma}_{22}^{-1}\mathbf{\Sigma}_{21} = \mathbf{\Sigma} - \mathbf{\Lambda}\mathbf{\Lambda}' = \mathbf{\Psi}$$

By considering a conditional model, when $\mathbf{x}$ is random, the least-squares solution given by (6.5.56) is applicable.

Thurstone (1935), Thompson (1934), and Anderson and Rubin (1956), among others, have proposed alternative solutions to the factor-score problem; however, Bartlett's procedure is now considered the preferred method. For a detailed discussion and comparison of several proposed methods for obtaining factor scores, see Harris (1967) and McDonald and Burr (1967).

We will use the correlation matrix analyzed by Di Vesta and Walls (given in Table 6.3.3 for eight variables and 292 words) to compare the unweighted least-squares and maximum-likelihood methods for acquiring $\hat{\mathbf{\Lambda}}$ and $\hat{\mathbf{\Psi}}$ discussed in this section. The computer program used in the analysis was written by Jöreskog and van Thillo (1971). The program is called UFABY3 and is distributed by the Education Testing Service (Princeton, N.J.).

Beginning with the unweighted least-squares method, it is necessary to determine the number of common factors to extract from **R**. By using Guttman's strongest lower-bound criterion, the number of nonnegative roots of

$$\mathbf{R}^* = \mathbf{R} - \hat{\mathbf{\Psi}}_0 \tag{6.5.59}$$

where $\hat{\mathbf{\Psi}}_0 = [\text{Dia } \mathbf{R}^{-1}]^{-1}$ is investigated. For **R** as defined in Table 6.3.3, the diagonal elements of **R*** are

$$\text{Dia } \mathbf{R}^* = [.926, .965, .971, .729, .425, .720, .489, .597]$$

since

$$\hat{\mathbf{\Psi}}_0 = \text{Dia } [.074, .035, .029, .271, .575, .280, .511, .403]$$

By determining the eigenvalues of **R***, three of the roots are nonnegative: $\hat{\delta}_1 = 4.57$, $\hat{\delta}_2 = 1.22$, and $\hat{\delta}_3 = .34$. That is, Guttman's strongest lower-bound criterion for Di Vesta and Walls' data is $k_G = 3$, so that the number of common factors extracted from **R** using the unweighted least-squares procedure is in the neighborhood of three, depending on the fit of **R** as measured by the residual matrix $\mathbf{R}_0 = \mathbf{R} - \hat{\mathbf{\Lambda}}\hat{\mathbf{\Lambda}}'$.

The three factors extracted from the correlation matrix given in Table 6.3.3, the loadings, and the unique variances are illustrated in Table 6.5.1.

TABLE 6.5.1. Summary of the Unweighted Least-Squares Solution Using 8 × 8 Correlation Matrix.

Variables	Factors 1	Factors 2	Factors 3	Unique Variances
1 (Friendly/Unfriendly)	.875	−.385	−.112	.074
2 (Good/Bad)	.898	−.398	−.090	.027
3 (Nice/Awful)	.892	−.418	−.151	.006
4 (Brave/Not Brave)	.853	.137	.131	.236
5 (Big/Little)	.521	.207	.441	.492
6 (Strong/Weak)	.832	.221	.297	.170
7 (Moving/Still)	.438	.520	−.112	.525
8 (Fast/Slow)	.614	.702	−.368	.000
Eigenvalue	4.627	1.355	.4911	
Percentage of total variance	57.8	16.9	6.1	
Cumulative percentage of total variance	57.8	74.7	80.8	

After setting the criterion for convergence of the loadings at the strict value .0005, considering the accuracy of the data, seven iterations are required for convergence. For the same data using iterative principal-factor analysis, convergence is not evidenced. The residual matrix $\mathbf{R}_3$ for the three-factor solution contained in Table 6.5.1 is given below.

$$\mathbf{R}_3 = \begin{bmatrix} 0 \\ .03 & 0 \\ .07 & -.09 & 0 \\ .01 & .00 & -.11 & 0 \\ .02 & .04 & -.03 & -.03 & 0 \\ -.08 & -.04 & .15 & .05 & .00 & 0 \\ .07 & -.06 & -.01 & -.00 & .05 & -.09 & 0 \\ -.03 & .03 & .00 & -.00 & -.02 & .03 & .01 & 0 \end{bmatrix}$$

Although correlations r_{63} and r_{43} do not fit as well as the others in $\mathbf{R}_3$, a fourth factor does not yield a significantly better fit. This is seen from the residual matrix $\mathbf{R}_4$ obtained by using four factors.

$$\mathbf{R}_4 = \begin{bmatrix} 0 \\ -.03 & 0 \\ -.01 & .09 & 0 \\ .04 & .01 & -.12 & 0 \\ -.00 & .01 & -.01 & -.00 & 0 \\ -.03 & -.03 & .11 & .00 & .00 & 0 \\ .03 & -.09 & .09 & -.02 & .00 & .02 & 0 \\ -.01 & .01 & .01 & .00 & .00 & -.00 & .00 & 0 \end{bmatrix}$$

The interpretation of the factors in Table 6.5.1 and the components in Table 6.3.4 are similar except that the magnitude of the loadings on factors 2 and 3 for variables 5 and 7 is lower than the loadings for the corresponding components. This is due to the large unique variances for variables 5 and 7, which are ignored in principal-component analysis. For a clearer interpretation of the factors in Table 6.5.1, the factors may be rotated by using some analytic criterion to a "simple" structure due to the indeterminacy in the common-factor-analysis model.

Given a $p \times k$ matrix $\hat{\mathbf{\Lambda}}$ of loadings, an orthogonal matrix $\mathbf{T}$ of order $k \times k$ is desired such that the elements of a new loading matrix $\mathbf{\Theta} = \hat{\mathbf{\Lambda}}\mathbf{T}$ are as close to Thurstone's simple-structure formulation as possible. A widely used analytic procedure procuring "simple" orthogonal factors is Kaiser's varimax criterion (Kaiser, 1958). Kaiser's raw varimax criterion for rotating to simple structure is to make the sum of the variances of the squared loadings within each column of the factor-loading matrix large. That is,

$$V_r = \sum_{j=1}^{k} \left[\frac{1}{p} \sum_{i=1}^{p} \theta_{ij}^4 - \frac{1}{p^2} \left(\sum_{i=1}^{p} \theta_{ij}^2 \right)^2 \right] \tag{6.5.60}$$

is maximized. However, the raw varimax criterion is overly influenced by variables with large communalities since each variable is weighted equally in equation (6.5.60). To adjust for this, the factors are normalized to unit length under the normal varimax criterion

$$V_N = \sum_{j=1}^{k} \left[\frac{1}{p} \sum_{i=1}^{p} \gamma_{ij}^4 - \frac{1}{p^2} \left(\sum_{i=1}^{p} \gamma_{ij}^2 \right)^2 \right] \tag{6.5.61}$$

where

$$\gamma_{ij} = \frac{\theta_{ij}}{\sum_{j=1}^{k} \lambda_{ij}^2}$$

and then returned to their original length. The criterion V_N is called simply Kaiser's varimax method in the literature since (6.5.60) is never used in practice.

Kaiser has shown that the angle θ in the orthogonal matrix

$$\mathbf{T} = \begin{bmatrix} \cos\theta & -\sin\theta \\ \sin\theta & \cos\theta \end{bmatrix}$$

which maximizes (6.5.61) for any two factors must satisfy the relationship

$$\tan 4\theta = \frac{2\left(p \sum_i u_i v_i - \sum_i u_i \sum_i v_i\right)}{p \sum_i (u_i^2 - v_i^2) - \left[\left(\sum_i u_i\right)^2 - \left(\sum_i v_i\right)^2\right]} \tag{6.5.62}$$

where $u_i = \gamma_{ij}^2 - \gamma_{ij'}^2$ and $v_i = 2\gamma_{ij}\gamma_{ij'}$, for $i = 1, \ldots, p$ and factors $j \neq j'$. For the second derivative of (6.5.61) with respect to Θ to be negative, it is essential that 4θ be in the correct quadrant according to the signs of the numerator and denominator of (6.5.62), as specified in Table 6.5.2. With more than two factors, (6.5.62) is applied to

TABLE 6.5.2. Quadrant of 4θ.

Numerator	Denominator	Quadrant of 4θ
+	+	I: $0° \leq 4\theta < 90°$
+	−	II: $90° \leq 4\theta < 180°$
−	−	III: $-180° \leq 4\theta < -90°$
−	+	IV: $-90° \leq 4\theta < 0°$

all $k(k-1)/2$ factors, where the complete set of $k(k-1)/2$ pairings constitute a *cycle*. Since the value of (6.5.61) is less than $(k-1)/k$, convergence is assured after a finite number of cycles.

By utilizing the varimax criterion, the pattern of loadings in Table 6.5.1 is transformed to that in Table 6.5.3. The interpretation given the factors in Table 6.5.3 is clearly evident and more nearly parallels Di Vesta and Walls' study, using principal-component analysis followed by a varimax rotation. The first factor is termed an evaluation factor and loads highest on the first three variables (friendly/unfriendly, good/bad, and nice/awful). The second factor is an activity factor, and loads highest on variable 8 (fast/slow) and slightly less on variable 7 (moving/still). The last factor

TABLE 6.5.3. Varimax Factor Pattern Using the Unweighted Least-Squares Method.

Variables	Factors 1	2	3
1 (Friendly/Unfriendly)	.925	.124	.236
2 (Good/Bad)	.944	.113	.263
3 (Nice/Awful)	.968	.124	.204
4 (Brave/Not Brave)	.537	.401	.560
5 (Big/Little)	.158	.167	.675
6 (Strong/Weak)	.427	.378	.710
7 (Moving/Still)	.062	.638	.251
8 (Fast/Slow)	.159	.975	.170

is termed a potency factory and loads highest on variables 5 and 6 (big/little and strong/weak).

By using Di Vesta and Walls' correlation matrix to obtain $\hat{\mathbf{\Lambda}}$ and $\hat{\mathbf{\Psi}}$ by the scale-free, maximum-likelihood method, a test of significance using equation (6.5.54) is available to help determine the number of factors, provided $p + k < (p - k)^2$. After setting the criterion for convergence of the loadings in the maximum-likelihood procedure as in the unweighted least-squares method, eight iterations are required for convergence. For the same data, two other scale-free procedures were tried, canonical factor analysis and alpha-factor analysis. For these two techniques, convergence was not evidenced. The unrotated factor pattern obtained by using Jöreskog and van Thillo's program for $k = 3$ factors is given in Table 6.5.4.

TABLE 6.5.4. Summary of Arbitrary Maximum-Likelihood Solution Using 8 × 8 Correlation Matrix.

Variables	Factors 1	2	3	Unique Variances
1 (Friendly/Unfriendly)	−.328	.907	−.008	.070
2 (Good/Bad)	−.338	.925	−.029	.029
3 (Nice/Awful)	−.339	.937	.029	.007
4 (Brave/Not Brave)	−.587	.527	−.386	.229
5 (Big/Little)	−.298	.244	−.614	.474
6 (Strong/Weak)	−.585	.454	−.522	.179
7 (Moving/Still)	−.679	−.039	−.119	.523
8 (Fast/Slow)	−.999	−.030	.001	.000
Eigenvalues	2.571	3.102	.814	

The first thing to notice about the solution given is that the result is not in canonical form. That is, the eigenvalues are not ordered from largest to smallest. To put the matrix of loadings in canonical form so that comparisons can be made between the maximum-likelihood and unweighted least-squares unrotated solutions, a matrix $\mathbf{T}$ needs to be found such that $\hat{\mathbf{\Lambda}}^* = \hat{\mathbf{\Lambda}}\mathbf{T}$, where the matrix $\hat{\mathbf{\Lambda}}^*$ is the canonical form of $\hat{\mathbf{\Lambda}}$. By letting $\mathbf{\Delta}$ denote the eigenvalues of $\hat{\mathbf{\Lambda}}'\hat{\mathbf{\Lambda}}$, there exists an orthogonal matrix **T** such that

$$\boldsymbol{\Delta} = \hat{\boldsymbol{\Lambda}}^{*\prime}\hat{\boldsymbol{\Lambda}}^{*} = \mathbf{T}'\hat{\boldsymbol{\Lambda}}'\hat{\boldsymbol{\Lambda}}\mathbf{T} \quad \text{and} \quad \hat{\boldsymbol{\Lambda}}^{*}\hat{\boldsymbol{\Lambda}}^{*\prime} = \hat{\boldsymbol{\Lambda}}\mathbf{T}\mathbf{T}'\hat{\boldsymbol{\Lambda}}' = \hat{\boldsymbol{\Lambda}}\hat{\boldsymbol{\Lambda}}'$$

For $\hat{\boldsymbol{\Lambda}}$ as defined in Table 6.5.4,

$$\hat{\boldsymbol{\Lambda}}'\hat{\boldsymbol{\Lambda}} = \begin{bmatrix} 2.57140 & & \\ -1.51900 & 3.10200 & \\ .79730 & -.59250 & .81440 \end{bmatrix}$$

where the eigenvalues for the matrix are $\delta_1 = 4.6249$, $\delta_2 = 1.3564$, and $\delta_3 = .5065$. The eigenvectors of $\hat{\boldsymbol{\Lambda}}'\hat{\boldsymbol{\Lambda}}$, represented columnwise, make up the matrix **T**:

$$\mathbf{T} = \begin{bmatrix} -.6364 & .6834 & -.3578 \\ .7308 & .6826 & .0038 \\ -.2468 & .2591 & .9338 \end{bmatrix}$$

By postmultiplying $\hat{\boldsymbol{\Lambda}}$ in Table 6.5.4 by **T**, the canonical form of the matrix $\hat{\boldsymbol{\Lambda}}$ is derived. The signs have been changed to agree with the unweighted least-squares solution summarized in Table 6.5.1.

$$\hat{\boldsymbol{\Lambda}}^{*} = \hat{\boldsymbol{\Lambda}}\mathbf{T} = \begin{bmatrix} .874 & -.393 & -.113 \\ .898 & -.393 & -.097 \\ .893 & -.415 & -.152 \\ .854 & .141 & .148 \\ .520 & .196 & .466 \\ .833 & .225 & .276 \\ .433 & .522 & .132 \\ .614 & .703 & .358 \end{bmatrix}$$

The residual matrix for the three-factor, maximum-likelihood solution is

$$\mathbf{R}_{ML} = \begin{bmatrix} 0 & & & & & & & \\ .00 & 0 & & & & & & \\ .00 & .00 & 0 & & & & & \\ .05 & .03 & -.03 & 0 & & & & \\ .03 & .05 & -.03 & -.06 & 0 & & & \\ -.07 & -.05 & .04 & .03 & .01 & 0 & & \\ .11 & -.06 & -.00 & .02 & .09 & -.07 & 0 & \\ -.00 & .00 & .00 & .00 & -.00 & .00 & .00 & 0 \end{bmatrix}$$

By comparing the factor loadings of the maximum-likelihood procedure with the unweighted least-squares method, there is very close agreement; however, the residual matrix for the maximum-likelihood solution is a little better than the unweighted least-squares case.

By calculating the chi-squared value for the maximum-likelihood solution to test the hypothesis

$$H_0: \mathbf{\Sigma} = \mathbf{\Lambda\Lambda}' + \mathbf{\Psi} \quad \text{or} \quad k = 3$$

the value of X_B^2, using equation (6.5.54), is 15.2351 or 7 degrees of freedom. The p value for the test is $\alpha = .033$. For four factors, the p value is .091. Thus, for the example considered, there is very close agreement between the unweighted least-squares procedure and the maximum-likelihood method. However, for the scale-free, maximum-likelihood method, it does not matter whether **R** or **S** is analyzed. This is not true for the unweighted least-squares procedure.

The overview of the factor-analytic model presented in this section was intended to alert researchers to "new" factoring procedures and to old problems associated with the common-factor model. It is hoped that many of the traditional factoring procedures in use today will be replaced by scale-free methods. First generation "little jiffy" is dead (see Kaiser, 1970).

EXERCISES 6.5

1. Using UFABY3 to analyze Harmon's 13 psychological tests, (Harmon, 1967, p. 178), the following four-factor solution with EPS = .0005 was obtained. Transform the solution to canonical form.

	Factors			
	1	2	3	4
1	.309	.395	−.236	−.482
2	.151	.262	−.097	−.282
3	.092	.359	−.128	−.502
4	.111	.443	−.179	−.325
5	.346	.716	.044	.130
6	.355	.734	.212	.037
7	.234	.806	.048	.129
8	.302	.646	−.114	−.026
9	.282	.773	.198	.093
10	.485	.141	−.507	.425
11	1.000	−.002	.001	−.000
12	.429	.124	−.647	.071
13	.536	.300	−.399	−.216

Unique variances

1	.461	5	.349	9	.276	13	.417
2	.820	6	.289	10	.307		
3	.595	7	.277	11	.000		
4	.654	8	.478	12	.378		

6.6 COVARIANCE-STRUCTURE ANALYSIS

Analysis of the structure of $\mathbf{\Sigma}$ for the common-factor-analysis model assumes that

(6.6.1) $$\mathbf{\Sigma} = \mathbf{\Lambda\Lambda}' + \mathbf{\Psi}$$

A very general model for the analysis-of-covariance structures, which includes factor analysis, MANOVA, path analysis, the Potthoff-Roy model, and so on, has been reviewed by Jöreskog (1970). The form of $\mathbf{\Sigma}$ is

$$\mathbf{\Sigma} = \mathbf{F}(\mathbf{\Lambda\Phi\Lambda}' + \mathbf{\Psi})\mathbf{F}' + \mathbf{\Theta} \tag{6.6.2}$$

where $E(\mathbf{Y}) = \mathbf{XBQ}$ and $V(\mathbf{Y}) = \mathbf{I} \otimes \mathbf{\Sigma}$, for an $N \times p$ data matrix $\mathbf{Y}$, where each row vector in $\mathbf{Y}$ has a multivariate normal distribution. The matrices $\mathbf{X}$, $\mathbf{B}$, $\mathbf{Q}$, $\mathbf{\Lambda}$, $\mathbf{\Phi}$, and $\mathbf{\Psi}$ have been previously defined. The matrices $\mathbf{F}$ and $\mathbf{\Theta}$ are parameter matrices where $\mathbf{\Theta}$ is diagonal. Ease in handling the model given by (6.6.2) is essential to future research in education and psychology. Special cases of the model given by (6.6.2) have been stressed and developed in this text. With these basics, it is hoped that applied researchers may better understand and implement the complex statistical tools at their disposal.

REFERENCES

Afifi, A., and R. M. Elashoff (1966). Missing observations in multivariate statistics I: Review of the literature, *Journal of the American Statistical Association*, **61**, 595–604.

———, and ——— (1967). Missing observations in multivariate statistics II: Point estimation in simple linear regression. *Journal of the American Statistical Association*, **62**, 10–29.

———, and ——— (1969). Missing observations in multivariate statistics III: Large sample analysis of simple linear regression. *Journal of the American Statistical Association*, **64**, 337–358.

Aitken, A. C. (1935). On least squares and linear combinations of observations. *Proceedings of the Royal Society of Edinburgh*, **55**, 42–48.

——— (1937. Studies in practical mathematics, II: The evaluation of the latent roots and latent vectors of a matrix. *Proceedings of the Royal Society of Edinburgh*, **57**, 269–304.

——— (1956). *Determinants and matrices.* (9th ed.) New York: Interscience.

Aitken, M. A. (1969). Some tests for correlation matrices. *Biometrika*, **56**, 443–446.

———, C. W. Nelson, and K. H. Reinfurt (1968). Tests for correlation matrices. *Biometrika*, **55**, 327–334.

Ammon, P., *Personal communication*. Berkeley: University of California.

Anderson, T. W. (1958). *An introduction to multivariate statistical analysis.* New York: John Wiley.

——— (1963). Asymptotic theory for principal components. *Annals of Mathematical Statistics*, **34**, 122–148.

———, S. Das Gupta, and G. P. H. Styan (1972). *A bibliography of multivariate statistical analysis*. New York: Halstead Press.

———, and H. Rubin (1956). Statistical inference in factor analysis. *Proceedings of the Third Berkeley Symposium on Mathematical Statistics and Probability*, **V**, 111–150 .

Andrews, D. F., R. Gnanadesikan, and J. L. Warner (1971). Transformations of multivariate data. *Biometrics*, **27**, 825–840.

Apostol, T. M. (1957). *Mathematical analysis*. Reading, Mass.: Addison-Wesley.

——— (1961). *Calculus*. Vol. I, New York: Blaisdell.

——— (1962). *Calculus*. Vol. II, New York: Blaisdell.

Bancroft, T. A. (1968). *Topics in intermediate statistical methods*. Vol. I. Ames, Iowa: Iowa University Press.

Banerjee, K. S. (1964). A note on idempotent matrices. *Annals of Mathematical Statistics*, **35**, 880–882.

Bargmann, R. E. (1957). A study of independence and dependence in multivariate normal analysis. *Mimeograph Series No. 186*. Chapel Hill: Institute of Statistics, University of North Carolina.

——— (1967). *MUDAID routine*. Atlanta: Department of Statistics, University of Georgia.

——— (1970). Interpretation and use of a generalized discriminant function. In R. C. Bose et al. (Eds), *Essays in probability and statistics*, pp. 35–60. Chapel Hill: University of North Carolina.

Barten, A. P. (1962). Note on unbiased estimation of the squared multiple correlation coefficients. *Statistica Neerlandica*, **16**, 151–163.

Barth, W., R. S. Morton, and J. H. Wilkinson (1967). Calculation of the eigenvalues of a symmetric tridiagonal matrix by the method of bisection. *Numerische Mathematik*, **9**, 386–393.

Bartlett, M. S. (1937a). Properties of sufficiency and statistical tests. *Proceedings of the Royal Society of London, Series A*, 160, 268–282.

——— (1937b). Some examples of statistical methods in research in agriculture and applied biology. *Journal of the Royal Statistical Society Supplement*, **4**, 137–183.

——— (1938a). Further aspects of the theory of multiple regression. *Proceedings of the Cambridge Philosophical Society*, **34**, 33–40.

——— (1938b). Methods of estimating mental factors. *Nature*, **141**, 609–610.

——— (1947). Multivariate analysis, *Journal of the Royal Statistical Society Supplement, Series B*, **9**, 176–197.

——— (1950). Tests of significance in factor analysis. *British Journal of Psychology (Statistics Section)*, **3**, 77–85.

——— (1951). The goodness of fit of a single hypothetical discriminant function in the case of several groups. *Annals of Eugenics*, **16**, 199–214.

——— (1954). A note on the multiplying factors for various χ^2 approximations. *Journal of the Royal Statistical Society, Series B*, **16**, 296–298.

———, and D. V. Rajalaksham (1953). Goodness of fit tests for simultaneous autoregressive series. *Journal of the Royal Statistical Society, Series B*, **15**, 107–124.

Bellman, R. (1960). *Introduction to matrix analysis*. New York: McGraw-Hill.

Bennett, B. M. (1951). Note on a solution of the generalized Behrens-Fisher problem. *Annals of the Institute of Statistical Mathematics*, **2**, 87–90.

Boardman, T. J. (1971). Graphical Monte Carlo type I error rates for multiple comparison procedures. *Biometrics*, **27**, 738–744.

Bock, R. D. (1963). Multivariate analysis of variance of repeated measures. In C. W. Harris (Ed.), *Problems of measuring change*, pp. 85–103. Madison: University of Wisconsin Press.

———, and E. A. Haggard (1968). The use of multivariate analysis of variance in behavioral research. In D. K. Whitla (Ed.), *Handbook of measurement and assessment in behavioral sciences*, pp. 100–112. Reading, Mass.: Addison-Wesley.

———, and L. V. Jones (1968). *The measurement and prediction of judgment and choice*, San Francisco: Holden-Day.

Bose, R. C., and S. N. Roy (1938). The distribution of studentized D^2-statistics. *Sankhyá*, **4**, 19–38.

Boullion, T. L., and P. L. Odell (1971). *Generalized inverse matrices*. New York: John Wiley.

Bowker, A. H. (1960). A representation of Hotelling's T^2 and Anderson's classification statistic. In I. Olkin et al. (Eds), *Contributions to probability and statistics*, pp. 142–149. Stanford, Calif.: Stanford University Press.

Box, G. E. P. (1949). A general distribution theory for a class of likelihood criteria. *Biometrika*, **36**, 317–346.

——— (1950). Problems in the analysis of growth and linear curves. *Biometrika*, **6**, 362–389.

——— (1953). Nonnormality and tests on variances. *Biometrika*, **40**, 318–335.

Brownlee, K. A. (1965). *Statistical theory and methodology in science and engineering*. (2nd ed.) New York: John Wiley.

Carroll, J. B. (1961). The nature of the data, or how to choose a correlation coefficient. *Psychometrika*, **26**, 347–372.

Chipman, J. J., and M. M. Rao (1964). The treatment of linear restrictions in regression analysis, *Econometrica*, **32**, 189–209.

Clyde, D. J. (1969). *Multivariate analysis of variance on large computers*. Miami: Clyde Computing Service.

Cochran, W. G. (1934). The distribution of quadratic forms in a normal system. *Proceedings of the Cambridge Philosophical Society*, **30**, 178–191.

Cooley, W. W., and P. R. Lohnes (1971). *Multivariate data analysis*. New York: John Wiley.

Cramer, C. M., and R. D. Bock (1966). Multivariate analysis. *Review of Educational Research*, **36**, 604–617.

Danford, M. B., and H. M. Hughes (1957). Mixed model analysis of variance, assuming equal variances and covariances. *Report 57–144*. School of Aviation Medicine, U.S.A.F.

———, H. M. Hughes, and R. C. McNee (1960). On the analysis of repeated experiments. *Biometrics*, **16**, 547–565.

Daniel, C., and F. S. Wood (1971). *Fitting equations to data*. New York: John Wiley.

Das Gupta, S. (1970). Step-down multiple decision rules. In R. C. Bose et al. (Eds), *Essay in probability and statistics*, pp. 229–250, Chapel Hill: University of North Carolina Press.

———, T. W. Anderson, and G. S. Mudholkar (1964). Monotonicity of the power functions of some tests of the multivariate linear hypothesis. *Annals of Mathematical Statistics*, **35**, 200–205.

Davis, A. W. (1970). Exact distributions of Hotelling's generalized T_0^2, *Biometrika*, **57**, 187–191.

Dempster, A. P. (1969). *Elements of continuous multivariate analysis*. Reading, Mass.: Addison-Wesley.

Di Vesta, F. J., and R. T. Walls (1970). Factor analysis of the semantic attributes of 487 words and some relationships to the conceptual behavior of fifth-grade children. *Journal of Educational Psychology*, **61**, 6, Pt. 2, December.

Dixon, W. J. (1967). *BMD biomedical computer programs*. Berkeley: University of California Press.

——— (1969). *BMD biomedical computer programs, X-series supplement*. Berkeley: University of California Press.

Draper, N. R., and H. Smith (1966). *Applied regression analysis*. New York: John Wiley.

Dunn, O. J. (1961). Multiple comparisons among means. *Journal of the American Statistical Association*, **56**, 52–64.

Dwyer, P. S. (1967). Some applications of matrix derivatives in multivariate analysis. *Journal of the American Statistical Association*, **62**, 607–625.

Eaton, M. L. (1969). Some remarks on Scheffé's solution to the Behrens-Fisher problem. *Journal of the American Statistical Association*, **64**, 1318–1322.

Eckart, C., and G. Young (1936). The approximation of one matrix by another of lower rank. *Psychometrika*, **1**, 211–218.

Edwards, A. L. (1968). *Experimental design in psychological research.* (3rd ed.) New York: Holt, Rinehart and Winston.

Elston, R. C., and J. E. Grizzle (1962). Estimation of time-response curves and their confidence bands. *Biometrics*, **18**, 148–159.

Faddeeva, V. N. (1959). *Computational methods of linear algebra.* New York: Dover.

Federer, W. T. (1955). *Experimental design.* New York: Macmillan.

Finn, J. (1972). *Multivariance: Univariate and multivariate analysis of variance, covariance and regression: Version V.* Ann Arbor: National Educational Resources.

Fisher, R. A. (1928). The general sampling distribution of the multiple correlation coefficient. *Proceedings of the Royal Society of London, Series A*, **121**, 654–673.

——— (1936). The use of multiple measurements in taxonomic problems. *Annals of Eugenics*, **7**, 179–188.

——— (1939). The sampling distribution of some statistics obtained from nonlinear linear equations. *Annals of Eugenics*, **9**, 238–249.

——— (1949). *Design of experiments.* (5th ed.) Edinburgh: Oliver & Boyd Ltd.

———, and F. Yates (1963). *Statistical tables for biological, agricultural, and medical research.* Edinburgh: Oliver & Boyd Ltd.

Forsythe, G. E. (1951). Theory of selected methods of finite matrix inversion and decomposition. *INA Report 52-5.* Washington, D.C.: National Bureau of Standards.

Fowlkes, E. B., and E. T. Lee (1971). *Computer programs for multivariate analysis of variance.* Murray Hill, N.J.: Bell Telephone Labs.

Fuller, E. L., and W. J. Hemmerle (1966). Robustness of the maximum-likelihood estimation procedure in factor analysis. *Psychometrika*, **31**, 255–266.

Gabriel, K. R. (1968). Simultaneous test procedures in multivariate analysis of variance. *Biometrika*, **55**, 489–504.

——— (1969). A comparison of some methods of simultaneous inference in MANOVA. In P. R. Krishnaiah (Ed.), *Multivariate analysis II*, pp. 67–86. New York: Academic Press.

———, and P. K. Sen (1968). Simultaneous test procedures for one-way ANOVA and MANOVA based on rank scores. *Sankhyá, Series A*, **30**, 303–312.

Gaito, J., and D. Wiley (1963). Univariate analysis of variance procedures in the measurement of change. In C. W. Harris (Ed.), *Problems in measuring change*, pp. 60–85. Madison: University of Wisconsin Press.

Geisser, S., and S. W. Greenhouse (1958). Extension of Box's results on the use of the *F*-distribution in multivariate analysis. *Annals of Mathematical Statistics*, **29**, 885–891.

Girshick, M. A. (1939a). The sampling distribution of some statistics obtained from nonlinear equations. *Annals of Eugenics*, **9**, 238–249.

——— (1939b). On the sampling theory of roots of determinantal equations. *Annals of Mathematical Statistics*, **10**, 203–224.

Givens, W. (1957). The characteristic value-vector problem. *Journal of the Association for Computing Machinery*. **4**, 298–307.

Glass, G. V. (1966). Alpha factor analysis of infallible variables. *Psychometrika*, **31**, 545–561.

———, and J. C. Stanley (1970). *Statistical methods in education and psychology.* Englewood Cliffs, N.J.: Prentice-Hall.

Gleser, L. J. (1968). Testing a set of correlation coefficients for equality. *Biometrika*, **34**, 513–517.

Gnanadesikan, R., and E. T. Lee (1970). Graphical techniques for internal comparisons amongst equal degrees of freedom groupings in multiresponse experiments. *Biometrika*, **57**, 229–237.

Goldberger, A. S. (1964). *Econometrics.* New York: John Wiley.

Goldstine, H., F. J. Murray, and J. von Neumann (1959). The Jacobi method for real symmetric matrices. *Journal of the Association for Computing Machinery*, **6**, 59–96.

Graybill, F. A. (1961). *An introduction to linear statistical models.* Vol. I. New York: McGraw-Hill.

——— (1969). *Introduction to matrices with application in statistics.* Belmont, Calif.: Wadsworth.

———, and G. Marsaglia (1957). Idempotent matrices and quadratic forms in the general linear hypothesis. *Annals of Mathematical Statistics*, **28**, 678–686.

Greenhouse, S. W., and S. Geisser (1959). On methods in the analysis of profile data. *Psychometrika*, **24**, 95–112.

Grizzle, J., and D. M. Allen (1969). Analysis of growth and dose response curves. *Biometrics*, **25**, 357–381.

Guttman, L. (1953). Image theory for the structure of quantitative variates. *Psychometrika*, **18**, 277–296.

——— (1954). Some necessary conditions for common-factor analysis. *Psychometrika*, **19**, 149–161.

——— (1956). Best possible systematic estimates of communalities. *Psychometrika*, **21**, 273–285.

Hahn, G. J., and S. S. Shapiro (1967). *Statistical methods in engineering.* New York: John Wiley.

Halmos, P. R. (1958). *Finite-dimensional vector spaces.* New York: Van Nostrand.

Hanumara, R. C., and W. A. Thompson, Jr. (1968). Percentage points of the extreme roots of a Wishart matrix. *Biometrika*, **55**, 505–512.

Harmon, H. H. (1967). *Modern factor analysis.* (2nd ed.) Chicago: The University of Chicago Press.

———, and W. Jones (1966). Factor analysis by minimizing residuals (minres). *Psychometrika*, **31**, 351–368.

Harris, C. W. (1962). Some Rao-Guttman relations. *Psychometrika*, **27**, 247–264.

——— (1963). Canonical factor models for the description of change. In C. W. Harris (Ed.), *Problems in measuring change*, pp. 138–155. Madison: University of Wisconsin Press.

——— (1964). Some recent developments in factor analysis. *Educational and Psychological Measurement*, **24**, 193–206.

——— (1967). On factors and factor scores. *Psychometrika*, **32**, 363–379.

——— (1971). Image analysis with a Spearman case. *Multivariate Behavioral Research*, **6**, 423–432.

———, and H. Kaiser (1964). Oblique factor analytic solutions by orthogonal transformations. *Psychometrika*, **29**, 347–362.

Hartley, H. O., and R. R. Hocking (1971). The analysis of incomplete data. *Biometrics*, **27**, 783–823.

Hays, W. L. (1963). *Statistics for psychologists.* New York: Holt, Rinehart and Winston.

Heck, D. L. (1960). Charts of some upper percentage points of the distribution of the largest characteristic root. *Annals of Mathematical Statistics*, **31**, 625–642.

Henderson, C. R. (1969). Design and analysis of animal husbandry experiments. In *Techniques and procedures in animal science research.* (2nd ed., chap. 1) Beltsville, Md.: American Society of Animal Science.

Hoel, P. G. (1971). *Introduction to mathematical statistics.* (4th ed.) New York: John Wiley.

Hogg, R. V., and A. T. Craig (1958). On the decomposition of certain χ^2 variables. *Annals of Mathematical Statistics*, **29**, 608–610.

Hohn, F. E. (1964). *Elementary matrix algebra.* (2nd ed.) New York: Macmillan.

Hotelling, H. (1931). The generalization of Student's ratio. *Annals of Mathematical Statistics*, **2**, 360–378.

——— (1933). Analysis of a complex of statistical variables into principal components. *Journal of Educational Psychology*, **24**, 417–441, 498–520.

——— (1936a). Relations between two sets of variates. *Biometrika*, **28**, 321–377.

——— (1936b). Simplified calculation of principal components. *Psychometrika*, **1**, 27–35.

——— (1951). A generalized T-test and measure of multivariate dispersion. *Proceedings of the Second Berkeley Symposium on Mathematics and Statistics*, 23–41.

Householder, A. S. (1964). *The theory of matrices in numerical analysis*. New York: Blaisdell.

Howe, W. G. (1955). Some contributions to factor analysis. *Report No. ORNL-1919*. Oak Ridge, Tenn.: Oak Ridge National Laboratory.

Hsu, P. L. (1938). Notes on Hotelling's generalized T^2. *Annals of Mathematical Statistics*, **9**, 231–243.

——— (1939). On the distribution of the roots of certain determinantal equations. *Annals of Eugenics*, **9**, 250–258.

Huang, D. S. (1970). *Regression and econometric methods*. New York: John Wiley.

Hughes, H. M., and M. B. Danford (1958). Repeated measurement designs, assuming equal variances and covariances. *Report 59-40*. Randolf AFB, Texas: Air University, School of Aviation Medicine, USAF.

Hummel, T. J., and J. R. Sligo (1971). Empirical comparison of univariate and multivariate analysis of variance procedures. *Psychological Bulletin*, **76**, 49–57.

Huynh, H., and L. S. Feldt (1970). Conditions under which mean square ratios in repeated measurements designs have exact *F*-distributions. *Journal of the American Statistical Association*, **65**, 1582–1589.

Ito, K. (1962). A comparison of powers of two multivariate analysis of variance tests. *Biometrika*, **49**, 455–462.

——— (1969). On the effect of heteroscedasticity and nonnormality upon some multivariate test procedures. In P. R. Krishnaiah (Ed.), *Multivariate analysis II*, pp. 87–120. New York: Academic Press.

———, and W. Schull (1964). On the robustness of the T_0^2 test in multivariate analysis of variance when variance-covariance matrices are not equal. *Biometrika*, **51**, 71–82.

James, A. T. (1964). Distributions of matrix variates and latent roots derived from normal samples. *Annals of Mathematical Statistics*, **35**, 475–501.

——— (1969). Tests of equality of latent roots of the covariance matrix. In P. R. Krishnaiah (Ed.), *Multivariate analysis II*, pp. 205–218. New York: Academic Press.

James, G. S. (1954). Tests of linear hypotheses in univariate and multivariate analysis when the ratios of the population variances are unknown. *Biometrika*, **41**, 19–43.

Jennrich, R. I., and S. M. Robinson (1969). A Newton-Raphson algorithm for maximum likelihood factor analysis. *Psychometrika*, **34**, 111–123.

Jensen, D. R., and R. B. Howe (1968). Tables of Hotelling's T^2-distribution. *Technical Report Number 9*. Virginia Polytechnic Institute.

John, J. A. (1970). Use of generalized inverse matrices in MANOVA. *British Journal of the Royal Statistical Society, Series B*, 137–143.

John, P. W. N. (1964). Pseudo-inverses in analysis of variance. *Annals of Mathematical Statistics*, **35**, 895–896.

Jones, L. V. (1966). Analysis of variance in its multivariate developments. In R. B. Cattell (Ed.), *Handbook of multivariate experimental psychology*, pp. 244–266. Chicago: Rand McNally.

Jöreskog, K. G. (1963). *Statistical estimation in factor analysis*. Stockholm: Almquist and Wiksell;

——— (1966). Testing a simple structure hypothesis in factor analysis. *Psychometrika*, **31**, 165–178.

——— (1967). Some contributions to maximum likelihood factor analysis. *Psychometrika*, **32**, 443–482.

——— (1969a). Efficient estimation in image factor analysis. *Psychometrika*, **34**, 51–76.

——— (1969b). A general approach to confirmatory maximum likelihood factor analysis. *Psychometrika*, **34**, 183–202.

——— (1970). A general method for analysis of covariance structures. *Biometrika*, **57**, 239–252.

———, and A. S. Goldberger (1971). Factor analysis by generalized least squares. *Research Bulletin 71-26*. Princeton, N.J.: Educational Testing Service. (*Psychometrika*, **37**, in press.)

———, and D. N. Lawley (1968). New methods in maximum likelihood factor analysis. *British Journal of Mathematical and Statistical Psychology*, **21**, 85–96.

———, and M. van Thillo (1971). New rapid algorithms for factor analysis by unweighted least squares, generalized least squares and maximum likelihood. *Research Memorandum 71-5*. Princeton, N.J.: Educational Testing Service.

Kaiser, H. F. (1958). The variance criterion for analytic rotation in factor analysis, *Psychometrika*, **23**, 187–200.

——— (1960). The application of electronic computers to factor analysis. *Educational and Psychological Measurement*, **20**, 141–151.

——— (1963). Image analysis. In C. W. Harris (Ed.), *Problems in measuring change*, pp. 156–166. Madison: University of Wisconsin Press.

——— (1970). A second generation little jiffy. *Psychometrika*, **35**, 401–415.

———, and J. Caffrey (1965). Alpha factor analysis. *Psychometrika*, **30**, 1–14.

Kendall, M. G. (1961). *A course in multivariate analysis*. New York: Hafner.

———, and A. Stuart (1961), *The advanced theory of statistics*. Vol. 2. New York: Hafner.

Khatri, O. G. (1966). A note on a MANOVA model applied to problems in growth curve. *Annals of the Institute of Statistical Mathematics*, **18**, 75–86.

Kirk, R. E. (1968). *Experimental design: Procedures for the behavioral sciences*. Monterey, Calif.: Brooks/Cole.

Koch, G. G. (1969). Some aspects of the statistical analysis of "split-plot" experiments in completely randomized layouts. *Journal of the American Statistical Association*, **64**, 485–505.

Korin, B. P. (1968). On the distribution of a statistic used for testing a covariance matrix. *Biometrika*, **55**, 171–178.

Kramer, C. Y., and D. R. Jensen (1969). Fundamentals of multivariate analysis. Part II: Inference about two treatments. *Journal of Quality Technology*, **1**, 189–204.

Krishnaiah, P. R. (1965). On the simultaneous ANOVA and MANOVA tests. *Annals of the Institute of Statistical Mathematics*, **17**, 35–53.

——— (1969). Simultaneous test procedures under general MANOVA models. In P. R. Krishnaiah (Ed.), *Multivariate analysis II*, 121–144. New York: Academic Press.

Kshirsagar, A. M. (1959). Bartlett decomposition and Wishart distributions. *Annals of Mathematical Statistics*, **30**, 239–241.

Kullback, S. (1959). *Information theory and statistics*. New York: John Wiley.

——— (1967). On testing correlation matrices. *Applied Statistics*, **16**, 80–85.

Lawley, D. N. (1938). A generalization of Fisher's z-test. *Biometrika*, **30**, 180–187.

——— (1940). The estimation of factor loadings by the method of maximum likelihood. *Proceedings of the Royal Statistical Society of Edinburgh, Section A*, **60**, 64–82.

——— (1956). Tests of significance on the latent roots of covariance and correlation matrices. *Biometrika*, **43**, 128–136.

——— (1963). On testing a set of correlation coefficients for equality. *Annals of Mathematical Statistics*, **34**, 149–151.

———, and A. E. Maxwell (1963). *Factor analysis as a statistical method.* London: Butterworth.

Layard, M. (1969). Asymptotically robust tests about covariance matrices. *Technical Report No. 37*. Stanford: Department of Statistics, Stanford University.

Lee, Y. H. K. (1970). Multivariate analysis of variance for analyzing trends in repeated observations. Unpublished doctoral dissertation, University of California, Berkeley.

Lehmann, E. L. (1959). *Testing statistical hypotheses*. New York: John Wiley.

Lohnes, P. R., and W. W. Cooley (1968). *Introduction to statistical procedures: With computer exercises*. New York: John Wiley.

Loynes, R. M. (1966). On idempotent matrices. *Annals of Mathematical Statistics*, **37**, 295–296.

Madow, W. (1940). The distribution of quadratic forms in noncentral normal random variables. *Annals of Mathematical Statistics*. **11**, 100–103.

Mahalanobis, P. C. (1936). On the generalized distance in statistics. *Proceedings of the National Institute of Science* (India), **12**, 49–55.

Marascuilo, L. (1969). Lecture notes in statistics, University of California, Berkeley. See also Marascuilo and Amster (1966). The effect of variety in children's concept learning. *California Journal of Educational Research*, Vol. XVII, No. 3.

Marcus, M. (1960). Basic theorems in matrix theory. *Applied Mathematics Series*, **57**, National Bureau of Standards. Washington, D.C.: U.S. Government Printing Office.

McDonald, L. L. (1971). On the estimation of missing data in the multivariate linear model. *Biometrics*, **27**, 535–543.

———, and G. A. Milliken (1971). Multivariate tests for nonadditivity: A general procedure. *Technical Report No. 20*. Manhattan, Kansas: Department of Statistics, Kansas State University.

McDonald, R. P. (1968). A unified treatment of the weighting problem. *Psychometrika*, **33**, 351–381.

———, and E. J. Burr (1967). A comparison of four methods of constructing factor scores. *Psychometrika*, **32**, 381–401.

———, and H. Swaminathan (1971). A simple matrix calculus with applications to structural models for multivariate data. Part I: Theory. Toronto: Ontario Institute for Studies in Education.

Mehta, J. S., and R. Srinivasan (1970). On the Behrens-Fisher problem. *Biometrika*, **57**, 649–656.

Miller, R. P. (1966). *Simultaneous statistical inference*. New York: McGraw-Hill.

——— (1968). Jackknifing variances. *Annals of Mathematical Statistics*, **39**, 567–582.

Milliken, G. A. (1971). Some results concerning restricted linear models. Paper presented at the Joint Statistical Meeting at Fort Collins, Colorado, August 1971.

———, and F. A. Graybill (1970). Extensions of the general linear hypothesis model. *Journal of the American Statistical Association*, **65**, 797–807.

———, and ——— (1971). Tests for interaction in the two-way model with missing data. *Biometrics*, **27**, 1079–1083.

Millman, J., and G. V. Glass (1967). Rules of thumb for writing the ANOVA table. *Journal of Educational Measurement*, **4**, 41–51.

Mood, A. M., and F. A. Graybill (1963). *Introduction to the theory of statistics*. (2nd ed.) New York: McGraw-Hill.

Myers, J. L. (1966). *Fundamentals of experimental design*. Boston: Allyn and Bacon.

Neudecker, H. (1968). The Kronecker matrix product and some of its applications in econometrics. *Statistica Neerlandia*, **22**, 69–82.

Neyman, J. (1937). Outline of a theory of statistical estimation based on the classical theory of probability. *Philosophical Transactions of the Royal Statistical Society, Series A*, **236**, 333–380.

———, and E. S. Pearson (1933). On the problem of the most efficient tests of statistical hypotheses. *Philosophical Transactions of the Royal Statistical Society, Series A*, **231**, 289–337.

Nobel, B. (1969). *Applied linear algebra*. Englewood Cliffs, N.J.: Prentice-Hall.

Nunnally, J. (1967). *Psychometric theory*. New York: McGraw-Hill.

Olson, C. L. (1973). A Monte Carlo investigation of the robustness of multivariate analysis of variance. Dissertation. Toronto, Canada: Department of Psychology, University of Toronto.

Ortega, J. M. (1960). On Strum sequences for tridiagonal matrices. *Journal of the Association for Computing Machinery*, **7**, 260–263.

Pearson, E. S., and H. O. Hartley (1951). Charts of the power function of the analysis of variance tests, derived from the noncentral *F*-distribution. *Biometrika*, **38**, 112–130.

———, and ——— (1966). *Biometrika tables for statisticians*. Vol. I. (3rd ed.) Cambridge, England: Cambridge University Press.

Pearson, K. (1934). *Tables of incomplete beta function*. Cambridge, England: Cambridge University Press.

Penrose, R. (1955). A generalized inverse for matrices. *Proceedings of the Cambridge Philosophical Society*, **51**, 406–413.

Petrinovich, L. F., and C. Hardyck (1969). Error ratio for multiple comparison methods; some evidence concerning the frequency of erroneous conclusions. *Psychological Bulletin*, **71**, 43–54.

Pillai, K. C. S. (1960). *Statistical tables for tests of multivariate hypotheses.* Manila: Statistical Center, University of the Philippines.

——— (1965). On the distribution of the largest characteristic root of a matrix in multivariate analysis. *Biometrika*, **52**, 405–414.

——— (1967). Upper percentage points of the largest root of a matrix in multivariate analysis. *Biometrika*, **54**, 189–193.

——— (1970). On the noncentral distributions of the largest roots of two matrices in multivariate analysis. In R. C. Bose et al. (Eds), *Essays in probability and statistics*, pp. 557–586. Chapel Hill: University of North Carolina Press.

———, and K. Jayachandran (1967). Power comparisons of tests of two multivariate hypotheses based on four criteria. *Biometrika*, **54**, 195–210.

———, and ——— (1968). Power comparisons of tests of equality of two covariance matrices based on four criteria. *Biometrika*, **55**, 335–342.

Pingel, L. A. (1966). Communality estimation using a modified canonical factor analysis. Unpublished master's thesis, University of Wisconsin.

Porebski, O. R. (1966a). Discriminatory and canonical analysis of technical college data. *British Journal of Mathematical and Statistical Psychology*, **19**, 215–236.

——— (1966b). On the interrelated nature of the multivariate statistics used in discriminatory analysis. *British Journal of Mathematical and Statistical Psychology*, **19**, 197–214.

Posten, H. O., and R. E. Bargmann (1964). Power of the likelihood-ratio test of the general linear hypotheses in multivariate analysis. *Biometrika*, **51**, 467–480.

Postovsky, V. A. (1970). Effects of delay in oral practice at the beginning of second language training. Unpublished doctoral dissertation, University of California, Berkeley.

Potthoff, R. F., and S. N. Roy (1964). A generalized multivariate analysis of variance model useful especially for growth curve problems. *Biometrika*, **51**, 313–326.

Puri, M. L., and P. K. Sen (1971). *Nonparametric methods in multivariate analysis.* New York: John Wiley.

Rao, B. R. (1969). Partial canonical correlations. *Trabajos de estudistica y de investigacion operative*, **20**, 211–219.

Rao, C. R. (1951). An asymptotic expansion of the distribution of Wilk's criterion. *Bulletin of the International Statistics Institute*, **33**, 177–180.

——— (1952). *Advanced statistical methods in biometric research.* New York: John Wiley.

——— (1954). Markoff's theorem with linear restrictions on parameters. *Sankhyá*, **7**, 16–19.

——— (1955a). Analysis of dispersion for multiple classified data with unequal numbers of cells. *Sankhyá*, **15**, 253–280.

——— (1955b). Estimation and tests of significance in factor analysis. *Psychometrika*, **20**, 93–111.

——— (1956). Analysis of dispersion with incomplete observations on one of the characters. *Journal of the Royal Statistical Society, Series B*, **18**, 259–264.

——— (1959). Some problems involving linear hypotheses in multivariate analysis. *Biometrika*, **46**, 49–58.

——— (1962). A note on a generalized inverse of a matrix with applications to problems in mathematical statistics. *Journal of the Royal Statistical Society, Series B*, **24**, 152–158.

——— (1965a). *Linear statistical inference and its applications.* New York: John Wiley.

——— (1965b). The theory of least squares when the parameters are stochastic and its application to the analysis of growth curves. *Biometrika*, **52**, 447–458.

——— (1966a). Covariance adjustment and related problems in multivariate analysis. In P. R. Krishnaiah (Ed.), *Multivariate analysis II*, pp. 87–103. New York: Academic Press.

——— (1966b). Generalized inverse for matrices and its application in mathematical statistics. In F. N. David (Ed.), *Research papers in statistics*, pp. 263–279. New York: John Wiley.

——— (1967). Least squares theory using an estimated dispersion matrix and its application to measurement of signals. *Proceedings of the 5th Berkeley Symposium on Mathematical Statistics*, **1**, 355–372.

———, and S. K. Mitra (1971). *Generalized inverse of matrices and its applications*. New York: John Wiley.

Robson, D. S. (1959). A simple method for constructing orthogonal polynomials when the independent variable is unequally spaced. *Biometrics*, **15**, 187–191.

Rohode, C. A. (1965). Generalized inverses of partitioned matrices. *Journal of the Society of Industrial and Applied Mathematics*, **13**, 1033–1035.

Rohwer, W. D. Personal communication.

Roy, J. (1958). Step-down procedure in multivariate analysis. *Annals of Mathematical Statistics*, **29**, 1177–1187.

——— (1966). Power of the likelihood-ratio test used in analysis of dispersion. In P. R. Krishnaiah (Ed.), *Multivariate analysis I*, pp. 105–127. New York: Academic Press.

Roy, S. N. (1939). *P*-statistics or some generalizations in the analysis of variance appropriate to multivariate problems. *Sankhyá*, **4**, 381–396.

——— (1953). On a heuristic method of test construction and its use in multivariate analysis. *Annals of Mathematical Statistics*, **24**, 220–238.

——— (1957). *Some aspects of multivariate analysis*. New York: John Wiley.

———, and R. C. Bose (1953). Simultaneous confidence interval estimation. *Annals of Mathematical Statistics*, **24**, 513–536.

———, R. Gnanadesikan, and J. N. Srivastava (1971). *Analysis and design of certain quantitative multiresponse experiments*. New York: Pergamon Press.

———, and A. E. Sarhan (1956). On inverting a class of patterned matrices. *Biometrika*, **43**, 227–231.

Rulon, P. J., D. V. Tiedeman, M. M. Tatsuoka, and C. R. Langmuir (1967). *Multivariate statistics for personnel classification*. New York: John Wiley.

Rummel, R. J. (1970). *Applied factor analysis*. Evanston, Ill.: Northwestern University Press.

Rutishauser, H. (1966). The Jacobi method for real symmetric matrices. *Numerische Mathematik*, **9**, 1–10.

Schatzoff, M. (1966). Sensitivity comparisons among tests of the general linear hypothesis. *Journal of the American Statistical Association*, **61**, 415–435.

Scheffé, H. (1943). On solutions of the Behrens-Fisher problem, based on the *t*-distribution. *Annals of Mathematical Statistics*, **14**, 35–44.

——— (1953). A method for judging all contrasts in the analysis of variance. *Biometrika*, **40**, 87–104.

——— (1959). *The analysis of variance*. New York: John Wiley.

——— (1970). Practical solutions of the Behrens-Fisher problem. *Journal of the American Statistical Association*, **65**, 1501–1508.

Seal, H. (1964). *Multivariate statistical analysis for biologists*. London: Methuen.

Searle, S. R. (1965). Additional results concerning estimable functions and generalized inverse matrices. *Journal of the Royal Statistical Society, Series B*, **27**, 486–490.

——— (1966). *Matrix algebra for the biological sciences*. New York: John Wiley.

——— (1968). Another look at Henderson's methods of estimating variance components. *Biometrics*, **24**, 749–787.

——— (1971). *Linear models*. New York: John Wiley.

Seber, G. A. F. (1966). *The linear hypothesis: A general theory*. New York: Hafner.

Shin, S. H. (1971). Creativity, intelligence and achievement: A study of the relationship between creativity and intelligence, and their effects upon achievement. Unpublished doctoral dissertation, University of Pittsburgh.

Spearman, C. (1904). General intelligence objectively determined and measured. *American Journal of Psychology*, **15**, 201–293.

Speed, F. M. (1969). A new approach to the analysis of linear models, *NASA Technical Memorandum, NASA TM X-58030*, June 1969.

Srivastava, J. N. (1964). On the monotonicity property of the three main tests for multivariate analysis of variance. *Journal of the Royal Statistical Society, Series B*, **26**, 77–81.

——— (1966). Some generalizations of multivariate analysis of variance. In P. R. Krishnaiah (Ed.), *Multivariate analysis I*, pp. 129–148. New York: Academic Press.

——— (1967). On the extensions of the Gauss–Markoff theorem to complex multivariate linear models. *Annals of the Institute of Statistical Mathematics*, **19**, 417–437.

Stewart, D. K., and W. A. Love (1968). A general canonical correlation index. *Psychological Bulletin*, **70**, 160–163.

Tan, W. K. (1972). A comparison of different procedures under the Potthoff-Roy generalized multivariate analysis of variance. Unpublished doctoral dissertation, University of Pittsburgh.

Tang, P. C. (1938). The power function of the analysis of variance tests with tables and illustrations of the use. *Statistical Research Memoirs*, **2**, 126–146.

Thomas, G. B. (1960). *Calculus and analytic geometry*. Reading, Mass.: Addison-Wesley.

Thompson, G. H. (1934). Hotelling's method modified to give Spearman's *g*. *Journal of Educational Psychology*, **25**, 366–374.

Thompson, W. A. (1962). Estimation of dispersion parameters. *Journal of the National Bureau of Standards*, **668**, 161–164.

Thurstone, L. L. (1931). Multiple factor analysis. *Psychological Review*, **38**, 406–427.

——— (1935). *The vectors of mind*. Chicago: The University of Chicago Press.

——— (1947). *Multiple factor analysis*. Chicago: The University of Chicago Press.

Timm, N. H. (1970). The estimation of variance-covariance and correlation matrices from incomplete data. *Psychometrika*, **35**, 417–437.

———, and J. E. Carlson (in press). Analysis of variance through full rank models. *Multivariate behavioral research*. Monograph.

Titku, M. L. (1967). Tables of the power of the *F* test. *Journal of the American Statistical Association*, **62**, 525–539.

Trawinski, I. M., and R. Bargmann (1964). Maximum-likelihood estimation with incomplete multivariate data. *Annals of Mathematical Statistics*, **35**, 647–657.

Tukey, J. W. (1949). One degree of freedom for nonadditivity. *Biometrics*, **5**, 232–242.

Urquhart, N. S., D. L. Weeks, and C. R. Henderson (1970). Estimation associated with linear models: A revisitation. *Paper BU-195*. Ithaca, N.Y.: Biometrics Unit, Cornell University.

Wald, A. (1943). Tests of statistical hypotheses concerning several parameters when the number of variables is large. *Transactions of the American Mathematical Society*, **54**, 426–482.

Wall, F. J. (1968). *The generalized variance ratio of the U-statistic*. Albuquerque: The Dikewood Corporation.

Welch, B. L. (1937). The significance of the difference between two means when the population variances are unequal. *Biometrika*, **29**, 350–362.

——— (1947). The generalization of Student's problem when several different population variances are involved. *Biometrika*, **34**, 28–35.

Welch, J. H. (1967). Certification of algorithm 254 [*F*-2]—eigenvalues and eigenvectors of a real symmetric matrix by the *QR* methods. *Journal of the Association for Computing Machinery*, **10**, 376–377.

Wijsman, R. A. (1957). Random orthogonal transformations and their use in some classical distribution problems in multivariate analysis. *Annals of Mathematical Statistics*, **29**, 415–423.

Wilk, M. B., and R. Gnanadesikan (1968). Probability plotting methods for the analysis of data. *Biometrika*, **55**, 1–17.

———, ———, and M. J. Huyett (1962). Probability plots for the gamma distribution, *Technometrics*, **4**, 1–20.

Wilkinson, J. H. (1962). Householder's method for symmetric matrices. *Numerische Mathematik*, **4**, 354–361.

——— (1965). *The algebraic eigenvalue problem*. London: Oxford University Press.

Wilks, S. S. (1932). Certain generalizations in the analysis of variance. *Biometrika*, **24**, 471–494.

Williams, E. J. (1959). *Regression analysis*. New York: John Wiley.

——— (1967). The analysis of association among many variables. *Journal of the Royal Statistical Society, Series B*, **29**, 199–242.

Winer, B. J. (1971). *Statistical principles in experimental design*. (2nd ed.) New York: McGraw-Hill.

Wishart, J. (1928). The generalized product moment distribution in samples from a normal multivariate population. *Biometrika*, **20A**, 32–52.

——— (1938). Growth rate determination in nutrition studies with bacon pig and their analysis. *Biometrika*, **30**, 16–28.

Yao, Y. (1965). An approximate degrees of freedom solution to the multivariate Behrens-Fisher problem. *Biometrika*, **52**, 139–147.

Yates, F. (1934). The analysis of multiple classifications with unequal numbers in the different classes. *Journal of the American Statistical Association*, **29**, 51–66.

TABLES

TABLE I. Upper Percentage Points of the Standard Normal Distribution, Z^{α}

Z^{α} denotes the upper α critical value for the standard normal distribution. If $X \sim N(0, 1)$ and $\alpha = .05$, the critical value Z^{α} such that the $P(X > Z^{\alpha}) = .05$ is read as $Z^{.05} = 1.645$ from the table.

TABLE II. Upper Percentage Points of the χ^2 Distribution, $\chi^2_{\alpha}(v)$

$\chi^2_{\alpha}(v)$ is the upper α critical value for a χ^2 distribution with v degrees of freedom. If $X^2 \sim \chi^2(v)$ and $\alpha = .05$ with $v = 10$, the critical constant such that the $P[X^2 > \chi^2_{\alpha}(v)] = .05$ is $\chi^2_{.05}(10) = 18.3070$.

TABLE III. Upper Percentage Points of Student's t Distribution, $t^{\alpha}(v)$

$t^{\alpha}(v)$ represents the upper α critical value for the t distribution with v degrees of freedom. If $T \sim t(v)$ and $\alpha = .05$ with $v = 10$, the $P[T > t^{\alpha}(v)] = .05$ is $t^{.05}(10) = 1.812$. For $v > 100$, the expressions $\chi^2_{\alpha} = v[1 - (2/9v) + X\sqrt{2/9v}]^3$ or $\chi^2_{\alpha} = [X + \sqrt{2v - 1}]^2/2$ may be used, with X defined in the last line of the table as an $N(0, 1)$ variable, depending on the degree of accuracy desired.

TABLE IV. Upper Percentage Points of the F Distribution, $F^{\alpha}(v_h, v_e)$

$F^{\alpha}(v_h, v_e)$ is the upper α critical value of the F distribution with v_h representing the numerator and v_e the denominator degrees of freedom. If $F \sim F^{\alpha}(v_h, v_e)$, the critical value for $\alpha = .05$, $v_h = 4$, and $v_e = 9$ such that the $P[F > F^{\alpha}(v_h, v_e)] = .05$ is $F^{.05}(4, 9) = 3.63$. To find the $P[F < F^{1-\alpha}(v_h, v_e)]$, the formula $F^{1-\alpha}(v_h, v_e) = 1/F^{\alpha}(v_e, v_h)$ is employed. Since $F^{\alpha}(9, 4) = 6.00$, the critical constant is $F^{1-\alpha}(v_h, v_e) = 1/6 = .167$.

TABLE V. Power Charts for the Noncentral F Distribution

These charts are entered with α, ν_1 (the degrees of freedom for the numerator), and ν_2 (the degrees of freedom for the denominator) of the noncentral F distribution, and the noncentrality parameter δ, which is related to ϕ by the formula

$$\phi = \left(\frac{\delta}{\nu_1 + 1}\right)^{1/2}$$

The value for the power of the test is then read from the margins of the charts.

TABLE VI. Upper Percentage Points of Hotelling's T^2 Distribution, $T^\alpha(p, \nu)$

Given the number of variables p, and the degrees of freedom for error ν, the critical values $T^\alpha(p, \nu)$ such that $P[T > T^\alpha(p, \nu_e)] = \alpha$ are tabled. If $T \sim T^2(p, \nu)$, the $P[T > T^\alpha(p, \nu)] = .05$, for $p = 5$ and $\nu = 50$, is $T^{.05}_{(5,50)} = 13.138$.

TABLE VII. Upper Percentage Points of Roy's Largest-Root Criterion, $\theta^\alpha(s, m, n)$

TABLE VIII. Heck's Charts for Roy's Largest-Root Criterion

The critical values $\theta^\alpha(s, m, n) \equiv X_\alpha$ for the distribution of the largest root of the determinantal equation $|\mathbf{Q}_h - \theta(\mathbf{Q}_h + \mathbf{Q}_e)| = 0$ may be obtained by using Table VII or the charts included in Table VIII and by employing the following notation,

$$\theta_s = \frac{\lambda_1}{1 + \lambda_1} \qquad m = \tfrac{1}{2}(|\nu_h - u| - 1)$$

$$n = \tfrac{1}{2}(\nu_e - u - 1) \qquad s = \min(\nu_h, u)$$

where u is equal to the number of variables after any transformation, λ_1 is the largest root of the equation $|\mathbf{Q}_h - \lambda\mathbf{Q}_e| = 0$, and $\mathbf{Q}_h$ and $\mathbf{Q}_e$ are the sum of squares and products matrices for the hypothesis and error based on ν_h and ν_e degrees of freedom, respectively. To determine the critical constant such that the $P[\theta_s > \theta^\alpha(s, m, n)] = \alpha$, for $\alpha = .05$, $s = 2$, $m = 0$, and $n = 5$, tables or charts are employed. From Table VII, $s = 2$ and $\alpha = .05$, $\theta^{.05}_{(2,0,5)} = .565$. For the charts in Table VIII, $m = -1/2, 0, 1, 2, \ldots, 10$ from the first curve line to the last within a set. The lower set of curves are extensions of the upper set, and the upper α percentage points are read from the bottom line labeled as $X_\alpha \equiv \theta^\alpha(s, m, n)$ by Heck. From Chart III, with $\alpha = .05$, $s = 2$, $m = 0$, and $n = 5$, the critical value is $\theta^{.05}_{(2,0,5)} \equiv X_{.05} = .565$ since the distance between each vertical grid line is .005 units. The tables and charts for Roy's criterion are included for $s = 2(1)6$. For percentage points for $6 \leq s \leq 10$ and $s = 14(2)20$, see Pillai (1965, 1967).

TABLE IX. Lower Percentage Points of Wilks' Lambda Criterion, $U^\alpha(p, q, n)$

By using the notation introduced to use Roy's test criterion in a MANOVA design and making the following correspondence in notation,

Walls' Table	This Text
p	u
q	ν_h
n	ν_e
$U^\alpha(p, q, n)$	$U^\alpha(u, \nu_h, \nu_e)$

Table IX gives the lower critical constants of the null distribution of the likelihood-ratio criterion: $\Lambda = |\mathbf{Q}_e|/|(\mathbf{Q}_h + \mathbf{Q}_e)|$. To find $U^\alpha(p, q, n)$ such that the $P[\Lambda < U^\alpha(p, q, n)] = .05$, for $p = u = 3$, $q = v_h = 4$, and $n = v_e = 20$, the value $U^{.05}_{(3,4,20)} = 0.347546$ is read from the table labeled $p = 3$ and $\alpha = .05$ within Table IX.

TABLE X. Upper Percentage Points of Hotelling and Lawley's Trace Criterion, $U_0^\alpha(s, m, n)$

For $\alpha = .05$ and .01, $s = 2(1)6$ and s, m, and n defined as in Tables VII and VIII, the upper critical values for the statistic

$$U^{(s)} = \text{Tr}(\mathbf{Q}_h\mathbf{Q}_e^{-1}) = \sum_i \lambda_i = \sum_i \frac{\theta_i}{1 - \theta_i}$$

have been tabulated. For $\alpha = .05$, $s = 2$, $n = 40$, and $m = 1.5$, the critical constant $U_0^\alpha(s, m, n)$ such that the $P[U^{(s)} > U_0^\alpha(s, m, n)] = \alpha$ is $U_0^{.05}(2, 1.5, 40) = .270$.

TABLE XI. Upper Percentage Points of Pillai's Trace Criterion, $V^\alpha(s, m, n)$

For $\alpha = .05$ and .01, $s = 2(1)6$, and s, m, and n defined above, the upper critical values for the statistic

$$V^{(s)} = \text{Tr}[\mathbf{Q}_h(\mathbf{Q}_h + \mathbf{Q}_e)^{-1}] = \sum_i \theta_i = \sum_i \frac{\lambda_i}{1 + \lambda_i}$$

are included. The critical constant $V^\alpha(s, m, n)$ such that the $P[V^{(s)} > V^\alpha(s, m, n)] = \alpha$, where $\alpha = .05$, $s = 2$, $m = 1.5$, and $n = 40$, is $V^{.05}_{(2,1.5,40)} = .230$.

TABLE XII. Upper Percentage Points for the Bonferroni-Dunn Procedure, $t^{\alpha/2k}(v)$

Upper $\alpha/2k$ (or two-tailed α/k) percentage points for the Student t distribution, where α is the size of the test, k is the number of comparisons, and v is the error degrees of freedom, are included in this table. Illustrated is control of the type I error α at $\alpha = .05$, while making $k = 5$ comparisons when the degrees of freedom for error $v = 10$. For Table XII, α denotes the $P[-t^{\alpha/2k}(v) \leq t \leq t^{\alpha/2k}(v)] \geq 1 - \alpha$, for k comparisons; since $k = 5$ and $v = 10$, the critical constant $t^{\alpha/2k}(v)$ such that the $P[t > t^{\alpha/2k}(v)] = .025$ is $t^{.025/5}(10) = t^{.005}(10) = 3.17$. This critical value is also given in Table III for this situation; however, Table XII allows us to readily obtain t-distribution critical values that are not included in standard Student t tables. For example, by increasing k to $k = 8$ when $\alpha = .05$, the upper $\alpha/2k$ critical value is $t^{\alpha/(2)(8)}(10) = t^{.025/8}_{(10)} = t^{.003125}_{(10)} = 3.45$.

TABLE XIII. Coefficients of Orthogonal Polynomials (Linear, Quadratic, and Cubic)

Orthogonal polynomial coefficients and orthogonal polynomials included in this table may be used to fit a polynomial model of degree three to n observations (X_i, Y_i) of equally spaced $X_i = x_i$ values. The curvilinear model $y_i = \beta_0 + \beta_1 x_i + \beta_2 x_i^2 + \beta_3 x_i^3 + \varepsilon_i$ is fitted by using the model $y_i = \bar{y} + \xi_1 f_1(x) + \xi_2 f_2(x) + \xi_3 f_3(x) + \varepsilon_i$, where the $f_j(x)$ are orthogonal polynomials in x of the jth degree. The values of $f_j(x_i)$, for $n = 3(1)12$ and $j = 1, 2, 3$, are summarized in Table XIII. The quantity $c_j = \Sigma_i [f_j(x_i)]^2$ is also given for each value of n, so that $\hat{\xi}_j = \Sigma_i y_i f_j(x_i)/c_j$. With the polynomials $f_j(x)$ given, we may easily transform to a model involving the β's. See Section 4.4 for examples of the technique.

TABLE XIV. Percentage Points of the Extreme Roots of a Wishart Matrix, $\ell(\alpha)$ and $u(\alpha)$

For $n = N - 1$, $n\mathbf{S} \sim W(n, \mathbf{\Sigma})$, for a sample size N and sample variance-covariance matrix $\mathbf{S}$. Letting c_1 and c_2 denote the largest and smallest roots of $n\mathbf{S}$, percentage points for c_1 and c_2 are given for the null case $\mathbf{\Sigma} = \mathbf{I}$. Constants α, ℓ, u_1, and u are calculated such that the $P(c_1 \leq u_1) = 1 - \alpha$ and the $P(c_2 \geq \ell) = 1 - \alpha$. For a fixed ℓ, u is determined such that

$$P[\ell(\alpha) \leq c_2 \leq c_1 \leq u(\alpha)] = 1 - 2\alpha$$

where $u(\alpha)$ and $\ell(\alpha)$ are the upper and lower α percentage points of c_1 and c_2, respectively, for the number of variables $p = 2(1)10$, for the value of $n = N - 1$ from 2 to 100, and for upper and lower α values equal to .005, .010, .025, and .050. As an example of the use of these tables, suppose $N = 101$ and $\mathbf{R}$ is a sample correlation matrix for $p =$ three variables, then, for the value $r_{13} = .60$ (say), a 95% confidence interval for ρ_{13} is obtained by using the values $u(\alpha/2) = 147.8$ and $\ell(\alpha/2) = 6.185$ from the table in formula (6.4.24). An approximate $1 - \alpha$ simultaneous confidence interval for ρ_{13} can be shown to be $.04 \leq \rho_{13} \leq .83$.

TABLE I. Upper Percentage Points of the Standard Normal Distribution, Z^{α}

α	Z^{α}	α	Z^{α}	α	Z^{α}
.50	0.00	.25	0.67	.050	1.645
.49	0.03	.24	0.71	.045	1.695
.48	0.05	.23	0.74	.040	1.751
.47	0.08	.22	0.77	.035	1.812
.46	0.10	.21	0.81	.030	1.881
.45	0.13	.20	0.84	.025	1.960
.44	0.15	.19	0.88	.020	2.054
.43	0.18	.18	0.92	.015	2.170
.42	0.20	.17	0.95	.010	2.326
.41	0.23	.16	0.99	.005	2.576
.40	0.25	.15	1.04	.004	2.652
.39	0.28	.14	1.08	.003	2.748
.38	0.30	.13	1.13	.002	2.878
.37	0.33	.12	1.17	.001	3.090
.36	0.36	.11	1.23		
.35	0.39	.10	1.28	.0005	3.291
.34	0.41	.09	1.34	.0001	3.719
.33	0.44	.08	1.41		
.32	0.47	.07	1.48	.00005	3.891
.31	0.50	.06	1.55	.00001	4.265
.30	0.52				
.29	0.55				
.28	0.58				
.27	0.61				
.26	0.64				

Abridged from Table 1, E. S. Pearson and H. O. Hartley (Eds.), *Biometrika Tables for Statisticians*, Vol. 1 (3rd ed.) New York: Cambridge, 1966. Reproduced by permission of the editors and trustees of *Biometrika*.

TABLE II. Upper Percentage Points of the χ^2 Distribution, $\chi^2_\alpha(\nu)$

df(ν) \ α	0·250	0·100	0·050	0·025	0·010	0·005	0·001
1	1·32330	2·70554	3·84146	5·02389	6·63490	7·87944	10·828
2	2·77259	4·60517	5·99146	7·37776	9·21034	10·5966	13·816
3	4·10834	6·25139	7·81473	9·34840	11·3449	12·8382	16·266
4	5·38527	7·77944	9·48773	11·1433	13·2767	14·8603	18·467
5	6·62568	9·23636	11·0705	12·8325	15·0863	16·7496	20·515
6	7·84080	10·6446	12·5916	14·4494	16·8119	18·5476	22·458
7	9·03715	12·0170	14·0671	16·0128	18·4753	20·2777	24·322
8	10·2189	13·3616	15·5073	17·5345	20·0902	21·9550	26·125
9	11·3888	14·6837	16·9190	19·0228	21·6660	23·5894	27·877
10	12·5489	15·9872	18·3070	20·4832	23·2093	25·1882	29·588
11	13·7007	17·2750	19·6751	21·9200	24·7250	26·7568	31·264
12	14·8454	18·5493	21·0261	23·3367	26·2170	28·2995	32·909
13	15·9839	19·8119	22·3620	24·7356	27·6882	29·8195	34·528
14	17·1169	21·0641	23·6848	26·1189	29·1412	31·3194	36·123
15	18·2451	22·3071	24·9958	27·4884	30·5779	32·8013	37·697
16	19·3689	23·5418	26·2962	28·8454	31·9999	34·2672	39·252
17	20·4887	24·7690	27·5871	30·1910	33·4087	35·7185	40·790
18	21·6049	25·9894	28·8693	31·5264	34·8053	37·1565	42·312
19	22·7178	27·2036	30·1435	32·8523	36·1909	38·5823	43·820
20	23·8277	28·4120	31·4104	34·1696	37·5662	39·9968	45·315
21	24·9348	29·6151	32·6706	35·4789	38·9322	41·4011	46·797
22	26·0393	30·8133	33·9244	36·7807	40·2894	42·7957	48·268
23	27·1413	32·0069	35·1725	38·0756	41·6384	44·1813	49·728
24	28·2412	33·1962	36·4150	39·3641	42·9798	45·5585	51·179
25	29·3389	34·3816	37·6525	40·6465	44·3141	46·9279	52·618
26	30·4346	35·5632	38·8851	41·9232	45·6417	48·2899	54·052
27	31·5284	36·7412	40·1133	43·1945	46·9629	49·6449	55·476
28	32·6205	37·9159	41·3371	44·4608	48·2782	50·9934	56·892
29	33·7109	39·0875	42·5570	45·7223	49·5879	52·3356	58·301
30	34·7997	40·2560	43·7730	46·9792	50·8922	53·6720	59·703
40	45·6160	51·8051	55·7585	59·3417	63·6907	66·7660	73·402
50	56·3336	63·1671	67·5048	71·4202	76·1539	79·4900	86·661
60	66·9815	74·3970	79·0819	83·2977	88·3794	91·9517	99·607
70	77·5767	85·5270	90·5312	95·0232	100·425	104·215	112·317
80	88·1303	96·5782	101·879	106·629	112·329	116·321	124·839
90	98·6499	107·565	113·145	118·136	124·116	128·299	137·208
100	109·141	118·498	124·342	129·561	135·807	140·169	149·449
X	+0·6745	+1·2816	+1·6449	+1·9600	+2·3263	+2·5758	+3·0902

For $\nu > 100$, the expressions $\chi^2_\alpha = \nu[1 - 2/9\nu + X\sqrt{2/9\nu}]^3$ or $\chi^2_\alpha = 1/2[X + \sqrt{(2\nu) - 1}]^2$ may be used, with X defined in the last line in the table, as a $N(0, 1)$ variable depending on the degree of accuracy desired. From Table 8, E. S. Pearson, and H. O. Hartley (Eds.), *Biometrika Tables for Statisticians*, Vol. 1 (3rd ed.) New York: Cambridge, 1966. Reproduced by permission of the editors and trustees of *Biometrika*.

TABLE III. Upper Percentage Points of Student's t Distribution, $t^{\alpha}(\nu)$

df(ν) \ α	0.250	0.100	0.050	0.025	0.010	0.005	0.001
1	1.000	3.078	6.314	12.706	31.821	63.657	318.31
2	0.816	1.886	2.920	4.303	6.965	9.925	22.327
3	0.765	1.638	2.353	3.182	4.541	5.841	10.214
4	0.741	1.533	2.132	2.776	3.747	4.604	7.173
5	0.727	1.476	2.015	2.571	3.365	4.032	5.893
6	0.718	1.440	1.943	2.447	3.143	3.707	5.208
7	0.711	1.415	1.895	2.365	2.998	3.499	4.785
8	0.706	1.397	1.860	2.306	2.896	3.355	4.501
9	0.703	1.383	1.833	2.262	2.821	3.250	4.297
10	0.700	1.372	1.812	2.228	2.764	3.169	4.144
11	0.697	1.363	1.796	2.201	2.718	3.106	4.025
12	0.695	1.356	1.782	2.179	2.681	3.055	3.930
13	0.694	1.350	1.771	2.160	2.650	3.012	3.852
14	0.692	1.345	1.761	2.145	2.624	2.977	3.787
15	0.691	1.341	1.753	2.131	2.602	2.947	3.733
16	0.690	1.337	1.746	2.120	2.583	2.921	3.686
17	0.689	1.333	1.740	2.110	2.567	2.898	3.646
18	0.688	1.330	1.734	2.101	2.552	2.878	3.610
19	0.688	1.328	1.729	2.093	2.539	2.861	3.579
20	0.687	1.325	1.725	2.086	2.528	2.845	3.552
21	0.686	1.323	1.721	2.080	2.518	2.831	3.527
22	0.686	1.321	1.717	2.074	2.508	2.819	3.505
23	0.685	1.319	1.714	2.069	2.500	2.807	3.485
24	0.685	1.318	1.711	2.064	2.492	2.797	3.467
25	0.684	1.316	1.708	2.060	2.485	2.787	3.450
26	0.684	1.315	1.706	2.056	2.479	2.779	3.435
27	0.684	1.314	1·703	2.052	2.473	2.771	3.421
28	0.683	1.313	1.701	2.048	2.467	2.763	3.408
29	0.683	1.311	1.699	2.045	2.462	2.756	3.396
30	0.683	1.310	1.697	2.042	2.457	2.750	3.385
40	0.681	1.303	1.684	2.021	2.423	2.704	3.307
60	0.679	1.296	1.671	2.000	2.390	2.660	3.232
120	0.677	1.289	1.658	1.980	2.358	2.167	3.160
∞	0.674	1.282	1.645	1.960	2.326	2.576	3.090

From Table 12, E. S. Pearson, and H. O. Hartley (Eds.), *Biometrika Tables for Statisticians*, Vol. 1 (3rd ed.) New York: Cambridge, 1966. Reproduced by permission of the editors and trustees of *Biometrika*.

TABLE IV. Upper Percentage Points of the *F* Distribution, $F^{\alpha}(v_h, v_e)$

df *for de- nomi- nator* (v_e)	α	df *for numerator* (v_h)											
		1	2	3	4	5	6	7	8	9	10	11	12
	.25	5.83	7.50	8.20	8.58	8.82	8.98	9.10	9.19	9.26	9.32	9.36	9.41
1	.10	39.9	49.5	53.6	55.8	57.2	58.2	58.9	59.4	59.9	60.2	60.5	60.7
	.05	161	200	216	225	230	234	237	239	241	242	243	244
	.25	2.57	3.00	3.15	3.23	3.28	3.31	3.34	3.35	3.37	3.38	3.39	3.39
2	.10	8.53	9.00	9.16	9.24	9.29	9.33	9.35	9.37	9.38	9.39	9.40	9.41
	.05	18.5	19.0	19.2	19.2	19.3	19.3	19.4	19.4	19.4	19.4	19.4	19.4
	.01	98.5	99.0	99.2	99.2	99.3	99.3	99.4	99.4	99.4	99.4	99.4	99.4
	.25	2.02	2.28	2.36	2.39	2.41	2.42	2.43	2.44	2.44	2.44	2.45	2.45
3	.10	5.54	5.46	5.39	5.34	5.31	5.28	5.27	5.25	5.24	5.23	5.22	5.22
	.05	10.1	9.55	9.28	9.12	9.01	8.94	8.89	8.85	8.81	8.79	8.76	8.74
	.01	34.1	30.8	29.5	28.7	28.2	27.9	27.7	27.5	27.3	27.2	27.1	27.1
	.25	1.81	2.00	2.05	2.06	2.07	2.08	2.08	2.08	2.08	2.08	2.08	2.08
4	.10	4.54	4.32	4.19	4.11	4.05	4.01	3.98	3.95	3.94	3.92	3.91	3.90
	.05	7.71	6.94	6.59	6.39	6.26	6.16	6.09	6.04	6.00	5.96	5.94	5.91
	.01	21.2	18.0	16.7	16.0	15.5	15.2	15.0	14.8	14.7	14.5	14.4	14.4
	.25	1.69	1.85	1.88	1.89	1.89	1.89	1.89	1.89	1.89	1.89	1.89	1.89
5	.10	4.06	3.78	3.62	3.52	3.45	3.40	3.37	3.34	3.32	3.30	3.28	3.27
	.05	6.61	5.79	5.41	5.19	5.05	4.95	4.88	4.82	4.77	4.74	4.71	4.68
	.01	16.3	13.3	12.1	11.4	11.0	10.7	10.5	10.3	10.2	10.1	9.96	9.89
	.25	1.62	1.76	1.78	1.79	1.79	1.78	1.78	1.78	1.77	1.77	1.77	1.77
6	.10	3.78	3.46	3.29	3.18	3.11	3.05	3.01	2.98	2.96	2.94	2.92	2.90
	.05	5.99	5.14	4.76	4.53	4.39	4.28	4.21	4.15	4.10	4.06	4.03	4.00
	.01	13.7	10.9	9.78	9.15	8.75	8.47	8.26	8.10	7.98	7.87	7.79	7.72
	.25	1.57	1.70	1.72	1.72	1.71	1.71	1.70	1.70	1.69	1.69	1.69	1.68
7	.10	3.59	3.26	3.07	2.96	2.88	2.83	2.78	2.75	2.72	2.70	2.68	2.67
	.05	5.59	4.74	4.35	4.12	3.97	3.87	3.79	3.73	3.68	3.64	3.60	3.57
	.01	12.2	9.55	8.45	7.85	7.46	7.19	6.99	6.84	6.72	6.62	6.54	6.47
	.25	1.54	1.66	1.67	1.66	1.66	1.65	1.64	1.64	1.63	1.63	1.63	1.62
8	.10	3.46	3.11	2.92	2.81	2.73	2.67	2.62	2.59	2.56	2.54	2.52	2.50
	.05	5.32	4.46	4.07	3.84	3.69	3.58	3.50	3.44	3.39	3.35	3.31	3.28
	.01	11.3	8.65	7.59	7.01	6.63	6.37	6.18	6.03	5.91	5.81	5.73	5.67
	.25	1.51	1.62	1.63	1.63	1.62	1.61	1.60	1.60	1.59	1.59	1.58	1.58
9	.10	3.36	3.01	2.81	2.69	2.61	2.55	2.51	2.47	2.44	2.42	2.40	2.38
	.05	5.12	4.26	3.86	3.63	3.48	3.37	3.29	3.23	3.18	3.14	3.10	3.07
	.01	10.6	8.02	6.99	6.42	6.06	5.80	5.61	5.47	5.35	5.26	5.18	5.11

From Table 18, E. S. Pearson, and H. O. Hartley (Eds.), *Biometrika Tables for Statisticians*, Vol. 1 (3rd ed.) New York: Cambridge, 1966. Reproduced by permission of the editors and trustees of *Biometrika*.

TABLE IV (*continued*).

df *for numerator* (v_h)													df *for de-nominator* (v_e)
15	20	24	30	40	50	60	100	120	200	500	∞	α	
9.49	9.58	9.63	9.67	9.71	9.74	9.76	9.78	9.80	9.82	9.84	9.85	.25	
61.2	61.7	62.0	62.3	62.5	62.7	62.8	63.0	63.1	63.2	63.3	63.3	.10	1
246	248	249	250	251	252	252	253	253	254	254	254	.05	
3.41	3.43	3.43	3.44	3.45	3.45	3.46	3.47	3.47	3.48	3.48	3.48	.25	
9.42	9.44	9.45	9.46	9.47	9.47	9.47	9.48	9.48	9.49	9.49	9.49	.10	2
19.4	19.4	19.5	19.5	19.5	19.5	19.5	19.5	19.5	19.5	19.5	19.5	.05	
99.4	99.4	99.5	99.5	99.5	99.5	99.5	99.5	99.5	99.5	99.5	99.5	.01	
2.46	2.46	2.46	2.47	2.47	2.47	2.47	2.47	2.47	2.47	2.47	2.47	.25	
5.20	5.18	5.18	5.17	5.16	5.15	5.15	5.14	5.14	5.14	5.14	5.13	.10	3
8.70	8.66	8.64	8.62	8.59	8.58	8.57	8.55	8.55	8.54	8.53	8.53	.05	
26.9	26.7	26.6	26.5	26.4	26.4	26.3	26.2	26.2	26.2	26.1	26.1	.01	
2.08	2.08	2.08	2.08	2.08	2.08	2.08	2.08	2.08	2.08	2.08	2.08	.25	
3.87	3.84	3.83	3.82	3.80	3.80	3.79	3.78	3.78	3.77	3.76	3.76	.10	4
5.86	5.80	5.77	5.75	5.72	5.70	5.69	5.66	5.66	5.65	5.64	5.63	.05	
14.2	14.0	13.9	13.8	13.7	13.7	13.7	13.6	13.6	13.5	13.5	13.5	.01	
1.89	1.88	1.88	1.88	1.88	1.88	1.87	1.87	1.87	1.87	1.87	1.87	.25	
3.24	3.21	3.19	3.17	3.16	3.15	3.14	3.13	3.12	3.12	3.11	3.10	.10	5
4.62	4.56	4.53	4.50	4.46	4.44	4.43	4.41	4.40	4.39	4.37	4.36	.05	
9.72	9.55	9.47	9.38	9.29	9.24	9.20	9.13	9.11	9.08	9.04	9.02	.01	
1.76	1.76	1.75	1.75	1.75	1.75	1.74	1.74	1.74	1.74	1.74	1.74	.25	
2.87	2.84	2.82	2.80	2.78	2.77	2.76	2.75	2.74	2.73	2.73	2.72	.10	6
3.94	3.87	3.84	3.81	3.77	3.75	3.74	3.71	3.70	3.69	3.68	3.67	.05	
7.56	7.40	7.31	7.23	7.14	7.09	7.06	6.99	6.97	6.93	6.90	6.88	.01	
1.68	1.67	1.67	1.66	1.66	1.66	1.65	1.65	1.65	1.65	1.65	1.65	.25	
2.63	2.59	2.58	2.56	2.54	2.52	2.51	2.50	2.49	2.48	2.48	2.47	.10	7
3.51	3.44	3.41	3.38	3.34	3.32	3.30	3.27	3.27	3.25	3.24	3.23	.05	
6.31	6.16	6.07	5.99	5.91	5.86	5.82	5.75	5.74	5.70	5.67	5.65	.10	
1.62	1.61	1.60	1.60	1.59	1.59	1.59	1.58	1.58	1.58	1.58	1.58	.25	
2.46	2.42	2.40	2.38	2.36	2.35	2.34	2.32	2.32	2.31	2.30	2.29	.10	8
3.22	3.15	3.12	3.08	3.04	3.02	3.01	2.97	2.97	2.95	2.94	2.93	.05	
5.52	5.36	5.28	5.20	5.12	5.07	5.03	4.96	4.95	4.91	4.88	4.86	.01	
1.57	1.56	1.56	1.55	1.55	1.54	1.54	1.53	1.53	1.53	1.53	1.53	.25	
2.34	2.30	2.28	2.25	2.23	2.22	2.21	2.19	2.18	2.17	2.17	2.16	.10	9
3.01	2.94	2.90	2.86	2.83	2.80	2.79	2.76	2.75	2.73	2.72	2.71	.05	
4.96	4.81	4.73	4.65	4.57	4.52	4.48	4.42	4.40	4.36	4.33	4.31	.01	

TABLE IV (*continued*).

df *for denominator* (v_e)	α	df *for numerator* (v_h) 1	2	3	4	5	6	7	8	9	10	11	12
	.25	1.49	1.60	1.60	1.59	1.59	1.58	1.57	1.56	1.56	1.55	1.55	1.54
10	.10	3.29	2.92	2.73	2.61	2.52	2.46	2.41	2.38	2.35	2.32	2.30	2.28
	.05	4.96	4.10	3.71	3.48	3.33	3.22	3.14	3.07	3.02	2.98	2.94	2.91
	.01	10.0	7.56	6.55	5.99	5.64	5.39	5.20	5.06	4.94	4.85	4.77	4.71
	.25	1.47	1.58	1.58	1.57	1.56	1.55	1.54	1.53	1.53	1.52	1.52	1.51
11	.10	3.23	2.86	2.66	2.54	2.45	2.39	2.34	2.30	2.27	2.25	2.23	2.21
	.05	4.84	3.98	3.59	3.36	3.20	3.09	3.01	2.95	2.90	2.85	2.82	2.79
	.01	9.65	7.21	6.22	5.67	5.32	5.07	4.89	4.74	4.63	4.54	4.46	4.40
	.25	1.46	1.56	1.56	1.55	1.54	1.53	1.52	1.51	1.51	1.50	1.50	1.49
12	.10	3.18	2.81	2.61	2.48	2.39	2.33	2.28	2.24	2.21	2.19	2.17	2.15
	.05	4.75	3.89	3.49	3.26	3.11	3.00	2.91	2.85	2.80	2.75	2.72	2.69
	.01	9.33	6.93	5.95	5.41	5.06	4.82	4.64	4.50	4.39	4.30	4.22	4.16
	.25	1.45	1.55	1.55	1.53	1.52	1.51	1.50	1.49	1.49	1.48	1.47	1.47
13	.10	3.14	2.76	2.56	2.43	2.35	2.28	2.23	2.20	2.16	2.14	2.12	2.10
	.05	4.67	3.81	3.41	3.18	3.03	2.92	2.83	2.77	2.71	2.67	2.63	2.60
	.01	9.07	6.70	5.74	5.21	4.86	4.62	4.44	4.30	4.19	4.10	4.02	3.96
	.25	1.44	1.53	1.53	1.52	1.51	1.50	1.49	1.48	1.47	1.46	1.46	1.45
14	.10	3.10	2.73	2.52	2.39	2.31	2.24	2.19	2.15	2.12	2.10	2.08	2.05
	.05	4.60	3.74	3.34	3.11	2.96	2.85	2.76	2.70	2.65	2.60	2.57	2.53
	.01	8.86	6.51	5.56	5.04	4.69	4.46	4.28	4.14	4.03	3.94	3.86	3.80
	.25	1.43	1.52	1.52	1.51	1.49	1.48	1.47	1.46	1.46	1.45	1.44	1.44
15	.10	3.07	2.70	2.49	2.36	2.27	2.21	2.16	2.12	2.09	2.06	2.04	2.02
	.05	4.54	3.68	3.29	3.06	2.90	2.79	2.71	2.64	2.59	2.54	2.51	2.48
	.01	8.68	6.36	5.42	4.89	4.56	4.32	4.14	4.00	3.89	3.80	3.73	3.67
	.25	1.42	1.51	1.51	1.50	1.48	1.47	1.46	1.45	1.44	1.44	1.44	1.43
16	.10	3.05	2.67	2.46	2.33	2.24	2.18	2.13	2.09	2.06	2.03	2.01	1.99
	.05	4.49	3.63	3.24	3.01	2.85	2.74	2.66	2.59	2.54	2.49	2.46	2.42
	.01	8.53	6.23	5.29	4.77	4.44	4.20	4.03	3.89	3.78	3.69	3.62	3.55
	.25	1.42	1.51	1.50	1.49	1.47	1.46	1.45	1.44	1.43	1.43	1.42	1.41
17	.10	3.03	2.64	2.44	2.31	2.22	2.15	2.10	2.06	2.03	2.00	1.98	1.96
	.05	4.45	3.59	3.20	2.96	2.81	2.70	2.61	2.55	2.49	2.45	2.41	2.38
	.01	8.40	6.11	5.18	4.67	4.34	4.10	3.93	3.79	3.68	3.59	3.52	3.46
	.25	1.41	1.50	1.49	1.48	1.46	1.45	1.44	1.43	1.42	1.42	1.41	1.40
18	.10	3.01	2.62	2.42	2.29	2.20	2.13	2.08	2.04	2.00	1.98	1.96	1.93
	.05	4.41	3.55	3.16	2.93	2.77	2.66	2.58	2.51	2.46	2.41	2.37	2.34
	.01	8.29	6.01	5.09	4.58	4.25	4.01	3.84	3.71	3.60	3.51	3.43	3.37
	.25	1.41	1.49	1.49	1.47	1.46	1.44	1.43	1.42	1.41	1.41	1.40	1.40
19	.10	2.99	2.61	2.40	2.27	2.18	2.11	2.06	2.02	1.98	1.96	1.94	1.91
	.05	4.38	3.52	3.13	2.90	2.74	2.63	2.54	2.48	2.42	2.38	2.34	2.31
	.01	8.18	5.93	5.01	4.50	4.17	3.94	3.77	3.63	3.52	3.43	3.36	3.30
	.25	1.40	1.49	1.48	1.46	1.45	1.44	1.43	1.42	1.41	1.40	1.39	1.39
20	.10	2.97	2.59	2.38	2.25	2.16	2.09	2.04	2.00	1.96	1.94	1.92	1.89
	.05	4.35	3.49	3.10	2.87	2.71	2.60	2.51	2.45	2.39	2.35	2.31	2.28
	.01	8.10	5.85	4.94	4.43	4.10	3.87	3.70	3.56	3.46	3.37	3.29	3.23

TABLE IV (*continued*).

df for numerator (v_h)												α	df for denominator (v_e)
15	20	24	30	40	50	60	100	120	200	500	∞		
1.53	1.52	1.52	1.51	1.51	1.50	1.50	1.49	1.49	1.49	1.48	1.48	.25	
2.24	2.20	2.18	2.16	2.13	2.12	2.11	2.09	2.08	2.07	2.06	2.06	.10	10
2.85	2.77	2.74	2.70	2.66	2.64	2.62	2.59	2.58	2.56	2.55	2.54	.05	
4.56	4.41	4.33	4.25	4.17	4.12	4.08	4.01	4.00	3.96	3.93	3.91	.01	
1.50	1.49	1.49	1.48	1.47	1.47	1.47	1.46	1.46	1.46	1.45	1.45	.25	
2.17	2.12	2.10	2.08	2.05	2.04	2.03	2.00	2.00	1.99	1.98	1.97	.10	11
2.72	2.65	2.61	2.57	2.53	2.51	2.49	2.46	2.45	2.43	2.42	2.40	.05	
4.25	4.10	4.02	3.94	3.86	3.81	3.78	3.71	3.69	3.66	3.62	3.60	.01	
1.48	1.47	1.46	1.45	1.45	1.44	1.44	1.43	1.43	1.43	1.42	1.42	.25	
2.10	2.06	2.04	2.01	1.99	1.97	1.96	1.94	1.93	1.92	1.91	1.90	.10	12
2.62	2.54	2.51	2.47	2.43	2.40	2.38	2.35	2.34	2.32	2.31	2.30	.05	
4.01	3.86	3.78	3.70	3.62	3.57	3.54	3.47	3.45	3.41	3.38	3.36	.01	
1.46	1.45	1.44	1.43	1.42	1.42	1.42	1.41	1.41	1.40	1.40	1.40	.25	
2.05	2.01	1.98	1.96	1.93	1.92	1.90	1.88	1.88	1.86	1.85	1.85	.10	13
2.53	2.46	2.42	2.38	2.34	2.31	2.30	2.26	2.25	2.23	2.22	2.21	.05	
3.82	3.66	3.59	3.51	3.43	3.38	3.34	3.27	3.25	3.22	3.19	3.17	.01	
1.44	1.43	1.42	1.41	1.41	1.40	1.40	1.39	1.39	1.39	1.38	1.38	.25	
2.01	1.96	1.94	1.91	1.89	1.87	1.86	1.83	1.83	1.82	1.80	1.80	.10	14
2.46	2.39	2.35	2.31	2.27	2.24	2.22	2.19	2.18	2.16	2.14	2.13	.05	
3.66	3.51	3.43	3.35	3.27	3.22	3.18	3.11	3.09	3.06	3.03	3.00	.01	
1.43	1.41	1.41	1.40	1.39	1.39	1.38	1.38	1.37	1.37	1.36	1.36	.25	
1.97	1.92	1.90	1.87	1.85	1.83	1.82	1.79	1.79	1.77	1.76	1.76	.10	15
2.40	2.33	2.29	2.25	2.20	2.18	2.16	2.12	2.11	2.10	2.08	2.07	.05	
3.52	3.37	3.29	3.21	3.13	3.08	3.05	2.98	2.96	2.92	2.89	2.87	.01	
1.41	1.40	1.39	1.38	1.37	1.37	1.36	1.36	1.35	1.35	1.34	1.34	.25	
1.94	1.89	1.87	1.84	1.81	1.79	1.78	1.76	1.75	1.74	1.73	1.72	.10	16
2.35	2.28	2.24	2.19	2.15	2.12	2.11	2.07	2.06	2.04	2.02	2.01	.05	
3.41	3.26	3.18	3.10	3.02	2.97	2.93	2.86	2.84	2.81	2.78	2.75	.01	
1.40	1.39	1.38	1.37	1.36	1.35	1.35	1.34	1.34	1.34	1.33	1.33	.25	
1.91	1.86	1.84	1.81	1.78	1.76	1.75	1.73	1.72	1.71	1.69	1.69	.10	17
2.31	2.23	2.19	2.15	2.10	2.08	2.06	2.02	2.01	1.99	1.97	1.96	.05	
3.31	3.16	3.08	3.00	2.92	2.87	2.83	2.76	2.75	2.71	2.68	2.65	.01	
1.39	1.38	1.37	1.36	1.35	1.34	1.34	1.33	1.33	1.32	1.32	1.32	.25	
1.89	1.84	1.81	1.78	1.75	1.74	1.72	1.70	1.69	1.68	1.67	1.66	.10	18
2.27	2.19	2.15	2.11	2.06	2.04	2.02	1.98	1.97	1.95	1.93	1.92	.05	
3.23	3.08	3.00	2.92	2.84	2.78	2.75	2.68	2.66	2.62	2.59	2.57	.01	
1.38	1.37	1.36	1.35	1.34	1.33	1.33	1.32	1.32	1.31	1.31	1.30	.25	
1.86	1.81	1.79	1.76	1.73	1.71	1.70	1.67	1.67	1.65	1.64	1.63	.10	19
2.23	2.16	2.11	2.07	2.03	2.00	1.98	1.94	1.93	1.91	1.89	1.88	.05	
3.15	3.00	2.92	2.84	2.76	2.71	2.67	2.60	2.58	2.55	2.51	2.49	.01	
1.37	1.36	1.35	1.34	1.33	1.33	1.32	1.31	1.31	1.30	1.30	1.29	.25	
1.84	1.79	1.77	1.74	1.71	1.69	1.68	1.65	1.64	1.63	1.62	1.61	.10	20
2.20	2.12	2.08	2.04	1.99	1.97	1.95	1.91	1.90	1.88	1.86	1.84	.05	
3.09	2.94	2.86	2.78	2.69	2.64	2.61	2.54	2.52	2.48	2.44	2.42	.01	

TABLE IV (*continued*).

df *for denominator* (v_e)	α	df *for numerator* (v_h) 1	2	3	4	5	6	7	8	9	10	11	12
	.25	1.40	1.48	1.47	1.45	1.44	1.42	1.41	1.40	1.39	1.39	1.38	1.37
22	.10	2.95	2.56	2.35	2.22	2.13	2.06	2.01	1.97	1.93	1.90	1.88	1.86
	.05	4.30	3.44	3.05	2.82	2.66	2.55	2.46	2.40	2.34	2.30	2.26	2.23
	.01	7.95	5.72	4.82	4.31	3.99	3.76	3.59	3.45	3.35	3.26	3.18	3.12
	.25	1.39	1.47	1.46	1.44	1.43	1.41	1.40	1.39	1.38	1.38	1.37	1.36
24	.10	2.93	2.54	2.33	2.19	2.10	2.04	1.98	1.94	1.91	1.88	1.85	1.83
	.05	4.26	3.40	3.01	2.78	2.62	2.51	2.42	2.36	2.30	2.25	2.21	2.18
	.01	7.82	5.61	4.72	4.22	3.90	3.67	3.50	3.36	3.26	3.17	3.09	3.03
	.25	1.38	1.46	1.45	1.44	1.42	1.41	1.39	1.38	1.37	1.37	1.36	1.35
26	.10	2.91	2.52	2.31	2.17	2.08	2.01	1.96	1.92	1.88	1.86	1.84	1.81
	.05	4.23	3.37	2.98	2.74	2.59	2.47	2.39	2.32	2.27	2.22	2.18	2.15
	.01	7.72	5.53	4.64	4.14	3.82	3.59	3.42	3.29	3.18	3.09	3.02	2.96
	.25	1.38	1.46	1.45	1.43	1.41	1.40	1.39	1.38	1.37	1.36	1.35	1.34
28	.10	2.89	2.50	2.29	2.16	2.06	2.00	1.94	1.90	1.87	1.84	1.81	1.79
	.05	4.20	3.34	2.95	2.71	2.56	2.45	2.36	2.29	2.24	2.19	2.15	2.12
	.01	7.64	5.45	4.57	4.07	3.75	3.53	3.36	3.23	3.12	3.03	2.96	2.90
	.25	1.38	1.45	1.44	1.42	1.41	1.39	1.38	1.37	1.36	1.35	1.35	1.34
30	.10	2.88	2.49	2.28	2.14	2.05	1.98	1.93	1.88	1.85	1.82	1.79	1.77
	.05	4.17	3.32	2.92	2.69	2.53	2.42	2.33	2.27	2.21	2.16	2.13	2.09
	.01	7.56	5.39	4.51	4.02	3.70	3.47	3.30	3.17	3.07	2.98	2.91	2.84
	.25	1.36	1.44	1.42	1.40	1.39	1.37	1.36	1.35	1.34	1.33	1.32	1.31
40	.10	2.84	2.44	2.23	2.09	2.00	1.93	1.87	1.83	1.79	1.76	1.73	1.71
	.05	4.08	3.23	2.84	2.61	2.45	2.34	2.25	2.18	2.12	2.08	2.04	2.00
	.01	7.31	5.18	4.31	3.83	3.51	3.29	3.12	2.99	2.89	2.80	2.73	2.66
	.25	1.35	1.42	1.41	1.38	1.37	1.35	1.33	1.32	1.31	1.30	1.29	1.29
60	.10	2.79	2.39	2.18	2.04	1.95	1.87	1.82	1.77	1.74	1.71	1.68	1.66
	.05	4.00	3.15	2.76	2.53	2.37	2.25	2.17	2.10	2.04	1.99	1.95	1.92
	.01	7.08	4.98	4.13	3.65	3.34	3.12	2.95	2.82	2.72	2.63	2.56	2.50
	.25	1.34	1.40	1.39	1.37	1.35	1.33	1.31	1.30	1.29	1.28	1.27	1.26
120	.10	2.75	2.35	2.13	1.99	1.90	1.82	1.77	1.72	1.68	1.65	1.62	1.60
	.05	3.92	3.07	2.68	2.45	2.29	2.17	2.09	2.02	1.96	1.91	1.87	1.83
	.01	6.85	4.79	3.95	3.48	3.17	2.96	2.79	2.66	2.56	2.47	2.40	2.34
	.25	1.33	1.39	1.38	1.36	1.34	1.32	1.31	1.29	1.28	1.27	1.26	1.25
200	.10	2.73	2.33	2.11	1.97	1.88	1.80	1.75	1.70	1.66	1.63	1.60	1.57
	.05	3.89	3.04	2.65	2.42	2.26	2.14	2.06	1.98	1.93	1.88	1.84	1.80
	.01	6.76	4.71	3.88	3.41	3.11	2.89	2.73	2.60	2.50	2.41	2.34	2.27
	.25	1.32	1.39	1.37	1.35	1.33	1.31	1.29	1.28	1.27	1.25	1.24	1.24
∞	.10	2.71	2.30	2.08	1.94	1.85	1.77	1.72	1.67	1.63	1.60	1.57	1.55
	.05	3.84	3.00	2.60	2.37	2.21	2.10	2.01	1.94	1.88	1.83	1.79	1.75
	.01	6.63	4.61	3.78	3.32	3.02	2.80	2.64	2.51	2.41	2.32	2.25	2.18

TABLE IV (*continued*).

df for numerator (v_h)													df for denominator (v_e)
15	20	24	30	40	50	60	100	120	200	500	∞	α	
1.36	1.34	1.33	1.32	1.31	1.31	1.30	1.30	1.30	1.29	1.29	1.28	.25	
1.81	1.76	1.73	1.70	1.67	1.65	1.64	1.61	1.60	1.59	1.58	1.57	.10	22
2.15	2.07	2.03	1.98	1.94	1.91	1.89	1.85	1.84	1.82	1.80	1.78	.05	
2.98	2.83	2.75	2.67	2.58	2.53	2.50	2.42	2.40	2.36	2.33	2.31	.01	
1.35	1.33	1.32	1.31	1.30	1.29	1.29	1.28	1.28	1.27	1.27	1.26	.25	
1.78	1.73	1.70	1.67	1.64	1.62	1.61	1.58	1.57	1.56	1.54	1.53	.10	24
2.11	2.03	1.98	1.94	1.89	1.86	1.84	1.80	1.79	1.77	1.75	1.73	.05	
2.89	2.74	2.66	2.58	2.49	2.44	2.40	2.33	2.31	2.27	2.24	2.21	.01	
1.34	1.32	1.31	1.30	1.29	1.28	1.28	1.26	1.26	1.26	1.25	1.25	.25	
1.76	1.71	1.68	1.65	1.61	1.59	1.58	1.55	1.54	1.53	1.51	1.50	.10	26
2.07	1.99	1.95	1.90	1.85	1.82	1.80	1.76	1.75	1.73	1.71	1.69	.05	
2.81	2.66	2.58	2.50	2.42	2.36	2.33	2.25	2.23	2.19	2.16	2.13	.01	
1.33	1.31	1.30	1.29	1.28	1.27	1.27	1.26	1.25	1.25	1.24	1.24	.25	
1.74	1.69	1.66	1.63	1.59	1.57	1.56	1.53	1.52	1.50	1.49	1.48	.10	28
2.04	1.96	1.91	1.87	1.82	1.79	1.77	1.73	1.71	1.69	1.67	1.65	.05	
2.75	2.60	2.52	2.44	2.35	2.30	2.26	2.19	2.17	2.13	2.09	2.06	.01	
1.32	1.30	1.29	1.28	1.27	1.26	1.26	1.25	1.24	1.24	1.23	1.23	.25	
1.72	1.67	1.64	1.61	1.57	1.55	1.54	1.51	1.50	1.48	1.47	1.46	.10	30
2.01	1.93	1.89	1.84	1.79	1.76	1.74	1.70	1.68	1.66	1.64	1.62	.05	
2.70	2.55	2.47	2.39	2.30	2.25	2.21	2.13	2.11	2.07	2.03	2.01	.01	
1.30	1.28	1.26	1.25	1.24	1.23	1.22	1.21	1.21	1.20	1.19	1.19	.25	
1.66	1.61	1.57	1.54	1.51	1.48	1.47	1.43	1.42	1.41	1.39	1.38	.10	40
1.92	1.84	1.79	1.74	1.69	1.66	1.64	1.59	1.58	1.55	1.53	1.51	.05	
2.52	2.37	2.29	2.20	2.11	2.06	2.02	1.94	1.92	1.87	1.83	1.80	.01	
1.27	1.25	1.24	1.22	1 21	1 20	1.19	1.17	1.17	1.16	1.15	1.15	.25	
1.60	1.54	1.51	1.48	1.44	1.41	1.40	1.36	1.35	1.33	1.31	1.29	.10	60
1.84	1.75	1.70	1.65	1.59	1.56	1.53	1.48	1.47	1.44	1.41	1.39	.05	
2.35	2.20	2.12	2.03	1.94	1.88	1.84	1.75	1.73	1.68	1.63	1.60	.01	
1.24	1.22	1.21	1.19	1.18	1.17	1.16	1.14	1.13	1.12	1.11	1.10	.25	
1.55	1.48	1.45	1.41	1.37	1.34	1.32	1.27	1.26	1.24	1.21	1.19	.10	120
1.75	1.66	1.61	1.55	1.50	1.46	1.43	1.37	1.35	1.32	1.28	1.25	.05	
2.19	2.03	1.95	1.86	1.76	1.70	1.66	1.56	1.53	1.48	1.42	1.38	.01	
1.23	1.21	1.20	1.18	1.16	1.14	1.12	1.11	1.10	1.09	1.08	1.06	.25	
1.52	1.46	1.42	1.38	1.34	1.31	1.28	1.24	1.22	1.20	1.17	1.14	.10	200
1.72	1.62	1.57	1.52	1.46	1.41	1.39	1.32	1.29	1.26	1.22	1.19	.05	
2.13	1.97	1.89	1.79	1.69	1.63	1.58	1.48	1.44	1.39	1.33	1.28	.01	
1.22	1.19	1.18	1.16	1.14	1.13	1.12	1.09	1.08	1.07	1.04	1.00	.25	
1.49	1.42	1.38	1.34	1.30	1.26	1.24	1.18	1.17	1.13	1.08	1.00	.10	∞
1.67	1.57	1.52	1.46	1.39	1.35	1.32	1.24	1.22	1.17	1.11	1.00	.05	
2.04	1.88	1.79	1.70	1.59	1.52	1.47	1.36	1.32	1.25	1.15	1.00	.01	

TABLE V. Power Charts for the Noncentral F Distribution

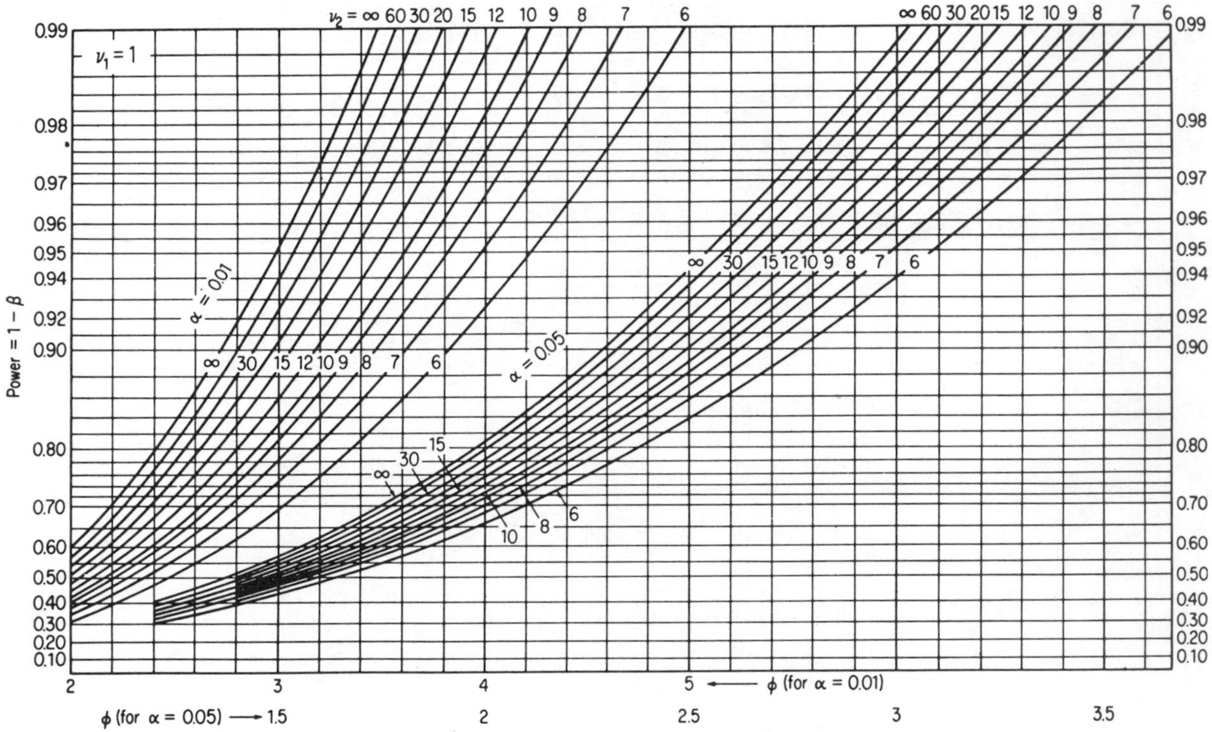

Reproduced with permission from E. S. Pearson and H. O. Hartley. Charts of the power function for analysis of variance tests, derived from the non-central F-distribution, *Biometrika*, 1951, **38**, 112–130.

TABLE V (*continued*).

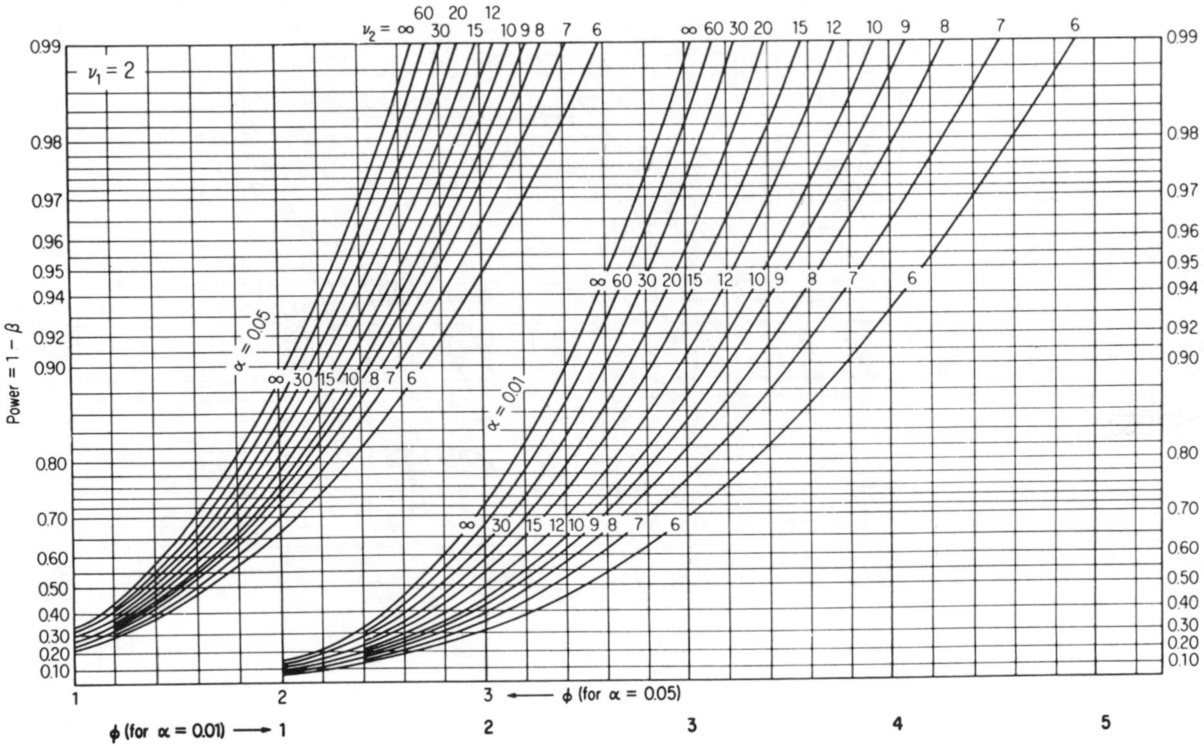

TABLE V (*continued*).

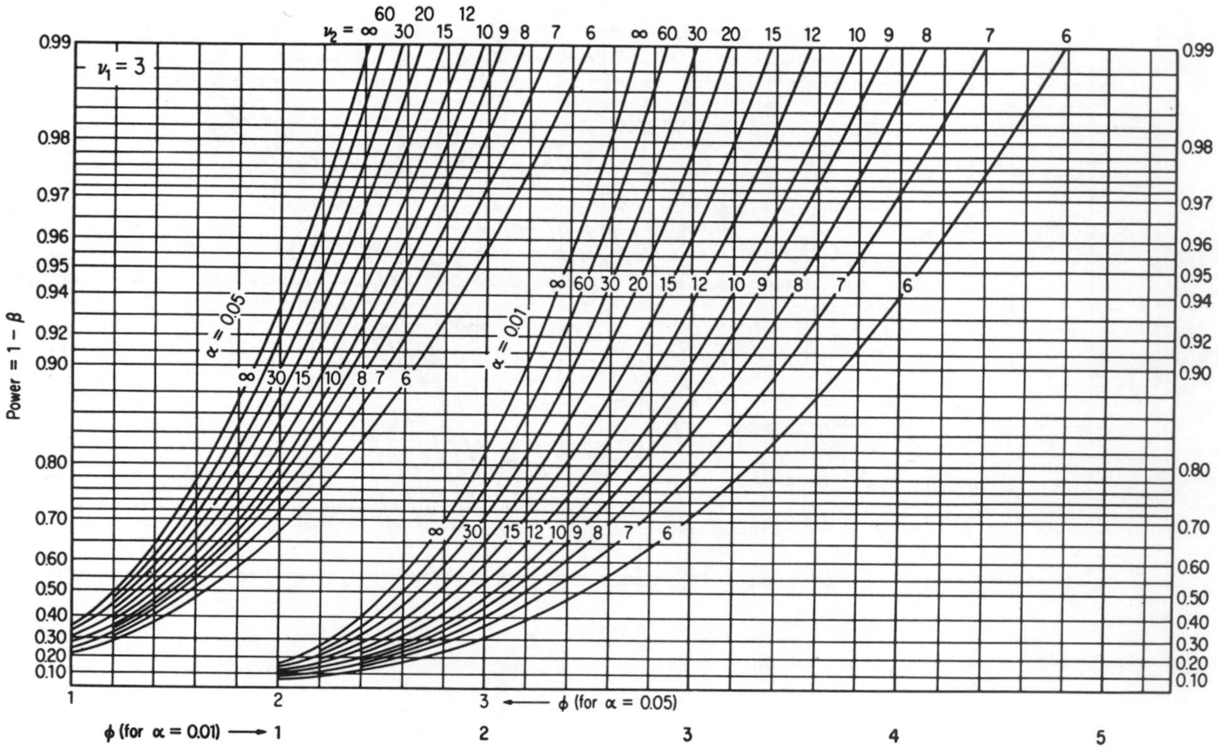

TABLE V (*continued*).

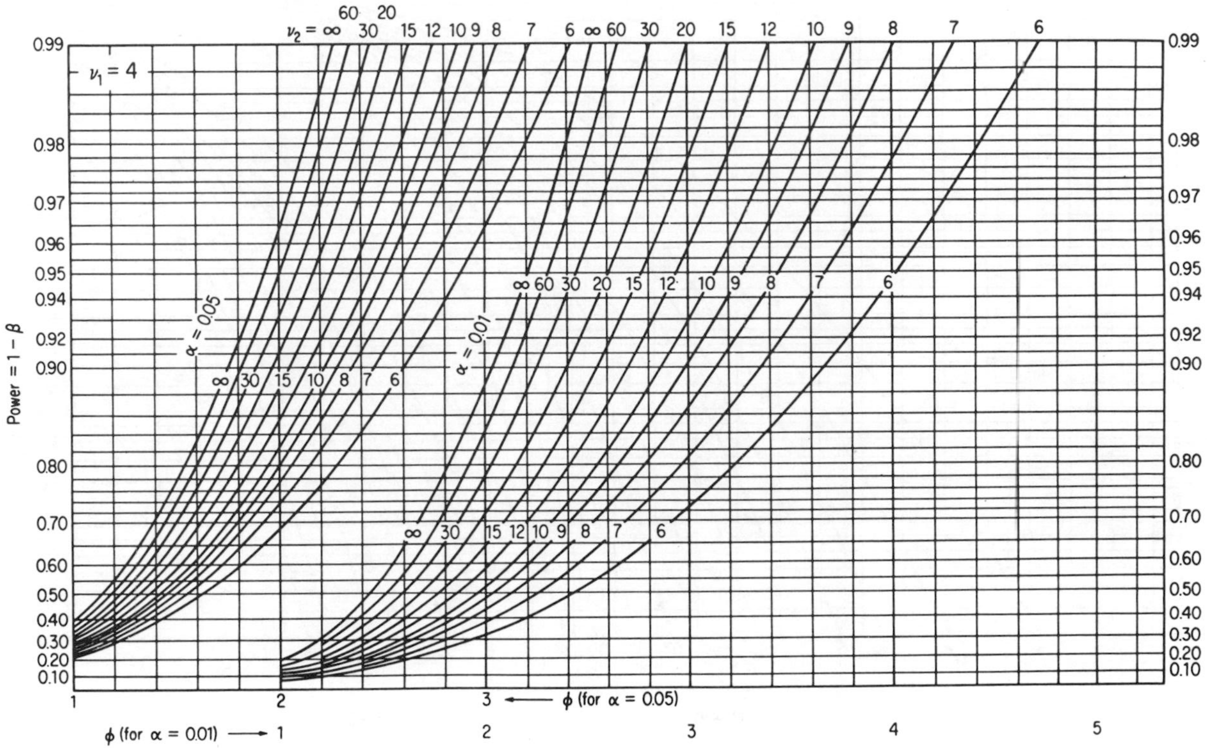

TABLE V (*continued*).

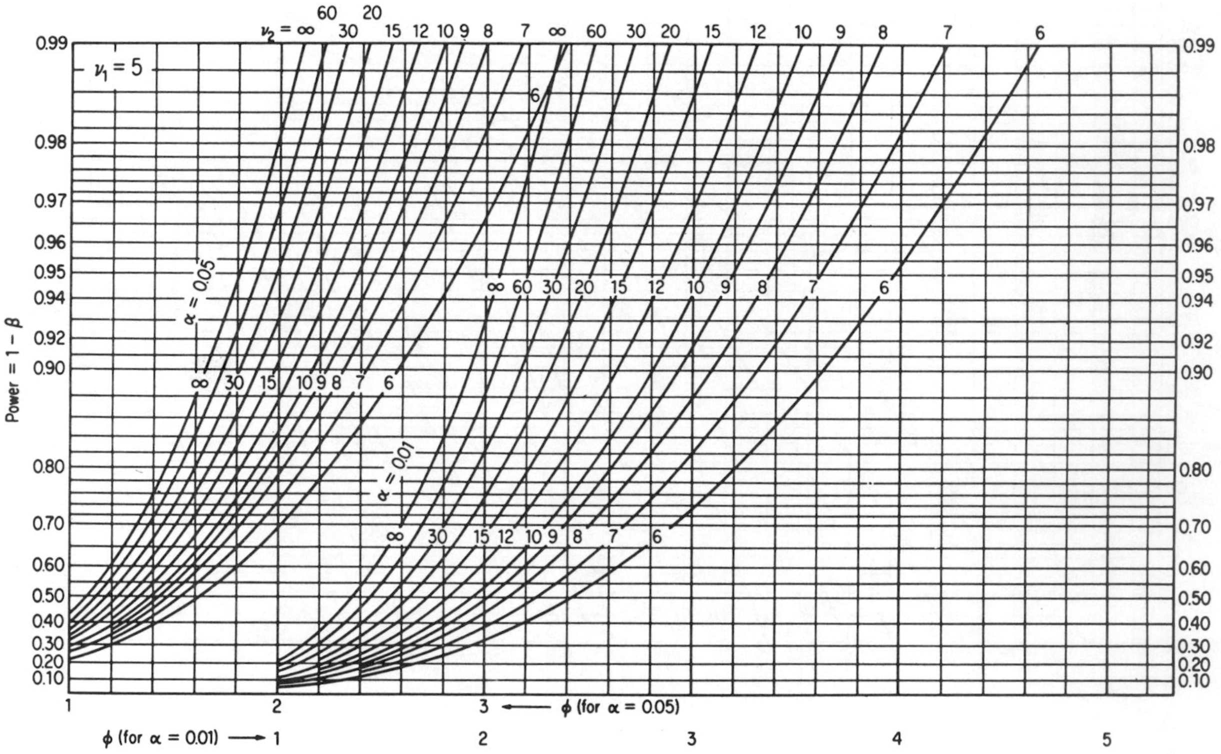

TABLE V (*continued*).

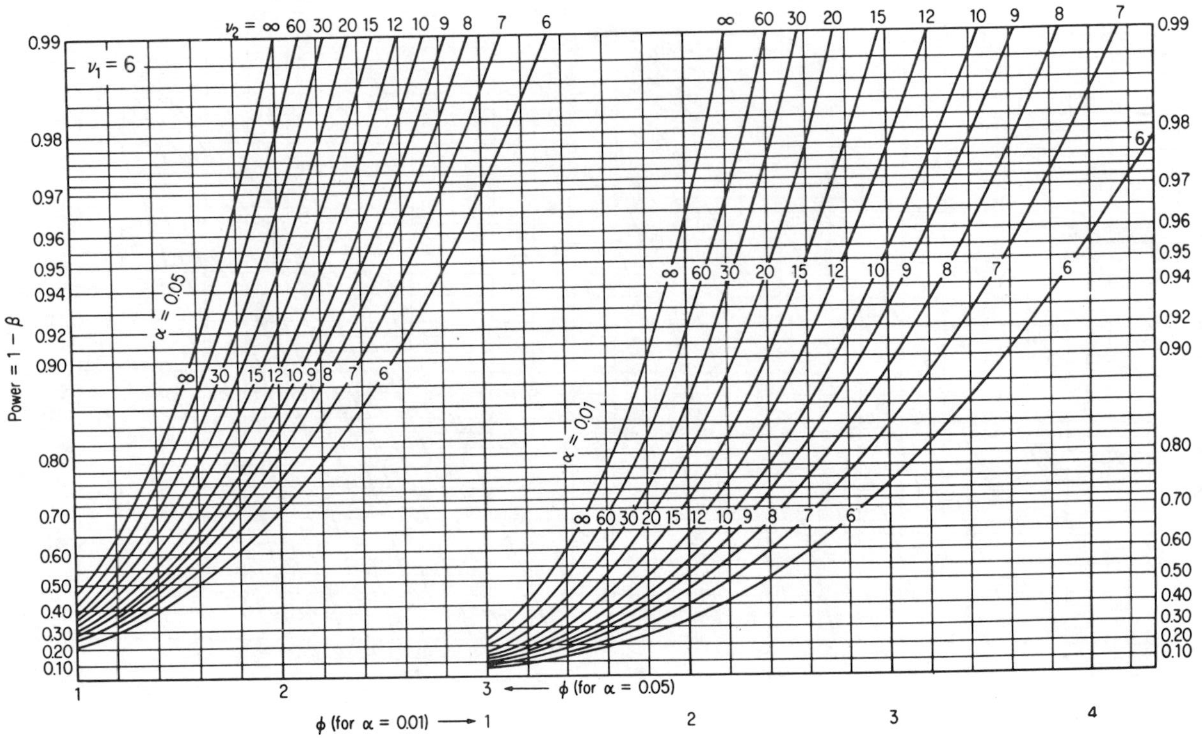

TABLE V (*continued*).

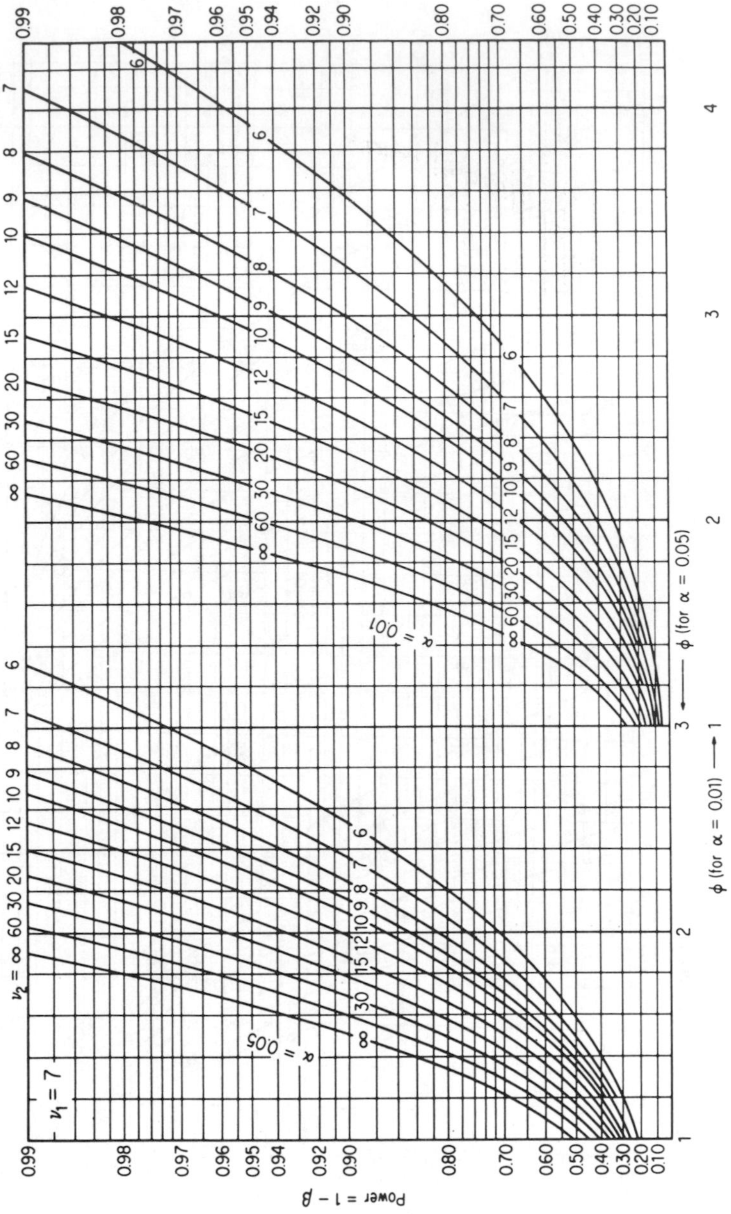

TABLE V (*continued*).

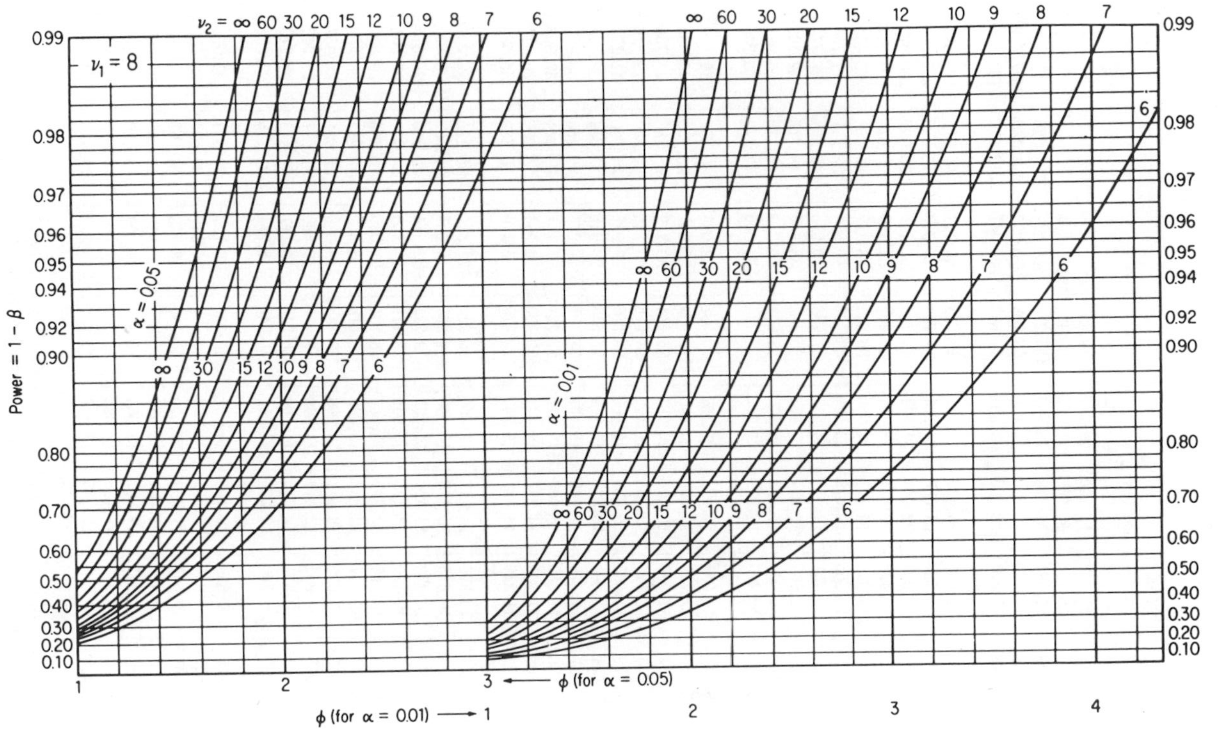

TABLE VI. Upper Percentage Points of Hotelling's T^2 Distribution: $T^\alpha(p, \nu)$, $\alpha = 0.01$

Degrees of Freedom, ν	p = 1	2	3	4	5	6	7	8	9	10
2	98.503									
3	34.116	297.000								
4	21.198	82.177	594.997							
5	16.258	45.000	147.283	992.494						
6	13.745	31.857	75.125	229.679	1489.489					
7	12.246	25.491	50.652	111.839	329.433	2085.984				
8	11.259	21.821	39.118	72.908	155.219	446.571	2781.978			
9	10.561	19.460	32.598	54.890	98.703	205.293	581.106	3577.472		
10	10.044	17.826	28.466	44.838	72.882	128.067	262.076	733.045	4472.464	
11	9.646	16.631	25.637	38.533	58.618	93.127	161.015	325.576	902.392	5466.956
12	9.330	15.722	23.588	34.251	49.739	73.969	115.640	197.555	395.797	1089.149
13	9.074	15.008	22.041	31.171	43.745	62.114	90.907	140.429	237.692	472.742
14	8.862	14.433	20.834	28.857	39.454	54.150	75.676	109.441	167.499	281.428
15	8.683	13.960	19.867	27.060	36.246	48.472	65.483	90.433	129.576	196.853
16	8.531	13.566	19.076	25.626	33.762	44.240	58.241	77.755	106.391	151.316
17	8.400	13.231	18.418	24.458	31.788	40.975	52.858	68.771	90.969	123.554
18	8.285	12.943	17.861	23.487	30.182	38.385	48.715	62.109	80.067	105.131
19	8.185	12.694	17.385	22.670	28.852	36.283	45.435	56.992	71.999	92.134
20	8.096	12.476	16.973	21.972	27.734	34.546	42.779	52.948	65.813	82.532
21	8.017	12.283	16.613	21.369	26.781	33.088	40.587	49.679	60.932	75.181
22	7.945	12.111	16.296	20.843	25.959	31.847	38.750	46.986	56.991	69.389
23	7.881	11.958	16.015	20.381	25.244	30.779	37.188	44.730	53.748	64.719
24	7.823	11.820	15.763	19.972	24.616	29.850	35.846	42.816	51.036	60.879
25	7.770	11.695	15.538	19.606	24.060	29.036	34.680	41.171	48.736	57.671
26	7.721	11.581	15.334	19.279	23.565	28.316	33.659	39.745	46.762	54.953
27	7.677	11.478	15.149	18.983	23.121	27.675	32.756	38.496	45.051	52.622
28	7.636	11.383	14.980	18.715	22.721	27.101	31.954	37.393	43.554	50.604
29	7.598	11.295	14.825	18.471	22.359	26.584	31.236	36.414	42.234	48.839
30	7.562	11.215	14.683	18.247	22.029	26.116	30.589	35.538	41.062	47.283
35	7.419	10.890	14.117	17.366	20.743	24.314	28.135	32.259	36.743	41.651
40	7.314	10.655	13.715	16.750	19.858	23.094	26.502	30.120	33.984	38.135
45	7.234	10.478	13.414	16.295	19.211	22.214	25.340	28.617	32.073	35.737
50	7.171	10.340	13.181	15.945	18.718	21.550	24.470	27.504	30.673	33.998
55	7.119	10.228	12.995	15.667	18.331	21.030	23.795	26.647	29.603	32.682
60	7.077	10.137	12.843	15.442	18.018	20.613	23.257	25.967	28.760	31.650
70	7.011	9.996	12.611	15.098	17.543	19.986	22.451	24.957	27.515	30.139
80	6.963	9.892	12.440	14.849	17.201	19.536	21.877	24.242	26.642	29.085
90	6.925	9.813	12.310	14.660	16.942	19.197	21.448	23.710	25.995	28.310
100	6.895	9.750	12.208	14.511	16.740	18.934	21.115	23.299	25.496	27.714
110	6.871	9.699	12.125	14.391	16.577	18.722	20.849	22.972	25.101	27.243
120	6.851	9.657	12.057	14.292	16.444	18.549	20.632	22.705	24.779	26.862
150	6.807	9.565	11.909	14.079	16.156	18.178	20.167	22.137	24.096	26.054
200	6.763	9.474	11.764	13.871	15.877	17.819	19.720	21.592	23.446	25.287
400	6.699	9.341	11.551	13.569	15.473	17.303	19.080	20.818	22.525	24.209
1000	6.660	9.262	11.426	13.392	15.239	17.006	18.713	20.376	22.003	23.600
∞	6.635	9.210	11.345	13.277	15.086	16.812	18.475	20.090	21.666	23.209

Abridged from D. R. Jensen, and R. B. Howe. Tables of upper percentage points of Hotelling's T^2 distribution. Technical Report No. 9, Virginia Polytechnic Institute, 1968. Reproduced by permission of the authors.

TABLE VI (*continued*). $\alpha = 0.05$

Degrees of Freedom, ν	p									
	1	2	3	4	5	6	7	8	9	10
2	18.513									
3	10.128	57.000								
4	7.709	25.472	114.986							
5	6.608	17.361	46.383	192.468						
6	5.987	13.887	29.661	72.937	289.446					
7	5.591	12.001	22.720	44.718	105.157	405.920				
8	5.318	10.828	19.028	33.230	62.561	143.050	541.890			
9	5.117	10.033	16.766	27.202	45.453	83.202	186.622	697.356		
10	4.965	9.459	15.248	23.545	36.561	59.403	106.649	235.873	872.317	
11	4.844	9.026	14.163	21.108	31.205	47.123	75.088	132.903	290.806	1066.774
12	4.747	8.689	13.350	19.376	27.656	39.764	58.893	92.512	161.967	351.421
13	4.667	8.418	12.719	18.086	25.145	34.911	49.232	71.878	111.676	193.842
14	4.600	8.197	12.216	17.089	23.281	31.488	42.881	59.612	86.079	132.582
15	4.543	8.012	11.806	16.296	21.845	28.955	38.415	51.572	70.907	101.499
16	4.494	7.856	11.465	15.651	20.706	27.008	35.117	45.932	60.986	83.121
17	4.451	7.722	11.177	15.117	19.782	25.467	32.588	41.775	54.041	71.127
18	4.414	7.606	10.931	14.667	19.017	24.219	30.590	38.592	48.930	62.746
19	4.381	7.504	10.719	14.283	18.375	23.189	28.975	36.082	45.023	56.587
20	4.351	7.415	10.533	13.952	17.828	22.324	27.642	34.054	41.946	51.884
21	4.325	7.335	10.370	13.663	17.356	21.588	26.525	32.384	39.463	48.184
22	4.301	7.264	10.225	13.409	16.945	20.954	25.576	30.985	37.419	45.202
23	4.279	7.200	10.095	13.184	16.585	20.403	24.759	29.798	35.709	42.750
24	4.260	7.142	9.979	12.983	16.265	19.920	24.049	28.777	34.258	40.699
25	4.242	7.089	9.874	12.803	15.981	19.492	23.427	27.891	33.013	38.961
26	4.225	7.041	9.779	12.641	15.726	19.112	22.878	27.114	31.932	37.469
27	4.210	6.997	9.692	12.493	15.496	18.770	22.388	26.428	30.985	36.176
28	4.196	6.957	9.612	12.359	15.287	18.463	21.950	25.818	30.149	35.043
29	4.183	6.919	9.539	12.236	15.097	18.184	21.555	25.272	29.407	34.044
30	4.171	6.885	9.471	12.123	14.924	17.931	21.198	24.781	28.742	33.156
35	4.121	6.744	9.200	11.674	14.240	16.944	19.823	22.913	26.252	29.881
40	4.085	6.642	9.005	11.356	13.762	16.264	18.890	21.668	24.624	27.783
45	4.057	6.564	8.859	11.118	13.409	15.767	18.217	20.781	23.477	26.326
50	4.034	6.503	8.744	10.934	13.138	15.388	17.709	20.117	22.627	25.256
55	4.016	6.454	8.652	10.787	12.923	15.090	17.311	19.600	21.972	24.437
60	4.001	6.413	8.577	10.668	12.748	14.850	16.992	19.188	21.451	23.790
70	3.978	6.350	8.460	10.484	12.482	14.485	16.510	18.571	20.676	22.834
80	3.960	6.303	8.375	10.350	12.289	14.222	16.165	18.130	20.127	22.162
90	3.947	6.267	8.309	10.248	12.142	14.022	15.905	17.801	19.718	21.663
100	3.936	6.239	8.257	10.167	12.027	13.867	15.702	17.544	19.401	21.279
110	3.927	6.216	8.215	10.102	11.934	13.741	15.540	17.340	19.149	20.973
120	3.920	6.196	8.181	10.048	11.858	13.639	15.407	17.172	18.943	20.725
150	3.904	6.155	8.105	9.931	11.693	13.417	15.121	16.814	18.504	20.196
200	3.888	6.113	8.031	9.817	11.531	13.202	14.845	16.469	18.083	19.692
400	3.865	6.052	7.922	9.650	11.297	12.890	14.447	15.975	17.484	18.976
1000	3.851	6.015	7.857	9.552	11.160	12.710	14.217	15.692	17.141	18.570
∞	3.841	5.991	7.815	9.488	11.070	12.592	14.067	15.507	16.919	18.307

TABLE VI (*continued*). $\alpha = 0.10$.

Degrees of Freedom, v	p									
	1	2	3	4	5	6	7	8	9	10
2	8.526									
3	5.538	27.000								
4	4.545	14.566	54.971							
5	4.060	10.811	26.954	92.434						
6	3.776	9.071	18.859	42.741	139.389					
7	3.589	8.081	15.202	28.751	61.940	195.836				
8	3.458	7.446	13.155	22.529	40.506	84.556	261.774			
9	3.360	7.005	11.857	19.085	31.077	54.132	110.590	337.204		
10	3.285	6.681	10.964	16.917	25.896	40.854	69.632	140.045	422.124	
11	3.225	6.434	10.314	15.435	22.655	33.600	51.866	87.009	172.920	516.536
12	3.177	6.239	9.820	14.361	20.448	29.082	42.202	64.114	106.263	209.216
13	3.136	6.081	9.432	13.548	18.854	26.016	36.204	51.706	77.601	127.396
14	3.102	5.951	9.119	12.912	17.651	23.808	32.146	44.025	62.113	92.327
15	3.073	5.842	8.862	12.401	16.713	22.145	29.229	38.840	52.547	73.423
16	3.048	5.750	8.648	11.981	15.960	20.850	27.036	35.120	46.102	61.772
17	3.026	5.670	8.465	11.631	15.344	19.814	25.331	32.329	41.486	53.933
18	3.007	5.600	8.309	11.335	14.830	18.966	23.969	30.161	38.026	48.326
19	2.990	5.539	8.173	11.081	14.396	18.261	22.857	28.431	35.343	44.219
20	2.975	5.485	8.053	10.860	14.023	17.665	21.931	27.020	33.203	40.877
21	2.961	5.437	7.948	10.667	13.701	17.154	21.150	25.847	31.459	38.286
22	2.949	5.394	7.854	10.497	13.419	16.713	20.482	24.857	30.011	36.175
23	2.937	5.355	7.770	10.345	13.170	16.327	19.904	24.012	28.790	34.424
24	2.927	5.320	7.695	10.210	12.949	15.987	19.400	23.281	27.747	32.949
25	2.918	5.288	7.626	10.088	12.752	15.685	18.955	22.643	26.846	31.690
26	2.909	5.259	7.564	9.977	12.574	15.415	18.561	22.082	26.061	30.603
27	2.901	5.232	7.507	9.877	12.414	15.173	18.209	21.584	25.370	29.655
28	2.894	5.207	7.455	9.785	12.268	14.954	17.893	21.140	24.757	28.821
29	2.887	5.184	7.407	9.701	12.135	14.755	17.607	20.741	24.211	28.083
30	2.881	5.163	7.363	9.624	12.013	14.573	17.348	20.380	23.720	27.424
35	2.855	5.007	7.184	9.316	11.530	13.862	16.343	19.001	21.866	24.972
40	2.835	5.013	7.054	9.095	11.190	13.369	15.657	18.074	20.642	23.382
45	2.820	4.965	6.957	8.930	10.937	13.006	15.158	17.408	19.773	22.269
50	2.809	4.927	6.880	8.802	10.743	12.729	14.779	16.907	19.125	21.446
55	2.799	4.896	6.818	8.699	10.588	12.510	14.481	16.516	18.623	20.814
60	2.791	4.871	6.768	8.616	10.462	12.332	14.242	16.202	18.223	20.312
70	2.779	4.831	6.690	8.487	10.270	12.062	13.880	15.731	17.625	19.568
80	2.769	4.802	6.632	8.392	10.129	11.867	13.619	15.394	17.200	19.042
90	2.762	4.780	6.588	8.320	10.023	11.719	13.422	15.141	16.882	18.651
100	2.756	4.762	6.553	8.263	9.939	11.603	13.268	14.944	16.635	18.348
110	2.752	4.747	6.524	8.217	9.871	11.509	13.145	14.786	16.438	18.107
120	2.748	4.735	6.501	8.179	9.815	11.432	13.043	14.657	16.278	17.911
150	2.739	4.708	6.449	8.096	9.694	11.266	12.826	14.380	15.934	17.493
200	2.731	4.682	6.399	8.015	9.576	11.105	12.614	14.112	15.603	17.092
400	2.718	4.643	6.324	7.895	9.403	10.870	12.309	13.727	15.131	16.523
1000	2.711	4.620	6.280	7.825	9.303	10.734	12.132	13.506	14.859	16.197
∞	2.706	4.605	6.251	7.779	9.236	10.645	12.017	13.362	14.684	15.987

TABLE VII. Upper Percentage Points of Roy's Largest-Root Criterion, $\theta^{\alpha}(s, m, n)$

$\alpha = 0.01 \qquad s = 2 \qquad \theta^{\alpha}(s, m, n)$

n \ m	0	1	2	3	4	5	7	10	15
5	0.675	0.745	0.787	0.817	0.839	0.8568	0.8821	0.9066	0.9306
10	0.470	0.544	0.597	0.638	0.670	0.6970	0.7391	0.7834	0.8309
15	0.357	0.425	0.476	0.517	0.551	0.5803	0.6279	0.6810	0.7418
20	0.288	0.347	0.394	0.433	0.467	0.4951	0.5435	0.5998	0.6670
25	0.240	0.293	0.336	0.372	0.403	0.4309	0.4782	0.5347	0.6045
30	0.207	0.254	0.293	0.326	0.355	0.3812	0.4266	0.4819	0.5521
40	0.161	0.200	0.232	0.261	0.286	0.3094	0.3503	0.4017	0.4697
60	0.1114	0.140	0.165	0.186	0.206	0.2244	0.2576	0.3008	0.3608
80	0.0852	0.1080	0.1273	0.1448	0.1609	0.1759	0.2035	0.2402	0.2925
100	0.0692	0.0878	0.1038	0.1184	0.1319	0.1446	0.1682	0.1999	0.2458
130	0.0539	0.0685	0.0813	0.0930	0.1039	0.1142	0.1331	0.1595	0.1983
160	0.0441	0.0562	0.0668	0.0765	0.0857	0.09430	0.1105	0.1328	0.1662
200	0.0355	0.0453	0.0540	0.0619	0.0694	0.07653	0.08994	0.1085	0.1366
300	0.0239	0.0305	0.0365	0.0419	0.0471	0.05202	0.06137	0.07446	0.1076
500	0.0144	0.0185	0.0221	0.0255	0.0287	0.03192	0.03753	0.04574	0.07415
1000	0.00725	0.00930	0.01114	0.01285	0.01448	0.01605	0.01904	0.02328	0.03811

$\alpha = 0.05 \qquad s = 2 \qquad \theta^{\alpha}(s, m, n)$

n \ m	0	1	2	3	4	5	7	10	15
5	0.565	0.651	0.706	0.746	0.776	0.7992	0.8337	0.8676	0.9011
10	0.374	0.455	0.514	0.561	0.598	0.6294	0.6787	0.7316	0.7889
15	0.278	0.348	0.402	0.445	0.483	0.5145	0.5671	0.6266	0.6954
20	0.222	0.281	0.329	0.369	0.403	0.4340	0.4855	0.5462	0.6197
25	0.183	0.236	0.278	0.314	0.346	0.3748	0.4239	0.4834	0.5580
30	0.157	0.203	0.241	0.274	0.303	0.3297	0.3760	0.4333	0.5070
40	0.121	0.158	0.190	0.218	0.243	0.2654	0.3064	0.3585	0.4282
60	0.0836	0.110	0.133	0.154	0.173	0.1909	0.2233	0.2661	0.3260
80	0.0638	0.0846	0.1027	0.1191	0.1345	0.1490	0.1756	0.2114	0.2630
100	0.0515	0.0686	0.0835	0.0972	0.1100	0.1221	0.1430	0.1753	0.2203
130	0.0400	0.0535	0.0652	0.0761	0.0864	0.09613	0.1141	0.1396	0.1771
160	0.0327	0.0473	0.0535	0.0626	0.0711	0.07925	0.09461	0.1159	0.1481
200	0.0263	0.0352	0.0432	0.0506	0.0576	0.06422	0.07687	0.09454	0.1215
300	0.0176	0.0237	0.0291	0.0342	0.0390	0.04356	0.05234	0.06471	0.09867
500	0.0106	0.0143	0.0176	0.0207	0.0237	0.02651	0.03194	0.03967	0.06110
1000	0.00535	0.00719	0.00888	0.01045	0.01195	0.01339	0.01618	0.02016	0.03547

From K. C. S. Pillai, *Statistical Tables for Tests of Multivariate Hypotheses*, The Statistical Center, University of the Phillipines, 1960. Reproduced by permission of the author and publisher.

TABLE VII (*continued*).

$\alpha = 0.01 \qquad s = 3 \qquad \theta^{\alpha}(s, m, n)$

n \ m	0	1	2	3	4	5	7	10	15
5	0.75816	0.8040	0.8344	0.8564	0.8730	0.8894	0.9056	0.9247	0.9437
10	0.55857	0.6164	0.6590	0.6923	0.7192	0.7415	0.7767	0.8141	0.8544
15	0.43751	0.4936	0.5374	0.5730	0.6029	0.6285	0.6703	0.7172	0.7708
20	0.35856	0.4104	0.4519	0.4867	0.5166	0.5428	0.5866	0.6376	0.6985
25	0.30343	0.3506	0.3893	0.4223	0.4511	0.4767	0.5203	0.5726	0.6370
30	0.26286	0.3058	0.3416	0.3726	0.3999	0.4245	0.4670	0.5189	0.5847
40	0.20727	0.2434	0.2742	0.3012	0.3256	0.3477	0.3869	0.4350	0.5001
60	0.14555	0.17271	0.19631	0.21751	0.23692	0.2549	0.2874	0.3298	0.3883
80	0.11212	0.13378	0.15282	0.17011	0.18608	0.2010	0.2285	0.2662	0.3166
100	0.09118	0.10916	0.12508	0.13963	0.15317	0.1659	0.1895	0.2211	0.2670
130	0.07120	0.08552	0.09829	0.11004	0.12104	0.1314	0.1508	0.1772	0.2162
160	0.05842	0.07030	0.08096	0.09079	0.10003	0.1088	0.1253	0.1479	0.1816
200	0.04713	0.05681	0.06554	0.07362	0.08123	0.08849	0.1021	0.1222	0.1540
300	0.03178	0.03839	0.04441	0.04997	0.05526	0.06034	0.06991	0.08332	0.1038
500	0.01924	0.02329	0.02696	0.03043	0.03370	0.03686	0.04285	0.05130	0.06441
1000	0.00969	0.01174	0.01362	0.01538	0.01706	0.01868	0.02547	0.03011	0.03304

$\alpha = 0.05 \qquad s = 3 \qquad \theta^{\alpha}(s, m, n)$

n \ m	0	1	2	3	4	5	7	10	15
5	0.66889	0.7292	0.7690	0.7994	0.8221	0.8456	0.8668	0.8934	0.9199
10	0.47178	0.5372	0.5862	0.6248	0.6564	0.6828	0.7246	0.7695	0.8185
15	0.36196	0.4218	0.4690	0.5078	0.5407	0.5690	0.6157	0.6687	0.7298
20	0.29310	0.3465	0.3898	0.4264	0.4582	0.4861	0.5333	0.5889	0.6559
25	0.24610	0.2937	0.3331	0.3671	0.3970	0.4237	0.4696	0.5251	0.5944
30	0.21201	0.2547	0.2907	0.3221	0.3500	0.3752	0.4192	0.4734	0.5429
40	0.16598	0.2013	0.2320	0.2584	0.2827	0.3050	0.3446	0.3950	0.4608
60	0.11566	0.14165	0.16440	0.18504	0.20400	0.2217	0.2538	0.2961	0.3550
80	0.08873	0.10925	0.12745	0.14404	0.15950	0.1740	0.2008	0.2382	0.2880
100	0.07199	0.08890	0.10403	0.11792	0.13091	0.1432	0.1660	0.1969	0.2421
130	0.05610	0.06951	0.08154	0.09268	0.10316	0.1131	0.1318	0.1574	0.1954
160	0.04594	0.05704	0.06705	0.07633	0.08510	0.09347	0.1092	0.1310	0.1637
200	0.03703	0.04604	0.05420	0.06182	0.06901	0.07591	0.08889	0.1079	0.1398
300	0.02491	0.03105	0.03665	0.04187	0.04684	0.05161	0.06072	0.07349	0.09332
500	0.01504	0.01882	0.02224	0.02546	0.02852	0.03149	0.03715	0.04517	0.05770
1000	0.00758	0.00947	0.01122	0.01286	0.01442	0.01594	0.01885	0.02741	0.02955

TABLE VII (*continued*).

$\alpha = 0.01 \qquad s = 4 \qquad \theta^{\alpha}(s, m, n)$

n \ m	0	1	2	3	4
5	0.8110	0.8436	0.8662	0.8830	0.8959
10	0.6247	0.6708	0.7057	0.7334	0.7560
15	0.5016	0.5490	0.5867	0.6177	0.6439
20	0.4175	0.4627	0.4997	0.5309	0.5579
25	0.3570	0.3992	0.4343	0.4645	0.4910
30	0.3116	0.3507	0.3837	0.4125	0.4380
40	0.2483	0.2819	0.3108	0.3364	0.35960
60	0.1763	0.2021	0.22486	0.24541	0.26429
80	0.1367	0.1575	0.17603	0.19300	0.20872
100	0.1115	0.12900	0.14459	0.15899	0.17241
130	0.0874	0.10139	0.11402	0.12572	0.13670
160	0.0719	0.08353	0.09411	0.10396	0.11323
200	0.0581	0.06764	0.07634	0.08444	0.09213
300	0.03928	0.04583	0.05185	0.05749	0.06285
500	0.02383	0.02787	0.03159	0.03508	0.03842
1000	0.01202	0.01407	0.01547	0.01777	0.01948

$\alpha = 0.05 \qquad s = 4 \qquad \theta^{\alpha}(s, m, n)$

n \ m	0	1	2	3	4
5	0.7387	0.7825	0.8131	0.8359	0.8537
10	0.5471	0.6004	0.6411	0.6736	0.7004
15	0.4307	0.4822	0.5235	0.5577	0.5869
20	0.3543	0.4017	0.4408	0.4741	0.5031
25	0.3006	0.3438	0.3802	0.4117	0.4395
30	0.2609	0.3004	0.3340	0.3635	0.3899
40	0.2063	0.2396	0.2685	0.2943	0.31767
60	0.1454	0.1704	0.19264	0.21284	0.23148
80	0.1122	0.1322	0.15014	0.16661	0.18196
100	0.0913	0.10795	0.12298	0.13686	0.14986
130	0.0714	0.08465	0.09672	0.10793	0.11849
160	0.0586	0.06963	0.07969	0.08909	0.09797
200	0.0473	0.05631	0.06454	0.07227	0.07959
300	0.03192	0.03808	0.04375	0.04909	0.05417
500	0.01934	0.02312	0.02661	0.02990	0.03306
1000	0.00974	0.01166	0.01344	0.01512	0.01674

TABLE VII (*continued*).

$\alpha = 0.01 \qquad s = 5 \qquad \theta^{\alpha}(s, m, n)$

n \ m	0	1	2	3	4
5	0.8478	0.8719	0.8892	0.9023	0.9126
10	0.6762	0.7136	0.7426	0.7659	0.7850
15	0.5544	0.5948	0.6274	0.6546	0.6777
20	0.4677	0.5074	0.5404	0,5685	0.5929
25	0.4038	0.4416	0.4735	0.5012	0.5255
30	0.3549	0.3904	0.4208	0.4475	0.4713
40	0.2854	0.3166	0.3438	0.3681	0.3963
60	0.2048	0.2293	0.2512	0.2710	0.2894
80	0.1597	0.1796	0.1977	0.2143	0.2296
100	0.1308	0.1476	0.1629	0.1771	0.1902
130	0.10284	0.11645	0.1289	0.1405	0.1514
160	0.08474	0.09615	0.10663	0.11641	0.12564
200	0.06862	0.07801	0.08665	0.09475	0.10244
300	0.04651	0.05300	0.05900	0.06466	0.07006
500	0.02828	0.03229	0.03602	0.03955	0.04292
1000	0.01429	0.01633	0.01825	0.02006	0.02180

$\alpha = 0.05 \qquad s = 5 \qquad \theta^{\alpha}(s, m, n)$

n \ m	0	1	2	3	4
5	0.7882	0.8210	0.8447	0.8626	0.8768
10	0.6069	0.6507	0.6848	0.7125	0.7354
15	0.4882	0.5327	0.5689	0.5993	0.6252
20	0.4072	0.4494	0.4847	0.5150	0.5413
25	0.3488	0.3881	0.4215	0.4506	0.4764
30	0.3049	0.3412	0.3726	0.4002	0.4250
40	0.2433	0.2746	0.3021	0.3267	0.3490
60	0.1732	0.1973	0.2188	0.2385	0.2567
80	0.1344	0.1538	0.1714	0.1877	0.2028
100	0.1098	0.1261	0.1409	0.1547	0.1676
130	0.08612	0.09918	0.1112	0.1224	0.1329
160	0.07084	0.08175	0.09179	0.10120	0.11012
200	0.05729	0.06622	0.07448	0.08224	0.08962
300	0.03875	0.04490	0.05061	0.05600	0.06115
500	0.02353	0.02731	0.03084	0.03418	0.03739
1000	0.01187	0.01380	0.01560	0.01732	0.01897

TABLE VII (*continued*).

$\alpha = 0.01 \qquad s = 6 \qquad \theta^{\alpha}(s, m, n)$

n \ m	0	1	2	3	4
5	0.8745	0.8929	0.9065	0.9169	0.9255
10	0.7173	0.7482	0.7724	0.7922	0.8086
15	0.5986	0.6334	0.6619	0.6858	0.7063
20	0.5111	0.5462	0.5757	0.6010	0.6231
25	0.4450	0.4790	0.5081	0.5335	0.5559
30	0.3936	0.4261	0.4542	0.4789	0.5011
40	0.3194	0.3484	0.3739	0.3969	0.4177
60	0.2315	0.2548	0.2757	0.2948	0.3125
80	0.1814	0.2006	0.2181	0.2342	0.2493
100	0.1491	0.1654	0.1803	0.1942	0.2072
130	0.11762	0.13091	0.14314	0.15457	0.16536
160	0.09713	0.10830	0.11901	0.12834	0.13754
200	0.07880	0.08803	0.09659	0.10466	0.11232
300	0.05355	0.05996	0.06594	0.07160	0.07701
500	0.03270	0.03661	0.04034	0.04388	0.04727
1000	0.01651	0.01855	0.02046	0.02229	0.02405

$\alpha = 0.05 \qquad s = 6 \qquad \theta^{\alpha}(s, m, n)$

n \ m	0	1	2	3	4
5	0.8246	0.8499	0.8685	0.8830	0.8945
10	0.6552	0.6917	0.7206	0.7442	0.7639
15	0.5371	0.5758	0.6077	0.6346	0.6577
20	0.4535	0.4912	0.5231	0.5505	0.5746
25	0.3918	0.4276	0.4583	0.4852	0.5091
30	0.3447	0.3782	0.4074	0.4332	0.4564
40	0.2775	0.3069	0.3329	0.3563	0.3776
60	0.1995	0.2225	0.2433	0.2624	0.2801
80	0.1556	0.1745	0.1916	0.2075	0.2224
100	0.1275	0.1434	0.1580	0.1716	0.1843
130	0.10036	0.11319	0.12504	0.13615	0.14666
160	0.08272	0.09348	0.10388	0.11284	0.12175
200	0.06702	0.07586	0.08409	0.09186	0.09926
300	0.04545	0.05156	0.05728	0.06281	0.06790
500	0.02765	0.03143	0.03498	0.03835	0.04160
1000	0.01397	0.01590	0.01772	0.01946	0.02113

TABLE VIII. Heck's Charts for Roy's Largest-Root Criterion

CHART I

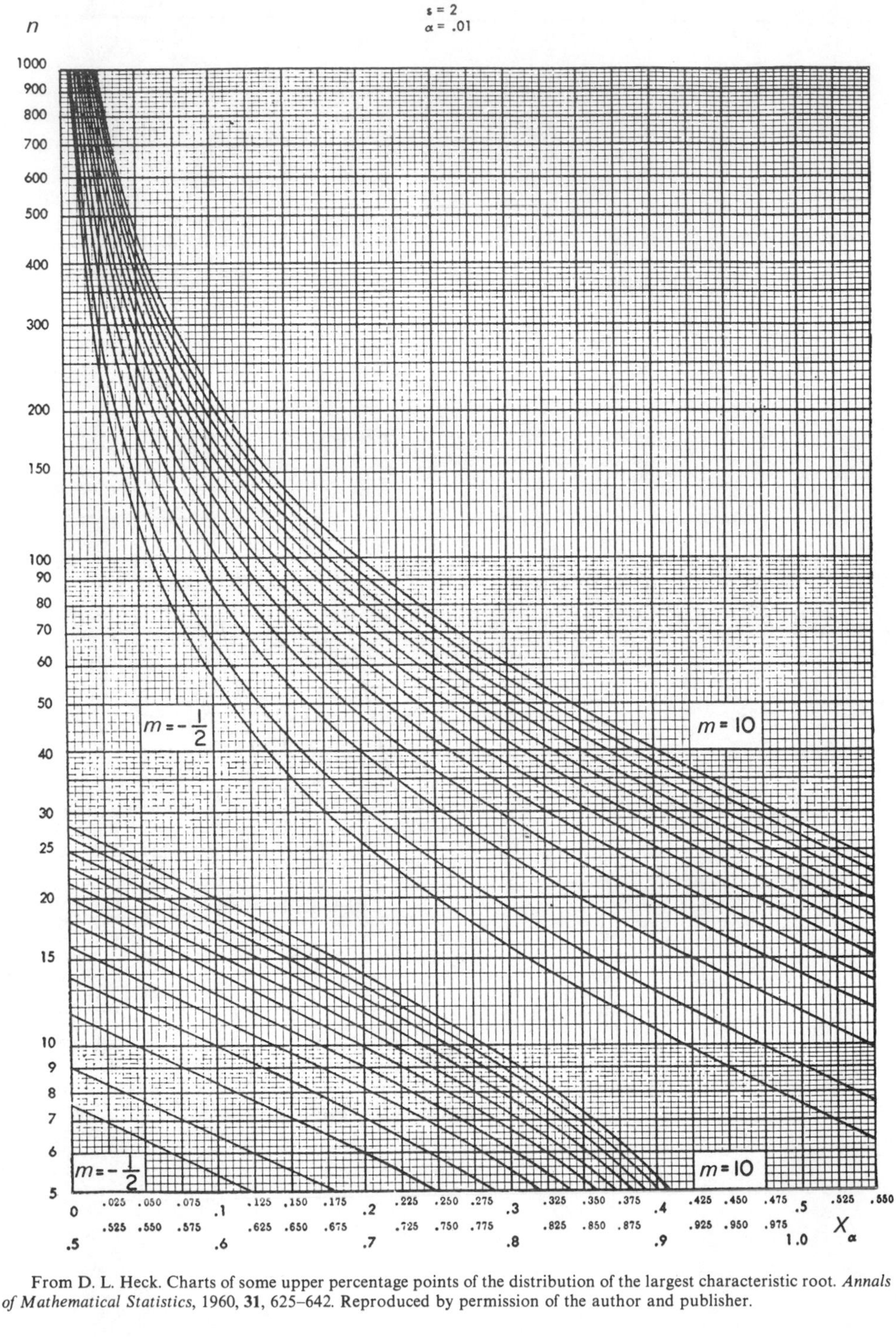

From D. L. Heck. Charts of some upper percentage points of the distribution of the largest characteristic root. *Annals of Mathematical Statistics*, 1960, **31**, 625–642. Reproduced by permission of the author and publisher.

TABLE VIII (*continued*).

CHART II

s = 2

α = .025

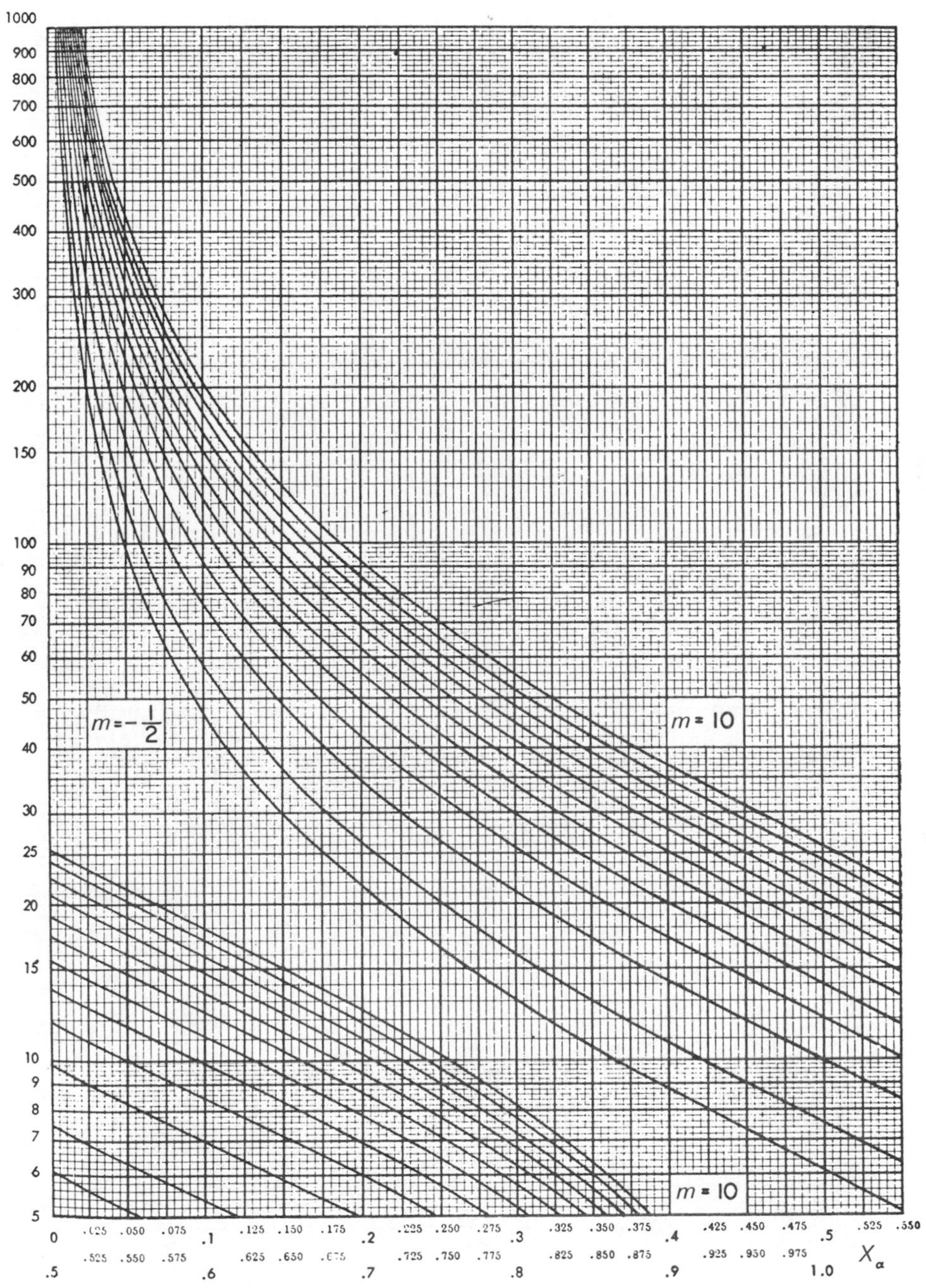

TABLE VIII (*continued*).

CHART III

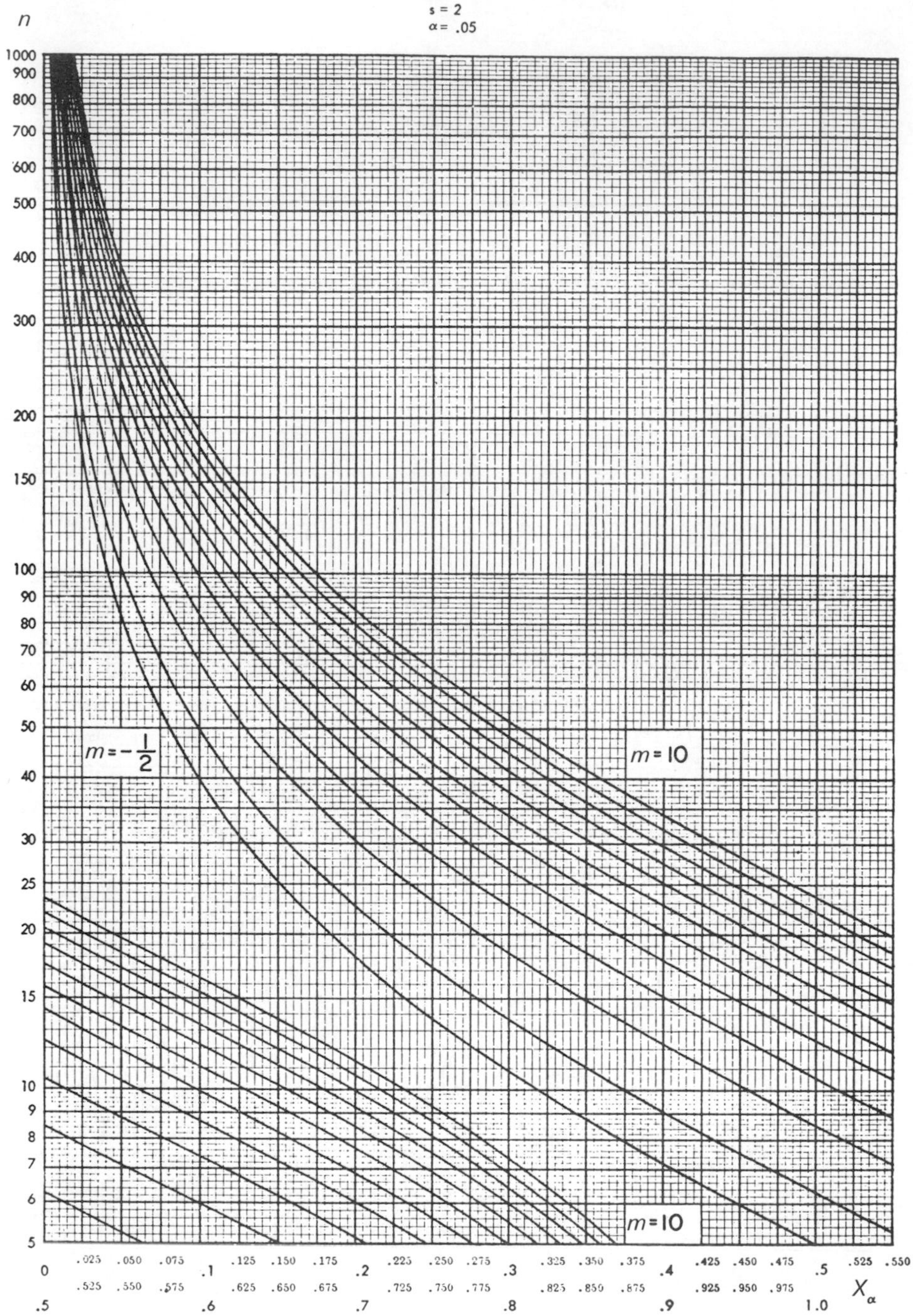

TABLE VIII (*continued*).

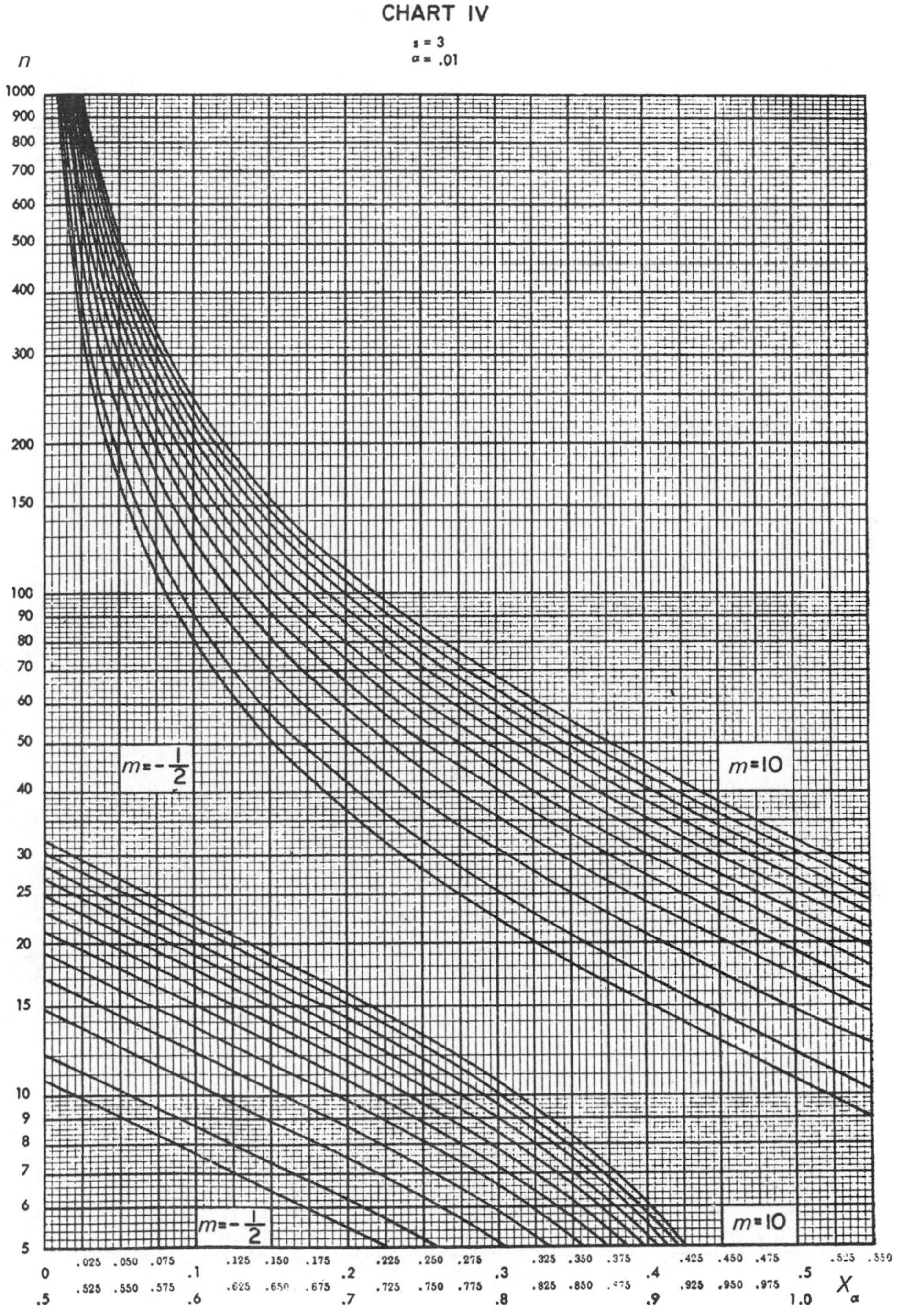

TABLE VIII (*continued*).

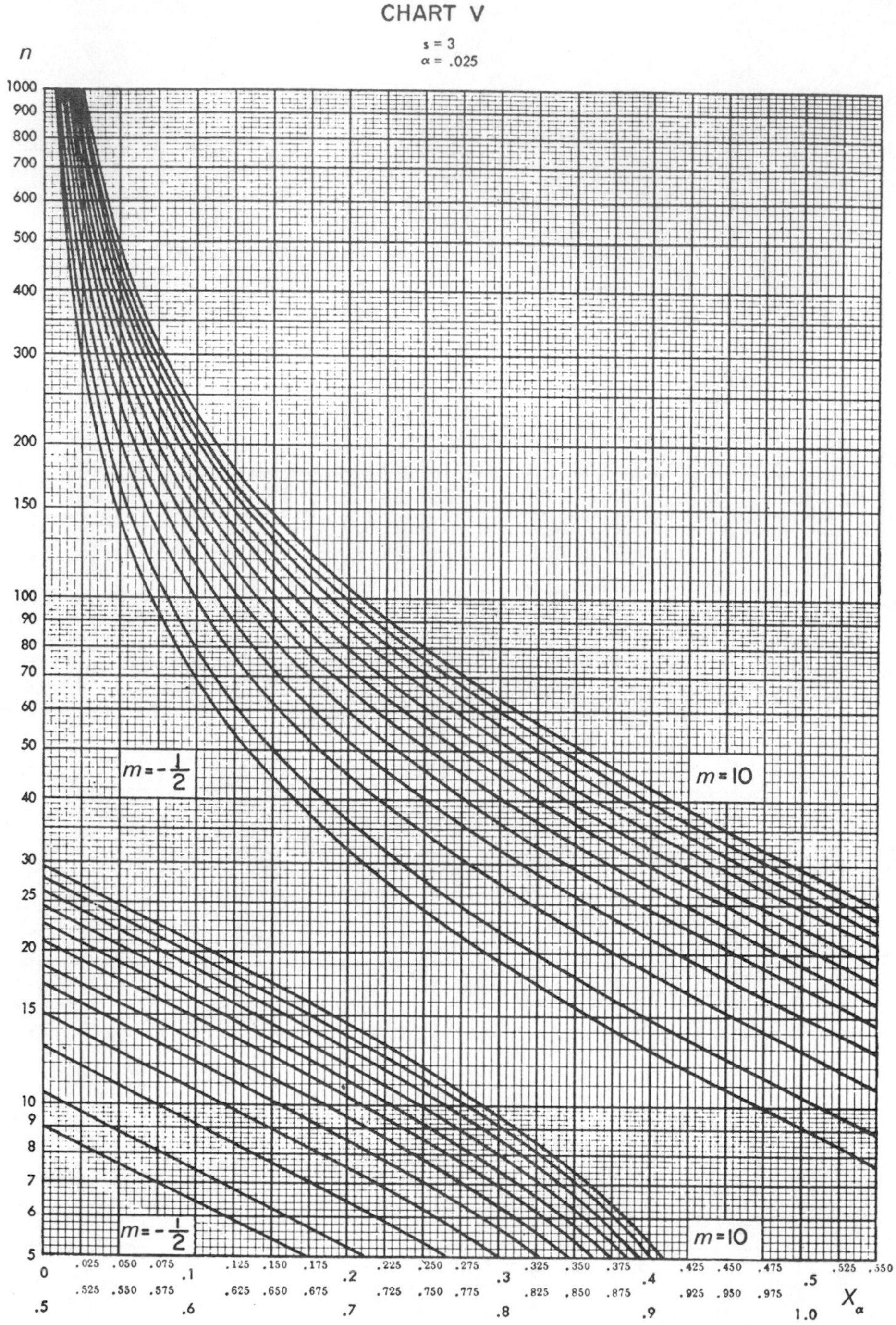

TABLE VIII (*continued*).

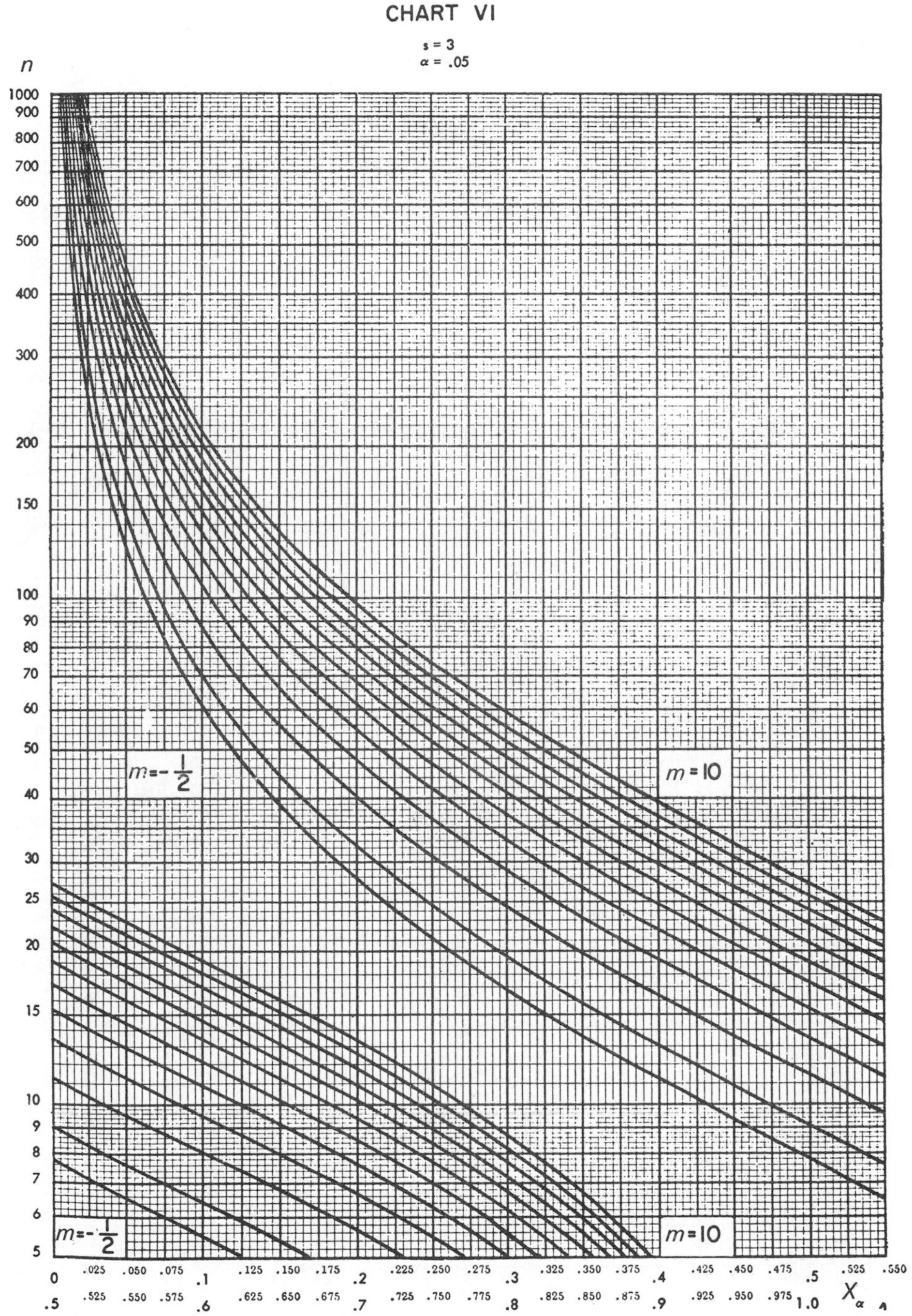

TABLE VIII (*continued*).

CHART VII

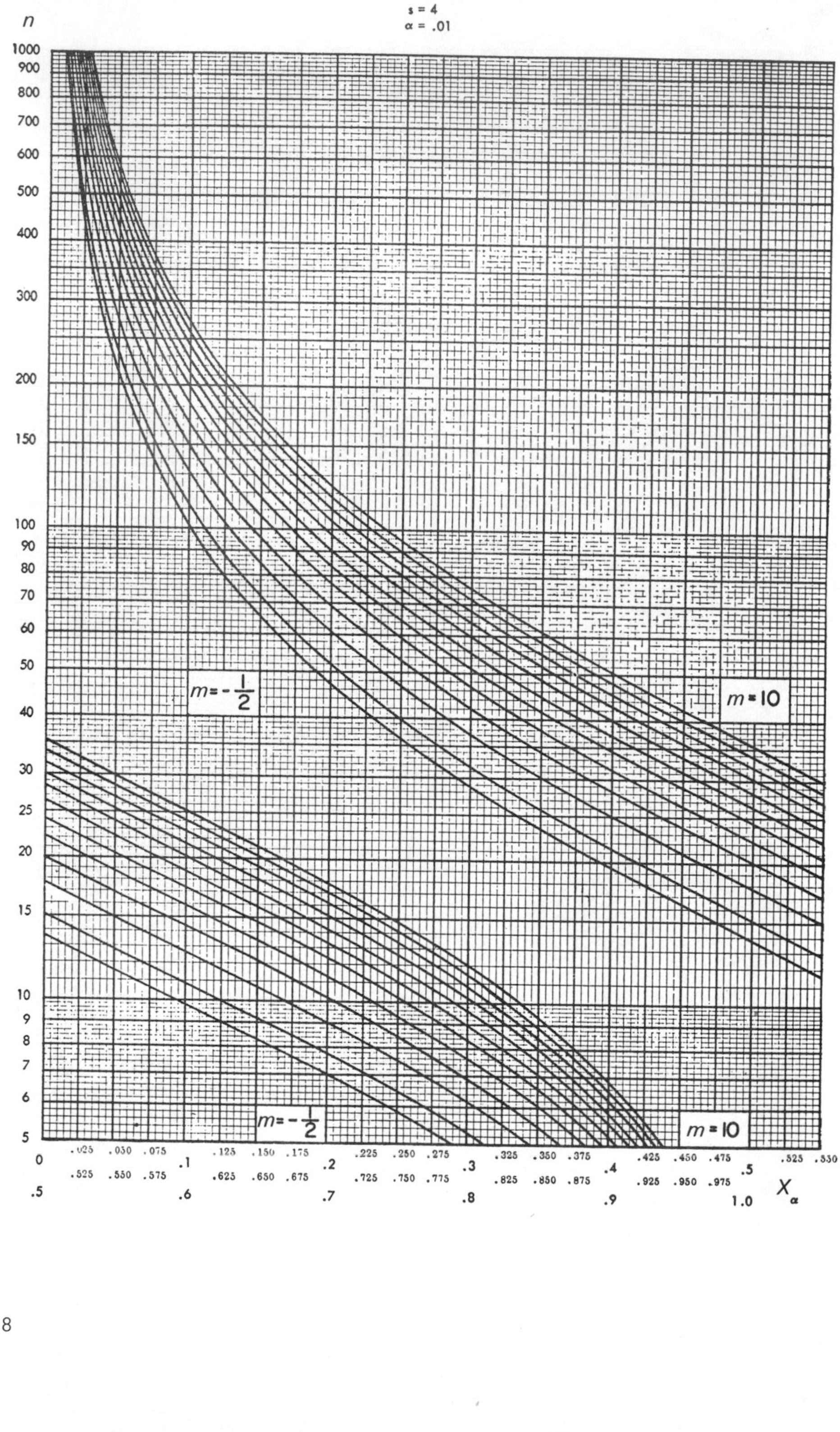

TABLE VIII (*continued*).

CHART VIII

s = 4
α = .025

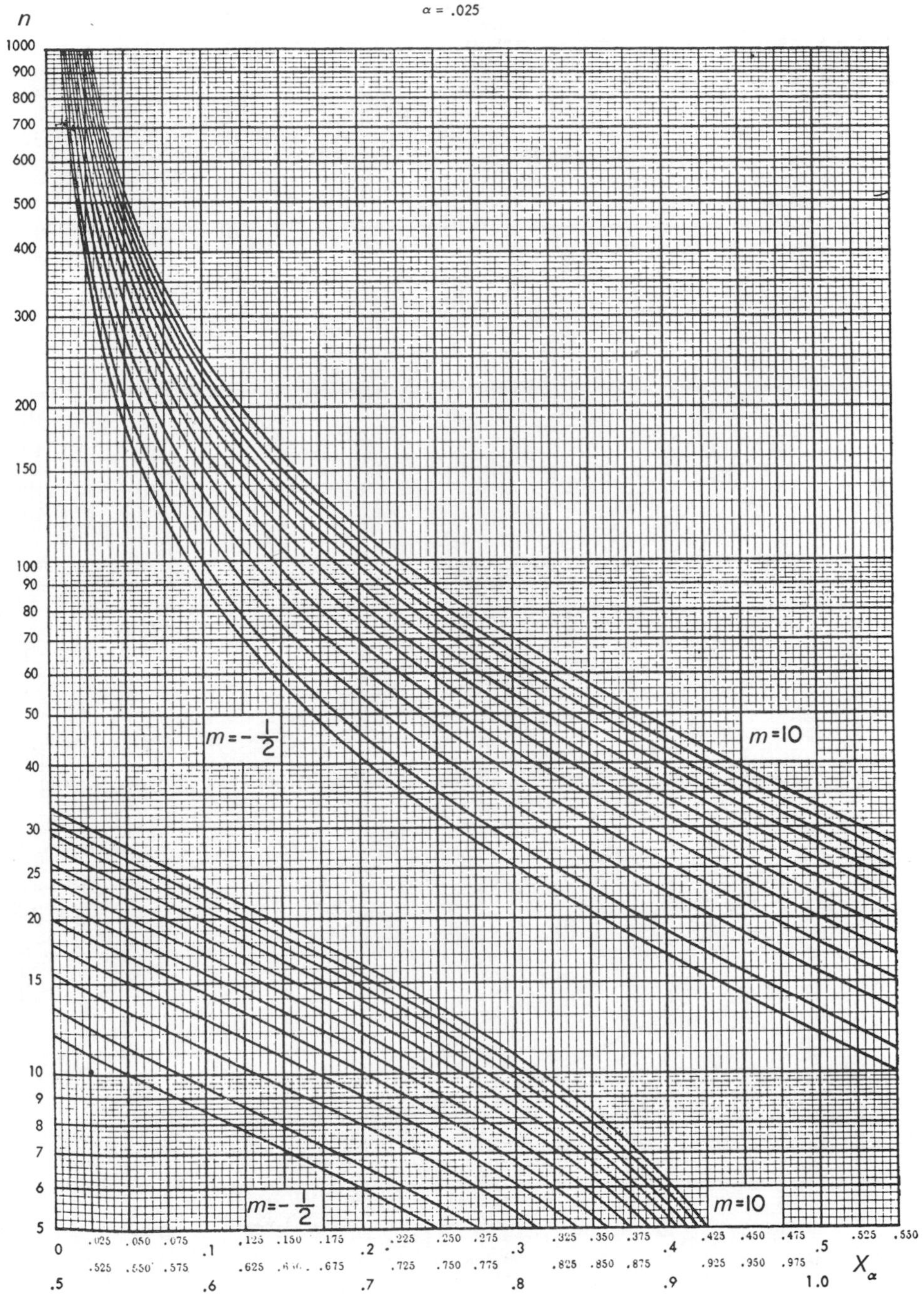

TABLE VIII (*continued*).

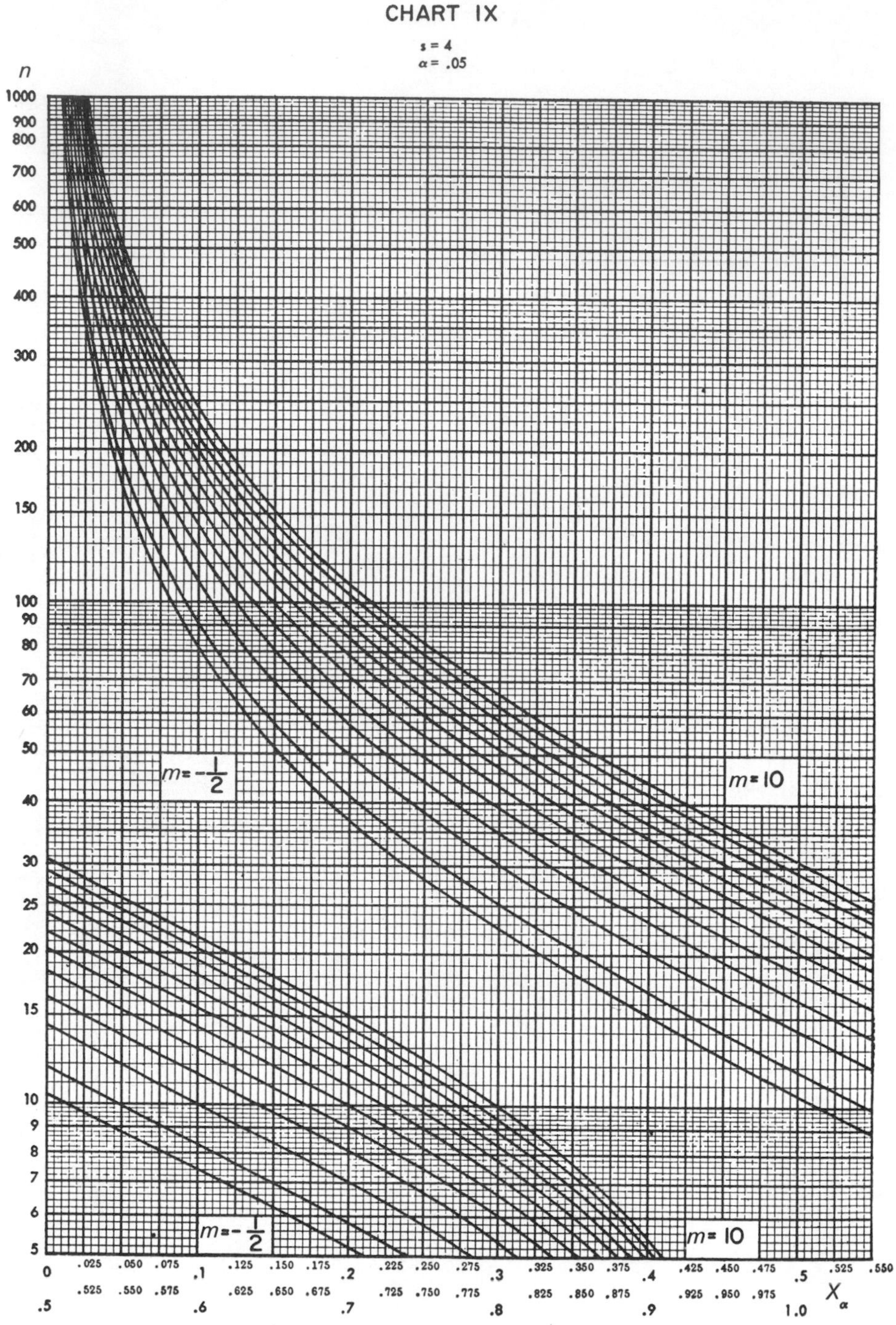

TABLE VIII (*continued*).

CHART X

s = 5
α = .01

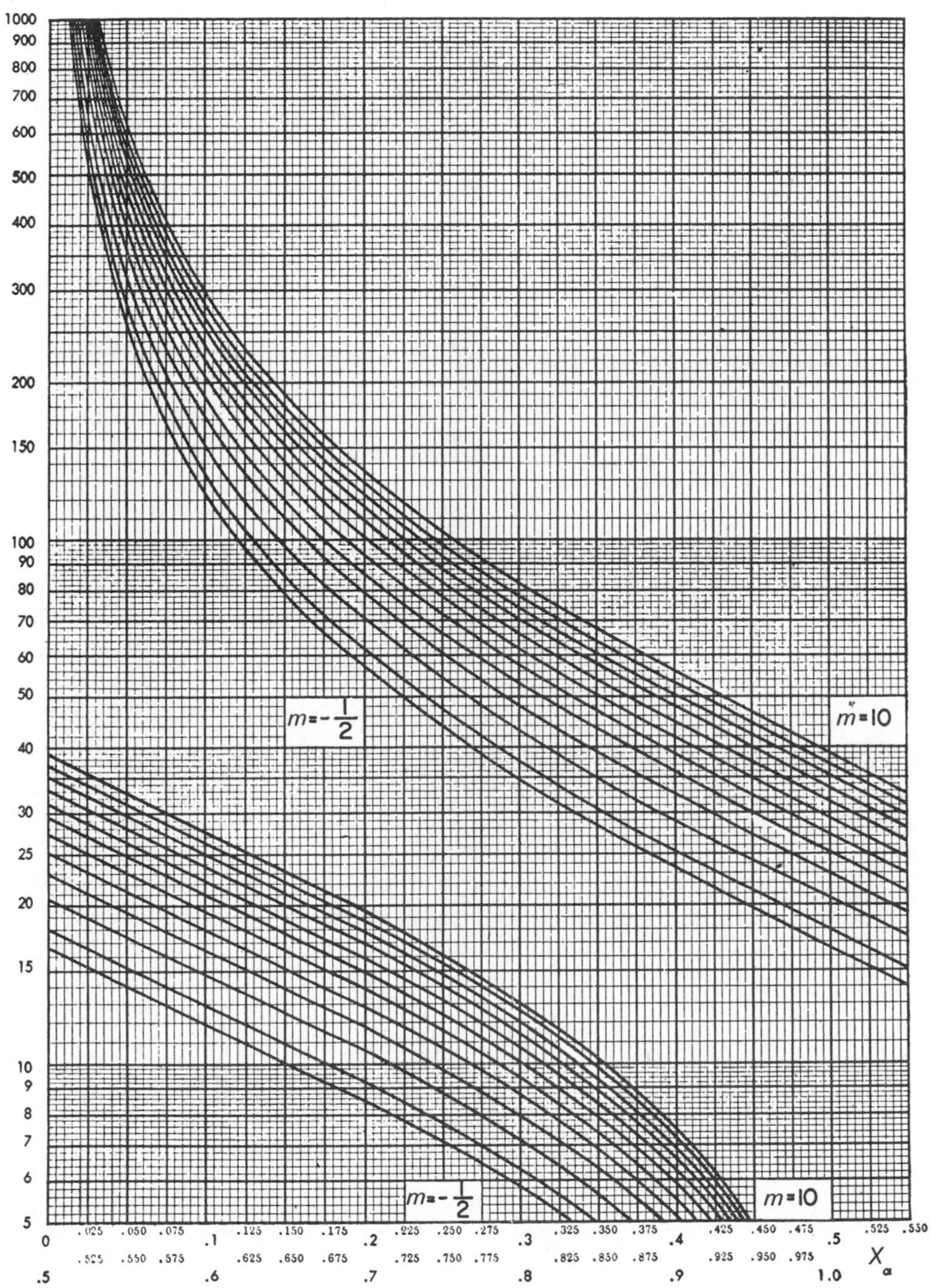

TABLE VIII (*continued*).

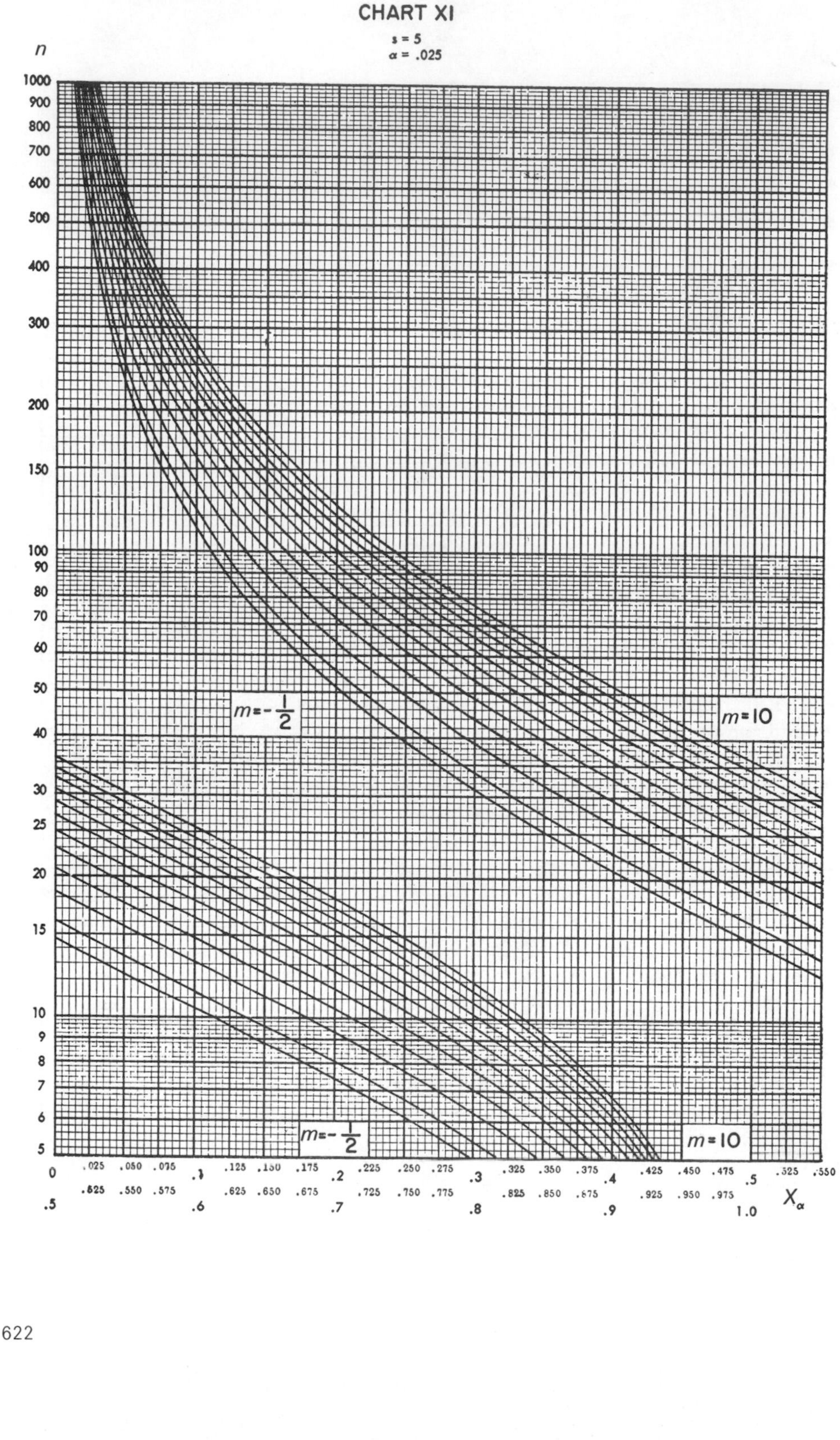

TABLE VIII (*continued*).

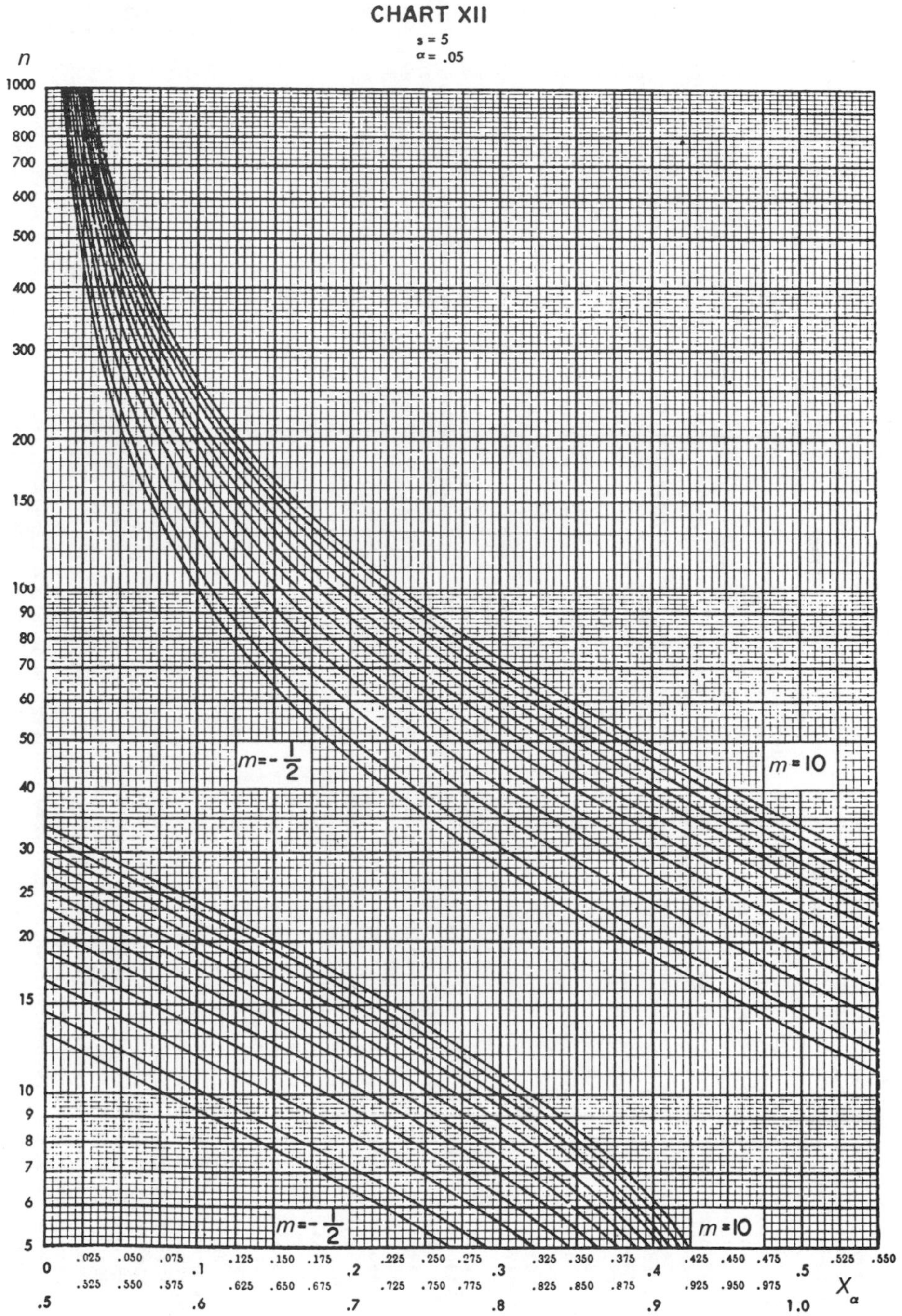

TABLE IX. Lower Percentage Points of Wilks' Lambda Criterion, $U^{\alpha}(p, q, n)$

$p = 1$ $\alpha = 0.01$ $U^{\alpha}(p, q, n)$ hypothesis df: q

n	1	2	3	4	5	6	7	8	9	10	11	12	n
1	0.006157	0.002501	0.001543	0.001112	0.000868	0.000712	0.000603	0.000523	0.000462	0.000413	0.000374	0.000341	1
2	0.097504	0.050003	0.033615	0.025322	0.020309	0.016953	0.014549	0.012741	0.011333	0.010208	0.009281	0.008512	2
3	0.228516	0.135712	0.097321	0.076019	0.062408	0.052963	0.046005	0.040672	0.036446	0.033020	0.030182	0.027794	3
4	0.341614	0.223602	0.168243	0.135345	0.113373	0.097610	0.085724	0.076446	0.068985	0.062851	0.057724	0.053375	4
5	0.430725	0.301697	0.235535	0.194031	0.165283	0.144073	0.127777	0.114822	0.104279	0.095505	0.088120	0.081787	5
6	0.500549	0.368408	0.295990	0.248596	0.214783	0.189255	0.169266	0.153168	0.139893	0.128754	0.119278	0.111114	6
7	0.555908	0.424896	0.349304	0.298096	0.260620	0.231811	0.208893	0.190186	0.174606	0.161423	0.150116	0.140289	7
8	0.600708	0.472870	0.396057	0.342590	0.302612	0.271332	0.246124	0.225311	0.207825	0.192902	0.180008	0.168747	8
9	0.637512	0.513916	0.437164	0.382446	0.340790	0.307770	0.280823	0.258362	0.239288	0.222931	0.208679	0.196182	9
10	0.668243	0.549286	0.473389	0.418213	0.375519	0.341248	0.313019	0.289246	0.268936	0.251373	0.235992	0.222443	10
11	0.694275	0.580017	0.505463	0.450317	0.407104	0.372040	0.342834	0.318054	0.296768	0.278229	0.261932	0.247467	11
12	0.716553	0.606964	0.534027	0.479309	0.435913	0.400299	0.370453	0.344940	0.322876	0.303528	0.286469	0.271240	12
13	0.735840	0.630737	0.559570	0.505524	0.462189	0.426361	0.396057	0.369995	0.347321	0.327362	0.309662	0.293823	13
14	0.752686	0.651825	0.582581	0.529327	0.486267	0.450348	0.419800	0.393372	0.370239	0.349823	0.331589	0.315247	14
15	0.767548	0.670715	0.603333	0.551025	0.508362	0.472534	0.441864	0.415222	0.391754	0.370941	0.352325	0.335541	15
16	0.780701	0.687653	0.622162	0.570862	0.528717	0.493103	0.462433	0.435638	0.411957	0.390869	0.371918	0.354797	16
17	0.792480	0.702972	0.639343	0.589081	0.547516	0.512176	0.481598	0.454742	0.430939	0.409637	0.390472	0.373077	17
18	0.803070	0.716858	0.655029	0.605835	0.564911	0.529907	0.499481	0.472687	0.448807	0.427368	0.408020	0.390411	18
19	0.812622	0.729553	0.669434	0.621307	0.581024	0.546448	0.516235	0.489502	0.465637	0.444138	0.424652	0.406891	19
20	0.821320	0.741135	0.682709	0.635651	0.596039	0.561890	0.531952	0.505341	0.481506	0.459991	0.440430	0.422546	20
21	0.829224	0.751770	0.694977	0.648941	0.610046	0.576355	0.546692	0.520264	0.496521	0.475006	0.455414	0.437469	21
22	0.836472	0.761597	0.706329	0.661316	0.623108	0.589905	0.560562	0.534332	0.510712	0.489258	0.469635	0.451660	22
23	0.843140	0.770660	0.716858	0.672867	0.635361	0.602631	0.573639	0.547638	0.524139	0.502762	0.483185	0.465179	23
24	0.849274	0.779083	0.726685	0.683655	0.646851	0.614609	0.585968	0.560211	0.536896	0.515594	0.496078	0.478088	24
25	0.854950	0.786896	0.735870	0.693771	0.657639	0.625900	0.597626	0.572128	0.548981	0.527817	0.508362	0.490402	25
26	0.860199	0.794189	0.744446	0.703278	0.667786	0.636566	0.608643	0.583435	0.560486	0.539459	0.520081	0.502167	26
27	0.865112	0.800995	0.752487	0.712189	0.677383	0.646637	0.619080	0.594147	0.571411	0.550537	0.531281	0.513428	27
28	0.869675	0.807373	0.760040	0.720612	0.686432	0.656174	0.628998	0.604370	0.581833	0.561127	0.541962	0.524200	28
29	0.873947	0.813339	0.767151	0.728546	0.694992	0.665222	0.638428	0.614075	0.591766	0.571228	0.552200	0.534515	29
30	0.877945	0.818970	0.773865	0.736053	0.703110	0.673798	0.647385	0.623321	0.601242	0.580872	0.561996	0.544418	30
40	0.907349	0.860886	0.824463	0.793274	0.765594	0.740539	0.717575	0.696365	0.676636	0.658188	0.640884	0.624603	40
60	0.937485	0.904968	0.878807	0.855911	0.835175	0.816055	0.798233	0.781494	0.765686	0.750702	0.736420	0.722809	60
80	0.952827	0.927841	0.907471	0.889450	0.872940	0.857590	0.843124	0.829437	0.816391	0.803925	0.791962	0.780464	80
100	0.962128	0.941845	0.925179	0.910324	0.896637	0.883835	0.871696	0.860153	0.849083	0.838455	0.828201	0.818314	100
120	0.968363	0.951297	0.937199	0.924578	0.912894	0.901916	0.891475	0.881501	0.871901	0.862660	0.853706	0.845045	120
140	0.972836	0.958107	0.945890	0.934921	0.924731	0.915131	0.905971	0.897200	0.888734	0.880563	0.872625	0.864929	140
170	0.977588	0.965370	0.955195	0.946025	0.937478	0.929401	0.921669	0.914245	0.907057	0.900101	0.893324	0.886738	170
200	0.980926	0.970487	0.961767	0.953893	0.946532	0.939564	0.932877	0.926443	0.920200	0.914149	0.908239	0.902486	200
240	0.984085	0.975345	0.968024	0.961396	0.955187	0.949296	0.943631	0.938171	0.932861	0.927705	0.922660	0.917740	240
320	0.988046	0.981451	0.975907	0.970876	0.966145	0.961649	0.957311	0.953121	0.949035	0.945058	0.941155	0.937344	320
440	0.991295	0.986475	0.982411	0.978715	0.975232	0.971914	0.968704	0.965599	0.962561	0.959604	0.956692	0.953846	440
600	0.993610	0.990064	0.987067	0.984337	0.981759	0.979301	0.976917	0.974611	0.972349	0.970144	0.967969	0.965842	600
800	0.995204	0.992539	0.990282	0.988225	0.986279	0.984422	0.982619	0.980873	0.979158	0.977487	0.975834	0.974218	800
1000	0.996161	0.994026	0.992216	0.990566	0.989003	0.987512	0.986062	0.984658	0.983276	0.981931	0.980598	0.979296	1000
INF	1.000000	1.000000	1.000000	1.000000	1.000000	1.000000	1.000000	1.000000	1.000000	1.000000	1.000000	1.000000	INF

Abridged from Francis J. Wall, Ph.D., *The Generalized Variance Ratio of the U-Statistic*, The Dikewood Corporation, 1968. Reproduced by permission of the author and the Dikewood Corporation.

TABLE IX (*continued*)..

$p = 1$ $\alpha = 0.05$ $U^{\alpha}(p, q, n)$ hypothesis df: q

n	1	2	3	4	5	6	7	8	9	10	11	12	*n*
1	0.000247	0.000100	0.000062	0.000044	0.000035	0.000028	0.000024	0.000021	0.000018	0.000017	0.000015	0.000014	1
2	0.019900	0.010000	0.006678	0.005013	0.004012	0.003344	0.002867	0.002509	0.002231	0.002008	0.001826	0.001674	2
3	0.080827	0.046416	0.032834	0.025458	0.020806	0.017599	0.015251	0.013458	0.012043	0.010898	0.009951	0.009157	3
4	0.158742	0.100000	0.073958	0.058903	0.049014	0.041999	0.036755	0.032682	0.029427	0.026763	0.024544	0.022665	4
5	0.235203	0.158489	0.121418	0.098877	0.083563	0.072430	0.063947	0.057265	0.051854	0.047390	0.043634	0.040434	5
6	0.303867	0.215443	0.169784	0.140867	0.120651	0.105640	0.094010	0.084728	0.077134	0.070801	0.065439	0.060839	6
7	0.363705	0.268270	0.216358	0.182355	0.158006	0.139585	0.125112	0.113416	0.103749	0.095628	0.088696	0.082716	7
8	0.415397	0.316227	0.259967	0.222073	0.194363	0.173070	0.156116	0.142270	0.130724	0.120944	0.112552	0.105261	8
9	0.460089	0.359381	0.300242	0.259453	0.229097	0.205430	0.186374	0.170658	0.157452	0.146189	0.136452	0.127957	9
10	0.498896	0.398108	0.337189	0.294313	0.261901	0.236323	0.215512	0.198202	0.183548	0.170965	0.160030	0.150442	10
11	0.532793	0.432877	0.370993	0.326670	0.292708	0.265602	0.243349	0.224692	0.208793	0.195061	0.183068	0.172501	11
12	0.562582	0.464159	0.401904	0.356635	0.321526	0.293230	0.269804	0.250027	0.233063	0.218338	0.205413	0.193976	12
13	0.588936	0.492388	0.430204	0.384373	0.348450	0.319237	0.294872	0.274166	0.256310	0.240729	0.226996	0.214791	13
14	0.612381	0.517948	0.456147	0.410058	0.373579	0.343685	0.318575	0.297116	0.278506	0.262202	0.247764	0.234893	14
15	0.633365	0.541170	0.479986	0.433866	0.397056	0.366662	0.340981	0.318908	0.299682	0.282756	0.267719	0.254259	15
16	0.652233	0.562342	0.501931	0.455967	0.418982	0.388257	0.362133	0.339583	0.319844	0.302404	0.286850	0.272887	16
17	0.669300	0.581709	0.522195	0.476513	0.439506	0.408565	0.382133	0.359198	0.339049	0.321175	0.305186	0.290785	17
18	0.684789	0.599484	0.540936	0.495647	0.458725	0.427677	0.401023	0.377807	0.357322	0.339101	0.322737	0.307970	18
19	0.698917	0.615848	0.558319	0.513499	0.476742	0.445681	0.418900	0.395470	0.374735	0.356217	0.339555	0.324463	19
20	0.711843	0.630958	0.574470	0.530184	0.493661	0.462657	0.435811	0.412241	0.391308	0.372565	0.355644	0.340290	20
21	0.723730	0.644947	0.589523	0.545805	0.509577	0.478683	0.451836	0.428178	0.407108	0.388183	0.371060	0.355477	21
22	0.734669	0.657933	0.603568	0.560456	0.524563	0.493830	0.467022	0.443332	0.422166	0.403108	0.385819	0.370054	22
23	0.744795	0.670019	0.616713	0.574221	0.538693	0.508161	0.481441	0.457752	0.436534	0.417377	0.399965	0.384048	23
24	0.754176	0.681292	0.629026	0.587173	0.552034	0.521736	0.495132	0.471485	0.450247	0.431029	0.413515	0.397487	24
25	0.762902	0.691831	0.640594	0.599381	0.564657	0.534611	0.508160	0.484576	0.463349	0.444097	0.426523	0.410397	25
26	0.771028	0.701704	0.651468	0.610905	0.576603	0.546834	0.520546	0.497064	0.475866	0.456613	0.438989	0.422804	26
27	0.778625	0.710971	0.661723	0.621798	0.587931	0.558452	0.532362	0.508986	0.487854	0.468607	0.450973	0.434734	27
28	0.785730	0.719685	0.671391	0.632109	0.598682	0.569507	0.543615	0.520379	0.499314	0.480110	0.462471	0.446211	28
29	0.792406	0.727896	0.680539	0.641884	0.608900	0.580037	0.554370	0.531274	0.510313	0.491149	0.473532	0.457257	29
30	0.798670	0.735642	0.689191	0.651161	0.618619	0.590076	0.564636	0.541702	0.520842	0.501748	0.484160	0.467894	30
40	0.845412	0.794328	0.755603	0.723155	0.694813	0.669500	0.646550	0.625549	0.606163	0.588188	0.571417	0.555726	40
60	0.894480	0.857696	0.828970	0.804330	0.782305	0.762272	0.743738	0.726513	0.710318	0.695108	0.680672	0.667012	60
80	0.919918	0.891251	0.868522	0.848784	0.830928	0.814526	0.799185	0.784809	0.771162	0.758235	0.745861	0.734069	80
100	0.935478	0.912011	0.893219	0.876803	0.861820	0.847989	0.834952	0.822679	0.810943	0.799788	0.789035	0.778749	100
120	0.945976	0.926119	0.910119	0.896070	0.883183	0.871238	0.859925	0.849237	0.838971	0.829183	0.819700	0.810616	120
140	0.953532	0.936329	0.922402	0.910129	0.898827	0.888325	0.878338	0.868886	0.859771	0.851065	0.842600	0.834476	140
170	0.961595	0.947263	0.935601	0.925292	0.915755	0.906869	0.898385	0.890335	0.882539	0.875082	0.867790	0.860800	170
200	0.967270	0.954993	0.944964	0.936079	0.927834	0.920134	0.912760	0.905757	0.898951	0.892434	0.886056	0.879907	200
240	0.972661	0.962351	0.953904	0.946399	0.939414	0.932883	0.926606	0.920640	0.914819	0.909247	0.903766	0.898492	240
320	0.979433	0.971628	0.965202	0.959483	0.954136	0.949127	0.944294	0.939692	0.935191	0.930865	0.926597	0.922489	320
440	0.985001	0.979285	0.974556	0.970342	0.966383	0.962678	0.959082	0.955661	0.952291	0.949067	0.945865	0.942783	440
600	0.988980	0.984767	0.981267	0.978151	0.975211	0.972459	0.969781	0.967231	0.964708	0.962302	0.959899	0.957590	600
800	0.991723	0.988553	0.985913	0.983561	0.981336	0.979256	0.977220	0.975291	0.973377	0.971545	0.969713	0.967956	800
1000	0.993372	0.990832	0.988711	0.986824	0.985033	0.983362	0.981722	0.980169	0.978621	0.977148	0.975664	0.974250	1000
INF	1.000000	1.000000	1.000000	1.000000	1.000000	1.000000	1.000000	1.000000	1.000000	1.000000	1.000000	1.000000	INF

TABLE IX (*continued*).

$p = 2$ $\alpha = 0.01$ $U^{\alpha}(p, q, n)$ hypothesis df: q

n	1	2	3	4	5	6	7	8	9	10	11	12	*n*
1	0.000000	0.000000	0.000000	0.000000	0.000000	0.000000	0.000000	0.000000	0.000000	0.000000	0.000000	0.000000	1
2	0.000100	0.000025	0.000011	0.000006	0.000004	0.000003	0.000002	0.000002	0.000001	0.000001	0.000000	0.000000	2
3	0.010000	0.003470	0.001764	0.001068	0.000716	0.000514	0.000386	0.000301	0.000241	0.000198	0.000165	0.000140	3
4	0.046416	0.019844	0.011160	0.007179	0.005013	0.003701	0.002846	0.002257	0.001834	0.001520	0.001280	0.001093	4
5	0.099999	0.049316	0.029953	0.020241	0.014627	0.011080	0.008688	0.006999	0.005760	0.004824	0.004099	0.003527	5
6	0.158490	0.086620	0.055849	0.039284	0.029229	0.022633	0.018059	0.014752	0.012283	0.010389	0.008903	0.007715	6
7	0.215443	0.127189	0.085984	0.062513	0.047671	0.037627	0.030485	0.025222	0.021222	0.018110	0.015639	0.013643	7
8	0.268270	0.168148	0.118119	0.088278	0.068750	0.055175	0.045317	0.037913	0.032206	0.027708	0.024097	0.021153	8
9	0.316228	0.207906	0.150743	0.115317	0.091448	0.074467	0.061901	0.052316	0.044821	0.038849	0.034006	0.030024	9
10	0.359382	0.245666	0.182908	0.142738	0.114989	0.094845	0.079687	0.067961	0.058684	0.051208	0.045092	0.040019	10
11	0.398107	0.281095	0.214051	0.169943	0.138805	0.115797	0.098225	0.084458	0.073448	0.064492	0.057100	0.050924	11
12	0.432876	0.314111	0.243868	0.196543	0.162496	0.136940	0.117163	0.101490	0.088832	0.078445	0.069807	0.062537	12
13	0.464159	0.344773	0.272209	0.222298	0.185786	0.157996	0.136231	0.118805	0.104603	0.092856	0.083018	0.074689	13
14	0.492388	0.373205	0.299027	0.247072	0.208495	0.178764	0.155227	0.136206	0.120576	0.107553	0.096574	0.087224	14
15	0.517947	0.399561	0.324338	0.270794	0.230506	0.199104	0.174003	0.153543	0.136603	0.122393	0.110341	0.100021	15
16	0.541170	0.424011	0.348190	0.293441	0.251751	0.218924	0.192450	0.170701	0.152570	0.137265	0.124211	0.112976	16
17	0.562341	0.446714	0.370654	0.315019	0.272197	0.238163	0.210492	0.187598	0.168387	0.152079	0.138096	0.126003	17
18	0.581709	0.467823	0.391807	0.335555	0.291830	0.256785	0.228078	0.204169	0.183988	0.166764	0.151924	0.139032	18
19	0.599484	0.487482	0.411733	0.355085	0.310657	0.274771	0.245174	0.220372	0.199322	0.181266	0.165639	0.152007	19
20	0.615848	0.505819	0.430515	0.373654	0.328695	0.292119	0.261761	0.236177	0.214352	0.195544	0.179196	0.164879	20
21	0.630957	0.522953	0.448231	0.391312	0.345965	0.308831	0.277829	0.251565	0.229052	0.209566	0.192561	0.177614	21
22	0.644947	0.538990	0.464956	0.408104	0.362497	0.324920	0.293378	0.266524	0.243403	0.223308	0.205707	0.190182	22
23	0.657933	0.554026	0.480762	0.424082	0.378320	0.340402	0.308412	0.281051	0.257394	0.236756	0.218614	0.202560	23
24	0.670019	0.568146	0.495715	0.439293	0.393467	0.355296	0.322939	0.295145	0.271020	0.249897	0.231267	0.214731	24
25	0.681292	0.581428	0.509875	0.453782	0.407970	0.369622	0.336971	0.308812	0.284279	0.262726	0.243657	0.226681	25
26	0.691831	0.593939	0.523299	0.467592	0.421860	0.383403	0.350522	0.322057	0.297172	0.275239	0.255776	0.238402	26
27	0.701704	0.605746	0.536039	0.480765	0.435169	0.396661	0.363607	0.334890	0.309703	0.287436	0.267621	0.249886	27
28	0.710971	0.616902	0.548144	0.493339	0.447927	0.409418	0.376242	0.347322	0.321877	0.299318	0.279190	0.261129	28
29	0.719686	0.627457	0.559655	0.505352	0.460162	0.421696	0.388443	0.359363	0.333703	0.310890	0.290483	0.272130	29
30	0.727896	0.637459	0.570615	0.516835	0.471903	0.433519	0.400227	0.371026	0.345187	0.322156	0.301504	0.282889	30
40	0.789652	0.714476	0.656673	0.608581	0.567185	0.530850	0.498541	0.469542	0.443323	0.419481	0.397694	0.377702	40
60	0.855467	0.799984	0.755573	0.717315	0.683328	0.652617	0.624558	0.598723	0.574795	0.552534	0.531746	0.512271	60
80	0.889953	0.846188	0.810436	0.779081	0.750765	0.724783	0.700697	0.678213	0.657114	0.637235	0.618444	0.600635	80
100	0.911163	0.875081	0.845239	0.818780	0.794644	0.772286	0.751373	0.731681	0.713048	0.695352	0.678495	0.662399	100
120	0.925522	0.894844	0.869263	0.846415	0.825431	0.805865	0.787452	0.770011	0.753414	0.737564	0.722385	0.707815	120
140	0.935886	0.909213	0.886840	0.866750	0.848208	0.830840	0.814421	0.798803	0.783877	0.769567	0.755808	0.742551	140
170	0.946959	0.924659	0.905836	0.888839	0.873070	0.858224	0.844122	0.830645	0.817710	0.805252	0.793223	0.781586	170
200	0.954772	0.935614	0.919375	0.904652	0.890941	0.877988	0.865642	0.853805	0.842407	0.831395	0.820731	0.810382	200
240	0.962196	0.946071	0.932346	0.919856	0.908184	0.897119	0.886539	0.876363	0.866534	0.857011	0.847760	0.838758	240
320	0.971540	0.959297	0.948819	0.939239	0.930247	0.921688	0.913469	0.905533	0.897838	0.890354	0.883057	0.875930	320
440	0.979238	0.970243	0.962512	0.955416	0.948732	0.942346	0.936194	0.930234	0.924437	0.918780	0.913248	0.907828	440
600	0.984741	0.978097	0.972369	0.967097	0.962118	0.957350	0.952745	0.948273	0.943913	0.939649	0.935470	0.931366	600
800	0.988539	0.983531	0.979204	0.975215	0.971440	0.967819	0.964316	0.960909	0.957581	0.954322	0.951123	0.947977	800
1000	0.990823	0.986804	0.983329	0.980120	0.977081	0.974162	0.971336	0.968584	0.965894	0.963257	0.960665	0.958115	1000
INF	1.000000	1.000000	1.000000	1.000000	1.000000	1.000000	1.000000	1.000000	1.000000	1.000000	1.000000	1.000000	INF

TABLE IX (*continued*).

$p = 2 \quad \alpha = 0.05 \quad U^{\alpha}(p, q, n) \quad$ hypothesis df: q

n	1	2	3	4	5	6	7	8	9	10	11	12	n
1	0.000000	0.000000	0.000000	0.000000	0.000000	0.000000	0.000000	0.000000	0.000000	0.000000	0.000000	0.000000	1
2	0.002500	0.000641	0.000287	0.000162	0.000104	0.000072	0.000053	0.000041	0.000032	0.000026	0.000022	0.000018	2
3	0.049998	0.018318	0.009528	0.005844	0.003950	0.002849	0.002152	0.001683	0.001352	0.001110	0.000928	0.000787	3
4	0.135725	0.061800	0.035817	0.023460	0.016578	0.012346	0.009555	0.007615	0.006212	0.005165	0.004362	0.003734	4
5	0.223605	0.117368	0.073621	0.050765	0.037211	0.028476	0.022507	0.018244	0.015092	0.012695	0.010826	0.009343	5
6	0.301715	0.174902	0.116450	0.083663	0.063188	0.049481	0.039834	0.032772	0.027440	0.023320	0.020068	0.017453	6
7	0.368405	0.229737	0.160239	0.118984	0.092129	0.073571	0.060172	0.050155	0.042465	0.036426	0.031600	0.027678	7
8	0.424876	0.280187	0.202813	0.154741	0.122376	0.099380	0.082397	0.069475	0.059404	0.051386	0.044908	0.039579	8
9	0.472866	0.325883	0.243151	0.189781	0.152779	0.125881	0.105643	0.089993	0.077615	0.067661	0.059515	0.052772	9
10	0.513885	0.367036	0.280802	0.223433	0.182643	0.152421	0.129282	0.111138	0.096610	0.084797	0.075044	0.066901	10
11	0.549280	0.404052	0.315720	0.255369	0.211592	0.178545	0.152898	0.132506	0.116013	0.102453	0.091177	0.081680	11
12	0.580029	0.437339	0.347988	0.285511	0.239373	0.203997	0.176155	0.153782	0.135511	0.120356	0.107656	0.096885	12
13	0.606971	0.467384	0.377744	0.313836	0.265838	0.228568	0.198874	0.174774	0.154909	0.138311	0.124284	0.112321	13
14	0.630737	0.494599	0.405216	0.340396	0.291016	0.252171	0.220930	0.195325	0.174061	0.156149	0.140923	0.127849	14
15	0.651851	0.519281	0.430564	0.365263	0.314363	0.274786	0.242249	0.215357	0.192837	0.173755	0.157442	0.143350	15
16	0.670711	0.541775	0.454003	0.388530	0.337412	0.296391	0.262763	0.234782	0.211185	0.191059	0.173755	0.158740	16
17	0.687662	0.562317	0.475724	0.410322	0.358763	0.316990	0.282502	0.253583	0.229036	0.208000	0.189807	0.173946	17
18	0.702981	0.581146	0.495888	0.430784	0.378964	0.336632	0.301430	0.271723	0.246366	0.224530	0.205530	0.188918	18
19	0.716866	0.598489	0.514629	0.449961	0.398041	0.355335	0.319573	0.289225	0.263169	0.240614	0.220915	0.203611	19
20	0.729531	0.614483	0.532092	0.467968	0.416109	0.373163	0.336951	0.306072	0.279429	0.256249	0.235936	0.218013	20
21	0.741124	0.629283	0.548399	0.484925	0.433211	0.390129	0.353609	0.322287	0.295146	0.271437	0.250565	0.232083	21
22	0.751776	0.643011	0.563622	0.500886	0.449429	0.406286	0.369555	0.337873	0.310325	0.286147	0.264800	0.245821	22
23	0.761598	0.655775	0.577893	0.515922	0.464800	0.421699	0.384810	0.352883	0.324978	0.300409	0.278639	0.259223	23
24	0.770680	0.667666	0.591286	0.530135	0.479373	0.436391	0.399429	0.367295	0.339116	0.314213	0.292087	0.272280	24
25	0.779088	0.678783	0.603884	0.543551	0.493227	0.450412	0.413436	0.381165	0.352775	0.327593	0.305127	0.285006	25
26	0.786893	0.689182	0.615752	0.556269	0.506409	0.463802	0.426867	0.394506	0.365946	0.340539	0.317798	0.297372	26
27	0.794192	0.698945	0.626937	0.568306	0.518951	0.476588	0.439744	0.407337	0.378645	0.353047	0.330095	0.309407	27
28	0.800992	0.708108	0.637517	0.579727	0.530891	0.488822	0.452093	0.419700	0.390911	0.365171	0.342019	0.321110	28
29	0.807354	0.716737	0.647497	0.590582	0.542291	0.500519	0.463948	0.431586	0.402753	0.376900	0.353591	0.332484	29
30	0.813343	0.724899	0.656961	0.600899	0.553155	0.511722	0.475325	0.443028	0.414182	0.388244	0.364802	0.343537	30
40	0.857594	0.786432	0.729818	0.681627	0.639419	0.601870	0.568076	0.537426	0.509476	0.483873	0.460296	0.438550	40
60	0.903437	0.852599	0.810662	0.773804	0.740586	0.710190	0.682157	0.656096	0.631804	0.609029	0.587643	0.567501	60
80	0.926967	0.887496	0.854347	0.824736	0.797636	0.772490	0.748974	0.726849	0.705927	0.686107	0.667279	0.649328	80
100	0.941272	0.909051	0.881684	0.856993	0.834186	0.812834	0.792697	0.773596	0.755405	0.738034	0.721395	0.705440	100
120	0.950898	0.923673	0.900382	0.879233	0.859569	0.841056	0.823491	0.806739	0.790700	0.775302	0.760485	0.746201	120
140	0.957812	0.934247	0.913983	0.895493	0.878224	0.861896	0.846339	0.831442	0.817125	0.803326	0.789999	0.777105	140
170	0.965169	0.945562	0.928606	0.913057	0.898465	0.884603	0.871338	0.858581	0.846267	0.834352	0.822797	0.811574	170
200	0.970341	0.953554	0.938982	0.925569	0.912940	0.900904	0.889349	0.878202	0.867412	0.856939	0.846755	0.836834	200
240	0.975243	0.961158	0.948887	0.937554	0.926848	0.916613	0.906758	0.897224	0.887968	0.878959	0.870174	0.861593	240
320	0.981393	0.970741	0.961415	0.952766	0.944563	0.936691	0.929082	0.921692	0.914493	0.907461	0.900579	0.893835	320
440	0.986445	0.978644	0.971788	0.965408	0.959337	0.953491	0.947824	0.942303	0.936908	0.931623	0.926435	0.921337	440
600	0.990047	0.984298	0.979233	0.974507	0.969998	0.965648	0.961420	0.957293	0.953251	0.949283	0.945380	0.941537	600
800	0.992529	0.988203	0.984384	0.980814	0.977404	0.974108	0.970900	0.967763	0.964686	0.961662	0.958683	0.955744	800
1000	0.994021	0.990552	0.987487	0.984620	0.981877	0.979224	0.976640	0.974110	0.971627	0.969184	0.966775	0.964397	1000
INF	1.000000	1.000000	1.000000	1.000000	1.000000	1.000000	1.000000	1.000000	1.000000	1.000000	1.000000	1.000000	INF

TABLE IX (*continued*).

$p = 3$ $\alpha = 0.01$ $U^{\alpha}(p, q, n)$ hypothesis df: q

n	1	2	3	4	5	6	7	8	9	10	11	12	n
1	0.000000	0.000000	0.000000	0.000000	0.000000	0.000000	0.000000	0.000000	0.000000	0.000000	0.000000	0.000000	1
2	0.000000	0.000000	0.000000	0.000000	0.000000	0.000000	0.000000	0.000000	0.000001	0.000002	0.000003	0.000004	2
3	0.000080	0.000021	0.000016	0.000015	0.000015	0.000017	0.000019	0.000021	0.000023	0.000026	0.000029	0.000031	3
4	0.006763	0.001829	0.000824	0.000484	0.000335	0.000258	0.000215	0.000188	0.000172	0.000161	0.000154	0.000149	4
5	0.032882	0.011211	0.005326	0.003037	0.001959	0.001383	0.001047	0.000837	0.000698	0.000602	0.000533	0.000483	5
6	0.073980	0.029981	0.015536	0.009229	0.006018	0.004211	0.003116	0.002414	0.001943	0.001614	0.001376	0.001200	6
7	0.121426	0.055863	0.031196	0.019423	0.013027	0.009244	0.006864	0.005293	0.004214	0.003450	0.002892	0.002474	7
8	0.169788	0.085991	0.051041	0.033146	0.022897	0.016575	0.012459	0.009664	0.007703	0.006286	0.005237	0.004444	8
9	0.216359	0.118124	0.073700	0.049627	0.035223	0.026018	0.019845	0.015549	0.012468	0.010203	0.008501	0.007198	9
10	0.259966	0.150745	0.098030	0.068087	0.049501	0.037260	0.028840	0.022851	0.018472	0.015200	0.012705	0.010772	10
11	0.300240	0.182909	0.123161	0.087852	0.065237	0.049951	0.039203	0.031408	0.025614	0.021217	0.017821	0.015158	11
12	0.337186	0.214052	0.148469	0.108380	0.081998	0.063758	0.050682	0.041037	0.033759	0.028161	0.023785	0.020317	12
13	0.370989	0.243868	0.173524	0.129256	0.099424	0.078384	0.063039	0.051550	0.042762	0.035922	0.030516	0.026188	13
14	0.401904	0.272209	0.198043	0.150167	0.117224	0.093576	0.076062	0.062771	0.052482	0.044385	0.037922	0.032700	14
15	0.430202	0.299027	0.221841	0.170888	0.135171	0.109125	0.089567	0.074542	0.062783	0.053438	0.045911	0.039779	15
16	0.456147	0.324338	0.244809	0.191257	0.153091	0.124859	0.103395	0.086724	0.073546	0.062977	0.054394	0.047349	16
17	0.479984	0.348191	0.266888	0.211160	0.170848	0.140644	0.117419	0.099197	0.084662	0.072908	0.063289	0.055338	17
18	0.501932	0.370654	0.288051	0.230524	0.188345	0.156370	0.131529	0.111859	0.096037	0.083144	0.072519	0.063680	18
19	0.522191	0.391807	0.308300	0.249300	0.205509	0.171955	0.145640	0.124624	0.107589	0.093611	0.082016	0.072312	19
20	0.540934	0.411734	0.327644	0.267462	0.222288	0.187333	0.159680	0.137421	0.119251	0.104243	0.091719	0.081178	20
21	0.558316	0.430515	0.346122	0.284999	0.238647	0.202457	0.173594	0.150193	0.130962	0.114983	0.101574	0.090228	21
22	0.574470	0.448231	0.363762	0.301910	0.254564	0.217287	0.187337	0.162890	0.142676	0.125784	0.111534	0.099418	22
23	0.589519	0.464956	0.380593	0.318203	0.270024	0.231801	0.200875	0.175473	0.154348	0.136602	0.121558	0.108708	23
24	0.603567	0.480761	0.396664	0.333888	0.285024	0.245977	0.214182	0.187912	0.165948	0.147403	0.131611	0.118064	24
25	0.616709	0.495715	0.412006	0.348987	0.299564	0.259807	0.227238	0.200180	0.177443	0.158158	0.141663	0.127455	25
26	0.629025	0.509875	0.426661	0.363513	0.313646	0.273282	0.240029	0.212260	0.188816	0.168840	0.151686	0.136855	26
27	0.640592	0.523299	0.440664	0.377492	0.327281	0.286401	0.252545	0.224135	0.200042	0.179430	0.161661	0.146240	27
28	0.651469	0.536040	0.454050	0.390942	0.340476	0.299164	0.264779	0.235795	0.211110	0.189910	0.171566	0.155593	28
29	0.661719	0.548144	0.466858	0.403887	0.353244	0.311575	0.276730	0.247231	0.222009	0.200265	0.181387	0.164895	29
30	0.671391	0.559656	0.479116	0.416348	0.365597	0.323637	0.288394	0.258438	0.232727	0.210485	0.191110	0.174132	30
40	0.744674	0.649620	0.577483	0.518712	0.469272	0.426891	0.390088	0.357822	0.329317	0.303979	0.281338	0.261014	40
60	0.823683	0.751990	0.694679	0.645816	0.602970	0.564801	0.530443	0.499282	0.470857	0.444810	0.420849	0.398737	60
80	0.865422	0.808282	0.761397	0.720482	0.683828	0.650513	0.619945	0.591715	0.565509	0.541095	0.518272	0.496881	80
100	0.891201	0.843804	0.804298	0.769332	0.737595	0.708389	0.681275	0.655947	0.632179	0.609801	0.588667	0.568664	100
120	0.908698	0.868241	0.834163	0.803715	0.775833	0.749959	0.725743	0.702949	0.681398	0.660958	0.641519	0.622994	120
140	0.921350	0.886074	0.856136	0.829206	0.804388	0.781216	0.759404	0.738756	0.719127	0.700413	0.682523	0.665388	140
170	0.934886	0.905306	0.880001	0.857075	0.835802	0.815812	0.796876	0.778843	0.761600	0.745064	0.729169	0.713861	170
200	0.944448	0.918986	0.897083	0.877137	0.858541	0.840986	0.824284	0.808309	0.792971	0.778202	0.763939	0.750169	200
240	0.953545	0.932073	0.913505	0.896514	0.880601	0.865514	0.851099	0.837256	0.823911	0.811012	0.798488	0.786374	240
320	0.965006	0.948662	0.934434	0.921336	0.909000	0.897239	0.885943	0.875039	0.864473	0.854210	0.844184	0.834449	320
440	0.974459	0.962428	0.951897	0.942156	0.932937	0.924109	0.915592	0.907335	0.899302	0.891467	0.883774	0.876280	440
600	0.981223	0.972324	0.964504	0.957246	0.950353	0.943732	0.937324	0.931093	0.925012	0.919063	0.913203	0.907480	600
800	0.985892	0.979179	0.973264	0.967760	0.962522	0.957478	0.952586	0.947819	0.943157	0.938588	0.934077	0.929662	800
1000	0.988702	0.983312	0.978556	0.974124	0.969900	0.965827	0.961872	0.958013	0.954234	0.950525	0.946859	0.943268	1000
INF	1.000000	1.000000	1.000000	1.000000	1.000000	1.000000	1.000000	1.000000	1.000000	1.000000	1.000000	1.000000	INF

TABLE IX (*continued*).

$p = 3$ $\alpha = 0.05$ $U^{\alpha}(p, q, n)$ hypothesis df: q

n	1	2	3	4	5	6	7	8	9	10	11	12	n
1	0.000000	0.000000	0.000000	0.000000	0.000000	0.000000	0.000000	0.000000	0.000000	0.000000	0.000000	0.000000	1
2	0.000000	0.000000	0.000000	0.000000	0.000000	0.000001	0.000002	0.000004	0.000005	0.000008	0.000010	0.000013	2
3	0.001698	0.000354	0.000179	0.000127	0.000105	0.000095	0.000091	0.000090	0.000091	0.000092	0.000095	0.000098	3
4	0.033740	0.009612	0.004205	0.002314	0.001479	0.001052	0.000809	0.000659	0.000562	0.000496	0.000449	0.000416	4
5	0.097355	0.035855	0.017521	0.010010	0.006357	0.004369	0.003195	0.002458	0.001971	0.001636	0.001397	0.001222	5
6	0.168271	0.073634	0.039672	0.024047	0.015792	0.011018	0.008067	0.006148	0.004849	0.003939	0.003281	0.002793	6
7	0.235525	0.116476	0.067711	0.043226	0.029433	0.021043	0.015642	0.012012	0.009485	0.007674	0.006345	0.005347	7
8	0.295976	0.160244	0.098932	0.065947	0.046378	0.033966	0.025706	0.019990	0.015911	0.012927	0.010697	0.008997	8
9	0.349276	0.202814	0.131378	0.090794	0.065660	0.049161	0.037855	0.029838	0.023995	0.019637	0.016323	0.013763	9
10	0.396084	0.243139	0.163846	0.116701	0.086448	0.066012	0.051643	0.041238	0.033514	0.027654	0.023135	0.019593	10
11	0.437147	0.280808	0.195556	0.142927	0.108110	0.083979	0.066659	0.053876	0.044225	0.036801	0.030993	0.026391	11
12	0.473377	0.315719	0.226090	0.168939	0.130131	0.102644	0.082534	0.067443	0.055894	0.046882	0.039757	0.034049	12
13	0.505452	0.347981	0.255220	0.194413	0.152159	0.121656	0.098973	0.081704	0.068298	0.057724	0.049278	0.042437	13
14	0.534017	0.377735	0.282849	0.219113	0.173959	0.140775	0.115736	0.096413	0.081246	0.069166	0.059407	0.051442	14
15	0.559570	0.405221	0.308951	0.242944	0.195322	0.159796	0.132619	0.111416	0.094593	0.081052	0.070028	0.060954	15
16	0.582577	0.430566	0.333588	0.265812	0.216138	0.178574	0.149493	0.126564	0.108178	0.093264	0.081026	0.070875	16
17	0.603338	0.454006	0.356777	0.287689	0.236338	0.197017	0.166236	0.141728	0.121917	0.105704	0.092299	0.081109	17
18	0.622168	0.475728	0.378631	0.308599	0.255858	0.215044	0.182762	0.156827	0.135693	0.118273	0.103768	0.091588	18
19	0.639337	0.495908	0.399223	0.328552	0.274710	0.232604	0.199009	0.171789	0.149446	0.130904	0.115361	0.102241	19
20	0.655028	0.514622	0.418629	0.347546	0.292843	0.249666	0.214918	0.186544	0.163097	0.143521	0.127018	0.113012	20
21	0.669437	0.532101	0.436898	0.365676	0.310304	0.266216	0.230467	0.201077	0.176620	0.156088	0.138689	0.123835	21
22	0.682712	0.548393	0.454182	0.382934	0.327083	0.282253	0.245626	0.215325	0.189969	0.168561	0.150321	0.134680	22
23	0.694960	0.563637	0.470473	0.399402	0.343191	0.297740	0.260397	0.229291	0.203123	0.180907	0.161896	0.145521	23
24	0.706310	0.577895	0.485889	0.415077	0.358665	0.312738	0.274743	0.242939	0.216044	0.193091	0.173370	0.156313	24
25	0.716875	0.591311	0.500491	0.430041	0.373523	0.327221	0.288709	0.256276	0.228718	0.205103	0.184720	0.167023	25
26	0.726681	0.603899	0.514336	0.444332	0.387790	0.341199	0.302238	0.269280	0.241137	0.216929	0.195944	0.177651	26
27	0.735837	0.615757	0.527453	0.457946	0.401488	0.354711	0.315386	0.281968	0.253300	0.228535	0.206998	0.188160	27
28	0.744404	0.626944	0.539914	0.470981	0.414658	0.367742	0.328131	0.294313	0.265188	0.239935	0.217899	0.198546	28
29	0.752437	0.637514	0.551741	0.483430	0.427307	0.380334	0.340477	0.306326	0.276805	0.251110	0.228615	0.208808	29
30	0.759984	0.647501	0.563023	0.495347	0.439474	0.392490	0.352461	0.318033	0.288158	0.262062	0.239155	0.218912	30
40	0.816139	0.723938	0.651355	0.590773	0.538846	0.493686	0.453976	0.418785	0.387401	0.359270	0.333940	0.311045	40
60	0.874843	0.807777	0.752424	0.704238	0.661334	0.622640	0.587440	0.555224	0.525598	0.498272	0.472957	0.449477	60
80	0.905160	0.852653	0.808266	0.768805	0.732964	0.700026	0.669520	0.641124	0.614572	0.589678	0.566281	0.544236	80
100	0.923660	0.880557	0.843610	0.810333	0.779746	0.751296	0.724666	0.699598	0.675935	0.653520	0.632235	0.611999	100
120	0.936178	0.899588	0.867973	0.839253	0.812632	0.787686	0.764150	0.741841	0.720623	0.700389	0.681054	0.662546	120
140	0.945137	0.913391	0.885776	0.860534	0.836998	0.814819	0.793780	0.773732	0.754565	0.736197	0.718557	0.701592	140
170	0.954680	0.928199	0.904999	0.883652	0.863624	0.844636	0.826518	0.809156	0.792465	0.776383	0.760857	0.745847	170
200	0.961395	0.938685	0.918687	0.900202	0.882782	0.866197	0.850307	0.835018	0.820262	0.805990	0.792160	0.778739	200
240	0.967765	0.948678	0.931793	0.916116	0.901281	0.887100	0.873459	0.860284	0.847521	0.835131	0.823081	0.811346	240
320	0.975762	0.961296	0.948422	0.936405	0.924971	0.913987	0.903369	0.893064	0.883033	0.873250	0.863692	0.854341	320
440	0.982336	0.971725	0.962235	0.953337	0.944835	0.936632	0.928671	0.920913	0.913332	0.905910	0.898630	0.891482	440
600	0.987028	0.979198	0.972173	0.965563	0.959229	0.953099	0.947133	0.941302	0.935589	0.929978	0.924461	0.919029	600
800	0.990261	0.984364	0.979060	0.974060	0.969257	0.964600	0.960057	0.955610	0.951243	0.946947	0.942713	0.938538	800
1000	0.992204	0.987475	0.983215	0.979193	0.975326	0.971571	0.967905	0.964310	0.960776	0.957296	0.953863	0.950473	1000
INF	1.000000	1.000000	1.000000	1.000000	1.000000	1.000000	1.000000	1.000000	1.000000	1.000000	1.000000	1.000000	INF

TABLE IX (*continued*).

$p = 4$ $\alpha = 0.01$ $U^{\alpha}(p, q, n)$ hypothesis df: q

n	1	2	3	4	5	6	7	8	9	10	11	12	n
1	0.000000	0.000000	0.000000	0.000000	0.000000	0.000000	0.000000	0.000000	0.000000	0.000000	0.000000	0.000000	1
2	0.000000	0.000000	0.000000	0.000000	0.000000	0.000000	0.000000	0.000000	0.000000	0.000000	0.000000	0.000000	2
3	0.000000	0.000000	0.000000	0.000000	0.000000	0.000000	0.000000	0.000000	0.000000	0.000000	0.000000	0.000000	3
4	0.000090	0.000026	0.000015	0.000011	0.000009	0.000008	0.000007	0.000007	0.000007	0.000007	0.000007	0.000007	4
5	0.005218	0.001224	0.000484	0.000250	0.000153	0.000106	0.000079	0.000063	0.000052	0.000045	0.000040	0.000037	5
6	0.025586	0.007345	0.003037	0.001538	0.000893	0.000574	0.000398	0.000293	0.000227	0.000183	0.000152	0.000130	6
7	0.058962	0.020352	0.009229	0.004891	0.002885	0.001846	0.001259	0.000906	0.000681	0.000531	0.000427	0.000353	7
8	0.098904	0.039349	0.019423	0.010860	0.006623	0.004315	0.002966	0.002131	0.001590	0.001225	0.000971	0.000789	8
9	0.140881	0.062551	0.033146	0.019474	0.012300	0.008211	0.005732	0.004155	0.003111	0.002396	0.001891	0.001527	9
10	0.182361	0.088300	0.049627	0.030445	0.019865	0.013591	0.009662	0.007095	0.005357	0.004145	0.003278	0.002644	10
11	0.222076	0.115330	0.068087	0.043357	0.029128	0.020392	0.014763	0.010993	0.008388	0.006539	0.005197	0.004202	11
12	0.259456	0.142746	0.087852	0.057777	0.039832	0.028478	0.020973	0.015836	0.012218	0.009607	0.007685	0.006242	12
13	0.294315	0.169948	0.108380	0.073308	0.051709	0.037679	0.028192	0.021570	0.016825	0.013349	0.010755	0.008785	13
14	0.326670	0.196546	0.129256	0.089607	0.064505	0.047814	0.036299	0.028118	0.022164	0.017742	0.014400	0.011834	14
15	0.356636	0.222301	0.150167	0.106392	0.077992	0.058712	0.045168	0.035392	0.028175	0.022747	0.018597	0.015378	15
16	0.384374	0.247074	0.170888	0.123435	0.091973	0.070212	0.054675	0.043297	0.034790	0.028315	0.023313	0.019396	16
17	0.410058	0.270796	0.191257	0.140556	0.106281	0.082172	0.064703	0.051742	0.041937	0.034394	0.028510	0.023860	17
18	0.433867	0.293442	0.211160	0.157615	0.120777	0.094469	0.075148	0.060641	0.049546	0.040928	0.034143	0.028739	18
19	0.455967	0.315021	0.230524	0.174505	0.135348	0.106995	0.085915	0.069912	0.057551	0.047861	0.040170	0.033996	19
20	0.476513	0.335555	0.249300	0.191144	0.149903	0.119660	0.096921	0.079482	0.065888	0.055142	0.046546	0.039596	20
21	0.495648	0.355086	0.267462	0.207474	0.164368	0.132389	0.108094	0.089287	0.074500	0.062720	0.053228	0.045503	21
22	0.513499	0.373655	0.284999	0.223450	0.178686	0.145118	0.119371	0.099266	0.083333	0.070548	0.060177	0.051683	22
23	0.530184	0.391312	0.301910	0.239044	0.192810	0.157796	0.130700	0.109371	0.092342	0.078585	0.067354	0.058103	23
24	0.545805	0.408104	0.318203	0.254237	0.206707	0.170381	0.142036	0.119555	0.101484	0.086790	0.074724	0.064730	24
25	0.560457	0.424083	0.333888	0.269015	0.220349	0.182837	0.153340	0.129781	0.110720	0.095129	0.082255	0.071537	25
26	0.574221	0.439293	0.348987	0.283374	0.233718	0.195137	0.164581	0.140015	0.120018	0.103570	0.089917	0.078495	26
27	0.587173	0.453781	0.363513	0.297314	0.246799	0.207259	0.175732	0.150228	0.129349	0.112085	0.097684	0.085579	27
28	0.599381	0.467592	0.377492	0.310838	0.259584	0.219187	0.186772	0.160396	0.138688	0.120648	0.105530	0.092768	28
29	0.610904	0.480765	0.390942	0.323953	0.272068	0.230906	0.197681	0.170499	0.148013	0.129238	0.113435	0.100038	29
30	0.621798	0.493340	0.403887	0.336665	0.284247	0.242408	0.208447	0.180518	0.157304	0.137834	0.121378	0.107372	30
40	0.704846	0.593044	0.510028	0.444079	0.390022	0.344862	0.306628	0.273929	0.245739	0.221270	0.199908	0.181164	40
60	0.795314	0.709205	0.641042	0.583746	0.534292	0.490946	0.452558	0.418305	0.387562	0.359837	0.334734	0.311927	60
80	0.843446	0.774138	0.717496	0.668503	0.625079	0.586056	0.550669	0.518371	0.488748	0.461472	0.436276	0.412939	80
100	0.873280	0.815461	0.767296	0.724909	0.686729	0.651890	0.619831	0.590158	0.562571	0.536836	0.512760	0.490184	100
120	0.893573	0.844036	0.802240	0.765030	0.731147	0.699908	0.670876	0.643745	0.618288	0.594325	0.571711	0.550325	120
140	0.908268	0.864962	0.828089	0.794989	0.764613	0.736396	0.709985	0.685132	0.661654	0.639410	0.618283	0.598178	140
170	0.924010	0.887597	0.856293	0.827943	0.801710	0.777147	0.753978	0.731988	0.711113	0.691172	0.672100	0.653829	170
200	0.935142	0.903739	0.876559	0.851792	0.828739	0.807033	0.786448	0.766791	0.748045	0.730070	0.712779	0.696132	200
240	0.945741	0.919212	0.896104	0.874924	0.855100	0.836335	0.818447	0.801269	0.784813	0.768957	0.753648	0.738836	240
320	0.959108	0.938870	0.921100	0.904693	0.889230	0.874494	0.860356	0.846684	0.833511	0.820740	0.808335	0.796267	320
440	0.970142	0.955217	0.942028	0.929777	0.918164	0.907036	0.896302	0.885863	0.875756	0.865908	0.856294	0.846895	440
600	0.978043	0.966989	0.957176	0.948021	0.939309	0.930927	0.922811	0.914887	0.907187	0.899657	0.892279	0.885041	600
800	0.983500	0.975154	0.967720	0.960765	0.954127	0.947724	0.941507	0.935422	0.929493	0.923681	0.917972	0.912357	800
1000	0.986785	0.980081	0.974098	0.968491	0.963130	0.957951	0.952914	0.947976	0.943158	0.938426	0.933773	0.929189	1000
INF	1.000000	1.000000	1.000000	1.000000	1.000000	1.000000	1.000000	1.000000	1.000000	1.000000	1.000000	1.000000	INF

TABLE IX (*continued*).

$p = 4 \quad \alpha = 0.05 \quad U^{\alpha}(p, q, n) \quad$ hypothesis df: q

n	1	2	3	4	5	6	7	8	9	10	11	12	n
1	0.000000	0.000000	0.000000	0.000000	0.000000	0.000000	0.000000	0.000000	0.000000	0.000000	0.000000	0.000000	1
2	0.000000	0.000000	0.000000	0.000000	0.000000	0.000000	0.000000	0.000000	0.000000	0.000000	0.000000	0.000000	2
3	0.000000	0.000000	0.000000	0.000000	0.000000	0.000001	0.000001	0.000001	0.000002	0.000002	0.000002	0.000003	3
4	0.001378	0.000292	0.000127	0.000075	0.000052	0.000040	0.000033	0.000029	0.000026	0.000025	0.000023	0.000022	4
5	0.025529	0.006091	0.002314	0.001128	0.000647	0.000416	0.000292	0.000218	0.000172	0.000141	0.000120	0.000105	5
6	0.076071	0.023604	0.010010	0.005073	0.002903	0.001818	0.001223	0.000872	0.000652	0.000508	0.000409	0.000338	6
7	0.135374	0.050839	0.024047	0.013014	0.007737	0.004938	0.003338	0.002365	0.001745	0.001333	0.001050	0.000848	7
8	0.194043	0.083695	0.043226	0.024857	0.015415	0.010129	0.006975	0.004994	0.003698	0.002819	0.002206	0.001766	8
9	0.248619	0.118995	0.065947	0.039919	0.025729	0.017408	0.012249	0.008907	0.006664	0.005112	0.004009	0.003208	9
10	0.298130	0.154758	0.090794	0.057378	0.038260	0.026586	0.019107	0.014130	0.010706	0.008288	0.006542	0.005254	10
11	0.342593	0.189778	0.116701	0.076502	0.052524	0.037385	0.027402	0.020589	0.015806	0.012365	0.009839	0.007948	11
12	0.382448	0.223411	0.142927	0.096664	0.068077	0.049495	0.036933	0.028170	0.021899	0.017314	0.013895	0.011302	12
13	0.418181	0.255376	0.168939	0.117377	0.084546	0.062632	0.047493	0.036731	0.028895	0.023075	0.018675	0.015303	13
14	0.450335	0.285511	0.194413	0.138286	0.101586	0.076537	0.058886	0.046115	0.036676	0.029572	0.024133	0.019917	14
15	0.479286	0.313829	0.219113	0.159131	0.118954	0.090983	0.070925	0.056188	0.045140	0.036722	0.030208	0.025101	15
16	0.505511	0.340400	0.242944	0.179688	0.136434	0.105779	0.083443	0.066806	0.054181	0.044440	0.036830	0.030804	16
17	0.529312	0.365253	0.265812	0.199832	0.153891	0.120780	0.096316	0.077856	0.063688	0.052645	0.043936	0.036980	17
18	0.551035	0.388530	0.287689	0.219490	0.171171	0.135856	0.109411	0.089236	0.073577	0.061263	0.051456	0.043568	18
19	0.570858	0.410325	0.308599	0.238570	0.188209	0.150905	0.122643	0.100843	0.083764	0.070213	0.059338	0.050514	19
20	0.589077	0.430766	0.328552	0.257052	0.204926	0.165853	0.135926	0.112607	0.094180	0.079441	0.067512	0.057782	20
21	0.605832	0.449947	0.347546	0.274909	0.221288	0.180626	0.149180	0.124462	0.104757	0.088877	0.075938	0.065315	21
22	0.621318	0.467988	0.365676	0.292142	0.237242	0.195197	0.162364	0.136342	0.115440	0.098474	0.084565	0.073068	22
23	0.635634	0.484922	0.382934	0.308765	0.252783	0.209511	0.175434	0.148204	0.126185	0.108191	0.093352	0.081008	23
24	0.648934	0.500883	0.399402	0.324767	0.267896	0.223535	0.188341	0.160009	0.136950	0.117977	0.102254	0.089100	24
25	0.661320	0.515918	0.415077	0.340175	0.282568	0.237277	0.201067	0.171726	0.147695	0.127818	0.111240	0.097305	25
26	0.672864	0.530124	0.430041	0.355004	0.296810	0.250710	0.213597	0.183333	0.158399	0.137656	0.120274	0.105608	26
27	0.683663	0.543561	0.444332	0.369254	0.310608	0.263809	0.225900	0.194794	0.169017	0.147483	0.129346	0.113968	27
28	0.693769	0.556262	0.457946	0.382979	0.323980	0.276602	0.237971	0.206105	0.179569	0.157274	0.138418	0.122368	28
29	0.703259	0.568303	0.470981	0.396197	0.336947	0.289051	0.249798	0.217241	0.189991	0.167006	0.147478	0.130784	29
30	0.712188	0.579734	0.483430	0.408914	0.349488	0.301188	0.261373	0.228198	0.200311	0.176673	0.156516	0.139205	30
40	0.778877	0.668158	0.582817	0.513297	0.455181	0.405867	0.363565	0.326959	0.295085	0.267163	0.242600	0.220888	40
60	0.849044	0.767047	0.700065	0.642556	0.592126	0.547349	0.507256	0.471148	0.438462	0.408771	0.381699	0.356960	60
80	0.885442	0.820705	0.766251	0.718260	0.675124	0.635912	0.600023	0.566986	0.536460	0.508176	0.481887	0.457414	80
100	0.907714	0.854312	0.808614	0.767700	0.730354	0.695928	0.663968	0.634166	0.606280	0.580111	0.555487	0.532298	100
120	0.922736	0.877325	0.838018	0.802443	0.769650	0.739118	0.710513	0.683595	0.658183	0.634132	0.611324	0.589657	120
140	0.933554	0.894066	0.859605	0.828175	0.798994	0.771635	0.745829	0.721386	0.698162	0.676045	0.654943	0.634778	140
170	0.945088	0.912072	0.883006	0.856283	0.831278	0.807661	0.785224	0.763821	0.743347	0.723717	0.704865	0.686733	170
200	0.953211	0.924848	0.899727	0.876499	0.854647	0.833900	0.814087	0.795095	0.776838	0.759251	0.742281	0.725885	200
240	0.960919	0.937047	0.915781	0.896012	0.877319	0.859481	0.842366	0.825881	0.809960	0.794554	0.779622	0.765130	240
320	0.970605	0.952477	0.936212	0.920990	0.906503	0.892593	0.879164	0.866153	0.853513	0.841211	0.829220	0.817517	320
440	0.978571	0.965253	0.953233	0.941922	0.931100	0.920655	0.910522	0.900654	0.891022	0.881602	0.872376	0.863330	440
600	0.984259	0.974422	0.965507	0.957084	0.948995	0.941160	0.933530	0.926075	0.918772	0.911606	0.904563	0.897634	600
800	0.988181	0.980767	0.974028	0.967644	0.961498	0.955529	0.949702	0.943994	0.938390	0.932877	0.927446	0.922092	800
1000	0.990538	0.984589	0.979173	0.974034	0.969078	0.964257	0.959545	0.954922	0.950376	0.945898	0.941481	0.937120	1000
INF	1.000000	1.000000	1.000000	1.000000	1.000000	1.000000	1.000000	1.000000	1.000000	1.000000	1.000000	1.000000	INF

TABLE IX (*continued*).

$p = 5$ $\alpha = 0.01$ $U^{\alpha}(p, q, n)$ hypothesis df: q

n	1	2	3	4	5	6	7	8	9	10	11	12	n
1	0.000000	0.000000	0.000000	0.000000	0.000000	0.000000	0.000000	0.000000	0.000000	0.000000	0.000000	0.000000	1
2	0.000000	0.000000	0.000000	0.000000	0.000000	0.000000	0.000000	0.000000	0.000000	0.000000	0.000000	0.000000	2
3	0.000000	0.000000	0.000000	0.000000	0.000000	0.000000	0.000000	0.000000	0.000000	0.000000	0.000000	0.000000	3
4	0.000000	0.000000	0.000000	0.000000	0.000000	0.000000	0.000000	0.000000	0.000000	0.000000	0.000000	0.000000	4
5	0.000164	0.000036	0.000015	0.000009	0.000006	0.000004	0.000003	0.000003	0.000003	0.000002	0.000002	0.000002	5
6	0.004668	0.000962	0.000335	0.000153	0.000084	0.000052	0.000035	0.000025	0.000020	0.000016	0.000013	0.000011	6
7	0.021333	0.005332	0.001959	0.000893	0.000472	0.000277	0.000177	0.000121	0.000088	0.000066	0.000052	0.000042	7
8	0.049302	0.014879	0.006018	0.002885	0.001557	0.000918	0.000582	0.000390	0.000275	0.000202	0.000154	0.000121	8
9	0.083710	0.029395	0.013027	0.006623	0.003709	0.002237	0.001432	0.000964	0.000677	0.000493	0.000371	0.000287	9
10	0.120729	0.047777	0.022897	0.012300	0.007165	0.004443	0.002899	0.001974	0.001394	0.001017	0.000763	0.000587	10
11	0.158044	0.068815	0.035223	0.019865	0.012007	0.007658	0.005103	0.003528	0.002519	0.001850	0.001392	0.001072	11
12	0.194389	0.091490	0.049501	0.029128	0.018203	0.011922	0.008113	0.005702	0.004122	0.003055	0.002315	0.001790	12
13	0.229107	0.115016	0.065237	0.039832	0.025645	0.017209	0.011946	0.008535	0.006252	0.004681	0.003576	0.002781	13
14	0.261911	0.138822	0.081993	0.051709	0.034186	0.023452	0.016584	0.012034	0.008929	0.006758	0.005207	0.004077	14
15	0.292711	0.162507	0.099424	0.064505	0.043665	0.030560	0.021981	0.016183	0.012158	0.009299	0.007228	0.005702	15
16	0.321529	0.185794	0.117224	0.077992	0.053925	0.038428	0.028076	0.020951	0.015926	0.012305	0.009648	0.007667	16
17	0.348449	0.208500	0.135171	0.091973	0.064813	0.046952	0.034797	0.026294	0.020207	0.015764	0.012465	0.009978	17
18	0.373583	0.230510	0.153091	0.106281	0.076195	0.056028	0.042071	0.032161	0.024971	0.019658	0.015670	0.012633	18
19	0.397053	0.251754	0.170848	0.120777	0.087948	0.065559	0.049825	0.038499	0.030179	0.023963	0.019247	0.015623	19
20	0.418988	0.272198	0.188345	0.135348	0.099970	0.075458	0.057989	0.045254	0.035793	0.028649	0.023178	0.018937	20
21	0.439504	0.291832	0.205509	0.149903	0.112170	0.085645	0.066497	0.052374	0.041771	0.033687	0.027441	0.022558	21
22	0.458721	0.310659	0.222288	0.164368	0.124471	0.096050	0.075288	0.059809	0.048073	0.039045	0.032012	0.026471	22
23	0.476739	0.328696	0.238647	0.178686	0.136809	0.106612	0.084307	0.067511	0.054660	0.044692	0.036865	0.030655	23
24	0.493662	0.345966	0.254564	0.192810	0.149131	0.117275	0.093504	0.075438	0.061497	0.050597	0.041977	0.035091	24
25	0.509575	0.362498	0.270024	0.206707	0.161392	0.127995	0.102837	0.083549	0.068546	0.056730	0.047322	0.039758	25
26	0.524560	0.378321	0.285024	0.220349	0.173556	0.138732	0.112264	0.091808	0.075777	0.063064	0.052876	0.044637	26
27	0.538691	0.393468	0.299564	0.233718	0.185593	0.149451	0.121752	0.100183	0.083160	0.069572	0.058617	0.049707	27
28	0.552034	0.407970	0.313646	0.246799	0.197480	0.160123	0.131271	0.108643	0.090666	0.076229	0.064523	0.054951	28
29	0.564655	0.421860	0.327281	0.259584	0.209196	0.170724	0.140795	0.117163	0.098272	0.083013	0.070573	0.060349	29
30	0.576601	0.435170	0.340476	0.272068	0.220728	0.181235	0.150300	0.125719	0.105955	0.089902	0.076748	0.065885	30
40	0.668249	0.542257	0.451107	0.380670	0.324502	0.278831	0.241173	0.209788	0.183401	0.161054	0.142005	0.125677	40
60	0.769057	0.669979	0.592611	0.528633	0.474345	0.427572	0.386853	0.351131	0.319602	0.291636	0.266722	0.244448	60
80	0.823038	0.742525	0.677232	0.621368	0.572440	0.529007	0.490107	0.455043	0.423278	0.394390	0.368029	0.343906	80
100	0.856606	0.789069	0.733048	0.684125	0.640446	0.600959	0.564970	0.531981	0.501610	0.473552	0.447557	0.423417	100
120	0.879484	0.821412	0.772515	0.729225	0.690073	0.654241	0.621196	0.590558	0.562038	0.535406	0.510473	0.487083	120
140	0.896071	0.845177	0.801862	0.763137	0.727789	0.695150	0.664792	0.636413	0.609782	0.584719	0.561076	0.538731	140
170	0.913861	0.870955	0.834024	0.800661	0.769907	0.741242	0.714313	0.688960	0.664941	0.642146	0.620465	0.599807	170
200	0.926453	0.889383	0.857222	0.827957	0.800792	0.775304	0.751191	0.728357	0.706614	0.685839	0.665962	0.646914	200
240	0.938451	0.907082	0.879663	0.854540	0.831068	0.808906	0.787800	0.767701	0.748450	0.729967	0.712175	0.695028	240
320	0.953594	0.929613	0.908455	0.888903	0.870485	0.852956	0.836125	0.819982	0.804406	0.789342	0.774746	0.760584	320
440	0.966104	0.948389	0.932642	0.917985	0.904086	0.890771	0.877902	0.865484	0.853429	0.841699	0.830267	0.819109	440
600	0.975067	0.961932	0.950192	0.939210	0.928746	0.918676	0.908897	0.899419	0.890178	0.881148	0.872309	0.863646	600
800	0.981261	0.971335	0.962430	0.954071	0.946080	0.938365	0.930849	0.923543	0.916398	0.909395	0.902519	0.895761	800
1000	0.984990	0.977013	0.969840	0.963094	0.956632	0.950381	0.944279	0.938337	0.932515	0.926799	0.921177	0.915641	1000
INF	1.000000	1.000000	1.000000	1.000000	1.000000	1.000000	1.000000	1.000000	1.000000	1.000000	1.000000	1.000000	INF

TABLE IX (*continued*).

$p = 5$ $\alpha = 0.05$ $U^{\alpha}(p, q, n)$ hypothesis df: q

n	1	2	3	4	5	6	7	8	9	10	11	12	n
1	0.000000	0.000000	0.000000	0.000000	0.000000	0.000000	0.000000	0.000000	0.000000	0.000000	0.000000	0.000000	1
2	0.000000	0.000000	0.000000	0.000000	0.000000	0.000000	0.000000	0.000000	0.000000	0.000000	0.000000	0.000000	2
3	0.000000	0.000000	0.000000	0.000000	0.000000	0.000000	0.000000	0.000000	0.000000	0.000000	0.000000	0.000000	3
4	0.000000	0.000000	0.000000	0.000000	0.000001	0.000001	0.000001	0.000001	0.000001	0.000001	0.000001	0.000001	4
5	0.001598	0.000291	0.000105	0.000052	0.000031	0.000021	0.000015	0.000012	0.000010	0.000008	0.000007	0.000007	5
6	0.021145	0.004391	0.001479	0.000647	0.000335	0.000197	0.000126	0.000087	0.000064	0.000049	0.000039	0.000032	6
7	0.062770	0.016898	0.006357	0.002903	0.001514	0.000872	0.000544	0.000361	0.000253	0.000185	0.000141	0.000110	7
8	0.113526	0.037390	0.015792	0.007737	0.004208	0.002479	0.001557	0.001032	0.000716	0.000516	0.000385	0.000296	8
9	0.165351	0.063279	0.029433	0.015415	0.008787	0.005348	0.003433	0.002304	0.001607	0.001159	0.000861	0.000657	9
10	0.214794	0.092191	0.046378	0.025729	0.015321	0.009639	0.006343	0.004335	0.003062	0.002225	0.001660	0.001267	10
11	0.260635	0.122403	0.065660	0.038260	0.023674	0.015360	0.010358	0.007216	0.005173	0.003802	0.002858	0.002192	11
12	0.302608	0.152793	0.086448	0.052524	0.033618	0.022418	0.015467	0.010980	0.007991	0.005946	0.004512	0.003486	12
13	0.340813	0.182662	0.108110	0.068077	0.044878	0.030680	0.021607	0.015611	0.011530	0.008685	0.006659	0.005187	13
14	0.375528	0.211602	0.130131	0.084546	0.057198	0.039965	0.028683	0.021061	0.015774	0.012024	0.009313	0.007317	14
15	0.407128	0.239373	0.152159	0.101586	0.070324	0.050117	0.036584	0.027266	0.020687	0.015949	0.012475	0.009884	15
16	0.435899	0.265851	0.173959	0.118954	0.084048	0.060965	0.045199	0.034145	0.026219	0.020428	0.016129	0.012885	16
17	0.462173	0.291015	0.195322	0.136434	0.098187	0.072367	0.054409	0.041618	0.032312	0.025427	0.020252	0.016307	17
18	0.486266	0.314859	0.216138	0.153891	0.112582	0.084178	0.064111	0.049602	0.038909	0.030904	0.024819	0.020133	18
19	0.508362	0.337418	0.236338	0.171171	0.127108	0.096308	0.074209	0.058024	0.045951	0.036810	0.029790	0.024339	19
20	0.528714	0.358776	0.255858	0.188209	0.141662	0.108634	0.084619	0.066805	0.053373	0.043100	0.035137	0.028896	20
21	0.547516	0.378956	0.274710	0.204926	0.156176	0.121083	0.095254	0.075885	0.061122	0.049724	0.040817	0.033782	21
22	0.564905	0.398037	0.292843	0.221288	0.170563	0.133590	0.106063	0.085203	0.069149	0.056652	0.046803	0.038962	22
23	0.581036	0.416105	0.310304	0.237242	0.184782	0.146095	0.116974	0.094699	0.077408	0.063832	0.053052	0.044411	23
24	0.596032	0.433216	0.327083	0.252783	0.198795	0.158544	0.127948	0.104337	0.085849	0.071231	0.059537	0.050103	24
25	0.610030	0.449429	0.343191	0.267896	0.212568	0.170898	0.138945	0.114058	0.094444	0.078809	0.066222	0.056005	25
26	0.623126	0.464800	0.358665	0.282568	0.226071	0.183129	0.149909	0.123843	0.103144	0.086536	0.073084	0.062103	26
27	0.635368	0.479382	0.373523	0.296810	0.239294	0.195207	0.160826	0.133657	0.111931	0.094385	0.080092	0.068358	27
28	0.646832	0.493247	0.387790	0.310608	0.252224	0.207116	0.171667	0.143454	0.120766	0.102328	0.087220	0.074761	28
29	0.657645	0.506421	0.401488	0.323980	0.264872	0.218828	0.182408	0.153240	0.129630	0.110336	0.094455	0.081283	29
30	0.667803	0.518945	0.414658	0.336947	0.277200	0.230347	0.193043	0.162971	0.138499	0.118393	0.101767	0.087901	30
40	0.744009	0.617178	0.521747	0.446045	0.384424	0.333492	0.290896	0.254963	0.224433	0.198322	0.175874	0.156480	40
60	0.824764	0.729155	0.652037	0.586878	0.530670	0.481578	0.438367	0.400085	0.365997	0.335520	0.308193	0.283593	60
80	0.866847	0.790730	0.727186	0.671775	0.622536	0.578316	0.538319	0.501966	0.468774	0.438392	0.410497	0.384827	80
100	0.892643	0.829563	0.775817	0.728040	0.684827	0.645343	0.609037	0.575508	0.544420	0.515540	0.488629	0.463515	100
120	0.910071	0.856267	0.809790	0.767957	0.729656	0.694256	0.661341	0.630608	0.601822	0.574793	0.549362	0.525395	120
140	0.922634	0.875748	0.834850	0.797705	0.763400	0.731431	0.701466	0.673268	0.646653	0.621477	0.597615	0.574968	140
170	0.936039	0.896748	0.862122	0.830370	0.800777	0.772953	0.746648	0.721687	0.697934	0.675284	0.653648	0.632953	170
200	0.945486	0.911680	0.881674	0.853973	0.827989	0.803406	0.780024	0.757705	0.736343	0.715856	0.696177	0.677251	200
240	0.954455	0.925960	0.900496	0.876838	0.854512	0.833264	0.812938	0.793426	0.774647	0.756540	0.739054	0.722148	240
320	0.965732	0.944055	0.924519	0.906224	0.888827	0.872146	0.856074	0.840535	0.825476	0.810855	0.796641	0.782805	320
440	0.975013	0.959064	0.944590	0.930949	0.917894	0.905302	0.893096	0.881226	0.869655	0.858357	0.847311	0.836500	440
600	0.981642	0.969850	0.959096	0.948912	0.939124	0.929642	0.920410	0.911396	0.902572	0.893921	0.885429	0.877084	600
800	0.986214	0.977320	0.969181	0.961450	0.953996	0.946753	0.939682	0.932756	0.925957	0.919273	0.912693	0.906209	800
1000	0.988963	0.981823	0.975277	0.969047	0.963029	0.957171	0.951441	0.945820	0.940292	0.934848	0.929480	0.924182	1000
INF	1.000000	1.000000	1.000000	1.000000	1.000000	1.000000	1.000000	1.000000	1.000000	1.000000	1.000000	1.000000	INF

TABLE IX (*continued*).

$p = 6$ $\alpha = 0.01$ $U^{\alpha}(p, q, n)$ hypothesis df: q

n	1	2	3	4	5	6	7	8	9	10	11	12	*n*
1	0.000000	0.000000	0.000000	0.000000	0.000000	0.000000	0.000000	0.000000	0.000000	0.000000	0.000000	0.000000	1
2	0.000000	0.000000	0.000000	0.000000	0.000000	0.000000	0.000000	0.000000	0.000000	0.000000	0.000000	0.000000	2
3	0.000000	0.000000	0.000000	0.000000	0.000000	0.000000	0.000000	0.000000	0.000000	0.000000	0.000000	0.000000	3
4	0.000000	0.000000	0.000000	0.000000	0.000000	0.000000	0.000000	0.000000	0.000000	0.000000	0.000000	0.000000	4
5	0.000000	0.000000	0.000000	0.000000	0.000000	0.000000	0.000000	0.000000	0.000000	0.000000	0.000000	0.000000	5
6	0.000295	0.000050	0.000017	0.000008	0.000004	0.000003	0.000002	0.000001	0.000001	0.000000	0.000000	0.000000	6
7	0.004608	0.000839	0.000258	0.000106	0.000052	0.000029	0.000018	0.000012	0.000008	0.000006	0.000005	0.000004	7
8	0.018808	0.004182	0.001383	0.000574	0.000277	0.000150	0.000089	0.000057	0.000038	0.000027	0.000020	0.000015	8
9	0.042762	0.011508	0.004211	0.001846	0.000918	0.000503	0.000297	0.000187	0.000124	0.000086	0.000062	0.000046	9
10	0.072861	0.022948	0.009244	0.004315	0.002237	0.001257	0.000754	0.000477	0.000317	0.000219	0.000156	0.000115	10
11	0.105882	0.037842	0.016575	0.008211	0.004443	0.002575	0.001578	0.001014	0.000678	0.000470	0.000336	0.000247	11
12	0.139723	0.055318	0.026013	0.013591	0.007658	0.004578	0.002873	0.001878	0.001272	0.000889	0.000639	0.000471	12
13	0.173151	0.074563	0.037260	0.020392	0.011922	0.007339	0.004715	0.003140	0.002158	0.001525	0.001105	0.000818	13
14	0.205478	0.094910	0.049951	0.028478	0.017209	0.010886	0.007151	0.004851	0.003384	0.002420	0.001769	0.001320	14
15	0.236354	0.115841	0.063758	0.037679	0.023452	0.015207	0.010199	0.007041	0.004984	0.003608	0.002664	0.002004	15
16	0.265622	0.136970	0.078384	0.047814	0.030560	0.020265	0.013856	0.009723	0.006980	0.005114	0.003815	0.002894	16
17	0.293243	0.158016	0.093576	0.058712	0.038428	0.026009	0.018099	0.012897	0.009383	0.006954	0.005240	0.004009	17
18	0.319245	0.178778	0.109125	0.070212	0.046952	0.032373	0.022896	0.016549	0.012192	0.009135	0.006950	0.005362	18
19	0.343692	0.199115	0.124859	0.082172	0.056028	0.039291	0.028206	0.020658	0.015398	0.011658	0.008952	0.006963	19
20	0.366666	0.218933	0.140644	0.094469	0.065559	0.046693	0.033983	0.025196	0.018987	0.014517	0.011245	0.008816	20
21	0.388259	0.238169	0.156370	0.106995	0.075458	0.054514	0.040181	0.030131	0.022939	0.017701	0.013827	0.010921	21
22	0.408567	0.256789	0.171955	0.119660	0.085645	0.062689	0.046751	0.035431	0.027233	0.021197	0.016688	0.013275	22
23	0.427679	0.274775	0.187333	0.132389	0.096050	0.071160	0.053649	0.041060	0.031843	0.024988	0.019819	0.015873	23
24	0.445681	0.292121	0.202457	0.145118	0.106612	0.079873	0.060830	0.046985	0.036744	0.029056	0.023208	0.018708	24
25	0.462657	0.308833	0.217287	0.157796	0.117275	0.088781	0.068253	0.053173	0.041911	0.033382	0.026841	0.021769	25
26	0.478684	0.324922	0.231801	0.170381	0.127995	0.097838	0.075880	0.059592	0.047319	0.037946	0.030703	0.025046	26
27	0.493829	0.340404	0.245977	0.182837	0.138732	0.107006	0.083676	0.066212	0.052941	0.042728	0.034778	0.028527	27
28	0.508160	0.355297	0.259807	0.195137	0.149451	0.116251	0.091609	0.073604	0.058756	0.047709	0.039052	0.032201	28
29	0.521737	0.369623	0.273282	0.207259	0.160123	0.125542	0.099650	0.079944	0.064740	0.052871	0.043508	0.036054	29
30	0.534611	0.383404	0.286401	0.219187	0.170724	0.134852	0.107773	0.087007	0.070872	0.058193	0.048132	0.040075	30
40	0.633971	0.495984	0.398981	0.326182	0.269778	0.225181	0.189401	0.160362	0.136572	0.116924	0.100582	0.086906	40
60	0.744292	0.633481	0.548313	0.479114	0.421438	0.372626	0.330877	0.294885	0.263658	0.236424	0.212564	0.191578	60
80	0.803733	0.712825	0.639827	0.578119	0.524751	0.477984	0.436634	0.399836	0.366925	0.337368	0.310733	0.286658	80
100	0.840806	0.764135	0.700945	0.646245	0.597871	0.554570	0.515501	0.480046	0.447731	0.418175	0.391064	0.366135	100
120	0.866117	0.799962	0.744489	0.695702	0.651910	0.612148	0.575775	0.542326	0.511446	0.482846	0.456292	0.431584	120
140	0.884492	0.826371	0.777033	0.733148	0.693333	0.656808	0.623066	0.591738	0.562544	0.535260	0.509701	0.485712	140
170	0.904218	0.855095	0.812851	0.774824	0.739929	0.707546	0.677354	0.649014	0.622338	0.597164	0.573355	0.550798	170
200	0.918192	0.875677	0.838782	0.805291	0.774313	0.745329	0.718117	0.692405	0.668017	0.644841	0.622774	0.601727	200
240	0.931516	0.895480	0.863940	0.835082	0.808188	0.782831	0.758861	0.736059	0.714299	0.693473	0.673511	0.654352	240
320	0.948345	0.920741	0.896322	0.873758	0.852531	0.832324	0.813059	0.794571	0.776777	0.759613	0.743028	0.726975	320
440	0.962259	0.941834	0.923609	0.906632	0.890537	0.875096	0.860269	0.845937	0.832046	0.818554	0.805427	0.792640	440
600	0.972232	0.957070	0.943458	0.930704	0.918546	0.906818	0.895499	0.884501	0.873788	0.863330	0.853106	0.843097	600
800	0.979127	0.967660	0.957321	0.947597	0.938291	0.929281	0.920553	0.912044	0.903725	0.895577	0.887583	0.879732	800
1000	0.983279	0.974060	0.965726	0.957869	0.950332	0.943020	0.935921	0.928984	0.922190	0.915520	0.908964	0.902511	1000
INF	1.000000	1.000000	1.000000	1.000000	1.000000	1.000000	1.000000	1.000000	1.000000	1.000000	1.000000	1.000000	INF

TABLE IX (*continued*).

$p = 6 \quad \alpha = 0.05 \quad U^{\alpha}(p, q, n) \quad$ hypothesis df: q

n	1	2	3	4	5	6	7	8	9	10	11	12	n
1	0.000000	0.000000	0.000000	0.000000	0.000000	0.000000	0.000000	0.000000	0.000000	0.000000	0.000000	0.000000	1
2	0.000000	0.000000	0.000000	0.000000	0.000000	0.000000	0.000000	0.000000	0.000000	0.000000	0.000000	0.000000	2
3	0.000000	0.000000	0.000000	0.000000	0.000000	0.000000	0.000000	0.000000	0.000000	0.000000	0.000000	0.000000	3
4	0.000000	0.000000	0.000000	0.000000	0.000000	0.000000	0.000000	0.000000	0.000000	0.000000	0.000000	0.000000	4
5	0.000007	0.000002	0.000001	0.000001	0.000001	0.000000	0.000000	0.000000	0.000000	0.000000	0.000000	0.000000	5
6	0.002045	0.000315	0.000095	0.000040	0.000021	0.000012	0.000008	0.000006	0.000004	0.000003	0.000003	0.000002	6
7	0.018804	0.003479	0.001052	0.000416	0.000197	0.000106	0.000063	0.000040	0.000027	0.000020	0.000015	0.000011	7
8	0.053911	0.012883	0.004369	0.001818	0.000872	0.000465	0.000270	0.000168	0.000111	0.000076	0.000055	0.000041	8
9	0.098038	0.028824	0.011018	0.004938	0.002479	0.001358	0.000798	0.000497	0.000325	0.000222	0.000157	0.000115	9
10	0.144274	0.049685	0.021043	0.010129	0.005348	0.003035	0.001826	0.001155	0.000762	0.000521	0.000369	0.000269	10
11	0.189355	0.073697	0.033966	0.017408	0.009639	0.005672	0.003507	0.002263	0.001514	0.001046	0.000744	0.000543	11
12	0.231866	0.099450	0.049161	0.026586	0.015360	0.009348	0.005940	0.003915	0.002664	0.001865	0.001338	0.000983	12
13	0.271356	0.125933	0.066012	0.037385	0.022418	0.014071	0.009172	0.006173	0.004273	0.003033	0.002200	0.001630	13
14	0.307797	0.152453	0.083979	0.049495	0.030680	0.019795	0.013205	0.009066	0.006381	0.004592	0.003370	0.002520	14
15	0.341285	0.178581	0.102644	0.062632	0.039965	0.026433	0.018012	0.012593	0.009005	0.006568	0.004877	0.003682	15
16	0.372033	0.204010	0.121656	0.076537	0.050117	0.033893	0.023544	0.016741	0.012147	0.008974	0.006740	0.005137	16
17	0.400304	0.228568	0.140775	0.090983	0.060965	0.042061	0.029737	0.021472	0.015794	0.011811	0.008966	0.006898	17
18	0.426364	0.252176	0.159796	0.105779	0.072367	0.050834	0.036522	0.026746	0.019924	0.015070	0.011554	0.008971	18
19	0.450349	0.274785	0.178574	0.120780	0.084178	0.060119	0.043825	0.032520	0.024510	0.018734	0.014503	0.011356	19
20	0.472562	0.296393	0.197017	0.135856	0.096308	0.069818	0.051576	0.038739	0.029518	0.022785	0.017796	0.014049	20
21	0.493091	0.316990	0.215044	0.150905	0.108634	0.079840	0.059715	0.045349	0.034906	0.027193	0.021418	0.017040	21
22	0.512181	0.336628	0.232604	0.165853	0.121083	0.090122	0.068178	0.052311	0.040646	0.031936	0.025354	0.020317	22
23	0.529913	0.355328	0.249666	0.180626	0.133590	0.100596	0.076899	0.059574	0.046695	0.036988	0.029581	0.023864	23
24	0.546452	0.373143	0.266216	0.195197	0.146095	0.111189	0.085836	0.067090	0.053016	0.042316	0.034078	0.027670	24
25	0.561889	0.390109	0.282253	0.209511	0.158544	0.121873	0.094944	0.074824	0.059586	0.047895	0.038825	0.031716	25
26	0.576348	0.406285	0.297740	0.223535	0.170898	0.132587	0.104168	0.082735	0.066362	0.053696	0.043795	0.035986	26
27	0.589899	0.421688	0.312738	0.237277	0.183129	0.143309	0.113485	0.090793	0.073318	0.059697	0.048977	0.040460	27
28	0.602633	0.436379	0.327221	0.250710	0.195207	0.153998	0.122849	0.098970	0.080420	0.065867	0.054339	0.045123	28
29	0.614602	0.450416	0.341199	0.263809	0.207116	0.164629	0.132250	0.107224	0.087654	0.072196	0.059866	0.049957	29
30	0.625896	0.463794	0.354711	0.276602	0.218828	0.175171	0.141648	0.115539	0.094994	0.078649	0.065542	0.054951	30
40	0.710937	0.569976	0.466792	0.387183	0.324162	0.273470	0.232192	0.198251	0.170132	0.146678	0.126985	0.110367	40
60	0.801604	0.693451	0.607528	0.536153	0.475641	0.423707	0.378774	0.339636	0.305361	0.275238	0.248638	0.225098	60
80	0.849063	0.762264	0.690479	0.628610	0.574313	0.526153	0.483144	0.444543	0.409736	0.378269	0.349725	0.323787	80
100	0.878218	0.805945	0.744748	0.690824	0.642495	0.598763	0.558956	0.522538	0.489125	0.458377	0.430004	0.403784	100
120	0.897944	0.836112	0.782919	0.735354	0.692128	0.652489	0.615927	0.582063	0.550602	0.521300	0.493955	0.468392	120
140	0.912172	0.858176	0.811198	0.768751	0.729786	0.693709	0.660119	0.628724	0.599296	0.571649	0.545628	0.521100	140
170	0.927365	0.882016	0.842092	0.805615	0.771775	0.740119	0.710350	0.682254	0.655667	0.630455	0.606507	0.583730	170
200	0.938078	0.899001	0.864313	0.832375	0.802523	0.774395	0.747758	0.722444	0.698328	0.675308	0.653300	0.632233	200
240	0.948255	0.915270	0.885761	0.858391	0.832628	0.808187	0.784886	0.762599	0.741229	0.720701	0.700953	0.681935	240
320	0.961056	0.935919	0.913212	0.891956	0.871772	0.852459	0.833892	0.815985	0.798676	0.781916	0.765666	0.749894	320
440	0.971597	0.953076	0.936212	0.920308	0.905097	0.890438	0.876249	0.862471	0.849063	0.835995	0.823242	0.810784	440
600	0.979129	0.965422	0.952870	0.940969	0.929528	0.918448	0.907669	0.897152	0.886868	0.876798	0.866923	0.857233	600
800	0.984325	0.973979	0.964469	0.955420	0.946688	0.938203	0.929921	0.921812	0.913858	0.906042	0.898354	0.890785	800
1000	0.987450	0.979142	0.971487	0.964187	0.957129	0.950256	0.943532	0.936937	0.930455	0.924073	0.917783	0.911578	1000
INF	1.000000	1.000000	1.000000	1.000000	1.000000	1.000000	1.000000	1.000000	1.000000	1.000000	1.000000	1.000000	INF

TABLE IX (*continued*).

$p = 7$ $\alpha = 0.01$ $U^{\alpha}(p, q, n)$ hypothesis df: q

n	1	2	3	4	5	6	7	8	9	10	11	12	n
1	0.000000	0.000000	0.000000	0.000000	0.000000	0.000000	0.000000	0.000000	0.000000	0.000000	0.000000	0.000000	1
2	0.000000	0.000000	0.000000	0.000000	0.000000	0.000000	0.000000	0.000000	0.000000	0.000000	0.000000	0.000000	2
3	0.000000	0.000000	0.000000	0.000000	0.000000	0.000000	0.000000	0.000000	0.000000	0.000000	0.000000	0.000000	3
4	0.000000	0.000000	0.000000	0.000000	0.000000	0.000000	0.000000	0.000000	0.000000	0.000000	0.000000	0.000000	4
5	0.000000	0.000000	0.000000	0.000000	0.000000	0.000000	0.000000	0.000000	0.000000	0.000000	0.000000	0.000000	5
6	0.000005	0.000001	0.000000	0.000000	0.000000	0.000000	0.000000	0.000000	0.000000	0.000000	0.000000	0.000000	6
7	0.000486	0.000068	0.000019	0.000007	0.000003	0.000002	0.000001	0.000001	0.000001	0.000000	0.000000	0.000000	7
8	0.004798	0.000782	0.000215	0.000079	0.000035	0.000018	0.000010	0.000006	0.000004	0.000003	0.000002	0.000002	8
9	0.017314	0.003481	0.001047	0.000398	0.000177	0.000089	0.000049	0.000029	0.000019	0.000012	0.000009	0.000006	9
10	0.038208	0.009312	0.003116	0.001259	0.000582	0.000297	0.000165	0.000098	0.000061	0.000040	0.000028	0.000020	10
11	0.064845	0.018560	0.006864	0.002966	0.001432	0.000754	0.000426	0.000255	0.000160	0.000105	0.000072	0.000051	11
12	0.094551	0.030855	0.012459	0.005732	0.002899	0.001578	0.000912	0.000555	0.000353	0.000233	0.000159	0.000112	12
13	0.125434	0.045575	0.019845	0.009662	0.005103	0.002873	0.001704	0.001057	0.000682	0.000455	0.000313	0.000221	13
14	0.156314	0.062081	0.028840	0.014763	0.008113	0.004715	0.002870	0.001817	0.001191	0.000804	0.000558	0.000397	14
15	0.186495	0.079813	0.039203	0.020973	0.011946	0.007151	0.004460	0.002881	0.001919	0.001314	0.000922	0.000661	15
16	0.215591	0.098313	0.050682	0.028192	0.016584	0.010199	0.006508	0.004286	0.002902	0.002013	0.001428	0.001034	16
17	0.243398	0.117225	0.063039	0.036299	0.021981	0.013856	0.009030	0.006056	0.004164	0.002928	0.002101	0.001535	17
18	0.269838	0.136276	0.076063	0.045168	0.028076	0.018099	0.012027	0.008203	0.005725	0.004077	0.002958	0.002182	18
19	0.294893	0.155259	0.089567	0.054675	0.034797	0.022896	0.015490	0.010733	0.007595	0.005476	0.004016	0.002991	19
20	0.318593	0.174026	0.103395	0.064703	0.042071	0.028206	0.019400	0.013641	0.009779	0.007133	0.005285	0.003972	20
21	0.340992	0.192467	0.117419	0.075148	0.049825	0.033983	0.023733	0.016917	0.012276	0.009053	0.006773	0.005135	21
22	0.362149	0.210506	0.131529	0.085915	0.057989	0.040181	0.028460	0.020544	0.015079	0.011235	0.008484	0.006487	22
23	0.382138	0.228088	0.145640	0.096921	0.066497	0.046751	0.033549	0.024505	0.018179	0.013677	0.010420	0.008031	23
24	0.401034	0.245182	0.159680	0.108094	0.075288	0.053649	0.038968	0.028778	0.021564	0.016372	0.012577	0.009768	24
25	0.418901	0.261767	0.173594	0.119371	0.084307	0.060830	0.044686	0.033340	0.025219	0.019311	0.014953	0.011698	25
26	0.435815	0.277834	0.187337	0.130700	0.093504	0.068253	0.050669	0.038170	0.029127	0.022486	0.017542	0.013817	26
27	0.451836	0.293382	0.200875	0.142036	0.102837	0.075880	0.056888	0.043242	0.033272	0.025883	0.020335	0.016123	27
28	0.467026	0.308414	0.214182	0.153341	0.112264	0.083676	0.063314	0.048535	0.037637	0.029490	0.023325	0.018609	28
29	0.481442	0.322941	0.227238	0.164581	0.121752	0.091609	0.069918	0.054026	0.042205	0.033296	0.026502	0.021269	29
30	0.495137	0.336973	0.240029	0.175732	0.131271	0.099650	0.076676	0.059694	0.046958	0.037285	0.029857	0.024096	30
40	0.601481	0.453452	0.352532	0.279077	0.223835	0.181405	0.148306	0.122167	0.101312	0.084528	0.070913	0.059797	40
60	0.720662	0.599212	0.507477	0.434307	0.374443	0.324703	0.282924	0.247545	0.217383	0.191527	0.169255	0.149991	60
80	0.785257	0.684670	0.604805	0.538151	0.481262	0.432066	0.389140	0.351441	0.318157	0.288649	0.262392	0.238958	80
100	0.825659	0.740370	0.670619	0.610809	0.558448	0.512056	0.470627	0.433418	0.399849	0.369460	0.341864	0.316744	100
120	0.853291	0.779445	0.717860	0.664095	0.616219	0.573107	0.533999	0.498338	0.465691	0.435710	0.408105	0.382634	120
140	0.873373	0.808339	0.753345	0.704715	0.660882	0.620947	0.584311	0.550538	0.519289	0.490293	0.463324	0.438191	140
170	0.894951	0.839849	0.792566	0.750184	0.711466	0.675794	0.642653	0.611744	0.582816	0.555676	0.530155	0.506117	170
200	0.910250	0.862477	0.821063	0.783589	0.749029	0.716923	0.686859	0.658572	0.631886	0.606648	0.582733	0.560038	200
240	0.924846	0.884289	0.848791	0.816382	0.786225	0.757986	0.731332	0.706070	0.682043	0.659149	0.637294	0.616401	240
320	0.943294	0.912166	0.884592	0.859136	0.835186	0.812531	0.790932	0.770257	0.750414	0.731325	0.712926	0.695173	320
440	0.958556	0.935488	0.914854	0.895630	0.877381	0.859974	0.843239	0.827091	0.811467	0.796319	0.781611	0.767310	440
600	0.969501	0.952359	0.936918	0.922439	0.908607	0.895334	0.882498	0.870040	0.857917	0.846097	0.834556	0.823272	600
800	0.977071	0.964096	0.952354	0.941294	0.930682	0.920457	0.910529	0.900854	0.891402	0.882150	0.873081	0.864181	800
1000	0.981630	0.971194	0.961722	0.952776	0.944171	0.935859	0.927767	0.919864	0.912124	0.904530	0.897070	0.889731	1000
INF	1.000000	1.000000	1.000000	1.000000	1.000000	1.000000	1.000000	1.000000	1.000000	1.000000	1.000000	1.000000	INF

TABLE IX (*continued*).

$p = 7 \quad \alpha = 0.05 \quad U^{\alpha}(p, q, n) \quad$ hypothesis df: q

n	1	2	3	4	5	6	7	8	9	10	11	12	n
1	0.000000	0.000000	0.000000	0.000000	0.000000	0.000000	0.000000	0.000000	0.000000	0.000000	0.000000	0.000000	1
2	0.000000	0.000000	0.000000	0.000000	0.000000	0.000000	0.000000	0.000000	0.000000	0.000000	0.000000	0.000000	2
3	0.000000	0.000000	0.000000	0.000000	0.000000	0.000000	0.000000	0.000000	0.000000	0.000000	0.000000	0.000000	3
4	0.000000	0.000000	0.000000	0.000000	0.000000	0.000000	0.000000	0.000000	0.000000	0.000000	0.000000	0.000000	4
5	0.000000	0.000000	0.000000	0.000000	0.000000	0.000000	0.000000	0.000000	0.000000	0.000000	0.000000	0.000000	5
6	0.000043	0.000006	0.000002	0.000001	0.000001	0.000000	0.000000	0.000000	0.000000	0.000000	0.000000	0.000000	6
7	0.002625	0.000350	0.000091	0.000033	0.000015	0.000008	0.000005	0.000003	0.000002	0.000002	0.000001	0.000001	7
8	0.017612	0.002953	0.000809	0.000292	0.000126	0.000063	0.000034	0.000020	0.000013	0.000009	0.000006	0.000005	8
9	0.047835	0.010329	0.003195	0.001223	0.000543	0.000270	0.000147	0.000086	0.000053	0.000035	0.000024	0.000017	9
10	0.086645	0.023060	0.008067	0.003338	0.001558	0.000798	0.000440	0.000259	0.000160	0.000104	0.000070	0.000049	10
11	0.128234	0.040186	0.015642	0.006974	0.003433	0.001826	0.001035	0.000619	0.000387	0.000252	0.000170	0.000119	11
12	0.169506	0.060396	0.025707	0.012249	0.006343	0.003508	0.002048	0.001252	0.000796	0.000525	0.000357	0.000249	12
13	0.209026	0.082538	0.037857	0.019109	0.010357	0.005940	0.003571	0.002234	0.001448	0.000967	0.000665	0.000468	13
14	0.246203	0.105734	0.051646	0.027402	0.015466	0.009172	0.005668	0.003628	0.002395	0.001625	0.001131	0.000804	14
15	0.280861	0.129346	0.066659	0.036933	0.021607	0.013206	0.008371	0.005476	0.003682	0.002537	0.001787	0.001285	15
16	0.313032	0.152929	0.082533	0.047494	0.028684	0.018013	0.011688	0.007801	0.005337	0.003733	0.002664	0.001936	16
17	0.342842	0.176179	0.098971	0.058884	0.036586	0.023544	0.015606	0.010611	0.007379	0.005235	0.003782	0.002778	17
18	0.370455	0.198894	0.115731	0.070921	0.045199	0.029736	0.020096	0.013900	0.009814	0.007057	0.005159	0.003829	18
19	0.396050	0.220944	0.132623	0.083445	0.054409	0.036520	0.025122	0.017653	0.012640	0.009204	0.006805	0.005102	19
20	0.419802	0.242252	0.149498	0.096315	0.064111	0.043824	0.030640	0.021845	0.015847	0.011676	0.008725	0.006605	20
21	0.441876	0.262777	0.166240	0.109415	0.074209	0.051579	0.036603	0.026450	0.019422	0.014469	0.010921	0.008342	21
22	0.462425	0.282503	0.182765	0.122645	0.084616	0.059717	0.042965	0.031435	0.023345	0.017571	0.013387	0.010314	22
23	0.481587	0.301432	0.199007	0.135923	0.095257	0.068177	0.049678	0.036769	0.027595	0.020971	0.016120	0.012521	23
24	0.499486	0.319577	0.214919	0.149181	0.106063	0.076901	0.056697	0.042416	0.032148	0.024653	0.019108	0.014956	24
25	0.516238	0.336959	0.230467	0.162364	0.116978	0.085838	0.063980	0.048346	0.036980	0.028599	0.022341	0.017614	25
26	0.531942	0.353606	0.245630	0.175429	0.127951	0.094941	0.071488	0.054525	0.042067	0.032794	0.025807	0.020487	26
27	0.546689	0.369546	0.260395	0.188339	0.138940	0.104168	0.079183	0.060924	0.047385	0.037217	0.029493	0.023565	27
28	0.560560	0.384810	0.274752	0.201068	0.149908	0.113482	0.087032	0.067514	0.052911	0.041851	0.033384	0.026838	28
29	0.573629	0.399430	0.288701	0.213591	0.160826	0.122851	0.095005	0.074268	0.058622	0.046678	0.037467	0.030296	29
30	0.585961	0.413438	0.302243	0.225894	0.171667	0.132247	0.103073	0.081161	0.064496	0.051680	0.041727	0.033928	30
40	0.679227	0.525996	0.417050	0.335433	0.272668	0.223571	0.184671	0.153533	0.128393	0.107940	0.091192	0.077392	40
60	0.779306	0.659576	0.566032	0.489695	0.426135	0.372561	0.327012	0.288026	0.254476	0.225471	0.200293	0.178361	60
80	0.831906	0.735024	0.655779	0.588321	0.529875	0.478709	0.433602	0.393626	0.358051	0.326284	0.297833	0.272287	80
100	0.864288	0.783251	0.715144	0.655689	0.602930	0.555673	0.513081	0.474521	0.439488	0.407570	0.378421	0.351744	100
120	0.886219	0.816680	0.757179	0.704361	0.656738	0.613420	0.573796	0.537400	0.503866	0.472893	0.444226	0.417647	120
140	0.902052	0.841199	0.788462	0.741086	0.697881	0.658148	0.621410	0.587314	0.555578	0.525974	0.498306	0.472408	140
170	0.918970	0.867751	0.822764	0.781839	0.744063	0.708913	0.676042	0.645194	0.616167	0.588800	0.562955	0.538514	170
200	0.930905	0.886705	0.847518	0.811553	0.778074	0.746666	0.717058	0.689053	0.662499	0.637274	0.613274	0.590412	200
240	0.942249	0.904887	0.871470	0.840546	0.811527	0.784091	0.758031	0.733197	0.709478	0.686784	0.665038	0.644178	240
320	0.956525	0.928004	0.902213	0.878097	0.855239	0.833417	0.812491	0.792362	0.772959	0.754224	0.736112	0.718583	320
440	0.968286	0.947243	0.928043	0.909937	0.892635	0.875985	0.859892	0.844294	0.829142	0.814403	0.800046	0.786051	440
600	0.976693	0.961103	0.946788	0.933208	0.920155	0.907522	0.895244	0.883276	0.871588	0.860157	0.848964	0.837994	600
800	0.982494	0.970720	0.959861	0.949517	0.939535	0.929836	0.920373	0.911114	0.902038	0.893128	0.884371	0.875758	800
1000	0.985983	0.976524	0.967778	0.959426	0.951346	0.943478	0.935782	0.928236	0.920822	0.913527	0.906342	0.899259	1000
INF	1.000000	1.000000	1.000000	1.000000	1.000000	1.000000	1.000000	1.000000	1.000000	1.000000	1.000000	1.000000	INF

TABLE IX (*continued*).

$p = 8$ $\alpha = 0.01$ $U^{\alpha}(p, q, n)$ hypothesis df: q

n	1	2	3	4	5	6	7	8	9	10	11	12	n
1	0.000000	0.000000	0.000000	0.000000	0.000000	0.000000	0.000000	0.000000	0.000000	0.000000	0.000000	0.000000	1
2	0.000000	0.000000	0.000000	0.000000	0.000000	0.000000	0.000000	0.000000	0.000000	0.000000	0.000000	0.000000	2
3	0.000000	0.000000	0.000000	0.000000	0.000000	0.000000	0.000000	0.000000	0.000000	0.000000	0.000000	0.000000	3
4	0.000000	0.000000	0.000000	0.000000	0.000000	0.000000	0.000000	0.000000	0.000000	0.000000	0.000000	0.000000	4
5	0.000000	0.000000	0.000000	0.000000	0.000000	0.000000	0.000000	0.000000	0.000000	0.000000	0.000000	0.000000	5
6	0.000000	0.000000	0.000000	0.000000	0.000000	0.000000	0.000000	0.000000	0.000000	0.000000	0.000000	0.000000	6
7	0.000021	0.000002	0.000001	0.000000	0.000000	0.000000	0.000000	0.000000	0.000000	0.000000	0.000000	0.000000	7
8	0.000738	0.000088	0.000021	0.000007	0.000003	0.000001	0.000001	0.000000	0.000000	0.000000	0.000000	0.000000	8
9	0.005130	0.000759	0.000188	0.000063	0.000025	0.000012	0.000006	0.000004	0.000002	0.000001	0.000001	0.000001	9
10	0.016457	0.003031	0.000837	0.000293	0.000121	0.000057	0.000029	0.000016	0.000010	0.000006	0.000004	0.000003	10
11	0.034984	0.007819	0.002414	0.000906	0.000390	0.000187	0.000098	0.000055	0.000033	0.000020	0.000013	0.000009	11
12	0.058795	0.015460	0.005293	0.002131	0.000964	0.000477	0.000255	0.000144	0.000086	0.000054	0.000035	0.000024	12
13	0.085701	0.025773	0.009664	0.004155	0.001974	0.001014	0.000555	0.000321	0.000194	0.000123	0.000080	0.000054	13
14	0.114026	0.038323	0.015549	0.007095	0.003528	0.001878	0.001057	0.000624	0.000384	0.000245	0.000162	0.000110	14
15	0.142659	0.052612	0.022851	0.010993	0.005702	0.003140	0.001817	0.001097	0.000687	0.000445	0.000296	0.000203	15
16	0.170909	0.068175	0.031408	0.015836	0.008535	0.004851	0.002881	0.001778	0.001134	0.000744	0.000502	0.000347	16
17	0.198367	0.084613	0.041037	0.021570	0.012034	0.007041	0.004286	0.002700	0.001752	0.001168	0.000797	0.000556	17
18	0.224804	0.101603	0.051550	0.028118	0.016183	0.009723	0.006056	0.003890	0.002567	0.001736	0.001200	0.000846	18
19	0.250102	0.118888	0.062771	0.035392	0.020951	0.012897	0.008203	0.005367	0.003600	0.002468	0.001727	0.001231	19
20	0.274219	0.136267	0.074542	0.043297	0.026294	0.016549	0.010733	0.007144	0.004864	0.003380	0.002393	0.001723	20
21	0.297154	0.153588	0.086724	0.051742	0.032161	0.020658	0.013641	0.009225	0.006371	0.004484	0.003210	0.002335	21
22	0.318934	0.170736	0.099197	0.060641	0.038499	0.025196	0.016917	0.011612	0.008127	0.005789	0.004189	0.003076	22
23	0.339603	0.187624	0.111859	0.069912	0.045254	0.030131	0.020544	0.014298	0.010133	0.007299	0.005336	0.003955	23
24	0.359212	0.204189	0.124624	0.079482	0.052374	0.035431	0.024505	0.017276	0.012388	0.009019	0.006658	0.004978	24
25	0.377817	0.220388	0.137421	0.089287	0.059809	0.041060	0.028778	0.020535	0.014887	0.010948	0.008156	0.006149	25
26	0.395477	0.236190	0.150193	0.099266	0.067511	0.046985	0.033340	0.024060	0.017624	0.013084	0.009833	0.007472	26
27	0.412247	0.251575	0.162890	0.109371	0.075438	0.053173	0.038170	0.027838	0.020590	0.015423	0.011686	0.008948	27
28	0.428183	0.266532	0.175473	0.119555	0.083549	0.059592	0.043242	0.031852	0.023775	0.017959	0.013714	0.010578	28
29	0.443335	0.281057	0.187912	0.129781	0.091808	0.066212	0.048535	0.036085	0.027167	0.020686	0.015914	0.012359	29
30	0.457755	0.295150	0.200180	0.140015	0.100183	0.073004	0.054026	0.040521	0.030756	0.023596	0.018280	0.014290	30
40	0.570452	0.414141	0.310972	0.238227	0.185181	0.145635	0.115665	0.092650	0.074781	0.060775	0.049703	0.040884	40
60	0.697943	0.566845	0.469642	0.393584	0.332536	0.282759	0.241719	0.207592	0.179013	0.154941	0.134560	0.117227	60
80	0.767444	0.657821	0.571844	0.501042	0.441431	0.390577	0.346803	0.308868	0.275817	0.246896	0.221493	0.199106	80
100	0.811035	0.717584	0.641821	0.577506	0.521783	0.472932	0.429764	0.391400	0.357150	0.326462	0.298881	0.274029	100
120	0.840895	0.759705	0.692428	0.634154	0.582694	0.536747	0.495425	0.458070	0.424165	0.393293	0.365111	0.339324	120
140	0.862619	0.790948	0.730631	0.677631	0.630186	0.587267	0.548177	0.512401	0.479541	0.449270	0.421319	0.395458	140
170	0.885982	0.825107	0.773033	0.726554	0.684388	0.645690	0.609966	0.576841	0.546023	0.517277	0.490407	0.465247	170
200	0.902559	0.849693	0.803953	0.762678	0.724868	0.689844	0.657190	0.626631	0.597943	0.570945	0.545489	0.521449	200
240	0.918384	0.873433	0.834123	0.798284	0.765147	0.734170	0.705043	0.677532	0.651483	0.626761	0.603257	0.580879	240
320	0.938398	0.903831	0.873198	0.844911	0.818448	0.793424	0.769627	0.746913	0.725172	0.704312	0.684271	0.664991	320
440	0.954965	0.929310	0.906326	0.884882	0.864628	0.845292	0.826734	0.808858	0.791594	0.774890	0.758703	0.742995	440
600	0.966853	0.947766	0.930536	0.914342	0.898941	0.884139	0.869838	0.855973	0.842497	0.829376	0.816581	0.804090	600
800	0.975077	0.960620	0.947501	0.935108	0.923267	0.911833	0.900736	0.889929	0.879380	0.869062	0.858958	0.849052	800
1000	0.980031	0.968398	0.957807	0.947773	0.938158	0.928847	0.919787	0.910939	0.902280	0.893788	0.885451	0.877255	1000
INF	1.000000	1.000000	1.000000	1.000000	1.000000	1.000000	1.000000	1.000000	1.000000	1.000000	1.000000	1.000000	INF

TABLE IX (*continued*).

$p = 8$ $\alpha = 0.05$ $U^{\alpha}(p, q, n)$ hypothesis df: q

n	1	2	3	4	5	6	7	8	9	10	11	12	n
1	0.000000	0.000000	0.000000	0.000000	0.000000	0.000000	0.000000	0.000000	0.000000	0.000000	0.000000	0.000000	1
2	0.000000	0.000000	0.000000	0.000000	0.000000	0.000000	0.000000	0.000000	0.000000	0.000000	0.000000	0.000000	2
3	0.000000	0.000000	0.000000	0.000000	0.000000	0.000000	0.000000	0.000000	0.000000	0.000000	0.000000	0.000000	3
4	0.000000	0.000000	0.000000	0.000000	0.000000	0.000000	0.000000	0.000000	0.000000	0.000000	0.000000	0.000000	4
5	0.000000	0.000000	0.000000	0.000000	0.000000	0.000000	0.000000	0.000000	0.000000	0.000000	0.000000	0.000000	5
6	0.000000	0.000000	0.000000	0.000000	0.000000	0.000000	0.000000	0.000000	0.000000	0.000000	0.000000	0.000000	6
7	0.000138	0.000015	0.000004	0.000001	0.000001	0.000000	0.000000	0.000000	0.000000	0.000000	0.000000	0.000000	7
8	0.003295	0.000393	0.000090	0.000029	0.000012	0.000006	0.000003	0.000002	0.000001	0.000001	0.000001	0.000000	8
9	0.017079	0.002632	0.000659	0.000218	0.000087	0.000040	0.000020	0.000011	0.000007	0.000004	0.000003	0.000002	9
10	0.043574	0.008626	0.002458	0.000872	0.000361	0.000168	0.000086	0.000047	0.000028	0.000017	0.000011	0.000008	10
11	0.078039	0.019031	0.006148	0.002365	0.001032	0.000497	0.000259	0.000144	0.000085	0.000052	0.000034	0.000023	11
12	0.115676	0.033314	0.012011	0.004993	0.002304	0.001155	0.000619	0.000351	0.000209	0.000130	0.000084	0.000056	12
13	0.153630	0.050518	0.019990	0.008908	0.004335	0.002263	0.001252	0.000727	0.000441	0.000278	0.000181	0.000122	13
14	0.190453	0.069716	0.029839	0.014129	0.007216	0.003915	0.002234	0.001331	0.000824	0.000527	0.000347	0.000235	14
15	0.225477	0.090151	0.041241	0.020590	0.010980	0.006173	0.003628	0.002215	0.001399	0.000910	0.000608	0.000416	15
16	0.258443	0.111245	0.053875	0.028171	0.015610	0.009065	0.005476	0.003422	0.002203	0.001457	0.000987	0.000683	16
17	0.289300	0.132575	0.067447	0.036729	0.021061	0.012594	0.007801	0.004982	0.003269	0.002197	0.001509	0.001057	17
18	0.318105	0.153836	0.081699	0.046115	0.027265	0.016740	0.010611	0.006915	0.004617	0.003151	0.002194	0.001555	18
19	0.344966	0.174814	0.096415	0.056185	0.034144	0.021472	0.013900	0.009228	0.006265	0.004339	0.003060	0.002194	19
20	0.370015	0.195359	0.111416	0.066805	0.041616	0.026747	0.017653	0.011923	0.008219	0.005771	0.004120	0.002987	20
21	0.393387	0.215374	0.126559	0.077857	0.049601	0.032519	0.021845	0.014991	0.010483	0.007456	0.005386	0.003946	21
22	0.415217	0.234796	0.141726	0.089233	0.058021	0.038737	0.026450	0.018419	0.013053	0.009397	0.006863	0.005078	22
23	0.435632	0.253588	0.156826	0.100843	0.066804	0.045350	0.031435	0.022192	0.015923	0.011593	0.008555	0.006390	23
24	0.454749	0.271732	0.171785	0.112606	0.075884	0.052311	0.036769	0.026287	0.019081	0.014041	0.010462	0.007885	24
25	0.472677	0.289225	0.186549	0.124457	0.085199	0.059573	0.042416	0.030685	0.022515	0.016733	0.012583	0.009565	25
26	0.489514	0.306072	0.201075	0.136338	0.094698	0.067091	0.048346	0.035361	0.026210	0.019663	0.014914	0.011428	26
27	0.505352	0.322285	0.215331	0.148203	0.104332	0.074826	0.054525	0.040293	0.030150	0.022818	0.017449	0.013472	27
28	0.520271	0.337880	0.229293	0.160010	0.114060	0.082739	0.060924	0.045457	0.034319	0.026189	0.020182	0.015694	28
29	0.534345	0.352879	0.242945	0.171728	0.123844	0.090796	0.067514	0.050831	0.038700	0.029764	0.023104	0.018089	29
30	0.547639	0.367302	0.256277	0.183330	0.133653	0.098967	0.074268	0.056394	0.043276	0.033529	0.026207	0.020651	30
40	0.648630	0.484826	0.371902	0.289857	0.228618	0.182082	0.146235	0.118316	0.096365	0.078964	0.065068	0.053897	40
60	0.757690	0.627279	0.527185	0.447009	0.381482	0.327255	0.281978	0.243910	0.211718	0.184362	0.161015	0.141011	60
80	0.815243	0.708843	0.622840	0.550577	0.488795	0.435425	0.388992	0.348380	0.312704	0.281253	0.253441	0.228779	80
100	0.850742	0.761330	0.686819	0.622411	0.565838	0.515687	0.470954	0.430871	0.394827	0.362322	0.332935	0.306310	100
120	0.874811	0.797857	0.732425	0.674791	0.623251	0.576764	0.534599	0.496197	0.461114	0.428982	0.399491	0.372376	120
140	0.892201	0.824719	0.766516	0.714559	0.667497	0.624521	0.585067	0.548712	0.515117	0.484002	0.455129	0.428296	140
170	0.910793	0.853874	0.804039	0.758920	0.717493	0.679163	0.643522	0.610267	0.579158	0.549999	0.522621	0.496881	170
200	0.923918	0.874725	0.831204	0.791410	0.754525	0.720081	0.687764	0.657345	0.628642	0.601508	0.575820	0.551470	200
240	0.936396	0.894758	0.857556	0.823223	0.791114	0.760867	0.732246	0.705079	0.679234	0.654605	0.631100	0.608645	240
320	0.952108	0.920269	0.891472	0.864586	0.839159	0.814944	0.791784	0.769570	0.748216	0.727659	0.707843	0.688722	320
440	0.965057	0.941534	0.920045	0.899793	0.880463	0.861889	0.843968	0.826629	0.809821	0.793502	0.777640	0.762209	440
600	0.974316	0.956873	0.940825	0.925599	0.910972	0.896826	0.883093	0.869724	0.856684	0.843948	0.831494	0.819306	600
800	0.980707	0.967524	0.955337	0.943721	0.932512	0.921624	0.911008	0.900629	0.890464	0.880494	0.870704	0.861084	800
1000	0.984551	0.973956	0.964134	0.954746	0.945661	0.936815	0.928167	0.919691	0.911367	0.903183	0.895127	0.887192	1000
INF	1.000000	1.000000	1.000000	1.000000	1.000000	1.000000	1.000000	1.000000	1.000000	1.000000	1.000000	1.000000	INF

TABLE X. Upper Percentage Points of Hotelling and Lawley's Trace Criterion, $U_0^{\alpha}(s, m, n)$

$\alpha = 0.01 \quad s = 2 \quad U_0^{\alpha}(s, m, n)$

m \ n	15	20	25	30	35	40	45	50	60	80	100	130	160	200
1	--	--	--	--	--	--	--	--	--	0.148	0.118	0.091	0.073	0.059
1.5	--	--	--	--	0.396	0.344	0.304	0.273	0.226	0.168	0.134	0.102	0.083	0.066
2	--	--	0.632	0.520	0.442	0.384	0.339	0.303	0.251	0.187	0.149	0.114	0.092	0.074
2.5	--	--	0.697	0.572	0.485	0.422	0.372	0.334	0.276	0.205	0.163	0.125	0.101	0.081
3	--	0.970	0.760	0.624	0.529	0.458	0.406	0.364	0.301	0.224	0.178	0.136	0.110	0.088
3.5	--	1.051	0.822	0.675	0.572	0.497	0.439	0.393	0.325	0.242	0.192	0.147	0.119	0.095
4	--	1.130	0.884	0.725	0.616	0.534	0.472	0.422	0.349	0.259	0.206	0.158	0.128	0.102
4.5	--	1.210	0.945	0.776	0.658	0.571	0.504	0.451	0.373	0.277	0.220	0.168	0.136	0.109
5	--	1.288	1.005	0.825	0.700	0.607	0.536	0.480	0.396	0.294	0.234	0.179	0.145	0.116
10	--	2.057	1.602	1.310	1.108	0.960	0.846	0.757	0.625	0.462	0.367	0.280	0.227	0.181
15	3.934	2.812	2.183	1.782	1.505	1.302	1.147	1.024	0.845	0.624	0.495	0.378	0.305	0.243
20	4.974	3.551	2.758	2.243	1.895	1.640	1.443	1.288	1.061	0.783	0.621	0.473	0.382	0.304
25	6.043	4.302	3.330	2.712	2.286	1.974	1.736	1.549	1.275	0.940	0.745	0.567	0.458	0.365
30	7.092	5.043	3.900	3.173	2.673	2.306	2.028	1.807	1.488	1.097	0.868	0.661	0.533	0.424
50	11.28	7.996	6.167	5.007	4.202	3.626	3.185	2.837	2.328	1.711	1.353	1.029	0.829	0.659
60	13.37	9.468	7.297	5.921	4.973	4.284	3.761	3.350	2.748	2.019	1.594	1.211	0.976	0.775
80	17.55	12.41	9.553	7.745	6.500	5.595	4.910	4.371	3.583	2.628	2.073	1.733	1.267	1.005
100	21.73	15.35	11.81	9.567	8.024	6.905	6.056	5.390	4.415	3.235	2.550	1.933	1.556	1.234
130	27.99	19.76	15.19	12.30	10.31	8.868	7.774	6.916	5.660	4.143	3.263	2.472	1.988	1.576
160	34.25	24.17	18.56	15.03	12.59	10.83	9.489	8.440	6.904	5.050	4.782	2.962	2.419	1.917
200	42.60	30.04	23.07	18.67	15.63	13.44	11.78	10.47	8.562	6.258	4.923	3.724	2.993	2.370
500	105	74.09	56.83	45.94	38.44	33.02	28.91	25.69	20.98	15.30	12.02	9.072	7.279	5.753
1000	210	147	113	91.38	76.45	65.64	57.46	51.04	41.66	30.36	23.83	17.97	14.41	11.38

From K. C. S. Pillai, *Statistical Tables for Tests of Multivariate Hypotheses*, The Statistical Center, University of the Phillipines, 1960. Reproduced by permission of the author and publisher.

TABLE X (*continued*).

$\alpha = 0.05 \qquad s = 2 \qquad U_0^\alpha(s, m, n)$

m \ n	15	20	25	30	35	40	45	50	60	80	100	130	160	200
1	--	--	--	--	--	--	--	--	--	0.116	0.092	0.071	0.058	0.046
1.5	--	--	--	--	0.310	0.270	0.239	0.215	0.179	0.133	0.106	0.081	0.066	0.053
2	--	--	0.496	0.410	0.350	0.305	0.270	0.242	0.202	0.150	0.120	0.092	0.075	0.060
2.5	--	--	0.551	0.456	0.389	0.339	0.300	0.269	0.224	0.167	0.133	0.102	0.083	0.066
3	--	0.768	0.607	0.502	0.428	0.373	0.330	0.296	0.246	0.183	0.146	0.112	0.091	0.073
3.5	--	0.838	0.662	0.547	0.466	0.406	0.360	0.323	0.268	0.200	0.159	0.122	0.099	0.079
4	--	0.907	0.717	0.592	0.505	0.439	0.389	0.349	0.290	0.216	0.172	0.132	0.107	0.085
4.5	--	0.976	0.771	0.637	0.542	0.472	0.418	0.375	0.311	0.232	0.185	0.142	0.115	0.092
5	--	1.045	0.824	0.681	0.580	0.505	0.447	0.401	0.332	0.248	0.197	0.151	0.123	0.098
10	--	1.718	1.352	1.115	0.947	0.823	0.728	0.652	0.541	0.402	0.320	0.245	0.199	0.159
15	3.263	2.381	1.870	1.539	1.306	1.134	1.003	0.899	0.744	0.553	0.440	0.337	0.273	0.217
20	4.171	3.039	2.383	1.958	1.661	1.442	1.275	1.142	0.945	0.701	0.558	0.426	0.345	0.275
25	5.077	3.692	2.894	2.374	2.014	1.749	1.545	1.383	1.143	0.848	0.674	0.515	0.417	0.332
30	5.981	4.345	3.403	2.792	2.366	2.053	1.813	1.622	1.341	0.994	0.790	0.603	0.488	0.389
50	9.590	6.947	5.430	4.449	3.767	3.266	2.881	2.577	2.127	1.573	1.248	0.953	0.770	0.613
60	11.39	8.246	6.442	5.275	4.466	3.870	3.413	3.051	2.516	1.861	1.476	1.126	0.910	0.724
80	14.99	10.84	8.463	6.926	5.860	5.074	4.475	3.998	3.295	2.435	1.930	1.580	1.188	0.945
100	18.59	13.43	10.48	8.575	7.253	6.280	5.534	4.944	4.072	3.007	2.382	1.814	1.465	1.165
130	23.99	17.32	13.51	11.05	9.340	8.084	7.122	6.361	5.237	3.864	3.059	2.328	1.879	1.494
160	29.39	21.21	16.54	13.52	11.43	9.888	8.709	7.777	6.400	4.721	3.735	2.810	2.292	1.822
200	36.58	26.40	20.57	16.81	14.21	12.29	10.82	9.663	7.951	5.861	4.635	3.525	2.842	2.258
500	90.54	65.27	50.83	41.51	35.06	30.31	26.68	23.80	19.57	14.40	11.38	8.640	6.959	5.521
1000	180	130	101	82.67	69.79	60.34	53.09	47.36	38.92	28.63	22.61	17.16	13.81	10.95

TABLE X (*continued*).

$\alpha = 0.01 \qquad s = 3 \qquad U_0^{\alpha}(s, m, n)$

m \ n	15	20	25	30	35	40	45	50	60	80	100	130	160	200
-0.5	--	--	--	--	--	--	--	--	--	--	--	0.084	0.068	0.055
0	--	--	--	0.464	0.394	0.341	0.303	0.271	0.225	0.167	0.133	0.102	0.083	0.066
.5	--	0.837	0.658	0.541	0.459	0.400	0.353	0.317	0.262	0.195	0.156	0.119	0.097	0.077
1	--	0.956	0.750	0.617	0.524	0.456	0.403	0.361	0.299	0.223	0.177	0.136	0.110	0.088
1.5	1.475	1.071	0.840	0.691	0.587	0.511	0.451	0.404	0.335	0.249	0.198	0.152	0.123	0.098
2	1.634	1.186	0.930	0.765	0.649	0.564	0.499	0.447	0.370	0.275	0.219	0.168	0.136	0.109
2.5	1.791	1.298	1.018	0.837	0.711	0.617	0.546	0.489	0.405	0.301	0.240	0.183	0.149	0.119
3	1.947	1.410	1.106	0.909	0.771	0.670	0.592	0.531	0.439	0.326	0.260	0.199	0.161	0.129
3.5	2.102	1.521	1.193	0.980	0.831	0.722	0.638	0.572	0.473	0.339	0.279	0.214	0.173	0.138
4	2.252	1.632	1.278	1.050	0.891	0.774	0.684	0.612	0.506	0.376	0.299	0.229	0.186	0.148
4.5	2.408	1.742	1.364	1.120	0.950	0.825	0.729	0.653	0.540	0.401	0.319	0.244	0.198	0.158
5	2.560	1.852	1.449	1.190	1.009	0.876	0.774	0.693	0.573	0.424	0.338	0.259	0.210	0.167
10	4.060	2.925	2.283	1.871	1.583	1.372	1.211	1.084	0.895	0.664	0.528	0.404	0.327	0.260
15	5.541	4.026	3.101	2.536	2.180	1.858	1.639	1.465	1.210	0.896	0.712	0.544	0.440	0.351
20	7.009	5.030	3.938	3.196	2.701	2.338	2.061	1.842	1.520	1.124	0.893	0.681	0.551	0.439
25	8.479	6.072	4.719	3.849	3.253	2.815	2.480	2.216	1.825	1.351	1.072	0.818	0.661	0.526
30	9.943	7.108	5.522	4.505	3.804	3.290	2.897	2.588	2.132	1.576	1.249	0.953	0.770	0.613
50	15.79	11.26	8.723	7.113	5.992	5.176	4.555	4.065	3.344	2.466	1.953	1.488	1.201	0.955
60	18.70	13.33	10.32	8.403	7.082	6.116	5.379	4.799	3.947	2.909	2.302	1.752	1.414	1.125
80	24.53	17.46	13.51	10.99	9.259	7.991	7.026	6.266	5.149	3.791	2.997	2.280	1.839	1.461
100	30.36	21.53	16.70	13.58	11.43	9.864	8.669	7.729	6.348	4.670	3.691	2.806	2.261	1.796
130	39.10	27.79	21.48	17.46	14.69	12.67	11.13	9.922	8.145	5.986	4.728	3.591	2.893	2.297
160	47.83	33.99	26.25	21.33	17.95	15.48	13.59	12.11	9.939	7.301	5.764	4.376	3.523	2.796
200	59.48	41.95	32.62	26.50	22.29	19.21	16.87	15.03	12.33	9.052	7.143	5.419	4.362	3.460
500	147	104	80.36	65.24	54.83	47.24	41.45	36.91	30.25	22.17	17.47	13.23	10.64	8.424
1000	292	207	160	130	109	93.93	82.41	73.36	60.09	44.01	34.66	26.23	21.08	16.68

TABLE X (*continued*).

$\alpha = 0.05 \quad s = 3 \quad U_0^{\alpha}(s, m, n)$

m \ n	15	20	25	30	35	40	45	50	60	80	100	130	160	200
-0.5	--	--	--	--	--	--	--	--	--	--	0.085	0.066	0.053	0.042
0	--	--	--	0.362	0.309	0.269	0.239	0.214	0.178	0.133	0.106	0.081	0.066	0.053
.5	--	0.658	0.520	0.431	0.368	0.321	0.284	0.255	0.212	0.158	0.126	0.097	0.079	0.063
1	--	0.761	0.603	0.499	0.425	0.371	0.329	0.295	0.245	0.183	0.146	0.112	0.091	0.073
1.5	1.172	0.864	0.683	0.566	0.482	0.420	0.372	0.334	0.278	0.207	0.165	0.127	0.103	0.082
2	1.310	0.964	0.763	0.632	0.538	0.469	0.416	0.373	0.310	0.231	0.184	0.141	0.115	0.092
2.5	1.446	1.065	0.843	0.696	0.594	0.517	0.458	0.411	0.341	0.254	0.203	0.156	0.126	0.101
3	1.581	1.164	0.921	0.761	0.649	0.565	0.500	0.449	0.372	0.278	0.222	0.170	0.138	0.110
3.5	1.717	1.263	0.998	0.825	0.703	0.612	0.542	0.486	0.403	0.293	0.240	0.184	0.149	0.119
4	1.852	1.362	1.076	0.889	0.757	0.659	0.583	0.524	0.434	0.324	0.258	0.198	0.161	0.128
4.5	1.986	1.459	1.153	0.952	0.810	0.706	0.625	0.561	0.465	0.347	0.277	0.212	0.172	0.137
5	2.120	1.557	1.229	1.015	0.864	0.752	0.666	0.598	0.496	0.369	0.295	0.226	0.183	0.146
10	3.438	2.517	1.982	1.634	1.391	1.210	1.071	0.961	0.796	0.593	0.472	0.362	0.293	0.234
15	4.742	3.494	2.725	2.245	1.933	1.661	1.469	1.316	1.090	0.811	0.646	0.495	0.401	0.320
20	6.039	4.407	3.478	2.852	2.424	2.107	1.862	1.668	1.381	1.026	0.817	0.626	0.507	0.404
25	7.336	5.345	4.197	3.455	2.935	2.550	2.253	2.018	1.670	1.241	0.987	0.755	0.612	0.488
30	8.628	6.282	4.930	4.057	3.445	2.991	2.643	2.367	1.958	1.454	1.156	0.884	0.716	0.571
50	13.79	10.02	7.853	6.455	5.475	4.750	4.194	3.754	3.102	2.300	1.827	1.396	1.129	0.900
60	16.37	11.89	9.312	7.651	6.487	5.627	4.967	4.445	3.671	2.721	2.161	1.650	1.335	1.063
80	21.52	15.62	12.23	10.04	8.509	7.378	6.510	5.824	4.808	3.561	2.826	2.157	1.744	1.389
100	26.67	19.34	15.14	12.43	10.53	9.127	8.052	7.201	5.943	4.399	3.490	2.663	2.152	1.713
130	34.39	24.93	19.51	16.01	13.56	11.75	10.36	9.266	7.644	5.654	4.483	3.419	2.762	2.198
160	42.12	30.52	23.87	19.59	16.58	14.37	12.67	11.33	9.344	6.909	5.476	4.175	3.371	2.682
200	52.41	37.98	29.70	24.36	20.62	17.86	15.75	14.08	11.61	8.580	6.799	5.181	4.183	3.327
500	130	93.85	73.34	60.12	50.87	44.05	38.82	34.69	28.58	21.10	16.70	12.71	10.25	8.147
1000	258	187	146	120	101	87.68	77.26	69.03	56.86	41.96	33.20	25.26	20.36	16.17

TABLE X (*continued*).

$\alpha = 0.01 \qquad s = 4 \qquad U_0^{\alpha}(s, m, n)$

m \ n	10	15	20	25	30	35	40	45	50	60	80	100	130	160	200
-0.5	--	--	0.875	0.687	0.565	0.480	0.418	0.370	0.331	0.274	0.204	0.163	0.125	0.101	0.081
0	--	1.413	1.028	0.807	0.664	0.565	0.491	0.434	0.389	0.322	0.240	0.191	0.146	0.119	0.095
.5	--	1.620	1.177	0.925	0.761	0.647	0.562	0.497	0.446	0.369	0.275	0.219	0.168	0.136	0.108
1	--	1.824	1.325	1.040	0.856	0.727	0.632	0.559	0.501	0.415	0.309	0.246	0.188	0.153	0.122
1.5	--	2.025	1.470	1.154	0.949	0.807	0.701	0.620	0.556	0.460	0.342	0.272	0.209	0.169	0.135
2	--	2.224	1.615	1.266	1.042	0.885	0.769	0.680	0.609	0.504	0.375	0.299	0.229	0.185	0.148
2.5	--	2.421	1.758	1.378	1.134	0.963	0.836	0.739	0.662	0.548	0.408	0.325	0.247	0.201	0.161
3	--	2.618	1.899	1.489	1.225	1.039	0.902	0.798	0.715	0.592	0.440	0.350	0.268	0.217	0.173
3.5	--	2.813	2.039	1.599	1.314	1.116	0.969	0.856	0.767	0.635	0.472	0.376	0.288	0.233	0.186
4	--	3.008	2.180	1.708	1.402	1.191	1.034	0.914	0.819	0.678	0.504	0.401	0.307	0.219	0.198
4.5	--	3.202	2.319	1.817	1.493	1.266	1.100	0.972	0.870	0.722	0.535	0.426	0.326	0.264	0.211
5	--	3.395	2.458	1.925	1.581	1.341	1.165	1.029	0.922	0.763	0.567	0.451	0.345	0.280	0.223
10	8.574	5.305	3.827	2.991	2.452	2.075	1.803	1.591	1.425	1.179	0.875	0.696	0.532	0.431	0.344
15	11.66	7.189	5.177	4.038	3.309	2.802	2.430	2.144	1.919	1.585	1.175	0.934	0.714	0.578	0.461
20	14.74	9.065	6.520	5.079	4.159	3.521	3.050	2.690	2.406	1.986	1.472	1.169	0.893	0.723	0.576
25	17.81	10.94	7.857	6.115	5.004	4.233	3.666	3.233	2.890	2.385	1.767	1.403	1.071	0.867	0.691
30	20.88	12.80	9.191	7.148	5.847	4.942	4.396	3.744	3.373	2.782	2.059	1.634	1.248	1.009	0.801
50	33.13	20.26	14.51	11.27	9.203	7.772	6.722	5.915	5.290	4.357	3.221	2.553	1.917	1.573	1.252
60	39.25	23.99	17.17	13.32	10.88	9.182	7.940	6.991	6.245	5.142	3.798	3.010	2.294	1.852	1.474
80	51.49	31.42	22.47	17.43	14.22	12.00	10.37	9.128	8.150	6.707	4.950	3.920	2.929	2.409	1.916
100	63.73	38.86	27.77	21.53	17.56	14.81	12.80	11.26	10.05	8.269	6.098	4.827	3.673	2.964	2.356
130	82.08	50.02	35.72	27.68	21.74	19.03	16.43	14.46	12.90	10.61	7.818	6.184	4.703	3.793	3.014
160	100	61.17	43.67	33.83	27.57	23.24	20.07	17.65	15.75	12.95	9.536	7.539	5.732	4.621	3.671
200	125	76.04	54.27	42.03	34.24	28.86	24.92	21.91	19.54	16.06	11.82	9.245	7.101	5.717	4.545
500	308	188	134	103	84.27	70.97	61.24	53.83	47.99	39.40	28.97	22.86	17.35	13.97	11.08
1000	614	373	266	206	168	141	122	107	95.39	78.29	57.52	45.38	34.41	27.69	21.95

TABLE X (*continued*).

$\alpha = 0.05$ $s = 4$ $U_0^2(s, m, n)$

m \ n	10	15	20	25	30	35	40	45	50	60	80	100	130	160	200
-0.5	--	--	0.691	0.547	0.453	0.387	0.337	0.299	0.268	0.223	0.167	0.133	0.102	0.083	0.066
0	--	1.122	0.827	0.655	0.542	0.463	0.403	0.357	0.321	0.267	0.199	0.159	0.122	0.099	0.079
.5	--	1.302	0.961	0.761	0.630	0.537	0.468	0.415	0.372	0.309	0.231	0.184	0.141	0.114	0.091
1	--	1.482	1.092	0.865	0.716	0.610	0.531	0.471	0.426	0.351	0.262	0.209	0.160	0.130	0.104
1.5	--	1.660	1.223	0.968	0.800	0.682	0.594	0.526	0.472	0.392	0.293	0.234	0.179	0.145	0.116
2	--	1.837	1.352	1.070	0.884	0.753	0.656	0.581	0.522	0.433	0.324	0.258	0.198	0.161	0.128
2.5	--	2.012	1.480	1.171	0.967	0.824	0.718	0.636	0.571	0.474	0.354	0.282	0.216	0.176	0.140
3	--	2.186	1.608	1.271	1.050	0.895	0.780	0.691	0.620	0.515	0.384	0.306	0.235	0.191	0.152
3.5	--	2.359	1.735	1.371	1.132	0.965	0.841	0.745	0.669	0.555	0.414	0.330	0.253	0.205	0.164
4	--	2.532	1.862	1.470	1.214	1.035	0.902	0.799	0.717	0.595	0.444	0.354	0.271	0.220	0.176
4.5	--	2.705	1.988	1.569	1.296	1.104	0.962	0.852	0.765	0.636	0.473	0.378	0.289	0.234	0.187
5	--	2.877	2.113	1.668	1.378	1.174	1.023	0.906	0.813	0.674	0.503	0.401	0.307	0.249	0.199
10	7.174	4.578	3.353	2.644	2.183	1.855	1.618	1.430	1.284	1.064	0.793	0.632	0.484	0.393	0.313
15	9.839	6.266	4.581	3.610	2.978	2.533	2.203	1.949	1.748	1.448	1.078	0.859	0.658	0.533	0.425
20	12.15	7.404	5.803	4.570	3.768	3.204	2.784	2.463	2.207	1.828	1.361	1.083	0.829	0.672	0.536
25	15.15	9.622	7.021	5.526	4.553	3.869	3.363	2.974	2.665	2.207	1.641	1.306	1.000	0.810	0.646
30	17.80	11.30	8.238	6.481	5.337	4.533	4.050	3.464	3.121	2.583	1.920	1.528	1.169	0.947	0.755
50	28.40	17.98	13.09	10.29	8.464	7.184	6.238	5.505	4.938	4.082	3.031	2.410	1.843	1.491	1.189
60	33.69	21.33	15.52	12.19	10.02	8.505	7.385	6.524	5.841	4.829	3.584	2.848	2.177	1.762	1.404
80	44.28	28.00	20.36	15.99	13.14	11.15	9.674	8.544	7.648	6.320	4.687	3.724	2.806	2.301	1.834
100	54.86	34.67	25.21	19.78	16.26	13.78	11.96	10.56	9.453	7.809	5.789	4.597	3.511	2.839	2.261
130	70.74	44.68	32.47	25.48	20.38	17.74	15.39	13.59	12.16	10.04	7.439	5.905	4.508	3.644	2.902
160	86.61	54.69	39.73	31.17	25.59	21.70	18.82	16.61	14.86	12.27	9.088	7.212	5.503	4.448	3.541
200	107	68.03	49.42	38.75	31.82	26.97	23.39	20.64	18.46	15.24	11.28	8.953	6.829	5.514	4.392
500	267	168	122	95.64	78.49	66.49	57.64	50.85	45.48	37.51	27.75	21.99	16.76	13.53	10.76
1000	531	335	243	190	156	132	115	101	90.48	74.62	55.18	43.72	33.30	26.88	21.36

TABLE X (*continued*).

$\alpha = 0.01 \qquad s = 5 \qquad U_0^{\alpha}(s, m, n)$

m \ n	10	15	20	25	30	35	40	45	50	60	80	100	130	160	200
-0.5	--	1.668	1.212	0.952	0.784	0.666	0.579	0.512	0.459	0.380	0.283	0.226	0.173	0.140	0.112
0	--	1.917	1.393	1.094	0.901	0.766	0.666	0.589	0.528	0.437	0.325	0.259	0.198	0.161	0.128
.5	--	2.162	1.572	1.234	1.016	0.863	0.750	0.664	0.595	0.492	0.366	0.292	0.223	0.181	0.144
1	--	2.405	1.748	1.372	1.129	0.959	0.834	0.737	0.661	0.547	0.407	0.324	0.248	0.201	0.161
1.5	--	2.646	1.922	1.509	1.241	1.054	0.916	0.810	0.726	0.601	0.447	0.356	0.273	0.221	0.176
2	--	2.885	2.095	1.644	1.352	1.148	0.998	0.882	0.790	0.654	0.487	0.388	0.297	0.241	0.192
2.5	5.005	3.123	2.267	1.778	1.463	1.241	1.078	0.953	0.854	0.708	0.526	0.419	0.321	0.260	0.208
3	5.387	3.360	2.438	1.912	1.572	1.334	1.159	1.025	0.918	0.760	0.566	0.450	0.345	0.279	0.223
3.5	5.768	3.596	2.607	2.045	1.680	1.426	1.239	1.095	0.982	0.813	0.604	0.481	0.369	0.299	0.238
4	6.148	3.831	2.776	2.177	1.788	1.520	1.319	1.166	1.045	0.865	0.643	0.512	0.392	0.318	0.253
4.5	6.527	4.064	2.945	2.308	1.896	1.609	1.400	1.236	1.107	0.916	0.682	0.543	0.415	0.337	0.268
5	6.906	4.297	3.113	2.439	2.003	1.701	1.477	1.305	1.170	0.968	0.720	0.573	0.439	0.355	0.284
10	10.66	6.604	4.774	3.731	3.064	2.598	2.255	1.992	1.783	1.475	1.096	0.872	0.667	0.540	0.431
15	14.39	8.891	6.412	5.008	4.108	3.481	3.020	2.666	2.386	1.973	1.464	1.164	0.890	0.721	0.575
20	18.11	11.17	8.042	6.278	5.145	4.358	3.778	3.334	2.984	2.465	1.829	1.453	1.111	0.899	0.717
25	21.81	13.44	9.667	7.541	6.178	5.229	4.532	3.999	3.578	2.954	2.191	1.740	1.330	1.076	0.858
30	25.52	15.70	11.29	8.802	7.207	6.098	5.284	4.662	4.169	3.442	2.551	2.026	1.548	1.252	0.997
35	29.22	17.96	12.91	10.06	8.235	6.966	6.034	5.322	4.759	3.927	2.909	2.310	1.764	1.426	1.136
40	32.93	20.23	14.53	11.32	9.261	7.832	6.783	5.981	5.348	4.412	3.267	2.593	1.980	1.601	1.275
45	36.63	22.49	16.14	12.57	10.29	8.696	7.531	6.639	5.935	4.896	3.624	2.876	2.195	1.774	1.413
50	40.32	24.75	17.76	13.83	11.31	9.560	8.278	7.297	6.522	5.379	3.981	3.158	2.409	1.947	1.551
60	47.72	29.26	20.99	16.34	13.36	11.29	9.770	8.610	7.694	6.343	4.692	3.720	2.836	2.293	1.825
80	62.51	38.29	27.45	21.35	17.44	14.74	12.75	11.23	10.03	8.268	6.111	4.842	3.691	2.981	2.372
100	77.30	47.32	33.90	26.36	21.53	18.18	15.73	13.85	12.37	10.19	7.526	5.961	4.542	3.667	2.917
130	99.47	60.86	43.58	33.87	27.64	23.34	20.19	17.77	15.87	13.07	9.646	7.636	5.815	4.693	3.732
160	122	74.40	53.25	41.37	33.77	28.50	24.65	21.70	19.37	15.95	11.76	9.309	7.086	5.717	4.545
200	151	92.44	66.15	51.38	41.93	35.38	30.59	26.92	24.03	19.78	14.58	11.54	8.778	7.081	5.627
500	373	228	163	126	103	86.96	75.14	66.10	58.99	48.50	35.72	28.23	21.45	17.29	13.72
1000	742	453	324	251	205	173	149	131	117	96.36	70.93	56.03	42.55	34.27	27.19

TABLE X (*continued*).

$\alpha = 0.05 \quad s = 5 \quad U_0^{\alpha}(s, m, n)$

m \ n	10	15	20	25	30	35	40	45	50	60	80	100	130	160	200
-0.5	--	1.345	0.992	0.786	0.651	0.555	0.484	0.429	0.385	0.319	0.238	0.190	0.146	0.118	0.095
0	--	1.568	1.156	0.915	0.757	0.645	0.562	0.498	0.447	0.371	0.277	0.221	0.170	0.138	0.110
.5	--	1.787	1.316	1.042	0.861	0.734	0.640	0.567	0.509	0.423	0.316	0.252	0.193	0.157	0.125
1	--	2.003	1.476	1.167	0.965	0.822	0.717	0.635	0.570	0.474	0.354	0.282	0.216	0.176	0.140
1.5	--	2.218	1.635	1.291	1.067	0.910	0.793	0.703	0.631	0.524	0.391	0.312	0.239	0.194	0.155
2	--	2.433	1.790	1.415	1.170	0.997	0.869	0.770	0.691	0.574	0.428	0.342	0.262	0.213	0.170
2.5	4.119	2.646	1.946	1.538	1.271	1.084	0.945	0.837	0.751	0.623	0.465	0.371	0.285	0.231	0.184
3	4.451	2.858	2.101	1.660	1.373	1.170	1.020	0.903	0.811	0.673	0.502	0.400	0.307	0.249	0.199
3.5	4.782	3.069	2.256	1.782	1.473	1.256	1.094	0.970	0.870	0.722	0.539	0.430	0.329	0.267	0.213
4	5.112	3.279	2.410	1.903	1.574	1.341	1.169	1.035	0.929	0.771	0.575	0.459	0.352	0.285	0.228
4.5	5.442	3.489	2.563	2.025	1.674	1.427	1.244	1.101	0.988	0.820	0.611	0.488	0.374	0.303	0.242
5	5.771	3.699	2.716	2.146	1.774	1.512	1.317	1.166	1.046	0.868	0.647	0.516	0.396	0.321	0.256
10	9.043	5.781	4.235	3.344	2.761	2.350	2.046	1.811	1.624	1.347	1.004	0.800	0.613	0.497	0.397
15	12.30	7.845	5.742	4.530	3.737	3.179	2.766	2.448	2.195	1.820	1.355	1.080	0.827	0.671	0.535
20	15.54	9.902	7.242	5.709	4.707	4.003	3.482	3.081	2.762	2.289	1.704	1.357	1.039	0.842	0.672
25	18.78	11.96	8.739	6.885	5.673	4.824	4.195	3.711	3.327	2.755	2.050	1.632	1.250	1.013	0.808
30	22.02	14.01	10.23	8.058	6.638	5.643	4.906	4.339	3.889	3.220	2.395	1.907	1.460	1.183	0.944
35	25.26	16.06	11.73	9.230	7.602	6.460	5.616	4.966	4.450	3.684	2.740	2.180	1.669	1.352	1.078
40	28.50	18.10	13.22	10.40	8.564	7.277	6.325	5.592	5.010	4.147	3.083	2.453	1.877	1.520	1.213
45	31.73	20.15	14.71	11.57	9.526	8.092	7.033	6.217	5.570	4.609	3.426	2.726	2.086	1.689	1.347
50	34.97	22.20	16.20	12.74	10.49	8.908	7.740	6.842	6.129	5.071	3.769	2.998	2.293	1.857	1.481
60	41.44	26.29	19.18	15.08	12.41	10.54	9.154	8.090	7.246	5.994	4.453	3.541	2.708	2.192	1.748
80	54.38	34.47	25.13	19.75	16.25	13.79	11.98	10.58	9.477	7.837	5.819	4.625	3.536	2.861	2.281
100	67.32	42.65	31.08	24.42	20.08	17.05	14.80	13.07	11.71	9.678	7.182	5.707	4.361	3.528	2.812
130	86.72	54.91	40.01	31.42	25.83	21.92	19.03	16.81	15.05	12.44	9.225	7.328	5.598	4.527	3.607
160	106	67.18	48.93	38.42	31.58	26.80	23.26	20.54	18.39	15.19	11.27	8.947	6.832	5.525	4.400
200	132	83.53	60.83	47.75	39.24	33.30	28.90	25.52	22.84	18.87	13.99	11.10	8.477	6.853	5.457
500	326	206	150	118	96.72	82.02	71.17	62.81	56.20	46.41	34.37	27.27	20.80	16.80	13.37
1000	649	411	299	234	193	163	142	125	112	92.30	68.33	54.19	41.32	33.37	26.54

TABLE X (*continued*).

$\alpha = 0.01 \qquad s = 6 \qquad U_0^{\alpha}(s, m, n)$

m \ n	10	15	20	25	30	35	40	45	50	60	80	100	130	160	200
-0.5	--	2.208	1.602	1.261	1.038	0.882	0.767	0.678	0.608	0.503	0.374	0.298	0.228	0.185	0.148
0	--	2.495	1.814	1.425	1.172	0.996	0.866	0.766	0.686	0.568	0.423	0.337	0.258	0.209	0.167
.5	4.445	2.780	2.020	1.586	1.305	1.109	0.963	0.852	0.764	0.632	0.471	0.375	0.287	0.233	0.186
1	4.899	3.062	2.225	1.746	1.437	1.220	1.060	0.937	0.840	0.696	0.518	0.413	0.316	0.256	0.204
1.5	5.350	3.343	2.427	1.905	1.567	1.330	1.156	1.022	0.916	0.759	0.565	0.450	0.345	0.279	0.223
2	5.799	3.621	2.628	2.063	1.696	1.440	1.252	1.107	0.992	0.821	0.611	0.487	0.373	0.302	0.241
2.5	6.246	3.897	2.828	2.219	1.824	1.549	1.347	1.191	1.067	0.884	0.658	0.524	0.401	0.325	0.259
3	6.692	4.173	3.028	2.374	1.952	1.658	1.441	1.274	1.141	0.945	0.703	0.560	0.429	0.347	0.277
3.5	7.136	4.447	3.226	2.529	2.080	1.766	1.535	1.357	1.216	1.007	0.749	0.596	0.457	0.370	0.295
4	7.580	4.721	3.422	2.683	2.209	1.873	1.628	1.439	1.290	1.068	0.794	0.632	0.484	0.392	0.313
4.5	8.022	4.995	3.621	2.837	2.333	1.981	1.721	1.521	1.363	1.129	0.839	0.668	0.512	0.415	0.331
5	8.464	5.268	3.819	2.991	2.459	2.088	1.814	1.603	1.437	1.189	0.884	0.704	0.539	0.437	0.349
10	12.85	7.971	5.763	4.512	3.706	3.144	2.729	2.411	2.160	1.789	1.328	1.057	0.809	0.655	0.523
15	17.20	10.65	7.689	6.014	4.936	4.185	3.631	3.207	2.872	2.375	1.764	1.403	1.073	0.869	0.693
20	21.55	13.32	9.607	7.508	6.159	5.219	4.527	3.997	3.578	2.957	2.195	1.746	1.335	1.081	0.862
25	25.88	15.98	11.52	8.997	7.376	6.248	5.417	4.783	4.280	3.537	2.624	2.086	1.594	1.290	1.029
30	30.22	18.64	13.43	10.48	8.591	7.275	6.308	5.566	4.980	4.114	3.051	2.424	1.850	1.499	1.195
35	34.55	21.30	15.33	11.97	9.803	8.299	7.194	6.347	5.678	4.689	3.477	2.761	2.109	1.706	1.360
40	38.88	23.95	17.24	13.45	11.02	9.322	8.079	7.127	6.380	5.265	3.901	3.097	2.366	1.914	1.525
45	43.21	26.61	19.14	14.93	12.23	10.34	8.964	7.906	7.071	5.837	4.325	3.433	2.621	2.120	1.689
50	47.54	29.26	21.04	16.41	13.43	11.37	9.847	8.684	7.766	6.410	4.748	3.768	2.877	2.326	1.853
60	56.19	34.56	24.85	19.37	15.85	13.41	11.61	10.24	9.155	7.553	5.591	4.436	3.386	2.737	2.180
80	73.49	45.17	32.45	25.28	20.68	17.48	15.14	13.34	11.93	9.836	7.270	5.769	4.401	3.556	2.831
100	90.79	55.77	40.04	31.18	25.50	21.55	18.66	16.44	14.70	12.12	8.956	7.099	5.413	4.373	3.480
130	117	71.67	51.43	40.04	32.73	27.66	23.94	21.09	18.84	15.53	11.47	9.091	6.928	5.595	4.451
160	143	87.56	62.83	48.89	39.96	33.76	29.21	25.73	22.99	18.94	13.99	11.08	8.440	6.814	5.420
200	177	109	78.01	60.69	49.59	41.89	36.24	31.92	28.52	23.49	17.34	13.73	10.46	8.439	6.710
500	437	268	192	149	122	103	88.96	78.32	69.94	57.57	42.44	33.58	25.54	20.60	16.36
1000	869	533	382	297	242	204	177	156	139	114	84.27	66.64	50.66	40.83	32.42

TABLE X (*continued*).

$\alpha = 0.05$ $s = 6$ $U^{\alpha}(s, m, n)$

m \ n	10	15	20	25	30	35	40	45	50	60	80	100	130	160	200
-0.5	--	1.829	1.347	1.066	0.882	0.752	0.655	0.580	0.521	0.433	0.323	0.258	0.198	0.161	0.128
0	--	2.086	1.537	1.216	1.005	0.857	0.747	0.662	0.594	0.494	0.369	0.294	0.226	0.183	0.146
.5	3.641	2.342	1.725	1.363	1.127	0.961	0.838	0.743	0.667	0.554	0.413	0.330	0.253	0.205	0.164
1	4.037	2.594	1.910	1.510	1.249	1.065	0.928	0.823	0,739	0.613	0.458	0.365	0.280	0.227	0.181
1.5	4.431	2.848	2.095	1.656	1.370	1.168	1.018	0.902	0.810	0.672	0.502	0.400	0.307	0.249	0.199
2	4.823	3.098	2.279	1.802	1.490	1.271	1.107	0.981	0.881	0.731	0.545	0.435	0.334	0.271	0.216
2.5	5.214	3.349	2.462	1.947	1.610	1.373	1.196	1.060	0.951	0.789	0.589	0.470	0.360	0.292	0.233
3	5.604	3.599	2.645	2.091	1.729	1.474	1.284	1.138	1.021	0.847	0.632	0.504	0.387	0.314	0.251
3.5	5.994	3.848	2.827	2.235	1.848	1.575	1.372	1.216	1.091	0.905	0.676	0.539	0.413	0.335	0.268
4	6.382	4.096	3.008	2.379	1.967	1.676	1.460	1.293	1.161	0.963	0.719	0.573	0.440	0.357	0.285
4.5	6.770	4.344	3.189	2.522	2.085	1.777	1.547	1.370	1.230	1.021	0.761	0.607	0.466	0.378	0.302
5	7.158	4.591	3.370	2.665	2.203	1.877	1.634	1.448	1.299	1.078	0.804	0.641	0.492	0.399	0.319
10	11.02	7.045	5.168	4.081	3.370	2.869	2.498	2.212	1.984	1.646	1.227	0.978	0.750	0.608	0.485
15	14.86	9.483	6.951	5.484	4.525	3.851	3.352	2.967	2.661	2.206	1.644	1.310	1.004	0.814	0.650
20	18.70	11.92	8.727	6.881	5.675	4.829	4.201	3.718	3.334	2.763	2.058	1.639	1.256	1.018	0.813
25	22.53	14.34	10.50	8.273	6.822	5.803	5.048	4.466	4.004	3.317	2.470	1.967	1.507	1.221	0.974
30	26.35	16.77	12.27	9.664	7.966	6.775	5.892	5.212	4.672	3.870	2.881	2.294	1.757	1.423	1.136
35	30.18	19.19	14.04	11.05	9.109	7.745	6.735	5.957	5.339	4.422	3.291	2.620	2.006	1.625	1.296
40	34.00	21.61	15.81	12.44	10.25	8.715	7.577	6.701	6.005	4.973	3.700	2.945	2.254	1.826	1.457
45	37.83	24.03	17.57	13.82	11.39	9.684	8.418	7.444	6.671	5.523	4.108	3.269	2.502	2.027	1.617
50	41.65	26.45	19.34	15.21	12.53	10.65	9.253	8.182	7.333	6.072	4.516	3.593	2.750	2.227	1.776
60	49.29	31.29	22.86	17.99	14.81	12.59	10.93	9.666	8.662	7.170	5.331	4.241	3.244	2.627	2.095
80	64.58	40.97	29.92	23.53	19.37	16.45	14.29	12.63	11.32	9.364	6.957	5.532	4.231	3.425	2.731
100	79.86	50.64	36.97	29.06	23.92	20.32	17.64	15.59	13.97	11.55	8.581	6.822	5.216	4.221	3.365
130	103	65.14	47.55	37.37	30.75	26.11	22.64	20.03	17.94	14.84	11.02	8.754	6.691	5.413	4.314
160	126	79.65	58.12	45.67	37.57	31.90	27.69	24.47	21.91	18.12	13.45	10.68	8.164	6.604	5.262
200	156	98.99	72.22	56.73	46.67	39.62	34.38	30.38	27.20	22.49	16.69	13.26	10.13	8.190	6.524
500	385	244	178	140	115	97.50	84.63	74.74	66.90	55.28	40.98	32.53	24.83	20.07	15.98
1000	767	486	354	278	229	194	168	149	133	110	81.45	64.64	49.32	39.85	31.71

TABLE XI. Upper Percentage Points of Pillai's Trace Criterion, $V^{\alpha}(s, m, n)$

$\alpha = 0.01 \quad s = 2 \quad V^{\alpha}(s, m, n)$

m \ n	5	10	15	20	25	30	35	40	45	50
.5	0.962	0.653	0.494	0.397	0.331	0.284	0.249	0.222	0.200	0.182
1	1.049	0.725	0.553	0.446	0.375	0.323	0.284	0.253	0.228	0.208
1.5	1.122	0.788	0.607	0.493	0.416	0.359	0.316	0.282	0.255	0.232
2	1.185	0.845	0.656	0.536	0.453	0.392	0.346	0.309	0.280	0.255
2.5	1.236	0.897	0.702	0.576	0.488	0.424	0.374	0.335	0.304	0.277
3	1.282	0.944	0.744	0.613	0.521	0.454	0.402	0.360	0.326	0.298
3.5	1.323	0.987	0.783	0.648	0.553	0.482	0.428	0.384	0.348	0.319
4	1.358	1.026	0.820	0.681	0.583	0.510	0.453	0.407	0.370	0.339
4.5	1.390	1.061	0.854	0.713	0.612	0.536	0.477	0.429	0.390	0.358
5	1.422	1.093	0.886	0.743	0.639	0.561	0.500	0.450	0.410	0.376
10	1.610	1.328	1.125	0.973	0.858	0.767	0.693	0.632	0.580	0.536
15	1.705	1.465	1.278	1.129	1.013	0.917	0.837	0.771	0.714	0.664
20	1.763	1.555	1.386	1.245	1.128	1.032	0.953	0.883	0.822	0.770
25	1.802	1.619	1.463	1.334	1.222	1.126	1.045	0.975	0.914	0.860
30	1.829	1.661	1.524	1.402	1.297	1.204	1.123	1.053	0.991	0.936
35	1.850	1.704	1.573	1.457	1.358	1.269	1.190	1.120	1.058	1.002
40	1.866	1.734	1.612	1.504	1.408	1.324	1.247	1.178	1.117	1.061
45	1.880	1.758	1.645	1.543	1.452	1.370	1.297	1.230	1.169	1.113
50	1.890	1.778	1.675	1.576	1.489	1.411	1.340	1.275	1.215	1.161

From K. C. S. Pillai, *Statistical Tables for Tests of Multivariate Hypotheses*, The Statistical Center, University of the Phillipines, 1960. Reproduced by permission of the author and publisher.

TABLE XI (*continued*).

$\alpha = 0.05 \qquad s = 2 \qquad V^{\alpha}(s, m, n)$

m \ n	5	10	15	20	25	30	35	40	45	50
-.5	0.567	0.357	0.259	0.204	0.168	0.143	0.124	0.110	0.098	0.089
0	0.699	0.451	0.333	0.263	0.218	0.186	0.162	0.144	0.129	0.117
.5	0.807	0.532	0.397	0.316	0.263	0.225	0.196	0.174	0.157	0.142
1	0.897	0.604	0.455	0.364	0.304	0.261	0.228	0.203	0.183	0.166
1.5	0.974	0.668	0.507	0.409	0.342	0.294	0.258	0.230	0.208	0.189
2	1.040	0.725	0.556	0.451	0.379	0.327	0.287	0.256	0.231	0.210
2.5	1.097	0.778	0.601	0.490	0.413	0.357	0.314	0.281	0.254	0.231
3	1.148	0.826	0.643	0.526	0.445	0.386	0.341	0.305	0.276	0.252
3.5	1.194	0.870	0.682	0.561	0.476	0.414	0.366	0.328	0.297	0.271
4	1.234	0.910	0.719	0.594	0.506	0.440	0.390	0.350	0.317	0.290
4.5	1.271	0.948	0.754	0.625	0.534	0.466	0.413	0.371	0.337	0.308
5	1.305	0.982	0.786	0.655	0.561	0.490	0.436	0.392	0.356	0.326
10	1.522	1.235	1.035	0.890	0.781	0.695	0.626	0.570	0.522	0.482
15	1.636	1.387	1.198	1.053	0.940	0.848	0.772	0.709	0.655	0.609
20	1.706	1.487	1.314	1.174	1.061	0.967	0.889	0.822	0.765	0.715
25	1.753	1.559	1.399	1.268	1.157	1.064	0.985	0.917	0.857	0.806
30	1.788	1.613	1.466	1.342	1.236	1.145	1.066	0.997	0.936	0.883
35	1.813	1.656	1.519	1.402	1.301	1.212	1.134	1.066	1.005	0.951
40	1.833	1.690	1.563	1.452	1.355	1.270	1.194	1.126	1.066	1.011
45	1.850	1.718	1.599	1.495	1.402	1.319	1.246	1.179	1.119	1.066
50	1.863	1.741	1.630	1.531	1.442	1.363	1.291	1.226	1.167	1.114

TABLE XI (*continued*).

$\alpha = 0.01 \quad s = 3 \quad V^{\alpha}(s, m, n)$

m \ n	5	10	15	20	25	30	35	40	45	50	60
-.5	1.052	0.710	0.535	0.429	0.359	0.308	0.269	0.240	0.216	0.196	0.166
0	1.212	0.829	0.630	0.508	0.426	0.366	0.321	0.286	0.258	0.235	0.199
.5	1.345	0.934	0.716	0.580	0.488	0.421	0.370	0.330	0.298	0.272	0.231
1	1.459	1.028	0.794	0.647	0.545	0.472	0.415	0.371	0.335	0.306	0.260
1.5	1.554	1.114	0.866	0.709	0.600	0.520	0.458	0.410	0.371	0.339	0.289
2	1.639	1.191	0.933	0.767	0.651	0.565	0.500	0.447	0.405	0.370	0.316
2.5	1.717	1.262	0.995	0.821	0.699	0.609	0.539	0.483	0.438	0.401	0.342
3	1.784	1.326	1.053	0.873	0.745	0.650	0.577	0.518	0.470	0.430	0.368
3.5	1.842	1.385	1.108	0.922	0.789	0.689	0.613	0.551	0.501	0.459	0.393
4	1.897	1.440	1.160	0.969	0.832	0.729	0.648	0.584	0.531	0.487	0.418
4.5	1.948	1.492	1.207	1.012	0.872	0.765	0.681	0.614	0.559	0.513	0.441
5	1.995	1.540	1.252	1.055	0.910	0.800	0.714	0.645	0.587	0.539	0.464
10	2.298	1.892	1.601	1.389	1.224	1.095	0.990	0.904	0.831	0.769	0.669
15	2.459	2.106	1.834	1.621	1.453	1.317	1.203	1.107	1.025	0.954	0.839
20	2.560	2.249	2.001	1.796	1.628	1.489	1.372	1.272	1.186	1.110	0.984
25	2.629	2.353	2.123	1.932	1.768	1.630	1.512	1.410	1.321	1.242	1.110
30	2.680	2.432	2.218	2.038	1.883	1.747	1.630	1.527	1.437	1.356	1.220
35	2.718	2.493	2.296	2.125	1.977	1.846	1.731	1.628	1.538	1.456	1.317
40	2.748	2.542	2.359	2.197	2.056	1.930	1.818	1.717	1.623	1.545	1.405
45	2.772	2.583	2.412	2.259	2.123	2.002	1.893	1.795	1.705	1.624	1.483
50	2.792	2.617	2.456	2.311	2.182	2.065	1.959	1.863	1.776	1.696	1.555
60	2.823	2.670	2.528	2.397	2.277	2.169	2.070	1.979	1.895	1.818	1.742

TABLE XI (*continued*).

$\alpha = 0.05 \quad s = 3 \quad V^{\alpha}(s, m, n)$

m \ n	5	10	15	20	25	30	35	40	45	50	60
-.5	0.886	0.583	0.434	0.345	0.287	0.245	0.214	0.190	0.171	0.155	0.131
0	1.041	0.697	0.524	0.420	0.350	0.300	0.263	0.234	0.210	0.191	0.162
.5	1.173	0.800	0.606	0.488	0.409	0.351	0.308	0.274	0.247	0.225	0.191
1	1.288	0.892	0.682	0.552	0.464	0.399	0.351	0.313	0.282	0.257	0.218
1.5	1.388	0.977	0.752	0.612	0.516	0.445	0.392	0.350	0.316	0.288	0.245
2	1.476	1.053	0.818	0.668	0.565	0.489	0.431	0.386	0.349	0.318	0.271
2.5	1.556	1.124	0.879	0.722	0.612	0.531	0.469	0.420	0.380	0.347	0.296
3	1.628	1.190	0.937	0.772	0.657	0.571	0.505	0.453	0.411	0.375	0.321
3.5	1.691	1.250	0.991	0.820	0.700	0.610	0.540	0.485	0.440	0.403	0.344
4	1.750	1.307	1.042	0.866	0.741	0.647	0.574	0.516	0.469	0.429	0.368
4.5	1.804	1.360	1.091	0.910	0.781	0.683	0.607	0.546	0.497	0.455	0.390
5	1.854	1.410	1.136	0.952	0.818	0.717	0.639	0.576	0.524	0.481	0.412
10	2.188	1.778	1.495	1.290	1.133	1.010	0.911	0.830	0.762	0.704	0.612
15	2.369	2.008	1.737	1.529	1.365	1.234	1.124	1.033	0.955	0.888	0.779
20	2.486	2.164	1.912	1.710	1.545	1.410	1.297	1.200	1.117	1.044	0.923
25	2.565	2.277	2.043	1.851	1.690	1.554	1.439	1.340	1.253	1.177	1.050
30	2.623	2.363	2.146	1.964	1.808	1.675	1.560	1.459	1.371	1.293	1.160
35	2.668	2.431	2.229	2.056	1.907	1.777	1.663	1.563	1.474	1.395	1.259
40	2.703	2.486	2.297	2.133	1.990	1.864	1.752	1.653	1.565	1.485	1.348
45	2.731	2.531	2.354	2.199	2.061	1.939	1.831	1.733	1.645	1.566	1.427
50	2.755	2.569	2.402	2.255	2.123	2.005	1.899	1.804	1.717	1.638	1.500
60	2.791	2.629	2.480	2.346	2.225	2.115	2.015	1.924	1.840	1.763	1.690

TABLE XI (*continued*).

$\alpha = 0.01 \qquad s = 4 \qquad V^{\alpha}(s, m, n)$

m \ n	5	10	15	20	25	30	35	40	45	50	60	80
.5	1.419	0.982	0.750	0.608	0.511	0.441	0.388	0.346	0.312	0.284	0.242	0.186
0	1.594	1.118	0.861	0.701	0.590	0.510	0.449	0.401	0.363	0.331	0.281	0.216
.5	1.744	1.241	0.962	0.786	0.665	0.576	0.508	0.454	0.411	0.372	0.319	0.246
1	1.875	1.352	1.056	0.866	0.735	0.638	0.563	0.504	0.457	0.417	0.356	0.275
1.5	1.992	1.454	1.143	0.941	0.801	0.696	0.616	0.553	0.501	0.458	0.391	0.302
2	2.098	1.547	1.224	1.013	0.864	0.753	0.667	0.599	0.543	0.497	0.425	0.330
2.5	2.189	1.632	1.301	1.080	0.924	0.807	0.716	0.643	0.584	0.535	0.458	0.356
3	2.273	1.712	1.373	1.144	0.981	0.858	0.763	0.687	0.624	0.572	0.490	0.382
3.5	2.350	1.787	1.440	1.206	1.036	0.908	0.808	0.729	0.663	0.608	0.522	0.407
4	2.420	1.857	1.504	1.264	1.089	0.956	0.853	0.769	0.701	0.643	0.553	0.431
4.5	2.486	1.922	1.564	1.319	1.140	1.003	0.895	0.808	0.737	0.677	0.583	0.456
5	2.547	1.984	1.622	1.371	1.188	1.048	0.937	0.847	0.773	0.711	0.612	0.479
10	2.956	2.440	2.071	1.801	1.591	1.425	1.290	1.173	1.084	1.004	0.875	0.696
15	3.185	2.729	2.378	2.104	1.888	1.713	1.567	1.443	1.336	1.246	1.096	0.883
20	3.331	2.924	2.601	2.336	2.120	1.940	1.788	1.659	1.547	1.449	1.285	1.048
25	3.432	3.068	2.768	2.518	2.306	2.126	1.973	1.841	1.724	1.622	1.450	1.196
30	3.507	3.178	2.898	2.662	2.459	2.283	2.130	1.996	1.878	1.773	1.595	1.329
35	3.564	3.264	3.003	2.780	2.586	2.415	2.264	2.131	2.012	1.906	1.724	1.448
40	3.610	3.334	3.091	2.879	2.693	2.527	2.380	2.248	2.130	2.024	1.840	1.558
45	3.666	3.392	3.164	2.963	2.784	2.625	2.481	2.352	2.236	2.129	1.945	1.658
50	3.677	3.441	3.227	3.035	2.864	2.710	2.571	2.444	2.329	2.224	2.040	1.750
60	3.725	3.518	3.327	3.153	2.995	2.851	2.720	2.600	2.489	2.387	2.190	1.914
80	3.787	3.622	3.465	3.319	3.183	3.058	2.941	2.833	2.732	2.637	2.465	2.180

TABLE XI (*continued*).

$\alpha = 0.05 \quad s = 4 \quad V^{\alpha}(s, m, n)$

m \ n	5	10	15	20	25	30	35	40	45	50	60	80
-.5	1.239	0.843	0.638	0.514	0.430	0.370	0.324	0.289	0.260	0.237	0.200	0.154
0	1.411	0.974	0.744	0.602	0.505	0.435	0.382	0.341	0.308	0.280	0.238	0.182
.5	1.560	1.094	0.842	0.684	0.576	0.497	0.438	0.391	0.353	0.322	0.273	0.210
1	1.693	1.203	0.932	0.761	0.643	0.556	0.490	0.438	0.396	0.362	0.308	0.237
1.5	1.812	1.304	1.017	0.834	0.707	0.613	0.541	0.484	0.438	0.401	0.341	0.264
2	1.919	1.396	1.097	0.903	0.768	0.667	0.590	0.529	0.479	0.438	0.374	0.289
2.5	2.015	1.482	1.172	0.969	0.826	0.720	0.637	0.572	0.519	0.475	0.406	0.314
3	2.103	1.563	1.243	1.032	0.882	0.770	0.683	0.614	0.557	0.510	0.437	0.339
3.5	2.183	1.638	1.311	1.092	0.935	0.818	0.727	0.654	0.595	0.545	0.467	0.363
4	2.257	1.709	1.374	1.149	0.987	0.865	0.770	0.694	0.631	0.579	0.497	0.387
4.5	2.325	1.776	1.435	1.204	1.037	0.910	0.811	0.732	0.666	0.612	0.526	0.410
5	2.389	1.839	1.493	1.257	1.085	0.954	0.852	0.769	0.701	0.644	0.554	0.433
10	2.829	2.312	1.952	1.689	1.488	1.329	1.201	1.095	1.007	0.932	0.810	0.643
15	3.079	2.615	2.268	2.000	1.789	1.619	1.479	1.359	1.258	1.171	1.028	0.827
20	3.241	2.825	2.499	2.238	2.026	1.850	1.702	1.577	1.468	1.374	1.216	0.990
25	3.355	2.979	2.675	2.425	2.216	2.040	1.890	1.761	1.648	1.548	1.382	1.136
30	3.439	3.098	2.814	2.576	2.373	2.200	2.049	1.918	1.803	1.701	1.527	1.269
35	3.503	3.191	2.926	2.700	2.505	2.335	2.186	2.055	1.939	1.835	1.658	1.389
40	3.555	3.267	3.019	2.804	2.616	2.451	2.305	2.175	2.059	1.955	1.775	1.499
45	3.610	3.330	3.097	2.892	2.712	2.552	2.409	2.281	2.166	2.062	1.881	1.599
50	3.631	3.383	3.164	2.969	2.795	2.640	2.501	2.376	2.262	2.158	1.977	1.692
60	3.685	3.468	3.271	3.094	2.933	2.788	2.656	2.53[illegible]	2.426	2.325	2.145	1.858
80	3.756	3.582	3.420	3.270	3.132	3.005	2.886	2.77[illegible]	2.676	2.581	2.410	2.127

TABLE XI (*continued*).

$\alpha = 0.01 \qquad s = 5 \qquad V^{\alpha}(s, m, n)$

m \ n	5	10	15	20	25	30	35	40	45	50	60	80	100	130	160	200
-0.5	1.806	1.276	0.994	0.812	0.686	0.594	0.524	0.468	0.424	0.387	0.329	0.254	0.206	0.164	0.135	0.108
0	1.991	1.431	1.117	0.915	0.776	0.673	0.595	0.532	0.482	0.440	0.375	0.290	0.236	0.185	0.152	0.122
0.5	2.156	1.568	1.231	1.013	0.862	0.748	0.662	0.593	0.538	0.492	0.420	0.324	0.265	0.207	0.170	0.138
1	2.304	1.692	1.337	1.104	0.941	0.820	0.726	0.652	0.591	0.541	0.463	0.358	0.293	0.229	0.189	0.152
1.5	2.436	1.807	1.437	1.191	1.017	0.888	0.788	0.709	0.643	0.589	0.504	0.391	0.320	0.251	0.206	0.167
2	2.553	1.917	1.530	1.275	1.093	0.955	0.848	0.763	0.693	0.636	0.545	0.423	0.346	0.272	0.224	0.181
2.5	2.660	2.013	1.618	1.353	1.162	1.018	0.906	0.816	0.742	0.681	0.584	0.455	0.373	0.293	0.241	0.195
3	2.759	2.107	1.702	1.428	1.230	1.079	0.962	0.867	0.790	0.725	0.623	0.486	0.398	0.316	0.258	0.209
3.5	2.850	2.195	1.781	1.499	1.294	1.139	1.016	0.917	0.836	0.768	0.660	0.516	0.423	0.334	0.275	0.223
4	2.935	2.271	1.857	1.567	1.356	1.195	1.068	0.966	0.881	0.810	0.697	0.546	0.448	0.353	0.292	0.237
4.5	3.013	2.353	1.928	1.633	1.416	1.250	1.109	1.004	0.925	0.851	0.733	0.575	0.473	0.373	0.308	0.250
5	3.087	2.426	1.997	1.696	1.474	1.303	1.167	1.058	0.967	0.891	0.769	0.603	0.497	0.392	0.324	0.263
10	3.595	2.978	2.538	2.212	1.960	1.758	1.594	1.458	1.343	1.245	1.086	0.866	0.720	0.574	0.478	0.390
15	3.887	3.338	2.915	2.584	2.322	2.108	1.930	1.780	1.651	1.531	1.356	1.094	0.918	0.739	0.618	0.507
20	4.079	3.585	3.191	2.871	2.607	2.388	2.203	2.045	1.908	1.788	1.588	1.297	1.096	0.889	0.748	0.618
25	4.214	3.767	3.401	3.096	2.838	2.619	2.431	2.269	2.127	2.001	1.791	1.479	1.259	1.029	0.870	0.721
30	4.315	3.909	3.567	3.277	3.028	2.813	2.625	2.462	2.317	2.189	1.970	1.625	1.405	1.159	0.984	0.820
35	4.393	4.021	3.701	3.426	3.187	2.977	2.792	2.629	2.483	2.353	2.130	1.790	1.544	1.279	1.092	0.914
40	4.454	4.112	3.812	3.551	3.321	3.118	2.937	2.776	2.630	2.500	2.274	1.926	1.670	1.392	1.194	1.003
45	4.505	4.188	3.906	3.657	3.437	3.240	3.064	2.905	2.761	2.631	2.404	2.050	1.787	1.498	1.290	1.088
50	4.547	4.252	3.986	3.749	3.538	3.347	3.176	3.020	2.878	2.749	2.522	2.165	1.896	1.598	1.381	1.169
60	4.612	4.353	4.115	3.899	3.704	3.526	3.364	3.215	3.079	2.953	2.729	2.369	2.093	1.781	1.550	1.321
80	4.700	4.491	4.297	4.113	3.944	3.788	3.643	3.509	3.383	3.266	3.053	2.700	2.420	2.093	1.844	1.591
100	4.755	4.581	4.414	4.257	4.109	3.971	3.841	3.719	3.604	3.495	3.296	2.957	2.680	2.349	2.091	1.823
130	4.808	4.668	4.533	4.403	4.279	4.160	4.048	3.942	3.840	3.744	3.564	3.249	2.985	2.659	2.396	2.118
160	4.842	4.725	4.611	4.501	4.394	4.291	4.193	4.099	4.008	3.922	3.759	3.468	3.218	2.903	2.644	2.361
200	4.872	4.777	4.682	4.590	4.500	4.413	4.329	4.247	4.169	4.093	3.949	3.687	3.456	3.159	2.908	2.629

TABLE XI (*continued*).

$\alpha = 0.05 \qquad s = 5 \qquad V^{\alpha}(s, m, n)$

m \ n	5	10	15	20	25	30	35	40	45	50	60	80	100	130	160	200
−0.5	1.617	1.128	0.871	0.708	0.595	0.514	0.453	0.404	0.365	0.333	0.283	0.217	0.177	0.140	0.114	0.092
0	1.800	1.277	0.989	0.807	0.681	0.590	0.520	0.465	0.420	0.383	0.326	0.251	0.204	0.160	0.131	0.106
0.5	1.964	1.410	1.099	0.901	0.763	0.662	0.584	0.523	0.473	0.432	0.368	0.284	0.231	0.181	0.149	0.120
1	2.112	1.533	1.203	0.990	0.841	0.731	0.646	0.579	0.525	0.479	0.409	0.316	0.258	0.202	0.166	0.134
1.5	2.245	1.647	1.301	1.074	0.915	0.797	0.706	0.633	0.574	0.525	0.449	0.348	0.284	0.222	0.183	0.148
2	2.366	1.753	1.393	1.155	0.987	0.861	0.764	0.686	0.623	0.570	0.488	0.378	0.309	0.242	0.200	0.161
2.5	2.476	1.853	1.480	1.232	1.055	0.922	0.819	0.737	0.670	0.614	0.526	0.409	0.334	0.262	0.216	0.175
3	2.578	1.947	1.563	1.306	1.121	0.982	0.874	0.787	0.716	0.656	0.563	0.438	0.359	0.284	0.232	0.188
3.5	2.672	2.035	1.642	1.376	1.184	1.040	0.926	0.835	0.760	0.698	0.599	0.467	0.383	0.301	0.248	0.201
4	2.759	2.116	1.717	1.444	1.246	1.095	0.977	0.882	0.804	0.738	0.635	0.496	0.407	0.320	0.264	0.214
4.5	2.840	2.196	1.789	1.509	1.305	1.149	1.020	0.922	0.846	0.778	0.670	0.524	0.430	0.339	0.280	0.227
5	2.916	2.271	1.858	1.572	1.362	1.202	1.075	0.973	0.888	0.817	0.704	0.552	0.453	0.358	0.296	0.240
10	3.454	2.838	2.408	2.090	1.847	1.653	1.497	1.367	1.258	1.165	1.015	0.807	0.670	0.534	0.444	0.362
15	3.769	3.212	2.793	2.470	2.213	2.005	1.833	1.688	1.564	1.450	1.281	1.032	0.864	0.695	0.580	0.476
20	3.977	3.474	3.078	2.762	2.503	2.289	2.109	1.955	1.821	1.705	1.512	1.233	1.040	0.843	0.708	0.584
25	4.125	3.668	3.298	2.993	2.739	2.524	2.340	2.181	2.042	1.920	1.715	1.413	1.201	0.980	0.828	0.686
30	4.236	3.819	3.473	3.181	2.933	2.721	2.537	2.376	2.234	2.108	1.895	1.564	1.347	1.109	0.911	0.783
35	4.322	3.939	3.614	3.336	3.097	2.889	2.706	2.545	2.403	2.275	2.056	1.724	1.485	1.228	1.048	0.876
40	4.390	4.036	3.730	3.467	3.236	3.033	2.854	2.694	2.551	2.423	2.201	1.860	1.611	1.341	1.148	0.964
45	4.446	4.118	3.830	3.578	3.356	3.159	2.984	2.826	2.684	2.556	2.332	1.985	1.728	1.446	1.244	1.048
50	4.493	4.186	3.915	3.675	3.461	3.270	3.098	2.944	2.803	2.676	2.452	2.101	1.837	1.546	1.335	1.129
60	4.566	4.295	4.052	3.833	3.635	3.456	3.293	3.144	3.008	2.883	2.661	2.306	2.035	1.729	1.503	1.280
80	4.663	4.445	4.243	4.058	3.886	3.728	3.582	3.446	3.320	3.203	2.991	1.641	2.363	2.042	1.797	1.549
100	4.724	4.542	4.370	4.210	4.059	3.919	3.787	3.664	3.548	3.439	3.239	2.901	2.627	2.300	2.045	1.782
130	4.784	4.637	4.497	4.364	4.237	4.117	4.003	3.895	3.793	3.695	3.514	3.199	2.935	2.612	2.352	2.077
160	4.822	4.700	4.581	4.467	4.358	4.254	4.154	4.058	3.967	3.879	3.714	3.423	3.173	2.859	2.601	2.322
200	4.856	4.756	4.658	4.563	4.471	4.382	4.296	4.213	4.133	4.056	3.910	3.647	3.415	3.118	2.868	2.591

TABLE XI (*continued*).

$\alpha = 0.01 \quad s = 6 \quad V^{\alpha}(s, m, n)$

m \ n	5	10	15	20	25	30	35	40	45	50	60	80	100	130	160	200
-0.5	2.210	1.606	1.260	1.037	0.881	0.765	0.677	0.607	0.550	0.503	0.429	0.332	0.271	0.212	0.174	0.141
0	2.406	1.764	1.394	1.150	0.980	0.854	0.756	0.678	0.615	0.563	0.481	0.373	0.304	0.238	0.196	0.158
0.5	2.582	1.911	1.518	1.258	1.075	0.938	0.832	0.747	0.679	0.621	0.531	0.412	0.337	0.264	0.218	0.176
1	2.739	2.047	1.636	1.361	1.165	1.018	0.905	0.814	0.739	0.678	0.581	0.451	0.369	0.290	0.239	0.193
1.5	2.880	2.174	1.745	1.458	1.252	1.096	0.975	0.878	0.799	0.732	0.628	0.489	0.400	0.315	0.259	0.210
2	3.013	2.293	1.849	1.550	1.334	1.171	1.043	0.940	0.856	0.786	0.675	0.526	0.431	0.339	0.280	0.227
2.5	3.133	2.405	1.949	1.638	1.414	1.243	1.109	1.001	0.912	0.838	0.720	0.563	0.461	0.364	0.300	0.243
3	3.244	2.510	2.043	1.723	1.490	1.312	1.173	1.060	0.967	0.888	0.765	0.598	0.491	0.387	0.320	0.259
3.5	3.347	2.607	2.133	1.804	1.564	1.380	1.234	1.117	1.020	0.938	0.808	0.633	0.521	0.411	0.339	0.275
4	3.444	2.699	2.219	1.882	1.635	1.445	1.294	1.172	1.071	0.986	0.851	0.668	0.549	0.434	0.359	0.291
4.5	3.533	2.786	2.301	1.957	1.703	1.508	1.352	1.226	1.121	1.033	0.893	0.702	0.578	0.457	0.378	0.307
5	3.618	2.869	2.379	2.030	1.770	1.569	1.409	1.278	1.170	1.079	0.933	0.735	0.606	0.480	0.397	0.323
10	4.217	3.512	3.003	2.625	2.331	2.097	1.904	1.743	1.608	1.492	1.304	1.041	0.867	0.693	0.577	0.472
15	4.571	3.936	3.446	3.062	2.755	2.505	2.296	2.120	1.968	1.840	1.620	1.310	1.100	0.886	0.742	0.610
20	4.809	4.234	3.774	3.400	3.096	2.835	2.618	2.432	2.270	2.129	1.893	1.549	1.311	1.065	0.896	0.740
25	4.984	4.455	4.026	3.668	3.366	3.109	2.888	2.697	2.530	2.382	2.132	1.763	1.503	1.230	1.041	0.863
30	5.111	4.627	4.226	3.885	3.592	3.339	3.119	2.926	2.755	2.603	2.345	1.957	1.678	1.383	1.176	0.980
35	5.206	4.765	4.388	4.064	3.782	3.535	3.317	3.125	2.953	2.799	2.535	2.133	1.840	1.526	1.304	1.093
40	5.286	4.878	4.523	4.215	3.944	3.704	3.490	3.299	3.128	2.974	2.706	2.294	1.990	1.661	1.424	1.197
45	5.350	4.972	4.638	4.344	4.083	3.850	3.642	3.454	3.284	3.130	2.861	2.442	2.130	1.787	1.539	1.298
50	5.404	5.051	4.735	4.455	4.205	3.979	3.776	3.592	3.424	3.271	3.002	2.579	2.259	1.905	1.648	1.395
60	5.489	5.178	4.895	4.639	4.407	4.195	4.003	3.826	3.664	3.515	3.249	2.822	2.494	2.124	1.849	1.577
80	5.602	5.351	5.117	4.899	4.699	4.513	4.340	4.180	4.030	3.891	3.638	3.218	2.885	2.496	2.200	1.899
100	5.673	5.464	5.265	5.077	4.900	4.737	4.580	4.434	4.297	4.167	3.930	3.526	3.197	2.803	2.495	2.176
130	5.743	5.575	5.412	5.256	5.107	4.966	4.832	4.705	4.584	4.468	4.253	3.878	3.562	3.173	2.773	2.528
160	5.788	5.648	5.510	5.377	5.249	5.126	5.009	4.896	4.788	4.684	4.489	4.141	3.843	3.467	3.068	2.820
200	5.829	5.714	5.600	5.488	5.380	5.276	5.175	5.077	4.983	4.892	4.719	4.405	4.130	3.774	3.474	3.141

TABLE XI (*continued*).

$\alpha = 0.05 \quad s = 6 \quad V^{\alpha}(s, m, n)$

m \ n	5	10	15	20	25	30	35	40	45	50	60	80	100	130	160	200
-0.5	2.015	1.446	1.127	0.923	0.782	0.678	0.598	0.536	0.485	0.443	0.377	0.291	0.237	0.185	0.152	0.123
0	2.208	1.601	1.256	1.033	0.877	0.762	0.674	0.604	0.547	0.500	0.427	0.330	0.269	0.210	0.173	0.140
0.5	2.382	1.745	1.377	1.137	0.968	0.843	0.747	0.670	0.608	0.556	0.475	0.368	0.300	0.235	0.193	0.156
1	2.541	1.880	1.491	1.237	1.056	0.921	0.817	0.734	0.666	0.610	0.522	0.405	0.331	0.259	0.213	0.173
1.5	2.684	2.006	1.600	1.332	1.140	0.996	0.885	0.796	0.723	0.663	0.568	0.441	0.361	0.283	0.233	0.189
2	2.818	2.124	1.704	1.422	1.221	1.069	0.951	0.856	0.779	0.714	0.612	0.477	0.390	0.307	0.253	0.205
2.5	2.941	2.236	1.802	1.509	1.298	1.139	1.015	0.915	0.833	0.764	0.656	0.512	0.419	0.330	0.272	0.220
3	3.054	2.341	1.895	1.593	1.374	1.207	1.077	0.972	0.886	0.813	0.699	0.546	0.447	0.353	0.291	0.236
3.5	3.160	2.439	1.985	1.673	1.446	1.273	1.137	1.028	0.937	0.861	0.741	0.580	0.476	0.375	0.310	0.251
4	3.258	2.532	2.070	1.750	1.516	1.337	1.196	1.082	0.987	0.908	0.783	0.613	0.504	0.398	0.328	0.267
4.5	3.350	2.621	2.153	1.825	1.584	1.399	1.253	1.135	1.037	0.954	0.823	0.646	0.531	0.420	0.347	0.282
5	3.436	2.705	2.231	1.897	1.650	1.460	1.309	1.186	1.084	0.999	0.863	0.678	0.559	0.442	0.365	0.297
10	4.063	3.362	2.864	2.495	2.210	1.983	1.798	1.645	1.516	1.405	1.226	0.978	0.813	0.649	0.540	0.441
15	4.443	3.800	3.315	2.938	2.639	2.394	2.191	2.020	1.874	1.749	1.539	1.242	1.041	0.838	0.701	0.576
20	4.697	4.112	3.652	3.282	2.982	2.728	2.516	2.334	2.177	2.039	1.811	1.479	1.249	1.014	0.853	0.704
25	4.882	4.346	3.913	3.556	3.258	3.005	2.789	2.602	2.437	2.293	2.050	1.692	1.440	1.177	0.995	0.825
30	5.020	4.529	4.122	3.780	3.489	3.239	3.022	2.832	2.665	2.516	2.264	1.885	1.615	1.329	1.129	0.940
35	5.125	4.675	4.292	3.966	3.684	3.439	3.224	3.034	2.865	2.714	2.455	2.061	1.776	1.471	1.255	1.051
40	5.212	4.795	4.435	4.123	3.851	3.611	3.400	3.211	3.042	2.890	2.627	2.223	1.926	1.605	1.375	1.155
45	5.283	4.894	4.555	4.257	3.995	3.762	3.554	3.368	3.200	3.048	2.784	2.371	2.065	1.730	1.489	1.255
50	5.342	4.979	4.658	4.373	4.121	3.895	3.692	3.508	3.342	3.191	2.926	2.509	2.195	1.849	1.597	1.351
60	5.435	5.114	4.825	4.565	4.330	4.118	3.924	3.748	3.587	3.439	3.176	2.754	2.431	2.067	1.798	1.532
80	5.560	5.300	5.060	4.838	4.635	4.446	4.272	4.111	3.961	3.822	3.570	3.154	2.824	2.441	2.149	1.853
100	5.640	5.421	5.216	5.024	4.845	4.677	4.520	4.373	4.235	4.105	3.868	3.466	3.138	2.749	2.445	2.131
130	5.717	5.541	5.373	5.213	5.062	4.918	4.782	4.654	4.531	4.414	4.198	3.823	3.508	3.122	2.724	2.483
160	5.766	5.619	5.477	5.341	5.210	5.085	4.965	4.851	4.742	4.637	4.440	4.091	3.793	3.418	3.021	2.777
200	5.811	5.690	5.572	5.458	5.347	5.241	5.138	5.039	4.944	4.851	4.677	4.361	4.085	3.730	3.431	3.099

TABLE XII. Upper Percentage Points for the Bonferroni-Dunn Procedure, $t^{\alpha/2k}(\nu)$

NUMBER OF CONTRASTS (k)	Error df (ν) 10	12	14	16	18	20	22	24	26	28	α = .10 30	35	40	45	50	60	80	100	250	500
1	1.812	1.782	1.761	1.746	1.734	1.725	1.717	1.711	1.706	1.701	1.697	1.690	1.684	1.679	1.676	1.671	1.664	1.660	1.651	1.648
2	2.228	2.179	2.145	2.120	2.101	2.086	2.074	2.064	2.056	2.048	2.042	2.030	2.021	2.014	2.009	2.000	1.990	1.984	1.970	1.965
3	2.466	2.403	2.360	2.328	2.304	2.285	2.270	2.258	2.247	2.238	2.231	2.215	2.204	2.195	2.188	2.178	2.165	2.158	2.140	2.134
4	2.634	2.560	2.510	2.473	2.445	2.423	2.405	2.391	2.379	2.368	2.360	2.342	2.329	2.319	2.311	2.299	2.284	2.276	2.255	2.248
5	2.764	2.681	2.624	2.583	2.552	2.528	2.508	2.492	2.479	2.467	2.457	2.438	2.423	2.412	2.403	2.390	2.374	2.364	2.341	2.334
6	2.870	2.779	2.718	2.673	2.639	2.613	2.591	2.574	2.559	2.546	2.536	2.515	2.499	2.487	2.477	2.463	2.445	2.435	2.410	2.402
7	2.960	2.863	2.796	2.748	2.712	2.683	2.661	2.642	2.626	2.613	2.601	2.579	2.562	2.549	2.539	2.524	2.505	2.494	2.467	2.459
8	3.038	2.934	2.864	2.813	2.775	2.744	2.720	2.700	2.684	2.669	2.657	2.633	2.616	2.602	2.591	2.575	2.555	2.544	2.516	2.507
9	3.107	2.998	2.924	2.870	2.829	2.798	2.772	2.751	2.734	2.719	2.706	2.681	2.663	2.648	2.637	2.620	2.600	2.587	2.558	2.549
10	3.169	3.055	2.977	2.921	2.878	2.845	2.819	2.797	2.779	2.763	2.750	2.724	2.704	2.690	2.678	2.660	2.639	2.625	2.596	2.586
11	3.225	3.106	3.025	2.967	2.923	2.888	2.861	2.838	2.819	2.803	2.789	2.762	2.742	2.726	2.714	2.696	2.674	2.660	2.629	2.619
12	3.277	3.153	3.069	3.008	2.963	2.927	2.899	2.875	2.856	2.839	2.825	2.797	2.776	2.760	2.747	2.729	2.705	2.692	2.659	2.649
13	3.324	3.196	3.109	3.047	3.000	2.963	2.933	2.909	2.889	2.872	2.857	2.828	2.807	2.791	2.778	2.758	2.734	2.720	2.687	2.676
14	3.368	3.236	3.146	3.082	3.034	2.996	2.965	2.941	2.920	2.902	2.887	2.857	2.836	2.819	2.805	2.785	2.761	2.747	2.712	2.701
15	3.409	3.273	3.181	3.115	3.065	3.026	2.995	2.970	2.949	2.930	2.915	2.885	2.862	2.845	2.831	2.811	2.786	2.771	2.736	2.725
16	3.448	3.308	3.214	3.146	3.095	3.055	3.023	2.997	2.975	2.957	2.941	2.910	2.887	2.869	2.855	2.834	2.809	2.793	2.758	2.746
17	3.484	3.341	3.244	3.175	3.123	3.082	3.049	3.023	3.000	2.981	2.965	2.933	2.910	2.892	2.877	2.856	2.830	2.815	2.778	2.766
18	3.518	3.371	3.273	3.202	3.149	3.107	3.074	3.046	3.024	3.004	2.988	2.955	2.931	2.913	2.898	2.877	2.850	2.834	2.797	2.785
19	3.551	3.401	3.300	3.228	3.173	3.131	3.097	3.069	3.046	3.026	3.009	2.976	2.952	2.933	2.918	2.896	2.869	2.853	2.815	2.803
20	3.581	3.428	3.326	3.252	3.197	3.153	3.119	3.091	3.067	3.047	3.030	2.996	2.971	2.952	2.937	2.915	2.887	2.871	2.832	2.820
25	3.716	3.550	3.438	3.358	3.298	3.251	3.214	3.183	3.158	3.136	3.118	3.081	3.055	3.034	3.018	2.994	2.964	2.946	2.905	2.892
30	3.827	3.649	3.530	3.444	3.380	3.331	3.291	3.258	3.231	3.208	3.189	3.150	3.122	3.100	3.083	3.057	3.026	3.007	2.964	2.949
35	3.922	3.733	3.607	3.517	3.450	3.398	3.356	3.322	3.293	3.269	3.249	3.208	3.178	3.155	3.137	3.111	3.078	3.058	3.013	2.998
40	4.005	3.807	3.675	3.581	3.510	3.455	3.412	3.376	3.346	3.321	3.300	3.258	3.227	3.203	3.184	3.156	3.122	3.102	3.054	3.039
45	4.078	3.871	3.734	3.636	3.563	3.506	3.461	3.424	3.393	3.367	3.345	3.301	3.269	3.244	3.225	3.196	3.161	3.140	3.091	3.075
50	4.144	3.930	3.787	3.686	3.610	3.552	3.505	3.467	3.435	3.408	3.385	3.340	3.307	3.282	3.261	3.232	3.195	3.174	3.123	3.107
60	4.259	4.031	3.880	3.773	3.692	3.630	3.581	3.540	3.507	3.479	3.454	3.407	3.372	3.345	3.324	3.293	3.254	3.232	3.179	3.161
70	4.357	4.117	3.958	3.846	3.762	3.697	3.645	3.603	3.567	3.538	3.513	3.463	3.426	3.398	3.376	3.344	3.304	3.280	3.225	3.207
80	4.442	4.192	4.026	3.909	3.822	3.754	3.700	3.656	3.620	3.589	3.563	3.511	3.473	3.444	3.421	3.388	3.346	3.322	3.265	3.246
90	4.518	4.258	4.086	3.965	3.874	3.804	3.749	3.703	3.666	3.634	3.607	3.553	3.514	3.484	3.461	3.426	3.383	3.358	3.299	3.280
100	4.587	4.318	4.140	4.015	3.922	3.850	3.792	3.745	3.707	3.674	3.646	3.591	3.551	3.520	3.496	3.460	3.416	3.391	3.330	3.310
250	5.202	4.847	4.616	4.454	4.334	4.242	4.168	4.109	4.060	4.018	3.983	3.914	3.864	3.825	3.795	3.750	3.696	3.664	3.589	3.565
500	5.694	5.263	4.985	4.791	4.648	4.539	4.452	4.382	4.324	4.276	4.234	4.153	4.094	4.050	4.014	3.962	3.899	3.862	3.776	3.748

Abridged from unpublished tables by C. M. Dayton, and W. D. Schafer. Reproduced by permission of the authors.

TABLE XII (*continued*).

NUMBER OF CONTRASTS (k)	Error df (ν) α = .05 10	12	14	16	18	20	22	24	26	28	30	35	40	45	50	60	80	100	250	500
1	2.228	2.179	2.145	2.120	2.101	2.086	2.074	2.064	2.056	2.048	2.042	2.030	2.021	2.014	2.009	2.000	1.990	1.984	1.970	1.965
2	2.634	2.560	2.510	2.473	2.445	2.423	2.405	2.391	2.379	2.368	2.360	2.342	2.329	2.319	2.311	2.299	2.284	2.276	2.255	2.248
3	2.870	2.779	2.718	2.673	2.639	2.613	2.591	2.574	2.559	2.546	2.536	2.515	2.499	2.487	2.477	2.463	2.445	2.435	2.410	2.402
4	3.038	2.934	2.864	2.813	2.775	2.744	2.720	2.700	2.684	2.669	2.657	2.633	2.616	2.602	2.591	2.575	2.555	2.544	2.516	2.507
5	3.169	3.055	2.977	2.921	2.878	2.845	2.819	2.797	2.779	2.763	2.750	2.724	2.704	2.690	2.678	2.660	2.639	2.626	2.596	2.586
6	3.277	3.153	3.069	3.008	2.963	2.927	2.899	2.875	2.856	2.839	2.825	2.797	2.776	2.760	2.747	2.729	2.705	2.692	2.659	2.649
7	3.368	3.236	3.146	3.082	3.034	2.996	2.965	2.941	2.920	2.902	2.887	2.857	2.836	2.819	2.805	2.785	2.761	2.747	2.712	2.701
8	3.448	3.308	3.214	3.146	3.095	3.055	3.023	2.997	2.975	2.957	2.941	2.910	2.887	2.869	2.855	2.834	2.809	2.793	2.758	2.746
9	3.518	3.371	3.273	3.202	3.149	3.107	3.074	3.046	3.024	3.004	2.988	2.955	2.931	2.913	2.898	2.877	2.850	2.834	2.797	2.785
10	3.581	3.428	3.326	3.252	3.197	3.153	3.119	3.091	3.067	3.047	3.030	2.996	2.971	2.952	2.937	2.915	2.887	2.871	2.832	2.820
11	3.639	3.480	3.374	3.297	3.240	3.195	3.159	3.130	3.106	3.085	3.067	3.033	3.007	2.987	2.972	2.948	2.920	2.903	2.864	2.851
12	3.691	3.527	3.417	3.339	3.279	3.233	3.196	3.166	3.141	3.120	3.102	3.066	3.039	3.019	3.003	2.979	2.950	2.933	2.892	2.879
13	3.740	3.571	3.458	3.377	3.316	3.268	3.230	3.199	3.174	3.152	3.133	3.096	3.069	3.048	3.032	3.007	2.977	2.960	2.918	2.904
14	3.785	3.611	3.495	3.412	3.349	3.301	3.262	3.230	3.204	3.181	3.162	3.124	3.096	3.075	3.058	3.033	3.003	2.984	2.942	2.928
15	3.827	3.649	3.530	3.444	3.380	3.331	3.291	3.258	3.231	3.208	3.189	3.150	3.122	3.100	3.083	3.057	3.026	3.007	2.964	2.949
16	3.867	3.684	3.562	3.475	3.410	3.359	3.318	3.285	3.257	3.234	3.214	3.174	3.145	3.123	3.106	3.080	3.048	3.029	2.984	2.970
17	3.904	3.717	3.593	3.504	3.437	3.385	3.344	3.310	3.282	3.258	3.237	3.197	3.168	3.145	3.127	3.101	3.068	3.049	3.003	2.989
18	3.939	3.749	3.621	3.531	3.463	3.410	3.368	3.333	3.304	3.280	3.259	3.218	3.188	3.165	3.147	3.120	3.087	3.067	3.021	3.006
19	3.973	3.778	3.649	3.556	3.487	3.433	3.390	3.355	3.326	3.301	3.280	3.239	3.208	3.185	3.166	3.139	3.105	3.085	3.038	3.023
20	4.005	3.807	3.675	3.581	3.510	3.455	3.412	3.376	3.346	3.321	3.300	3.258	3.227	3.203	3.184	3.156	3.122	3.102	3.054	3.039
25	4.144	3.930	3.787	3.686	3.610	3.552	3.505	3.467	3.435	3.408	3.385	3.340	3.307	3.282	3.261	3.232	3.195	3.174	3.123	3.107
30	4.259	4.031	3.880	3.773	3.692	3.630	3.581	3.540	3.507	3.479	3.454	3.407	3.372	3.345	3.324	3.293	3.254	3.232	3.179	3.161
35	4.357	4.117	3.958	3.846	3.762	3.697	3.645	3.603	3.567	3.538	3.513	3.463	3.426	3.398	3.376	3.344	3.304	3.280	3.225	3.207
40	4.442	4.192	4.026	3.909	3.822	3.754	3.700	3.656	3.620	3.589	3.563	3.511	3.473	3.444	3.421	3.388	3.346	3.322	3.265	3.246
45	4.518	4.258	4.086	3.965	3.874	3.804	3.749	3.703	3.666	3.634	3.607	3.553	3.514	3.484	3.461	3.426	3.383	3.358	3.299	3.280
50	4.587	4.318	4.140	4.015	3.922	3.850	3.792	3.745	3.707	3.674	3.646	3.591	3.551	3.520	3.496	3.460	3.416	3.391	3.330	3.310
60	4.706	4.422	4.234	4.102	4.004	3.928	3.867	3.818	3.777	3.743	3.714	3.656	3.614	3.582	3.556	3.519	3.473	3.446	3.383	3.362
70	4.809	4.510	4.314	4.175	4.073	3.994	3.931	3.879	3.837	3.801	3.771	3.711	3.667	3.634	3.607	3.568	3.521	3.493	3.427	3.406
80	4.898	4.587	4.383	4.239	4.133	4.051	3.985	3.932	3.888	3.851	3.820	3.758	3.713	3.678	3.651	3.611	3.561	3.532	3.465	3.443
90	4.977	4.655	4.444	4.296	4.186	4.101	4.034	3.979	3.934	3.896	3.863	3.799	3.753	3.171	3.689	3.648	3.597	3.567	3.498	3.475
100	5.049	4.716	4.499	4.346	4.233	4.146	4.077	4.021	3.974	3.935	3.902	3.836	3.788	3.752	3.723	3.681	3.629	3.598	3.527	3.504
250	5.694	5.263	4.985	4.791	4.648	4.539	4.452	4.382	4.324	4.276	4.234	4.153	4.094	4.050	4.014	3.962	3.899	3.862	3.776	3.748
500	6.211	5.695	5.364	5.134	4.966	4.837	4.736	4.655	4.587	4.531	4.483	4.389	4.321	4.269	4.229	4.169	4.096	4.054	3.956	3.925

TABLE XII (*continued*).

Error df (ν) $\alpha = .01$

NUMBER OF CONTRASTS (k)	10	12	14	16	18	20	22	24	26	28	30	35	40	45	50	60	80	100	250	500
1	3.169	3.055	2.977	2.921	2.878	2.845	2.819	2.797	2.779	2.763	2.750	2.724	2.704	2.690	2.678	2.660	2.639	2.626	2.596	2.586
2	3.581	3.428	3.326	3.252	3.197	3.153	3.119	3.091	3.067	3.047	3.030	2.996	2.971	2.952	2.937	2.915	2.887	2.871	2.832	2.820
3	3.827	3.649	3.530	3.444	3.380	3.331	3.291	3.258	3.231	3.208	3.189	3.150	3.122	3.100	3.083	3.057	3.026	3.007	2.964	2.949
4	4.005	3.807	3.675	3.581	3.510	3.455	3.412	3.376	3.346	3.321	3.300	3.258	3.227	3.203	3.184	3.156	3.122	3.102	3.054	3.039
5	4.144	3.930	3.787	3.686	3.610	3.552	3.505	3.467	3.435	3.408	3.385	3.340	3.307	3.282	3.261	3.232	3.195	3.174	3.123	3.107
6	4.259	4.031	3.880	3.773	3.692	3.630	3.581	3.540	3.507	3.479	3.454	3.407	3.372	3.345	3.324	3.293	3.254	3.232	3.179	3.161
7	4.357	4.117	3.958	3.846	3.762	3.697	3.645	3.603	3.567	3.538	3.513	3.463	3.426	3.398	3.376	3.344	3.304	3.280	3.225	3.207
8	4.442	4.192	4.026	3.909	3.822	3.754	3.700	3.656	3.620	3.589	3.563	3.511	3.473	3.444	3.421	3.388	3.346	3.322	3.265	3.246
9	4.518	4.258	4.086	3.965	3.874	3.804	3.749	3.703	3.666	3.634	3.607	3.553	3.514	3.484	3.461	3.426	3.383	3.358	3.299	3.280
10	4.587	4.318	4.140	4.015	3.922	3.850	3.792	3.745	3.707	3.674	3.646	3.591	3.551	3.520	3.496	3.460	3.416	3.391	3.330	3.310
11	4.649	4.372	4.189	4.060	3.964	3.890	3.831	3.783	3.744	3.710	3.681	3.625	3.584	3.553	3.528	3.491	3.446	3.420	3.358	3.338
12	4.706	4.422	4.234	4.102	4.004	3.928	3.867	3.818	3.777	3.743	3.714	3.656	3.614	3.582	3.556	3.519	3.473	3.446	3.383	3.362
13	4.759	4.467	4.275	4.140	4.039	3.962	3.900	3.850	3.808	3.773	3.743	3.685	3.642	3.609	3.583	3.545	3.498	3.470	3.406	3.385
14	4.809	4.510	4.314	4.175	4.073	3.994	3.931	3.879	3.837	3.801	3.771	3.711	3.667	3.634	3.607	3.568	3.521	3.493	3.427	3.406
15	4.855	4.550	4.349	4.208	4.104	4.023	3.959	3.907	3.864	3.827	3.796	3.735	3.691	3.657	3.630	3.590	3.542	3.513	3.446	3.425
16	4.898	4.587	4.383	4.239	4.133	4.051	3.985	3.932	3.888	3.851	3.820	3.758	3.713	3.678	3.651	3.611	3.561	3.532	3.465	3.443
17	4.939	4.622	4.414	4.268	4.160	4.077	4.010	3.956	3.912	3.874	3.842	3.779	3.733	3.698	3.671	3.630	3.580	3.550	3.482	3.460
18	4.977	4.655	4.444	4.296	4.186	4.101	4.034	3.979	3.934	3.896	3.863	3.799	3.753	3.717	3.689	3.648	3.597	3.567	3.498	3.475
19	5.014	4.687	4.472	4.322	4.210	4.124	4.056	4.000	3.955	3.916	3.883	3.818	3.771	3.735	3.707	3.665	3.613	3.583	3.513	3.490
20	5.049	4.716	4.499	4.346	4.233	4.146	4.077	4.021	3.974	3.935	3.902	3.836	3.788	3.752	3.723	3.681	3.629	3.598	3.527	3.504
25	5.202	4.847	4.616	4.454	4.334	4.242	4.168	4.109	4.060	4.018	3.983	3.914	3.864	3.825	3.795	3.750	3.696	3.664	3.589	3.565
30	5.329	4.955	4.712	4.542	4.416	4.320	4.243	4.181	4.130	4.086	4.049	3.977	3.925	3.885	3.853	3.807	3.750	3.717	3.639	3.614
35	5.438	5.048	4.794	4.617	4.486	4.386	4.306	4.242	4.188	4.144	4.105	4.031	3.976	3.935	3.902	3.854	3.795	3.761	3.681	3.655
40	5.533	5.128	4.865	4.682	4.547	4.443	4.361	4.294	4.239	4.193	4.154	4.077	4.021	3.978	3.944	3.895	3.834	3.799	3.716	3.690
45	5.618	5.199	4.928	4.739	4.600	4.493	4.409	4.341	4.284	4.237	4.196	4.117	4.060	4.016	3.981	3.930	3.869	3.832	3.748	3.721
50	5.694	5.263	4.985	4.791	4.648	4.539	4.452	4.382	4.324	4.276	4.234	4.153	4.094	4.050	4.014	3.962	3.899	3.862	3.776	3.748
60	5.827	5.375	5.084	4.881	4.731	4.617	4.527	4.454	4.393	4.343	4.300	4.216	4.154	4.108	4.071	4.017	3.951	3.913	3.824	3.795
70	5.942	5.471	5.168	4.957	4.802	4.683	4.590	4.514	4.452	4.400	4.355	4.268	4.205	4.157	4.119	4.063	3.995	3.956	3.864	3.835
80	6.042	5.554	5.241	5.023	4.863	4.741	4.645	4.567	4.503	4.449	4.403	4.313	4.248	4.199	4.160	4.103	4.033	3.993	3.898	3.869
90	6.131	5.628	5.305	5.081	4.917	4.792	4.693	4.613	4.547	4.492	4.445	4.353	4.287	4.236	4.196	4.138	4.067	4.025	3.929	3.899
100	6.211	5.695	5.364	5.134	4.966	4.837	4.736	4.655	4.587	4.531	4.483	4.389	4.321	4.269	4.229	4.169	4.096	4.054	3.956	3.925
250	6.940	6.291	5.881	5.599	5.393	5.237	5.114	5.016	4.934	4.867	4.809	4.697	4.616	4.554	4.506	4.435	4.350	4.300	4.185	4.151
500	7.528	6.766	6.287	5.960	5.723	5.544	5.403	5.290	5.198	5.121	5.055	4.928	4.836	4.767	4.713	4.633	4.537	4.481	4.354	4.319

TABLE XIII. Coefficients of Orthogonal Polynomials (Linear, Quadratic, and Cubic)

$n = 3$	$f_1(x_i)$	$f_2(x_i)$
	-1	1
	0	-2
	1	1
c_j	2	6
$f_1(x)$	$x - 2$	
$f_2(x)$	$3x^2 - 12x + 10$	
$f_3(x)$	...	

$n = 4$	$f_1(x_i)$	$f_2(x_i)$	$f_3(x_i)$
	-3	1	-1
	-1	-1	3
	1	-1	-3
	3	1	1
c_j	20	4	20
$f_1(x)$	$2x - 5$		
$f_2(x)$	$x^2 - 5x + 5$		
$f_3(x)$	$\frac{1}{3}(10x^3 - 75x^2 + 167x - 105)$		

$n = 5$	$f_1(x_i)$	$f_2(x_i)$	$f_3(x_i)$
	-2	2	-1
	-1	-1	2
	0	-2	0
	1	-1	-2
	2	2	1
c_j	10	14	10
$f_1(x)$	$x - 3$		
$f_2(x)$	$x^2 - 6x + 7$		
$f_3(x)$	$\frac{1}{6}(5x^3 - 45x^2 + 118x - 84)$		

$n = 6$	$f_1(x_i)$	$f_2(x_i)$	$f_3(x_i)$
	-5	5	-5
	-3	-1	7
	-1	-4	4
	1	-4	-4
	3	-1	-7
	5	5	5
c_j	70	84	180
$f_1(x)$	$2x - 7$		
$f_2(x)$	$\frac{1}{2}(3x^2 - 21x + 28)$		
$f_3(x)$	$\frac{1}{6}(10x^3 - 105x^2 + 317x - 252)$		

$n = 7$	$f_1(x_i)$	$f_2(x_i)$	$f_3(x_i)$
	-3	5	-1
	-2	0	1
	-1	-3	1
	0	-4	0
	1	-3	-1
	2	0	-1
	3	5	1
c_j	28	84	6
$f_1(x)$	$x - 4$		
$f_2(x)$	$x^2 - 8x + 12$		
$f_3(x)$	$\frac{1}{6}(x^3 - 12x^2 + 41x - 36)$		

$n = 8$	$f_1(x_i)$	$f_2(x_i)$	$f_3(x_i)$
	-7	7	-7
	-5	1	5
	-3	-3	7
	-1	-5	3
	1	-5	-3
	3	-3	-7
	5	1	-5
	7	7	7
c_j	168	168	264
$f_1(x)$	$2x - 9$		
$f_2(x)$	$x^2 - 9x + 15$		
$f_3(x)$	$\frac{1}{3}(2x^3 - 27x^2 + 103x - 99)$		

$n = 9$	$f_1(x_i)$	$f_2(x_i)$	$f_3(x_i)$
	-4	28	-14
	-3	7	7
	-2	-8	13
	-1	-17	9
	0	-20	0
	1	-17	-9
	2	-8	-13
	3	7	-7
	4	28	14
c_j	60	2772	990
$f_1(x)$	$x - 5$		
$f_2(x)$	$3x^2 - 30x + 55$		
$f_3(x)$	$\frac{1}{6}(5x^3 - 75x^2 + 316x - 330)$		

$n = 10$	$f_1(x_i)$	$f_2(x_i)$	$f_3(x_i)$
	-9	6	-42
	-7	2	14
	-5	-1	35
	-3	-3	31
	-1	-4	12
	1	-4	-12
	3	-3	-31
	5	-1	-35
	7	2	-14
	9	6	42
c_j	330	132	8580
$f_1(x)$	$2x - 11$		
$f_2(x)$	$\frac{1}{2}(x^2 - 11x + 22)$		
$f_3(x)$	$\frac{1}{6}(10x^3 - 165x^2 + 761x - 858)$		

$n = 11$	$f_1(x_i)$	$f_2(x_i)$	$f_3(x_i)$
	-5	15	-30
	-4	6	6
	-3	-1	22
	-2	-6	23
	-1	-9	14
	0	-10	0
	1	-9	-14
	2	-6	-23
	3	-1	-22
	4	6	-6
	5	15	30
c_j	110	858	4290
$f_1(x)$	$x - 6$		
$f_2(x)$	$x^2 - 12x + 26$		
$f_3(x)$	$\frac{1}{6}(5x^3 - 90x^2 + 451x - 546)$		

$n = 12$	$f_1(x_i)$	$f_2(x_i)$	$f_3(x_i)$
	-11	55	-33
	-9	25	3
	-7	1	21
	-5	-17	25
	-3	-29	19
	-1	-35	7
	1	-35	-7
	3	-29	-19
	5	-17	-25
	7	1	-21
	9	25	-3
	11	55	33
c_j	572	12012	5148
$f_1(x)$	$2x - 13$		
$f_2(x)$	$3x^2 - 39x + 91$		
$f_3(x)$	$\frac{1}{3}(2x^3 - 39x^2 + 211x - 273)$		

From Owen, D. B., *Handbook of Statistical Tables*, 1962, Addison-Wesley, Reading, Mass. Courtesy of U.S. Atomic Energy Commission.

TABLE XIV. Percentage Points of the Extreme Roots of a Wishart Matrix, $\ell(\alpha)$ and $u(\alpha)$

n \ α	Lower 100α % points, $\ell(\alpha)$				Upper 100α % points, $u(\alpha)$			
	0·005	0·010	0·025	0·050	0·005	0·010	0·025	0·050
				$p = 2$				
2	0·0^{4}1518	0·0^{4}6287	0·0^{3}3858	0·0^{2}1500	13·66	12·16	10·15	8·594
3	·0^{2}5012	·0^{2}1005	·02532	·05129	16·16	14·57	12·42	10·74
4	·04047	·06477	·1216	·1980	18·40	16·73	14·46	12·68
5	·1264	·1812	·2948	·4314	20·48	18·73	16·36	14·49
6	·2659	·3573	·5340	·7333	22·45	20·64	18·17	16·21
7	·4550	·5858	·8278	1·090	24·33	22·47	19·91	17·88
8	·6880	·8595	1·167	1·489	26·15	24·23	21·59	19·49
9	·9597	1·172	1·544	1·926	27·92	25·95	23·24	21·06
10	1·265	1·518	1·953	2·392	29·65	27·63	24·84	22·60
15	3·184	3·629	4·358	5·059	37·83	35·59	32·48	29·96
20	5·558	6·177	7·166	8·094	45·51	43·08	39·69	36·94
25	8·233	9·009	10·23	11·37	52·86	50·27	46·63	43·67
30	11·13	12·05	13·49	14·80	59·99	57·24	53·39	50·24
40	17·38	18·56	20·38	22·03	73·76	70·75	66·50	63·02
50	24·07	25·48	27·65	29·60	87·08	83·84	79·24	75·46
60	31·07	32·70	35·18	37·39	100·1	96·72	91·72	87·66
70	38·31	40·14	42·91	45·37	112·9	109·2	104·0	99·70
80	45·75	47·76	50·79	53·48	125·4	121·6	116·1	111·6
90	53·34	55·52	58·81	61·71	137·8	133·8	128·1	123·4
100	61·06	63·40	66·93	70·04	150·1	145·9	140·0	135·0
				$p = 3$				
3	0·0^{5}9820	0·0^{4}3927	0·0^{3}2454	0·0^{3}9817	18·96	17·18	14·90	13·11
4	0·0^{2}3342	0·0^{2}6701	·01688	·03420	21·26	19·50	17·12	15·24
5	0·02844	·04550	·08538	·1390	23·45	21·66	19·18	17·22
6	·09224	·1322	·2149	·3142	25·55	23·69	21·13	19·09
7	·1997	·2682	·4004	·5492	27·56	25·64	22·99	20·88
8	·3495	·4497	·6346	·8339	29·49	27·52	24·80	22·62
9	·5383	·6719	·9106	1·160	31·37	29·34	26·55	24·31
10	·7625	·9300	1·223	1·522	33·19	31·12	28·26	25·96
15	2·301	2·638	3·191	3·724	41·79	39·52	36·36	33·80
20	4·338	4·833	5·623	6·364	49·82	47·37	43·95	41·18
25	6·710	7·350	8·356	9·285	57·49	54·89	51·24	48·27
30	9·326	10·10	11·31	12·41	64·90	62·15	58·30	55·15
40	15·08	16·10	17·66	19·07	79·18	76·18	71·96	68·50
50	21·33	22·57	24·45	26·14	92·95	89·73	85·18	81·44
60	27·93	29·37	31·55	33·48	106·4	102·9	98·09	94·09
70	34·81	36·43	38·88	41·04	119·5	115·9	110·8	106·5
80	41·90	43·69	46·39	48·77	132·4	128·6	123·2	118·8
90	49·16	51·12	54·06	56·64	145·2	141·2	135·6	130·9
100	56·57	58·69	61·85	64·63	157·7	153·6	147·8	143·0
				$p = 4$				
4	0·0^{5}7074	0·0^{4}2830	0·0^{3}1769	0·0^{3}7085	23·78	21·97	19·49	17·52
5	·0^{2}2506	·0^{2}5025	·01266	·02565	26·11	24·24	21·67	19·63
6	·02197	·03514	·06595	·1073	28·31	26·39	23·74	21·62
7	·07289	·1045	·1698	·2481	30·41	28·43	25·71	23·53
8	·1607	·2158	·3220	·4414	32·43	30·41	27·61	25·37
9	·2854	·3671	·5177	·6798	34·39	32·32	29·45	27·15
10	·4451	·5552	·7519	·9574	36·29	34·18	31·25	28·90
15	1·675	1·935	2·365	2·781	45·25	42·94	39·73	37·13
20	3·435	3·839	4·488	5·096	53·58	51·10	47·64	44·84
25	5·555	6·095	6·946	7·730	61·51	58·88	55·21	52·22
30	7·939	8·607	9·645	10·59	69·16	66·40	62·53	59·37

From R. C. Hanumara, and W. A. Thompson, Jr. Percentage points of the extreme roots of a Wishart Matrix. *Biometrika*, 1968, **55**, 505–512. Reproduced by permission of the author and the editors and trustees of *Biometrika*.

TABLE XIV (*continued*).

$n \backslash \alpha$	Lower 100α % points, $\ell(\alpha)$ 0·005	0·010	0·025	0·050	Upper 100α % points, $u(\alpha)$ 0·005	0·010	0·025	0·050
				$p = 4$ (*cont.*)				
40	13·28	14·18	15·56	16·79	83·86	80·86	76·64	73·18
50	19·16	20·27	21·95	23·45	98·00	94·79	90·25	86·53
60	25·43	26·73	28·69	30·43	111·8	108·3	103·5	99·55
70	31·99	33·47	35·69	37·65	125·2	121·6	116·5	112·3
80	38·80	40·44	42·90	45·06	138·4	134·7	129·3	124·9
90	45·79	47·59	50·27	52·63	151·4	147·5	142·0	137·4
100	52·94	54·89	57·79	60·33	164·3	160·2	154·4	149·7
				$p = 5$				
5	0·0^{5}5521	0·0^{4}2209	0·0^{3}1381	0·0^{3}5527	28·85	26·62	23·97	21·85
6	·0^{2}2005	·0^{2}4020	·01013	·02052	31·01	28·86	26·13	23·95
7	·01791	·02865	·05377	·08750	33·11	31·00	28·21	25·96
8	·06035	·08648	·1405	·2054	35·17	33·05	30·19	27·88
9	·1347	·1809	·2698	·3698	37·17	35·04	32·11	29·75
10	·2418	·3110	·4383	·5754	39·14	36·98	33·98	31·57
15	1·210	1·411	1·746	2·073	48·40	46·05	42·79	40·15
20	2·728	3·063	3·602	4·109	56·99	54·49	50·99	48·14
25	4·629	5·092	5·820	6·493	65·15	62·51	58·80	55·78
30	6·811	7·394	8·301	9·128	73·01	70·23	66·34	63·16
40	11·79	12·59	13·82	14·93	88·09	85·08	80·84	77·37
50	17·34	18·35	19·87	21·23	102·6	99·34	94·81	91·08
60	23·32	24·51	26·30	27·88	116·6	113·2	108·4	104·4
70	29·61	30·97	33·01	34·81	130·3	126·8	121·7	117·5
80	36·16	37·68	39·95	41·94	143·8	140·1	134·8	130·4
90	42·91	44·58	47·08	49·25	157·0	153·2	147·6	143·1
100	49·84	51·66	54·36	56·71	170·1	166·1	160·4	155·6
				$p = 6$				
6	0·0^{5}4590	0·0^{4}1835	0·0^{3}1148	0·0^{3}4596	33·22	31·19	28·39	26·14
7	·0^{2}1671	·0^{2}3350	·0^{2}8440	·01710	35·48	33·40	30·54	28·23
8	·01512	·02420	·04540	·07389	37·65	35·53	32·60	30·24
9	·05153	·07383	·1200	·1753	39·75	37·59	34·60	32·19
10	·1161	·1558	·2325	·3185	41·79	39·59	36·54	34·08
15	·8580	1·012	1·272	1·529	51·33	48·96	45·64	42·96
20	2·162	2·440	2·889	3·313	60·16	57·63	54·09	51·21
25	3·865	4·264	4·893	5·475	68·53	65·87	62·13	59·07
30	5·866	6·379	7·178	7·907	76·58	73·79	69·86	66·66
40	10·51	11·24	12·35	13·25	92·00	88·98	84·72	81·24
50	15·78	16·69	18·09	19·33	106·8	103·6	99·00	95·27
60	21·49	22·58	24·23	25·69	121·1	117·7	112·9	108·9
70	27·53	28·79	30·69	32·35	135·1	131·5	126·4	122·3
80	33·85	35·27	37·39	39·24	148·7	145·0	139·7	135·4
90	40·39	41·95	44·28	46·32	162·2	158·3	152·8	148·3
100	47·12	48·82	51·35	53·56	175·5	171·5	165·8	161·1
				$p = 7$				
7	0·0^{5}3835	0·0^{4}1534	0·0^{4}9592	0·0^{3}3841	37·82	35·70	32·76	30·40
8	·0^{2}1432	·0^{2}2872	·0^{2}7234	·01466	40·05	37·89	34·90	32·48
9	·01309	·02095	·03930	·06395	42·22	40·02	36·96	34·49
10	·04498	·06444	·1047	·1530	44·31	42·07	38·96	36·45
15	·5909	·7071	·9057	1·105	54·11	51·70	48·34	45·61
20	1·701	1·931	2·306	2·662	63·16	60·60	57·02	54·10

TABLE XIV (*continued*).

n \ α	Lower 100α % points, ℓ(α)				Upper 100α % points, u(α)			
	0·005	0·010	0·025	0·050	0·005	0·010	0·025	0·050
				p = 7 (*cont.*)				
25	3·224	3·569	4·114	4·620	71·72	69·03	65·26	62·17
30	5·058	5·513	6·220	6·867	79·95	77·13	73·18	69·95
40	9·403	10·06	11·07	11·98	95·67	92·64	88·36	84·86
50	14·40	15·24	16·53	17·66	110·7	107·5	102·9	99·18
60	19·86	20·88	22·41	23·77	125·3	121·9	117·1	113·1
70	25·69	26·36	28·63	30·18	139·5	135·9	130·9	126·7
80	31·79	33·12	35·11	36·84	153·4	149·7	144·4	140·0
90	38·13	39·60	41·80	43·71	167·1	163·2	157·7	153·2
100	44·67	46·28	48·67	50·74	180·5	176·5	170·8	166·1
				p = 8				
8	0·0^{5}3327	0·0^{4}1331	0·0^{4}8318	0·0^{3}3332	42·35	40·15	37·10	34·63
9	·0^{2}1253	·0^{2}2513	·0^{2}6330	·01283	44·57	42·33	39·22	36·71
10	·01154	·01847	·03465	·05638	46·72	44·45	41·28	38·72
15	·3902	·4753	·6235	·7744	56·76	54·32	50·91	48·15
20	1·323	1·513	1·824	2·122	66·01	63·43	59·81	56·86
25	2·681	2·979	3·452	3·892	74·76	72·04	68·23	65·12
30	4·361	4·764	5·392	5·968	83·15	80·31	76·33	73·07
40	8·426	9·026	9·945	10·77	99·16	96·11	91·81	88·29
50	13·17	13·95	15·14	16·19	114·5	111·2	106·7	102·9
60	18·41	19·36	20·78	22·04	129·3	125·8	121·0	117·0
70	24·02	25·12	26·78	28·23	143·7	140·1	135·0	130·9
80	29·93	31·18	33·05	34·69	157·8	154·0	148·8	144·4
90	36·09	37·48	39·55	41·35	171·6	167·8	162·3	157·8
100	42·45	43·97	46·23	48·20	185·3	181·3	175·6	170·9
				p = 9				
9	0·0^{5}2936	0·0^{4}1175	0·0^{4}7343	0·0^{3}2941	46·84	44·57	41·40	38·84
10	·0^{2}1114	·0^{2}2234	·0^{2}5626	·01140	49·05	46·74	43·52	40·91
15	·2426	·3024	·4090	·5202	59·32	56·85	53·39	50·58
20	1·014	1·169	1·425	1·672	68·76	66·14	62·48	59·50
25	2·218	2·475	2·884	3·267	77·67	74·93	71·09	67·94
30	3·753	4·111	4·670	5·183	86·21	83·35	79·34	76·05
40	7·557	8·104	8·944	9·698	102·5	99·43	95·11	91·57
50	12·07	12·79	13·89	14·86	118·0	114·8	110·2	106·4
60	17·09	17·98	19·31	20·49	133·1	129·6	124·8	120·8
70	22·50	23·54	25·10	26·47	147·7	144·1	139·0	134·8
80	28·25	29·41	31·18	32·72	162·0	158·2	152·9	148·6
90	34·21	35·53	37·49	39·20	176·0	172·1	166·6	162·1
100	40·41	41·86	44·01	45·88	189·8	185·8	180·1	175·4
				p = 10				
10	0·0^{5}2628	0·0^{4}1051	0·0^{4}6573	0·0^{3}2632	51·32	48·98	45·73	43·12
15	·1382	·1777	·2503	·3284	61·78	59·28	55·78	52·94
20	·7608	·8863	1·095	1·298	71·41	68·77	65·07	62·05
25	1·821	2·042	2·396	2·728	80·48	77·72	73·84	70·67
30	3·221	3·538	4·035	4·493	89·17	86·29	82·24	78·93
40	6·777	7·278	8·047	8·738	105·7	102·6	98·28	94·72
50	11·07	11·74	12·76	13·66	121·5	118·2	113·6	109·8
60	15·89	16·72	17·97	19·07	136·7	133·3	128·4	124·4
70	21·11	22·09	23·56	24·85	151·5	147·9	142·8	138·7
80	26·66	27·78	29·46	30·92	166·0	162·2	157·0	152·6
90	32·48	33·74	35·60	37·23	180·2	176·3	170·8	166·3
100	38·52	39·90	41·95	43·74	194·1	190·1	184·5	179·8

PARTIAL SOLUTIONS TO SELECTED EXERCISES

EXERCISES 1.2

1. a. $\begin{bmatrix} 8 \\ 2 \\ -1 \end{bmatrix}$ b. $\begin{bmatrix} s_1 + 2s_2 \\ s_1 \\ s_1 - s_2 \end{bmatrix}$ c. $\begin{bmatrix} 1/4 \\ -3/4 \\ -5/4 \end{bmatrix}$

6. Linearly independent. 7. Yes.

EXERCISES 1.3

3. a. $\|\mathbf{y}_1\| = 3$, $\|\mathbf{y}_2\| = \sqrt{10}$, $\|\mathbf{y}_1 - \mathbf{y}_2\| = 3$, $\theta = \cos^{-1}(5/3\sqrt{10}) = 58°$

   b. $[3/\sqrt{2}, -3/\sqrt{2}]$

4. a. $[1, 9, -7]$ b. If $\mathbf{y} \perp V$, then $P_V\mathbf{y} = \mathbf{0}$.

5.
$$\mathbf{x}_1 = \begin{bmatrix} 1 \\ 1 \\ 1 \end{bmatrix}, \quad \mathbf{x}_2 = \begin{bmatrix} -1 \\ 0 \\ 1 \end{bmatrix}, \quad \mathbf{x}_3 = \begin{bmatrix} 1/6 \\ -2/6 \\ 1/6 \end{bmatrix}$$

6. a. $(s\mathbf{x}, \mathbf{y} - s\mathbf{x}) = 0 \Rightarrow s(\mathbf{x}, \mathbf{y}) - s^2(\mathbf{x}, \mathbf{x}) = 0$
$\Rightarrow s = 0$ or $(\mathbf{x}, \mathbf{y}) - s(\mathbf{x}, \mathbf{x}) = 0$
$\Rightarrow s = 0$ or $s = (\mathbf{x}, \mathbf{y})/(\mathbf{x}, \mathbf{x})$, if $(\mathbf{x}, \mathbf{x}) \neq 0$
($s = 0$ should be discarded. Why?)

b.

$$\begin{bmatrix} y_{11} - \bar{y}_1 \\ y_{12} - \bar{y}_1 \\ y_{21} - \bar{y}_2 \\ y_{22} - \bar{y}_2 \end{bmatrix}$$

c. $P_{\mathbf{1}}\mathbf{y} = \dfrac{(\mathbf{y}, \mathbf{1})\mathbf{1}}{\|\mathbf{1}\|^2} = \left[\dfrac{\sum_i y_i}{n}\right]\mathbf{1} = \bar{y}\mathbf{1}$

9. a. $\left\{\dfrac{1}{\sqrt{5}}\begin{bmatrix} 0 \\ 1 \\ -2 \end{bmatrix}, \dfrac{1}{\sqrt{105}}\begin{bmatrix} -5 \\ 8 \\ 4 \end{bmatrix}\right\} = V_2$

b. $\dfrac{1}{\sqrt{6}}\begin{bmatrix} -1 \\ 1 \\ 2 \end{bmatrix} = V_1$ c. $\dfrac{1}{\sqrt{14}}\begin{bmatrix} 1 \\ -3 \\ 2 \end{bmatrix} = W_1$

d.

$$\begin{aligned} E_3 &= \left\{\frac{1}{\sqrt{21}}\begin{bmatrix} 4 \\ 2 \\ 1 \end{bmatrix}\right\} \oplus \{V_2\} \\ &= \left\{\frac{1}{\sqrt{21}}\begin{bmatrix} 4 \\ 2 \\ 1 \end{bmatrix}\right\} \oplus V_1 \oplus W_1 \\ &= \left\{\frac{1}{\sqrt{3}}\begin{bmatrix} 1 \\ 1 \\ 1 \end{bmatrix}, \frac{1}{\sqrt{5}}\begin{bmatrix} 2 \\ 0 \\ -1 \end{bmatrix}\right\} \oplus \frac{1}{\sqrt{14}}\begin{bmatrix} 1 \\ -3 \\ 2 \end{bmatrix} \end{aligned}$$

11. The dimension of V is 3.

15. a. $\mathbf{1} = \{\mathbf{v}_1\}$, $A/\mathbf{1} = \{\mathbf{v}_2 - \mathbf{v}_3\}$, $B/\mathbf{1} = \{\mathbf{v}_4 - \mathbf{v}_5\}$, $AB/(A + B) = \{\mathbf{v}_6 - \mathbf{v}_7 - \mathbf{v}_8 + \mathbf{v}_9\}$. The dimension of each subspace is 1.

b. Letting $y_{ij.} = \sum_{k=1}^{2} y_{ijk}/2$, $y_{i..} = \sum_{j=1}^{2} y_{ij.}/2$, $y_{.j.} = \sum_{i=1}^{2} y_{ij.}/2$, and $y_{...} = \sum_{i=1}^{2}\sum_{j=1}^{2} y_{ij.}/4$, the

$$\|P_{A/\mathbf{1}}\mathbf{y}\|^2 = \|P_A\mathbf{y} - P_{\mathbf{1}}\mathbf{y}\|^2 = 4\sum_{i=1}^{2}(y_{i..} - y_{...})^2$$

$$\|P_{B/\mathbf{1}}\mathbf{y}\|^2 = \|P_B\mathbf{y} - P_{\mathbf{1}}\mathbf{y}\|^2 = 4\sum_{j=1}^{2}(y_{.j.} - y_{...})^2$$

$$\begin{aligned} \|P_{AB/(A+B)}\mathbf{y}\|^2 &= \|P_{AB}\mathbf{y} - P_{A+B}\mathbf{y}\|^2 = \|P_{AB}\mathbf{y} - P_{A/\mathbf{1}}\mathbf{y} - P_{B/\mathbf{1}}\mathbf{y} - P_{\mathbf{1}}\mathbf{y}\|^2 \\ &= \|P_{AB}\mathbf{y} - P_A\mathbf{y} - P_B\mathbf{y} + P_{\mathbf{1}}\mathbf{y}\|^2 \\ &= 2\sum_{i=1}^{2}\sum_{j=1}^{2}(y_{ij.} - y_{i..} - y_{.j.} + y_{...})^2 \end{aligned}$$

EXERCISES 1.4

2.
$$\mathbf{AB} = \begin{bmatrix} 7 & -2 & 3 \\ 4 & -2 & 22 \\ 3 & 0 & 13 \end{bmatrix} \neq \mathbf{BA} = \begin{bmatrix} 3 & 6 & 7 \\ 4 & 8 & 4 \\ 5 & -2 & 7 \end{bmatrix}$$

EXERCISES 1.5

2. $|\mathbf{A}| = 3$ 6. $\mathbf{A}^{-1} = \frac{1}{5}\begin{bmatrix} 0 & 5 & 0 \\ 1 & -7 & 3 \\ 1 & 3 & -2 \end{bmatrix}$

7.
$$\mathbf{A}^{-1} = \frac{1}{55}\left[\begin{array}{cc|cc} 8 & 10 & 17 & -11 \\ -25 & 10 & -5 & 0 \\ \hline 7 & -5 & 8 & 11 \\ 47 & -10 & -17 & 11 \end{array}\right]$$

8. a.
$$\mathbf{P} = \begin{bmatrix} 1 & 0 & 0 & 0 \\ -3 & 1 & 0 & 0 \\ 0 & 0 & 0 & 1 \\ 1 & -2 & 1 & 0 \end{bmatrix} \quad \text{and} \quad \mathbf{Q} = \begin{bmatrix} 1 & 0 & -2 \\ 0 & 1 & 1 \\ 0 & 0 & 1 \end{bmatrix}$$

we have that $\mathbf{A} = \mathbf{P}^{-1}\begin{bmatrix} \mathbf{I}_3 \\ \mathbf{0} \end{bmatrix}\mathbf{Q}^{-1} = \mathbf{P}_1\mathbf{Q}_1$ where

$$\mathbf{P}_1 = \begin{bmatrix} 1 & 0 & 0 \\ 3 & 1 & 0 \\ 5 & 2 & 0 \\ 0 & 0 & 1 \end{bmatrix} \quad \text{and} \quad \mathbf{Q}_1 = \begin{bmatrix} 1 & 0 & 2 \\ 0 & 1 & -1 \\ 0 & 0 & 1 \end{bmatrix}$$

b. For $\mathbf{P} = \begin{bmatrix} 1 & 0 & 0 \\ -1/2 & 1 & 0 \\ -4 & 6 & 1 \end{bmatrix}$, $\mathbf{PAP}' = \mathbf{\Lambda} = \begin{bmatrix} 2 & 0 & 0 \\ 0 & -1/2 & 0 \\ 0 & 0 & 0 \end{bmatrix}$ so that

$$\mathbf{P}_1 = \mathbf{P}^{-1}\mathbf{\Lambda}^{1/2} = \begin{bmatrix} 1 & 0 & 0 \\ 1/2 & 1 & 0 \\ 1 & -6 & 1 \end{bmatrix}\begin{bmatrix} \sqrt{2} & 0 & 0 \\ 0 & \sqrt{-1/2} & 0 \\ 0 & 0 & 0 \end{bmatrix}$$

EXERCISES 1.6

1. a. Inconsistent system, no solution exists.
   b. Consistent system, unique solution, $x = -1$, $y = 7$, and $z = 1$.
   c. Consistent system, unique trivial solution, $x = y = 0$.
   d. Consistent system, infinite nontrivial solutions, $x = 0$ and $y = z$.
   e. Consistent system, infinite solutions, $y = 2$ and $x + z = 4$.

2. a.
$$\begin{bmatrix} \mu \\ \alpha_1 \\ \alpha_2 \end{bmatrix} = \begin{bmatrix} 2y_{..} - y_{1.} \\ -2y_{..} + 2y_{1.} \\ 0 \end{bmatrix} = \begin{bmatrix} y_{2.} \\ y_{1.} - y_{2.} \\ 0 \end{bmatrix}$$

b. Side condition is not appropriate, the $R\begin{bmatrix} \mathbf{A} \\ \mathbf{R} \end{bmatrix} = 2 < 3$.

3. a. $\begin{bmatrix} x \\ y \\ z \end{bmatrix} = \begin{bmatrix} 2 \\ -2 \\ 2 \end{bmatrix}$ d. $\begin{bmatrix} x \\ y \\ z \end{bmatrix} = \begin{bmatrix} 0 \\ -2 \\ 4 \end{bmatrix}$

4. a. $\begin{bmatrix} \mu + \alpha_1 \\ \mu + \alpha_2 \end{bmatrix} = \begin{bmatrix} y_{1.} \\ y_{2.} \end{bmatrix}$ b. $R\begin{bmatrix} \mathbf{A} \\ \mathbf{C} \end{bmatrix} = 3 > R(A) = 2$

c. $\begin{bmatrix} \mu + \alpha_1 \\ \mu + \alpha_2 \end{bmatrix} = \begin{bmatrix} y_{1.} \\ y_{1.} - y_2 \end{bmatrix}$ d. same as b.

5. a. $\begin{bmatrix} y \\ x + z \end{bmatrix} = \begin{bmatrix} -2 \\ 4 \end{bmatrix}$

6. a. $\mathbf{A}^- = [1 \quad 0 \quad 0]$ b. $\mathbf{A}^- = \begin{bmatrix} 0 \\ 1 \\ 0 \end{bmatrix}$

c. $\mathbf{A}^- = \begin{bmatrix} 1 & 2 & 0 \\ 0 & -1 & 0 \end{bmatrix}$ d. $\mathbf{A}^- = \begin{bmatrix} 1/4 & -1/4 & 0 \\ -1/4 & 1/2 & 0 \\ 0 & 0 & 0 \end{bmatrix}$

e.
$$\mathbf{A}^- = \begin{bmatrix} 0 & 0 & 0 & 0 \\ 0 & -5/3 & 2/3 & 0 \\ 0 & 4/3 & -1/2 & 0 \\ 0 & 0 & 0 & 0 \end{bmatrix}$$

8. $\begin{bmatrix} x \\ y \\ z \end{bmatrix} = \begin{bmatrix} 4 - z_3 \\ -2 \\ z_3 \end{bmatrix}$ where two linearly independent solutions are $\begin{bmatrix} 4 \\ -2 \\ 0 \end{bmatrix}$ and $\begin{bmatrix} 3 \\ -2 \\ 1 \end{bmatrix}$

9. Setting $\beta_2 = \alpha_1 = 0$

$$(\mathbf{A}'\mathbf{A})^- = \begin{bmatrix} 3/8 & -1/4 & 0 & -1/4 & 0 \\ -1/4 & 1/2 & 0 & 0 & 0 \\ 0 & 0 & 0 & 0 & 0 \\ -1/4 & 0 & 0 & 1/2 & 0 \\ 0 & 0 & 0 & 0 & 0 \end{bmatrix}$$

$$\mathbf{H} = \begin{bmatrix} 1 & 0 & 1 & 0 & 1 \\ 0 & 1 & -1 & 0 & 0 \\ 0 & 0 & 0 & 0 & 0 \\ 0 & 0 & 0 & 1 & -1 \\ 0 & 0 & 0 & 0 & 0 \end{bmatrix}$$

and

$$\begin{bmatrix} \mu \\ \alpha_1 \\ \alpha_2 \\ \beta_1 \\ \beta_2 \end{bmatrix} = \begin{bmatrix} y_{...} \\ y_{1..} - y_{2..} \\ 0 \\ y_{.1.} - y_{.2.} \\ 0 \end{bmatrix} + \begin{bmatrix} 0 & 0 & -1 & 0 & -1 \\ 0 & 0 & 1 & 0 & 0 \\ 0 & 0 & 1 & 0 & 0 \\ 0 & 0 & 0 & 0 & 1 \\ 0 & 0 & 0 & 0 & 1 \end{bmatrix} \begin{bmatrix} z_1 \\ z_2 \\ z_3 \\ z_4 \\ z_5 \end{bmatrix}$$

$$= \begin{bmatrix} y_{...} - z_3 - z_5 \\ y_{1..} - y_{2..} + z_3 \\ z_3 \\ y_{.1.} - y_{.2.} + z_5 \\ z_5 \end{bmatrix}$$

is a solution set. Setting $z_3 = z_5 = 0$, the form of unique solutions for an arbitrary vector $\mathbf{t}' = [t_0, t_1, t_2, t_3, t_4]$ is

$$\mathbf{p}'\mathbf{x} = \mathbf{t}'\mathbf{H}\mathbf{x} = t_0(\mu + \alpha_2 + \beta_2) - t_1(\alpha_1 - \alpha_2) + t_2(\beta_1 - \beta_2)$$

with solution given by

$$\mathbf{p}'\hat{\mathbf{x}} = \mathbf{t}'\mathbf{H}\mathbf{A}^-\mathbf{y} = t_0 y_{...} - t_1(y_{1..} - y_{2..}) + t_2(y_{.1.} - y_{.2.})$$

An alternative choice of $(\mathbf{A}'\mathbf{A})^-$ is

$$(\mathbf{A}'\mathbf{A})^- = \begin{bmatrix} -1/8 & 0 & 0 & 0 & 0 \\ 0 & 1/4 & 0 & 0 & 0 \\ 0 & 0 & 1/4 & 0 & 0 \\ 0 & 0 & 0 & 1/4 & 0 \\ 0 & 0 & 0 & 0 & 1/4 \end{bmatrix}$$

12. *Hint*: $\mathbf{P}_1\mathbf{y} = \mathbf{1}(\mathbf{1}'\mathbf{1})^{-1}\mathbf{1}'\mathbf{y} \equiv \mathbf{P}_1\mathbf{y}$

$$\mathbf{P}_2\mathbf{y} = [\mathbf{A}(\mathbf{A}'\mathbf{A})^-\mathbf{A}' - \mathbf{1}(\mathbf{1}'\mathbf{1})^{-1}\mathbf{1}']\mathbf{y} \equiv P_{A/1}\mathbf{y}$$

$$\mathbf{P}_3\mathbf{y} = [\mathbf{X}(\mathbf{X}'\mathbf{X})^-\mathbf{X}' - \mathbf{A}(\mathbf{A}'\mathbf{A})^-\mathbf{A}']\mathbf{y} \equiv P_{(A+B)/A}\mathbf{y} = P_{B/1}\mathbf{y}$$

$$\mathbf{P}_4\mathbf{y} = [\mathbf{I} - \mathbf{X}(\mathbf{X}'\mathbf{X})^-\mathbf{X}']\mathbf{y} \equiv P_{(A+B)^\perp}\mathbf{y}$$

where $\mathbf{I} = \mathbf{P}_1 + \mathbf{P}_2 + \mathbf{P}_3 + \mathbf{P}_4$

EXERCISES 1.7

1. a.

$$\mathbf{A} = \begin{bmatrix} 3 & 0 & 1 \\ 0 & 1 & 0 \\ 1 & 0 & 2 \end{bmatrix}, \quad \mathbf{B} = \begin{bmatrix} 1 & 1 & 0 \\ 1 & 5 & -2 \\ 0 & -2 & 1 \end{bmatrix}$$

b.

$$\mathbf{P}' = \begin{bmatrix} 1 & 0 & 0 \\ 0 & 1 & 3 \\ -1 & 0 & 3 \end{bmatrix}, \quad \mathbf{P}'\mathbf{A}\mathbf{P} = \begin{bmatrix} 3 & 0 & 0 \\ 0 & 1 & 0 \\ 0 & 0 & 15 \end{bmatrix}, \quad 3x_1^2 + x_2^2 + 15x_3^2$$

$$\mathbf{P}' = \begin{bmatrix} 1 & 0 & 0 \\ -1 & 1 & 0 \\ -1 & 1 & 2 \end{bmatrix}, \quad \mathbf{P}'\mathbf{B}\mathbf{P} = \begin{bmatrix} 1 & 0 & 0 \\ 0 & 4 & 0 \\ 0 & 0 & 0 \end{bmatrix}$$

c. $R(\mathbf{A}) = 3$ and $\mathbf{A}$ is p.d. $R(\mathbf{B}) = 2$ and $\mathbf{B}$ is p.s.d.

d. $\mathbf{A} = [(\mathbf{P}')^{-1}\mathbf{\Lambda}^{1/2}][\mathbf{\Lambda}^{1/2}\mathbf{P}^{-1}] = \mathbf{FF}'$ where for $\mathbf{P}'$ in part b is

$$\mathbf{F} = \begin{bmatrix} \sqrt{3} & 0 & 0 \\ 0 & 1 & 0 \\ \sqrt{3}/3 & 0 & \sqrt{5/3} \end{bmatrix}$$

e.

$$Q = [\sqrt{3}y_1 + \sqrt{3}y_3/3]^2 + y_2^2 + 5y_3^2/3$$

$$\mathbf{T} = \begin{bmatrix} \sqrt{3} & 0 & 0 \\ 0 & 1 & 0 \\ \sqrt{3/3} & 0 & \sqrt{5/3} \end{bmatrix} \qquad \mathbf{T}^{-1} = \begin{bmatrix} \sqrt{3}/3 & 0 & 0 \\ 0 & 1 & 0 \\ -\sqrt{15}/15 & 0 & \sqrt{3/5} \end{bmatrix}$$

$$\mathbf{A}^{-1} = (\mathbf{T}')^{-1}\mathbf{T}^{-1} = \frac{1}{5}\begin{bmatrix} 2 & 0 & -1 \\ 0 & 1 & 0 \\ -1 & 0 & 3 \end{bmatrix}$$

2.

$$\mathbf{A}^- = \frac{1}{2}\begin{bmatrix} 2 & -1 & 0 \\ 0 & 1 & 0 \\ 0 & 0 & 0 \end{bmatrix}\frac{1}{2}\begin{bmatrix} 2 & 0 & 0 \\ -1 & 1 & 0 \\ 0 & 0 & 0 \end{bmatrix} = \frac{1}{4}\begin{bmatrix} 5 & -1 & 0 \\ -1 & 1 & 0 \\ 0 & 0 & 0 \end{bmatrix}$$

3. a. $\mathbf{P} = \begin{bmatrix} 1/\sqrt{2} & -1/\sqrt{2} \\ 1/\sqrt{2} & 1/\sqrt{2} \end{bmatrix}$, $\lambda_1 = \frac{5}{2}$, $\lambda_2 = \frac{3}{2}$, $Q = \frac{5}{2}x_1^2 + \frac{3}{2}x_2^2$

b. major axis $-2\sqrt{2}$; minor axis $-2\sqrt{30}/5$ c. $\lambda_1^* = 2/3$, $\lambda_2^* = 2/5$

4. a. For $\mathbf{A}$, $\lambda_1 = 3$ and $\lambda_2 = 1$. For $\mathbf{B}$, $\lambda_1 = 6$, $\lambda_2 = 1$, and $\lambda_3 = 0$.

b. For $\mathbf{A}$, $\mathbf{P} = \begin{bmatrix} 1/\sqrt{2} & 1/\sqrt{2} \\ 1/\sqrt{2} & -1/\sqrt{2} \end{bmatrix}$

For $\mathbf{B}$, $\mathbf{P} = \begin{bmatrix} 1/\sqrt{30} & 2/\sqrt{5} & 1/\sqrt{6} \\ 5/\sqrt{30} & 0 & -1/\sqrt{6} \\ -2/\sqrt{30} & 1/\sqrt{5} & -2/\sqrt{6} \end{bmatrix}$

5.

$$\mathbf{B}^8\mathbf{x}_0 = \begin{bmatrix} 223950 \\ 1119744 \\ -447897 \end{bmatrix}, \qquad \mathbf{X}_8 = \begin{bmatrix} .20 \\ 1.00 \\ .40 \end{bmatrix}$$

Normalizing,

$$\mathbf{p}_1 = \begin{bmatrix} .20/\sqrt{1.20} \\ 1.00/\sqrt{1.20} \\ -.40/\sqrt{1.20} \end{bmatrix}, \qquad \lambda_1 = \mathbf{p}_1'\mathbf{Bp} \doteq 6.0$$

6. The roots of the matrix are $\lambda_1 = 1 + (n - r)r$ with the characteristic vector $\mathbf{1}_n' = [1, \ldots, 1]$ and $\lambda_i = 1 - r$ for $i = 2, 3, \ldots, n$ with $(n - 1)$ vectors orthogonal to $\mathbf{1}_n'$.

7. $\lambda_1 = .7342$ and $\lambda_2 = .0014$.

EXERCISES 1.8

1. a. $\dfrac{dy}{dx} = 15x^4 + \sin x$ b. $\dfrac{dy}{dx} = e^x(\sin x + \cos x)$ c. $\dfrac{dy}{dx} = \dfrac{-4}{(x-1)^2}$

   d. $\dfrac{dy}{dx} = 2x\, e^{x^2} \cos(e^{x^2})$ e. $\dfrac{dy}{dx} = \dfrac{-\sin x}{\cos x}$

   f. The function in part a at $x = 0$ is a maximum, and the slope of the tangent line at $x = 0$ is parallel to the x-axis. For the function in part b, the slope of the tangent line at $x = 0$ is equal to 1.

2. Minimum $\begin{cases} x = 0 \\ y = 0 \end{cases}$ Maximum $\begin{cases} x = -2/3 \\ y = 4/27 \end{cases}$

   Point of inflection $\begin{cases} x = -2/6 \\ y = 2/27 \end{cases}$

3. a. $$\frac{\partial y}{\partial \mathbf{x}} = \begin{bmatrix} 1 & x_2 \\ 0 & x_1 \\ 1 & 0 \end{bmatrix}$$

   b. $$\frac{\partial f}{\partial \mathbf{x}} = \begin{bmatrix} 2 & 4 \\ 4 & 2 \end{bmatrix}\begin{bmatrix} x_1 \\ x_2 \end{bmatrix} = \begin{bmatrix} 2x_1 + 4x_2 \\ 4x_1 + 2x_2 \end{bmatrix}$$

   $$\frac{\partial f}{\partial \mathbf{A}} = \begin{bmatrix} x_1^2 & 2x_1x_2 \\ 2x_1x_2 & x_2^2 \end{bmatrix} \qquad \frac{\partial f}{\partial \mathbf{B}} = \begin{bmatrix} x_1^2 & x_1x_2 \\ x_1x_2 & x_2^2 \end{bmatrix}$$

   c. If $\mathbf{A}$ is not symmetric, the

   $$\frac{\partial|\mathbf{A}|^{-1/2}}{\partial \mathbf{A}} = -\frac{1}{2}|\mathbf{A}|^{-3/2} \text{ adjoint } \mathbf{A}' = -\frac{1}{2}|\mathbf{A}|^{-3/2} \frac{\text{adjoint } \mathbf{A}'}{|\mathbf{A}'|}|\mathbf{A}'|$$

   $$= -\frac{1}{2}|\mathbf{A}|^{-1/2}(\mathbf{A}^{-1})'$$

   When $\mathbf{A}$ is symmetric, the

   $$\frac{\partial|\mathbf{A}|^{-1/2}}{\partial \mathbf{A}} = -\frac{1}{2}|\mathbf{A}|^{-1/2}[2\mathbf{A}^{-1} - \text{Dia } \mathbf{A}^{-1}]$$

4. a. Critical points: $x_1 = 1$ $x_2 = 0$

   $x_1 = 0$ $x_2 = 1$

   $x_1 = 2/5$ $x_2 = 3/5$

   b. Critical points: $x_1 = (2/3)\sqrt[5]{9/4}$, $x_2 = \sqrt[5]{9/4}$

   c. There is no relationship between parts a and b.

EXERCISES 2.3

1. a. $\mu = 2$ and $\sigma = 1$

   b. $Y \sim N(4, 20)$

$$P(|Y-3|>1) = P(Y-3>1) + P(Y-3<-1) = P(Y>4) + P(Y<2) = .83$$

2. a. $X + Y \sim N(3, 12)$ b. $X - Y \sim N(-1, 12)$ c. $3X - 2Y \sim N(-1, 68)$

EXERCISES 2.4

1. a. Let $\mathbf{y} = \begin{bmatrix} y_1 \\ y_2 \end{bmatrix}, \boldsymbol{\mu}_1 = \begin{bmatrix} 1 \\ 2 \end{bmatrix}, \boldsymbol{\Sigma}_{11} = \begin{bmatrix} 3 & 1 \\ 1 & 4 \end{bmatrix}, \mathbf{x} = \begin{bmatrix} y_2 \\ y_4 \end{bmatrix}, \boldsymbol{\mu}_2 = \begin{bmatrix} 2 \\ 4 \end{bmatrix}, \boldsymbol{\Sigma}_{22} = \begin{bmatrix} 4 & 0 \\ 0 & 20 \end{bmatrix}.$

$$f(\mathbf{y}) = \frac{1}{2\pi|\boldsymbol{\Sigma}_{11}|} \exp -\frac{1}{2}(\mathbf{y}-\boldsymbol{\mu}_1)'\boldsymbol{\Sigma}_{11}^{-1}(\mathbf{y}-\boldsymbol{\mu}_1)$$

$$f(\mathbf{x}) = \frac{1}{2\pi|\boldsymbol{\Sigma}_{22}|} \exp -\frac{1}{2}(\mathbf{y}-\boldsymbol{\mu}_2)'\boldsymbol{\Sigma}_{22}^{-1}(\mathbf{y}-\boldsymbol{\mu}_2)$$

   b. $\rho_{12} = \sqrt{3}/6, \quad \rho_{24} = 0$

   c. length of semimajor axis is $\sqrt{100(\lambda_1)} = 20\sqrt{5}$

2.
$$\boldsymbol{\Sigma}^{-1} = \begin{bmatrix} 2 & 1 & 1 \\ 1 & 1 & 0 \\ 1 & 0 & 3 \end{bmatrix}; \quad f(\mathbf{y}) = \frac{1}{(2\pi)^{3/2}|\boldsymbol{\Sigma}|} e^{-1/2}\mathbf{y}'\boldsymbol{\Sigma}^{-1}\mathbf{y}$$

5. a.

X \ Y	0	1
−1	0	1/3
0	1/3	0
1	0	1/3

   b. (See Barr, D. R., and P. W. Zehna, (1971), *Probability*, Brooks/Cole, page 180, for a simple example.)

EXERCISES 2.5

1. $\rho_{12\cdot3} = \sqrt{15}/5, \quad \rho_{23\cdot1} = -\sqrt{5}/5, \quad \rho_{13\cdot2} = \sqrt{3}/3$

2. a. $\boldsymbol{\mu}_{1\cdot2} = \boldsymbol{\mu}_1 + \boldsymbol{\Sigma}_{12}\boldsymbol{\Sigma}_{22}^{-1}(\mathbf{Y}_2 - \boldsymbol{\mu}_2)$

$$= 1 + [1, 0, 0]\begin{bmatrix} 1/4 & 0 & 0 \\ 0 & 5 & -1 \\ 0 & -1 & 1/4 \end{bmatrix}\begin{bmatrix} y_2 - 2 \\ y_3 - 3 \\ y_4 - 4 \end{bmatrix}$$

$$= \frac{y_2}{4} + \frac{1}{2}$$

$$\boldsymbol{\Sigma}_{1\cdot2} = \boldsymbol{\Sigma}_{11} - \boldsymbol{\Sigma}_{12}\boldsymbol{\Sigma}_{22}^{-1}\boldsymbol{\Sigma}_{21} = 11/4$$

   b. $\rho_{13\cdot2} = 0$ c. $\rho_{1\cdot234} = 3/6$

EXERCISES 3.4

1. a. $T^2 = 51.43$, $\Lambda = .1628$, $\theta_s = .8372$, $U^{(s)} = 5.1432$ with all multivariate criteria being significant.

   b. Confidence intervals:

T^2	$43.00 \le \mu_1 \le 71.91$	Bonferroni	$46.83 \le \mu_1 \le 68.08$
	$7.82 \le \mu_2 \le 66.91$		$15.65 \le \mu_2 \le 59.08$
	$24.62 \le \mu_3 \le 65.56$		$30.05 \le \mu_3 \le 60.14$

EXERCISES 3.5

1. a.

$$\mathbf{X'X} = \begin{bmatrix} 3 & 0 & 2 & 0 \\ 0 & 2 & 0 & 2 \\ 2 & 0 & 2 & 0 \\ 0 & 2 & 0 & 2 \end{bmatrix} \qquad (\mathbf{X'X})^- = \begin{bmatrix} 1 & 0 & -1 & 0 \\ 0 & 1/2 & 0 & 0 \\ -1 & 0 & 3/2 & 0 \\ 0 & 0 & 0 & 0 \end{bmatrix}$$

$$\mathbf{H} = \begin{bmatrix} 1 & 0 & 0 & 0 \\ 0 & 1 & 0 & 1 \\ 0 & 0 & 1 & 0 \\ 0 & 0 & 0 & 0 \end{bmatrix} \qquad \hat{\boldsymbol{\beta}} = \begin{bmatrix} y_2 \\ (y_3 - y_1)/2 \\ (y_1 - 2y_2 + y_3)/2 \\ 0 \end{bmatrix}$$

   c. $\beta_0 + \beta_2$ and $\beta_1 + \beta_3$ are both estimable.

   d. The general form of estimable functions is

$$\psi = \mathbf{c}'\boldsymbol{\beta} = \mathbf{t}'\mathbf{H}\boldsymbol{\beta} = t_0\beta_0 + t_1(\beta_1 + \beta_3) + t_2\beta_2$$

   where the form of the estimators is

$$\hat{\psi} = t_0 y_2 + t_1(y_3 - y_1)/2 + t_2(y_1 - 2y_2 + y_3)/2$$

2. a.

$$\mathbf{X'X} = N\begin{bmatrix} 1 & X \\ X & X^2 \end{bmatrix} \qquad (\mathbf{X'X})^- = \begin{bmatrix} 1/N & 0 \\ 0 & 0 \end{bmatrix}$$

$$\mathbf{H} = \begin{bmatrix} 1 & X \\ 0 & 0 \end{bmatrix} \qquad \hat{\boldsymbol{\beta}} = \begin{bmatrix} \bar{y} \\ 0 \end{bmatrix}$$

   b. No.

   c. The general form of estimable functions is

$$\psi = \mathbf{c}'\boldsymbol{\beta} = \mathbf{t}'\mathbf{H}\boldsymbol{\beta} = t_0(\mu + \beta x)$$

   where the form of the estimators is $\hat{\psi} = t_0\bar{y}$.

   d. $V(\hat{\psi}) = \sigma^2/N$

3. b. $\delta = 2J\beta^2/\sigma^2$

c. ANOVA Table

Source	df	SS	E(MS)
β	1	$\dfrac{(\sum y_{2j} - \sum y_{0j})^2}{2J}$	$\sigma_2 + 2J\beta^2$
Error	$3J - 2$	$\sum\sum y_{ij}^2 - \dfrac{(\sum\sum y_{ij})^2}{3J} - \dfrac{(\sum y_{2j} - \sum y_{0j})^2}{2J}$	σ^2
"Total"	$3J - 1$	$\sum y_{ij}^2 - \dfrac{(\sum\sum y_{ij})^2}{3J}$	

5. a. Letting $\bar{x} = \Sigma_i w_i x_i / \Sigma_i w_i$,

$$\hat{\beta}_1 = \frac{\sum_i w_i(x_i - \bar{x})y_i}{\sum_i w_i(x_i - \bar{x})^2} \quad \text{and} \quad \hat{\beta}_0 = \frac{\sum_i w_i y_i}{\sum_i w_i - \hat{\beta}_1 \bar{x}}$$

EXERCISES 3.6

1. b. $\mathbf{X}\hat{\mathbf{B}}_\omega$ and $\mathbf{Q}_t$ are statistically independent.

EXERCISES 3.7

2. a. $F = \dfrac{(t_1 y_{1.} + t_2 y_{2.})^2}{s_p^2(t_1^2/N_1 + t_2^2/N_2)}$ b. $H_0: \mu = 0$ c. $H_0: \dfrac{t_1\mu_1 + t_2\mu_2}{t_1 + t_2} = 0$

EXERCISES 3.9

1. a. $T^2 = 25.96, \Lambda = .3957, \theta_s = .6043, U^{(s)} = 1.53$
   b. The correlation between the discriminant function and each variable is

$$\hat{\boldsymbol{\rho}} = \begin{bmatrix} -.40 \\ .05 \\ .70 \end{bmatrix}$$

For $\alpha = .05$, $T^{.05}_{(3,17)} = 11.177$ and $c_0 = 1.553$ so that

$$-42.38 \le \mu_{11} - \mu_{21} \le 10.54$$
$$-30.68 \le \mu_{12} - \mu_{22} \le 35.91$$
$$1.66 \le \mu_{13} - \mu_{23} \le 47.27$$

   c. For the Bonferroni method with $\alpha = .05$ and $k = 3$ variables, $c_0 = 2.67$. Corresponding confidence intervals are

$$-37.06 \le \mu_{11} - \mu_{21} \le 5.22$$
$$-23.98 \le \mu_{12} - \mu_{22} \le 29.21$$
$$6.25 \le \mu_{13} - \mu_{23} \le 42.68$$

EXERCISES 3.12

1. a. For the overall test, $T^2 = 13.78$, $\Lambda = .6621$, $\theta_s = .3379$, and $U^{(s)} = .5104$. The correlation between the discriminant function and each variable is

$$\hat{\boldsymbol{\rho}} = \begin{bmatrix} -.07 \\ -.67 \\ -.46 \\ -.67 \end{bmatrix}$$

EXERCISES 3.14

1. a. $T^2 = 232.93$; $T^{.05}_{(3,7)} = 22.72$
   b. For T^2, $c_0 = 4.77$. For the Bonferroni procedure, $c_0 = 3.76$.

EXERCISES 3.16

1. a. For the test of parallelism, $T^2 = 1.21 < T^{.05}_{(3,15)} = 11.806$.
   b. To test H_{02}, $T^2 = 7.65$. To test H^*_{03}, $F = 6.29$ and a confidence interval for $\mu_1 - \mu_2$ is $1.92 \leq \mu_1 - \mu_2 \leq 23.57$.

EXERCISES 3.17

1. For Bartlett, $X^2_B = 11.3$; for Box, $F = 1.85$
2. $X^2_I \doteq 390$
3. $X^2_{BX} = 7.66$ for Bartlett; for Box, $F = 1.87$
4. $X^2_s = 7.8$

EXERCISES 4.3

1. b. $\hat{\beta}_0 = 15.883$; $\hat{\beta}_1 = .49286$ c. $F = 10.5863$ d. $\hat{\sigma}_{\hat{\beta}_1} = .15148$
3. *Hint:* Plot residuals against fitted values, plot residuals against time, and make an appropriate transformation.

EXERCISES 4.4

1. a. Matrix of Orthogonalized Polynomials

$$\begin{bmatrix} .577350 & -.707107 & .408248 \\ .577350 & .000000 & -.816497 \\ .577350 & .707107 & .408248 \end{bmatrix}$$

   b. Matrix of Orthogonalized Polynomials

$$\begin{bmatrix} .577350 & -.617213 & .534522 \\ .577350 & -.154303 & -.801784 \\ .577350 & .771517 & .267261 \end{bmatrix}$$

EXERCISES 4.5

1.

Source	df	SS
ξ_0	1	28,037.025
$\xi_1 \mid \xi_0$	1	5712.200
$\xi_2 \mid \xi_1, \xi_0$	1	217.286
$\xi_3 \mid \xi_2, \xi_1, \xi_0$	1	255.613
$\xi_4 \mid \xi_3, \xi_2, \xi_1, \xi_0$	1	58.502
Residual	35	1210.374
Total	40	

EXERCISES 4.9

1.

Source	df	SSP
$\xi_1 \mid \xi_0$	1	$\begin{bmatrix} 3{,}675.000 & \\ 4{,}410.000 & 5{,}292.000 \end{bmatrix}$
$\xi_2 \mid \xi_1, \xi_0$	1	$\begin{bmatrix} 173.571 & \\ 192.857 & 214.286 \end{bmatrix}$
$\xi_3 \mid \xi_2, \xi_1, \xi_0$	1	$\begin{bmatrix} 1{,}200.000 & \\ -240.000 & 48.000 \end{bmatrix}$
$\xi_4 \mid \xi_3, \xi_2, \xi_1, \xi_0$	1	$\begin{bmatrix} 207.429 & \\ -414.857 & 829.714 \end{bmatrix}$

EXERCISES 4.11

1. For the test of parallelism, $F = .26$

EXERCISES 4.15

1. Matrix of Partial Intercorrelations (with the effects of IQ removed)

	2	3	4	5	6	7	8	9	10	11	12	13
2	1.00											
3	.70	1.00										
4	.08	.05	1.00									
5	.30	.37	.37	1.00								
6	.24	.26	.18	.24	1.00							
7	.16	−.02	.37	.34	.22	1.00						
8	.04	.09	.08	.16	.09	.15	1.00					
9	.10	.06	.00	.13	.19	−.01	.35	1.00				
10	−.01	−.05	−.05	.01	−.05	−.02	.11	.47	1.00			
11	.00	−.03	−.06	−.00	−.00	.04	.32	.47	.46	1.00		
12	.06	.10	.36	.38	.33	.34	.24	.27	.15	.36	1.00	
13	.05	.08	.32	.20	.14	.18	.20	.11	.06	.18	.54	1.00

EXERCISES 5.4

1. Group 1 with group 2, 6.4 standard deviations. Group 2 with group 3, 6.2 standard deviations.

2. a. MANOVA Table

Source	df	SSP
Groups	3	$\begin{bmatrix} 17{,}101.137 & \\ 12{,}860.000 & 27{,}882.000 \end{bmatrix}$
Within error	76	$\begin{bmatrix} 8926.050 & \\ 3926.750 & 5411.000 \end{bmatrix}$
"Total"	79	$\begin{bmatrix} 26{,}027.187 & \\ 16{,}786.750 & 33{,}293.000 \end{bmatrix}$

Multivariate criteria: $\Lambda = .5623$, $\theta_s = .8492$, $U^{(s)} = 7.3120$, and $V^{(s)} = 1.4767$.

d. There are two significant discriminant functions for this problem; the correlations between the functions and the variables are as follows:

$$\boldsymbol{\rho}_{L_1} = \begin{bmatrix} .24 \\ .94 \end{bmatrix} \qquad \boldsymbol{\rho}_{L_2} = \begin{bmatrix} .97 \\ .35 \end{bmatrix}$$

EXERCISES 5.5

3. Using the hypothesis test matrix

$$\mathbf{C} = \begin{bmatrix} 1 & -1 & -1 & 1 \\ 1 & 1 & -1 & -1 \\ 1 & -1 & 1 & -1 \end{bmatrix}$$

to test for a significant treatment effect, there are significant differences in the effects of morning versus afternoon and nonsense words versus letter words; however, there is no difference for the training versus no training comparison.

EXERCISES 5.7

2. a. MANOVA Table

Source	df	MSSP
Method	1	$\begin{bmatrix} 2440.167 & & \\ 2379.667 & 2320.667 & \\ 312.583 & 304.833 & 40.042 \end{bmatrix}$
Time	2	$\begin{bmatrix} 354.667 & & \\ 144.333 & 6515.167 & \\ 33.667 & 181.208 & 7.452 \end{bmatrix}$
Linear	1	$\begin{bmatrix} 4.000 & & \\ -225.000 & 12{,}656.250 & \\ -5.500 & 309.370 & 7.560 \end{bmatrix}$
Quadratic	1	$\begin{bmatrix} 705.333 & & \\ 513.667 & 374.083 & \\ 72.833 & 53.042 & 7.521 \end{bmatrix}$
Method × time	2	$\begin{bmatrix} 2.667 & & \\ 2.667 & 164.667 & \\ -.667 & 3.833 & .2917 \end{bmatrix}$
Within error	18	$\begin{bmatrix} 8.806 & & \\ 6.111 & 36.278 & \\ 2.181 & 12.986 & 15.236 \end{bmatrix}$

b. ANOVA Table

Source	df		
Between schools	3		
G			1
$(G)S$ (error 1)			2
Within schools	60		
$(SG)V$		4	
V			1
GV			1
$(G)SV$			2
$(SG)F$		4	
F			1
GF			1
$(G)SF$			2
$(SG)VF$		4	
VF			1
GVF			1
$(G)SVF$			2
$(G)SVF$ (error 2)		48	
"Total"	63		

EXERCISES 5.9

1. MANOVA Table

Source	df		SSP
A Programs	1		$\begin{bmatrix} 21{,}939.13 & \\ -11{,}335.22 & 5856.53 \end{bmatrix}$
$(A)B$ Schools w. programs	4		$\begin{bmatrix} 4774.61 & \\ -198.78 & 25{,}184.03 \end{bmatrix}$
$(A_1)B$		2	$\begin{bmatrix} 1069.91 & \\ -4379.13 & 19{,}670.17 \end{bmatrix}$
$(A_2)B$		2	$\begin{bmatrix} 3704.70 & \\ 4180.35 & 5513.86 \end{bmatrix}$
Within error	132		$\begin{bmatrix} 48{,}852.87 & \\ 18{,}806.26 & 60{,}852.43 \end{bmatrix}$
"Total"	137		

	Tests		
Criteria	A	$(A_1)B$	$(A_2)B$
Λ	.5612	.8849	.6871
θ_s	.4389	.1067	.3118
$U^{(s)}$	.7817	.1289	.4546
$V^{(s)}$	.4388	.1150	.3134

For each of the three tests, there is only one significant discriminant function. The correlations follow.

$$\hat{\boldsymbol{\rho}}_A = \begin{bmatrix} .76 \\ -.36 \end{bmatrix} \qquad \hat{\boldsymbol{\rho}}_{(A_1)B} = \begin{bmatrix} .78 \\ .85 \end{bmatrix} \qquad \hat{\boldsymbol{\rho}}_{A_2(B)} = \begin{bmatrix} -.22 \\ .85 \end{bmatrix}$$

EXERCISES 5.11

1. MANOVA Table

Source	df	SSP
Machines (M)	3	$\begin{bmatrix} 855.24 & \\ 26.76 & 92.25 \end{bmatrix}$
Ability groups (A)	3	$\begin{bmatrix} 71.25 & \\ -7.74 & 2.25 \end{bmatrix}$
Treatments (T)	3	$\begin{bmatrix} 6.75 & \\ -14.76 & 113.25 \end{bmatrix}$
Within error	6	$\begin{bmatrix} 128.50 & \\ 123.50 & 260.00 \end{bmatrix}$
"Total"	15	

Criteria	Tests		
	M^*	A	T
Λ	.0561	.4657	.4697
θ_s	.9241	.5320	.5171
$U^{(s)}$	12.5352	1.1416	1.0988
$V^{(s)}$	1.1852	.5368	.5445
*Significant			

EXERCISES 5.13

1. H_{01}: Parallelism ($m = 0, n = 4$)

$$\Lambda = .2179, \theta_s = .7797, U^{(s)} = 3.5510, V^{(s)} = .7906$$

H_{02}: Conditions ($m = -.5, n = 4$)

$$\Lambda = .1664, \theta_s = .8315, U^{(s)} = 4.9469, V^{(s)} = .8443$$

H_{03}: Groups ($m = .5, n = 3.5$)

$$\Lambda = .1414, \theta_s = .8315, U^{(s)} = 5.2309, V^{(s)} = .9773$$

EXERCISES 5.15

1. Multivariate Tests of Parallelism ($\Lambda = .7059$)

MANCOVA Table

Source	df	SSP
Treatments	2	$\begin{bmatrix} 5.1892 & \\ 2.4901 & 2.0231 \end{bmatrix}$
Error	25	$\begin{bmatrix} 2.0475 & \\ .0297 & 3.7757 \end{bmatrix}$
"Total"	27	

Criteria: $\Lambda = .2170, \theta_s = .7407, U^{(s)} = 3.0515, V^{(s)} = .9036$. The significant standardized discriminant function is

$$L = 3.2629(R_2) + .8955(M_2)$$

The correlation between L and each dependent variable is

$$\hat{\boldsymbol{\rho}}_L = \begin{bmatrix} .94 \\ .36 \end{bmatrix}$$

EXERCISES 5.16

1. a. Parallelism hypothesis: (nonsignificant)

$$\Lambda = .2429, \theta_s = .6538, U^{(s)} = 2.3138, V^{(s)} = .9522$$

Coincidence hypothesis: (significant)

$$\Lambda = .0736, \theta_s = .8642, U^{(s)} = 7.2083, V^{(s)} = 1.3219$$

For $p = q = 4$ and

$$\mathbf{Q} = \begin{bmatrix} .50000 & .50000 & .50000 & .50000 \\ -.67082 & -.22361 & .22361 & .67082 \\ .50000 & -.50000 & -.50000 & .50000 \\ -.22361 & .67082 & -.67082 & .22361 \end{bmatrix}$$

$$\hat{\mathbf{B}} = \begin{bmatrix} 21.500 & -4.472 & -2.500 & 0.000 \\ 26.125 & -3.969 & -2.375 & -2.068 \\ 15.750 & -1.118 & .250 & 0.000 \end{bmatrix}$$

For $p = q = 4$ and

$$\mathbf{Q} = \begin{bmatrix} 1 & 1 & 1 & 1 \\ 1 & 2 & 3 & 4 \\ 1 & 4 & 9 & 16 \\ 1 & 8 & 27 & 64 \end{bmatrix}$$

$$\hat{\mathbf{B}} = \begin{bmatrix} 9.500 & 4.250 & -1.250 & 0.000 \\ 27.750 & -21.583 & 10.375 & -1.542 \\ 9.750 & -1.125 & .125 & 0.000 \end{bmatrix}$$

b. For quadratic growth curves, with

$$\mathbf{Q} = \begin{bmatrix} 1 & 1 & 1 & 1 \\ 1 & 2 & 3 & 4 \\ 1 & 4 & 9 & 16 \end{bmatrix} \quad \text{and} \quad \mathbf{G} = \mathbf{S}$$

$$\hat{\mathbf{B}} = \begin{bmatrix} 9.500 & 4.250 & -1.250 \\ 8.350 & 9.278 & -2.221 \\ 9.750 & -1.125 & .125 \end{bmatrix}$$

For the test of parallelism, $\Lambda = .3469$, $\theta_s = 1.6799$, $U^{(s)} = 1.7556$. For the test of coincidence, $\Lambda = .0817$, $\theta_s = 5.8668$, $U^{(s)} = 6.6501$.

2. a. Using

$$\mathbf{Q} = \begin{bmatrix} 1 & 1 & 1 & 1 & 1 & 0 & 0 & 0 & 0 & 0 \\ 1 & 2 & 3 & 4 & 5 & 0 & 0 & 0 & 0 & 0 \\ 1 & 4 & 9 & 16 & 25 & 0 & 0 & 0 & 0 & 0 \\ 1 & 8 & 27 & 64 & 125 & 0 & 0 & 0 & 0 & 0 \\ 1 & 16 & 81 & 256 & 625 & 0 & 0 & 0 & 0 & 0 \\ 0 & 0 & 0 & 0 & 0 & 1 & 1 & 1 & 1 & 1 \\ 0 & 0 & 0 & 0 & 0 & 1 & 2 & 3 & 4 & 5 \\ 0 & 0 & 0 & 0 & 0 & 1 & 4 & 9 & 16 & 25 \\ 0 & 0 & 0 & 0 & 0 & 1 & 8 & 27 & 64 & 125 \\ 0 & 0 & 0 & 0 & 0 & 1 & 16 & 81 & 256 & 625 \end{bmatrix}$$

$$\hat{\mathbf{B}} = \begin{bmatrix} 14.647 & 1.212 & .586 & -.324 & .036 & 36.594 & -16.661 & 7.839 & -1.573 & .115 \\ 3.975 & 10.215 & -4.502 & .871 & -.062 & 28.000 & 19.721 & -6.616 & .638 & .007 \end{bmatrix}$$

For the test of parallelism, $\Lambda = .6549$ and $\theta_s = .5268$. For the test of coincidence, $\Lambda = .2139$ and $\theta_s = 3.6751$.

EXERCISES 5.17

1. a. For the test of $H_0: \alpha_1 = \alpha_2, F = 1.667$. For the test of $H_0: \beta_1 = \beta_2, F = 4.667$.

   b. For the test of $H_0: \alpha_i + \frac{\sum_{j=1}^{2} \gamma_{ij}}{2}$, $F = 2.1857$ (equal weights).

   For the test of $H_0: \alpha_i + \frac{1}{N_{i+}} \sum_{j=1}^{2} N_{ij}(\beta_j + \gamma_{ij})$, $F = 2.8167$ (unequal weights).

   Similarly, for the test of B, we have $F = .5714$ (equal weights) and $F = 1.0667$ (unequal weights).

2. Model with Interaction Term

Test	Λ criterion	p value
A (equal weights)	.9080	.1940
A (unequal weights)	.9194	.2399
B (equal weights)	.8604	.0777
B (unequal weights)	.8693	.0925

Additive Model

Test	Λ criterion	p value
A	.9064	.1791
A (unequal weights)	.9201	.2327
B	.8572	.0676
B (unequal weights)	.8695	.0866

EXERCISES 6.2

1. b. Eigenvalues: 46.6182, 6.5103, 5.4144
2. b. Eigenvalues: 1.5978, 1.0794, .9258, .7579, .6390

   c. Components

	1	2
1	$-.07$	.85
2	.59	.45
3	.52	.16
4	.76	$-.08$
5	.62	$-.36$

EXERCISES 6.3

1. Eigenvalues: 3.4701, .2430, .1601, .1267

EXERCISES 6.4

1. a. $\hat{\lambda}_1 = 46.6182$

INDEX